# PRINCIPLES OF
# VIROLOGY
## Molecular Biology, Pathogenesis, and Control

# PRINCIPLES OF

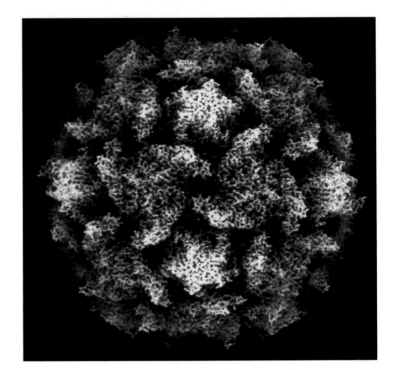

# VIROLOGY

## Molecular Biology, Pathogenesis, and Control

**S. J. Flint**
Department of Molecular Biology
Princeton University
Princeton, New Jersey

**L. W. Enquist**
Department of Molecular Biology
Princeton University
Princeton, New Jersey

**R. M. Krug**
Institute for Cellular and
    Molecular Biology
University of Texas at Austin
Austin, Texas

**V. R. Racaniello**
Department of Microbiology
College of Physicians & Surgeons
Columbia University
New York, New York

**A. M. Skalka**
Institute for Cancer Research
Fox Chase Cancer Center
Philadelphia, Pennsylvania

ASM
PRESS

Washington, D.C.

*Front cover illustration:* A model of the atomic structure of the poliovirus type 1 Mahoney strain. The model has been highlighted by radial depth cuing so that the portions of the model that are farthest from the center are bright. Prominent surface features include a star-shaped mesa at each of the fivefold axes and a propeller-shaped feature at each of the threefold axes. A deep cleft or canyon surrounds the star-shaped feature. This canyon is the receptor-binding site. Courtesy of Robert Grant, Stéphane Crainic, and James Hogle (Harvard Medical School).

*Back cover illustration:* Progress in the global eradication of poliomyelitis has been striking, as illustrated by maps showing areas of known or probable circulation of wild-type poliovirus in 1988 and 1998. Blue indicates either the presence of virus or insufficient surveillance to exclude the possibility of virus circulation. In 1988, virus was present on all continents except Australia. By 1998, the Americas were free of wild-type poliovirus, and transmission was interrupted in the western Pacific region (including the People's Republic of China) and in the European region (with the exception of southeastern Turkey). There are currently three major foci of transmission: southern Asia (Afghanistan, Pakistan, and India), West Africa (mainly Nigeria), and central Africa (mainly the Democratic Republic of Congo). Source: World Health Organization.

**Library of Congress Cataloging-in-Publication Data**
Principles of virology: molecular biology, pathogenesis, and control
/ S. J. Flint ... [et al.].
    p.   cm.
  Includes bibliographical references and index.
  ISBN 1-55581-127-2 (hardcover)
  1. Viruses. 2. Virology. I. Flint, S. Jane.
  [DNLM: 1. Viruses. 2. Virology—methods.        QW 160 P957 2000]
QR360.P697  2000
579.2—dc21
DNLM/DLC                                                             99-14697
for Library of Congress                                                  CIP

10 9 8 7 6 5 4 3 2 1

*Illustrations and illustration concepting:* J/B Woolsey Associates
*Cover and interior design:* Susan Brown Schmidler

*We dedicate this book to the students,*
*current and future scientists, and physicians*
*for whom it was written.*
*We kept them ever in mind.*

*We also dedicate it to our families:*
Jonn, Gethyn, and Amy Leedham
Kathy and Brian
Marjorie, Dana, Nora, and Stephen
Doris, Aidan, Devin, and Nadia
Rudy, Jeanne, and Chris


*Oh, be wiser thou!*
*Instructed that true knowledge leads to love.*

WILLIAM WORDSWORTH
*Lines Left upon a Seat in a Yew-Tree*
*1888*

# Contents

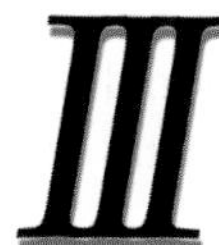

# Acknowledgments

This book could not have been completed without help and contributions from many individuals. Indeed, this venture would not have begun were it not for the initial encouragement we received from our colleagues in virology. We are especially indebted to Patrick Fitzgerald (formerly Director, ASM Press) for his enthusiastic support when for the project was little more than a rough plan and for his sound advice during its development. The creative energy and ideas of Priscilla Schaffer (University of Pennsylvania) had a substantial impact on the initial planning. Our sincere thanks also go to our colleagues who took considerable time and effort to review this text in its evolving manifestations. Their expert knowledge and advice on issues ranging from teaching virology to organization of individual chapters and style were invaluable, even when orthogonal to our approach, and are inextricably woven into the final form of the book.

We thank the following individuals for their reviews and comments on multiple chapters or the entire book: Nicholas Acheson (McGill University), Clifford W. Bond (Montana State University), Martha Brown (University of Toronto Medical School), Teresa Compton (University of Wisconsin), Stephen Dewhurst (University of Rochester Medical Center), Mary K. Estes (Baylor College of Medicine), Ronald Javier (Baylor College of Medicine), Richard Kuhn (Purdue University), Muriel Lederman (Virginia Polytechnic Institute and State University), Richard Moyer (University of Florida College of Medicine), Leonard Norkin (University of Massachusetts), Martin Petric (University of Toronto Medical School), Marie Pizzorno (Bucknell University), Nancy Roseman (Williams College), David Sanders (Purdue University), Dorothea Sawicki (Medical College of Ohio), Bert Semler (University of California, Irvine), and Bill Sugden (University of Wisconsin).

We also are very grateful to those who gave so generously of their time to serve as expert reviewers of individual chapters or topics: Edward Arnold (Center for Advanced Biotechnology and Medicine, Rutgers University),

Albert Bendelac (Princeton University), Susan Berget (Baylor College of Medicine), Jim Broach (Princeton University), Michael J. Buchmeier (The Scripps Research Institute), Bruce Chesebro (Rocky Mountain Laboratories, National Institute of Allergy and Infectious Diseases), Marie Chow (University of Arkansas Medical Center), Michael Cordingly (Bio-Mega/Boehringer Ingelheim), Ted Cox (Princeton University), Robert Doms (University of Pennsylvania), Emilio Emini (Merck Sharp & Dohme Research Laboratories), Charles Grose (Iowa University Hospital), Peter Howley (Harvard Medical School), James Hoxie (University of Pennsylvania), Frederick Hughson (Princeton University), Tom Kelly (Johns Hopkins University School of Medicine), Elliott Kieff (Harvard Medical School), Arnold Levine (The Rockefeller University), Michael Malim (University of Pennsylvania), William Mason (Fox Chase Cancer Center), Bernard Moss (National Institutes of Health), Roger Pomerantz (Thomas Jefferson University), Glenn Rall (Fox Chase Cancer Center), Barry Rouse (University of Tennessee College of Veterinary Medicine), Priscilla Schaffer (University of Pennsylvania), Robert Schneider (New York University School of Medicine), Christoph Seeger (Fox Chase Cancer Center), Aaron Shatkin (Center for Advanced Biotechnology and Medicine, Rutgers University), Thomas Shenk (Princeton University), John M. Taylor (Fox Chase Cancer Center), Alice Telesnitsky (University of Michigan Medical School), Peter Vogt (The Scripps Research Institute), M. Gerard Waters (Princeton University), Robin Weiss (Institute of Cancer Research, Chester Beatty Laboratories), Lindsay Whitton (The Scripps Research Institute), and Eckard Wimmer (State University of New York at Stony Brook). Their rapid responses to our requests for details and checks on accuracy, as well as their assistance in simplifying complex concepts, were invaluable. All remaining errors or inconsistencies are entirely ours.

We thank our colleagues who provided expert advice on specific topics or information prior to publication: James Alwine (University of Pennsylvania), Carl Baker (National Institutes of Health), Amiya Banerjee (Cleveland Clinic Foundation), Silvia Barabino (University of Basel), Barklie Clements (University of Glasgow), Don Coen (Harvard Medical School), Ron Desrosiers (Harvard Medical School), Bert Flanagan (University of Florida), Samuel Gunderson (European Molecular Biology Laboratory), Walter Keller (University of Basel), Ihor Lemischka (Princeton University), Michael Malim (University of Pennsylvania), James Manley (Columbia University), William Mason (Fox Chase Cancer Center), Bernard Moss (National Institutes of Health), Peter O'Hare (Marie Curie Research Institute), Radhakris Padmanabhan (University of Kansas Medical Center), Stuart Peltz (University of Medicine and Dentistry of New Jersey—Robert Wood Johnson Medical School), Jack Rose (Yale University), Rozanne Sandri-Goldin (University of California, Irvine), Bert Semler (University of California, Irvine), Thomas Shenk (Princeton University), Joan Steitz (Yale University), Wesley Sundquist (University of Utah), Heinz-Jürgen Thiel (Institut für Virologie, Giessen, Germany), Sandra Weller (University of Connecticut Health Center), and Eckard Wimmer (State University of New York at Stony Brook).

Since the inception of this work, we have believed that the illustrations must complement and enrich the text. Execution of this plan would have not been possible without the support of Jeff Holtmeier (Director, ASM Press) and the technical expertise and craft of our illustrators, J/B Woolsey Associates. The illustrations are an integral part of the exposition of the information and ideas discussed, and credit for their execution goes to the knowledge, insight, and enthusiasm of Patrick Lane of J/B Woolsey Associates. Many of the figures could not have been completed without the help and generosity of our many colleagues (mentioned in the figure legends) who provided original images. Special thanks go out to those who crafted figures tailored specifically to our needs or provided multiple pieces: Mark Andrake (Fox Chase Cancer Center), Edward Arnold (Rutgers University), Bruce Banfield (Princeton University), Christopher Basler and Peter Palese (Mount Sinai School of Medicine), Amy Brideau (Princeton University), Roger Burnett (Wistar Institute), Rajiv Chopra and Stephen Harrison (Harvard University), Marie Chow (University of Arkansas Medical Center), Richard Compans (Emory University), Friedrich Frischknecht (European Molecular Biology Laboratory), Wade Gibson (Johns Hopkins University School of Medicine), Ramon Gonzalez (Princeton University), Thomas Leitner (Los Alamos National Laboratory), Paul Masters (New York State Department of Health), Rolf Menzel (Small Molecule Therapeutics), B. V. Venkataram Prasad (Baylor College of Medicine), Wesley Sundquist (University of Utah), Jose Varghese (Commonwealth Scientific and Industrial Research Organisation), Robert Webster (St. Jude Children's Research Hospital), Thomas Wilk (European Molecular Biology Laboratory), and Alexander Wlodawer (National Cancer Institute).

We thank all those who guided and assisted in the preparation and production of the book: Jeff Holtmeier (Director, ASM Press) for steering us through the complexities inherent in a team effort, Ken April (Production Editor, ASM Press) for keeping us on track during production, Greg Payne (Senior Editor, ASM Press) for organizing the reviews of the text, and Susan Schmidler (Susan Schmidler Graphic Design) for her elegant and creative designs for the layout and cover. We are also grateful for expert secretarial and administrative support from Norma Caputo (Princeton University), Helen Pirrello (Rutgers University), Josephina de la Rosa (Columbia University), and Mary Estes and Rose Walsh (Fox Chase Cancer Center) that facilitated preparation of this text, and we thank Amy Leedham for assistance with editing.

We owe a very special debt of gratitude to the Rockefeller Foundation for the award of a residency for our team at the Bellagio Study and Conference Center, Bellagio, Italy, in June 1997. The extraordinary beauty of the Center and the gracious hospitality of its staff made that an unforgettable as well as rewarding experience. The unique opportunity for study and extended consultation with each other and for interaction with resident scholars from many countries and disciplines had a major impact on the delineation of the purpose, approach, and scope of this textbook. ASM Press generously provided the financial support for travel to Bellagio as well as for our many meetings.

This often consuming enterprise was made possible by the emotional, intellectual, and logistical support of our families, to whom the book is dedicated.

Our acknowledgments close with a special tribute to the person who has done more than any other to support the author team, Trisha Barney (Princeton University), who has carried responsibilities too numerous to list since the very beginning of the project. However, we are especially grateful for her patience and good humor and particularly for collecting all the permissions necessary for preparation of the figures. Trisha's dedication, attention to detail, and problem-solving skills were critically important to our progress.

# Organization of This Book

*The enduring goal of scientific endeavor, as of all human enterprise, I imagine, is to achieve an intelligible view of the universe. One of the great discoveries of modern science is that its goal cannot be achieved piecemeal, certainly not by the accumulation of facts. To understand a phenomenon is to understand a category of phenomena or it is nothing. Understanding is reached through creative acts.*

A. D. Hershey
*Carnegie Institution Yearbook 65*

The major goal of this book is to illustrate why and how animal viruses are studied. In so doing, we highlight unifying principles that emerge from these studies that may be used for investigation of questions still outstanding and of viral systems about which our knowledge is still quite limited. We have been selective in our choice of topics, viruses, and examples. The purpose was not to be comprehensive, for it is not possible for one reasonably sized text to cover every facet of animal virology. *Fields Virology* (1996), in two encyclopedic volumes, has made the best attempt to be all-inclusive and is recommended as a resource for detailed reviews of specific virus families.

For purposes of readability, references are generally omitted from the text, but each chapter ends with a list of relevant books, review articles, and selected research papers, for readers who wish to pursue specific topics. If an experiment is featured in a chapter, one or more references will be listed to provide more detailed information. In many cases, the figures and their legends also provide important detail whose inclusion in the text might interrupt the flow of thought. Text boxes include basic background information (boxes with aqua and pale green backgrounds); detailed discussions of specific topics (purple and greenish beige); definitions of terms, explanations of nomenclature, and clarifications of common sources of confusion in terminology (maroon and gray); discussions of specific experiments (blue and yellow); and cautionary information (orange). The text itself follows the general strategy used by all viruses and covers the principles of critical steps in virus replication in the following sections.

### Part I: The Science of Virology

Chapter 1 provides a general introduction to the field with historical perspectives and a discussion of how viruses are classified. It summarizes the unifying principles that are the foundations of virology. In chapter 2, we describe general methods by which animal viruses are studied in the laboratory as an entrée into the molecular biology section.

### Part II: Molecular Biology

The next section, consisting of chapters 3 to 13, focuses on molecular processes that take place in an infected host cell. Chapter 3 provides a description of the extracellular virus particle (a virion). Virion architecture is discussed in the context of providing both protection and delivery of the viral genome in a single vehicle. Chapters 4 through 13 describe the broad spectrum of molecular processes that characterize the common steps of the reproductive cycle of viruses in a single cell, from decoding genetic information to genome replication. The intention is to describe how these common steps are accomplished by diverse but representative viruses, while illuminating principles applicable to all.

### Part III: Pathogenesis and Control

This section moves away from events in the single cell and addresses issues related to the interplay between viruses and their host organisms. Chapter 14 provides a detailed description of the components of the immune system and strategies that viruses have evolved for coping with it, while chapter 15 describes the different types of relationships (some short and others of long duration) that viruses can establish with their hosts. Chapter 16 describes the roles of viruses in cell transformation and in oncogenesis, the formation of tumors. As viruses are of paramount importance as disease agents, chapters 17 to 19 address issues of pathogenesis and the principles involved in developing methods of treatment and control of infection. An entire chapter (chapter 18) is devoted to the AIDS virus, not only because it is the causative agent of a serious, worldwide epidemic but also because of its unique and informative interactions with the human immune defenses. Chapter 20, the final chapter, is a foray into the future and a consideration of viral evolution, the ways in which new viruses emerge, and the overall implications for virus and host survival.

### Appendix

The appendix provides a brief description and a general scheme for the reproductive cycles in single cells of viruses that serve as important examples in several chapters of the text. This section is intended to be a reference resource when one reads individual chapters and a convenient means by which specific topics may be related to the overall infectious cycles of the selected viruses. On the opening page of each chapter, therefore, is a set of icons, which are based on the diagrams of virus structures in the appendix and which represent the selected viruses described in that chapter.

### Reference

Fields, B. N., D. M. Knipe, and P. M. Howley (ed. in chief). 1996. *Fields Virology*, 3rd ed. Raven Press, New York, N.Y.

# *I* The Science of Virology

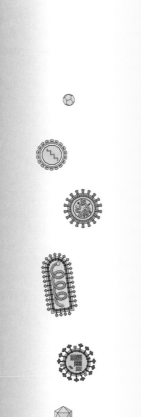

*1*

# Foundations of Virology

## Luria's Credo

Forty-five years have passed since Salvador Luria wrote the following credo in the introduction to the classic textbook he wrote with James Darnell, *General Virology:* "There is an intrinsic simplicity of nature and the ultimate contribution of science resides in the discovery of unifying and simplifying generalizations, rather than in the description of isolated situations—in the visualization of simple, overall patterns rather than in the analysis of patchworks."

Despite an explosion of information in biology since Luria wrote these words, his vision of unity in diversity is as relevant now as it was then. That such first principles exist may not be obvious considering the often bewildering array of viruses, genes, and proteins recognized in modern virology. Indeed, new viruses are being described regularly (at least 50 since 1988), and viral diseases like AIDS, hepatitis, and influenza continue to defy our efforts to control them. Yet, as discussed below, Luria's credo still stands: all viruses follow a simple three-part strategy to ensure their survival. This insight has been hard won over many years of observation, research, and debate; the history of virology is rich and instructive.

## Virus Prehistory

Viruses have been known as distinct biological entities for little more than a century. Consequently, efforts to understand and control these important agents of disease are phenomena of the 20th century. Nevertheless, evidence of viral infection can be found among the earliest recordings of human activity, and methods for combatting viral disease were practiced long before the first virus was recognized.

## Viral Infections in Antiquity

Reconstruction of the prehistoric past to provide a plausible account of when or how viruses established themselves in human populations is a challenging task. However, extrapolating from current knowledge, we can deduce that

**3**

some modern viruses undoubtedly were associated with the earliest precursors of mammals and coevolved with humans. Others entered human populations only recently. It is instructive to consider the last 10,000 years of human development, a time of radical change for humans and our viruses: animals were domesticated, the human population increased dramatically, large population centers appeared, and commerce drove interactions among unprecedented numbers of humans. We can infer from scattered glimpses of ancient history that viruses have long been a part of human experience.

Some viruses that eventually established themselves in human populations were transmitted to early humans from animals, much as still happens today. The initial human experience with viruses from nonhuman sources probably depended on many parameters. For example, early human groups that domesticated and lived with their animals were undoubtedly more exposed to different viruses than were nomadic hunter societies whose exposure to animals was only transient. Similarly, as many different viruses are endemic in the tropics, human societies in that environment must have been exposed to a greater variety of viruses than societies established in temperate climates. When nomadic bands met groups with domesticated animals or tropical cultures mingled with those from cooler climates, human-to-human contact could have provided new avenues for virus interaction and spread. Even so, it seems unlikely that highly virulent viruses, such as measles virus and smallpox virus, could have entered a permanent relationship with these small groups of early humans. Such virulent viruses either kill their hosts or induce life-long immunity in the survivors. Consequently, they can survive only where large, interacting host populations are available for their continued propagation. Such viruses could not have been established in human populations until large, settled communities appeared. Thus, less virulent viruses that enter into a more benign, long-term relationship with their hosts were more likely to be the first to become adapted to replication in the earliest human populations. These viruses include the modern retroviruses, herpesviruses, and papillomaviruses.

Evidence of several viral diseases can be found in ancient records (Fig. 1.1). The Greek poet Homer characterizes Hector as "rabid" in *The Iliad*. Mesopotamian laws describing the responsibilities of the owners of rabid dogs date from before 1,000 B.C. Their existence indicates that the communicable nature of this viral disease was already well known by that time, as it was to other ancient civilizations. Egyptian hieroglyphs illustrating what appear to be the consequences of poliovirus infection (a withered leg typical of poliomyelitis) or pustular lesions characteristic of smallpox also date from that period. Smallpox, which was

probably endemic in the Ganges river basin by the 5th century B.C. and subsequently spread to other parts of Asia and Europe, has played an important part in human history. Its introduction to the previously unexposed native peoples of Central and South America by colonists in the 15th century led to lethal epidemics, which are considered an important factor in the conquests achieved by a small number of European soldiers. Other viral diseases known in ancient times include mumps and, perhaps, influenza. Yellow fever has been described since the discovery of Africa by Europeans, and it has been suggested that this scourge of the tropical trade was the basis for legends about ghost ships, such as the *Flying Dutchman*, in which an entire ship's crew perished mysteriously.

Humans have not only been subject to viral disease throughout much of their history but have also manipulated these agents, albeit unknowingly, for much longer than might be imagined. One classic example is provided by efforts to cultivate marvelously patterned tulips, which were of enormous value in 17th century Holland. Such cultivation included deliberate spread of a virus (tulip breaking virus or tulip mosaic virus) that we now know causes the striping of tulip petals so highly prized at that time (Fig. 1.2). Efforts to control viral disease have an even more venerable history.

## The First Vaccines

Measures to control one viral disease have been used with some success for the last millennium. The disease is smallpox (Fig. 1.3), and the practice is called **variolation**, inoculation of healthy individuals with material from a smallpox pustule into a scratch made on the arm. Variolation was widespread in China and India by the 11th century and was based on the recognition that survivors of smallpox infection were protected against subsequent bouts of the disease. Variolation subsequently spread to Asia Minor, where its value was recognized by Lady Mary Wortley Montague, wife of the British ambassador to the Ottoman Empire. She introduced this practice into England in 1721, where it became quite widespread following successful inoculation of children of the royal family. George Washington is said to have introduced variolation among Continental Army soldiers in 1776. However, the consequences of variolation were unpredictable and never pleasant. Serious skin lesions invariably developed at the site of inoculation and also might be accompanied by more generalized rash and disease, with a fatality rate of 1 to 2%. From the comfortable viewpoint of a developed country in the 20th century, such a high death rate seems unacceptable. However, in the 18th century, variolation was perceived as a much better alternative than natural smallpox, a disease associated with a fatality rate of 25%

A

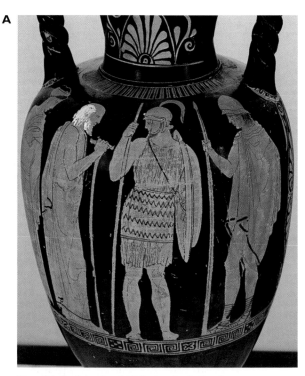

B

*Here this firebrand, rabid Hector, leads the charge.*
HOMER, *The Iliad*,
translated by Robert Fagels
(Viking Penguin)

**Figure 1.1 References to viral diseases abound in the ancient literature.** (A) An image of Hector from an ancient Greek vase. Courtesy of the University of Pennsylvania Museum (object 30-44-4). (B) An Egyptian stele, or stone tablet, from the 18th dynasty (1580–1350 B.C.) depicting a man with a withered leg and the "drop foot" syndrome characteristic of polio.

in the whole population and 40% in babies and young children.

In the 1790s, Edward Jenner, an English country physician, recognized the principle on which modern methods of viral immunization are based even though viruses themselves were not to be identified for another 100 years. Jenner himself was variolated as a boy and also practiced this procedure. Thus, he was undoubtedly familiar with its effects and risks. Perhaps this experience spurred his great insight, which was that milkmaids who had been exposed to cowpox (a mild disease in humans) were protected against smallpox. Jenner followed up this astute observation with direct experiments. In 1794–1796, he demonstrated that inoculation of a young boy with extracts of cowpox lesions induced only mild symptoms but protected him against the far more dangerous smallpox. It is from these experiments with cowpox that we derive the term **vaccination** (*vacca* = cow in Latin). This term was first given its general meaning by Louis Pasteur in 1881 to honor Jenner's accomplishments.

Initially, the only way to propagate and maintain cowpox vaccine was by serial infection of human subjects. This method was eventually banned, as it was associated with transmission of other diseases, such as syphilis and hepatitis, at significant frequencies. By 1860, the vaccine had been passaged in cows; later, sheep and water buffaloes were also used. While Jenner's original vaccine was based on cowpox virus, sometime during the human-to-human and cow-to-cow transfers, the cowpox virus was replaced unknowingly with the poxvirus now called vaccinia virus. Vaccinia virus is the basis for the modern smallpox vaccine, but its origins are not known. It exhibits little genetic similarity to cowpox virus, smallpox virus, or many of the known poxviruses. One popular idea is that vaccinia virus

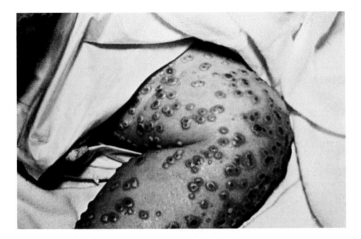

**Figure 1.3 Characteristic smallpox lesions in a young smallpox victim.** Illustrations like these were used to track down individuals infected with smallpox during the World Health Organization campaign to eradicate this disease. Photo courtesy of the Immunization Action Coalition (original source: Centers for Disease Control and Prevention).

**Figure 1.2 *Three Broken Tulips.*** A painting by Nicolas Robert (1624–1685), now in the collection of the Fitzwilliam Museum, Cambridge, England. Striping patterns (color breaking) in tulips were described in 1576 in western Europe and were caused by a virus infection. This beautiful image depicts the remarkable consequences of infection with the tulip mosaic virus. Courtesy of the Fitzwilliam Museum, University of Cambridge.

and an influenza virus vaccine was available in 1936. These advances became possible only with radical changes in our knowledge of living organisms and of the causes of disease during the second half of the previous century.

## Microorganisms as Pathogenic Agents

The 19th century was a period of revolution in scientific thought, particularly in ideas about living things and their origins. The publication of Charles Darwin's *The Origin of Species* in 1859 crystallized startling (and to many people, shocking) new ideas about the origin of diversity in plants and animals, until then generally attributed directly to the hand of God. These insights permanently undermined the belief that humans were somehow set apart from all other members of the animal kingdom. From the point of view of the science of virology, the most important changes were in ideas about the causes of disease.

The diversity of macroscopic organisms has been appreciated and cataloged since the dawn of recorded human history. A vast new world of organisms too small to be visible to the naked eye was revealed with the invention of the microscope by Antony van Leeuwenhoek (1632–1723). Among van Leeuwenhoek's vivid and enthusiastic descriptions of living microorganisms as "wee animalcules," seen in such ordinary materials as rain or seawater, are examples of what we now know as protozoa, algae, and bacteria. By the early 19th century, the scientific community had accepted the existence of microorganisms and turned to the question of their origin—a topic of fierce debate. Some believed that microorganisms arose sponta-

originated in horses. Scientists have recovered the smallpox vaccine used in New York in 1876 and have verified that it is vaccinia virus and not cowpox virus. Speculation about when and how the switch occurred has produced some fascinating scenarios.

The first deliberately attenuated viral vaccine was made by Pasteur in 1885. To accomplish this, he infected rabbits with the rabies agent and then used aqueous suspensions of their dried spinal cords to infect other rabbits. After several such passages, the resulting virus preparations caused mild disease (i.e., were **attenuated**) yet produced effective immunity against subsequent infection by the virulent rabies virus. Safer and more efficient methods for the production of larger quantities of these first vaccines awaited the recognition of viruses as distinctive biological entities and of their parasitism in the cells of their hosts. Indeed, it took almost 50 years to discover the next antiviral vaccines: a vaccine for yellow fever virus disease appeared in 1935, neously, for example, in decomposing matter, in which they were especially abundant. Others held the view that all were generated by the reproduction of like microorganisms, as were macroscopic organisms. The death knell of the spontaneous generation hypothesis was sounded with the famous experiments of Pasteur. He demonstrated that boiled (i.e., sterilized) medium remained free of microorganisms as long as it was maintained in special flasks with curved, narrow necks designed to prevent entry of airborne microbes (Fig. 1.4). Pasteur also established that particular microorganisms were associated with specific processes, for example, the production of alcohol, lactic acid, or acetic acid (vinegar) by fermentation. This idea was crucial in the development of modern explanations of the causes of disease.

From the earliest times, poisonous air (miasma) was generally invoked to account for epidemics of contagious diseases, and there was little recognition of the differences

**Figure 1.4 Four experiments to challenge spontaneous generation of life.** The first step in each experiment was to boil the broth medium very thoroughly to destroy all living organisms. Air was then admitted to the flasks to satisfy the believers in spontaneous generation, who insisted that oxygen must be present for life to originate. However, the air admitted to the broth medium first was freed of living organisms in several ingenious ways. Under these conditions the broths remained perfectly sterile; no microorganisms appeared in them, showing that living things could not be generated spontaneously from the lifeless liquid. (A) Sterilizing air by chemical treatment. The set of bulbs next to the person's face contained alkali, and the other set concentrated acid. Air was drawn in through the acid to inactivate microbes before it reached the broth. (B) Sterilizing air by heat. Air could enter the flask of broth only by pasing through the coiled glass tube kept hot by the flame. (C) Sterilizing air by physically trapping particles. The aspirating bottle drew air into the flask through the tube containing cotton at the right. The cotton filtered out the microbes in the air, just as the cotton or foam plugs now used in bacteriological culture tubes protect the culture from air contamination. (D) Pasteur's famous U-neck flasks provided passive exclusion of microbes from the sterilized broth. Although the flask was freely open to the air, the broth remained sterile so long as the microbe-bearing dust that collected in the U did not reach the liquid.

**A**

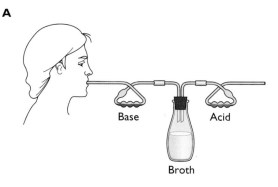

**B**

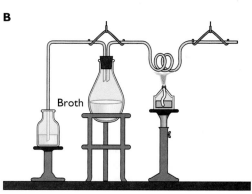

**C**

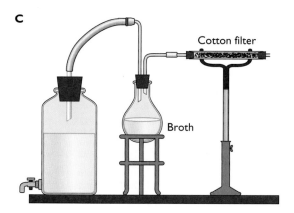

**D**

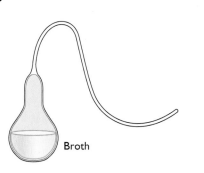

among their causative agents. The association of particular microorganisms, initially bacteria, with specific diseases can be attributed to the ideas of the German physician Robert Koch. He developed and applied a set of criteria for identification of the agent responsible for a specific disease (a **pathogen**). These criteria, **Koch's postulates**, are still applied in the identification of those pathogens that can be propagated in the laboratory and tested in an appropriate animal model. The postulates are as follows.

- The organism must be regularly associated with the disease and its characteristic lesions.
- The organisms must be isolated from the diseased host and grown in culture.
- The disease must be reproduced when a pure culture of the organism is introduced into a healthy, susceptible host.
- The same organism must again be reisolated from the experimentally infected host.

By applying these principles, Koch demonstrated that anthrax, a common disease of cattle, was caused by a specific bacterium (designated *Bacillus anthracis*) and that a second,

distinct bacterial species caused tuberculosis in humans. Armed with these principles and methods for the sterile culture and isolation of pure preparations of bacterial species developed by Pasteur, Joseph Lister, and Koch, scientists working during the last part of the 19th century identified and classified many pathogenic bacteria (as well as yeasts and fungi) (Fig. 1.5). Thus, understanding of the causes of infectious disease was placed on a secure scientific foundation, the first step toward rational treatment and ultimately control. During the last decade of the 19th century, failures of the paradigm that bacterial or fungal agents are responsible for disease led to the identification of a new class of infectious agents—submicroscopic pathogens that came to be called **viruses**.

## The Discovery of Viruses

### The First Viruses

The first report of a pathogenic agent smaller than any known bacterium appeared in 1892. The Russian scientist Dimitrii Ivanowsky observed that the causative agent of tobacco mosaic disease was not retained by the unglazed porcelain filters used at that time to remove bacteria from

**Figure 1.5 The pace of early discovery of new infectious agents of humans.** Koch's introduction of efficient bacteriological techniques spawned an explosion of new discoveries of bacterial agents in the early 1880s. Similarly, Ivanowsky's discovery of filterable agents launched the field of virology in the early 1900s. Despite an early surge of virus discovery, only 19 distinct human viruses had been discovered by 1935. TMV, tobacco mosaic virus. Adapted from K. L. Burdon, *Medical Microbiology* (MacMillan Co. New York, N.Y., 1939), with permission.

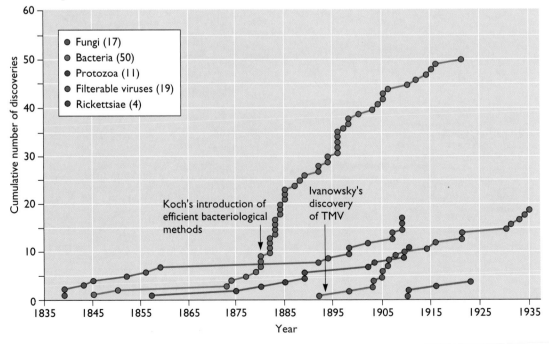

extracts and culture media. Six years later, Martinus Beijerinck made the same observation independently in Holland. More importantly, Beijerinck made the conceptual leap that the pathogen responsible for tobacco mosaic dis-

ease must be a distinctive agent so small that it could pass through filters that trapped all known bacteria (Fig. 1.6).

The same year (1898), the German scientists Friedrich Loeffler and Paul Frosch, both former students and assis-

**Figure 1.6 Filter systems used to characterize viruses.** Filters for making liquids free of cultivatable organisms were instrumental in the first identification of viruses. Several types of filters were used in the early days of virus research, and five are illustrated here. (A) The Seitz filter was used as a prefilter to remove large particulate material. Berkefeld filters (B) and Chamberland filters (C) are typical of the "candle" style of filter comprising diatomaceous earth pressed into the shape of a hollow candle open at one end. The pore size could be controlled with some accuracy: V pores were 8 to 12 μm in diameter, N pores were 5 to 7 μm in diameter, and W pores were 3 to 4 μm in diameter. Only N and W pore sizes retained bacteria and allowed viruses to pass through. For the Chamberland filter (C), the fluid to be filtered was introduced into the open end by means of a funnel, and filtration occurred when liquid passed from within the candle to the outside. The filter is cut away to illustrate the hollow center. Filters similar to these were probably used by Ivanowsky, Loeffler, and Frosch to isolate the first plant and animal viruses. (D) Jenkins filter with a portion cut away to show the square porcelain block which is the actual filtering material. A second filter or a collection tube could be added in the receiving flask. (E) Example of a collodion membrane "ultrafilter." The thin membrane is the filtering surface and is stretched over a filter plate. (F) A Mudd filtration apparatus illustrating the typical setup for filtering fluids. All the filters were designed to operate under negative pressure by connection with a suction pump or aspirator. Positive pressure was also used. The trap (on the right) was essential to catch back flow of liquid when the suction was released. The manometer (in the middle) provided an accurate means for measuring and regulating pressure. The filtrate was collected in a graduated buret from which measured samples were collected and tested.

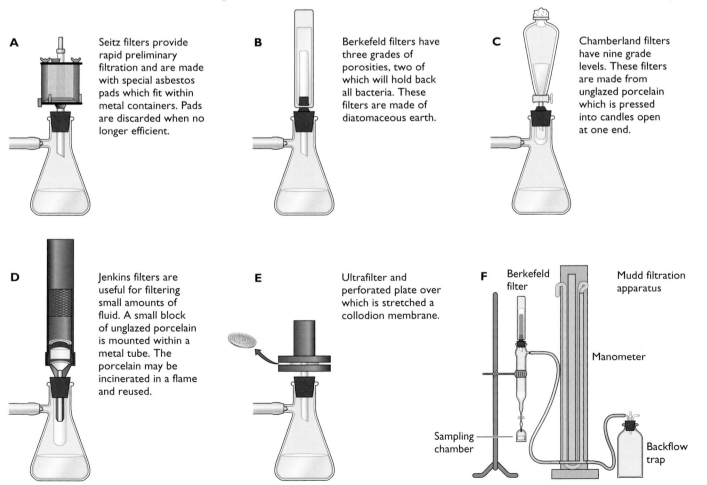

**A**  Seitz filters provide rapid preliminary filtration and are made with special asbestos pads which fit within metal containers. Pads are discarded when no longer efficient.

**B**  Berkefeld filters have three grades of porosities, two of which will hold back all bacteria. These filters are made of diatomaceous earth.

**C**  Chamberland filters have nine grade levels. These filters are made from unglazed porcelain which is pressed into candles open at one end.

**D**  Jenkins filters are useful for filtering small amounts of fluid. A small block of unglazed porcelain is mounted within a metal tube. The porcelain may be incinerated in a flame and reused.

**E**  Ultrafilter and perforated plate over which is stretched a collodion membrane.

**F**  Berkefeld filter

Mudd filtration apparatus

Manometer

Sampling chamber

Backflow trap

tants of Koch, observed that the causative agent of foot-and-mouth disease, a widespread, devastating infection of cattle and other livestock, was also filterable. Not only were tobacco mosaic and foot-and-mouth disease pathogens much smaller than any previously recognized microorganism, but they also replicated only in their host organisms. For example, extracts of an infected tobacco plant diluted into sterile solution produced no additional infectious agents until introduced into leaves of healthy plants, which subsequently developed tobacco mosaic disease. The serial transmission of tobacco mosaic and foot-and-mouth diseases by diluted extracts established that these diseases were not caused by a bacterial toxin present in the original preparations derived from infected tobacco plants or cattle. The failure of both pathogens to multiply in solutions that readily supported growth of bacteria, and their dependence on host organisms for reproduction, further distinguished these new agents from pathogenic bacteria. Beijerinck termed the submicroscopic agent responsible for tobacco mosaic disease *contagium vivium fluidum* to emphasize its infectious nature and distinctive reproductive and physical properties. Agents that pass through filters that retain bacteria came to be called ultra-filterable viruses. This term eventually evolved simply to viruses, appropriating the term **virus** from the Latin for "poison."

The discovery of the first virus, tobacco mosaic virus, is often dated to the work of Ivanowsky in 1892. However, Ivanowsky did not identify the tobacco mosaic disease pathogen as a distinctive agent, nor was he convinced that its passage through bacterial filters was not the result of some technical failure. It is therefore probably more appropriate to attribute the founding of the field of virology to the astute insights of Beijerinck, Loeffler, and Frosch, who recognized the distinctive nature of the plant and animal pathogens they were studying 100 years ago.

The pioneering work on tobacco mosaic virus and foot-and-mouth disease virus was followed by the identification of viruses associated with specific diseases in many other organisms. Important landmarks from this early period include the identification of viruses that cause leukemias or solid tumors in chickens by Vilhelm Ellerman and Olaf Bang in 1908 and by Peyton Rous in 1911, respectively. The study of viruses associated with cancers in chickens, particularly the Rous sarcoma virus, eventually led to an understanding of the molecular basis of neoplastic disease.

Bacterial viruses were first described by Frederick Twort in 1915 and Felix d'Hérelle in 1917. d'Hérelle named them **bacteriophages** because of their ability to lyse bacteria on the surface of agar plates (**phage** is derived from the Greek for "eating"). In an interesting twist of serendipity, Twort made his discovery of bacterial viruses while testing smallpox vaccine virus to see if it would grow on simple media. He found bacterial contaminants, some of them exhibiting a "glassy transformation" due to lysis by a bacteriophage. Investigation of bacteriophages provided the foundations for the field of molecular biology as well as important general insights into interactions of viruses with their host cells.

## Identification of Human Viruses

The first human virus to be identified, in 1901, was that responsible for yellow fever. The complicated transmission and life cycle of this virus made this achievement all the more remarkable. Yellow fever is known to have been widespread in tropical countries since the 15th century. It was responsible for devastating epidemics associated with such high rates of mortality (e.g., 28% in the New Orleans epidemic of 1853) that normal life became impossible. This disease was not, however, directly contagious, and an infectious agent could not be demonstrated in yellow fever patients. These puzzling properties defeated efforts to establish the origin of yellow fever until the Cuban physician Carlos Juan Finlay proposed in 1880 that a blood-sucking insect, most likely a mosquito, played a part in the transmission of this disease. A commission to study the etiology of yellow fever was established by the U.S. Army under Colonel Walter Reed in 1899, in part because of the high incidence of the disease among soldiers who were occupying Cuba at that time. Dr. Jesse Lazear, a member of Reed's commission, provided leadership and ultimately gave his life to demonstrate that yellow fever was transmitted by mosquitoes. He was the first experimentally infected person who died from the disease in Reed's studies. The results of the Reed Commission's study proved conclusively that mosquitoes are the disease vector. They also demonstrated that human blood is the source of infection for mosquitoes. For example, in 1901, Reed and James Carroll injected diluted, filtered serum from an experimentally infected yellow fever patient into three nonimmune individuals. Two subsequently developed yellow fever. Reed and Carroll concluded that a filterable virus was the cause of the disease. In the same year, Juan Guiteras, a professor of pathology and tropical medicine at the University of Havana, attempted to produce immunity to yellow fever by exposing volunteers to mosquitoes carrying yellow fever virus. Of 19 volunteers, 8 contracted yellow fever and 3 died. One of the dead was Clara Louise Maass, a U.S. Army nurse from New Jersey. There could be little doubt that mosquitoes transmitted yellow fever virus. Yellow fever had been a constant scourge of Havana for 150 years, but the conclusions of Reed and his colleagues were a revelation. Rapid introduction of mosquito control, by William Gorgas, dramatically reduced the incidence of disease by 1902, and mosquito control remains to this day an important method for control of yellow fever.

Other human viruses were identified during the early decades of this century (Fig. 1.5). However, the pace of discovery was slow, not least because of the dangers and difficulties associated with experimental manipulation of human viruses so amply illustrated by the experience with yellow fever virus. Consequently, agents of some important human diseases were not identified for many years, and then only with some good luck. A classic case in point is the virus responsible for influenza, a name derived from Italian in the mid-1700s to indicate that a disease resulted from the "influence" of miasma (bad air) and other astrological signs. The human disease is thought to have arisen as a result of the transfer of virus among humans and other animals following human domestication of animals about 6,000 years ago. Worldwide epidemics, called **pandemics**, of influenza have been documented in humans for well over 100 years. These pandemics were typically associated with mortality among the very young and the very old, but the 1918–1919 pandemic following the end of World War I was especially devastating. Over 20 million people died during this pandemic, more than were killed in the preceding war. Despite many efforts, a human influenza virus was not isolated until 1933. The virus that causes human influenza was first identified by Wilson Smith, Christopher Andrewes, and Patrick Laidlaw only because they found a host suitable for its propagation. They infected ferrets with human throat washings and isolated the prototype strain of human influenza A virus. The use of ferrets seems to be an unusual choice of animal host, and in fact the success achieved with these animals was serendipitous: Laidlaw was using ferrets in studies on canine distemper virus, and they were available in his laboratory. Subsequently, influenza A virus was shown to infect adult mice and chicken embryos. The latter proved to be an especially valuable host system, for vast quantities of the virus are produced in the allantoic sac. Indeed, chicken eggs are still used today to produce influenza vaccines.

## The Definitive Properties of Viruses

Throughout the early period of virology, in which many viruses of plants, animals, and bacteria were cataloged, the origin and nature of these distinctive infectious agents were controversial. For example, arguments centered on whether viruses represented parts of a cell that had somehow acquired a new kind of existence or were built from virus-specific components. Little progress was made toward resolving these issues and establishing the definitive properties of viruses until new techniques that allowed their visualization or propagation in cultured cells were developed.

## The Structural Simplicity of Viruses

Dramatic confirmation of the structural simplicity of viruses came in 1935 when crystals of tobacco mosaic virus were obtained by Wendell Stanley. At that time, nothing was known of the structural organization of biologically important macromolecules such as proteins and DNA. Indeed the crucial role of DNA as genetic material had not even been recognized. The ability to obtain an infectious agent in crystalline form, a state that is more generally associated with inorganic material, therefore created much wonder and speculation about whether a virus is truly a life form. In retrospect, it is obvious that the relative ease with which tobacco mosaic virus could be crystallized was a direct result of its structural simplicity and the ability of many particles to associate in regular arrays.

The same decade saw introduction of the instrument that rapidly revolutionized virology, the electron microscope. The great magnifying power of this microscope (eventually over 100,000-fold) allowed direct visualization of virus particles for the first time. It has always been an exciting experience for investigators to obtain images of viruses, especially as they often prove to be remarkably elegant in appearance (see the appendix for examples). Images of many different viruses confirmed that these agents are very small (Fig. 1.7) and far simpler in structure than any cellular organism. Many appeared as regular helical or spherical particles. The description of the morphology of virus particles made possible by electron microscopy opened the way for the first rational classification of viruses.

## The Intracellular Parasitism of Viruses

### Organisms as Hosts

The fundamental characteristic of viruses is their absolute dependence on a living host organism for reproduction: they are **obligate parasites**. All early studies on viruses relied on the interactions with host organisms.

Transmission of plant viruses such as tobacco mosaic virus can be readily achieved, for example, by applying extracts of an infected plant to a scratch made on the leaf of a healthy plant. Furthermore, as a single infectious particle of many plant viruses induces the characteristic lesion (Fig. 1.8), the concentration of the infectious agent could be

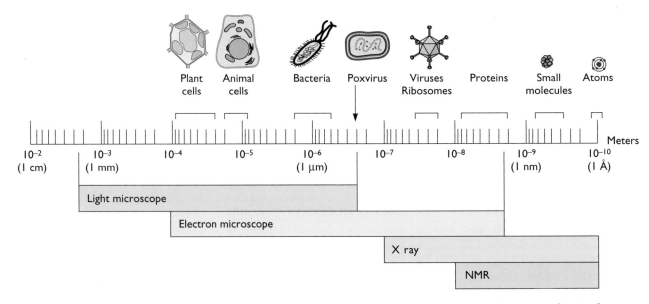

**Figure 1.7 The small size of viruses is illustrated with a logarithmic, metric scale.** Sizes of animal and plant cells, bacteria, viruses, proteins, molecules, and atoms are indicated. The resolving powers of various techniques used in virology, including light microscopy, electron microscopy, X-ray crystallography, and nuclear magnetic resonance (NMR) spectroscopy, are indicated below the logarithmic size scale. Viruses, which are within the resolving power of the electron microscope, are about two orders of magnitude smaller than the smallest bacterium. The units commonly used in descriptions of virus particles or their components are the nanometer (nm [$10^{-9}$ m]) and the angstrom (Å [$10^{-10}$ m]).

**Figure 1.8 Lesions induced by tobacco mosaic virus on *Nicotiana glutinosa* leaves.** In 1886 Adolph Mayer first described the characteristic patterns of light and dark-green areas on the leaves of tobacco plants infected with tobacco mosaic virus. He demonstrated that the mosaic lesions could be transmitted from an infected plant to a healthy plant by aqueous extracts derived from infected plants. The number of local necrotic lesions that result is directly proportional to the number of infectious particles in the preparation. Reprinted with permission from S. E. Luria et al., *General Virology*, 3rd ed. (John Wiley & Sons, New York, N.Y., 1978).

readily measured. Plant viruses were therefore the first to be studied in detail. Viruses of animals, such as the chicken viruses described in the previous section, could also be propagated under experimental conditions, and methods were developed to quantify these agents by determining the lethal dose. Human volunteers were initially used in the study of human viruses, but as this practice had potentially lethal consequences, laboratory animals were used if they could be infected. For example, yellow fever virus was transmitted to mice by Max Theiler in 1930, an achievement that led to the isolation of an attenuated vaccine strain still considered perhaps the safest and most effective vaccine ever produced.

After specific viruses and host organisms were identified, it became possible to produce sufficient quantities of many viruses for investigation of their physical and chemical properties. Scientists also were able to determine the consequences of infection for the host. Parameters such as the incubation period, gross symptoms of infection, and effects on specific tissues and organs were investigated. Animal hosts remain an essential tool in investigations of the pathogenesis of viruses that cause disease in humans and other animals. However, real progress toward understanding the mechanisms of virus replication was made only with the development of tissue and cell culture systems. Among the simplest, but of crucial importance in both virology and molecular biology, were cultures of bacterial cells.

### Lessons from Bacteriophages

In the late 1930s and early 1940s, bacteriophages, or "phages," came to receive increased attention as a result of a controversy centering on how they were formed. d'Hérelle, one of the discoverers of bacteriophages, favored the hypothesis that there was only one phage that attacked all bacteria. This hypothesis was disproved by F. Macfarlane Burnet, who showed that there were a great variety of phages with different physical and biological properties. Nevertheless, the precise nature of these agents remained elusive. John Northrup, a biochemist at the Rockefeller Institute in Princeton, N.J., championed the theory that a phage was a metabolic product of a bacterium. Phage formation was said to be analogous to the autocatalytic formation of enzymes from inactive precursors, a direct rejection of the proposal by d'Hérelle that a phage was a living organism. On the other hand, Max Delbrück, in his work with Emory Ellis and later with Luria, regarded phages as autonomous, stable, self-replicating entities characterized by heritable traits. In this paradigm, phages were seen as ideal tools with which to understand genes and heredity. Probably the most critical early contribution of Delbrück and Ellis was the perfection of the one-step growth technique for synchronization of the replication of phage. This achievement allowed analysis of a single cycle of phage growth in a population of bacteria. The new technique introduced highly quantitative studies to biology as well as a rigor of analysis that was unprecedented in virology. The first experiments showed that phages indeed multiplied in the bacterial host and were liberated in a "burst" by lysis of the cell.

Delbrück was a zealot for phage research and recruited talented scientists (the "phage school") to pursue the fundamental issues of what is now known as the field of molecular biology. The list of Nobel laureates who were trained as phage workers is a testament to Delbrück's insight. Progress was rapid, primarily because of the simplicity of the phage system. The phage life cycle is measured in minutes, whereas the infectious cycles of animal and plant viruses occupy hours or days. Phages replicate in bacterial hosts that can be readily obtained by overnight culture, as opposed to eukaryotic cells, which take much longer to grow to concentrations sufficient for infection. By the mid-1950s, it was evident that viruses from bacteria, animals, and plants share many fundamental properties. However, bacteriophages provided a far more tractable experimental system for analysis. Consequently, their study had a profound impact on the development of virology.

One critical lesson came from the definitive experiments of Alfred Hershey and Martha Chase establishing that viral nucleic acid carries the genetic information. It was known from studies on the "transforming principle" of *Pneumoccocus* by Oswald Avery, Colin MacLeod, and Maclyn McCarty (1944) that nucleic acid was both necessary and sufficient for the transfer of genetic traits. However, in the early 1950s, viral protein was still suspected to be an important component of viral heredity. In a brilliantly simple experiment that included the use of a common kitchen item, a Waring blender, Hershey and Chase showed that the viral protein formed merely the vehicle of delivery. By differentially labeling the nucleic acid and protein components of virus particles with radioactive phosphorus ($^{32}$P) and radioactive sulfur ($^{35}$S), respectively, they showed that the protein coat of the infecting virus could be removed soon after infection by agitating the bacteria for a few minutes in the blender. Conversely, $^{32}$P-labeled bacteriophage DNA entered and remained associated with the bacterial cells; such cells nonetheless produced a normal burst of new virus particles. This elegant experiment established two additional enduring principles: that the first step in viral growth is the liberation of nucleic acid from a proteinaceous coat (a process now known as uncoating) and that nucleic acid alone can be infectious.

Bacteriophages were originally thought to be lethal agents, killing their hosts after infection. In the early 1920s, a previously unknown interaction in which the host not only survived the infection but also stably inherited the genetic information of the virus was described. It was also observed that certain bacterial strains not known to be infected could lyse spontaneously and produce bacteriophages after a period of growth in culture. Such strains were called **lysogenic**, and the phenomenon was called **lysogeny**. Studies on lysogeny identified many previously unrecognized features of virus-host cell interactions (Box 1.1).

Recognition of these principles resulted from the work of many scientists, but it began with the elegant experiments of André Lwoff and colleagues at the Institut Pasteur in Paris. Lwoff demonstrated the existence of the viral genome in lysogenic cells as a specific genetic element called the **prophage**. This element determined the ability of lysogenic bacteria to produce infectious bacteriophage. Subsequent studies of *Escherichia coli* bacteriophage lambda established a paradigm for one of the many mechanisms of lysogeny, the integration of a circular phage genome into a specific site on the bacterial chromosome. Other mechanisms were soon found (Box 1.1).

Bacteriophages became inextricably associated with the new field of molecular biology (Table 1.1). Their study also established many fundamental principles. For example,

**BOX I.I**

*Studies of lysogeny established several general principles in virology*

Some viruses can enter into either destructive (**lytic**) or relatively benign (**lysogenic**) relationships with their host cells. Such bacteriophages were called "temperate."

In a lysogenic bacterial cell, viral genetic information persists, but viral gene expression is restricted. Such cells are called **lysogens**, and the quiescent viral genome is called a **prophage.**

For some bacteriophages like lambda and Mu, prophage DNA is integrated into the host genome of lysogens and passively replicated by the host. Virally encoded enzymes, known as **integrase** (lambda) and **transposase** (Mu), mediate the covalent insertion of viral DNA into the chromosome of the host bacterium, establishing it as a prophage.

Bacteriophage Mu (Mu stands for mutator) inserts its genome into many random locations on the host chromosome, causing numerous mutations. This process is called **insertional mutagenesis.**

The mechanism of genome **transposition** by Mu establishes an integrated prophage when repressed but also replicates the viral genome during the lytic cycle.

The prophage DNA of other bacteriophages, such as P1, is a **plasmid**, a self-replicating, autonomous chromosome in the cytoplasm of a lysogen.

Viral gene expression in lysogens is turned off by the action of viral proteins called **repressors**. Viral gene expression can be turned on when repressors are inactivated (a process called **induction**).

Viruses can pick up cellular genes and deliver them to new cells (a process known as **transduction**).

The process of picking up any gene segment from the host chromosome by a virus is called **generalized transduction.**

**Specialized transduction** is the acquisition of a specific segment of host DNA by a virus. This process is limited to phage that integrate into the host DNA. For example, occasional mistakes in excision of the lambda prophage after induction result in production of unusual progeny phage that have lost some of their own DNA but have acquired the bacterial DNA adjacent to the prophage.

**Table 1.1**  Bacteriophages: landmarks in molecular biology[a]

| Year | Discovery |
|------|-----------|
| 1939 | One-step growth of viruses (Ellis and Delbrück) |
| 1946 | Mixed phage infection leads to genetic recombination (Delbrück) |
| 1947 | Mutation and DNA repair (multiplicity reactivation) (Luria) |
| 1952 | Transduction of genetic information (Zinder and Lederberg) |
| 1952 | DNA, not protein, found to be the genetic material (Hershey and Chase) |
| 1952 | Restriction and modification of DNA (Luria) |
| 1955 | Definition of a gene (*cis-trans* test) (Benzer) |
| 1958 | Mechanisms of control of gene expression by repressors and activators are established (Pardee, Jacob, and Monod) |
| 1958 | Definition of the episome (Jacob and Wollman) |
| 1961 | Discovery of mRNA (Brenner, Jacob, and Meselson) |
| 1961 | Elucidation of the triplet code by genetic analysis (Crick, Barnett, Brenner, and Watts-Tobin) |
| 1961 | Genetic definition of nonsense codons as stop signals for translation (Campbell, Epstein, and Bernstein) |
| 1964 | Colinearity of the gene with the polypeptide chain (Sarabhai, Stretton, and Brenner) |
| 1966 | Pathways of macromolecular assembly (Edgar and Wood) |
| 1974 | Vectors for recombinant DNA technology (Murray and Murray, Thomas, Cameron, and Davis) |

[a]Sources: T. D. Brock, *The Emergence of Bacterial Genetics* (Cold Spring Harbor Laboratory Press, Cold Spring Harbor, N.Y., 1990); K. Denniston and L. Enquist, *Recombinant DNA* (*Benchmark Papers in Microbiology*, vol. 15) (Dowden, Hutchinson and Ross, Inc., Stroudsburg, Pa., 1981); and C. K. Mathews, E. Kutter, G. Mosig, and P. Berget, *Bacteriophage T4* (American Society for Microbiology, Washington, D.C., 1983).

control of the decision to enter a lysogenic or a lytic pathway was clearly encoded in the genome of the virus. The first mechanisms for the genetic control of gene expression, exemplified by the elegant operon theory of Nobel laureates François Jacob and Jacques Monod, were deduced in part from studies of bacteriophage lambda lysogeny. The biology of bacteriophage lambda provided a fertile ground for work on gene regulation, but study of nontemperate, virulent T phages (T1 to T7, where T stands for type) of *E. coli* paved the way for many other important advances (Table 1.1). As we shall see, these systems also provided an extensive preview of phenomena to be discovered in studies of animal viruses (Box 1.2).

## Animal Cells as Hosts

The culture of animal cells in the laboratory was initially more of an art than a science, restricted to cells that grew out of organs or tissues maintained in nutrient solutions under sterile conditions. The finite life span of such **primary** cells, their dependence for growth on natural media such as lymph, plasma, or chicken embryo extracts, and the technical demands of sterile culture prior to the discovery of antibiotics, made reproducible experimentation very difficult. By 1955 the work of many investigators had led to a series of important methodological advances. These included the development of defined media optimal for growth of mammalian cells, incorporation of antibiotics into tissue culture media, and development of im-

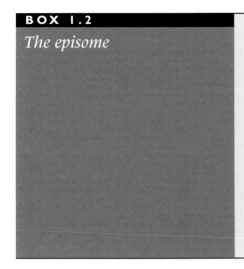

In 1958, François Jacob and Elie Wollman realized that lambda prophage, the *E. coli* F sex factor, and the colicinogenic factor had many common genetic properties. Their remarkable insight led to the concept of the episome.

An **episome** is an exogenous genetic element not necessary for cell survival. Its defining characteristic is the ability to reproduce in two alternative states by integrating into the host chromosome or by autonomous replication.

Nowadays this term is often applied to genomes that can be maintained in cells by autonomous replication and never integrate, for example, certain viral DNA genomes.

Jacob and Wollman immediately understood that the episome had value in understanding larger problems, including cancer, as revealed by this quotation: "Thus in the no man's land between heredity and infection, between physiology and pathology at the cellular level, episomes provide a new link and a new way of thinking about cellular genetics in bacteria, and perhaps in mice, men and elephants" (F. Jacob and E. Wollman, *Viruses and Genes: Readings from Scientific American* [W. H. Freeman & Co., New York, N.Y., 1961]).

mortal cell lines such as the mouse L and human HeLa cells which are still in widespread use. These advances allowed growth of animal cells in culture to become a routine, reproducible exercise.

The availability of well-characterized cell cultures had several important consequences for virology. It allowed the discovery of several new human viruses, such as adenovirus, measles virus, and rubella virus, for which animal hosts were not available. In 1949 John Enders and colleagues used cell cultures to propagate poliovirus, a feat that led to development of the Sabin vaccine a few years later. Tissue culture technology revolutionized the ability to investigate the replication of viruses in host cells. Viral infectious cycles could be studied under precisely controlled conditions by employing the analog of the one-step growth cycle of bacteriophages and simple methods for quantification of infectious particles. Our current understanding of the molecular basis of parasitism by viruses, the focus of the initial chapters of this book, is based almost entirely on analyses of one-step growth cycles with cells in culture. Such studies established that viruses are **molecular parasites**: they reproduce by exploiting their host cell's biosynthetic machinery for synthesis of the components from which they are built. Thus, in contrast to cells, viruses do not reproduce by growth and division. Rather, the infecting genome contains viral information necessary to redirect cellular systems to the production of all the components needed to assemble new virus particles.

## Viruses Defined

Advances in knowledge of the structure of virus particles and the mechanisms by which viruses reproduce in their host cells have been accompanied by increasingly accurate definitions of these unique agents. The earliest pathogenic agents, distinguished by their small size and dependence on a host organism for reproduction, emphasized the importance of viruses as agents of disease. We can now provide a much more precise definition of viruses, elaborating their relationship with the host cell and the important features of virus particles. The definitive properties of viruses are summarized as follows:

- A virus is a very small, infectious, obligate intracellular (molecular) parasite.
- The virus genome comprises either DNA or RNA.
- Within an appropriate host cell, the viral genome is replicated and directs the synthesis, by cellular systems, of other virion components.
- Progeny virions are formed by de novo assembly from newly synthesized components within the host cell.
- A progeny virion assembled during the infectious cycle is the vehicle for transmission of the viral genome to the next host cell or organism, where its disassembly leads to the beginning of the next infectious cycle.

With these properties in mind, we can accurately place viruses within the evolutionary continuum of biological agents. They are far simpler than even the smallest microorganisms and lack the complex energy-generating and biosynthetic systems necessary for independent existence. On the other hand, viruses are not the simplest biologically active agents: even the smallest virus, built from a very limited genome and a single type of protein, is significantly more complex than other parasitic molecular pathogens. Some of these minimalist molecular pathogens, **viroids**, which are infectious agents of a variety of economically important plants, comprise a single small molecule of RNA. Others, termed **prions**, are believed to be single molecules of protein.

## Cataloging of Diverse Animal Viruses

### Many Sizes and Shapes Produced by Evolution

Around 1960 virus classification was a subject of colorful and quite heated controversy. New viruses were being discovered and studied by electron microscopy. The virus world consisted of a veritable zoo of particles with different sizes, shapes, and compositions (Fig. 1.9). Very strong opinions were advanced concerning classification and nomenclature, and opposing camps developed, as in any controversy involving individuals who tend to focus on differences and those who look for similarities (conventionally known as **splitters** and **lumpers**). Splitters pointed to the inability to infer, from the known properties of viruses, anything about their evolutionary origin or their relationships to one another—the major goal of clas-

sical taxonomy. Lumpers maintained that, despite such limitations, there were significant practical advantages in grouping isolates with similar properties. Furthermore, it seemed likely that a good classification might actually stimulate fruitful investigation. A major sticking point, however, was finding agreement on the properties that should be considered most important in constructing a scheme for virus classification.

### *The Classical System of Classification*

In 1962 Lwoff, R. W. Horne, and P. Tournier advanced a comprehensive scheme for the classification of all viruses (bacterial, plant, and animal) under the classical Linnaean hierarchical system consisting of phylum, class, order, family, genus, and species. Although a subsequently formed international committee on the nomenclature of viruses did not adopt this system in toto, many principles embodied in the system were accepted, and its designation of families, genera, and species was used by this committee in the classification of animal viruses.

One of the most important principles embodied in the system advanced by Lwoff and his colleagues was that viruses should be grouped according to *their* shared prop-

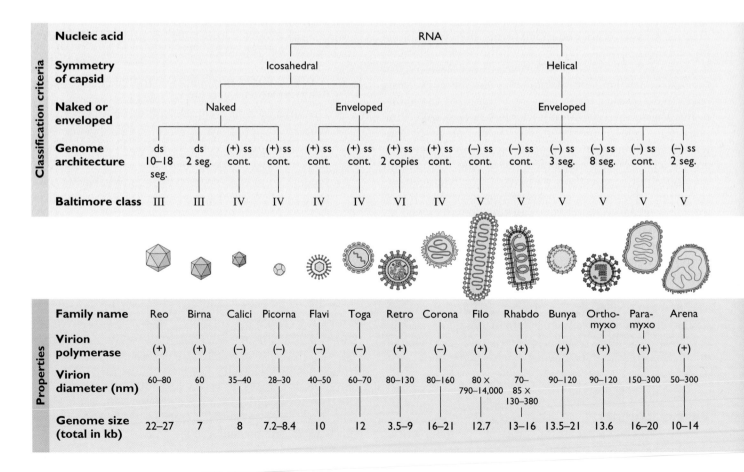

erties rather than the properties of the cells or organisms they infect. A second was a focus on the nucleic acid genome as the primary criterion for classification. The importance of the genome had become clear when it was inferred from the Hershey-Chase experiment that viral nucleic acid alone can be infectious. The system delineated four characteristics to be used in the classification of all viruses:

1. Nature of the nucleic acid in the virion (DNA or RNA)
2. Symmetry of the capsid
3. Presence or absence of an envelope
4. Dimensions of the virion and capsid.

These and a few additional properties are summarized in Fig. 1.9, which lists 21 families of viruses that infect humans and other mammals.

The order in which families are aligned in the figure is arbitrary, as the evolutionary relationships among families are still largely unknown. However, it is clear that the morphological distinctions highlighted in the scheme belie some significant similarities, for example, in the organization of the picornavirus and flavivirus genomes and of the rhabdovirus and paramyxovirus genomes. **Genomics**, the elucidation of evolutionary relationships by analyses of nucleic acid and protein sequence similarities, increasingly is being used to order members within a virus family, to classify newly isolated viral nucleic acids, and to predict the functions of genes. For example, human herpesvirus 8 was placed in the subfamily *Gammaherpesvirinae* on the basis of sequence analysis of a small segment of its DNA, and hepatitis C virus was classified as a member of the family *Flaviviridae* from the sequence of a cloned DNA copy of its genome. Indeed, as our knowledge of molecular properties of viruses and their replication has increased, it has become apparent that *any* two-dimensional comparison (e.g., virus versus size, or virus versus capsid morphology) is somewhat misleading. For example, *Hepadnaviridae*, *Retroviridae*, and some plant viruses are classified as different families on the basis of the nature of their genomes, but all use reverse transcription in their reproductive cycles. Moreover, the viral polymerases that perform this task exhibit similarities in amino acid sequence.

As of the latest report (1995) of the International Committee on Taxonomy of Viruses (ICTV) approximately

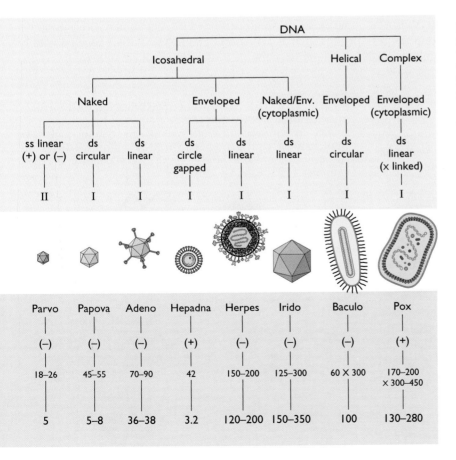

**Figure 1.9 Classification schemes for animal viruses.** Summary of the major characteristics of 21 representative families of viruses that infect vertebrates and 1 that infects insects. Not all virus families are shown in the figure; the one insect virus family (*Baculoviridae*) is included because it has become an important tool in biotechnology. Adapted from G. P. Martelli et al. (ed.), *Virus Taxonomy Classification and Nomenclature of Viruses: Sixth Report of the International Committee on Taxonomy of Viruses* (Springer-Verlag, Vienna, Austria, 1995).

4,000 viruses that infect bacteria, plants, and animals have been assigned to one of 71 recognized families, 9 subfamilies, and 164 genera—with yet others assigned provisionally. As few new virus families have been identified in recent years, it seems likely that a significant fraction of all existing virus families are now known. However, as we learn more and more about the genes, proteins, and reproduction strategies of these families, other relationships will certainly be revealed. Thus, classification refinements can be expected to continue in the future. The ICTV report also includes descriptions of subviral agents (satellites, viroids, and prions) and a list of viruses for which information is still insufficient to make assignments. **Satellites** are composed of nucleic acid molecules that depend for their productive multiplication on coinfection of a host cell with a helper virus. However, they are not related to their helper virus. When a satellite encodes the coat protein in which its nucleic acid is encapsidated, it is referred to as a **satellite virus** (e.g., hepatitis delta virus is a satellite virus).

Several years ago, the bacterial virologists who were members of the ICTV agreed to coin similar Latinized family names for the different types of bacteriophages. However, this nomenclature never really took hold, and it has not been widely used by those who do research with bacteriophages. Plant virologists do not classify their viruses into families and genera. Instead, they use group names derived from the prototype virus of each group. For animal viruses, however, the ICTV nomenclature has proved quite useful and, as it has been applied widely in both the scientific and medical literature, we adopt it in this text. In this nomenclature, the Latinized virus family names are recognized as starting with capital letters and ending with *-viridae*, as, for example, in the family name *Parvoviridae*. These names are used interchangeably with their common derivatives as, for example, **parvoviruses**.

### The Baltimore Classification System

The past 25 years or so have brought an enormous increase in knowledge of the molecular biology of viruses and cells. Among the most significant advances has been the elucidation of pathways by which information encoded in genomes is expressed. We know that the genes of cells are encoded in their DNA but that cells express this information via messenger RNAs (mRNAs) which are translated in the cytoplasm by ribosomes and associated machinery. This is the so-called central dogma conceptualized by Francis Crick.

$$DNA \rightarrow mRNA \rightarrow proteins$$

All viruses must direct the synthesis of mRNA to produce proteins, as they are completely dependent on the cell's translational machinery. It was appreciation of the central role of the translational machinery and of the importance of mRNA molecules in the programming of viral protein synthesis that inspired an alternative classification scheme. Proposed by David Baltimore (Fig. 1.10), this classification is based on the genetic system of each virus and describes the obligatory relationship between the viral genome and its mRNA.

By the molecular biologist's convention, mRNA is defined as a **positive [(+)] strand** because it contains immediately translatable information. In the Baltimore classification, a strand of DNA that is of equivalent polarity is also designated a (+) strand. The RNA and DNA complements of (+) strands are designated **negative [(−)] strands**. The Baltimore classification is superimposed on the ICTV nomenclature in Fig. 1.9. It is apparent that these two schemes are not mutually exclusive but rather complement one another. In the case of animal viruses, the Baltimore designations can be seen as simply providing additional defining characteristics. The Baltimore class number designations are rarely used in the animal virus literature. However, the principles embodied in the classification have proved to be most valuable, especially for viruses with single-stranded RNA genomes. Thus, designations such as (+) strand and (−) strand are applied widely to describe genomes, as will be illustrated in this text.

**Figure 1.10 The Baltimore classification.** All viruses must produce mRNA that can be translated by cellular ribosomes. In this classification system, the unique pathways from various viral genomes to mRNA define specific virus classes on the basis of the nature and polarity of their genomes.

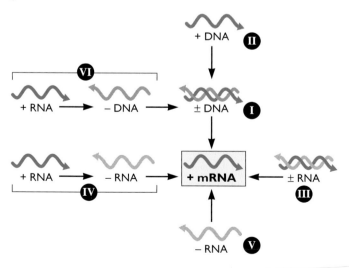

Knowledge of these designations provides virologists with immediate insight into the steps that must take place during replication and expression of the genome.

### The Molecular Biologist's Focus

Because the viral genome carries the entire blueprint for virus propagation, molecular virologists have long considered it the most important characteristic for classification purposes. Thus, this feature is placed at the top of our hierarchy of classification (Fig. 1.9).

## Genetic Content

### DNA Viruses

As shown in Fig. 1.9, the members of all but one family of DNA viruses contain double-stranded genomes (Baltimore's class I). The family *Parvoviridae* (class II) is the lone exception. Most of the DNA genomes are not simple linear molecules. For example, papovavirus genomes are covalently closed, double-stranded circles, and hepadnavirus genomes are duplex circles in which one strand contains a nick and the other contains a gap. The ends of the double-stranded poxvirus genome are covalently joined, whereas herpesvirus genomes contain terminal and internal duplications.

### RNA Viruses

In contrast to DNA viruses, most RNA viruses have single-stranded genomes. Those RNA viruses with continuous (+) strand genomes (Baltimore's class IV) include the morphologically diverse picornaviruses and caliciviruses. Retroviruses (the sole members of Baltimore's class VI) also contain (+) strand RNA genomes, but two copies of genomic RNA are included in each virion. This is a unique feature of the family *Retroviridae*, as all other viruses contain only one copy of their genome.

RNA viruses with (−) strand genomes (Baltimore's class V) are all enveloped by helical capsids. Three members of this class, rhabdoviruses, paramyxoviruses, and filoviruses, are sufficiently related to be organized into the order Mononegavirales. As suggested by the name, the genomes of all viruses in this order consist of a single (−) strand RNA molecule. The other (−) strand RNA viruses have segmented genomes: orthomyxovirus genomes comprise seven or eight separate RNA molecules, and arenavirus and bunyavirus genomes comprise two and three RNA molecules, respectively. The genomes of arenaviruses and some bunyaviruses are **ambisense**, i.e., not only are their complements mRNAs, but the 5' ends of some of the constituent genomic RNAs also function as mRNAs.

Two families comprise the class of viruses with double-stranded RNA genomes (Baltimore's class III). Reovirus genomes consist of 10 to 12 double-stranded linear RNA molecules that encode different genes; birnavirus genomes comprise two double-stranded RNA molecules.

## Capsid Symmetry

Viral genomes are encased in protective protein structures called **capsids** or **nucleocapsids**. As first pointed out by Francis Crick and James Watson in 1956, most viruses appeared to be rod shaped or spherical under the electron microscope. Because of the limited coding capacity of viral genomes, they suggested that it was most economical to construct capsids from a small number of protein subunits by using helical symmetry (rod-shaped viruses) or the symmetry of Platonic polyhedra (e.g., tetrahedron, octahedron, or icosahedron) (spherical viruses). Indeed, capsids are most often formed by the assembly of identical structural units in ways that provide maximal contact between them. The requirement for proper fit excludes inappropriately folded protein subunits from being incorporated, thus ensuring the integrity of the capsid structure.

### Helical Symmetry

Helical capsids are described by the number of protein subunits per turn of the helix and the rise per subunit. Such structures are formed by stacking repeated components with a constant relationship of amplitude and pitch to one another. The nucleic acid is found inside the helical capsid. One can visualize a helix by forming a two-dimensional lattice and rolling it into a cylindrical structure of a diameter that accommodates the viral nucleic acid (Fig. 1.11). Helical capsids may be rigid or flexible, depending on the arrangement of the protein subunits with respect to one another. As summarized in Fig. 1.9, all animal viruses with filamentous nucleocapsids in a helical conformation also possess envelopes.

### Icosahedral Symmetry

The most economical way to build a symmetric shell of maximal internal volume with nonsymmetric protein molecules is to arrange the proteins in identical equilateral triangular structures that can be joined to form an icosahedron (Fig. 1.11B). In the simplest case, each triangular face comprises 3 identically placed proteins, there are 20 faces (by definition), and the total number of protein subunits in the capsid is 60. The volume of such a capsid depends on the size of the protein but is quite small and consequently can contain only small genomes. Larger genomes require larger capsids, which therefore must include more than 60 small proteins. However, as will be described in chapter 3, they retain the defining features of icosahedral symmetry: characteristic two-, three-, and fivefold rotational axes (illustrated in Fig. 1.11).

### Complex Particles

Some viruses fit into neither the helical nor the icosahedral category. They form complex capsids with structures that are still not well understood. Members of the family *Poxviridae* and some members of the *Retroviridae* represent prime examples among animal viruses.

## Enveloped or Naked Particles

Many animal viruses are surrounded by a lipid bilayer which they acquire when the nucleocapsid buds through cell membranes. The envelopes of all such viruses contain virus-encoded proteins that are required for infectivity. Viruses that are not enveloped are referred to as **naked viruses.**

## Virion Enzymes

Many viruses contain virus-encoded enzymes within mature, infectious virions (Fig. 1.9). The most important of these are polymerases, which some viruses use to copy all or part of their genomes, often immediately on entry into a newly infected cell. These enzymes are essential for viruses with single (−) strand RNA genomes because host cells lack the RNA-dependent RNA polymerase necessary to produce viral mRNAs. However, the first viral polymerase was discovered in a DNA virus, the poxvirus vaccinia virus. This is a special case, though, because unlike most other DNA viruses, poxviruses replicate in the cytoplasm and lack access to the cellular transcription machinery for production of their mRNA.

## Unifying Principles

### A Common Strategy

The basic thesis of this textbook is that *all* viruses follow a simple **three-part general strategy** to ensure their survival:

- All viruses package their genomes inside a *particle* that mediates transmission of the viral genome from host to host.
- The viral genome contains the information for initiating and completing an *infectious cycle* within a susceptible, permissive cell. An infectious cycle includes attachment and entry of the particle, decoding of genome information, translation of viral mRNA by host ribosomes, genome replication, and assembly and release of particles containing the genome.
- All viruses are able to *establish themselves in a host population* so that virus survival is ensured.

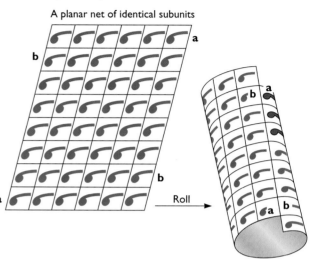

**A Helical symmetry**

A planar net of identical subunits

Roll →

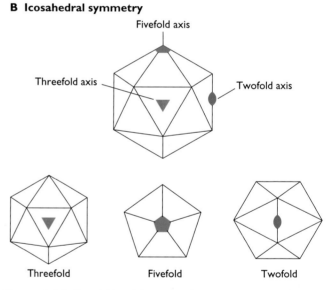

**B Icosahedral symmetry**

Fivefold axis

Threefold axis

Twofold axis

Threefold        Fivefold        Twofold

**Figure 1.11 Capsid architecture is based on principles of helical and icosahedral symmetry.** (A) A planar net of identical subunits and the same planar net rolled to form a helix. Many virions are rod-shaped (e.g., tobacco mosaic virus) with the viral RNA inside bound to the capsid protein subunits arranged on the outside with helical symmetry. (B) An icosahedron comprises 20 equilateral triangles with characteristic positions of rotational symmetry. The three views at the bottom highlight these positions.

Modern virology is both exciting and challenging because of the many and varied tactics used by individual virus families to put this simple strategy into effect. As viruses are obligate molecular parasites, every tactical solution must of necessity tell us something about the host as well as the virus. The intellectual satisfaction of discovering and understanding new principles is as rewarding as the practical consequences of providing solutions to problems of disease.

## Highlights of Virus Strategies Outlined in This Book

### Making Viral Proteins

All viral RNAs must be decoded by the host's protein synthesizing machinery; all viruses are parasites of *translation*. Unless the viral genome can function directly as an mRNA, this requirement means that strategies must exist for synthesizing mRNAs soon after the viral genome enters the cell. Even when the genome is a functional mRNA, successful replication necessarily includes production of many new copies of the genome. Analysis of the various means of producing mRNA have taught us much about the cell itself.

### Making Viral Genomes

Many viral genomes are copied by the cell's synthetic machinery in cooperation with viral proteins. The cell provides nucleotide substrates, energy, and other components. In all cases, the internal compartmentalization of the cell must be reckoned with because essential components are found only in the nucleus, are restricted to the cytoplasm, or are present in cellular membranes. Study of the intramolecular trafficking strategies required and the mechanisms of viral genome replication has illuminated fundamental principles of cell biology and nucleic acid replication and control.

### Forming Progeny Virions

The various components of a virion, the nucleic acid genome, capsid protein(s), and in some cases envelope proteins, are often synthesized in different cellular compartments. Their trafficking through and among the cell's compartments and organelles requires that they be equipped with the proper homing signals. Virion components must be assembled at some central location, and the information for assembly must be preprogrammed in the component molecules.

### Overcoming Host Defenses

Organisms have evolved many physical barriers to protect themselves from dangers in their environment, including invading parasites. In addition, vertebrates possess an effective immune system to defend themselves against anything recognized as "nonself" or dangerous. Studies on the interactions among viruses and the immune system are particularly instructive because of the many counter-measures used by viruses to defeat this system. Understanding these measures teaches us much about the basis of immunity. Furthermore, many of the symptoms and pathologies associated with viral disease are manifestations of the immune system's attempts to destroy the viral invaders.

## References

### Books

Brock, T. D. 1990. *The Emergence of Bacterial Genetics.* Cold Spring Harbor Laboratory Press, Cold Spring Harbor, N.Y.

Brothwell, D., and A. T. Sandison (ed.). 1967. *Diseases in Antiquity.* Charles C Thomas, Publisher, Springfield, Ill.

Cairns, J., G. S. Stent, and J. D. Watson (ed.). 1966. *Phage and the Origins of Molecular Biology.* Cold Spring Harbor Laboratory for Quantitative Biology, Cold Spring Harbor, N.Y.

Denniston, K., and L. Enquist. 1981. *Recombinant DNA. Benchmark Papers in Microbiology*, vol. 15. Dowden, Hutchinson and Ross, Inc., Stroudsburg, Pa.

Fields, B. N., D. M. Knipe, and P. M. Howley (ed.). 1996. *Fields Virology.* Raven Press, New York, N.Y.

Hughes, S. S. 1977. *The Virus: A History of the Concept.* Heinemann Educational Books, London, United Kingdom.

Karlen, A. 1996. *Plague's Progress, a Social History of Man and Disease.* Indigo, Guernsey Press Ltd., Guernsey, Channel Islands.

Levine, A. J. 1992. *Viruses.* Scientific American Library. W. H. Freeman & Co., New York, N.Y.

Luria, S. E., and J. E. Darnell, Jr. 1953. *General Virology*, 2nd ed. John Wiley & Sons, Inc., New York, N.Y.

Fiennes, R. 1978. *Zoonoses and the Origins and Ecology of Human Disease.* Academic Press, Inc., New York, N.Y.

Martelli, G. P., M. A. Mayo, and M. D. Summers (ed.). 1995. *Virus Taxonomy Classification and Nomenclature of Viruses: Sixth Report of the International Committee on Taxonomy of Viruses.* Springer-Verlag, Vienna, Austria.

Oldstone, M. B. A. 1998. *Viruses, Plagues and History.* Oxford University Press, New York, N.Y.

Stent, G. S. 1960. *Papers on Bacterial Viruses.* Little, Brown & Co., Boston, Mass.

Stent, G. S. 1965. *Papers on Bacterial Viruses*, 2nd ed. Little, Brown & Co., Boston, Mass.

Murphy, F. A., C. M. Fauquet, D. H. L. Bishop, S. A. Ghabrial, A. W. Jarvis, and N. Rasmussen. 1997. *Picture Control: the Electron Microscope and the Transformation of Biology in America 1940–1960.* Stanford University Press, Stanford, Calif.

Waterson, A. P., and L. Wilkinson. 1978. *An Introduction to the History of Virology.* Cambridge University Press, London, United Kingdom.

### Papers of Special Interest

Baltimore, D. 1971. Expression of animal virus genomes. *Bacteriol. Rev.* **35:**235–241.

Brown, F., J. Atherton, and D. Knudsen. 1989. The classification and nomenclature of viruses: summary of results of meetings of the International Committee on Taxonomy of Viruses. *Intervirology* **30**:181–186.

Burnet, F. M. 1953. Virology as an independent science. *Med. J. Aust.* **40**:842.

Crick, F. H. C., and J. D. Watson. 1956. Structure of small viruses. *Nature* 177:473–475.

Lustig, A., and A. J. Levine. 1992. One hundred years of virology. *J. Virol.* **66**:4629.

Van Helvoort, T. 1993. *Research Styles in Virus Studies in the Twentieth Century: Controversies and the Formation of Consensus.* Doctoral dissertation.

*Web Sites of Interest to Virologists*

http://www.tulane.edu/~dmsander/garryfavweb.html  *Has links to almost all the virology sites on the Web.*

http://boris.qub.ac.uk/vsd/vsd6d.html

http://hivgenome.hjf.org/

http://inet.uni-c.dk/~iaotb/bse.htm

http://life.anu.edu.au:80/viruses/indxvir2.htm *A list of ICTV-approved virus names and links to virus databases.*

http://life.anu.edu.au:80/./viruses/virus.html

http://www.bocklabs.wisc.edu/welcome.html

http://www.hs.state.az.us/lab/virus.html

http://www.mni.uwo.ca/Bio221a/intro97.html

http://www.outbreak.org

http://www.res.bbsrc.ac.uk/plantpath/

http://www.uct.ac.za/depts/mmi/stannard/linda.html

# 2

# Virus Cultivation, Detection, and Genetics

## Introduction

The knowledge of virus replication and pathogenesis discussed in subsequent chapters of this book had its origin in many research and diagnostic virology laboratories. A variety of experimental techniques have been employed to acquire the results used to test hypotheses and to formulate new models of virus replication. The purpose of this chapter is to explain how some of these methods work. Understanding how experiments are done not only is necessary for understanding why certain conclusions are made but also gives the reader the tools to formulate new approaches to unsolved problems in virology. The technology for manipulating viral genomes, which came with the development of recombinant DNA methods, will also be explored. These techniques facilitate the study of viral gene function and have made viruses invaluable tools as vectors for the delivery of genes into cells and organisms. This chapter should be read in sequence with the remainder of the book and should also be consulted as a reference source for other chapters.

## Cultivation of Viruses

### Cell Culture

#### Types of cell culture

Although human and other animal cells were first cultured in the early 1900s, contamination with bacteria, mycoplasmas, fungi, and yeast initially made routine work on such systems extremely difficult. For this reason, most viruses were grown in laboratory animals. In 1949 John Enders, Thomas Weller, and Frederick Robbins made the discovery that poliovirus could multiply in cultured cells not of neuronal origin. As noted in chapter 1, this revolutionary finding, for which these three investigators were awarded the Nobel Prize in physiology or medicine in 1954, led the way to the propagation of many other viruses in cultured cells, the discovery of new viruses, and the development of viral vaccines such as those against poliomyelitis, measles, and rubella. The ability to infect cultured cells synchronously permitted studies of the biochemistry and molecular biology of viral replication. The growth and purification of viruses on a large scale allowed studies on the composition of virus particles, leading to the solution of high-resolution three-dimensional structures of some viruses, as discussed in chapter 3.

**Cell culture** is still the most common method for the propagation of viruses. To prepare a cell culture, tissues are dissociated into a single-cell suspension by mechanical disruption followed by treatment with proteolytic enzymes. The cells are then suspended in culture medium and placed in plastic flasks or plates. As the cells divide, they cover the plastic surface. Epithelioid and fibroblastic cells attach to the plastic and form a **monolayer,** whereas blood cells such as lymphocytes settle but do not adhere. The cells are grown in a chemically defined medium consisting of an isotonic solution of salts, glucose, vitamins, coenzymes, and amino acids buffered to a pH between 7.2 and 7.4 and often supplemented with antibiotics to inhibit microbial growth. Serum is usually added to the cell culture medium to provide a source of growth factors required by the cells. Commonly used cell lines double in 24 to 48 h in such media.

There are three main kinds of cell cultures (Fig. 2.1). **Primary cell cultures** are prepared from animal tissues as described above. They consist of several cell types and have a limited life span, usually no more than 5 or 10 cell divisions. The most commonly used primary cell cultures are derived from monkey kidney, human embryonic amnion, kidney, or foreskin, and chicken or mouse embryos. Such cells are used for experimental virology when the state of cell differentiation is important or when appropriate cell lines are not available. They are also used in vaccine production—for example, live attenuated poliovirus vaccine strains are propagated in primary monkey kidney cells. Primary cell cultures were mandated for the growth of human vaccines to avoid contamination of the product with potentially oncogenic DNA from continuous cell lines (see below). Some viral vaccines are now prepared in **diploid cell strains,** which consist of a homogeneous population of a single type and can divide up to 100 times before dying. Despite the numerous divisions, these cell strains retain the diploid chromosome number. The most widely used diploid cells are those established from human embryos, such as the WI-38 strain derived from human embryonic lung. **Continuous cell lines** consist of a single cell type that can be propagated indefinitely in culture. These immortal lines are usually derived from tumor tissue or by treating a primary cell culture or a diploid strain

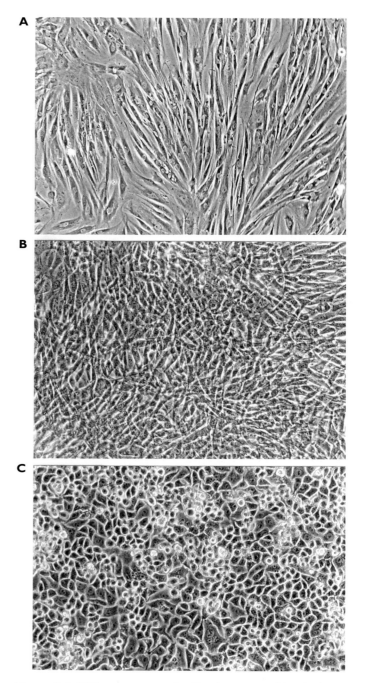

**Figure 2.1 Different types of cell cultures used in virology.** Confluent cell monolayers photographed by low-power light microscopy. (A) Primary human foreskin fibroblasts; (B) established line of mouse fibroblasts (3T3); (C) continuous line of human epithelial cells (HeLa). Note the differences in morphology among the primary (human foreskin fibroblasts), established (i.e., immortalized [3T3]), and transformed (HeLa) cells, as well as the ability of transformed HeLa cells to overgrow one another. This property is the result of a loss of contact inhibition. Courtesy of R. Gonzalez, Princeton University.

with a mutagenic chemical or a tumor virus. Such cell lines often do not resemble the cell of origin; they are less differentiated (having lost the morphology and biochemical features that they possessed in vivo), often abnormal in chromosome morphology and number (**aneuploid**), and can be tumorigenic (e.g., they produce tumors when inoculated into nude mice). Examples of commonly used continuous cell lines include HEp-2 and HeLa cells, derived from human carcinomas; Vero cells from African green monkey kidney; L and 3T3 cells derived from mice, and BHK-21 cells derived from hamster kidneys. Continuous cell lines have the advantage that they can be propagated indefinitely and, like diploid cell strains, retain viability after being frozen at low temperatures ($-70$ to $-196°C$). They provide a uniform population of cells that can be infected synchronously for growth curve analyses (see "The One-Step Growth Cycle" below) or biochemical studies on virus replication.

In contrast to cells that grow in monolayers on plastic dishes, others can be maintained in **suspension cultures,** in which cells are continuously stirred by a spinning magnet. The advantage of suspension cultures is that a large number of cells can be grown in a relatively small volume. This culture method is thus well suited for applications that require large quantities of virus particles, such as X-ray crystallography.

Initially it was believed that viruses, as obligatory intracellular parasites, could not replicate in any cell-free medium, no matter how complex. This dictum was nullified in 1991 by the demonstration that infectious poliovirus could be synthesized in a cell-free extract of human cells incubated with viral RNA. Despite this finding, most work on viruses is done in cultured cells, embryonated eggs, or laboratory animals.

### Evidence of Viral Growth in Cultured Cells

Some viruses kill the cells in which they replicate, and the infected cells may eventually detach from the plastic cell culture plate. As more cells are infected, the changes become visible and are called **cytopathic effects** (Table 2.1). Many types of cytopathic effects can be seen with a simple light or phase-contrast microscope at low power without fixing or staining the cells. These changes include the rounding up and detachment of cells from the plastic culture dish, cell lysis, swelling of nuclei, and formation of fused cells called **syncytia** (Fig. 2.2). Other changes, which require observation by high-power microscopy, are also forms of cytopathic effect. These include the development of intracellular masses of virions or unassembled viral components in the nucleus and/or cytoplasm (**inclusion bodies**), formation of crystalline arrays of virus proteins,

**Table 2.1** Some examples of cytopathic effects of viral infection of animal cells

| Cytopathic effect(s) | Virus(es) |
|---|---|
| **Morphological alterations** | |
| Nuclear shrinking (pyknosis), proliferation of membrane | Picornaviruses |
| Proliferation of nuclear membrane | Alphaviruses, herpesviruses |
| Vacuoles in cytoplasm | Papovaviruses |
| Syncytia (cell fusion) | Paramyxoviruses, coronaviruses |
| Margination and breaking of chromosomes | Herpesviruses |
| Rounding up and detachment of tissue culture cells | Herpesviruses, rhabdoviruses, adenoviruses, picornaviruses |
| **Inclusion bodies** | |
| Virions in nucleus | Adenoviruses |
| Virions in the cytoplasm (Negri bodies) | Rabies virus |
| "Factories" in the cytoplasm (Guarnieri bodies) | Poxviruses |
| Clumps of ribosomes in virions | Arenaviruses |
| Clumps of chromatin in nucleus | Herpesviruses |

membrane blebbing, duplication of membranes, and fragmentation of organelles. The time required for the development of cytopathic effects varies greatly among animal viruses. For example, depending on the size of the inoculum, enteroviruses and herpes simplex virus can cause cytopathic effects in 1 to 2 days and destroy the cell monolayer in 3 days. In contrast, cytomegalovirus, rubella virus, and some adenoviruses may not produce cytopathic effects for several weeks.

The development of cytopathic effects in infected cell cultures is frequently monitored in diagnostic virology when viruses are being isolated from specimens of infected patients or animals. However, cytopathic effect is also of value in the research laboratory: it can be used to monitor the progress of an infection, and it is often one of the phenotypic traits used to characterize mutant viruses.

Many viruses multiply in cells without causing an obvious cytopathic effect. For example, many members of the families *Arenaviridae, Paramyxoviridae,* and *Retroviridae* do not cause obvious damage to cultured cells. The growth of such viruses in cells must therefore be assayed using alternative methods, as described in "Detection of Viruses in the Host" below.

### Embryonated Eggs

Before the advent of cell culture, many viruses were propagated in embryonated chicken eggs (Fig. 2.3). At 5 to 14

**A**

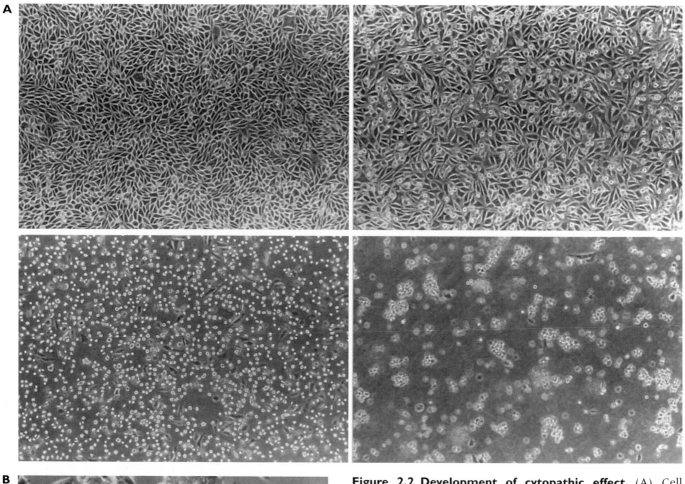

**B**

**Figure 2.2 Development of cytopathic effect.** (A) Cell rounding and lysis during poliovirus infection. Cells were infected with poliovirus type 1 and incubated at 37°C. (Upper left) Uninfected cells. (Upper right) At 5 1/2 h after infection some cells have begun to "round up." (Lower left) At 8 h after infection, more cells have rounded up and detached from the culture dish. (Lower right) At 24 h after infection all cells have detached and formed clumps as they lyse. (B) Syncytium formation by murine leukemia virus. The field shows a mixture of individual small cells and syncytia, indicated by arrows, which are large, multinucleate cells. Courtesy of R. Compans, Emory University School of Medicine.

days after fertilization, a hole is drilled in the shell and virus is injected into the site appropriate for its replication. This method of virus propagation is now used only for influenza viruses. The robust yield of this virus from embryonated chicken eggs has led to their widespread use in research laboratories and for vaccine production.

## Laboratory Animals

In the early 1900s, when viruses were first isolated, freezers and cell cultures were not available and it was neces-

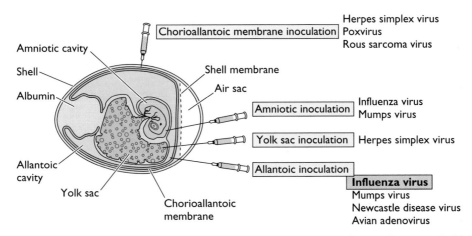

**Figure 2.3 Growth of viruses in embryonated eggs.** Cutaway view of an embryonated chicken egg showing the different compartments in which viruses may grow. The different routes by which viruses are inoculated into eggs are indicated. Adapted from F. Fenner et al., *The Biology of Animal Viruses* (Academic Press, New York, N.Y., 1974), with permission.

sary to maintain virus stocks by continuous passage of virus from animal to animal. This practice was not only inconvenient but, as we shall see in chapter 20, led to the selection of viral mutants. For example, monkey-to-monkey intracerebral passage of poliovirus selected a mutant that could no longer infect chimpanzees by the oral route, the natural primary site of infection.

Cell culture has largely supplanted the use of animals for propagating viruses, but some viruses cannot be grown in cell culture. For example, hepatitis C virus replicates only in chimpanzees. Many viruses that cause human gastroenteritis, such as Norwalk virus, also cannot be grown in cell culture.

Experimental infection of laboratory animals has always been, and will continue to be, obligatory for studying the processes by which viruses cause disease, a field of study known as **viral pathogenesis.** The studies of poliomyelitis, the paralytic disease caused by poliovirus, in monkeys led to an understanding of the basis of the disease caused by the virus and were instrumental in the development of a successful vaccine. Similarly, the development of vaccines against hepatitis B virus would not have been possible without experimental studies with chimpanzees. Understanding how the immune system or any complex organ system reacts to a virus cannot be achieved without research on living animals. The development of viral vaccines, antiviral drugs, and diagnostic tests for veterinary use has also benefited from research on viral diseases in experimental animals.

## Detection of Viruses in the Host

The ability to recognize viral infection in the host is of prime importance in a research or clinical virology laboratory. The approaches for detecting viruses fall into three categories. Because viruses were first recognized by their **infectivity,** the earliest assays focused on this property, which remains the most sensitive and informative. In viral **serology,** the specific interaction of viral antigens with antibodies is used to detect viruses and the immune response against them. Viral proteins in infected cells or tissues can be detected with the use of antibodies; alternatively, the assay can be designed to identify antibodies produced in the host in response to viral infection. More recently, **molecular diagnostics,** which allow the detection of viral nucleic acid sequences in infected cells or tissues, have been developed. Examples of each class of assay are discussed in this section.

### Measurement of Infectious Units

One of the most important procedures in virology is measuring the concentration of a virus in a sample, or the **virus titer.** This parameter is determined by inoculating serial dilutions of virus into host cell cultures, chick embryos, or laboratory animals and monitoring for evidence of virus multiplication. The response of the host may be quantitative (as in assays for plaques, fluorescent foci, infectious centers, and transformation) or all-or-none, in which the presence or absence of infection is measured (as in an endpoint dilution assay).

#### Plaque Assay

In 1952, Renato Dulbecco modified the plaque assay developed to determine the titers of bacterial virus stocks for use in animal virology. The plaque assay was rapidly adopted for reliable determination of the titers for a wide variety of viruses. The importance of this development was recognized by the award of the Nobel Prize in physiology or medicine to Dulbecco in 1975. In this assay, monolayers of cultured cells are incubated with a preparation of virus to allow the virus to adsorb to cells. After removal of the inoculum, the cells are covered with nutrient medium containing a supplement that results in the formation of a gel, most commonly agar. When the original infected cells release new progeny viruses, their spread to neighboring uninfected cells in the culture is restricted by the gel. As a result, each infectious particle produces a circular zone of infected cells, or plaque. If the infected cells are damaged, the plaque can be distinguished from the surrounding monolayer. In time the plaque becomes large enough to see with the naked eye (Fig. 2.4). The plaque assay can be used to determine titers only of viruses that cause visible cell damage.

In some cases, viral plaques can be seen without further manipulation of the monolayer. To enhance the contrast between the plaque and the surrounding monolayer, the cells may be stained with a vital dye, neutral red or crystal violet. Living cells absorb the stain, and plaques appear clear against a red (neutral red) or purple (crystal violet) background of healthy cells (Fig. 2.4). When crystal violet is used, the agar overlay is first removed. However, neutral red staining, which is not toxic and can be done through the agar overlay, allows the recovery of infectious virus from a plaque (see below).

For the majority of animal viruses, there is a linear relationship between the number of infectious virus particles and the plaque count (Fig. 2.5). One infectious particle is therefore sufficient to initiate infection, and the virus is said to infect cells with **one-hit kinetics.** Some examples of **two-hit kinetics,** in which two different types of virus particles must infect a cell to ensure replication, have been recognized. For example, the genomes of some (+) strand RNA viruses of plants consist of two RNA molecules that are separately encapsidated. Both RNAs are required for infectivity. The dose-response curve in plaque assays for these viruses is parabolic rather than linear (Fig. 2.5).

When a single infectious virus is sufficient to form a plaque, the titer of the virus stock can be calculated in **plaque-forming units (PFUs)** per milliliter, according to the dilution of the sample and the number of plaques observed. To minimize error in calculating the virus titer, a series of dilutions is used for the plaque assay and only those plates containing between 20 and 100 plaques are counted, depending on the area of the cell culture vessel. According to statistical principles, when 100 plaques are counted, the sample titer varies ±10%. Plates with more than 100 plaques are generally not used because the plaques may overlap, making counting inaccurate.

When one infectious virus particle initiates a plaque, the viral progeny within the plaque are clones, and virus stocks prepared from a single plaque are known as **plaque**

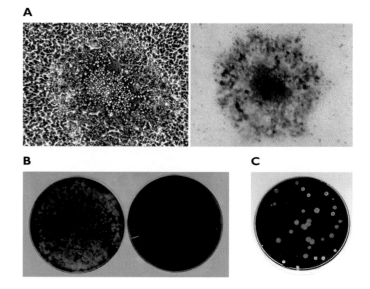

**Figure 2.4 Plaques formed by different animal viruses.** Plaque sizes reflect the life cycle of a virus in a particular cell type. (A) Photomicrograph of a single plaque formed by pseudorabies virus in Georgia bovine kidney cells. (Left) Unstained cells. Detached cells and syncytia are evident. (Right) Cells were stained with the chromogenic substrate X-Gal. The virus used carries the β-galactosidase gene, which converts X-Gal to a blue product. Infected and uninfected cells can be clearly distinguished. Courtesy of B. Banfield, Princeton University. (B) Different plaque morphology of influenza C virus strains. Monolayers were stained with crystal violet. One virus produces small, clear plaques (left), and the other produces small, turbid plaques that are difficult to see (right). (C) Plaques formed by poliovirus on human HeLa cells stained with crystal violet. Note the well-defined plaques.

**purified.** The tip of a small pipette is plunged into the overlay above the plaque, and the plug of agar containing the virus is recovered. The virus within the agar plug is eluted into buffer and used to prepare virus stocks. To ensure purity, this process is usually repeated at least one more time. Plaque purification is used widely in virology to establish clonal virus stocks.

### Fluorescent-Focus Assay

The fluorescent-focus assay, a modification of the plaque assay, is useful in determining the titers of viruses that do not kill cells. The initial procedure is the same as in the plaque assay. However, after a period sufficient for adsorption and propagation, cells are permeabilized with methanol or acetone and incubated with an antibody raised against the virus. A second antibody that recognizes the first antibody is then added. This second antibody is usually conjugated to an indicator, such as fluorescein. The cells are then examined under a microscope with ultraviolet illumination; foci of infected cells fluoresce against the

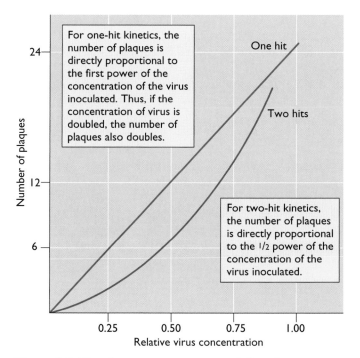

For one-hit kinetics, the number of plaques is directly proportional to the first power of the concentration of the virus inoculated. Thus, if the concentration of virus is doubled, the number of plaques also doubles.

One hit

Two hits

For two-hit kinetics, the number of plaques is directly proportional to the 1/2 power of the concentration of the virus inoculated.

**Figure 2.5 The dose-response curve of the plaque assay.** The number of plaques produced by a virus with one-hit kinetics (line) or two-hit kinetics (curve) is plotted versus the relative concentration of the virus. In two-hit kinetics, there are two classes of uninfected cells, those receiving one particle and those receiving no particles. The Poisson distribution can be used to determine the proportion of cells in each class: they are $e^{-m}$ and $me^{-m}$ (Box 2.2). Since one particle is not sufficient for infection, $P(0) = e^{-m}(1 + m)$. At a very low multiplicity of infection, this equation becomes $P(i) = (1/2)m^2$, which gives a parabolic curve. Adapted from B. D. Davis et al., *Microbiology* (J. B. Lippincott Co., Philadelphia, Pa., 1980), with permission.

dark background. When each fluorescent focus is initiated by a single virus particle, the titer of the virus can be expressed in fluorescent-focus-forming units per milliliter.

### Infectious Center Assay

Another modification of the plaque assay, the infectious center assay, is used to determine the fraction of cells in a culture infected with a virus. Monolayers of cells infected with a virus are suspended before progeny viruses are produced. Dilutions of a known number of infected cells are then plated on monolayers of susceptible cells and covered with an agar overlay. The number of plaques that form on the indicator cells is a measure of the number of cells infected in the original culture. The fraction of infected cells in the culture can therefore be easily determined. A typical use of the infectious center assay is to measure the proportion of infected cells in **persistently infected** cultures, which produce infectious virus in the absence of cell destruction for extended periods.

### Transformation Assay

The transformation assay is useful for determining titers of some retroviruses that do not form plaques on cells. For example, Rous sarcoma virus transforms chick embryo cells. As a result they lose their contact inhibition (the property that governs whether cultured cells grow as a single monolayer [see chapter 16]) and become heaped upon one another. The transformed cells form small piles, or **foci,** that can be easily distinguished from the rest of the monolayer (Fig. 2.6). Infectivity is expressed in focus-forming units per milliliter.

**Figure 2.6 Transformed-focus assay.** Chicken cells transformed by two different strains of Rous sarcoma virus. Loss of contact inhibition causes cells to pile up rather than grow as a monolayer. One focus is seen in panel A, and three foci are seen in panel B. Courtesy of H. Hanafusa, Rockefeller University.

**A**

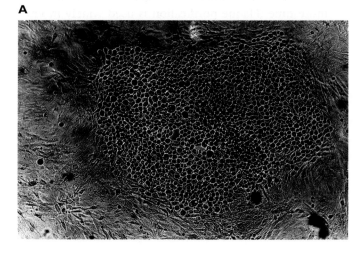

**B**

### Endpoint Dilution Assay

The endpoint dilution assay was used extensively before development of the plaque assay. It is still used for certain viruses that do not form plaques or for determining the virulence of a virus in animals. Serial dilutions of a virus stock are inoculated into replicate **test units** (typically 8 to 10), which can be cell cultures, embryonated eggs, or animals.

The number of test units that have become infected is then determined for each virus dilution. When cell culture is used, infection can be gauged by the development of cytopathic effect, and in embryonated eggs or animals by death or disease. An example of an endpoint dilution assay involving cell cultures is shown in Box 2.1. At high dilutions, none of the cell cultures are infected because the dilution

---

**BOX 2.1**

*Endpoint dilution assays*

In the first example, 10 monolayer cell cultures were infected with each virus dilution. After the incubation period, plates that displayed cytopathic effect were scored +. Fifty percent of the cell cultures displayed cytopathic effect at the $10^{-5}$ dilution, and therefore the virus stock contains $10^5$ $ID_{50}$ units.

| Virus dilution | Cytopathic effect | | | | | | | | | |
|---|---|---|---|---|---|---|---|---|---|---|
| $10^{-2}$ | + | + | + | + | + | + | + | + | + | + |
| $10^{-3}$ | + | + | + | + | + | + | + | + | + | + |
| $10^{-4}$ | + | + | − | + | + | + | + | + | + | + |
| $10^{-5}$ | − | + | + | − | + | − | − | + | − | + |
| $10^{-6}$ | − | − | − | − | − | − | + | − | − | − |
| $10^{-7}$ | − | − | − | − | − | − | − | − | − | − |

In practice, the 50% endpoint frequently falls between dilutions. The endpoint must therefore be interpolated by statistical methods, such as that of Reed and Muench. The use of this approach is illustrated in the following example, in which the lethality of poliovirus in mice is the endpoint. Eight mice were inoculated per dilution.

| Dilution | Alive | Dead | Total alive | Total dead | Mortality ratio | Mortality (%) |
|---|---|---|---|---|---|---|
| $5 \times 10^{-2}$ | 0 | 8 | 0 | 40 | 0/40 | 100 |
| $5 \times 10^{-3}$ | 0 | 8 | 0 | 32 | 0/32 | 100 |
| $5 \times 10^{-4}$ | 1 | 7 | 1 | 24 | 1/25 | 96 |
| $5 \times 10^{-5}$ | 0 | 8 | 1 | 17 | 1/18 | 94 |
| $5 \times 10^{-6}$ | 2 | 6 | 3 | 9 | 3/12 | 75 |
| $5 \times 10^{-7}$ | 5 | 3 | 8 | 3 | 8/11 | 27 |

In the method of Reed and Muench, the results are pooled, as shown in the table. The interpolated value of the 50% endpoint, known as *I*, which falls between the fifth and sixth dilutions, is calculated by the following formula:

$$I = h\left(\frac{\% \text{ of animals affected at dilution above } 50\% - 50\%}{\% \text{ of animals affected at dilution above } 50\% - \% \text{ of animals affected at dilution below } 50\%}\right)$$

where *h* is the dilution factor, in this case 10. For this example *I* is 0.524. The endpoint is then calculated as follows:

$$50\% \text{ endpoint titer} = 10^{\log \text{ total dilution above } 50\% - (I \times \log h)}$$

which in our example is $10^{-5.301 - (0.524 \times 1.0)} = 10^{-5.825}$. The $LD_{50}$ is calculated as the concentration of the stock virus in plaque-forming units per milliliter times the 50% endpoint titer. In the example shown, the $LD_{50}$ is $1.5 \times 10^3$.

is so high that no infectious particles are delivered to the cells in the test unit; at low dilutions, every culture is infected. The endpoint of a limit dilution assay is the dilution of virus that affects 50% of the test units. This number can be calculated from the data and expressed as 50% infectious dose ($ID_{50}$) per milliliter. The preparation illustrated in Box 2.1 contains $10^5$ $ID_{50}$ per ml.

In most cases the 50% endpoint does not always fall on a dilution tested as shown in the example in Box 2.1; for this reason, various statistical procedures have been developed to calculate the endpoint of the titration. One popular method is that of L. J. Reed and H. Muench, in which the dilution containing the $ID_{50}$ is identified by interpolation between two dilutions on either side of this value (Box 2.1). The assumption is made that the location of the 50% endpoint varies linearly with the log of the dilution. Because the number of test units used at each dilution is usually small, the accuracy of this method is relatively low. For example, if six test units are used at each 10-fold dilution, differences of only 50-fold or more in virus titer can be reliably detected.

When the endpoint dilution assay is used to assess the virulence of a virus or its capacity to cause disease, as defined in chapter 17, the result of the assay can be expressed in terms of 50% lethal dose ($LD_{50}$) per ml or 50% paralytic dose ($PD_{50}$), endpoints of death and paralysis, respectively. If the virus titer can be determined separately by plaque assay, the 50% endpoint determined in an animal host can be related to the virus titer. In this way, the effects of route of inoculation or specific mutations on viral virulence can be quantified.

### Efficiency of Plating

The term efficiency of plating (often abbreviated eop) was coined in studies on bacteriophages to assign a value to the plaque count determined with the same bacteriophage for different strains of bacteria. The absolute efficiency of plating was later defined as the plaque titer divided by the number of virus particles in the sample. The **particle to plaque-forming unit ratio,** a term more commonly used today, is the inverse value (Table 2.2). For many bacteriophages, the particle to plaque-forming unit ratio approaches 1. However, for animal viruses this value can be much higher, ranging from 1 to 10,000. These high values have complicated the study of animal viruses. For example, when the particle to plaque-forming unit ratio is high, it is never certain whether properties measured biochemically are in fact due to the infectious or the noninfectious component.

Although the linear nature of the dose-response curve indicates that a single particle is *capable* of initiating an infection (one-hit kinetics) (Fig. 2.5), the high particle to plaque-forming unit ratio of many viruses demonstrates

**Table 2.2** Particle/plaque-forming unit ratios of some animal viruses

| Virus | Particle/PFU ratio |
|---|---|
| *Picornaviridae* | |
| Poliovirus | 30–1,000 |
| *Alphaviridae* | |
| Semliki Forest virus | 1–2 |
| *Orthomyxoviridae* | |
| Influenza virus | 20–50 |
| *Reoviridae* | |
| Reovirus | 10 |
| *Papovaviridae* | |
| Polyomavirus | 38–50 |
| Simian virus 40 | 100–200 |
| Papillomavirus | 10,000 |
| *Adenoviridae* | 20–100 |
| *Poxviridae* | 1–100 |
| *Herpesviridae* | |
| Herpes simplex virus | 50–200 |

that not all virions are *successful* in infecting a cell. The high particle to plaque-forming unit ratio of animal viruses may sometimes be caused by the presence of noninfectious particles whose genomes harbor lethal mutations or which have been damaged during growth or purification. An alternative explanation is that all viruses in a preparation are in fact capable of initiating infection, but that all do not succeed because of the complexity of the infectious cycle. Virus replication entails an obligate sequence of steps—attachment, entry, replication, and assembly—and failure at any one of these prevents completion of the infectious cycle.

### Measurement of Virus Particles and Their Components

Although the numbers of virus particles and infectious units are often not equal, assays for particle number are frequently used to approximate the quantity of virus present in a sample. Physical assays are usually more rapid and easier to carry out than assays for infectivity, which may be cumbersome or not available. Assays for subviral components also provide information on particle number if their stoichiometry in the virus particle is known.

### Electron Microscopy

With the exception of members of the *Poxviridae*, virus particles are too small to be observed by light microscopy. However, they can be seen readily by electron microscopy. If a sample contains only one type of virus, the particle count can be determined. First, the viruses are usually mixed with a known concentration of latex beads. The numbers of virus particles and beads are then counted, allowing the concentration of the virus in the sample to be determined by comparison.

### *Hemagglutination*

Many viruses contain proteins that can bind to erythrocytes, or red blood cells; these viruses can link multiple cells, resulting in a lattice. This property is called **hemagglutination.** For example, influenza viruses contain an envelope glycoprotein called hemagglutinin (HA), which binds to *N*-acetylneuraminic acid-containing glycoproteins on erythrocytes. In practice, twofold serial dilutions of the virus stock are prepared, mixed with a defined quantity of red blood cells, and added to the small wells of a plastic tray (Fig. 2.7). Unadsorbed red blood cells tumble to the bottom of the well and form a sharp dot or button. In contrast, agglutinated red blood cells form a lattice that coats the well. Because the assay is rapid (30 min), it is often used as a quick indicator for the relative quantities of virus particles. However, it is not sensitive for determining small numbers of particles.

### *Measurement of Viral Enzyme Activity*

Some animal virus particles contain nucleic acid polymerases which can be assayed by mixing permeabilized particles with radioactively labeled precursors and calculating the incorporation of radioactivity into nucleic acid. Assays for nucleic acid polymerase are most frequently used for retroviruses, many of which do not transform cells or form plaques. In one example, the reverse transcriptase incorporated into the virus particle is assayed by mixing cell culture supernatants with a mild detergent (to permeabilize the viral envelope), a poly(rC) template and an oligo(dG) primer, and [α-$^{32}$P]dGTP. If reverse transcriptase is present, a radioactive product will be produced by priming on the poly(rC) template. This product can be detected by precipitation or bound to a filter and quantified. This assay cannot be used to calculate the number of infectious or noninfectious particles. However, since the

**Figure 2.7 Hemagglutination assay.** Samples of different influenza viruses were diluted, and a sample from each dilution was mixed with a suspension of chicken red blood cells and added to the wells. After 30 min at 4°C, the wells were photographed. Sample A causes hemagglutination through a dilution of 1:256 and has an HA titer of 256. Many viruses that hemagglutinate red blood cells also contain a neuraminidase in the virus particle. The presence of this enzyme, which cleaves *N*-acetylneuraminic acid from glycoprotein receptors, elutes bound viruses from red blood cells. As a result, the lattice formed by hemagglutination is reversed. An example of elution of the virus from red blood cells can be seen in column 1, rows D and E. To minimize elution, hemagglutination assays should not be incubated for more than 30 min at 4°C. Courtesy of C. Basler and P. Palese, Mt. Sinai School of Medicine of the City University of New York.

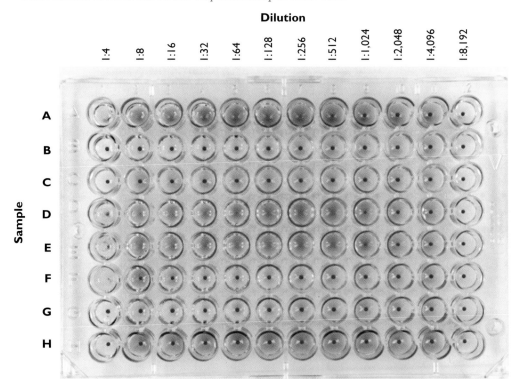

amount of enzymatic activity is proportional to particle number, the assay does allow rapid tracking of virus production in the course of an infection.

### Serological Methods

Many virological techniques take advantage of the specificity of the antibody-antigen reaction. Assays have been designed to detect the presence of viral antigens with defined antibodies, or antibodies against a virus, with the use of a defined antigen. Some of the techniques, such as immunostaining, immunoprecipitation, immunoblotting, and enzyme-linked immunosorbent assay, are by no means limited to the detection of viruses and viral proteins. All these approaches have been used extensively to study the structures and functions of uninfected cell proteins.

**Virus neutralization.** When viruses are inoculated into an animal, an array of antibodies against the virus are produced. Some of these antibodies can bind virus particles and neutralize their infectivity, as discussed in chapter 14. Virus neutralization assays are usually conducted by mixing dilutions of serum or monoclonal antibody with virus, incubating them, and assaying for remaining infectivity with cultured cells, embryonated eggs, or animals. The endpoint is defined as the highest dilution of antiserum that inhibits the development of cytopathic effect in cultured cells or virus replication in the eggs or animals.

Neutralizing antibodies define **type-specific antigens** on the virus particle. For example, the three **serotypes** of poliovirus are distinguished on the basis of neutralization tests; type 1 poliovirus is neutralized by antiserum to type 1 virus but not antiserum to type 2 or 3 poliovirus, and so forth. Neutralization tests have therefore been invaluable for virus classification.

Knowledge of the antigenic structure of a virus is useful in understanding the immune response to viruses and in designing novel vaccination strategies. The use of **monoclonal antibodies** in neutralization assays permits mapping of neutralization antigenic sites on a virus particle or of the amino acid sequences that give rise to neutralizing antibodies. Each monoclonal antibody specifically binds to a small number (8 to 12) of amino acids that fit into the antibody-combining site. This amino acid sequence, which may be linear or nonlinear, is known as an **epitope.** Direct methods for identification of neutralization antigenic sites include cross-linking of the monoclonal antibody to the virus and determining which protein is the target of the antibody, and assessing the abilities of monoclonal antibodies to bind synthetic peptides representing viral protein sequences. When the monoclonal antibody recognizes a linear epitope, it may react with the protein in Western blot analysis (see "Immunoprecipitation and immunoblotting" below), facilitating direct identification of the viral protein harboring the antigenic site. The most elegant understanding of antigenic structures has come from the isolation and study of variant viruses that have become resistant to neutralization with specific monoclonal antibodies (called monoclonal antibody-resistant or *mar* variants). By identifying the amino acid change responsible for the *mar* phenotype, the antibody-binding site can be located and, together with three-dimensional structural information, can provide detailed information on the nature of neutralization antigenic sites.

**Hemagglutination inhibition.** Antibodies against viral proteins with hemagglutination activity can block the ability of virus to bind red blood cells. In this assay, dilutions of serum are incubated with virus, and erythrocytes are added as outlined above. After incubation, the hemagglutination inhibition titer is read as the highest dilution of serum that inhibits hemagglutination. This test is sensitive, simple, inexpensive, and rapid and is the method of choice for assaying antibodies to any virus that causes hemagglutination. The hemagglutination inhibition assay can be used to detect antibodies to viral hemagglutinin in animal and human sera or to identify the origin of the hemagglutinin of influenza viruses produced in cells coinfected with two parent viruses (see "Mapping Mutations" below).

**Complement fixation.** The complement fixation assay can be used to determine whether antibodies against a virus are present in serum. The interaction of viral antigen and antibody can cause complement fixation, which leads to membrane lysis, as discussed in chapter 14. Red blood cells are used as targets because lysis of their membranes is readily observed. In the assay for complement-fixing antibodies, samples of patient sera are heated to inactivate endogenous complement and then incubated with preparations of viral antigen and a standardized quantity of guinea pig complement. If an antibody-antigen reaction takes place, complement fixation will occur. The latter is detected by adding sheep red blood cells that have been coated with rabbit anti-red blood cell antibodies. If complement has been fixed (complexed) by binding of viral antigen to antibody, the red blood cells will remain intact; if complement is not fixed, the red blood cells will be lysed by the action of free complement.

Because crude preparations of viral antigens are often used for the complement fixation test, it is not highly specific. The antigen preparations may include proteins or specific epitopes shared by related groups of viruses. As a result, the test detects most serotypes within a given group of viruses and is therefore said to recognize **group-specific antigens.** For example, the three serotypes of poliovirus share a complement-fixing group-specific antigen. Complement fixation is not as sensitive as a neutralization or hemagglutination inhibition test. Nevertheless, it is often the first assay performed on sera from infected patients to identify the group to which the infecting virus belongs.

**Immunostaining.** Antibodies can be used to visualize viral proteins in infected cells or tissues. In **direct immunostaining,** an antibody directed against a viral antigen is directly coupled to an indicator such as a fluorescent dye or an enzyme (Fig. 2.8). A more sensitive approach is **indirect immunostaining,** in which a second antibody is coupled to the indicator. The second antibody recognizes a general epitope on the virus-specific antibody. Multiple second-antibody molecules bind to the first antibody, resulting in an increased signal from the indicator compared to that obtained with direct immunostaining. This method provides a single indicator-coupled second-antibody reagent for use in many assays, avoiding the need to purify and couple an indicator to many first antibodies.

In practice, virus-infected cells are fixed with acetone or methanol and incubated with polyclonal or monoclonal antibodies directed against viral antigen. Excess antibody is washed away, and in direct immunostaining, cells are examined for the indicator by microscopy. For indirect immunostaining, the second antibody is added before examination of the cells by microscopy. Commonly used indicators include fluorescein and rhodamine, which fluoresce on exposure to ultraviolet light. Filters are placed between the specimen and the eyepiece to remove blue and ultraviolet light so that the field is dark except for cells to which the antibody has bound, which emit green (fluorescein) (Fig. 2.8) or red (rhodamine) light. Antibodies can also be coupled to molecules other than fluorescent indicators, including enzymes such as alkaline phosphatase, horseradish peroxidase, and β-galactosidase, a bacterial enzyme that in a test system converts the chromogenic substrate X-Gal (5-bromo-4-chloro-3-indolyl-β-D-galactopyranoside) to a blue product. After excess antibody is washed away, a suitable chromogenic substrate is added, and the presence of the indicator antibody is revealed by development of a color in or on the cells that can be visualized with a light microscope.

Immunostaining has been applied widely in the research laboratory in determining the subcellular localization of proteins in cultured cells, monitoring the expression of viral proteins, determining the effects of mutation on protein expression (Fig. 2.8), and investigating the sites of virus replication in animal hosts. It is the basis of the fluorescent-focus assay, which can be used in conjunction with the plaque assay to calculate the ratio of focus-forming units to plaque-forming units. When the fluorescent-focus assay is performed early in infection to detect the expression of early genes before genome amplification, it can serve as an indirect measure of how many genomes have entered cells. The ratio of focus-forming units to plaque-forming units provides information about what fraction of a virus stock can complete an infectious cycle.

Immunostaining of viral antigens in smears of clinical specimens is also used to diagnose viral infections. For example, direct and indirect immunofluorescence assays with nasal swabs or washes are routinely utilized to diagnose infections with respiratory syncytial virus, influenza virus, parainfluenza virus, measles virus, and adenovirus.

**Figure 2.8 Direct and indirect methods for antigen detection.** (A) The sample (tissue section, smear, or bound to a solid phase) is incubated with a virus-specific antibody. In direct immunostaining, the antibody is linked to an indicator such as fluorescein. In indirect immunostaining, a second antibody coupled to an indicator is used which is directed against a general epitope on the virus-specific antibody. (B) Use of immunofluorescence to demonstrate the intracellular localization of a viral protein and the effect of mutation on protein localization. Plasmids directing the expression of wild-type herpes simplex virus ICP0 protein or an altered ICP0 were introduced into Vero cells. Forty-eight hours later, the cells were fixed and reacted with a polyclonal antibody against ICP0, followed by fluorescein-conjugated secondary antibodies. Cells were photographed using a microscope with ultraviolet illumination. Wild-type ICP0 is distributed in discrete areas of the nucleus, whereas the altered protein is expressed in a diffuse pattern. Courtesy of E. Lium and S. Silverstein, Columbia University.

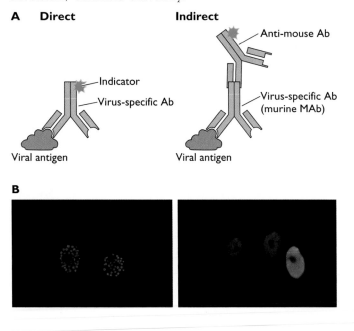

**A    Direct**            **Indirect**

Anti-mouse Ab

Indicator
Virus-specific Ab

Virus-specific Ab
(murine MAb)

Viral antigen          Viral antigen

**B**

**Immunoprecipitation and immunoblotting.** Immunoprecipitation depends on the interaction of specific antiviral antibodies with viral proteins in solubilized extracts of infected cells or tissues (Fig. 2.9). The antibody-protein complexes are isolated, and the viral proteins, usually metabolically labeled with a radioactive amino acid, are fractionated by electrophoresis in polyacrylamide gels. Immunoprecipitation can establish many properties of viral proteins, including the types present in virions and infected cells, molecular mass, kinetics of synthesis and subcellular localization, and whether they associate with other viral or cellular proteins.

Immunoblotting, also known as Western blot analysis, depends on the reaction of an antibody with a viral protein but in a different format—immobilized on a thin membrane (Fig. 2.10). Proteins are fractionated by electrophoresis in a polyacrylamide gel and then transferred to a thin synthetic membrane that has a strong affinity for proteins. Proteins bound to the membrane are detected by immunostaining. Because proteins bound to the membrane are denatured, antibodies that recognize nonlinear determinants are usually not suitable for their detection. The main advantage of this technique is that it does not require labeling of proteins and therefore can be conducted with tissues and organs as well as with cultured cells. However, this assay does not provide information about the kinetics of protein synthesis. Western blot analysis can detect only relatively stable proteins, not those that are translated and then rapidly degraded. When used alone, it cannot reveal whether a viral protein associates with other proteins. However, such information can be obtained if proteins are first isolated by immunoprecipitation and then subjected to Western blot analysis with a different antibody.

Western blot analysis is used in the clinical laboratory to identify antibodies to human immunodeficiency virus in clinical specimens and in donated blood. The sample is incubated with membranes to which viral proteins separated by gel electrophoresis have been transferred. If antibodies against human immunodeficiency virus are present, one or more of the viral proteins on the membrane will be stained.

**Enzyme-linked immunosorbent assay.** Detection of viral antigen or antibody can be accomplished by solid-phase methods in which viral antibody or protein is adsorbed to a plastic surface. To detect viral antigens in serum or clinical samples, a "capture" antibody, directed against the virus, is linked to a solid support, a plastic dish or bead (Fig. 2.11). The specimen is added to the plastic support, and if viral antigens are present, they will be captured by the bound antibody. Bound viral antigen is detected by using a second antibody linked to an enzyme. A chromogenic molecule is then added and is converted by the enzyme to an easily detectable product. The use of an enzyme amplifies the signal because a single enzyme molecule can generate many products. To detect immunoglobulin G (IgG) antibodies to viruses, viral antigen is first linked to the plastic support, and then the specimen is added (Fig. 2.11). If antibodies against the virus are present in the specimen, they will bind to the immobilized antibody. The bound antibodies are then detected by using a second antibody directed against a general epitope on

**Figure 2.9 Visualization of proteins by immunoprecipitation.** In the example shown, cell proteins are labeled during translation by incubating cells with radioactive amino acids. Cells are lysed with a detergent to solubilize the proteins. Antibodies that have been coupled to plastic beads are added to the cell lysate. The beads are removed by centrifugation and washed free of proteins not bound by the antibody. The bound proteins can then be fractionated by gel electrophoresis and visualized by autoradiography. SDS-PAGE, sodium dodecyl sulfate polyacrylamide gel electrophoresis.

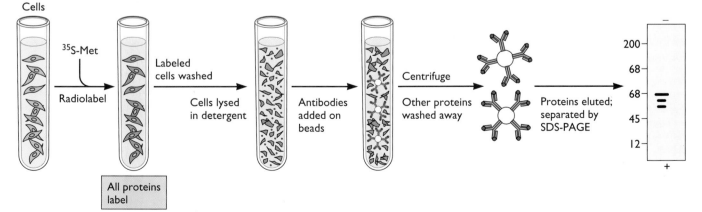

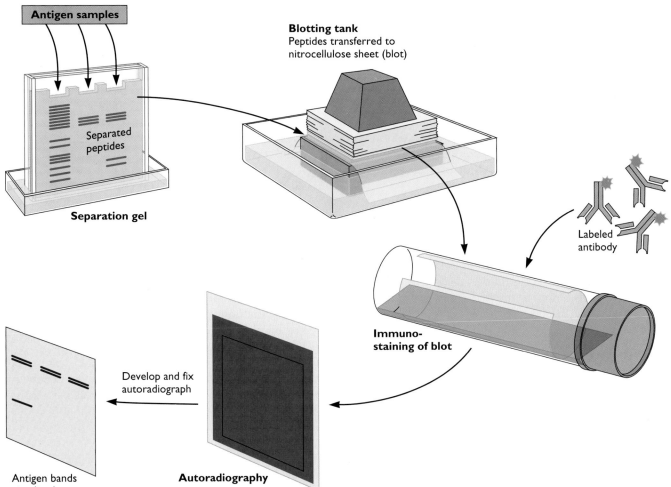

**Figure 2.10 Visualization of proteins by Western blot analysis.** The sample, which can be cells, tissues, virus, or partially purified protein, is solubilized with detergent. Proteins are fractionated by electrophoresis in a polyacrylamide gel. A thin membrane is placed over the gel, and the fractionated proteins are transferred. In the illustration, protein transfer is accomplished by capillary transfer in a blotting tank; transfer may also be achieved by an electric current. The membrane, to which the proteins are tightly bound, is incubated with a specific antibody coupled to an enzyme such as horseradish peroxidase. Bound antibody is detected by the addition of a substrate that is converted to a luminescent compound after reaction with the enzyme, followed by autoradiography.

antibody molecules. Like other detection methods, enzyme-linked immunosorbent assays are used in both diagnostic and experimental virology.

### Nucleic Acid Detection

The development of methods to detect viral nucleic acids among the vast excess of cellular sequences has revolutionized the study of viruses. In **Southern blot hybridization analysis** (developed by Edward Southern), DNA preparations are digested with a restriction endonuclease, fractionated by gel electrophoresis, denatured

within the gel, and transferred to a membrane to which they bind strongly. The bound DNAs are then detected by hybridization with an isotopically labeled viral nucleic acid sequence. **Northern blot hybridization analysis** is used to detect RNA in a similar manner. The name of this technique, like that of Western blot analysis, was not derived from the inventor but is a play on Dr. Southern's name.

In one variation of Southern blot analysis, DNAs are concentrated onto a small region of the filter, generally by a vacuum. This method, called slot blot or dot blot analysis depending on the shape of the manifold wells, provides

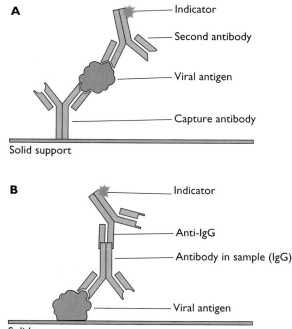

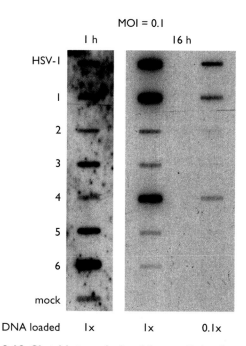

**Figure 2.11 Detection of viral antigen or antibodies against viruses by enzyme-linked immunosorbent assay.** (A) To detect viral proteins in a sample, antibodies specific for the virus are immobilized on a solid support such as a plastic well. The sample is placed in the well, and viral proteins are "captured" by the immobilized antibody. After washing to remove unbound proteins, a second antibody against the virus is added, which is linked to an indicator. Another wash is done to remove unbound second antibody. If viral antigen has been captured by the first antibody, the second antibody will bind and will be detected by the indicator. (B) To detect antibodies to a virus in a sample, viral antigen is immobilized on a solid support. The sample is placed in the well, and viral antibodies bind the immobilized antigen. After washing to remove unbound antibodies, a second antibody, directed against a general epitope on the first antibody, is added. Another wash is done to remove unbound second antibody. If viral antibodies are bound by the immobilized antigen, the second antibody will bind and will be detected by the indicator.

**Figure 2.12 Slot blot analysis of intracellular herpes simplex virus DNA.** Cells were mock-infected or infected with wild-type herpes simplex virus type 1 (HSV-1) or six different mutants at a low (0.1) multiplicity of infection. Total infected-cell DNAs were prepared at 1 and 16 h after infection, and undiluted (1×) and 10-fold dilutions (0.1×) of DNAs were applied to nylon membranes. The infected-cell DNAs were hybridized with herpes simplex virus type 1-specific DNA. The viral DNA detected 1 h after infection, before genome amplification, represents the incoming viral genomes. By determining the amount of radioactivity in each slot, the relative levels of input DNA can be found for each mutant. This information, together with the virus titer, can be used to calculate a relative genome to plaque-forming unit ratio for each virus. This ratio provides information similar to that obtained from the particle to plaque-forming unit ratio. For example, although the multiplicity of infection was the same for all viruses, it is clear that cells infected with mutant 6 received more viral DNA than the others. The mutation may exert this effect by reducing the number of defective particles or by increasing the efficiency of the early steps in infection. Figure courtesy of E. Lium and S. Silverstein, Columbia University.

a rapid means for determining relative concentrations of different nucleic acids in the same sample or a specific viral nucleic acid in different samples. In the example shown in Fig. 2.12, slot blot analysis is used to calculate the relative genome/plaque-forming unit ratio of a DNA virus. This approach is effective when many viral mutants are analyzed and it is not practical to determine particle numbers by electron microscopy.

Nucleic acid hybridization probes can also be used to locate viral genomes or transcripts in sections of cells, tissues, or even whole animals by a technique known as **in situ hybridization.** The **polymerase chain reaction** has also found wide application in the detection of viral nucleic acids. Specific oligonucleotides are used to amplify viral DNA sequences from infected cells or tissues; in situ methods involving the polymerase chain reaction have also been developed. Polymerase chain reaction techniques have also allowed the identification of new viruses associated with human diseases, as described in chapter 20. Clinical virology laboratories also employ these methods to diagnose viral infections, except that radioactive nucleic acids are not used because of their short half-lives and the hazards posed by particle emissions. Other methods for labeling nucleic acids have been developed. In one version, the nucleic acid hybridization probes contain biotin. The biotinylated nucleic acids are detected by incubation with streptavidin, which binds with high affinity to biotin and which has been linked to an indicator such as β-galactosidase. These nonradioactive approaches have proven so sensitive that they have found their way back into research laboratories.

## The One-Step Growth Cycle

One important objective of research in virology is to understand how viruses enter individual cells, replicate, and assemble new infectious viruses. These studies are usually carried out with cell cultures rather than with animals, because cell cultures provide a much simpler and more homogeneous experimental system. More importantly, cell cultures can be infected in such a way as to ensure that only a single replication cycle occurs synchronously in every infected cell. The idea that one-step growth analysis can be used to study the single-cell life cycle of viruses was first set forth by Emory Ellis and Max Delbrück in 1939 in their studies with bacteriophages described in chapter 1. Synchronous infection, the key to the one-step growth cycle, is accomplished by infecting cells with a sufficient number of virus particles to ensure that most of the cells

are infected rapidly. Most one-step growth experiments are conducted at a **multiplicity of infection** (Box 2.2) of 5 to 10 plaque-forming units per cell to ensure that almost all cells receive at least one infectious unit.

One-step growth curve analysis begins with removal of the cell culture medium from the cell monolayer and addition of virus in a small volume to promote rapid adsorption. After approximately 1 h, the unadsorbed inoculum is removed, the cells are washed, and fresh medium is added. At different times after infection, samples of the cell culture supernatant are collected, and the number of viruses present is determined. The kinetics of intracellular virus production can be monitored by removing the medium containing released viruses, scraping the cells into fresh medium, and lysing the cells by repeated cycles of freeze-thawing. A cell-free extract is prepared after removal of cellular debris by centrifugation, and the virus titer in the extract is measured.

When the results of a one-step growth experiment are plotted graphically, a number of important features about virus replication are revealed. In the example shown in Fig. 2.13A, the first 11 h after infection constitute the **eclipse period,** during which the viral nucleic acid is uncoated from its protective capsid and no infectious virus

## BOX 2.2
### Multiplicity of infection (moi)

Infection depends on the random collision of cells and virus particles. When susceptible cells are mixed with a suspension of virus, some cells are uninfected and other cells receive one, two, three, etc., particles. The distribution of virus particles per cell is best described by the **Poisson distribution:**

$$P(k) = \frac{e^{-m}m^k}{k!}$$

In this equation, $P(k)$ is the fraction of cells infected by $k$ virus particles. The multiplicity of infection, $m$, is calculated from the proportion of uninfected cells, $P(0)$, which can be determined experimentally. If $k$ is made 0 in the above equation, then

$$P(0) = e^{-m} \quad \text{and} \quad m = -\ln P(0)$$

The fraction of cells receiving 0, 1, and more than one virus particle in a culture of $10^6$ cells infected with an moi of 10 can be determined as follows.

Fraction of cells that receive 0 particles:

$$P(0) = e^{-10} = 4.5 \times 10^{-5}$$

and in a culture of $10^6$ cells this equals 45 uninfected cells.

Fraction of cells that receive 1 particle:

$$P(1) = 10 \times 4.5 \times 10^{-5} = 4.5 \times 10^{-4}$$

and in a culture of $10^6$ cells, 450 cells receive 1 particle.

Fraction of cells that receive >1 particle:

$$P(>1) = 1 - e^{-m}(m + 1)^* = 99.95\%$$
of cells receive more than 1 particle

For moi = 0.001:

$$P(0) = 99.99\%$$
$$P(1) = 0.0999\% \text{ (for } 10^6 \text{ cells, } 10^4 \text{ are infected)}$$
$$P(>1) = 10^{-6}$$

The moi required to infect 99% of the cells in a cell culture dish:

$$P(0) = 1\% = 0.01$$
$$m = -\ln (0.01) = 4.6 \text{ plaque-forming units per cell}$$

*Obtained by subtracting from 1 (the sum of all probabilities for any value of $k$) the probabilities $P(0)$ and $P(1)$.

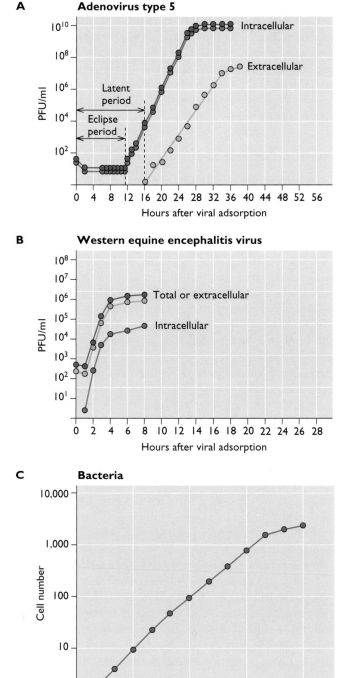

**Figure 2.13 One-step growth curves.** (A) Growth of a nonenveloped virus, adenovirus type 5. The virus titer is plotted on the $y$ axis, and the number of hours after adsorption is plotted on the $x$ axis for extracellular virus (e.g., released from the cell), intracellular virus, and the sum of both (purple dots). The burst size (plaque-forming unit per cell) can be calculated by subtracting the residual infectivity observed during the eclipse period from the total number of infectious viruses produced and dividing this number by the number of cells in the culture. (B) Growth of an enveloped virus, western equine encephalitis virus, a member of the *Togaviridae*. The total replication cycle is short, as are the eclipse, latent, and synthetic periods. This virus acquires infectivity after maturation at the plasma membrane, and therefore little intracellular virus can be detected (the small amounts observed at each time point likely represent released virus contaminating the cell extract). (C) Growth curve for a bacterium. Time is plotted on the $x$ axis, and number of bacteria is plotted on the $y$ axis. One bacterium is added to the culture at time zero; after a brief lag, the bacterium begins to divide. The number of bacteria doubles every 20 min until nutrients in the medium are depleted and the growth rate decreases. (A and B) Adapted from B. D. Davis et al., *Microbiology* (J. B. Lippincott Co., Philadelphia, Pa., 1980), with permission. (C) Adapted from B. Voyles, *The Biology of Viruses* (McGraw-Hill, New York, N.Y., 1993), with permission.

**phase,** during which new virus particles are assembled. During the **latent period** no extracellular virus can be detected. At 18 h after adsorption, virus is released from cells and found in the extracellular medium. Ultimately virus production plateaus as the cells become metabolically and structurally unable to support additional replication.

The **burst size,** or the yield of infectious virus per cell, can be calculated from the data collected during a one-step growth experiment (Fig. 2.13). This value varies widely among different viruses and with different virus-host cell combinations. For many viruses, increasing the multiplicity of infection above a certain point does not increase the burst size. Thus, cells have a finite capacity to produce new virus particles.

The nature of the one-step growth curve can vary dramatically among different viruses. The curve shown in Fig. 2.13A illustrates the pattern observed for a DNA virus with long latent and synthetic phases typical of many DNA viruses, some retroviruses, and reovirus. For small RNA viruses, the entire growth curve is complete within 6 to 8 h, and the latent phases are correspondingly shorter. Latent phases can be as short as 2 h (poliovirus and vesicular stomatitis virus) and as long as 12 to 14 h (polyomavirus and adenovirus). The synthetic phase may occupy from 2 to 6 h after infection (poliovirus and vesicular stomatitis virus) to 12 to 48 h after infection (polyomavirus). Polyomavirus, which has one of the smallest genomes of the DNA viruses, has a latent period as long as that of the other DNA viruses. This similarity reflects the fact that DNA synthesis of small DNA viruses requires that

can be detected inside cells. The low level of infectivity detected during this period probably results from adsorbed virus that was not uncoated. Beginning at 12 h after adsorption, the quantity of intracellular infectious virus begins to increase, marking the onset of the **synthetic** cells enter S phase (chapter 9), which takes time. Furthermore, enveloped viruses that mature by budding from the plasma membrane, as discussed in chapter 13, generally become infectious only as they leave the cell, and therefore little intracellular infectious virus can be detected (Fig. 2.13B).

The first one-step growth curves of viruses were prepared for bacteriophages, and the results surprised scientists who had expected that they would resemble the growth curves of bacteria or cultured cells. After a short lag, cell growth becomes exponential (e.g., each progeny cell is capable of dividing) and follows a straight line (Fig. 2.13C). Exponential growth continues until the nutrients in the medium are exhausted. The one-step growth curves of viruses are different: they begin with a longer lag period (eclipse) during which no virus growth is observed, followed by the sudden appearance of new infectious virus. We now know that during the eclipse period the components of new virus particles are being synthesized and assembled.

One-step growth curve analysis can provide quantitative information about different virus-host systems. It is frequently employed to study mutant viruses to determine what parts of the replication cycle are affected by a particular genetic lesion. It is also valuable for studying the multiplication of a new virus or viral replication in a new virus-host cell combination.

When cells are infected at a low multiplicity of infection, several cycles of viral multiplication may occur. Growth curves established under these conditions can also provide useful information. For example, if a mutation fails to have an obvious effect on viral multiplication in a one-step growth curve, the defect may become obvious following a low-multiplicity infection. Because the effect of a mutation in each cycle is multiplied over several cycles, a small effect in a one-step growth cycle can be amplified after several cycles. Defects in the ability of viruses to spread from cell to cell may also be revealed when multiple cycles of replication occur.

## Genetic Analysis of Viruses

The use of genetic analysis to study the structure and function of animal virus genes and proteins began with development of the plaque assay by Dulbecco in 1952. This assay permitted the preparation of clonal stocks of virus, the measurement of virus titers, and a convenient system for studying viruses with conditional lethal mutations. Although a limited repertoire of classical genetic methods was available for the study of animal viruses, the mutants that were isolated were invaluable in elucidating many aspects of the infectious cycles of viruses, as well as of cell transformation by viruses. Modern methods of genetic analysis based on recombinant DNA technology, developed subsequently, confer an essentially unlimited scope for genetic manipulation; in principle, any viral gene of interest can be mutated, and the precise nature of the mutation can be predetermined by the investigator. Much of the large body of information about viruses and their lifestyles that we now possess can be attributed to the power of these methods.

## Classical Genetic Methods

### Spontaneous and Induced Mutations

In the early days of experimental virology, mutant viruses could be isolated only by screening virus stocks for interesting phenotypes, for none of the tools we now take for granted, such as restriction endonucleases, efficient DNA sequencing methods, or molecular cloning procedures, were developed until the mid- to late 1970s. RNA virus stocks usually contain a high proportion of mutants, and it is only a matter of devising the appropriate selection conditions (e.g., high or low temperature or drugs that inhibit viral growth) to select mutants with the desired phenotype from the total population. For example, the live attenuated poliovirus vaccine strains developed by Albert Sabin are mutants that were selected from a virulent virus stock. RNA viral mutants resistant to neutralization with monoclonal antibodies are often isolated from wild-type stocks at a frequency of 1 in 100,000 plaque-forming units. RNA viruses have mutation frequencies on the order of 1 misincorporation in $10^4$ to $10^5$ nucleotides polymerized, compared to 1 misincorporation in $10^8$ to $10^{11}$ nucleotides incorporated for DNA viruses. This difference has been attributed to a lack of proofreading abilities in enzymes that replicate RNA, a property that results in lower-fidelity replication.

The low spontaneous mutation rate of DNA viruses necessitated random mutagenesis of the virus by exposure to a chemical **mutagen.** In vitro mutagens such as nitrous acid, hydroxylamine, and alkylating agents chemically modify the nucleic acid in the virus particle, resulting in changes in base pairing during subsequent replication and the substitution of an incorrect nucleotide. In vivo mutagens such as base analogs, intercalating agents, and ultraviolet light are applied to the infected cell. Such agents introduce mutations more or less at random. Some mutations are lethal under all conditions, while others have no effect and are said to be silent. To increase the chances of obtaining viruses with a single mutation, mutants were sought after exposure to the lowest concentration of mutagen that produced a useful number of mutants. Under such conditions, many of the virus particles in the mutagenized population contain wild-type genomes, making identification of mutants laborious. Some investigators have used chemicals to mutagenize RNA viral genomes despite their high mutation rate.

To facilitate identification of mutants, the population must be screened for a phenotype that can be easily identified in a plaque assay. One such phenotype is temperature-sensitive growth of the virus. Viral mutants with this phenotype reproduce well at low temperatures but poorly or not at all at high temperatures. The permissive and nonpermissive temperatures are typically 39 and 33°C, respectively, for viruses that replicate in mammalian cells. Other commonly sought phenotypes were changes in plaque size or morphology, cold-sensitive growth, drug resistance, antibody escape, and host range (e.g., loss of ability to multiply in certain hosts or host cells). The nomenclature used to identify viral and cellular genes, essential for describing mutations, is explained in Box 2.3.

## BOX 2.3

### Genetic nomenclature

From the earliest days of genetic analysis, it has been customary to use abbreviations to designate gene names. Unfortunately, no standard nomenclature has been established for the genes of mammalian cells and their viruses. To facilitate our discussion of genes and mutations, we shall use the following conventions. Abbreviations for cellular genes will be given in lowercase, italicized letters (e.g., *pvr* for poliovirus receptor gene). In abbreviations for proteins, the first letter will be capitalized, and the remainder will be lowercase (e.g., Pvr for poliovirus receptor protein). We will use the names of viral genes and proteins that have become firmly entrenched in the scientific literature, as they have come to have a life of their own. Sadly, this nomenclature is not consistent among viruses, and in some cases the gene and protein names do not relate to one another (e.g., the VP1 capsid protein of poliovirus is encoded by the 1A gene, and the membrane glycoprotein gD of herpes simplex virus is encoded by the US6 gene). The reader should consult the appendix to avoid confusion about such names.

## Mapping Mutations

Before the era of recombinant DNA, it was extremely difficult for investigators to determine the locations of mutations in viral genomes. The **marker rescue** technique (described in "Introducing Mutations into the Viral Genome" below) was the solution to this problem, but before it was developed, other less satisfactory approaches were exploited.

**Recombination mapping** can be applied to both DNA and RNA viruses. Recombination results in genetic exchange between genomes within the infected cell. For viruses with unimolecular genomes, the recombination frequency between two mutations increases with the physical distance separating them. In practice, cells are coinfected with two mutants, and the frequency of recombination is calculated by dividing the titer of phenotypically wild-type virus obtained under restrictive conditions (e.g., high temperature) by the titer obtained under permissive conditions (e.g., low temperature). The recombination frequency is determined between pairs of mutants, which can then be placed on linear maps. Although a location can be assigned for each mutation relative to others, this approach does not result in a **physical map** of the actual location of the base change in the genome.

For RNA viruses with segmented genomes, the technique of **reassortment** allowed the localization of mutations to specific genome segments. When cells are coinfected with both mutant and wild-type viruses, the progeny includes reassortants that inherit RNA segments from either parental virus. The origins of the RNA segments can be deduced from their migration patterns after gel electrophoresis (Fig. 2.14) or by nucleic acid hybridization. When a panel of such reassortants is analyzed, the segment responsible for the phenotype can be identified. When the protein products of each RNA segment are later identified, the mutation can be unambiguously assigned to a specific polypeptide. However, the physical location of the mutation on each RNA segment cannot be determined by this approach.

## Functional Analysis

**Complementation** describes the ability of gene products from two different viruses to interact functionally in the same cell, permitting virus replication. In a complementation assay, cells are coinfected with two mutants. If the mutations are in separate genes, each virus will supply a functional gene product, allowing both viruses to grow. If both viruses carry mutations in the same gene, no growth will occur. In this way, the members of collections of mutants obtained by chemical mutagenesis were initially organized into **complementation groups** defining

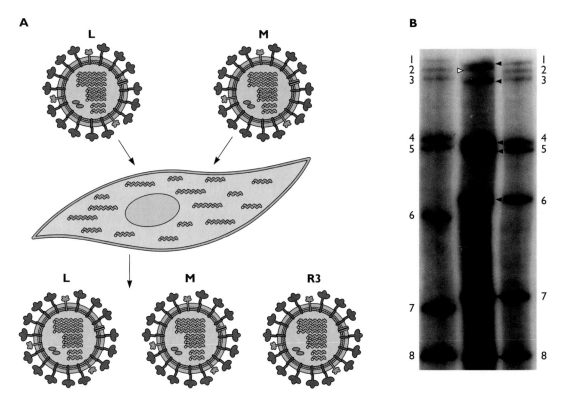

**Figure 2.14 Reassortment of influenza virus RNA segments.** (A) Schematic diagram of influenza virus reassortment. Cells are coinfected with two influenza virus strains, L and M. In infected cells the eight (−) strand RNAs of both viruses replicate. Progeny viruses include both parents and viruses that derived RNA segments from both parents. Recombinant R3 has inherited segment 2 from the L strain and the remaining seven segments from the M strain. (B) $^{32}$P-labeled influenza virus RNAs were fractionated on a polyacrylamide gel and detected by autoradiography. Migration differences of parental virus RNAs (M and L) permitted identification of the origin of RNA segments in the progeny virus R3. Solid arrowheads next to the RNAs of recombinant viruses indicate RNA segments derived from the M parent virus, and the open arrowhead indicates an RNA segment derived from the L parent virus. This analysis indicates that reassortant virus R3 derives all RNA segments from parent M except RNA 2, which is derived from virus L. Panel B is reprinted from V. R. Racaniello and P. Palese, *J. Virol.* **29:**361–373, 1979, with permission.

separate viral functions. In theory, there can be as many complementation groups as genes.

Complementation can be distinguished from recombination or reassortment by examining the progeny produced by coinfected cells. True complementation yields both parental mutants, while wild-type genomes result from recombination or reassortment.

## Engineering Mutations into Viruses

### Infectious DNA Clones

Recombinant DNA techniques have made it possible to introduce any kind of mutation anywhere in the genome of most animal viruses, whether the genome contains DNA or RNA. The holy grail of virology today is the **infectious DNA clone**—a double-stranded DNA copy of the viral genome that is carried on a bacterial plasmid. In-

fectious DNA clones, or in vitro transcripts derived from them, can be introduced into cultured cells by **transfection** (Box 2.4) to recover infectious virus. This approach is a modern validation of the Hershey-Chase experiment described in chapter 1. The availability of site-specific bacterial restriction endonucleases, DNA ligases, and an array of methods for mutagenesis has made it possible to manipulate these infectious clones at will. Infectious viral DNA clones also provide a stable repository of the virus, which is particularly important for vaccine strains.

**DNA viruses.** Current genetic methods for the study of most viruses with DNA genomes are based on the infectivity of viral DNA. When deproteinized viral DNA molecules are introduced into permissive cells by transfection or other methods, they generally initiate a complete infectious cycle, although the infectivity (i.e., plaques per mi-

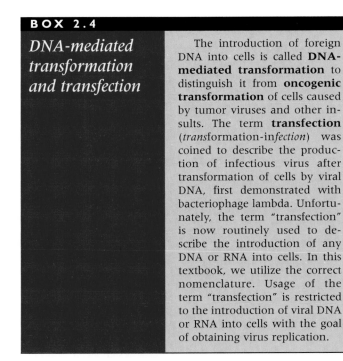

*DNA-mediated transformation and transfection*

The introduction of foreign DNA into cells is called **DNA-mediated transformation** to distinguish it from **oncogenic transformation** of cells caused by tumor viruses and other insults. The term **transfection** (*trans*formation-in*fection*) was coined to describe the production of infectious virus after transformation of cells by viral DNA, first demonstrated with bacteriophage lambda. Unfortunately, the term "transfection" is now routinely used to describe the introduction of any DNA or RNA into cells. In this textbook, we utilize the correct nomenclature. Usage of the term "transfection" is restricted to the introduction of viral DNA or RNA into cells with the goal of obtaining virus replication.

crogram of DNA) may be low. For example, the infectivity of deproteinized human adenoviral DNA is between 10 and 100 plaque-forming units per microgram. When the genome is isolated by procedures that do not degrade the covalently attached terminal protein, infectivity is increased to $10^3$ to $10^4$ plaque-forming units per μg, probably because this protein participates in the assembly of initiation complexes on the viral origins of replication.

The complete genomes of papovaviruses and adenoviruses can be cloned in plasmid vectors, and such DNA is infectious under appropriate conditions. The DNA genomes of herpesviruses are too large to insert into conventional bacterial plasmid vectors, but they can be cloned in vectors that can accept larger insertions (e.g., cosmids and bacmids), and these cloned DNAs are infectious. Among the DNA-containing viruses, poxvirus DNA is not infectious, and consequently other approaches to manipulating these genomes have been developed (see "Introducing Mutations into the Viral Genome" below).

**RNA viruses.**    *(+) strand RNA viruses.*    The first infectious DNA clones of RNA viruses were those of (+) strand RNA viruses, whose genomic RNA is infectious. To make an infectious clone of poliovirus, the single-stranded genomic RNA was reverse transcribed by using purified retroviral enzyme to form a complementary strand of DNA (cDNA). The cDNA was then made double stranded with DNA-dependent DNA polymerase and cloned into a plas-

mid vector. Introduction of the plasmid containing the cloned poliovirus DNA into cultured mammalian cells resulted in the production of progeny virus (Fig. 2.15A). The mechanism by which cloned poliovirus DNA initiates infection is not known, but it has been suggested that the DNA is transported to the nucleus, where it is transcribed by cellular DNA-dependent RNA polymerase from cryptic, promoterlike elements on the plasmid. Although the resulting RNA transcripts are longer than poliovirus RNA, they can be translated and replicated. During replication, the extra terminal nucleotide sequences must be removed or ignored because the viruses that are produced contain RNA with the authentic 5' and 3' termini.

The genomic RNA of poliovirus has a higher specific infectivity ($10^6$ plaque-forming units per μg) than cloned DNA ($10^3$ plaque-forming units per μg). By incorporating promoters for bacteriophage SP6 or T7 DNA-dependent RNA polymerase in plasmids containing poliovirus DNA, full-length (+) strand RNA transcripts can be synthesized in vitro. The specific infectivity of such RNA transcripts resembles that of genomic RNA. Consequently, most infectious clones of (+) strand RNA viruses are transcribed into infectious RNA in vitro. Infectious DNA clones have been constructed for many (but not all) (+) strand RNA viruses, including members of the *Picornaviridae*, *Caliciviridae*, *Togaviridae*, and *Flaviviridae*, and arteriviruses.

The genomic RNA of retroviruses is copied into a double-stranded DNA form by the viral enzyme reverse transcriptase early during infection, a process described in chapter 7. Such viral DNA is infectious when introduced into cells, as are molecularly cloned forms inserted into bacterial plasmids.

*(−) strand RNA viruses.*    Purified genomic RNA of (−) strand RNA viruses is not infectious because it can be neither translated nor copied into (+) strand RNA by host cell RNA polymerases, as discussed in chapter 6. This limitation delayed the isolation of infectious DNA clones of these viruses, as it was necessary to reconstitute functional **ribonucleoprotein complexes,** a technically difficult task. Two different experimental approaches have been used to achieve this goal (Fig. 2.15B and C).

A system for isolation of influenza viruses containing RNA segments produced from genetically modified DNA has been developed. However, it is not yet possible to reconstitute the entire viral genome from cloned DNAs. The influenza virus genome consists of eight single-stranded RNA segments, each associated with four viral proteins (NP, PA, PB1, and PB2) in a ribonucleoprotein that is infectious when introduced into cells by transfection. RNA segments produced by in vitro transcription of a cloned DNA copy of one of the genomic segments are therefore

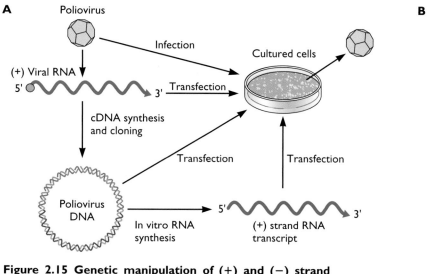

**A**

Poliovirus

Infection

Cultured cells

(+) Viral RNA

5' ◯ ～～～～～～～ 3'    Transfection

cDNA synthesis
and cloning

Transfection    Transfection

Poliovirus
DNA

In vitro RNA
synthesis

5' ～～～～～～ 3'

(+) strand RNA
transcript

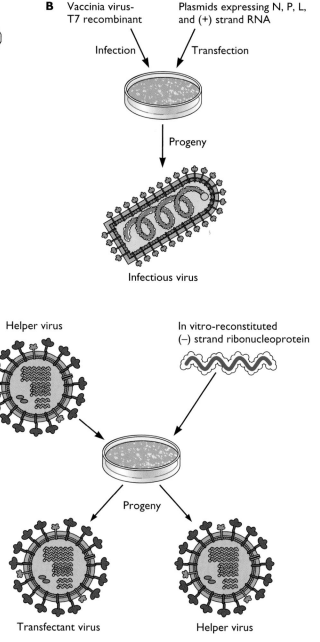

**B**    Vaccinia virus-        Plasmids expressing N, P, L,
        T7 recombinant         and (+) strand RNA

Infection              Transfection

Progeny

Infectious virus

**C**    Helper virus            In vitro-reconstituted
                                (−) strand ribonucleoprotein

Progeny

Transfectant virus        Helper virus

**Figure 2.15 Genetic manipulation of (+) and (−) strand RNA viruses.** (A) Recovery of infectivity from cloned DNA of (+) strand RNA genomes as exemplified by poliovirus, whose genomic RNA is infectious when introduced into cultured cells by transfection. A complete DNA clone of the viral RNA, carried in the plasmid pBR322, is also infectious in cultured cells, as are RNAs derived by in vitro transcription (using bacteriophage T7 RNA polymerase) from the full-length DNA. (B) Recovery of infectivity from cloned DNA of viruses with a (−) strand RNA genome. Cells are infected with a vaccinia virus recombinant that expresses T7 RNA polymerase and transformed with plasmids that express (+) strand RNA and proteins required for viral RNA synthesis. The (+) strand RNA is copied into (−) strands, which in turn are used as templates for mRNA synthesis and genome replication. Infectious progeny viruses are produced. The example shown is for viruses with a single (−) strand RNA genome (e.g., rhabdoviruses and paramyxoviruses). A similar approach has been demonstrated for Bunyamwera virus, whose genome consists of three (−) strand RNAs. (C) Method for introducing DNA-derived RNA segments into the genome of influenza virus. Cells are infected with a helper virus and transformed with a ribonucleoprotein complex reconstituted in vitro from a (−) strand RNA segment transcribed from cloned DNA. Replication and reassortment of genome segments occurs in the cell. Progeny viruses include helper virus ($\sim 10^5$ plaque-forming units per ml) and transfectant virus in which one RNA segment has been replaced by a DNA-derived transcript ($\sim 10$ plaque-forming units per ml). The few transfectant viruses can be isolated with appropriate selection techniques. For example, if the RNA synthesized in vitro encodes drug resistance, transfectant viruses can be enriched in the presence of drug.

mixed with NP, PA, PB1, and PB2 proteins to form a ribonucleoprotein. When this complex is introduced into cells infected with a **helper virus,** the RNA segment made in vitro is replicated by the helper virus. Progeny viruses that contain the in vitro-transcribed RNA can be identified if an appropriate selection system is available. For example, if the RNA transcript corresponds to a segment con-

taining a drug resistance marker, viruses containing this segment can be selected in the presence of the drug. In this way, mutations introduced into the transcript can be incorporated into virus. More recently, NP, PA, PB1, and PB2 proteins have been provided to the cell with plasmid-based expression vectors, bypassing the need for in vitro reconstitution of the ribonucleoprotein.

A helper virus-free system has been developed for Bunyamwera virus, obviating the need for selection methods for isolating recombinant viruses. Members of the *Bunyaviridae* contain three segmented (−) strand RNAs. When plasmids containing full-length DNA copies of the three RNA segments, under the control of the bacteriophage T7 promoter, are introduced into cells producing T7 RNA polymerase and Bunyamwera virus proteins, infectious virus is produced.

For (−) strand RNA viruses possessing a single RNA strand, such as vesicular stomatitis virus (a rhabdovirus), the approach to obtaining an infectious DNA has been to transcribe a DNA clone of the entire genome in vitro to produce a full-length (+) strand RNA (Fig. 2.15). This RNA is introduced into cells that express the vesicular stomatitis virus nucleocapsid protein, phosphoprotein, and polymerase from separate plasmids. The (+) strand RNA is copied into (−) strand RNAs, which initiate an infectious cycle leading to the production of new virus particles. For unknown reasons, transfection of a full-length (−) strand RNA into cells producing the nucleocapsid, phosphoprotein, and polymerase fails to initiate an infectious cycle.

### Types of Mutations

Recombinant DNA techniques allow the introduction of many kinds of mutations at any desired site in cloned DNA. Indeed, provided that the sequence of the segment of the viral genome to be mutated is known, there is little restriction on the type of mutation that can be introduced. **Deletion mutations** can be used to remove an entire gene to assess its role in replication, to produce truncated gene products, or to assess the functions of specific segments of a coding sequence. Noncoding regions can be deleted to identify and characterize regulatory elements such as promoters. **Insertion mutations** can be made by the addition of unrelated sequences or sequences derived from a closely related virus. **Substitution mutations,** which can correspond to one or more nucleotides, are often made in coding or noncoding regions. Included in the latter class are **nonsense mutations,** in which a termination codon is introduced, and **missense mutations,** in which a single base or a codon is changed, resulting in the expression of a protein with a single amino acid substitution. The introduction of a termination codon is frequently exploited to cause truncation of a membrane protein so that it is secreted or to eliminate the synthesis of a protein without changing the size of the viral genome or mRNA. Point mutations are used to assess the roles of specific nucleotides in regulatory sequences or of single amino acids in protein function, such as polymerase activity or binding of viral protein to a cell receptor. An example of a longer substitution is the exchange of nucleic acid

**BOX 2.5**

*Operations on DNA and on protein*

A mutation is a change in DNA or RNA comprising base changes, nucleotide additions, deletions, and rearrangements. When mutations occur in open reading frames, they can be manifested as changes in the expressed proteins. For example, one or more base changes in a specific codon may produce a single amino acid substitution, a truncated protein, or no protein. The terms "mutation" and "deletion" are often used incorrectly or ambiguously to describe alterations in proteins. In this textbook, these terms are used to describe genetic changes, and the terms "amino acid substitution" and "truncation" are used to describe protein alterations.

sequences encoding a specific protein sequence among virus strains that differ in host range. The construction of such hybrid viruses can be used to assign a biological function to a specific gene product and to determine whether it functions independently of the viral genetic background.

A large variety of experimental methods can be used to construct deletion, insertion, and substitution mutations, and it is beyond the scope of this book to cover all of them. Much of what can be achieved with mutagenesis today depends on the development of two important factors: low-cost chemical synthesis of specific **oligonucleotides** and the polymerase chain reaction for amplifying selected regions of DNA by using synthetic oligonucleotides. Figure 2.16 illustrates how the polymerase chain reaction can be used to introduce a deletion, insertion, or substitution mutation into DNA.

### Introducing Mutations into the Viral Genome

Mutations can be introduced rapidly into a viral genome when it is cloned in its entirety, in either bacterial plasmids or cosmids. However, mutagenesis is usually carried out on cloned subfragments of the viral genome, which are then substituted into full-length cloned DNA using recombinant DNA techniques. The final step is introduction of the mutagenized DNA into cultured cells by transfection. This approach has been applied to cloned DNA copies of RNA viral genomes and smaller viral DNA genomes and has recently been extended to the genomes of human adenoviruses and herpesviruses.

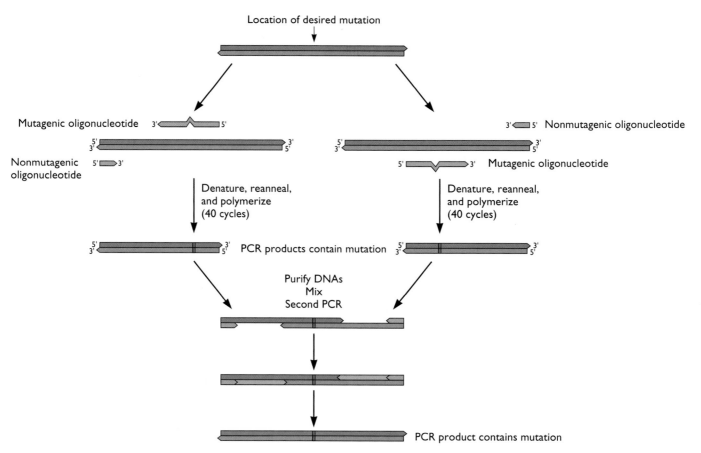

**Figure 2.16 Mutagenesis by polymerase chain reaction.** To introduce a mutation into DNA, two polymerase chain reaction mixtures are assembled with oligonucleotides complementary to either the left or right ends of the DNA to be mutagenized (which may be part of a larger sequence) and a mutagenic oligonucleotide. The products of the first set of reactions are two fragments that overlap in the area to be mutagenized. Next, the two DNAs are combined and polymerase chain reaction (PCR) is done with both the left- and right-end oligonucleotides. The double-stranded DNAs are denatured, and a fraction anneals such that the two different products overlap in the area of the mutation. Oligonucleotide-primed DNA synthesis results in the formation of completely double-stranded molecules containing the mutation at the desired site. Depending on the sequence of the oligonucleotide, this strategy may be used to introduce point mutations, deletions, or insertions into DNA.

A less direct, but effective, method for introducing mutations into a viral DNA genome is **marker transfer** (a mutation is marked or traced by its phenotype). This method was originally developed to locate mutations within the viral genome in mutant viruses isolated by classical methods. For DNA viruses, site-specific bacterial restriction endonucleases were used to generate physical maps of viral DNA on which the reference points are the sites at which such enzymes cleave. These endonucleases also allowed the preparation of discrete subgenomic fragments of viral DNA genomes. In the marker rescue mapping technique, the discrete fragments representing the complete genome of the virus are individually introduced into host cells together with a mutant virus. Only recombination with the wild-type fragment that spans the site of the mutation generates wild-type progeny virus, allowing the mutation to be mapped to a specific interval on the physical map. This method of mutant genome rescue is now largely of historical interest as a procedure for mapping mutations as mutations can be introduced at predetermined sites within the viral genome. However, this technique is widely used to introduce mutations into viral genomes and is now known as marker transfer.

The marker transfer method is applicable to any DNA virus provided the target sequence is available as a DNA fragment in a vector that allows cloning and amplification in *Escherichia coli*. As in the marker rescue method, cells are transfected with the viral genome together with a DNA fragment containing the desired mutation. Recombination between the cotransfected DNAs produces viral genomes containing the desired mutation. The use of marker transfer to introduce mutations into the herpes simplex UL54 gene is shown in Fig. 2.17. A similar approach is used to introduce mutations into the genome of poxviruses, whose genomic DNA is not infectious. In marker transfer with adenoviruses, overlapping subgenomic fragments, one of which contains the desired mutation, are introduced into cells (Fig. 2.18). Overlap recombination gives

**Figure 2.17 Marker transfer in herpes simplex virus.** (A) Schematic drawing of the viral DNA genome with the region containing the UL54 gene expanded. The viral gene to be mutated, the herpes simplex UL54 gene, is first replaced in the virus with the *lacZ* gene, which encodes a bacterial enzyme (β-galactosidase) that converts the chromogenic substrate X-Gal to a blue product. The substitution is achieved by co-transforming cells with viral DNA and a plasmid containing the *lacZ* gene surrounded by genome sequences that flank the UL54 gene. Recombination between the homologous sequences in these DNAs leads to replacement of the wild-type UL54 gene by *lacZ*. Recombinant viral progeny can be detected because they form blue plaques in the presence of X-Gal. Because these viruses lack the UL54 gene, they must be grown in cells that produce its gene product, the α27 protein. (B) Next, the UL54 gene on another plasmid is mutagenized, and the mutant versions are introduced into cells with viral DNA containing the *lacZ* substitution. In the presence of X-Gal, white plaques are formed by viruses in which the *lacZ* gene has been replaced by UL54 mutant alleles by recombination. Courtesy of R. Soliman and S. Silverstein, Columbia University.

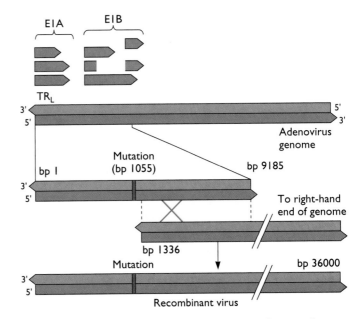

**Figure 2.18 Marker transfer in adenovirus by overlap recombination.** To introduce mutations into the adenovirus genome, two overlapping DNA fragments which together constitute the entire viral genome are prepared. In the example shown, a mutation is introduced into the E1A region. The left-end fragment is approximately 9 kb long and contains the mutation at bp 1055 (indicated by a red bar). The right-end fragment contains all but 1.3 kb from the left-hand end of the genome. Alone, neither DNA is infectious, but when they are introduced together into cells, homologous recombination generates a full-length genomic DNA that contains the desired mutation. (Top) Schematic drawing of the adenovirus genome showing the location of E1A and E1B RNA transcripts. Courtesy of C. S. H. Young, Columbia University.

rise to mutant viruses, and because no wild-type genomes are present, the only viruses produced are mutants. A procedure similar to the marker transfer method can be used to introduce mutations into the large (26 to 31-kb) (+) strand RNA genome of coronaviruses, for which an infectious clone has not been developed. Targeted RNA recombination is used to introduce mutations into the viral genome, as illustrated in Fig. 2.19.

Introduction of mutagenized viral nucleic acid into cultured cells by transfection may produce one of several possible results. The mutation might have no effect on virus replication; it might have a subtle effect that is discovered only on subsequent study; it might impart a readily detectable phenotype, which may be constitutive or conditionally lethal; or it might prevent the production of infectious viruses. In the last-mentioned case, it is necessary to complement these defects in cell lines that have been engineered to produce specific viral proteins. For example, herpes simplex viruses containing the *lacZ* gene in

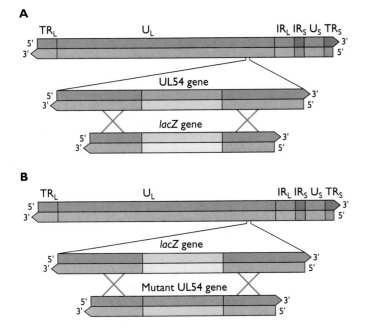

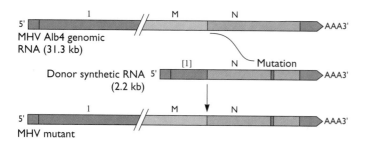

**Figure 2.19 Introduction of defined mutations into a coronavirus, mouse hepatitis virus (MHV), by targeted RNA recombination.** Site-specific mutations are engineered into the viral genome by RNA-RNA recombination between a recipient virus and donor RNA synthesized in vitro. In the example shown, the recipient virus is a mouse hepatitis virus mutant, Alb4, that is thermolabile because of a deletion in the nucleocapsid protein (N) gene and forms tiny plaques at the nonpermissive temperature. Only genes 1, M, and N are shown. The donor synthetic RNA contains the entire (undeleted) N gene and the 3' untranslated region of the viral genome, as well as sequences from the 5' end of the genome that allow it to replicate in the presence of Alb4 as a helper virus. Only a fragment of gene 1 is present (1). The donor N gene has also been constructed with a site-specific mutation of interest (indicated by a red bar). Recombinants resulting from transfection of donor RNA into Alb4-infected cells are selected as progeny virus capable of forming large plaques at the nonpermissive temperature following heat treatment to kill parental Alb4 virus. These selected progeny are then screened for the presence of the mutation of interest, which is contained in the recombinants that were generated by a crossover event at any point in the RNA to the left of the mutation. Figure courtesy of P. Masters, New York State Department of Health.

place of the UL54 gene must be propagated in complementing cell lines that contain and express the viral gene targeted for mutagenesis (Fig. 2.17).

When the sequences of interest are those that control viral gene expression or replication, rather than coding sequences, complementation is not possible. This limitation is not as prohibitive as it might appear. One approach that can be used to study control regions is to introduce mutations that impair but do not eliminate virus replication. For example, certain mutations introduced into the major late promoter of subgenus C adenoviruses reduce virus yield more than 30-fold. An alternative approach is to place the viral promoter at an **ectopic site** (a second site at which they are not present in the wild-type genome) linked to a reporter gene whose product is easily assayed (e.g., bacterial chloramphenicol transferase or β-galactosidase, firefly luciferase, or green fluorescent protein from the jellyfish *Aequorea victoria*). The ectopic site can be located within a nonessential region of the same genome or that of another viral genome altogether. An ectopic site commonly used for the insertion of reporter genes with

putative control elements is the E3 region of human adenovirus (Fig. 2.20). The E3 gene products are not necessary for growth of the virus in cells in culture, for their function is to counteract the host immune defenses, as described in chapter 14. This approach permits the study of severely inhibitory mutations of transcriptional control elements when introduction of the same mutations into the control elements in their normal context would generate mutants that would not be viable.

### Reversion Analysis

The phenotypes caused by mutation can **revert** in one of two ways: by reversion of the original mutation to the wild-type sequence or by mutation at a second site, either in the same gene or in a different gene. Phenotypic reversion caused by second-site mutation is known as **suppression,** or **pseudoreversion,** to distinguish it from reversion at the original site of mutation. Reversion has been studied since the beginnings of classical genetic analysis. When suspected pseudorevertants are crossed with wild-type viruses, some viruses with the mutant phenotype should be observed in the progeny as a result of segregation of the

**Figure 2.20 Genetic analysis of control sequences placed at an ectopic site in the adenovirus genome.** To study lethal mutations in a region that controls the synthesis of RNAs (e.g., a promoter region), a second copy of the control region is inserted into the E3 region of the adenovirus genome. This region can be deleted, as it is nonessential for growth in cultured cells. The effect of different mutations on the expression of a reporter gene can then be examined in infected cells and compared to that obtained with the wild-type control sequence. The control region may be derived from adenovirus, in which case the ectopic copy results in a duplication. This approach can also be used to study control regions of other viral and even cellular genes.

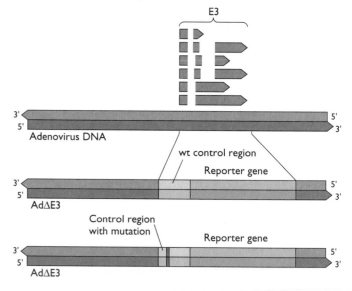

original mutation from the suppressor mutation. In the modern era of genetics, cloning and sequencing techniques can be used to demonstrate suppression and to identify the nature of the suppressor mutation (see below). The identification of suppressor mutations is a powerful tool for studying protein-protein and protein-nucleic acid interactions. Some suppressor mutations complement changes made at several sites, whereas **allele-specific** suppressors complement only a specific change. The allele specificity of second-site mutations provides evidence for physical interactions among proteins and nucleic acids.

Phenotypic revertants can be isolated either by propagating the mutant virus under restrictive conditions or, with mutants exhibiting nonconditional phenotypes, by searching for wild-type properties. For DNA viruses, chemical mutagenesis may be required to generate revertants, but this is not necessary for RNA viruses, which spawn mutants at a higher frequency. Next, nucleotide sequence analysis is used to determine whether the original mutation is still present in the genome of the revertant. The presence of the original mutation(s) indicates that reversion has occurred by second-site mutation. Even in the smallest viral genomes, it is not practical to sequence the entire genome to identify the suppressor mutation. Therefore, a form of marker rescue is used to identify the region that includes the suppressor. For DNA viruses, restriction fragments derived from different parts of the genome of phenotypic revertants can be substituted in the original mutant viral genome by ligation in vitro or recombination in vivo. The objective is to identify a fragment that contains the suppressor mutation and is small enough (~1 kb) so that its nucleotide sequence can be determined. A similar analysis can be carried out for RNA viruses, except that DNA clones must first be derived from the genomes of phenotypic suppressors. The final step is introduction of the suspected suppressor mutation into the genome of the original mutant virus to confirm its effect. Several specific example of suppressor analysis are provided below.

Some mutations within the origin of replication (Ori) of simian virus 40 inhibit viral DNA replication, with the formation of small viral plaques. Pseudorevertants of Ori mutants were isolated by random, local mutagenesis of mutant viral DNA followed by introduction into cultured cells and selection of viruses that form large plaques. The second-site mutations that suppressed the replication defects were localized to a specific region within the gene for large T antigen by in vitro recombination or marker rescue experiments. These results indicated that a specific domain of large T antigen interacts with the Ori signal during virus replication.

The 5' untranslated region of the poliovirus genome contains extensive RNA secondary-structure features that are important for RNA translation and replication, as discussed in chapters 6 and 11. Disruption of such secondary structure by insertion of an eight-nucleotide sequence produces a virus that is poorly infectious and readily gives rise to pseudorevertants (Fig. 2.21). Nucleotide sequence analysis of two pseudorevertants demonstrated that they contain base changes that restore the disrupted secondary structure. These results confirm that the RNA secondary structure is important for the biological activity of this untranslated region.

Mutations in the viral genome may also suppress defects in a host cell protein. An example of such extragenic suppression is the selection of virus mutants that utilize altered cell receptors (Fig. 2.22). Amino acid changes were introduced into the poliovirus receptor. Wild-type virus cannot bind to cells that express the altered receptors. Nevertheless, viruses that bind to the altered receptors can be selected. Some of the selected viruses contain allele-specific suppressor mutations that allow the virus to bind to a receptor containing a single specific amino acid change. These capsid protein alterations are believed to define regions of the viral capsid that bind the cell receptor.

## Engineering Viral Genomes: Viral Vectors

Naked DNA can be introduced into cultured animal cells by a variety of methods, including the formation of complexes with calcium chloride or lipid-based reagents or by electroporation. Such DNA can support expression of its gene products transiently or stably from an integrated or episomal copy of the DNA. Introduction of DNA into cells is a routine method in virological research and is also employed for certain clinical applications, such as the production of a therapeutic protein or a vaccine. However, this approach is not suitable for certain applications. For example, one goal of gene therapy is to deliver a gene to patients who either lack the gene or carry defective versions of it. An approach to this problem is to create a cell line that expresses the gene product. After infusion into patients, the cells become permanently established in tissues. If the primary cells to be used are limiting in a culture (e.g., stem cells), it is not practical to select and amplify the rare cells that receive naked DNA. Recombinant viruses carrying foreign genes can infect a greater percentage of cells and thus facilitate generation of the desired cell lines. These viral vectors have also found widespread use in the research laboratory. While the subject of this textbook is not gene therapy, some clinical applications will be discussed. The reader should note that a complete understanding of the structure and function of viral vectors requires a knowledge of viral genome replication, which is discussed in subsequent chapters for selected viruses and is summarized in the appendix.

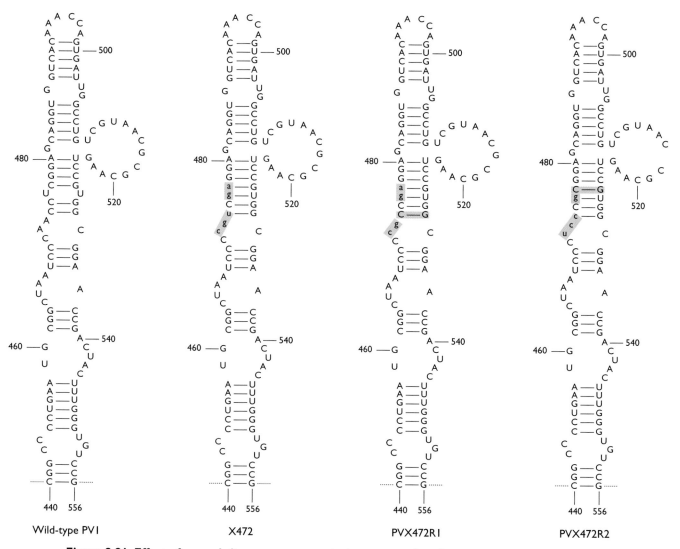

**Figure 2.21 Effect of second-site suppressor mutations on predicted secondary structure in the 5′ untranslated region of poliovirus (+) strand RNA.** A diagram of the region between nucleotides 448 and 556 is shown, which corresponds to stem-loop V (chapter 6). The sequences of wild-type poliovirus type 1 (PV1), X472 (a mutant containing the nucleotide changes highlighted in pink), and PVX472R1 and PVX472R2 (two phenotypic revertants) are shown. Two C·G base pairs present in the wild-type parent and destroyed by the mutation are restored by second-site reversion (blue shading). Adapted from A. A. Haller et al., *J. Virol.* **70**:1467–1474, 1996, with permission.

The design requirements for viral vectors include (1) the inclusion of appropriate promoter elements, (2) maintenance of genome size within the packaging limit of the virus, and (3) elimination of viral virulence, the capacity of the virus to cause disease. Expression of foreign genes in viral vectors may be controlled by homologous or heterologous promoters and enhancers chosen to support efficient transcription (e.g., the human cytomegalovirus immediate-early transcriptional control region) or tissue-specific expression, depending on the goals of the experiment. Such genes can be built directly into the viral

genome or introduced by recombination in cells, as described previously. However, the recipient viral genome is generally not the wild type but rather carries deletions and sometimes additional mutations. Deletion of some viral sequences is often required to overcome the limitations on the size of viral genomes that can be packaged in virions when the foreign gene(s) is introduced. For example, adenoviral DNA molecules more than 105% the length of the wild type are packaged very poorly. As this limitation would allow only 1.8 kbp of exogenous DNA to be inserted into the genome, adenovirus vectors often include

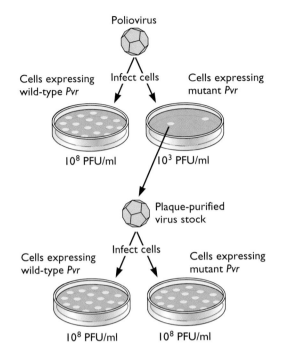

**Figure 2.22 Mutations in a virus that suppress host cell receptor defects.** Cell lines that express wild-type and altered versions of the poliovirus receptor, Pvr, are established. Poliovirus cannot efficiently bind to altered Pvr molecules, and as a result virus replication is impaired. The titer of poliovirus on cells expressing wild-type Pvr is approximately $10^8$ plaque-forming units per ml, and that on cells expressing altered Pvr is $10^3$ plaque-forming units per ml. Viruses selected after growth on cells expressing altered Pvr have similar titers ($10^8$ plaque-forming units per ml) on both cell lines. These selected viruses contain mutations that allow them to bind to the defective receptor (in addition to the wild-type receptor), restoring their ability to grow on cells expressing mutant Pvr. Some of the selected viruses contain allele-specific suppressors which allow the virus to utilize receptors with a specific amino acid change. Other selected viruses contain mutations that can suppress the defects in all altered Pvr molecules.

deletions of the E3 gene (which is not essential for growth in cells in culture) and of the E1A and E1B transcription units, which encode proteins that can be provided in complementing cell lines.

When viral vectors are designed for therapeutic purposes, it is essential to prevent replication of the vector within and hence destruction of the target host cells. The deletions necessary to accommodate a foreign gene may contribute to such disabling of the vector. For example, the E1A protein-coding sequences that are invariably deleted from adenovirus vectors are necessary for efficient transcription of viral early genes; in their absence, viral yields from cells in culture are reduced by about three to six orders of magnitude (depending on the cell type). Thus, removal of E1A coding sequences from adenovirus vectors is

doubly beneficial, although not sufficient to ensure that the vector cannot replicate or induce damage in a host animal. As discussed in detail in chapter 17, this goal can be more difficult to achieve.

A summary of viral vectors is presented in Table 2.3, and selected examples are discussed below.

### DNA Virus Vectors

One goal of gene therapy is to introduce genes into terminally differentiated cells. Such cells normally do not divide, they cannot be cultured in vitro, and the organs of which they are a part cannot be infused with virus-infected cells. DNA virus vectors have been developed to overcome some of these problems.

Adenovirus vectors were originally developed for the treatment of cystic fibrosis because of the tropism of the virus for the respiratory epithelium. Adenovirus can infect terminally differentiated cells, but only transient expression is provided, as this viral DNA is not integrated into the chromosome. Adenoviruses carrying the cystic fibrosis transmembrane regulator gene, which is defective in patients with this disease, have been used in clinical trials. Many other gene products with therapeutic potential have been expressed from adenovirus vectors, and these have been expressed in a wide variety of cell types.

Vaccinia virus vectors have the advantages of a wide host range, a genome that accepts very large fragments, high expression, and relative ease of preparation. Foreign DNA is usually inserted into the viral genome by homologous recombination, with an approach similar to that described for marker transfer. Because of the relatively low pathogenicity of the virus, vaccinia virus recombinants expressing foreign proteins have been considered candidates for human and animal vaccines.

### RNA Virus Vectors

A number of RNA viruses have also been used as vectors for foreign gene expression (Table 2.3). Because poliovirus is such an excellent mucosal immunogen, it has been an attractive candidate for the delivery of antigens to these surfaces. Alphavirus vectors have been shown to be capable of expressing foreign genes. These (+) strand RNA animal viruses produce subgenomic mRNAs from which efficient gene expression can be obtained. In one type of vector, the foreign gene replaces those encoding the viral structural proteins. As a result, this vector requires the presence of a helper virus to provide the missing proteins so that the vector genome can be packaged into virus particles. An alternative arrangement is to place the foreign gene under the control of a second, subgenomic RNA promoter, leaving intact the promoter that controls synthesis

**Table 2.3** Viral vectors

| Virus | Insert size | Persistence | Advantages | Disadvantages |
|---|---|---|---|---|
| Poliovirus | ~300 bp for helper-free virus, ~3 kb for defective virus | Transient | Excellent mucosal immunity | Limited capacity, reversion to neurovirulence |
| Alphavirus | ~5 kb | Transient | Broad host range, high-level expression | Virulence |
| Rhabdovirus | Unknown | Transient | High-level expression, rapid cell killing | Virulence, highly cytopathic |
| Influenza virus | Unknown | Transient | Strong immune response | Virulence |
| Retrovirus | 1–8 kb | Permanent | Stable integration | May rearrange genome, may require cell division (human immunodeficiency virus type 1 is one exception) |
| Adeno-associated virus | ~4 kb | Transient; wild type integrates | Prolonged transient | Limited capacity, current vectors nonintegrating |
| Adenovirus | 2–35 kb | Transient | Efficient gene delivery | Transient, immunogenic |
| Herpes simplex virus | Unknown | Stable in neurons | Neurotropic, large capacity | Virulence, persistence in neurons |
| Vaccinia virus | At least ~25 kb, probably ~75–100 kb | Transient | Wide host range, ease of isolation, large capacity, high expression | Transient, immunogenic |

of the structural proteins. This vector replicates without a helper and produces infectious particles. Both vectors can be used to express proteins in eukaryotic cells.

The ability to manipulate the genome of influenza virus has been exploited in the use of this virus as a vector, particularly for vaccine purposes. For example, immunization of mice with a recombinant influenza virus that expresses specific antigens of *Plasmodium yoelii* resulted in a protective immune response against the malaria parasite *Plasmodium falciparum*. Similarly, mice immunized with recombinant influenza viruses that express epitopes from human immunodeficiency virus generated human immunodeficiency virus-specific serum antibodies and secretory IgA in the nasal, vaginal, and intestinal mucosa. These results suggest that influenza viruses may be engineered for use as live vaccines for protection against other pathogens.

The development of an infectious DNA clone of vesicular stomatitis virus, a (−) strand RNA virus, has led to a potential approach to therapy for human immunodeficiency virus type 1 infection. Recombinant vesicular stomatitis virus particles that lack the viral glycoprotein gene used for attachment to host cells have been produced. In its place are two cell proteins required for the attachment of human immunodeficiency virus type 1 to cells (CD4 and Cxcr4) (chapter 5). The recombinant vesicular stomatitis viruses cannot infect normal cells. However, they can infect and lyse cells infected with human immunodeficiency virus type 1 because these cells express viral glyco-

proteins on the surface. This approach has the potential of limiting the number of infected cells in the host.

Retroviruses have enjoyed great popularity as vectors because their replication cycle includes the integration of a double-stranded DNA copy of viral RNA into the cell genome, a topic of chapter 7. The integrated provirus remains permanently in the cell's genome and is passed on to progeny cells during cell division. This feature of retroviral vectors results in permanent modification of the genome of the infected cell. The minimal viral sequences required for retroviral vectors are 5'- and 3'-terminal sequences that control gene expression and packaging of the RNA genome (Fig. 2.23). The foreign gene is cloned between the terminal sequences, and the DNA is introduced into cell lines that have been modified to synthesize viral proteins. As no wild-type viral RNA is present, the result is the formation of retrovirus particles containing only the recombinant vector genome and no wild-type particles.

A problem with the use of retroviruses in correcting genetic deficiencies is that only a few cell types can be infected by the commonly used retroviral vectors, which can integrate their DNA only in actively dividing cells. Often the cells that are targets of gene therapy, bone marrow stem cells, hepatocytes, and muscle cells, do not divide. This problem can be circumvented if ways can be found to induce such cells to divide before infection with the retrovirus. An alternative approach is to use viral vectors containing sequences from human immunodeficiency virus type 1, which can infect nondividing cells.

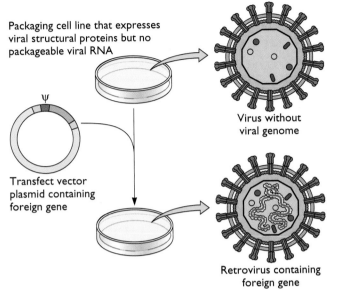

Packaging cell line that expresses viral structural proteins but no packageable viral RNA

Ψ

Transfect vector plasmid containing foreign gene

Virus without viral genome

Retrovirus containing foreign gene

**Figure 2.23 Retrovirus vectors.** The minimal viral sequences required for retroviral vectors are 5′- and 3′-terminal sequences that control gene expression (yellow boxes) and packaging (green box labeled Ψ) of the RNA genome. The foreign gene (red box) is cloned between the viral sequences. The DNA is introduced into packaging cell lines that synthesize viral structural proteins. No wild-type viral RNA is present in these packaging cell lines, which produce virus particles without viral genomes. When retrovirus vector DNA is introduced into packaging cell lines, the result is the formation of retrovirus particles that contain only the recombinant vector genome and no wild-type particles.

## Perspectives

The development of new technologies is essential for scientific progress. Methods are not developed for their own sake, but rather to drive a field forward and open new vistas. For example, the development of methods for propagating animal viruses in cell culture in the 1950s led to an explosion of activity in the field of viral genetics. The techniques of DNA sequencing, molecular cloning, site-directed mutagenesis, and monoclonal antibody production introduced in the 1970s and 1980s made it possible to isolate, characterize, mutagenize, and express individual genes, the foundations of today's genetic analysis. In recognition of the importance of technology to scientific progress, some Nobel Prizes have been bestowed for the development of new methods (Table 2.4). We can only guess what the next blockbuster method might be. Just when it appeared that it would be difficult to top recombinant DNA methodology, the polymerase chain reaction was developed, and it is now used in nearly every molecular biology laboratory. The beauty of a great method is not only its universal utility, but its ability to invoke in all of us the same thought: "I *knew* that would work!"

**Table 2.4** Some Nobel Prizes awarded for the development of new technologies

| Yr | Prize | Recipient(s) | Discovery |
|----|-------|--------------|-----------|
| 1993 | Chemistry | Kary Mullis | Polymerase chain reaction |
| | | Michael Smith | Oligonucleotide-directed mutagenesis |
| 1984 | Medicine | Niels K. Jerne, Georges Köhler, Cesar Milstein | Monoclonal antibodies |
| 1980 | Chemistry | Paul Berg | Recombinant DNA |
| | | Walter Gilbert, Frederick Sanger | DNA sequencing |
| 1978 | Medicine | Werner Arber, Daniel Nathans, Hamilton Smith | Restriction endonucleases |
| 1977 | Medicine | Rosalyn Yalow | Radioimmunoassay |
| 1975 | Medicine | David Baltimore, Renato Dulbecco, Howard Temin | Reverse transcriptase |
| 1958 | Chemistry | Frederick Sanger | Protein sequencing |
| 1954 | Medicine | John Enders, Thomas Weller, Frederick Robbins | Growth of poliovirus in cell culture |

## References

### Textbooks

Ausubel, F. M., R. Brent, R. E. Kingston, D. D. Moore, J. G. Seidman, J. A. Smith, and K. Struhl. 1998. *Current Protocols in Molecular Biology.* John Wiley & Sons, Inc., New York, N.Y.

Coligan, J. E., B. M. Dunn, H. C. Ploegh, D. W. Speicher, and P. T. Wingfeld. 1998. *Current Protocols in Protein Science.* John Wiley & Sons, Inc., New York, N.Y.

Harlow, E., and D. Lane. 1988. *Antibodies: a Laboratory Manual.* Cold Spring Harbor Laboratory Press, Cold Spring Harbor, N.Y.

Sambrook, J., E. F. Fritsch, and T. Maniatis. 1989. *Molecular Cloning: a Laboratory Manual,* 2nd ed. Cold Spring Harbor Laboratory Press, Cold Spring Harbor, N.Y.

Suzuki, D. T., A. J. F. Griffiths, J. H. Miller, and R. C. Lewontin. 1985. *An Introduction to Genetic Analysis.* W. H. Freeman & Co., New York, N.Y.

### Review articles

Coen, D. M., and R. F. Ramig. 1996. Viral genetics, p. 113–152. *In* B. N. Fields, D. M. Knipe, P. M. Howley, R. M. Chanock, J. L. Melnick, T. P. Monath, B. Roizman, and S. E. Straus (ed.), *Fields Virology.* Lippincott-Raven, New York, N.Y.

Conzelmann, K.-K., and G. Meyers. 1996. Genetic engineering of animal RNA viruses. *Trends Microbiol.* **4:**386–393.

Maramorosch, K., and H. Koprowski. *Methods in Virology.* Academic Press, New York, N.Y. A series of volumes begun in the 1960s that contain review articles on virological methods.

*Papers of Special Interest*

Bridgen, A., and R. M. Elliott. 1996. Rescue of a segmented negative-strand RNA virus entirely from cloned complementary DNAs. *Proc. Natl. Acad. Sci. USA.* **93:**15400–15404.

Goff, S. P., and P. Berg. 1976. Construction of hybrid viruses containing SV40 and lambda phage DNA segments and their propagation in cultured monkey cells. *Cell* **9:**695–705.

Graham, F. L., J. Rudy, and P. Brinkley. 1989. Infectious circular DNA of human adenovirus type 5: regeneration of viral DNA termini from molecules lacking terminal sequences. *EMBO J.* **8:**2077–2085.

Luytjes, W., M. Krystal, M. Enami, J. D. Pavin, and P. Palese. 1989. Amplification, expression, and packaging of a foreign gene by influenza virus. *Cell* **59:**1107–1113.

Mackett, M., G. L. Smith, and B. Moss. 1982. Vaccinia virus: a selectable eukaryotic cloning and expression vector. *Proc. Natl. Acad. Sci. USA.* 79:7415–7419.

Palese, P., and J. L. Schulman. 1976. Mapping of the influenza virus genome: identification of the hemagglutinin and the neuraminidase genes. *Proc. Natl. Acad. Sci. USA* **73:**2142–2146.

Post, L. E., and B. Roizman. 1981. A generalized technique for deletion of specific genes in large genomes: alpha gene 22 of herpes simplex virus 1 is not essential for growth. *Cell* **25:** 227–232.

Racaniello, V. R., and D. Baltimore. 1981. Cloned poliovirus complementary DNA is infectious in mammalian cells. *Science* **214:**916–919.

Saiki, R. K., D. H. Gelfand, S. Stoffel, S. J. Scharf, R. Higuchi, G. T. Horn, K. B. Mullis, and H. A. Erlich. 1988. Primer-directed enzymatic amplification of DNA with a thermostable DNA polymerase. *Science* **239:**487–491.

Schnell, M. J., T. Mebatsion, and K. K. Conzelmann. 1994. Infectious rabies viruses from cloned cDNA. *EMBO J.* **13:** 4195–4203.

The monthly journal *Biotechnology* contains articles on new methods in cell and molecular biology.

# Ⅱ   Molecular Biology

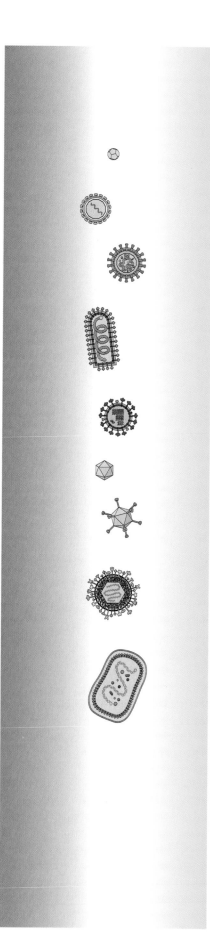

# 3

# Virus Structure

## Introduction

Virus particles are elegant assemblies of viral, and occasionally cellular, macromolecules. They are marvelous examples of architecture on the molecular scale, with form beautifully adapted to the functions of the virion. Virus particles come in many sizes and shapes (see Fig. 1.9) and vary enormously in the number and nature of the molecules from which they are built. Nevertheless, they fulfill common functions and are built according to general principles that apply to all of them. These properties are described in subsequent sections, in which we also discuss some examples of the architectural detail characteristic of different virus families.

## Functions of the Virion

Virus particles are designed for effective transmission of the nucleic acid genome from one host cell to another within a single animal or among host organisms. A primary function of the virion is therefore protection of the genome, which can be damaged irreversibly by a break in the nucleic acid or by mutation during passage through hostile environments. During its travels, a virus particle may encounter a variety of potentially lethal chemical and physical agents, including proteolytic and nucleolytic enzymes, extremes of pH or temperature, and various forms of natural radiation. In all virus particles, the nucleic acid is sequestered within a sturdy barrier formed by extensive interactions among the viral proteins comprising the protein coat. Such protein-protein interactions maintain surprisingly stable capsids: many virus particles composed of only protein and nucleic acid survive exposure to large variations in the temperature, pH, or chemical composition of their environment. Some, such as certain picornaviruses, are even resistant to strong detergents like sodium dodecyl sulfate. The highly folded nature of coat proteins and their dense packing within virions render them largely inaccessible to proteolytic enzymes. Some viruses also possess an envelope, derived from cellular membranes, into which viral glycoproteins have been inserted in the infected cell.

**59**

The envelope adds not only a protective lipid membrane but also an external layer of protein and sugars formed by the glycoproteins. Like the cellular membranes from which they are derived, viral envelopes are impermeable and block entry of chemicals or enzymes in aqueous solution into the virus particle.

The protective function of virus particles demands that these particles be stable structures. However, virions also mediate attachment of the virus to an appropriate host cell and delivery of the genome to the interior of that cell, where the particle is at least partially disassembled. The functions of virus particles therefore depend on very stable intermolecular interactions among their components during assembly, egress from the virus-producing cell, and transmission, but also ready reversal of these interactions during entry and uncoating in a new host cell. The molecular mechanisms by which these apparently paradoxical requirements are met are understood in only a few cases. Nevertheless, it is clear that contact of a virus with the appropriate cell-surface receptor or exposure to a specific intracellular environment can trigger substantial conformational changes in the virion. Virus particles are therefore **metastable structures**, which have not yet attained the minimum free energy conformation. The latter state can be gained only when an unfavorable energy barrier is surmounted following induction of irreversible conformational transitions associated with attachment and entry. Thus, virions are not simply inert structures. Rather, they are molecular machines that play an active role in delivery of the nucleic acid genome to the appropriate host cell and initiation of the reproductive cycle.

Specific functions of virion proteins associated with protection and delivery of the genome are summarized in Table 3.1.

**Table 3.1**  Functions of virion proteins

Protection of the genome

  Assembly of a stable, protective protein shell

  Specific recognition and packaging of the nucleic acid genome

  In many viruses, interaction with host cell membranes to form envelope

Delivery of the genome

  Specific binding to external receptors of the host cell

  Transmission of specific signals that induce uncoating of the genome

  Fusion with host cell membranes

  Interactions with internal components of the infected cell allowing transport of the genome to the appropriate site

Other interactions with the host

  With cellular components to ensure an efficient infectious cycle

  With cellular components for transport to intracellular sites of assembly

  With the host immune system

## Nomenclature

Virus architecture is described in terms of structural units of increasing complexity, from the smallest biochemical unit (the **polypeptide chain**) to the infectious particle (or **virion**). These terms, which will be used throughout this chapter, are defined in Table 3.2. Although virus particles are complex assemblies of macromolecules exquisitely designed for protection and delivery of viral genomes, they are constructed according to the general principles of biochemistry and protein structure. In this and subsequent chapters, we describe structural features of viral proteins using the nomenclature summarized in Box 3.1.

**Table 3.2**  Nomenclature used in description of virus structure

| Term | Synonym | Definition |
|---|---|---|
| Subunit (protein subunit) | None | Single, folded polypeptide chain |
| Structural unit | Protomer, asymmetric unit | Unit from which capsids or nucleocapsids are built; may comprise one protein subunit or multiple different protein subunits |
| Morphological unit | Capsomere | Surface structures (e.g., knobs, projections, clusters) seen by electron microscopy; because these structures do not necessarily correspond to structural units (see text), the term is restricted to descriptions of electron micrographs of viruses |
| Capsid | Coat | Protein shell surrounding the nucleic acid genome |
| Nucleocapsid | Core | Nucleic acid-protein assembly packaged within the virion; used when this complex is a discrete substructure of a complex particle |
| Envelope | Viral membrane | Lipid bilayer carrying viral glycoproteins |
| Virion | | Complete infectious virus particle |

## BOX 3.1

### *The description of protein structure*

The **primary structure** (A) of a polypeptide chain is its amino acid sequence, conventionally given in the N- to C-terminal direction. Throughout this text, amino acid sequences are given in the one-letter code, summarized at the right. The **secondary structure** (B) describes local regular structures formed in proteins. As illustrated, these are stabilized by numerous hydrogen bonds. The α-helix (left) and β-sheet (right) are by far the most common secondary structures. The former is formed by amino acids that are consecutive in the primary sequence. However, individual β-strands in a β-sheet are connected by sharp turns or larger loops, and a β-sheet can therefore be formed with amino acids that are not close neighbors in the primary sequence. β-Sheets may be antiparallel (shown) or parallel, when all strands possess the same orientation. In the simplified diagrams of protein structure that have become the convention in the field, α-helices are represented by cylinders or coiled ribbons and the individual strands of a β-sheet are shown as arrows drawn in the N- to C-terminal direction of the strand. These allow the **tertiary structure** (C), the folded structure of the

polypeptide chain, to be summarized in a clear, accessible fashion. The large number of high-resolution protein structures that has now been obtained has established that tertiary structure is based on a finite number of structural **motifs**. These are combinations of a limited number of secondary-structure units in a specific geometric arrangement, as illustrated with the Greek key motif. This motif is represented in a two-dimensional **topology diagram**, which summarizes the orientations of and connections among secondary-structure units. In general, several such motifs interact in the folded protein to form globular structures termed **domains**. A domain is a part of a polypeptide chain that can fold into a stable tertiary structure independently of other parts of the chain. For example, the β-barrel jelly roll domain, which is found in many viral capsid proteins, is formed from two Greek key motifs linked to one another by four connections across both the top and bottom of the barrel-shaped structure formed by the β-sheets. Simple proteins contain a single domain, but many proteins are multidomain structures in which each domain is an independent structural unit and performs a

*(continued)*

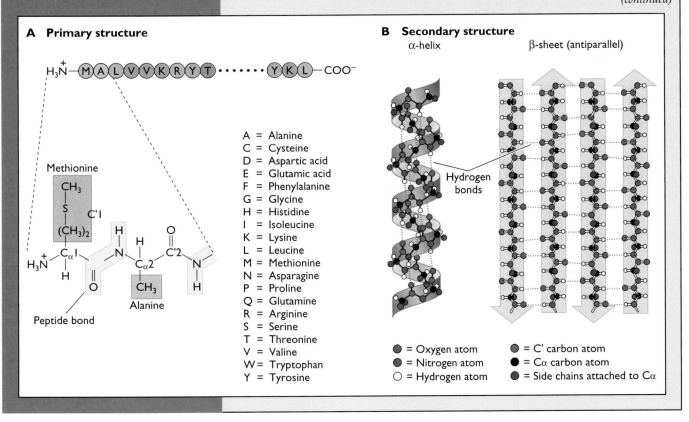

**A Primary structure**

$H_3\overset{+}{N}$—(M)(A)(L)(V)(V)(K)(R)(Y)(T)······(Y)(K)(L)—COO⁻

Methionine

Alanine

Peptide bond

A = Alanine
C = Cysteine
D = Aspartic acid
E = Glutamic acid
F = Phenylalanine
G = Glycine
H = Histidine
I = Isoleucine
K = Lysine
L = Leucine
M = Methionine
N = Asparagine
P = Proline
Q = Glutamine
R = Arginine
S = Serine
T = Threonine
V = Valine
W = Tryptophan
Y = Tyrosine

**B Secondary structure**

α-helix          β-sheet (antiparallel)

Hydrogen bonds

● = Oxygen atom       ● = C' carbon atom
● = Nitrogen atom     ● = Cα carbon atom
○ = Hydrogen atom     ● = Side chains attached to Cα

**BOX 3.1**    *(continued)*

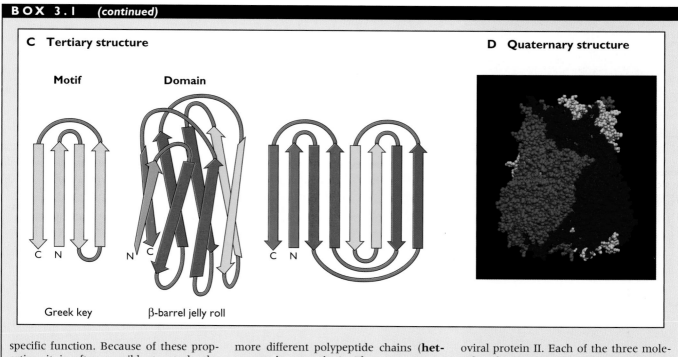

**C Tertiary structure**

Motif          Domain

Greek key      β-barrel jelly roll

**D Quaternary structure**

specific function. Because of these properties, it is often possible to study the functions of individual domains of complex proteins. Many functional proteins comprise a single folded polypeptide chain. However, many others function only as multimeric structures containing several copies of the same polypeptide chain (**homomeric proteins**) or two or more different polypeptide chains (**heteromeric proteins**). The **quaternary structure** (D) of a protein describes the arrangement of and interactions among the individual folded polypeptide chains (**subunits**) in such multimeric proteins. Shown is the human adenovirus type 2 hexon, the major structural unit of the capsid, which is a homotrimer of adenoviral protein II. Each of the three molecules of protein II in the hexon (shown in different colors) contains two β-barrel jelly roll domains. Most other virion proteins described in this chapter are also multimeric. Panel C is adapted from C. Branden and J. Tooze, *An Introduction to Protein Structure* (Garland Publishing, Inc., New York, N.Y., 1991), with permission.

## Methods for Studying Virus Structure

The method most widely used in the examination of virus structure and morphology is electron microscopy. Detailed, often beautiful images (see the appendix) are obtained when purified virus particles are negatively stained with an electron-dense material such as uranyl acetate or phosphotungstate. This technique, which has been applied to viruses since the middle decades of this century, provided the first rational basis for the classification of these agents. Because the greatest contrast between virus particle and stain occurs where portions of the folded protein chain protrude from the virion surface, such structures, termed **morphological units**, are the main features identified by negative-contrast electron microscopy. However, these surface knobs or projections are often formed by protein domains from neighboring structural units in the virion, and so their organization does not necessarily correspond to that of the individual proteins that form the capsid shell. Even when virus structure is well preserved and a high degree of contrast can be achieved, the minimal size of an object that can be distinguished by this technique (the **resolution**) is limited to 50 to 75 Å, making detailed structural interpretation impossible. Cryoelectron microscopy, in which samples are rapidly frozen and examined at very low temperatures in a hydrated, vitrified (noncrystalline, glasslike) state, preserves native structure and allows direct visualization of the contrast inherent in the virus particle rather than that between virus and stain. When combined with computerized mathematical methods of image analysis and three-dimensional reconstruction (Fig. 3.1), cryoelectron microscopy can improve resolution to 10 to 20 Å. Indeed, sufficient resolution (less than 10 Å) was attained in recent applications of these techniques to core particles of hepatitis B virus to identify secondary-structure features of the core protein and to determine how it is folded. Within the past decade, this approach has provided a wealth of previously inaccessible information about the external and internal structures of many viruses, as described below.

The diameter of an α-helix in a protein is on the order of 10 Å (Box 3.1), so even the highest resolution attained by these sophisticated techniques cannot reveal the mole-

Scanned micrograph

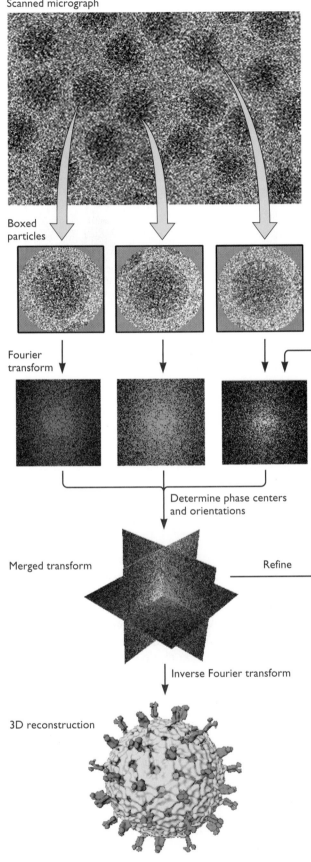

Boxed
particles

Fourier
transform

Determine phase centers
and orientations

Merged transform

Refine

Inverse Fourier transform

3D reconstruction

Concentrated preparations of purified virus particles are prepared for cryoelectron microscopy by rapid freezing on an electron microscope grid so that a glasslike, noncrystalline (vitrified) water layer is produced. This procedure avoids damage to the sample that can be caused by crystallization of the water as well as alterations that may be caused by chemical modification or dehydration during conventional negative-contrast electron microscopy. The sample is maintained at or below −160°C during all subsequent operations, including microscopy at low electron doses and recording of images. Fields containing sufficient numbers of vitrified virus particles are identified by transmission electron microscopy at low magnification (to minimize sample damage from the electron beam) and photographed at high resolution. Because the inherent contrast of unstained virus particles is low, defocusing of the objective lens of the microscope is used to generate phase contrast. Images may be recorded at different levels of defocus to enhance features of different sizes in the sample.

These electron micrographs can be treated as two-dimensional projections (Fourier transforms) of the particles. Three-dimensional structures can be reconstructed from such two-dimensional projections by mathematically combining the information given by different views of the particles, which may be provided by particles lying in different orientations or by tilting the sample in the electron microscope. For the purpose of reconstruction, the images of different particles are therefore treated as different views of the same structure.

For reconstruction, micrographs are first digitized for computer processing. Each particle to be analyzed is then framed and centered inside a box, and its orientation is determined by application of programs that orient the particle on the basis of its icosahedral symmetry. The parameters that define the orientation of the particle must be determined with a high degree of accuracy, for example, to within 1° for even a low-resolution reconstruction (>40 Å). These parameters are improved in accuracy (**refined**) by comparison of different views (particles) to identify common data.

Once the orientations of a number of particles sufficient to represent all parts of the asymmetric unit of the icosahedron have been determined, a low-resolution three-dimensional reconstruction is calculated from the initial set of two-dimensional projections by using computerized algorithms.

This reconstruction is refined by including data from additional views (particles). The number of views required depends on the size of the particle and the resolution sought. The reconstruction is initially interpreted in terms of the external features of the virus particle. Various computational and computer graphics procedures have been developed to facilitate interpretation of internal features (e.g., Fig. 3.21). Courtesy of B. V. V. Prasad, Baylor College of Medicine.

*And is it not true that even the small step of a glimpse through the microscope reveals to us images that we should deem fantastic and over-imaginative if we were to see them somewhere accidentally, and lacked the sense to understand them.*
Paul Klee, *On Modern Art*, translated by Paul Findlay (London, United Kingdom, 1948)

**Figure 3.1 Cryoelectron microscopy and image reconstruction illustrated with images of rotavirus.**

**A**

X-ray
source

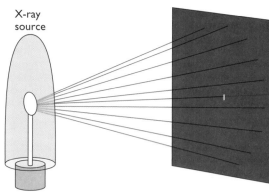

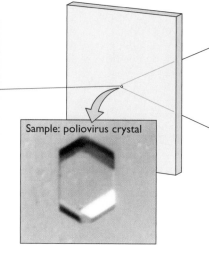

Sample: poliovirus crystal

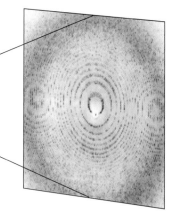

A picture of a section of the diffraction pattern generated by the poliovirus crystal.

**B**

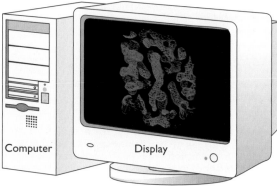

Computer     Display

A section of the poliovirus electron density map showing part of the region around the fivefold axis of symmetry.

**C**

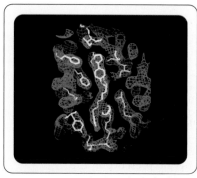

The same section of the map through one plane of the virus particle with segments of the structure built to fit the electron density (left). The built structure without the electron density is shown below. The C–C bonds and atoms are white, blue indicates nitrogen, and red indicates oxygen. The double-ring structure with the long aliphatic chain is an antiviral drug (Win51711) that is bound to the poliovirus particles in the crystal.

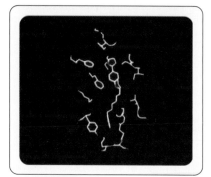

**D**

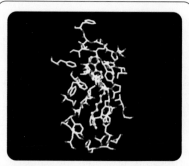

A portion of the completely built virus structure as a ball-and-stick model. The region displayed is that surrounding the drug binding pocket. A portion of this structure corresponds to the electron density shown in panel C. The antiviral drug is shown in green.

**E**

A portion of the virus structure shown as a ribbon diagram, with the three proteins that form the surface, VP1, VP2, and VP3, colored blue, yellow, and red, respectively.

cular interactions that cement structural units in the virion. Such information can be obtained by X-ray crystallography (Fig. 3.2), provided that the virus yields crystals suitable for X-ray diffraction. It has been known for more than 60 years that simple plant viruses can be crystallized, and the first high-resolution virus structure determined was that of tomato bushy stunt virus. Since this feat was accomplished in 1978, high-resolution structures of a number of smaller animal viruses have been determined, placing our understanding of the principles of capsid structure on a firm molecular foundation. Although they cannot capture dynamic processes, atomic-level descriptions of specific viruses, or of individual viral proteins, have also greatly improved our knowledge of the attachment of virions to host cell receptors and of mechanisms of entry and virus assembly and have provided new opportunities for the design of antiviral drugs.

Unfortunately, more complex viruses cannot yet be examined directly by X-ray crystallography: they generally do not form suitable crystals, and the larger viruses lie beyond the power of the current procedures by which X-ray diffraction spots are converted into a structural model (Fig. 3.2). Nevertheless, their component proteins can be examined by this method, an approach that has been particularly important in illuminating mechanisms of attachment and entry of enveloped viruses. The structures of individual viral proteins in solution can also be determined by multidimensional **nuclear magnetic resonance (NMR)** methods. These techniques allow structural models to be constructed from knowledge of the distances between specific atoms in a polypeptide chain (Fig. 3.3). At present, NMR methods can be applied to only relatively small proteins (20 to 30 kDa), but their power is being expanded rapidly. When combined with information collected by cryoelectron microscopy of the intact virus, the high-resolution structure of a single virion protein can reveal previously inaccessible information about the architecture of virus particles.

## Building a Protective Coat

Regardless of their structural complexity, all virions contain at least one protein coat, the capsid or nucleocapsid (Table 3.2) that encases and protects the nucleic acid genome. As discussed in chapter 1, genetic economy dictates that such structures be built from identical copies of a small number of viral proteins. These protein molecules are arranged to provide maximal contact and noncovalent bonding among subunits and structural units. The repetition of such interactions among a limited number of pro-

◁ **Figure 3.2 Determination of virus structure by X-ray diffraction.** A virus crystal is composed of virus particles arranged in a well-ordered three-dimensional lattice. When the crystal is bombarded with a monochromatic X-ray beam, each atom within the virus particle scatters the radiation. Interactions of the scattered rays with each other form a diffraction pattern that is recorded on a series of X-ray films or phosphor image plates (A). Each spot contains information about the position and the identity of all atoms in the crystal. The locations and intensities of the spots are digitized and stored electronically. Determination of the three-dimensional structure of the virus from the diffraction pattern requires information that is lost in the X-ray diffraction experiment. This missing information (the phases of the diffracted rays) can be retrieved by collecting the diffraction information from isomorphous crystals of the same sample in which the phases have been systematically perturbed by the introduction of heavy metal atoms at known positions; heavy metal atoms scatter X-rays more strongly because of their high electron density. Comparison of the two diffraction patterns, by using computer programs to calculate how spot intensities at each position have been altered by the presence of the heavy metal, yields the phases. This process is called **multiple isomorphous replacement**. Alternatively, if the structure of a related molecule is known, the diffraction pattern generated from the crystal can be interpreted by using the phases from the known structure as a starting point and by subsequently using computer algorithms to iteratively calculate the actual values of the phases. This method is known as **molecular replacement**. Once the phases are known, the intensities and spot positions from the diffraction pattern are used to calculate the locations of the atoms within the crystal, again by using computer programs. The product of this mathematical analysis is an electron density map (B), which is a map of the location of electron-dense atoms (typically C, N, O, and S) as well as of the covalent bonds within and between amino acids in the virus. A model of the peptide backbone linking amino acids within proteins is built by tracing the electron density through each section of the map and specifying the location of each atom (C and D). The amino acids are identified by the shapes of the electron density of the side chains and by using the primary sequence of the proteins. When the entire structure is built, the completed images can be reassembled and visualized in various representations (D and E). Courtesy of J. Hogle, Harvard Medical School, and M. Chow, University of Arkansas for Medical Sciences.

teins results in a regular structure whose symmetry describes the spatial patterns of the interactions. The protein coats of all but the most complex animal viruses display helical or icosahedral symmetry.

## Helical Structures

The nucleocapsids of some enveloped animal viruses, as well as certain plant viruses and bacteriophages, are rod-like or filamentous structures with helical symmetry. Helical symmetry is described by the number of structural units per turn of the helix, $\mu$, the axial rise per unit, $\rho$, and the pitch of the helix, $P$, given by the formula

$$P = \mu \times \rho.$$

A characteristic feature of a helical structure is that any volume can be enclosed simply by varying the length of the helix. Such a structure is said to be **open**. In contrast, capsids with icosahedral symmetry (described below) are **closed** structures of fixed internal volume.

From a structural point of view, the best understood helical nucleocapsid is that of tobacco mosaic virus, the first virus to be identified. The virus particle comprises a single molecule of (+) strand RNA, about 6.4 kb in length, enclosed within a helical protein coat (Fig. 3.4A). The coat is built from a single protein which folds into an extended clog-shaped structure. Repetitive interaction of coat protein subunits forms disks that have been likened to lock washers, which in turn assemble as a long, rodlike, right-handed helix with 16.3 coat protein molecules per turn. In the interior of the helix, each coat protein molecule binds three nucleotides of the RNA genome. Thus, the coat protein molecules engage in identical, equivalent interactions with one another and with the genome, allowing construction of a large, stable structure from a single protein subunit.

The virions of several families of animal viruses with (−) strand RNA genomes, including paramyxoviruses, rhabdoviruses, and orthomyxoviruses, contain internal structures with helical symmetry encased within an envelope. In all cases, these structures contain an RNA molecule, many copies of an RNA packaging protein (designated NP or N), and the viral RNA polymerase and associated enzymes responsible for synthesis of mRNA at the beginning of the infectious cycle. Despite common helical symmetry and similar composition, the internal components of these (−) strand RNA viruses exhibit considerable diversity in morphology and organization. Like tobacco mosaic virus particles, the nucleocapsids of paramyxoviruses and rhabdoviruses contain a single molecule of RNA (of about 15 and 11 kb, respectively) tightly associated with the nucleocapsid protein. Those of paramyxoviruses, such as Sendai virus, are long, filamentous

structures in which the RNA and NP protein form a left-handed helix with a hollow core (Fig. 3.4B). Nucleocapsids of vesicular stomatitis viruses (and other rhabdoviruses) are quite different in shape (Fig. 3.4C). Furthermore, an additional virion protein is essential to maintain their organization. Vesicular stomatis virus nucleocapsids released from within the virion envelope retain the dimensions and morphology observed in intact particles but become highly extended and filamentous if the matrix (M) protein is also removed.

The internal components of influenza A virus particles differ more radically. In the first place, they comprise not a single nucleocapsid but rather multiple ribonucleoproteins, one for each molecule of the segmented RNA genome (appendix, Fig. 7) present in the virion. The viral NP protein organizes each RNA into a helical structure analogous to the nucleocapsids of viruses with nonsegmented (−) strand RNA genomes. However, each influenza A virus ribonucleoprotein is further folded into a compact, circular conformation as a result of the binding of the virion enzyme complex (P proteins) to specific sequences conserved at the 5′ and 3′ ends of each (−) RNA molecule (Fig. 3.5). Finally, apart from the terminal sequences bound by P proteins, the RNA in these ribonucleoproteins is fully accessible to solvent. This property suggests that the RNA is wound around a helical core formed by the NP protein, rather than sequestered on the inside of a helical structure.

## Capsids or Nucleocapsids with Icosahedral Symmetry

### General Principles

**Icosahedral symmetry.** An icosahedron is a solid with 20 triangular faces and 12 vertices related by two-, three-, and fivefold axes of rotational symmetry (Fig. 1.11). Many virions possess closed protein shells (capsids or nucleocapsids) based on icosahedral symmetry. Such particles do not always appear icosahedral, for example, because the most prominent surface structures or the viral glycoproteins in the envelope do not conform to the underlying symmetry of the capsid shell. Nevertheless, the symmetry with which the structural units of the capsid or nucleocapsid interact is that of an icosahedron. As mentioned in chapter 1, icosahedral symmetry allows formation of a closed shell with the smallest number (60) of identical subunits. In such an arrangement, these subunits are related to one another by the two-, three-, and fivefold rotational axes that define icosahedral symmetry (Fig. 3.6A).

**Large capsids and quasiequivalent bonding.** In the simple icosahedral packing arrangement shown in Fig. 3.6A, each of the 60 subunits (or asymmetric units) consists of a single molecule in a structurally identical environment. Consequently, all subunits interact with their neighbors in an identical (or equivalent) manner, just like the subunits of a helical nucleocapsid such as that of tobacco mosaic virus. However, construction of a protein shell according to such a simple icosahedral design severely restricts the size of the genome that can be packaged. In fact, the capsids or nucleocapsids of the majority of animal viruses are built from structural units composed of several copies of the same protein or of several different proteins and can accommodate quite large genomes. In 1962 Donald Caspar and Aaron Klug developed a theoretical framework accounting for the structural properties of larger particles with icosahedral symmetry. This theory has had enormous influence on the way virus structures are described and interpreted.

A cornerstone of the theory developed by Caspar and Klug was the proposition that when a capsid contains more than 60 subunits, each subunit occupies a **quasiequivalent position**; that is, the bonding properties of subunits in different structural environments are similar (but not identical, as in the case of the simplest, 60-subunit structure). This property is illustrated in Fig. 3.6B for a particle with 180 identical subunits. In the small, 60-subunit structure, 5 subunits make fivefold symmetric contact at each of the 12 vertices (Fig. 3.6A). In the larger structure with 180 subunits this arrangement of protein subunits is retained at the 12 vertices, but the additional subunits, arranged with sixfold symmetry, are interposed between the fivefold symmetric clusters (Fig. 3.6B). In such a structure, each subunit can be present in one of three different structural environments, designated A, B, or C in Fig. 3.6B. Nevertheless, all subunits bond to their neighbors in similar ways, for example, via head-head and tail-tail interactions (Fig. 3.6B).

**The triangulation number, *T*.** A second important idea introduced by Caspar and Klug was that of **triangulation**, the description of the triangular face of a large icosahedral structure in terms of its subdivision into smaller triangles termed **facets**. As shown in Fig. 3.7, combining several triangular facets allows assembly of a larger face of an icosahedrally symmetric structure from the same structural unit, in this case a homotrimer. This process is described by the **triangulation number, *T***, which gives the number of structural units per face. The total number of subunits in the structure is 60*T*. As in the *T* = 3 structure illustrated in Fig. 3.6B, the packing interactions of the 240 (60 × 4) protein subunits in the *T* = 4 structure shown in Fig. 3.7 cannot be identical, but according to the original theory of Caspar and Klug they are very similar, or quasiequivalent.

Icosahedrally symmetric structures can be formed from quasiequivalent subunits for only certain values of *T* (Box 3.2). Virus structures corresponding to various values of *T*, some very large, have been described (Table 3.3). The capsids of a number of small (+) strand RNA viruses, such as that of tomato bushy stunt virus (*T* = 3), show quasiequivalent bonding of their identical subunits: the local contacts made by the A, B, and C monomers that occupy distinctive packing environments (Fig. 3.6B) are largely similar, and the corresponding parts of each engage in similar interactions. However, an important determinant of which type of packing interaction each subunit adopts in such *T* = 3 structures is whether a flexible arm of the protein is ordered or disordered. This property indicates that interactions among chemically identical protein subunits must be regulated during assembly, with appropriate conformational switching of the flexible segment. Such large conformational differences between small regions of chemically identical (and otherwise quasiequivalent) subunits were not anticipated in early considerations of virus structure. This is not surprising, for these principles were formulated when little was known about the structural properties of proteins.

The triangulation number and quasiequivalent bonding among subunits describe the structural properties of many simple viruses with icosahedral symmetry. However, as we discuss in the next sections, the structural properties of both small and more complex viruses can depart radically from the constraints imposed by quasiequivalent bonding. For example, the capsid of the small papovavirus simian virus 40 is built from 360 subunits, corresponding to the *T* = 6 triangulation number excluded by the rules formulated by Caspar and Klug. Conversely, the single protein that forms the innermost of three protein shells of bluetongue virus, a member of the *Reoviridae*, is organized as a *T* = 2 structure, also forbidden by these rules (Box 3.2). Our current view of icosahedrally symmetric virus structures is thus one that includes greater diversity in the mechanisms by which stable capsids can be formed than anticipated by the pioneers in this field.

### Structurally Simple Capsids

Several nonenveloped animal viruses are small enough to be amenable to high-resolution structural studies by X-ray crystallography. We have chosen two examples, the picornavirus poliovirus and the papovavirus simian virus 40, to illustrate molecular foundations of icosahedral architecture.

**Structure of poliovirus.** As their name implies, the picornaviruses are among the smallest and simplest of animal viruses. The approximately 300-Å-diameter poliovirus

**Figure 3.3 Information obtained from multidimensional NMR spectra.** (A) Example of a two-dimensional $^1$H$^{15}$N heteronuclear, single quantum coherence (HSQC) nuclear magnetic resonance (NMR) spectrum of the human immunodeficiency virus type 1 matrix (MA) protein. The horizontal and vertical axes show the radiation absorbed by covalently linked protons and nitrogen atoms expressed relative to reference signals and termed the **chemical shifts**. These signals are observed when the protein is placed in a strong magnetic field and exposed to radiofrequency pulses. The chemical shift is different for each atom and is determined in part by the molecular environment of its nucleus. Thus, the different atoms in a complex molecule like the MA protein exhibit different chemical shifts in this kind of experiment. Modern multidimensional NMR spectroscopic techniques allow structural determinations of moderate-sized proteins (up to ~30 kDa), although proteins at the upper end of this size range typically require enrichment with the "NMR-active" atomic nuclei $^{15}$N and $^{13}$C (in addition to protons). The initial problem is to resolve and assign the very large number of signals generated by even a small protein (~1,000 for a protein of 15 kDa). This is typically achieved in multidimensional experiments, which allow individual protons to be resolved on the basis of several different chemical shift indices ($^{15}$N and $^1$H in the spectrum shown). (B) Proton-proton correlations obtained from $^1$H,$^1$H total correlation spectroscopy (TOCSY) and nuclear Overhauser effect spectroscopy (NOESY) NMR spectra. TOCSY reveals correlations between protons that are covalently connected via only one or two additional atoms. A $^1$H,$^1$H TOCSY experiment identifies interactions among protons within the *same* amino acid residue in a protein. Hydrogen atoms in adjacent amino acids are connected by at least three other atoms (Box 3.1). A TOCSY experiment therefore generates a characteristic set of linked signals associated with each amino acid, which differs for the different amino acids because each bears a unique side chain. NOESY reveals correlations between protons that are closer than 5 Å in space regardless of whether they are closely linked in the primary sequence of the protein. Thus, secondary structure elements, such as α-helices, generate characteristic sets of NOE signals. Because NOE signals require only that protons be close in space, they also provide information about the tertiary structure of the protein. NOE signals place constraints on the distances between specific pairs of hydrogen atoms in the protein. When sufficient NOE constraints have been collected and assigned, it is possible to generate structural models of the protein (C) that are consistent with these constraints, as shown here for the human immunodeficiency virus type 1 MA protein. (A and C) Courtesy of M. F. Summers, University of Maryland, Baltimore County, and W. I. Sundquist, University of Utah. (B) Adapted from C. Branden and J. Tooze, *An Introduction to Protein Structure* (Garland Publishing, Inc., New York, N.Y., 1991), with permission.

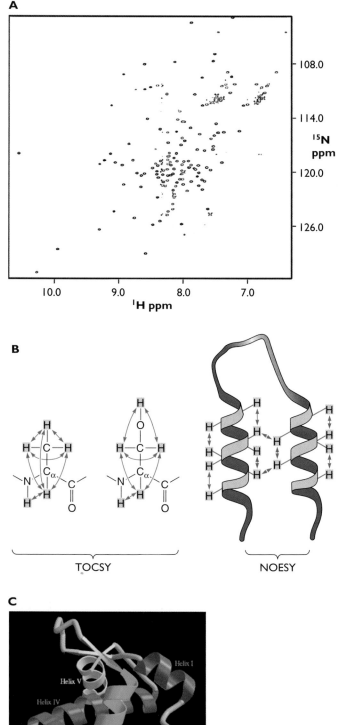

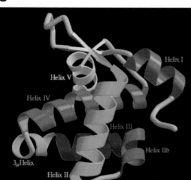

particle is composed of 60 copies each of four virally encoded structural proteins, VP1, VP2, VP3, and VP4 (VP = virion protein), that form a closed capsid encasing the (+) strand RNA genome of about 7.5 kb and its covalently attached 5'-terminal protein, VPg (appendix, Fig. 1). Our

**A   Tobacco mosaic virus**

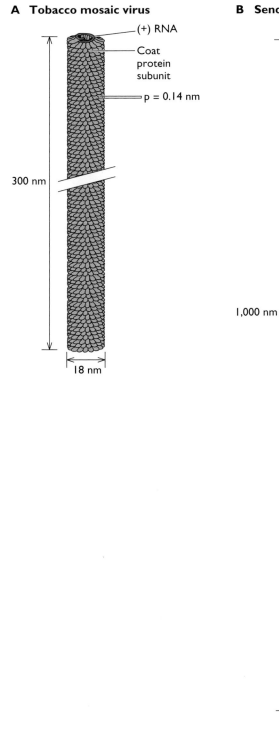

(+) RNA

Coat protein subunit

p = 0.14 nm

300 nm

18 nm

**B   Sendai virus nucleocapsid**

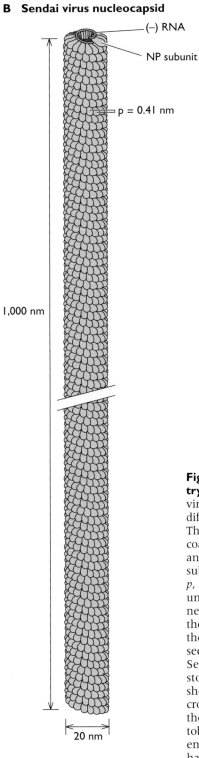

(−) RNA

NP subunit

p = 0.41 nm

1,000 nm

20 nm

**C   Vesicular stomatitis virus nucleocapsid**

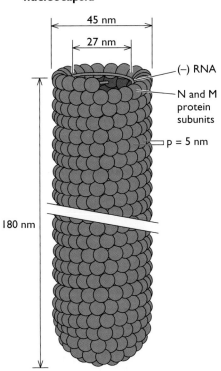

45 nm

27 nm

(−) RNA

N and M protein subunits

p = 5 nm

180 nm

**Figure 3.4 Virus structures with helical symmetry.** (A) Tobacco mosaic virus. The structure of this virus has been determined at high resolution by X-diffraction of fibers or oriented gels of the particles. The single genomic RNA molecule and the single coat protein of tobacco mosaic virus particles form an extended right-handed helix with $16\frac{1}{3}$ protein subunits per turn, $\mu$, and an axial rise per residue, $p$, of 0.14 nm, for a pitch, $P$, of 2.3 nm. Each subunit engages in the same interactions with its neighbors, and each binds to three nucleotides in the RNA. The broken disk-shaped organization of the proteins in a single turn of the helix can be seen at the top of the model. The structures of Sendai virus (a paramyxovirus) and vesicular stomatitis virus (a rhabdovirus) nucleocapsids shown in panels B and C are based on electron microscopy of virions or nucleocapsids released from them. The Sendai virus nucleocapsid resembles the tobacco mosaic virus particle in diameter and presence of a hollow core but is in the form of a left-handed helix with distinctive helical parameters ($\mu = 13$, $p = 0.41$ nm, $P = 5.3$ nm). The vesicular stomatitis virus nucleocapsid is also hollow and comprises about 30 turns of uniform diameter, followed by 5 or 6 turns of decreasing diameter. Helical parameters have not been determined.

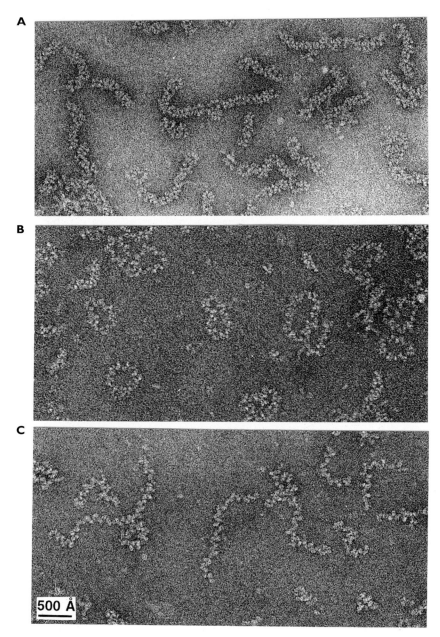

**Figure 3.5 Circular conformation of influenza A virus ribonucleoproteins.** The electron micrographs show negatively stained ribonucleoproteins as purified from virions (A and B) or following removal of P proteins by exposure to detergent and density gradient centrifugation (C). The samples shown in panels B and C were incubated in low salt prior to electron microscopy, a treatment that unwinds the compact native ribonucleoproteins (A) to circular structures (B). In the absence of the P proteins, only linear structures are observed on unwinding (C). Therefore, the P proteins, which protect both 5'- and 3'-terminal sequences of RNA segments against chemical modification, hold the virion ribonucleoproteins in a compact, circular conformation. From K. Klumpp et al., *EMBO J.* **16:**1248–1257, 1997, with permission. Courtesy of R. W. H. Ruigrok, European Molecular Biology Laboratory Grenoble Outstation, Grenoble, France.

understanding of the structure of picornaviruses took a quantum leap in 1985 with the determination of high-resolution structures of human rhinovirus 14 (genus *Rhinovirus*) and poliovirus (genus *Enterovirus*).

The heteromeric structural unit of the poliovirus capsid contains one copy each of VP1, VP2, VP3, and VP4 (Fig. 3.8A). VP4 lies on the inner surface of the protein shell formed by VP1, VP2, and VP3. Although the three latter proteins are not related in amino acid sequence, all three contain a central β-sheet structure termed a β-barrel jelly roll (or a Swiss-roll β-barrel or an eight-stranded antiparallel β-barrel). The organization of this β-barrel domain is illustrated schematically in Fig. 3.8B for comparison with

the actual structures of VP1, VP2, and VP3. It is a wedge-shaped structure comprising two antiparallel β-sheets. One of these forms one wall of the wedge, while the second, sharply twisted β-sheet forms both the second wall and the floor (Fig. 3.8B). The β-barrel domains of VP1, VP2, and VP3 are folded in the same way; that is, they possess the same **topology**. The differences among these proteins are therefore largely restricted to the loops that connect β-strands and to the N- and C-terminal segments that extend from the central β-barrel domains (Fig. 3.8B).

The β-barrel jelly rolls of picornavirus structural proteins are similar in structure to the core domains of capsid proteins of a number of plant, insect, and vertebrate (+)

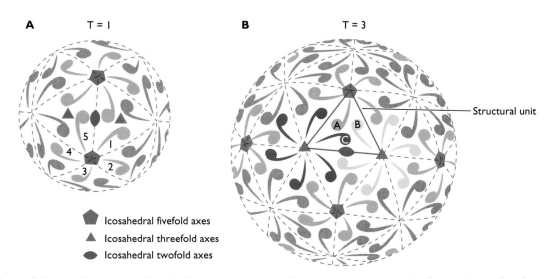

**A** T = 1

**B** T = 3

Structural unit

Icosahedral fivefold axes

Icosahedral threefold axes

Icosahedral twofold axes

**Figure 3.6 Icosahedral packing in simple structures.** A comma represents a single protein molecule. In the simplest and ideal case, $T = 1$ (A), the protein molecule forms the structural unit and each of the 60 molecules is related to its neighbors by the two-, three-, and fivefold rotational axes that define a structure with icosahedral symmetry. At a twofold rotational axis, rotation of an object by 180° results in an appearance identical to that at the starting position; at a threefold axis, rotations of 120° and 240° restore the starting appearance; and rotations of 72°, 144°, 216°, and 288° around a fivefold axis restore identity. In such a simple icosahedral structure, the interactions of all molecules with their neighbors are identical. In the $T = 3$ structure (B) with 180 identical protein subunits, there are three modes of packing of a subunit shown in orange, yellow, and purple because the **structural unit** (outlined in blue) is now the **asymmetric unit** whose replication according to 60-fold icosahedral symmetry generates the complete structure. The orange subunits are present in pentamers, formed by tail-to-tail interactions, and interact in rings of three (back to back) with purple and yellow subunits and in pairs (head to head) with a purple or a yellow subunit. The latter subunits are arranged in rings of six molecules (also by tail-to-tail interactions) that alternate in the particle. Despite these packing differences, the bonding interactions in which each subunit engages are similar: for example, all engage in tail-tail and head-head interactions. This type of packing arrangement is seen in small plant RNA viruses, such as tomato bushy stunt virus, in which the variation in subunit-subunit interactions needed for the three types of packing is accommodated by whether an N-terminal arm on each molecule adopts an ordered (purple subunits) or a disordered (orange and yellow subunits) conformation. In other small icosahedral viruses, such as the nodavirus black beetle virus, interaction of internal surfaces of adjacent subunits at the twofold axes with ordered segments of the RNA genome contributes to the variation in packing of the subunits. Adapted from S. C. Harrison et al., *in* B. N. Fields et al. (ed.), *Fundamental Virology* (Lippincott-Raven, New York, N.Y., 1995), with permission.

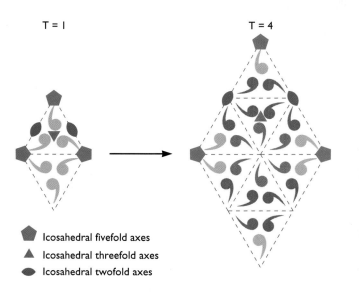

T = 1

T = 4

Icosahedral fivefold axes

Icosahedral threefold axes

Icosahedral twofold axes

**Figure 3.7 $T = 4$ triangulation of an icosahedral face.** Two faces of a $T = 1$ icosahedron, each composed of three protein subunits, are shown flattened into a single plane on the left. The two-, three-, and fivefold rotation axes that define icosahedral symmetry are indicated. The combination of four such identical faces (now facets) to form a $T = 4$ face is shown at the right. The icosahedral symmetry axes are in the same relative positions as in the $T = 1$ face. In such a $T = 4$ structure, the size of the face, which is defined by the length of one of its sides, is twice that of the $T = 1$ structure. Adapted from S. Casjens, *in* S. Casjens (ed.), *Virus Structure and Assembly* (Jones and Bartlett Publishers Inc., Boston, Mass., 1985), with permission.

*The triangulation number,*
*T, and how it is determined*

In developing their theories about virus structure, Caspar and Klug used graphic illustrations of capsid subunits, such as the net of flat hexagons shown at the top of (A) and (B). Each hexagon represents a hexamer whose identical subunits are shown as equilateral triangles. When all subunits assemble into such hexamers, the result is a flat sheet, or lattice, which can never form a closed structure. However, curvature can be introduced into the hexagonal net by converting hexamers to pentamers. As an icosahedron has 12 axes of fivefold symmetry, 12 pentamers must be introduced to form a closed structure with icosahedral symmetry. If 12 adjacent hexamers are converted to pentamers (A), an icosahedron of the minimum size possible for the net is formed. This structure is built from 60 equilateral triangle asymmetric units and corresponds to a $T = 1$ icosahedron. Larger structures with icosahedral symmetry are built by converting 12 **nonadjacent** hexamers to pentamers at appropriately spaced and regular intervals. To illustrate this, we use nets in

which an origin ($O$) is fixed and the positions of all other hexamers are defined by the coordinates along the axes labeled $h$ and $k$, where $h$ and $k$ are any integers greater than or equal to zero. The hexamer ($h$, $k$) is therefore defined as that reached from the origin ($O$) by $h$ steps in the direction of the $h$ axis and $k$ steps in the direction of the $k$ axis. In the $T = 1$ structure, $h = 1$ and $k = 0$, or $h = 0$ and $k = 1$, and adjacent hexamers are converted to pentamers (A, center). To construct a model of an icosahedron when $h = 1$ and $k = 1$ (B), we generate one of its faces in the net by conversion of the origin ($O$) and (1, 1) hexamers to pentamers. The lattice point (1, 1) shown is reached by one step in each direction. The third hexamer replaced ($-1$, 1) is that identified by threefold symmetry to complete the large, marked equilateral triangle (black) representing the face of an icosahedron formed when $h = 1$ and $k = 1$. The resulting quasiequivalent lattice (B, center) and the $T = 1$ lattice (A, center) can be folded to form the icosahedron shown, by excision from the sheet

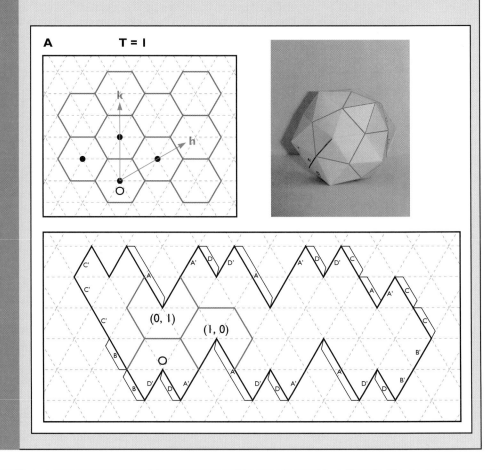

and sequential joining of the edges marked A to D with those marked A′ to D′, respectively.

The triangulation number, *T*, is the number of asymmetric units per face of icosahedra constructed in this way. A geometric derivation of *T* in terms of *h* and *k*, developed by Caspar, is shown in panel C. In the general case, it is the points (*h*, *k*) in the hexamer lattice that are converted to pentamers to create the new face, represented by the large equilateral triangle (purple) with sides of length *S*. Because each small equilateral triangle represents a subunit of the smallest icosahedron, to calculate *T* we need to know how many such small triangles can be accommodated in the large triangle of sides *S* that represents the face of the larger (*h*, *k*) icosahedron. Thus,

$$T = \frac{\text{area of equilateral triangle of sides } S \text{ (the new face)}}{\text{area of small equilateral triangle representing 1 subunit}}$$

The area of an equilateral triangle with sides of length *S* is $(\sqrt{3}/4)S^2$. By definition, the small triangle, which represents one subunit, has sides of unit length.

$$\therefore T = \frac{(\sqrt{3}/4)S^2}{(\sqrt{3}/4)(1)^2}$$

As shown on the right, we can draw additional triangles on the lattice to derive *S* in terms of *h* and *k*. Here *k* is the hypotenuse of the 30-60-90 triangle that connects the lattice point (*h*, *k*) to the *h* axis. The other sides of this triangle are therefore $\sqrt{3}k/2$ and *k*/2, as shown. *S* is therefore the hypotenuse of a right-angled triangle whose sides are (*h* + *k*/2) and $\sqrt{3}k/2$. By the Pythagorean theorem,

$$S^2 = \left(h + \frac{k}{2}\right)^2 + \left(\frac{\sqrt{3}k}{2}\right)^2$$

$$= h^2 + hk + k^2$$

$$\therefore T = h^2 + hk + k^2$$

Therefore, when both *h* and *k* are 1 (B), *T* = 3, and each face of the icosahedron contains three asymmetric units. The total number of units, which must be 60*T* (see text), is 180.

The integers h and k describe the spacing and spatial relationships of pentamers in the lattice and of fivefold vertices in the corresponding icosahedron. For example, when h 5 0 and k 5 0, pentamers are in

*(continued)*

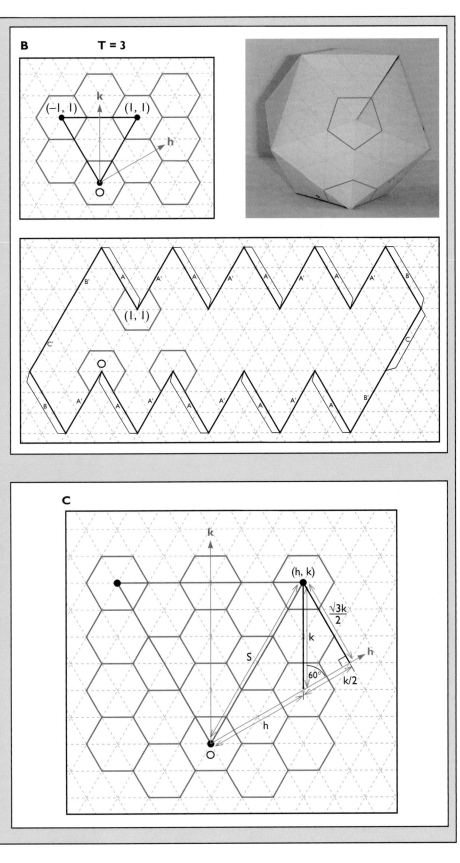

**BOX 3.2** (continued)

direct contact with other pentamers (e.g., Fig. 3.6A), but when $h = 1$ and $k = 1$, each pentamer is separated from neighboring pentamers by one hexamer (Fig. 3.6B). The values of $h$ and $k$ are therefore determined by inspection of electron micrographs of virus particles or their constituents (D). For example, in the phage p22 capsid (top), one pentamer is separated from another by two steps (from one structural unit to the next) along the $h$ axis and one step along the $k$ axis, as illustrated for the bottom left pentamer shown. Hence, $h = 2$, $k = 1$, and $T = h^2 + hk + k^2 = 7$. In contrast, pentamers of the herpes simplex virus type 1 (HSV-1) nucleocapsid (bottom) are separated by four and zero steps, respectively, in the directions of the $h$ and $k$ axes. Thus, $h = 4$, $k = 0$, and $T = 16$. Cryoelectron micrographs of phage P22 and herpes simplex virus type 1 courtesy of B. V. V. Prasad and W. Chiu (Baylor College of Medicine), respectively. (A and B) Adapted from Fig. 2 of J. E. Johnson and A. J. Fisher, in R. G. Webster and A. Granoff (ed.), *Encyclopedia of Virology,* 3rd ed. (Academic Press, London, England, 1994), with permission. (C) Adapted from S. Casjens, in S. Casjens (ed.), *Virus Structure and Assembly* (Jones and Bartlett Publishers, Inc., Boston, Mass., 1985), with permission.

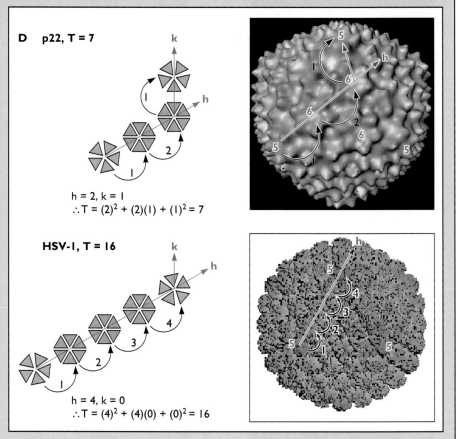

**D   p22, T = 7**

$h = 2, k = 1$
$\therefore T = (2)^2 + (2)(1) + (1)^2 = 7$

**HSV-1, T = 16**

$h = 4, k = 0$
$\therefore T = (4)^2 + (4)(0) + (0)^2 = 16$

strand RNA viruses, such as tomato bushy stunt virus and black beetle virus, a property that was entirely unanticipated. Even more remarkably, this relationship is not restricted to small RNA viruses: the major capsid proteins of the DNA-containing papovaviruses and adenoviruses also contain such β-barrel domains (see below). One of the most important principles established by high-resolution structural studies of proteins is the ability of proteins with

**Table 3.3**   $T$ numbers of representative viruses

| Triangulation number ($T$)[a] | Family | Member(s) | 60$T$ | Protein(s) that forms the capsid or nucleocapsid shell |
|---|---|---|---|---|
| 1 | *Parvoviridae* | Canine parvovirus | 60 | 60 copies of VP2 |
| 3 | *Nodaviridae* | Black beetle virus | 180 | 180 copies of coat protein |
| Pseudo-3 | *Picornaviridae* | Poliovirus, human rhinovirus | 180 | 60 copies each of VP1, VP2, VP3, and VP4 |
| 4 | *Alphaviridae* | Ross River virus | 240 | 240 copies of C protein |
| 7 | *Papovaviridae* | Simian virus 40, polyomavirus | 420 | 360 copies of VP1 |
| 13 | *Reoviridae* | Reovirus, outer shell | 780 | 60 copies of λ2, 600 copies of μ1 |
| 16 | *Herpesviridae* | Herpes simplex virus type 1 | 960 | 960 copies of VP5 |
| 25 | *Adenoviridae* | Adenovirus type 2 | 1,500 | 720 copies of protein II, 60 copies of protein III |
| 189–217 | *Iridoviridae* | Ranavirus | 11,340; 13,020 | ? |

[a]The triangulation number, $T$, is a description of the face of an icosahedron indicating the number of equilateral triangles into which each face is divided following the laws of solid geometry (Box 3.2). The total number of subunits is equal to 60$T$ because a minimum of 60 subunits is required to form a closed shell with icosahedral symmetry. Thus, the simplest assembly is $T = 1$. A $T = 3$ icosahedron requires 180 subunits in sets of three; a $T = 4$ icosahedron requires 240 subunits; a $T = 7$ icosahedron requires 420 subunits; and so forth. The 60 structural units of picornavirus capsids each contain four different polypeptides (VP1, VP2, VP3, and VP4). Hence the capsid is described as a pseudo $T = 3$ structure. Note that papovaviruses, reoviruses, and adenoviruses do not contain the number of subunits predicted.

largely unrelated primary sequences to form the same three-dimensional structure. This property makes it possible to speak of topological families of proteins, whose members are also often functionally related. For example, there is very little identity of amino acid sequences among all known globins or lysozymes, and the residues that are conserved are those important for the functions of the protein. Nevertheless, all globins possess a common three-dimensional structure based on eight α-helices. All lysozymes also adopt a similar folded structure, but one that is quite distinct from the "globin fold." Thus, the three-dimensional structures of the folded proteins have been highly conserved, presumably to sustain critical functions. The common occurrence of the β-barrel jelly roll domain in virion proteins of different viruses therefore suggests that modern viruses that seem quite unrelated to one another (e.g., picornaviruses and adenoviruses) may share some portion of their evolutionary history. Alternatively, this structural unit may represent one of a limited number that allow proteins to pack to form a sphere and thus may be an example of convergent evolution.

The overall similarity in shape of the β-barrel domains of poliovirus VP1, VP2, and VP3 facilitates both their interaction with one another to form the 60 structural units of the capsid and the packing of these structural units in the poliovirus capsid. How well these interactions are tailored to form a protective shell is illustrated by the space-filling model of the capsid shown in Fig. 3.9A and B: the extensive interactions among the β-barrel domains of adjacent proteins form a dense, rigid protein shell around a central cavity in which the genome resides. The packing of the β-barrel domains is reinforced by a network of protein-protein contacts on the inside of the capsid. These contacts are particularly extensive about the fivefold axes (Fig. 3.9C).

One of several important lessons learned from high-resolution structures of picornaviruses is that their design does not strictly conform to the principle of quasiequivalence. For example, despite the topological identity and geometric similarity of the central domains of the poliovirus proteins that form the capsid shell, the subunits do not engage in quasiequivalent bonding: interactions among VP1 molecules around the fivefold axes are not chemically or structurally equivalent to those in which VP2 or VP3 engages (Fig. 3.9A and B). These differences account for the characteristic features of the surface of the capsid. For example, the interaction of five VP1 molecules unique to the fivefold axes results in a prominent protrusion extending to about 25 Å from the capsid shell (Fig. 3.9A and B). The resulting structure appears as a steep plateau encircled by a valley or cleft. In human rhinovirus 14, these depressions are so deep that they have been termed **canyons** and are the receptor binding sites. The exposed loops, which connect the β-strands in the β-barrel domains (Fig. 3.8B), contain the antigenic sites of the virion recognized by host antibodies.

**Structure of simian virus 40.**    The capsids of the small DNA papovaviruses, simian virus 40 and polyomavirus, about 500 Å in diameter, are organized according to a rather different design. The structural unit is a pentamer of the major structural protein, VP1 (Fig. 3.10A). The capsid is built from 72 such structural units engaged in one of two kinds of interaction, which are described by the number of neighbors that surround any specific pentamer. Twelve structural units occupy the 12 positions of fivefold rotational symmetry, in which each is surrounded by five neighbors. Each of the remaining 60 structural units is surrounded by six neighbors at positions of sixfold rotational symmetry in the capsid (Fig. 3.10A). Thus, in contrast to the 60 structural units of poliovirus, the 72 pentamers of simian virus 40 (or polyomavirus) do not engage in identical interactions but occupy a number of different local environments in the capsid because of differences in packing around the five- and sixfold axes (Fig. 3.10A and C).

Like the three poliovirus proteins that form the capsid shell, simian virus 40 VP1 contains a large central β-barrel domain, in this case with an N-terminal arm and a long C-terminal extension. Furthermore, the topology of the VP1 β-barrel domain (Fig. 3.10B) is the same as that of the β-barrels of poliovirus capsid proteins. However, the arrangement and packing of VP1 molecules in the simian virus 40 capsid bears little resemblance to the organization of poliovirus capsid proteins. In the first place, the VP1 β-barrels in each pentamer project outward from the surface of the capsid to a distance of about 50 Å (Fig. 3.10A), in sharp contrast to those of the poliovirus capsid proteins, which tilt along the surface of the capsid shell. As a result, the surface of simian virus 40 is much more "bumpy" than that of poliovirus (compare Fig. 3.10A and 3.9B). Also, the VP1 molecules present in adjacent pentamers in the simian virus 40 capsid do not make extensive contacts via the surfaces of their β-barrel domains. Rather, stable interactions among pentamers are made by their N-terminal and C-terminal arms (Fig. 3.10B and C). The packing of VP1 pentamers in both pentameric and hexameric arrays in the simian virus 40 capsid requires different contacts among these structural units, depending on their local environment. In fact, just three kinds of interpentamer contact are needed (Fig. 3.10C). The different contacts are the result of alternative conformations and arrangements of the long C-terminal arms of VP1 molecules (Fig. 3.10B and C).

Along its long axis, each VP1 pentamer containing a tapering cavity that is narrowest toward the outer surface of

**A**

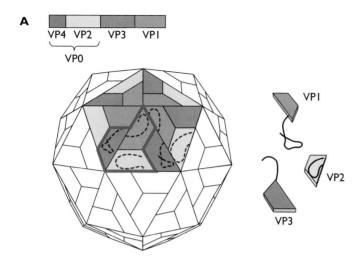

**B**

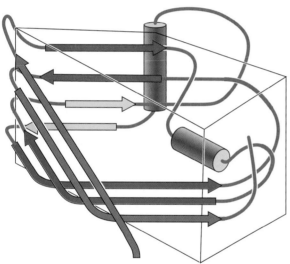

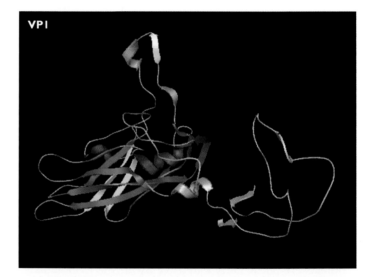

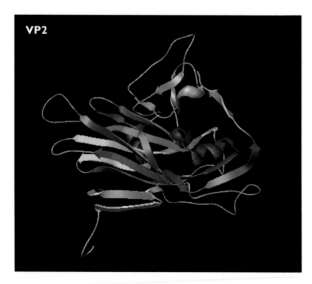

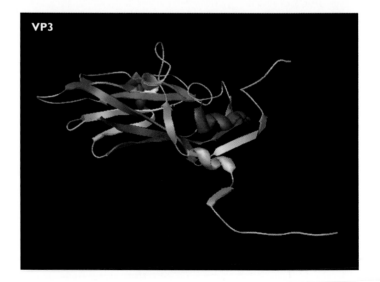

the virion. This aperture contains a common C-terminal sequence of either VP2 or VP3 which associates tightly with the VP1 pentamer. Both these proteins are internal and make no contribution to the outer surface of the virion. VP2 plays an important role during entry, but why VP3 is present in virions is not known. This protein is identical in sequence to the C-terminal segment of the larger VP2 protein (appendix, Fig. 13). It is possible that the virus particle is simply too small to accommodate 72 molecules of VP2 in its interior. VP3 could serve the same structural function as the C-terminal portion of VP2, presumably stabilizing VP1 pentamers, but occupy less internal space.

Simian virus 40 and poliovirus capsids differ in their surface appearance, in the number of structural units, and in the ways in which the structural units interact. Despite such differences, these virions share important features, including modular organization of proteins that form the capsid shell and a common β-barrel domain as the capsid building block. Neither poliovirus nor simian virus 40 capsids conform to strict quasiequivalent construction: all contacts made by all protein subunits are not similar, and in the case of simian virus 40 the majority of VP1 pentamers are packed in hexameric arrays. Nevertheless, close packing with icosahedral symmetry is achieved by limited variations of the contacts either among topologically and geometrically similar, but chemically distinct, surfaces (poliovirus) or made by a flexible arm (simian virus 40).

The capsids of nonenveloped viruses such as picornaviruses and papovaviruses must perform both the protective and the delivery functions of the virion. Thus, as mentioned above and described in detail in chapter 4, their surfaces contain binding sites for specific cellular receptors. Moreover, their tightly knit protective protein shells (Fig. 3.9A and 3.10B) are designed for disassembly when appropriate conditions are encountered, for example, contact with and specific binding to the appropriate cell surface receptor.

**Structurally simple capsids or nucleocapsids in more complex virions.** Several viruses that are structurally more sophisticated than the picornaviruses or papovaviruses nevertheless possess simple protein coats built from one or a few structural proteins. In such viruses, the complexity comes from the additional protein and lipid layers in which the capsid is enclosed. For example, togaviruses, such as Sindbis virus and Ross River virus (genus *Alphavirus*), contain a $T = 4$ icosahedral capsid built from a single protein (Table 3.3) within an envelope. A more dramatic illustration of this type of virion architecture is provided by herpesviruses. The virions of members of this family, such as herpes simplex virus type 1, contain far more proteins than any virus described so far: over half of the more than 80 genes of this virus encode proteins found in virus particles (Table 3.4), which are correspondingly large, about 2,000 Å in diameter. These proteins form three distinct substructures in the virus particle, the nucleocapsid surrounding the DNA genome, a protein layer encasing this structure called the **tegument**, and the external envelope from which glycoprotein spikes project (Fig. 3.11A). Despite such complexity, a single protein, the major capsid protein VP5, forms the structural units of the $T = 16$ icosahedral capsid (Fig. 3.11B). Like papovaviral VP1, herpes simplex virus type 1 VP5 forms both the pentons at the 12 vertices of the nucleocapsid and the hexons from which the remainder of the shell is built.

◁ **Figure 3.8 Packing and structures of poliovirus proteins.** (A) The packing of the 60 VP1-VP2-VP3 structural units, represented by wedge-shaped blocks corresponding to their β-barrel domains. Note that the structural unit (outlined in blue) contributes to two adjacent faces of an icosahedron rather than corresponding to a facet. The wedge-shaped β-barrels of VP1, VP2, and VP3 facilitate tight packing of the structural units, with VP1 proteins from five adjacent structural units clustered around the 12 fivefold axes of the virus particle and the thin ends of VP2 and VP3 alternating around the threefold axes (see also Fig. 3.9). (B) The topology of the polypeptide chain in a β-barrel jelly roll is shown at the top left. The β-strands, indicated by arrows, form two antiparallel sheets juxtaposed in a wedgelike structure. The two α-helices (purple cylinders) that surround the open end of the wedge are also conserved in location and orientation in these proteins. The VP1, VP2, and VP3 proteins each contain a central β-barrel jelly roll domain. However, the loops that connect the β-strands in this domain of the three proteins vary considerably in length and conformation, particularly at the top of the β-barrel domain, which, as represented here, corresponds to the outer surface of the capsid. The N- and C-terminal segments of the proteins, which extend from their β-barrel cores, also vary in length and structure. The very long N-terminal extension of VP3 has been truncated in this representation. Adapted from J. M. Hogle et al., *Science* **229**:1358–1365, 1985, with permission.

**A**

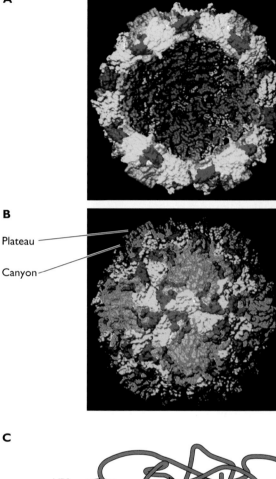

**Figure 3.9 Interactions among the proteins of the poliovirus capsid.** (A) Space-filling representation of the particle, with four pentamers removed from the capsid shell and VP1 in blue, VP2 in yellow, VP3 in red, and VP4 in green. Note the large central cavity in which the RNA genome resides, the dense protein shell formed by packing of the VP1, VP2, and VP3 β-barrel domains, and the interior location of VP4, which decorates the inner surface of the capsid shell. (B) Space-filling representation of the exterior surface showing the packing of the β-barrel domains of VP1, VP2, and VP3. Interactions among the loops connecting the upper surface of the β-barrel domains of these proteins create the surface features of the virion, the plateaus at the fivefold axes (arrow), and the knobs and ridges surrounding the threefold axes. For example, the outward tilt of the upper surface of each VP1 β-barrel domain and of the loops that connect the β-strands at the upper (outer) surface (Fig. 3.8B) of each of the five VP1 molecules around the fivefold axes form the plateaus, which are encircled by a deep cleft or canyon. The virion is also stabilized by numerous interactions among the proteins on the inner side of the capsid. As illustrated in panel C, these internal contacts are most extensive around the fivefold axes, where the N termini of five VP3 molecules are arranged in a tubelike, parallel β-sheet. The N-termini of VP4 molecules carry chains of the fatty acid myristate (gray), which are added to the protein posttranslationally. The lipids mediate interaction of the β-sheet formed by VP3 N termini with a second β-strand structure containing strands contributed by both VP4 (green) and VP1 (blue) molecules. This internal structure is not completed until the final stages of, or after, assembly of virus particles, when proteolytic processing liberates VP2 and VP4 from their precursor, VP0. This reaction therefore stabilizes the capsid and also primes it for entry into a new host cell. (A and B) From J. M. Hogle et al., *Science* **229**:1358–1365, 1985, with permission. Courtesy of M. Chow, University of Arkansas for Medical Sciences, and J. M. Hogle, Harvard Medical School.

**B**

Plateau

Canyon

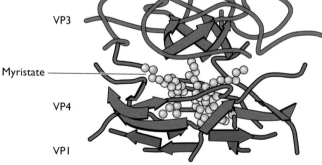

**C**

VP3

Myristate

VP4

VP1

## Structurally Complex Capsids

Some naked viruses are considerably larger and more elaborate than the small RNA and DNA viruses described in the previous section. The characteristic feature of such virus particles is the presence of proteins devoted to specialized structural or functional roles. Despite such complexity, reasonably detailed pictures of the organization of this type of virus particle can be constructed by using a combination of biochemical, structural, and genetic methods. We have chosen the well-studied human adenovirus type 2 to exemplify this approach.

**Adenovirus.** The most striking morphological features of the adenovirus particle (maximum diameter, 1,500 Å) are the well-defined icosahedral appearance of the capsid and the presence of long projections, the fibers, at the 12 vertices (Fig. 3.12A). Each fiber, which terminates in a distal knob that binds to cell surface receptors, is attached to one of the 12 penton bases located at positions of fivefold symmetry in the capsid. The remainder of the shell is built from 240 additional subunits, the hexons, each of which is a trimer of viral protein II (Table 3.5). Formation of this capsid therefore depends on nonequivalent

**A**

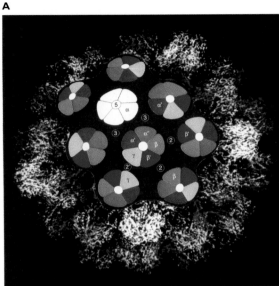

**C**

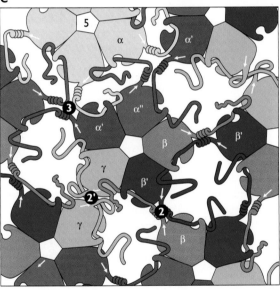

**B**

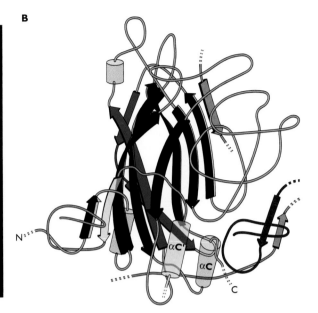

**Figure 3.10 Structural features of the simian virus 40 (SV40) virion.** The structure of simian virus 40 (and other papovaviruses) was long viewed as an enigma because organization of these particles cannot be explained by the theoretical principles elaborated by Caspar and Klug. In the first place, the 360 subunits of the virion would correspond to $T = 6$, a triangulation number that is not permitted according to the rules developed by these authors. Furthermore, the hexameric packing of 60 VP1 pentamers represents a direct violation of the principle of quasiequivalence. The high-resolution view shown here therefore made an important contribution to advancing our understanding of the molecular principles of virus structure. (A) View of the SV40 virion showing the organization of VP1 pentamers. The twelve 5-coordinated pentamers are shown in white, and the 60 pentamers present in hexameric arrays are colored. The three types of interpentameric clustering that allow 5-coordinated and 6-coordi-

nated clustering of pentamers are shown in the schematic overlay. The VP1 subunits shown in white, purple, and green form a threefold cluster, designated 3. Those shown in red (β) and blue (β') engage in one kind of twofold interaction, designated 2 , and the yellow subunits (γ) form a second kind of twofold cluster, labeled 2'. (B) The topology of the VP1 protein shown in a ribbon diagram. The strands of the β-barrel jelly roll are shown as in Fig. 3.8B. This β-barrel domain is radial to the capsid surface rather than organized tangentially as in poliovirus VP1, VP2, and VP3 proteins. Two C-terminal arms, which include αC helices, are shown in orange and light gray. The former is the C-terminal arm of the VP1 subunit shown, which invades a neighboring pentamer (not shown). The second is the invading arm from a different neighboring pentamer (not shown), which is clamped in place by extensive interactions of its β-strand with the N-terminal segment of the subunit shown. The subunit shown also interacts with the N-terminal arm from its anticlockwise neighbor in the same pentamer (dark gray) and with the C-terminal arm of the pentamer that invades that neighbor (black). Thus, the pentamers interact in the capsid shell via extensive contacts among N- and C-terminal arms of VP1 subunits in the same and in neighboring pentamers. (C) The interpentamer contacts of C-terminal arms, with pentamers and their subunits represented and color coded as in panel A, with C-terminal extensions shown as lines and with coils representing the αC-helices in panel B. The three types of interpentamer clustering shown in panel A, 3, 2, and 2', are accommodated by variation in interpentamer contacts made by the C-terminal extensions. For example, in the threefold cluster, the C-terminal α-helices from the three interacting VP1 molecules present in different pentamers form a three-stranded α-helical bundle. A two-chain α-helical bundle is formed at one kind of twofold cluster (2, between red and blue subunits), whereas at the second kind of twofold cluster (2', yellow subunits) the subunits are packed so closely that there is no space to accommodate an α-helical structure. (A and B) From R. C. Liddington et al., *Nature* **354:**278–284, 1991, with permission. Courtesy of S. C. Harrison, Harvard University.

**Table 3.4** Some herpes simplex virus type 1 virion proteins

| Location | Protein | Gene | Properties or function |
|---|---|---|---|
| Nucleocapsid | VP5 | UL19 | Major capsid protein; forms hexamers and pentamers of the icosahedral shell |
| | VP19C | UL38 | Heterotrimer ("triplex") connecting VP5 subunits on surface |
| | VP23 | UL18 | Heterotrimer ("triplex") connecting VP5 subunits on the surface |
| | VP24 | UL26 | Synthesized as a precursor linked to scaffolding protein VP21; protease |
| | VP26 | UL35 | Associated with VP5 hexons but not pentons |
| Tegument | VP1–3 | UL36 | Very large, ~273 kDa |
| | VP16 | UL48 | Abundant structural protein; activator of immediate-early gene transcription |
| | VP18.8 | UL13 | Protein kinase |
| | VP22 | UL49 | Not known |
| | Vhs | UL41 | Host shutoff protein; binds to VP16 |
| | US11 | US11 | Binds to 60S ribosomal subunit |
| Envelope | gB | UL27 | Binds heparan sulfate; fusion; binds tegument proteins including VP16 |
| | gC | UL44 | Attachment protein; binds heparan sulfate |
| | gD | US6 | Binds cell surface receptors; entry |
| | gE/gI | US8/US7 | Heterodimer; binds Fc domain of immunoglobulin G; cell-cell spread |
| | gH/gL | UL22/UL1 | Heterodimer; fusion |
| | gG | US4 | Not known |
| | gJ | US5 | Not known |
| | gK | UL53 | Virus-induced cell fusion; egress |
| | gM/gN | UL10/UL49.5 | Heterodimer; function not known |
| | US9 | US9 | No large external domain |
| | UL20 | UL20 | Probably multiple membrane-spanning domains; egress |

interactions among subunits: the hexons that surround pentons occupy a different bonding environment than those surrounded entirely by other hexons. Furthermore, the 780 total protein subunits ($12 \times 5$ in pentons plus $3 \times 240$ in hexons) that form the closed capsid surface are not arranged to correspond to the $T = 13$ (780/60) icosahedron. Rather, determination of the trangulation number of the adenovirus capsid by electron microscopy (Box 3.2, Fig. D) yielded a value of $T = 25$.

The adenovirus particle contains at least seven additional proteins that play important structural roles (Table 3.5), as well as several small polypeptides produced by proteolytic cleavage of precursor proteins during maturation of newly assembled virus particles. The large size of the virion and the presence of so many proteins have made elucidation of adenovirus structure a challenging problem. One approach that has proved generally useful in the study of complex viruses is the isolation and characterization of discrete subassemblies or subviral particles following exposure of purified virions to appropriate chemicals or enzymes. For example, adenovirus particles can be dissociated into a core structure which contains the DNA genome, groups-of-nine hexons, and pentons (Fig. 3.13). Analysis of the composition of such subassemblies, together with identification of virion proteins that contact one another (by cross-linking methods), identified two classes of virion proteins in addition to the major capsid proteins described previously. One comprises the proteins present in the core, such as protein VII, the major DNA-binding protein (Table 3.5). The remaining proteins are associated with either individual hexons or groups-of-nine hexons (Table 3.5), suggesting that they stabilize the structure. Indeed, protein IX has been clearly identified as capsid "cement": a mutant virus that lacks the protein IX coding sequence produces the typical yield of virions, but they are much more heat sensitive than wild-type particles, indicating that this protein makes an important contribution to virion stability. It has been proposed that the several stabilizing proteins resolve the paradoxical requirements for assembly of this large particle and its transmission. During assembly, interactions among hexons and other major structural proteins must be relatively weak so

**A**

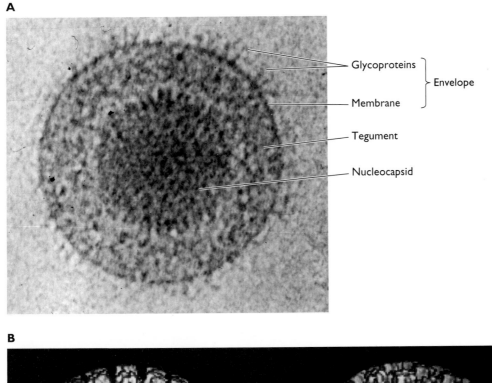

Glycoproteins ⎤
                ⎬ Envelope
Membrane    ⎦

Tegument

Nucleocapsid

**B**

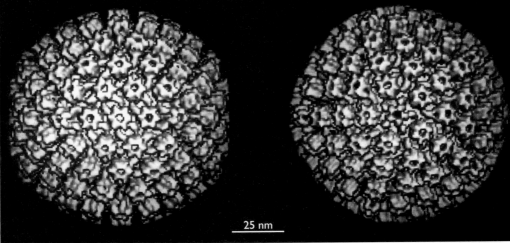

25 nm

**Figure 3.11 Structure of herpesvirus particles.** (A) Electron micrograph of a frozen, hydrated HSV-1 virion showing the surface glycoproteins, lipid membrane, tegument, and nucleocapsid. Note the regular appearance not only of the capsid but also of the tegument and envelope; the traditional view of the tegument as an amorphous, irregular structure was the result of distortion during negative staining. From F. J. Rixon, *Semin. Virol.* **4:**135–144, 1993, with permission. Courtesy of F. Rixon, Institute of Virology, Glasgow, United Kingdom, and W. Chiu, Baylor College of Medicine. (B) Three-dimensional reconstruction of the surface of the herpes simplex virus type 1 nucleocapsid viewed along the twofold (left) and fivefold (right) axes, made from cryoelectron micrographs and computed to about 25-Å resolution. Note the hexameric (left) and pentameric (right) clustering of the major capsid protein, VP5. These structural units are connected at the floor of the outer surface by a heterotrimer of two minor virion proteins (Table 3.4). These triplexes can also be seen in the reconstruction of the nucleocapsid shown in panel D of the figure in Box 3.2. From W. M. Newcomb et al., *J. Mol. Biol.* **232:**499–511, 1993, with permission. Courtesy of J. C. Brown, University of Virginia Health Sciences Center.

**Figure 3.12 Structural features of adenovirus particles.** (A) Surface view of the adenovirus type 2 virion, obtained by cryoelectron microscopy and image reconstruction. This view is oriented along an icosahedral threefold axis and has a nominal resolution of 30 Å. This is sufficient to visualize individual hexons, as well as their general shape, and to distinguish pentons. Only the virion-proximal portions of the projecting fibers are observed. From P. L. Stewart et al., *Cell* **67:**145–154, 1991, with permission. (B and C) Structure of the hexon. The monomer (B) is shown as a Richardson diagram, with gaps indicating regions that were not defined in the X-ray crystal structure at 2.9-Å resolution, and the trimer (C) is shown as a space-filling model with each monomer in a different color. The monomer contains two β-barrel jelly roll domains colored green and blue. These form the base of the hexon trimer (C), from which rise three towers formed by loops emanating from the β-barrel domains of each monomer. The trimers are stabilized by extensive interactions within both the base and the towers. The shape of the base facilitates packing of hexons in hexameric arrays to form the capsid shell. From M. M. Roberts et al., *Science* **232:**1148–1151, 1986, and F. K. Athappilly et al., *J. Mol. Biol.* **242:**430–455, 1994, with permission. Courtesy of J. Rux, S. Benson, and R. M. Burnett, The Wistar Institute.

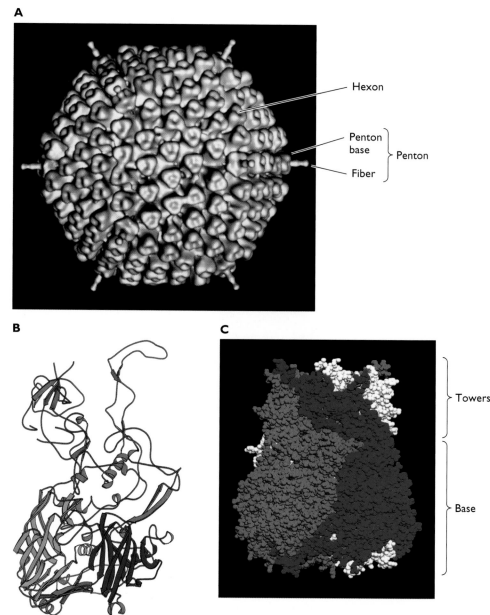

that incorrect associations can be reversed and corrected. However, the assembled virion must be stable enough to survive passage from one host to another. The incorporation of stabilizing proteins like protein IX allows both requirements to be met.

The intact adenovirus particle has not been studied by X-ray crystallography, but high-resolution structures have been determined for the knob of the fiber and the hexon. Each hexon subunit (a molecule of protein II) (Table 3.5) contains two β-barrel domains, each with the topology of the β-barrels of the simpler RNA and DNA viruses described in the previous section (Fig. 3.12B). As the two β-barrels are very similar in structure, the hexon trimer exhibits pseudohexagonal symmetry, a property that facilitates its close packing in the capsid. In the trimer, the β-barrel domains of the three monomers are packed together in a hollow base from which rise three towers formed by intertwining loops from each monomer (Fig. 3.12C). The interactions among monomers are very extensive, particularly in the tower, where loops from the

**Table 3.5**  Specialization of adenovirus type 2 structural proteins[a]

| Protein | Molecular mass (kDa) | Number of copies | Location | Function |
|---------|---------------------|------------------|----------|----------|
| II | 109,677 | 720 | Hexon (trimer) | Formation of capsid shell |
| III | 63,296 | 60 | Penton base (pentamer) | Formation of capsid shell |
| IV | 61,960 | 36 | Fiber (trimer) | Attachment to host cell |
| IIIa | 63,287 | $74 \pm 1$ | Edges between hexon faces | Stabilization of capsid |
| VI | 23,449 | $342 \pm 4$ | Hexon associated or interior surface | Stabilization of capsid |
| VIII | 14,539 | $211 \pm 2$ | Hexon associated | Stabilization of capsid |
| IX | 14,339 | $247 \pm 2$ | Outer surface of groups-of-nine hexons | Stabilization of capsid |
| V | 41,631 | $157 \pm 1$ | Core, outer surface | Packaging of DNA genome |
| VII | 19,412 | $835 \pm 20$ | Core, bound to DNA | Packaging of DNA genome |
| μ | 2,441 | $120 \pm 15$ | Core | Packaging of DNA genome |

[a]In addition to the structural proteins listed, the adenovirus virion contains two copies of pTP, one covalently linked to each 5′ end of the viral genome, the viral L3 protease, and several small proteins generated upon cleavage of proteins that enter the virion as precursors. Only one of these, the C-terminal extension from pVI, which activates the L3 protease, has been ascribed a function.

three subunits are intertwined with one another. Consequently, once the trimer has formed, the monomers cannot be easily dissociated, and the hexon is an extremely stable structure. Because of its large central cavity, which is open at the base but closed at the top, the trimer occupies a large volume for its mass, an advantageous property for a structural unit that forms a large protein shell.

The high-resolution structure of the hexon has allowed the resolution gap between X-ray crystallography and electron microscopy to be bridged by the technique of **difference imaging**. This procedure (Fig. 3.14) essentially subtracts the hexon image obtained by X-ray crystallography (at appropriate resolution) from an electron microscopic image. As illustrated in Fig. 3.14 for the groups-of-nine hexons described above, this subassembly contains 12 molecules of protein IX, each extending along a hexon-hexon interface like mortar between bricks in a wall. This arrangement neatly accounts for the stabilizing function of protein IX. Difference imaging has yielded previously inaccessible structural information about adenoviruses, although not the molecular details of protein-protein or protein-nucleic acid interactions. Its application to other complex viruses is proving equally valuable (see Fig. 3.17, for example).

**Other large icosahedral viruses.** The structural complexity of adenovirus particles is the result of the relatively large number of proteins with which the single protein shell of these virions is built and stabilized. Other icosahedral virions that contain a similar number of proteins exhibit quite different architectures. One of the most elegant is that of reovirus, naked particles 800 Å in diameter containing the 10 segments of double-stranded RNA that comprise the genome. The eight proteins of reovirus particles are organized in two concentric shells with spikes projecting from the inner layer through and beyond the outer layer at each of the 12 vertices (Fig. 3.15A). Rotavirus particles are also constructed from nested protein layers, in this case three layers punctuated by 60 spikes.

The structure of the inner core of the reovirus bluetongue virus has recently been determined by X-ray crystallography. This structure, which contains the genome, virion enzymes, and structural proteins and is transcriptionally active, represents the largest yet to be determined at high resolution. Furthermore, this structure has revealed a new mechanism of packing protein structural units in a closed shell with icosahedral symmetry. The core comprises an outer "bristly" layer of trimers of a single protein (VP7) organized into a classical $T = 13$ icosahedral lattice (Fig. 3.15C, left). Indeed, the interactions among the VP7 structural units that form pentamers and hexamers in this shell are very similar. This outer layer lies upon a thin inner shell built from monomers of the VP3 protein (Fig. 3.15C, right). The architecture of this layer is radically different from that predicted from the principles of quasiequivalence. It contains 120 monomers that occupy one of two icosahedrally unique environments (Fig. 3.15C, right), an arrangement excluded in the classic theory of Caspar and Klug (Box 3.2). However, substantial differences in the conformations of the VP3 molecules that occupy the two structurally distinct environments allow close packing in the equivalent of a $T = 2$ lattice. Because the distortion necessary for formation of the VP3 shell precludes closely similar contacts among "structural units" (one monomer of each type), this organization has been termed **geometrical quasiequivalence**.

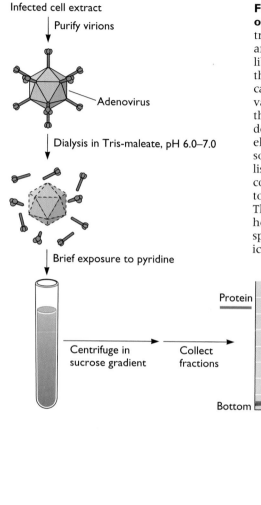

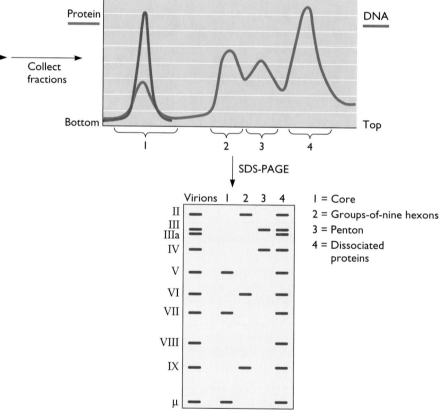

**Figure 3.13 Analysis of subassemblies generated by controlled dissociation of the adenovirus particle.** Adenovirus particles are readily purified from extracts prepared from infected cells by repeated cycles of freezing and thawing and low-speed centrifugation to remove large cellular debris, followed by equilibrium centrifugation in CsCl gradients. Dialysis of such purified virions under the conditions shown liberates the pentons and renders the resulting pentonless capsid susceptible to further dissociation. Brief exposure to pyridine generates a variety of subassemblies that can be separated by velocity centrifugation. Of these, only the fastest sedimenting also contains viral DNA and can therefore be defined as the nucleoprotein core of the virion. Examination of fractions in the electron microscope (not shown) and analysis of their protein composition by sodium dodecyl sulfate-polyacrylamide gel electrophoresis (SDS-PAGE) established the identity and composition of the subassemblies illustrated. The core contains the internal DNA-associated proteins, V, VII and μ, whereas the penton fractions contain the proteins that form the fiber (IV) and penton base (III). The subassembly designated "groups-of-nine" comprises nine copies of the hexon and hexon-associated proteins, such as protein IX. This structure corresponds to the nine hexons that form the center of each of the 20 faces of the icosahedral capsid. The separation of these structures is idealized in the figure.

## Complex Viruses

Some large viruses contain no structural elements that conform simply to helical or icosahedral symmetry. For example, the cores of the complex poxviruses, such as vaccinia virus, appear as smooth rectangles encased within a membrane and surface protein layer (Fig. 3.16). This shape may be the result of the combination of icosahedrally symmetric terminal caps with a tubular structure with matching helical symmetry, as in the elongated heads of certain bacteriophages.

**A** Groups-of-nine hexons

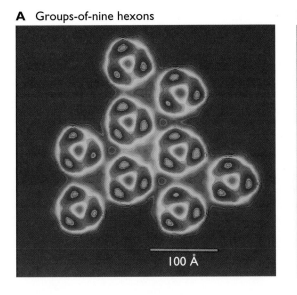

100 Å

**B** Single hexon

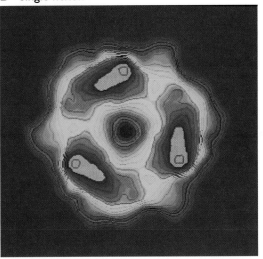

**C** Artificial image of 9 hexons placed as in A

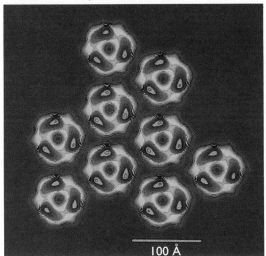

100 Å

**D** Difference image, A minus C

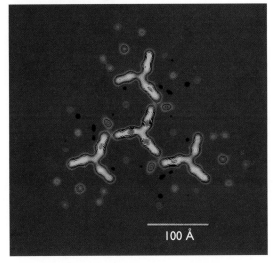

100 Å

**Figure 3.14 Difference imaging of adenovirus groups-of-nine hexons.** The images used in this process are shown as two-dimensional contour plots. (A) The average of 57 images of groups-of-nine hexons obtained by scanning transmission electron microscopy (STEM) of these structures with a resolution of 15 to 17 Å. (B) Two-dimensional image of a single hexon trimer at the same resolution, constructed from the high-resolution hexon structure determined by X-ray diffraction. (C) Artificial image of the groups-of-nine hexons created by rotational and translational alignment of nine copies of the X-ray-based hexon image (B) with the STEM image of the assembly (A). The very good agreement of hexon separation in this artificial image with that observed by electron microscopy established the accuracy of the image shown. (D) Difference image formed by subtracting the artificial image of the placement of hexons in groups of nine (C) from the STEM image of natural groups of nine containing both hexons and protein IX (A). This procedure reveals 12 molecules of protein IX arranged in four clusters of 3 molecules in the hexon-hexon interfaces. More recently, three-dimensional difference imaging of the intact particle has located protein IIIa to the edges between icosahedral faces, consistent with its necessary role in virion assembly, and provided insights into the packing of trimeric hexons around the penton base pentamer at each vertex. From P. S. Furcinitti et al., *EMBO J.* **8:**3563–3570, 1989, with permission. Courtesy of J. Rux and R.M. Burnett, The Wistar Institute.

**A**

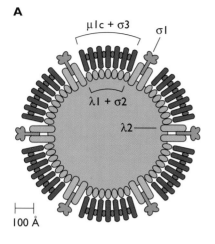

μlc + σ3   σl

λl + σ2

λ2

100 Å

**B**

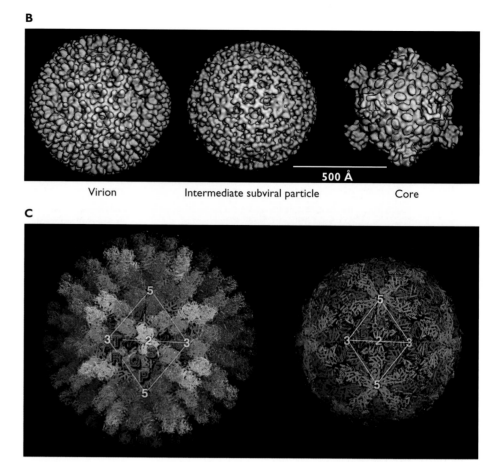

500 Å

Virion          Intermediate subviral particle          Core

**C**

Core particle                                Inner layer

**Figure 3.15  Structure of members of the *Reoviridae*.** (A) Organization of mammalian reovirus particles is shown schematically to indicate the locations of the virion proteins deduced from the protein composition of virions and of subviral particles that can be readily isolated from them. (B) Structures determined by cryoelectron microscopy and image reconstruction of these particles. Exposure of virions (left) to proteases generates infectious (intermediate) subviral particles (center), with release of the σ3 protein and cleavage of μ1, and cores (right) containing only the inner capsid proteins and a conformationally altered form of λ2. Infectious subviral particles correspond closely to particles generated by cellular proteases during entry of reovirus into host cells (see chapter 5). Cores are not infectious but are active for transcription of the double-stranded RNA genome by viral enzymes. The surface projections of the virion are formed by σ3 protein. On protease treatment, this protein is cleaved and removed from the particle. The surface of the subviral particle is formed of clusters of μ1, and the σ1 fibers that project from pentameric λ2 spikes at each of the 12 vertices are clearly visible, although they are not in intact virions. Thus, σ1 undergoes a large conformational change when infectious subviral particles are formed. Removal of μ1 and σ1 in the core particles (right) reveals the prominent λ2 spikes projecting from the inner capsid, which is formed of σ2 protein. From K. A. Dryden et al., *J. Cell Biol.* **122:**1023–1041, 1993, with permission. Courtesy of T. S. Baker, Purdue University. (C) X-ray crystal structure of the core of bluetongue virus, a member of the orbivirus genus of the *Reoviridae*, showing the core particle (left) and the inner scaffold (right). The VP7 trimers project radially from the outer layer. As indicated, each icosahedral asymmetric unit, the triangular areas defined by the icosahedral symmetry axes, contains 13 copies of VP7 arranged as five trimers designated P, Q, R, S, and T and colored red, orange, green, yellow, and blue, respectively. Trimer T, which is located at the threefold axes of icosahedral symmetry, contributes only a monomer to an asymmetric unit. Thus, the outer layer is organized with classical $T = 13$ icosahedral symmetry. The VP3 molecules of the inner layer are red and green. As shown on the right, this scaffold is built from VP3 monomers that occupy one of two completely different structural environments designated A and B and colored green and red, respectively. A-type monomers span the icosahedral twofold axes and interact around the icosahedral fivefold axes. In contrast, B-type monomers are organized as triangular "plugs" around the threefold axes. Distortion in the interactions among monomers at different positions allows close packing to form the closed shell in a $T = 2$ structure (see text). As might be anticipated, VP7 trimers in pentameric or hexameric arrays in the outer layer make completely different contacts with the two classes of VP3 monomer in the inner layer. Nevertheless, each type of interaction is extensive and, in total, these contacts compensate for the symmetry mismatch between the two layers of the core. The details of these contacts suggest that the inner shell both defines the size of the virus particle and provides a template for assembly of the outer $T = 13$ structure. From J. M. Grimes et al., *Nature* **395:**470–478, 1998, with permission. Courtesy of D. I. Stuart, University of Oxford.

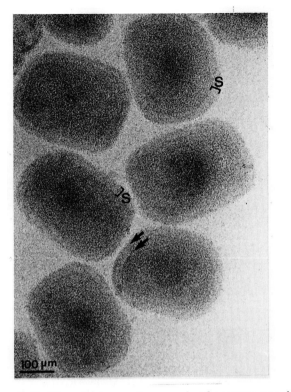

**Figure 3.16 Cryoelectron micrograph of mature vaccinia virus particles.** These particles comprise a 300-Å-thick surface domain (*S*) bounded by the membranes marked by the arrows and surrounding a homogeneous core. The latter is studded with 200-Å spikes and has the shape of a rounded rectangle. From J. Dubochet et al., *J. Virol.* **68:**1935–1941, 1994, with permission. Courtesy of M. Adrian and J. Dubochet, Université de Lausanne.

**Table 3.6**  Mechanisms of packaging the viral genome

| Mechanism | Examples | Protein(s) contacting nucleic acid |
|---|---|---|
| By direct contact with nucleocapsid or capsid proteins | Parvoviruses | Ca |
| | Herpesviruses | ? |
| | Caliciviruses | ? |
| | Alphaviruses | C |
| | Reoviruses | λ1? |
| | Picornaviruses | VP4? |
| By specialized viral nucleic acid-binding proteins | Coronaviruses | N |
| | Filoviruses | NP |
| | Rhabdoviruses | N |
| | Paramyxoviruses | NP |
| | Orthomyxoviruses | NP |
| | Retroviruses | NC |
| | Adenoviruses | VII, V, μ |
| | Poxviruses | L4R, F17R? |
| By cellular DNA-binding proteins | Papovaviruses | Histones H2A, H2B, H3, and H4 |

## Packaging the Nucleic Acid Genome

A definitive property of a virion is the presence of the nucleic acid genome. Incorporation of the genome requires its discrimination from a large population of cellular nucleic acid and its packaging in the interior of the virion. Three mechanisms for condensing, and presumably organizing, nucleic acid molecules within the capsid or nucleocapsid can be distinguished (Table 3.6).

### Direct Contact of the Genome with Proteins of the Outer Shell

The simplest genome packaging arrangement is direct contact of viral nucleic acid with the protein(s) that forms the protective shell of the particle (Table 3.6). The structural units of the icosahedral capsids of many small RNA viruses, including poliovirus and other picornaviruses, contain proteins that interact with the viral genome at the inner surface of the capsid shell. As we have seen above, the interior of the poliovirus capsid can be described in detail. Nevertheless, we possess no structural information about the arrangement of the RNA genome, for the nucleic acid was not visible in the X-ray structure. This property indicates that the RNA genome lacks the symmetry of the virion and does not adopt the identical conformation in every virus particle. In contrast, in other small viruses with icosahedral symmetry, segments of the RNA or DNA genome are highly ordered and can be seen in direct and specific contact with internal surfaces of capsid proteins in X-ray crystal structures. For example, in the *T* = 3 nodavirus black beetle virus (Table 3.3), duplex segments of the RNA genome bind to subunits of adjacent structural units at the icosahedral twofold axes (equivalent to the C subunits in Fig. 3.6B). Such position-specific interaction of the RNA genome with capsid protein molecules is an important regulator of the quasiequivalent interactions necessary to form the *T* = 3 capsid.

Use of the same protein or proteins both to package the genome and to build a capsid allows efficient utilization of limited genetic capacity. It is therefore an advantageous arrangement for viruses with small genomes. However, this mode of genome packing is also characteristic of some more complex viruses, notably rotaviruses and herpesviruses. The genome of rotaviruses (members of the family *Reoviridae*) comprises 11 segments of double-stranded RNA located within the innermost of the three protein shells of the virion. About 25% of the RNA (more than 4,000 bp) is highly ordered, forming a dodecahedral structure in which RNA helices are in close contact with the interior surface of the inner nucleocapsid built from VP2 (Fig. 3.17). At present, rotaviruses hold the record for the fraction of the genome that is icosahedrally ordered within the virion.

One of the most surprising properties of the large herpesvirus nucleocapsid is the absence of internal proteins associated with viral DNA: despite intense efforts, no such core proteins have been identified. It is therefore probable that the double-stranded DNA genome is packaged in a liquid crystalline state in which the DNA is present at a high concentration (typical of crystals) but not organized into a specific structure (like the molecules in a liquid). Such packing must require neutralization of the negative charges of the sugar-phosphate backbone, either by proteins that form the inner surface of the capsid or by incorporation of small, positively charged cellular molecules like spermine and spermidine.

## Packaging by Specialized Virion Proteins

In many other virus particles, the genome is associated with specialized nucleic acid binding proteins, such as the nucleocapsid proteins of (−) strand RNA viruses and retroviruses or the core proteins of adenoviruses (Table 3.6). Although these virion nucleic acid binding proteins were recognized many years ago, relatively few details of the nucleoprotein structures they form within virions have been established. As described above, the ribonucleoproteins of (−) strand RNA viruses are helical structures. Electron microscopy of adenovirus particles or of the core structures released from them suggested that the internal nucleoprotein is also organized in some regular fashion. This organization has proved difficult to study in detail, not least because the compact nucleoprotein released from virions rapidly unfolds into an elongated structure whose appearance depends on experimental conditions. However, the fundamental DNA packaging unit is a multimer of protein VII (Table 3.5), which appears as beads on a string of adenoviral DNA when other core proteins are removed. Protein V probably occupies a more external position in the nucleoprotein core, where it makes contact with capsid proteins. These, and the third core protein, μ, are basic, as would be expected for proteins that bind to a negatively charged DNA molecule without sequence specificity.

Retrovirus virions contain approximately 2,000 molecules of nucleocapsid (NC) protein that bind to the two copies of the encapsidated (+) strand RNA genome (appendix, Fig. 9). The NC protein is also responsible for encapsidation of the genome during assembly, by recognition of a specific packaging signal in the RNA. Like other viral

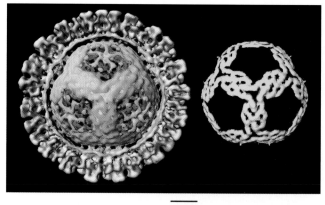

**100 Å**

**Figure 3.17 Structural organization of double-stranded genomic RNA within rotavirus.** This structure was determined by cryoelectron microscopy, image reconstruction, and difference imaging of infectious double-layered particles and various virus-like particles formed from virion proteins expressed from baculovirus vectors. The 19-Å-resolution structure of the double-layered particle is shown at the left, with the outer VP6 capsid of the double-layered particle (middle capsid in the intact virion) in blue and the inner VP2 capsid in green. A complex of the virion RNA-dependent RNA polymerase and RNA capping enzyme (VP1 and VP3), shown in red, is associated with the inner surface of the VP2 nucleocapsid at the axes of fivefold rotational symmetry. The double-stranded RNA, which is packed around the structures formed by VP1 and VP3, is shown in yellow. The dodecahedral shell of ordered RNA, in which each strand is about 20 Å in diameter as expected for RNA in double-stranded helices, is shown separately at the right. From B. V. V. Prasad et al., *Nature* **382:** 471–473, 1996, with permission. Courtesy of B. V. V. Prasad, Baylor College of Medicine.

genome binding proteins, NC is positively charged. It contains at least one copy of a well-characterized nucleic acid binding motif which is defined by the presence of cysteine and histidine residues (described in more detail in chapters 12 and 13). Retroviral ribonucleoproteins are encased within a protein shell built from the capsid (CA) protein to form an internal core. Although they contain the same components, these cores vary considerably in morphology (Fig. 3.18). Recent studies of in vitro assembly of human immunodeficiency virus type 1 CA indicate that this protein determines the conical shape of the core of this virus. However, the structural basis for the morphological differences among retroviral cores is not yet well understood.

## Packaging by Cellular Proteins

The final mechanism of packaging the viral genome, exploitation of cellular proteins, is unique to papovaviruses, such as simian virus 40. The circular, double-stranded DNA genomes of these viruses are coiled around cellular

**A**

**B**

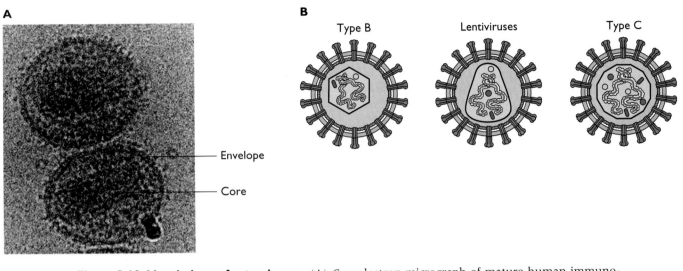

Envelope

Core

**Figure 3.18 Morphology of retroviruses.** (A) Cryoelectron micrograph of mature human immunodeficiency virus type 1 (HIV-1) showing the elongated internal cores of these particles. Courtesy of T. Wilk, European Molecular Biology Laboratory. (B) Variations in the morphology of retroviruses shown schematically. Although retrovirus particles are assembled from the same components, some contain roughly spherical cores, whereas the cores of lentiviruses like HIV-1 are elongated and conical. The structure of the N-terminal, protease-resistant domain of the CA protein of HIV-1 has been determined by NMR methods (chapter 5). However, whether the striking differences in shape among retroviral cores is the result of variation in CA structure, of differences in the structure of the internal ribonucleoprotein, of the ways in which CA proteins interact with NC proteins, or of a combination of such differences, is not known.

nucleosomes containing the four core histones H2A, H2B, H3, and H4. Thus, this viral genome is organized within the virion (and in infected cells) like cellular DNA in chromatin to form a minichromosome. This is an elegant packaging mechanism with two major advantages to the virus: none of its limited genomic information needs to be devoted to DNA-binding proteins, and the viral genome, which is transcribed by cellular RNA polymerase II, enters the infected cell nucleus as a nucleoprotein closely resembling the cellular templates for this enzyme. The 20 or so nucleosomes present in the virion, which package all sequences of the viral genome, condense the DNA by a factor of approximately 7. Within the virion, the minichromosome must be further condensed, presumably as a result of its interactions with the internal proteins of the capsid, VP2 and VP3, and perhaps the N-terminal arms of VP1 that lie on the interior surface of the capsid. Both VP2 and VP3 can bind nonspecifically to DNA as well as to cellular histones, and all three proteins associate with the minichromosome during virion assembly. The internal nucleoprotein is not organized with icosahedral symmetry matching that of the capsid, but neither its structure nor how it interacts with capsid proteins has been determined.

Although three different ways of condensing and organizing genomic nucleic acids within virions can be readily distinguished (Table 3.6), none of these packaging arrangements is understood in detail. High-resolution structural descriptions of virion interiors, or of the nucleoproteins that reside within them, would undoubtedly provide important insights into mechanisms of condensation of the nucleic acid and the structural (symmetry) relationships between nucleic acid-containing structures and the protein shells that surround them. They might also improve our understanding of the advantages conferred by the several packaging mechanisms described above.

## Viruses with Envelopes

Many types of animal virus contain structural elements in addition to the capsids or nucleocapsids described previously. All such virus particles possess an envelope formed by a viral glycoprotein-containing membrane derived from the host cell, but they vary considerably in size, morphology, complexity, and nature of the membrane. Typical features of viral envelopes and their proteins are described in the next section to set the stage for consideration of the structures of envelope proteins and the mechanisms by which they interact with internal components of the virion.

## Viral Envelope Components

The foundation of all viral envelopes is a lipid membrane acquired from the host cell during assembly. This membrane consists of a bilayer of phospholipids, molecules containing negatively charged head groups attached to long chains of hydrophobic fatty acids. As discussed in more detail in chapter 4, the apolar fatty acyl chains are arranged in the interior with the polar head groups on the outside, facing the aqueous environment. However, the precise lipid composition is variable, for different viruses capture their envelopes from different kinds of cellular membranes. Embedded in the membrane are viral proteins, the great majority of which are **glycoproteins** that carry covalently linked sugar chains, or **oligosaccharides** (Fig. 3.19). As described in more detail in chapter 12, sugars are added to the proteins posttranslationally during transport to the cellular membrane at which progeny virions assemble. Intra- or interchain disulfide bonds, another common chemical feature of these proteins, are also acquired during transport to assembly sites. These covalent bonds stabilize the tertiary or quaternary structures of viral glycoproteins (Table 3.7).

### Envelope Glycoproteins

Viral glycoproteins are **integral membrane proteins** firmly embedded in the lipid bilayer by a short **membrane-spanning domain** (Fig. 3.19) or less often by two or more transmembrane segments. Membrane-spanning domains of both viral and cellular proteins are hydrophobic α-helices of sufficient length to span the lipid bilayer. The membrane-spanning α-helices of viral glycoproteins

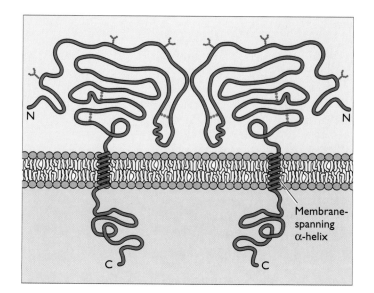

**Figure 3.19 Structural and chemical features of a typical viral envelope glycoprotein shown schematically.** The protein is inserted into the lipid bilayer via a single membrane-spanning domain during its entry into the cellular secretory pathway. This segment separates a larger external domain, decorated with N-linked oligosaccharides (purple) added as the protein travels the secretory pathway and containing disulfide bonds (green), from a smaller internal domain. Viral glycoproteins are oligomers (Table 3.7) whose external portions form the spikes or other projections characteristic of the outer surfaces of enveloped viruses.

generally separate a large external domain, the region decorated with oligosaccharides, from a smaller internal domain (Fig. 3.19). The former contains binding sites for cell surface virus receptors, major antigenic determinants, and sequences that mediate fusion of viral with cellular membranes during entry. Internal domains, which make con-

**Table 3.7** Oligomeric structures of some viral membrane proteins[a]

| Virus | Protein | Quaternary structure |
|---|---|---|
| Alphavirus | | |
|    Semliki Forest virus | E1, E2, E3 | $(E1E2E3)_3$ |
|    Ross River virus | E1, E2 | $(E1E2)_3$ |
| Rhabdovirus (vesicular stomatitis virus) | G | $(G)_3$ |
| Orthomyxovirus (influenza virus) | HA | $(HA1\text{-}HA2)_3$ |
| | NA | $(NA\text{-}NA)_2$ |
| | M2 | $(M2\text{-}M2)_2$ |
| Retrovirus (avian sarcoma virus) | Env | $(SU\text{-}TM)_3$ |
| Herpesvirus (herpes simplex virus type 1) | gH | $(gHgL)_?$ |
| | gL | |

[a]The best predicted oligomeric structures of the viral membrane proteins listed are shown in column 3, with hyphens indicating disulfide-bonded protein chains. The proteins that comprise many of the heteromeric subunits listed (e.g., influenza virus HA, retroviral Env) are produced by proteolytic processing of precursors during transport to the cell surface.

tact with internal components of the virion (see below), are often essential for virus assembly.

With few if any exceptions, the structures formed by viral membrane glycoproteins are oligomeric. These oligomers vary considerably in composition. Some comprise multiple copies of a single protein, but in many cases each subunit contains two or more protein chains (Table 3.7). The subunits are generally held together by noncovalent interactions. On the exterior of the virion, these oligomers form spikes or other surface projections, the structures by which viruses attach to and initiate entry into their host cells. Because of their critical roles in initiating infection, the structures of several viral glycoproteins have been determined. Indeed, the hemagglutinin (HA) protein of human influenza A virus was the first glycoprotein for which high-resolution structural information was obtained.

The HA protein is a trimer of disulfide-linked HA1 and HA2 molecules. Its structure is illustrated in Fig. 3.20 and described in detail in chapters 4 and 5 in the context of the well-understood attachment and entry functions of this protein. The HA protein contains a globular head whose top surface is projected about 135 Å from the viral membrane by a long stalk (Fig. 3.20). The latter is formed by the coiling of α-helices present in each monomer, the major interaction that stabilizes the HA trimer. The membrane-distal globular domain contains both the binding site for the sialic acid virus receptor and the sites of antigenic variation.

Thus, these important functional regions are located more than 100 Å away from the lipid membrane of influenza virus particles. Other viral glycoproteins that mediate cell attachment and entry, such as the E protein of the flavivirus tick-borne encephalitis virus are quite different in structure; the external domain of E protein is a flat, elongated dimer that would lie on the surface of the viral membrane rather than projecting from it (Fig. 3.20). Despite their lack of common structural features, both the influenza A virus HA protein and the tick-borne encephalitis virus E protein are primed for dramatic conformational change to allow entry of internal virion components into a host cell.

The particles of the majority of other enveloped viruses contain one or two glycoproteins with typical properties. For example, the surfaces of all retroviruses are covered by a dense array of spikes (Fig. 3.18) formed by the Env proteins, TM (transmembrane) and SU (surface unit). However, some enveloped viruses contain a larger collection of glycoproteins. One of the several remarkable features of the virions of herpesviruses is the large number of envelope proteins: so far, more than 10 viral glycoproteins have been identified (Table 3.4). As expected, some of them are important for attachment of the virus to host cells and for entry.

The high-resolution viral glycoprotein structures mentioned above are those of the large external domains of the proteins cleaved from the viral envelope by proteases. This treatment facilitated crystallization but of course precluded

**Figure 3.20 Structures of extracellular domains of viral glycoproteins and their orientation with respect to the lipid membrane of the viral envelope illustrated schematically (not to scale).** The influenza virus HA trimer and the tick-borne encephalitis virus E protein dimer are the envelope proteins that mediate attachment to and entry of these virus particles into a host cell. Nevertheless, these proteins are quite different in their tertiary and quaternary structures and orientation with respect to the membrane of the viral envelope.

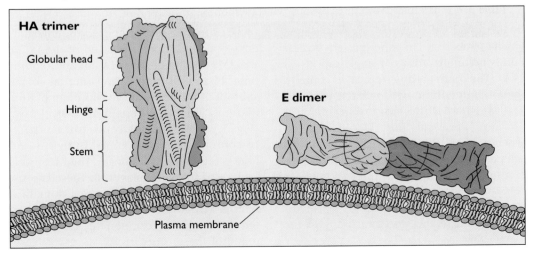

determination of the structure of membrane-spanning or internal segments of these proteins, which also play important structural or functional roles. Interactions among membrane-spanning domains can contribute to the stability of oligomeric glycoproteins, as in influenza virus HA. These segments may also transmit signals from the exterior to the interior of the virion, for example, conformational change induced by specific contact of the external portion with a cellular receptor. The internal domains of viral glycoproteins can participate in anchoring the envelope to internal virion structures, as discussed below.

### Other Envelope Proteins

The envelopes of some more complex viruses, including orthomyxoviruses and poxviruses, contain integral membrane proteins that lack large external domains or possess multiple membrane-spanning segments (see, for example, Table 3.4). Among the best characterized of these is the influenza A virus M2 protein. This small (97-amino-acid) protein is a minor component of virions. Estimates of its concentration range from 14 to 68 copies per virion. In the viral membrane, two disulfide-linked M2 dimers associate to form a noncovalent tetramer that functions as an ion channel. The M2 ion channel is the target of the influenza virus inhibitor amantadine. The effects of this drug, as well as of mutations in the M2 coding sequence, indicate that M2 can play important roles in entry and assembly by controlling the pH of the local environments in which HA molecules and virus particles find themselves.

### Simple Enveloped Viruses: Direct Contact of External Proteins with the Capsid or Nucleocapsid

In the simplest enveloped viruses, exemplified by (+) strand RNA togaviruses such as Sindbis and Ross River viruses, the envelope directly abuts the protein shell that packages and protects the RNA genome. These virus particles can therefore be thought of as comprising two protein shells separated by a membrane. The inner protein layer is a $T = 4$ icosahedral shell built from the single capsid (C) protein (Fig. 3.21A). The outer glycoprotein layer contains 240 copies of a heterodimer of the two envelope proteins, E1 and E2 (Table 3.7), whose clustering in trimers imparts a characteristic surface appearance (Fig. 3.21A). The glycoproteins are also organized into a $T = 4$ icosahedral shell (Fig. 3.21A). Such congruence of the inner and outer protein shells is the result of binding of the internal portions of the glycoprotein spikes to the C-protein subunits of the underlying nucleocapsid (Fig. 3.21B). The glycoproteins are thus directly anchored to the structural units of the nucleocapsid.

### Enveloped Viruses with Additional Protein Layers

Larger enveloped viruses contain additional proteins that mediate interactions of the innermost, genome-containing structure with the viral envelope. In the simplest case, a single viral structural protein, termed the **matrix protein**, welds an internal ribonucleoprotein or capsid to the envelope. This arrangement is found in several groups of ($-$) strand RNA viruses as well as in retroviruses (appendix, Fig. 5, 7, and 9). In contrast, more complex enveloped viruses can contain several proteins in their intermediate layer. As we have seen, the herpesvirus nucleocapsid is surrounded by a regular, spherical structure called the tegument (Fig. 3.11A), which contains at least eight viral proteins (Table 3.4). Because the internal capsids or nucleocapsids of these viruses are not in direct contact with the envelope, the organization and symmetry of the internal structure are not necessarily evident from the external appearance of the surface glycoprotein layer. For example, the outer surface of all retroviruses appears as a dense, roughly spherical array of projecting knobs or spikes, regardless of whether the internal core is also spherical or cone shaped (Fig. 3.18). Likewise, influenza virus particles, which contain helical nucleocapsids, are generally roughly spherical particles 800 to 1,200 Å in diameter (appendix, Fig. 7), although long, filamentous forms are common in fresh virus isolates.

In general, neither the interior architecture of these more complex enveloped viruses nor the molecular connections among internal structures and external protein and protein-plus-lipid layers can be described in detail, for example, because these viruses do not crystallize well (or at all). Nevertheless, there is good evidence, from both mutational analyses and in vitro binding assays, for binding of the proteins that abut the envelope to internal domains of envelope glycoproteins and for their direct association with the lipid membrane. The former interaction is exemplified by the binding of the vesicular stomatitis virus matrix (M) protein to the internal domain of the envelope glycoprotein (G) and of the influenza virus matrix (M1) protein to the HA and neuraminidase (NA) proteins of the envelope. The internal segment of the TM subunit of the envelope glycoprotein of human immunodeficiency virus type 1 and some B- and D-type retroviruses (Fig. 3.18) also directly contacts the matrix (MA) protein. In contrast, in many simple retroviruses, the viral envelope is tethered by binding of the MA protein to the inner surface of the lipid bilayer. In these cases, deletion of the coding sequence for the internal domain of TM perturbs neither virion integrity nor incorporation of the glycoprotein (or any other component) into virions. Furthermore, many of these retroviruses can assemble functional virus

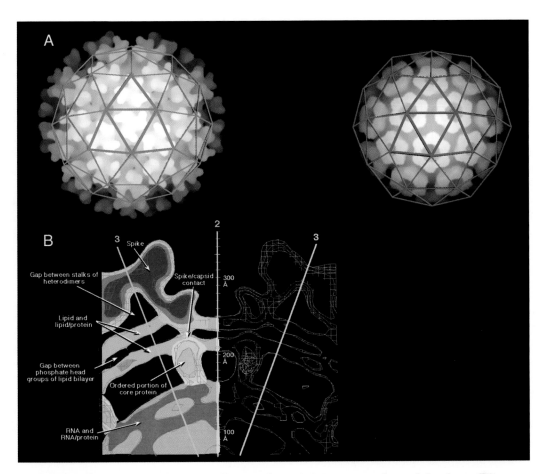

**Figure 3.21 Structural features of a togavirus.** (A) Reconstructions of the three-dimensional structure of the surface (left) and internal nucleocapsid (right) of Ross River virus viewed along a threefold axis. The dispositions of the surface spikes (left) and structural units of the nucleocapsid (right) relative to the icosahedral lattice (blue and red, respectively) are indicated. The glycoprotein E1-E2 dimers and the single capsid (C) protein form $T = 4$ icosahedral shells. The latter encases the (+) strand RNA genome. However, the three-dimensional structure of the C protein bears no resemblance to those of the capsid proteins of naked viruses described in the text: the C protein has a basic N-terminal segment (probably involved in RNA packaging) and a C-terminal domain with a chymotrypsinlike structure. The latter is a functional protease responsible for the liberation of the C protein from a polyprotein translation product. Thus, while the β-barrel domain is a common structural feature of viral capsid proteins, it is not universal. (B) Detailed view of an equatorial slice of the Ross River virus image reconstruction. This view includes one twofold and two threefold axes, indicated in yellow. Distances from the center of the particle, in angstroms, are listed along the twofold axis. Blue, purple, and red indicate low, medium, and high densities in the reconstructed image. In the left half, the different components of the virus particle are shown in different colors. Note the transmembrane contact of the viral glycoproteins in the surface spikes with C protein in the nucleocapsid. From R. H. Cheng et al., *Cell* **80:**621–630, 1995, with permission. Courtesy of T. S. Baker, Purdue University.

particles containing glycoproteins of other, unrelated viruses or even cellular glycoproteins. The matrix proteins of several viruses, including influenza A virus and human immunodeficiency virus type 1, interact with both glycoprotein and lipid components of the envelope.

Internal proteins that mediate contact with the viral envelope are not embedded within the lipid bilayer but rather bind to its inner face and to the cytoplasmic surface of the cellular membrane (such as the plasma membrane) from which the envelope is derived. These viral proteins are targeted to and interact with membranes via specific signals, some of which have been identified. For example, specific segments of the influenza A virus M1 protein are required for its binding to lipid vesicles in vitro and to membranes within infected cells. A posttranslationally added fatty acid chain is important for membrane binding of the MA proteins of most retroviruses. These matrix proteins and their Gag polyprotein precursors (appendix, Fig. 9) carry the 14-carbon, saturated fatty acid myristate covalently linked to N-terminal glycine residues. In some cases, this lipid is necessary and sufficient for binding of MA to the membrane, but the human immunodeficiency virus type 1 MA protein contains a bipartite membrane-binding signal: the myristoylated N-terminal segment and a largely basic sequence located a short distance downstream mediate high-affinity binding of MA and of the Gag polyprotein to the lipid bilayer by hydrophobic interactions with membrane lipids and ionic bonds with negatively charged head groups of membrane phospholipids, respectively. The MA protein of this complex retrovirus was the first viral peripheral membrane protein whose structure was determined, initially by NMR methods (Fig. 3.3). Subsequent analysis of the crystal structure of the protein established that MA is a trimer (Fig. 3.22). Each MA molecule comprises a compact, globular domain of α-helices capped by a β-sheet containing the positively charged amino acids necessary for membrane binding. As illustrated in the model of MA oriented on a membrane shown in Fig. 3.22, the basic residues form a positively charged surface positioned for interaction with phospholipid head groups on the inner surface of the envelope membrane.

The common functions of envelope glycoproteins in attachment and entry of the virus at the beginning of an infectious cycle are well established, as is the common mechanism by which these proteins reach the appropriate cellular membrane. In contrast, high-resolution structural information has been collected for only a few of these proteins, and in no case do the structural models include the transmembrane or internal domains. Similarly, we possess few detailed views of the structure, or organization within virus particles, of internal proteins that bind to viral envelopes. Consequently, interactions among internal and envelope components, which must impart stability to the virion and influence its morphology, are well understood only for the simplest enveloped viruses in which envelope glycoproteins are directly tethered to an internal nucleocapsid (e.g., Fig. 3.21). It is to be hoped that the methods of image reconstruction that have generated such clear views of this method of anchoring the envelope will be equally informative when applied to more complex enveloped viruses. However, the lack of symmetry characteristic of some of these viruses (e.g., influenza virus) imposes a huge hurdle for such methods.

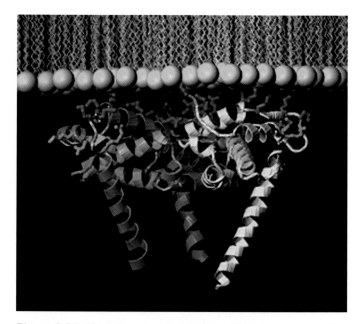

**Figure 3.22 Model of the interaction of human immunodeficiency virus type 1 MA protein with the membrane.** This model is based on the X-ray crystal structure of recombinant MA protein synthesized in *Escherichia coli* and consequently lacking the N-terminal myristate chain normally added in human cells. The three monomers in the MA trimer are shown in different colors. Basic residues in the β-sheet that caps the globular α-helical domain are magenta or green. Substitution of those shown in magenta impairs replication of the virus in cells in culture. From C. P. Hill et al., *Proc. Natl. Acad. Sci. USA* **93**:3099–3104, with permission. Courtesy of C. P. Hill and W. L. Sundquist, University of Utah.

## Other Components of Virions

Some virus particles comprise only the nucleic acid genome and structural proteins, or proteins plus a lipid membrane, necessary for protection and delivery into a host cell. However, many contain additional viral proteins or other components, which are generally present at much lower concentrations than the structural proteins described in previous sections but play essential or important roles in establishing an efficient infectious cycle (Table 3.8).

### Virion Enzymes

Many types of virus particle contain enzymes necessary for synthesis of viral nucleic acids within an infected cell. Such enzymes generally catalyze synthetic reactions unique to viruses, such as synthesis of viral mRNA from an RNA template or of viral DNA from an RNA template. These reactions, which have no cellular counterparts, are catalyzed by RNA-dependent RNA polymerases present in the nucleocapsids of (−) strand RNA viruses and retroviral

**Table 3.8**  Some virion enzymes and other components necessary for efficient viral reproduction

| Virus | Component | Function(s) |
|---|---|---|
| Viral enzymes | | |
| Rhabdovirus (vesicular stomatitis virus) | L | RNA-dependent RNA polymerase, synthesis of viral mRNA and genome RNA |
| Orthomyxovirus (influenza A virus) | P proteins | RNA-dependent RNA polymerase, synthesis of viral mRNA and genome RNA; cap-dependent endonuclease, initiation of mRNA synthesis |
| Reovirus (reovirus type 1) | λ2 | Guanyl transferase |
| | λ3 | Double-stranded RNA-dependent RNA polymerase |
| Retrovirus (avian sarcoma virus, human immunodeficiency virus type 1) | RT | Reverse transcriptase, proviral DNA synthesis |
| | IN | Integrase, integration of viral DNA into the cellular genome |
| | PR | Protease, production of infectious virions |
| Adenovirus (human adenovirus type 2) | L3 23k | Protease, production of infectious virions; uncoating |
| Herpesvirus (herpes simplex virus type 1) | VP24 | Protease, nucleocapsid maturation for genome encapsidation |
| | VP18.8 | Protein kinase |
| Poxvirus (vaccinia virus[a]) | Eight RPO proteins | DNA-dependent RNA polymerase |
| | Poly(A) polymerase (two proteins) | Synthesis of poly(A) on viral mRNA |
| | Capping enzyme (two proteins) | Addition of 5′ caps to viral pre-mRNA |
| | DNA topoisomerase | Sequence-specific nicking of viral DNA |
| Other viral proteins | | |
| Retrovirus (human immunodeficiency virus type 1) | Vpr, Vif | Required for efficient infection in some cell types? |
| Herpesviruses | | |
| Herpes simplex virus type 1 | VP16 | Structural protein, activation of immediate-early gene transcription |
| | Vhs | Induction of degradation of cellular mRNA |
| Human cytomegalovirus | pp65 | Inhibition of presentation of viral immediate-early proteins as antigens |
| Poxvirus (vaccinia virus) | VETF | Binds to early promoters, essential for their transcription |
| | RAP94 | Associated with virion RNA polymerase, specificity factor |
| Cellular components | | |
| Retroviruses | | |
| Avian sarcoma virus, human immunodeficiency virus type 1 | tRNA | Primer for synthesis of first viral DNA strand |
| Human immunodeficiency virus type 1 | Cyclophilin A | Not fully known; an early step in the infectious cycle |

[a]Vaccinia virions contain at least 11 enzymes, only a few of which are listed.

reverse transcriptase, respectively. In contrast, virions of the poxvirus vaccinia virus contain a DNA-dependent RNA polymerase, analogous to cellular RNA polymerases, as well as several enzymes that modify viral RNA transcripts (Table 3.8). This complement of enzymes is necessary because transcription of the viral double-stranded DNA genome takes place in the cytoplasm of infected cells, whereas cellular DNA-dependent RNA polymerases and the RNA-processing machinery are restricted to the nucleus. Other types of enzyme present in virions include retroviral integrase, influenza A virus cap-dependent en-

donuclease, and the proteases of several virus families (Table 3.8). The latter, which eliminate covalent connections among specific proteins from which virions assemble, are necessary for the production of infectious particles, and some may also participate in uncoating and liberation of the viral genome.

## Other Viral Proteins

More complex virions may also contain additional viral proteins that are not necessarily enzymes but are nevertheless important for an efficient infectious cycle (Table

3.8). Among the best characterized examples of this class of virion proteins are several tegument proteins of herpesviruses, such as herpes simplex virus type 1 (Table 3.8). The VP16 protein activates transcription of viral immediate-early genes to initiate the viral program of gene expression. In contrast to the majority of viral proteins discussed in this section, VP16 is present at high concentration in virions and is necessary for their assembly. Other herpesvirus tegument proteins (Table 3.8) induce degradation of cellular mRNA or block the cellular mechanism by which viral proteins are presented to the host's immune system. Complex retroviruses like human immunodeficiency virus type 1 also contain additional proteins required for efficient viral replication in certain cell types. At least two such proteins, Vpr and Vif, which are present in the internal core, have been identified. These proteins are discussed in chapter 18.

### Cellular Macromolecules

Virus particles can also contain cellular macromolecules that play important roles during the infectious cycle, such as the cellular histones that package papovavirus DNA within virions described above. Because of the budding mechanisms by which they form, enveloped viruses can readily incorporate cellular proteins and other macromolecules. Cellular glycoproteins may not be excluded from the membrane from which the viral envelope is derived. Moreover, as a bud enlarges and pinches off during virus assembly, internal cellular components may be trapped within it. In addition, enveloped viruses are generally more difficult to purify than naked viruses. As a result, preparations of these viruses may be contaminated with vesicles formed from cellular membranes during extraction of the virus from infected cells. Because of these properties, it can be difficult to distinguish cellular components specifically incorporated into enveloped virus particles from those trapped randomly or copurifying with the virus. Nevertheless, in some cases it is clear that cellular molecules are important components of virus particles: these molecules are reproducibly observed in virions at a specific stoichiometry and can be shown to play essential or important roles in the infectious cycle. The best-characterized examples of the capture of cellular components are provided by retroviruses.

The primer for initiation of synthesis of the first strand of proviral DNA during reverse transcription is invariably a cellular tRNA incorporated into virions by virtue of its binding to a specific sequence in the RNA genome and to reverse transcriptase. Some retroviruses make more radical use of cellular components: one of the most unusual properties of human immunodeficiency virus type 1 virions is the presence of cellular cyclophilin A, a **chaperone**, which assists or catalyzes protein folding. This protein is a member of a ubiquitous family of peptidyl-prolyl isomerases, which catalyze the intrinsically slow cis-trans isomerization of peptide bonds preceding proline residues in newly synthesized proteins. Cyclophilin A is the major cytoplasmic member of this family. It is incorporated within human immunodeficiency virus type 1 particles (but not those of even closely related retroviruses) via specific interactions with the central portion of CA in multimers of the Gag polyprotein from which particles assemble. Although incorporation of cyclophilin A is not a prerequisite for assembly, particles that lack this cellular chaperone are poorly infectious because they fail to complete some early step in the infectious cycle. Recent experiments controlling for contamination of virus particles by cellular membrane vesicles indicate that cellular membrane proteins (e.g., Icam-1) (chapter 4) can also be incorporated in the viral envelope and can contribute to attachment and entry of human immunodeficiency virus type 1 particles.

It is clear from these examples that, despite the limited coding capacity of their genomes and the finite small volumes of their particles, viruses deliver components with a surprisingly broad repertoire of molecular functions to their host cells (Table 3.8). This repertoire is undoubtedly larger than we presently appreciate, for the precise molecular functions and/or roles in the infectious cycles of many of these important virion components have yet to be established.

### Perspectives

Much has been learned about the architecture of viruses and the molecular principles of their design since investigators first considered the problem of constructing these particles from a limited number of components. Despite great variety in size and morphology, all virus particles contain one protein shell that encases and protects the nucleic acid genome. With few exceptions, these capsids or nucleocapsids are built from a small number of protein subunits arranged with helical or icosahedral symmetry. The detailed views of nonenveloped virions provided by X-ray crystallography emphasize just how well such protein shells are designed for protection of the genome during passage from one host cell or organism to another. They have also identified several mechanisms by which identical or nonidentical subunits can interact to form icosahedrally symmetric structures. As we shall see, atomic-level descriptions of virus particles, or of the individual proteins that contact surface components of the host cell, can provide critical insights into the mechanisms by which virus particles attach to host cells and deliver the genome.

More complex virus particles, which may contain additional protein layers, a lipid envelope carrying viral proteins, and enzymes or other proteins necessary to initiate the infectious cycle, cannot be described with the exquisite detail provided by a crystal structure. Indeed, for many years we possessed only schematic views of their structures deduced from negative-contrast electron microscopy and biochemical or genetic methods of analysis. Within the past decade, the development and refinement of cryoelectron microscopy and techniques of image reconstruction have revolutionized structural studies of larger and more complex viruses. These methods have yielded previously inaccessible information about the structural features and the roles of individual proteins of both naked and enveloped viruses, as well as remarkable views of genome organization, or interactions among envelope and internal proteins. The recent improvement in the resolution that can be achieved by these techniques to <10 Å, the power of difference imaging methods, and the increasing ability to produce large quantities of specific structural elements assembled from recombinant viral proteins promise many new insights into the structure, molecular organization, and function of virus particles. Such information seems likely to improve our understanding, presently quite limited, of mechanisms of packaging of nucleic acid genomes within virus particles and of tethering envelopes (when present) to internal structures. Some surprises are undoubtedly in store. But we can predict with some confidence that future studies will reinforce and elaborate the general principles of virus structure described here.

# References

## Chapters in Books

Baker, T. S., and J. E. Johnson. 1997. Principles of virus structure determination, p. 38–79. *In* W. Chiu, R. M. Burnett, and R. L. Garcea (ed.), *Structural Biology of Viruses*. Oxford University Press, New York, N.Y.

Chow, M., R. Basavappa, and J. M. Hogle. 1997. The role of conformational transitions in poliovirus pathogenesis, p. 157–187. *In* W. Chiu, R. M. Burnett, and R. L. Garcea (ed.), *Structural Biology of Viruses*. Oxford University Press, New York, N.Y.

Garcea, R. L., and R. C. Liddington. 1997. Structural biology of polyomaviruses, p. 187–208. *In* W. Chiu, R. M. Burnett, and R. L. Garcea (ed.), *Structural Biology of Viruses*. Oxford University Press, New York, N.Y.

Harrison, S. C., J. J. Skehel, and D. C. Wiley. 1996. Virus structure, p. 59–99. *In* B. N. Fields, D. M. Knipe, and P. K. Howley (ed.). *Fundamental Virology*, Lippincott-Raven, New York, N.Y.

## Reviews

Chapman, M. S., V. L. Giranda, and M. G. Rossmann. 1990. The structures of human rhinovirus and mengo virus: relevance to function and drug design. *Semin. Virol.* **1:**413–427.

Rixon, F. J. 1993. Structure and assembly of herpesviruses. *Semin. Virol.* **4:**135–144.

Rux, J. J., and R. M. Burnett. 1998. Spherical viruses. *Curr. Opin. Struct. Biol.* **1:**142–149.

Stabbs, G. 1990. Molecular structures of viruses from the tobacco mosaic virus group. *Semin. Virol.* **1:**405–512.

Stewart, P. L., and R. M. Burnett. 1990. The structure of adenovirus. *Semin. Virol.* **1:**477–487.

## Papers of Special Interest

### Theoretical Foundations

Caspar, D. L. D., and A. Klug. 1962. Physical principles in the construction of regular viruses. *Cold Spring Harbor Symp. Quant. Biol.* **27:**1–22.

Crick, F. H. C., and J. D. Watson. 1956. Structure of small viruses. *Nature* **177:**473–475.

### Structures of Virions

Brenner, S., and R. W. Horne. 1959. A negative staining method for high resolution electron microscopy of viruses. *Biochim. Biophys. Acta* **34:**103–110.

Cheng, R. H., R. G. Kuhn, N. H. Olson, M. G. Rossmann, H.-K. Choi, T. J. Smith, and T. S. Baker. 1995. Nucleocapsid and glycoprotein organization in an enveloped virus. *Cell* **80:**621–630.

Dryden, K. A., G. Wang, M. Yeager, M. L. Nibert, K. M. Coombs, D. B. Furlong, B. N. Fields, and T. S. Baker. 1993. Early steps in reovirus infection are associated with dramatic changes in supramolecular structure and protein conformation: analysis of virions and subviral particles by cryoelectron microscopy and image reconstruction. *J. Cell. Biol.* **122:**1023–1041.

Harrison, S. C., A. Olson, C. E. Schutt, F. K. Winkler, and G. Bricogne. 1978. Tomato bushy stunt virus at 2.9Å resolution. *Nature* **276:**368–373.

Hogle, J. M., M. Chow, and D. J. Filman. 1985. Three-dimensional structure of poliovirus at 2.9Å resolution. *Science* **229:**1358–1365.

Liddington, R. C., Y. Yan, H. C. Zhao, R. Sahli, T. L. Benjamin, and S. C. Harrison. 1991. Structure of simian virus 40 at 3.8Å resolution. *Nature* **354:**278–284.

Paredes, A. M., D. T. Brown, R. Rothnagel, W. Chiu, R. J. Schoepp, R. E. Johnston, and B. V. V. Prasad. 1993. Three-dimensional structure of a membrane containing virus. *Proc. Natl. Acad. Sci. USA* **90:**9095–9099.

Prasad, B. V. V., R. Rothnagel, C. Q.-Y. Zeng, J. Jakana, J. A. Lawton, W. Chiu, and M. K. Estes. 1996. Visualization of ordered genomic RNA and localization of transcriptional complexes in rotavirus. *Nature* **382:**471–473.

Rossman, M. G., E. Arnold, and J. W. Erickson. 1985. Structure of a common cold virus and functional relationship to other picornaviruses. *Nature* **317:**145–153.

Wery, J.-P., V. S. Reddy, M. V. Hosur, and J. E. Johnson. 1994. The refined three-dimensional structure of an insect virus at 2.8Å resolution. *J. Mol. Biol.* **235:**565–586.

## Structures of Virion Proteins

Böttcher, B., S. A. Wynne, and R. A. Crowther. 1997. Determination of the fold of the core protein of hepatitis B virus by electron cryomicroscopy. *Nature* **386:**88–91.

Chen, X. S., T. Stehe, and S. Harrison. 1998. Interaction of the polyomavirus internal protein VP2 with the capsid protein VP1 and implications for participation of VP2 in viral entry. *EMBO J.* **37:**3233–3240.

Hill, C. P., D. Worthylake, D. P. Bancroft, A. M. Christensen, and W. L. Sundquist. 1996. Crystal structures of the trimeric human immunodeficiency virus type 1 matrix protein: implications for membrane association and assembly. *Proc. Natl. Acad. Sci. USA* **93:**3099–3104.

Massiak, M. A., M. R. Stanick, C. Paschall, M. F. Summers, A. M. Christensen, and W. L. Sundquist. 1994. The three-dimensional structure of the human immunodeficiency virus type 1 matrix protein. *J. Mol. Biol.* **244:**198–223.

Rey, F. A., F. X. Heinz, C. Mandl, C. Kunz, and S. C. Harrison. 1995. The envelope glycoprotein from tick-borne encephalitis virus at 2Å resolution. *Nature* **375:**291–298.

Wilson, J. A., T. S. Skehel, and D. C. Wiley. 1981. Structure of the haemagglutinin membrane glycoprotein of influenza virus at 3Å resolution. *Nature* **289:**366–373.

*4*

# Virus Attachment to Host Cells

## Introduction

To initiate replication in a host cell, viruses must first adhere to the plasma membrane, an interaction mediated by binding to a specific receptor. Until the 1980s, our understanding of attachment lagged far behind that of other steps in viral replication because technical difficulties prevented scientists from identifying viral receptors. As a result of the development of monoclonal antibody and recombinant DNA methods, this field exploded with the identification of receptors for a number of medically important viruses, including human immunodeficiency virus type 1 and poliovirus. Viral receptors are now being identified at a dizzying pace, and with their isolation comes the ability to understand the virus-receptor interaction at a molecular level, learn how the interaction leads to uncoating of the viral genome, design novel antiviral therapies, and produce new transgenic mouse models for studying viral diseases and disease prevention. In this chapter we will discuss the nature of the virus-receptor interaction, how receptors are isolated, and what can be learned from their study.

Our understanding of the earliest interactions of viruses with cells comes almost exclusively from analysis of synchronously infected cells in culture (with a one-step growth curve), as it is difficult to examine virus-receptor interactions in infected animals. In our discussion, we will for the moment ignore the many steps that occur before attachment, including how viruses enter the environment, the initial contact with a host organism, and bypassing host defense mechanisms. These steps are very important parts of the total picture of viral infection and are discussed in chapters 14 and 17 to 19.

A viral receptor is defined as the cellular molecule to which a virus attaches to initiate replication. For some viruses, the receptor is the only cell surface molecule required for entry into cells. For other viruses, binding to a cellular receptor is not sufficient for infection. In such cases, an additional cell surface molecule, or **coreceptor,** is required for entry of the virus into cells. The roles of coreceptors in viral replication are discussed in chapter 5.

**101**

## Protection and Release of Genetic Material

Successful infection of a new host cell requires, at a minimum, the introduction of an intact viral genome. However, to survive in the extracellular environment, the genome must be encapsidated in a protective coat that will shield viral nucleic acid from the variety of potentially harsh conditions that a virion may meet in the environment. For example, mechanical shearing, UV irradiation from sunlight, extremes of pH (in the gastrointestinal tract), dehydration (in the air), and enzymatic attack (in body fluids) are all capable of damaging viral nucleic acids. The chemically stable, protective protein shells and lipid envelopes that encase viral genomes are designed to perform this function. However, once in the host cell, this protective structure must become sufficiently unstable to release the genetic material. Thus, viral particles cannot be viewed only as entirely passive vehicles. They must be able to undergo structural transformations that are important for attachment and entry into a new host cell and for the subsequent disassembly required for viral replication. Examples of some of these structural transformations are described in chapter 5.

## Finding the Right Cells To Infect

Viral infection is initiated by a collision between the virus and the cell, a process that is governed by chance. Therefore, a higher concentration of virus increases the probability of infection. However, a virus is not able to infect every cell it encounters. It must come in contact with the cells and tissues in which it has evolved to replicate. Such cells are normally recognized by means of a specific virion-cell surface receptor interaction, a process that can be either promiscuous or highly selective, depending on the virus and distribution of the cell receptor. The presence of such receptors determines whether the cell will be **susceptible** to the virus. However, whether a cell is **permissive** for the replication of a particular virus depends on other, intracellular components found only in certain cell types. Thus, cells must be both susceptible *and* permissive if an infection is to be successful. The predilection of a virus to invade and replicate in a particular cell type (its biological niche) is known as its cell or tissue **tropism**. In this chapter we will focus on the determinants of cell susceptibility and the mechanisms by which viruses attach to specific cell surface molecules. The cellular and viral genes and products that determine whether a cell is permissive will be discussed in the following chapters that describe later events in the replication cycle.

In general, viruses have no means of locomotion, but their small size facilitates their diffusion and Brownian movement. Their propagation is dependent on essentially random encounters with potential hosts and host cells. In this sense they can be thought of as tiny, opportunistic "Darwinian machines." For such machines, features that increase the probability of favorable encounters are very important. Thus, viral propagation is critically dependent on the production of large numbers of progeny virions with surfaces composed of many copies of structures that enable the virions to attach to susceptible cells.

## The Architecture of Cell Surfaces

The initiation and spread of viral infection most often follow encounters at the epithelial surfaces of the body exposed to the environment. These surfaces are covered by cells, and the part of these cells exposed to the environment is called the **apical** surface. Conversely, the **basolateral** surfaces of such cells are in contact with adjacent or underlying cells or tissues. These cells exhibit a differential (**polar**) distribution of proteins and lipids in the plasma membranes that creates the two distinct surface domains. As illustrated in Fig. 4.1, these cell layers can differ in thickness and organization. Movement of macromolecules between the cells in the epithelium is prevented by **tight junctions,** which circumscribe the cells at the apical edges of their lateral membranes (Fig. 4.1). Many viruses initiate infection by entering epithelial or endothelial cells at their exposed apical surfaces, often by attaching to cell surface molecules specific for this domain. Viruses that both enter and are released at apical membranes can be transmitted laterally from cell to cell without ever transversing the epithelial or endothelial layers; they generally cause localized infections. In other cases, newly replicated viruses are transported to the basolateral surface and released into the underlying cells and tissues, which facilitates viral spread to other sites of replication.

There are also more specialized pathways by which viruses find susceptible cells. For example, some epithelial tissues contain M cells, specialized cells that overlie the collections of lymphoid cells in the gut known as Peyer's patches. M cells function in the transport of intestinal contents to Peyer's patches, and certain viruses, such as poliovirus and human immunodeficiency virus type 1, can also be transported through them by a mechanism called **transcytosis.** These viruses can gain access to underlying tissues by this mechanism.. Mobile cells, such as dendritic cells, which are present throughout the body in various tissues, can also be infected by human immunodeficiency virus type 1, and the virus is carried across epithelial layers by this mechanism. These specialized pathways of invasion will be considered in chapters 17 and 18. Below we describe briefly the structures that surround cells and tissues, the membrane components that mediate the interaction of cells with their environment, and the ways in which viruses attach to cells by means of these components.

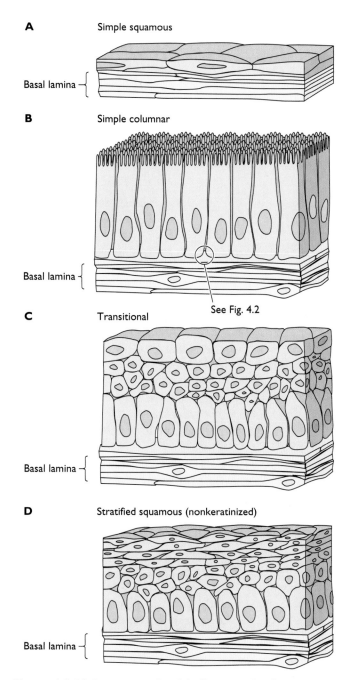

**A**    Simple squamous

Basal lamina

**B**    Simple columnar

Basal lamina

See Fig. 4.2

**C**    Transitional

Basal lamina

**D**    Stratified squamous (nonkeratinized)

Basal lamina

**Figure 4.1 Major types of epithelia.** (A) Simple squamous epithelium made up of thin cells such as those lining blood vessels and many body cavities. (B) Simple columnar epithelium. In the lining of the stomach and the cervical tract, this layer includes mucus-secreting cells. In the lining of the small intestine, it includes absorptive cells. (C) Transitional epithelium is composed of several layers with cells of different shapes. This layer lines cavities, such as the urinary bladder, that are subject to expansion and contraction. (D) Stratified, nonkeratinized epithelium lining surfaces such as the mouth and vagina. These layers resist abrasion. Adapted from H. Lodish et al., *Molecular Cell Biology*, 3rd ed. (W. H. Freeman & Co., New York, N.Y., 1995), with permission.

## Extracellular Matrix: Components and Biological Importance

Extracellular matrices hold the cells and tissues of the body together. The extracellular matrix is made up of two main classes of macromolecules (Fig. 4.2). The first class comprises **glycosaminoglycans** (such as heparan sulfate and chondroitin sulfate), which are unbranched polysaccharides made up of linked disaccharides. Glycosaminoglycans are usually linked to proteins to form **proteoglycans.** The second class of macromolecules in the extracellular matrix consists of fibrous proteins with structural (**collagen** and **elastin**) and adhesive (**fibronectin** and **laminin**) functions. The proteoglycan molecules in the matrix form hydrated gels in which the fibrous proteins are embedded. The gel provides resistance to compression and allows the diffusion of nutrients between blood and tissue cells. Fibrous proteins provide strength and resilience to the matrix. The extracellular matrix of each cell type is specialized for the particular function required, varying in thickness, strength, and degree of adhesion.

Most organized groups of cells, like epithelial cells of the skin (Fig. 4.1 and 4.3), are bound tightly on their basal surface to a thin layer of extracellular matrix called the **basal lamina.** This matrix is linked to the basolateral membrane by specific receptor proteins called **integrins** (which are discussed in "Cell Membrane Proteins" below). Integrins are anchored to the intracellular structural network (the **cytoskeleton**) at the inner surface of the cell membrane. The basal lamina is attached to collagen and other material in the underlying loose connective tissue found in many organs of the body (Fig. 4.3). Capillaries, glands, and specialized cells are embedded in this connective tissue. Several viruses are known or thought to gain access to susceptible cells by attaching specifically to components of the extracellular matrix, including some cell adhesion proteins, fibronectins, proteoglycans, and integrins (Table 4.1).

## Properties of the Plasma Membrane

The plasma membrane of every cell type is composed of a similar phospholipid/glycolipid bilayer, but different sets of membrane proteins and glycolipids allow the cells of different tissues to carry out their specialized functions. The lipid bilayer consists of molecules that possess both hydrophilic and hydrophobic portions; they are known as amphipathic molecules, from the Greek word *amphi* (on both sides) (Fig. 4.4). They form a sheetlike structure whose polar head groups face the aqueous environment of the cell's cytoplasm (inner surface) or the surrounding environment (outer surface). The polar heads of the inner and outer leaflets bear side chains with different lipid compositions

A

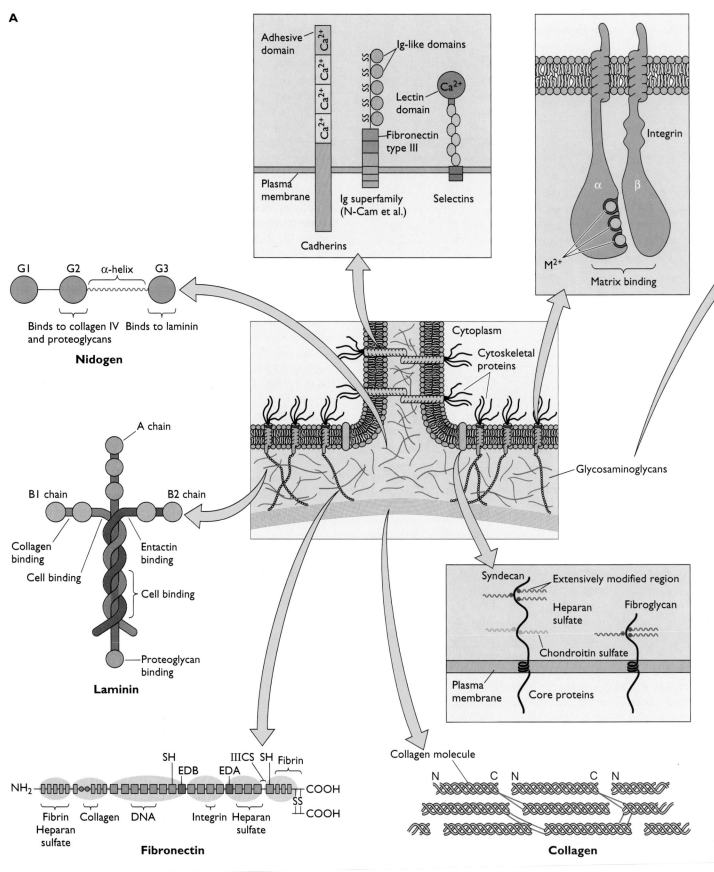

Adhesive domain

Ig-like domains

Lectin domain

Ca²⁺

Fibronectin type III

Plasma membrane

Ig superfamily (N-Cam et al.)

Selectins

Cadherins

Integrin

α    β

M²⁺

Matrix binding

G1    G2    α-helix    G3

Binds to collagen IV    Binds to laminin
and proteoglycans

**Nidogen**

Cytoplasm

Cytoskeletal proteins

Glycosaminoglycans

A chain

B1 chain    B2 chain

Collagen binding    Entactin binding

Cell binding

Cell binding

Proteoglycan binding

**Laminin**

Syndecan    Extensively modified region

Heparan sulfate    Fibroglycan

Chondroitin sulfate

Plasma membrane    Core proteins

SH    IIICS  SH    Fibrin
EDB    EDA

NH₂    COOH
SS
COOH

Fibrin    Collagen    DNA    Integrin    Heparan
Heparan                                      sulfate
sulfate

**Fibronectin**

Collagen molecule

N    C    N    C    N

**Collagen**

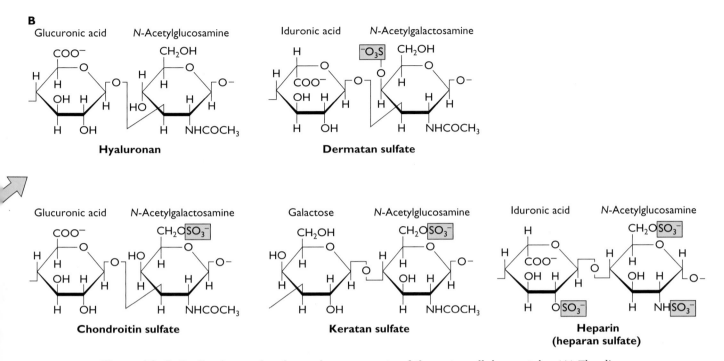

**Figure 4.2 Cell adhesion molecules and components of the extracellular matrix.** (A) The diagram in the center (expanded from Fig. 4.1B) illustrates the variety of cell surface components that play a role in cell-cell adhesion and attachment to the extracellular matrix. Additional details are provided for each major molecular class. $M^{2+}$, divalent cation. (B) Chemical structures of glycosaminoglycans. Adapted from H. Lodish et al., *Molecular Cell Biology*, 3rd ed. (W. H. Freeman & Co., New York, N.Y. 1995), and G. M. Cooper, *The Cell: a Molecular Approach* (ASM Press, Washington, D.C., and Sinauer Associates, Sunderland, Mass., 1997), with permission.

(Fig. 4.5). The fatty acyl side chains form a continuous hydrophobic interior about 3 nm thick. Hydrophobic interactions are the driving force for formation of the bilayer. However, hydrogen bonding and electrostatic interactions among the polar groups and water molecules or membrane proteins also stabilize the structure.

Thermal energy permits the phospholipid and glycolipid molecules comprising natural cell membranes to rotate freely around their long axes and diffuse laterally. If unencumbered, a lipid molecule can diffuse the length of an animal cell in only 20 s at 37°C. In most cases, phospholipids and glycolipids do not flip-flop from one side of a bilayer to the other, and the outer and inner leaflets of the bilayer remain separate. However, membrane proteins not anchored to the extracellular matrix and/or the underlying structural network of the cell can diffuse rapidly, moving laterally like icebergs in this fluid bilayer. In this way, certain membrane proteins can form functional aggregates. Intracellular organelles such as the nucleus, endoplasmic reticulum, compartments of the Golgi apparatus, mitochondria, lysosomes, and other internal vesicles are also enclosed in lipid bilayers, although their composition and physical properties differ.

## Cell Membrane Proteins

Membrane proteins are classified into two broad categories, **integral** and **peripheral,** names that describe the nature of their interactions with the plasma membrane (Fig. 4.4).

Integral proteins are embedded in the lipid bilayer. Many integral membrane proteins contain one or more **membrane-spanning** domains and portions that protrude out into the exterior and interior of the cell. A membrane-spanning domain consists of an α-helix typically 3.7 nm long (Fig. 4.6). It includes 20 to 25 generally hydrophobic or uncharged residues embedded in the membrane with the hydrophobic side chains protruding outward to form contacts with the fatty acyl chains of the lipid bilayer. The polar carbonyl (C=O) and amide (NH) groups are contained within the helix. The first and last residues are often positively charged amino acids (lysine or arginine) that can interact with the negatively charged polar head groups of the phospholipids to stabilize the membrane-spanning domain. Extracellular portions of the protein may be organized so that they can perform specialized functions such as binding specific molecules, known generally as **ligands.** Proteins with membrane-

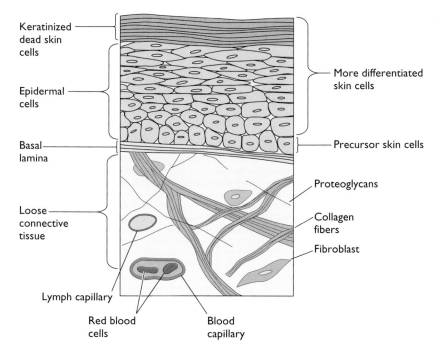

Keratinized dead skin cells

Epidermal cells

Basal lamina

Loose connective tissue

More differentiated skin cells

Precursor skin cells

Proteoglycans

Collagen fibers

Fibroblast

Lymph capillary

Red blood cells

Blood capillary

**Figure 4.3 Cross section through skin.** In this diagram of skin from a pig the precursor epidermal cells rest on a thin layer of extracellular matrix called the basal lamina. Underneath is loose connective tissue consisting mostly of extracellular matrix. Fibroblasts in the connective tissue synthesize the connective tissue proteins, hyaluronan, and proteoglycans. Blood and lymph capillaries are also located in the loose connective tissue layer. Adapted from H. Lodish et al., *Molecular Cell Biology,* 3rd ed. (W. H. Freeman & Co., New York, N.Y., 1995), with permission.

spanning domains enable the cell to respond to signals from outside the cell. These membrane proteins are designed to bind specific ligands (e.g., hormones, cytokines, or membrane proteins on the same cell or on other cells) and to signal the occurrence of such interactions to molecules in the interior of the cell. Some proteins with multiple membrane-spanning domains (Fig. 4.6) form critical components of molecular pores or pumps, which mediate the internalization of required nutrients or expulsion of undesirable material from the cell, or maintain homeostasis with respect to cell volume, pH, and ion concentration.

In many cases, the external portions of membrane proteins are decorated by complex or branched **carbohydrate chains** linked to the peptide backbone through nitrogen (**N linked**) in the side chain of asparagine residues or through oxygen (**O linked**) in the side chains of serine or threonine residues. Such membrane **glycoproteins,** as they are called, appear to be used quite frequently as viral receptors. Many viruses attach specifically to one or more of the external components of glycoproteins.

Some membrane proteins do not span the lipid bilayer but are anchored in the inner or outer leaflet by covalently attached hydrocarbon chains. These lipid-bound proteins fall into three classes according to the nature of the anchoring hydrocarbons (Fig. 4.7). They include amide-linked **myristoyl** and thioether-linked **farnesyl** chains by which proteins may be bound to the inner cytoplasmic face of the plasma membrane. Several viruses (e.g., retroviruses) also use a myristoyl group to anchor viral proteins

to the cell membrane during virion assembly. The third class of nonembedded integral membrane proteins are bound to the outer portion of the cell membrane via amide-linked glycosylphosphatidylinositol.

**Peripheral membrane** proteins are bound to the plasma membrane lipid bilayer indirectly, either by interacting with integral membrane proteins or with the charged sugars of the glycolipids within the membrane lipid. Fibronectin, a protein in the extracellular matrix that binds to integrins (Fig. 4.2), is an example.

## Interaction of Viruses with Cell Receptors

### General Principles

Many different cell surface molecules can serve as receptors for the attachment of viruses (Table 4.1). Evolutionary forces appear to have selected for use of molecules that are either abundant or unique on the surface of permissive cells. In some cases, receptor molecules are proteins that are expressed only on the surfaces of differentiated cells. These viruses have restricted tropism, as is seen, for example, with hepatitis B virus for liver cells and human immunodeficiency virus type 1 for CD4+ T cells or macrophages. Some viruses may be able to attach to more than one receptor, and in some cases viruses from different families utilize the same receptor molecule, as do human immunodeficiency virus type 1 and human herpesvirus 7, and adenovirus and coxsackievirus B.

**Table 4.1**   Viral receptors and coreceptors[a]

| Virus | Receptor | Type of molecule | Coreceptor |
|---|---|---|---|
| *Picornaviridae* | | | |
| Foot-and-mouth disease virus (cell culture adapted) | Heparan sulfate | Glycosaminoglycan | |
| Foot-and-mouth disease virus | $\alpha_v\beta_3$ (vitronectin receptor) | Integrin | |
| Encephalomyocarditis virus | Vcam-1 | Ig-like | |
| | Sialylated glycophorin A (for hemag-glutination only) | Carbohydrate | |
| Poliovirus type 1 to 3 | Pvr | Ig-like | |
| Coxsackieviruses A13, A18, A21 | Icam-1 | Ig-like | |
| Coxsackievirus A21 | Decay-accelerating protein (CD55) | SCR-like (complement cascade) | Icam-1 |
| Coxsackievirus A9 | $\alpha_v\beta_3$ | Integrin | |
| Coxsackieviruses B1 to B6 | Car (coxsackievirus-adenovirus receptor) | Ig-like | |
| Coxsackieviruses B1, B3, B5 | CD55 | SCR-like (complement cascade) | $\alpha_v\beta_6$ integrin |
| Echoviruses 1 and 8 | $\alpha_2\beta_1$ integrin (Vla-2) | Integrin | $\beta_2$ microglobulin |
| Echovirus 22 | $\alpha_v\beta_3$ (vitronectin receptor) | Integrin | |
| Echoviruses 3, 6, 7, 11 to 13, 20, 21, 24, 29, 33 | CD55 | SCR-like (complement cascade) | $\beta_2$ microglobulin |
| Enterovirus 70 | CD55 | SCR-like (complement cascade) | |
| Bovine enterovirus | Sialic acid | Carbohydrate | |
| Hepatitis A virus | HAVCr-1 | Ig-like, mucin-like | |
| Major group rhinoviruses (91 serotypes) | Icam-1 | Ig-like | |
| Minor group rhinovirus (10 serotypes) | Low-density lipoprotein receptor protein family | Signaling receptor | |
| Rhinovirus 87 | Sialic acid | Carbohydrate | |
| *Coronaviridae* | | | |
| Mouse hepatitis virus | Bgp (biliary glycoprotein) | Ig-like | |
| Human coronavirus 229E | Aminopeptidase N | Protease | |
| Transmissible gastroenteritis virus | Aminopeptidase N | Protease | |
| Human coronavirus OC43 | Sialic acid | Carbohydrate | |
| Bovine coronavirus | Sialic acid | Carbohydrate | |
| *Togaviridae* | | | |
| Semliki Forest virus | Major histocompatibility class I molecule | Ig-like | |
| Sindbis virus | High-affinity laminin receptor | Integrin | |
| | Heparan sulfate | Glycosaminoglycan | |
| Dengue virus | Heparan sulfate | Glycosaminoglycan | |
| *Rhabdoviridae* | | | |
| Rabies virus | Nicotinic acetylcholine receptor | Neurotransmitter receptor | |
| | Neural cell adhesion molecule CD56 | Ig-like | |
| | Low-affinity nerve growth factor receptor | Tnf receptor protein superfamily | |
| *Paramyxoviridae* | | | |
| Measles virus | Membrane cofactor protein, CD46 | Complement-regulating protein | |
| Sendai virus | Sialic acid | Carbohydrate | |
| | Asialoglycoprotein receptor Gp-2 | Transport protein (receptor-mediated endocytosis) | |
| *Orthomyxoviridae* | | | |
| Influenza A and B viruses | Sialic acids (*N*-acetyl neuraminic acid) | Carbohydrate | |
| Influenza C virus | Sialic acids (9-*O*-acetyl neuraminic acid) | Carbohydrate | |
| *Arenaviridae* | | | |
| Lymphocytic choriomeningitis virus | $\alpha$-Dystroglycan | Laminin receptor | |
| Lassa virus | $\alpha$-Dystroglycan | Laminin receptor | |

*(continued)*

**Table 4.1** Viral receptors and coreceptors[a] *(continued)*

| Virus | Receptor | Type of molecule | Coreceptor |
|---|---|---|---|
| *Reoviridae* | | | |
| Reovirus | Sialic acids | Carbohydrate | |
| Group A porcine rotavirus | Sialic acids | Carbohydrate | |
| *Retroviridae* | | | |
| Human immunodeficiency virus type 1 | CD4 | Ig-like | Chemokine receptors (Ccr5, Cxcr4, Ccr3) |
| | Galactosylceramide | Glycolipid | |
| Human immunodeficiency virus type 2 | CD4 | Ig-like | Chemokine receptors |
| | Cxcr4 | 7-transmembrane superfamily | |
| Simian immunodeficiency virus | CD4 | Ig-like | Chemokine receptors |
| Gibbon ape leukemia virus | Glvr1 | Sodium-dependent phosphate transport protein | |
| Feline leukemia virus B | Glvr1 | Sodium-dependent phosphate transport protein | |
| Amphotropic murine leukemia virus | Ram-1 | Sodium-dependent phosphate transport protein | |
| Ecotropic murine leukemia virus | Cat | Cationic amino acid transport protein | |
| Subgroup A avian leukosis and sarcoma virus | Tva | Low-density lipoprotein receptor protein family | |
| Subgroup B and D avian leukosis and sarcoma viruses | Car1 | Tnf receptor family protein superfamily | |
| Bovine leukemia virus | BLVRcp 1 | Unknown | |
| Feline immunodeficiency virus | Cxcr4 | 7-transmembrane superfamily | |
| Visna virus | Major histocompatibility complex class II molecule | Ig-like | |
| *Parvoviridae* | | | |
| Bovine parvovirus | Sialic acids | Carbohydrate | |
| Adeno-associated virus type 2 | Heparan sulfate | Glycosaminoglycan | $\alpha_v\beta_5$ integrin |
| *Papovaviridae* | | | |
| Simian virus 40 | Major histocompatibility class I molecule | Ig-like | |
| *Adenoviridae* | | | |
| Adenovirus subgroups A, C, D, E, F | Car | Ig-like | $\alpha_v$ integrins |
| Adenovirus type 5 (subgroup C) | Major histocompatibility class II molecule | Ig-like | $\alpha_v$ integrins |
| Adenovirus type 2 (subgroup C) | $\alpha_M\beta_2$ | Integrin | $\alpha_v$ integrins |
| Adenovirus type 9 (subgroup D) | $\alpha_v$ integrins | Integrin | |
| *Herpesviridae* | | | |
| Herpes simplex type 1 | Heparan sulfate | Glycosaminoglycan | HveA, Prr1 |
| Herpes simplex type 2 | Heparan sulfate | Glycosaminoglycan | HveA, Prr1, Prr2 |
| Pseudorabies virus | Heparan sulfate | Glycosaminoglycan | Pvr, Prr1, Prr2 |
| Bovine herpesvirus 1 | Heparan sulfate | Glycosaminoglycan | Pvr, Prr1 |
| Human herpesvirus 7 | CD4 | Ig-like | |
| Epstein-Barr virus | Complement receptor Cr2 (CD21) | SCR-like (complement cascade) | |
| Human cytomegalovirus | Heparan sulfate | Glycosaminoglycan | Aminopeptidase N (CD13) |
| *Poxviridae* | | | |
| Vaccinia virus | Heparan sulfate | Glycosaminoglycan | |
| | Epidermal growth factor receptor | Signaling receptor | |

[a]The name of the receptor and the type of molecule are listed for selected viruses. When coreceptors have been identified, they are listed; a blank in the coreceptor column indicates that none have been identified to date. Abbreviations: Vcam, vascular cell adhesion molecule; Prr1, Prr2, Pvr-related proteins 1 and 2; SCR, short consensus repeat; Ig, immunoglobulin; Tnf, tumor necrosis factor; Car1, cytopathic avian leukosis and sarcoma virus receptor; Car, coxsackievirus and adenovirus receptor; HveA, herpesvirus entry mediator.

**A**

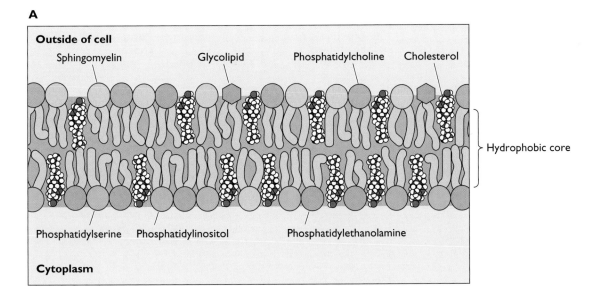

**B**

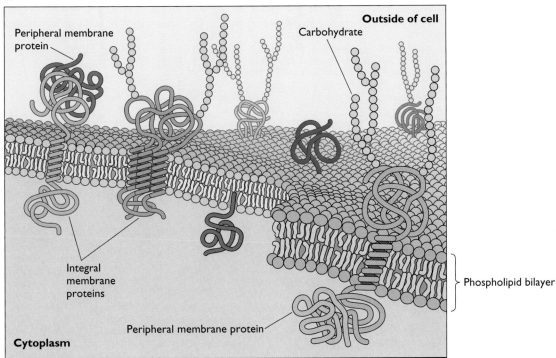

**Figure 4.4  The plasma membrane.** (A) Lipid components of the plasma membrane. The membrane consists of two layers (leaflets) of phospholipid and glycolipid molecules. Their fatty acid tails converge to form the hydrophobic interior of the bilayer; the polar hydrophilic head groups (shown as balls) line both surfaces. Major components of the outer leaflet include phosphatidylcholine, sphingomyelin, and glycolipids; the inner leaflet is made up largely of phosphatidylethanolamine, phosphatidylserine, and phosphatidylinositol. Cholesterol is found in both leaflets. The head groups of phosphatidylserine and phosphatidylinositol have a net negative charge. (B) Fluid mosaic model of the plasma membrane. Integral and peripheral membrane proteins are illustrated. Some integral membrane proteins are transmembrane proteins and are exposed on both sides of the bilayer. Proteins on the external leaflet are usually glycosylated. Adapted from G. M. Cooper, *The Cell: a Molecular Approach* (ASM Press, Washington, D.C., and Sinauer Associates, Sunderland, Mass., 1997), with permission.

**A**

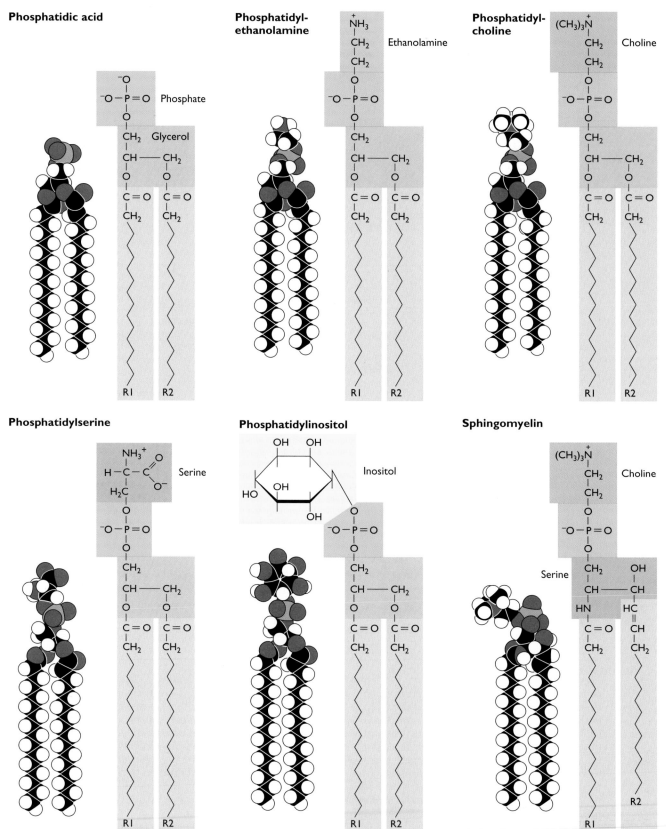

**Phosphatidic acid**

**Phosphatidyl-ethanolamine**

**Phosphatidyl-choline**

**Phosphatidylserine**

**Phosphatidylinositol**

**Sphingomyelin**

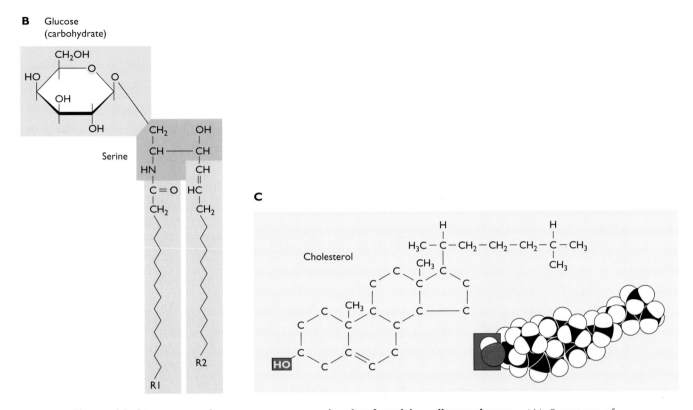

**Figure 4.5 Structures of some common molecules found in cell membranes.** (A) Structure of phospholipids. Glycerol phospholipids contain two fatty acids (R1 and R2) linked to glycerol. The fatty acyl side chains can be saturated or can contain one or more double bonds. Addition of a small polar molecule to phosphatidic acid produces the different phospholipids. Sphingomyelin is the only non-glycerol phospholipid found in membranes. (B) Structure of glycolipids. The polar head group, composed of serine linked to a carbohydrate, is joined to two hydrocarbon chains. (C) Structure of cholesterol. This molecule is amphipathic because of the presence of the polar hydroxyl group. Reprinted from G. M. Cooper, *The Cell: a Molecular Approach*, p. 44–45 (ASM Press, Washington, D.C., and Sinauer Associates, Sunderland, Mass., 1997), with permission.

The **host range** of a virus, its ability to infect a particular animal or cell culture, can be determined by cell receptors, but they are not the only determinants. For example, poliovirus infects primates and cultured primate cells but not mice or mouse cell cultures. Mouse cells express a protein that is homologous to the poliovirus receptor but is sufficiently different so that poliovirus cannot attach to it. In this example, the poliovirus receptor is the determinant of poliovirus host range. However, expression of the virus receptor in a particular cell type does not ensure that virus replication will occur. Thus, some primate cell cultures express the poliovirus receptor but cannot be infected. The restricted host range of the virus in such cells is most likely due to a block in viral replication beyond the attachment step. Cell receptors can also be determinants of tissue tropism, but there are many other determinants as well. For example, the sialic acid residues on membrane glycoproteins or glycolipids, which are receptors for influenza virus, are found on many tissues, yet viral replication in

the host is restricted. This topic is discussed more extensively in chapter 17.

Cell culture systems have made it possible to study the general properties of virus attachment. The initial interactions between virions and cells probably involve electrostatic forces, as they are sensitive to low pH or high concentrations of salt. Subsequent high-affinity binding relies mainly on hydrophobic and other short-range forces whose strength and specificity are governed primarily by the conformations of the interacting viral and cellular interfaces. Virion binding can usually occur at 0°C (even though entry does not) as well as at body temperature (e.g., 37°C). Thus, infection of cultured cells can be synchronized by allowing binding to take place at a low temperature (4°C) and then shifting the cells to a physiological temperature to allow the initiation of subsequent steps.

The first steps in virus attachment are governed largely by the probability that a virion and a cell will collide, which depends on the concentration of free virions and

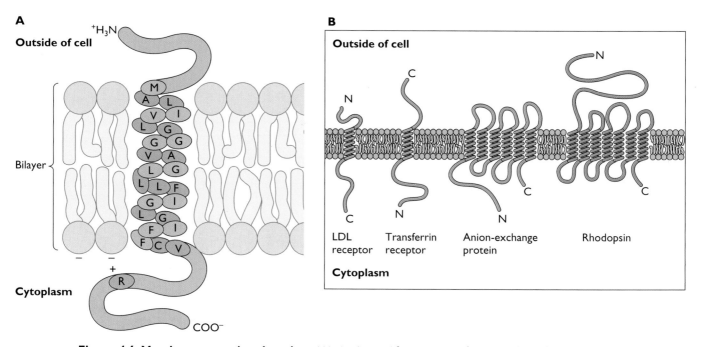

**Figure 4.6 Membrane-spanning domains.** (A) Amino acid sequence of a typical single membrane-spanning domain, in this case for the CD4 receptor of human immunodeficiency virus type 1 and several other primate lentiviruses. Essential features include an α-helical stretch of 25 hydrophobic residues bounded on the C-terminal edge by a hydrophilic positively charged (R) residue. (B) Different arrangements of N- and C-termini and membrane-spanning domains in a variety of integral membrane proteins. Adapted from L. H. Stryer, *Biochemistry,* 4th ed. (W. H. Freeman & Co., New York, N.Y., 1995), with permission.

host cells. The rate of attachment can be described by the equation $dA/dt = k[V][H]$, where $[V]$ and $[H]$ are the concentrations of virions and host cells and $k$ is a rate constant. Values of $k$ for animal viruses vary greatly from a maximal value that represents the limits of diffusion to one that is as much as five orders of magnitude slower. It can be seen from this equation that if a mixture of viruses and cells is diluted after virions have been allowed to attach, subsequent binding will be greatly reduced. For example, a 100-fold dilution of the mixture will reduce the attachment rate 10,000-fold (i.e., $1/100 \times 1/100$). Thus, dilution can be used to prevent subsequent virus adsorption and hence to synchronize an infection.

## Identification of Viral Receptors on Cells

### Experimental Approaches

How were viral receptors first discovered? Early experimentalists made use of a variety of enzymes to characterize components on the cell surface that are required for virus attachment. The first viral receptor discovered, sialic acid, which binds influenza virus, was identified because the enzyme neuraminidase removes this carbohydrate

from cells and blocks virus attachment. In a similar way, experiments with proteases showed that many receptors are proteins. These types of analysis provided the first clues concerning the chemical nature of cell surface components to which virions become attached. Early cell culture experiments also made it possible to determine whether receptors are shared by different viruses. If a particular cell type could bind two different viruses, one could determine if saturating cells with one virus prevented binding of the second. If significant competition was observed, then it was surmised that the same receptor molecule was probably used by both. This method was first used to establish different viral subgroups among avian retroviruses and to type cells from individual animals for the presence of genetic determinants of susceptibility to each subgroup.

Despite these approaches, identification of cell receptors for viruses languished because biochemical purification of viral receptors proved difficult. As late as 1985, only one viral receptor, the sialic acid receptor of influenza viruses, had been unequivocally identified. The development of two crucial technologies rapidly changed this situation. Monoclonal antibodies provide a powerful means for iso-

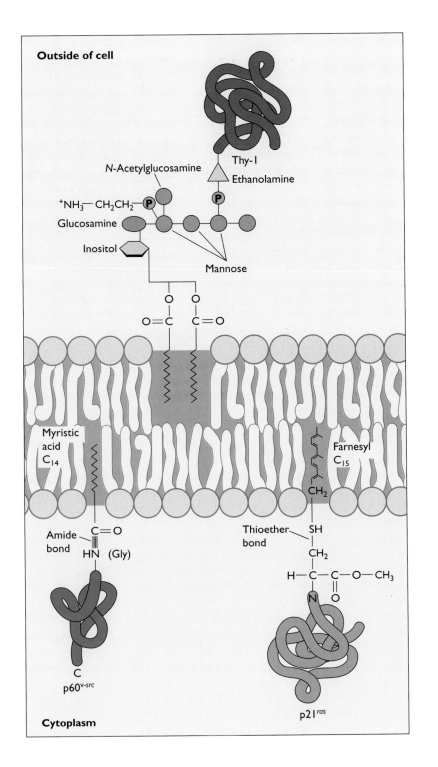

**Figure 4.7 Lipid anchors for some integral membrane proteins.** Myristyl- and farnesyl-anchored proteins are found only on the cytoplasmic face of the plasma membrane. They include various proteins involved in cell growth, as well as a number of viral proteins. Glyco-sylphosphatidylinositol-anchored proteins are found only on the extracellular face.

lating and characterizing individual cell surface proteins. In practice, hybridoma cell lines that secrete monoclonal antibodies to cell surface proteins are obtained after immunizing mice with intact cells. Hybridoma cell lines are identified that secrete an antibody which blocks virus attachment. Finally, the antibody can be used to purify the receptor protein by immunoprecipitation or affinity chromatography. A second crucial technology that advanced the viral receptor field was the development of DNA-mediated transformation and molecular cloning technologies.

The strategy for isolating receptor DNA is to introduce DNA from susceptible cells into nonsusceptible cells by DNA-mediated transformation (Fig. 4.8). Cells that acquire DNA encoding the receptor and express the corresponding protein on their surface will be able to bind virus specifically. Clones of such cells are recognized (and selected) by the binding of receptor-specific monoclonal antibodies or following infection with genetically altered viruses carrying a drug-selectable marker or an indicator molecule. The receptor genes can then be cloned from these selected cells. Examples of the use of these approaches to identify specific receptor molecules are provided in subsequent sections.

The availability of cloned receptor genes has made it possible to investigate the details of receptor interaction with viruses by site-directed mutagenesis. Parts of the gene can be altered or deleted, and the protein can be expressed in bacterial or animal cells with appropriate vectors. Receptor protein can be expressed and purified, and its properties can be studied in vitro. Animal cells expressing the altered receptor proteins can be used to test the effects of the mutations on virus attachment. Because of their hydrophobic membrane-spanning domains, many of these cell surface receptor proteins are relatively insoluble and difficult to work with. In a number of cases, soluble forms have been obtained by the expression of truncated receptor proteins lacking the membrane-spanning domain. Soluble forms of receptors have been instrumental in structural studies of receptor-virus interactions. Cloned receptor genes have also been used to produce transgenic animals that express receptor proteins. Such transgenic animals can serve as useful models in the study of human viral diseases.

The receptors for a number of viruses have now been identified (Table 4.1), but many more remain to be discovered. Criteria for identifying receptors are discussed in Box 4.1. Despite their obvious importance to our understanding of viral biology and pathogenesis, the actual host cell functions of few of the receptors are known. We know even less about the molecular mechanisms mediating virus attachment and entry. Among the best studied are the receptors for members of the picorna-, orthomyxo-, and retrovirus families.

### A Single Type of Receptor Molecule Mediates Virus Binding and Entry

**The rhinovirus receptor, intercellular adhesion molecule 1.** Rhinoviruses are unstable below pH 5 to 6 and multiply primarily in the upper respiratory tract. Up to 50% of all human common colds are caused by members of the major group of rhinoviruses. The cell surface receptor for the major group of human rhinoviruses (~90 serotypes) was identified by screening monoclonal antibodies against human cell surface components for their ability to block rhinovirus infection. When such a monoclonal antibody was identified, it was used to isolate a 95-kDa cell surface glycoprotein by affinity chromatography. Amino acid sequence analysis of the purified protein, which bound to rhinovirus in vitro, identified it as the integral membrane protein **intercellular adhesion molecule 1 (Icam-1).** The murine homolog of human Icam-1 is sufficiently different that it cannot bind major group rhinoviruses. This difference allowed molecular cloning of the Icam-1 gene from mouse cells that had been transformed with human DNA. Mouse cells that expressed the receptor protein were identified with a monoclonal antibody specific for human Icam-1.

Icam-1 (Fig. 4.9 and 4.10) is a member of the immunoglobulin (Ig) superfamily. It has five domains that are homologous to one another and to the constant domains of antibody molecules. As in antibody molecules, each domain is stabilized by intrachain disulfide bonds. Icam-1 is found on the surface of many tissues, including nasal epithelium, the normal entry site for rhinoviruses. The natural ligand of Icam-1 is another integral membrane glycoprotein, the integrin known as **lymphocyte function-associated antigen 1 (Lfa-1).** Icam-1 is also the receptor on endothelial cells for erythrocytes infected with the malaria parasite *Plasmodium falciparum.* The normal function of Icam-1 is to bind Lfa-1 on the surface of lymphocytes and promote a wide variety of immunological and inflammatory functions. Increased expression of Icam-1 is induced by mediators of the inflammatory response. Therefore, the initial reaction of host defenses to rhinoviruses, which leads to the production of such mediators, might actually induce the appearance of its receptor in nearby cells and thereby enhance subsequent spread of the virus.

Mutational analyses of cloned Icam-1 DNA established that the binding sites for rhinovirus, *P. falciparum*-infected erythrocytes, and Lfa-1 are located in domains 1 and 2, with critical contact points in both cases mapping to domain 1, the most membrane distal and accessible (Fig. 4.10). However, although the binding sites for rhinovirus and Lfa-1 partially overlap, they are distinct. The three-dimensional atomic structure of the two amino-terminal domains of a soluble form of Icam-1 has been determined (Fig. 4.10). This structure was then modeled into a reconstruction of the rhinovirus–Icam-1 complex made by cryo-electron microscopy. The residues of Icam-1 that contact the virus are located in the BC, CD, DE, and FG loops of the first amino-terminal domain (Fig. 4.10). Amino acids identified by mutational analyses as crucial for virus binding are located in these loops.

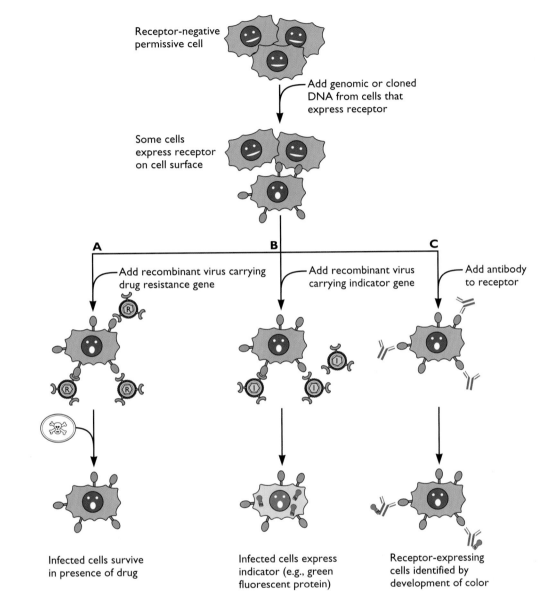

Receptor-negative permissive cell

Add genomic or cloned DNA from cells that express receptor

Some cells express receptor on cell surface

**A**

Add recombinant virus carrying drug resistance gene

**B**

Add recombinant virus carrying indicator gene

**C**

Add antibody to receptor

Infected cells survive in presence of drug

Infected cells express indicator (e.g., green fluorescent protein)

Receptor-expressing cells identified by development of color

**Figure 4.8 Experimental strategies for identification and isolation of cellular genes encoding viral receptors.** Genomic DNA or pools of DNA clones (e.g., prepared by reverse transcription of mRNAs) from cells known to express the receptor are introduced into receptor-negative permissive cells. A small number of recipient cells express the receptor. Three different strategies are outlined for identifying such rare receptor-expressing cells. (A) The cells are infected with a virus that has been engineered so that it carries a gene encoding drug resistance. Cells that express the receptor will become resistant to the drug. The receptor DNA can then be isolated from these cells. This strategy works only for viruses that persist in cells without killing them. (B) For lytic viruses, an alternative is to engineer the virus to express an indicator, such as green fluorescent protein or β-galactosidase. Cells that express the correct receptor and become infected with such viruses can be distinguished by a color change, such as green in the case of green fluorescent protein. (C) The third approach depends on the availability of an antibody directed against the receptor, which binds to cells that express the receptor. Bound antibodies can be detected by an indicator molecule. After one of these assays has been conducted, one of two general approaches is used to isolate receptor DNA. When the donor DNA is genomic DNA, the presence of species-specific repetitive sequences allows it to be distinguished from the DNA of the recipient cell. Genomic DNA from the susceptible cell is cloned into an appropriate vector, and the repetitive sequence can be used as a hybridization probe to identify the cloned receptor DNA. When cDNA cloned in a plasmid is used as the donor DNA, pools of individual clones (usually 10,000 clones per pool) are prepared and introduced individually into cells. The specific DNA pool that yields receptor-expressing cells is then subdivided, and the screening process is repeated until a single receptor-encoding DNA is identified.

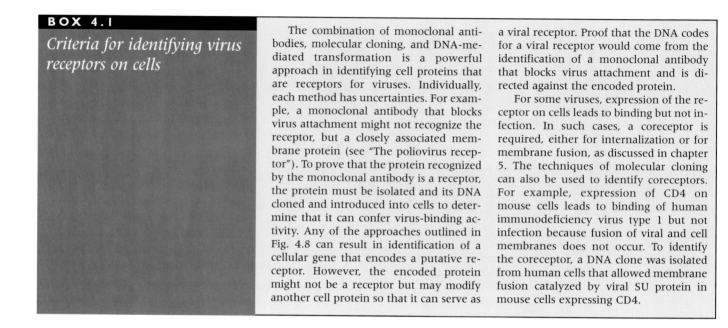

**BOX 4.1**

*Criteria for identifying virus receptors on cells*

The combination of monoclonal antibodies, molecular cloning, and DNA-mediated transformation is a powerful approach in identifying cell proteins that are receptors for viruses. Individually, each method has uncertainties. For example, a monoclonal antibody that blocks virus attachment might not recognize the receptor, but a closely associated membrane protein (see "The poliovirus receptor"). To prove that the protein recognized by the monoclonal antibody is a receptor, the protein must be isolated and its DNA cloned and introduced into cells to determine that it can confer virus-binding activity. Any of the approaches outlined in Fig. 4.8 can result in identification of a cellular gene that encodes a putative receptor. However, the encoded protein might not be a receptor but may modify another cell protein so that it can serve as a viral receptor. Proof that the DNA codes for a viral receptor would come from the identification of a monoclonal antibody that blocks virus attachment and is directed against the encoded protein.

For some viruses, expression of the receptor on cells leads to binding but not infection. In such cases, a coreceptor is required, either for internalization or for membrane fusion, as discussed in chapter 5. The techniques of molecular cloning can also be used to identify coreceptors. For example, expression of CD4 on mouse cells leads to binding of human immunodeficiency virus type 1 but not infection because fusion of viral and cell membranes does not occur. To identify the coreceptor, a DNA clone was isolated from human cells that allowed membrane fusion catalyzed by viral SU protein in mouse cells expressing CD4.

**Figure 4.9 Cell receptors for picornaviruses.** A schematic diagram of the cell proteins that function as receptors for different picornaviruses is shown. The different domains (Ig-like, short consensus repeat-like [SCR-like], low density lipoprotein-like [LDL-like], and threonine/serine/proline [T/S/P]) are labeled. Car, coxsackievirus and adenovirus receptor; Vcam-1, vascular cell adhesion molecule 1; Ldlr, low-density lipoprotein receptor. Adapted from D. J. Evans and J. W. Almond, *Trends Microbiol.* **6:**198–202, 1998, with permission.

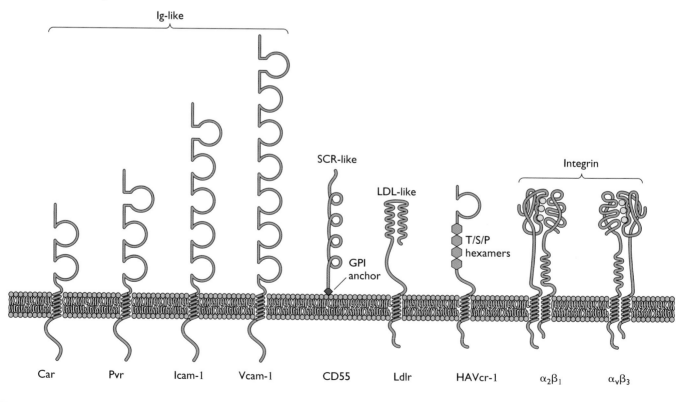

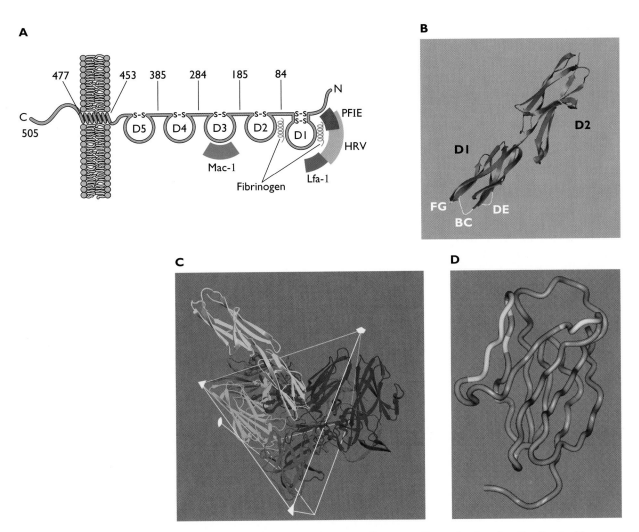

**Figure 4.10  Structures of Icam-1 and Pvr.** (A) Schematic diagram of Icam-1, showing the five Ig-like domains, glycosylation sites (ball and stick), and binding sites for Lfa-1, Mac-1, rhinoviruses (HRV), fibrinogen, and malaria-infected erythrocytes (PFIE). (B) Model of domains 1 and 2 of Icam-1 derived from X-ray crystallographic data. The loops of domain 1 that contact rhinovirus are white and are labeled BC, DE, and FG. (C) Interaction of Icam-1 (yellow) in the canyon of rhinovirus 16. Viral proteins VP1, VP2, and VP3 are blue, green, and red. Two symmetry-related copies of VP1 and VP3 are included. Five-, three-, and twofold axes of symmetry are indicated; white lines indicate the boundary of an icosahedral symmetric unit. (D) Model of domain 1 (V-type) of Pvr. The locations of mutations that reduce poliovirus binding (white and green) are shown. Amino acids in white are believed to form part of the virus contact region on Pvr. Residues in green are not believed to contact poliovirus; rather, mutations at these sites probably cause global changes in the structure of the receptor. (A to C) Adapted from J. Bella et al., *Proc. Natl. Acad. Sci. USA* **95:**4140–4145, 1998, with permission. (C) Courtesy of Michael Rossmann, Purdue University.

The results of competition assays had indicated that coxsackievirus A21 and human rhinovirus 14 bind to the same cellular receptor. It was subsequently found that coxsackieviruses A13, A18, and A21 could bind to mouse cells that express human Icam-1. Thus, Icam-1 serves as a receptor for coxsackievirus A13, A18, and A21 and the major group rhinoviruses (Table 4.1).

**The poliovirus receptor.** The genus *Enterovirus* includes human polioviruses, coxsackieviruses, echoviruses, and enteroviruses. These viruses are distinguished from rhinoviruses by their stability at acid pH and their ability to multiply in the gastrointestinal tract. However, they can also replicate in other tissues, such as nerve, heart, and muscle. The poliovirus receptor (Pvr) was identified at about the same time as the rhinovirus receptor with the DNA transformation and cloning strategy outlined in Fig. 4.8. It was well known that mouse cells cannot be infected with poliovirus. However, introduction of infectious poliovirus RNA into mouse tissue culture cells leads to poliovirus replication, indicating that there is no intracellular block to virus multiplication. The nonsusceptible state of mouse cells is therefore due to lack of a receptor. Cloning of the human gene from such receptor-positive mouse cells revealed that Pvr glycoprotein is also a member of the immunoglobulin superfamily, in this case possessing three Ig-like domains (Fig. 4.9). The first of these domains is essential and sufficient for receptor function, but efficient (wild-type) binding requires all three. A molecular model of domain 1 of Pvr derived from structural data from crystals of the homologous Ig variable domain is shown in Fig. 4.10. Mutational studies indicate that Pvr contacts poliovirus through at least two β-strands and adjacent loops. A different region has been mapped for contacts with rhinovirus in domain 1 of Icam-1 (Fig. 4.10). These findings demonstrate that interactions may differ even among structurally related viruses and receptors.

As mouse cells are permissive for poliovirus replication and susceptibility appeared to be limited only by the absence of Pvr, a small-animal model for the disease was made by producing transgenic mice expressing this receptor. Inoculation of Pvr transgenic mice with poliovirus by various routes produces paralysis, tremors, and death, as is observed in human poliomyelitis. Pvr-expressing mice were the first new animal model created by transgenic technology for the study of viral disease. As new viral receptors are identified, similar approaches should lead to animal models for a variety of viral diseases.

The observation that even some human cells that express Pvr are resistant to poliovirus infection prompted the search for a second cell protein that might regulate receptor activity. One candidate was the lymphocyte homing receptor, CD44. This molecule was thought to be a coreceptor for poliovirus because a monoclonal antibody against this protein blocks poliovirus binding to cells. This multifunctional 100-kDa membrane glycoprotein is normally involved in directing the migration of lymphocytes to the lymph nodes and in the regulation of lymphocyte adhesion and other functions. However, CD44 is not a receptor for poliovirus, nor is it required for poliovirus infection of cells expressing Pvr. It is believed that Pvr and CD44 are associated in the cell membrane and that anti-CD44 antibodies may block poliovirus attachment by sterically hindering the poliovirus-binding site on Pvr. These results emphasize that monoclonal antibodies that block virus binding do not necessarily identify the virus receptor on cells. Why certain cells that

express Pvr are resistant to poliovirus infection remains unknown.

The fact that the major group rhinoviruses, poliovirus, and group B coxsackieviruses use immunoglobulin superfamily members as cell surface receptors might lead one to believe that this is a general property of the family *Picornaviridae*. However, this notion is quickly dispelled by inspection of Fig. 4.9 and Table 4.1. The minor group of human rhinoviruses use a different type of receptor (Ldlr), and various members of the other genera in the picornavirus family bind specifically to yet other types of cell surface molecules. For example, several use specific integrins as receptors. Most members of the integrin superfamily are known to recognize a specific sequence, RGD, common in a variety of extracellular matrix proteins that are the normal ligands of integrins. As we shall see in our discussion of viral attachment sites, this motif is also found in the capsid proteins of both picornaviruses and adenoviruses that attach to integrins.

**The influenza virus receptor, sialic acid.** The family *Orthomyxoviridae* comprises the three genera of influenza viruses, A, B, and C. Influenza A viruses are the best studied members of this family. The surfaces of influenza viruses were shown in the early 1940s to contain receptor-destroying enzymes that, paradoxically, remove the receptors for attachment from the surface of cells. Later these enzymes were identified as virus-encoded envelope proteins, neuraminidases (NA), which cleave the glycoside linkages of sialic acids (Fig. 4.11). Influenza A virus binds to negatively charged terminal sialic acid moieties present in a number of the oligosaccharide chains that are covalently attached to cell surface glycoproteins or glycolipids. The presence of sialic acid on most cell surfaces accounts for the ability of influenza virus to attach to many types of cells. The interaction of influenza virus with individual sialic acid moieties is of low affinity, but the opportunity for multiple interactions among the numerous hemagglutinin (HA) molecules on the surface of the virion and multiple terminal sialic acid residues on cellular glycoproteins and glycolipids results in a high overall avidity of the virus for the cell surface.

## Receptors and Coreceptors Required for Infection

**The human immunodeficiency virus type 1 receptor, CD4.** Animal retroviruses have long been of interest to virologists because of their ability to cause a variety of serious diseases, especially cancers (caused by oncogenic retroviruses) and neurological disorders (caused by lentiviruses). When retroviruses infect a cell, they exhibit

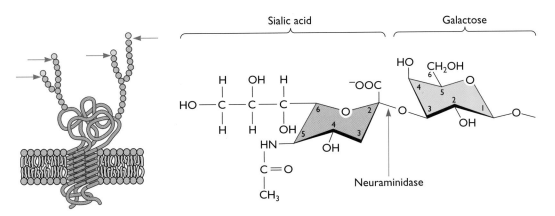

**Figure 4.11 Sialic acid receptors for influenza viruses.** The diagram on the left represents an integral membrane glycoprotein; the arrows point to terminal sialic acid units that are attachment sites for influenza virus. The diagram on the right illustrates the structure of a terminal sialic acid moiety that is recognized by the viral envelope protein, hemagglutinin. Sialic acid is attached to galactose by an α(2,3) linkage in the example shown; certain HA subtypes preferentially bind to molecules with an α(2,6) linkage. The site of cleavage by the influenza virus envelope protein neuraminidase is indicated. The sialic acid shown is *N*-acetylneuraminic acid, which is the preferred receptor for influenza A and B viruses. These viruses do not bind to 9-*O*-acetyl-*N*-neuraminic acid, the receptor for influenza C viruses. Adapted from E. Wimmer, *Cellular Receptors for Animal Viruses* (Cold Spring Harbor Laboratory Press, Cold Spring Harbor, N.Y., 1994), with permission.

a phenomenon called **interference,** a block to subsequent infection by a related retrovirus that uses the same receptor. The mechanism of interference is not well understood but is presumed to be the result of occupation of receptor molecules by the envelope proteins produced in the infected cells. A study of receptor interference by 20 different mammalian retroviruses that can infect human cells in culture indicated that these viruses recognize no fewer than eight different receptors (Table 4.2). Although this is a seemingly large number, it is still considerably smaller than the number of retroviruses. Receptors for four of the groups listed in the table have been identified. At least one receptor is also used by members of another viral family (e.g., CD4 and human herpesvirus 7 [Table 4.1]).

The worldwide AIDS epidemic has focused enormous attention on the lentivirus human immunodeficiency virus type 1, its close relative human immunodeficiency virus type 2, and the primate lentiviruses, simian immunodeficiency viruses. Thus, although this group (group 8 in Table 4.2) comprises the most recently discovered retroviruses, their interactions with cell surface receptors have been among the most intensively studied and are currently the best understood.

Early immunological studies showed that sensitivity to human immunodeficiency virus type 1 correlated with the presence of the T-helper cell surface marker CD4 and suggested that CD4 might be important for virus attachment. Subsequently, the observation that certain monoclonal antibodies against CD4 could block infection and cell fu-

sion (syncytium formation) caused by human immunodeficiency virus type 1 provided strong evidence that this integral membrane glycoprotein was the major cell receptor for the virus. Molecular cloning of CD4 DNA, and the demonstration that nonsusceptible human cells engineered to express CD4 became susceptible to infection, confirmed the importance of this receptor for human immunodeficiency virus type 1. The CD4 protein, a 55-kDa rodlike molecule, is a member of the Ig superfamily containing four Ig-like domains. The three-dimensional structure of the first two Ig-like domains of CD4 was solved by X-ray crystallography. The site of interaction with human immunodeficiency virus type 1 was mapped by site-directed mutagenesis, monoclonal antibody studies, and X-ray crystallographic studies on a complex of CD4 and SU (surface), the viral attachment protein (Fig. 4.12). Residues that contact SU are located between amino acids 25 and 64 of CD4.

As can be seen by comparison with Fig. 4.10, the interaction site for SU in domain 1 of CD4 is in a region analogous to the site in Pvr that binds to poliovirus. Because two viruses with entirely different virion architectures bind to analogous surfaces of these Ig-like domains, it can be hypothesized that something about this region is especially advantageous for virus attachment.

One of the first strategies to be considered for the treatment of AIDS was the use of soluble CD4 protein, which lacks the membrane-spanning domain, to inhibit viral infection. The rationale for such treatment was that soluble

**Table 4.2**　Receptor interference groups of mammalian retroviruses that can infect human cells[a]

| Receptor group | Virus[b] | Receptor | Type | Species from which isolated | Origin[c] |
|---|---|---|---|---|---|
| 1 | RD114 | Unknown | C | Domestic cat | Endogenous |
|  | BaEV |  | C | Baboon | Endogenous |
|  | SRV-1 |  | D | Rhesus monkey | Exogenous |
|  | SRV-2 |  | D | Black macaque | Exogenous |
|  | MPMV (SRV-3) |  | D | Rhesus monkey | Endogenous |
|  | SRV-4 |  | D | Cynomolgus monkey | Exogenous |
|  | SRV-5 |  | D | Rhesus monkey | Exogenous |
|  | PO-1-Lu |  | D | Spectacled langur | Endogenous |
|  | SMRV |  | D | Squirrel monkey | Endogenous |
| 2 | MLV-A | Ram-1/Glvr2 | C | Wild mouse | Exogenous |
| 3 | MLV-X | Unknown | C | Laboratory mouse | Endogenous |
| 4 | FeLV-C | Unknown | C | Domestic cat | Exogenous |
| 5 | FeLV-B | Glvr1 | C | Domestic cat | Exogenous |
|  | SSAV |  | C | Woolly monkey | Exogenous |
|  | GALV |  | C | Gibbon | Exogenous |
| 6 | BLV | Blvr | C (complex) | Cattle | Exogenous |
| 7 | HTLV-1 | Unknown | C (complex) | Human | Exogenous |
|  | HTLV-2 |  | C (complex) | Human | Exogenous |
|  | ChTLV |  | C (complex) | Chimpanzee | Exogenous |
|  | STLV |  | C (complex) | Bonnet macaque | Exogenous |
| 8 | HIV-1 | CD4 | Lentivirus | Human | Exogenous |
|  | HIV-2 |  | Lentivirus | Human | Exogenous |
|  | SIV$_{mac}$ |  | Lentivirus | Rhesus monkey | Exogenous |
|  | SIV$_{smm}$ |  | Lentivirus | Sooty mangabey | Exogenous |
|  | SIV$_{agm}$ |  | Lentivirus | Green monkey | Exogenous |

[a]Data from M. A. Sommerfelt and R. A. Weiss, *Virology* **176**:58–69, 1990, with permission. The family *Retroviridae* has recently been reorganized into seven genera, based on knowledge of molecular structure and probable evolutionary relatedness. The first four of these include known oncoretroviruses with simple genomes (appendix). They comprise mammalian type B, mammalian type C, and mammalian type D viruses that were originally distinguished by the type of morphology that characterized their virion cores and their sites of assembly. Members of the fourth genus, avian type C, do not replicate in human cells and were not included in this survey. Genera of the *Retroviridae* that comprise viruses with more complex genomes include bovine (bovine lentivirus) and human (human T-cell leukemia virus) retroviruses that have cores with type C morphology, lentiviruses that have cone-shaped cores and produce diseases with long or intermittent clinical latency, and spumaviruses (not included in this survey) that cause no known disease. Receptors for members of four of these genera, representing receptor groups 2, 5, 6, and 8, have been identified.

[b]RD114, cat endogenous retrovirus; BaEV, baboon endogenous virus: SRV-1 to -5, simian retrovirus types 1 to 5; MPMV, Mason-Pfizer monkey virus; Po-1-Lu, langur endogenous retrovirus; SMRV, squirrel monkey retrovirus; MLV-A and -X, amphotropic and xenotropic murine leukemia virus; FeLV-B and -C, feline leukemia virus subgroups B and C; SSAV, simian sarcoma-associated virus; GALV, gibbon ape leukemia virus; BLV, bovine leukemia virus; HTLV-1 and -2, human T-cell leukemia virus types 1 and 2; ChTLV, chimpanzee T-cell leukemia virus; STLV, simian T-cell leukemia virus; HIV-1 and -2, human immunodeficiency virus types 1 and 2; SIV, simian immunodeficiency virus.

[c]Endogenous origin refers to proviruses that are integrated into the germ line cells of the indicated species.

CD4 should bind virus and block its attachment to CD4 on host cell surfaces. Unfortunately, although inhibition of viral infectivity could be demonstrated in cell culture experiments, clinical trials gave disappointing results. This failure can be attributed in part to the fact that each viral envelope includes many copies ($\approx$30) of the structure that binds CD4 such that relatively high serum concentrations of soluble CD4 are required in order to block all of them. This problem was further compounded by the short half-life of soluble CD4 in the blood. Furthermore, human im-

munodeficiency virus can also be spread by cell fusion, a process that is not readily blocked by circulating soluble CD4.

Although the presence of CD4 is sufficient to render *human* cells susceptible to human immunodeficiency virus infection, expression of CD4 on the surface of *mouse* cells allows the virus to attach but not to enter. These results suggested that one or more human coreceptors are required for steps in infection subsequent to attachment. The experimental strategies used to isolate these coreceptors, their identification as **chemokine receptors,** and their role in promoting cell fusion are described in chapter 5. Certain isolates of human immunodeficiency virus type 2 and simian immunodeficiency virus can utilize chemokine receptors to enter cells independently of CD4. The relevance of these observations for pathogenesis remains to be determined. Given the large number of members of the chemokine receptor family and the ability of human immunodeficiency virus type 1 to interact with these proteins, it is possible that CD4-independent, chemokine receptor-mediated infection may occur at some frequency. If so, then the host range of the virus would be broader than previously suspected, thereby changing our view of the pathogenesis of AIDS.

It is noteworthy that both CD4 and human immunodeficiency virus type 1 coreceptor molecules are coupled via their cytoplasmic domains to intracellular signaling pathways. For example, the normal role of CD4 is to participate with the T-cell receptor in the recognition of major histocompatibility class II molecules during the presentation of antigens (Fig. 4.13) and to signal such interactions by contact with a protein kinase (p56$^{lck}$) attached to the inner leaflet of the plasma membrane. The chemokine receptors also signal interaction with their ligand, affecting cellular gene expression. One might speculate that contact with viral envelope proteins could activate these signaling pathways and produce cellular responses that enhance virus propagation and/or affect pathogenesis. Interactions between SU and CD4 have been implicated in virus-induced cell killing. One example is the induction of apoptosis by cross-linking of SU bound to human CD4$^+$ T cells, followed by signaling through the T-cell receptor. However, little is known about the relevance of such responses by the receptors of human immunodeficiency virus type 1. It has also been reported that interactions between SU and chemokine receptors on neuronal cells induces apoptosis. The destruction of cytotoxic T cells by macrophages has also been attributed to such interactions. Such effects could help to explain the neurological disorders and loss of cytotoxic T cells that are symptoms of AIDS.

**Other picornavirus receptors.** Decay-accelerating protein (CD55), a member of the complement cascade, has been identified as the cell receptor for many enteroviruses (Fig. 4.9; Table 4.1). For most of these viruses, binding to decay-accelerating protein is not sufficient for infection, which requires the presence of a coreceptor. For example, coxsackievirus A21 can bind to cell surface decay-accelerating protein, but this interaction does not lead to infection unless Icam-1 is also present. Virus binding to Icam-1 alone is sufficient for infection. However, when both

Icam-1 and decay-accelerating protein are present in the same cell, complete blockade of virus binding and infection requires the use of antibodies against both receptors. In such cells, decay-accelerating protein and Icam-1 form a receptor-coreceptor complex required for infection. It is believed that virus first binds to decay-accelerating protein and then interacts with Icam-1 to allow cell entry. A similar mechanism with different receptors and coreceptors is likely to occur during infection with other enteroviruses (Table 4.1).

When decay-accelerating protein is cross-linked with antibodies in the absence of Icam-1, it can serve as a receptor mediating infection with coxsackievirus A21. In host cells, cross-linking of decay-accelerating protein might be a consequence of binding of a cellular ligand. Such binding might alter the structure of the receptor, exposing a binding site for the virus. Depending on the cell type, decay-accelerating protein, Icam-1, or a complex of both molecules may serve as a productive receptor for coxsackievirus A21.

**Adenovirus receptors and coreceptors.** The results of competition experiments (see "Experimental Approaches") indicated that members of two different virus families, group B coxsackieviruses and adenoviruses, share a cell receptor. This receptor has been identified as a 46-kDa member of the immunoglobulin superfamily and is called Car (coxsackievirus and adenovirus receptor) (Table 4.1). Adenoviruses of subgroups A, C, D, E, and F bind to Car. An alternative receptor has been identified for adenovirus types 2 and 5, members of subgroup C. Adenovirus type 2 can bind to major histocompatibility class I molecules on the surface of human epithelial and B lymphoblastoid cells, while adenovirus type 5 binds to the integrin $\alpha_M\beta_2$. However, binding to Car, major histocompatibility class I molecules, or $\alpha_M\beta_2$ is not sufficient for infection by most adenoviruses. Interaction with a coreceptor, identified as the $\alpha_v$ integrins $\alpha_v\beta_3$ and $\alpha_v\beta_5$, is required for uptake of the capsid into the cell by receptor-mediated endocytosis. An exception is adenovirus type 9, which apparently can infect hematopoietic cells after binding directly to $\alpha_v$ integrins.

**Alphaherpesvirus receptors and coreceptors.** The alphaherpesvirus subfamily of the *Herpesviridae* includes herpes simplex virus types 1 and 2, pseudorabies virus, and bovine herpesvirus. Initial contact of these viruses with the cell surface is first made by low-affinity binding to glycosaminoglycans (preferentially heparan sulfate), abundant components of the extracellular matrix (see "Extracellular Matrix: Components and Biological Importance" above). This initial interaction concentrates viruses

**A**

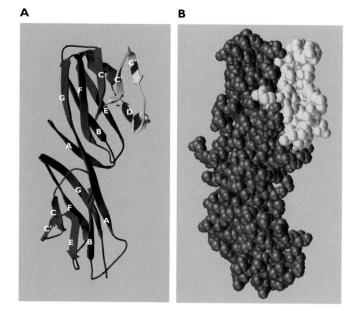

**B**

**C**

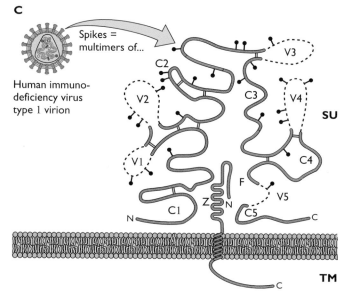

Human immuno-
deficiency virus
type 1 virion

Spikes =
multimers of...

C2

V3

V2

C3

V4

SU

V1

C4

F

V5

C1

Z

N

C5

N

C

TM

**D**

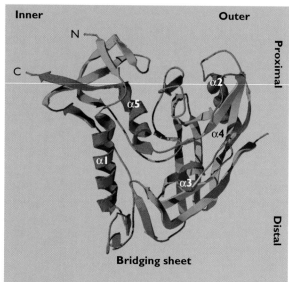

Inner

Outer

N

C

Proximal

α2

α5

α4

α1

α3

Distal

Bridging sheet

**E**

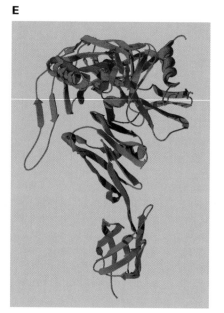

**F**

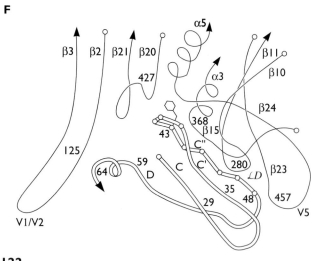

β3   β2   β21   β20

α5

β11

427

α3

β10

368

β24

43

β15

C''

280

V1/V2

125

64

59

C   C'

LD

β23

29

35   48

457

D

V5

**G**

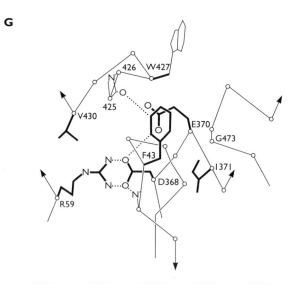

426   W427

N

425

O

V430

O

E370

G473

F43

I371

D368

R59

N

◁ **Figure 4.12 Interaction of human immunodeficiency virus type 1 SU with its cell receptor, CD4.** (A) Ribbon model of the backbone carbons of domains 1 and 2 (residues 1 to 182), derived from X-ray crystallography. The β-strands are labeled with letters; strand directions are indicated by arrowheads, and connecting loops are shown as strings. (B) A space-filling model of CD4 shows the positions of residues (yellow) that interact with human immunodeficiency virus type 1 during attachment, as revealed by mutagenesis. (C) Schematic diagram of a monomer of human immunodeficiency virus type 1 envelope glycoprotein. Conserved (C) and variable (V) regions of SU are indicated. Lines show the position of disulfide bridges, and lollipops the positions of glycosylation. F is the fusogenic peptide, and Z is a leucine zipper-like domain in the transmembrane protein, TM. Regions in the C domains of SU that are known to contact the CD4 receptor are highlighted. (D) Ribbon diagram of SU, derived from X-ray crystallographic data. The structure was determined for a core of SU lacking 52 and 19 amino acids from the N and C termini, respectively, and containing Gly-Ala-Gly in place of 67 V1/V2 loop residues and 32 V3 loop residues. This modified SU binds CD4 with affinity comparable to that of the native protein. α-helices are red, β-strands are magenta, and β-strand 15 is yellow. The last forms an antiparallel β-sheet with strand C″ of CD4. Loops are gray, except for loop V4 (dashed line), which is disordered. Selected features of the molecule are labeled. (E) Ribbon diagram depicting the SU-CD4 interaction, derived from X-ray crystallographic data. SU is red, and CD4 is brown. (F) Schematic diagram of the interface between SU and CD4. Solid black lines represent the six segments of SU that interact with CD4, which is shown by double lines. Secondary structure elements are labeled, and some residues of SU that interact with CD4 are indicated. The directions of the main chains are indicated by arrows. An important residue of CD4 identified by mutagenesis, Phe43, is shown penetrating into SU. Phe43 makes 23% of the interatomic contacts between CD4 and SU. This residue is at the center of the interface and appears to stabilize the entire complex. (G) Contacts in the area of CD4 amino acids Phe43 and Arg59. SU residues in direct contact with these two CD4 amino acids are labeled. Electrostatic interactions are shown as dashed lines. There are hydrophobic interactions between Phe43 of CD4 and Trp427, Glu370, Gly473, and Ile371 of SU and between Arg59 of CD4 and Val430 of SU. Adapted from J. Wang et al., *Nature* **348:**411–418, 1990 (A and B); B. N. Fields et al. (ed.), *Fields Virology,* 3rd ed. (Lippincott-Raven Publishers, Philadelphia, Pa., 1996) (C); and P. D. Kwong et al., *Nature* **393:**648–659, 1998 (D to G), with permission.

near the cell surface and facilitates subsequent attachment to a membrane coreceptor, which is required for entry of virus into the cell. Members of at least two different protein families serve as entry coreceptors for alphaherpesviruses (Table 4.1). One of these families comprises the poliovirus receptor Pvr and related proteins, another example of receptors shared by different viruses.

### *Alternative Receptors*

Some examples of the use of alternative cell surface molecules as receptors for the same virus have already been discussed. Two additional examples illustrate how receptor usage depends on the nature of the virus isolate or the cell line. Infection of cells with foot-and-mouth disease virus type A12 requires the RGD-binding integrin $\alpha_v\beta_3$ (Table 4.1). However, the O strain of foot-and-mouth disease virus, which has been extensively passaged in cell culture, cannot use this integrin as a receptor but instead binds to cell surface heparan sulfate. High concentrations of heparin (a derivative of heparan sulfate proteoglycan) (Fig. 4.2) adsorb to the virus and block its attachment to cells. On the other hand, the type A12 strain cannot infect cells that lack $\alpha_v\beta_3$ even if heparan sulfate is present. In a similar way, adaptation of Sindbis virus to cultured cells has led to the selection of variants that use heparan sulfate

as a receptor. Thus, laboratory adaptation of viruses to cells that express low levels of the virus receptor can lead to the use of alternative receptors.

## How Virions Attach to Receptors

Animal viruses have multiple receptor binding sites on their surfaces. Of necessity, nonenveloped viruses must rely on specific interactions of the cell receptor with one or more of their capsid proteins or spikes. Receptor binding sites for enveloped viruses are surface glycoproteins that have been incorporated into their cell-derived membranes. Some receptors are "hooks" that hold virus particles to cells, while uncoating is triggered by low pH or proteases. Other receptors are "unzippers": their binding produces dramatic conformational changes in the structures of virions that facilitate subsequent steps of infection. Although the details vary among viruses, most use one of several attachment mechanisms to attach to cell receptors, as illustrated by the best studied examples described below.

### *Nonenveloped Viruses*

**Attachment via surface features: canyons and loops.** RNA genomes of picornaviruses are protected by capsids

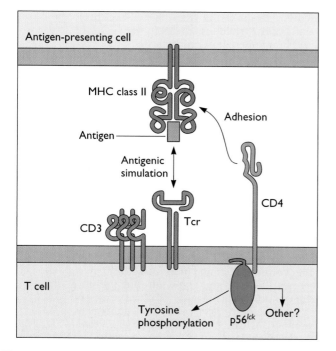

**Figure 4.13 Function of CD4 in lymphocytes.** CD4 interacts with a conserved region in major histocompatibility class II molecules on antigen presenting cells and stabilizes the interaction between it and the T-cell receptor complex. The interaction triggers an intracellular signaling pathway through tyrosine phosphorylation by p56$^{lck}$. The T-cell receptor complex also stimulates intracellular signaling. Binding of CD4 by human immunodeficiency virus type 1 prevents its interaction with major histocompatibility class II molecules and presumably dampens the T-cell response. Adapted from M. Foti et al., *J. Lab. Clin. Med.* **126**:233–239, 1995, with permission.

made up of four virus-encoded proteins, VP1, VP2, VP3, and VP4, arranged with icosahedral symmetry (Fig. 3.8A). Three-dimensional structures have been determined by X-ray crystallography for at least one member of all five picornavirus genera. The picornavirus capsid is built from 60 subunits arranged as 12 pentamers (Fig. 3.8A). Each subunit contains the four capsid proteins in an identical arrangement, with portions of the first three exposed on the surface. Although similar in arrangement, the surface architecture of the three exposed proteins varies among the family members, a property that accounts for the different serotypes and the different modes of interaction that can take place with cell receptors. For example, the capsids of rhinoviruses and some enteroviruses like poliovirus have deep canyons surrounding the 12 fivefold axes of symmetry in the pentamers (Fig. 4.14A), whereas cardioviruses and aphthoviruses lack this feature.

The canyons in the capsids of rhinoviruses and enteroviruses are generally accepted to be the sites of interaction with cell surface receptors (Fig. 4.14A). Consistent

with this view, amino acids that line the canyons are more conserved than any others on the viral surface, and their substitution can alter the affinity of binding to cells. A human rhinovirus bound to a receptor fragment containing Icam-1 domains 1 and 2 has been visualized in reconstructed images from cryoelectron microscopy. The three-dimensional atomic structure of Icam-1, determined by X-ray crystallography, was then modeled into the virus receptor model. The results indicate that Icam-1 binds to the central portion of the canyon in an orientation perpendicular to the surface of the virion. The BC, DE, and FG loops of Icam-1 domain 1 penetrate deep into the canyon, while the CD loop lies against one wall (Fig. 4.10). There are extensive charge-charge interactions between rhinovirus and Icam-1. Certain amino acids of the rhinovirus capsid that are crucial for receptor binding (previously identified by mutagenesis) all fall within the footprint of the receptor. Comparison of the amino acid sequences of major and minor group rhinoviruses (~10 serotypes) in the areas that contact Icam-1 do not reveal why the latter bind a different receptor (Table 4.1). In fact, it appears that both major and minor group rhinoviruses should be able to bind Icam-1 to some extent.

The canyons of picornaviruses were first believed to be too deep and narrow to permit entry of antibody molecules with adjacent Ig domains. It was hypothesized that this physical barrier would allow amino acids crucial for receptor interactions to be hidden from the immune system. Recent X-ray crystallographic analyses of a specific rhinovirus-antibody complex have shown that this is not the case (Fig. 4.14A): the antibody penetrates deep into the canyon, as does Icam-1. Therefore, the shape of the canyon is not likely to play a role in immune escape.

Beneath the canyon floor is a curved tunnel lined with hydrophobic side chains that opens at the base of the canyon and extends towards the fivefold axis of symmetry. In poliovirus and most rhinoviruses, the tunnel appears to contain a fatty acid or related compound (Fig. 4.14B). In poliovirus, this lipid is believed to be sphingosine. Some antiviral drugs (WIN compounds, named after Sterling-Winthrop, the pharmaceutical company at which they were discovered) can irreversibly replace the lipid, thereby inhibiting uncoating, and in some cases attachment, of picornaviruses. Early studies of human rhinovirus 14 demonstrated that drug binding causes conformational changes in the canyon that deform the receptor-binding site and prevent attachment. However, drug binding to human rhinoviruses 1A and 16 and poliovirus induces much smaller structural changes in the capsid. Because WIN compounds cause minor changes in the canyon of rhinovirus 16, it is believed that the drugs block attachment of this virus not by altering the receptor binding site

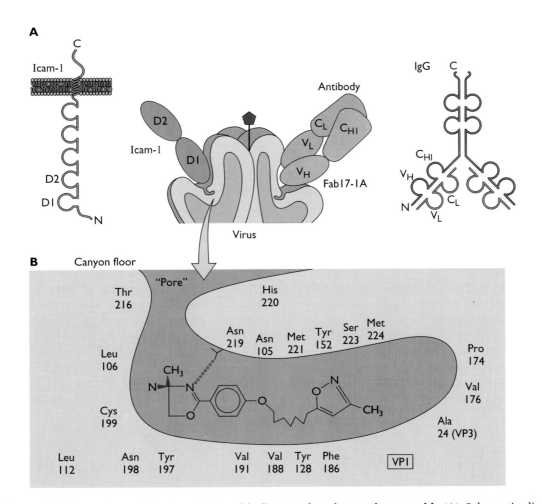

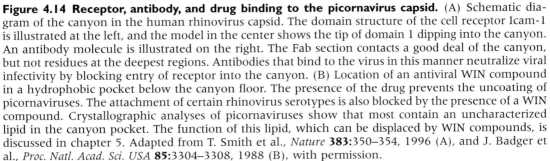

**Figure 4.14 Receptor, antibody, and drug binding to the picornavirus capsid.** (A) Schematic diagram of the canyon in the human rhinovirus capsid. The domain structure of the cell receptor Icam-1 is illustrated at the left, and the model in the center shows the tip of domain 1 dipping into the canyon. An antibody molecule is illustrated on the right. The Fab section contacts a good deal of the canyon, but not residues at the deepest regions. Antibodies that bind to the virus in this manner neutralize viral infectivity by blocking entry of receptor into the canyon. (B) Location of an antiviral WIN compound in a hydrophobic pocket below the canyon floor. The presence of the drug prevents the uncoating of picornaviruses. The attachment of certain rhinovirus serotypes is also blocked by the presence of a WIN compound. Crystallographic analyses of picornaviruses show that most contain an uncharacterized lipid in the canyon pocket. The function of this lipid, which can be displaced by WIN compounds, is discussed in chapter 5. Adapted from T. Smith et al., *Nature* **383**:350–354, 1996 (A), and J. Badger et al., *Proc. Natl. Acad. Sci. USA* **85**:3304–3308, 1988 (B), with permission.

but by preventing conformational changes that may be required for receptor binding. Poliovirus uncoating, but not binding, is inhibited by these drugs.

For picornaviruses whose capsids do not have prominent canyons, including coxsackievirus A and foot-and-mouth disease virus, attachment is accomplished through VP1 surface loops that include RGD motifs recognized by their integrin receptors. Alteration of the RGD sequence in VP1 of foot-and-mouth disease virus blocks

virus binding. However, changes in the same sequence of coxsackievirus A9 does not abolish virus binding to cells, suggesting that these viruses can bind to an alternative receptor.

**Attachment via protruding fibers.** The nonenveloped DNA-containing adenoviruses are much larger than picornaviruses, and their icosahedral capsids are more complex, comprising at least 10 different polypeptides. By electron microscopy adenovirus particles can be seen to have fibers protruding from each pentamer (Fig. 3.12A). The fibers are composed of homotrimers of the adenovirus fiber protein and are anchored in the pentameric penton base protein; both proteins have roles to play in virus attachment and uptake.

For many adenovirus serotypes, attachment via the fibers is necessary but not sufficient for infection. A region comprising the N-terminal 40 amino acids of each subunit of the fiber protein is bound noncovalently to the penton base. The central shaft region is composed of repeating motifs of approximately 15 amino acids; the length of the shaft in different serotypes is determined by the number of these repeats. The three constituent shaft regions appear to form a rigid triple-helical structure in the trimeric fiber. The C-terminal 180 amino acids of each subunit interact to form a terminal knob. Genetic analyses and competition experiments indicate that determinants for the initial, specific attachment to host cell receptors reside in this knob. The crystal structure of this receptor binding domain has been determined for adenovirus type 5. The three-dimensional model shows a deep surface depression on the top of the knob, centered on a threefold axis (Fig. 4.15). As the amino acids that line the depression are conserved in adenovirus types 5 and 2, serotypes that share the same cell receptor, the depression seems likely to be the site of attachment to fiber receptors described above.

The observation that soluble fiber protein can bind to cells but is not endocytosed suggested that adenoviruses require a coreceptor for uptake into the cell. A second interaction has been identified between the penton base and cellular integrins in what should now be recognized as a recurrent theme: as in certain members of the picornaviruses (echoviruses, group A coxsackieviruses, and foot-and-mouth disease virus) (Table 4.1), attachment is mediated by RGD sequences that mimic the normal ligands of cell surface integrins. The same RGD sequences in each of the five subunits of the adenovirus penton base bind to the $\alpha_v\beta_3$ and $\alpha_v\beta_5$ vitronectin-binding integrins and perhaps other integrin family members. This interaction is critical, because adenovirus infection is blocked by competition with RGD-containing peptides or by substitutions in the integrin-binding region of the penton base protein. However, with the exception of adenovirus type 9 (see "Adenovirus receptors and coreceptors" above), while necessary, in most cases the integrin interaction is not sufficient to support infection in the absence of specific attachment via the fiber protein. Thus, attachment and uptake is hypothesized to be a two-step process, with the presumed role of the second receptor being to trigger receptor-mediated endocytosis of virus captured by interaction with the receptor.

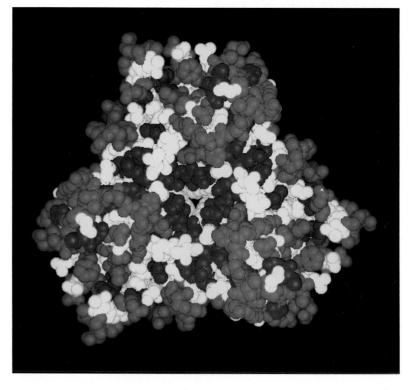

**Figure 4.15 Space-filling model of the trimeric knob of adenovirus type 5 fiber protein.** A deep surface depression is centered on the threefold axis and is surrounded by shallower valleys. Red, amino acids that differ between adenovirus types 5 and 2; yellow, amino acids that are identical in adenovirus types 5 and 2; blue, residues that are identical in adenovirus types 5 and 2 but different from residues that are identical in adenovirus types 7 and 3. The main conserved regions are in the central surface depression and on the floor of the valley. From D. Xia et al., *Curr. Biol.* **2:**1259–1270, 1994, with permission. Courtesy of J. Deisenhofer, University of Texas.

### Enveloped Viruses Attach to Receptors by means of Viral Membrane Glycoproteins

As noted above, enveloped viruses acquire their lipid membranes from the cells they infect. The process of virion assembly includes insertion into membranes of specific viral proteins that carry membrane-spanning domains analogous to those of cellular integral membrane proteins. Attachment sites (i.e., viral ligands) on one or more of these envelope proteins bind to specific cell receptors. The two best studied examples of enveloped virus attachment and its consequences are the interactions of influenza A virus and the retrovirus human immunodeficiency virus type 1 with their cell receptors.

**The HA of influenza virus.**   Influenza virus HA is the viral glycoprotein that binds to the cell receptor, sialic acid (Fig. 4.16). The HA monomer is synthesized as a precursor that is glycosylated and subsequently cleaved to form HA1 and HA2 subunits. HA is anchored in the viral envelope by a single membrane-spanning domain at the C terminus of HA2. The two subunits are held together by a disulfide bond and many noncovalent bonds. The N-terminal tip of HA2 protein formed by cleavage contains a sequence known as a "fusion" peptide. This peptide mediates fusion of viral and cellular membranes, as described in chapter 5. Each glycoprotein spike seen by electron microscopy is a trimer of HA. The extracellular domain of the HA ho-

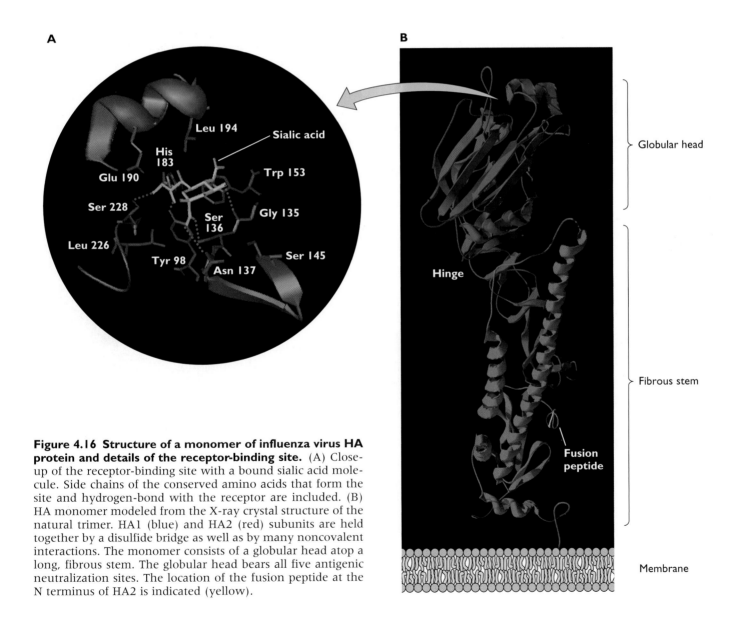

**Figure 4.16 Structure of a monomer of influenza virus HA protein and details of the receptor-binding site.** (A) Close-up of the receptor-binding site with a bound sialic acid molecule. Side chains of the conserved amino acids that form the site and hydrogen-bond with the receptor are included. (B) HA monomer modeled from the X-ray crystal structure of the natural trimer. HA1 (blue) and HA2 (red) subunits are held together by a disulfide bridge as well as by many noncovalent interactions. The monomer consists of a globular head atop a long, fibrous stem. The globular head bears all five antigenic neutralization sites. The location of the fusion peptide at the N terminus of HA2 is indicated (yellow).

motrimer, released from the membrane with the protease bromelain, was the first membrane protein whose structure was solved by X-ray crystallography. Each HA monomer consists of a long, helical stalk anchored in the membrane by HA2 and topped by a large HA1 globule that includes the sialic acid-binding pocket and a number of highly antigenic surface loops (Fig. 4.16). The trimer is held together largely by hydrophobic interactions among the helices.

While attachment of all influenza A virus strains requires sialic acid, strains vary in their affinities for different sialyloligosaccharides. For example, avian virus strains prefer sialic acids attached to galactose with an α(2,3) linkage (Fig. 4.11), while human virus strains prefer the α(2,6) linkage. Amino acid differences at positions 226 and 228 of certain HA subtypes account for these differences in sugar linkage specificity. Changes at these residues, which are in the receptor-binding pocket in HA (Fig. 4.16), influence its contact with sialic acid. These amino acids can also influence the host range of the virus. Avian strains of influenza virus replicate in the intestine, while human strains do not. A recombinant duck virus, containing an H3 (human) HA gene, does not replicate in duck intestine. However, introduction of two mutations into the H3 HA gene—an L-to-Q change at amino acid 226 and an S-to-G change at amino acid 228—allow the recombinant virus to replicate in duck intestine. These findings indicate that recognition of different receptors by HA is an important determinant of virus replication in the bird intestine.

**The envelope glycoprotein of human immunodeficiency virus type 1.** When examined by electron microscopy, the envelopes of human immunodeficiency virus type 1 and other retroviruses appear to be studded with "spikes" (Fig. 4.12). These structures are composed of oligomers (probably trimers) of the single viral envelope glycoprotein. The spikes bind the cell receptor, CD4 (Fig. 4.12). The monomers of the spike protein are synthesized as heavily glycosylated precursors that are cleaved by a cellular protease to form SU and TM. The latter is anchored in the envelope by a single membrane-spanning domain and remains bound to SU by numerous noncovalent bonds. The extensive antigenic variability among different isolates of human immunodeficiency virus type 1 maps to five discrete regions called V1 to V5. They are separated by relatively conserved regions, C1 to C5. Sequences important for binding to the major receptor, CD4, are included in regions C2 to C4.

The atomic structure of a complex of human immunodeficiency virus type 1 SU, a two-domain fragment of CD4, and a neutralizing antibody against SU has been determined by X-ray crystallography (Fig. 4.12). The polypeptide of SU is folded into an inner and an outer domain linked by an antiparallel four-stranded "bridging sheet." A depression at the interface of the outer and inner domains and the bridging sheet forms the binding site for CD4. Interatomic contacts are made between 22 amino acids of CD4 and 26 amino acids of SU (Fig. 4.12). Although the structure of SU in the absence of CD4 is not known, certain features of the structure of the complex indicate that receptor binding induces a conformational change in SU. For example, CD4 amino acid Phe43 is located in a cavity in SU lined with hydrophobic amino acids. It is not likely that hydrophobic amino acids would be exposed in the pocket in the absence of CD4. The conformational changes induced in SU by CD4 binding expose binding sites on SU for the chemokine receptors. Interaction with chemokine receptors is required for fusion of viral and cell membranes, a process described in the next chapter.

## Perspectives

The receptors and viral attachment strategies provided as examples in this chapter represent only a small subset of those that have been studied to some degree or are still to be explored. Receptors for fewer than 100 viruses are listed in Table 4.1, a fraction of the ~4,000 known viruses. Although many more viral receptors await identification, studies of a small number of viral receptors have provided fundamental concepts about how receptors function during virus attachment and entry into cells. A general conclusion that can be drawn is that the interactions governing virus attachment are complex, consisting of reciprocal changes in the virion and the cell. For some viruses, such as poliovirus, rhinovirus, and influenza virus, a single type of cellular molecule is sufficient for virus binding. In other cases, including human immunodeficiency virus type 1, adenovirus, and alphaherpesviruses, one type of cellular molecule is required for attachment and a coreceptor is necessary for virus entry into the cell. There is a growing list of viruses that can recognize multiple receptors; which one is used depends on the virus strain, the type of cultured cell, the host species, and even the site of replication within the host. The notion that a virus uses one receptor to enter cells, influenced by early studies with influenza viruses, has been shattered.

A perusal of Table 4.1 yields two striking conclusions about receptor utilization by viruses. The molecules that function as cellular receptors for viruses are a small subset of the large number of cell surface components. It is also surprising that many different viruses share the same cell receptor. Perhaps as additional receptors for viruses are identified, these patterns will no longer be so obvious. It seems more likely that viruses have evolved to utilize certain types of molecules because interaction induces cellular processes that are beneficial for replication. Possible consequences of virus-receptor interactions include enhanced virus entry, more efficient replication, recruitment of susceptible cells to the site of infection, or inhibition of cellular defense mechanisms. Induction of cell death by receptor interaction (see "The human immunodeficiency virus type 1 receptor, CD4" above) may serve to facilitate virus release from cells. The triggering of such events by virus binding to receptors may mimic interaction of the receptor with its cellular ligand. Future research is likely to focus more intensely on the question of whether virus-receptor interactions lead to intracellular signaling. The results should clarify why viruses have evolved to utilize certain receptors and may also lead to new ways to prevent viral disease.

## References

### Reviews

Berger, E. A. 1997. HIV entry and tropism: the chemokine receptor connection. *AIDS* **11**:S3–S16.

Evans, D. J., and J. W. Almond. 1998. Cell receptors for picornaviruses as determinants of cell tropism and pathogenesis. *Trends Microbiol.* **6**:198–202.

Harrison, S. C. 1995. Virus structures and conformational rearrangements. *Curr. Opin. Struct. Biol.* **5**:157–164.

Norkin, L. C. 1995. Virus receptors: implications for pathogenesis and the design of antiviral agents. *Clin. Microbiol. Rev.* **8**: 293–315.

Weiss, R. A., and C. S. Tailor. 1995. Retrovirus receptors. *Cell* **82**:531–533.

Wimmer, E. 1994. *Cellular Receptors for Animal Viruses.* Cold Spring Harbor Press, Cold Spring Harbor, N.Y.

### Papers of Special Interest

#### Picornaviruses

Bella, J., P. R. Kolatkar, C. W. Marlor, J. M. Greve, and M. G. Rossmann. 1998. The structure of the two amino-terminal domains of human ICAM-1 suggests how it functions as a rhinovirus receptor and as an LFA-1 integrin ligand. *Proc. Natl. Acad. Sci. USA* **95**:4140–4145.

Bergelson, J. M., J. A. Cunningham, G. Droguett, E. A. Kurt-Jones, A. Krithivas, J. S. Hong, M. S. Horwitz, R. L. Crowell, and R. W. Finberg. 1997. Isolation of a common receptor for coxsackie B viruses and adenoviruses 2 and 5. *Science* **275**:1320–1323.

Bergelson, J. M., M. P. Shepley, B. M. Chan, M. E. Hemler, and R. W. Finberg. 1992. Identification of the integrin VLA-2 as a receptor for echovirus 1. *Science* **255**:1718–1720.

Berinstein, A., M. Roivainen, T. Hovi, P. W. Mason, and B. Baxt. 1995. Antibodies to the vitronectin receptor (integrin $\alpha_v\beta_3$) inhibit binding and infection of foot-and-mouth disease virus to cultured cells. *J. Virol.* **69**:2664–2666.

Casasnovas, J. M., T. Stehle, J. H. Liu, J. H. Wang, and T. A. Springer. 1998. A dimeric crystal structure for the N-terminal two domains of intercellular adhesion molecule-1. *Proc. Natl. Acad. Sci. USA* **95**:4134–4139.

Greve, J. M., G. Davis, A. M. Meyer, C. P. Forte, S. C. Yost, C. W. Marlor, M. E. Kamarck, and A. McClelland. 1989. The major human rhinovirus receptor is ICAM-1. *Cell* **56**:839–847.

Karnauchow, T. M., D. L. Tolson, B. A. Harrison, E. Altman, D. M. Lublin, and K. Dimock. 1996. The HeLa cell receptor for enterovirus 70 is decay-accelerating factor (CD55). *J. Virol.* **70**:5143–5152.

Mendelsohn, C. L., E. Wimmer, and V. R. Racaniello. 1989. Cellular receptor for poliovirus: molecular cloning, nucleotide sequence and expression of a new member of the immunoglobulin superfamily. *Cell* **56**:855–865.

Shafren, D. R. 1998. Viral cell entry induced by cross-linked decay-accelerating factor. *J. Virol.* **72**:9407–9412.

Staunton, D. E., V. J. Merluzzi, R. Rothlein, R. Barton, S. D. Marlin, and T. A. Springer. 1989. A cell adhesion molecule, ICAM-1, is the major surface receptor for rhinoviruses. *Cell* **56**:849–853.

Tomassini, J. E., D. Graham, C. M. DeWitt, D. W. Lineberger, J. A. Rodkey, and R. J. Colonno. 1989. cDNA cloning reveals that the major group rhinovirus receptor on HeLa cells is intercellular adhesion molecule 1. *Proc. Natl. Acad. Sci. USA* **86**:4907–4911.

#### Coronaviruses

Dveksler, G. S., C. W. Dieffenbach, C. B. Cardellichio, K. McCuaig, M. N. Pensiero, G. S. Jiang, N. Beauchemin, and K. V. Holmes. 1993. Several members of the mouse carcinoembryonic antigen-related glycoprotein family are functional receptors for the coronavirus mouse hepatitis virus-A59. *J. Virol.* **67**:1–8.

Yeager, C. L., R. A. Ashmun, R. K. Williams, C. B. Cardellichio, L. H. Shapiro, A. T. Look, and K. V. Holmes. 1992. Human aminopeptidase N is a receptor for human coronavirus 229E. *Nature* **357**:420–422.

#### Togaviruses

Klimstra, W. B., K. D. Ryman, and R. E. Johnston. 1998. Adaptation of Sindbis virus to BHK cells selects for use of heparan sulfate as an attachment receptor. *J. Virol.* **72**:7357–7366.

Wang, K. S., R. J. Kuhn, E. G. Strauss, S. Ou, and J. H. Strauss. 1992. High-affinity laminin receptor is a receptor for Sindbis virus in mammalian cells. *J. Virol.* **66**:4992–5001.

#### Rhabdoviruses

Thoulouze, M. I., M. Lafage, M. Schachner, U. Hartmann, H. Cremer, and M. Lafon. 1998. The neural cell adhesion molecule is a receptor for rabies virus. *J. Virol.* **72**:7181–7190.

## Paramyxoviruses

Dorig, R. E., A. Marcil, A. Chopra, and C. D. Richardson. 1993. The human CD46 molecule is a receptor for measles virus (Edmonston strain). *Cell* **75**:295–305.

Naniche, D., G. Varior-Krishnan, F. Cervoni, T. F. Wild, B. Rossi, C. Rabourdin-Combe, and D. Gerlier. 1993. Human membrane cofactor protein (CD46) acts as a cellular receptor for measles virus. *J. Virol.* **67**:6025–6032.

## Influenza Virus

Rogers, G. N., G. Herrler, J. C. Paulson, and H. D. Klenk. 1986. Influenza C virus uses 9-*O*-acetyl-*N*-acetylneuraminic acid as a high affinity receptor determinant for attachment to cells. *J. Biol. Chem.* **261**:5947–5951.

Vines, A., K. Wells, M. Matrosovich, M. R. Castrucci, T. Ito, and Y. Kawaoka. 1998. The role of influenza A virus hemagglutinin residues 226 and 228 in receptor specificity and host range restriction. *J. Virol.* **72**:7626–7631.

Weis, W., J. H. Brown, S. Cusack, J. C. Paulson, J. J. Skehel, and D. C. Wiley. 1988. Structure of the influenza virus hemagglutinin complexed with its receptor, sialic acid. *Nature* **333**:426–431.

## Arenaviruses

Cao, W., M. D. Henry, P. Borrow, H. Yamada, J. H. Elder, E. V. Ravkov, S. T. Nichol, R. W. Compans, K. P. Campbell, and M. B. A. Oldstone. 1998. Identification of α-dystroglycan as a receptor for lymphocytic choriomeningitis virus and Lassa fever virus. *Science* **282**:2079–2081.

## Retroviruses

### Human immunodeficiency virus

Choe, H., M. Farzan, Y. Sun, N. Sullivan, B. Rollins, P. D. Ponath, L. Wu, C. R. Mackay, G. LaRosa, W. Newman, N. Gerard, C. Gerard, and J. Sodroski. 1996. The β-chemokine receptors CCR3 and CCR5 facilitate infection by primary HIV-1 isolates. *Cell* **85**:1135–1148.

Deng, H., R. Liu, W. Ellmeier, S. Choe, D. Unutmaz, M. Burkhart, P. DiMarzio, S. Marmon, R. E. Sutton, C. M. Hill, C. B. Davis, S. C. Peiper, T. J. Schall, D. R. Littman, and N. R. Landau. 1996. Identification of a major co-receptor for primary isolates of HIV-1. *Nature* **381**: 661–666.

Doranz, B.J., J. Rucker, Y. Yi, R. J. Smyth, M. Samson, S. C. Peiper, M. Parmentier, R. G. Collman, and R. W. Doms. 1996. A dual-tropic primary HIV-1 isolate that uses fusin and the β-chemokine receptors CKR-5, CKR-3 and CKR-2b as fusion cofactors. *Cell* **85**:1149–1158.

Dragic, T., V. Litwin, G. P. Allaway, S. R. Martin, Y. Huang, K. A. Nagashima, C. Cayanan, P. J. Maddon, R. A. Koup, J. P. Moore, and W. A. Paxton. 1996. HIV-1 entry into CD4+ cells is mediated by the chemokine receptor CC-CKR-5. *Nature* **381**:667–673.

Feng, Y., C. C. Broder, P. E. Kennedy, and E. A. Berger. 1996. Human immunodeficiency virus type 1 entry cofactor: functional cDNA cloning of a seven-transmembrane, G protein-coupled receptor. *Science* **272**:872–877.

Herbein, G., U. Mahlknecht, F. Batliwalla, P. Gregersen, T. Pappas, J. Butler, W. A. O'Brien, and E. Verdin. 1998. Apop-

tosis of CD8+ T cells is mediated by macrophages through the interaction of human immunodeficiency virus gp120 with chemokine receptor CXCR4. *Nature* **395**:189–194.

Hesselgesser, J., D. Taub, P. Basker, M. Greenberg, J. Hoxie, D. L. Kolson, and R. Horuk. 1998. Neuronal apoptosis induced by human immunodeficiency virus-1 gp120 and the chemokine Sdf-1α is mediated by the chemokine receptor CXCR4. *Curr. Biol.* **8**:595–598.

Kwong, P. D., R. Wyatt, J. Robinson, R. W. Sweet, J. Sodroski, and W. A. Hendrickson. 1998. Structure of an HIV gp120 envelope glycoprotein in complex with the CD4 receptor and a neutralizing human antibody. *Nature* **393**:648–659.

Maddon, P. J., A. G. Dalgleish, J. S. McDougal, P. R. Clapham, R. A. Weiss, and R. Axel. 1986. The T4 gene encodes the AIDS virus receptor and is expressed in the immune system and the brain. *Cell* **47**:333–348.

Ryu, S.-E., P. D. Kwong, A. Truneh, T. G. Porter, J. Arthos, M. Rosenberg, X. Dai, N.-H. Xuong, R. Axel, R. W. Sweet, and W. A. Hendrickson. 1990. Crystal structure of an HIV-binding recombinant fragment of human CD4. *Nature* **348**: 419–426.

Wang, J., Y. Yan, T. P. J. Garrett, J. Liu, D. W. Rodgers, R. L. Garlick, G. E. Tarr, Y. Husain, E. L. Reinherz, and S. C. Harrison. 1990. Atomic structure of a fragment of human CD4 containing two immunoglobulin-like domains. *Nature* **348**: 411–418.

### Other retroviruses

Albritton, L. M., L. Tseng, D. Scadden, and J. M. Cunningham. 1989. A putative murine ecotropic retrovirus receptor gene encodes a multiple membrane-spanning protein and confers susceptibility to virus infection. *Cell* **57**:659–666.

Bates, P., J. A. Young, and H. E. Varmus. 1993. A receptor for subgroup A Rous sarcoma virus is related to the low density lipoprotein receptor. *Cell* **74**:1043–1051.

Brojatsch, J., J. Naughton, M. M. Rolls, K. Zingler, and J. A. Young. 1996. Car1, a TNFR-related protein, is a cellular receptor for cytopathic avian leukosis-sarcoma viruses and mediates apoptosis. *Cell* **87**:845–855.

## Parvoviruses

Summerford, C., and R. J. Samulski. 1998. Membrane-associated heparan sulfate proteoglycan is a receptor for adeno-associated virus type 2 virions. *J. Virol.* **72**:1438–1445.

Summerford, C., J. S. Bartlett, and R. J. Samulski. 1999. α$_v$β$_5$ integrin: a coreceptor for adeno-associated virus type 2 infection. *Nat. Med.* **5**:78–82.

## Papovaviruses

Breau, W. C., W. J. Atwood, and L. C. Norkin. 1992. Class I major histocompatibility proteins are an essential component of the simian virus 40 receptor. *J. Virol.* **66**:2037–2045.

## Adenoviruses

Bergelson, J. M., J. A. Cunningham, G. Droguett, E. A. Kurt-Jones, A. Krithivas, J. S. Hong, M. S. Horwitz, R. L. Crowell, and R. W. Finberg. 1997. Isolation of a common receptor for coxsackie B viruses and adenoviruses 2 and 5. *Science* **275:**1320–1323.

Huang, S., T. Kamata, Y. Takada, Z. M. Ruggeri, and G. R. Nemerow. 1996. Adenovirus interaction with distinct integrins mediates separate events in cell entry and gene delivery to hematopoietic cells. *J. Virol.* **70:**4502–4508.

Liebermann, H., R. Mentel, U. Bauer, P. Pring-Akerblom, R. Dolling, S. Modrow, and W. Seidel. 1998. Receptor binding sites and antigenic epitopes on the fiber knob of human adenovirus serotype 3. *J. Virol.* **72:**9121–9130.

Roelvink, P. W., I. Kovesdi, and T. J. Wickham. 1996. Comparative analysis of adenovirus fiber-cell interaction: adenovirus type 2 (Ad2) and Ad9 utilize the same cellular fiber receptor but use different binding strategies for attachment. *J. Virol.* **70:**7614–7621.

Roelvink, P. W., A. Lizonova, J. G. Lee, Y. Li, J. M. Bergelson, R. W. Finberg, D. E. Brough, I. Kovesdi, and T. J. Wickham. 1998. The coxsackievirus-adenovirus receptor protein can function as a cellular attachment protein for adenovirus serotypes from subgroups A, C, D, E, and F. *J. Virol.* **72:**7909–7915.

Tomko, R. P., R. L. Xu, and L. Philipson. 1997. HCAR and MCAR: the human and mouse cellular receptors for subgroup C adenoviruses and group B coxsackieviruses. *Proc. Natl. Acad. Sci. USA* **94:**3352–3356.

Wickham, T. J., P. Mathias, D. A. Cheresh, and G. R. Nemerow. 1993. Integrins alpha v beta 3 and alpha v beta 5 promote adenovirus internalization but not virus attachment. *Cell* **73:**309–319.

## Herpesviruses

Geraghty, R. J., C. Krummenacher, G. H. Cohen, R. J. Eisenberg, and P. G. Spear. 1998. Entry of alphaherpesviruses mediated by poliovirus receptor-related protein 1 and poliovirus receptor. *Science* **280:**1618–1620.

Montgomery, R. I., M. S. Warner, B. J. Lum, and P. G. Spear. 1996. Herpes simplex virus-1 entry into cells mediated by a novel member of the TNF/NGF receptor family. *Cell* **87:**427–436.

Shieh, M. T., D. WuDunn, R. I. Montgomery, J. D. Esko, and P. G. Spear. 1992. Cell surface receptors for herpes simplex virus are heparan sulfate proteoglycans. *J. Cell Biol.* **116:**1273–1281.

Warner, M. S., R. J. Geraghty, W. M. Martinez, R. I. Montgomery, J. C. Whitbeck, R. Xu, R. J. Eisenberg, G. H. Cohen, and P. G. Spear. 1998. A cell surface protein with herpesvirus entry activity (HveB) confers susceptibility to infection by mutants of herpes simplex virus type 1, herpes simplex virus type 2, and pseudorabies virus. *Virology* **246:**179–189.

## Poxviruses

Chung, C. S., J. C. Hsiao, Y. S. Chang, and W. Chang. 1998. A27L protein mediates vaccinia virus interaction with cell surface heparan sulfate. *J. Virol.* **72:**1577–1585.

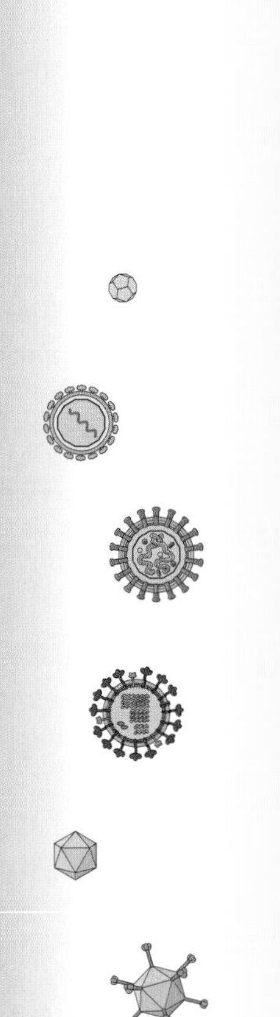

# 5

# Virus Entry into Cells

## Introduction

Once a virus has attached to a cellular receptor, it must enter the host cell so that genome replication can begin. Early studies of virus entry into the host cell, from the 1950s until the late 1970s, led to the view that viruses enter cells by an entirely passive process: virus particles attach to the cell surface, are taken up into the cell, release their genomes, and begin to replicate. No active role for the receptor in release of the viral genome into the cell, a process called uncoating, was envisioned. In these studies, investigation of cell entry comprised largely photographing sections of virus-infected cells at high magnification. While these findings showed that viruses did indeed cross the plasma membrane, the mechanisms of virus entry remained unknown. Indeed, in a popular virology textbook published in 1974, virus entry was described in only a few paragraphs. Beginning in the 1980s, the techniques of cellular, molecular, and structural biology were applied to elucidate the earliest events in viral infection. It is now understood that virus entry into cells is not a passive process but rather relies on viral usurpation of normal cellular processes, including endocytosis, membrane fusion, vesicular trafficking, and transport into the nucleus.

Successful entry of a virus into a host cell requires that the virus cross the plasma membrane, and in some cases the nuclear membrane as well, that the virus particle disassemble to make the viral genome accessible to the cytoplasm, and that the nucleic acid be targeted to the correct cellular compartment. These are not simple processes. Cell membranes are not permeable to particles as large as viruses. To cross such barriers, viruses utilize the specific transport mechanisms by which large molecules normally enter cells. Virions have evolved to be metastable vehicles for transporting their cargo, the viral genome, from cell to cell; the elegant products of viral assembly dissociate, at least in part, to release the viral genome from its protective shell. Viral disassembly often begins with conformational changes in the viral capsid or nucleocapsid or integral membrane proteins. In some cases, interaction of the virus with a cell receptor initiates conformational change that primes the capsid for uncoating. In others, the cell receptor is nothing more than a tether that holds the virus to the cell surface where it can be taken into the cell. Conformational changes are then triggered by low pH or by the action of proteases. Finally, the viral genome must be targeted to the correct cellular compartment, either the nucleus or the cytoplasm. Viruses that replicate in the nucleus include all DNA-containing viruses (except poxviruses), retroviruses, influenza viruses, and Borna disease virus. All other RNA-containing viruses replicate in the cytoplasm. Because all viruses first enter the cytoplasm, those destined for the nucleus must undergo an additional trafficking step.

Before we consider how viruses enter cells, we will briefly discuss how cells take up macromolecules. These are the same pathways by which viruses enter cells.

## Uptake of Macromolecules by Cells

### The Plasma Membrane Is a Semipermeable Barrier

The plasma membrane, the limiting membrane of the cell, permits nutrient molecules to enter and waste molecules to leave, thereby ensuring a constant internal environment. The semipermeable nature of this barrier results from a combination of the physical properties of the lipid bilayer and the existence of specific transport processes. Water, gases, and small hydrophobic molecules such as ethanol can freely traverse the lipid bilayer, but most metabolites, including glucose, ATP, nucleosides, amino acids and proteins, and certain ions ($Ca^{2+}$, $H^+$, $K^+$, and $Na^+$), cannot cross the membrane. These essential components

enter the cell by specific transport processes. Integral membrane proteins known as **permeases** are responsible for the transport of ions, sugars, and amino acids, while proteins and large particles (such as some viruses and bacteria) are taken into the cell by phagocytosis and endocytosis.

### Cells Take Up Macromolecules by Phagocytosis and Endocytosis

**Phagocytosis** is the process by which cells engulf large particles such as bacteria or cell fragments (Fig. 5.1). **Pinocytosis** is the process whereby small amounts of fluid, including dissolved molecules or suspended small particles, are taken up by vesicles approximately 0.1 μm in diameter (Fig. 5.1). Both processes are nonspecific, which means that any particle or molecule can be taken into the cell. Specific molecules are selectively taken into cells from the extracellular fluid by **receptor-mediated endocytosis** (Fig. 5.1 and 5.2). Molecules (**ligands**) in the extracellular medium bind to cells via specific plasma membrane receptor proteins. The receptor-ligand complex diffuses along the membrane until it reaches an invagination of the plasma membrane which is coated on its cytoplasmic surface by a cagelike lattice composed of the fibrous protein **clathrin** and is referred to as a **clathrin-coated pit** (Fig. 5.2B). Clathrin-coated pits can comprise as much as 2% of the surface area of a cell, and some receptors are clustered over these areas even in the absence of their ligands. Following the accumulation of receptor-ligand complexes, the clathrin-coated pit invaginates and then pinches off, forming a **clathrin-coated vesicle** containing the ligand-receptor complex. Within a few seconds, the clathrin coat is lost and the vesicles fuse with small, smooth-walled vesicles located near the cell surface called

**Figure 5.1 Mechanisms for the uptake of macromolecules from extracellular fluid.** During phagocytosis, large particles such as bacteria or cell fragments that come in contact with the cell surface are engulfed by extensions of the plasma membrane. The extensions fuse to form a 1- to 2-μm-diameter closed vesicle, or *phagosome,* containing the particle. Phagosomes ultimately fuse with lysosomes, resulting in degradation of the material within the vesicle. Macrophages use phagocytosis to ingest bacteria and destroy them. Endocytosis involves the invagination and pinching off of small regions of the plasma membrane, resulting in the nonspecific internalization of molecules (**pinocytosis** or **fluid-phase endocytosis**) or the specific uptake of molecules bound to cell surface receptors (**receptor-mediated endocytosis**). Adapted from J. Darnell et al., *Molecular Cell Biology* (Scientific American Books, New York, N.Y., 1986), with permission.

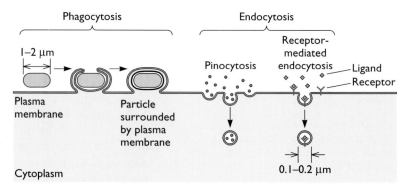

**A**

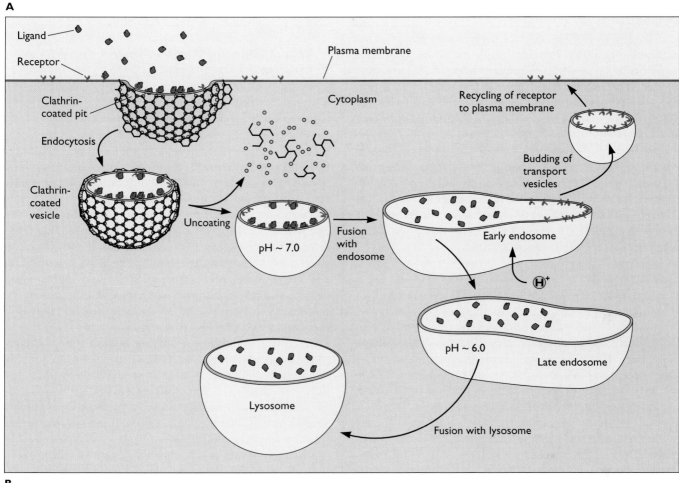

**B**

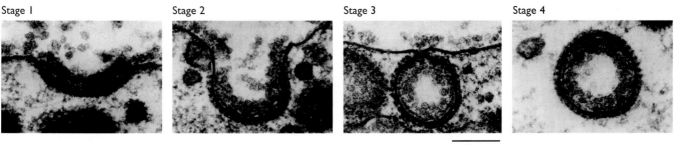

Stage 1          Stage 2          Stage 3          Stage 4

0.2 μm

**Figure 5.2 Receptor-mediated endocytosis.** (A) This process begins when a ligand binds a specific cell surface receptor. The ligand-receptor complex diffuses into an invagination of the plasma membrane coated with the protein clathrin on the cytosolic side (clathrin-coated pits; see below). The coated pit invaginates and pinches off, forming a coated vesicle. The coated vesicle then fuses with an early endosome. The early endosome is acidic, as a result of the activity of vacuolar proton ATPases. Receptors may be recycled to the cell surface from the early endosome by the budding off of transport vesicles. Transport of the contents of early endosomes to the late endosomes occurs by a microtubule-mediated process involving multivesicular bodies. The late endosomes then fuse with lyso-somes. Whether a receptor is recycled from early endosomes or is degraded in the lysosome varies according to the particular function of the receptor. For example, the low-density lipoprotein (Ldl) receptor is dissociated from its ligand, Ldl, in the early endosome and is recycled to the plasma membrane. Ldl is then carried to lysosomes. In contrast, the epidermal growth factor (Egf)-Egf receptor complex is transported to lysosomes, where both are degraded. (B) Clathrin-coated pits and formation of clathrin-coated vesicles. Electron micrographs of four sequential stages in the formation of clathrin-coated vesicles from clathrin-coated pits are shown. Adapted from G. M. Cooper, *The Cell: a Molecular Approach* (ASM Press, Washington, D.C., and Sinauer Associates, Sunderland, Mass., 1997).

early **endosomes.** The lumen of early endosomes is acidic, a result of energy-dependent transport of protons into the interior of the vesicles by a proton pump in the endosome membrane. Some ligands dissociate from their receptors in the acidic environment of the endosome, and the receptors are recycled to the cell surface by transport vesicles that bud from the endosome and fuse with the plasma membrane. The contents of the early endosome are then transported to the late endosomes located close to the nucleus. Late endosomes in turn fuse with **lysosomes**, which are vesicles that contain a variety of enzymes that degrade sugars, proteins, nucleic acids, and lipids. Ligands that reach the lysosomes are degraded by enzymes for further use of their constituents in the cell. In some cases, the entire ligand-receptor complex travels to the lysosomal compartment, where it is degraded.

## Cell Membranes Fuse during Endocytosis

The formation of vesicles during the process of endocytosis requires the fusion of cell membranes. For example, during endocytosis, fusion produces the intracellular vesicle following invagination of a small region of the plasma membrane (Fig. 5.2). Similar fusion reactions occur during phagocytosis and pinocytosis (Fig. 5.1). Membrane fusion also takes place during many other cellular processes, such as cell division, myoblast fusion, and exocytosis.

### Cell Membrane Fusion Is Regulated

Membrane fusion must be regulated in order to maintain the integrity of the cell and its intracellular compartments. Therefore, membrane fusion does not occur spontaneously but proceeds by specialized mechanisms mediated by proteins. The two membranes must first come into close proximity. This reaction is mediated by interactions of integral membrane proteins that protrude from the lipid bilayer, a targeting protein on one membrane and a docking protein on the other. The next step, fusion, requires an even closer approach of the membranes to within 1.5 nm of each other. This step depends on the removal of water molecules from the membrane surfaces, an energetically unfavorable process. In mammalian cells, a multisubunit protein complex called a **fusion protein machine** is thought to provide the energy required for such a close approach. The complex formed by targeting and docking proteins recruits additional protein components that lead to fusion of the two membranes. In mammalian cells, these proteins have been identified as *N*-ethylmaleimide-sensitive fusion protein (Nsf) and soluble Nsf attachment proteins (Snaps). Targeting and docking proteins are called Snap receptors, or Snares. After fusion occurs, the complex dissociates until needed once again. The individual components of the fusion protein

machine lack fusion activity. As a result, fusion can be regulated by assembling and disassembling the complex.

### Viral Membrane Fusion Proteins Have Been Identified

The precise mechanism by which lipid bilayers fuse is not well characterized, but the action of fusion proteins is believed to result in the formation of an opening called a **fusion pore,** allowing exchange of material across the membranes. Much of our understanding about membrane fusion reactions comes from studies using viral fusion proteins. Individual viral proteins that promote membrane fusion have been identified. Membrane fusion reactions catalyzed by these viral fusion proteins appear to be less complex than those mediated by cellular proteins, for in most cases a single viral gene product is sufficient. The sites at which viruses enter cells, the plasma membrane and the endosomal membrane, are relatively homogeneous. As a result, it is not necessary to accommodate extensive diversity in the fusion process. In contrast, fusion mediated by cellular proteins occurs in many different cellular compartments, necessitating a more flexible, complex fusion machinery. In addition, viral fusion proteins need not be recycled, as abundant quantities will be produced in the virus-infected cell. Cell fusion proteins are far less abundant and must be recycled, a requirement that is best accomplished by assembling and disassembling a multisubunit protein complex.

The membranes of enveloped viruses fuse with those of the cell as a first step in delivery of the viral nucleic acid into a host cell. Viral fusion may occur either at the plasma membrane or from within an endosome. The membranes of the virus and the cell are first brought into close contact by interaction of a viral glycoprotein with a cell receptor. The same viral glycoprotein, or a different viral integral membrane protein, then catalyzes fusion of the juxtaposed membranes. As we will see in the following sections, virus-mediated fusion must be regulated to prevent viruses from aggregating or to ensure that fusion does not occur in the incorrect cellular compartment. In some cases, fusogenic potential is masked until the fusion protein interacts with other integral membrane proteins. In others, low pH is required to expose fusion domains. The activity of fusion proteins may also be regulated by cleavage of a precursor. This requirement probably prevents activation of fusion potential during virus assembly: if fusion proteins were to become cleaved before assembly is complete, the mildly acidic compartments of the transport pathway might induce aberrant fusion reactions. Cleavage also generates the metastable states of viral glycoproteins that can subsequently undergo conformational rearrangements required for fusion activity.

## Mechanisms of Virus Entry into Cells

The entry of viruses into cells begins with attachment to a cell surface receptor, a process that has been discussed in chapter 4. The next step, **uncoating,** is the release of viral nucleic acid from its protective protein coat or lipid envelope, although in most cases the liberated nucleic acid is still complexed with viral proteins. In some cases, uncoating is relatively simple and is accomplished upon fusion of the viral membrane with the plasma membrane (Fig. 5.3). For other viruses, uncoating is a multistep process that may include passage through the endocytic pathway and even docking at the nuclear membrane (Fig. 5.3). In the following sections, we will consider examples of uncoating at different cellular locations to illustrate molecular mechanisms of entry and the viral and cellular proteins that regulate these processes.

## Uncoating at the Plasma Membrane

### Separate Attachment and Fusion Proteins

Many enveloped virus particles, including members of the family *Paramyxoviridae,* such as Sendai virus, measles virus, and respiratory syncytial virus, fuse directly with the plasma membrane at neutral pH. These virions bind to cell surface receptors via a viral integral membrane protein (Fig. 5.4A). Once the viral and cell membranes have been closely juxtaposed by this receptor-ligand interaction, fusion is induced by a second viral glycoprotein known as fusion protein (F), and the viral nucleocapsid is released into the cell cytoplasm.

F protein is a type I integral membrane glycoprotein (the amino terminus lies outside the viral membrane) which is synthesized as a precursor called F0. During tran-

**Figure 5.3 Three uncoating strategies.** Examples of genome uncoating at the plasma membrane (top), from within endosomes (middle), and at the nuclear membrane (bottom) are illustrated. In the first two cases, the viral genome is released into the cytoplasm, and in the last case, the genome is released into the nucleus. Not shown are viruses such as reovirus that penetrate the cytosol at the plasma membrane, after activation by proteases.

**Uncoating at the plasma membrane**

**Uncoating within endosomes**

**Uncoating at the nuclear membrane**

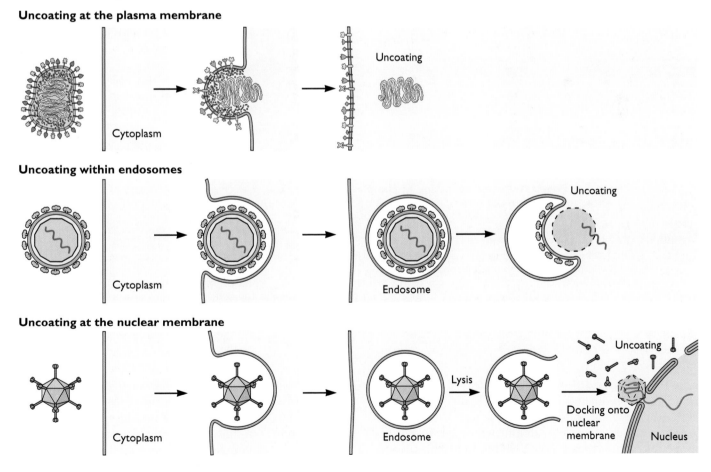

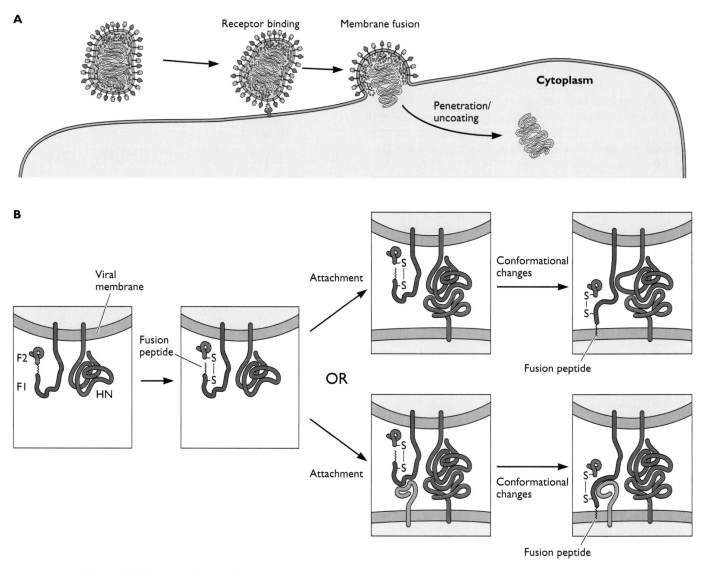

**Figure 5.4 Penetration and uncoating at the plasma membrane.** (A) Overview. Members of the *Paramyxoviridae* bind to cell surface receptors via HN, H, or G glycoprotein. The fusion protein (F) then catalyzes membrane fusion at the cell surface at neutral pH. The viral nucleocapsid, a ribonucleoprotein complex, is then released into the cytoplasm, where RNA synthesis begins. The mechanism by which contacts between the viral nucleocapsid and the M protein, which forms a shell beneath the lipid bilayer, are broken to facilitate release of the nucleocapsid into the cytoplasm is not known. (B) Models for F-protein-mediated membrane fusion. In one model, binding of HN to the cell receptor (red) induces conformational changes in HN which in turn induce conformational changes in the F protein, moving the fusion peptide from a buried position nearer to the cell membrane. In another model, interaction of HN with a cell receptor leads to the binding of F to a putative receptor (orange), which induces conformational changes in F, exposing the fusion peptide.

sit to the cell surface, this precursor is cleaved by a host cell protease to produce two subunits, F1 and F2, held together by disulfide bonds. The newly formed N-terminal 20 amino acids of the F1 subunit, which are highly hydrophobic, form a region called the fusion peptide because it is believed to insert into target membranes to initiate fu-

sion. Viruses containing the uncleaved F0 precursor can be produced in cells lacking the protease responsible for its cleavage. Such virus particles are noninfectious; they bind to target cells, but the viral genome does not enter. Thus, cleavage of the F0 precursor is necessary for fusion, presumably because the fusion peptide becomes available for insertion into the plasma membrane, and to generate the metastable state of the protein that can undergo conformational rearrangements needed for fusion (see "Acid-Catalyzed Membrane Fusion" below).

Because cleaved F-protein-mediated fusion can occur at neutral pH, it must be controlled, both to ensure that virus particles fuse with only the appropriate cell and to prevent aggregation of newly assembled virions. The fusion peptide of F1 is probably buried in the native protein. Conformational changes in F protein might move the fusion peptide from a buried position in the molecule to an exposed one that can contact the target membrane (Fig. 5.4B). Such movement of the fusion peptide in influenza virus hemagglutinin (HA) in response to acid pH has been described at the molecular level (see "Acid-Catalyzed Membrane Fusion" below). The results of experiments in which hemagglutinin-neuraminidase (HN) and F glycoproteins are expressed in cultured mammalian cells indicate that F protein lacks fusion activity if HN glycoprotein is not present. It has therefore been hypothesized that an interaction between HN and F proteins is essential for fusion. It is believed that binding of HN protein to its cellular receptor induces conformational changes in this protein, which in turn trigger conformational change in F protein, exposing the fusion peptide and making the protein fusion competent (Fig. 5.4B). The requirement for HN protein in F fusion activity has been observed only with certain paramyxoviruses, including Newcastle disease virus, human parainfluenza virus type 3, and mumps virus. In contrast, F protein of the paramyxovirus simian virus 5 apparently does not require HN protein for fusion activity. Conformational changes in simian virus 5 F glycoprotein needed to expose the fusion peptide might occur as the protein contacts the cell membrane or perhaps as it interacts with an as yet unidentified F protein receptor on the plasma membrane (Fig. 5.4B).

As a result of fusion of the viral and plasma membranes, the viral nucleocapsid, which is a ribonucleoprotein (RNP) consisting of the ($-$) strand viral RNA genome associated with the viral proteins L, NP, and P, is released into the cytoplasm (Fig. 5.4A). Once in the cytoplasm, L, NP, and P proteins begin the synthesis of viral mRNAs, a process that will be discussed in chapter 6. Because members of the *Paramyxoviridae* replicate in the cytoplasm, fusion of the viral and plasma membranes achieves uncoating and transport of the viral genome to the correct cellular compartment in a single step.

### A Single Viral Glycoprotein That Mediates Attachment and Fusion

Members of the *Retroviridae* also enter cells by fusion of the viral envelope with the plasma membrane, and there are interesting similarities to and differences from the *Paramyxoviridae*. Members of the subgroup A avian sarcoma and leukosis viruses (ASLVs) bind to the cellular receptor Tva via the viral envelope glycoprotein (Env-A) (Fig. 5.5A). The SU subunit of Env-A binds the cell receptor. The TM subunit, which is embedded in the viral envelope at its carboxy terminus, contains a fusion peptide at its amino terminus. The fusion of ASLVs differs from the fusion of paramyxoviruses in that the cell-binding and fusion activities of ASLVs are present in the same viral glycoprotein. However, both F protein of paramyxoviruses and Env-A protein of ASLVs must undergo proteolytic cleavage so that conformational rearrangements can occur, leading to exposure of the fusion peptide.

The cleavage of Env-A into SU and TM subunits, which occurs in the trans-Golgi network during virus maturation, is required for fusion. The fusion peptide is probably buried within the Env-A molecule. When Env-A binds Tva, conformational rearrangements of the glycoprotein lead to exposure of the fusion peptide, which can then catalyze fusion of the viral and cellular membranes (Fig. 5.5A). A similar mechanism is presumed to mediate fusion of other simple retroviruses.

### Virus-Cell Membrane Fusion That Requires a Cellular Coreceptor

The interaction of Env-A and Tva is believed to be sufficient to lead to virus-cell fusion at the plasma membrane. In contrast, fusion of human immunodeficiency virus type 1 with the plasma membrane requires participation not only of the cell receptor CD4 but also of an additional cellular protein. When human CD4 protein is expressed in cultured mouse cells, human immunodeficiency virus type 1 binds to the receptor but cannot fuse with the plasma membrane because mouse cells lack a cellular coreceptor required for fusion. The experimental strategies used to identify the first of these coreceptors started with cultured mouse cells engineered to express CD4; DNA that included candidate coreceptor genes was then introduced. Cells that expressed both CD4 and the coreceptor were selected on the basis of their ability to fuse (form syncytia) with other cells that expressed the viral envelope protein on their surface. The coreceptor gene was then cloned from the selected cells. Other, related human immunodeficiency virus type 1 coreceptors were identified and cloned within a few weeks of this first report. All the coreceptors are cell surface receptors for small molecules that many cells can produce to attract and stimulate immune defense cells at sites of infection; hence these small molecules are called **chemotactic cytokines** or **chemokines.** The chemokine receptors on such immune cells comprise a large family of proteins with seven membrane-spanning domains and are coupled to intracellular signal transduction pathways.

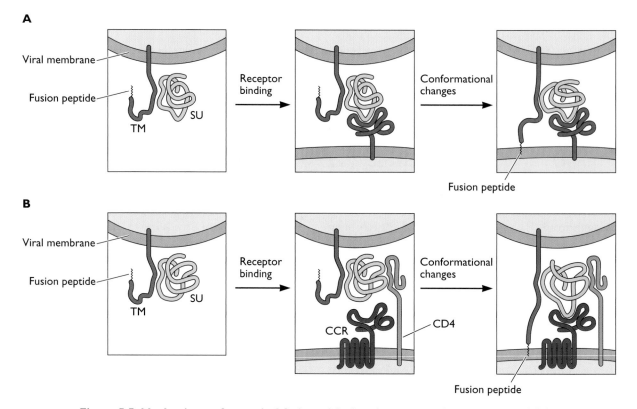

**Figure 5.5 Mechanisms of retroviral fusion with the plasma membrane.** (A) Model for fusion mediated by envelope (Env) glycoprotein of retroviruses. Env consists of two proteins that are cleaved from a single precursor: SU (yellow), which contains the receptor-binding site, and TM (olive), which contains the fusion peptide and the viral membrane-spanning segment. The fusion peptide is buried near the membrane surface. When SU binds the cell receptor (red), conformational changes are transmitted to TM, resulting in exposure of the fusion peptide. (B) Model of the role of chemokine receptors in human immunodeficiency virus type 1 fusion at the plasma membrane. For simplicity, the envelope glycoprotein is shown as a monomer, although trimer and tetramer forms have been reported. Binding of SU to CD4 is reported to be followed by cleavage of the V3 region by a cellular protease, a reaction that could effect a structural change. Binding of SU to CD4 exposes a high-affinity chemokine receptor-binding site on SU. The SU-chemokine receptor interaction leads to conformational changes in TM that expose the fusion peptide and permit it to insert into the cell membrane, catalyzing fusion in a manner similar to that proposed for influenza virus (cf. Fig. 5.7 and 5.8).

The first coreceptor for human immunodeficiency virus type 1 to be isolated was initially called **fusin** because it facilitated fusion of the viral envelope with the cell membrane. Its predicted structure (Fig. 5.6) resembles that of receptors for CXC chemokines (i.e., chemokines characterized as having their first two cysteines separated by a single amino acid). However, when first identified, fusin, which appears to be involved in lymphocyte migration, was considered an **orphan receptor**, as its natural ligand was unknown. This ligand was subsequently identified as stromal cell-derived protein (Sdf1); fusin was then given the "family" name of CXC chemokine receptor 4 (Cxcr4). Cxcr4 appears to be a specific coreceptor for human immunodeficiency virus type 1 strains that preferentially infect T cells and are thus T-cell-tropic.

Hints about the identity of a second coreceptor came from earlier observations that a specific subgroup of CC chemokines (CC chemokines are distinguished from CXC chemokines by having the first two cysteines adjacent) appeared to inhibit infection by macrophage-tropic strains of human immunodeficiency virus type 1. As a chemokine receptor that responded to this same chemokine subgroup

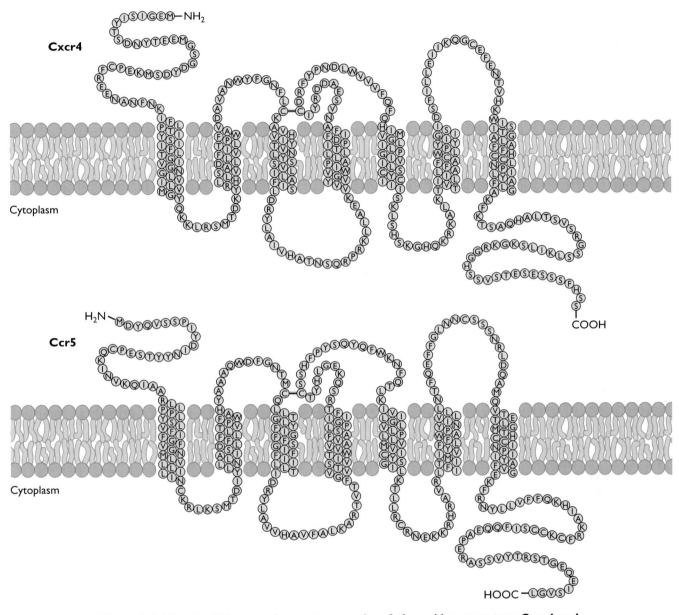

**Figure 5.6 Models of the membrane topography of chemokine receptors Cxcr4 and Ccr5.** Amino acids are shown as circles and identified by the one-letter code. The first amino-terminal extracellular domain of Ccr5 (bottom) is crucial for its function as an human immunodeficiency virus type 1 entry coreceptor. Adapted from J. F. Berson et al., *J. Virol.* **70:**6288–6295, 1996, with permission.

was already known, namely, the receptor Ccr5, its cloned DNA was tested directly and shown to express a coreceptor for the macrophage-tropic strains of the virus. The chemokines that bind to this receptor activate both T cells and macrophages, and the receptor is found on both types of cells. Interestingly, individuals who are homozygous for deletions in the Ccr5 gene and produce nonfunctional coreceptors appear to be resistant to infection with human

immunodeficiency virus type 1 but have no discernible immune function abnormality. Even heterozygous individuals seem to be somewhat resistant to the virus. Other members of the CC chemokine receptor family (i.e., Ccr2b and Ccr3) were subsequently found to serve as coreceptors for human immunodeficiency virus type 1.

Clues about how the chemokine receptors facilitate human immunodeficiency virus type 1 fusion come from experiments demonstrating that SU can bind to Ccr5 but that the binding affinity is far greater when SU is first bound to CD4. Attachment to CD4 appears to create a high-affinity binding site on SU for Ccr5. Resolution of the atomic structure of a complex of SU and CD4 revealed that the conformational changes induced in SU by CD4 binding expose binding sites on SU for chemokine receptors (Fig. 4.12). Studies of Ccr5 have shown that the first amino-terminal extracellular domain is crucial for cofactor function (Fig. 5.6), suggesting that this sequence might interact with SU.

Like subgroup A avian sarcoma and leukosis viruses, human immunodeficiency virus type 1 TM is likely to mediate envelope fusion with the cell membrane. The high-affinity SU-Ccr5 interaction may produce conformational changes in TM that expose the fusion peptide, placing it near the cell membrane, where it can catalyze fusion (Fig. 5.5B). Such changes might be similar to those that influenza virus hemagglutinin (HA) undergoes on exposure to low pH. X-ray crystallographic analysis of fusion-active human immunodeficiency virus type 1 TM reveals that its structure is strikingly similar to that of the low-pH fusogenic form of HA (see "Acid-Catalyzed Membrane Fusion" below).

## Virus Entry via the Endocytic Pathway

### Acid-Catalyzed Membrane Fusion

In contrast to fusion at plasma membranes, which is characteristic of many viral infections, other enveloped viruses undergo fusion within an endosomal compartment. The entry of influenza virus from the endosomal pathway is one of the best understood viral entry mechanisms. At the cell surface, the virus attaches to sialic acid-containing receptors via the viral HA glycoprotein (Fig. 4.16 and 5.7). The virus-receptor complex is then internalized by the clathrin-dependent receptor-mediated endocytic pathway. When the vesicular pH reaches approximately 5.0, in the late endosome, HA undergoes an acid-catalyzed conformational rearrangement, exposing a fusion peptide. The viral and endosome membranes then fuse, allowing penetration of the viral ribonucleoprotein (vRNP) into the cytoplasm.

The fusion reaction mediated by the influenza virus HA protein is a remarkable event when viewed at atomic resolution. The viral HA on the virion surface is a homotrimer in which each subunit consists of two disulfide-linked polypeptides, HA1 and HA2 (Fig. 5.8). These polypeptides are produced by proteolytic cleavage of the HA precursor during its transit to the plasma membrane (see chapter 12). The globular head (HA1) contains the sialic acid-binding pocket and is perched atop a fibrous

stem consisting largely of HA2. The fusion peptides, part of HA2, are buried in the subunit interface in the trimer stem, about 100 Å from the globular head of the molecule.

A comparison of the crystal structures of the native and low-pH forms of influenza virus HA reveals the dramatic molecular changes induced by acidic conditions in late endosomes (Fig. 5.8). In native HA, the fusion peptide is joined to the three-stranded coiled-coil core by a 28-amino-acid sequence that forms an extended loop structure buried deep inside the molecule. In contrast, in the low-pH HA structure, this loop region is transformed into a three-stranded coiled coil. In addition, the long α-helices of the coiled coil bend upward and away from the viral membrane. The result is that the fusion peptide has moved a great distance toward the endosome membrane (Fig. 5.8). Despite these dramatic changes, HA remains trimeric, and the globular heads can still bind sialic acid. In this conformation, HA holds the viral and endosome membranes 100 Å apart, too distant for the fusion reaction to occur. To bring the viral and cellular membranes closer, the top of the acid-induced coiled coil is believed to splay apart, spreading into the lipid bilayer (Fig. 5.7). The stems of the HA probably tilt with respect to the membrane, further enabling the membranes to come into close contact.

Cleavage of HA is a prerequisite for the low-pH-induced conformational change because it not only produces the free N terminus of HA2 containing the fusion peptide but also converts HA to a metastable state that can undergo the structural rearrangements that promote fusion. In contrast to cleaved HA, the precursor HA is stable at low pH and cannot undergo structural changes. How does cleavage of HA produce a metastable protein? Cleavage of the covalent bond between HA1 and HA2 might simply allow movement of the fusion peptide, which is restricted in the uncleaved molecule. Another possibility is suggested by the observation that cleavage of HA is accompanied by movement of the fusion peptide into the cavity in HA (Fig. 5.8). This movement buries ionizable residues of the fusion peptide, perhaps setting the low-pH "trigger." It should be emphasized that after cleavage the N terminus of HA2 is tucked into the hydrophobic interior of the trimer (Fig. 5.8). This rearrangement presumably buries the fusion peptide so that newly synthesized virions do not aggregate and lose infectivity.

When the core structure of influenza virus HA is compared with those of the Env proteins of two retroviruses, human immunodeficiency virus type 1 and Moloney murine leukemia virus, remarkable similarities emerge despite the diversity in mechanisms of fusion in these viruses (Fig. 5.9). In all three cases, the fusion peptides are presented to membranes on top of a three-stranded coiled coil. Such a scaffold might be a common feature of viral

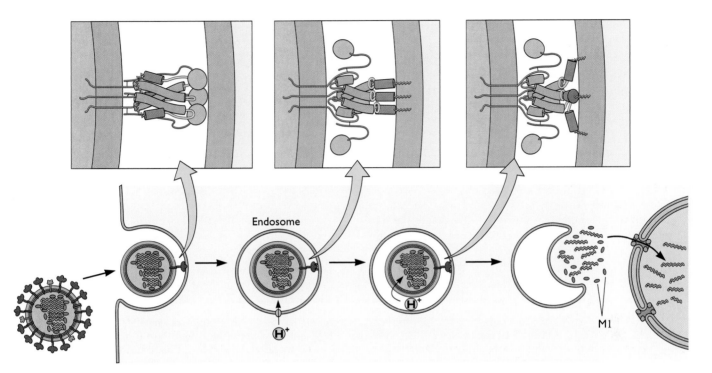

**Figure 5.7 Influenza virus entry.** The globular heads of native HA mediate binding of the virus to sialic acid-containing cell receptors. The virus-receptor complex is endocytosed, and import of H⁺ ions into the endosome acidifies the interior. Upon acidification, the viral HA undergoes a conformational rearrangement that produces a fusogenic protein. The loop region of native HA (yellow) becomes a coiled coil, moving the fusion peptides (red) to the top of the molecule near the cell membrane. At the viral membrane, the long α-helix (pink) packs against the trimer core, pulling the globular heads to the side. The long coiled coil splays into the cell membrane, bringing it closer to the viral membrane so that fusion can occur. Not shown is the tilting of HA that occurs. To allow release of vRNP into the cytoplasm, the H⁺ ions in the acidic endosome are pumped into the virion interior by the M2 ion channel. As a result, vRNP is primed to dissociate from M1 after fusion of the viral and endosome membranes. The released vRNPs are imported into the nucleus through the nuclear pore complex via a nuclear localization signal-dependent mechanism (see "Import of Influenza Virus RNP into the Cell Nucleus"). Adapted from C. M. Carr and P. S. Kim, *Science* **266:**234–236, 1994, with permission.

membrane fusion proteins: many have a region of high α-helical content and a 4-3 heptad repeat of hydrophobic amino acids, characteristic of coiled coils, next to the amino-terminal fusion peptide.

### Membrane Fusion Proceeds through a Hemifusion Intermediate

Because its fusogenic activity can be controlled by pH, the HA polypeptide has become a model for studying how proteins mediate membrane fusion. Fusion is believed to proceed through a hemifusion intermediate in which the outer leaflets of two opposing bilayers fuse (Fig. 5.10), followed by fusion of the inner leaflets and the formation of a fusion pore. Direct evidence that fusion involves a hemi-

fusion intermediate has been obtained with influenza virus HA (see Figure 5.10 for details). This experiment also demonstrates that the transmembrane domain of the HA polypeptide plays a role in the fusion process. The extracellular portion of the HA is believed to move horizontally in the cell membrane, perhaps in response to low pH (Fig. 5.10). The transmembrane domain also reorients, and this movement generates stress within the single lipid bilayer that has resulted from hemifusion. The stress is then relieved by the formation of a fusion pore. When HA attached to the membrane by linkage with glycosylphosphatidylinositol, instead of its normal transmembrane segment, is exposed to low pH, no stress occurs, and the area of the single lipid bilayer expands (Fig. 5.10).

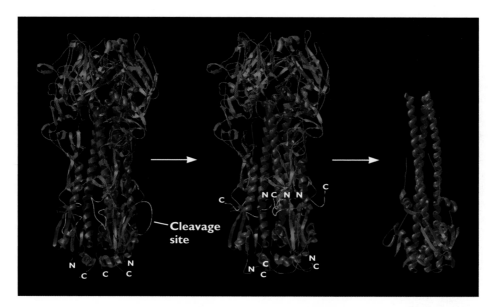

**Figure 5.8 Cleavage- and low-pH-induced structural changes in the influenza virus HA.** (Left) Structure of the uncleaved HA0 precursor. HA1 subunits are blue, HA2 subunits are red, residues 323 of HA1 to 12 of HA2 are yellow, and locations of some of the N and C termini are indicated. The viral membrane is at the bottom, and the globular heads are at the top. The transmembrane segment is not present because the HA was cleaved from cells expressing it with the protease trypsin. The cleavage site between HA1 and HA2 is in a loop adjacent to a deep cavity. (Middle) Structure of the HA0 trimer at neutral pH. The transmembrane segment is not present because the HA was cleaved from the virus with the protease bromelain. Cleavage of HA0 generates new N and C termini, which are separated by 20 Å. The N and C termini visible in this model are labeled. The cavity is now filled with residues 1 to 10 of HA2, part of the fusion peptide. (Right) Structure of the low-pH trimer. The protein used for crystallization was prepared from bromelain HA by further treatment with thermolysin and trypsin, and therefore the HA1 subunit and the fusion peptide are not present. This treatment is necessary to prevent aggregation of HA at low pH. At neutral pH the fusion peptide is close to the viral membrane, linked to a short α-helix, and at acid pH this α-helix is reoriented toward the cell membrane, carrying with it the fusion peptide. The structures are aligned on a central α-helix that is unaffected by the conformational change. Adapted from J. Chen et al., *Cell* **95**:409–417, 1998, with permission.

## Release of vRNP from the Viral Envelope

The genomes of many enveloped RNA viruses are present as vRNP in the virus particle. One mechanism for release of vRNP from the viral membrane during virus entry has been revealed by studies of influenza virus. Fusion of the influenza virus envelope with the endosome membrane exposes vRNPs to the cytoplasm. Each influenza virus vRNP is composed of a segment of the RNA genome complexed with nucleoprotein (NP) molecules at about 10- to 15-nucleotide intervals and the virion RNA polymerase. This complex interacts with viral M1 protein, an abundant virion protein that underlies the viral envelope and provides rigidity (Fig. 5.7). M1 protein is also believed to contact the internal tails of HA and NA proteins in the viral envelope. This arrangement presents two problems.

Unless M1-vRNP interactions are disrupted, vRNPs might not be released into the cytoplasm. Furthermore, the vRNPs must enter the nucleus, where transcription and replication occur. However, vRNP cannot enter the nucleus if M1 protein is bound because this protein masks a nuclear localization signal (see "Import of Influenza Virus RNP into the Cell Nucleus" below).

The influenza virus M2 protein, the first viral protein discovered to be an ion channel, probably provides the solution to both problems. The virion envelope contains a small number (14 to 68) of molecules of M2 protein, whose native form is a homotetramer. When expressed in *Xenopus laevis* oocytes or in cultured mammalian cells, the M2 protein can transport monovalent cations into the cell interior, a property of proteins with ion channel activity.

**Influenza virus**    **Human immunodeficiency virus type I**    **Moloney murine leukemia virus**

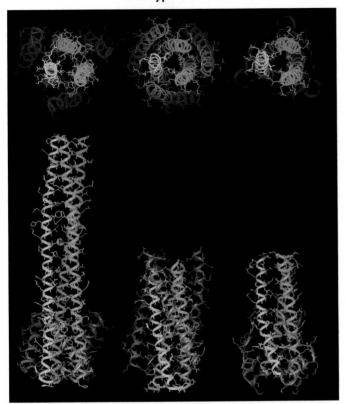

**Figure 5.9 Structures of influenza virus HA, human immunodeficiency virus type I TM, and Moloney murine leukemia virus TM.** (Top) View from the top of the three structures. (Bottom) Side view. The three monomers that form the central coiled coil of each are shown in yellow, green, and blue, and the supporting structures are purple. The structure shown for HA is the low-pH, or fusogenic form. Structures of retrovirus TM proteins are derived from interacting human immunodeficiency virus type 1 peptides and a peptide from Moloney murine leukemia virus and are presumed to represent the fusogenic forms because of structural similarity to HA. In all three molecules, fusion peptides would be located at the membrane-distal portion (the tops of the molecules in the bottom view). Despite the differences in the mechanism of membrane fusion among the three viruses (HA-mediated fusion is acid catalyzed, that of human immunodeficiency virus type 1 requires a cellular coreceptor, and that of Moloney murine leukemia virus occurs at the plasma membrane), they all present fusion peptides to cells on top of a central three-stranded coiled coil supported by carboxy-terminal structures. Adapted from D. C. Chan et al., *Cell* **89**:263–273, 1997, with permission.

Furthermore, when purified M2 was reconstituted into synthetic lipid bilayers, ion channel activity was also observed, indicating that this property requires only M2 protein. The M2 protein channel is structurally much simpler than other ion channels, which generally possess 24 membrane-spanning domains, and is the smallest channel discovered to date.

The M2 ion channel is activated by the low pH of the endosome before HA-catalyzed membrane fusion occurs. As a result of M2 ion channel activity, protons enter the interior of the virus. It has been suggested that the reduced pH of the virion interior leads to conformational changes in M1 protein, thereby disrupting M1-vRNP interactions. When fusion between the viral envelope and the endosomal membrane subsequently occurs, vRNPs are released into the cytosol free of M1 and can then be imported into the nucleus (Fig. 5.7). Support for this model comes from studies with the anti-influenza virus drug **amantadine,** which specifically inhibits M2 ion channel activity. In the presence of this drug, influenza virus can bind to cells, enter endosomes, and undergo HA-mediated membrane fusion, but vRNPs are not released from the endosomal membrane.

### Uncoating in the Cytoplasm by Ribosomes

By allowing protons to enter the virus as it passes through acidic endosomes, the M2 ion channel of influenza viruses promotes release of vRNPs from the endosomal membrane into the cytoplasm. Other enveloped RNA-containing viruses, such as Semliki Forest virus, contain nucleocapsids that must be disassembled in the cytoplasm to release viral RNA. The nucleocapsid of this virus is an icosahedral shell composed of a single viral protein, C protein, which encloses the positive-stranded viral RNA. This structure is surrounded by an envelope containing viral glycoproteins called E1 and E2. These glycoproteins are arranged as heterodimers clustered into groups of three, each cluster forming a spike on the virus surface.

After binding to its cell surface receptor, Semliki Forest virus is internalized by receptor-mediated endocytosis (Fig. 5.11). The low pH in endosomes induces conformational changes in the spike glycoprotein to promote virus fusion. Semliki Forest virus-mediated fusion exhibits several unusual features. For example, this process requires the presence of cholesterol in the cell membrane, whereas cholesterol is not required for fusion mediated by proteins of other viruses. Why cholesterol is needed for fusion is not understood. One notion is that it might interact directly with the E1 subunit. At low pH, the E1-E2 heterodimer spikes present in the viral envelope dissociate and the E1 protein molecules reassociate as trimers, which may be the fusion-active form. In contrast to the situation with other viruses, proteolytic cleavage of E1 is not required to produce a fusogenic protein. However, protein processing is used in another way to control fusion potential. In the endoplasmic reticulum E1 protein is associated

**Figure 5.10 Glycosylphosphatidylinositol-anchored influenza virus HA induces hemifusion.** (Left) Model of the steps of fusion mediated by wild-type HA. Cultured mammalian cells expressing wild-type HA are fused with erythrocytes containing two different types of fluorescent dye, one in the cytoplasm and one in the lipid membrane. Upon exposure to low pH, HA undergoes conformational change and the fusion peptide is inserted into the erythrocyte membrane. The green dye is transferred from the lipid bilayer of the erythrocyte to the bilayer of the cultured cell. The HA trimers tilt, causing reorientation of the transmembrane domain and generating stress within the hemifusion diaphragm. Fusion pore formation relieves the stress. The red dye within the cytoplasm of the erythrocyte is then transferred to the cytoplasm of the cultured cell. (Right) An altered form of HA was produced lacking the transmembrane and cytoplasmic domains, with membrane anchoring provided by linkage to a glycosylphosphatidylinositol (GPI) moiety. Upon exposure to low pH, the HA fusion peptide is inserted into the erythrocyte membrane, and green dye is transferred to the membranes of the mammalian cell. When the HA trimers tilt, no stress is transmitted to the hemifusion diaphragm because no transmembrane domain is present, and the diaphragm becomes larger. Fusion pores do not form, and there is no mixing of the contents of the cytoplasm, indicating that complete membrane fusion has not occurred. These results prove that hemifusion, or fusion of only the inner leaflet of the bilayer, can occur among whole cells. Adapted from G. B. Melikyan et al., *J. Cell Biol.* **131:**679–691, 1995, with permission.

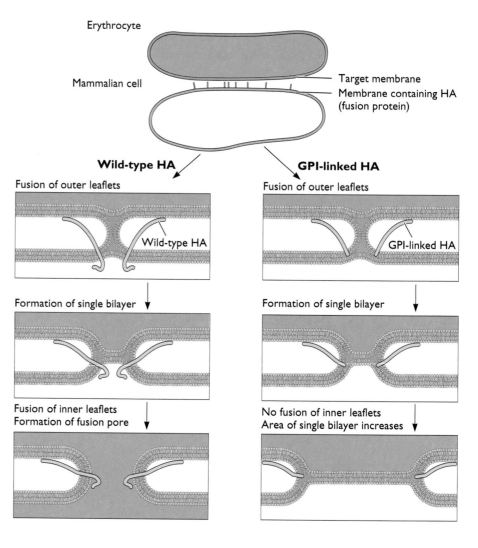

with the precursor of E2, called p62. In this heterodimer form, p62-E1, E1 protein cannot be activated for fusion by mildly acidic conditions. Only after p62 has been cleaved to E2 can low pH induce disruption of E1-E2 heterodimers and formation of fusion-active E1 homotrimers.

Fusion of the viral envelope and endosomal membrane exposes the nucleocapsid to the cytoplasm (Fig. 5.11). The viral RNA within this structure is sensitive to digestion with RNase, suggesting that the nucleocapsid is permeable. Crystallographic studies of the nucleocapsid of Sindbis virus, an alphavirus closely related to Semliki Forest virus, confirm the presence of holes ranging in diameter from 30

to 60 Å. These holes permit the entry of small proteins such as RNase (25 to 40 Å in diameter) into the nucleocapsid, but not of ribosomes, whose individual subunits are approximately 150 Å in diameter. To begin translation of (+) strand viral RNA, the nucleocapsid must be disassembled, a process mediated by an abundant cellular component—the ribosome. Each ribosome binds from three to six molecules of C protein, causing them to detach from the nucleocapsid. This process occurs while the nucleocapsid is attached to the cytoplasmic side of the endosome membrane (Fig. 5.11) and ultimately results in disassembly. The uncoated viral RNA remains associated with cel-

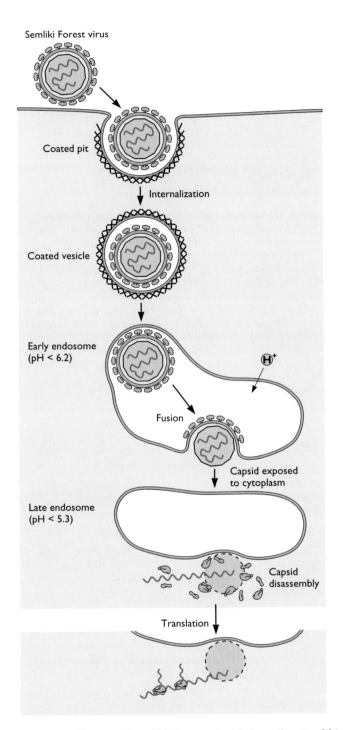

**Figure 5.11 Entry of Semliki Forest virus into cells.** Semliki Forest virus enters cells by clathrin-dependent receptor-mediated endocytosis, and membrane fusion is catalyzed by acidification of endosomes. Fusion results in exposure of the viral nucleocapsid to the cytoplasm, although the nucleocapsid remains attached to the cytosolic side of the endosome membrane. Cellular ribosomes then bind the capsid, disassembling it and distributing the capsid protein throughout the cytoplasm. The viral RNA is then accessible to ribosomes, which initiate translation. Adapted from M. Marsh and A. Helenius, *Adv. Virus Res.* **36:**107–151, 1989, with permission.

lular membranes, at which translation and replication begin. The mechanism by which the replication complex is anchored to the vesicle membrane is not known.

## Entry of Nonenveloped Viruses into Cells

Enveloped virus components enter cells by crossing cellular membranes when viral and cellular membranes fuse. How nonenveloped viruses enter cells is more difficult to envision because it is not obvious how naked viral capsids can cross the plasma membrane. Information on this step obtained from study of the members of several virus families has established that naked viruses gain entry to the cell by one of several mechanisms.

### Disrupting the Endosome Membrane

Adenoviruses are composed of a double-stranded DNA genome packaged in an icosahedral capsid made up of at least 10 structural proteins, as described in chapter 3. At the start of infection, viral fiber proteins attach to an integrin or immunoglobulin (Ig)-like cell surface receptor. Next, the penton base, to which the fiber is attached, interacts with a second cell receptor, the integrins $\alpha_v\beta_3$ and $\alpha_v\beta_5$. For most adenovirus serotypes, this interaction is required for subsequent internalization of the virus by receptor-mediated endocytosis. As the virus is transported through the endosomes from the cell surface toward the nuclear membrane, it undergoes multiple uncoating steps by which structural proteins are sequentially eliminated (Fig. 5.12). In the endosome, the fiber, protein IIIa, and protein VIII are dissociated from the capsid. As the endosome becomes acidified, the low pH causes release of the penton base and the facet-stabilizing protein IX. The penton base is believed to mediate lysis of the endosome membrane, thereby releasing the remainder of the virus into the cytoplasm. Protein VI is proteolytically degraded in the acidic endosome, probably by the viral protease L3/p23. This protease, which is present in the core of the virus particle, is inactive in the virion, possibly because of intrachain disulfide bonds that form in the oxidizing extracellular environment. The protease is activated on entry into the reducing environment of the endosome. Furthermore, interaction of the penton base with integrin receptors is believed to lead to a conformational change in the penton that exposes L3/p23 cleavage sites on protein VI. The capsid, now devoid of most of its components except viral DNA, the core proteins, and a small amount of hexon protein, then docks onto the nuclear pore complex (Fig. 5.12). The nucleoprotein is released into the nucleus, where viral transcription begins.

The entire process of adenovirus entry, from the beginning of endocytosis to the release of DNA into the nucleus, takes approximately 30 min at 37°C (Fig. 5.12). During this

**A**

Cell surface binding

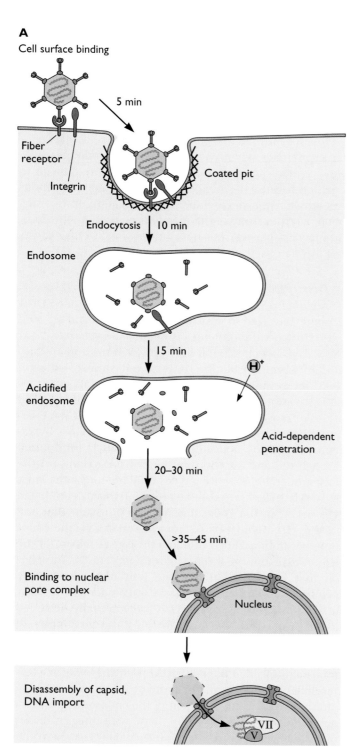

**B**

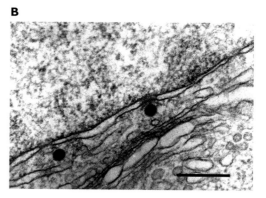

**Figure 5.12 Stepwise uncoating of adenovirus during cell entry.** (A) Adenoviruses bind a cell receptor via the fiber protein. Interaction of the penton base with an integrin receptor leads to internalization by endocytosis. Viral capsid proteins are sequentially removed in the endosome and the acidified endosome, leading to penetration of the capsid into the cytoplasm. The capsid then docks onto the nuclear pore complex, and the viral DNA is delivered into the nucleus. Adapted from U. F. Greber et al., *Cell* **75:**477–486, 1993, with permission. (B) Electron micrograph of adenovirus type 2 bound to the cytoplasmic face of the nuclear pore complex. Bar = 200 nm. Reprinted from U. F. Greber et al., *Trends Microbiol.* **2:**52–56, 1994, with permission. Courtesy of Ari Helenius, Urs Greber, and Paul Webster, University of Zurich.

time, the virus undergoes a series of obligatory sequential changes in which the stabilizing interactions of the capsid are broken down. Some of these changes are induced by the low pH of the endosome; some are caused by the action of the viral protease; and others, such as the release of the fiber, might result from interaction of the capsid with cell receptors. Adenovirus crosses the cell membrane by using an acid-activated membrane lytic protein, probably the penton base, to disrupt endosomal membranes.

### Forming a Pore in the Cell Membrane

A different mechanism is used to transfer the genome of the nonenveloped picornaviruses across the cell membrane. Structural information at the atomic level, and complementary genetic and biochemical data obtained from studies of cell entry, are available for these viruses. The interaction of poliovirus with its immunoglobulinlike cell receptor, Pvr, leads to major conformational rearrangements in the virus particle (Fig. 5.13A). The resulting altered (A) particles have lost the internal capsid protein VP4 and have externalized the normally internal amino terminus of capsid protein VP1. As a result of the latter change, altered particles are hydrophobic and possess an increased affinity for membranes compared to the native virus particle. It is believed that the exposed lipophilic amino terminus of VP1 inserts into the cell membrane, forming a pore that allows transport of viral RNA into the cytoplasm (Fig. 5.13B). In support of this model, ion channel activity can be detected when A particles are added to lipid bilayers.

The fate of VP4 is not known, but the study of a virus containing a mutation in VP4 indicates that this protein is required for an early stage of cell entry. Mutant virus par-

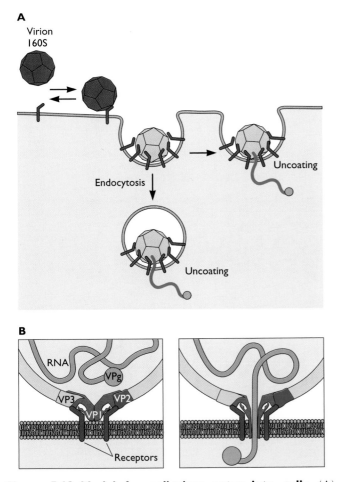

**Figure 5.13 Model for poliovirus entry into cells.** (A) Overview. The native virion (160S) binds to its cell receptor, Pvr, and at temperatures higher than 33°C undergoes a receptor-mediated conformational transition resulting in the formation of altered (A) particles. The A particles have lost the internal viral protein VP4, and the hydrophobic amino terminus of VP1, also normally internal, is displaced to the virion surface. The A particle may be an intermediate in the uncoating of the viral RNA. It is not known whether the viral RNA, shown as a curved green line, leaves the capsid at the plasma membrane or from within endosomes. (B) Model of the formation of a pore in the cell membrane after poliovirus binding. The interaction of poliovirus with Pvr leads to the formation of A particles in which the hydrophobic N termini of VP1 are extruded. These termini might form a pore in the cell membrane through which the RNA is released from the capsid into the cytosol. Note the location of the hydrophobic pocket, which may be occupied by sphingosine in poliovirus type 1. Adapted from B. N. Fields et al. (ed.), *Fields Virology* (Lippincott-Raven Publishers, Philadelphia, Pa., 1996), with permission.

ticles can bind to target cells and convert to altered particles but are blocked at a subsequent, unidentified step in virus entry. During poliovirus assembly, VP4 and VP2 are part of the precursor VP0, which remains uncleaved until the viral RNA has been encapsidated. The cleavage of VP0

during poliovirus assembly therefore primes the capsid for uncoating by separating VP4 from VP2.

Whether the altered particle is an obligatory uncoating intermediate (Fig. 5.13A) has been called into question by the result of experiments showing that poliovirus can replicate at 25°C, a temperature at which conversion to altered particles does not occur. These findings suggest the possibility that the altered particle detected at higher temperatures may be a stable end product that is an exaggerated form of the uncoating intermediate. The latter may be short-lived and consequently difficult to detect.

It is not known whether the release of poliovirus RNA occurs at the plasma membrane or from within endosomes (Fig. 5.13A). Bafilomycin A1, a drug that prevents the acidification of endosomes by inhibiting the transport of protons into the vesicles, has no effect on poliovirus infection. In contrast, this drug completely inhibits infection of cells with influenza virus, Semliki Forest virus, and vesicular stomatitis virus, all of which penetrate the cytosol from acidified endosomes. If poliovirus does enter the cell by endocytosis, this pathway is not sufficient to trigger uncoating. For example, antibody-poliovirus complexes can bind to cells that express Fc receptors but cannot infect them. As the Fc receptor is known to be endocytosed, these results suggest that interaction of poliovirus with Pvr is required to induce conformational changes in the particle required for uncoating.

Some picornaviruses enter cells by a pH-dependent pathway, but precisely how acidic conditions lead to uncoating of these viruses in the cytoplasm is not understood. For example, foot-and-mouth disease virus is believed to enter cells by receptor-mediated endocytosis. Uncoating is then triggered by acidification of the endosome. Consistent with this entry mechanism, antibody-coated foot-and-mouth disease virus can bind to and infect cells that express Fc receptors, in contrast to findings with poliovirus. This result suggests that the cell receptor for foot-and-mouth disease virus is a "hook": it serves only to tether the virus to the cell and bring it into the endocytic pathway and does not induce uncoating-related changes in the virus particle.

### Lysosomal Uncoating

Most viruses that enter cells by receptor-mediated endocytosis leave the pathway before the vesicles reach the lysosomal compartment. This is not surprising, for lysosomes contain proteases and nucleases that would degrade virus particles. However, members of the *Reoviridae* take advantage of these enzymes in their uncoating process, which takes place in lysosomes.

Reovirus is a naked icosahedral virus containing a double-stranded RNA genome of 10 segments. The viral capsid

**Figure 5.14 Entry of reovirus into cells.** (A) The different stages in cell entry of reovirus. After the attachment of σ1 protein to the cell receptor, the virus particle enters the cell by receptor-mediated endocytosis. Proteolysis in the late endosome produces the infectious subviral particle (ISVP), which may then cross the lysosome membrane and enter the cytoplasm as a core particle. The intact virion is composed of two concentric, icosahedrally organized protein capsids. The outer capsid is made up largely of σ3 and μ1 and stabilized by σ3-μ1, λ2-σ3, and λ2-μ1 interactions. The dense core shell is formed mainly by λ1 and σ2. In the ISVP, 600 σ3 subunits have been released by proteolysis, and the σ1 protein changes from a compact form to an extended flexible fiber, possibly as a result of the release of σ3. The μ1 protein is believed to mediate interaction of the ISVP with membranes. This protein is present in virions as two cleaved fragments, μ1N and μ1C (see schematic of μ1 in panel B). The amino terminus of μ1N is modified with myristate, suggesting that the protein functions in the penetration of membranes. A pair of amphipathic α-helices flank a carboxy-terminal trypsin/chymotrypsin cleavage site at which μ1C is cleaved by lysosomal proteases. Such cleavage may release the helices to facilitate membrane penetration. The membrane-penetrating potential of μ1C in the virion may be masked by σ3; release of the σ3 in ISVPs might then allow μ1C to interact with membranes. The core is produced by the release of 12 σ1 fibers and 600 μ1 subunits. In the transition from ISVP to core, domains of λ2 rotate upward and outward to form a turretlike structure. (Insets) Close-up views of the emerging turretlike structure as the virus progresses through the ISVP and core stages. This structure may facilitate the entry of nucleotides into the core and the exit of newly synthesized viral mRNAs. (B) Schematic of the σ protein, showing locations of myristate, protease cleavage sites, and amphipathic α-helices. Virus images reprinted from K. A. Dryden et al., *J. Cell Biol.* **122:**1023–1041, 1993, with permission. Courtesy of Norm Olson and Tim Baker, Purdue University.

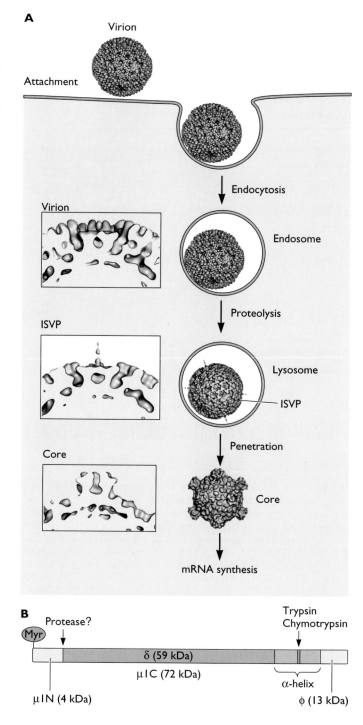

is a double-shelled structure composed of eight different structural proteins. These viruses bind to cell receptors via protein σ1 and are internalized into cells via endocytosis (Fig. 5.14A). Infection of cells by reoviruses is sensitive to bafilomycin A1, indicating that these viruses require acidification of endosomes to enter cells. Low pH activates lysosomal proteases, which then modify several virion proteins, enabling the virus to cross the vesicle membrane. One viral outer capsid protein is cleaved and another is removed from the particle, resulting in what is known as an infectious subviral particle. These subviral particles penetrate the lysosome membrane and escape into the cytosol by a mechanism that is not yet understood. Isolated infectious subviral particles cause cell membranes to become permeable to toxins and produce pores in artificial membranes. These particles can initiate an infection by penetrating the plasma membrane, entering the cytoplasm directly. Their infectivity is not sensitive to bafilomycin A1, further supporting the idea that these particles are primed for membrane entry and do not require further acidification for this process. The core particles generated from in-

fectious subviral particles after penetration into the cytoplasm mediate viral mRNA synthesis.

Cryoelectron microscopy has established the structural basis for some of the entry steps of reovirus (Fig. 5.14A). These studies indicate that this virus is designed to undergo a series of sequential disassembly steps in which layers of the capsid are removed, ultimately resulting in entry of the core into the cytoplasm and initiation of viral mRNA synthesis. The strategy is similar to that used by adenovirus, another nonenveloped virus, although the molecular mechanisms differ significantly.

### Endocytosis via Caveolae to the Endoplasmic Reticulum

The majority of simian virus 40 particles enter cells by endocytosis into small, uncoated vesicles, although a small fraction are taken up into more conventional large, coated vesicles. Approximately one-third of cell-associated virus then enters the endoplasmic reticulum. The small vesicles resemble **caveolae,** which are membranous vesicles that are believed to pinch off from the plasma membrane and migrate to the endoplasmic reticulum. Caveolae can be distinguished from clathrin-coated vesicles by their size and shape. They contain a marker protein called caveolin. According to this hypothesis, simian virus 40 might enter cells via a novel pathway for targeting molecules from the cell surface to the endoplasmic reticulum.

If simian virus 40 entry via the endoplasmic reticulum results in productive infection, an interesting question is how virus particles reach the nucleus, the site of replication of this virus. In the endoplasmic reticulum, virus particles would not have access to cytoplasmic import proteins or the cytoplasmic side of the nuclear pore complex (see "The Cellular Pathway of Protein Import into the Nucleus" below). There may be a specific mechanism by which virus crosses either the membrane of the endoplasmic reticulum or the outer nuclear membrane to enter the cytoplasm so that it can gain access to the nuclear import machinery. Alternatively, virus might cross the inner nuclear membrane and enter the nucleoplasm directly. This mechanism would be unique, for such a pathway of entry into the nucleus has never been described.

## Cellular Molecules That Participate in Virus Uncoating

### Uncoating of Human Immunodeficiency Virus Type 1 Requires Cyclophilin A

There is evidence that members of the cyclophilin family of proteins play an important role in the changes in the human immunodeficiency virus type 1 core that occur after it is released at the plasma membrane. Members of this family are chaperones, facilitating protein folding and assembly, and possess peptidylprolyl isomerase activity. Cyclophilin A is incorporated into human immunodeficiency virus type 1 virions at a stoichiometry of approximately 1 molecule to 10 molecules of the capsid protein CA. When cyclophilin A is excluded from virions, for ex-

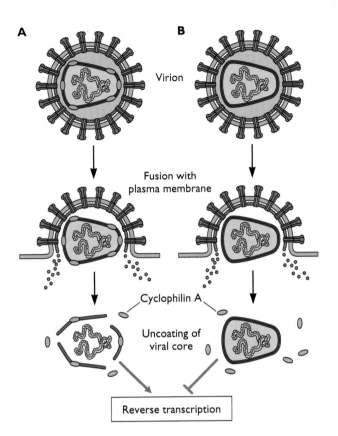

**Figure 5.15 Model of the role of cyclophilin A in human immunodeficiency virus type 1 uncoating.** The pathways of infection for mature virus particles containing (A) and lacking (B) cyclophilin A are depicted. After fusion at the plasma membrane, the viral nucleocapsid is released into the cytoplasm. Cyclophilin A (orange) packaged into the virus particle destabilizes the nucleocapsid, leading to release of CA and the initiation of reverse transcription. Adapted from J. L. Luban, *Cell* **87:**1–20, 1996, with permission.

ample, by changing the sequence of CA that binds to cyclophilin A, the virus particles made are morphologically and biochemically normal and can enter cells. However, they are not infectious because of a block early in replication before reverse transcription takes place (Fig. 5.15). The presence of cyclophilin A in infected cells does not rescue infectivity of the mutant virus.

A model has been proposed for the function of cyclophilin A in human immunodeficiency virus type 1 replication. Shortly after the virus fuses with the plasma membrane to release the core into the cytoplasm, this structure may need to expand to facilitate reverse transcription. This expansion may be caused by the release of some of the CA subunits. It has been suggested that cyclophilin A acts as a chaperone, a protein that facilitates the folding of other proteins, and assists in the release of CA. When cyclophilin A is not present in the virus particle, the core does not expand, a property that may explain why reverse transcription cannot begin (Fig. 5.15). If correct, this mechanism would identify a unique, novel function for chaperones. These proteins are best known for their roles in facilitating protein folding or assembly, not disassembly.

Examination of the crystal structure of cyclophilin bound to the amino-terminal 151 residues of the CA protein (called $CA_{151}$) has identified features consistent with a function for cyclophilin in uncoating of the human immunodeficiency virus type 1 core. Human immunodeficiency virus type 1 $CA_{151}$ is a triangular pyramid with a high content of α-helices (Fig. 5.16A). One of the few parts of $CA_{151}$ that does not form an α-helix is a loop believed to be exposed on the surface of the virion core. This loop contains a proline-rich region required for binding cyclophilin A and is the only part of CA protein that contacts cyclophilin A in cocrystals of the two molecules. Substitution of residues G89 and P90 of CA abrogates binding to cyclophilin A. These amino acids are located at the top of the loop in $CA_{151}$, buried deep within the hydrophobic pocket of cyclophilin A. A model for the organization of CA within the virion core suggests that "strips" composed of two rows of CA molecules interact side to side to form a two-dimensional surface (Fig. 5.16B). In this model, the N termini of all the CA molecules point upward toward the viral membrane. The presence of a molecule of cyclophilin A for each 10 molecules of CA protein might introduce minor dislocations between neighboring strips of CA molecules, thereby destabilizing the core and facilitating disassembly. This mechanism of disassembly would be specific for human immunodeficiency virus type 1 because other retroviruses do not contain cyclophilin A.

### Cellular Lipids Regulate Poliovirus Uncoating

A critical regulator of the receptor-induced structural transitions of poliovirus appears to be a hydrophobic tunnel located below the surface of each structural unit (Fig. 4.14). In poliovirus type 1, each tunnel is occupied by a natural ligand believed to be a molecule of sphingosine (Fig. 5.17). Rhinoviruses 1A and 16 each also contain a lipid, although they are shorter than the sphingosine of poliovirus. Because of the symmetry of the capsid, each virion may contain up to 60 lipid molecules.

The lipids are believed to contribute to the stability of the native virus particle, possibly by locking the capsid in a stable conformation and preventing conformational changes. Removal of the lipid is probably necessary to endow the particle with sufficient flexibility to permit the RNA to leave the shell. These conclusions come from the study of antiviral drugs known as WIN compounds that displace the lipid and fit tightly in the hydrophobic tunnel (Fig. 4.14). Polioviruses containing bound WIN compounds can bind to

the cell receptor, but altered particles are not produced. WIN compounds may therefore inhibit poliovirus infectivity by preventing the receptor-mediated conformational alterations required for uncoating. The properties of poliovirus mutants that cannot replicate in the absence of WIN compounds further support the role of the lipids in uncoating. These drug-dependent mutants spontaneously convert to altered particles at 37°C, in the absence of the cell receptor, probably because they do not contain lipid in the hydrophobic pocket. The lipids are therefore viewed as switches whose presence or absence determines whether the virus is stable or will uncoat. The interaction of the virus with receptor probably initiates structural changes in the virion that lead to the release of lipid. Consistent with this hypothesis is the observation that Pvr docks onto the poliovirus capsid just above the hydrophobic pocket (see section in chapter 4 entitled "Attachment via surface features: canyons and loops").

## Transport of the Viral Genome into the Nucleus

Most DNA viruses, retroviruses, influenza viruses, Borna disease virus, and hepatitis delta satellite virus begin replication in the cell nucleus. The genomes of these viruses must therefore be transported from the cytoplasm across the nuclear membrane. This movement is accomplished via the cellular pathway for protein import into the nucleus.

Although the discussions in this chapter concern use of the nuclear pore complex to enter the nucleus, it should be noted that this portal is also used for export of cellular and viral mRNAs. For viruses that replicate in the cell nucleus, the nuclear pore complexes mediate the transport of nucleoproteins into the cytoplasm for the final steps of assembly. These events are discussed in chapters 10 and 12.

## The Cellular Pathway of Protein Import into the Nucleus

### Nuclear Localization Signals

Proteins that reside within the nucleus are characterized by the presence of specific nuclear targeting signals. Such signals, termed **nuclear localization signals,** are both necessary for nuclear localization of the proteins in which they are present and sufficient to direct heterologous nonnuclear proteins to enter the nucleus. Nuclear localization signals identified by these criteria share a number of common properties: they are generally fewer than 20 amino acids in length; they are not found at the same position within all nuclear proteins, they are not removed after entry of the protein into the nucleus, and they are usually rich in positively charged (basic) amino

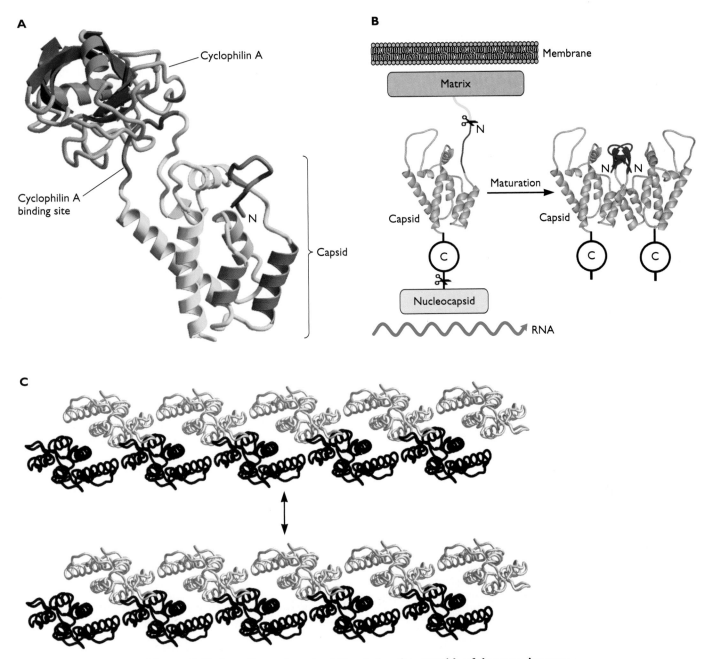

**Figure 5.16 Location of cyclophilin A in the capsid of human immuno-deficiency virus type 1.** (A) Interaction between CA$_{151}$ and cyclophilin A. A ribbon diagram of the CA$_{151}$/cyclophilin A complex is shown. Cyclophilin A is at the upper left, bound to the loop between helices IV and V of CA. Cyclophilin A β-strands are in red, and α-helices are in green. (B and C) Model of CA$_{151}$ interactions in the human immunodeficiency virus type 1 particle. (B) Schematic of the arrangements of MA, CA, NC, and the viral RNA. The viral membrane is shown at the top. The structure of CA$_{151}$ is shown as a ribbon diagram; the remainder of the molecule, whose structure has not been determined is labeled C. (Right) The association of two CA$_{151}$ molecules. (C) Higher-order interaction of CA to form a shell beneath the viral envelope. The cyclophilin A-binding loop of CA projects upward from the page. Adapted from T. R. Gamble et al., *Cell* **87:**1285–1294, 1996, with permission. Courtesy of Theresa Gamble, Wesley Sundquist, and Christopher Hill, University of Utah.

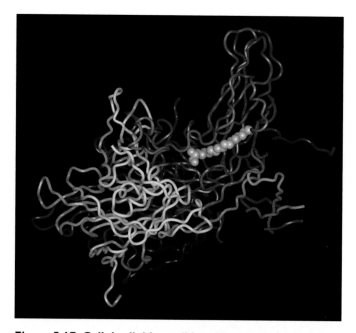

**Figure 5.17 Cellular lipid, possibly sphingosine, bound in the hydrophobic pocket of poliovirus type 1.** A single structural unit, composed of one copy each of VP1 (blue), VP2 (yellow), VP3 (red), and VP4 (green), is shown from the side. The RNA interior is below the structural unit. Shown as gray spheres is a molecule which is found in the hydrophobic tunnel in poliovirus type 1 and which has been modeled as the lipid sphingosine. This lipid is located just beneath the canyon floor.

ularly well-characterized example of a simple nuclear localization signal is that of simian virus 40 large T antigen, which comprises five contiguous basic residues flanked by a single hydrophobic amino acid (Fig. 5.18). This sequence is sufficient to relocate the enzyme pyruvate kinase, normally found in the cytoplasm, to the nucleus. Many other viral and cellular nuclear proteins contain short, basic, nuclear localization signals, but these signals are not identical in primary sequence to the T-antigen nuclear localization signal (Table 5.1). The presence of a nuclear localization signal is all that is needed to target a macromolecular substrate for import into the nucleus. Even gold particles with diameters as large as 26 nm are readily imported following their microinjection into the cytoplasm as long as they are coated with proteins or peptides containing a nuclear localization signal.

### The Nuclear Pore Complex

The nuclear envelope is composed of two typical lipid bilayers separated by a luminal space (Fig. 5.19A). Like all other cellular membranes, it is impermeable to macromolecules such as proteins. However, the nuclear pore complexes that stud the nuclear envelopes of all eukaryotic cells (Fig. 5.19A) provide aqueous channels that span both the inner and outer nuclear membranes for exchange of small molecules, macromolecules, and macromolecular assemblies between nuclear and cytoplasmic compartments. Numerous experimental techniques, including direct visualization of gold particles attached to proteins or RNA molecules as they are transported, have established that nuclear proteins enter and RNA molecules exit the nucleus by transport through the nuclear pore complex. The func-

acids. In spite of these similarities, no consensus nuclear localization sequence can be defined.

Most nuclear localization signals belong to one of two classes, simple or bipartite sequences (Fig. 5.18). A partic-

**Figure 5.18 Nuclear localization signals.** The general form and a specific example of simple and bipartite nuclear localization signals are shown in the one-letter amino acid code, where X is any residue. The sequences shown were defined by the criteria described in the text. Furthermore, specific substitutions, such as of K129 of the simian virus 40 (SV40) T antigen nuclear localization signal, prevent nuclear accumulation of the protein. As the name implies, bipartite nuclear targeting signals are defined by the presence of two clusters of positively charged amino acids separated by a spacer region of variable sequence. Both clusters of basic residues, which often resemble the simple targeting sequences of proteins like SV40 T antigen, are required for efficient import of the proteins in which they are found. The subscript indicates either length (3–7) or composition (e.g., 3/5 means at least 3 residues out of 5 are basic).

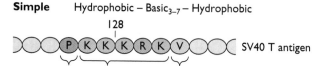

**Simple**   Hydrophobic – Basic$_{3-7}$ – Hydrophobic

SV40 T antigen

**Bipartite**   Basic$_{2/2}$ – X$_{10}$ – Basic$_{3/5}$

Nucleoplasmin

**Table 5.1** Nuclear localization signals of some viral proteins

| Virus | Protein | Nuclear targeting signal | |
|---|---|---|---|
| | | Position (residues)[a] | Sequence |
| **Structural proteins** | | | |
| Simian virus 40 | VP1 | 1–11 | APTKRK[6] |
| | VP2[b] | 316–324 | GPNKKKRKL |
| Adenovirus type 2 | Fiber | 1–11 | MKRARPSEDTF |
| | Pre-TP | 362–373 | RLPVRRRRRRVP |
| Influenza virus | M1 | 101–105 | RKLKR |
| | NP | 327–445 | AAFEDLRVRS[345] |
| | | | |
| **Nonstructural proteins** | | | |
| Simian virus 40 | Large T antigen | 126–132 | PKKKRKV |
| Adenovirus type 2 | EIA 13S | 285–289 | KRPRP |
| | DNA polymerase | 1–48 | Bipartite[c] |
| Epstein-Barr virus | EBNA-1 | 379–386 | LKRPRSPSS |
| Human immunodeficiency virus type 1 | Tat | 48–52 | GRKKR |
| | Rev | 38–45 | RRNRRRRW |
| Influenza virus | NS1 | | SXGTKRSYXXM and TKRSXXXM |

[a]Position indicates the region of the protein shown to contain a nuclear localization signal. When the sequences listed in column 4 are part of a large region shown in column 3, they were generally identified by similarity to another nuclear localization signal rather than actually demonstrated to be a nuclear localization signal.

[b]This sequence is also present in the VP3 protein.

[c]Bipartite nuclear localization signals contain two basic sequences (Fig. 5.18).

tions of the nuclear pore complex in both protein import and RNA export are far from completely understood, not least because this important cellular machine is large (molecular mass, approximately $124 \times 10^6$ Da in vertebrates), built from many different proteins (Table 5.2), and architecturally complex (Box 5.1 and Fig. 5.19B). In comparison, ribosomes, which consist of ~82 proteins and four RNA molecules, have a molecular mass of $4.2 \times 10^6$ Da.

### The Nuclear Import Pathway

Import of a protein into the nucleus occurs in two distinct, and experimentally separable, steps. A protein containing a nuclear localization signal first binds to a soluble cytoplasmic receptor protein. This complex then engages with the cytoplasmic surface of the nuclear pore complex, in a reaction often called docking, and is translocated through the nuclear pore complex into the nucleus. The former process is energy-independent, but the latter consumes energy. In the nucleus, the complex is disassembled, releasing the cargo into the nucleus. Our current, but still incomplete, model of the nuclear import pathway and its components is summarized in Figure 5.20.

Different groups of proteins are imported into the nucleus by specific nuclear import receptor systems. In the classical system of import, proteins containing basic nuclear localization signals are initially recognized by a cytoplasmic nuclear localization signal receptor. The best-studied receptor is **importin-α, or karyopherin α** (Fig. 5.20). This complex then binds **importin-β (karyopherin β)**, which mediates docking with the nuclear pore complex, by binding to members of a family of nucleoporins (Table 5.2). Some of these nucleoporins are found in the cytoplasmic filaments of the nuclear pore complex (Fig. 5.19B), with which import substrates are seen to be associated in the electron microscope. These properties suggest that cytoplasmic filaments are the site of the initial association of receptor-substrate complexes with the nuclear pore complex. A monomeric receptor called **transportin** mediates import of heterogeneous nuclear RNA-binding proteins that contain glycine- and arginine-rich nuclear localization signals. Transportin is related to importin β, as are other monomeric receptors involved in the nuclear import of ribosomal proteins.

Translocation requires additional transport proteins, in most cases including a small GTP-binding protein termed Ran (Fig. 5.20) and nucleoporins (Table 5.1). How these components work together to move the import substrate through the channel of the nuclear pore complex, a distance of more than 100 nm, is not yet understood. Such movement may be the result of repeated cycles of binding of the cargo-receptor complex to, and dissociation from, nucleoporins, as it "walks" across the nuclear pore. It is believed that guanosine nucleotide concentration differences, with GTP most abundant in the nucleus and GDP

**A**

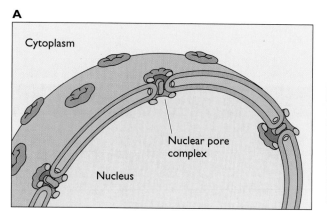

Cytoplasm

Nuclear pore complex

Nucleus

**B** Cytoplasmic surface view    Cross-sectional view

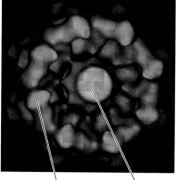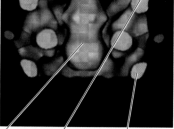

Cytoplasmic ring    Central plug    Cytoplasmic ring    Nuclear ring

**C**

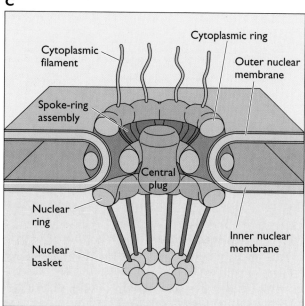

Cytoplasmic filament

Cytoplasmic ring

Spoke-ring assembly

Outer nuclear membrane

Nuclear ring

Central plug

Nuclear basket

Inner nuclear membrane

**D**

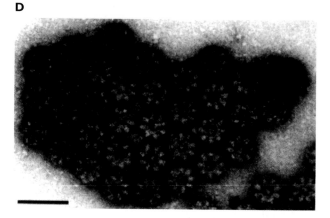

most abundant in the cytoplasm, drives GDP-GTP exchange in the nucleus and provides the driving force and directionality for transport (Fig. 5.20).

Not all proteins imported into the nucleus contain a nuclear localization signal. For example, β-catenin, a protein that mediates a late step in signal transduction in the fruit fly *Drosophila,* localizes to the nucleus without a nuclear localization signal. It has been shown that β-catenin binds directly to the nuclear pore complex without first binding to a cytoplasmic receptor. A similar nuclear localization signal-independent mechanism might account for the import of some viral proteins into the nucleus, such as the remarkable VP22 protein of herpes simplex virus type 1 (Box 5.2).

## Import of DNA Genomes

The capsids of many DNA-containing viruses are larger than 26 nm and cannot be imported into the nucleus from

**Figure 5.19 Structural model of the nuclear pore complex.** (A) Overview of the nuclear membrane, showing the topology of the nuclear pore complexes. (B) Three-dimensional image reconstructions from cryoelectron microscopy of a cytoplasmic surface view and a cross section of the nuclear pore complex. Reprinted from C. W. Akey and M. Radermacher, *J. Cell Biol.* **122:**1–19, 1993, with permission. (C) Schematic drawing of the nuclear pore complex, showing the spoke-ring assembly at the waist of the nuclear complex and its attachment to cytoplasmic filaments and the nuclear basket. The latter comprises eight filaments, extending 50 to 100 nm from the central structure and terminating in globular structures that form a distal annulus. The nuclear pore channel is shown containing a central plug, which is generally believed to be the pore complex transporter but may represent material being transported or peripheral structures that collapsed during the preparation of samples for microscopy: the identity of the central plug is presently controversial. This model is based on electron microscopy (D) and image analysis of nuclear pore complexes (B), generally of *Xenopus* oocytes, in which the nuclear envelope is packed with a paracrystalline array of nuclear pore complexes. It represents a composite of features observed in different experiments, for no single method of sample preparation preserves all the structural elements shown in the figure. Adapted from Fig. 8.7B of G. M. Cooper, *The Cell: a Molecular Approach* (ASM Press, Washington, D.C., and Sinauer Associates, Sunderland, Mass., 1997). (D) Face-on view of negatively stained nuclear pore complexes in an electron microscope. The very abundant nuclear complexes, which are about 150 nm in diameter, clearly exhibit eightfold symmetry. A central channel is visible and in some cases lacks a plug. Bar, 0.2 μm. Courtesy of Ron Milligan (Scripps Research Institute).

**Table 5.2**  Some metazoan nucleoporins[a]

| Protein | Mass (kDa) | Location | Properties |
|---------|-----------|----------|------------|
| Ran BP | 135 | Cytoplasmic filaments; perhaps also nuclear face | Binds Ran; required for import; contains zinc fingers |
| Nup250 | 180 | Cytoplasmic rings and filaments | Contains FG repeats |
| Nup180 | 180 | Cytoplasmic rings and filaments | |
| Nup153 | 190 | Distal annulus of nuclear basket | Contains zinc fingers |
| p62 | 62 | Spoke-ring assembly; central plug[b] | p62 contains FG repeats; p62-p58-p54 complex required for protein docking |
| p58 | 58 | Spoke-ring assembly; central plug[b] | p62-p58-p54 complex required for protein docking |
| p54 | 54 | Spoke-ring assembly; central plug[b] | p62-p58-p54 complex required for protein docking |
| Pom121 | 145 | Pore membrane | Contains FG repeats; bulk of protein exposed to spoke-ring assembly |

[a]The locations and properties of some of the metazoan nucleoporins are listed. A subset of the nucleoporins are characterized by the presence of repeated sequences, such as the repeats containing phenylalanine (F) and glycine (G) listed. Zinc fingers are structural motifs which are stabilized by binding of $Zn^{2+}$, associated with nucleic acid-binding proteins.

[b]It is not known whether the central plug is composed of an intrinsic component of the nuclear pore complex, is an artifact of the isolation procedure, or represents molecules being transported through the channel.

the cytoplasm. These viruses have evolved a different mechanism for crossing the nuclear membrane in which capsids are released into the cytoplasm, dock onto the nuclear pore complex, and deliver their DNA into the nucleus. Adenoviral and probably herpesviral DNAs are transported into the nucleus by a similar strategy, although the viral proteins that mediate this activity have not been identified.

## Import of the Retroviral Genome into the Nucleus

As we have seen, fusion of retroviral and plasma membranes releases the viral core into the cytoplasm. The retroviral core consists of the viral RNA genome, coated with NC protein, and the enzymes, RT and IN, enclosed by CA protein. Retroviral DNA synthesis commences in the cytoplasm, within the nucleocapsid core, and after 4 to 8 h of DNA synthesis the preintegration complex, comprising viral DNA, IN, and perhaps other proteins, localizes to the nucleus. There the viral DNA is integrated into a cellular chromosome and viral transcription begins. The mechanism of nuclear import of the preintegration complex is poorly understood, but it is quite clear that this structure is too large (ca. 60S) to pass through the nuclear pore complex. Moloney murine leukemia virus can infect only dividing cells. This observation may indicate that the breakdown of the nuclear membrane that occurs during

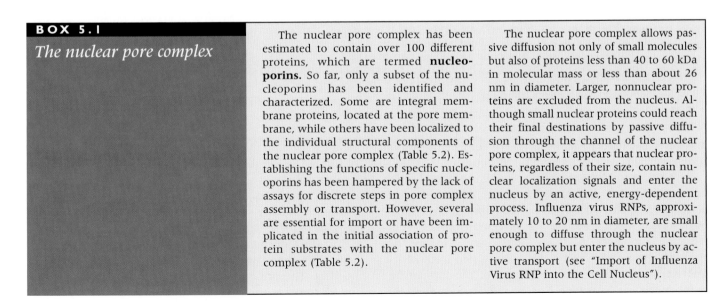

**BOX 5.1**

*The nuclear pore complex*

The nuclear pore complex has been estimated to contain over 100 different proteins, which are termed **nucleoporins**. So far, only a subset of the nucleoporins has been identified and characterized. Some are integral membrane proteins, located at the pore membrane, while others have been localized to the individual structural components of the nuclear pore complex (Table 5.2). Establishing the functions of specific nucleoporins has been hampered by the lack of assays for discrete steps in pore complex assembly or transport. However, several are essential for import or have been implicated in the initial association of protein substrates with the nuclear pore complex (Table 5.2).

The nuclear pore complex allows passive diffusion not only of small molecules but also of proteins less than 40 to 60 kDa in molecular mass or less than about 26 nm in diameter. Larger, nonnuclear proteins are excluded from the nucleus. Although small nuclear proteins could reach their final destinations by passive diffusion through the channel of the nuclear pore complex, it appears that nuclear proteins, regardless of their size, contain nuclear localization signals and enter the nucleus by an active, energy-dependent process. Influenza virus RNPs, approximately 10 to 20 nm in diameter, are small enough to diffuse through the nuclear pore complex but enter the nucleus by active transport (see "Import of Influenza Virus RNP into the Cell Nucleus").

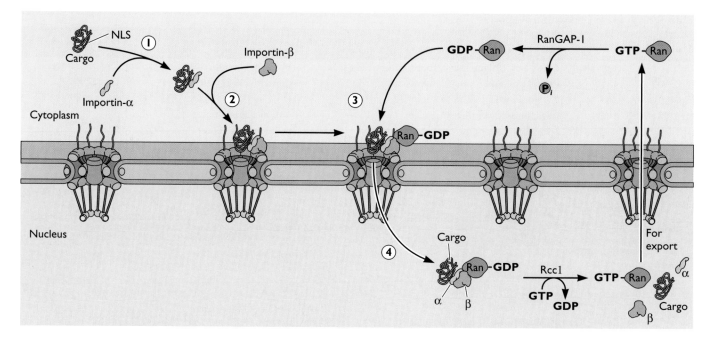

**Figure 5.20 The protein import pathway.** This pathway is illustrated schematically from left to right. Cytoplasmic and nuclear compartments are shown separated by the nuclear envelope studded with nuclear pore complexes. In step 1, a nuclear localization signal on the cargo (red) is recognized by importin-α (karyopherin α). Although the simple simian virus 40 and some bipartite nuclear localization signals (Table 5.2) bind to the same nuclear localization signal (NLS) receptor, metazoan cells contain multiple related, but distinct, importin-α proteins which do not appear to fulfill identical functions. In step 2, importin-β binds the cargo–importin-α complex and docks onto the nucleus, probably by associating initially with nucleoporins present in the cytoplasmic filaments of the nuclear pore complex. As shown, this interaction would bring the nuclear localization signal-containing protein to the cytoplasmic face of the nuclear pore complex. Translocation of the substrate into the nucleus (step 4) requires additional soluble proteins, including the guanine nucleotide-binding protein Ran. A Ran-specific guanine nucleotide exchange protein (Rcc1) and a Ran-GTPase-activating protein (RanGAP-1) are localized in the nucleus and cytoplasm, respectively. The action of RanGAP-1 maintains cytoplasmic Ran in the GDP-bound form. When Ran is in the GTP-bound form, nuclear import cannot occur. Ran binds specifically to certain nucleoporins. It has been suggested that translocation of the import substrate is achieved by repeated cycles of docking and release of substrate-importin complexes with specific nucleoporins. However, the precise mechanism of translocation remains to be established. Following import, the complexes are dissociated when Ran-GDP is converted to Ran-GTP by Rcc1. Ran-GTP participates in export from the nucleus. When Ran is in the GDP-bound form, nuclear export cannot occur. Thus, hydrolysis of Ran-GTP in the cytoplasm and GTP-GDP exchange in the nucleus maintain a gradient of Ran-GTP/Ran-GDP. The asymmetric distribution of RanGAP-1 and Rcc1 allows for the formation of such a gradient. This gradient is believed to provide the driving force and directionality for nuclear transport.

mitosis is essential to allow entry of the preintegration complex of this retrovirus into the nucleus.

In contrast to Moloney murine leukemia virus, human immunodeficiency virus type 1 DNA can replicate in nondividing cells. This virus, and probably other lentiviruses, must have a mechanism for the active transport of preintegration complexes into the nucleus. Some evidence suggests that Vpr and a minor, phosphorylated form of the matrix protein direct nuclear import of the human immunodeficiency virus type 1 preintegration complex. Nu-

clear localization signals have been found in the IN protein of human immunodeficiency virus type 1, and they may also contribute to such import.

Avian sarcoma and leukosis viruses, like Moloney murine leukemia virus, do not replicate in nondividing cells. IN of these viruses contains a nuclear localization signal in the C-terminal domain which, when fused to heterologous, cytoplasmic proteins, can direct them to the nucleus. The presence of a nuclear localization signal in the IN protein suggests that the block to replication of

**BOX 5.2**

*The remarkable VP22 protein of herpes simplex virus type 1*

The region of the herpes simplex virion between the capsid and the envelope, called the tegument, contains several proteins. One of these, a protein of 38-kDa molecular mass called VP22, is distinguished by its trafficking properties. Despite lacking a signal sequence, VP22 is exported from infected cells and can then reenter surrounding cells. It does so with such high efficiency that in cell cultures infected at a low multiplicity of infection, the protein can spread to nearly every cell. Its uptake into cells does not require the endocytic pathway. However, drugs that disrupt the actin cytoskeleton interfere with the uptake of VP22 into cells. It is therefore believed that VP22 binds to a cell surface receptor linked to actin microfilaments and enters cells via a novel pathway involving the cytoskeleton. After entering cells, VP22 is imported into the nucleus, despite lacking a nuclear localization signal.

Other proteins can be delivered into cells by fusing their coding regions with that of VP22. The unique properties of VP22 make it an ideal vector for the therapeutic delivery of proteins into cells.

**Elliott, G., and P. O'Hare.** 1997. Intercellular trafficking and protein delivery by a herpesvirus structural protein. *Cell* **88:**223–233.

these viruses in nondividing cells may occur at some step after nuclear localization.

## Import of Influenza Virus RNP into the Cell Nucleus

Influenza virus is unusual among RNA-containing viruses in that it replicates in the cell nucleus. After vRNPs separate from M1 and are released into the cytosol, they are rapidly imported into the nucleus (Fig. 5.7). Such transport is carried out by the cell machinery responsible for the import of proteins into the nucleus. The import of vRNP into the nucleus depends on the presence of a karyophilic signal—a nuclear localization signal—in the NP protein, a component of vRNPs. Naked viral RNA does not dock onto the nuclear pore complex nor is it taken up into the nucleus, but in the presence of NP the viral RNA can enter the nucleus. Therefore, NP contains nuclear import signals that enable the viral RNA to enter the nucleus.

Two nuclear localization signals have been identified in NP protein and are called nonconventional because they do not resemble sequences identified in the simian virus 40 T antigen or nucleoplasmin (Table 5.1). In experimental systems NP binds to karyopherin α, and studies with permeabilized cells have shown that the influenza virus NP protein docks at the nuclear envelope in the presence of karyopherin and is imported into the nucleus in an energy-dependent reaction when Ran and p10 (Fig. 5.20) are present.

## Perspectives

Although viruses are structurally diverse, the mechanisms by which they enter cells appear to be surprisingly limited. Depending on whether the virus is enveloped, and whether the viral genome must reach the cytoplasm or the nucleus, common entry themes are evident. For example, all enveloped viruses must fuse with a membrane, either at the cell surface or from within an intracellular vesicle. Studies of this entry step have revealed that fusion is catalyzed by a viral glycoprotein whose fusogenic potential may be regulated by proteolytic cleavage. Fusion may be triggered by a reduction in pH or by interaction with coreceptors, which cause structural changes in the viral glycoprotein and expose the fusion peptide. Although the structures of some protein intermediates in the fusion reaction have been solved, "snapshots" of others will be required to better understand this process. We know even less about the precise mechanism by which lipid bilayers fuse.

Once fusion has occurred, the genome can enter the cytoplasm. Release of the influenza virus RNP into the cytoplasm requires disruption of interactions with the M1 protein. A similar step may be required during uncoating of other (−) strand viruses, but few have been studied. In some enveloped viruses, the genome is surrounded by a capsid or nucleocapsid, which must be disassembled to allow the RNA access to the cytoplasm. At least one togavirus nucleocapsid is disassembled in the cytoplasm by ribosomes; whether this is a universal mechanism is not yet known.

There are large gaps in our understanding of how nonenveloped viruses enter cells. One strategy of entry is exemplified by adenoviruses and reoviruses, which become structurally altered as they pass through the endocytic pathway, resulting in particles that can escape from endosomes. Poliovirus may form a pore at the cell surface, but crucial aspects of this hypothesis must be tested. Structural studies of intermediates in the cell entry of these three viruses should reveal the underlying mechanisms.

Much less is known about how other nonenveloped viruses enter cells. Studies of these viruses may reveal novel mechanisms of uncoating.

The genomes of many viruses replicate in the nucleus. Incoming viral genomes enter this cellular compartment by transport through the nuclear pore complex. Two distinct themes have emerged from the study of this step in infection. In one approach, exemplified by influenza virus, vRNP, which has been released into the cytoplasm, is imported into the nucleus. In the other approach, seen with DNA viruses, the viral capsid docks onto the nuclear pore complex, discharging the genome into the nucleus. There are many unanswered questions about nuclear import, e.g., whether components of the nuclear pore complex play a role in capsid disassembly and how molecules pass through the nuclear pore. Studies of viral entry into cells have been invaluable in clarifying aspects of normal cellular processes such as membrane fusion and endocytosis, and similar benefits are likely to come from studying how viral macromolecules enter the nucleus.

# References

## Reviews

Bates, P. 1996. Chemokine receptors and HIV-1: an attractive pair? *Cell* **86:**1–3.

Berger, E. A. 1997. HIV entry and tropism: the chemokine receptor connection. *AIDS* **11:**S3–S16.

Cole, C. N., and C. M. Hammell. 1998. Nucleocytoplasmic transport: driving and directing transport. *Curr. Biol.* **8:**R368–R372.

Greber, U. F., I. Singh, and A. Helenius. 1994. Mechanisms of virus uncoating. *Trends Microbiol.* **2:**52–56.

Kasamatsu, H., and A. Nakanishi. 1998. How do animal DNA viruses get to the nucleus? *Annu. Rev. Microbiol.* **52:**627–686.

Luban, J. 1996. Absconding with the chaperone: essential cyclophilin-gag interaction in HIV-1 virions. *Cell* **87:**1–20.

Marsh, M., and A. Helenius. 1989. Virus entry into animal cells. *Adv. Virus Res.* **36:**107–151.

Perez, L., and L. Carrasco. 1994. Involvement of the vacuolar H-ATPase in animal virus entry. *J. Gen. Virol.* **75:**2595–2606.

Whittaker, G. 1998. Squeezing through pores: how microorganisms get into and out of the nucleus. *Trends Microbiol.* **6:**178–179.

Whittaker, G., M. Bui, and A. Helenius. 1996. The role of nuclear import and export in influenza virus infection. *Trends Cell Biol.* **6:**67–71.

## Papers of Special Interest

### Adenovirus

Greber, U. F., M. Suomalainen, R. P. Stidwill, K. Boucke, M. W. Ebersold, and A. Helenius. 1997. The role of the nuclear pore complex in adenovirus DNA entry. *EMBO J.* **16:**5998–6007.

Greber, U. F., P. Webster, J. Weber, and A. Helenius. 1996. The role of the adenovirus protease in virus entry into cells. *EMBO J.* **15:**1766–1777.

Greber, U. F., M. Willetts, P. Webster, and A. Helenius. 1993. Stepwise dismantling of adenovirus 2 during entry into cells. *Cell* **75:**477–486.

### Retroviruses

Chan, D. C., D. Fass, J. M. Berger, and P. S. Kim. 1997. Core structure of gp41 from the HIV envelope glycoprotein. *Cell* **89:**263–273.

Feng, Y., C. C. Broder, P. E. Kennedy, and E. A. Berger. 1996. HIV-1 entry cofactor: functional cDNA cloning of a seven-transmembrane, G protein-coupled receptor. *Science* **272:**872–876.

Gallay, P., T. Hope, D. Chin, and D. Trono. 1997. HIV-1 infection of nondividing cells through the recognition of integrase by the importin/karyopherin pathway. *Proc. Natl. Acad. Sci. USA* **94:**9825–9830.

Gallay, P., S. Swingler, C. Aiken, and D. Trono. 1995. HIV-1 infection of nondividing cells: C-terminal tyrosine phosphorylation of the viral matrix protein is a key regulator. *Cell* **80:**379–388.

Gamble, T. R., F. F. Vajdos, S. Yoo, D. K. Worthylake, M. Houseweart, W. I. Sundquist, and C. P. Hill. 1996. Crystal structure of human cyclophilin A bound to the amino-terminal domain of HIV-1 capsid. *Cell* **87:**1285–1294.

Gilbert, J. M., L. D. Hernandez, J. W. Balliet, P. Bates, and J. M. White. 1995. Receptor-induced conformational changes in the subgroup A avian leukosis and sarcoma virus envelope glycoprotein. *J. Virol.* **69:**7410–7415.

Popov, S., M. Rexach, G. Zybarth, N. Reiling, M. A. Lee, L. Ratner, C. M. Lane, M. S. Moore, G. Blobel, and M. Bukrinsky. 1998. Viral protein R regulates nuclear import of the HIV-1 pre-integration complex. *EMBO J.* **17:**909–917.

### Influenza Virus

Bullough, P. A., F. M. Hughson, J. J. Skehel, and D. C. Wiley. 1994. Structure of influenza hemagglutinin at the pH of membrane fusion. *Nature* (London) **371:**37–43.

Chen, J., K. H. Lee, D. A. Steinhauer, D.J. Stevens, J. J. Skehel, and D. C. Wiley. 1998. Structure of the hemagglutinin precursor cleavage site, a determinant of influenza pathogenicity and the origin of the labile conformation. *Cell* **95:**409–417.

Martin, K., and A. Helenius. 1991. Nuclear transport of influenza virus ribonucleoproteins: the viral matrix protein (M1) promotes export and inhibits import. *Cell* **67:**117–130.

Melikyan, G. B., J. M. White, and F. S. Cohen. 1995. GPI-anchored influenza hemagglutinin induces hemifusion to both red blood cell and planar bilayer membranes. *J. Cell Biol.* **131:**679–691.

O'Neill, R. E., R. Jaskunas, G. Blobel, P. Palese, and J. Moroianu. 1995. Nuclear import of influenza virus RNA can be mediated by viral nucleoprotein and transport factors required for protein import. *J. Biol. Chem.* **270:**22701–22704.

Pinto, L. H., L. J. Holsinger, and R. A. Lamb. 1992. Influenza virus M2 protein has ion channel activity. *Cell* **69:**517–528.

Tatulian, S. A., P. Hinterdorfer, G. Baber, and L. K. Tamm. 1995. Influenza hemagglutinin assumes a tilted conformation during membrane fusion as determined by attenuated total reflection FTIR spectroscopy. *EMBO J.* **14:**5514–5523.

## Semliki Forest Virus

Kielian, M. C., and A. Helenius. 1984. Role of cholesterol in fusion of Semliki Forest virus with membranes. *J. Virol.* **52:**281–283.

Marsh, M., E. Bolzau, and A. Helenius. 1983. Penetration of Semliki Forest virus from acidic prelysosomal vacuoles. *Cell* **32:**931–940.

Marsh, M., and A. Helenius. 1980. Adsorptive endocytosis of Semliki Forest virus. *J. Mol. Biol.* **142:**439–454.

Salminen, A., J. M. Wahlberg, M. Lobigs, P. Liljestrom, and H. Garoff. 1992. Membrane fusion process of Semliki Forest virus. II. Cleavage-dependent reorganization of the spike protein complex controls virus entry. *J. Cell Biol.* **116:**349–357.

Singh, I., and A. Helenius. 1992. Role of ribosomes in Semliki Forest virus nucleocapsid uncoating. *J. Virol.* **66:**7049–7058.

Wahlberg, J. M., and H. Garoff. 1992. Membrane fusion process of Semliki Forest virus. I. Low pH-induced rearrangement in spike protein quaternary structure precedes virus penetration into cells. *J. Cell Biol.* **116:**339–348.

White, J., and A. Helenius. 1980. pH-dependent fusion between the Semliki Forest virus membrane and liposomes. *Proc. Natl. Acad. Sci. USA* **77:**3273–3277.

## Picornaviruses

Badger, J., I. Minor, M. J. Kremer, M. A. Oliveira, T. J. Smith, J. P. Griffith, D. M. Guerin, S. Krishnaswamy, M. Luo, M. G. Rossmann, et al. 1988. Structural analysis of a series of antiviral agents complexed with human rhinovirus 14. *Proc. Natl. Acad. Sci. USA* **85:**3304–3308.

Curry, S., M. Chow, and J. M. Hogle. 1996. The poliovirus 135S particle is infectious. *J. Virol.* **70:**7125–7131.

Dove, A. W., and V. R. Racaniello. 1997. Cold-adapted poliovirus mutants bypass a postentry replication block. *J. Virol.* **71:**4728–4735.

Fricks, C. E., and J. M. Hogle. 1990. Cell-induced conformational change in poliovirus: externalization of the amino terminus of VP1 is responsible for liposome binding. *J. Virol.* **64:**1934–1945.

Mason, P. W., B. Baxt, F. Brown, J. Harber, A. Murdin, and E. Wimmer. 1993. Antibody-complexed foot-and-mouth disease virus, but not poliovirus, can infect normally insusceptible cells via the Fc receptor. *Virology* **192:**568–577.

Moscufo, N., A. G. Yafal, A. Rogove, J. Hogle, and M. Chow. 1993. A mutation in VP4 defines a new step in the late stages of cell entry by poliovirus. *J. Virol.* **67:**5075–5078.

Mosser, A. G., and R. R. Rueckert. 1993. WIN 51711-dependent mutants of poliovirus type 3: evidence that virions decay after release from cells unless drug is present. *J. Virol.* **67:**1246–1254.

Perez, L., and L. Carrasco. 1993. Entry of poliovirus into cells does not require a low-pH step. *J. Virol.* **67:**4543–4548.

Tosteson, M. T., and M. Chow. 1997. Characterization of the ion channels formed by poliovirus in planar lipid membranes. *J. Virol.* **71:**507–511.

Zeichhardt, H., M. J. Otto, M. A. McKinlay, P. Willingmann, and K. O. Habermehl. 1987. Inhibition of poliovirus uncoating by disoxaril (WIN 51711). *Virology* **160:**281–285.

## Reovirus

Dryden, K. A., G. Wang, M. Yeager, M. L. Nibert, K. M. Coombs, D. B. Furlong, B. N. Fields, and T. S. Baker. 1993. Early steps in reovirus infection are associated with dramatic changes in supramolecular structure and protein conformation: analysis of virions and subviral particles by cryoelectron microscopy and image reconstruction. *J. Cell Biol.* **122:**1023–1041.

Keljo, D. J., M. Kuhn, and A. Smith. 1988. Acidification of endosomes is not important for the entry of rotavirus into the cell. *J. Pediatr. Gastroenterol. Nutr.* **7:**257–263.

Martinez, C. G., R. Guinea, J. Benavente, and L. Carrasco. 1996. The entry of reovirus into L cells is dependent on vacuolar proton-ATPase activity. *J. Virol.* **70:**576–579.

Suzuki, H., S. Kitaoka, T. Konno, T. Sato, and N. Ishida. 1985. Two modes of human rotavirus entry into MA 104 cells. *Arch. Virol.* **85:**25–34.

Tosteson, M. T., M. L. Nibert, and B. N. Fields. 1993. Ion channels induced in lipid bilayers by subvirion particles of the nonenveloped mammalian reoviruses. *Proc. Natl. Acad. Sci. USA* **90:**10549–10552.

## Papovaviruses

Anderson, H. A., Y. Chen, and L. C. Norkin. 1996. Bound simian virus 40 translocates to caveolin-enriched membrane domains, and its entry is inhibited by drugs that selectively disrupt caveolae. *Mol. Biol. Cell* **7:**1825–1834.

Chen, X. S., T. Stehle, and S. C. Harrison. 1998. Interaction of polyomavirus internal protein VP2 with the major capsid protein VP1 and implications for participation of VP2 in viral entry. *EMBO J.* **17:**3233–3240.

## Herpesviruses

Elliott, G., and P. O'Hare. 1997. Intercellular trafficking and protein delivery by a herpesvirus structural protein. *Cell* **88:**223–233.

Sodeik, B., M. W. Ebersold, and A. Helenius. 1997. Microtubule-mediated transport of incoming herpes simplex virus 1 capsids to the nucleus. *J. Cell Biol.* **136:**1007–1021.

## Nuclear Import

Fagotto, F., U. Gluck, and B. M. Gumbiner. 1998. Nuclear localization signal-independent and importin/karyopherin-independent nuclear import of beta-catenin. *Curr. Biol.* **8:**181–190.

Kalderon, D., B. L. Roberts, W. D. Richardson, and A. E. Smith. 1984. A short amino acid sequence able to specify nuclear location. *Cell* **39:**499–509.

Nakielny, S., and G. Dreyfuss. 1998. Import and export of the nuclear protein import receptor transportin by a mechanism independent of GTP hydrolysis. *Curr. Biol.* **8:**89–95.

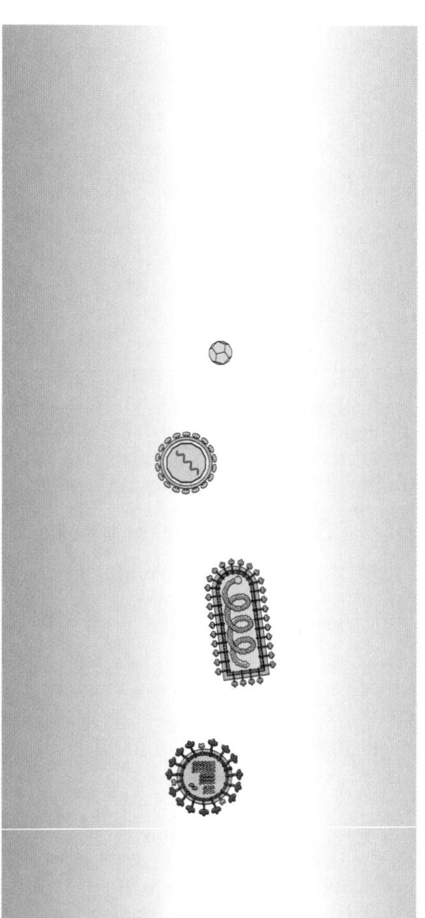

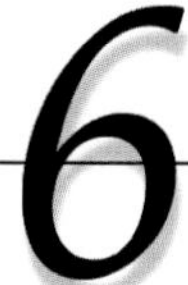

# 6

# Genome Replication and mRNA Production by RNA Viruses

## Introduction

The genomes of RNA viruses come in a number of configurations, including unimolecular or segmented, single stranded of (+) or (−) polarity, double stranded, and circular (Table 6.1). These structurally diverse viral RNA genomes share a common requirement: they must be efficiently copied within the infected cell to provide both genomes for assembly into progeny virions and mRNAs for the synthesis of viral proteins. The synthesis of these RNA molecules by RNA viruses is a unique process that has no parallel in the cell, which lacks the machinery to carry out RNA-directed RNA synthesis. All RNA viruses except retroviruses (see below) encode an **RNA-dependent RNA polymerase** (Box 6.1) to catalyze the synthesis of new genomes and mRNAs.

RNA viruses with (−) strand and double-stranded RNA genomes must carry the RNA polymerase in the viral particle because the incoming viral RNA can be neither translated nor copied by the cellular machinery. The deproteinized RNA genomes of (−) strand and double-stranded RNA viruses are therefore noninfectious. In contrast, (+) strand viral particles lack a virion polymerase; the deproteinized RNAs of these viruses are infectious because they are translated in cells to produce, among other viral proteins, the viral RNA polymerase.

The mechanisms by which viral mRNA is made and the RNA genome replicated in cells infected by RNA viruses (Fig. 6.1) appear even more diverse than the structure and organization of viral RNA genomes (Table 6.1). For example, picorna- and alphaviruses both possess single molecules of (+) strand genomic RNA but have evolved quite different strategies for the production of viral RNA (Fig. 6.1). Nevertheless, each of the mechanisms of viral RNA synthesis summarized in Fig. 6.1 meets two essential requirements common to the infectious cycles of all these viruses. During replication, the RNA genome must be copied from one end to the other with no loss of nucleotide sequence. The second requirement is for the production of viral mRNAs that can be efficiently translated by the cellular protein synthetic machinery.

**163**

**Table 6.1** Representative RNA-containing viruses

| Virus | Genome structure | RNA polymerase[a] | Viral accessory proteins involved in RNA synthesis | Host factors |
|---|---|---|---|---|
| Poliovirus | Single-stranded linear (+) strand | 3D$^{pol}$ | 3AB: membrane-bound VPg donor; cofactor for 3D$^{pol}$ | Poly(rC)-binding protein |
| | | | 3B (VPg): primer for RNA synthesis; linked to 5′ end of (+) and (−) strand RNAs | |
| | | | 3CD: precursor of 3C and 3D$^{pol}$; binds cloverleaf in the (+) strand | |
| | | | 3C: viral proteinase that cleaves 3CD and 3AB | |
| | | | 2C: NTPase; possible helicase; anchors RNA polymerase to cellular membranes | |
| Reovirus | Double-stranded linear, 10–11 segments | λ3 | λ2: guanylyltransferase; methyltransferase | None identified |
| | | | λ1: binds double-stranded RNA; function unknown | |
| | | | μNS: role in mRNA synthesis | |
| Sindbis virus | Single-stranded linear (+) strand | nsP4 | nsP1: methyltransferase | Required but not identified |
| | | | nsP2: NTPase; possible RNA helicase | |
| | | | nsP3: unknown function | |
| | | | nsP1,2,3,4: form a multisubunit polymerase | |
| Influenza virus | Single-stranded linear (−) strand, 7–8 segments | PB1 | PA: role in genome RNA synthesis | None identified |
| | | | PB2: cap-binding subunit | |
| | | | NP: nucleocapsid protein | |
| Vesicular stomatitis virus | Single-stranded linear (−) strand | L | N: nucleocapsid protein | β-Tubulin, actin |
| | | | P: phosphoprotein; binds L; required for polymerase activity | |
| Arenavirus | Single-stranded linear (−) strand, 2 segments | L | Z: regulation of mRNA synthesis | None identified |
| | | | NP: nucleocapsid protein | |
| Hepatitis delta virus | Single-stranded circular, rodlike (−) strand | Cellular DNA-dependent RNA polymerase II | Delta antigen | Cellular general initiation proteins? |

[a] The catalytic subunit; accessory proteins may be required for enzyme activity.

In this chapter we will consider the mechanisms of viral RNA synthesis, the ways viruses switch from mRNA production to genome replication, and how the process of RNA-directed RNA synthesis can result in genetic diversity. Much of our understanding of viral RNA synthesis comes from in vitro experiments. Because it is possible that events proceed differently in infected cells, the results of in vitro studies are used to build hypothetical models for the different steps in RNA synthesis. While many models exist for each step in RNA synthesis, those presented in this chapter were selected because they are consistent with experimental results obtained in different laboratories.

The general principles of RNA synthesis will be illustrated with a few viruses, including influenza virus, vesicular stomatitis virus, poliovirus, alphaviruses, and arenaviruses, as examples. Summaries of the genome organization of these viruses are presented in the appendix. Although retroviruses are RNA viruses, they will be discussed in chapter 7. Retroviruses encode an RNA-dependent DNA polymerase, and therefore mRNA synthesis and genome replication are very different from those of other RNA viruses.

## Mechanisms of Viral RNA Synthesis

### RNA-Dependent RNA Polymerases

#### Identification

The first evidence for a viral RNA-dependent RNA polymerase emerged in the early 1960s from studies with mengovirus and poliovirus, both (+) strand RNA viruses. In these experiments, extracts were prepared from virus-infected cells and incubated with the four ribonucleoside triphosphates (ATP, UTP, CTP, and GTP), one of which was radioactively labeled. The incorporation of nucleoside monophosphate into RNA was then measured. Infection with mengovirus or poliovirus resulted in the appearance of a cytoplasmic enzyme that could synthesize viral RNA in the presence of actinomycin D, a drug that was known

**BOX 6.1**

*What should we call RNA polymerases and the processes they catalyze?*

Viral RNA polymerases are frequently multisubunit assemblies consisting of several viral gene products. One protein is the catalytic subunit that polymerizes the RNA chain, while accessory proteins have other functions. In this book, the term **RNA-dependent RNA polymerase** refers to the protein assembly required to carry out RNA synthesis, not just the polypeptide that catalyzes chain elongation.

Historically, viral RNA-dependent RNA polymerases have been given two different names depending on their activities during infection. The term **replicase** is used to describe the enzyme that copies the viral RNA to produce additional genomes, while the enzyme that produces mRNA is called **transcriptase**. In some cases this terminology indicates true differences in the enzymes that carry out synthesis of functionally different RNAs. However, these terms can also be inaccurate and misleading, and so they will not be used here. For example, for some RNA viruses, genomic replication and mRNA synthesis are the same reaction. For double-stranded RNA viruses, mRNA synthesis produces templates that can also be used for genomic replication.

The production of mRNAs from viral RNA templates is often designated **transcription**. However, this term refers to a specific process, the copying of genetic information carried in DNA into RNA. Consequently, it will not be used here to describe synthesis of the mRNAs of viruses with RNA genomes. Similarly, we will reserve use of the term **promoter** to designate sequence elements in DNA at which transcription begins.

to inhibit cellular DNA-directed RNA synthesis. Lack of sensitivity of the enzyme activity to actinomycin D suggested that it was virus-specific and could copy RNA from an RNA template and not from a DNA template. This enzyme was presumed to be an RNA-dependent RNA polymerase. Several years later, similar assays were used to demonstrate that the virions of (−) strand viruses and of double-stranded RNA viruses contain an RNA-dependent RNA polymerase that synthesizes mRNAs from the (−) strand RNA present in the virions.

The initial discovery of a putative RNA polymerase in poliovirus-infected cells was followed by attempts to purify the enzyme and show that it can copy viral RNA. Because polioviral genomic RNA contains a 3′ poly(A) tail, polymerase activity was measured with a poly(A) template and an oligo(U) primer. After several fractionation steps, a poly(U) polymerase that could copy polioviral genomic RNA in the presence of an oligo(U) primer was purified from poliovirus-infected cells. Poly(U) polymerase activity coincided with a single polypeptide, now known to be the polioviral RNA polymerase 3D$^{pol}$ (see the appendix and chapter 11 for a description of this nomenclature). The 3D$^{pol}$ RNA polymerase purified either from infected human cells or following expression of the viral coding sequence in bacteria or insect cells cannot copy polioviral genomic RNA in the absence of an oligo(U) primer. The 3D$^{pol}$ is therefore a template- and primer-dependent enzyme.

Because many (+) strand viral RNA-dependent RNA polymerases are tightly associated with membranes or nucleocapsids in virus-infected cells, it has been difficult to purify these enzymes and study their properties. Indeed,

polioviral 3D$^{pol}$, vesicular stomatitis virus L protein, and retroviral reverse transcriptase (to be discussed in the following chapter) are the only animal virus RNA-dependent nucleic acid polymerases that have been purified and characterized. Nevertheless, many viral RNA polymerases have been identified. Assays for the RNA polymerase activity described in the previous paragraph have been used to demonstrate virus-specific enzymes in virions or in extracts of infected cells. Amino acid sequence alignments (see "Sequence Relationships among Polymerases" below) can be used to identify the viral protein with motifs characteristic of RNA-dependent RNA polymerases. These approaches have been used to determine that the L proteins of paramyxo- and bunyaviruses, the PB1 protein of influenza viruses, and the nsP4 protein of alphaviruses are likely to be RNA polymerases. When expressed in mammalian cells, the putative RNA polymerases of influenza virus and alphaviruses copy viral RNA templates.

RNA-directed RNA synthesis follows a set of universal rules that differ slightly from those followed by DNA-dependent DNA polymerases. RNA synthesis begins and ends (**initiation** and **termination**) at specific sites in the template and is catalyzed by virus-encoded polymerases, but viral accessory proteins and even host cell proteins may also be required. Like cellular DNA-dependent RNA polymerases, most RNA-dependent RNA polymerases can initiate RNA synthesis de novo. Some RNA-dependent RNA polymerases require a primer with a free 3′-OH end to which nucleotides complementary to the template strand are added. Some RNA primers are protein-linked, and others contain a 5′ cap structure (described in chapter 10).

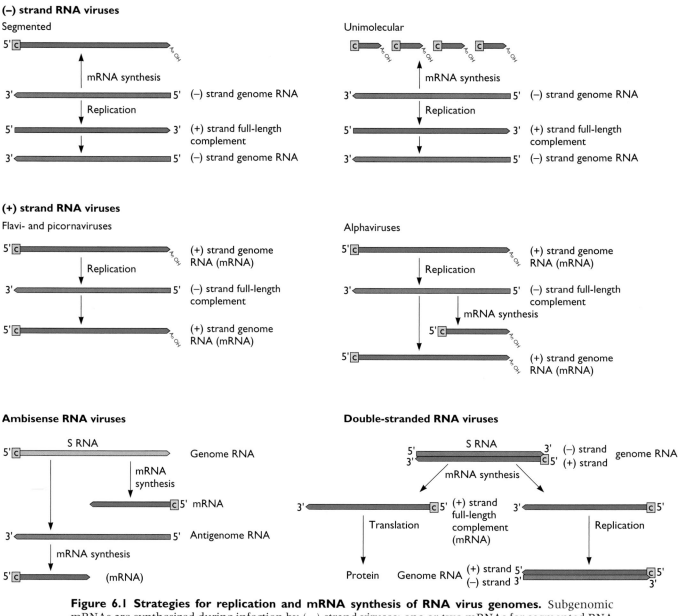

**Figure 6.1 Strategies for replication and mRNA synthesis of RNA virus genomes.** Subgenomic mRNAs are synthesized during infection by (−) strand viruses: one or two mRNAs for segmented RNA viruses or multiple mRNAs for viruses with unimolecular genomes. As these mRNAs are incomplete copies of the genomic RNA, they cannot be used as replicative intermediates. Antitermination mechanisms result in the synthesis of full-length (+) strands, which can then be copied to produce additional (−) genomic RNAs. For (+) strand viruses, the genomic RNA is also an mRNA. A subgenomic mRNA is also produced in alphavirus-infected cells. The segmented genomes of ambisense RNA viruses are copied to form one mRNA; a second mRNA is synthesized from the complement, or antigenome. In cells infected with double-stranded RNA viruses, the mRNAs first synthesized can either be translated into protein or serve as templates for the synthesis of (−) strands, resulting in double-stranded genomic RNA. Polioviral genomic RNA is linked to VPg at the 5′ end. The (+) genomic RNA of some flaviviruses does not contain poly(A).

RNA is usually synthesized by template-directed stepwise incorporation of ribodeoxynucleoside monophosphates from ribodeoxynucleoside triphosphate substrates into the 3′-OH end of the growing RNA chain, which is **elongated** in the 5′ → 3′ direction. However, some examples of nontemplated synthesis of viral RNA have been described.

Viral RNA synthesis is highly efficient. For example, from the single molecule of polioviral RNA that enters a cell, approximately 50,000 copies are produced during the course of an 8-h infection.

### Sequence Relationships among Polymerases

The amino acid sequences of viral RNA polymerases have been compared to identify conserved regions essential for polymerase function and to provide information about the evolution of these enzymes. Despite wide variation in genomic configuration, and very different amino acid sequences, four common motifs (A to D) have been identified in the sequences of all polymerases (Fig. 6.2). One of these motifs (C) includes a Gly-Asp-Asp sequence conserved in the RNA-dependent RNA polymerases of all (+) strand RNA viruses that have been studied. As a result, it was suggested that the Gly-Asp-Asp sequence is part of the active site of the enzyme. In support of this hypothesis, mutagenesis of the first Asp in polioviral 3D$^{pol}$ produces an inactive polymerase. Evidence that a viral protein is an RNA polymerase is considerably strengthened when this three-amino-acid sequence is found.

The Asp-Asp sequence is also conserved in RNA-dependent DNA polymerases of retroviruses and in RNA-dependent RNA polymerases of double-stranded RNA and segmented (−) strand viruses. The RNA polymerases of nonsegmented (−) strand viruses contain Gly-Asp-Asn instead of Gly-Asp-Asp. Mutational studies have shown that in the L protein of vesicular stomatitis virus, this sequence is essential for RNA synthesis.

When the sequences of DNA-dependent DNA and RNA polymerases were included in comparisons, it became evident that two of the motifs (A and C) identified in RNA-dependent RNA polymerases were present (Fig. 6.2).

Furthermore, the entire polymerase domain of DNA-dependent DNA and RNA-dependent DNA polymerases of retroviruses could be aligned. These comparisons suggested that all four classes of nucleic acid polymerases might share a common topology.

### Three-Dimensional Structure of 3D$^{pol}$

The crystal structures of three of the four types of polymerases, DNA-dependent DNA polymerase, DNA-dependent RNA polymerase, and RNA-dependent DNA polymerase (reverse transcriptase), are often characterized as being analogous to a right hand consisting of a palm, fingers, and a thumb, with the active site of the enzyme located in the palm subdomain. The first structure of an RNA-dependent RNA polymerase obtained was that of polioviral 3D$^{pol}$, which was determined at 2.6-Å resolution by X-ray crystallography (Fig. 6.3). The enzyme has the same overall shape as other polymerases, although the fingers and thumb are distinctive.

The palm subdomain of 3D$^{pol}$ is structurally similar to that of other polymerases and contains the four motifs (A to D) conserved in RNA-dependent polymerases (Fig. 6.3 and 6.4A). The fifth motif, which is present in RNA-dependent but not in DNA-dependent polymerases, packs between the palm and thumb subdomains. Determination of the 3D$^{pol}$ structure has allowed the realignment of the amino acid sequences of the five motifs for all four classes of polymerases (Fig. 6.4B). This realignment has revealed new functional relationships between individual amino acids. For example, in motif A, Tyr115 of human immunodeficiency virus 1 reverse transcriptase is aligned with Asp238 of 3D$^{pol}$. These two residues are nearly completely conserved in these two

**Figure 6.2 Protein domain alignments for the four categories of nucleic acid polymerases.** Numbers at the top are from the 3D$^{pol}$ amino acid sequence. Sequence motifs in each polymerase category are enclosed with solid lines and labeled, and the alignments between the different polymerases are shown with dashed lines. Motifs A to E are found in RNA-dependent RNA and DNA polymerases; of these, only motifs A and C are also found in DNA-dependent DNA and RNA polymerases. Motif B* found in DNA-dependent DNA and RNA polymerases is not the same as motif B in the other two enzymes. The structure and sequence motifs that resulted from comparing the crystal structure of 3D$^{pol}$ are labeled A to E and are shown in green. This comparison allowed the identification of motifs B and D in DNA-dependent DNA and RNA polymerases. Adapted from J. L. Hansen et al., *Structure* **5:**1109–1127, 1997, with permission. Courtesy of S. Schultz, University of Colorado, Boulder.

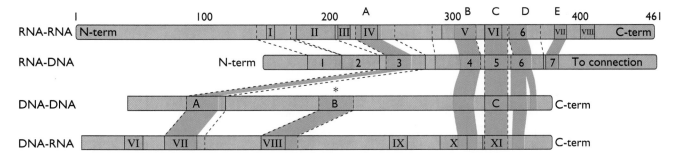

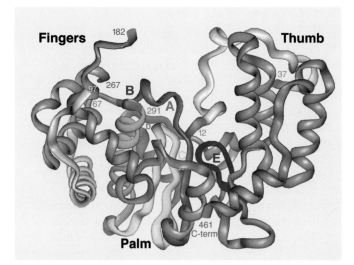

**Figure 6.3 Structure of polioviral 3Dpol.** The thumb, palm, and fingers subdomains are labeled. Structure/sequence motifs are labeled and colored as follows: A, red; B, green; C, yellow; D, purple; E, dark blue. The top of the finger subdomain is disordered in the crystals; the two N-terminal regions that are ordered are in white, and amino acids flanking disordered regions are numbered. The two white regions have no counterpart in other categories of polymerases. Because the region connecting the two white sequences (residues 33 to 66) is disordered, it is not known how they are connected. It seems unlikely that residues 33 to 66 stretch across the active cleft of the polymerase, and therefore it has been suggested that residues 14 to 35 are from a separate molecule. Such interactions could lead to oligomerization of 3D$^{pol}$ (Fig. 6.7). Adapted from J. L. Hansen et al., *Structure* **5:**1109–1127, 1997, with permission. Courtesy of S. Schultz, University of Colorado, Boulder.

polymerases, and it has been suggested that they control the discrimination between ribonucleoside triphosphates and deoxyribonucleoside triphosphates, a major functional difference between the two enzymes. Motif C of 3D$^{pol}$ contains the Asp-Asp sequence conserved in RNA-dependent polymerases; the first Asp is also conserved in DNA-dependent polymerases. All nucleic acid polymerases require a metal ion cofactor for activity, and the two Asp residues of motif C and the conserved Asp of motif A form a cluster that may coordinate such metals.

Based on the similarity between polioviral 3D$^{pol}$ and members of the other three classes of polymerases, it is likely that the structures of RNA polymerases of other RNA viruses also resemble a right hand.

## Accessory Proteins in RNA-Dependent RNA Synthesis

### Viral Proteins

In addition to the RNA-dependent RNA polymerase, other viral proteins participate in RNA synthesis. Such ac-

cessory proteins can play several roles: they may direct the polymerase to the correct intracellular site, facilitate recognition of the initiation site on the RNA template, or stimulate the activity of the polymerase. In addition, (+) strand RNA viruses with large genomes encode a **helicase**, an enzyme that unwinds double-stranded RNA structures by disrupting the hydrogen bonds between the two strands.

**Directing RNA polymerase to the correct cellular site.** Viral RNA synthesis does not occur at random locations in the cell but takes place either in the nucleus or on cytoplasmic membranes (see "Cellular Sites of Viral RNA Synthesis" below). Because influenza virus RNA synthesis occurs in the cell nucleus, the viral RNA polymerase must be transported there from the cytoplasm. As described in chapter 5, nuclear import of influenza virus RNA polymerase is mediated by nuclear localization sequences on the NP protein. Polioviral RNA synthesis occurs on membranous vesicles, and two polioviral polypeptides, 2C and 3AB, bring the RNA polymerase to these membranes. Low concentrations of guanidine hydrochloride inhibit polioviral RNA synthesis, and it is believed that this compound disrupts the interaction between the RNA polymerase and cell membranes. Mutations responsible for resistance of poliovirus to guanidine hydrochloride map to the coding sequence for the 2C protein, which contains an RNA-binding domain. It therefore appears that protein 2C anchors viral RNA to membranes during replication. As discussed later, polioviral protein 3AB may anchor the viral protein primer for RNA synthesis, VPg, in cell membranes.

**Targeting RNA polymerase to the correct site on RNA templates.** RNA-dependent RNA polymerases are template-specific enzymes; that is, they copy viral but not cellular RNAs. Such template specificity may reside in part in the recognition of structures at the 5′ and 3′ ends of viral RNAs by viral proteins (see "The Nature of the RNA Template" below). Polioviral 3CD protein plays an important role in viral RNA synthesis by participating in the formation of a ribonucleoprotein at the 5′ end of the (+) strand RNA. The 3CD protein is a precursor of the 3C$^{pro}$ protease and 3D$^{pol}$. Protein 3CD, together with a cellular protein known as poly(rC)-binding protein 2, binds to a cloverleaflike structure in the viral RNA (Fig. 6.5D). The RNA-binding domain of 3CD is contained within the 3C portion of the protein, and mutations within this domain abolish complex formation and RNA synthesis without affecting viral protein processing. Protein 3AB-3CD also binds the 3′-terminal sequence of polioviral RNA and may direct the polymerase to that site for the initiation of (−) strand RNA synthesis.

**A**

| Klenow | T7 RNAP | HIV-1 RT | Poliovirus polymerase |

**B**

**Motif A**

| Poliovirus | 225 | From fingers | M E E K L F A F **D** Y T G Y D A S L | Fingers 241 |
|---|---|---|---|---|
| HIV-1 RT | 102 | From fingers | K K K S V T V L **D** V G D A Y F S V _____ | Fingers 118 |
| Klenow | 697 | From thumbs | E D Y V I V S A **D** Y S Q I E L R I M A H L S R | Fingers 719 |
| T7 RNAP | 529 | From thumbs | N C S L P L A F **D** G S C S G I Q H F S A M L R | Fingers 551 |

β β β β β β β β β    h h h h h h h h h h

**Motif B**

| Poliovirus | 278 | K T Y C V K G G M **P S G** C S G T S I F N S M I N N L I I R T L L L K T | 312 |
|---|---|---|---|
| HIV-1 RT | 141 | G I R Y Q Y N V L **P Q G** W K G S P A I F Q S S M T K I L E P F K K Q N | 175 |
| Klenow | 833 | G A R R A A A E R A A I N A P M Q G T A A D I I K R A M I A V D A W L Q A E | 870 |
| T7 RNAP | 168 | E I D A H K Q E S G I A P N F V H S Q D G S H I R K T V V W A H E K Y G | 803 |

α α α α α α α α α α α α α α α α α α α α α α α α

**Motif C**

| Poliovirus | 313 | Y K G I D L D H L K M   I A **Y G D D** V I A S Y P H E | 337 |
|---|---|---|---|
| HIV-1 RT | 176 | P D I V I   Y Q **Y M D D** L Y V G S D L E | 194 |
| Klenow | 871 | Q P R V R M I M Q **V H D E** L V F E V H K D | 891 |
| T7 RNAP | 804 | I E S F   A L **I H D S** F G T I | 817 |

β β β    β β β T T β β β β β β

**Motif D**

| Poliovirus | 338 | V D A S L L A Q S G K D Y **G** L T M T P A D **K** S A T | 362 |
|---|---|---|---|
| HIV-1 RT | 195 | I G Q H R T K I E E L R Q H L L R W **G** L T T P D · **K** K H Q K | 223 |
| Klenow | 892 | D V D A V A K Q I H Q L M E N C * L L V E V G S G E N W D Q A H | C-term 928 |
| T7 RNAP | 818 | P A D A A N L F K A V R E T α continues | to C-term 831 |

α α α α α α α α α α α α α α α α α α α α α    β β β β

**Motif E**

| Poliovirus | 363 | F E T V T W E N V T F L K R F F R A   To thumb | 380 |
|---|---|---|---|
| HIV-1 RT | 224 | E P P F L W M G Y E L H P   To thumb | 236 |

β β β T T β β β β

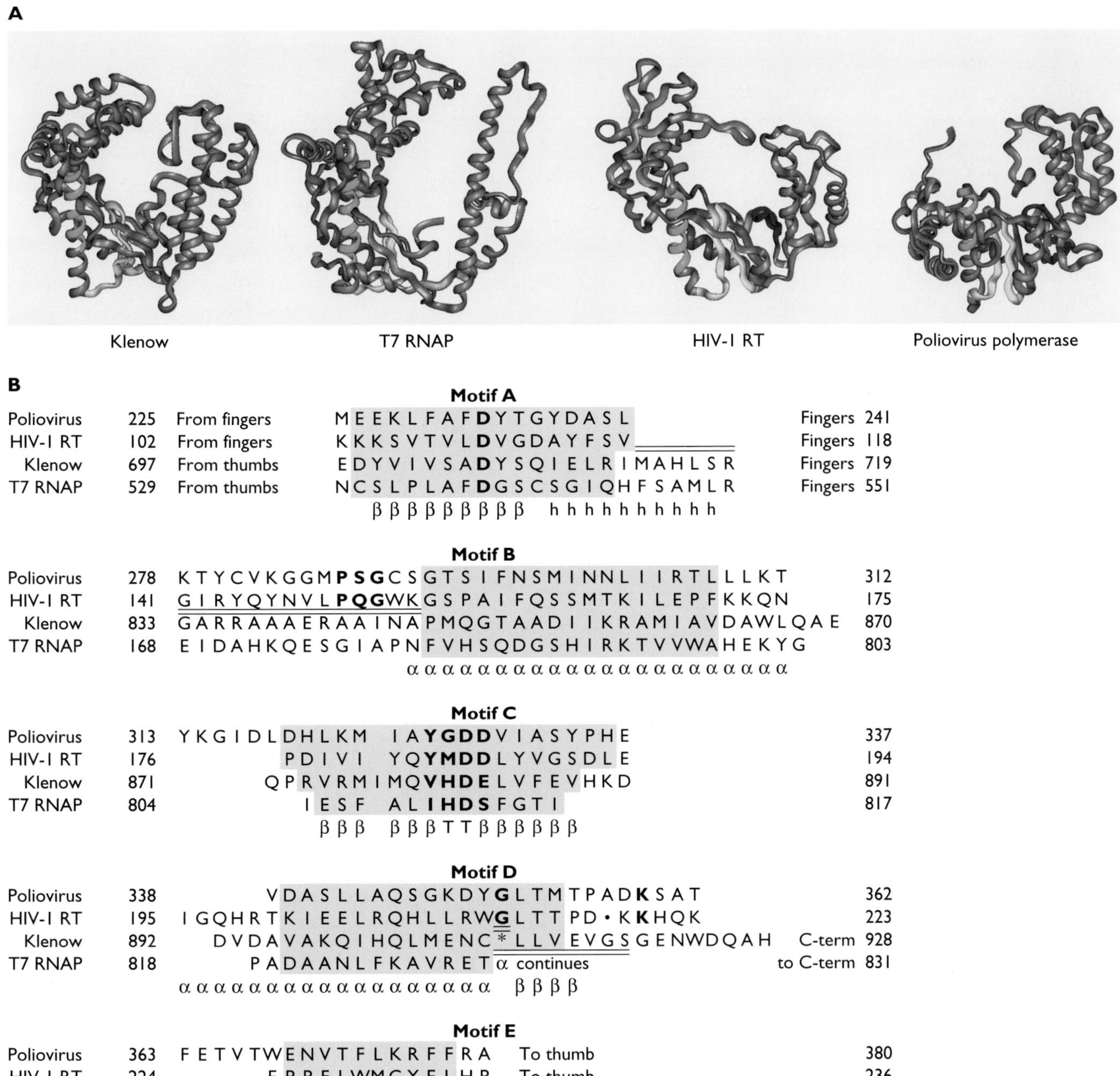

**Figure 6.4 Polymerase structure and sequence motifs in representative structures of each of the four types of nucleic acid polymerases.** (A) Ribbon diagrams of the polymerase domain of the large (Klenow) fragment of *Escherichia coli* DNA polymerase I, a DNA-dependent DNA polymerase; T7 RNA polymerase (T7 RNAP), a DNA-dependent RNA polymerase; human immunodeficiency virus type 1 reverse transcriptase (HIV-1 RT), an RNA-dependent DNA polymerase; and polioviral 3D$^{pol}$, an RNA-dependent RNA polymerase. The thumb subdomain is at the right, and the fingers subdomain is at the left. The structure/sequence motifs A, B, C, D, and E are labeled and are red, green, yellow, purple, and dark blue, respectively. (B) Sequence alignments based on the structures of 3D$^{pol}$, human immunodeficiency virus type 1 reverse transcriptase, the large (Klenow) fragment of DNA polymerase I, and T7 RNA polymerase. Motif B is the green portion in panel A, not motif B* of DNA-dependent polymerases (Fig. 6.2). α-helical and β-strand regions are labeled. Purple boxes indicate the extent of structural agreement. The most conserved amino acids are in bold type. Double lines are regions in which structures differ. Abbreviations: C-term, C terminus; h, helix; T, turn. Adapted from J. L. Hansen et al., *Structure* **5**:1109–1127, 1997, with permission. Courtesy of S. Schultz, University of Colorado, Boulder.

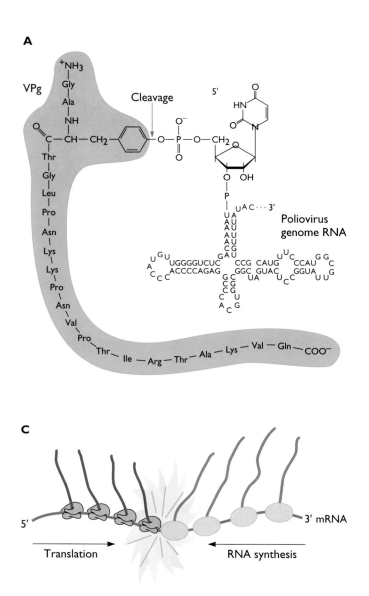

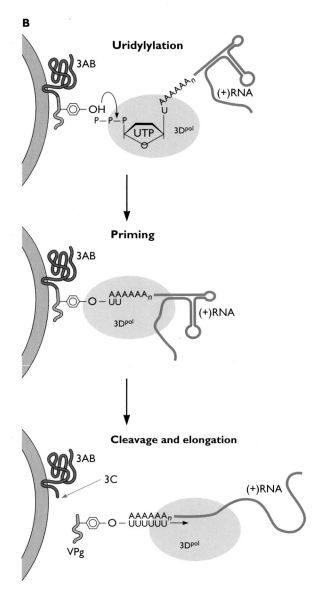

**Figure 6.5 Polioviral RNA synthesis.** (A) Linkage of VPg to polioviral genomic RNA. Polioviral RNA is linked to the 22-amino-acid-long VPg (orange) via an O4-(5′-uridylyl)-tyrosine linkage. This phosphodiester bond is cleaved at the indicated site by a cellular enzyme to produce the viral mRNA containing a 5′-terminal Up. Adapted from N. Kitamura et al., *Nature* **291:**547–553, 1981, with permission. (B) Model of VPg priming of polioviral RNA synthesis. A precursor of VPg, known as 3AB, is likely to participate in the initiation reaction. 3AB is a membrane-bound polypeptide and therefore an ideal candidate to act as a VPg donor in membranous replication complexes. The 3AB protein contains a hydrophobic domain. A virus carrying a mutation that reduces the hydrophobicity of this 3AB domain is defective in initiation of RNA synthesis, in vitro uridylylation of VPg, and in vivo synthesis of (+) strand viral RNAs. However, because no protein larger than VPg linked to nascent RNA strands has been detected, it is likely that 3AB is rapidly cleaved by the protease 3C to form VPg-RNA. Uridylylation of membrane-bound 3AB is carried out by 3D$^{pol}$ and is dependent on the presence of an RNA template.

VPg-pUpU is then cleaved from 3A by the viral protease 3C and serves as a primer for elongation by the viral polymerase 3D$^{pol}$. (C) Ribosome-RNA polymerase collisions. A strand of viral RNA is shown, with ribosomes translating in the 5′ → 3′ direction and RNA polymerase copying the RNA chains in the 3′ → 5′ direction. Ribosome-polymerase collisions would occur in cells infected with (+) strand RNA viruses unless mechanisms exist to avoid simultaneous translation and replication. Adapted from R. Andino et al., *EMBO J.* **12:**3587–3598, 1993, with permission. (D) Model of the synthesis of polioviral (−) strand RNA. (a) Schematic of the structure of the ribonucleoprotein complex that forms on a cloverleaflike structure within the first 108 nucleotides of (+) strand RNA. Viral protein 3CD and cellular poly(rC)-binding protein are shown interacting with different loops of the cloverleaf. Formation of this ribonucleoprotein complex is required for synthesis of (+) and (−) strand RNA. (b) Three steps in the initiation of (−) strand synthesis. The top panel shows a (+) strand on which the ribonucleoprotein is formed, interacting with membrane-bound 3AB. The viral (+) strand is shown with the 5′ clover-

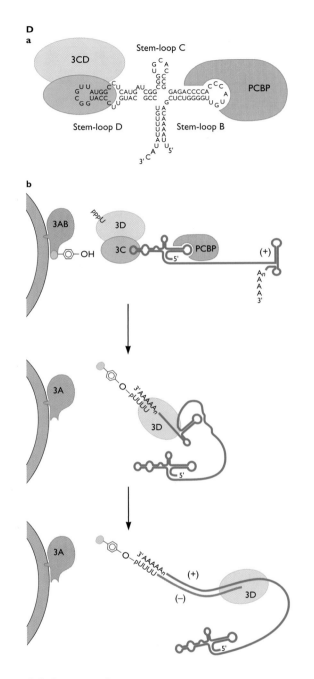

**Figure 6.5** *(continued)*

leaf, 3' pseudoknot, and poly(A). In the next panel, 3AB is uridylylated and cleaved to release VPg-pUpU, and 3CD is cleaved. All three activities are proposed to be carried out by 3CD. The released 3D$^{pol}$ primes synthesis at the 3' end of the (+) strand with VPg-pUpU as a primer. Recognition of the 3' end is believed to be facilitated by a pseudoknot structure. The last panel shows the elongation of a new (−) strand RNA. Once the (−) strand is complete, a cloverleaf forms at its 5' end, with the formation of a new ribonucleoprotein in anticipation of a round of (+) strand synthesis. Many initiation reactions can occur on the original (+) strand RNA while (−) strands are being elongated. A role for protein 2C is not shown, nor is a role for the head-to-tail oligomerization of 3D$^{pol}$.

Another way that the RNA polymerase may be directed to the RNA template is by protein-protein interactions. The vesicular stomatitis virus RNA polymerase consists of the P protein, an RNA-binding protein required for viral mRNA synthesis, and the L protein, the catalytic subunit. The C-terminal region of the P protein binds to the L protein, while the N-terminal region associates with the RNA template. In this way the P protein brings the L polymerase to the correct location for initiating mRNA synthesis. Cellular general initiation proteins have a similar function in bringing RNA polymerase II to the correct site to initiate transcription of DNA templates.

**Stimulating polymerase activity.** Although purified polioviral 3D$^{pol}$ can copy RNA in the presence of a primer, addition of purified 3AB to an in vitro reaction greatly stimulates polymerase activity. Protein 3AB interacts with 3D$^{pol}$ and 3CD in infected cells, and mutations in 3D$^{pol}$ that disrupt its interaction with 3AB result in viruses with defects in viral RNA synthesis. These results indicate that 3AB is a stimulatory cofactor for 3D$^{pol}$. Its postulated role in viral RNA synthesis is described in the model shown in Fig. 6.5.

**RNA helicases.** Base-paired regions in viral RNA must be disrupted to permit copying by RNA-dependent RNA polymerase. RNA helicases, which are encoded in the genomes of many RNA viruses (see Table 6.1 for examples), are believed to be required to melt the genomes of double-stranded RNA viruses and the secondary structures in template RNAs and to prevent extensive base pairing between template RNA and the nascent complementary strand. The RNA helicase of hepatitis C virus, an important human pathogen, has been studied extensively because the protein is a potential target for chemotherapeutic intervention. To facilitate the development of new agents that inhibit the hepatitis C virus helicase, its three-dimensional structure has been determined by X-ray crystallography.

### Cell Proteins in Viral RNA Synthesis

It has been known for over 30 years that host cell components are required for viral RNA synthesis. Such components were initially called "host factors" because nothing was known about their chemical composition. Evidence that host cell proteins are essential components of a viral RNA polymerase came first from studies of the bacteriophage Qβ. The RNA-dependent RNA polymerase of this bacteriophage is a multisubunit enzyme, consisting of a 65-kDa virus-encoded protein and three host proteins: ribosomal protein S1 and the translation elongation proteins EF-Tu and EF-Ts. The 65-kDa viral protein exhibits no RNA polymerase activity in the absence of the host proteins but has sequence similarity with known RNA-dependent RNA polymerases.

Polioviral RNA synthesis also requires host cell proteins. When purified polioviral RNA is incubated with a cytoplasmic extract prepared from cultured permissive cells, the genomic RNA is translated and the viral RNA polymerase is made. If guanidine hydrochloride is included in the reaction, the polymerase assembles on the viral genome, but RNA synthesis does not initiate. The RNA polymerase-template assembly can be isolated free of guanidine, but RNA synthesis cannot begin unless a new cytoplasmic extract is added, indicating that soluble cellular proteins are required for initiation. A similar conclusion comes from studies in which polioviral RNA was injected into oocytes derived from the African clawed toad *Xenopus laevis*. Polioviral RNA cannot replicate in *Xenopus* oocytes unless it is coinjected with a cytoplasmic extract from human cells.

Host cell poly(rC)-binding proteins are required for polioviral RNA synthesis. As described above, these proteins bind to a cloverleaflike secondary structure that forms in the first 108 nucleotides of (+) strand RNA (Fig. 6.5). Formation of a ribonucleoprotein composed of the 5′ cloverleaf, 3CD, and poly(rC)-binding proteins is essential for the initiation of viral RNA synthesis. Binding of poly(rC)-binding proteins to the cloverleaf facilitates the binding of viral protein 3CD to the opposite side of the same cloverleaf. A model of how these interactions lead to viral RNA synthesis is presented in Fig. 6.5.

Another candidate for a host protein that is essential for polioviral RNA synthesis is Sam68 (Src-associated in mitosis, 68 kDa), which is physically associated with 3D$^{pol}$ in poliovirus-infected cells. Sam68 is normally located in the cell nucleus but becomes redistributed to the cytoplasm, the site of viral RNA synthesis, during polioviral infection. These properties implicate Sam68 in polioviral RNA synthesis, although no additional evidence for such a role has been obtained.

Host cell cytoskeletal proteins participate in paramyxoviral RNA synthesis. Measles virus and Sendai virus mRNA synthesis in vitro is stimulated by tubulin, the major structural component of microtubules of the cell's cytoskeleton, and inhibited by anti-β-tubulin antibodies. In contrast, mRNA synthesis of human parainfluenza virus type 3 and respiratory syncytial virus requires cellular actin, which forms the microfilaments of the cytoskeletal network. Actin and tubulin might serve as anchoring sites on the cytoskeleton for the viral RNA polymerase. In support of this hypothesis, it has been shown that the incoming viral ribonucleoproteins associate with the cytoskeletal framework, where they are active in RNA synthesis. Assembling the RNA polymerase on the cytoskeleton or on cell membranes may ensure high local concentrations of

replication components and hence increase the rates or efficiencies of replication reactions.

## The Nature of the RNA Template

### Naked or Nucleocapsid RNA

The genomes of (−) strand viruses are organized into nucleocapsids in which protein molecules, including the RNA-dependent RNA polymerase and accessory proteins, are bound to the genomic RNAs at regular intervals. The RNA polymerases of (−) strand viruses copy viral RNAs only when they are present in the nucleocapsid and not when they are naked RNAs. For example, vesicular stomatitis virus genomic RNA is a template for RNA polymerase only when it is bound to the nucleocapsid protein, N. The function of the bound N protein molecules is likely to maintain the RNA in a conformation suitable for initiation of mRNA synthesis. This suggestion is based on the observation that influenza virus RNA is wound around the nucleoprotein with the bases exposed to the outside of the complex. In contrast, the genomes of (+) strand RNA viruses are not coated with viral proteins in the virion. The main reason for this structural difference may be that the genomes of (−) strand RNA viruses must be ready to begin mRNA synthesis upon cell entry, whereas the genomes of (+) strand RNA viruses are translated.

The viral nucleoproteins are cooperative single-stranded RNA-binding proteins, as are the single-stranded nucleic acid-binding proteins required during DNA-directed DNA and RNA synthesis. Their function during replication is to keep the RNA single-stranded and prevent base pairing between the template and product so that additional rounds of RNA synthesis can occur. Many (+) strand RNA viruses encode helicases that serve a similar function. Interestingly, in addition to its enzymatic activity, polioviral 3D$^{pol}$ is also a cooperative single-stranded RNA-binding protein and can unwind RNA duplexes without the hydrolysis of ATP characteristic of helicase-mediated unwinding. Polioviral RNA polymerase is therefore functionally similar to the RNA-binding nucleoproteins of (−) strand viruses. Consistent with this hypothesis, two regions of polymerase-polymerase interaction have been observed in 3D$^{pol}$ protein crystals (Fig. 6.6). Interface I is extensive and consists of more than 23 amino acid side chains on two different surfaces of the protein. This interaction occurs in a head-to-tail fashion and results in long fibers of polymerase molecules. When RNA is modeled into the 3D$^{pol}$ structure according to the location of DNA in the human immunodeficiency virus type 1 reverse transcriptase-DNA cocrystal, it falls along the fibers formed by interface I interactions (Fig. 6.7). Interface II is formed by N-terminal polypeptide segments, perhaps from different polymerase molecules, which in

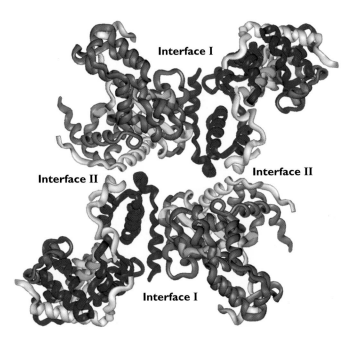

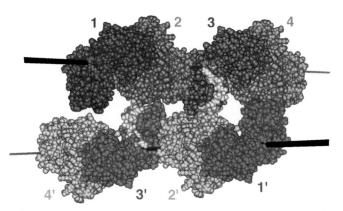

**Figure 6.7 Model of oligomerization of the polioviral RNA-dependent RNA polymerase 3Dpol.** Two fibers formed by interface I interactions are shown; each contains four molecules of 3D$^{pol}$ (labeled 1, 2, 3, 4, and 1′, 2′, 3′, 4′). The two fibers are shown interacting along interface II. The fibers cross at 120° angles; as a result, 1 and 1′ come out of the page, and 4 and 4′ go into the page. The axis of each fiber is shown in black. Shown in white are the N-terminal strands of the thumb subdomains that might be contributed by a molecule in the other fiber. Only one set of interface II interactions is shown. Adapted from J. L. Hansen et al., *Structure* **5:**1109–1127, 1997, with permission. Courtesy of S. Schultz, University of Colorado, Boulder.

**Figure 6.6 Interactions between 3Dpol molecules in crystals of the polymerase.** Four 3D$^{pol}$ molecules are shown; the structure and sequence motifs and N-terminal regions are color-coded as in Fig. 6.3. The two interfaces between polymerase molecules, called interface I and interface II, are labeled. Interface I is formed by interactions between the front of the thumb subdomain of one molecule and the back of the palm subdomain of another. This interaction leads to head-to-tail fibers of polymerase. Interface II is formed by interactions between the two N-terminal regions of the polymerase, residues 12 to 37 and residues 67 to 97 (shown in white). Interface II is shown formed by the contact of polymerase molecules in neighboring fibers formed by interface I. Because of the lack of order at the N-terminal regions, the origin of these regions among interacting 3D$^{pol}$ molecules is not known. However, it has been suggested that residues 14 to 35 might derive from an adjacent polymerase molecule. The consequences of such a model are shown in Fig. 6.7. Adapted from J. L. Hansen et al., *Structure* **5:**1109–1127, 1997, with permission. Courtesy of S. Schultz, University of Colorado, Boulder.

combination with interactions at interface I might lead to a network of polymerase fibers. Because the N terminus of 3D$^{pol}$ is required for enzyme activity, interface II interactions seem likely to be functionally important.

### Secondary Structures in Viral RNA

RNA molecules are not simple linear chains but can form secondary structures that have important functions. Viral RNA genomes contain secondary-structure elements such as base-paired **stem regions, hairpin loops, bulge loops, interior loops**, and **multibranched loops** (Fig. 6.8). An **RNA pseudoknot** is formed when a single-stranded loop region base pairs with a complementary sequence outside the loop (Fig. 6.8). These structures are important in RNA viral translation, assembly, and RNA synthesis.

The first step in identifying a structural feature in RNA is to scan the nucleotide sequence with computer programs designed to fold the RNA into energetically stable structures. Comparative sequence analysis can provide evidence that RNA secondary structures exist. For example, comparison of the RNA sequences of several related viruses might reveal that the structure but not the sequence of a stem-loop is conserved. The best evidence for RNA structure comes from experiments in which RNAs are treated with enzymes or chemicals that specifically attack single- or double-stranded regions. The results of such analyses confirm whether predicted stem regions are base-paired, while loop regions are unpaired.

As mentioned previously, the template specificity of RNA-dependent RNA polymerases may in part reside in the recognition of structures at the 5′ and 3′ ends of viral RNAs. Polioviral RNA polymerase 3D$^{pol}$ copies only viral RNAs in infected cells. On the other hand, the purified enzyme copies any polyadenylated RNA in vitro if provided with an oligo(U) primer. This result has led to the suggestion that the specificity of the polymerase in cells must derive from its interaction with sequence elements in the viral RNA. The 3′ noncoding region of polioviral genomic RNA contains an RNA pseudoknot structure that is conserved among picornaviruses and was identified by chemical, enzymatic, and thermodynamics-based structure analysis. Disruption of the pseudoknot by mutagenesis impairs viral RNA synthesis, suggesting that the structure functions in the synthesis of (−) strand RNA. A model of

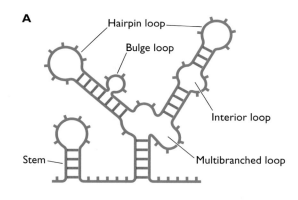

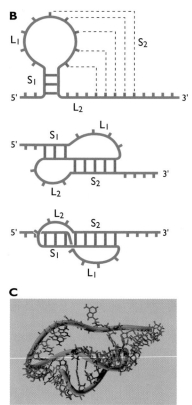

**Figure 6.8 RNA secondary structure.** (A) Schematic of different structural motifs in RNA. Black bars indicate base pairs, and open bars, unpaired nucleotides. (B) Schematic of a pseudoknot. (Top) Stem 1 ($S_1$) is formed by base pairing in the stem-loop structure, and stem 2 ($S_2$) is formed by base paring of nucleotides in the loop with nucleotides outside the loop. (Middle) A different view of the formation of stems $S_1$ and $S_2$. (Bottom) Coaxial stacking of $S_1$ and $S_2$ resulting in a quasi-continuous double helix. (C) Structure of a pseudoknot as determined by nuclear magnetic resonance. The sugar backbone is highlighted with a tube. Stacking of the bases in the areas of $S_1$ and $S_2$ can be seen. Adapted from C. W. Pleij, *Trends Biochem. Sci.* **15:**143–147, 1990, with permission.

how the 3′ end of polioviral RNA participates in the initiation of RNA synthesis is shown in Fig. 6.5. The 5′ end of the poliovindal (+) strand contains a cloverleaflike structure necessary for (+) strand RNA synthesis. A model of the function of this structure is depicted in Fig. 6.5D. Similarly, a stem-loop structure found at the 3′ end of (−) strand alphaviral RNA is believed to serve as the initiation site for (+) strand RNA synthesis.

RNA stem-loop structures not only provide binding sites for proteins involved in viral RNA synthesis but may suppress this process under certain conditions. In cells infected with arenaviruses, only the 3′ half of the genomic RNAs is copied to produce mRNAs (Fig. 6.9). Continued mRNA synthesis may be blocked by a predicted stem-loop structure in the two genomic RNA segments. It is not known how this block is overcome to produce full-length copies of the genomic RNA during replication. A stem-loop structure in full-length antigenomic RNAs also limits mRNA production to the 3′ ends.

## Initiation and Elongation

### Initiation Mechanisms

The requirement for a primer in the initiation step of nucleic acid synthesis varies among the different classes of polymerases. Both DNA- and viral RNA-dependent DNA polymerases are primer-dependent enzymes, while DNA-dependent RNA polymerases can initiate RNA synthesis de novo. Most of the RNA-dependent RNA polymerases of RNA viruses probably initiate RNA synthesis without a primer, but there are three notable exceptions, the enzymes of picornaviruses, influenza viruses, and bunyaviruses. These RNA polymerases use one of two unusual priming mechanisms.

**Protein priming.** Polioviral RNA replication is probably primed by a genome-linked protein, a mechanism that occurs during adenoviral DNA replication, as described in chapter 9. As we have seen, polioviral 3D^pol is a primer-dependent enzyme which will not copy polioviral RNA in vitro without an oligo(U) primer. The finding that poliovindal genomic RNA, as well as newly synthesized (+) and (−) strand RNAs, are covalently linked at their 5′ ends to the 22-amino-acid-long protein VPg (Fig. 6.5A) suggested that VPg might function as a primer for RNA synthesis. This hypothesis was supported by the discovery of a uridylylated form of the protein, VPg-pUpU, in infected cells. If VPg is first uridylylated in vitro and then added to a polymerase reaction containing 3D^pol and a poly(A) template, the uridy-

**Genomic segments**

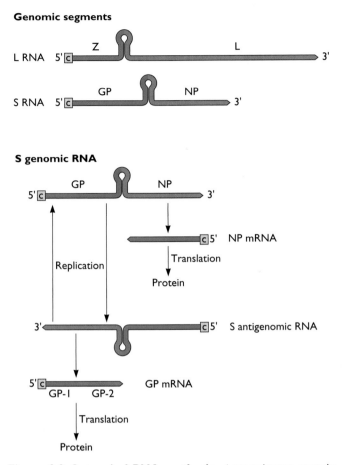

**Figure 6.9 Arenaviral RNA synthesis.** Arenaviruses contain two genomic RNA segments, S (small) and L (large) (top). At early times after infection, only the 3' region of each of these segments is copied to form mRNA: the N mRNA from the S genomic RNA and the L mRNA from the L genomic RNA. Copying of the remainder of the S and L genomic RNAs may be blocked by a stem-loop structure in the genomic RNAs. After the S and L genomic RNAs are copied into full-length (+) strands, the 3' regions of the (+) RNAs are copied to produce mRNAs: the glycoprotein precursor (GP) mRNA from S (+) RNA and the Z mRNA (encoding a protein of unknown function) from the L (+) RNA. Termination of mRNA synthesis on the (+) strands may also occur at a stem-loop structure. Only RNA synthesis from the S RNA is shown in detail. Adapted from B. N. Fields et al. (ed.), *Fields Virology* (Lippincott-Raven Publishers, Philadelphia, Pa., 1996), with permission.

lylated VPg is extended to form poly(U). A model of VPg priming of polioviral RNA synthesis is depicted in Fig. 6.5B.

**Priming by capped RNA fragments.** Influenza virus mRNA synthesis occurs in the host cell nucleus and is blocked by the fungal toxin α-amanitin, which inhibits cellular DNA-dependent RNA polymerase II. This dependence on nuclear function is a consequence of the requirement by the viral RNA polymerase for capped RNA primers derived by cleavage of host cell RNA polymerase

II transcripts. Such transcripts do not accumulate in the nucleus because they are either degraded or rapidly processed into mature mRNAs and transported to the cytoplasm. Consequently, viral mRNA synthesis is inhibited by α-amanitin because a continuous supply of newly synthesized cellular RNA polymerase II transcripts is needed to provide primers for influenza virus mRNA synthesis.

In a process called **cap snatching**, nuclear RNA polymerase II transcripts are cleaved 10 to 13 nucleotides from their 5' ends by a virus-encoded, cap-dependent endonuclease that is part of the RNA polymerase (Fig. 6.10A and B). The resulting capped fragments serve as primers for the initiation of viral mRNA synthesis. Priming does not require hydrogen bonding between the cellular capped RNA fragment and the 3' end of the genomic RNA templates. However, a 5'-methylated cap structure is necessary, indicating that the cap of the RNA primer is recognized by one or more viral proteins.

Small quantities of the three P proteins (PB1, PB2, and PA) are associated with viral nucleocapsids, which are the templates for mRNA synthesis. During mRNA synthesis, the PB2 protein binds to the cap of the RNA primer, and the PB1 protein catalyzes nucleotide addition during RNA chain elongation. The PA protein does not function during mRNA synthesis but plays an active role during genomic RNA replication (see "Antitermination and Different RNA Polymerases" below).

Bunyaviral mRNA synthesis is also primed with capped fragments of cellular RNAs. In contrast to that of influenza virus, bunyaviral mRNA synthesis is not inhibited by α-amanitin because it occurs in the cytoplasm, in which there are many capped cellular RNAs. Therefore, active RNA polymerase II is not required to provide a continuous supply of capped RNA primers.

### Polyadenylation

The mRNAs synthesized during infection by most RNA viruses (exceptions are arena- and reoviruses) contain a 3' poly(A) sequence, as do the vast majority of cellular mRNAs. For (+) strand viruses whose genome is also mRNA, the poly(A) sequence is encoded in the viral RNA. For example, both polioviral genomic RNA and mRNA contain a 3' stretch of poly(A), approximately 62 nucleotides in length, which is required for infectivity. The (−) strand RNA contains a 5' stretch of poly(U), which is copied to form poly(A) of the (+) strand.

Another mechanism for poly(A) addition is reiterative copying of, or "stuttering" at, a short U sequence in the (−) strand template. After initiation, vesicular stomatitis virus mRNAs are elongated until the RNA polymerase reaches a stretch of seven uridine (U) residues located in each intergenic region (Fig. 6.11C). Poly(A) is added by re-

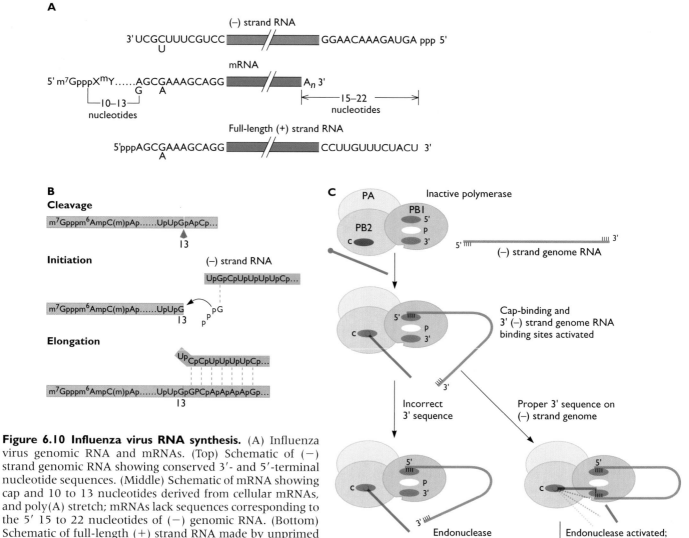

**A**

(−) strand RNA

3' UCGCUUUCGUCC▓▓▓▓//▓▓▓▓GGAACAAAGAUGA ppp 5'
        U

mRNA

5' m⁷GpppXᵐY......AGCGAAAGCAGG▓▓▓▓//▓▓▓▓Aₙ 3'
              G      A
    └─10–13─┘                    ├─── 15–22 ───┤
    nucleotides                      nucleotides

Full-length (+) strand RNA

5'pppAGCGAAAGCAGG▓▓▓▓//▓▓▓▓CCUUGUUUCUACU 3'
                A

**B**

**Cleavage**

m⁷Gpppm⁶AmpC(m)pAp......UpUpGpApCp...
                          ▲
                          13

**Initiation**

                        (−) strand RNA

                    UpGpCpUpUpUpUpCp...

m⁷Gpppm⁶AmpC(m)pAp......UpUpG        pG
                          13      p   p

**Elongation**

                    Up  CpCpUpUpUpUpCp...

m⁷Gpppm⁶AmpC(m)pAp......UpUpGpGPCpApApApApGp...
                          13

**Figure 6.10 Influenza virus RNA synthesis.** (A) Influenza virus genomic RNA and mRNAs. (Top) Schematic of (−) strand genomic RNA showing conserved 3'- and 5'-terminal nucleotide sequences. (Middle) Schematic of mRNA showing cap and 10 to 13 nucleotides derived from cellular mRNAs, and poly(A) stretch; mRNAs lack sequences corresponding to the 5' 15 to 22 nucleotides of (−) genomic RNA. (Bottom) Schematic of full-length (+) strand RNA made by unprimed end-to-end copying of (−) strand genomic RNA. (B) Mechanism of cap snatching, capped RNA-primed initiation of influenza virus mRNA synthesis. The first step is the cleavage of capped RNA fragments from the 5' ends of cellular nuclear RNAs. These fragments then serve as primers for viral mRNA synthesis. The 10 to 13 nucleotides in these primers do not need to hydrogen-bond to the common sequence found at the 3' ends of the influenza virus genomic RNA segments. The first nucleotide added to the primer is a G residue templated by the penultimate C residue of the genomic RNA segment, followed by elongation of the mRNA chains. The terminal U residue of the genomic RNA segment does not direct the incorporation of an A residue. The 5' ends of the viral mRNAs therefore comprise 10 to 13 nucleotides plus a cap structure snatched from host nuclear RNAs. Adapted from S. J. Plotch et al., *Cell* **23:**847–858, 1981, with permission. (C) Activation of the influenza virus RNA polymerase by specific virion RNA sequences. The RNA polymerase, comprising the three P proteins, lacks activity until it binds to the 5'-terminal sequence of genomic RNA. Binding of the PB1 protein to this RNA sequence induces the PB2 protein to bind to the cap of a cellular RNA. When the 3'-terminal sequence of genomic RNA binds to a second amino acid sequence in the PB1 protein, the polymerase acquires the activity to cleave the capped cellular RNA 10 to 13 nucleotides from the cap. The RNA polymerase can then carry out initiation and elongation of mRNAs. 5' and 3' indicate the binding sites for the 5' and 3' ends, respectively, of (−) strand genomic RNA. Capped RNA is green; the 5' cap structure is a circle at one end; p is the polymerase active site. Blue indicates an inactive site, and red indicates an active site. The p site is not colored because it is not known if it is constitutively active or requires activation. The mature polymerase is bound to both the 5' and 3' ends of the genomic RNA, with the capped RNA primer asso-

**C**

PA  PB1  PB2    Inactive polymerase

        5'
        p
        3'

        5' ⅢⅢ                    ⅢⅢ 3'
              (−) strand genome RNA

Cap-binding and 3' (−) strand genome RNA binding sites activated

Incorrect 3' sequence

Proper 3' sequence on (−) strand genome

Endonuclease inactive

Endonuclease activated; initiation and elongation occur

UUUUUU

Elongation

AAAA
UUUUUU

Polyadenylation

**D**

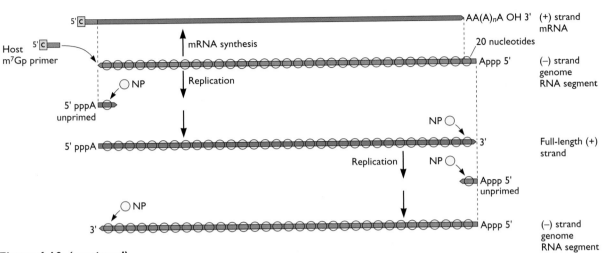

**Figure 6.10** *(continued)*

ciated with the PB2 protein. During RNA synthesis, the polymerase remains bound to the 5′ end of the genomic RNA, and the 3′ end of the genomic RNA is threaded through (or along the surface of) the polymerase as the PB1 protein catalyzes each nucleotide addition to the growing mRNA chain. This threading process continues until the mRNA reaches a position on the genomic RNA that is close to the binding site of the polymerase. At this point the polymerase itself blocks further mRNA synthesis, and reiterative copying of the adjacent U7 tract occurs. After about 150 A residues are added to the 3′ end of the mRNA, mRNA synthesis terminates. The polymerase specifically binds to the sequence AGCAAAGCAGG, which is found in all viral mRNAs near their 5′ ends, immediately downstream of the sequence snatched from host cell capped RNAs. Such binding blocks access of the viral cap-dependent endonuclease to the viral mRNAs. The 5′ ends of host cell mRNAs are not protected in this way and are therefore efficiently cleaved by the viral endonuclease. (Right) Activated endonuclease; (left) influenza virus mRNA bound to the RNA polymerase and protected from cleavage by the endonuclease. Adapted from B. N. Fields et al. (ed.), *Fields Virology*, 3rd ed. (Lippincott-Raven Publishers, Philadelphia, Pa., 1996), with permission. (D) The switch from influenza virus mRNA synthesis to genomic RNA replication is regulated by both the number of "free" nucleocapsid (NP) protein molecules and the acquisition by the viral RNA polymerase of the ability to catalyze initiation without a primer. Binding of the NP protein to elongating (+) strands enables the polymerase to read all the way to the 5′ end of genomic RNA. (E) Functionally different forms of the viral RNA polymerase carry out mRNA synthesis and genomic replication. During mRNA synthesis, the PB2 cap-binding protein is required for the initiation of mRNA chains by host cell capped RNA primers, and the PA protein subunit does not have a role. In contrast, during replication the PB2 protein would have no role, and the PA protein is postulated to be required for initiation of both (+) and (−) strands without a primer. The three virus-specific RNA sequences highlighted in different colors activate the polymerase for mRNA synthesis and genomic replication. One of the virus-specific sequences (shown in red) binds to the RNA polymerase for mRNA synthesis or genomic replication and is found in both viral mRNAs and full-length (+) strand RNAs.

**E**

**RNA polymerase for mRNA synthesis**

Viral mRNA synthesis

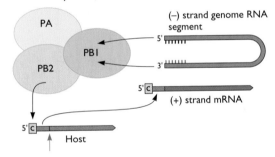

**RNA polymerase for genome replication**

Synthesis of full-length (+) strands

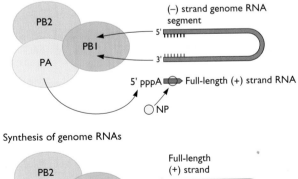

Synthesis of genome RNAs

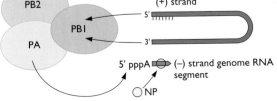

**A**

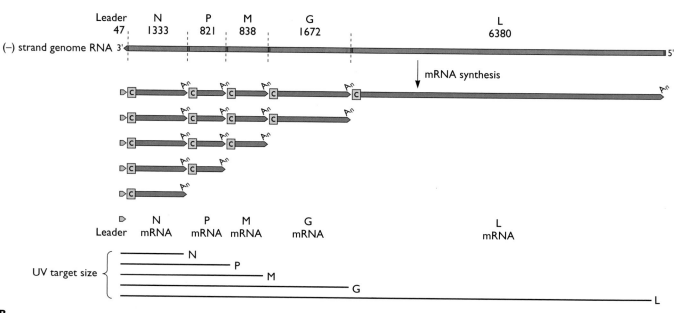

**B**

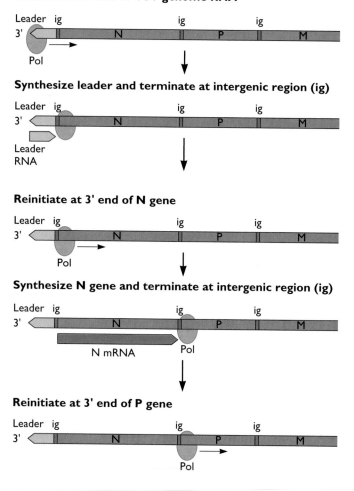

**Initiation at 3' end of VSV genome RNA**

**Synthesize leader and terminate at intergenic region (ig)**

**Reinitiate at 3' end of N gene**

**Synthesize N gene and terminate at intergenic region (ig)**

**Reinitiate at 3' end of P gene**

**Figure 6.11 Vesicular stomatitis virus RNA synthesis.** (A) Vesicular stomatitis virus mRNA map and UV map. The genome is shown as a thick green line at the top, and the N, P, M, G, and L genes and their relative sizes are indicated. The 47-nucleotide leader RNA is encoded at the 3' end of the genomic RNA. The leader and intergenic regions are shown in red. The RNAs encoded at the 3' end of the genome are made in larger quantities than the RNAs encoded at the 5' end of the genome. UV irradiation experiments determined the size of the vesicular stomatitis virus genome (UV target size) required for synthesis of each of the viral mRNAs. The UV target size of each viral mRNA corresponded to the size of the genomic RNA sequence encoding the mRNA plus all of the genomic sequence 3' to this coding sequence. (B) Stop-start model of mRNA synthesis. The RNA polymerase (Pol) initiates RNA synthesis at the 3' end of the genome RNA. The leader RNA is elongated, and synthesis terminates at the intergenic region (ig). RNA synthesis reinitiates at the 3' end of the N gene, and after synthesis of the N mRNA, RNA synthesis terminates at the intergenic region, followed by reinitiation at the 3' end of the P gene. This process continues until all five mRNAs are synthesized. Reinitiation does not occur after the last mRNA (the L mRNA) is synthesized, and as a consequence the 59 5'-terminal nucleotides of the vesicular stomatitis virus genomic RNA are not copied. Only a fraction of the polymerase molecules successfully make the transition from termination to reinitiation of mRNA synthesis at each intergenic region. (C) Functions of the RNA polymerase at an intergenic region. After the U7 sequence at the end of the mRNA-encoding sequence has been copied, the resulting A7 sequence in the mRNA slips off the U7 genomic sequence, which is then recopied. This process continues until approximately 200 A residues are added to the 3' end of the mRNA. Termination then occurs, completing the synthesis of the mRNA, followed by initiation and capping of the next mRNA. The dinucleotide NA in the genomic RNA is not copied. The

**C**

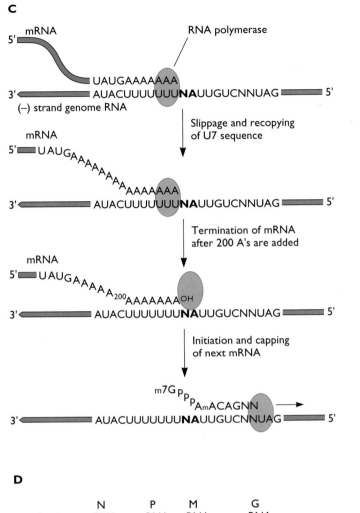

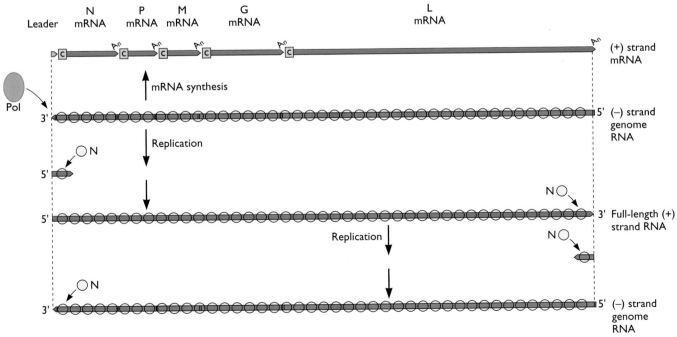

**Figure 6.11** *(continued)*

genomic RNA being copied is bound to viral N protein molecules at regular intervals (15 to 20 nucleotides). (D) The switch from mRNA synthesis to genomic RNA replication is regulated by the viral nucleocapsid (N) protein. When concentrations of N protein are low, mRNA synthesis is favored (top). When the N protein concentration is higher, it binds to nascent (+) strands and allows the RNA polymerase to read through the intergenic junctions (ig) at which polyadenylation and termination take place during mRNA synthesis.

**D**

iterative copying of the U stretch, followed by termination. Polyadenylation is achieved by a similar mechanism during influenza virus mRNA synthesis (Fig. 6.10D).

Conventional models for RNA synthesis involve a stationary template and moving polymerase. If we consider a model for influenza virus mRNA synthesis in which the enzyme is stationary and the template moves, a mechanism becomes apparent for reiterative copying of the five to seven U residues that produces the poly(A) sequence (Fig. 6.10D). Because the RNA polymerase specifically binds the 5′ end of (−) strand RNA, it has been postulated that it remains bound at this site of each genomic RNA segment throughout mRNA synthesis. The genomic RNAs would therefore be threaded through the polymerase in a 3′ → 5′ direction as mRNA synthesis proceeds. Eventually the polymerase itself would act as a physical barrier to mRNA synthesis, resulting in reiterative copying of the U residues.

### Protein Processing and Conformational Changes in the Regulation of RNA Synthesis

Viral RNA polymerases can be modified so that they are adapted to the specific requirements of mRNA synthesis and genomic replication. Alterations such as proteolytic processing and conformational changes may influence template recognition and initiation.

For example, in alphavirus-infected cells three different multisubunit RNA polymerases are produced from a single polyprotein precursor, P1234 (Fig. 6.12A). Cleavage at different sites in the precursor produces enzymes that contain the same protein sequences in different polypeptides and recognize different initiation sites for RNA synthesis (Figure 6.12B). These cleavages most likely cause conformational changes that alter template specificity (see "Regulation of Initiation: Different Polymerase Specificities" below).

On the other hand, in poliovirus-infected cells, two different cleavage reactions occur during the initiation of RNA synthesis that activate the replication proteins for nucleic acid synthesis (Fig. 6.5). The first is cleavage of uridylylated VPg from its membrane anchor, 3AB, releasing it so that it can be used for priming. In the second, the precursor 3CD, which is bound to the 5′ cloverleaf structure via 3C, is cleaved to 3C and 3D$^{pol}$. As 3CD does not possess polymerase activity, proteolysis of 3CD activates the poly-

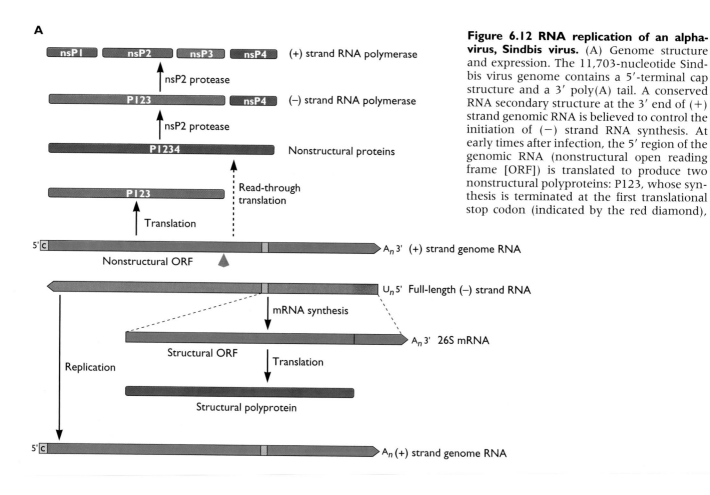

**A**

**Figure 6.12 RNA replication of an alphavirus, Sindbis virus.** (A) Genome structure and expression. The 11,703-nucleotide Sindbis virus genome contains a 5′-terminal cap structure and a 3′ poly(A) tail. A conserved RNA secondary structure at the 3′ end of (+) strand genomic RNA is believed to control the initiation of (−) strand RNA synthesis. At early times after infection, the 5′ region of the genomic RNA (nonstructural open reading frame [ORF]) is translated to produce two nonstructural polyproteins: P123, whose synthesis is terminated at the first translational stop codon (indicated by the red diamond),

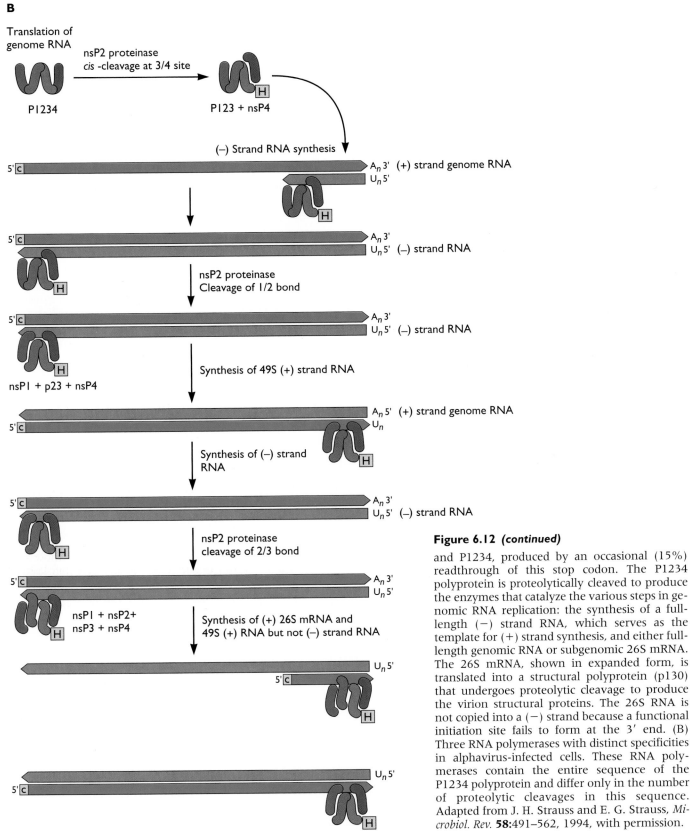

**Figure 6.12** *(continued)*

and P1234, produced by an occasional (15%) readthrough of this stop codon. The P1234 polyprotein is proteolytically cleaved to produce the enzymes that catalyze the various steps in genomic RNA replication: the synthesis of a full-length (−) strand RNA, which serves as the template for (+) strand synthesis, and either full-length genomic RNA or subgenomic 26S mRNA. The 26S mRNA, shown in expanded form, is translated into a structural polyprotein (p130) that undergoes proteolytic cleavage to produce the virion structural proteins. The 26S RNA is not copied into a (−) strand because a functional initiation site fails to form at the 3′ end. (B) Three RNA polymerases with distinct specificities in alphavirus-infected cells. These RNA polymerases contain the entire sequence of the P1234 polyprotein and differ only in the number of proteolytic cleavages in this sequence. Adapted from J. H. Strauss and E. G. Strauss, *Microbiol. Rev.* **58:**491–562, 1994, with permission.

merase once it has been targeted to the correct location on the viral RNA via the 3C portion.

The cap-binding and endonuclease functions of the influenza virus RNA polymerase are activated, probably by conformational changes, via sequential interactions with two viral RNA sequences (Fig. 6.10C). When coexpressed, the three P proteins form a multisubunit assembly that can neither bind to capped primers nor synthesize mRNAs. Addition of a sequence corresponding to the 5'-terminal 11 nucleotides of the viral RNA, which is almost fully conserved in all eight genome segments, activates the cap-binding activity of the P proteins. Such activation appears to occur by an allosteric mechanism, for it is the PB1 protein, rather than the cap-binding PB2 subunit, that interacts with this RNA sequence. Concomitantly with activation of its cap-binding activity, the P proteins acquire the ability to bind to a conserved sequence at the 3' ends of genomic RNA segments. This second interaction activates the endonuclease that cleaves capped host cell RNAs, enabling the P proteins to synthesize viral mRNA.

### Imbalanced (−) and (+) Strand Synthesis

One of the unexplained curiosities of RNA virus infectious cycles is the imbalance in the synthesis of genomes and their full-length complements in infected cells. For example, in cells infected with poliovirus, genomic RNA is produced at 100-fold higher concentrations than its complement. Assuming that RNA genomes and their complementary strands have similar stabilities, the simplest explanation for this difference is that the two strands are synthesized by different mechanisms which vary in efficiency. In alphavirus-infected cells, (−) strand RNAs are synthesized only for a short time early in infection, while the synthesis of (+) strands continues for much longer and leads to accumulation of excess mRNA and (+) strand genomic RNA.

In cells infected with (−) strand RNA viruses, there is an excess of full-length (−) over full-length (+) strands. In the case of influenza virus, this difference can probably be attributed to the synthesis of full-length (+) strands for only a short period at early times after infection. These (+) strands, which are in the form of nucleocapsids, remain in the nucleus throughout infection, where they serve as templates for genomic RNA synthesis. The mechanism for the temporal regulation of full-length (+) strand synthesis is unknown.

### The Problem of Ribosomes Encountered by (+) Strand RNA Virus Polymerases

The genomic RNA of (+) strand viruses can be translated in the cell, and the translation products include the viral RNA polymerase. At a certain point in infection, the RNA polymerase copies the RNA in a 3' → 5' direction, while ribosomes traverse it in a 5' → 3' direction (Fig. 6.5C), raising the question of how the viral polymerase avoids collisions with ribosomes translating in the opposite direction. In vitro experiments with inhibitors of protein synthesis demonstrate that when ribosomes are frozen on polioviral RNA, its replication is inhibited. In contrast, when ribosomes are released, replication of the RNA increases. These results suggest that ribosomes must be cleared from viral RNA before it can serve as a template for (−) strand RNA synthesis; in other words, replication and translation cannot occur simultaneously.

It has been suggested that the interaction of viral and cellular proteins with the cloverleaf structure in the polioviral 5' untranslated region might determine whether the RNA genome is translated or replicated. In this model, interaction of cellular poly(rC)-binding protein with the cloverleaf stimulates early translation in infected cells. Once the 3CD precursor has been synthesized, it binds to the cloverleaf structure, repressing translation and promoting RNA synthesis (Fig. 6.5D). However, this model cannot account for the observation that viral RNA translation continues in infected cells for several hours after the onset of RNA replication.

Even if mechanisms exist for controlling whether the polioviral genome is translated or replicated, some ribosome-RNA polymerase collisions do occur. This conclusion is drawn from the isolation of a polioviral mutant whose genome contains an insertion of a 15-nucleotide sequence from 28S rRNA. After colliding with a ribosome, the RNA polymerase apparently copied 15 nucleotides of rRNA before returning to the viral RNA template.

### Discrimination of Viral and Cellular RNAs

How do viral RNA polymerases selectively copy viral RNA in the midst of hundreds of thousands of polyadenylated cellular mRNAs? It has been suggested that conserved RNA structural features in the terminal noncoding region of viral RNAs are responsible for the specific recognition by viral RNA polymerases. For example, the RNA pseudoknot that appears to be important for the initiation of polioviral (−) strand synthesis may play a role in the specificity of copying by 3D$^{pol}$. The 3D$^{pol}$ and 3CD proteins cannot bind to the 3' end of polioviral RNA unless 3AB is present. The 3AB-3D$^{pol}$ may determine the specificity of binding to the 3' pseudoknot. The template specificity imparted by such terminal structures might be of greater importance early in infection when concentrations of viral proteins are low and few replication sites have been assembled. This suggestion originates from the observation that polioviruses from which the entire 3' noncoding region of the viral RNA has been removed by mutagenesis can replicate. Primary determinants of template selection by the polymerase might include the membrane association of the RNA polymerase, a property that would lead to increased local concentrations of essential components, as well as the fact that translation of 3D$^{pol}$, the most 3′-terminal gene, would position the polymerase at the 3′ end of the genome, ready for initiation. Early in infection, the 3′ pseudoknot structure might facilitate template selection when few viral polymerase molecules are available and membrane association has not yet provided highly concentrated areas of replication components.

The specific RNA-binding activity of the influenza virus P proteins explains the selective snatching of caps only from cellular RNAs. If the 5′ ends of newly synthesized viral mRNAs were also cleaved and used as primers, then no net synthesis of viral mRNAs would occur. The P protein binds not only to the cap of a nascent viral mRNA, via its PB2 subunit, but also to a (+) strand RNA sequence located immediately downstream of the capped cellular RNA primer (Fig. 6.10E). Such binding to two sites in the nascent viral mRNA blocks access of a second P protein and thus selectively protects newly synthesized viral mRNA from endonucleolytic cleavage by P proteins.

### Synthesis of Nested Subgenomic mRNAs

An unusual pattern of mRNA synthesis occurs in cells infected with members of the families *Coronaviridae* and *Arteriviridae*, in which subgenomic mRNAs that form a 3′-coterminal nested set with the viral genome are synthesized (Fig. 6.13). These viral families were recently combined into the order *Nidovirales* to reflect this property (*nidus* is Latin for "nest").

The subgenomic mRNAs of these viruses are composed of a leader and a body which are synthesized from noncontiguous sequences in the viral (+) strand genome. The leader is encoded in the 5′ end of the viral genome, and the mRNA bodies are encoded in the 3′ end. The leader and body are separated by a conserved junction sequence encoded both in the 3′ end of the leader and in the 5′ end of the mRNA body. The complementarity between the (+) strand junction sequence at the 3′ end of the leader and the (−) strand junction sequence at the start of each mRNA body led to the leader-primed model for mRNA synthesis (Fig. 6.13). This model was initially supported by the failure to detect, in virus-infected cells, subgenome-length (−) strands, either free or as part of replicative intermediates. Recently, both these structures have been detected in infected cells, leading to a new model of discontinuous mRNA synthesis. Subgenome-length (−) strand RNAs are produced when the viral RNA polymerase copies the (+) strand genome and "jumps" from the junction sequence to the leader sequence at the 5′ end (Fig.

6.13). Such (−) strand subgenome-length RNAs then serve as templates for mRNA synthesis.

## Cellular Sites of Viral RNA Synthesis

### Nucleus

mRNA synthesis and genome replication of influenza virus and Borna disease virus, a paramyxovirus with a nonsegmented (−) strand RNA genome, are unique among the nonretroid RNA viruses because they take place in the cell nucleus. The use of the nucleus by influenza virus cannot be dictated by the need to produce capped primers from cellular mRNAs; such mRNAs are found in the cytoplasm and are used by bunyaviruses in a similar process of cap snatching. Rather, influenza virus RNA synthesis may take place in the nucleus to allow production of spliced viral mRNAs. The mRNA and genomic RNAs of the viral satellite agent hepatitis delta virus are also made in the nucleus, where these processes are carried out by host cell DNA-dependent RNA polymerase II, a nuclear enzyme.

### Cytoplasm

mRNA and genome synthesis of most RNA viruses occurs in the cytoplasm of the cell, and the RNA polymerase may be membrane associated. For example, alphaviral RNA synthesis occurs on the cytoplasmic surface of endosomes and lysosomes, and polioviral RNA polymerase is located on the surfaces of small, membranous vesicles. Cells infected with poliovirus accumulate these vesicles as a result of the inhibition of membrane vesicle transport from the endoplasmic reticulum to the cell surface. Polioviral infection therefore induces the membrane vesicles important for replication by inhibition of an intracellular protein transport pathway. Because poliovirus is nonenveloped, its replication does not require the protein glycosylation and transport functions of the endoplasmic reticulum and Golgi apparatus.

Why viral replication complexes should be membrane-associated is not known. However, tight membrane association may ensure high local concentrations of replication components and hence increase the rates or efficiencies of replication reactions. As we have seen, this property may contribute to the specificity of polioviral 3D$^{pol}$ for viral RNA templates. Membrane association of replication complexes may also allow efficient packaging of progeny RNA into virions.

The replication of some RNA viruses, such as reovirus, occurs in the cytoplasm in **inclusion bodies** that are not bound by membranes. These bodies contain viral RNA, proteins, and incomplete particles and probably serve the same role as membranes in RNA synthesis of other viruses,

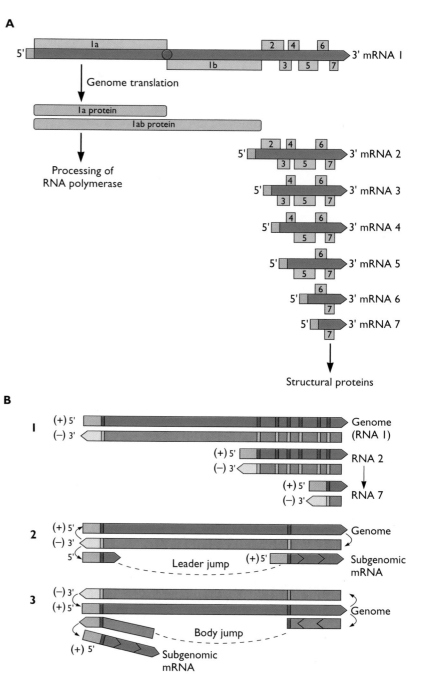

**Figure 6.13 Nidoviral genome organization and expression.** (A) Organization of open reading frames. The (+) strand viral RNA is shown at the top, with open reading frames as boxes. The genomic RNA is translated to form proteins 1a and 1ab, which are processed to form the RNA polymerase. Structural proteins are encoded by six nested mRNAs that share a common 5′ leader sequence (orange box). (B) Models of the synthesis of nested mRNAs. (1) Overview of (+) and (−) strand RNAs. (2) Discontinuous transcription was proposed to occur during (+) strand RNA synthesis by leader priming on a full-length (−) strand RNA template. Leader RNA (orange box) is joined to the mRNAs by a junction sequence also found adjacent to the leader RNA at the 5′ end of the viral RNA. The (+) strand junction sequence at the 3′ end of the leader can base-pair with the (−) strand junction sequence of the region that encodes each mRNA. (3) After a nested set of subgenomic replicative intermediates were discovered in infected cells [i.e., containing both (+) and (−) strand RNAs], a new model was proposed in which a (−) strand is synthesized on the (+) strand genomic RNA and then translocated to the 5′ end of the genomic RNA to complete synthesis of the leader RNA. The resulting (−) strand RNAs, with leader sequences at the 3′ ends, are then copied to form mRNAs. The nested set of mRNAs would be produced by translocation signaled by the individual junction sequences between mRNA coding regions. Adapted from E. J. Snijder et al., *J. Gen. Virol.* **79:**961–979, 1998, with permission.

to focus the replication machinery on a discrete location in the cell.

## Unique Mechanisms of mRNA and Genome Synthesis of the Hepatitis Delta Satellite Virus

The strategy for replication and mRNA synthesis of the hepatitis delta virus satellite genome is quite different from that of any other virus. The viral RNAs are synthesized by host cell DNA-dependent RNA polymerase II, and the hepatitis delta virus RNAs are RNA catalysts, or **ribozymes** (Box 6.2). The genome of hepatitis delta virus is a 1,700-nucleotide (−) strand circular RNA, the only RNA with this structure that has been found in animal cells. As approximately 70% of the nucleotides are base-paired, the viral RNA is folded into an unusual rodlike structure (Fig. 6.14A). Surprisingly, all hepatitis delta virus-specific RNA

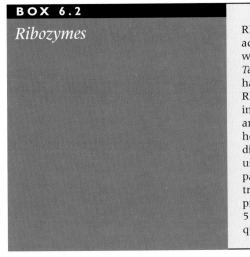

A **ribozyme** is an enzyme in which RNA, not protein, carries out the catalytic activity. The first ribozyme discovered was the group I intron of the ciliate *Tetrahymena thermophila*. Other ribozymes have since been discovered, including RNase P of bacteria, group II self-splicing introns, hammerhead RNAs of viroids and satellite RNAs, and the ribozyme of hepatitis delta virus. Ribozymes are very diverse in size, sequence, and mechanism used for catalysis. For example, the hepatitis delta virus ribozyme catalyzes a transesterification reaction that produces products with 2′,3′-cyclic phosphate and 5′-OH termini. Only an 85-nucleotide sequence is required for activity of this ri-bozyme, which can cleave optimally with as little as a single nucleotide 5′ to the site of cleavage. By joining the 85-nucleotide fragment to other upstream sequences, accurate 3′ ends of heterologous RNA transcripts synthesized in vitro can be obtained. Ribozymes can also be used to target the cleavage of other RNAs and are therefore being widely explored as antiviral and antitumor agents.

**Kruger, K., P. J. Grabowski, A. J. Zaug, J. Sands, D. E. Gottschling, and T. R. Cech.** 1982. Self-splicing RNA: autoexcision and auto-cyclization of the ribosomal RNA intervening sequence of *Tetrahymena*. *Cell* **31:**147–157.

**Westhof, E., and F. Michel.** Ribozyme architectural diversity made visible. *Science* **282:**251–252.

synthesis is catalyzed by cellular DNA-dependent RNA polymerase II, which has not been known to utilize RNA as a template. Genomic RNA replication requires reactions catalyzed by ribozymes residing in both the hepatitis delta virus genomic (−) RNA and its full-length (+) RNA copy. Both these hepatitis delta virus RNAs catalyze self-cleavage in the absence of any protein and may also catalyze self-ligation.

All hepatitis delta virus RNAs, including (−) genomic RNAs, (+) strand RNAs, and a single polyadenylated mRNA that encodes the delta antigen, are synthesized in the nucleus by RNA polymerase II. The current model for hepatitis delta virus RNA synthesis is summarized in Fig. 6.14B. The switch from mRNA synthesis to the production of full-length (+) RNA is controlled by suppression of a poly(A) signal (see "Suppression of Polyadenylation" below). Full-length (−) and (+) RNAs are copied by a rolling-circle mechanism, and ribozyme self-cleavage releases linear monomers. Subsequent ligation of the two termini produces a monomeric circular RNA. The hepatitis delta virus ribozymes are therefore needed to process the intermediates of rolling-circle RNA replication.

## The Switch from mRNA Production to Genome RNA Synthesis

The mRNAs of most RNA viruses are not complete copies of the viral RNA but are subgenomic. However, to assemble infectious viral particles, exact replicas of the RNA genome must be made. Viruses that utilize subgenomic mRNAs must therefore engineer a switch from mRNA synthesis to the production of full-length genomes. The mechanism used may be reversible, to ensure that subgenomic mRNAs can be made at the expense of genomic RNA when structural protein levels fall as new virions are assembled.

In this section we will consider examples of different mechanisms for the switch from mRNA synthesis to the production of full-length genomes. As might be expected, the majority of these mechanisms involve regulation at either the initiation or termination step of RNA synthesis. For example, initiation of RNA synthesis in alphavirus-infected cells is regulated by the production of RNA polymerases of different specificities. The switch from mRNA synthesis to genomic RNA synthesis is regulated at the termination or polyadenylation step of mRNA synthesis in cells infected with rhabdoviruses, paramyxoviruses, influenza viruses, arenaviruses, and hepatitis delta virus.

When the only viral mRNA made in infected cells is identical to the (+) strand genome, an RNA synthesis strategy exemplified by picornaviruses (Fig. 6.1), no switch from mRNA to genomic RNA synthesis is needed. However, newly synthesized polioviral (+) strand RNA molecules can serve as RNA templates for further genomic RNA replication, as mRNAs for the synthesis of viral proteins, or as genomic RNAs to be packaged into progeny virions. The finding that only newly synthesized genomic RNAs are encapsidated suggests a hypothetical mechanism for the assignment of new genomic RNAs to one of these three roles. When capsid protein concentrations in the cell are high, the majority of newly synthesized genomic RNA molecules would be encapsidated (Fig. 6.15) and thus unavailable for replication or translation. Most newly synthesized genomic RNA molecules would not be encapsidated when concentrations of capsid proteins are low. Unencapsidated genomic RNAs retained at membranous RNA replication sites would serve as templates for genomic RNA replication (see "Cellular Sites of RNA Synthesis" above for a description of replication sites), whereas (+) strand RNA molecules released from RNA replication sites would function as mRNAs. If packaging is

## A   RNAs of HDV

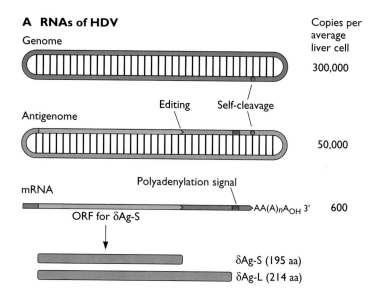

Genome

Editing   Self-cleavage

Antigenome

Polyadenylation signal

mRNA

ORF for δAg-S

AA(A)nAOH 3'

δAg-S (195 aa)

δAg-L (214 aa)

Copies per
average
liver cell

300,000

50,000

600

**Figure 6.14 Hepatitis delta satellite virus RNA synthesis.** (A) Schematic of the forms of hepatitis delta virus (HDV) RNA and delta antigen found in infected cells. aa, amino acids. (B) Overview of hepatitis delta virus mRNA and genomic RNA synthesis. In steps 1 to 3, RNA synthesis is initiated, most likely by host RNA polymerase II, at the indicated position on the (−) strand genomic RNA. The polymerase passes the poly(A) signal (purple box) and the self-cleavage domain (red circle). In steps 4 and 5, the 5′ portion of this RNA is processed by cellular enzymes to produce delta mRNA with a 3′ poly(A) tail, while RNA synthesis continues beyond the cleavage site and the RNA undergoes self-cleavage (step 6). RNA synthesis continues until at least one unit of the (−) genomic RNA template is copied. The poly(A) signal is ignored in this full-length (+) strand. In steps 7 to 10, after self-cleavage to release a full-length (+) strand, self-ligation produces a (+) strand circular RNA. In steps 11 to 20, mRNA synthesis initiates on the full-length (+) strand to produce (−) strands by a rolling-circle mechanism. Unit-length genomes are released by the viral ribozyme (step 15) and self-ligated to form (−) strand circular genomic RNAs. Adapted from J. M. Taylor, *Curr. Top. Microbiol. Immunol.* **239:**107–122, 1999, with permission.

**B**

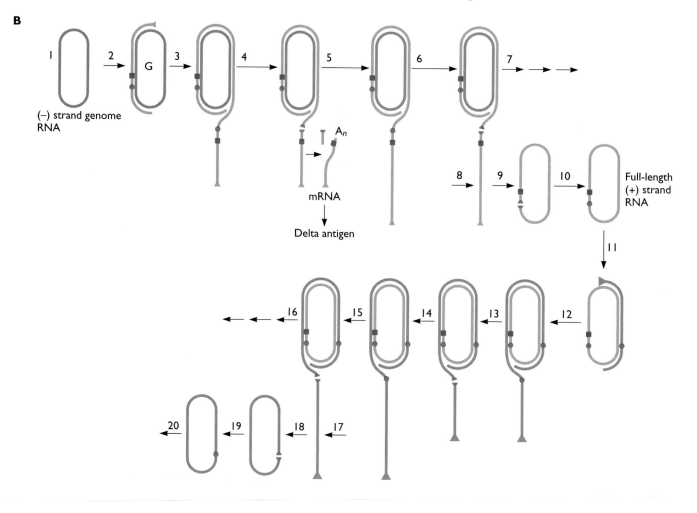

(−) strand genome
RNA

G

mRNA

Delta antigen

An

Full-length
(+) strand
RNA

**A  Low capsid protein**

**B  High capsid protein**

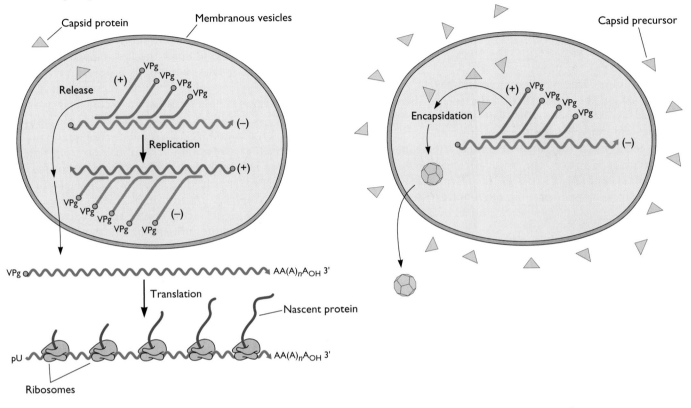

**Figure 6.15 Hypothetical model of how polioviral (+) strand RNAs play different roles in infected cells.** (+) strand RNA synthesis occurs on membranous vesicles. (A) When the concentration of capsid proteins is low, encapsidation of genomic RNA does not occur. Some of the newly synthesized genomic RNAs are retained on the vesicles and participate in further genomic RNA replication. The genomic RNAs released from the vesicles can serve as viral mRNAs. These mRNAs, when found on polyribosomes, lack the 5′-terminal VPg protein. (B) When the concentration of capsid proteins is high, encapsidation of genomic RNA is favored, and RNAs do not enter the replication/translation pathways.

directly linked to RNA replication, the level of virion capsid proteins in the infected cell would regulate the quantity of viral RNA available for translation or replication.

Because picornaviral mRNA is identical in sequence to the viral RNA genome, all RNAs needed for the reproduction of these viruses can be made by a simple set of RNA synthesis reactions (Fig. 6.5). Such simplicity comes at a price because there can be no regulation of individual gene products at the level of mRNA synthesis. However, polioviral gene expression can be controlled by the rate and extent of polyprotein processing. For example, the viral protease 3C$^{pro}$, which cleaves at certain Gln-Gly amino acid pairs in the polyprotein, does not efficiently process the capsid protein precursor VP0. The 3C$^{pro}$ precursor, 3CD, is required for the processing of VP0. By regulating the amount of 3CD that is produced, the level of capsid

protein processing can be controlled. However, as discussed previously 3CD is not an RNA polymerase and must be cleaved for RNA replication to occur.

## Regulation of Initiation: Different Polymerase Specificities

Other (+) strand RNA viruses have evolved a different mechanism of mRNA synthesis, one that allows structural and nonstructural proteins (generally needed in greater and lesser quantities, respectively) to be separately produced. The latter are expressed from full-length (+) strand (genomic) RNA, while structural proteins arise from subgenomic mRNA(s). This strategy is used by corona-, calici-, and alphaviruses. For example, translation of the Sindbis virus (+) strand RNA genome yields the nonstructural proteins that synthesize a full-length (−) strand (Fig.

6.12A). Such RNA molecules contain not only a 3'-terminal sequence for initiation of (+) strand RNA synthesis but also an internal initiation site. The initiation of (+) strand synthesis at this internal site produces a 26S subgenomic mRNA which is translated into the precursor of the structural proteins.

Regulation of alphaviral genome and mRNA synthesis is achieved by the sequential production of three RNA polymerases with different template specificities. All three enzymes are derived from the nonstructural polyprotein P1234 and contain the complete amino acid sequence of this precursor. However, the covalent connections among the segments (P1, P2, P3, P4) of the polyprotein are successively broken, with ensuing alterations in RNA polymerase specificity. The first RNA polymerase detected in alphavirus-infected cells, which recognizes only the initiation site for (−) strand RNA synthesis at the 3' end of genomic RNA (Fig. 6.12B), is cleaved only between the P3 and P4 sequences. Subsequent severing of the covalent link between the P1 and P2 portions of the polyproteins produces an RNA polymerase that specifically recognizes the initiation site for synthesis of full-length (+) strands at the 3' end of (−) strand templates and continues to recognize the initiation site for synthesis of full-length (−) strand RNA. A final proteolytic cleavage between P2 and P3 sequences imparts the ability to recognize the internal sequence for initiation of 26S subgenomic (+) mRNA and at the same time shuts down the ability to synthesize additional (−) strand RNA. This form of the polymerase can also synthesize full-length (+) strands but can no longer initiate (−) strand synthesis. It seems likely that each proteolytic cleavage induces a conformational change in the polymerase that alters its template specificity.

## Suppression of Intergenic Stop-Start Reactions by Nucleocapsid Protein

The genes of RNA viruses with a single (−) strand RNA genome are expressed by the production of subgenomic mRNAs in infected cells. However, replication requires synthesis of a single full-length (+) strand copy of the genome. The switch from subgenomic mRNA to full-length (+) strand synthesis is by suppression of intergenic stop-start reactions that produce individual mRNAs. This strategy is exemplified by vesicular stomatitis virus (Fig. 6.11). The genome of this virus, a single (−) strand RNA species, encodes five proteins, each of which is made by translation of a separate capped and polyadenylated subgenomic mRNA.

A **stop-start mechanism** (Fig. 6.11B) has been proposed to explain how vesicular stomatitis virus mRNAs are synthesized. The viral RNA polymerase, which consists of two subunits, the L and P proteins, initiates synthesis at a single entry site at or very near the 3' end of the genomic RNA. A 47-nucleotide leader RNA is produced, and RNA synthesis terminates in the **intergenic region**. RNA synthesis resumes, without dissociation of the polymerase from the template, to produce the first capped mRNA. As discussed above, poly(A) addition is a result of reiterative copying of a sequence of seven U residues present in each intergenic region. Chain termination occurs after approximately 200 A residues have been added to the mRNA, which is then released. The RNA polymerase then initiates synthesis of the next mRNA. The transition from reiterative copying and termination to initiation is not perfect, and only about one-third to one-half of the polymerase molecules accomplish this transition at each intergenic region. Such inefficiency accounts for the observation that 3'-proximal mRNAs are more abundant than 5'-proximal mRNAs. The stop-start mechanism also accounts for the observation that 3'-proximal genes must be copied before downstream genes (Box 6.3).

To copy the entire (−) strand RNA into a full-length (+) strand RNA, the stop-start reactions at intergenic regions during mRNA synthesis are suppressed (Fig. 6.11D). Suppression occurs on synthesis of the N protein, which is directed to nascent RNA chains through an interaction with P protein and binds to leader RNA in a sequence-specific manner. Additional N protein molecules then associate with the (+) RNAs as they are elongated and eventually bind to the A7 sequence in the intergenic region. This interaction prevents the A7 sequences from slipping backward along the genomic RNA template and thus blocks reiterative copying of the genomic U7 sequence by the L-P protein. Consequently, RNA synthesis continues through the intergenic region. The number of N protein molecules in infected cells therefore regulates the relative efficiencies of mRNA synthesis and genome RNA replication because full-length (+) strand RNAs can be made only when sufficient N protein is present. Once replication has begun, the levels of free N protein are reduced because each full-length genome requires approximately 2,600 N monomers to coat the chain.

The copying of full-length (+) strand RNAs to (−) strand genomic RNAs also requires the addition of N protein to elongating RNA molecules. Thus, newly synthesized (−) strand RNAs are produced as nucleocapsids that can be readily packaged into progeny viral particles.

The (−) strand RNA genome of paramyxoviruses is efficiently copied only when its length in nucleotides is a multiple of 6. This requirement, called the **rule of six**, is probably a consequence of the association of each N monomer with exactly six nucleotides. Assembly of the nucleocapsid begins with the first nucleotide at the 5' end of the RNA and continues until the 3' end is reached. The

**BOX 6.3**

*Mapping gene order by UV irradiation*

The effects of UV irradiation provide insight into the mechanism of vesicular stomatitis virus mRNA synthesis. In these experiments, vesicular stomatitis virus particles are irradiated with UV light, and the effect on the synthesis of individual mRNAs is determined. In principle, larger genes require less UV irradiation to inactivate mRNA synthesis and have a greater **target size**. The dosage of UV irradiation needed to inactivate synthesis of the N mRNA corresponded to the predicted size of the N gene, but this was not the case for the other viral mRNAs. The target size

of each other mRNA was the sum of its size plus the size of other genes located 3' of the gene. For example, the UV target size of the L mRNA is the size of the entire vesicular stomatitis virus genome (Fig. 6.11A). These results indicate that these mRNAs are synthesized sequentially, in the 3' → 5' order in which their genes are arranged in the vesicular stomatitis virus genome: N-P-M-G-L.

**Ball, L. A., and C. N. White.** 1976. Order of transcription of genes of vesicular stomatitis virus. *Proc. Natl. Acad. Sci. USA* **73:**442–446.

efficiency of initiation of RNA synthesis at the 3' end depends on the position of the RNA polymerase relative to the NP monomer, which is determined by the total number of nucleotides in the genome.

## Antitermination and Different RNA Polymerases

The switch from viral mRNA to full-length (+) strand synthesis during the influenza virus infectious cycle is regulated in part by the inhibition of poly(A) addition by NP protein as described for vesicular stomatitis virus (Fig. 6.11D). However, the activities of the viral RNA polymerase also appear to be regulated.

As discussed previously, the cap-binding subunit of the RNA polymerase is essential during influenza virus mRNA synthesis because it produces RNA primers for initiation. Even in the presence of NP protein, this form of polymerase synthesizes only subgenomic polyadenylated mRNAs. In contrast, the enzyme responsible for replication initiates the synthesis of full-length (+) and (−) strand RNAs de novo, that is, without a primer, and, in the presence of NP protein, reads through the polyadenylation and termination signals used for mRNA production (Fig. 6.10D). It is believed that the PA subunit of the RNA polymerase, which plays no known role in mRNA synthesis, is important for replication, but exactly how the initiation properties of the enzyme and its responsiveness to NP protein are regulated is not yet clear. The NP protein is believed to bind nascent (+) strand transcripts and block poly(A) addition by a mechanism analogous to that described for vesicular stomatitis virus N protein. Copying of (+) strand RNAs into (−) strand RNAs also requires NP protein. Intracellular concentrations of NP protein are therefore an important determinant of whether mRNAs or full-length (+) strands are synthesized.

## Suppression of Termination Caused by a Stem-Loop Structure

Certain (−) strand RNA viruses, such as arenaviruses, produce subgenomic mRNAs that terminate at a stem-loop structure in the viral RNA. The suppression of termination, by an unknown mechanism, results in the synthesis of full-length (+) RNAs.

Although arenaviruses are considered (−) strand RNA viruses, their genomic RNA is in fact **ambisense** because mRNAs are produced both from (−) strand genomic RNA and from complementary full-length (+) strands. The arenavirus genome comprises two RNA segments, S (small) and L (large) (Fig. 6.9). Shortly after entering cells, a virion-associated RNA polymerase synthesizes mRNAs from the 3' region of the S and L genomic RNA segments. Synthesis of each mRNA terminates at a stem-loop structure that is believed to block further RNA synthesis and production of full-length RNAs (Fig. 6.9). These mRNAs, which are translated to produce the nucleocapsid (N) protein and RNA polymerase (L) protein, respectively, are the only viral RNAs produced during the first several hours of infection. Later in infection, the block imposed by the stem-loop structure is somehow overcome, permitting the synthesis of full-length S and L (+) RNAs. The 3' regions of these RNAs serve as templates for the production of mRNAs that encode the envelope glycoprotein precursor and the Z protein. A stem-loop structure also restricts mRNA synthesis to the 3' ends of (+) strand RNAs (Fig. 6.9). Such temporal regulation of mRNA synthesis is a property of few RNA viruses.

## Different Templates Used for mRNA Synthesis and Genome Replication

A distinctive feature of the infectious cycle of double-stranded RNA viruses such as reovirus is the production of mRNAs and genomic RNAs from different templates in different viral particles. Because the viral genomes are double stranded, they cannot be translated by the cell. Therefore, the first step in infection is the production of mRNAs from each viral RNA segment by the virion-associated RNA polymerase (Fig. 6.16). Reoviral mRNAs contain 5' cap structures but lack 3' poly(A) tails.

The viral mRNAs are extruded from the viral particle into the cytoplasm, in which they may be translated or copied to form double-stranded RNA. Examination of the structure of actively transcribing rotavirus, a member of the *Reoviridae*, has allowed a three-dimensional visualization of how mRNAs are released from the particle (Box 6.4). Viral (+) strand RNAs that will serve as templates for (−) strand RNA synthesis are first packaged into newly assembled subviral particles (Fig. 6.16). Each (+) strand RNA is copied just once within the subviral particle to produce double-stranded RNA. This two-step replication process of double-stranded RNA bears little resemblance to the simultaneous copying of both strands that generally occurs during the replication of double-stranded DNA.

## Suppression of Polyadenylation

Another mechanism for switching from mRNA synthesis to genomic replication is to suppress a poly(A) addition signal, and an example occurs in hepatitis delta satellite virus-infected cells. Host RNA polymerase II initiates delta mRNA synthesis at a position on the genome near the beginning of the delta antigen-coding region (Fig. 6.14B). Once the polymerase has moved past the polyadenylation signal and the self-cleavage domain, the 3' poly(A) end of delta mRNA is produced by host cell enzymes. The RNA downstream of the poly(A) site is not degraded, in contrast to that of other mRNA precursors made by RNA polymerase II, but is elongated around the circular genome until a complete full-length (+) strand is made. The poly(A) site in this full-length (+) RNA is not used. The delta antigen bound to the rodlike RNA may block access of cellular enzymes to the poly(A) signal, thereby inhibiting polyadenylation.

## Origins of Diversity in RNA Viruses

### Misincorporation of Nucleotides

All nucleic acid polymerases make errors during chain elongation, resulting in insertion of an incorrect nucleotide. DNA-directed DNA polymerases have **proofreading** capabilities in the form of **exonuclease** activities that can correct such mistakes. RNA-dependent RNA polymerases do not possess proofreading ability. The result is that error frequencies in RNA replication can be as high as one misincorporation per $10^3$ to $10^4$ nucleotides polymerized, whereas the errors in DNA replication are about

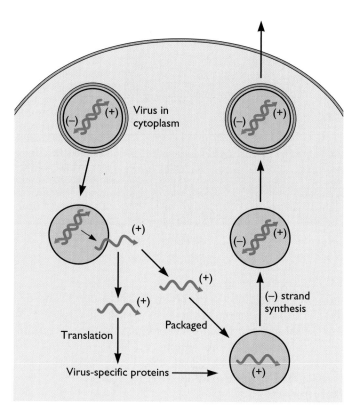

**Figure 6.16 mRNA synthesis and replication of double-stranded RNA genomes.** These processes occur in subviral particles containing the RNA templates and enzymes required for these reactions. During cell entry, the virus passes through the lysosomal compartment, and proteolysis of viral capsid proteins activates the RNA synthetic machinery. Single-stranded (+) viral mRNAs, which are synthesized in parental subviral particles, are extruded into the cytoplasm, where they serve either as mRNAs or as templates for the synthesis of (−) RNA strands. The latter viral mRNAs are first packaged into newly assembled subviral particles in which the synthesis of (−) RNAs to produce double-stranded RNAs occurs. These subviral particles become infectious particles. Only one double-stranded RNA is shown, whereas the reoviral genome consists of 10 to 12 double-stranded RNA segments.

1,000-fold lower. Many of these polymerization errors cause amino acid changes, and the viruses may not be infectious, while other mutations may appear in the genomes of infectious virions. This phenomenon has led to the realization that RNA virus populations are **quasispecies**, or mixtures of many different genome sequences. The errors introduced during RNA replication have important consequences for viral pathogenesis and evolution, as discussed in chapter 20. Because RNA virus stocks are mixtures of genotypically different viruses, specific viral mutants may be isolated directly from such stocks. For example, live attenuated poliovirus vaccine strains are viral mutants that were isolated from an unmutagenized stock of wild-type virus.

**BOX 6.4**

*Release of mRNA from rotavirus particles*

Rotaviruses, the most important cause of gastroenteritis in children, are large icosahedral viruses made of a three-shelled capsid containing 11 double-stranded RNA segments. Resolution of the structure of this virus by cryoelectron microscopy revealed that a large portion of the viral genome (~25%) is ordered within the particle and forms a dodecahedral structure (Fig. 3.17). In this structure, the RNAs interact with the inner capsid layer and pack around the RNA polymerase located at the fivefold axis of symmetry. Further analysis of rotavirus particles in the process of synthesizing mRNA has revealed that newly synthesized molecules are extruded through the capsid through several channels located at the fivefold axes (see figure). Multiple mRNAs are released at the same time from such particles. On the basis of these observations, it has been suggested that each double-stranded genomic RNA segment is copied by an RNA polymerase located at a fivefold axis of symmetry. This model may explain why no double-stranded RNA virus with more than 12 genomic segments—the maximum number of fivefold axes—has ever been identified.

Lawton, J. A., M. K. Estes, and B. V. Prasad. 1997. Three-dimensional visualization of mRNA release from actively transcribing rotavirus particles. *Nat. Struct. Biol.* **4**:118–121.

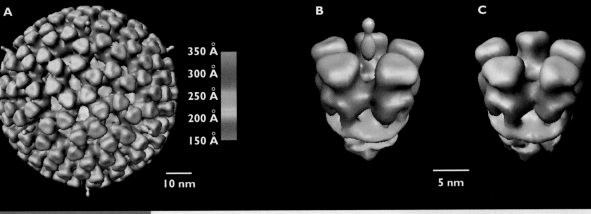

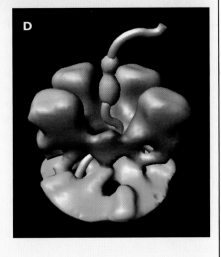

**Three-dimensional visualization of mRNA release from rotavirus particles synthesizing mRNA.** (A) Three-dimensional structure of rotavirus in the process of synthesizing mRNA. The capsid is depth-cued according to the color chart. Parts of newly synthesized mRNA that are ordered, and therefore structurally visible, are shown in pink at the fivefold axes of symmetry. (B) Close-up view of the channel at the fivefold axis and the visible mRNA. The mRNA is surrounded by five trimers of capsid protein VP6. (C) Close-up view of the channel at the fivefold axis of a particle not in the process of synthesizing mRNA. (D) Model of the pathway of mRNA transit through the capsid. One VP6 trimer has been omitted for clarity. The green protein is VP2, and the mRNA visible in the structure is shown in pink. The gray tube represents the possible path of an mRNA molecule passing through the VP2 and VP6 layers through the channel. Courtesy of B. V. V. Prasad, Baylor College of Medicine. Reprinted from J. A. Lawton et al., *Nat. Struct. Biol.* **4**:118–121, 1997, with permission.

## Segment Reassortment and RNA Recombination

**Reassortment** is the exchange of entire RNA molecules between genetically related viruses with segmented genomes. In cells coinfected with two different influenza viruses, the eight genome segments of each virus replicate. When new progeny viruses are assembled, they can package eight different RNA segments from either parental virus. Because reassortment involves the simple exchange of RNA segments, it can occur at high frequencies. Reassortment between human and animal strains of influenza virus produces antigenic diversity that is responsible for new outbreaks of the disease each year.

In contrast to reassortment, recombination is the exchange of nucleotide sequences among different genomic RNA molecules (Fig. 6.17, top). Recombination is a feature of many RNA viruses; it is an important mechanism for producing new genomes with selective growth advantages and has shaped the RNA virus world by rearranging genomes or moving functional parts of RNA molecules among different viruses. RNA recombination was first discovered in cells infected with poliovirus and was subsequently found to occur with other (+) and (−) strand RNA viruses. The frequency of recombination can be relatively high: it has been estimated that 10 to 20% of polioviral genomic RNAs recombine in one growth cycle. RNA recombination is not limited to the research laboratory. For example, recombinants among the three serotypes of poliovirus present in live virus (Sabin) vaccines are readily isolated from the intestines of vaccinated individuals. The significance of such recombinants is unknown, but it has been suggested that such viruses possess an improved ability to replicate in the human alimentary tract and thus have a selective advantage over the parental viruses.

Polioviral recombination is predominantly **base pairing dependent**: it occurs between nucleotide sequences of the two parental genomic RNA strands that have a high percentage of nucleotide identity (Fig. 6.17, bottom). Other viruses undergo **base-pairing-independent** recombination between very different nucleotide sequences. RNA recombination is believed to be coupled with the process of genomic RNA replication: it occurs by template exchange during (−) strand synthesis, as first demonstrated in poliovirus-infected cells. The RNA polymerase first copies the 3' end of one parental (+) strand and then exchanges one template for another at the corresponding position on a second parental (+) strand (Fig. 6.17, bottom). Template exchange in poliovirus-infected cells occurs predominantly during (−) strand synthesis because the concentration of (+) strands available for template exchange is 100-fold higher than (−) strands. This template exchange mechanism of recombination is also known as **copy choice**. The exact mechanism of template exchange is not known, but it might be triggered by pausing of the polymerase during chain elongation. In the **breakage-induced template exchange** model, it is postulated that breakage of the RNA template that is being copied promotes exchange of one template for another. It has been suggested that recombinant RNAs can be formed by a **breakage-and-ligation** mechanism similar to that which occurs during some DNA recombination. Although these last two models are plausible, they have not yet been experimentally demonstrated.

## RNA Recombination Leading to the Production of Pathogenic Viruses

A remarkable property of pestiviruses, members of the *Flaviviridae*, is that RNA recombination produces viruses that cause disease. Bovine viral diarrhea virus causes a usually fatal gastrointestinal disease. Infection of a fetus with this virus during the first trimester is noncytopathic, but RNA recombination produces a cytopathic virus that causes severe gastrointestinal disease after the animal is born.

Pathogenicity of bovine viral diarrhea virus is associated with the expression of a nonstructural protein, NS3, encoded by the recombinant cytopathic virus (Fig. 6.18). The NS3 protein cannot be produced in cells infected by the noncytopathic parental virus because its precursor, the NS2-3 protein, is not proteolytically processed. In contrast, NS3 is synthesized in cells infected by cytopathic bovine viral diarrhea virus because base-pairing-independent RNA recombination reactions that give rise to virus (Fig. 6.17, bottom) add an extra protease cleavage site in the viral polyprotein, precisely at the N terminus of the NS3 protein. This cleavage site can be created in several ways (see also "RNA Recombination Leading to the Production of Defective-Interfering RNAs" below). One of the most frequent is insertions of a cellular RNA sequence coding for **ubiquitin**, which targets cellular proteins to a degradative pathway. Insertion of ubiquitin at the N terminus of NS3 permits cleavage of NS2-3 by any member of a widespread family of cellular proteases. This recombination event provides a selective advantage because pathogenic viruses outgrow nonpathogenic viruses. Why cytopathogenicity is associated with release of the NS3 protein, which is thought to be part of the enzymatic system for genomic RNA replication, is not known.

Retroviruses acquire cellular genes by recombination, and the resulting viruses have lethal disease potential, as described in chapter 20.

## RNA Recombination Leading to the Production of Defective-Interfering RNAs

If the RNA polymerase skips sequences during template switching, deletion of viral RNA sequences may occur. The resulting subgenomic RNAs can replicate if they contain the appropriate signals for the initiation of RNA synthesis and are provided with RNA polymerase. These subgenomic RNAs are smaller than full-length genomic RNAs, replicate more rapidly, and therefore compete for the components of the RNA synthesis machinery. Because of this

## RNA recombination

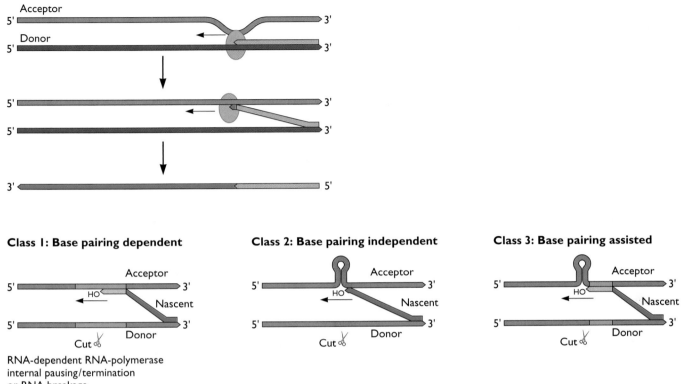

**Figure 6.17 RNA recombination.** (Top) Schematic representation of RNA recombination occurring during template switching by RNA polymerase, or copy choice. Two parental genomes are shown. The RNA polymerase (purple oval) has copied the 3' end of the donor genome and is switching to the acceptor genome. The resulting recombinant molecule is shown. (Bottom) Three classes of RNA recombination. All three classes involve RNA polymerase-mediated template exchange. Events that occur after the exchange are shown by an arrow. The hairpin symbolizes various RNA features required for class 2 and 3 recombination. In base-pairing-dependent recombination, substantial sequence similarity between parental RNAs is required and is the major determinant of recombination. In base-pairing-independent recombination, sequence similarity is not required but may be present. Recombination may be determined by other RNA features, such as RNA polymerase-binding sites, secondary structures, and heteroduplex formation between parental RNAs. Base-pairing-assisted recombination combines features of class 1 and class 2 recombination. Sequence similarity influences the frequency or site of recombination, but additional RNA features are required. Adapted from P. D. Nagy and A. E. Simon, *Virology* **235:**1–9, 1997, with permission.

property they are called **defective-interfering RNAs**. Such RNAs lack a full complement of viral genes and are packaged into viral particles only in the presence of a **helper** virus that provides viral proteins.

Defective-interfering viruses accumulate during the replication of both (+) and (−) strand RNA viruses. Production of these viruses requires a high multiplicity of infection, a condition that is readily achieved in the laboratory but rarely in nature. It is not known whether defective-interfering viruses generally play a role in viral pathogenesis. However, some recombination reactions that lead to the appearance of cytopathic bovine viral di-

arrhea viruses delete viral RNA sequences rather than inserting cellular RNA sequences. Such deletions create a new protease cleavage site at the N terminus of the NS3 protein, and the defective-interfering viruses also cause severe gastrointestinal disease in livestock.

## RNA Editing

Diversity in RNA viral genomes is also achieved by RNA editing. Viral mRNAs can be edited either by insertion of a nontemplated nucleotide during synthesis or by altering the base after synthesis. Examples of RNA editing have been documented in members of the *Paramyxoviridae* and

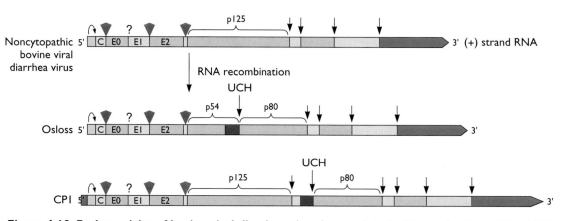

**Figure 6.18 Pathogenicity of bovine viral diarrhea virus is associated with production of the NS3 nonstructural protein by a recombinant cytopathic virus.** During infection of cells with noncytopathic bovine viral diarrhea virus, the NS3 protein is not produced but remains part of the precursor polypeptide NS2-3 (shown as p125). During replication, RNA recombination leads to precise insertion of the cellular ubiquitin gene into the polyprotein. Cleavage of the polyprotein at the N terminus of the NS3 protein is catalyzed by any member of a widespread family of cellular proteases that cleave ubiquitin in uninfected cells. Because of the release of this NS3 protein, the resulting viral strains are pathogenic. Two cytopathic viruses, Osloss and CP1, in which the ubiquitin sequence (UCH) has been inserted at different sites are shown. In Osloss, UCH has been inserted into the NS2-3 precursor, and NS3 (shown as p80) is produced. In CP1, a duplication has also occurred such that an additional copy of NS3 is present after the UCH sequence. Adapted from B. N. Fields et al. (ed.), *Fields Virology*, 3rd ed., (Lippincott-Raven Publishers, Philadelphia, Pa., 1996), with permission.

*Filoviridae* and in hepatitis delta satellite virus. This process is described in chapter 10.

## Perspectives

Our understanding of genomic replication and mRNA synthesis of RNA viruses has expanded considerably since the discovery of RNA-dependent RNA polymerase in the early 1960s. From early studies of infected cells and of partially purified enzymes, the field has progressed enormously with the development of recombinant DNA methods. These procedures have allowed the production and purification of RNA polymerases and their associated proteins, and the production of altered proteins and RNA templates, to better understand the mechanisms of RNA synthesis. As outlined in this chapter, many questions still remain. Our knowledge is rudimentary compared to that on reverse transcriptase, to which the entire next chapter is devoted. We expect that future research will provide insight into the nature and function of host proteins required for viral RNA synthesis and will help us to understand why RNA synthesis occurs in certain cellular compartments and how translation and replication are regulated on individual RNA molecules. Resolution of the crystal structure of polioviral 3D^pol hints at the beautiful pictures of other viral RNA polymerases and their associated proteins that will be forthcoming and how they complex with RNA. Because

our current understanding of RNA-directed RNA synthesis indicates that it is unique to viruses, this process is a good target for therapeutic intervention.

## References

### Reviews

**Deiman, B. A. L. M., and C. W. A. Pleij.** 1997. Pseudoknots: a vital feature in viral RNA. *Semin. Virol.* **8:**166–175.

**Horikami, S. M., and S. A. Moyer.** 1995. Structure, transcription, and replication of measles virus. *Curr. Top. Microbiol. Immunol.* **191:**35–50.

**Kolakofsky, D., T. Pelet, D. Garcin, S. Hausmann, J. Curran, and L. Roux.** 1998. Paramyxovirus RNA synthesis and the requirement for hexamer genome length: the rule of six revisited. *J. Virol.* **72:**891–899.

**Nagy, P. D., and A. E. Simon.** 1997. New insights into the mechanisms of RNA recombination. *Virology* **235:**1–9.

**Nowakowski, J., and I. Tinoco.** 1997. RNA structure and stability. *Semin. Virol.* **8:**153–165.

**Pleij, C. W.** 1990. Pseudoknots: a new motif in the RNA game. *Trends Biochem. Sci.* **15:**143–147.

**Sawicki, D. L., and S. G. Sawicki.** 1998. Role of the nonstructural polyproteins in alphavirus RNA synthesis. *Adv. Exp. Med. Biol.* **440:**187–198.

**Sawicki, S. G., and D. L. Sawicki.** 1998. A new model for coronavirus transcription. *Adv. Exp. Med. Biol.* **440:**215–219.

Xiang, W. K., A. V. Paul, and E. Wimmer. 1997. RNA signals in entero- and rhinovirus genome replication. *Semin. Virol.* **8:**256–273.

## Papers of Special Interest

### Poliovirus

Andino, R., G. E. Rieckhof, P. L. Achacoso, and D. Baltimore. 1993. Poliovirus RNA synthesis utilizes an RNP complex formed around the 5′-end of viral RNA. *EMBO J.* **12:**3587–3598.

Baltimore, D., and R. M. Franklin. 1963. Properties of the mengovirus and poliovirus RNA polymerases. *Cold Spring Harbor Symp. Quant. Biol.* **28:**105–108.

Barton, D. J., and J. B. Flanegan. 1997. Synchronous replication of poliovirus RNA: initiation of negative-strand RNA synthesis requires the guanidine-inhibited activity of protein 2C. *J. Virol.* **71:**8482–8489.

Charini, W. A., S. Todd, G. A. Gutman, and B. L. Semler. 1994. Transduction of a human RNA sequence by poliovirus. *J. Virol.* **68:**6547–6552.

Dasgupta, A., M. H. Baron, and D. Baltimore. 1979. Poliovirus replicase: a soluble enzyme able to initiate copying of poliovirus RNA. *Proc. Natl. Acad. Sci. USA* **76:**2679–2683.

Gamarnik, A. V., and R. Andino. 1998. Switch from translation to RNA replication in a positive-stranded RNA virus. *Genes Dev.* **12:**2293–2304.

Gamarnik, A. V., and R. Andino. 1996. Replication of poliovirus in *Xenopus* oocytes requires two human factors. *EMBO J.* **15:**5988–5998.

Giachetti, C., and B. L. Semler. 1991. Role of a viral membrane polypeptide in strand-specific initiation of poliovirus RNA synthesis. *J. Virol.* **65:**2647–2654.

Hansen, J. L., A. M. Long, and S. C. Schultz. 1997. Structure of the RNA-dependent RNA polymerase of poliovirus. *Structure* **5:**1109–1122.

Kirkegaard, K., and D. Baltimore. 1986. The mechanism of RNA recombination in poliovirus. *Cell* **47:**433–443.

Maynell, L. A., K. Kirkegaard, and M. W. Klymkowsky. 1992. Inhibition of poliovirus RNA synthesis by brefeldin A. *J. Virol.* **66:**1985–1994.

McBride, A. E., A. Schlegel, and K. Kirkegaard. 1996. Human protein Sam68 relocalization and interaction with poliovirus RNA polymerase in infected cells. *Proc. Natl. Acad. Sci. USA* **93:**2296–2301.

McBride, A. E., S. J. Taylor, D. Shalloway, and K. Kirkegaard. 1998. KH domain integrity is required for wild-type localization of Sam68. *Exp. Cell. Res.* **241:**84–95.

Molla, A., A. V. Paul, and E. Wimmer. 1993. In vitro synthesis of poliovirus. *Dev. Biol. Stand.* **78:**39–53.

Novak, J. E., and K. Kirkegaard. 1994. Coupling between genome translation and replication in an RNA virus. *Genes Dev.* **8:**1726–1737.

Parsley, T. B., J. S. Towner, L. B. Blyn, E. Ehrenfeld, and B. L. Semler. 1997. Poly(rC) binding protein 2 forms a ternary complex with the 5′-terminal sequences of poliovirus RNA and the viral 3CD proteinase. *RNA* **3:**1124–1134.

Paul, A. V., J. H. van Boom, D. Filippov, and E. Wimmer. 1998. Protein-primed RNA synthesis by purified poliovirus RNA polymerase. *Nature* (London) **393:**280–284.

Todd, S., J. S. Towner, D. M. Brown, and B. L. Semler. 1997. Replication-competent picornaviruses with complete genomic RNA 3′ noncoding region deletions. *J. Virol.* **71:**8868–8874.

Tolskaya, E. A., L. I. Romanova, M. S. Kolesnikova, A. P. Gmyl, A. E. Gorbalenya, and V. I. Agol. 1994. Genetic studies on the poliovirus 2C protein, an NTPase: a plausible mechanism of guanidine effect on the 2C function and evidence for the importance of 2C oligomerization. *J. Mol Biol.* **236:**1310–1323.

Towner, J. S., T. V. Ho, and B. L. Semler. 1996. Determinants of membrane association for poliovirus protein 3AB. *J. Biol Chem.* **271:**26810–26818.

### Alphaviruses

Lemm, J. A., A. Bergqvist, C. M. Read, and C. M. Rice. 1998. Template-dependent initiation of Sindbis virus RNA replication in vitro. *J. Virol.* **72:**6546–6553.

Lemm, J. A., T. Rumenapf, E. G. Strauss, J. H. Strauss, and C. M. Rice. 1994. Polypeptide requirements for assembly of functional Sindbis virus replication complexes: a model for the temporal regulation of minus- and plus-strand RNA synthesis. *EMBO J.* **13:**2925–2934.

Sawicki, D. L., and S. G. Sawicki. 1994. Alphavirus positive and negative strand RNA synthesis and the role of polyproteins in formation of viral replication complexes. *Arch. Virol. Suppl.* **9:**393–405.

Wang, Y. F., S. G. Sawicki, and D. L. Sawicki. 1994. Alphavirus nsP3 functions to form replication complexes transcribing negative-strand RNA. *J. Virol.* **68:**6466–6475.

### Paramyxoviruses

Barik, S. 1992. Transcription of human respiratory syncytial virus genome RNA in vitro: requirement of cellular factor(s). *J. Virol.* **66:**6813–6818.

Grosfeld, H., M. G. Hill, and P. L. Collins. 1995. RNA replication by respiratory syncytial virus (RSV) is directed by the N, P, and L proteins; transcription also occurs under these conditions but requires RSV superinfection for efficient synthesis of full-length mRNA. *J. Virol.* **69:**5677–5686.

Gupta, S., B. P. De, J. A. Drazba, and A. K. Banerjee. 1998. Involvement of actin microfilaments in the replication of human parainfluenza virus type 3. *J. Virol.* **72:**2655–2662.

Horikami, S. M., S. Smallwood, and S. A. Moyer. 1996. The Sendai virus V protein interacts with the NP protein to regulate viral genome RNA replication. *Virology* **222:**383–390.

### Rhabdoviruses

Das, T., M. Mathur, A. K. Gupta, G. M. Janssen, and A. K. Banerjee. 1998. RNA polymerase of vesicular stomatitis virus specifically associates with translation elongation factor-1 alphabetagamma for its activity. *Proc. Natl. Acad. Sci. USA* **95:**1449–1454.

Pattnaik, A. K., L. Hwang, T. Li, N. Englund, M. Mathur, T. Das, and A. K. Banerjee. 1997. Phosphorylation within the amino-terminal acidic domain I of the phosphoprotein of vesicular stomatitis virus is required for transcription but not for replication. *J. Virol.* **71**:8167–8175.

Stillman, E. A., and M. A. Whitt. 1998. The length and sequence composition of vesicular stomatitis virus intergenic regions affect mRNA levels and the site of transcript initiation. *J. Virol.* **72**:5565–5572.

Testa, D., P. K. Chanda, and A. K. Banerjee. 1980. Unique mode of transcription in vitro by vesicular stomatitis virus. *Cell* **21**:267–275.

### Influenza Virus

Beaton, A. R., and R. M. Krug. 1986. Transcription antitermination during influenza viral template RNA synthesis requires the nucleocapsid protein and the absence of a 5′ capped end. *Proc. Natl. Acad. Sci. USA* **83**:6282–6286.

Cianci, C., L. Tiley, and M. Krystal. 1995. Differential activation of the influenza virus polymerase via template RNA binding. *J. Virol.* **69**:3995–3999.

Hagen, M., T. D. Chung, J. A. Butcher, and M. Krystal. 1994. Recombinant influenza virus polymerase: requirement of both 5′ and 3′ viral ends for endonuclease activity. *J. Virol.* **68**:1509–1515.

Klumpp, K., R. W. Ruigrok, and F. Baudin. 1997. Roles of the influenza virus polymerase and nucleoprotein in forming a functional RNP structure. *EMBO J.* **16**:1248–1257.

Li, M. L., B. C. Ramirez, and R. M. Krug. 1998. RNA-dependent activation of primer RNA production by influenza virus polymerase: different regions of the same protein subunit constitute the two required RNA-binding sites. *EMBO J.* **17**:5844–5852.

Palese, P., M. B. Ritchey, and J. L. Schulman. 1977. P1 and P3 proteins of influenza virus are required for complementary RNA synthesis. *J. Virol.* **21**:1187–1195.

Plotch, S. J., M. Bouloy, I. Ulmanen, and R. M. Krug. 1981. A unique cap(m7GpppXm)-dependent influenza virion endonuclease cleaves capped RNAs to generate the primers that initiate viral RNA transcription. *Cell* **23**:847–858.

Pritlove, D. C., L. L. Poon, E. Fodor, J. Sharps, and G. G. Brownlee. 1998. Polyadenylation of influenza virus mRNA transcribed in vitro from model virion RNA templates: requirement for 5′ conserved sequences. *J. Virol.* **72**:1280–1286.

### Reovirus

Lawton, J. A., M. K. Estes, and B. V. Prasad. 1997. Three-dimensional visualization of mRNA release from actively transcribing rotavirus particles. *Nat. Struct. Biol.* **4**:118–121.

Lawton, J. A., C. Q. Zeng, S. K. Mukherjee, J. Cohen, M. K. Estes, and B. V. Prasad. 1997. Three-dimensional structural analysis of recombinant rotavirus-like particles with intact and amino-terminal-deleted VP2: implications for the architecture of the VP2 capsid layer. *J. Virol.* **71**:7353–7360.

Prasad, B. V., R. Rothnagel, C. Q. Zeng, J. Jakana, J. A. Lawton, W. Chiu, and M. K. Estes. 1996. Visualization of ordered genomic RNA and localization of transcriptional complexes in rotavirus. *Nature* **382**:471–473.

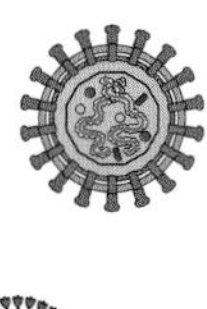

# 7

# Reverse Transcription and Integration: Hallmarks of the Retroid Viruses

## Retroviral Reverse Transcription

## Introduction

### Discovery

In 1970, back-to-back reports in the scientific journal *Nature* from the laboratories of Howard Temin and David Baltimore provided the first concrete evidence for the existence of an RNA-directed DNA polymerase activity in retrovirus particles. The pathways to this discovery were quite different for the two laboratories. In Temin's case, the discovery came about through attempts to understand how this group of RNA-containing viruses could permanently alter the heredity of cells, as they do in the process of oncogenic transformation. Temin proposed that retroviral RNA genomes become integrated into the host cell's chromatin in a DNA form, an idea supported by the observation that cellular DNA polymerases can use RNA as a template under certain in vitro conditions. Furthermore, studies on bacterial viruses such as bacteriophage lambda had established a precedent for viral DNA integration into host DNA. However, with the technology available at the time, it was a difficult hypothesis to test, and attempts by Temin and others to demonstrate such a phenomenon in infected cells were generally met with skepticism. Baltimore's entrée into the problem of reverse transcription came from his interest in virion-associated polymerases, in particular one that he had just discovered to be present in the virion of vesicular stomatitis virus, a virus that possesses a ($-$) strand RNA genome. It occurred to Baltimore and Temin independently that retroviruses might also contain the sought-after RNA-dependent DNA polymerase. As it turned out, this was indeed the case, and the retrovirus particles themselves yielded up the enzyme activity that had earlier eluded Temin. In 1975, Temin and Baltimore were awarded the Nobel Prize in physiology or medicine for their independent discoveries of retroviral reverse transcriptase (RT).

## *Impact*

The immediate impact of the discovery of RT was to amend the then-accepted "central dogma" of molecular biology, that the transfer of genetic information is unidirectional: DNA → RNA → protein. It was now apparent that there could also be a *retro*grade flow of information from RNA to DNA, and the name **retroviruses** eventually came to replace the earlier designation of RNA tumor viruses. In the years following this revision of dogma, many additional reverse transcription reactions have been discovered. Furthermore, as Temin hypothesized, study of RT has an important role in our understanding of cancer. As described in chapter 16, the study of oncogenic retroviruses has provided the framework for some current concepts of the genetic basis of this disease. Study of reverse transcription (and integration) has also allowed us to understand the persistence of retroviral infections and other aspects of the pathogenesis of AIDS. Finally, RT itself, first purified from virions and now expressed in bacteria, has become an indispensable tool in molecular biology, allowing experimentalists to capture cellular mRNAs as complementary DNAs, which can then be amplified, cloned, and expressed by well-established methodologies. For such reasons, we devote an entire chapter to these very important reactions.

## The Pathway of Retroviral Reverse Transcription

Significant insight into the mechanism of reverse transcription can be obtained by comparing the amino acid sequences of RTs with those of other enzymes that catalyze similar reactions. For example, RTs share certain sequence motifs with the RNA and DNA polymerases of both prokaryotes and eukaryotes in regions known to include critical active site residues (Fig. 6.4B). Thus, it is hypothesized that these enzymes employ similar catalytic mechanisms for nucleic acid polymerization reactions. Like DNA polymerases, viral RTs cannot initiate DNA synthesis de novo but require a specific primer to get started. Retroelements (see Box 7.2) other than retroviruses that employ RTs use a wide variety of primers as well as sites and modes of initiation on their RNA templates. In this chapter, we provide a detailed description of priming and reverse transcription for retroviruses and hepadnaviruses. But it should be remembered that even as arcane and distinct from each other as these two systems may appear, they do not exhaust the repertoire that has evolved for reverse transcription reactions in nature.

Much of what has been learned about reverse transcription in retroviruses comes from the analysis of intermediates in the reaction pathway that have been identified in infected cells. Reverse transcription intermediates have also been analyzed from **endogenous reactions**, which take place within purified virus particles, using the encapsidated viral RNA template. It was amazing to discover that intermediates and products virtually identical to those made in infected cells can actually be synthesized in purified virions; all that is required is the application of mild detergents to permeabilize the envelope and addition of the metal cofactor and deoxyribonucleoside triphosphate substrates. The efficiency of the endogenous reaction suggests that the reverse transcription system is poised to begin as soon as the virus enters the cell on infection. Retroviral reverse transcription intermediates have also been analyzed in totally reconstituted reactions with purified enzymes and model RNA templates.

Viral RT is the only protein required to accomplish all the diverse steps in the pathway described below. However, the reactions that take place inside cells are significantly more efficient than those observed in either endogenous or reconstituted systems. Thus, not all the essential features have been reproduced in these systems.

## *Essential Components*

**Genomic RNA.** Retrovirus particles contain two copies of the RNA genome held together by multiple regions of base pairing. Such RNA sediments as a 70S complex, as expected for a dimer of 35S genomes. Partial denaturation and electron microscopic analyses of the 70S complex indicate that the strongest pairing is located near the 5′ ends of the two genomes (Fig. 7.1). Thus, retroviral genomes can be thought of as being annealed head to head, an arrangement that may discourage the encapsidation of multiples larger than two (i.e., concatemers). The 70S RNA complex also includes two molecules of a specific cellular tRNA that serves as a primer for the initiation of reverse transcription (discussed below).

Despite the fact that two genomes are encapsidated, generally only one provirus is detected after infection with single virions. Therefore, retrovirions are considered **pseudodiploid** in character. Why should such a feature have been selected in evolution? One popular notion is that the availability of two RNA templates can help retroviruses survive extensive damage to their genomes. At least parts of both genomes can and typically *are* used during the reverse transcription process, accounting for the high rates of genetic recombination in these viruses. Presumably, being able to patch together one complete DNA provirus from two randomly damaged RNA genomes would provide survival value, as might the genetic recombination produced during the process. However, genetic experiments have shown that the use of two RNA templates is not an essential feature of the reverse transcrip-

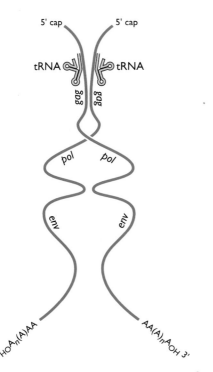

**Figure 7.1 The diploid retroviral genome.** The diploid genome includes the following, from 5' to 3': the m⁷Gppp capping group, the primer tRNA, the coding regions for viral structural proteins and enzymes, *gag, pol,* and *env,* and the 3' poly(A) sequence. From J. M. Coffin, p. 1767–1848, *in* B. N. Fields et al. (ed.), *Fields Virology,* 3rd ed. (Lippincott-Raven, Philadelphia, Pa., 1996), with permission.

tion process; one genome can suffice for all the steps outlined.

Like (−) strand genome RNA viruses, retroviral genome RNA is coated along its length with a viral nucleocapsid protein (NC), with approximately one molecule for every 10 nucleotides. This small basic protein can bind nonspecifically to both RNA and DNA and can promote the annealing of nucleic acids. Biochemical experiments suggest that NC may perform a role in reverse transcription similar to that of prokaryotic single-stranded binding protein in the synthesis of DNA catalyzed by bacterial DNA polymerases, namely, enhancing processivity and facilitating template exchanges. The ability of NC first to organize RNA genomes within the virion and then to facilitate reverse transcription within the infected cell may account for some of the differences in efficiency observed when comparing reactions reconstituted in vitro with those that take place in a natural infection.

**Primer tRNA.**   It was surprising to find that retrovirions contain a collection of cellular RNAs in addition to the

viral genome. These include approximately 100 copies of a nonrandom collection of tRNAs, some 5S rRNA, 7S RNA, and traces of cellular mRNAs. We do not know how most of these cellular RNAs become incorporated into virus particles, and most have no obvious function. However, one particular tRNA molecule does have a critical role, that of serving as primer for the initiation of reverse transcription. The tRNA primer is positioned on the template genome during virus assembly and maturation via interactions with both viral RNA and RT which are probably facilitated by NC. The primer tRNA is partially unwound and hydrogen-bonded near the 5' end of each RNA genome (Fig. 7.1 and 7.2) in a region called the **primer binding site (pbs).** (Lowercase designations are used throughout to refer to RNA sequences; uppercase is used to identify the same or complementary sequences in DNA.) Of the numerous retroviruses studied to date, all use one of just a few classes of cellular tRNAs as primers for reverse transcription. For example, most mammalian viruses use tRNA^Pro, tRNA^Lys3, or tRNA^Lys1,2.

In addition to the 3'-terminal 18 nucleotides that anneal to the pbs, other regions in the tRNA primer contact the RNA template and modulate reverse transcription. The template-primer interaction has been studied most extensively in reconstituted reactions with RNA and RT of avian sarcoma/leukosis virus (ASLV). In these in vitro analyses, the ability of the viral RNA to form stem-loop structures, and specific interactions between the primer tRNA^Trp and one of these loops, appear to be critical (Fig. 7.2). As similar interactions can be modeled for other retroviral RNAs and their primers, they may be biologically significant. We do not yet know how RTs recognize these structural features in the RNA template-primer complex.

**Reverse transcriptase.**   Retrovirions contain approximately 50 to 100 molecules of RT. Viral RTs probably function as dimers, but the number of dimers in each virion that are actually engaged in reverse transcription is not known. As noted above, studies of endogenous reverse transcription in purified virions suggest that viral DNA synthesis can begin as soon as the viral envelope is removed and deoxynucleotide triphosphates are made available. With the three viral systems studied most extensively (avian sarcoma/leukosis virus, murine leukemia virus [MLV], and human immunodeficiency virus), reverse transcription takes place mainly in the cytoplasm before viral DNA enters the nucleus to be integrated. Enzymes of these three viruses will be used as examples throughout this chapter.

Retroviral RTs are complex molecular machines with moving parts and multiple activities. Four distinct catalytic activities brought into play at various stages in the path-

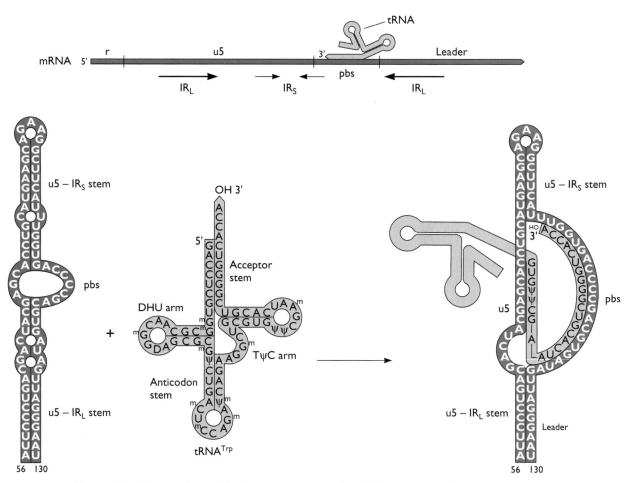

**Figure 7.2 Primer tRNA binding to the retroviral RNA genome.** (Top) Linear representation of the 5′ terminus of retroviral RNA. The r, u5, and leader regions are indicated. A tRNA primer is shown schematically annealed to the pbs. Two inverted repeat sequences (IR) that flank the pbs are represented by arrows. (Bottom) Avian sarcoma/leukosis virus RNA can form an extended hairpin structure around the pbs in the absence of primer tRNA (left). (Middle) Primer tRNA^Trp in the cloverleaf structure. Modified bases are indicated. (Right) Viral RNA annealed to tRNA^Trp. The u5-leader and u5-IR stem structures are indicated. An interaction with the TΨC arm of the primer and u5 RNA is also shown. (Bottom diagram) From J. Leis et al., p. 33–47, *in* A. M. Skalka and S. P. Goff (ed.), *Reverse Transcriptase* (Cold Spring Harbor Laboratory Press, Cold Spring Harbor, N.Y., 1993), with permission.

way of reverse transcription are illustrated in Fig. 7.3. These include RNA-directed and DNA-directed DNA polymerase activities, a DNA-unwinding activity (helicase-like), and RNase H activity. In many other systems these activities would require separate enzymes, but in RT they are all combined, the first three within a single polymerase domain. The RNase H activity, contained in a separate domain of RT, catalyzes the hydrolysis of RNA in RNA·DNA heteroduplexes. In digesting such RNA, this domain can function as both an exonuclease and an endonuclease, producing oligoribonucleotides 2 to 15 nucleotides in length. These activities are employed as the RNase H of RT degrades the genomic RNA after it has been copied into complementary DNA, forms the primer for (+) strand DNA synthesis from the genomic RNA, and finally removes this primer and the tRNA primer from the 5′ ends of the viral DNA duplex.

## Critical Reactions in Reverse Transcription

**RNA priming and template strand exchange.** Based on our understanding of DNA synthesis, the simplest way of transcribing an RNA template into DNA would be to start at its 3′ end and finish at its 5′ end. It was therefore somewhat of a shock to early researchers to discover that reverse transcription in fact started near the 5′ end only to run out of template after little more than about 100 nucleotides, at which point most DNA synthesis stopped. As we will see later, this counterintuitive strategy for initiation of DNA synthesis allows the duplication and translocation of critical transcription and integration signals encoded in both the 5′ and 3′ ends of the genome RNA (called u5 and u3, respectively). The short (ca. 100-nucleotide) product of this first reaction (Fig. 7.3, steps 1 to 3) accumulated in large amounts in the endogenous and reconstituted systems, and thus it was called **(−) strong-stop DNA.** In the next step, the 3′ end of the same or the second RNA genome is engaged as template so that DNA synthesis can continue. This reaction, which we call the **first template exchange**, corresponds to the substitution of one end of the RNA for another to be copied by the RT "machine" (Box 7.1).

The first template exchange (Fig. 7.3, middle) requires the presence of direct repeats, r, at the ends of the viral RNA. After it is copied to form (−) strong-stop DNA, the 5′ end of the genome RNA is digested by the RNaseH activity of RT. The R complement in (−) strong-stop DNA can then anneal to the complementary r sequence upstream of the poly(A) tail at the 3′ end of either the same or the second RNA genome in the virion (Fig. 7.3, steps 3 and 4). It has been suggested that transient melting, facilitated by NC or by RT helicase activity, may allow this exchange to take place without complete digestion of the RNA and that some special properties of the u3-r junction may facilitate the reaction. As (−) strong-stop DNA is barely detectable in infected cells, this first template exchange must be quite efficient in vivo.

Once the 3′ end of the genome RNA is engaged, the RNA-directed DNA polymerase activity of RT can proceed with copying all the way to the 5′ end of that template (Fig. 7.3, steps 4 to 7). This (−) strand DNA product (also called complementary DNA or cDNA) terminates at the primer binding site on the 5′ end of the RNA template (Fig. 7.3, step 7). It comprises an entire genome's equivalent of DNA, but in permuted order. This is, of course, an idealized picture. In actuality, analyses of reaction intermediates show that the enzyme pauses along the way, presumably at some sequences (or possibly breaks) that impede copying. If a break is encountered in the RNA template, synthesis can be completed by utilization of the second RNA genome. Such *internal* template exchanges, known to occur even in the absence of such breaks, probably proceed via the same steps outlined for the first template exchange. Internal exchanges that take place during RNA-directed DNA synthesis are estimated to be the source of at least half of the genetic recombination that occurs in retroviruses; the mechanism is known as **copy choice** (see Fig. 7.5).

**Production of primers for (+) strand DNA synthesis by RNase H digestion of copied viral RNA.** Viral RNA that has been copied into DNA is digested following (−) strand DNA synthesis. In order to initiate synthesis of the second, (+) strand of DNA and commence DNA-directed DNA polymerization, another primer is needed by RT. In contrast to the tRNA primer for (−) strand DNA synthesis, this second primer is not derived from the cell but is a specific product of RNase H digestion of the viral RNA genome.

The (+) strand primer comprises a **polypurine tract (ppt)** approximately 11 nucleotides long. It is presumed to be produced through specific recognition of the polypurine sequence located near the 3′ end of the genome RNA. The DNA-directed DNA polymerase activity of RT begins as soon as the polypurine tract primer is formed, usually before (−) strand DNA synthesis is completed. Synthesis of (+) strand DNA proceeds to the nearby end of the (−) strand DNA template and terminates after copying the first 18 nucleotides of the primer tRNA when it encounters a modified base that cannot be copied by RT. This product is called (+) strong-stop DNA, and after removal of the tRNA primer by the RNase H activity of RT, its 3′ end becomes available to anneal to the PBS sequence at the 3′ end of the (−) strand DNA (Fig. 7.3, step 8). This process is called the **second template exchange**.

Other oligoribonucleotide products formed by RNaseH digestion of the RNA template can also serve as primers for DNA-directed DNA polymerization by RT. Human immunodeficiency virus contains a second, specific internal polypurine tract for priming (+) strand synthesis located near the center of the genome. With both the human virus and avian sarcoma/leukosis viruses, (+) strand DNA synthesis can also be initiated at several additional locations; this is the situation illustrated in Fig. 7.3 (steps 6 and 7). On the other hand, for murine leukemia virus there is little or no evidence of (+) strand initiation at sites other than the 3′ polypurine tract.

**Strand displacement synthesis and formation of long terminal repeats.** The product of the second template

**Initiation**

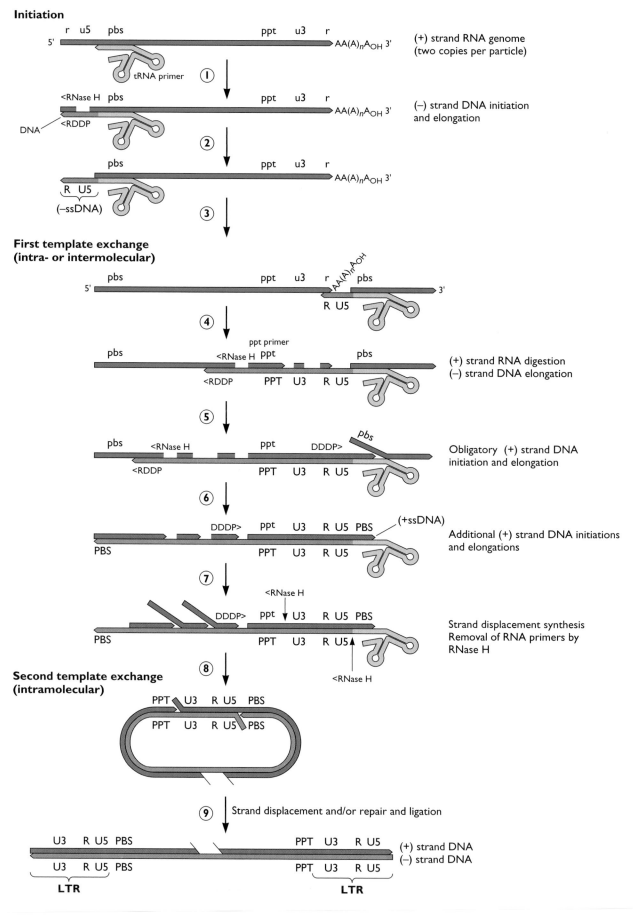

**First template exchange (intra- or intermolecular)**

**Second template exchange (intramolecular)**

The exchange of one template for another to be copied by either DNA or RNA polymerases is sometimes referred to as enzyme "jumping." This inappropriate term comes from a too literal reading of cartoon illustrations of the process, in which the templates to be exchanged may be on opposite ends of the nucleic acid or on different nucleic acid molecules. In actuality, such enzyme movement is quite improbable, and use of this terminology can cloud thinking about these processes. In almost all cases these polymerases are components of large complexes with architecture designed to bring different parts of the template or different templates close to each other. Although some dynamic changes must occur for the proteins to accommodate template exchanges, it seems likely that most of the "movement" is done by the flexible nucleic acid molecules.

exchange is a circular DNA molecule with overlapping 5′ ends that border the U3-R-U5 region (Fig. 7.3, bottom). Were reverse transcription to stop at this point, repair enzymes in the nucleus of the cell could produce a covalently closed circular DNA molecule with a single genome's equivalent of DNA. Such molecules have been isolated from infected cells and are most likely by-products of incomplete reverse transcription produced by such repair.

However, perhaps owing to its helicase activity, RT has the ability to displace the annealed DNA strand so that synthesis is usually continued to the end of the DNA template strand. Such **strand displacement synthesis** by RT produces a blunt-ended, linear duplex product with **long terminal repeats (LTRs)** (Fig. 7.3, step 9). Sequences at the ends of the LTRs are essential for DNA integration, the next step in the viral replication cycle.

◁ **Figure 7.3 The reverse transcription process.** Two copies of the capped 7- to 10-kb RNA genome with a poly(A) tail are present in each retrovirus particle. Critical *cis*-acting sequences at the ends of the RNA genome are indicated in lowercase letters. They are represented in uppercase letters in the DNA products. (−) strand DNA synthesis starts near the 5′ end of the (+) strand RNA genome with a specific host tRNA as primer. The partially unfolded tRNA is annealed via at least 18 nucleotides at its 3′ end to a site on the RNA genome called the primer-binding site (pbs). Step 1: Synthesis proceeds to the 5′ end of the RNA genome through the u5 region, ca. 100 nucleotides, ending after the r region, which is terminally redundant in the RNA, forming the (−) strand strong-stop DNA (−ssDNA). Step 2: The RNA portion of the RNA-DNA hybrid product is digested by the RNaseH activity of RT, resulting in a single-strand DNA product. Step 3: This exposure facilitates hybridization with the r region at the 3′ end of the same or the second RNA genome, a reaction that produces the first template exchange for reverse transcriptase (RT), and allows continuation of RNA-dependent DNA polymerization (RDDP). Step 4: When (−) strand elongation passes a specific region near the 3′ end of the RNA genome, characterized by a polypurine tract (ppt), the complementary sequence of the RNA template escapes digestion by RNase H and serves as a primer for (+) strands in DNA-dependent DNA polymerization (DDDP). (+) Strand synthesis then continues back to the U5 region with the (−) strand DNA as a template. Step 5: Meanwhile, (−) strand synthesis continues along (+) strand RNA template, after which the RNA template is removed via RNase H activity. Step 6: The RNase digestion products formed can in some cases provide additional primers for (+) strand synthesis at internal locations along the (−) strand DNA. Step 7: ppt-initiated (+) strand DNA synthesis stops after the annealed portion of the tRNA has been copied, generating the (+) strand DNA form of the PBS, and forming the (+) strand strong stop product, denoted +ssDNA. The tRNA is then removed by RNase H activity of RT. Step 8: The exposed PBS facilitates annealing to its complement on the (−) strand DNA, allowing the second template exchange. DNA synthesis then continues. Step 9: Strand displacement synthesis by RT to the PBS and PPT ends produces a linear molecule with long terminal repeats (LTRs). Internal discontinuities in the (+) strand, indicated by the gap, are probably repaired by host cell enzymes. Adapted from R. A. Katz and A. M. Skalka, *Annu. Rev. Biochem.* **63:**133–173, 1994, with permission.

Retroviral reverse transcription has been called "destructive replication," as there is no net gain of genomes but rather a substitution of one double-stranded DNA for two molecules of single-stranded RNA. However, by this rather complex but elegant pathway, RT not only generates a linear DNA copy of the retroviral genome to be integrated but also produces the LTRs that include signals necessary for transcription of the integrated DNA, which is called the **provirus**. Integration ensures subsequent replication of the provirus via the host's DNA synthesis machinery as the cell divides. In addition, the promoter in the *upstream* LTR is now in the appropriate location for synthesis of progeny RNA genomes and viral mRNAs by host cell RNA polymerase II.

**Structure of retroviral DNA products.** Unintegrated linear retroviral DNA extracted from infected cells appears to have continuous (−) strands and discontinuous (+) strands. For example, the (+) strands of murine retroviruses (mouse mammary tumor virus and murine leukemia virus) contain large gaps (Fig. 7.4). In contrast, avian sarcoma/leukosis virus (+) strands contain single-stranded tails generated by strand displacement synthesis across internal sites of initiation throughout the DNA template. Human immunodeficiency virus DNA also has a uniquely positioned tail generated by strand displacement synthesis across the internal polypurine tract. This apparent abundance of single-stranded DNA, opposite gaps and in tails, could provide additional opportunities for recom-

**Figure 7.4 Differences in the structure of retroviral DNAs extracted from infected cells.** (A) Schematic representations. Like all retroviral DNAs, most unintegrated murine leukemia virus (MLV) DNA molecules extracted from infected cells (top) have a full-length, continuous DNA (−) strand but only short fragments of the (+) strand. Human immunodeficiency virus type 1 (HIV-1) DNA (middle) has two classes of (+) strands. In the left half of the molecules (+) strands are continuous to a point just beyond the internal PPT, where they appear to encounter a block to further elongation. In the right half there appear to be many possible initiation sites, generating a collection of (+) strands of varying length. The (+) strands of avian sarcoma/leukosis virus (ASLV) DNA can be initiated at several (apparently random) locations. These are elongated past adjacent initiation sites to generate long, single-stranded tails. (B) An electron micrograph of ASLV DNA formed in an endogenous reaction. Arrows point to single-stranded tails (about three or four per molecule). The image shows two molecules of full length joined by a single-stranded tail as proposed in the strand displacement-assimilation recombination model. From R. P. Junghans et al., *Cell* **30**:53–62, 1982, with permission.

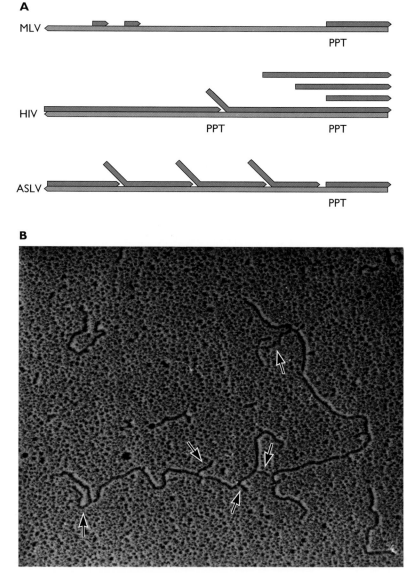

bination. It is estimated that approximately half of the genetic recombination that occurs during retroviral reverse transcription takes place during (+) strand synthesis. One proposed mechanism, the **strand displacement-assimilation** model, is illustrated in Fig. 7.5.

It is not known how or when linear DNA products with (+) strand discontinuities or branches are processed to complete the covalent joining that produces an integrated provirus in the host chromosome. From in vitro studies it appears unlikely that strand displacement synthesis by RT can completely displace all downstream (+) strand segments. It may be that cellular repair enzymes in the nucleus participate in the final process. Theoretically, such repair could occur even after integration, as viral DNA is competent for integration as soon as the duplex LTR ends are formed. Indeed, the first step in the integration reaction, processing of the blunt ends of the LTRs, takes place in the cytoplasm before the viral DNA enters the nucleus, and nucleoprotein complexes isolated from the cytoplasmic fraction of human immunodeficiency virus type 1 (HIV-1)-infected cell extracts are competent to integrate viral DNA into exogenously supplied target DNA despite the fact that their (+) strands are not yet continuous.

Beginning approximately 8 to 12 h after the infection of cells in culture, a portion of the newly synthesized viral DNA becomes tightly associated with the nucleus. At about the same time, integration events are also first detected. Such nucleus-associated viral DNA comprises both linear molecules and a small but variable proportion of covalently closed circular molecules containing one or two LTRs. It is now clear that linear viral DNA molecules are the substrates for integration (see below). The covalently closed circles appear to be dead-end products. However, as joined LTRs are easy to detect by polymerase chain reaction techniques, the two LTR circles have been used as markers for the nuclear transport of viral DNA.

## General Properties and Structure of Retroviral Reverse Transcriptases

### Domain Structure and Subunit Organization

The RTs of all retroviruses are encoded in *pol* genes. Despite the sequence homologies and similar organization of coding sequences in their *pol* regions, retroviruses from different vertebrate species show considerable diversity with respect to the subunit composition and organization of their RTs (Fig. 7.6). Each of the three enzymes studied most extensively, those of avian sarcoma/leukosis, murine leukemia, and human immunodeficiency viruses, has a distinctive arrangement. Virtually nothing is known about the subunit compositions of RTs from most other retroviruses. In the absence of more structural data, it is diffi-

cult to know the significance of this apparent structural diversity. However, it seems likely that in the confines of the subviral nucleoprotein complex in which reverse transcription takes place in vivo, Pol-derived proteins all work together and in similar ways.

### Catalytic Properties

**DNA polymerization.** The biochemical properties of RT have been studied with enzymes purified from virions or expressed in bacteria acting on model templates and primers. Kinetic analyses have revealed an ordered reaction pathway for DNA polymerization similar to that of other polymerases. Like cellular polymerases and nucleases, RTs require divalent cations ($Mg^{2+}$ or $Mn^{2+}$) as cofactors. The in vitro rate of elongation by RT on natural RNA templates is 1 to 1.5 nucleotides per s, approximately 1/10 the rate of other eukaryotic DNA polymerases. The relatively long time required to generate a genome's equivalent of retroviral DNA after infection, approximately 4 h for ca. 9,000 nucleotides, supports the view that even in vivo reverse transcription is a relatively slow process.

In reconstructed systems, the rate of enzyme-template/primer dissociation decreases considerably after addition of the first nucleotide, suggesting that initiation and elongation are distinct steps in reverse transcription, as is the case in DNA synthesis by DNA-dependent DNA polymerases. However, in contrast to such polymerases, retroviral RTs do not remain attached to their template/primers during a large number of successive deoxyribonucleotide additions; this property is described as "poor processivity."

**Fidelity.** Retroviral genes, like those of other RNA viruses, mutate at much higher rates than cellular genes. RTs lack an editing activity (resembling the $3' \rightarrow 5'$ exonuclease of *Escherichia coli* DNA polymerase I, which is capable of excising mispaired nucleotides) and have been shown to be error-prone in vitro. Therefore, RTs are presumed to contribute to the high in vivo mutation rate of retroviruses.

In vitro, RT errors have been shown to include not only misincorporations but also rearrangements such as deletions and additions (Fig. 7.7). Misincorporations by human immunodeficiency virus type 1 RT can occur as frequently as 1 per 70 copies at some positions and as infrequently as 1 per $10^6$ copies at others. Many types of genetic experiments have been conducted in attempts to determine the in vivo error rates of RTs in a single replication cycle. The general conclusion is that such rates are also quite high, with reported misincorporations in the range of 1 per $10^4$ to $10^6$ nucleotides polymerized, in contrast to 1 per $10^7$ to $10^{11}$ for cellular DNA replication. As

**A Recombination during minus-strand synthesis: copy choice**

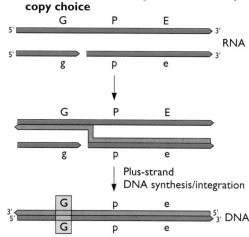

Heterozygous particle with random breaks in one RNA genome

Minus (first) strand DNA synthesis starts on one genome and switches to the second at a break point, pause site, or random location.

Plus-strand DNA synthesis/integration

Copying of the resultant recombined minus strand produces a double-strand homoduplex

**B Recombination during plus-strand synthesis: strand assimilation**

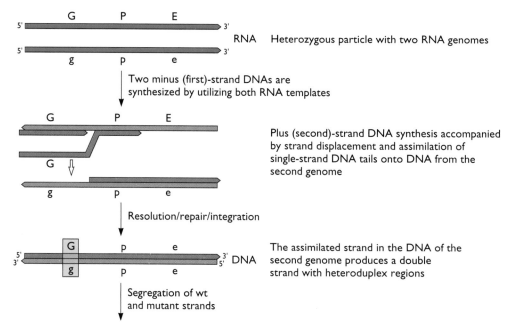

Heterozygous particle with two RNA genomes

Two minus (first)-strand DNAs are synthesized by utilizing both RNA templates

Plus (second)-strand DNA synthesis accompanied by strand displacement and assimilation of single-strand DNA tails onto DNA from the second genome

Resolution/repair/integration

The assimilated strand in the DNA of the second genome produces a double strand with heteroduplex regions

Segregation of wt and mutant strands

**Figure 7.5 Two models proposed to explain genetic recombination during reverse transcription.** Virtually all retroviral recombination occurs between coencapsidated genomes at the time of reverse transcription. The copy choice model (A) postulates a mechanism for genetic recombination during first, or (−) strand, DNA synthesis; the strand displacement-assimilation model (B) proposes a mechanism for recombination during (+) strand DNA synthesis. These two models are not mutually exclusive, and there is experimental support for both. Viral genes (*gag*, *pol*, and *env* alleles, indicated as G/g, P/p, and E/e, respectively) are shown to illustrate recombination. Single recombinations are shown for simplicity, arbitrarily focusing on the *gag* gene (G/g). However, multiple crossover events are frequently observed. The yellow boxes highlight the regions of recombination. (A) Recombination during (−) strand synthesis. A break in the RNA is represented as a gap. Exchanges between RNA templates may occur at RNA breaks, as shown, but such breaks may not be required. Introducing intentional breaks, e.g., by gamma irradiation, does not seem to increase the frequency of such recombination. The copy choice mechanism predicts only one DNA homoduplex (G/G) product from two RNA molecules. (B) Recombination during (+) strand synthesis. (+) Strand DNA synthesis initiates at internal sites with RNA primers generated by partial RNase H digestion. The strand assimilation model requires that (+) strand DNA synthesis take place on two templates and predicts formation of a heteroduplex (G/g) as a consequence of recombination of (+) strands. One DNA molecule is shown for simplicity. Adapted from R. Katz and A. M. Skalka, *Annu. Rev. Genet.* **24:**409–445, 1990, with permission.

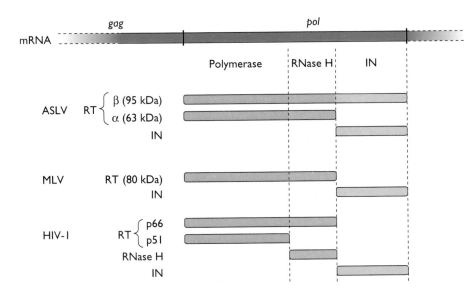

**Figure 7.6 Domain and subunit relationships of RTs from different retroviruses.** The RT of ASLV includes sequences corresponding to integrase in its largest subunit, β. Both subunits of this enzyme include RNase H sequences. MLV RT functions as a homodimer; both subunits contain the RNase H domain, but neither includes IN sequences. Like that of ASLV, the HIV-1 RT is a heterodimer. However, in this case only one subunit includes the RNase H domain, and neither includes IN sequences. Adapted from R. A. Katz and A. M. Skalka, *Annu. Rev. Biochem.* **63**:133–173, 1994, with permission.

retroviral genomes are approximately $1 \times 10^4$ nucleotides in length, this rate can translate to approximately one lesion per retroviral genome per replication cycle. This high mutation rate explains in part the difficulties inherent in treating AIDS patients with RT inhibitors, as a large population of mutant viruses preexist in every chronically infected individual, some encoding drug-resistant RTs. These mutants can replicate in the presence of a drug and quickly comprise the bulk of the population (see chapters 18 and 19).

Several unique activities of RTs are also likely to contribute to their high error rates. Avian sarcoma/leukosis virus and human immunodeficiency virus RTs are both proficient at extending mismatched terminal base pairs such as those that result from nontemplated addition (Fig. 7.7). This process facilitates incorporation of the mismatched nucleotide into the RT product. RTs also seem to allow a certain type of slippage within homopolymeric runs, in which one or more bases are extruded on the template strand (Fig. 7.7B). Mispairing occurs after the next deoxyribonucleotide is added and the product strands attempt to realign with the template. Deletions can also be produced by this mechanism (Fig. 7.7D). Both deletions and insertions are also known to occur when template exchanges are made in vivo, apparently because these exchanges can take place within short sequence repeats (e.g., four or five nucleotides) that are not in homologous locations on the two RNA templates. This mechanism is of major importance in the capture of oncogenes.

**RNase H.**   The RNase H activity of RT also requires a divalent cation, $Mg^{2+}$ or $Mn^{2+}$. Like the other RNase H enzymes present in all prokaryotic and eukaryotic cells, the

RNase H of RT digests only RNA that is annealed to complementary DNA. RNase H cleaves phosphodiester bonds to produce $5'\text{-PO}_4$ and $3'\text{-OH}$ ends; the latter can be extended directly by the polymerase activity of RT.

### Structure of RT

Although RTs from avian and murine retroviruses have been studied extensively in vitro, the potential of human immunodeficiency virus type 1 RT as a target for drugs for treating AIDS has focused intense interest and resources

**Figure 7.7 Mutational intermediates for base substitution and frameshift errors.** Misincorporated deoxynucleotides are indicated in red. Slippage and dislocations are presumed to be mediated by looping out of nucleotides in the template. Only single-nucleotide dislocations are shown here, but large dislocations leading to deletions are also possible. From K. Bebenek and T. A. Kunkel, p. 85–102, *in* A. M. Skalka and S. P. Goff (ed.), *Reverse Transcriptase* (Cold Spring Harbor Laboratory Press, Cold Spring Harbor, N.Y., 1993), with permission.

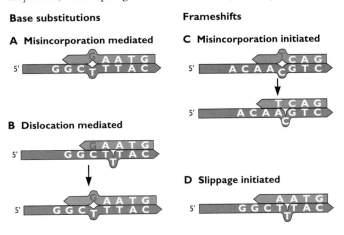

on this enzyme. As a consequence, we know more about the human virus protein than about any other RT. Figure 7.8 shows a linear map of RT subunits of the human virus indicating evolutionarily conserved regions in both the polymerase and RNase H domains and acidic amino acids that have been shown by mutational studies to be required for catalysis. The aspartic acid residues highlighted in the polymerase region are included in conserved motifs in a large number of polymerases (Fig. 6.4B) and are believed to coordinate the required metal ions and contribute to binding deoxyribonucleoside triphosphates and catalysis.

The first portion of human immunodeficiency virus type 1 RT to be crystallized was the independently folded RNase H domain obtained as a proteolytic digestion product. As illustrated by the ribbon models in Fig. 7.9, the general structure of this domain is almost superimposable on that of *E. coli* RNase H, with which it shows only 24% amino acid identity. Included in the regions of structural homology are seven residues that are components of conserved motifs present in all retroviral and bacterial RNases H, clustered at what is believed to be the active site at which the metal ion cofactors are bound. Site-directed mutagenesis studies have confirmed that many of the conserved residues are important for the activities of both *E. coli* and human retroviral RNase H. Although the fragment used to obtain crystals of the RNase H domain of the human immunodeficiency virus RT is enzymatically inactive, activity can be reconstituted by addition of the smaller subunit of this RT, p51, which may provide missing determinants or stabilize the structure.

The first high-resolution structure of the intact human immunodeficiency virus type 1 RT (p66-p51) heterodimer

was obtained for complexes with the nonnucleoside inhibitor nevirapine. Subsequently, structures were determined for the enzyme bound to a short 18- to 19-mer DNA duplex (Fig. 7.10) and with no bound inhibitors. The most surprising finding to come from the analysis of these RT crystals was the **structural asymmetry** in the subunits. Although the primary sequence of p51 is the same as that of p66, minus the RNase H domain (Fig. 7.8), the analogous portions of these subunits were found to be arranged quite differently. Consistent with this observation, biochemical studies showed that they also perform different functions in the heterodimeric enzyme. The catalytic functions are contributed by the larger subunit, p66; the smaller p51 subunit's role appears to be mainly structural. In the crystal structures, the two subunits are nestled one on top of the other (Fig. 7.10) with an extensive subunit interface. As noted in the previous chapter, analogous to the schematization proposed for the topology of the polymerase domain of the *E. coli* DNA polymerase I Klenow fragment, the p66 polymerase domain was divided into three subdomains denoted "finger," "palm," and "thumb" (Fig. 6.4A). A fourth subdomain lies between the remainder of the polymerase and the RNase H domain and is therefore called the "connection" domain. It contains the major contacts between the two subunits of RT. The extended thumb of p51 contacts the RNase H domain of p66, an interaction that appears to be required for RNase H activity. Not only are human immunodeficiency virus type 1 RT and *E. coli* DNA polymerase similar topologically, but this retroviral RT can actually substitute for the bacterial enzyme in *E. coli* cells that carry a temperature-sensitive mutation in the DNA polymerase I gene. Such genetic complementation has been used to develop a

**Figure 7.8 Linear map showing conserved residues in HIV-1 RT.** The most highly conserved residues, which are thought to be components of the two active sites, are indicated by dots. The sequences included in the two subunits are shown by the lines below. In the polymerase domain, the colored blocks in the blue background indicate regions of homology within the RT family. The more intensely colored areas indicate regions of homology for all polymerases. In the RNase H domain, the intensely colored areas indicate regions with homology to *E. coli* RNase H. The arrowheads show sites of complete (red) or partial (pink) processing by the retroviral protease (PR). The dashed line indicates partial processing. Adapted from R. A. Katz and A. M. Skalka, *Annu. Rev. Biochem.* **63:**133–173, 1994, with permission.

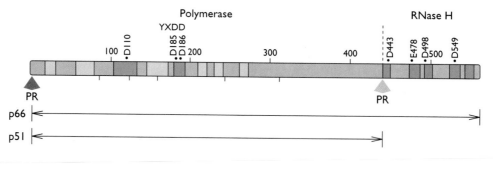

**E. coli RNase H**     **HIV-1 RNase H**

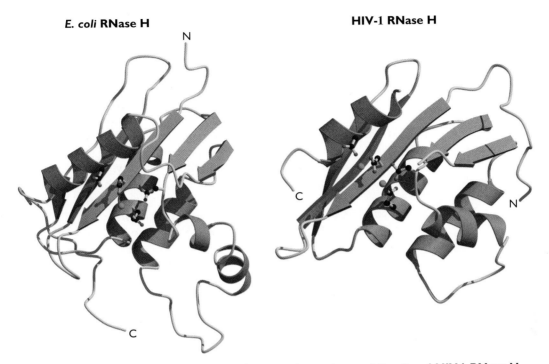

**Figure 7.9 Ribbon diagrams based on the crystal structures of *E. coli* and HIV-1 RNase H.** The highly conserved acidic residues are shown as ball-and-stick models, with metal ion complexes indicated by the green spheres. Courtesy of Mark Andrake, Fox Chase Cancer Center.

screen in *E. coli* for nucleoside analogs that block the human retroviral RT activity.

Analysis of the various structural models predicts highly dynamic interactions among human immunodeficiency virus type 1 RT, its template/primer, and deoxyribonucleoside triphosphate substrates. The substrates are bound in a defined order: the template/primer first and then the deoxyribonucleoside triphosphate to be added. On binding the template/primer, the thumb is moved away from the fingers, allowing contacts between the fingers and the 5' extension of the template. Another conformational change takes place when the deoxyribonucleoside triphosphate is bound. This substrate interacts directly with two fingertip residues; this interaction may induce closure of its binding pocket. This conformational change facilitates attack of the 3' OH of the primer on the α-phosphate of the incoming deoxyribonucleotide. After this addition, the fingertips may resume their open position so that the diphosphate product can be released and the template/primer can be translocated to accept the next deoxyribonucleoside triphosphate. Such translocation may be driven in part by the energy released on hydrolysis of the previous deoxyribonucleoside triphosphate substrate.

Figure 7.11 shows a predicted mechanism for formation of new phosphodiester bonds at the catalytic center of the polymerase domain, emphasizing the presumed roles of residues known to be critical from both structural and biochemical data. The chemical mechanism proposed for the polymerase reaction of both *E. coli* DNA polymerase I and the human retroviral RT (and RNase H) is based on the two-metal mechanism of catalysis deduced for the 3'-5' exonuclease reaction of the DNA polymerase I Klenow fragment. Figure 7.12 is a schematic drawing of an RNA·DNA heteroduplex bound in the cleft region of the human immunodeficiency virus type 1 RT. As illustrated in the structure shown in Fig. 7.10, the cleft is flanked on one end by the polymerase active site and on the other end by the RNase H active site. This arrangement is consistent with biochemical evidence for polymerase-linked RNase H activity; it can account for the length (ca. 18 nucleotides) of the terminal RNase H oligoribonucleotide product. Cross-linking experiments suggest that the p51 subunit may perform a unique function in the RT heterodimer, that of binding the tRNA primer.

Production of the p51 subunit that possesses identical amino acid sequences but has both structure and function that are distinct from the p66 subunit appears to be a unique example of the economical use of viral coding capacity by human immunodeficiency virus. The C terminus of the p51 subunit at the end of the connection domain is

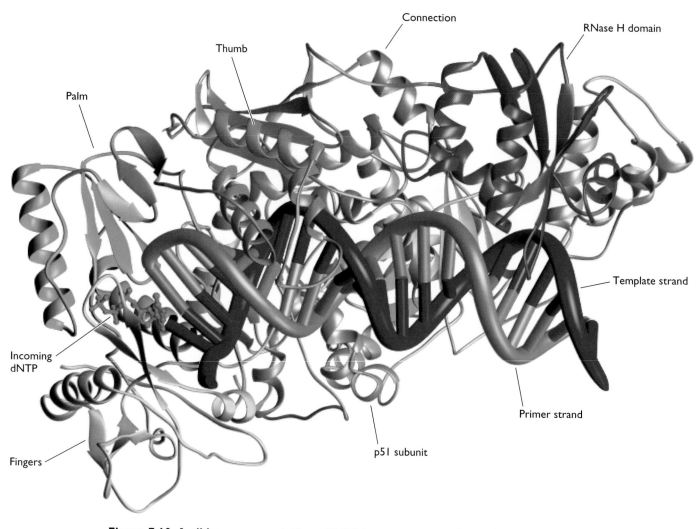

**Figure 7.10 A ribbon representation of HIV-1 reverse transcriptase in complex with DNA template/primer.** dNTP, deoxyribonucleoside triphosphate. Courtesy of Rajiv Chopra and Stephen Harrison, Harvard University.

buried within the N-terminal β-sheet of the RNase H domain of p66 in the crystal structure. This arrangement suggests a model for proteolytic processing in which a p66 homodimer intermediate in the reaction is also arranged asymmetrically and the RNase H domain of the subunit destined to become p51 is unfolded. This mechanism would account for asymmetric cleavage by the viral protease. Without comparable structural data for other retroviral RTs, we can only wonder if a similar economy of coding and an analogous processing strategy have been adopted by other retroviruses. The variability in subunit composition of the three best studied systems, illustrated in Fig. 7.6, makes this an interesting question.

## Other Examples of Reverse Transcription

When it was first discovered, RT was thought to be a peculiarity of retroviruses. We now know that other animal viruses, such as the hepadnaviruses, and some plant viruses, the caulimoviruses, also replicate by producing duplex genomic DNA via an mRNA intermediate. All are thus classified as **retroid viruses**. In fact, the discovery of RT activity in some strains of myxobacteria and *E. coli* places the evolutionary origin of this enzyme before the separation of prokaryotes and eukaryotes. As it is now widely held that the evolving biological world was initially based on RNA as both catalysts and genomes, the development of the modern (DNA) stage of evolution would

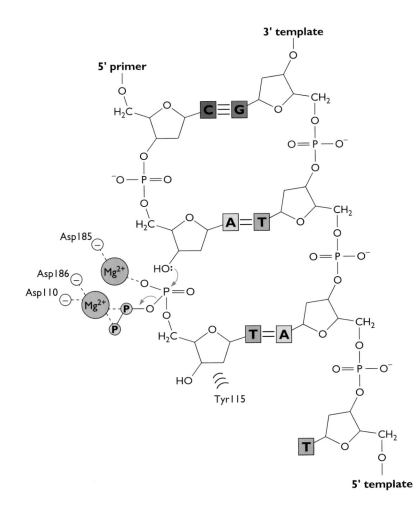

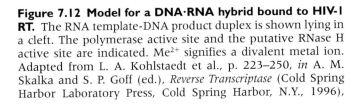

**Figure 7.11 Schematic representation of the proposed mechanism of phosphodiester bond formation by HIV-1 RT.** The reaction is proposed to be metal-mediated, requiring two divalent cations separated by ~4 Å ($Mg^{2+}$ in the model can also be replaced by $Mn^{2+}$ or $Zn^{2+}$). One of the metal ions is shown coordinating Asp-110 and Asp-186 with the β- and γ-phosphates of the incoming deoxynucleoside triphosphate, while the other ion is chelated by Asp-185 and the α-phosphate. The second metal ion is in position to participate directly in the nucleophilic attack step by contributing to the electropositive character of the α-phosphate. The resulting $PP_i$ may transiently bind to Asp-110 and Asp-186 via $Mg^{2+}$ coordination prior to release. Tyr-115 is believed to stabilize binding of the incoming deoxyribonucleotide by base-stacking interactions symbolized by ellipses. Red arrows indicate the net movement of electrons. From P. H. Patel et al., *Biochemistry* **34**:5351–5363, 1995, with permission.

have required an RT activity. If so, the retroid viruses may also be viewed as living fossils, shedding the first dim light into an ancient evolutionary passageway from the primordial world.

Since its discovery in retroviruses, additional evidence for reverse transcription and RT-related sequences has been found in cellular genomes. Some of these cellular constituents, known generally as **retroelements**, are derived from retroviruses. During the retroviral life cycle (appendix, Fig. 10), the double-stranded DNA molecule synthesized by reverse transcription is integrated into the genomes of infected animal cells by a second retroviral enzyme, integrase (discussed in the following section). In some cases, retroviruses can infect and integrate their DNA in germ line cells of the host. These integrated retroviral DNAs are then passed on to future generations in Mendelian fashion as **endogenous proviruses**. Endogenous proviruses are often replication-defective, a property that may facilitate coexistence with their hosts. A

**Figure 7.12 Model for a DNA·RNA hybrid bound to HIV-1 RT.** The RNA template-DNA product duplex is shown lying in a cleft. The polymerase active site and the putative RNase H active site are indicated. $Me^{2+}$ signifies a divalent metal ion. Adapted from L. A. Kohlstaedt et al., p. 223–250, *in* A. M. Skalka and S. P. Goff (ed.), *Reverse Transcriptase* (Cold Spring Harbor Laboratory Press, Cold Spring Harbor, N.Y., 1996),

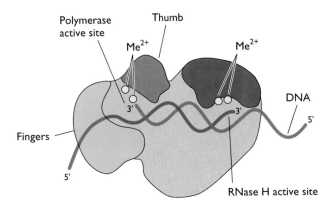

surprisingly large fraction of the mammalian genome (an estimated 10%) comprises endogenous proviruses and other retroelements that have been accumulated during evolution. Three other types of reverse-transcribed elements that are common residents in the genomes of animal cells include retrotransposons, retroposons, and processed pseudogenes (Box 7.2). Chromosomal telomeres are also formed by reverse transcription mediated by an enzyme known as telomerase. The family tree shown in Fig. 7.13 was constructed by using computer algorithms to align the predicted RT-related amino acid sequences of representative retroelements. Such comparisons provide clues about the relatedness of the various retroelements and hints of their possible origin.

## Retroviral DNA Integration

### Introduction

Retroviruses and the related retrotransposons are unique in the eukaryotic world for their possession of **integrase (IN)**, the enzyme that catalyzes specific and efficient integration of the DNA product of RT into the genome of the host cell. Some retroposons also contain IN-related sequences (Fig. 7.13). Integration is a critical step in the retroviral life cycle, as it ensures a stable association between viral DNA and the host cell chromosome. Integrated proviral DNA is transcribed by cellular RNA polymerase II to produce the viral RNA genome and the mRNAs required to complete the replication cycle.

IN is encoded in the 3′ region of the retroviral *pol* gene (Fig. 7.7), and the mature protein is generated by viral protease (PR)-mediated processing, in most cases, of the Gag-Pol polyprotein precursor. During progeny virus assembly, all three viral enzymes (PR, RT, and IN) are incorporated into the virion core. Virus particles contain equimolar concentrations of RT and IN (ca. 50 to 100 molecules per virion). As the double-stranded DNA product of RT is the direct substrate for IN, it seems likely that these enzymes function in concert within a nucleoprotein subvirion complex in the cytoplasm of the infected cell. As already noted, IN sequences are actually included in one of the subunits of avian sarcoma/leukosis virus RT, and gentle extraction of murine leukemia virus particles yields RT-IN complexes. However, as with RT, virtually nothing is known about the molecular organization of IN within retrovirions.

The first insights into the mechanism of retroviral integration came in the early 1980s, when it was established that proviral DNA is flanked by LTRs and is colinear with the unintegrated linear viral DNA and the RNA genome (Fig. 7.14). Nucleotide sequencing of cloned retroviral DNAs and host-viral DNA integration junctions revealed several unique features of the process. Both viral and cellular DNAs undergo characteristic changes as a result of integration. Viral DNA is shortened, usually by 2 bp from each end, and a short (4- to 6-bp) duplication of host DNA flanks the provirus on either end. Finally, the proviral ends of all retroviruses comprise the same dinucleotide: 5′-TG . . . CA-3′. This dinucleotide is often embedded in an extended, imperfect inverted repeat that can be as long as 20 bp for some viruses. The fact that the length of the host cell DNA duplication is characteristic of the virus provided the first clue that a viral protein and not a host cell protein was responsible for integration.

The inverted repeat, conserved terminal dinucleotide sequence, and flanking direct repeats of host DNA are strikingly reminiscent of features observed in a number of bacterial insertion sequences and an *E. coli* bacteriophage (Mu). Predicted amino acid sequences homologous to a portion of the retroviral IN are also found in the transposases of certain bacterial insertion sequences, suggesting that, like RT, IN probably evolved before the divergence of prokaryotes and eukaryotes. These similarities suggest common mechanisms for retroviral DNA integration and DNA transposition.

### Pathway of Retroviral Integration

A generally accepted model for the retroviral DNA integration reaction has been developed on the basis of results from many different types of experiments, including in vivo analyses and studies with reconstituted systems. The two ends of viral DNA are recognized specifically, nicked, and then joined covalently to the host DNA in seemingly random locations, at staggered nicks also introduced by IN. The enzyme accomplishes these multiple activities as a multimer in a large nucleoprotein complex, probably assisted by other viral components and perhaps by other proteins in the infected cell. As illustrated in Fig. 7.15, the reaction catalyzed by IN is known to occur in two biochemically and temporally distinct steps.

### *The Processing Step*

The first step in integration is a processing reaction which can take place only when synthesis of the ends of unintegrated linear viral DNA is complete; it requires a specific sequence in duplex DNA and close proximity of this sequence to a terminus (Fig. 7.15). It has been demonstrated that the processing step can occur in the cytoplasm of an infected cell before viral DNA enters the nucleus. Endonucleolytic nicking adjacent to the conserved dinucleotide near each DNA end results in the removal of a terminal dinucleotide, with formation of a new, recessed 5′ . . . CA$_{OH}$-3′ end that will be joined to target DNA in the second step of the reaction. Site-directed mutagenesis of

## BOX 7.2

### *Retroelements*

**Retrotransposons** are dispersed widely in nature. Their gene content and arrangements are similar to those of retroviruses. Like retroviral proviruses, integrated retrotransposon DNAs have LTRs that include signals for transcription by cellular RNA polymerase II and short direct repeats of host DNA at their borders. RT sequence comparisons (Fig. 7.13) have allowed classification of these elements into two families, *Pseudoviridae* and *Metaviridae* (classification by S. Sandmeyer and J. Boeke for the seventh report of the International Committee on Taxonomy of Viruses, in preparation). Most retrotransposons are distinguished from retroviruses by lack of an extracellular phase; they have no *env* gene, and hence the viruslike particles formed intracellularly are noninfectious. However, in the family *Metaviridae*, members of one genus do include open reading frames corresponding to *env*, and at least one of these elements, the *Drosophila* gypsy, produces infectious extracellular particles. Because of these similarities, retrotransposons may be thought of as retroviral progenitors. An alternative possibility, but with less phylogenetic support, is that retrotransposons are degenerate forms of retroviruses.

**Retroposons**, also called non-LTR retrotransposons, are dispersed widely in the DNA of human cells. Like retroviral proviruses and retrotransposons, they are

flanked by direct repeats of host DNA at their boundaries. Although they lack LTRs, they contain internal promoters for transcription by cellular RNA polymerase III. All retroposons have A-rich stretches at one terminus, presumed to be derived by reverse transcription of the 3'-poly(A) tails in their RNA intermediates. Retroposons comprise fairly long stretches of related sequences (up to 6 kbp). Thus, these long interspersed repeat elements are commonly called **LINEs**. Most LINEs encode RT-related sequences, but they often contain large deletions and translational stop codons and thus are defective. Recently, LINEs that are capable of retrotransposition in the human genome and encode an active RT have been discovered.

**Retrosequences**, another type of reverse-transcribed sequence, are found as short interspersed repeat elements on the order of 300 bp long. These sequences, called **SINEs**, are even more abundant than LINEs. SINEs have no known open reading frames. The mechanism by which they are retrotransposed is not yet known.

**Processed pseudogenes** have no introns (hence "processed"), and their sequences are related to exons in functional genes that map elsewhere in the genome. Like retroposons, they include long A-rich stretches at one end and lack LTRs. However, they contain no internal promoter for transcription, and no RT sequences have been found associated with processed pseudogenes. Generally, they are thought to arise from rare reverse transcription of cellular mRNAs catalyzed by the RTs of retroviruses or nondefective LINEs. The following table summarizes the defining characteristics of the major classes of retroelements.

**Retroelements table with diagrams.** Boxes labeled LTR show the position of long terminal repeat sequences. *gag*, *pol*, and *env* are retroviral genes (appendix, Fig. 9). Pol II and Pol III refer to host RNA polymerases I and II. (A)$_n$ refers to poly(A) sequences of variable length. Arrows flanking the elements represent direct repeats.

---

### Characteristics of retroelements resident in eukaryotic genomes

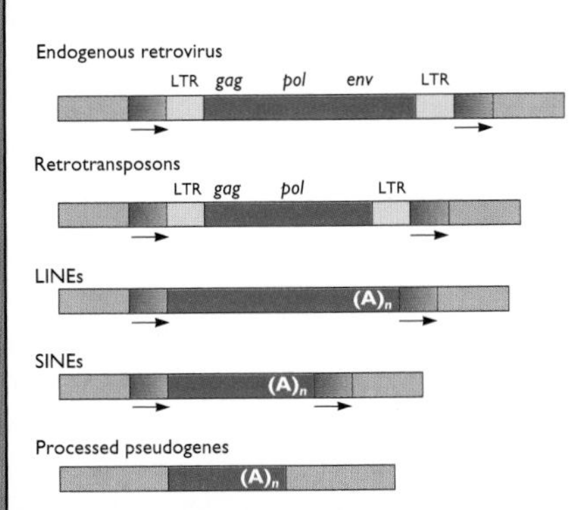

| Designation | Characteristic | Example | Copy number |
|---|---|---|---|
| Endogenous retroviruses | RT, LTR (internal Pol II promoter), and *env* | HERVs (human) | $1–10^2$ |
| Retrotransposons | RT, LTR (internal Pol II promoter) | Ty3 (yeast) | $10^2–10^4$ |
| Retroposons (LINEs) | RT, internal Pol III promoter A-rich sequence at end | LINE 1 (human) | $10^4–10^5$ |
| Retrosequences (SINEs) | A-rich sequence at end, internal Pol III promoter, but no RT | *Alu* (human) | $10^5–10^6$ |
| Processed pseudogenes | A-rich sequence at end, no internal promoter, no RT | β-Tubulin (human) | $1–10^2$ |

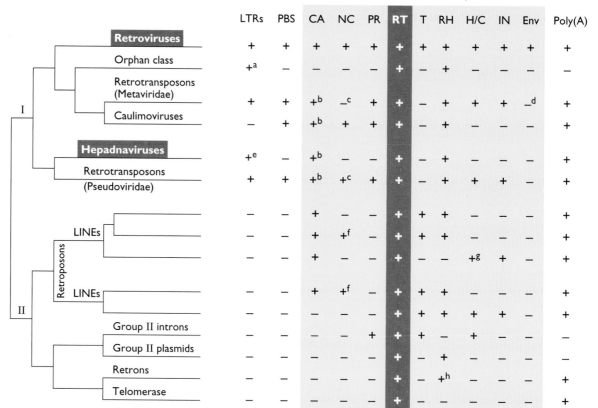

Deduced amino acid sequence

| | LTRs | PBS | CA | NC | PR | RT | T | RH | H/C | IN | Env | Poly(A) |
|---|---|---|---|---|---|---|---|---|---|---|---|---|
| **Retroviruses** | + | + | + | + | + | + | + | + | + | + | + | + |
| Orphan class | +[a] | − | − | − | − | + | − | + | − | − | − | − |
| Retrotransposons (Metaviridae) | + | + | +[b] | −[c] | + | + | − | + | + | + | −[d] | + |
| Caulimoviruses | − | + | +[b] | + | + | + | − | + | − | − | − | + |
| **Hepadnaviruses** | +[e] | − | +[b] | − | − | + | − | + | − | − | − | + |
| Retrotransposons (Pseudoviridae) | + | + | +[b] | +[c] | + | + | − | + | + | + | − | + |
| LINEs | − | − | + | − | − | + | + | + | − | − | − | + |
| | − | − | + | +[f] | − | + | + | + | − | − | − | + |
| | − | − | + | − | − | + | − | − | +[g] | + | − | + |
| LINEs | − | − | + | +[f] | − | + | + | + | − | − | − | + |
| | − | − | − | − | − | + | + | + | + | + | − | + |
| Group II introns | − | − | − | − | + | + | + | − | + | − | − | − |
| Group II plasmids | − | − | − | − | − | + | − | + | − | − | − | − |
| Retrons | − | − | − | − | − | + | − | +[h] | − | − | − | + |
| Telomerase | − | − | − | − | − | + | − | − | − | − | − | + |

**Figure 7.13 Schematic representation of the eukaryotic retroid family phylogeny.** The unrooted family tree derived from comparison of deduced RT amino acid sequences is displayed at the left. It is separated into two major branches (I and II) based primarily on the presence of long terminal repeats (LTRs) that contain important regulatory sequences in the DNA versions of some of its members. The table compares DNA sequences or deduced amino acid sequences encoded in all family members with those characteristic of retroviruses. The region that defines membership in the retroid family, the RT, is highlighted. Superscript letters in the figure indicate the following. (a) Some members have inverted LTR-like sequences. (b) Members of the retrotransposon, caulimovirus, and hepadnavirus lineages have capsid proteins (CA), but convincing similarity to retroviral capsid sequences has not been demonstrated. (c) The presence of nucleocapsid protein (NC) varies among different members of the retrotransposon lineages. (d) Some members of this retrotransposon lineage have a region that encodes an envelopelike protein (Env). (e) The direct repeats, DR1 and DR2, of hepadnaviruses are not *long* terminal repeats but may be related to retroviral LTRs. (f) Among the retroposons, the NC relationship consists only of alignment to the essential conserved $CX_2CX_4HX_4C$ motif of the retroviral NC. This $Zn^{2+}$-binding motif, found in an upstream open reading frame, is variable in copy number and position. (g) In one of the LINEs (R2Bm), this region is highly divergent. (h) Among the retrons, RNase H activity and sequence similarity are variable. Gene key: LTRs, long terminal repeats flanking the retroviral genome; PBS, tRNA primer-binding site; CA, capsid protein; NC, nucleocapsid; PR, protease; RT, reverse transcriptase domain; T, "tether" region connecting the RT and RNase H domains; RH, RNase H domain; H/C, histidine and cysteine motif in the amino-terminal portion of the integrase; IN, integrase; Env, envelope protein; poly(A), 3' poly(A) tract in RNA or DNA form. From M. A. McClure, *in* A. M. Skalka and S. P. Goff (ed.), *Reverse Transcriptase* (Cold Spring Harbor Laboratory Press, Cold Spring Harbor, N.Y., 1996), with permission.

viral DNA ends has shown that the conserved 5' . . . $CA_{OH}$-3' dinucleotide is essential for correct and efficient integration. Indeed, mutation of this sequence at only one DNA terminus is sufficient to impair processing of both

viral DNA ends. Sequences upstream also seem to affect the efficiency of the reaction (Fig. 7.16). However, although there is strong evidence for sequence specificity in vivo, attempts to demonstrate sequence-specific binding of

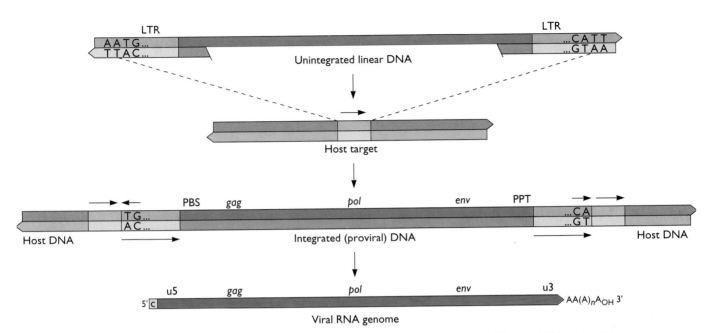

**Figure 7.14 Characteristic features of retroviral integration.** Unintegrated linear DNA of the "simple" avian retrovirus avian leukosis virus (top) after reverse transcription has produced two blunt-ended LTRs (Fig. 7.3). The break in the bottom strand indicates that this strand can include discontinuities, whereas the top strand is continuous (Fig. 7.3). Two base pairs (AA·TT) are removed from both termini on integration. A 6-bp "target site" in host DNA (indicated by an arrow) is duplicated on either side of the proviral DNA during integration. The integrated proviral DNA (middle) includes 15-bp, imperfect, inverted repeats at its termini, which end with the conserved 5'-TG . . . CA-3' sequence; these inverted repeats are embedded in the LTR, which is itself a direct repeat. The gene order is identical in unintegrated and proviral DNA and is colinear with that in the viral RNA genome (bottom), for which a provirus serves as a template (described more fully in chapter 8).

purified IN protein to retroviral DNA in vitro have generally not been fruitful. This failure may be a manifestation of the ability of IN to recognize host DNA in a sequence-independent manner in the joining step.

### The Joining Step

The second step in integration is a concerted cleavage and ligation reaction in which the two newly processed 3' viral DNA ends are joined to staggered (4- to 6-bp) phosphates at the target site in host DNA (Fig. 7.15). The product of the joining step is a *gapped* intermediate in which the 5'-$PO_4$ ends of the viral DNA are not linked to the 3'-OH ends of host DNA. These gaps are subsequently repaired, producing the characteristic flanking direct repeats of host DNA and a net loss of 2 bp from the ends of the viral DNA. This final repair step may be accomplished by host cell enzymes; however, a role for RT and possibly IN in this process can also be envisioned.

### The Preintegration Complex and Nuclear Entry

Rapidly sedimenting nucleoprotein complexes have been isolated from the cytoplasm of avian sarcoma/leuko-

sis virus-, murine leukemia virus-, and human immunodeficiency virus type 1-infected cells. These complexes appear to be compact, but the viral DNA associated with them is accessible to nucleases. The nucleoproteins contain IN and viral DNA in a form that can be joined to exogenously provided plasmid or bacteriophage DNA. Such ex vivo integrants exhibit all the features expected for products of authentic integration. Other viral proteins, among them RT, capsid, and nucleocapsid proteins, have also been reported to be present in the preintegration complexes of MLV and HIV-1, but no clearly defined role in the integration process has yet been assigned to any of these proteins. How nucleoprotein preintegration complexes gain access to target sites in host DNA is largely unknown. Some recent models are described in chapter 5.

### Specificity of the Integration Reaction

The integration reaction is site-specific with respect to viral DNA; it always occurs at the ends of LTRs. Each retroviral IN recognizes a specific sequence at its cognate viral DNA ends that includes the conserved 5' . . . CA-3' dinucleotide and a short stretch of adjacent sequences that are

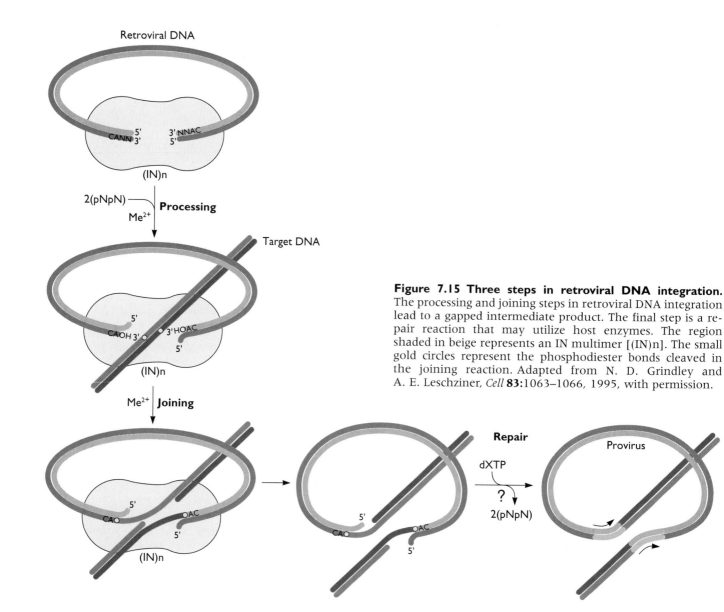

**Figure 7.15 Three steps in retroviral DNA integration.** The processing and joining steps in retroviral DNA integration lead to a gapped intermediate product. The final step is a repair reaction that may utilize host enzymes. The region shaded in beige represents an IN multimer [(IN)n]. The small gold circles represent the phosphodiester bonds cleaved in the joining reaction. Adapted from N. D. Grindley and A. E. Leschziner, *Cell* **83**:1063–1066, 1995, with permission.

often part of an imperfect inverted repeat (Fig. 7.16). While specificity for viral DNA end sequences can be demonstrated under certain conditions in vitro, this recognition is clearly less efficient than that which occurs in vivo. Thus, it seems likely that some structural features or interactions among components within the preintegration complex help to place IN at its site of action near the viral DNA termini in living cells.

Many sites in host DNA can be targets for retroviral integration. Furthermore, comparison of cellular sequences flanking a number of independent integrations have identified no clear target consensus sequence. In vitro studies with purified enzymes and naked target DNA have failed to show any DNA sequence specificity for joining, although the patterns exhibited by different retroviral IN with the same DNA target are not identical. The most extensive in vivo study to date examined the integration of avian sarcoma/leukosis virus DNA into the DNA of cultured turkey or quail embryo fibroblasts. The results suggest that virtually any region can be a target for integration. Although there was some clustering of integrations at particular sites within some genes, no gene was targeted with significant preference over any other.

In the absence of discernible sequence or gene specificity, should we expect any selectivity for retroviral DNA integration into host chromosomes? The available data

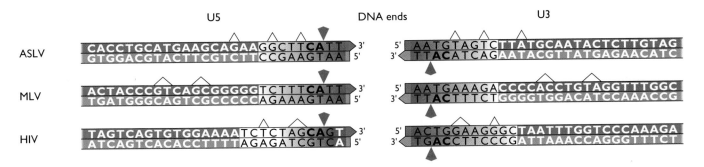

**Figure 7.16 Nucleotide sequences at the termini of retroviral DNAs.** Sequences are arranged with juxtaposed termini. Red-shaded sections are important for integration. The most intensely shaded regions appear to be most critical, and the conserved 5′ . . . CA . . . 3′ is essential for efficient and accurate integration. Red arrowheads indicate the location of the nick produced during processing.

suggest yes. The IN protein of a yeast retrotransposon (Ty3) appears to interact with RNA polymerase III transcription initiation proteins during target site selection. A specific interaction between human immunodeficiency virus type 1 IN and a cellular protein (called IN-interacting protein 1 [Ini-1]), which is a component of a large complex that binds to chromatin and activates transcription, has also been observed. In vitro studies have established a preference for retroviral DNA integration into DNA sequences that are bent because of intrinsic properties or underwound as a consequence of being wrapped around a nucleosome. Thus, these results suggest that interactions with particular chromatin-associated proteins may guide retroviral integration into sites where the DNA conformation has been favorably altered by protein binding.

### Model Reactions In Vitro

The development of simple in vitro assays for the processing and joining steps in retroviral integration marked an important turning point for investigation of the biochemistry of this reaction (Fig. 7.17). These assays include as substrates short duplex oligodeoxynucleotides (15 to 25 bp) whose sequences correspond to one retroviral DNA terminus. The use of model DNA substrates also showed that IN can catalyze an apparent reversal of the joining reaction, which has been called **disintegration** (Fig. 7.17). The disintegration reaction is relatively inefficient, but it appears to be the least demanding in terms of the viral DNA sequence or (as discussed in the next section) enzyme determinants required. With simple in vitro assays it was discovered that retroviral IN protein is both necessary and sufficient for catalysis, that no source of energy (ATP or an ATP-generating system) is needed, and that the only required cofactor is a divalent cation, $Mn^{2+}$ or $Mg^{2+}$.

Use of these simple substrates and other biochemical analyses with purified IN protein helped to delineate the sequence and structural requirements for DNA recognition. They also showed that while the processing, joining, and disintegration reactions produce different products, their underlying mechanisms are the same. These reactions all comprise nucleophilic attack on a phosphorus atom by the oxygen in an OH group and result in cleavage of a phosphodiester bond in the DNA backbone (Fig. 7.17). In the processing step, the OH comes from a water molecule. In the joining step, the OH is derived from the processed 3′ end of the viral DNA, and the result is a direct transesterification. In the disintegration reaction, also a direct transesterification, a 3′-OH end in the interrupted duplex attacks an adjacent phosphorus atom, forming a new phosphodiester bond and releasing the overlapping DNA.

Assays with single-end model substrates have been invaluable in elucidation of the catalytic mechanisms of IN. Nonetheless, they are limited in that the major products represent "half-reactions" in which only one viral end is processed and joined to a target. More recently, conditions for efficient *concerted* processing and joining have been described with a variety of specially designed, short model DNA substrates that include *two* viral DNA end sequences. These studies demonstrate that IN alone is also sufficient to accomplish the concerted reactions, an impressive feat for this relatively small protein, even acting as a multimer.

### Integrase Structure and Mechanism

#### Domain Structure

Retroviral IN proteins are approximately 300 amino acids in length. The IN monomer contains three domains (Fig. 7.18). Attempts to determine the three-dimensional

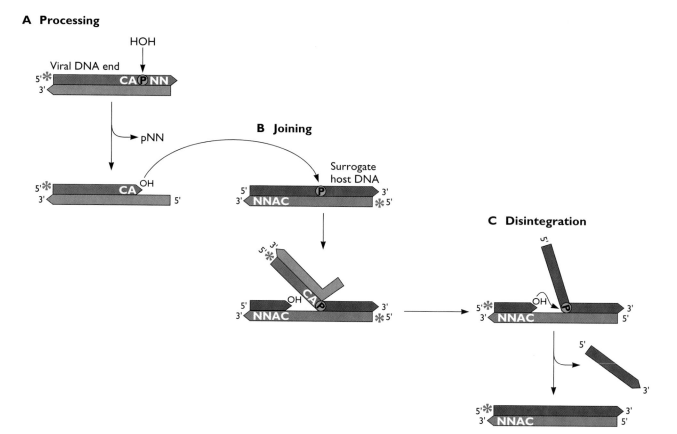

**A Processing**

HOH

Viral DNA end

**B Joining**

Surrogate host DNA

**C Disintegration**

**Figure 7.17 Principles of simple in vitro assays for IN activity with oligodeoxynucleotides.** (A) The substrate representing viral DNA comprises an oligodeoxynucleotide duplex whose sequence matches one end of the viral DNA. In the simplest form of the assay, the 5′ end of the strand to be cleaved is radioactively labeled with ³²P, shown here as a red asterisk. The processed product is then detected on a denaturing sequencing gel as a single chain two nucleotides shorter than the substrate. Production of the other product (pNN) can also be monitored if one or both of the 3′-terminal phosphates in the substrate are labeled. (B) In the joining reaction, a collection of products that are longer than the substrate are generated. These molecules can also be detected in a denaturing gel. If the target DNA is a much larger molecule, such as a bacterial plasmid or phage DNA, the products can be detected by electrophoretic separation in agarose gels. (C) Only the simplest version of the disintegration assay is shown here. A variety of disintegration substrates, in which all strands are duplex and some are continuous, have been used.

structure of a full-length retroviral IN by X-ray crystallography have so far proved unsuccessful, perhaps because the linkages between the domains are too flexible. However, considerable insight has been obtained from analyses of the separate domains studied as isolated polypeptides.

**N-terminal domain.** The N-terminal domain, comprising approximately the first 50 amino acids, is characterized by two invariant histidines and cysteines, HHCC. A Zn²⁺ ion bound by these residues stabilizes a helix-turn-helix structural motif (Fig. 7.18, left, green ball in the ribbon diagram)

that is almost identical in topology to the DNA-binding domain of the bacterial *trp* repressor protein. However, the isolated N-terminal domain does not bind tightly to DNA although there is evidence that it may interact with DNA in the context of the intact protein. The isolated N-terminal domain is a dimer in solution. As with the C-terminal domain, this property is consistent with its proposed role in promoting the multimerization of full-length IN.

**Catalytic core domain.** The catalytic domain (Fig. 7.18, center), included in a central region of approxi-

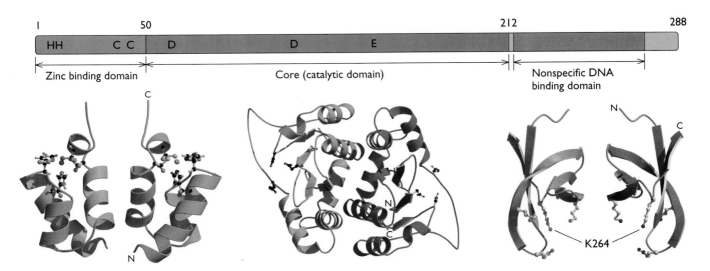

**Figure 7.18 Linear map of human immunodeficiency virus type 1 IN showing the three independently folding domains.** Numbers at the top indicate amino acid residues starting with 1 at the N terminus. Evolutionarily conserved amino acids are indicated in the single-letter code. (Left) Ribbon diagram showing the nuclear magnetic resonance (NMR) structure of a dimer of the N-terminal domain of HIV-1 IN (residues 1 to 46). The structure of the single subunit is shown in pale green, the zinc atom is displayed as a green ball, and the coordinating cysteines and histidine residues are shown in ball-and-stick representations. Coordinates were kindly provided by Marius Clore (National Institutes of Health). (Center) Ribbon model of the structure of a dimer of the catalytic core domain determined by X-ray crystallography. The structure of the single subunit is shown in dark green. Side chains of conserved acidic residues of the active site are shown in a red ball-and-stick representation. The primary contacts between subunits at this interface are between α-helix 1 of one subunit and α-helix 5′ of the other subunit. Coordinates were kindly provided by Alex Wlodawer, National Cancer Institute-Frederick Cancer Research and Development Center. (Right) Ribbon model of the NMR-determined structure of a dimer of the C-terminal domain (residues 220 to 270). The structure of a single subunit is shown in red. The positions of several basic amino acids that extend into the saddlelike groove and are proposed to bind DNA are indicated in ball-and-stick representation. Coordinates were kindly provided by Marius Clore; the figure was prepared by Mark Andrake (Fox Chase Cancer Center).

mately 150 amino acids, is characterized by a D,D(35)E motif: a constellation of three invariant acidic amino acids, the last two separated by 35 residues. There is also significant sequence conservation in amino acids surrounding the invariant triad, including two nearby lysine residues that have been implicated in viral DNA binding. Site-directed mutagenesis experiments have shown that the three invariant acidic amino acids are required for all catalytic functions of IN. This finding indicates that there is one catalytic center for both processing and joining. The isolated catalytic cores of human immunodeficiency virus type 1 IN and avian sarcoma/leukosis virus IN do not have processing and joining activity. However, both catalyze the disintegration reaction, albeit at low efficiency, and the avian virus IN core possesses efficient endonuclease activity. Thus, this domain clearly contains the catalytic center.

Solution of the crystal structures of the catalytic core domains of the human and avian IN proteins was a signif-

icant milestone in the study of IN. These structures established that IN proteins are members of a large superfamily of nucleases and recombinases that includes the RNase H domain of the human virus RT. In overall topology, the core domains of these two retroviral INs are quite similar. Specific metal complexes are formed when the core domain crystals are soaked in solutions of divalent metals. The orientation of the side chains and the positions of the metals in such complexes are superimposable on those of metal ion complexes obtained with the RNase H domain of the human immunodeficiency virus type 1 RT (Fig. 7.19). The binding of two metal ions at the active sites of this RNase H and avian retrovirus IN is consistent with a two-metal mechanism of catalysis by this superfamily, as proposed for the human retroviral RT polymerase activity (Fig. 7.11). The core domains of the human and avian virus INs both crystallize as dimers, and determinants for multimerization have been mapped to the core domains of both proteins by independent methods.

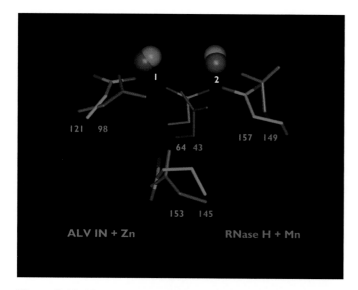

**Figure 7.19 Metal complexes in the HIV-1 RNase H and the ASLV IN catalytic core domain.** In both structures, the side chains of critical acidic amino acids form a tripod that complexes two divalent cations in almost superimposable positions. The ASLV IN core also binds a single $Mn^{2+}$ or $Mg^{2+}$ ion in position 1 with no change in the position of the side chains of D64 and D121.

**C-terminal domain.** The sequence of the 80- to 100-amino-acid C-terminal domain is not highly conserved among IN proteins from different retroviral genera. Nevertheless, mutational analyses of both avian sarcoma/leukosis virus IN and human immunodeficiency virus IN indicate that this domain contains nonspecific DNA binding activity and multimerization determinants. The structure of a 50-amino-acid fragment (corresponding to residues 220 to 270) from the C-terminal domain of the human retrovirus IN has been determined by nuclear magnetic resonance; it exists as a dimer in solution (Fig. 7.18, right). It has been suggested that a DNA-binding site may be formed by the dimer in a saddlelike groove that includes favorably positioned basic side chains and could accommodate a duplex DNA molecule. In support of this model, substitution of conserved basic amino acids significantly reduces the DNA-binding activity of this domain of IN.

### Higher-Order Structure in the Integration Complex

Retroviral DNA integration is an intricate, dynamic reaction in which two separate DNA duplexes are bound to the enzyme, cut, and reorganized. We have yet to understand completely how the substrates and protein are organized and acted on in any recombination reaction. Because of the medical importance of retroviruses, the higher-order structure of this integration complex is not only of scientific interest but also of practical significance.

**DNA organization.** In most models of retroviral integration the viral DNA ends are shown to be adjacent to one another, as presumed to be necessary for their concerted processing and joining to a target site. The presence of inverted repeats at the viral DNA ends (Fig. 7.16) suggests that they are recognized by a multimeric protein with a twofold axis of symmetry. In vitro experiments with model substrates including two viral DNA end sequences indicate that the invariant 5′ . . . CA-3′ dinucleotides are bound to IN in a nonlinear arrangement and are at a fixed distance from each other during the concerted processing reaction. The optimal distance between them is related to the distance between the staggered cuts in target DNA made during joining (i.e., equivalent to six deoxyribonucleotides for the avian virus and five deoxyribonucleotides for human immunodeficiency virus type 1). Experiments with these and other model substrates also indicate that the viral DNA ends do not maintain B-DNA conformation when bound at the active site of the enzyme but that the two strands are partially unwound and distorted. As noted above, partially unwound, bent DNA is a preferred target for joining. Thus, there appears to be a certain symmetry in the conformations of viral DNA ends and target DNA during joining.

**Protein organization.** Biochemical and in vitro complementation studies indicate that IN functions as a multimer. A reversible equilibrium among monomeric, dimeric, and tetrameric forms of IN is observed in the absence of DNA substrate. A conservative estimate of the intracapsid concentration of IN gives a value ($\sim$150 $\mu$M) high enough for most of the IN protein in virions to be in the form of dimers and tetramers. While it is clear that all three isolated domains dimerize independently and together facilitate tetramerization, we know little about how the different domains of IN interact within a multimer. Nevertheless, as a mutation in the 5′ . . . CA-3′ dinucleotide of one viral DNA end impairs processing at both ends, it seems certain that IN monomers do indeed "talk" to each other by protein-protein interactions.

On the basis of an analogy with Mu transposase and other recombinases, a tetrameric IN structure seems to be a reasonable functional unit. In the case of IN, one active site in each monomer subunit might catalyze each of the four nucleophilic attacks required in the concerted processing and joining reactions. However, as illustrated in the simple models in Fig. 7.20, the sequential action of two monomers can also be envisioned. On the other hand, other multimeric recombination complexes include some monomers that appear to play only a structural role, and we have learned from analysis of the human immunodeficiency virus type 1 RT heterodimer that proteins with almost identical primary sequences can be put to very different uses in a multimeric retroviral enzyme. A recent model suggested by DNA cross-linking studies with human retrovirus IN proposes that at least eight monomers are required for concerted integration of two viral DNA ends into a target DNA. Only two of the subunits function catalytically; the other subunits help to position the substrates and stabilize the complex. This provocative model should provide a framework for further investigation of the higher-order structure of the integration complex.

**Role of host proteins.** The viral DNA within preintegration complexes is not itself a target for integration; such autointegration reactions would be suicidal in vivo. In murine leukemia virus preintegration complexes isolated from infected mouse cell lines, this restriction was shown to depend on the presence of a previously unknown, 89-amino-acid cellular protein called **barrier-to-autointegration protein.** If the cellular protein is removed from such complexes by salt extraction, the complexes will catalyze only autointegration. Subsequent addition of purified barrier-to-autointegration protein reconstitutes the block to autointegration and restores normal integration of viral DNA into exogenously added target DNA. The barrier-to-autointegration protein forms dimers in solution, binds to DNA, and causes it to become compacted, presumably through formation of intermolecular bridges. It has been proposed that the protein prevents autointegration because such compacted DNA is unavailable for integration. As purified murine leukemia virus particles do not contain this cellular protein, it must be acquired after the

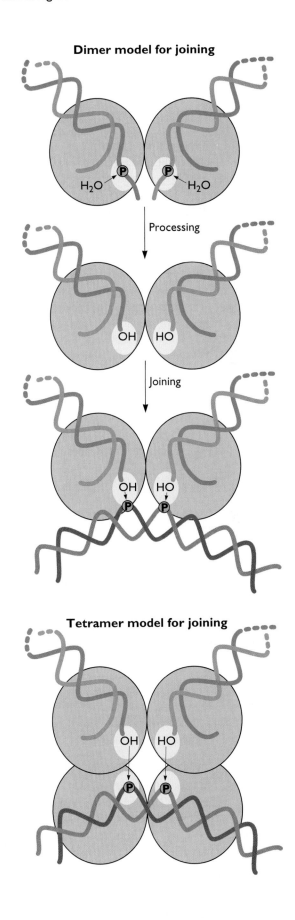

**Figure 7.20 Two models for functional IN multimers.** In both models the ends of the viral DNA are shown in a nonlinear arrangement with the termini partially denatured and distorted as the strands are bound to the protein. The dimer model assumes sequential reactions at two active sites in which the viral DNA ends are first processed. The target DNA is then "captured," perhaps by binding in the region vacated by the dinucleotide clipped from the viral 3' end during processing, and the joining reaction ensues. The tetramer model shown assumes a symmetry in the binding of viral and target DNAs. Processing occurs at the active sites of two symmetrically arranged monomers. The new 3'-OH ends act as nucleophiles in the attack on phosphorus atoms on both chains of the target DNA. The attacked phosphates are situated across from each other in the major groove of target DNA which is held at the active sites of two other monomers in an arrangement similar to that of the viral DNA.

virus enters the cytoplasm of its host cell and viral DNA is synthesized. It is currently unknown whether a similar mechanism blocks autointegration in other retroviral preintegration complexes.

In the best understood prokaryotic recombination systems, such as integration of bacteriophage lambda DNA and transposition of Mu DNA, DNA-bending host proteins are required to establish the proper configuration of DNA substrates. The same may be true for retroviral integration. Addition of eukaryotic high-mobility-group proteins, which are nonhistone DNA-bending chromosomal proteins, stimulates the concerted integration of two viral DNA ends into a target DNA by purified avian sarcoma/leukosis virus IN and human immunodeficiency virus type 1 IN. Furthermore, one such high-mobility-group protein, HMG(I)Y, was found to be present in preintegration complexes isolated from cells infected with the human retrovirus. Salt extraction of such complexes abolishes integration of the viral DNA into exogenously provided target DNA, but this activity can be restored partially by the addition of purified HMG(I)Y. Like the barrier-to-integration protein, the high-mobility-group protein is presumed to be acquired after the virion enters the cell. It is not yet clear exactly how these two host proteins are acquired, whether their activities are related, or how important each is to the organization of functional viral DNA-IN complexes in vivo.

## Hepadnaviral Reverse Transcription

### Introduction

#### Another Discovery

The concept that a virus with an RNA genome can replicate through a DNA intermediate was followed with another big surprise: a DNA virus that used RNA as an intermediate in its replication. Early hints that a mechanism other than semiconservative DNA synthesis was responsible for hepadnaviral replication came from the discovery of asymmetries in the genomic DNA and in the endogenous DNA polymerase reaction in virions. The viral DNA comprises one full-length (−) strand and an incomplete complementary (+) strand (Fig. 7.21), and the endogenous polymerase reaction can extend only the (+) strand. The replication intermediates isolated from infected cells were also unusual, comprising mainly (−) strands of less than unit length, few of which were associated with (+) strands. To Jesse Summers and William Mason, all this seemed suspiciously like retroviral reverse transcription, and in 1982 they published the results of their landmark studies on the unique features of hepadnaviral replication. Studying subviral particles from cytoplasmic extracts of duck hepatitis B virus-infected liver, they found that, un-

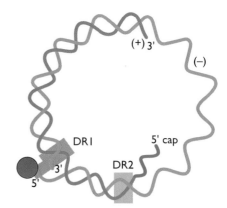

**Figure 7.21 Hepadnaviral DNA.** The DNA in extracellular hepadnavirions is a relaxed circular molecule of approximately 3 kb whose circularity is maintained by overlapping 5′ ends. The (−) strand is slightly longer than unit length, and the polymerase, shown as a blue ball, is attached to its 5′ end. The (+) strand has a capped RNA of ∼18 nucleotides at its 5′ end and is less than unit length. The 5′ ends include short (10- to 12-bp) direct repeats called DR1 and DR2. As in retroviruses, these repeat sequences play an important role in facilitating critical template transfers during reverse transcription. Details of the genetic content are provided in Fig. 11 of the appendix.

like the endogenous reaction typical of extracellular virions, these particles incorporated deoxynucleotide triphosphates into both (+) and (−) strands. As with RNA-dependent DNA polymerization in retroviruses, synthesis of the (−) strand was resistant to the DNA-intercalating drug actinomycin D, whereas synthesis of the (+) strand was inhibited by this compound. Furthermore, a portion of the newly synthesized (−) strand DNA sedimented with the density of RNA·DNA hybrids. These and related findings marked an important turning point in our understanding of hepadnaviruses and extended our knowledge of reverse transcription.

#### Role of Reverse Transcription in the Hepadnaviral Life Cycle

An abbreviated model for the single-cell replication cycle of hepadnaviruses is shown in Fig. 7.22. After the virion core enters the cytoplasm, viral DNA is imported into the nucleus and repaired to produce a covalently closed circular molecule by processes that are largely unknown. Unlike retroviral DNA, hepadnaviral DNA is not normally integrated into the host's genome; hepadnaviruses encode no IN. However, covalently closed circular DNA persists in the nucleus as an autonomous episome from which viral RNAs are transcribed by the host cell RNA polymerase II. The 3.5-kb **pregenome mRNA** is exported to the cytoplasm, where it serves as the template

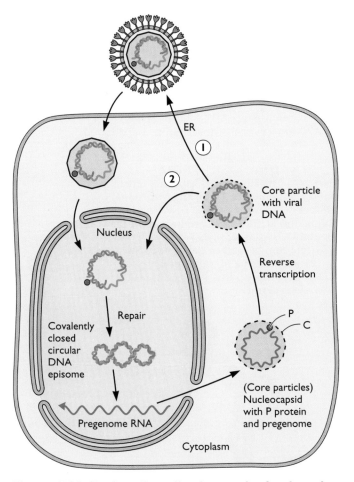

**Figure 7.22 Single-cell replication cycle for hepadnaviruses.** Pathway 2 provides additional copies of covalently closed circular episomal DNA. ER, endoplasmic reticulum.

Hepadnaviral covalently closed circular DNA is not replicated by the host's semiconservative DNA synthesis machinery. Thus, *all* hepadnaviral DNA is produced by reverse transcription. This contrasts with the situation in retroviruses, in which the integrated nuclear form, the provirus, is replicated along with the host DNA. Both retroviruses and hepadnaviruses cause persistent infections by maintaining their nuclear DNAs for the life of their host. But, as we see, they do so in quite different ways.

The analysis of reverse transcription in hepadnaviruses has been difficult for a number of technical reasons. For one, suitable tissue culture systems were not available until 1986, when hepatoma cell lines permissive for viral replication following transfection with cloned viral DNA were identified. Furthermore, mutational studies are confounded by the compact coding organization of the DNA. The tiny genome (ca. 3.2 kbp) is used very efficiently, with more than half of its nucleotides translated in more than one reading frame. This makes it difficult to produce mutations that change only one protein. Finally, as in retroviruses, reverse transcription takes place in subviral particles, but until quite recently it had not been possible to prepare enzymatically active protein to study the reaction free from these hepadnaviral particles (commonly called **cores**). Despite these difficulties, details currently available reveal fascinating analogies but also striking differences in the reverse transcription of hepadnaviruses and retroviruses.

## Pathway of Reverse Transcription

### *Essential Components*

**Pregenome RNA.** The **pregenome** RNA molecule that serves as the template for production of hepadnaviral genomic DNA is capped and polyadenylated and is also an mRNA for both C and P proteins. The transcription of pregenome RNA from covalently closed circular DNA in the nucleus is initiated at a position approximately 6 bp upstream of one copy of a short direct repeat called DR1 (Fig. 7.21). Transcription then proceeds along the entire DNA molecule, past the initiation site, to terminate after a polyadenylation signal just downstream of DR1. Thus, pregenome RNA is longer than its template DNA. Because the region from the transcription initiation site to the polyadenylation site is copied twice, there is a longer direct repeat (R) at either end of the RNA. This longer repeat includes DR1 and a structural element called **epsilon (ε)** (Fig. 7.23 and 7.26). The ε structure at the 5' end provides both the site for initiation of (−) strand synthesis and the signal for encapsidation into core particles.

for reverse transcription. This process takes place in a subviral particle that includes core and polymerase proteins (products of the C and P genes). P protein provides enzymatic and other activities required for reverse transcription. Following reverse transcription, the DNA-containing nascent core particles can follow one of two pathways. Late in infection, when the cisternae of the endoplasmic reticulum contain an abundance of viral envelope glycoprotein, they can bud into the endoplasmic reticulum and eventually be secreted as progeny virions. Conversely, if they do not become enveloped, the core particles can recycle to the nucleus, where they give rise to additional copies of covalently closed circular DNA. The amplification pathway predominates at early times after infection when there is little envelope protein available. Eventually, ca. 10 to 20 covalently closed circles accumulate in the nucleus, and a steady-state balance of DNA and cytoplasmic virion protein components is maintained.

There is a marked preference for reverse transcription of the pregenome from which P protein is translated. The basis of such *cis* selectivity is unknown; C protein, which is also translated from the pregenome and has RNA- and DNA-binding domains, appears to function perfectly well in *trans*. It is possible that the nascent P polypeptide binds to its own mRNA cotranslationally or to mRNAs that are in close proximity soon after translation. In any case, the virus may benefit from selection for genomes that express functional P protein.

Unlike those of retroviruses, the hepadnaviral cores in which reverse transcription occurs include only one genomic RNA and no cellular RNAs (Table 7.1). This selectivity is determined by the presence of the encapsidation signal(s) at the 5′ end of the hepadnaviral pregenome RNA. After reverse transcription, hepadnaviruses contain

**Figure 7.23 *cis*-acting signals on pregenome RNA.** (A) The RNA bears terminal repetitions (R) that contain copies of the packaging signal (ε), but only the 5′ copy has functional activity in vivo. Indicated are positions of the 5′ ends of (−) and (+) strand DNAs, and the 5′-UUAC-3′ motifs in duck hepatitis B virus within ε and at DR1, that are important for (−) strand DNA synthesis. Adapted from C. Seeger and W. S. Mason, p. 815–832, *in* E. De-Pamphilis (ed.), *DNA Replication in Eukaryotic Cells* (Cold Spring Harbor Laboratory Press, Cold Spring Harbor, N.Y., 1996), with permission. (B) Sequences and proposed base pairing in ε of duck hepatitis B virus (DHBV) and human hepatitis B virus (HBV). From J. R. Pollack and D. Ganem, *J. Virol.* **68:**5579–5587, 1994, with permission.

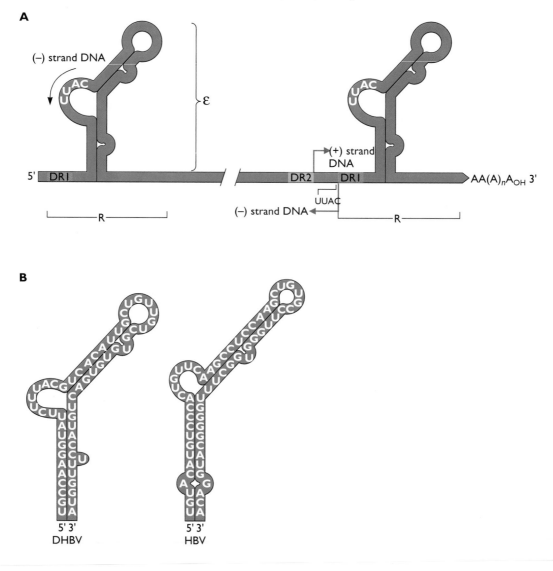

**Table 7.1**    Comparison of retroviral and hepadnaviral reverse transcription

| Parameter | Retroviruses[a,b] | Hepadnaviruses[a] |
|---|---|---|
| Viral genome | RNA (pseudodiploid) | DNA (incomplete duplex) |
| Template RNA also: | Genome RNA, *mRNA (gag, pol)* | Pregenome RNA, *mRNA (C and P proteins)* |
| DNA intermediate | *Circular DNA with 5′ overlaps* | *Circular DNA with 5′ overlaps* |
| Virus-encoded enzyme | RT | P protein |
| Molecules per core | 50–100 | 1 |
| Functions | *DNA polymerase, RNase H*, helicase (strand displacement) | *DNA polymerase, RNase H*, protein priming, template RNA encapsidation |
| Primer, first (−) DNA strand | tRNA (host) | Viral P protein |
| Site of initiation | *Near 5′ end of genome* | *Near 5′ end of pregenome* |
| First DNA product | (−) strong-stop DNA, ca. 100 nucleotides | Four nucleotides copied from bulge in 5′ ε |
| First template exchange | *To complementary sequence in repeated sequence, r, at 3′ end of template RNA* | *To complementary sequence in repeated sequence, R, at 3′ end of template RNA* |
| Primer, second (+) DNA strand | *Derived from template RNA*, internal RNase H product (ppt) | *Derived from template RNA*, 5′ cap, terminal RNase H product |
| Site of initiation | *Near 5′ end of (−) strand DNA* | *Near 5′ end of (−) strand DNA* |
| Time of initiation | Before completion of (−) strand | After completion of (−) strand |
| Type of priming | Priming in situ | Primer translocated |
| Second template exchange | *To the 3′ end of (−) strand DNA via complementary PBS sequence* | *To the 3′ end of (−) strand DNA via complementary sequence* |
| Reverse transcribing nucleo-protein complex | *Subviral "core" particles*, deposited in the *cytoplasm* on viral entry | *Nascent subviral "cores," cytoplasmic* intermediates in viral assembly |
| Final product(s) | Double-stranded linear DNA | Circular viral DNA or covalently closed episomal DNA |
| DNA maintained in the nucleus | Integrated into host genome, proviral DNA | Nonintegrated episome in host nucleus |

[a]Italics indicate similarities.

[b]ppt, polypurine tract.

only one copy of viral DNA. Quantitation of P protein in cytoplasmic core particles gives a P protein-to-viral DNA molecule molar ratio of approximately 1:1.

**Primers.**    The primers for hepadnavirus RT remain attached to the 5′ ends of the viral DNA strands. They are for (−) strand synthesis, the P protein itself, and for (+) strand synthesis, a capped RNA fragment derived from the pregenome. A protein priming mechanism was first described for adenovirus DNA replication and later for the bacteriophage φ29. Priming by a viral protein, VPg, is also used for the synthesis of poliovirus RNAs. Hepadnaviral reverse transcription is distinguished from these three viral systems by the fact that the protein primer and the polymerase are components of the same polypeptide.

**P protein.**    P protein is translated from an internal AUG in the pregenome approximately 550 to 650 nucleotides downstream from the start of the overlapping C protein reading frame (appendix, Fig. 11B). The translation product is a 90-kDa protein whose C-terminal enzymatic domains were first identified by amino acid

sequence alignment with the retroviral RTs (Fig. 7.24). Site-directed mutagenesis studies have verified that the highly conserved residues in the homologous domains are essential for hepadnaviral reverse transcription. Hepadnaviral P protein also contains an N-terminal domain separated from the RT region by a **spacer**, so called because it is refractory to mutation and apparently not necessary for any known activity of the protein. The N-terminal domain, referred to as the **terminal protein region**, includes the tyrosine residue utilized for priming (−) strand DNA synthesis (see Fig. 7.26). In addition to its other functions, P protein is also required for encapsidation of viral RNA, another departure from the retroviral strategy, in which the NC protein sequences in the Gag polyprotein are used for this purpose (Table 7.1).

An important breakthrough in the study of hepadnaviral reverse transcription was achieved with the 1992 demonstration by Guang-Hua Wang and Christoph Seeger that enzymatically active P protein can be produced on translation of P mRNA from duck hepatitis B virus in a cell-free rabbit reticulocyte lysate. Essential components in this reaction include the ε sequence and host proteins that

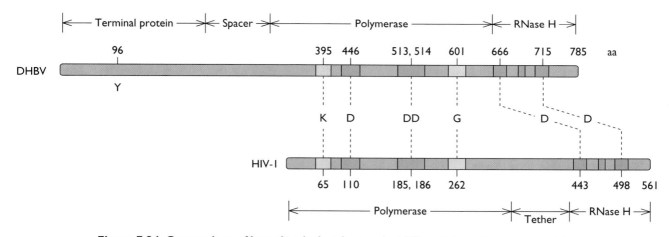

**Figure 7.24 Comparison of hepadnaviral and retroviral RT proteins.** Linear maps of the duck hepatitis B virus (DHBV) and human immunodeficiency virus (HIV) *pol* gene products. The maps were aligned relative to amino acids that are generally conserved among all RTs. Position 1 on the DHBV protein corresponds to the AUG codon at position 170 on the DHBV genome. See also Fig. 7.8. Adapted from C. Seeger and W. S. Mason, p. 815–832, *in* E. DePamphilis (ed.), *DNA Replication in Eukaryotic Cells* (Cold Spring Harbor Laboratory Press, Cold Spring Harbor, N.Y., 1996), with permission.

assist in the formation of an active nucleoprotein complex. Subsequent studies showed that a short, three- or four-nucleotide DNA fragment copied from the bulge in ε was covalently attached to a tyrosine residue in the newly synthesized polymerase. As described below, this short DNA product serves as a primer for further RNA-dependent DNA synthesis following the first template exchange, and P protein remains covalently attached to this DNA during all subsequent steps in reverse transcription (see Fig. 7.26).

**Other components.** Binding of P protein to the ε sequence at the 5′ end of pregenome RNA is required both for its encapsidation and for the initiation of reverse transcription. Furthermore, reverse transcription in vivo takes place in core particles. Thus, it was valid to ask if C protein participates in the initiation of hepadnaviral reverse transcription. Because C protein was absent in the cell-free lysates described above and in other studies in which P protein was expressed in yeast cells, it was concluded that P protein is the only viral protein required for DNA synthesis to be initiated.

Binding of P protein to ε may be required for folding of the protein into an active conformation. If the ε region is not present during synthesis of P protein in yeast, the enzyme is inactive even if the ε region is supplied later. Furthermore, P protein synthesized in the presence of the ε region is more resistant to proteolysis. Recent studies suggest that host cell proteins may also affect P-protein folding. Synthesis of active P protein in the cell-free system

requires the presence of cellular chaperone proteins and a source of energy (ATP). Furthermore, these host cell proteins appear to be incorporated into viral cores in a polymerase-dependent manner. It has been proposed that chaperones are needed to maintain viral P protein in a conformation that is competent to bind to ε and prime DNA synthesis (Fig. 7.25).

### Critical Steps in Reverse Transcription

**Initiation and the first template exchange.** Synthesis of the (−) strand of DNA is initiated by the polymerization of three or four nucleotides primed by the OH group of a tyrosine residue in the terminal protein domain of P protein. The template for this reaction is a sequence in the bulge of ε at the 5′ end of the pregenome, 3′-CAUU-5′ for DHBV. Why this copy and not the ε at the 3′ end is normally chosen for initiation by P protein remains a mystery, but this point has been verified by mutational studies. This initial synthesis is followed by a template exchange in which another CAUU sequence is engaged at the edge of DR1 at the 3′ end of the pregenome (Fig. 7.23 and 7.26, step 1). Although the sequences are complementary, the sequence is not unique (Fig. 7.23). Furthermore, mutation of the normal acceptor sequence leads to the synthesis of (−) strands whose 5′ ends map to other sites in the vicinity of the 3′ DR1 that apparently can serve as alternative acceptors. Thus, it seems reasonable to assume that selection of the normal site is guided by specific organization of pregenomic RNA in core particles, which places the accep-

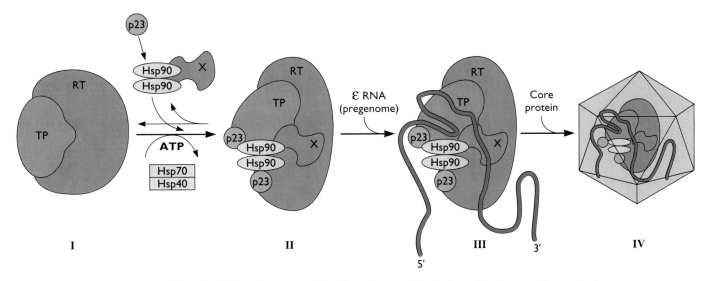

**Figure 7.25 Model for the assembly of nucleocapsids in hepadnaviruses.** P protein is translated to form an apostructure (I) that provides a substrate for the Hsp90 pathway. TP denotes the N-terminal domain of the P protein (the so-called terminal protein), which is unique to hepadnaviral RTs and harbors the primer tyrosine residue for the initiation of reverse transcription. RT indicates the polymerase plus RNase H domains which are conserved among all known RTs. The heat shock protein 90 (Hsp90) complex is depicted together with the chaperone partner p23 and possible additional proteins (X) that may be components of the complex. The function of Hsp90 also depends on the ATPase Hsp70 and its partner Hsp40. Interaction of the polymerase with the chaperone complex induces a conformational change (II) that allows binding of P protein to $\epsilon$ RNA (III), which in turn provides the signal for nucleocapsid assembly and initiation of viral DNA synthesis (IV). From J. Hu et al., *EMBO J.* **16:**59–68, 1997, with permission.

tor sequence close to the site of initiation (Fig. 7.27). There is evidence from genetic studies with woodchuck hepadnavirus that sequences 3' of DR1 may help to specify the acceptor site. During the first template exchange and all subsequent steps, P protein remains covalently attached to the 5' end of the (−) strand.

**Elongation and RNase H digestion of the RNA template.** Following the first template exchange, (−) strand DNA synthesis continues all the way to the 5' end of the pregenome RNA template (Fig. 7.26, steps 2 and 3). Because synthesis is initiated in the 3' DR1, a short repeat of nine nucleotides (r) is generated at the end of this elongation step when the 5' DR1 sequence is copied (Fig. 7.26, step 3). The reader should note this fact, as it will become important in the final step of viral DNA synthesis. The RNA template is degraded by the RNase H activity of P protein as (−) strand synthesis proceeds. Unlike those of retroviruses, none of these RNase H products are used as primers for (+) strand DNA synthesis in situ, and (+) strand DNA synthesis does not start until the (−) strand has been completed (Table 7.1). The final product of RNase

H digestion is a short RNA molecule, corresponding to the capped end of the pregenome, which includes the 5' DR1. This 15- to 18-nucleotide product is similar in length to the limit product of polymerase-associated RNase H digestion by human immunodeficiency virus type 1 RT. In the human retrovirus protein, such a product provides a measure of the distance between the polymerase and RNase H active sites on the enzyme (Fig. 7.12); a similar relationship may apply to hepadnaviral P protein (Table 7.1).

The short, capped RNA could serve as a primer in situ for synthesis of the (+) strand of DNA (Fig. 7.26, step 4). In 5 to 10% of reverse transcription reactions, such priming does take place, and double-stranded linear genomes are produced. These linear DNAs can be converted to covalently closed circular DNA after transfection into permissive cells. However, most of these circular DNA molecules are formed by base-pairing-independent recombination between their linear ends, a process in which genetic information is lost. Thus, infectious virus particles are not normally produced after such transfections. What happens most of the time is a translocation of the capped RNA

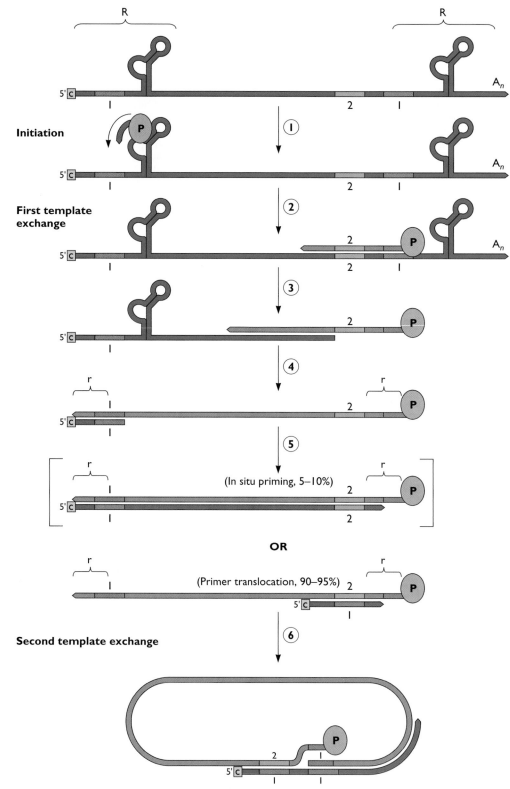

Initiation

First template
exchange

(In situ priming, 5–10%)

**OR**

(Primer translocation, 90–95%)

Second template exchange

◁ **Figure 7.26 Critical steps in the pathway of hepadnavirus reverse transcription.** The terminally redundant pregenomic RNA (top line) is capped and polyadenylated and encapsidated into a core particle. Boxes 1 and 2 represent DR1 and DR2, respectively. The horizontal bracket labeled R represents the terminally redundant region of the RNA. (−) Strand DNA synthesis is initiated at ε at the 5′ end of the RNA and is primed by P protein. Following the first template exchange, step 2, DNA synthesis continues, using the 3′ copy of DR1 as the template (step 3). As synthesis proceeds, the newly copied RNA template is degraded by the associated RNase H activity of the polymerase (step 4). Elongation of (−) strand DNA is finished on complete copying of the pregenomic RNA template. The product is a terminally redundant complete (−) strand DNA species. The redundancies (8 to 10 nucleotides) are labeled r. At this time, the primer for (+) strand synthesis is generated from the 5′-terminal 15 to 18 nucleotides of the pregenomic RNA. The primer is capped and includes the sequence to the 3′ end of DR1. At a low frequency (5 to 10%), instead of being translocated, the (+) strand primer is extended in situ (step 5; the structure set off in vertical brackets). Elongation of this (+) strand results in a duplex linear genome. In the majority of cases, the primer is translocated to base pair with the DR2 sequence near the 5′ end of (−) strand DNA (step 5). (+) Strand synthesis is initiated, and then elongation begins. On reaching the 5′ end of (−) strand DNA, an intramolecular template exchange occurs, resulting in a circular DNA genome (step 6). This exchange is facilitated by the short terminal redundancy, r, in (−) strand DNA. (+) strand DNA synthesis then continues for a variable distance, resulting in the relaxed circular form of the genome found in mature virions. Adapted from J. E. Tavis, *Viral Hepatitis Rev.* **2:**205–218, 1996, with permission.

primer to another location on the (−) strand DNA template, DR2.

**Translocation of the primer for (+) strand DNA synthesis.** Translocation of the primer for (+) strand DNA synthesis is facilitated by the homology between DR1 and DR2 (Fig. 7.26, step 5). The capped RNA primer, which includes DR1 sequences, can anneal to both. How the primer is induced to dissociate from DR1 and associate with DR2 is another mystery, and, as in the first template exchange, specific organization of the template in the core

particles has been invoked (Fig. 7.28). If the potential for the primer to hybridize with DR2 is disrupted by mutation, the pathway leading to formation of linear duplex DNA molecules predominates.

Thus, we have a situation similar to that producing the strong-stop DNAs in retroviral reverse transcription. The (+) strand DNA synthesis begins near the 5′ end at DR2 and soon runs out of (−) strand template. As in the retroviral case, this problem is solved by a second template exchange, in this instance facilitated by the short repeat, r, generated during synthesis of the (−) strand (Fig. 7.26, step 3).

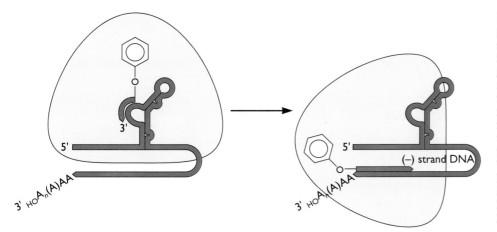

**Figure 7.27 Model for the protein-priming reaction.** Priming of DNA synthesis occurs at the RNA packaging signal ε near the 5′ end of pregenomic RNA, using a tyrosine residue of the polymerase polypeptide. After the polymerization of four nucleotides, the nascent DNA strand is transferred to the 3′ end of the RNA template, where (−) strand DNA synthesis continues. From C. Seeger and W. S. Mason, p. 815–832, *in* E. De-Pamphilis (ed.), *DNA Replication in Eukaryotic Cells* (Cold Spring Harbor Laboratory Press, Cold Spring Harbor, N.Y., 1996), with permission.

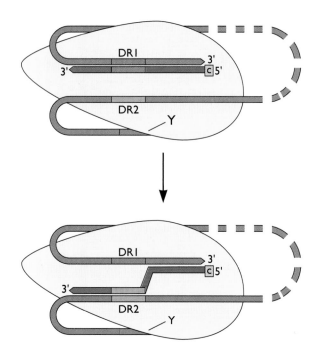

**Figure 7.28 Model for (+)-strand priming.** (−) Strand DNA is arranged to facilitate transfer of the capped RNA primer to DR1 at the 3′ end. The polymerase is covalently linked to the 5′ end of the (−) strand DNA through a tyrosine residue. From C. Seeger and W. S. Mason, p. 815–832, *in* E. DePamphilis (ed.), *DNA Replication in Eukaryotic Cells* (Cold Spring Harbor Laboratory Press, Cold Spring Harbor, N.Y., 1996), with permission.

**The second template exchange forms a circle.** The structural requirements for the next step must be complicated, as P protein is still attached to the 5′ end of the (−) strand that must be displaced to allow template transfer. It is difficult to envision how a single protein accommodates all three DNA ends at once and catalyzes polymerization while still attached to one. Nevertheless, this exchange does occur with high efficiency in vivo, and subsequent elongation of the (+) strand produces the relaxed, partially duplex circle that comprises virion DNA (Fig. 7.26, step 6).

It is not clear what causes premature termination of the (+) strand of hepadnaviral DNA. Synthesis of (+) strand DNA is affected by mutations in C protein, consistent with the idea that C protein is somehow involved in this reaction. It has been proposed that DNA synthesis induces a change in the outer surface of the core and that envelopment is regulated by interaction of the envelope proteins with the altered core. Once the cores are enveloped, DNA synthesis stops, presumably because deoxyribonucleoside triphosphate substrates can no longer enter the particle.

In mammalian hepadnaviruses, the (+) strand comprises only about half of the genome, but in avian hepadnaviruses, it is nearly full length. However, even in avian viral DNA the capped RNA primer remains attached to the (−) strand. The primer is not removed by the P protein RNase H, as is the case during retroviral replication, and strand displacement synthesis, an essential activity of retroviral RT, is not observed with P protein (Table 7.1).

## Perspectives

### Comparison of Systems

The description of critical steps in hepadnaviral reverse transcription reveals interesting points of similarity and contrast with retroviral systems, as summarized in Table 7.1. Amino acid sequences and functions are conserved among retroviral RT and hepadnaviral P proteins, and both enzymes use terminal nucleic acid repeats to mediate template exchanges. However, the mechanisms by which their templates are reverse transcribed are quite distinct. Differences in the form and function of the final products of the two pathways are especially striking. With retroviruses, a relaxed DNA circle with overlapping 5′ ends is an *intermediate* in the formation of the final product of the reaction, a linear duplex DNA. Repair of this relaxed-circle intermediate is an aberrant reaction, and the covalently closed circle thus formed is a dead-end product. In contrast, linear hepadnaviral DNA is an aberrant product of its reverse transcription, and the covalently closed circle is the functional form.

The single-cell replication cycles of retroviruses and hepadnaviruses are in a sense permutations of one another. In comparing them, it is instructive to interpose cauliflower mosaic virus, a plant retroid virus that seems to combine some features of both animal viruses (Fig. 7.29). This plant virus has a relaxed circular DNA genome and directs synthesis of a covalently closed episomal form, but its pathway of reverse transcription and priming mechanism is quite analogous to that of retroviruses and retrotransposons. On the other hand, as with hepadnaviruses, all, or remnants, of the RNA primers remain attached to the 5′ ends of cauliflower mosaic virus DNA. Thus, as suggested by the family tree of retroelements (Fig. 7.13), retroid viruses represent a continuum in evolution and remind us of the varied combinations of strategies that exist in nature for replicating viral genomes and related genetic elements.

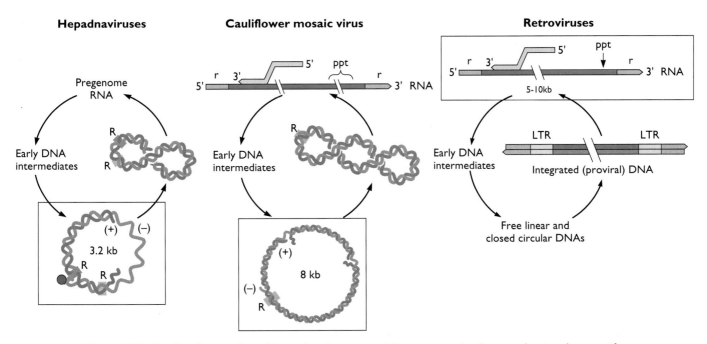

**Figure 7.29 Replication cycles of hepadnaviruses, cauliflower mosaic virus, and retroviruses.** The double-stranded DNA circle found in cauliflower mosaic virus particles contains three interruptions. At each interruption there is a short 5′ overlap of DNA as if formed by strand displacement synthesis. Ribonucleotides are often found attached to the 5′ ends. The (−) strand starts with either a ribo- or a deoxyriboadenosine. The 5′ ends of the (+) strand each contain 8 to 10 purine-rich matches to the viral DNA at the same location, suggesting a primer function. r, short sequence at both ends of viral RNA; R, same sequence in DNA. Updated from H. E. Varmus, *Nature* **304**:116–117, 1983, with permission.

# References

## Reviews

Arts, E. J., and S. F. Le Grice. 1998. Interaction of retroviral reverse transcriptase with template-primer duplexes during replication. *Prog. Nucleic Acid Res. Mol. Biol.* **58**:339–393.

Brown, P. O. 1997. Integration, p. 161–204. *In* J. M. Coffin, S. H. Hughes, and H. E. Varmus (ed.), *Retroviruses.* Cold Spring Harbor Laboratory Press, Cold Spring Harbor, N.Y.

Ganem, D. 1996. Hepadnaviridae and their replication, p. 2703–2727. *In* B. N. Fields, D. M. Knipe, and P. M. Howley (ed.), *Fields Virology,* 3rd ed. Lippincott-Raven, Philadelphia, Pa.

Seeger, C., and J. Hu. 1997. Why are hepadnaviruses DNA and not RNA viruses? *Trends Microbiol.* **5**:447–450.

Seeger, C., and W. S. Mason. 1996. Replication of the hepatitis virus genome, p. 815–832. *In* E. DePamphilis (ed.), *DNA Replication in Eukaryotic Cells.* Cold Spring Harbor Laboratory Press, Cold Spring Harbor, N.Y.

Skalka, A. M. (ed.). 1999. *Advances in Virus Research*, vol. 52. *Seminars in Virology: Retroviral DNA Integration.* Academic Press, New York, N.Y.

Skalka, A. M., and S. Goff (ed.). 1993. *Reverse Transcriptase.* Cold Spring Harbor Laboratory Press, Cold Spring Harbor, N.Y.

Steitz, T. A. 1993. DNA- and RNA-dependent DNA polymerases. *Curr. Opin. Struct. Biol.* **3**:31–38.

Tavis, J. E. 1996. The replication strategy of the hepadnaviruses. *Viral Hepatitis Rev.* **2**:205–218.

Telesnitsky, A., and S. P. Goff. 1997. Reverse transcriptase and the generation of retroviral DNA, p. 121–160. *In* J. M. Coffin, S. H. Hughes, and H. E. Varmus (ed.), *Retroviruses.* Cold Spring Harbor Laboratory Press, Cold Spring Harbor, N.Y.

## Papers of Special Interest

### Retroviral Reverse Transcription

Davies, J. F., Jr., Z. Hostomska, Z. Hostomsky, S. R. Jordan, and D. A. Matthews. 1991. Crystal structure of the ribonuclease H domain of HIV-1 reverse transcriptase. *Science* **252**:88–95.

Hsiou, Y., J. Ding, K. Das, A. D. Clark, Jr., S. H. Hughes, and E. Arnold. 1996. Structure of unliganded HIV-1 reverse transcriptase at 2.7 Å resolution: implications of conformational changes for polymerization and inhibition mechanisms. *Structure* **4**:853–860.

Huang, H., R. Chopra, G. L. Verdine, and S. Harrison. 1998. Structure of a covalently trapped catalytic complex of HIV-1 reverse transcriptase: implications for drug resistance. *Science* **282**:1669–1675.

Jacobo-Molina, A., J. Ding, R. G. Nanni, A. D. Clark, Jr., X. Lu, C. Tantillo, R. L. Williams, G. Kamer, A. L. Ferris, P. Clark, A. Hizi, S. H. Hughes, and E. Arnold. 1993. Crystal structure of human immunodeficiency virus type 1 reverse transcriptase complexed with double-stranded DNA at 3.0 Å resolution shows bent DNA. *Proc. Natl. Acad. Sci. USA* **90:**6320–6324.

Kim, B., and L. Loeb. 1995. Human immunodeficiency virus reverse transcriptase substitutes for DNA polymerase I in *Escherichia coli. Proc. Natl. Acad. Sci. USA* **92:**684–688.

Kim, B., and L. Loeb. 1995. A screen in *Escherichia coli* for nucleoside analogs that target human immunodeficiency virus (HIV) reverse transcriptase and herpes simplex virus thymidine kinase. *J. Virol.* **69:**6563–6566.

Kohlstaedt, L. A., J. Wang, J. M. Friedman, P. A. Rice, and T. A. Steitz. 1992. Crystal structure at 3.5 Å resolution of HIV-1 reverse transcriptase complexed with an inhibitor. *Science* **256:**1783–1790.

Patel, P. H., A. Jacobo-Molina, J. Ding, C. Tantillo, A. D. Clark, Jr., R. Raag, R. G. Nanni, S. H. Hughes, and E. Arnold. 1995. Insights into DNA polymerization mechanisms from structure and function analysis of HIV-1 reverse transcriptase. *Biochemistry* **34:**5351–5363.

Rodgers, D. W., S. J. Gamblin, B. A. Harris, S. Ray, J. S. Culp, B. Hellmig, D. J. Woolf, C. Debouck, and S. C. Harrison. 1995. The structure of unliganded reverse transcriptase from the human immunodeficiency virus type 1. *Proc. Natl. Acad. Sci. USA* **92:**1222–1226.

Temin, H. M., and S. Mizutani. 1970. RNA-dependent DNA polymerase in virions of Rous sarcoma virus. *Nature* **226:**1211–1213.

Topping, R., M.-A. Demoitie, N. H. Shin, and A. Telesnitsky. 1998. Cis-acting elements required for strong stop acceptor template selection during Moloney murine leukemia virus reverse transcription. *J. Mol. Biol.* **281:**1–15.

## Retroviral DNA Integration

Aiyar, A., P. Hindmarsh, A. M. Skalka, and J. Leis. 1996. Concerted integration of linear retroviral DNA by the avian sarcoma virus integrase in vitro: dependence on both long terminal repeat termini. *J. Virol.* **70:**3571–3580.

Baltimore, D. 1970. RNA-dependent DNA polymerase in virions of RNA tumour viruses. *Nature* **226:**1209–1211.

Bujacz, G., M. Jaskólski, J. Alexandratos, A. Wlodawer, G. Merkel, R. A. Katz, and A. M. Skalka. 1995. High-resolution structure of the catalytic domain of the avian sarcoma virus integrase. *J. Mol. Biol.* **253:**333–346.

Craigie, R., T. Fujiwara, and F. Bushman. 1990. The IN protein of Moloney murine leukemia virus processes in the viral DNA ends and accomplishes their integration *in vitro. Cell* **62:**829–837.

Dyda, F., A. B. Hickman, T. M. Jenkins, A. Engelman, R. Craigie, and D. R. Davies. 1994. Crystal structure of the catalytic domain of HIV-1 integrase: similarity to other polynucleotidyl transferases. *Science* **266:**1981–1986.

Eijkelenboom, A. P. A. M., R. A. Puras Lutzke, R. Boelems, R. H. A. Plasterk, R. Kaptein, and K. Hård. 1995. The DNA-binding domain of HIV-1 integrase has an SH3-like fold. *Nat. Struct. Biol.* **2:**807–810.

Farnet, C. M., and F. D. Bushman. 1997. HIV-1 cDNA integration: requirement of HMG 1(Y) protein for function of preintegration complexes *in vitro. Cell* **88:**483–492.

Heuer, T. S., and P. O. Brown. 1998. Cross-linking studies suggest a model for the architecture of an active human immunodeficiency virus type 1 integrase-DNA complex. *Biochemistry* **37:**6667–6678.

Jones, J. S., R. W. Allan, and H. W. Temin. 1994. One retroviral RNA is sufficient for synthesis of viral DNA. *J. Virol.* **68:**207–216.

Katz, R. A., G. Merkel, J. Kulkosky, J. Leis, and A. M. Skalka. 1990. The avian retroviral IN protein is both necessary and sufficient for integrative recombination *in vitro. Cell* **63:**87–95.

Kukolj, G., and A. M. Skalka. 1995. Enhanced and coordinated processing of synapsed viral DNA ends by retroviral integrases *in vitro. Genes Dev.* **9:**2556–2567.

Lee, M. S., and R. Craigie. 1998. A previously unidentified host protein protects retroviral DNA from autointegration. *Proc. Natl. Acad. Sci. USA* **95:**1528–1533.

Lodi, P. J., J. A. Ernst, J. Kuszewski, A. B. Hickman, A. Engelman, R. Craigie, G. M. Clore, and A. M. Gronenborn. 1995. Solution structure of the DNA binding domain of HIV-1 integrase. *Biochemistry* **34:**9826–9833.

Murphy, J. E., T. De Los Santos, and S. P. Goff. 1993. Mutational analysis of the sequences at the termini of the Moloney murine leukemia virus DNA required for integration. *Virology* **195:**432–440.

Withers-Ward, E. S., Y. Kitamura, J. P. Barnes, and J. M. Coffin. 1994. Distribution of targets for avian retrovirus DNA integration *in vivo. Genes Dev.* **8:**1473–1487.

Yang, W., and T. A. Steitz. 1995. Recombining the structures of HIV integrase, RuvC and RNase H. *Structure* **3:**131–134.

## Hepadnaviral Reverse Transcription

Hu, J., D. O. Toft, and C. Seeger. 1997. Hepadnavirus assembly and reverse transcription require a multi-component chaperone complex which is incorporated into nucleocapsids. *EMBO J.* **16:**59–68.

Summers, J., and W. S. Mason. 1982. Replication of the genome of a hepatitis B-like virus by reverse transcription of an RNA intermediate. *Cell* **29:**403–415.

Wang, G.-H., and C. Seeger. 1992. The reverse transcriptase of hepatitis B virus acts as a protein primer for viral DNA synthesis. *Cell* **71:**663–670.

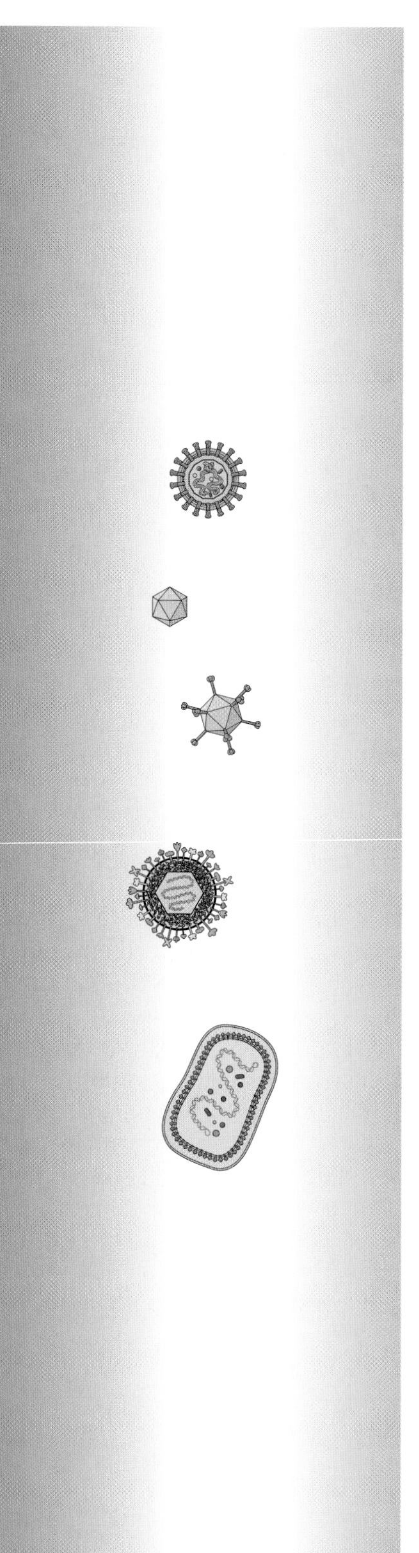

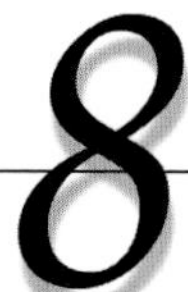

8 ———————————————————————————————

# Transcription Strategies: DNA Templates

It is possible that nature invented DNA for the purpose of achieving regulation at the transcriptional rather than at the translational level. The level of control in small DNA viruses and large RNA viruses should be interesting to study.

A. CAMPBELL, 1967

## Introduction

In cells infected by viruses that encapsidate DNA, synthesis of viral mRNA must precede production of proteins. The majority of DNA viruses, including the parvo-, papova-, hepadna-, adeno-, and herpesviruses, depend on cellular RNA polymerase II, the enzyme that produces cellular mRNA. The retroviruses, which establish themselves as DNA proviruses early in their infectious cycles, also use the cellular RNA polymerase II system. The signals that control expression of the genes of these viruses are therefore similar in organization to those of cellular genes. In fact, much of our current understanding of the mechanisms of cellular transcription stems from study of the transcription of viral DNA templates. Some of these viruses, notably simple retroviruses, depend exclusively on the cellular transcriptional machinery for expression of their genetic information. The genomes of poxviruses are transcribed by a viral DNA-dependent RNA polymerase and accessory viral proteins, which control the recognition of viral promoters. However, this strategy is an exception, not the rule.

A hallmark of cells infected by DNA viruses is the expression of viral genes in a strictly defined, reproducible sequence. In general, viral enzymes and regulatory proteins are made during the initial period of infection, whereas structural proteins from which virus particles are built are synthesized only after viral DNA synthesis begins. Such orderly expression of viral genes is primarily the result of transcriptional regulation by viral proteins. Thus, viral genes are transcribed in a fixed sequence, and transcription of specific genes is activated during specific periods in the infectious cycle. This pattern of gene expression is quite different from that characteristic of the majority of RNA viruses. The molecular strategies that ensure orderly (and efficient) transcription of the genes of these DNA viruses therefore lay the foundations for the infectious cycle. As discussed in this chapter, the elucidation of these strategies, one of the major goals of studies of DNA viruses, has identified a number of common mechanisms executed in virus-specific fashion. An unanticipated dividend has been new insights into cellular mechanisms that control progression through the cell cycle.

## Properties of Cellular and Viral Transcription Systems

### Cellular Transcription Systems Used by Viruses

#### The Three Transcriptional Systems of Eukaryotes

A general feature of eukaryotic cells is the division of transcriptional labor among three DNA-dependent RNA polymerases: these enzymes, designated RNA polymerases I, II, and III, synthesize different kinds of cellular RNA (Table 8.1). RNA polymerase II makes mRNA species or their precursors, while RNA polymerases I and III produce stable RNAs, such as ribosomal RNAs (rRNAs) and transfer RNAs (tRNAs), respectively. Synthesis of these abundant "housekeeping" RNAs must be adjusted to match the rates of cell growth and division. But regulation of mRNA synthesis is crucial for orderly development and differentiation in eukaryotes, as well as for the responses of cells to their environment. The evolution of RNA polymerases with distinct transcriptional responsibilities appears to represent a device for maximizing opportunities for regulation of mRNA synthesis while maintaining a constant supply of the RNA species essential for the metabolism of all cells.

Although the three eukaryotic RNA polymerases transcribe different classes of genes and must meet different regulatory requirements, several of the 12 to 16 subunits of these large enzymes are identical, and others are related in sequence. Such conservation of sequence, which extends to the proteins of prokaryotic and viral DNA-dependent RNA polymerases, can be attributed to the common biochemical capabilities of the enzymes. These activities include binding of ribonucleoside triphosphate substrates, binding to template DNA and to product RNA, and catalysis of phosphodiester bond formation.

Transcription of cellular and viral genes by either cellular or viral RNA polymerases requires not only template-directed synthesis of RNA but also correct interpretation of DNA punctuation signals that mark the sites at which transcription must start and stop. The point at which copying of the DNA should begin, the **initiation site**, must be identified by the transcription machinery. Such recognition together with synthesis of the first few phosphodiester bonds of RNA completes **initiation** (Fig. 8.1). During the **elongation phase**, nucleotides are added rapidly to the 3' end of the nascent RNA as the transcriptional machinery reads the sequence of a gene. When **termination sites** are encountered in the DNA, both the RNA product and the RNA polymerase are released from the template (Fig. 8.1). Purified RNA polymerases I, II, and III perform the elongation reactions in vitro but, despite their complexity, are incapable of specific initiation of transcription without the assistance of additional, polymerase-specific proteins. This requirement for additional proteins during initiation is a universal property of DNA-dependent RNA polymerases, including those encoded by viruses.

### Transcription of Viral Templates by Cellular RNA Polymerases II and III

The dependence of many DNA viruses that reproduce in eukaryotic cells on the cellular enzyme that synthesizes pre-mRNA or mRNA (RNA polymerase II) was expected because prokaryotic RNA polymerases transcribe many bacteriophage genes. In contrast, the discovery that some of these viruses also make use of cellular RNA polymerase III was not anticipated. This phenomenon was initially observed in human cells infected by adenovirus, but RNA polymerase III transcription units are also present in the genomes of other viruses. Viruses can also take advantage of the cellular transcriptional machinery in unique ways. Perhaps the most dramatic example of such virus-specific mechanisms is provided by (−) strand RNA viruses, like influenza A viruses, that acquire the 5' caps for their mRNAs from the cellular pre-mRNAs or mRNAs synthesized by RNA polymerase II.

Because the host cells' RNA polymerase II transcriptional machinery is responsible for transcription of many viral DNA templates, this cellular transcription system is described in some detail below. We then consider the regulatory mechanisms that ensure appropriate expression of genetic information of DNA viruses and retroviruses by this cellular system.

### Production of Viral DNA Templates Transcribed by Cellular RNA Polymerases

All viral templates that are transcribed by cellular RNA polymerases must enter the infected cell nucleus. However, there is considerable variation in the reactions needed to produce templates that can be recognized by the

**Table 8.1** Eukaryotic RNA polymerases synthesize different classes of cellular and viral RNA

| Enzyme | RNAs synthesized[a] | |
| --- | --- | --- |
| | Cellular | Viral |
| RNA polymerase I | Pre-rRNA | None known |
| RNA polymerase II | Pre-mRNA, snRNAs | Pre-mRNA, mRNA |
| RNA polymerase III | Pre-tRNAs, 5S rRNA, U6 snRNA | Ad2 VA-RNAs, EBV EBER RNAs |

[a] snRNA, small nuclear RNA; Ad2, adenovirus type 2; EBV, Epstein-Barr virus.

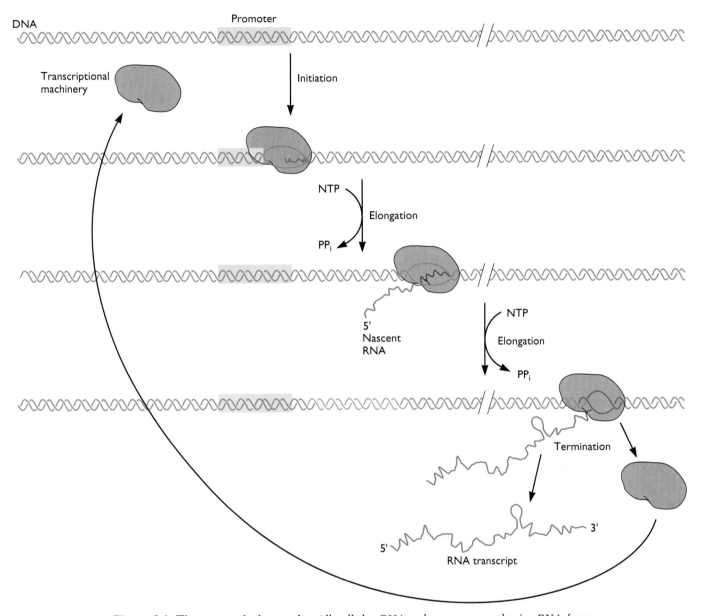

**Figure 8.1  The transcription cycle.** All cellular RNA polymerases synthesize RNA from DNA templates by a common set of reactions. The transcriptional machinery binds to the promoter and induces local unwinding of the double-stranded DNA template so that copying of the DNA into RNA can begin. These reactions comprise **initiation** of transcription. The transcribing complex progressively unwinds the template as it reads the DNA sequence and adds nucleotides to the 3′ end of the nascent RNA chain. These **elongation** reactions continue until a termination signal is encountered in the template. Transcription then ceases, and both the RNA transcript and the transcriptional machinery are released from the DNA template during **termination**.

cellular transcriptional machinery from genomic nucleic acids (Table 8.2). Some of these viral genomes, like the cellular genome, are double-stranded DNA molecules that can be transcribed as soon as they reach the nucleus. In the case of papovaviruses, the resemblance to cellular tem-

plates for transcription is even greater because the circular viral genome enters cells as a "minichromosome" in which the DNA is packaged by cellular nucleosomes. Transcription of specific genes is therefore the first biosynthetic reaction in cells infected by papova-, adeno-, and her-

**Table 8.2** Prerequisites for transcription of viral genes by RNA polymerase II

| Requirement | Virus | Form of the intracellular template |
|---|---|---|
| Entry of genome into nucleus | Papovavirus: simian virus 40 | Closed circular double-stranded DNA organized by cellular nucleosomes |
| | Adenovirus: human adenovirus type 2 | Linear double-stranded DNA associated with viral protein VII |
| Entry of viral genome plus a virion protein (e.g., VP16) into the nucleus | Herpesvirus: herpes simplex virus type 1 | Circular double-stranded DNA formed by ligation of genomic DNA |
| Production of double-stranded DNA from genome and entry into nucleus | Hepadnavirus: hepatitis B virus | Closed circular DNA synthesized from genomic molecules with gap in the (+) strand and nick in the (−) strand |
| | Parvovirus: adeno-associated virus | Linear double-stranded DNA synthesized from single-stranded DNA genome during replication |
| Reverse transcription, entry into nucleus, and integration | Retrovirus: avian sarcoma virus | Linear double-stranded DNA integrated into the cellular genome produced by action of viral reverse transcriptase and integrase |

pesviruses. Although viral DNA synthesis is not required prior to transcription, this process dramatically increases the concentration of templates and, as we shall see, is crucial for transcription of specific viral genes.

Other viral DNA genomes must be converted from the form in which they enter the cells to double-stranded molecules that serve as transcriptional templates (Table 8.2). The hepadnaviral genome, an incomplete circular DNA molecule with a large gap in one strand (appendix, Fig. 11B), appears as fully double-stranded DNA in the nucleus. During the initial period of the hepadnaviral infectious cycle, the supply of templates is supplemented when pregenomic RNA transcribed by RNA polymerase II is copied into double-stranded DNA by the viral DNA polymerase. The prerequisites for expression of retroviral genetic information are even more demanding, for the (+) strand RNA genome must be both converted into viral DNA and, following import into the nucleus, integrated into the cellular genome. The former reactions, catalyzed by reverse transcriptase, create both an appropriate double-stranded DNA template and the signals needed for its recognition by components of the cellular transcriptional machinery. Once the viral DNA becomes integrated into cellular DNA, it is indistinguishable in all its properties, including intranuclear location and organization in chromatin, from cellular genes.

## Transcription by RNA Polymerase II

Accurate initiation of transcription of a gene by RNA polymerase II is directed by specific DNA sequences located near the site of initiation, called the **promoter** (Fig. 8.2). The promoter and the set of DNA sequences that control transcription make up the **transcriptional control region** (Fig. 8.2). Transcriptional control regions of DNA viruses and retroviruses were among the first to be exam-

ined experimentally. For example, the human adenovirus type 2 major late promoter was the first from which accurate initiation of transcription was achieved in vitro. As viral genes transcribed by RNA polymerase II must be recognized by the cellular transcriptional machinery, their transcriptional control regions are built from DNA sequences typical of those of cellular genes. Consequently, the study of viral transcription has yielded fundamental information about transcriptional control signals and the mechanisms by which RNA polymerase II transcription is initiated and regulated.

Biochemical studies of model transcriptional control regions, such as the adenoviral major late promoter, established that initiation of transcription by RNA polymerase II is a multistep process. The initiation reactions include **promoter recognition**, formation of an **open initiation complex**, in which the two strands of the DNA template in the vicinity of the initiation site are unwound, and **promoter clearance**, movement of the transcribing complex away from the promoter (Fig. 8.3). At least 40 proteins, which comprise RNA polymerase II itself and auxiliary initiation proteins, are needed to complete the intricate process of initiation. Our understanding of the functions of these proteins and of the DNA sequences that control initiation of transcription of both viral and cellular genes is largely based on in vitro systems or simple assays for detecting gene expression within cells (Fig. 8.4). Application of these methods has identified a very large number of transcriptional control sequences. Fortunately, all of them can be assigned to one of the three functionally distinct regions identified in Fig. 8.2.

**Core promoters** of viral and cellular genes contain all the information necessary for recognition of the site of transcription initiation. They direct assembly of an RNA

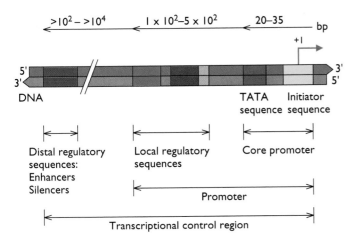

**Figure 8.2 RNA polymerase II transcriptional control elements.** The **core promoter** contains the minimal sequence necessary to specify accurate initiation of transcription. The site of initiation is represented by the jointed red arrow drawn in the direction of transcription on the nontranscribed DNA strand, a convention used throughout this chapter. The **TATA sequence** is the binding site for TfIId (Box 8.2) and conforms to a well-conserved consensus sequence. The core promoter often also contains a sequence sufficient to specify initiation at a unique site. Such autonomous **initiator sequences** are more variable than TATA sequences, do not conform to a single consensus, and are recognized by a variety of different proteins. Although TATA sequences are considered a characteristic feature of RNA polymerase II promoters, they are absent from many promoters (e.g., Fig. 8.5). The activity of the core promoter is modulated by the **local regulatory sequences** found within a few hundred base pairs of the initiation site. The constellation of local regulatory sequences upstream of the TATA sequence shown is a common arrangement. However, such sequences can also lie within transcribed sequences (e.g., Fig. 8.5). Distant regulatory sequences that stimulate (**enhancers**) or repress (**silencers**) transcription are present in a large number of RNA polymerase II **transcriptional control regions**. Such sequences can function over distances of greater than 10,000 bp in the DNA and when located upstream of the promoter, within an intron, or downstream of the gene. Enhancers play critical roles in regulation of transcription during development and differentiation or in response to environmental stimuli. Both local regulatory and enhancer sequences are binding sites for sequence-specific DNA-binding proteins.

polymerase II-containing complex competent to begin transcription when supplied with ribonucleotide triphosphates. Such **preinitiation complexes** (Fig. 8.3) contain a common set of general initiation proteins and RNA polymerase II in an assembly precisely organized to support transcription initiation at the appropriate site(s). A hallmark of core RNA polymerase II promoters is the presence of a TA-rich **TATA sequence** 20 to 35 bp upstream of the site of initiation (Fig. 8.2 and 8.5A and B). Short sequences, termed **initiators**, which specify accurate (but

inefficient) initiation of transcription in the absence of any other promoter sequences, are also commonly found (Fig. 8.2 and 8.5). Either an initiator or an appropriately positioned TATA element can be sufficient to establish specific initiation, but both may be present (Fig. 8.5). In vitro studies of formation of preinitiation complexes on core promoters elaborated a stepwise assembly pathway in which initiation proteins (with nomenclature explained in Box 8.1) and RNA polymerase II enter the assembling complex in a defined sequence (Box 8.2). Recognition of the promoter is by the TATA-binding protein (Tbp) subunit of TfIId, which binds to the TATA sequence. Other initiation proteins and RNA polymerase II bind to DNA-bound TfIId and to each other (Box 8.2). In the cell, many of the interactions among components of the transcriptional machinery take place before a promoter is encountered, for RNA polymerase II is present in extremely large assemblies that contain initiation proteins as well as other proteins that are essential for transcription or participate in regulation of transcription. Such assemblies, termed **holoenzymes**, appear to be poised to initiate transcription as soon as they are recruited to a promoter, for example, by binding of one or more constituent proteins to DNA-bound TfIId.

### Regulation of RNA Polymerase II Transcription

**Patterns of transcriptional regulation.** Numerous patterns of gene expression are necessary for eukaryotic life: some RNA polymerase II transcription units must be expressed in all cells, whereas others are transcribed only during specific developmental stages or in specialized differentiated cells. Many others must be maintained in a silent ground state from which they can be rapidly activated in response to specific stimuli and to which they can be readily returned. Transcription of viral genes is also regulated during the single-cell life cycles of most of the viruses considered in this chapter. Large quantities of viral proteins for assembly of progeny virions must be made within a finite (and often short) infectious cycle. Consequently, some viral genes must be transcribed at high rates. As discussed previously, the genes comprising the majority of viral DNA templates are transcribed in a specific temporal sequence. Such regulated transcription is achieved in part by means of cellular control mechanisms, for example, mechanisms that transmute specific environmental stimuli into a transcriptional response or cellular proteins that repress transcription. However, viral proteins are generally critical components of the circuits that establish efficient and orderly transcription of viral genes by cellular components.

**Local and distant regulatory sequences and their recognition.** Both local and distant sequences (Fig. 8.2) can control transcription from core promoters. However,

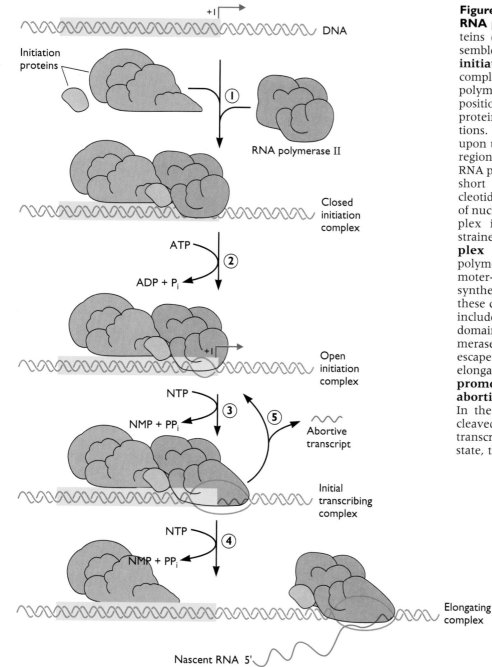

**Figure 8.3 Initiation of transcription by RNA polymerase II.** General initiation proteins (Box 8.2) and RNA polymerase II assemble at the promoter into the **closed initiation complex** (step 1). The initiation complex is a very stable entity, in which RNA polymerase II is firmly anchored at a specific position on the DNA template by numerous protein-protein and protein-DNA interactions. The **open initiation complex** forms upon unwinding of the DNA template in the region spanning the site of initiation (step 2). RNA polymerase II then synthesizes an initial short transcript (less than 10 to 15 nucleotides) by template-directed incorporation of nucleotides (step 3). The transcribing complex is believed to be conformationally strained in the **initial transcribing complex** formed in this way, because RNA polymerase II remains in contact with promoter-bound general initiation proteins as it synthesizes initial transcripts. The severing of these contacts, by mechanisms that probably include phosphorylation of the C-terminal domain of the largest subunit of RNA polymerase II, allows the transcribing complex to escape from the promoter and proceed with elongation (step 4). On many promoters, this **promoter clearance** step is inefficient, with **abortive initiation** (step 5) predominating. In the latter process, initial transcripts are cleaved and released from the complex as the transcribing machinery returns to its ground state, the open initiation complex.

in many cases, local sequences are sufficient for proper transcriptional regulation. These local regulatory sequences are binding sites for sequence-specific DNA-binding proteins (Fig. 8.6), a property first demonstrated with the simian virus 40 early promoter. An enormous number of sequence-specific proteins that regulate transcription are now known, many first identified through analyses of viral promoters (Fig. 8.6). Unfortunately, the nomenclature applied to these regulatory proteins presents serious difficulties for both writer and reader, for it is unsystematic and whimsical. The names of some transcriptional regulators and the DNA elements they recognize indicate their function (e.g., the glucocorticoid receptor [Gr] and the glucocorticoid response element [GRE]). Many other

**A**

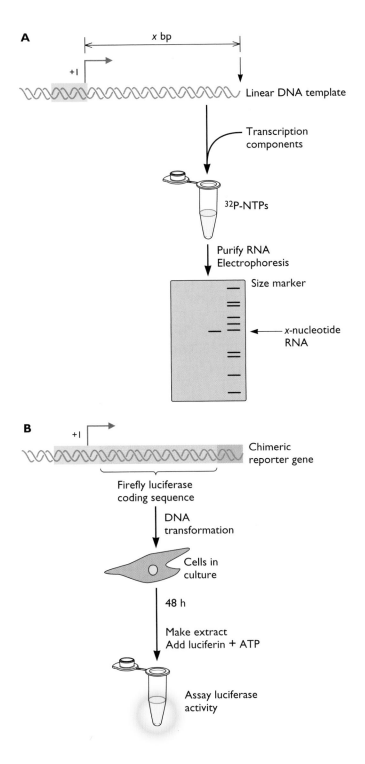

**B**

**Figure 8.4 Assays for the activity of RNA polymerase II promoters.** (A) In vitro transcription assay. In this simple assay, linear DNA templates are prepared by restriction endonuclease cleavage (black arrow at right end of $x$-bp sequence) of plasmid DNAs, a known distance, $x$ bp, downstream of position +1 of the promoter of interest. When the template is incubated with the transcriptional machinery and nucleoside triphosphate (NTP) substrates, transcription initiated at position +1 continues until the transcribing complex "runs off" the linear template. Specific transcription is therefore assayed as the production of $^{32}$P-labeled RNA $x$ nucleotides in length. This runoff transcription assay is convenient and can be used to assess both specificity and efficiency of transcription. However, it can be applied only to linear DNA templates and often suffers from high background, for example "end-to-end" transcription of the linear DNA template. These problems can be circumvented by the use of circular templates and an indirect assay for specific transcripts. Such in vitro transcription assays differ from transcription under normal intracellular conditions in several important ways. In in vitro assays, transcription templates are typically provided as naked circular or linear DNA molecules, physical states that may not resemble those of cellular or viral genes within cells. When components of the transcriptional machinery are extracted from cells, some of the enzymes or other essential proteins may be recovered more efficiently than others. Another important parameter that is altered in in vitro transcription systems is therefore the relative, as well as the absolute, concentrations of the proteins necessary for transcription. (B) Transient-expression assay. For transient-expression assays, a segment of DNA containing the transcriptional control region of interest is ligated to the coding sequence (orange) of an enzyme not expressed in the recipient cells to be used, and RNA processing signals, such as those specifying polyadenylation, shown by the green box at the end of the coding sequence. In this example, the firefly luciferase coding sequence is shown as the **reporter gene**. Plasmids containing such chimeric reporter genes are introduced into cells in culture by any one of several methods, including coprecipitation with calcium phosphate and electroporation. The proportion of cells that will take up foreign DNA varies with a number of parameters, including cell type, but is generally low: uptake of DNA by 10% of cells is considered excellent in this kind of assay. Within a cell that does take up the reporter gene, the DNA enters the nucleus, where the transcriptional control region directs transcription of chimeric RNA. The RNA is exported from the nucleus following processing and translated on cytoplasmic polyribosomes. The activity of the luceriferase enzyme is then assayed, generally 48 h after introduction of the reporter gene, as a measure of the activity of the transcriptional control region. Note that this indirect measure of transcription assumes that it is *only* the activity of the transcriptional control region that determines the intracellular level of the enzyme whose activity is measured. Alternatively, the concentration of the chimeric reporter RNA can be measured.

names, however, originate from the promoters in which the protein's binding sites were first identified (e.g., adenovirus E2 transcription factor [E2f]) or report some very general property (e.g., upstream stimulatory factor [Usf]). The universal use of abbreviations and the subsequent recognition that many "factors" in fact comprise families of closely related proteins compound this difficulty. A glossary of sequence-specific regulatory proteins described in this and other chapters is therefore provided in Table 8.3.

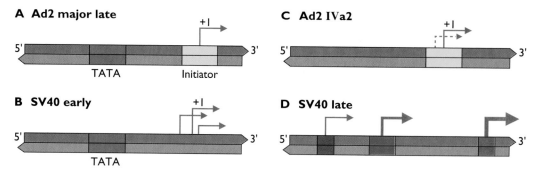

**A  Ad2 major late**

TATA    Initiator

**B  SV40 early**

TATA

**C  Ad2 IVa2**

**D  SV40 late**

**Figure 8.5  Variations in core RNA polymerase II promoter architecture.** Some of the variability in the DNA sequence elements of RNA polymerase II core promoters is illustrated with four viral promoters represented as in Fig. 8.2. The TATA or initiator sequences of the different promoters are not identical in DNA sequence. In the case of the simian virus 40 (SV40) late transcription unit (D), each of the several sites of initiation is included within a DNA sequence resembling an initiator of another promoter. It has not been shown experimentally that all actually function as autonomous initiator sequences. The relative frequency with which different initiation sites are used in a single promoter is indicated by the thickness of the arrow. Ad2, adenovirus type 2.

---

**BOX 8.1**

*Nomenclature of general initiation proteins*

The idealized profile of elution of human (HeLa cell) general initiation proteins from a phosphocellulose column illustrates the origin of the nomenclature of these proteins. The fractions essential for transcription by purified RNA polymerase II in vitro were designated A to D in order of their elution, and the initiation proteins present in them were termed transcription factor (Tf) IIa, b, and so on.

The fraction designated C in the original experiments proved not to be an essential component. Further chromatography showed that some fractions contained more than one essential protein, as summarized in the figure. Homologs of all the general initiation proteins shown are present in the yeast *Saccharomyces cerevisiae*. They are therefore considered universal components of eukaryotic cells.

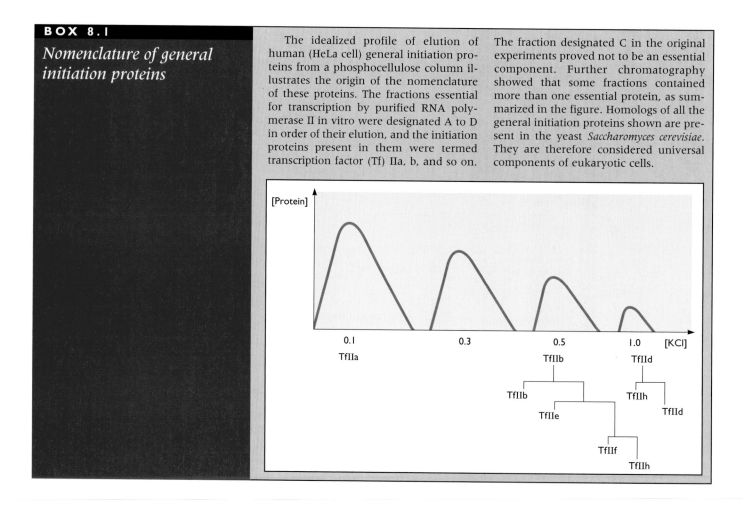

## BOX 8.2

*Stepwise assembly of the RNA polymerase II closed initiation complex*

The assembly of general initiation proteins and RNA polymerase II into the closed initiation complex (Fig. 8.3) is shown for a promoter that contains both a TATA and an initiator sequence (e.g., the adenovirus major late promoter). The first protein to bind, TfIId (step 1), contains a subunit that recognizes the TATA sequence (TATA-binding protein [Tbp]) and 8 to 10 additional subunits, termed Tbp-associated proteins (Tafs). Binding of Tbp to the TATA sequences bends the DNA sharply, as indicated. One popular hypothesis is that such bending facilitates interaction among proteins bound to local regulatory sequences located upstream of the TATA sequence and the basal transcriptional machinery. TfIId is required for transcription from all RNA polymerase II promoters. It can recognize those that lack TATA sequences (e.g., Fig. 8.5) by binding of a Taf to an initiator sequence, or it can be recruited to the promoter by interactions with proteins bound to specific sequences near the initiation site.

The TfIIa protein is shown entering the complex at step 1 because it can facilitate binding of Tbp to DNA and bind directly to one of the Tbp-associated proteins present in TfIId. Binding of TfIIb to the TfIId-promoter complex (step 2) is necessary for delivery of RNA polymerase II by TfIIf. TfIIb binds directly to Tbp, RNA polymerase II, and TfIIf, and TfIIf also interacts with the enzyme. The complex containing TfIId, TfIIb, and TfIIf RNA polymerase II is competent to initiate transcription when the DNA template is supercoiled, a state that can be considered energetically activated. Such a complex therefore contains the minimal set of components necessary for promoter recognition. The assembly process is completed by the sequential association of TfIIe and TfIIh. Protein components of TfIIe and TfIIh interact with one another and with RNA polymerase II. TfIIh contains at least nine different proteins. Its largest subunit, which also plays an important role in nucleotide excision repair of damaged DNA, supplies DNA-dependent ATPase and helicase activities essential for transcription. Two others are components of a kinase that can phosphorylate a C-terminal sequence of the largest subunit of RNA polymerase II. Although assembly of initiation complexes within cells is probably via a much-truncated pathway (see text), elucidation of this in vitro pathway has identified protein-DNA and protein-protein interactions important in the assembly and stabilization of the closed initiation complex, as well as reactions that can be regulated.

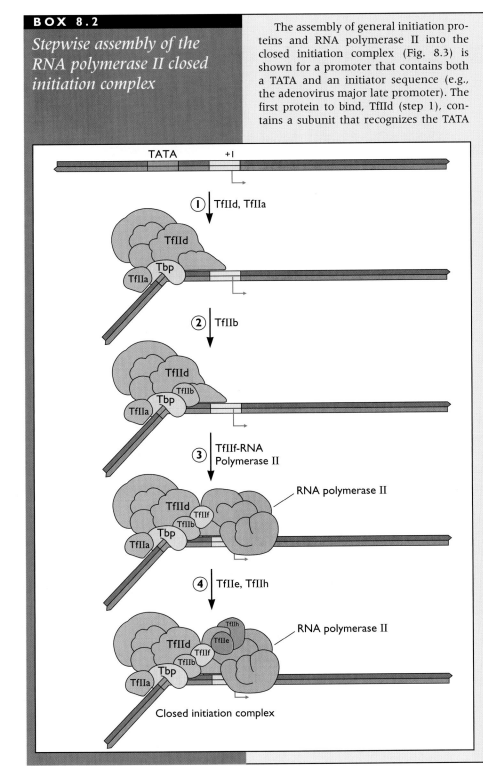

Proper regulation of transcription of many viral and cellular genes requires not only the promoter but also more distant regulatory sequences in the DNA template (Fig. 8.2), whose properties were entirely unanticipated. The first example was discovered in the genome of the papovavirus simian virus 40 and was termed an **enhancer** because it stimulated transcription to a large degree. Enhancers are defined by their position- and orientation-in-

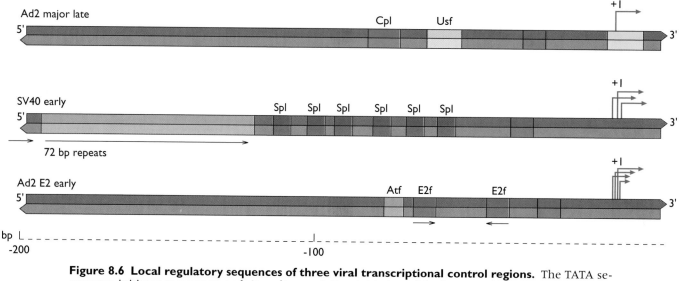

**Figure 8.6 Local regulatory sequences of three viral transcriptional control regions.** The TATA sequences, initiator sequences, and sites of transcription initiation of the human adenovirus type 2 (Ad2) major late and E2 early and the simian virus 40 (SV40) early transcriptional control regions are depicted as in Fig. 8.5. The local regulatory sequences of each promoter, which are recognized by the cellular DNA-binding proteins listed, are drawn to the scale shown at the bottom. The 72-bp repeats that lie upstream of the Sp1-binding sites of the simian virus 40 early transcriptional control region include the viral enhancer (Fig. 8.17).

dependent stimulation of transcription of homologous and heterologous genes over distances as great as 10,000 bp in the genome. Despite these unusual properties, enhancers are built with binding sites for the proteins that recognize local promoter sequences.

All viral templates transcribed by the cellular RNA polymerase II system contain transcriptional control regions that are directly recognized by cellular regulatory proteins (e.g., Fig. 8.6). Indeed, the transcriptional programs of the simple retroviruses are executed by the cellular transcriptional machinery alone. However, many viruses also encode transcriptional regulatory proteins, some of which resemble cellular proteins that recognize local promoter or more distant enhancer sequences. Other viral proteins that regulate transcription do so by means of such cellular proteins. Thus, cellular, sequence-specific, transcriptional regulatory proteins play pivotal roles in transcription from viral promoters and in the regulation of viral gene expression. These roles are illustrated in Fig. 8.7 by the many ways in which viruses depend on the cellular regulator nuclear factor κb (Nf-κb).

**Properties of sequence-specific proteins that regulate transcription.** The cellular DNA-binding proteins necessary for transcription from viral DNA templates, as well as viral proteins that bind to specific transcriptional control sequences, share a number of common properties. Their

most characteristic feature is modular organization: they are built from discrete functional and structural units termed **domains** (Fig. 8.8). The basic modules are a DNA-binding domain and an activation domain, which function as independent units. For the experimenter, this arrangement affords the opportunity to build artificial transcriptional regulators by mixing and matching DNA-binding and activation domains from different proteins. Other common properties of these regulators of transcription are binding to DNA as dimers and membership in families of related proteins that share the same types of DNA-binding and dimerization domains (Fig. 8.8).

Once a regulatory protein is bound to a specific sequence in a transcriptional control region, it must effect a change in the rate of initiation (the speed at which the reactions summarized in Fig. 8.3 are completed), in the number of initiation cycles, or in other transcription reactions. Although the molecular mechanisms responsible for regulation of transcription are not fully understood, it is clear that even a single regulatory protein can act by multiple mechanisms. This property is discussed in more detail below.

Regulation of transcription by sequence-specific DNA-binding proteins usually requires the participation of additional proteins termed **coactivators** (or corepressors) and **mediators**. In general, coactivators cannot bind specifically to DNA, nor can they modulate transcription on their

**Table 8.3**  Some cellular transcriptional regulators that recognize specific DNA sequences

| Abbreviation | Name | Characteristics | Viral or cellular promoters/enhancers recognized |
|---|---|---|---|
| Ap-1 | Activator protein 1 | Dimers of various basic-leucine zipper proteins, including c-Fos and c-Jun; regulated by dimerization and phosphorylation; mediate transcriptional responses to a variety of extracellular stimuli | Simian virus 40 enhancer |
| Atf | Activating transcription factor 2 | Member of large Atf/Creb family; basic-leucine zipper proteins, related to Ap-1 family; mediate transcriptional responses to second messenger cyclic AMP | Adenovirus type 2 E2E, E3, and E4 promoters; virus-responsive enhancer of human β-interferon gene |
| Cp-1 | CCAAT-binding protein 1 | Heterodimer | Adenovirus type 2 major late promoter |
| Creb | CAMP response element (CRE)-binding protein | Member of the large Atf/Creb family (see above) | Human cytomegalovirus major immediate-early enhancer |
| E2f | E2 factor | Member of family of dimeric activators; activity regulated by association with Rb and Rb-related proteins; important in cell cycle regulation | Adenovirus type 2 E2E promoter |
| Ebps, A1/Ebp, C/Ebp-α | CCAAT/enhancer-binding protein | Members of basic-leucine zipper family; C/Ebpα enriched in hepatic cells | Rous sarcoma virus enhancer; hepatitis B virus enhancer |
| Ets-1 | E twenty-six 1 | Defined by conserved DNA-binding domain; expressed in T cells; expression and activity regulated by multiple mechanisms | Human immunodeficiency virus type 1 enhancers; herpes simplex virus type 1 ICP4 promoter |
| Gata-3 | GATA (binding protein) 3 | Member of family defined by conserved Zn finger DNA-binding domain; expressed in T cells and specific neurons of the central nervous system | Human immunodeficiency virus type 1 enhancers |
| Ibp | Initiator binding protein | Member of nuclear hormone receptor superfamily; binds via Zn fingers as dimers | Simian virus 40 major late promoter |
| Irf | Interferon-regulatory factor 1 | Expression regulated by virus infection and mitogens | Human beta interferon enhancer |
| Nf-κb | Nuclear factor κb | Dimer of p50-p65 Rel family proteins; activity regulated by sequestration in cytoplasm (Fig. 8.7) | Human immunodeficiency virus type 1 core enhancer; simian virus 40 enhancer |
| NfI | Nuclear factor I | Binds as dimer via novel DNA-binding domain to viral and cellular promoters | BK virus late promoter; mouse mammary tumor virus enhancer; adenovirus type 2 origins of replication (chapter 9) |
| Nf-Il6 | Nuclear factor for interleukin-6 expression | Member of basic-leucine zipper protein family | Rous sarcoma virus and human immunodeficiency virus type 1 enhancers |
| Obps | Octamer-binding proteins | Defined by binding to an octameric DNA sequence and POU homeodomain DNA-binding motif; some ubiquitous, some restricted in expression | Simian virus 40 enhancer |
| Oct-1 | Octamer-binding protein 1 | Ubiquitously expressed member of Obp family | Herpes simplex virus type 1 immediate-early promoters, in complex with VP16 and Hcf (see text) |
| Spl | Stimulatory protein 1 | Binds as monomer via Zn finger DNA-binding domain; first sequence-specific transcription factor identified | Simian virus 40 early promoter |
| Srf | Serum response factor | Member of Mads box DNA-binding domain family; mediates transcriptional response to serum growth factors | Rous sarcoma virus enhancer/promoter; interacts with human T-cell leukemia virus type 1 Tax |
| Tef-1 | Transcription enhancer factor 1 | Complex DNA-binding domains; no independent activation domain identified; requires limiting coactivator | Simian virus 40 enhancer; human papillomavirus type 16 E6/E7 promoter |
| Tef-2 | Transcription enhancer factor 2 | Ubiquitous | Simian virus 40 enhancer; human papillomavirus type 16 enhancer |
| Usf | Upstream stimulatory factor | Contains basic, helix-loop-helix, and leucine zipper; domains; binds as dimer | Adenovirus type 2 major late promoter |

**Figure 8.7 The cellular transcriptional regulator Nf-κb and its exploitation by different viruses.** (A) Nf-κb and the Rel protein family. The members of this family are defined by the presence of the Rel homology region (orange segment), which contains DNA-binding and dimerization motifs and a nuclear localization signal (red). The p50-p65 Nf-κb heterodimer contains proteins from each of the two groups into which the Rel protein family can be divided. One group (right) contains p50 (Nf-κb1) and p52 (Nf-κb2). As illustrated for p50, they are synthesized as precursors, which include a C-terminal segment closely related in sequence to members of the Iκb family of inhibitory proteins. Proteolytic processing of p105 at the site indicated by the vertical arrow generates mature p50. The members of the second group of Rel proteins, which includes c-Rel, the *Drosophila* protein Dorsal, and p65 (Rel A), also contain an acidic activation domain (blue) at their C termini. The v-*rel* gene of the avian retrovirus reticuloendotheliosis virus strain T is a transduced version (see chapter 16) of c-*rel* expressed as a fusion with the viral *env* gene. It is responsible for the oncogenic properties of this retrovirus. The X-ray crystal structure of a p50-p65 heterodimer bound specifically to DNA is shown below. The structure is viewed down the helical axis of DNA with the two strands in purple and yellow, with the p53 and p65 subunits in green and red, respectively. The dimer makes extensive contact with DNA via protein loops, and the DNA-binding residues of the protein and those mediating contact between the dimers are in close proximity. From F. E. Chen et al., *Nature* (London) **391**:410–413, 1998. Courtesy of G. Ghosh, University of California, San Diego. (B) Activation of Nf-κb. Two forms of inactive Nf-κb are found in the cytoplasm (e.g., of unstimulated T cells). One consists of mature Rel protein heterodimers, p50-p65, associated with an inhibitory protein such as Iκb-α (left). Other members of the inhibitor family, which are characterized by the presence of a C-terminal region containing seven ankyrin repeats, include Iκb-β, the Bcl-3 oncogene product, and the *Drosophila* protein Cactus. The second inactive form of Nf-κb is the p65 protein dimerized with a Rel precursor, such as the p105 precursor of p50 (right). The C-terminal segment of p105 functions like Iκb, with which it shares sequences, to block nuclear localization signals and retain Nf-κb in the cytoplasm. Exposure of the cells to any one of several cytokines or mitogens (chapter 14), indicated by tumor necrosis factor α (Tnf-α), interleukin-1 (Il-1), and phorbol esters, results in activation (green arrows) of protein kinases that phosphorylate specific residues of Iκb-β or p105. Upon phosphorylation, Iκb-α dissociates from the complex and is recognized by the system of enzymes that adds branched chains of ubiquitin (Ub) to proteins targeted for degradation. It is then degraded by the proteasome, a multicatalytic protease that degrades polyubiquitinated proteins. Specific cleavage of p105 (A) generates the p50-p65 dimer Nf-κb. The free Nf-κb generated by either mechanism can translocate to the nucleus because its nuclear localization signals (A) are now accessible. Within the nucleus of uninfected cells, Nf-κb binds to specific promoter sequences to stimulate transcription via the p65 activation domain (A). Viral transcriptional control regions to which Nf-κb binds and some of the viral proteins that induce activation (green arrows) of Nf-κb are indicated. HCMV, human cytomegalovirus; HIV-1, human immunodeficiency virus type 1; HTLV-1, human T-cell leukemia virus type 1; SV40, simian virus 40; EBV, Epstein-Barr virus.

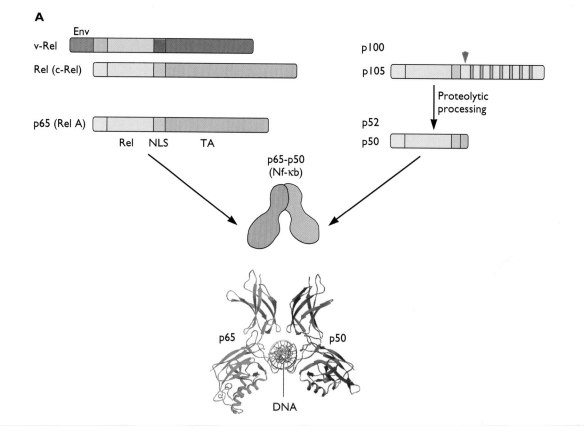

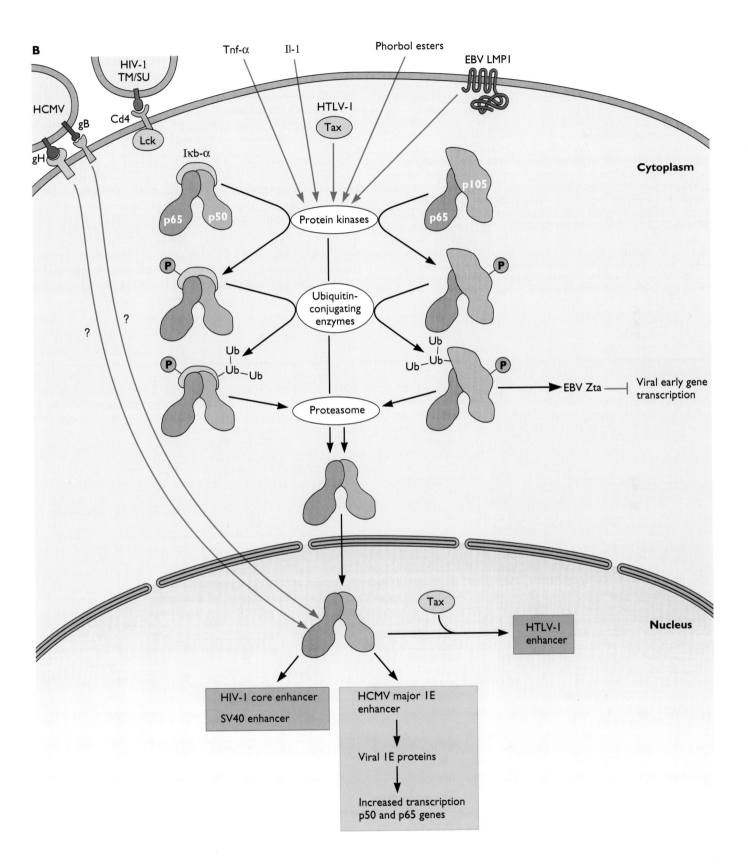
B
HIV-1 TM/SU
HCMV
gB
gH
Cd4
Lck
Tnf-α
Il-1
Phorbol esters
EBV LMP1
HTLV-1
Tax
Iκb-α
Cytoplasm
p65
p50
p105
p65
Protein kinases
P
P
Ubiquitin-conjugating enzymes
?
?
Ub
Ub
Ub
Ub
Ub
Ub
P
P
EBV Zta
Viral early gene transcription
Proteasome
Tax
HTLV-1 enhancer
Nucleus
HIV-1 core enhancer
SV40 enhancer
HCMV major IE enhancer
Viral IE proteins
Increased transcription p50 and p65 genes

| DNA binding | Dimer formation | Activation |
|---|---|---|
| Zn finger | | Acidic |
| Helix-turn-helix | | Glutamine rich |
| Basic | Leucine | Proline rich |
| | zipper | Isoleucine rich |
| Rel homology region | | |

**Figure 8.8 Modular organization of sequence-specific transcriptional activators.** Common functional domains of eukaryotic proteins that regulate transcription by binding specifically to local regulatory regions of promoters are listed. Some of the types of each domain are listed below. DNA-binding and activation domains are defined by their structure (e.g., Zn finger or helix-turn-helix) and chemical makeup (e.g., acidic, glutamine-rich), respectively. As discussed in the text, such DNA-binding and activation domains can function as independent units. Transcriptional activators are often more complex than illustrated here. They can contain more than a single activation domain and regulatory domains such as ligand-binding domains.

own. However, they dramatically augment (or damp) the transcriptional responses induced by sequence-specific regulators. Among the best characterized coactivators are the Swi/Snf complex (named for the products of the yeast mating type switch [*SWI*] and sucrose nonfermentable [*SNF*] genes present in the first such complex identified) and the highly related proteins Cbp (cyclic AMP response element-binding protein-binding protein) and p300. The latter was discovered by virtue of its binding to adenoviral E1A proteins, which stimulate transcription of both viral and cellular genes. Each of these coactivators can cooperate with multiple sequence-specific transcriptional activators and stimulate transcription from many promoters, but neither is required globally. Many questions about the mechanism of action of coactivators remain to be addressed. However, it is now well established that many coactivators can alter the organization of the chromatin template to facilitate binding of TfIId and sequence-specific activators.

*Synergistic stimulation of RNA polymerase II transcription by promoter-bound proteins*

Panel A of the figure illustrates the definition of synergy. The synergistic stimulation of transcription is illustrated for a hypothetical case in which a promoter, with a basal transcription rate defined as 1.0, contains binding sites for the two stimulatory proteins X and Y. Binding of X alone or Y alone stimulates the rate of transcription 10- or 5-fold, respectively. If X and Y targeted different reactions in the initiation sequence and functioned *independently* of one another, the 10-fold stimulation by protein X would be amplified a further 5-fold by protein Y, for a total 50-fold activation. **Synergism** is the ability of one protein to assist the action of the second when both are bound to the promoter, in this case resulting in a greater-than-50-fold increase in the rate of transcription. The synergistic action of two proteins is therefore defined as an increase in transcription when both are pre-

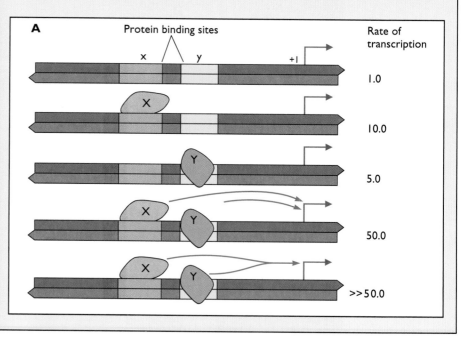

**BOX 8.3** *(continued)*

sent that is greater than the *product* of the increases induced by each protein on its own.

Panel B illustrates mechanisms of synergistic stimulation of transcription. Cooperativity in binding of the two proteins X and Y to their respective promoter elements (model 1) is one simple mechanism of synergistic activation. The binding of one protein (Y) can stabilize the interaction of the other (X) with a nearby sequence in the promoter via direct contact between the two proteins. Such synergism in binding of proteins to a promoter may be particularly important on nucleosomal templates when access to specific sequences is limited by interaction of the DNA with the histones that comprise nucleosomes. However, synergism among transcriptional activators cannot often be explained by cooperative DNA binding of the proteins to the promoter. Other possible mechanisms are illustrated with effects mediated by TfIId. Because individual Tbp-associated proteins present in TfIId interact with activators representative of different classes (acidic, isoleucine-rich, and so on) (Fig. 8.8), it has been proposed that this general initiation protein functions as a "central processing" unit for regulatory signals. In model 2, binding of protein X to TfIId is unfavorable because the conformation of TfIId proteins is not suitable for its binding to this protein. However, when protein Y binds to its recognition sequence, it interacts with one of the Tbp-associated proteins to induce a conformational change in TfIId. As a result, protein X (once bound to its promoter sequence) can now bind directly to TfIId. Thus, as in model 1, the primary effect is on recruitment of the sequence-specific DNA-binding proteins to the transcriptional complex. Alternatively, it is possible that the two proteins function cooperatively to set the "readout" from TfIId. As illustrated in model 3, binding of the two proteins simultaneously to the promoter and to different Tbp-associated proteins of TfIId might be required to induce a conformational change in TfIId, resulting in an increased rate of transcription. Although TfIId was chosen for these examples, other general transcription factors, notably TfIIb, are capable of binding to numerous transcriptional activators and could therefore also mediate a synergistic response to two or more promoter-bound activating proteins.

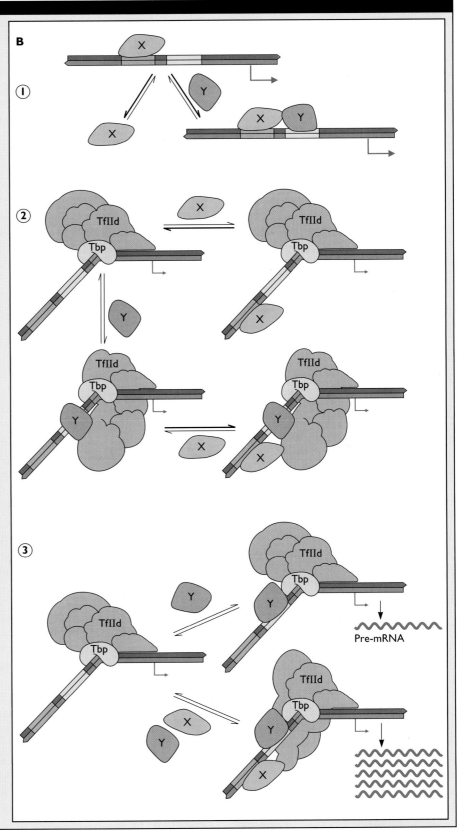

The variety in the nature of core RNA polymerase II promoters and in the constellations of sequence-specific proteins and coactivators that regulate their activity contribute to the plasticity of the RNA polymerase II system. Equally important is the power of the transcriptional machinery to integrate signals from multiple promoter-bound regulators (Box 8.3). The transcriptional machinery must also be able to sense and respond to developmental and environmental cues. Thus, transcriptional regulators are themselves frequently regulated by mechanisms that determine their activity, availability, or intracellular concentration. These mechanisms include regulation of the phosphorylation of specific amino acids, which can determine how well a protein binds to DNA, its oligomerization state, or the properties of its regulatory domain(s). In some cases, the intracellular location of a sequence-specific DNA-binding protein or its association with inhibitory proteins within the nucleus is controlled. Autoregulation of transcription of the genes encoding transcriptional regulators is also common. This brief summary illustrates the varied repertoire of regulatory mechanisms available to viruses that depend on the cellular transcriptional machinery. Not surprisingly, viruses appear to have sampled all the items on the menu and have also added the zest of virus-specific mechanisms.

## Transcription of Viral DNA Templates by the Cellular Machinery Alone

Transcription of the DNA of several major virus families is controlled by DNA sequence signals organized just like those of cellular genes transcribed by RNA polymerase II. In the simplest case, the components of the cellular transcriptional machinery described in the previous section complete the viral transcriptional program without the assistance of viral proteins. Many simple retroviruses employ just such a mechanism. The proviral DNA created by reverse transcription and integration (Table 8.2) comprises a single RNA polymerase II transcription unit organized into chromatin like the cellular templates for transcription.

The transcriptional strategy adopted by simple retroviruses suits their parasitic life style. Integrated proviral DNA is a permanent resident in the cellular genome. Because these retroviruses do not encode transcriptional regulators, the rate at which proviral DNA is transcribed is determined by the constellation of cellular transcription proteins present in an infected cell. This rate may be influenced by the nature and growth state of the infected cell as well as by the organization of cellular chromatin around the site of proviral DNA integration. Nevertheless, transcription of viral genetic information occurs throughout the lifetime of the host cell, indeed even in descendants of the initially infected cell. Thus, retroviral transcription is

characterized by continuous production of a single viral transcript, which serves as both the genome for assembly of progeny virions and the source of viral mRNA species. This strategy for transcription of viral DNA is exemplified by avian sarcoma and leukosis viruses such as Rous-associated viruses.

The long terminal repeat (LTR) of these proviral DNAs contains a compact enhancer located immediately upstream of the viral promoter (Fig. 8.9). Although enhancers can stimulate transcription over large distances and generally do so when regulating the activity of cellular genes, the close proximity of the avian proviral LTR enhancer to promoter sequences is typical of viral genomes (see, for example, Fig. 8.12 and 8.17). The avian and mammalian serum response proteins that bind to the enhancer also bind to a specific sequence in the promoter (Fig. 8.9). This arrangement emphasizes the fact that enhancer and promoter sequences cannot be distinguished by the kinds of proteins that recognize them. The other proteins that bind to this viral enhancer are all members of the basic-leucine zipper family of proteins (Fig. 8.8). Such proteins share a "leucine zipper" dimerization motif located immediately adjacent to a DNA-binding domain rich in basic-residues (Fig. 8.10). As with many other transcriptional regulators, dimerization is essential for DNA binding, in this case to align the adjacent basic regions of the monomers to form the DNA-binding surface (Fig. 8.10).

The most remarkable property of the avian retroviral transcriptional control region is that it is active in many different cell types of both the natural avian hosts and mammals. This unusual feature can be explained by the widespread distribution of the cellular proteins that bind to it. For example, both avian and mammalian members of the very large basic-leucine zipper family of transcriptional regulators are present in many types of cells. Thus, these enhancer- and promoter-binding proteins, presumably assembled on the LTR transcriptional control region in different combinations in different cell types, allow efficient transcription of the provirus in both avian and mammalian cells. This property of the LTR enhancers/promoters of avian retroviruses has been exploited in the development of viral vectors used both in the laboratory and for gene therapy.

Because the LTRs are direct repeats of one another (Fig. 8.9), transcription directed by the 3' LTR cannot contribute to the expression of retroviral genetic information. In fact, the transcriptional control region of the 3' LTR is normally inactivated by **promoter occlusion**. The passage of transcribing complexes initiating at the 5' LTR through the 3' LTR prevents recognition of the latter transcriptional control signals by enhancer- and promoter-binding proteins.

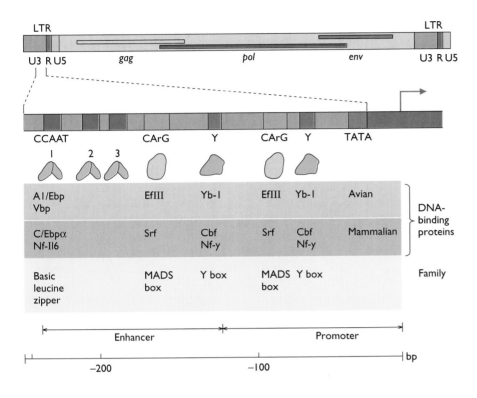

**Figure 8.9 The transcriptional control region of an avian retrovirus.** The proviral DNA of an avian leukosis virus is shown at the top. The enhancer and promoter present in the U3 regions of the LTRs are drawn to scale below. Each of the multiple CCAAT, CArG, and Y-box sequences present in this transcriptional control region, which are required for maximally efficient transcription, is recognized by both avian and mammalian proteins. All those recognizing the CCAAT enhancer sequences are members of the basic-leucine zipper family of proteins described in the text (see also Fig. 8.10). The chicken protein EfIII that binds to the CArG sequences appears to be the avian homolog of mammalian serum response factor (Srf). This protein plays an important role in the activation of transcription in response to serum growth factors, hence its name. These proteins are members of a large, widespread family defined by a conserved sequence motif within the DNA-binding domain (the MADS box) and named for four of the originally identified members.

Absolute dependence on the cellular transcriptional machinery for the expression of viral genetic information, which avoids the need to devote limited viral genetic information to transcriptional regulatory proteins, appears to be an effective strategy, for simple retroviruses reproduce in a very wide range of organisms and cell types. Nevertheless, such a strategy is not the rule, even among retroviruses.

## Viral Proteins That Regulate Transcription by RNA Polymerase II

### Patterns of Regulation by Viral Proteins

Many viruses with DNA templates transcribed by the RNA polymerase II machinery rely on the synthesis of large quantities of viral transcripts (in some cases, more than $10^5$ copies of individual mRNA species per cell) in relatively short periods. Such bursts of transcription of viral genetic information are elicited by transcriptional regulatory proteins encoded in the viral genomes. Viral proteins that stimulate RNA polymerase II transcription establish one of two kinds of regulatory circuit. The first is a **positive autoregulatory loop** (Fig. 8.11A), epitomized by the complex retrovirus human immunodeficiency virus type 1. A viral activating protein stimulates the rate of transcription of proviral DNA but does not alter the nature of viral proteins made in infected cells. The second is a

**transcriptional cascade** in which different viral transcription units are activated in a fixed sequence (Fig. 8.11B). The latter mechanism, which ensures that different classes of viral proteins are made during different periods in the infectious cycle, is characteristic of viruses with DNA genomes, including papova-, adeno-, and herpesviruses. The participation of viral regulatory proteins confers a measure of control lacking when the transcriptional program is executed solely by cellular components. We can infer that control by viral proteins is advantageous because most viruses that rely on transcription of DNA templates by RNA polymerase II have evolved such regulatory strategies. The following sections describe some well-studied examples of regulatory circuits established by viral proteins.

### Autoregulation by the Human Immunodeficiency Virus Type 1 Tat Protein

Like those of their simpler cousins, the proteins of human immunodeficiency virus type 1 and other complex retroviruses are encoded in a single proviral transcription unit whose expression is controlled by an LTR enhancer and promoter. However, in addition to the common structural proteins and enzymes, complex retroviruses encode auxiliary proteins, including transcriptional regulators. Some of these regulatory proteins, such as the Tax protein of human T-cell leukemia virus type 1, resemble activators of

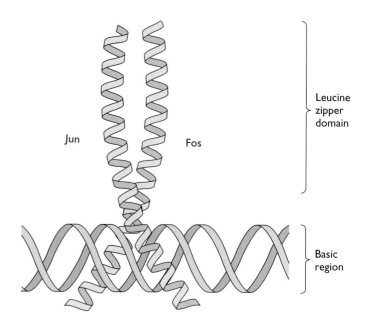

**Figure 8.10 Model of the structure of a basic-leucine zipper domain bound specifically to DNA.** The model of the leucine zipper and adjacent basic region of the Jun-Fos heterodimer (Ap1) bound to a specific recognition sequence in DNA is based on the structure determined by X-ray crystallography. The leucine zipper forms an α-helical coiled coil, both in crystals and in solution. In the protein-DNA complex, the basic (DNA-binding) region is also α-helical, but in solution it is disordered. Thus, DNA binding induces a major conformational change in proteins of this class. The angle between the two leucine zipper region α-helices in the dimer increases slightly as the coiled coil approaches the DNA. As a result, the α-helical regions containing the basic amino acids that make specific contact with DNA fit snuggly into successive major grooves. Thus, the basic-leucine zipper dimerization and DNA-binding domains have been likened to DNA "forceps." Adapted from J. N. M. Glover and S. C. Harrison, *Nature* (London) **373:**257–261, 1995, with permission.

other virus families and stimulate transcription from a wide variety of viral and cellular promoters (Table 8.4). Others, exemplified by the transactivator of transcription (Tat) of human immunodeficiency virus type 1, are unusual sequence-specific activators of transcription that recognize an RNA element in nascent transcripts.

In outline, the positive feedback loop (Fig. 8.11A), established once a sufficient concentration of Tat has been

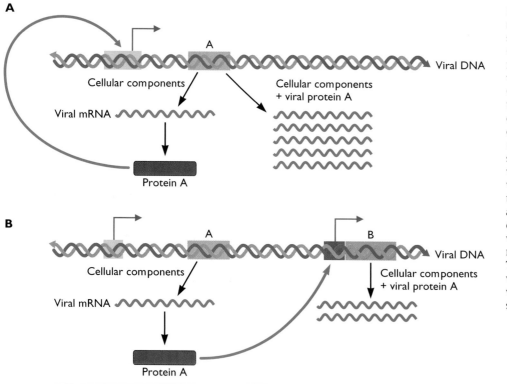

**Figure 8.11 General mechanisms of stimulation of transcription by viral proteins.** Cellular transcriptional components acting alone transcribe the viral gene for viral protein A. Once synthesized and returned to the nucleus, viral protein A can stimulate transcription either of the same transcription unit (A) or of a different viral transcription unit (B). In either case, viral protein A acts in concert with components of the cellular transcriptional machinery. The autoregulatory strategy (A) is exemplified by complex retroviruses such as human immunodeficiency virus type 1. The transcriptional programs of viruses with DNA genomes are variations of the sequential transcriptional cascade (B).

**Table 8.4** Viral enhancers and their recognition

| Virus | Enhancer location | Enhancer-binding proteins Cellular | Viral | Enhancer properties and functions |
|---|---|---|---|---|
| **Retroviruses** | | | | |
| Rous sarcoma virus | LTR enhancer, −250 to −130 of viral transcription unit | bZip proteins, Srf, Y-box-binding proteins | None | Broad cell-type specificity; active in many avian and mammalian cells because recognized by ubiquitous cellular proteins |
| Human immuno-deficiency virus type 1 | Core enhancer, −95 to −50 of viral transcription unit | Nf-κb, Ets-1 | None | Active only in cells in which Nf-κb is activated (e.g., T cells exposed to various mitogens or cytokines); necessary for efficient human immunodeficiency virus type 1 transcription in vitro |
| | Upstream enhancer region, 5′ to −100 of viral transcription unit | Ets-1, Gata 3, Lef, Nf-Il6 | None | Active in T cells and hematopoetic cells containing Ets-1 and Gata 3 (i.e., narrow cell-type specificity) |
| **Hepadnavirus** | | | | |
| Hepatitis B virus | Enhancer I, adjacent to X gene promoter (appendix, Fig. 8.11B) | Nf-κb, Ap-1, Nf-1, bZip protein, C/ebp, hepatocyte nuclear factor 3 (Hnf3) Hnf4 | None | Strong specificity for hepatic cells because C/ebp, Hnf3, and Hnf4 are specific for or enriched in these cells; some activity in nonhepatic cells; stimulates transcription from all viral promoters in infected cells; activity may be increased by the viral X protein |
| | Enhancer II, adjacent to pregenomic RNA promoter (appendix, Fig. 8.11B) | Hnf1, Hnf3, nuclear receptor superfamily members | None | Hepatic cell specific; stimulates transcription from pregenomic and pre-S promoters in infected cells |
| **Papovaviruses** | | | | |
| Simian virus 40 | Between early and late promoters; tandem copies of a 72-bp repeat sequence | Nf-κb, Ap-1, Tef-1, octamer family members | None | Broad cell type specificity; active in many mammalian cell types; activated by signal transduction pathways that converge on Ap-1 |
| Human papilloma-virus type 16 | "Constitutive" enhancer in the long control region between coding sequences for early and late proteins | Nf-κb, Ap-1, Tef-1, Tef-2, Oct-1, Nf-1 | None | Epithelial cell specific, at least in part because of the presence of a specific subset of Nf-1 family members in these cells; Ap-1 binding sites confer responsiveness to epidermal growth factor |
| | E2 protein-dependent enhancers formed by pairs of E2-binding sites present in the long control region | None | E2 protein dimers | Active only in cells expressing the viral E2 protein; stimulates transcription of all early genes |
| **Adenovirus** | | | | |
| Human adeno-virus type 2 | Enhancer 1, sequences repeated at −300 and −200 of E1A transcriptional control region (appendix, Fig. 15B) | E2f, Ets family members | None | Broad cell type specificity; active in a variety of cell types; negatively regulated in undifferentiated cells by Zn finger containing cellular repressor |
| | Enhancer II, between repeated sequences of enhancer 1 | ? | None | Independently of enhancer 1, stimulates transcription from all immediate-early and early promoters |
| **Herpesvirus** | | | | |
| Human cyto-megalovirus | Immediate-early proximal enhancer, −613 to −70 of major immediate-early transcription unit | Nf-κb, Ap-1, Creb, Srf, Ets family member Elk1, Rxr | None | Can function as a strong basal enhancer in established cell lines in culture and as an inducible enhancer activated by signal transduction pathways that converge on Nf-κb, Creb, and Srf + Ets; activity believed to be crucial for entry into lytic cycle and reactivation from latency |

attained in a human immunodeficiency virus type 1-infected cell, is simplicity itself. Cellular proteins initially direct transcription of the proviral DNA in infected cells at some basal rate; among the processed products of viral primary transcripts are spliced mRNA species from which the Tat protein is synthesized; this protein is imported into the nucleus, where it stimulates transcription of the proviral template upon binding to its RNA recognition site in nascent viral transcripts. However, the molecular mechanisms that establish this autostimulatory loop in human immunodeficiency virus type 1-infected cells are sophisticated and unusual. Their elucidation has been an important area of research because the Tat protein is essential for virus propagation and represents a good target for antiviral therapy.

### Cellular Proteins and the Human Immunodeficiency Virus Type 1 LTR Transcriptional Control Region

The LTR of human immunodeficiency virus type 1 proviral DNA, like the retroviral transcriptional control regions described previously, contains an enhancer and promoter that are recognized by cellular proteins. These regulatory sequences support some proviral transcription before Tat is made in infected cells and are also necessary for Tat-dependent transcription. In contrast to the simple retrovirus Rous sarcoma virus, human immunodeficiency virus type 1 propagates efficiently in only a few cell types, notably CD4$^+$ T lymphocytes and cells of the monocytic lineage. Viral reproduction (i.e., transcription) in infected T cells in culture is stimulated by T-cell mitogens or activators, indicating that viral transcription requires cellular components available only in activated T cells. Indeed, the failure of the virus to propagate efficiently in unstimulated T cells can be correlated with the absence of active forms of specific enhancer-binding proteins. Thus, for both simple and complex retroviruses, the distribution of cellular enhancer-binding proteins is an important determinant of host range. The difference is that the avian sarcoma viruses depend on proteins that are widely distributed, whereas human immunodeficiency virus type transcription requires proteins found in only a few cell types or whose activity is regulated.

Within the LTR, the promoter is immediately preceded by two important regulatory regions (Fig. 8.12). The promoter-proximal enhancer core is essential for enhancer

**Figure 8.12 Protein-binding sites in the human immunodeficiency virus type 1 transcriptional control region.** The organization of the U3 region of the viral LTR is shown to scale. The proteins that bind to the core enhancer or promoter discussed in the text are shown below the DNA sequence. As shown above the line, the core enhancer also contains binding sites for other proteins. The upstream enhancer region, which stimulates human immunodeficiency virus type 1 transcription in infected cells, also contains binding sites for several proteins that are enriched in cells in which the virus reproduces. For example, the Lef protein is lymphocyte-specific. Upon binding, this protein induces a large (130°) bend in DNA. Similarly, Ets-1 and Gata-3 are enriched in and restricted to, respectively, T lymphocytes. The cellular enhancer-binding proteins listed have been shown to modulate the efficiency of transcription from the viral promoter in experimental situations (usually in transient-expression assays). However, with the exception of Nf-κb (see text), their contributions to human immunodeficiency virus type 1 transcription in infected cells are generally not known.

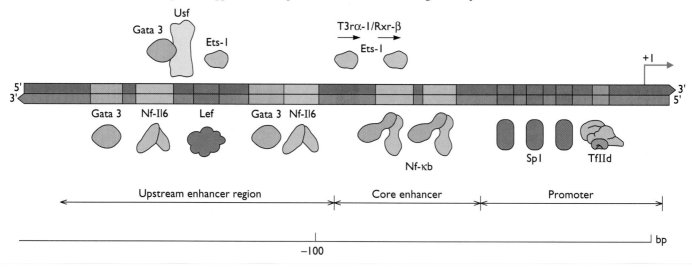

function in transient-expression assays (Fig. 8.4B) and virus reproduction in T cells. The upstream enhancer region is also necessary for efficient viral transcription in both peripheral blood lymphocytes and certain T-cell lines. Both the core and the upstream enhancers are densely packed with binding sites for cellular proteins, many of which are enriched in the types of cell in which human immunodeficiency virus type 1 can propagate. For example, the human Gata-3 and Ets-1 proteins, which can bind within the upstream enhancer region (Fig. 8.12) and stimulate viral transcription in transient-expression assays, are restricted to cells of the T-lymphocyte and hematopoietic lineages, respectively. However, it is not known whether these proteins are important for transcription of the provirus in infected cells. Transient-expression assays in which a reporter gene is introduced together with a sequence encoding the protein of interest (Gata 3 or Ets-1 in this case) do not reproduce physiological conditions (Box 8.4). Thus, a positive result in this kind of assay establishes only that a certain protein *can* stimulate transcription, not that it normally does so.

The mechanisms by which viral transcription is regulated by cellular pathways are illustrated by the cellular transcription factor Nf-κb, which plays a critical role in

human immunodeficiency virus type 1 propagation in T cells. Unstimulated T cells display no Nf-κb activity, because the protein is retained in inactive form in the cytoplasm by binding of inhibitory proteins of the Iκb family (Fig. 8.7B). This interaction masks the nuclear localization signal of Nf-κb, preventing its translocation to the nucleus. Upon T-cell activation by treatment with any one of several cytokines or mitogens, Iκb proteins rapidly disappear from the cytoplasm, with the concomitant appearance of active Nf-κb in the nucleus. Phosphorylation of Iκb at specific sites, the result of signaling pathways activated by cytokines or mitogens at the cell surface, targets the inhibitor for degradation by the cytoplasmic multiprotease complex (the **proteasome**). Thus, Nf-κb is free for transit to the nucleus, where it can bind to its recognition sites within the viral LTR core enhancer (Fig. 8.12). This pathway can account for the induction of human immunodeficiency virus type 1 transcription observed when T cells are stimulated. The importance of activation of Nf-κb in the infectious cycle of this virus is emphasized by the severe (in some cases complete) inhibition of virus reproduction (transcription) in normal human CD4$^+$ lymphocytes by mutations in the enhancer core sequence that prevent binding of this cellular protein. Nevertheless, Nf-κb and

---

**BOX 8.4**

*Caution: transient-coexpression assays do not reproduce conditions within virus-infected cells*

Transient-coexpression assays, in which the transcriptional control region of interest linked to a reporter gene is introduced into cells in culture with a vector expressing a viral (or cellular) regulator of transcription, provide a powerful, efficient way to investigate control of transcription. Advantages include the following:

Simplicity and sensitivity of assays for reporter gene activity (Fig. 8.4B)

Ready analysis of mutated promoters to identify DNA sequences needed for the action of the regulatory protein

Application with chimeric fusion proteins and synthetic promoters to avoid transcriptional responses due to endogenous cellular proteins

Simplification of complex transcriptional regulatory circuits to focus on the activity of a single protein

Despite these advantages, transient assays do not necessarily tell us how transcription is regulated in virus-infected cells, for they do not reproduce intracellular conditions. Important differences include the following:

Association of transcriptional templates with different proteins: the exogenous DNA may associate with cellular histones, but viral DNA may be packaged by virus-specific proteins.

Abnormally high concentrations of exogenous template DNA: concentrations of reporter genes as high as 10$^6$ copies per cell are not unusual, a value significantly greater than even maximal concentrations of viral DNA molecules attained toward the end of an infectious cycle.

Abnormally high concentrations of the regulatory protein as a result of its deliberate overexpression.

Because of these high concentrations of template and protein, potential interactions of the viral regulator with template or components of the cellular transcriptional machinery that would not take place in an infected cell.

Absence of viral components that might negatively or positively modulate the activity of the viral protein under investigation.

the other cellular proteins that act via LTR enhancer or promoter binding sites are not sufficient to support efficient expression of human immunodeficiency virus type 1 genes. This process depends on synthesis of the viral Tat protein.

### Regulation of Transcription by the Tat Protein

**Recognition of an RNA transcriptional control sequence by Tat.** Stimulation of human immunodeficiency virus type 1 transcription by Tat requires an LTR sequence, termed the *trans*-**activation response** (**TAR**) sequence, which lies within the transcription unit (Fig. 8.13A). This sequence is active only in the sense orientation and only when located close to the initiation site of the promoter, properties that distinguish it from enhancer elements. The observation that mutations that disrupted the secondary structure predicted for the RNA copy of the TAR sequence inhibited stimulation of transcription by Tat initially indicated that the TAR sequence is recognized as RNA. Indeed, the Tat protein binds specifically to a trinucleotide bulge and adjacent base pairs in the stem of the TAR RNA stem-loop structure (Fig. 8.13B). Such specific binding requires an arginine-rich basic region and adjacent sequences of Tat (Fig. 8.13B). Binding of Tat to this region of TAR induces a local conformational rearrangement in the RNA, resulting in formation of a more stable and compact structure (Fig. 8.13C) and is therefore energetically favorable. This mechanism of recognition of a transcriptional control sequence by a regulatory protein remains unique to Tat.

**Mechanism of stimulation of human immunodeficiency virus type 1 transcription.** Binding of Tat to TAR RNA stimulates production of viral RNA as much as 100-fold. In contrast to many cellular and viral proteins that stimulate transcription by RNA polymerase II, the Tat protein has little effect on initiation. Rather, it greatly improves elongation. Complexes initiating viral transcription in the absence of Tat elongate poorly, and many terminate transcription within 60 bp of the initiation site (Fig. 8.14A). Consequently, in the absence of Tat, full-length transcripts of the proviral genome account for no more than 10% of the total. This property resolves the paradox of why the human immunodeficiency virus type 1 LTR enhancer and promoter do not support productive viral RNA synthesis: they direct efficient initiation, but the transcription complexes formed are not competent for sustained transcription of the proviral genome over long distances from the initiation site (Fig. 8.14A). The Tat protein overcomes such poor **processivity** of elongating complexes to allow efficient production of full-length human immunodeficiency virus type 1 transcripts. The molecular mechanism by which Tat stimulates transcriptional elongation upon bind-

ing to TAR RNA proved to be unprecedented and is more complex than initially anticipated. The current model, which is based on a variety of experimental approaches, emphasizes the importance of regulation of elongation by RNA polymerase II. Importantly, it also suggests new targets for inhibitors of Tat function and human immunodeficiency virus type 1 replication.

Cellular proteins that bind specifically to the N-terminal region of Tat of human immunodeficiency virus type 1 (Fig. 8.13B), and other complex retroviruses including human immunodeficiency virus type 2, mediate stimulation of transcription by the viral proteins. A search for such Tat-binding cellular proteins identified a human Ser/Thr kinase called Tat-associated kinase (Tak). This protein, which proved to correspond to a protein that stimulates elongation (positive-acting transcription factor b), is necessary for efficient stimulation of human immunodeficiency virus type 1 transcription by Tat. The Tat-associated kinase comprises a regulatory subunit termed a cyclin (because the first members of the family to be identified accumulate during specific periods of the cell cycle) and a cyclin-dependent kinase (Cdk9). This kinase phosphorylates an unusual domain at the C terminus of the largest subunit of RNA polymerase II, which is essential for Tat-dependent stimulation of viral transcription. This domain comprises multiple tandem copies of a heptapeptide sequence rich in Ser and Thr residues. The C-terminal domain of RNA polymerase II present in preinitiation complexes is unphosphorylated, and its hyperphosphorylation has been implicated in successful transcriptional elongation by the enzyme both in vitro and in vivo. Tat-associated kinase is required for efficient transcriptional elongation in vitro, suggesting that it mediates stimulation of transcription upon binding to Tat. Indeed, the kinase activity of this cellular protein is essential for Tat-dependent stimulation of processive human immunodeficiency virus type 1 transcription both in vitro and in vivo. Furthermore, compounds identified in a large-scale screen for inhibitors of Tat function inhibit this cellular kinase.

As mentioned above, Tat binds specifically to the bulge region in the stem of TAR RNA. Although the viral protein makes no contact with residues in the TAR RNA loop (Fig. 8.13B), these sequences are crucial for the stimulation of transcription by Tat. This apparent paradox has been resolved with the discovery that the activation domain of Tat binds directly to the cyclin subunit of the Tat-associated kinase (cyclin T). The viral-cellular protein complex makes contact with the RNA loop (Fig. 8.14B) and binds to TAR RNA with higher affinity and greater specificity than does Tat alone. Thus, the binding of Tat to cyclin T by which the cellular kinase is recruited to the nascent RNA is a very favorable process. It appears that Tat, and presumably the kinase bound to it, then becomes associated with

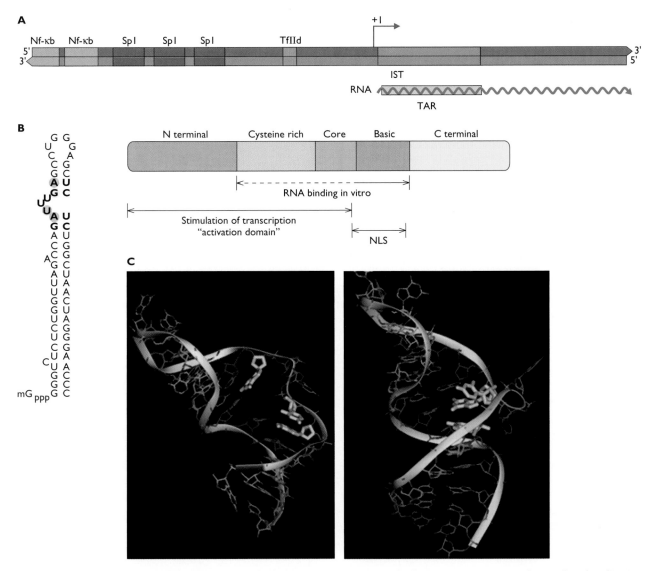

**Figure 8.13 Human immunodeficiency virus type 1 intragenic regulatory sequences.** (A) Locations of regulatory sequences. The region of the viral genome spanning the site of transcriptional initiation drawn to scale, with the core enhancer and promoter depicted as in Fig. 8.12. The DNA sequence lying just downstream of the initiation site (pink) controls synthesis of short transcripts from the human immunodeficiency virus type 1 promoter and has therefore been termed the inducer of short transcripts (IST). Transcription of the proviral DNA produces nascent transcripts that contain the TAR sequence (yellow box). (B) TAR RNA and the Tat protein. The TAR RNA hairpin (left) extends from position +1 to position +59 in viral nascent RNA and includes the minimal TAR sequence (+19 to +42) necessary and sufficient to support Tat-dependent stimulation of transcription in vivo. Sequences important for recognition of TAR RNA by the Tat protein are colored. Optimal stimulation of transcription by Tat requires not only this binding site in TAR but also the terminal loop. The Tat protein (right) is made from several different, multiply spliced mRNAs (appendix, Fig. 9B) and therefore varies in length (at its C terminus) from 82 to 101 amino acids. The regions of the protein are named for the nature of their sequences (Basic, Cysteine rich) or greatest conservation among lentiviral Tat proteins (Core). Experiments with fusion proteins containing various segments of Tat and a heterologous RNA-binding domain identified the N-terminal segment as sufficient to stimulate tran-

scription. The basic region contains the nuclear localization signal (NLS). Analysis of the RNA-binding properties of short peptides indicated that the basic region, which is very rich in arginine residues, can bind specifically to RNA containing the bulge characteristic of TAR RNA (left). However, high-affinity binding, effective discrimination of wild-type TAR from mutated sequences in vitro, and RNA-dependent stimulation of transcription within cells require additional, N-terminal regions of the protein, shown by the dashed arrow. Thus, the sequences required for the stimulation of transcription and specific recognition of TAR RNA in cells do not form discrete domains of the protein. (C) Conformational change in TAR RNA induced by binding of the Tat peptide containing residues from the core and basic regions of the protein. Major groove view of structures of a free TAR RNA corresponding to the apical stem and loop regions (but with a truncated stem) (left) and of the same RNA when bound to the Tat peptide (right) were determined by nuclear magnetic resonance methods. The residues shown in yellow are A22, U23, and G26 (from bottom to top in the left panel), which are colored lavender, green, and yellow, respectively, in panel B. Note the substantial conformational change in the trinucleotide bulge region on binding of the Tat peptide. From F. Aboul-ela et al., *J. Mol. Biol.* **253**:313–332, 1995, and F. Aboul-ela et al., *Nucleic Acids Res.* **24**:3974–3981, 1996, with permission. Courtesy of M. Afshar, RiboTargets, and J. Karn, MRC Laboratory of Molecular Biology.

**Figure 8.14 Mechanism of stimulation of transcription by the human immunodeficiency virus type I Tat protein.** (A) Summary of the experimentally observed differences in human immunodeficiency virus type 1 transcription in the presence and absence of the Tat protein. Before Tat is made in infected cells, 9 out of 10 proviral transcripts synthesized by the cellular proteins described in Fig. 8.12 and the text are terminated, most within 60 bp of the initiation site. Thus, synthesis of full-length human immunodeficiency virus type 1 RNA is inefficient. Production of the Tat protein upon translation of mRNAs spliced from rare, full-length transcripts allows all transcriptional complexes to pass through the elongation blocks to synthesize full-length viral RNA. Current evidence indicates that these two patterns result from opposing effects of the DNA sequences IST and Tat. (B) Model for stimulation of transcription of human immunodeficiency virus type 1 proviral DNA by Tat. Initiation of transcription by the RNA polymerase II machinery results in the synthesis of short transcripts containing the TAR sequence. In the absence of Tat, transcription complexes are poorly processive, and the great majority terminate transcription, releasing transcription components and short transcripts. Binding of Tat to TAR in conjunction with the cyclin T subunit of Tak (see text) leads to stimulation of phosphorylation (P) of the C-terminal domain of the largest subunit of RNA polymerase II by the TfIIh-associated kinase and allows phosphorylation of this domain by Tak. It is believed that the two kinases act sequentially (see text). As a result of modification of this C-terminal domain of the enzyme subunit, the transcriptional complexes become competent to carry out highly processive transcription.

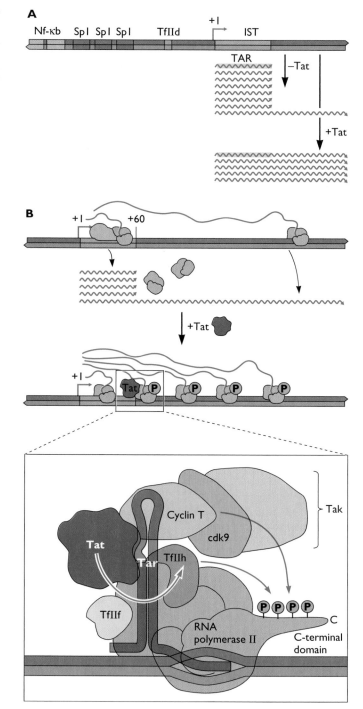

transcription complexes, which would be converted to the form competent for processive elongation following phosphorylation of the C-terminal domain of RNA polymerase II (Fig. 8.14B).

In this mechanism, the sole function of Tat is to ensure association of the C-terminal domain kinase (Tak) with the transcriptional complex (Fig. 8.14B). However, the viral protein stimulates the activity of a second C-terminal domain kinase present in the general initiation protein TfIIh. The Tat protein binds directly to TfIIh. Furthermore, inhibition of the TfIIh kinase impairs the ability of Tat to stimulate processive human immunodeficiency virus type 1 transcription. As we have seen, TfIIh is a component of the preinitiation complex and acts during initiation and promoter clearance. However, it is released from the transcriptional complex when nascent RNA chains of 30 to 50 nucleotides have been made. In contrast, the Tat-associated kinase modulates the activity of transcriptional complexes at a later stage, when nascent RNAs are 15 to about 500 nucleotides in length. It is therefore believed that phosphorylation of the C-terminal domain of RNA polymerase II to allow efficient and processive elongation is a two-step process, with TfIIh and Tat-associated kinase acting sequentially. Thus, according to the current model, Tat

stimulates human immunodeficiency virus type 1 transcriptional elongation so effectively because it facilitates and integrates the action of both cellular kinase-containing complexes (Fig. 8.14B).

At this juncture, it is difficult to appreciate the value of the intricate transcriptional program of human immunodeficiency virus type 1. Many other viruses successfully employ transcriptional regulators that operate, directly or indirectly, via specific DNA sequences, and Tat can stimulate transcription effectively when made to bind to DNA experimentally. Why, then, have human immunodeficiency virus type 1 and its relatives evolved a unique RNA-dependent mechanism of bringing their transcriptional activators to the promoter? Binding of Tat to nascent viral RNA close to the site at which many transcriptional complexes pause or stall (Fig. 18.14) could provide a particularly effective way to recruit the cellular proteins that stimulate processive transcription. A deeper understanding of the normal mechanism of action of Tat-associated kinase may shed light on this issue. Or it may be that the use of a nascent RNA sequence in transcriptional regulation is a legacy from retroviral ancestors that operated in the RNA world.

It is clear that the efficient production of progeny human immunodeficiency virus type 1 genomes by transcription of the provirus ultimately depends on the constellation or activation state of cellular enhancer-binding proteins. Transcription of even small quantities of full-length viral transcripts in response to these cellular proteins allows the synthesis of Tat mRNA and Tat protein and consequently activation of the positive autoregulatory loop. The provirus can be considered dormant in cells that do not contain the necessary enhancer-binding proteins or in which these proteins are inactive, such as unstimulated T cells that contain inactive cytoplasmic Nf-κb. However, it is important to keep in mind that such transcriptional inactivity is not the cause of the clinical latency characteristic of human immunodeficiency virus type 1 infections. In clinical latency, few, if any, symptoms are manifest. Nevertheless, virus is produced continuously (see chapter 18). Thus, the positive autoregulatory loop starts to work at full blast from the moment that appropriate conditions for LTR enhancer-dependent transcription are encountered by the virus.

## The Transcriptional Cascades of DNA Viruses

The transcriptional strategies of viruses with DNA genomes exhibit a number of common features. The most striking is the transcription of viral genes in a reproducible and precise sequence. Prior to initiation of viral DNA synthesis, during **immediate-early** and **early phases** in the reproductive cycle, infected cells are devoted to the production of viral proteins necessary for viral DNA synthesis, efficient expression of viral genes, or other regulatory functions. Transcription of the **late genes**, most of which encode viral structural proteins, requires viral DNA synthesis (Fig. 8.15). This property ensures coordinated production of the DNA genomes and structural proteins from which progeny virus particles are assembled. Another common feature is the use of viral proteins, as well as replication of viral DNA, to control the transitions from one transcriptional stage to the next (Fig. 8.15).

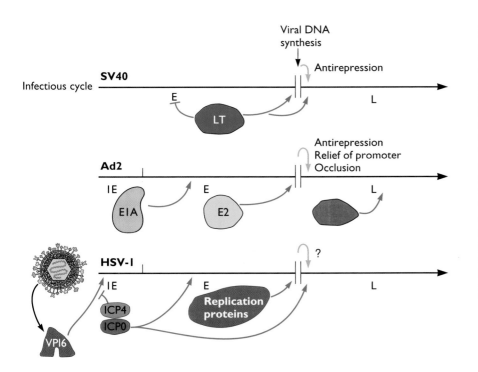

**Figure 8.15 Important features of the simian virus 40 (SV40), human adenovirus type 2 (Ad2), and herpes simplex virus type 1 (HSV-1) transcriptional programs.** The transcriptional programs of these three viruses are depicted by the horizontal time lines, on which the onset of viral DNA synthesis is indicated by vertical lines. For comparative purposes only, the three reproductive cycles are represented by lines of equal length. The immediate-early (IE), early (E), and late (L) transcriptional phases are indicated, as are viral proteins that participate in regulation of transcription. Stimulation of transcription by these proteins and effects contingent on viral DNA synthesis in infected cells are indicated by green and orange arrows, respectively. Red bars indicate negative regulation of transcription. LT, large T antigen.

The transcriptional activating proteins encoded by different viruses exhibit distinctive molecular properties. However, all cooperate with cellular transcriptional components to stimulate transcription of specific sets of viral genes expressed inefficiently in their absence. In addition, many (including simian virus 40 large T antigen and the herpes simplex virus type 1 ICP4 protein) can repress transcription of their own genes. Such autoregulation presumably circumvents unfavorable consequences of too-high concentrations of these viral proteins, for example, unnecessary use of the resources of the infected cell or inappropriate transcription of cellular genes. Thus, the transcriptional programs of these viruses closely resemble those regulating many developmental processes in animals, in both the transcription of individual genes in a predetermined sequence and the sequential action of viral proteins that regulate the transcription of different sets of viral genes.

### The Transcriptional Programs of Different DNA Viruses

**Sequential transcription of viral genes.** DNA viruses with small genomes establish simple transcriptional programs in the infected cell. For example, the genome of the papovavirus simian virus 40 contains only two transcription units (appendix, Fig. 13B), each of which encodes more than one protein. This type of organization reduces the genetic information that must be devoted to transcription punctuation marks and regulatory sequences, a significant advantage when genome size is limited by packaging constraints. The price for such a transcriptional strategy is heavy dependence on the host cell's RNA processing systems to generate multiple mRNAs by differential polyadenylation and/or splicing of a single primary transcript. Expression of the simian virus 40 early transcription unit by the cellular enhancer- and promoter-binding proteins described below leads to the synthesis of large T antigen in infected cells. This multifunctional viral early protein both induces initiation of viral DNA synthesis and activates late transcription (Fig. 8.15). Thus, T-antigen synthesis in a permissive host cell inevitably induces progression to the late phase of infection, when the segment of the viral genome encoding structural proteins is transcribed.

Adenovirus genomes also contain a minimal number of transcriptional regulatory sequences in their genomes, for the sequences encoding the more than 40 viral proteins are organized into only eight RNA polymerase II transcription units (appendix, Fig. 15B). In contrast to that of papovaviruses, the initial period of the adenoviral infectious cycle comprises two distinct transcriptional phases (Fig.

8.15). Upon entry of the viral genome into the nucleus, a single viral gene (E1A) is transcribed under the control of typical enhancers. If functional E1A proteins cannot be made, progression of the infectious cycle beyond this immediate-early phase is seriously impaired. The E1A proteins are necessary for efficient transcription of all viral early transcription units (Fig. 8.15). These proteins regulate transcription by multiple mechanisms. The viral early genes include the E2 gene, which encodes the replication proteins required for initiation of viral DNA synthesis.

The adenovirus late phase is marked not only by transcription of late phase-specific genes but also by an increased rate of initiation from the major late promoter, which is active during the early phase of infection. Thus, late genes of the more complex DNA viruses are defined as those that attain their maximal rates of transcription following viral DNA synthesis. However, no known activator of late transcription is included among the adenoviral early gene products. Instead, the synthesis of progeny adenoviral DNA molecules is indirectly coordinated with production of the protein components that will encapsidate them in virions (Fig. 8.15). In this respect, adenovirus can be distinguished from both the simpler papovavirus simian virus 40 and the more complex herpesviruses, such as herpes simplex virus type 1.

The organization of the herpes simplex virus type 1 genome demands an entirely different transcriptional strategy, for the more than 70 known genes of this virus are, with few exceptions, expressed as individual transcription units. Furthermore, splicing of herpes simplex virus type 1 primary transcripts is the exception rather than the rule. In contrast to those of simian virus 40 or human adenoviruses, herpes simplex virus type 1 genes encoding proteins that participate in the same process, for example, viral DNA synthesis, are not clustered but are scattered throughout the genome (appendix, Fig. 17B). The basic distinction of early and late phases is maintained in the herpes simplex virus type 1 transcriptional program (Fig. 8.15), but temporal control of the activity of more than 70 viral promoters is obviously more complicated. In fact, the potential for finely tuned regulation is clearly much greater when the viral genome comprises a large number of independent transcription units.

Initial expression of both papovavirus and adenovirus genes is directed by enhancers that operate efficiently with cellular proteins. In contrast, a viral activating protein is imported into cells infected by herpes simplex virus type 1 (Fig. 8.15). This virion structural protein, VP16, is necessary for efficient transcription of viral immediate-early genes. Although this simple device might seem to guarantee transcription of these genes in all infected cells, this is not necessarily the case, for VP16 can operate only in conjunction with specific cellular proteins. Stimulation of immediate-early gene transcription by VP16 is of considerable importance for the success of the viral productive cycle. For example, viruses encoding VP16 that is defective for activation of transcription grow well only at high multiplicities of infection. Mutant adenoviruses lacking E1A genes exhibit the same phenotype. Thus, neither of these transcriptional regulatory proteins is absolutely essential for virus reproduction, at least in cells in culture.

Herpesvirus immediate-early gene products resemble adenovirus E1A proteins in performing a number of regulatory functions. The regulatory scheme of herpes simplex virus type 1 is, however, considerably more complex, for this viral genome carries five such genes (appendix, Fig. 17B). The products of at least two, ICP4 and ICP0, are transcriptional regulators. The ICP4 protein, which is essential for progression beyond the immediate-early phase of infection, is regarded as the major herpes simplex virus type 1 transcriptional activator. It stimulates transcription of both early and late genes and also acts as a repressor of immediate-early gene transcription. As for the simpler DNA viruses, transcription of herpesviral late genes is governed by viral DNA replication (Fig. 8.15). Herpes simplex virus type 1 late genes can be divided into those that are not transcribed prior to viral DNA synthesis and those that are transcribed at maximal rates following initiation of viral DNA synthesis. More subtle distinctions among the large number of late genes may be made as their transcriptional regulation becomes better understood.

This summary of DNA virus transcriptional programs illustrates the increasing sophistication of the circuits controlling transcription with increasing genome size, a feature of little surprise. Underlying the diversity of the specific mechanisms employed by these viruses are the common themes of the central role of virus-encoded transcriptional regulators and the coordination of transcriptional control with viral DNA synthesis. As we shall see in considering these topics in more detail, the molecular devices by which these viruses activate transcription often also optimize the cellular environment for virus reproduction. Regulation of initiation is the mainstay of the transcriptional control programs of these (and other) DNA viruses. Nevertheless, expression of some genes is controlled at other steps in the transcription cycle, such as termination.

**Entry into one of two alternative transcriptional programs.**  The herpesviruses are unique among the viruses described in this section in their ability to establish **latent infections** in specific cell types. This state is characterized by both lack of efficient expression of all viral genes transcribed during productive infection and activation of a unique latent-phase transcriptional program. For example, during latent infection of neurons by herpes simplex virus type 1, a set of specific RNAs, termed the latency-associated transcripts (LATs) (Fig. 8.16), are the only viral products synthesized in large quantities. The 2.0- and 1.5-kb latency-associated transcripts (Fig. 8.16) accumulate to 40,000 to 100,000 copies within a latently infected neuron, lack poly(A) tails, are restricted to the nucleus, and are not linear molecules. Indeed, all properties established to date indicate that they are stable introns produced by splicing of precursor RNA, as discussed in chapter 10. Regulatory mechanisms that determine whether the transcriptional cascade described above is activated or repressed and that control transcription of LAT genes must be of paramount importance in determining the outcome of herpes simplex virus infection. Reentry into the lytic cycle from the latent state, a process called **reactivation**, must also be governed by transcriptional regulatory mechanisms.

The mechanisms that mediate latency and reactivation are by no means fully understood. However, the available evidence indicates that the latency-associated transcripts themselves are important for establishing a latent infection. For example, stable expression of the abundant latency-associated transcripts in neuronal cells suppresses replication of the virus and production of the immediate-early gene products needed for progression through the productive infectious cycle. The intriguing question of how viral RNAs with the properties of introns excised from pre-mRNAs might repress expression of these viral genes cannot yet be answered. The products of other RNAs made in smaller quantities in latently infected neurons, such as the L/ST RNAs (Fig. 8.16), can block the ability of the immediate-early ICP4 protein to bind to promoter sequences. These observations suggest that once latency-associated genes are expressed in an infected neuron, the transcriptional cascade characteristic of lytic infection is repressed. Much less clear at present are the mechanisms that control transcription of these genes. As discussed below, the regulatory circuits that govern entry into the lytic or latent transcriptional programs of other herpesviruses, notably the gammaherpesvirus Epstein-Barr virus, appear to be considerably simpler and are better understood.

### The Simian Virus 40 Enhancer: a Model for Viral and Cellular Enhancers

The majority of viral DNA templates described in this chapter contain enhancers of transcription that are recognized by cellular DNA-binding proteins. The simian virus 40 enhancer has been studied more intensively than other viral enhancers, in part because it was the first such regu-

**A**

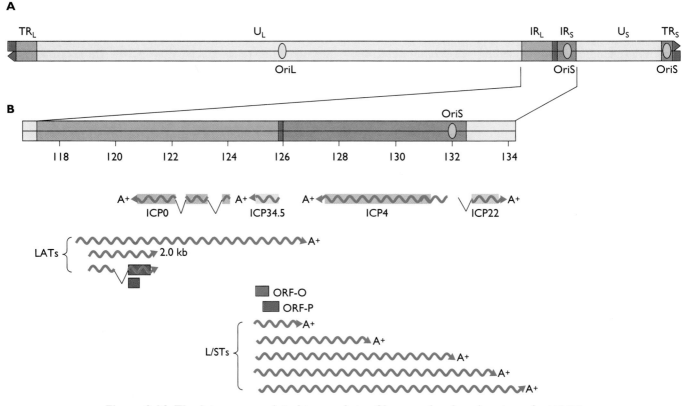

**Figure 8.16 The latency-associated transcripts of herpes simplex virus type 1.** (A) Diagram of the herpes simplex virus type 1 genome, showing the unique long and short segments, U$_L$ and U$_S$, respectively, the terminal repeat (TRL, TRS) and internal repeat (IRL, IRS) sequences, and the origins of replication, OriL and OriS. (B) Expanded map of the IRL region, with the scale in kilobase pairs. This region encodes immediate-early proteins, ICP0, ICP4, and ICP22, which play important roles in establishing a productive infection. Below are shown the locations of sequences encoding the LATs and the transcripts named L/STs. The arrow indicates the direction of transcription and A+ 3′ poly(A) sequences where known. Open reading frames, including ORF-O and ORF-P, are shown as boxes. The evidence currently available suggests that ORF-P (and perhaps ORF-O) may control LAT RNA synthesis. This region of the herpes simplex virus genome therefore contains both *cis*-acting sequences and a variety of genes that are important in establishing a lytic or a latent infection.

latory sequence to be identified. Its properties and mechanism of action are considered characteristic of many enhancers, whether of viral or cellular origin.

The simian virus 40 enhancer displays the hierarchical organization typical of these regulatory regions (Box 8.5). It is built from three units, termed **enhancer elements**, which are subdivided into smaller sequences recognized by DNA-binding proteins (Fig. 8.17). The DNA-binding proteins that interact with this viral enhancer are differentially expressed in different cell types. For example, Nf-κb and certain members of the octamer-binding protein (Obp) family (Fig. 8.17) are specific to cells of lymphoid origin, and their binding sites are necessary for enhancer activity in these cells. Similarly, the DNA sequences bound by transcriptional enhancer factor 1 (Tef-1) (Fig. 8.17) are required for enhancer activity only in cells in which this protein is present, such as human lymphoid cells. Yet other elements of the simian virus 40 enhancer, such as the activator protein 1 (Ap-1) binding sites, confer responsiveness to cellular signaling pathways. This combination of DNA sequences ensures the activity of the enhancer (hence initiation of the viral life cycle) in many different circumstances in which the virus may find itself. This property is exhibited by several other viral enhancers, including that of Rous sarcoma virus discussed previously (Table 8.4). In contrast, some viral templates for RNA

**BOX 8.5**

*Typical properties of the simian virus 40 enhancer*

An enhancer is composed of multiple units, termed **enhancer elements** (e.g., the B1, B2, and A elements of the simian virus 40 enhancer) (Fig. 8.17).

Enhancer elements operate synergistically: a single element has little activity, but the complex set stimulates transcription more than 100-fold.

Multimerization of a single inactive enhancer element can also generate an enhancer active in infected or transfected cells.

Enhancer activity is relatively insensitive to the orientation or position of individual enhancer elements: stimulation of transcription generally depends on the presence of multiple elements rather than on the exact way in which these are arranged.

Enhancer elements comprise multiple sequences recognized by sequence-specific DNA-binding proteins that also bind to promoter sequences (Fig. 8.9, 8.12, and 8.17).

**Figure 8.17 Organization of the simian virus 40 enhancer.** The positions of the 72-bp repeat region of the simian virus 40 genome containing the enhancer elements relative to the early promoter are shown at the top. Shown to scale below are functional DNA sequence units, and the proteins that bind to them, of the early promoter-distal 72-bp repeat and its 5'-flanking sequence, which forms part of enhancer element B. All the protein-binding sites shown between the expansion lines are repeated in the promoter-proximal 72-bp repeat. Thus, the complete enhancer contains one copy of the enhancer element B1 and two directly repeated copies of the enhancer elements B2 and A. The definition of enhancer elements is discussed in the text. Some enhancer elements are built from repeated binding sites for a single sequence-specific protein. The Sph-II sequences, which are recognized by the cellular protein Tef-1, are representative of this class, as formation of a functional enhancer element requires cooperative binding of Tef-1 to the two Sph-II sequences. Such cooperative binding, which renders enhancer activity sensitive to small changes in the concentration of a single protein, represents a regulatory device of great potential sensitivity. A second class of enhancer elements comprises sequences bound by two different proteins, as illustrated by enhancer element B. Although the binding of Tef-1 and Tef-2 to the GT-IIC and GT-I sequences, respectively, is not cooperative, these proteins interact once bound to DNA to generate an active enhancer element. A third kind of enhancer element contains neither a precisely spaced duplicate (to permit cooperative binding) nor a second protein-binding site. Rather, oligomerization of a single sequence is sufficient to generate an active enhancer element. The TC-II sequence recognized by Nf-κb functions in this manner.

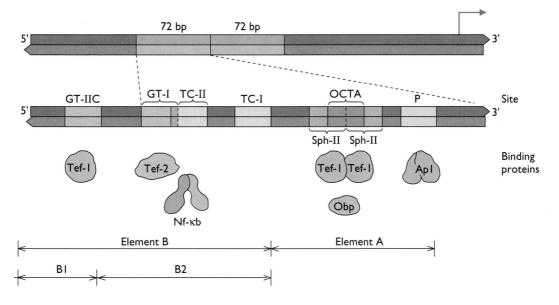

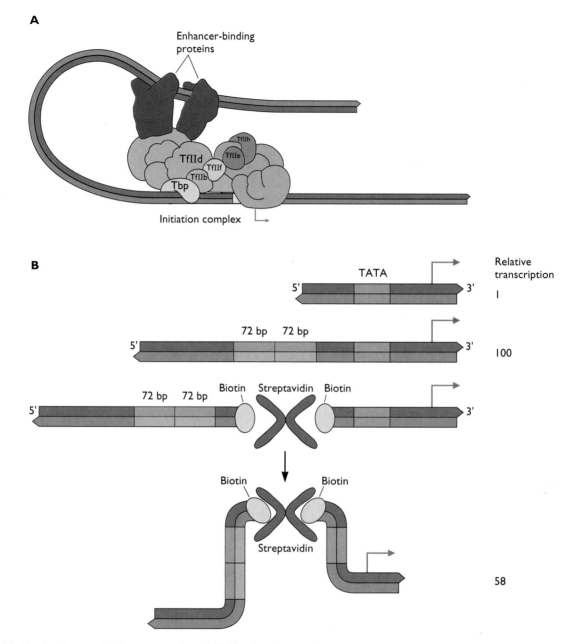

**Figure 8.18 Mechanisms of enhancer action.** (A) The DNA looping model postulates that proteins bound to a distant enhancer, here shown upstream of a gene, interact directly with components of the transcription initiation complex, with the intervening DNA looped out. These interactions of proteins bound to a distal enhancer are analogous to those that can take place when the enhancer is located close to the promoter. Such interactions might stabilize the initiation complex and thus stimulate transcription. (B) An enhancer noncovalently linked to a promoter via a protein bridge is functional. When placed immediately upstream of the rabbit β-globin gene promoter, the simian virus 40 enhancer stimulates specific in vitro transcription from circular plasmids by a factor of 100. In the experiment summarized here, the enhancer and promoter were separated by restriction endonuclease cleavage. Under this condition, the enhancer cannot stimulate transcription (not shown). The protein biotin was added to the ends of each DNA fragment by incorporation of biotinylated UTP. Biotin binds the protein streptavidin nonco-valently, but with extremely high affinity ($Kd$, $10^{-15}$ M). Because streptavidin can bind four molecules of biotin, its addition to the biotinylated DNA fragments allows formation of a noncovalent protein "bridge" linking the enhancer and the promoter. Under these conditions, the viral enhancer stimulates in vitro transcription from the rabbit β-globin promoter almost as efficiently as when present in the same DNA molecule, as summarized at the right. Because this result indicates that an enhancer can stimulate transcription when present in a separate DNA molecule (i.e., in *trans*), it rules out models in which enhancers are proposed to serve as entry sites for RNA polymerase II (or other components of the transcription machinery): such a mechanism requires that RNA polymerase II slide along the DNA from the enhancer to the promoter, a passage that would be blocked by the protein bridge. Rather, this result suggests that enhancer function requires close proximity (but not covalent linkage) to the promoter. Thus, the results of this experiment are consistent with the looping model shown in panel A.

polymerase II transcription contain enhancers whose activity is highly tissue-specific (for example, hepatitis B virus) or strictly dependent on the presence of inducers or viral proteins (Table 8.4).

The simian virus 40 enhancer is located within 200 bp of the transcription initiation site. More typically, enhancers are found hundreds, thousands, or tens of thousands of base pairs up- or downstream of the promoters that they regulate. The most popular model of the mechanism by which enhancers exert remote control of transcription, the DNA looping model, invokes interactions among enhancer-bound proteins and the transcriptional components assembled at the promoter, with the intervening DNA looped out (Fig. 8.18A). Experimental evidence in favor of this model has been cleverly collected by using the simian virus 40 enhancer (Fig. 8.18B). Enhancers can also facilitate access of the transcriptional machinery to chromatin templates. For example, the simian virus 40 enhancer lies within, and contains DNA sequences necessary for formation of, a nucleosome-free region of the viral genome in infected cells. Thus, enhancers may stimulate RNA polymerase II transcription by multiple molecular mechanisms. The primary effect of these mechanisms is to increase the probability that the gene to which an enhancer is linked will be transcribed, because enhancers ensure that a stable, active transcriptional template will be formed, rather than to increase the rate of transcription per se.

### Transcriptional Regulators Encoded by DNA Viruses

Virus-encoded proteins that stimulate transcription are the linchpins of the molecular circuits that govern transcription in cells infected by viruses with DNA genomes. Consequently, they have received much attention. In this section, we focus on a few selected viral regulators to illustrate general principles of their operation or fundamental insights into cellular processes that have been gained through their study.

Some viral transcriptional regulators (Table 8.5) are close relatives of well-studied cellular proteins that bind to specific DNA sequences from which promoters or enhancers are built. Not every viral DNA-binding protein listed in Table 8.5 has been studied in detail, but all their known properties emphasize this resemblance. These viral proteins possess discrete DNA-binding domains, some with sequences that conform to motifs characteristic of cellular DNA-binding proteins, and activation domains that interact with general initiation proteins (Table 8.5). These properties are described in more detail for one such viral protein, the Epstein-Barr virus Zta protein, in the next section. Sequence-specific DNA-binding proteins play

ubiquitous roles in the transcription of cellular genes by RNA polymerase II, so it is hardly surprising that viruses relying on this enzyme have evolved (or captured) genes for analogous proteins. However, viral transcriptional regulators that possess no intrinsic ability to bind specifically to DNA are equally common (Table 8.6). The preponderance of such viral proteins, quite unexpected when they were first characterized, was a strong indication that host cells also contain proteins that control transcription without themselves binding to DNA. Many such proteins (e.g., coactivators or mediators) have now been recognized. Two examples, the herpes simplex virus type 1 VP16 and adenovirus E1A proteins, illustrate the diversity of mechanisms by which this class of viral proteins regulate transcription.

**Multiple mechanisms of regulation of the activity and availability of the Epstein-Barr virus Zta protein.** When the gammaherpesvirus Epstein-Barr virus infects B lymphocytes, a latent state characterized by expression of a limited number of viral genes is established. The products of these viral genes maintain the viral genome via replication from a latent-phase-specific origin of replication, allow infected cells to evade the immune system, and alter the growth properties of the cells. Treatment of such latently infected lymphocytic cells in culture with a variety of agents induces the viral replicative (or lytic) cycle. Epstein-Barr virus also enters the lytic cycle upon infection of epithelial cells. Virus reproduction begins with expression of three viral proteins that regulate gene expression. However, expression of just one of these, the transcriptional regulator Zta (also known as ZEBRA, Z, or EB-1) in latently infected cells is sufficient to interrupt latency and induce entry into the lytic cycle.

The Zta protein exhibits many properties of the cellular proteins that recognize local promoter sequences. It is a modular, sequence-specific DNA-binding protein that belongs to the basic-leucine zipper family (Table 8.5). This dimerization/DNA-binding domain mediates binding of Zta to viral promoters that contain its recognition sequence. Its discrete activation domain, which can bind directly to cellular initiation proteins, such as Tbp and Tbp-associated proteins of TfIIId, is believed to facilitate the assembly of preinitiation complexes. The Zta protein acts synergistically (Box 8.3) with a second viral activating protein (Rta) expressed at the beginning of the lytic cycle to stimulate transcription from viral early promoters. The availability or activity of Zta is regulated by numerous mechanisms. Expression of the viral gene encoding Zta is controlled by several cellular DNA-binding proteins that stimulate or repress its transcription (Fig. 8.19). Many of the cellular proteins that stimulate transcription from the

**Table 8.5** Some viral transcriptional regulators that bind to specific DNA sequences

| Virus | Protein | Properties | Functions |
|---|---|---|---|
| Papovavirus | | | |
| Bovine papilloma-virus type 1 | E2 | Typical domain organization; binds as dimers; truncated forms lack activation domains; can bind TfIIb and TfIId | Stimulation of transcription from several viral promoters as enhancer-binding protein; truncated forms compete for DNA binding and inhibit transcriptional stimulation by full-length E2 protein; replication of viral DNA in vivo |
| Adenovirus | | | |
| Adenovirus type 2 | IVa2 | Binds intragenic sequence in the major late promoter; late phase-specific | In conjunction with a second, infected cell-specific protein that binds intragenic sequences, stimulates transcription from major late promoter |
| Herpesviruses | | | |
| Herpes simplex virus type 1 | ICP4 | Typical domain organization; one domain related in sequence to a domain of a cellular coactivator; binds to a degenerate consensus sequence | Stimulates transcription from early and late promoters by a mechanism that is not well understood but may not require binding to a specific DNA sequence; represses transcription from its own and other immediate-early promoters by binding to promoter sequences |
| Epstein-Barr virus | Zta | Basic-leucine zipper viral protein; expression and activity regulated by multiple mechanisms (see text) | Essential activator of early gene transcription; commits to lytic infection (see text) |
| Baculovirus | | | |
| *Autographa californica* nuclear polyhedrosis virus | IE-1 | Typical domain organization; binds as dimer to 28-bp palindromic sequences | Stimulates transcription from early promoters, including its own, by binding to enhancer elements |
| Poxvirus | | | |
| Vaccinia virus | VETF | Binds to TA-rich early promoter sequences as heterodimer; DNA-dependent ATPase | Essential for recognition of and transcription from early promoters by the viral RNA polymerase (Fig. 8.27) |

Zta gene promoter are targets of signal transduction cascades that induce entry of Epstein-Barr virus into the lytic cycle in latently infected B cells (Fig. 8.19). In addition, the Zta protein is itself a positive autoregulator of transcription. The availability of Zta mRNA for translation in the cytoplasm of infected cells is also regulated, in part by annealing of Zta pre-mRNA to the complementary transcripts of the viral EBNA-1 gene (Fig. 8.19). Finally, phosphorylation of Zta modulates its DNA-binding activity, whereas binding of cellular proteins, including the p53 tumor suppressor protein and the p65 subunit of Nf-κb (Fig. 8.7), prevents Zta from stimulating transcription.

The net effect of these regulatory mechanisms, which depends on the type, growth state, differentiation state, and environment of the Epstein-Barr virus-infected cell, determines whether active Zta protein is available and thus whether infection leads to latent or lytic replication. Entry into the lytic cycle appears to be an inevitable consequence of production of active Zta. This protein stimulates transcription not only from the promoter of its own gene but also from that of the viral gene encoding the Rta protein, with which Zta cooperates to activate early gene transcription. Furthermore, Zta plays an important role in replication from the lytic origins of Epstein-Barr virus replication.

**Sequence-specific transcriptional regulation via cellular DNA-binding proteins: the herpes simplex virus type 1 VP16 protein.** The herpesviral VP16 protein, which enters infected cells in the virion (Fig. 8.15), has taught us much about mechanisms by which transcription by RNA polymerase II can be stimulated. Furthermore, its unusual mode of promoter recognition illustrates the importance of conformational change during formation of DNA-bound complexes.

Although it lacks a DNA-binding domain, the herpes simplex virus type 1 VP16 protein has been widely used to investigate mechanisms of stimulation of transcription, for its C-terminal acidic activation domain is one of the most potent known. Chimeric proteins in which this acidic acti-

**Table 8.6** Some viral transcriptional regulators that lack sequence-specific DNA-binding activity

| Virus | Protein | Functions and properties |
|---|---|---|
| **Papovavirus** | | |
| Simian virus 40 | Large T antigen | Stimulates late gene transcription; can stimulate transcription from many simple promoters in experimental assays; can bind to Tbp, TfIIb, and Tef-1 via different segments; contains no activation domain; may stabilize initiation complexes |
| **Adenovirus** | | |
| Adenovirus type 2 | E1A proteins | Larger stimulates early gene transcription in infected cells via CR3; both bind to the Rb protein, releasing cellular E2f, and to the cellular coactivator p300 (see text); both repress enhancer-dependent transcription |
| **Herpesviruses** | | |
| Herpes simplex virus type 1 | VP16 | Stimulates transcription from IE promoters via potent acidic activation domain; achieves promoter specificity by interaction with the cellular Oct-1 protein (see text) |
| | ICP0 | Cooperates with ICP4 to stimulate transcription of E and L genes in infected cells; important for reactivation from latency in model systems; stimulates transcription from many promoters in experimental assays; can stimulate transcription synergistically with ICP4 in such assays and bind to ICP4 |
| Epstein-Barr virus | EBNA-2 | Essential for B-cell transformation; induces transcription of viral LMP-1 and LMP-2 and of some cellular genes via specific DNA sequence; sequence recognition achieved by interaction with the cellular recombination binding protein Jκ; contains an acidic activation domain that binds to Tbp and a Tbp-associated protein of TfIId |
| **Hepadnavirus** | | |
| Hepatitis B virus | X | Required for efficient virus reproduction in vivo; promiscuous activator of transcription by RNA polymerase II and III in experimental assays; stimulation of RNA polymerase II transcription depends on cellular sequence-specific activators; may be a general coactivator; may also stimulate transcription indirectly via activation of cellular signaling pathways |
| **Retrovirus** | | |
| Human T-cell leukemia virus type 1 | Tax | Stimulates transcription from the viral LTR transcriptional control region; in vitro, binds directly to the basic region of basic-leucine zipper proteins to stabilize the dimers and increase their affinity for DNA; binds to free and DNA-bound Nf-κb in vitro; also regulates transcription indirectly, by inducing degradation of Iκb (Fig. 8.7) |

vation domain are fused to heterologous DNA-binding domains strongly stimulate transcription from promoters that contain the appropriate binding sites. When part of such fusion proteins, the VP16 acidic activation domain can stimulate several initiation reactions, consistent with its ability to bind directly to multiple general initiation proteins (Fig. 8.20A). It can also increase the rate of transcriptional elongation and stimulate transcription from chromatin templates (Fig. 8.20B). These properties established that a single protein can regulate RNA polymerase II transcription by multiple molecular mechanisms.

The herpes simplex virus type 1 VP16 protein is the founding member of a class of viral regulators that activate transcription from promoters containing a specific consensus sequence yet possess no sequence-specific DNA-binding activity. The 5'-flanking regions of herpes simplex virus type 1 immediate-early genes contain at least one copy of a consensus sequence, 5'TAATGARAT3' (where R is a purine), which is necessary for VP16-dependent acti-

vation of their transcription. This sequence is bound by VP16 only in association with at least two cellular proteins, octamer-binding protein 1 (Oct-1) and host cell factor (Hcf) (Fig. 8.21). The Oct-1 protein is a ubiquitous transcriptional activator named for its recognition of a consensus DNA sequence termed the octamer motif. This protein and VP16 can associate to form a ternary (three-component) complex on the 5'TAATGARAT3' sequence, but the second cellular protein, Hcf, is necessary for stable, high-affinity binding. The VP16 protein and Hcf form a heteromeric complex in the absence of Oct-1 or DNA. The primary function of Hcf is therefore believed to be induction of a conformational change in VP16, to allow its stable binding to Oct-1 on the 5'TAATGARAT3'-containing sequences of herpes simplex virus type 1 immediate-early genes (Fig. 8.21).

One of the most remarkable features of the mechanism by which the VP16 protein is recruited to viral immediate-early promoters is its specificity for Oct-1. This protein is a

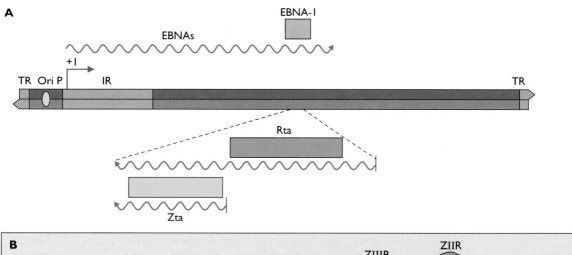

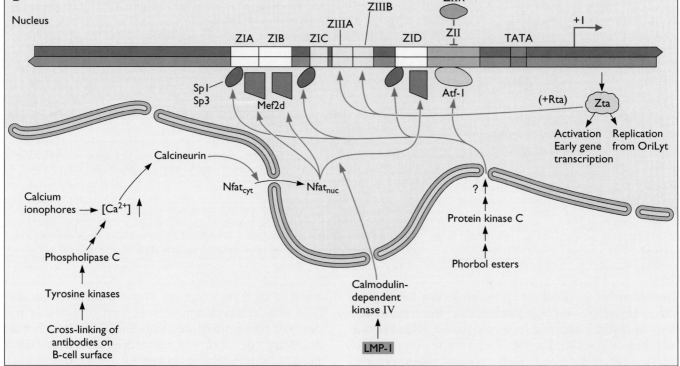

**Figure 8.19 Organization and regulation of the Epstein-Barr virus Zta gene promoter Zp.** (A) Organization of the transcription units that contain the coding sequence for the Epstein-Barr virus nuclear antigen (EBNA) proteins (an ~100-kb transcription unit) and Zta and Rta. The locations of the genomic terminal (TR) and internal (IR) repeat sequences, the origin of replication for plasmid maintenance (OriP), and the coding sequences for the EBNA-1, Zta, and Rta proteins are indicated. The last two proteins are synthesized from spliced mRNAs processed from the primary transcripts shown. (B) Sequences that regulate transcription from the Zp promoter, shown to scale, and the cellular or viral proteins that operate through them are shown. The Atf-1 and Mef-2d proteins are members of the basic-leucine zipper and Mads families, respectively. The activities of the cellular transcriptional regulators are controlled by several pathways that transmit signals from the surface of the cell to the transcriptional machinery. Such signals, which include cross-linking of antibodies on the B-cell surface by anti-immunoglobulin antibodies and phorbol esters, induce entry into the lytic cycle, presumably as a result of activation of transcription from the Zp promoter. Expression of viral latent membrane protein 1 (LMP-1) can also activate transcription, as does synthesis of Zta itself, which binds to the ZIII sites in the promoter. In addition to the ZII sequence, which negatively regulates transcription from the Zp promoter, several negative elements lie upstream of position −221. Transcripts from the upstream promoter of the Rta gene (A) yield processed bicistronic mRNAs that encode Zta and Rta. The Zta protein, a sequence-specific DNA-binding protein belonging to the basic-leucine zipper family, activates both transcription of viral genes and replication from the lytic origin, OriLyt.

member of a family of related transcriptional regulators defined by a common DNA-binding motif called the POU homeodomain. The VP16 protein distinguishes Oct-1 from all other members of this family, including Oct-2, which binds to exactly the same DNA sequence as Oct-1. In fact, VP16 detects a single amino acid difference in the exposed surfaces of DNA-bound Oct-1 and Oct-2 homeodomains. Such discrimination is not only remarkable but also of biological importance, because only the Oct-1 protein is ubiquitously expressed.

The GARAT segment of TAATGARAT-containing elements of immediate-early genes is indispensable for assembly of VP16 into the DNA-bound complex and therefore defines promoters recognized by the viral VP16 protein. Nevertheless, the homeodomain of Oct-1, not VP16, recognizes the GARAT motif. The Oct-1 homeodomain appears to adopt an altered conformation to bind to this sequence, and it is only this conformation that is recognized by VP16 (Fig. 8.21). Thus, the GARAT DNA sequence is the crucial effector of assembly of the quaternary complex. The mechanism by which VP16, in conjunction with Hcf and Oct-1, recognizes herpesviral immediate-early promoters illustrates the important contributions of induced conformational changes to transcriptional specificity.

The VP16 protein interacts with cellular proteins via its amino-terminal region. Its C-terminal region contains the acidic activation domain discussed previously, which acts from sites far from a promoter when fused to a GAL4 DNA-binding domain. In contrast, when present in the ternary complex with cellular proteins, VP16 stimulates transcription only if the complex is bound close to the promoter. This difference indicates that the activation domain does not function in the same manner in natural complexes and artificial chimeras. Which, if any, of the biochemical activities exhibited by the VP16 acidic activation domain when fused to a heterologous DNA-binding domain (e.g., Fig. 8.20) pertain to the mechanism of stimulation of immediate-early gene transcription in infected cells has yet to be determined.

The incorporation of the VP16 protein into herpes simplex virus type 1 virions at the end of one infectious cycle is an effective way to ensure transcription of viral genes and initiation of a new cycle once a new host cell has been infected. Nevertheless, some features of this mechanism are puzzling, in particular the location of VP16 within virions. As a component of the tegument, this protein is interposed between the capsid and the envelope (appendix, Fig. 17A) and is not close to the genome carrying its target genes. The significance of such an arrangement is not yet clear. Similarly, the benefits conferred by the indirect mechanism by which VP16 recognizes immediate-early promoters are not fully appreciated. One advantage over

direct DNA binding may be an opportunity to monitor the growth state of the host cell provided by the requirement for binding to Hcf, a protein that plays a role in cell cycle progression in uninfected cells. And the strictly local action of VP16 imposed by its association with Oct-1 and Hcf (see above) might well be necessary to prevent untimely transcription of other viral genes by the potent acidic activation domain of this viral transcriptional regulator.

**Viral proteins with multiple regulatory activities: adenovirus E1A proteins.** Two E1A proteins, expressed from differentially spliced mRNAs (Fig. 8.22), are synthesized within an hour or two of adenovirus infection. These two proteins share all sequences except for an internal segment (conserved region 3 [CR3]) that is unique to the larger protein (Fig. 8.22). Nevertheless, they differ considerably in their regulatory potential because the CR3 segment is primarily responsible for stimulation of transcription of viral early genes in infected cells. As the larger E1A protein neither binds specifically to DNA nor depends on a specific promoter sequence, it is often considered the prototypical example of viral proteins that stimulate transcription by indirect mechanisms, some of which are listed in Table 8.6.

Several properties suggest that the larger E1A protein may be a general coactivator. In transient-expression assays (Fig. 8.4B), stimulation of transcription by this E1A protein is remarkably insensitive to the nature of the promoter. The protein can stimulate transcription by RNA polymerase III as well as expression from the great majority of the many RNA polymerase II promoters that have been tested. This E1A protein can interact physically with RNA polymerase II initiation proteins, such as Tbp and Tbp-associated proteins, and stimulate transcription via a large variety of sequence-specific transcriptional regulators. The E1A protein can thus be defined as a general coactivator, which allows sequence-specific DNA-binding proteins to stimulate transcription more effectively. However, it is not known how CR3, a region of only 49 amino acids, can cooperate with so many different components of the cellular transcriptional machinery. When human adenovirus type 2 infects a permissive cell, the larger E1A protein does not stimulate transcription globally, as might be expected from the experimental observations just described. Instead, the rates of transcription of only viral early genes and a small number of cellular genes are increased. The reasons for this difference are not understood. However, it is important to keep in mind that transcription in transient-expression assays differs in several important respects from transcription in infected cells (Box 8.4).

Adenoviral E1A proteins activate transcription by a second mechanism, which was discovered more recently but

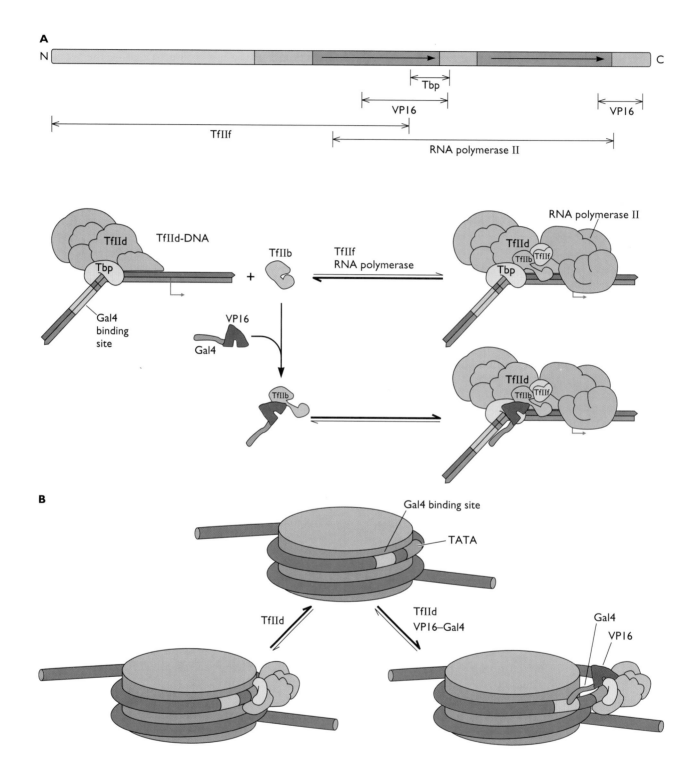

is better understood. This function, mediated by the conserved regions of the N-terminal exon, CR1 and CR2 (Fig. 8.22), was elucidated through studies of transformation and immortalization by E1A gene products. The CR1 or CR2 segments interact with several cellular proteins. Two

of these, Rb and p300 (Fig. 8.23), are of special relevance to transcriptional regulation.

The Rb protein is the product of the cellular retinoblastoma susceptibility gene, a **tumor suppressor**, whose crucial role in cell cycle checkpoint control is described in

◁ **Figure 8.20 Models for transcriptional activation by the acidic activation domain of herpes simplex virus type I VP16 protein.** (A) Induction of conformational change in TfIIb. The C-terminal domain of human TfIIb protein (top) contains two directly repeated sequences and sequences required for binding to Tbp, the acidic activation domain of VP16 and RNA polymerase II. The N-terminal domain is necessary for the binding of TfIIf to TfIIb. In native TfIIb, these two domains interact with one another such that the segments of the protein that interact with TfIIf-RNA polymerase II are blocked. Thus, entry of TfIIb into an assembling initiation complex and recruitment of TfIIf-RNA polymerase II to the complex are unfavorable reactions that require conformational change in TfIIb. Binding of the acidic activation domain of VP16, as a chimera with the DNA-binding domain of the yeast protein Gal4, disrupts the intermolecular association of the N- and C-terminal domains of TfIIb, exposing its binding sites for TfIIf and RNA polymerase II. Consequently, formation of the preinitiation complex containing TfIIf and RNA polymerase II is now more favorable. When the VP16 activation domain is bound to the promoter, here through interaction of the Gal4 DNA-binding domain with its recognition site, its binding to TfIIb can also facilitate interaction of this protein with the TfIId-DNA complex. Thus, two sequential reactions in the assembly sequence (Box 8.2) are driven in the forward direction by interaction of the VP16 acidic activation domain with TfIIb. However, this activation domain can also bind directly to proteins of TfIId. These interactions, or those with TfIIf or TfIIh, may also favor formation of the closed initiation complex by directly stabilizing assembling complexes or by inducing favorable conformational changes. Or they may be related to the ability of the VP16 acidic activation domain to stimulate other reactions, such as promoter clearance or elongation of the transcriptional machinery through sequences in the DNA template that can cause premature termination of transcriptions. Other acidic activating domains have also been shown to facilitate formation of the open initiation complex. (B) Alleviation of transcriptional repression by nucleosomes. Many activators, including the acidic activation domain of VP16, stimulate transcription from nucleosomal DNA templates to a much greater degree than they do transcription from naked DNA. This property is the result of their ability to alleviate repression of transcription by nucleosomes, an activity therefore termed **antirepression**. The association of DNA with a nucleosome, an octamer of two copies each of the histones H2A, H2B, H3, and H4, can block access of proteins to their DNA-binding sites, as illustrated for binding of TfIId to a TATA sequence (left). Interaction of the DNA template with the histone proteins may directly prevent binding of the protein, or the conformation adopted by the DNA to wrap around the nucleosome may be incompatible with its recognition by the protein. Binding of the acidic activation domain of VP16 to the template, in this example via the heterologous Gal4 DNA-binding domain, alters the interaction of the DNA with the nucleosome to allow TfIId access to the TATA sequence (right). The mechanism by which VP16 and other activators might remodel nucleosomal DNA has not been established but might involve their interactions with nucleosomal histones or with certain coactivators. These coactivators are multicomponent complexes that, in ATP-dependent reactions, dramatically alter the association of nucleosomes with DNA (but do not displace them) and in so doing stimulate the binding of Tbp and other DNA-binding proteins. Other coactivators stimulate acetylation of nucleosomal histones, a modification long known to be associated with transcriptionally active chromatin. Although many activators alter nucleosomal DNA templates in this way, some can prevent binding of nucleosomes to promoters once bound to their recognition sequences.

chapter 16. The discovery that Rb binds to adenovirus E1A proteins, as well as to transforming proteins of papovaviruses, led to elucidation of the mechanism of action of this important cellular protein and of one mechanism by which E1A proteins regulate transcription. In uninfected cells, Rb binds to cellular E2f proteins, sequence-specific transcriptional activators discovered by virtue of binding sites in the human adenovirus type 2 E2 early promoter (Fig. 8.6). Such E2f-Rb complexes possess the specific DNA-binding activity characteristic of E2f alone, but Rb represses transcription (Fig. 8.23). The CR1 and CR2 regions of the E1A proteins bind to the same segments of Rb as do E2f heterodimers. Thus, these viral proteins release E2f and sequester Rb. Free E2f proteins bind to sequences in the viral E2 early promoter (Fig. 8.6) that are essential for efficient E2 transcription during the early phase of infection. As the E2 transcription unit encodes the proteins required for viral DNA synthesis, the availability of free E2F controls the synthesis of viral replication proteins and progression into the late phase of infection (Fig. 8.15).

The N-terminal sequences common to the two E1A proteins also bind to a second cellular protein that participates in regulation of transcription, the p300 protein (Fig. 8.22). This protein is one of a family of coactivators which bind to and mediate stimulation of transcription by the basic-leucine zipper protein Creb and steroid/thyroid hormone receptors. The function of p300 appears to be to modify the chromatin structure of the promoter, for it

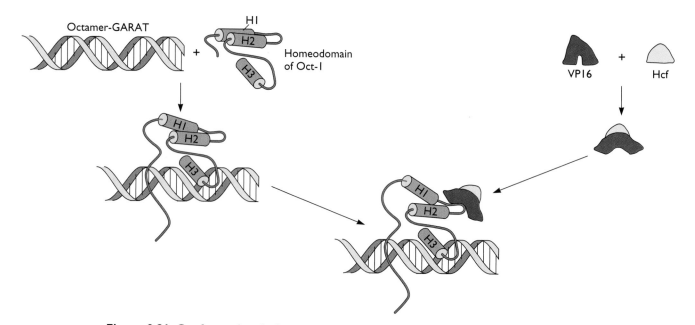

**Figure 8.21 Conformational changes are necessary for Oct-1 protein-dependent recruitment of VP16 to herpes simplex virus type 1 promoters.** The VP16 protein, which enters herpes simplex virus type 1-infected cells in the virion tegument, must interact with the host cell protein Hcf to adopt a conformation that allows recognition of the cellular Oct-1 protein bound to DNA. The homeodomain segment of the bipartite DNA-binding domain of Oct-1 also undergoes a conformational change necessary for its recognition by VP16. When the binding site (Octamer-GARAT) contains the GARAT sequence, the conformation of the DNA-bound homeodomain of the Oct-1 protein is altered from the structure observed when the protein is bound to the octamer sequence alone, which has been determined by X-ray crystallography. The novel conformation adopted on binding sites that contain the GARAT sequence has not been determined and is illustrated schematically at the left. The conformationally altered VP16 protein can then recognize the unique form of the Oct-1 homeodomain bound to a site containing the GARAT sequence. This mechanism ensures that the VP16 protein is recruited only to promoters that contain the GARAT sequence, viral immediate-early promoters. Thus, binding of VP16 to the many cellular promoters that contain the octamer sequence, interactions that would be of little benefit to the virus, is avoided.

both possesses histone acetylase activity and binds to other histone acetylases. Modification of nucleosomal histones by acetylation has long been known to correlate with activation of transcription. The E1A proteins, which bind directly to p300, both disrupt interaction of p300 with Creb and compete for binding of p300 to the histone acetylase. This activity of E1A proteins has been implicated in repression of enhancer-dependent transcription and, as discussed in chapter 16, is required for induction of cell proliferation and transformation.

The multiplicity of mechanisms by which the E1A proteins engage with components of the cellular transcriptional machinery is one of their most interesting features. Regulation by multiple mechanisms may prove to be a definitive property of viral proteins that cannot bind directly to viral promoters (Table 8.6). For example, the human T-cell leukemia virus type 1 Tax protein stimulates transcription by both binding to specific cellular members of the basic-leucine zipper family and activation of Nf-κb. The latter mechanism includes activation of the signaling cascade leading to degradation of Iκb and translocation of Nf-κb from the cytoplasm to the nucleus (Fig. 8.7), a pathway illustrating the potential for intervention of viral regulators in cellular pathways well "upstream" of any transcriptional activator.

### Coordination of Transcription of Late Genes with Viral DNA Synthesis

The structural proteins of viruses must be made in stoichiometric quantities sufficient to ensure assembly of the maximal number of progeny viruses. Such high concentration of foreign (viral) proteins is undoubtedly injurious

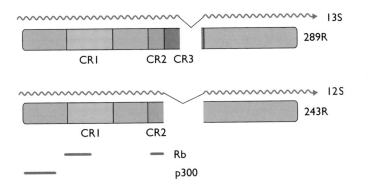

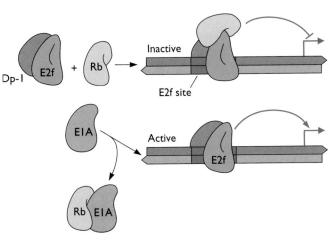

**Figure 8.22 Structural and functional features of the adenovirus type 2 E1A proteins.** Primary transcripts of the immediate-early E1A gene are alternatively spliced to generate the abundant 13S and 12S mRNAs. As such splicing does not change the translational reading frame, the E1A proteins synthesized from these mRNAs are identical except for an internal segment of 46 amino acids unique to the larger protein. Comparison of the E1A sequences of different human adenovirus serotypes identified the three most highly conserved regions, designated CR1, CR2, and CR3. The CR3 segment includes the unique sequence of the larger E1A protein, which is required for stimulation of transcription of viral early genes in infected cells. Conserved regions CR1 and CR2 present in both E1A proteins are necessary for the mitogenic activity of E1A gene products and for transformation (see chapter 16). A large number of cellular proteins bind to these regions, as well as to N-terminal sequences, of E1A proteins. The regions of the E1A proteins necessary for interaction with two of these, the Rb protein and p300, are indicated.

**Figure 8.23 Model of competition between E1A proteins and E2f for binding to Rb protein.** The E2f transcriptional activators are heterodimers of a member of the E2f protein family described in chapter 16 and a Dp protein, such as Dp-1. The binding of E2f to its recognition sites in specific promoters is not inhibited by association with the Rb protein. However, when present in a ternary complex, Rb can repress transcription from E2f-dependent promoters (top). Adenoviral E1A proteins made in infected (or transformed) cells bind to Rb and disrupt the E2f-Rb interaction. Thus, binding of E1A proteins to Rb releases E2f, which can then stimulate transcription.

to the host cell and could therefore interfere with cellular processes on which viruses rely. In the case of the viruses under consideration here, synthesis of structural proteins is restricted to the late phase of infection, and this potential problem is avoided. Because replication of viral DNA genomes is in full swing by the time late genes are expressed, synthesis of structural proteins and virion assembly are the only steps then necessary to complete the reproductive cycle. Thus, side effects of structural protein synthesis at this time pose no problem for the virus. This optimal pattern of viral structural protein synthesis is a result of the dependence of late gene transcription on viral DNA replication: drugs or mutations that inhibit viral DNA synthesis in infected cells also block expression of late genes. Despite their importance, the mechanisms by which activation of transcription can be integrated with viral DNA synthesis have been investigated only recently and remain incompletely understood.

**Regulation of transcription initiation.** The most obvious consequence of genome replication in cells infected by any DNA virus is the large increase in intranuclear concentration of viral DNA molecules. Even in experimental

situations, infected cells contain a relatively small number of copies of the viral genome during the early phase of infection, typically 5 to 100 copies per cell depending on the multiplicity of infection. As soon as viral DNA synthesis begins, this number rises geometrically to values as high as hundreds of thousands of viral DNA molecules per infected cell nucleus. At such high concentrations, viral promoters can compete effectively for components of the cellular transcription machinery. The increase in DNA template concentration also titrates cellular repressing proteins that bind to specific sequences of certain viral late promoters. For example, the simian virus 40 major late promoter contains multiple binding sites for a cellular repressor that belongs to the steroid/thyroid hormone receptor superfamily (Fig. 8.24A). Viral DNA replication increases the concentration of the late promoter above that at which every copy can be bound by the cellular repressor and thus allows this promoter to become active (Fig. 8.24B).

Although titration of the cellular repressor allows transcription from the simian virus 40 major late promoter, maximally efficient late transcription requires the viral early gene product large T antigen. This protein controls simian virus 40 late transcription (Fig. 8.15) both directly, as an activating protein, and indirectly, as a result of its essential functions in viral DNA synthesis. Among the more

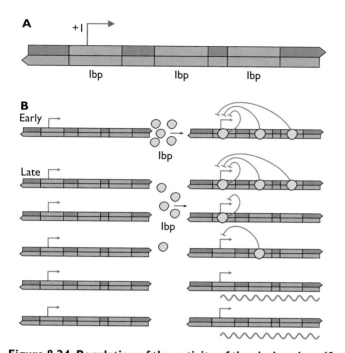

**Figure 8.24 Regulation of the activity of the simian virus 40 late promoter by cellular repressors.** (A) The sequence surrounding the simian virus 40 major late initiation site (the thickest arrow in Fig. 8.5D) contains three binding sites for the cellular proteins termed initiator-binding proteins (Ibps), which are members of the steroid/thyroid receptor superfamily. (B) "Antirepression" resulting from synthesis of viral DNA synthesis in infected cells. During the early phase of infection, the concentration of Ibps relative to that of the viral major late promoter is sufficiently high to allow all Ibp-binding sites in the viral genomes to be occupied. Consequently, the late promoter is transcriptionally silent because the Ibps are repressors of transcription. The concentration of Ibps does not change during the course of simian virus 40 infection. However, as viral DNA synthesis takes place in the infected cell, the concentration of the major late promoter becomes sufficiently high that not all Ibp-binding sites can be occupied. The major late promoter therefore becomes accessible to cellular transcription components. Therefore, although we generally speak of "activation" of late gene transcription, this DNA replication-dependent mechanism is one of escape from repression.

complex DNA viruses, the molecular coupling of replication to transcription of late genes can be more indirect. For example, late phase-specific transcription of the adenoviral IVa2 gene, like simian virus 40 major late transcription, appears to be controlled by viral DNA synthesis-dependent titration of a cellular repressor. However, the IVa2 protein is itself a sequence-specific activator of transcription. Once synthesized in infected cells, it cooperates with a second, as yet unidentified protein to stimulate the rate of initiation of transcription from the major late promoter at least 20-fold. Thus, synthesis of adenoviral DNA initiates a tran-

scriptional cascade in which late promoters are activated sequentially (Fig. 8.15).

The increase in intracellular concentration of viral DNA can also alleviate promoter occlusion. The pIX gene of adenovirus is completely contained within the early E1B transcription unit (appendix, Fig. 15B) and transcription from the pIX promoter is initially inhibited by transcription from the upstream E1B promoter. As the concentration of viral DNA molecules increases following viral DNA synthesis, the concentration of E1B promoters exceeds that of the cellular transcriptional components necessary for initiation of E1B transcription. Consequently, some viral DNA molecules would carry transcriptionally inactive E1B genes in which the pIX promoter is accessible for recognition by cellular transcription proteins.

Newly replicated DNA molecules must be partitioned for such fates as additional replication cycles, service as templates for transcription, or assembly into virions. This process must be regulated so that different fates predominate at different times in the infectious cycle. Although transcription of all viral DNA molecules made in infected cells seems to be a simple mechanism to ensure efficient transcription of late genes, no more than 5 to 10% of the large numbers that accumulate are transcriptionally active. In the case of simian virus 40, transcriptional activity can be ascribed to establishment of the open chromatin region in minichromosomes described above. For the more complex viruses, it is not known whether a subset of viral DNA molecules is also marked for transcriptional activity by some unknown molecular device, or indeed whether all viral late genes can be concurrently transcribed from a single DNA molecule. However, one important parameter governing the concentration of transcriptional templates must be the relative concentrations of viral DNA molecules and the proteins that package them for assembly into virions because packaging and transcription are mutually exclusive.

**Regulation of termination.** Viral DNA synthesis can also affect transcriptional termination, a regulatory device illustrated by the human adenovirus type 2 major late transcription unit. During the early phase of infection, major late transcription terminates within a region in the middle of the transcription unit such that no elongating complexes reach the L4 poly(A) addition site (Fig. 8.25). As discussed in chapter 10, such restricted transcription is coupled with preferential utilization of specific RNA processing signals to produce a single major late mRNA and protein during the early phase of infection. Viral DNA synthesis is necessary, and probably sufficient, to induce transcription of distal segments of the transcription unit to a termination site close to the right-hand end of the viral

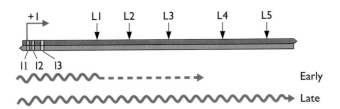

**Figure 8.25 DNA replication-dependent alterations in termination of human adenovirus type 2 major late transcription.** The segment of the viral genome from 15 to 100 map units containing the major late transcription unit is shown to scale (top). The five sites at which primary major late transcripts are polyadenylated, designated L1 to L5, are indicated by the vertical arrows. During the early phase of infection, major late transcripts synthesized in response to the stimulatory activity of the 289R E1A protein terminate at multiple sites beyond the L1 site, as indicated by the dashed line. No transcript extends beyond the position indicated by the arrowhead. The mechanisms responsible for this termination, which takes place within a DNA segment of 11 to 12 kb, are not understood. Following synthesis of both viral DNA and late proteins, all regions of the major late transcription unit to termination sites near the right-hand end of the genome are transcribed with equal efficiency.

genome (Fig. 8.25). The inefficient termination of early transcription at many sites spread over about 12 kb of DNA and the fact that only replicated viral DNA molecules can support complete transcription of the major late transcription unit are unusual features suggesting a novel regulatory mechanism. In virions, adenoviral DNA is packaged by viral DNA-binding proteins (appendix, Fig. 15A), at least one of which is retrained when the genome enters the infected cell nucleus. Because these viral DNA proteins are products of late genes, viral DNA synthesis must necessarily be accompanied by reorganization of the viral nucleoproteins. Indeed, the templates for major late transcription during the late phase of infection appear protein-free in the electron microscope. Such alterations in template structure upon viral DNA synthesis may play an important role in regulation of termination of major late transcription.

## A Viral DNA-Dependent RNA Polymerase

The DNA of viruses considered in preceding sections is replicated in the nucleus of infected cells and thus is well positioned to utilize the cellular transcriptional machinery. In contrast, poxviruses such as vaccinia virus reproduce exclusively in the cytoplasm of their host cells. This feat is possible because these viruses encode the components of a transcription and RNA processing system that produces viral mRNAs with all the hallmarks [5' caps and 3' poly(A) tails] of cellular mRNAs. This system, which is carried into infected cells within vaccinia virus virions, includes a DNA-dependent RNA polymerase responsible for transcription of all vaccinia virus genes. A striking feature of this viral enzyme is its structural and functional resemblance to cellular RNA polymerases.

The vaccinia virus RNA polymerase, like RNA polymerase II and the other eukaryotic RNA polymerases, is a complex, multisubunit enzyme built from the products of at least eight viral genes. The amino acid sequences of several of these subunits, including the two largest and the smallest, are clearly related to subunits of RNA polymerase II. The viral enzyme transcribes all classes of vaccinia virus genes expressed at different times in the infectious cycle (early, intermediate, and late). Like the cellular enzymes, vaccinia virus RNA polymerase achieves promoter recognition by cooperation with additional proteins. For example, formation of transcriptionally competent initiation complexes on vaccinia virus early promoters is mediated by two viral proteins, VETF and RAP14, which are responsible for the recognition of promoter sequences and recruitment of the RNA polymerase, respectively (Fig. 8.26). Thus, these viral proteins are functional analogs of the cellular RNA polymerase II general initiation proteins TfIId and TfIIf (Box 8.2). However, the vaccinia virus transcriptional machine is not analogous to its cellular counterpart in every respect. Transcription of the majority of viral early genes terminates at discrete sites 20 to 50 bp downstream of specific T-rich sequences in the template. Termination requires the viral termination protein, which is also the viral mRNA capping enzyme. Thus, the 3' ends of the viral mRNAs correspond to sites of transcription termination. In contrast, cellular RNA polymerase II can transcribe far beyond the sites at which the 3' ends of mature cellular or viral mRNAs are produced by processing of the primary transcript and does not terminate transcription at simple, discrete sites.

The viral RNA polymerase and proteins necessary for transcription of early genes enter host cells within vaccinia virus particles. Subsequently, viral gene expression is controlled by the ordered synthesis of viral transcriptional regulatory proteins that permit sequential recognition of intermediate and late promoters. For example, transcription of intermediate genes requires synthesis of the viral RP030 gene product, while late transcription depends on synthesis of several intermediate gene products. Activation of intermediate and late gene transcription also requires viral DNA replication. Thus, transcription of vaccinia virus genetic information is regulated by mechanisms similar to those operating in cells infected by other DNA viruses, even though the transcriptional machinery is viral in origin.

Surprisingly, the vaccinia virus transcription system is not entirely self-contained, for a cellular protein is essen-

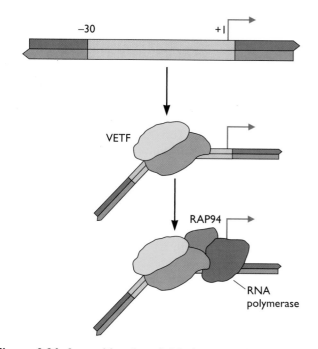

**Figure 8.26 Assembly of an initiation complex on a vaccinia virus early promoter.** Vaccinia virus early promoters are characterized by the presence of an AT-rich sequence (tan) immediately upstream of the site of initiation of transcription. Vaccinia virus RNA polymerase cannot recognize these (or any other) viral promoters in the absence of other viral proteins. One of these, the vaccinia virus early transcription protein (VETF), is necessary for early promoter recognition and must bind before the viral RNA polymerase. This is a heteromeric protein composed of subunits of 77 and 82 kDa which binds specifically to early promoters and induces DNA bending. This protein also possesses DNA-dependent ATPase activity. It, and the second protein necessary for early promoter specificity, RAP94, enter infected cells in virions. Some 40% of the virion RNA polymerase molecules are associated with RAP94, and it is only this RNA polymerase population that can transcribe early promoters in vitro. The RAP97-RNA polymerase complex associates with early promoter-bound VETF, an interaction that probably includes direct binding of RAP94 to VETF, to form a functional initiation complex. Thus, assembly of these vaccinia virus initiation complexes is analogous to, although simpler than, formation of RNA polymerase II initiation complexes (Box 8.2).

tial for transcription of viral intermediate genes. This protein, termed Vitf2, is located in the nucleus of uninfected cells but is present in both the cytoplasm and the nucleus of vaccinia virus-infected cells. As vaccinia virus devotes a significant number of its genes to proteins necessary for transcription, such dependence on a cellular factor must confer some special advantage. An attractive possibility is that interaction of the viral transcriptional machinery with a cellular protein serves to integrate the virus's reproduc-

tive cycle with the growth state of its host cell, just as simian virus 40 and human adenovirus type 2 "sense" and modulate cell cycling and growth by the interactions of large T-antigen and E1A proteins, respectively, with Rb. The molecular identity of Vitf2 and elucidation of its function in uninfected cells should shed light on the physiological significance of the limited but essential participation of cellular components in the transcription of vaccinia virus genetic information.

## Transcription of Viral Genes by RNA Polymerase III

Several of the viruses considered in this chapter contain RNA polymerase III transcription units in their genomes. This enzyme transcribes cellular genes that encode functional RNA species that participate in such essential processes as translation (tRNAs and 5S rRNA), protein trafficking (the 7S RNA component of signal recognition particles), and pre-mRNA splicing (U6 snRNA). In view of the important roles played by small cellular RNA species, it is not surprising that some viruses also exploit such molecules and thus cellular RNA polymerase III (Table 8.7).

The first, and still best understood, example of such a viral gene is that encoding the major RNA polymerase III transcription product of human adenovirus type 1, virus-associated RNA I (VA-RNA I). The VA-RNA I gene, which specifies an RNA product that ameliorates the effects of a host cell defense mechanism, contains a typical intragenic promoter that has been widely used in studies of the mechanism by which RNA polymerase III initiates transcription.

**Table 8.7**  Viral RNA polymerase III transcription units

| Virus | RNA polymerase III transcript | Function |
|---|---|---|
| Adenovirus | | |
|   Human adenovirus type 2 | VA-RNA I | Prevention of activation of RNA-dependent protein kinase |
| | VA-RNA II | Not known |
| Herpesviruses | | |
|   Epstein-Barr virus | EBER 1, EBER 2 | Not known |
|   Herpesvirus saimiri | HVSR 1–5 | Degradation of certain cellular mRNAs? |
| Retrovirus | | |
|   Moloney murine leukemia virus | Let | Stimulation of transcription of specific cellular genes (e.g., *CD4* and major histocompatibility complex class I genes) |

## Transcription of the Adenoviral VA-RNA Genes

### The VA-RNA I Promoter

The human adenovirus type 2 genome contains two VA-RNA genes located very close to one another in the viral chromosome (appendix, Fig. 15B). The VA-RNA I promoter is described here, for it has been more thoroughly characterized than the closely related VA-RNA II promoter. Transcription of the VA-RNA I gene depends on two intragenic sequences, the A and B boxes (Fig. 8.27A). Sequences upstream of sites of initiation are not necessary for promoter activity but do modulate the specificity or efficiency of initiation. As in the RNA polymerase II system, the essential promoter sequences are binding sites for accessory proteins necessary for promoter recognition by the enzyme. The internal sequences are specifically recognized

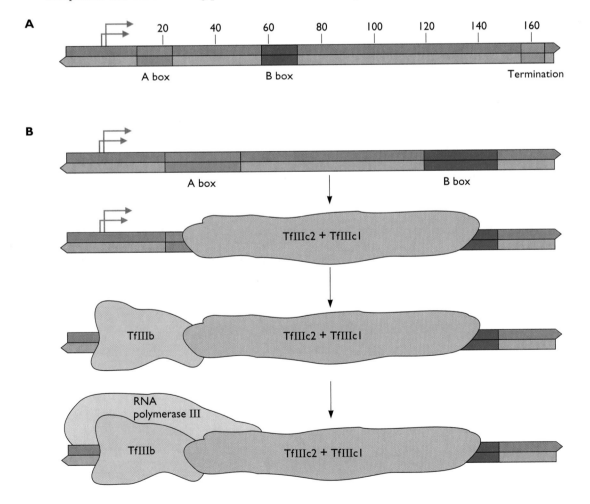

**Figure 8.27 Organization and recognition of the human adenovirus type 2 VA-RNA I RNA polymerase III promoter.** (A) The VA-RNA I gene is depicted to scale. The intragenic A and B box sequences are essential for efficient VA-RNA I transcription by RNA polymerase III. The sequences of the A and B boxes of the VA-RNA gene are closely related to the consensus A and B sequences of cellular tRNA genes. The VA-RNA termination site sequences are also typical of those of cellular genes transcribed by RNA polymerase III. Numbers at the top are base pairs. (B) Assembly of an RNA polymerase III initiation complex on the VA-RNA I promoter is initiated by binding of TfIIIc2 to the intragenic promoter sequences. This reaction is stimulated by TfIIIc1, which makes contact with both the A box and sequences spanning the termination site. The TfIIIc-DNA complex is recognized by TfIIIb, which in turn allows recruitment of RNA polymerase III to the promoter. Although TfIIIb does not possess sequence-specific DNA-binding activity, it does contact DNA in the vicinity of the site of initiation. All mammalian RNA polymerase III initiation proteins are large, multiprotein complexes. The TATA-binding protein is an essential component of TfIIIb.

by the RNA polymerase III-specific initiation protein TfIIIc, whose binding to the promoter begins assembly of an initiation complex that also contains TfIIIb and RNA polymerase III (Fig. 8.27B). The pathway of initiation of VA-RNA I transcription by RNA polymerase III shown in Fig. 8.27B was elucidated by in vitro assays. We can be reasonably confident that this same pathway operates in adenovirus-infected cells, for there is excellent agreement between the effects of A and B box mutations on VA-RNA I transcription by RNA polymerase III in vitro and on VA-RNA I synthesis in mutant virus-infected cells.

### Regulation of VA-RNA Gene Transcription During Adenovirus Infection

The two adenoviral VA-RNA genes are initially transcribed at similar rates, but during the late phase of infection VA-RNA I, but not VA-RNA II, is produced at a greatly accelerated rate. Such preferential transcription of the VA-RNA I gene is the result of competition of its promoter with the intrinsically much weaker transcriptional control region of the VA-RNA II gene for a limiting component of the RNA polymerase III transcriptional machinery. Repression of VA-RNA I transcription may account for the similar rates at which the two genes are transcribed during the early phase. The control of transcription of VA-RNA genes during the adenoviral infectious cycle emphasizes the fact that transcription by RNA polymerase III can, and must, be regulated, although the mechanisms are probably less elaborate than those that govern transcription by RNA polymerase II. The properties of other viral RNA polymerase III transcription units illustrate the kinship of the RNA polymerase II and III systems (Fig. 8.28).

### Virus-Specific Applications of Cellular Transcription Components

In the preceding sections, we concentrated on the close relationships among the mechanisms by which viral DNA is transcribed in cells infected by a wide range of viruses to their counterparts responsible for transcription of cellular genes. Even though every mechanism by which these viruses regulate expression of their genes by the host cell's RNA polymerase II or RNA polymerase III cannot be described in molecular detail, it seems clear that the majority are not unique to viral systems. In most cases, viral regulatory mechanisms not only rely on cellular components but also mimic those of their host cells. Such a resemblance is to be expected, for this very dependence on the host's transcriptional machinery constrains the structural and functional properties of virus-encoded molecules that participate in the regulation of transcription. It is thus an axiom of molecular virology that every mechanism by which viral transcription units are expressed, or by which their activity is regulated, will prove directly analogous to a normal cellular mechanism. However, this is not to say that viruses cannot evolve novel virus-specific uses of cellular transcription components.

One example of the latter kind of mechanism is the transcription of hepatitis delta satellite virus RNA from an *RNA* template by RNA polymerase II, described in chapter 6. Such RNA-dependent RNA synthesis by RNA polymerase II from a specific RNA promoter is one of the most remarkable interactions of a virus with the cellular transcriptional machinery. No cellular analog of this reaction is yet known. Although RNA polymerase II copies a novel template, hepatitis delta satellite virus RNA, it is nevertheless still carrying out transcription. More divergent applications of cellular transcriptional components by viruses are illustrated by the participation of a human chromosome remodeling complex protein (a coactivator) in the integration of human immunodeficiency virus type 1 proviral DNA and of the RNA polymerase III initiation proteins TfIIIb and TfIIIc in integration of the yeast retroid element Ty3 (see chapter 7). Given the large repertoire of molecular and biochemical activities displayed by components of the eukaryotic transcriptional machinery, it seems

**Figure 8.28 Some viral RNA polymerase III genes contain both external and internal promoter elements.** The 5' end of the Epstein-Barr virus EBER-2 transcription unit is shown to scale. This gene, which is transcribed by RNA polymerase III in Epstein-Barr virus-infected cells, contains intragenic A and B box sequences, which resemble those of other genes transcribed by this enzyme in both their sequences and their positions. However, efficient transcription by RNA polymerase III also depends on the 5'-flanking sequence, which includes binding sites for the RNA polymerase II stimulatory proteins Sp1 and Atf. It has not been determined whether these proteins stimulate RNA polymerase III transcription from the EBER-2 promoter in infected cells. The TATA-like sequence is important for efficient transcription and essential for specifying transcription by RNA polymerase III.

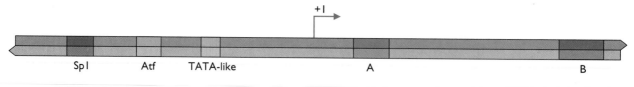

likely that viruses will prove to have evolved other unusual uses for these cellular proteins.

## Inhibition of the Cellular Transcription Machinery in Virus-Infected Cells

Inhibition of cellular transcription in virus-infected cells offers several advantages to a virus. Cellular resources, such as substrates for RNA synthesis, can be devoted to the exclusive production of viral mRNA (and in many cases, genome RNA), and competition from cellular mRNAs during translation is reduced. The retroviruses and most DNA viruses, which depend absolutely on cellular transcriptional systems for expression of genetic information, cannot inactivate this machinery. Nevertheless, the larger DNA viruses achieve selective expression of viral genes. Herpes simplex virus type 1 infection induces inhibition of transcription of the majority of cellular genes by RNA polymerase II. Such selective transcription of viral genes depends on the expression of multiple viral immediate-early genes, including ICP4 and ICP27, and has been correlated with loss from infected cells of RNA polymerase II in which the C-terminal domain of the largest subunit is hyperphosphorylated. However, the mechanism by which transcription of cellular genes is inhibited is not yet known. The poxviruses, whose genomes encode a viral transcription system, induce rapid inhibition of synthesis of all classes of cellular RNA in infected cells. Such inhibition requires viral proteins, but these have not been identified. The redistribution of RNA polymerase II from the nucleus to the cytoplasm observed in vaccinia virus-infected cells undoubtedly contributes to inhibition of transcription by this enzyme. However, RNA polymerase II transcription must also be inactivated by more direct mechanisms because whole-cell extracts prepared from vaccinia virus-infected cells do not support transcription in vitro.

The majority of viruses with RNA genomes also inhibit cellular mRNA synthesis in infected cells, for they require neither the cellular transcriptional machinery nor its RNA products. Among the best-characterized examples is the inhibition of transcription by RNA polymerase II characteristic of poliovirus-infected cells. The poliovirus protease 3C$^{pro}$ is both necessary and sufficient for such inactivation of this cellular transcription system. The defect in RNA polymerase II transcription in extracts of poliovirus-infected cells can be explained by the fact that 3C$^{pro}$ cleaves the Tbp subunit of TfIId at several specific sites. This modification eliminates the DNA-binding activity of Tbp and hence the ability of TfIId to support initiation of transcription by RNA polymerase II. As TfIId is required for initiation of transcription from all promoters (Box 8.2), its inactivation is likely to be primarily responsible for the in-

hibition of RNA polymerase II transcription observed in poliovirus-infected cells.

The TATA-binding protein is also a subunit of initiation proteins that function with RNA polymerase III (TfIIIb) and RNA polymerase I. Apart from some subunits shared by the three RNA polymerases, it is the only protein known to be common to the three cellular transcription systems. Its cleavage by 3C$^{pro}$ in poliovirus-infected cells therefore appears to be a very efficient way to inhibit all cellular transcriptional activity. Indeed, TfIIIb activity is reduced in poliovirus-infected cells, as is that of TfIIIc2 (Fig. 8.28B), whose largest subunit is also cleaved by the polioviral protease.

Two gene products of the rhabdovirus vesicular stomatitis virus have been implicated in the inhibition of RNA polymerase II transcription, emphasizing the advantages associated with disruption of cellular mRNA synthesis. The first is the viral leader RNA described in chapter 6, which inhibits transcription by both RNA polymerase II and RNA polymerase III in vitro and is primarily responsible for the rapid inhibition of cellular RNA synthesis in vesicular stomatitis virus-infected cells. Following synthesis in the cytoplasm, the leader RNA enters the nucleus. The question of how short RNA molecules inhibit DNA-dependent RNA transcription cannot yet be answered, although in vitro experiments suggest that binding of a cellular protein to specific sequences within the RNA may be important. The viral M protein is also a potent inhibitor of transcription by RNA polymerase II, even in the absence of other viral gene products. This activity may become important later in infection, when replication of genome RNA predominates over transcription such that less leader RNA is produced.

## Perspectives

Because the majority of viruses that rely on synthesis of mRNA or pre-mRNA from a DNA template depend on the cellular RNA polymerase II transcription system, their study has yielded pioneering insights into the molecular biology of eukaryotic transcription. Indeed, it is difficult to exaggerate the importance of viral systems in the elucidation of mechanisms of transcription and its regulation in eukaryotic cells. The organization of RNA polymerase II promoters considered archetypal was first described for viral transcriptional control regions; enhancers, which operate in a manner that had no precedent when these regulatory sequences were identified, were first discovered in viral genomes, and many important cellular regulators of transcription (e.g., Table 8.3) were discovered by virtue of their specific binding to viral transcriptional control regions. Perhaps even more importantly, efforts to elucidate the molecular basis of transcriptional regulatory circuits crucial to viral infectious cycles have established general principles of eukaryotic transcriptional regulation. These include the importance of proteins that do not recognize transcriptional control regions directly and the ability of a single activating protein to act by means of multiple components of the transcriptional machinery and to affect several reactions in the transcription cycle. The recent insights into regulation of elongation by RNA polymerase II gained from investigation of the virus-specific, RNA-dependent activation of human immunodeficiency virus type 1 transcription by Tat emphasize the intimate relationship of viral regulatory proteins with cellular transcriptional components that make viral systems such rich resources for the investigation of eukaryotic transcription.

The identification of cellular and viral proteins necessary for transcription of specific viral transcription units has allowed many circuits that control viral gene expression to be traced. For example, the tissue distribution, or the regulation of the availability or activity, of particular cellular activators that bind to specific viral DNA sequences can neatly account for the tropism of individual viruses or conditions under which different transcriptional programs (e.g., latent or lytic) can be established. Furthermore, the mechanisms that allow sequential expression of viral genes are quite well established. Regardless of whether regulatory circuits are constructed of largely cellular (most viruses with DNA genomes) or mostly viral (poxviruses) proteins, these transcriptional cascades share such mechanistic features as sequential production of viral activators in infected cells and integration of transcription of late genes encoding structural proteins with synthesis of viral DNA in the infected cell. Our ability to describe these regulatory circuits in some detail represents enormous progress. Nevertheless, it is important to appreciate that greater challenges lie ahead.

In very few cases have the models for individual regulatory processes crucial in viral infectious cycles that were developed in convenient and powerful experimental systems actually been tested in virus-infected cells. This dearth must be remedied if we are to gain a better understanding of the regulation of transcription of viral DNA templates. It is clear that such experimental systems (e.g., in vitro transcription reactions and transient-expression assays) do not reproduce all features characteristic of infected cells, even in the simplest cases, and cannot address such issues of how transcription of specific genes can be coupled with replication of the viral genome. It is more difficult to elucidate the molecular functions and mechanisms of action of transcriptional components in vivo. Many viral regulatory proteins perform multiple functions, a property that can confound genetic analysis. Moreover, the analysis of individual reactions, such as

binding of a protein to a specific promoter sequence, formation of preinitiation complexes, or elongation, is technically demanding. Nevertheless, viral cis-acting transcriptional control region sequences and regulatory proteins remain far more amenable to genetic analyses of their function in the natural context than their cellular counterparts. In conjunction with the increasingly powerful and sensitive molecular methods for examining intracellular processes now available, continued efforts to exploit such genetic manipulability will eventually establish how transcription of viral DNA templates is mediated and regulated within infected cells. Such information will not only address outstanding virological questions but should also allow current models of the fundamental cellular process of transcription by RNA polymerase II to be assessed and refined.

# References

## Books
McKnight, S. L., and K. Yamamoto (ed.). 1992. *Transcriptional Regulation.* Cold Spring Harbor Laboratory Press, Cold Spring Harbor, N.Y.

## Book Chapters
Cole, C. N. 1996. Polyomavirinae: the viruses and their replication, p. 917–946. *In* B. N. Fields, D. M. Knipe, P. M. Howley, R. M. Chanock, J. L. Melnick, T. P. Monath, B. Roizman, and S. E. Straus (ed.), *Fundamental Virology.* Lippincott-Raven, Philadelphia, Pa.

Kieff, E. 1996. Epstein-Barr virus and its replication, p. 1109–1162. *In* B. N. Fields, D. M. Knipe, P. M. Howley, R. M. Chanock, J. L. Melnick, T. P. Monath, B. Roizman, and S. E. Straus (ed.), *Fundamental Virology.* Lippincott-Raven, Philadelphia, Pa.

Roizman, B., and A. E. Sears. 1996. Herpes simplex viruses and their replication, p. 1043–1108. *In* B. N. Fields, D. M. Knipe, P. M. Howley, R. M. Chanock, J. L. Melnick, T. P. Monath, B. Roizman, and S. E. Straus (ed.), *Fundamental Virology.* Lippincott-Raven, Philadelphia, Pa.

Shenk, T. 1996. Adenoviridae: the viruses and their replication, p. 979–1016. *In* B. N. Fields, D. M. Knipe, P. M. Howley, R. M. Chanock, J. L. Melnick, T. P. Monath, B. Roizman, and S. E. Straus (ed.), *Fundamental Virology.* Lippincott-Raven, Philadelphia, Pa.

## Reviews
Block, T. M., and J. M. Hill. 1997. The latency-associated transcripts (LAT) of herpes simplex virus: still no end in sight. *J. Neurovirol.* **3:**313–321.

Flint, J., and T. Shenk. 1997. Viral transactivating proteins. *Annu. Rev. Genet.* **31:**177–212.

Gait, M. J., and J. Karn. 1993. RNA recognition by the human immunodeficiency virus Tat and Rev proteins. *Trends Biochem. Sci.* **18:**255–259.

Greenblatt, J. 1997. RNA polymerase II holoenzyme and transcriptional activation. *Curr. Opin. Cell Biol.* **9:**310–319.

Hiscott, J., L. Petropoulos, and J. Lacoste. 1995. Molecular interactions between HTLV-1 Tax protein and the Nf-κB/IκB complex. *Virology* **214:**3–11.

Jones, K. A. 1997. Taking a new TAK on Tat transactivation. *Genes Dev.* **11:**2593–2599.

Kadonaga, J. T. 1998. Eukaryotic transcription: an interlaced network of transcription factors and chromatin-modifying machines. *Cell* **92:**307–313.

Khoury, G., and P. Grüss. 1983. Enhancer elements. *Cell* **33:**313–314.

Maniatis, T., S. Goodbourn, and J. F. Fischer. 1987. Regulation of inducible and tissue-specific gene expression. *Science* **236:**1237–1245.

Moss, B., B.-Y. Ahn, B. Amegadzie, P.-D. Gershon, and J. G. Keck. 1991. Cytoplasmic transcription system encoded by vaccinia virus. *J. Biol. Chem.* **266:**1355–1358.

O'Hare, P. 1993. The virion transactivator of herpes simplex virus. *Semin. Virol.* **4:**145–155.

Pabo, C. O., and R. T. Sauer. 1992. Transcription factors: structural families and principles of DNA recognition. *Annu. Rev. Biochem.* **61:**1053–1095.

Reinberg, D., and L. Zawel. 1995. Common themes in assembly and function of eukaryotic transcription complexes. *Annu. Rev. Biochem.* **64:**533–561.

Roeder, R. G. 1996. The role of general initiation factors in transcription by RNA polymerase II. *Trends Biochem. Sci.* **21:**327–335.

Ruddell, A. 1995. Transcription regulatory elements of the avian retroviral long terminal repeat. *Virology* **206:**1–7.

Wade, P. A., D. Pruss, and A. P. Wolffe. 1997. Histone acetylation: histones in action. *Trends Biochem. Sci.* **22:**128–132.

## *Papers of Special Interest*

### Viral RNA Polymerase II Promoters and the Cellular Transcriptional Machinery

Buratowski, S., S. Hahn, L. Guanrente, and P. A. Sharp. 1989. Five intermediate complexes in transcription initiation by RNA polymerase II. *Cell* **56:**549–561.

Dynan, W. S., and R. Tjian. 1983. The promoter-specific transcription factor Sp1 binds to upstream sequences in the SV40 early promoter. *Cell* **35:**79–87.

Grüss, P., R. Dhar, and G. Khoury. 1981. Simian virus 40 tandem repeated sequences as an element of the early promoter. *Proc. Natl. Acad. Sci. USA* **78:**943–947.

Hu, S. L., and J. L. Manley. 1981. DNA sequences required for initiation of transcription *in vitro* from the major late promoter of adenovirus 2. *Proc. Natl. Acad. Sci. USA* **78:**820–824.

Matsui, T., J. Segall, P. A. Weil, and R. G. Roeder. 1980. Multiple factors required for accurate initiation of transcription by purified RNA polymerase II. *J. Biol. Chem.* **255:**11992–11996.

Weil, P. A., D. S. Luse, J. Segall, and R. G. Roeder. 1979. Selective and accurate initiation of transcription at the Ad2 major late promoter in a soluble system dependent on purified RNA polymerase II and DNA. *Cell* **18:**469–484.

Yamamoto, T., B. de Crombrugghe, and I. Pastan. 1980. Identification of a functional promoter in the long terminal repeat of Rous sarcoma virus. *Cell* **22:**787–797.

### Viral Transcriptional Enhancers

Banerji, J., S. Rusconi, and W. Schaffner. 1981. Expression of the α-globin gene is enhanced by remote SV40 DNA sequences. *Cell* **27:**299–308.

Fromental, C., M. Kanno, H. Nomiyama, and P. Chambon. 1988. Cooperativity and hierarchical levels of functional organization in the SV40 enhancer. *Cell* **54:**943–953.

Moreau, P., R. Hen, B. Wasylyk, R. Everett, M. P. Gaub, and P. Chambon. 1981. The SV40 72 base repeat has a striking effect on gene expression both in SV40 and other chimeric recombinants. *Nucleic Acids Res.* **9:**6047–6068.

Müller, H. P., J. M. Sogo, and W. Schaffner. 1989. An enhancer stimulates transcription in *trans* when attached to the promoter via a protein bridge. *Cell* **58:**767–777.

Nabel, G. J., and D. Baltimore. 1987. An inducible transcription factor that activates expression of human immunodeficiency virus in T cells. *Nature* (London) **326:**711–713.

Ross, E. K., A. J. Buckler-White, A. B. Rabson, G. Englund, and M. A. Martin. 1991. Contributions of Nf-κB and Sp1 binding motifs to the replicative capacity of human immunodeficiency virus type 1: distinct patterns of viral growth are determined by T-cell types. *J. Virol.* **65:**4350–4358.

Spalholz, P. A., Y.-C. Yang, and P. M. Howley. 1985. Transactivation of a bovine papillomavirus transcriptional regulatory element by the E2 gene product. *Cell* **42:**183–191.

Yee, J. 1989. A liver-specific enhancer in the core promoter of human hepatitis B virus. *Science* **246:**658–670.

### Regulation of Transcription of the Genomes of DNA Viruses by Viral Proteins

Ace, C. I., T. A. McKee, M. Ryan, J. C. Cameron, and C. M. Preston. 1989. Construction and characterization of a herpes simplex virus type 1 mutant unable to transinduce immediate-early gene expression. *J. Virol.* **63:**2260–2269.

Bandara, L. R., and N. B. La Thangue. 1991. Adenovirus E1A prevents the retinoblastoma gene product from complexing with a cellular transcription factor. *Nature* (London) **351:**494–497.

Batterson, W., and B. Roizman. 1983. Characterization of the herpes simplex virion-associated factor responsible for the induction of alpha genes. *J. Virol.* **46:**371–377.

Berk, A. J., F. Lee, T. Harrison, J. F. Williams, and P. A. Sharp. 1979. Pre-early adenovirus 5 gene product regulates synthesis of early viral messenger RNAs. *Cell* **17:**935–944.

Chellappan, S. P., S. Hiebert, M. Mudryj, J. M. Horowitz, and J. R. Nevins. 1991. The E2F transcription factor is a cellular target for the Rb protein. *Cell* **65:**1053–1061.

Chi, T., and M. Carey. 1993. The ZEBRA activation domain: modular organization and mechanism of action. *Mol. Cell. Biol.* **13:**7045–7055.

Gerster, T., and R. G. Roeder. 1988. A herpes virus trans-activating protein interacts with transcription factor Oct-1 and other cellular proteins. *Proc. Natl. Acad. Sci. USA* **85:**6347–6351.

Hardy, S., D. Engle, and T. Shenk. 1989. An adenovirus early region 4 gene product is required for induction of the infection-specific form of cellular E2F activity. *Genes Dev.* **3:**1062–1074.

Jones, N., and T. Shenk. 1979. An adenovirus type 5 early gene function regulates expression of other early viral genes. *Proc. Natl. Acad. Sci. USA* **76:**3665–3669.

Keller, J. M., and J. C. Alwine. 1984. Activation of the SV40 late promoter: direct effects of T antigen in the absence of viral DNA replication. *Cell* **36:**381–389.

Lai, J.-S., M. A. Cleary, and W. Herr. 1992. A single amino acid exchange transfers VP16-induced positive control from the Oct-1 to the Oct-2 homeo domain. *Genes Dev.* **6:**2058–2065.

Lee, W. S., C. C. Kao, G. O. Bryant, X. Liu, and A. J. Berk. 1991. Adenovirus E1A activation domain binds the basic repeat in the TATA box transcription factor. *Cell* **67:**365–376.

Lillie, J. W., and M. R. Green. 1989. Transcription activation by the adenovirus E1A protein. *Nature* (London) **338:**39–44.

Mador, N., D. Goldenberg, O. Cohen, A. Panet, and I. Steiner. 1998. Herpes simplex virus type 1 latency-associated transcripts suppress viral replication and reduce immediate-early gene mRNA levels in a neuronal cell line. *J. Virol.* **72:**5067–5075.

Pereira, D. J., D. M. McCarty, and N. Muzyczka. 1997. The adeno-associated virus (AAV) Rep protein acts as both a repressor and an activator to regulate AAV transcription during a productive infection. *J. Virol.* **71:**1079–1088.

Tribouley, C., P. Lutz, A. Staub, and C. Kédinger. 1994. The product of the adenovirus intermediate gene IVa$_2$ is a transcriptional activator of the major late promoter. *J. Virol.* **68:**4450–4457.

Vales, L. D., and J. E. Darnell. 1989. Promoter occlusion prevents transcription of adenovirus polypeptide IX mRNA until after DNA replication. *Genes Dev.* **3:**49–59.

Walker, S., S. Hayes, and P. O'Hare. 1994. Site-specific conformational alteration of the Oct-1 POU domain-DNA complex as the basis for differential recognition by Vmw65 (VP16). *Cell* **79:**841–852.

Wiley, S. R., R. J. Kraus, F. Zuo, E. E. Murray, K. Loritz, and J. E. Mertz. 1993. SV40 early-to-late switch involves titration of cellular transcriptional repressors. *Genes Dev.* **7:**2206–2219.

Wilson, A. C., R. W. Freeman, H. Goto, T. Nishimoto, and W. Herr. 1997. VP16 targets an amino-terminal domain of HCF involved in cell cycle progression. *Mol. Cell. Biol.* **17:**6139–6146.

## RNA-Dependent Stimulation of Human Immunodeficiency Virus Type 1 Transcription by the Tat Protein

Dingwall, C., J. Ernberg, M. J. Gait, S. M. Green, S. Heaphy, J. Karn, M. Singh, and J. J. Shinner. 1990. HIV-1 Tat protein stimulates transcription by binding to a U-rich bulge in the stem of the TAR RNA structure. *EMBO J.* **9:**4145–4153.

Feinberg, M. B., D. Baltimore, and A. D. Frankel. 1991. The role of Tat in the human immunodeficiency virus life cycle indicates a primary effect on transcriptional elongation. *Proc. Natl. Acad. Sci. USA* **88:**4045–4049.

Fisher, A. G., M. B. Feinberg, S. F. Josephs, M. E. Harper, L. M. Marselle, G. Reyes, M. A. Gonda, A. Aldovini, C. Debouk, R. C. Gallo, and F. Wong-Staal. 1986. The *trans*-activator gene of HTLV-III is essential for virus replication. *Nature* (London) **320:**367–371.

Kao, S., A. F. Calman, P. A. Luciw, and B. M. Peterlin. 1987. Anti-termination of transcription within the long terminal repeat of HIV-1 by the Tat gene product. *Nature* (London) **330:**489–493.

Mancebo, H. S. Y., G. Lee, J. Flygare, J. Tomassini, P. Luu, Y. Zhu, J. Peng, C. Blau, D. Hazuda, D. Price, and O. Flores. 1997. P-TEFb kinase is required for HIV Tat transcriptional activation *in vivo* and *in vitro*. *Genes Dev.* **11:**2633–2644.

Parada, C. A., and R. G. Roeder. 1996. Enhanced processivity of RNA polymerase II triggered by Tat-induced phosphorylation of its carboxy-terminal domain. *Nature* (London) **384:**375–378.

Rosen, C. A., J. G. Sodroski, and W. A. Haseltine. 1985. The location of *cis*-acting regulatory sequences in the human T-cell lymphotropic virus type III (HTLV-III/LAV) long terminal repeat. *Cell* **41:**813–823.

Wei, P., M. F. Garber, S.-M. Fang, W. H. Fischer, and K. A. Jones. 1998. A novel CDK9-associated C-type cyclin interacts directly with HIV-1 Tat and mediates its high-affinity, loop-specific binding to TAR RNA. *Cell* **92:**451–462.

## The Poxviral Transcriptional System

Hagler, J., and S. Shuman. 1992. A freeze-frame view of eukaryotic transcription during elongation and capping of nascent RNA. *Science* **255:**983–986.

Kates, J. R., and B. R. McAuslan. 1967. Poxvirus DNA-dependent RNA polymerase. *Proc. Natl. Acad. Sci. USA* **58:**134–141.

Lu, J., and S. S. Broyles. 1993. Recruitment of vaccinia virus RNA polymerase to an early gene promoter by the viral early transcription factor. *J. Biol. Chem.* **268:**2773–2780.

Rosales, R., G. Sutter, and B. Moss. 1994. A cellular factor is required for transcription of vaccinia viral intermediate-stage genes. *Proc. Natl. Acad. Sci. USA* **91:**3794–3798.

## Transcription of Viral Genes by RNA Polymerase III

Fowlkes, D. M., and T. Shenk. 1980. Transcriptional control regions of the adenovirus VA1 RNA gene. *Cell* **22:**405–413.

Hoeffler, W. K., R. Kovelman, and R. G. Roeder. 1988. Activation of transcription factor IIIC by the adenovirus E1A protein. *Cell* **53:**907–920.

Howe, E., and M.-D. Shu. 1993. Upstream basal promoter element important for exclusive RNA polymerase III transcription of EBER2 gene. *Mol. Cell. Biol.* **13:**2656–2665.

Huang, W., R. Pruzan, and S. J. Flint. 1994. *In vivo* transcription from the adenovirus E2 early promoter by RNA polymerase III. *Proc. Natl. Acad. Sci. USA* **91:**1265–1269.

$\mathcal{9}$

---

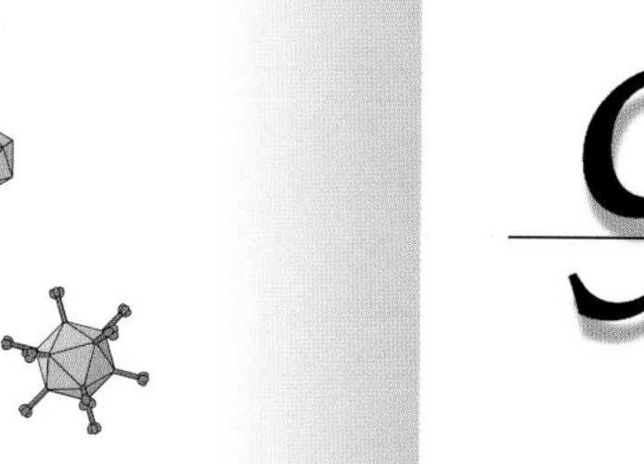

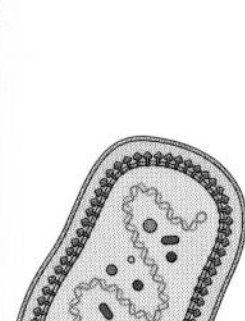

# Genome Replication Strategies: DNA Viruses

## Introduction

The genomes of DNA viruses come in a considerable variety of sizes and shapes, from small molecules of single-stranded DNA to large double-stranded molecules that may be linear or circular (Table 9.1). Whatever their physical nature, viral DNA molecules must be replicated efficiently within an infected cell to provide genomes for assembly into progeny virions. This important step in DNA virus reproduction invariably requires the synthesis of at least one viral protein and often expression of several viral genes. Consequently, viral DNA synthesis cannot begin immediately on arrival of the genome at the appropriate intracellular site but rather is delayed until viral replication proteins have attained a sufficient concentration within an infected cell. Initiation of viral DNA synthesis typically leads to many cycles of replication and the accumulation of very large numbers of newly synthesized viral DNA molecules in the infected cell. However, more long-lasting virus-host cell interactions, such as latent infections, are also common both in nature and in the laboratory. In these circumstances, both the number of viral DNA molecules made in an infected cell and the timing of their replication are strictly controlled.

Replication of DNA in all organisms, from the simplest virus to the most complex vertebrate cell, follows a set of universal rules: (1) DNA is always synthesized by template-directed, stepwise incorporation of deoxynucleoside monophosphates (dNMPs) from deoxynucleoside triphosphate (dNTP) substrates into the 3' OH end of the growing DNA chain, which is elongated in the 5' → 3' direction; (2) each parental strand of a duplex DNA template is copied by base pairing to produce two daughter molecules identical to one another and to their parent (**semiconservative replication**); (3) replication of DNA begins and ends at specific sites in the template, termed **origins** and **termini**, respectively; and (4) DNA synthesis is catalyzed by DNA-dependent DNA polymerases, but many accessory proteins are required for initiation or elongation. In contrast to all DNA-dependent and many RNA-dependent RNA polymerases,

**Table 9.1**    Cellular and viral proteins used in the synthesis of viral DNA

| Virus | Genome | Viral origin-binding protein(s) | DNA polymerase(s) | Other replication proteins | Effects on cellular DNA synthesis |
|---|---|---|---|---|---|
| **Parvovirus** | | | | | |
| Adeno-associated virus type 2 | 4,680 bp, linear, single (−) or (+) strands | None | Cellular | Many cellular; viral Rep 68/78, a site-specific endonuclease and ATP-dependent helicase | None |
| **Papovavirus** | | | | | |
| Simian virus 40 | 5,243 bp, closed, double-stranded circle | LT | Cellular, DNA polymerases α and γ | Many cellular; no viral, but LT supplies ATP-dependent helicase activity | Induction of entry into S phase and cellular DNA synthesis |
| **Adenovirus** | | | | | |
| Human adenovirus type 2 | 35,937 bp, linear, double-stranded | DNA polymerase-preterminal protein complex | Viral | Cellular transcription proteins (Nf-1, Oct-1) and topoisomerase; viral preterminal protein primer; viral single-stranded DNA-binding protein | Induction of entry into S phase in quiescent cells; inhibition of cellular DNA synthesis in actively growing cells |
| **Herpesviruses** | | | | | |
| Herpes simplex virus type 1 (HSV-1) | ~150 kbp, linear, double-stranded, 4 isomers | UL9 | Viral | Cellular protein requirement not known; at least 7 viral, including a primase complex | Inhibition of cellular DNA synthesis |
| Epstein-Barr virus | ~172 kbp, linear, double-stranded | | | | |
| OriP | | EBNA-1 | Cellular | Many cellular; no viral | Immortalization of latently infected B cells |
| OriLyt | | Zta | Viral | Several viral; analogous to HSV-1 replication proteins | Induction of cell cycle arrest and inhibition of cellular DNA synthesis by Zta |
| **Poxvirus** | | | | | |
| Vaccinia virus | ~200 kbp, linear, double-stranded with closed terminal loops | None | Viral | Several viral; no cellular | Inhibition of cellular DNA synthesis |

no DNA polymerase can initiate template-directed DNA synthesis de novo. All require a **primer** with a free 3′ OH end to which dNMPs complementary to those of the template strand are added. As discussed below, this property of DNA polymerases has profound consequences for the mechanisms of DNA synthesis.

Viruses with RNA genomes must encode enzymes that catalyze RNA-dependent RNA or DNA synthesis. In contrast, the genomes of DNA viruses can be replicated by the cellular machinery. Indeed, the smaller DNA viruses, such as parvo- and papovaviruses (Table 9.1), contribute only a single replication protein, and the great majority of reactions necessary for synthesis of viral DNA are carried out by cellular replication proteins. This strategy avoids the need to devote limited viral genetic coding capacity to enzymes and other proteins needed for DNA synthesis. However, the genomes of all larger DNA viruses encode DNA polymerases and additional replication proteins (Table 9.1). In the extreme case, exemplified by poxviruses, the virus supplies a complete DNA synthesis system and replicates in the cytoplasm of its host cells.

There is also considerable variety in the mechanism of priming of viral DNA synthesis. In some cases, DNA synthesis is initiated with RNA primers, the mechanism used during replication of cellular genomes, but in others, unusual structural features of the genome or a viral protein serve as primers. Despite such distinctions, the replication mechanisms employed by different viruses are based on common molecular principles. For example, the genomes of papovaviruses and herpesviruses, which are quite different in size and structure (Table 9.1), are replicated by the cellular replication machinery and viral replication proteins, respectively. Nevertheless, these two types of virus use the same priming mechanism, and the virally encoded replication machinery of herpesviruses carries out the same biochemical reactions as host proteins exploited by papovaviruses such as simian virus 40.

## DNA Synthesis by the Cellular Replication Machinery: Lessons from Simian Virus 40

Our current understanding of the intricate reactions by which both strands of a typical double-stranded DNA template are completely copied during replication in a eukaryotic cell is largely based on detailed studies of simian virus 40 DNA synthesis in in vitro reactions. In this section, we focus on the cellular replication machinery and the molecular functions of its components established by using this important viral system. However, we first briefly discuss general features of eukaryotic DNA replication and

why simian virus 40 proved to be a priceless resource for those seeking to understand this process.

## Eukaryotic Replicons

### General Features

The complete replication of large eukaryotic genomes within the lifetime of an actively growing cell depends on their division into smaller units of replication termed **replicons** (Fig. 9.1): at the maximal rate of replication observed in vivo, a typical human chromosome could not be replicated as a single unit in less than 10 days! Each eukaryotic chromosome therefore contains many replicons, ranging in length from about 20 to 300 kbp. No viral DNA genome is larger than the upper limit reported for such cellular replicons. Nevertheless, all but the smallest viral DNA molecules are replicated from two or three origins (see "Viral Replication Origins and Their Recognition" below).

Each replicon contains a single origin at which replication begins. The sites at which nascent DNA chains are being synthesized, the ends of "bubbles" seen in the electron microscope (Fig. 9.1A), are termed **replication forks**. In **bidirectional replication**, two replication forks

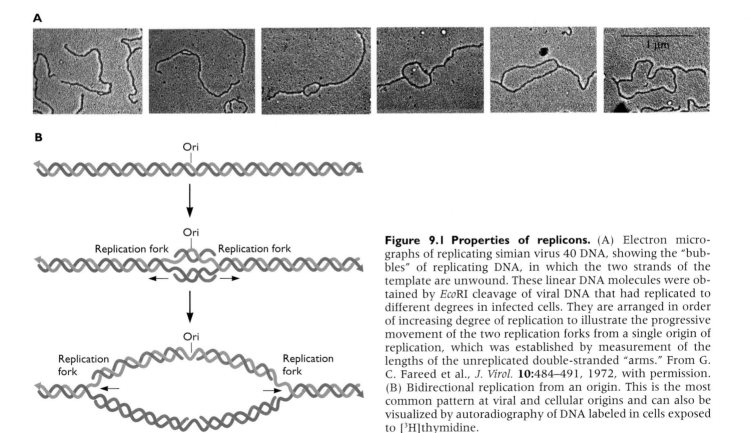

**Figure 9.1 Properties of replicons.** (A) Electron micrographs of replicating simian virus 40 DNA, showing the "bubbles" of replicating DNA, in which the two strands of the template are unwound. These linear DNA molecules were obtained by *Eco*RI cleavage of viral DNA that had replicated to different degrees in infected cells. They are arranged in order of increasing degree of replication to illustrate the progressive movement of the two replication forks from a single origin of replication, which was established by measurement of the lengths of the unreplicated double-stranded "arms." From G. C. Fareed et al., *J. Virol.* **10:**484–491, 1972, with permission. (B) Bidirectional replication from an origin. This is the most common pattern at viral and cellular origins and can also be visualized by autoradiography of DNA labeled in cells exposed to [³H]thymidine.

**A**

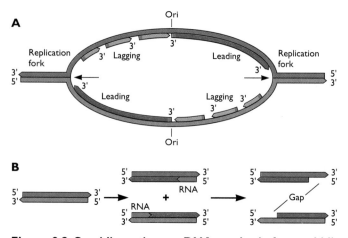

**B**

**Figure 9.2 Semidiscontinuous DNA synthesis from a bidi-rectional origin.** (A) Semidiscontinuous synthesis of the daughter strands. Synthesis of the RNA primers (green) at the origin allows initiation of continuous copying of one of the two strands on either side of the origin in the replication bubble. The second strand cannot be made in the same way because DNA can be synthesized only in the $5' \rightarrow 3'$ direction. The nascent DNA population, identified by pulse-labeling with [³H]thymidine, contains many small molecules termed **Okazaki fragments** in honor of the investigator who first described them. The presence of short segments of RNA at the 5' ends of Okazaki fragments indicated that the primers necessary for DNA synthesis are molecules of RNA. With increasing time of replication, these small fragments are incorporated into long DNA molecules, indicating that they are precursors. It was therefore deduced that the second nascent DNA strand is synthesized discontinuously, also in the $5' \rightarrow 3'$ direction. Because synthesis of this strand cannot begin until the replication fork has moved some distance from the origin, it is called the **lagging strand**, while the strand synthesized continuously is termed the **leading strand**. Because the lagging strand is initially made as short pieces, its complete replication requires not only an enzyme to make the RNA primer (a **primase**) and a DNA polymerase but also enzymes that can remove RNA primers, repair the gaps thus created, and ligate the individual DNA fragments to produce a continuous copy of the template strand. (B) Incomplete synthesis of the lagging strand. When a DNA molecule is linear, synthesis of the leading strand can be completed simply by copying of its template to the end of the molecule. In contrast, removal of the terminal RNA primer from the 5' end of the lagging strand creates a gap that cannot be repaired by any DNA polymerase. Thus, in the absence of specialized mechanisms for maintaining the ends of linear DNA molecules, successive cycles of replication would progressively eliminate genetic information.

are established at a single origin and move away from it as the new DNA strands are synthesized (Fig. 9.1B). However, as DNA can be synthesized only in the $5' \rightarrow 3'$ direction, only one of the two parental strands can be copied continuously from a primer deposited at the origin. The long-standing conundrum of how the second strand is synthesized was solved with elucidation of the discontinu-

ous mechanism of synthesis of this strand (Fig. 9.2A). Primers for DNA synthesis are synthesized at multiple sites such that this new DNA strand is initially made as short discontinuous segments.

The discontinuous mechanism of DNA synthesis creates a special problem at the ends of linear DNA molecules, where excision of the terminal primer creates a gap at the 5' end of the daughter DNA molecules (Fig. 9.2.B). In the absence of a mechanism for completing synthesis of termini, discontinuous DNA synthesis would lead to an intolerable loss of genetic information. In chromosomal DNA, specialized DNA elements, called **telomeres**, at the ends of each chromosome prevent loss of DNA upon discontinuous DNA synthesis. These structures comprise simple repeated sequences maintained by reverse transcription of an RNA template, which is an essential component of the ribonucleoprotein enzyme **telomerase**. Viruses with linear DNA genomes (Table 9.1) have evolved a variety of elegant mechanisms to solve the "end" problem and ensure complete replication of all sequences of their genomes.

### Origins of Cellular Replication

Identification of replication origins of eukaryotes (other than simple yeasts) has proved to be one of the most frustrating endeavors of modern molecular biology. A large body of evidence indicates that replication initiates at specific sites in vivo. With the availability of cloned genomic DNA segments, these origins can be identified and mapped with some precision. The difficulty lies in the functional identification of origins. Yeast origins can be assayed readily because they support replication of small plasmids and hence their maintenance as episomes. All yeast origins behave as such autonomously replicating sequences. In contrast, this simple functional assay has generally failed to identify analogous mammalian sequences even when applied to DNA segments containing origins mapped in vivo. Conversely, many long DNA fragments (including prokaryotic DNAs) support replication of plasmids in mammalian cells or *Xenopus* eggs but apparently direct random initiation of DNA replication. The paradox of why specific DNA sequences that support plasmid replication are so difficult to find when replication initiates at specific sites in vivo has yet to be resolved. Current ideas focus on the long-range organization of the genome, features imposed by the organization of the genome into chromatin, and the integration of replication and transcription.

### The Origin of Simian Virus 40 DNA Replication

Because the identification of functional origins of most eukaryotic genomes has been so difficult, progress in elucidation of the mechanisms of origin-dependent DNA syn-

**BOX 9.1**

*Mapping of the simian virus 40 origin of replication by analysis of the distribution of radioactivity in pulse-labeled DNA molecules that completed replication during a brief labeling period*

As illustrated in panel A (left), exposure of simian virus 40-infected monkey cells to [³H]thymidine ([³H]dT) for a period less than the time required to complete one round of replication (e.g., 5 min) results in labeling of the growing points of replicating DNA. If replication proceeds from a specific origin (Ori) to specific termination sites (T), the DNA replicated last will be preferentially labeled in the population of completely replicated DNA molecules (panel A, right). The distribution of [³H]thymidine among the fragments of completely replicated viral DNA generated by cleavage with *Hin*dII and *Hin*dIII is shown in panel B. The circular simian virus 40 genome is represented as cleaved within the G fragment, and relative distances are given with respect to the junction of the A and C fragments. The observation of two decreasing gradients of labeling that can be extrapolated (dashed lines) to the same region of the genome confirmed that simian virus 40 replication is bidirectional (Fig. 9.1A) and allowed location of the origin on the physical map of the viral genome. Modified from K. J. Danna and D. Nathans, *Proc. Natl. Acad. Sci. USA* **69:**3097–3100, 1972, with permission.

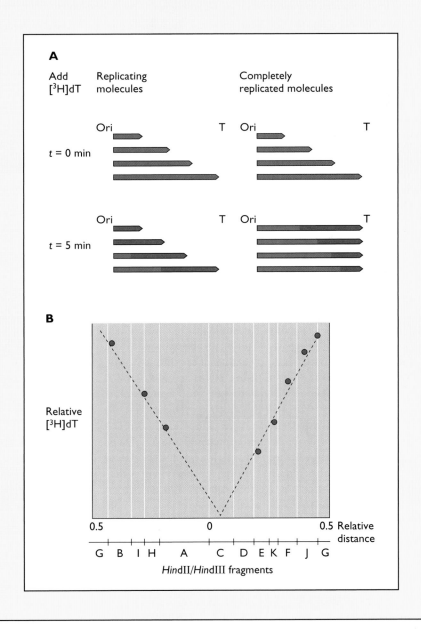

thesis came through study of viral origins, notably that of the papovavirus simian virus 40 (SV40). The simian virus 40 origin of replication, which supports bidirectional replication, was the first viral control sequence to be located on a physical map on which the reference points are restriction endonuclease cleavage sites (Fig. 9.1; Box 9.1). We now possess a detailed picture of this origin (Fig. 9.3) and of the binding sites for the simian virus 40 origin recognition protein, large T antigen (LT).

A 64-bp sequence, the **core origin**, which lies between the initiation sites of early and late transcription, is sufficient for initiation of simian virus 40 DNA synthesis in infected cells. This sequence contains four copies of a pentanucleotide binding site for LT flanked by an AT-rich element and a 10-bp imperfect palindrome (Fig. 9.3). Additional DNA sequences within this busy control region of the viral genome modulate the efficiency of initiation of DNA synthesis from the core origin. A second LT-binding site (site I in Fig. 9.3) stimulates in vivo replication modestly, whereas the simian virus 40 enhancer and Spl-binding sites of the nearby early promoter increase the efficiency of viral DNA replication in infected cells at least 10-fold.

## Cellular Replication Proteins and Their Functions during Simian Virus 40 DNA Synthesis

### Eukaryotic DNA Polymerases

It has been known for about 40 years that eukaryotic cells contain DNA-dependent DNA polymerases that catalyze the incorporation of deoxyribonucleoside monophosphates into DNA when provided with an appropriate template-primer. Mammalian cells contain four nuclear enzymes, designated DNA polymerases α, β, δ, and ε (Table 9.2). These enzymes are distinguished by such properties as sensitivities to various inhibitors, whether they are associated with $3' \rightarrow 5'$ exonuclease activity, and degree of **processivity**, the number of nucleotides incorporated into a nascent DNA chain per initiation from a primer carrying a 3′ OH group (Table 9.2). These characteristics can be readily assayed in in vitro reactions with artificial template-primers such as gapped or nicked DNA molecules. However, distinguishing their roles in DNA replication was impossible until simian virus 40 origin-dependent DNA synthesis was reproduced in 1984 (see next section). The requirements for viral DNA synthesis in vitro and genetic analyses (performed largely with yeast) identify DNA polymerases α, δ, and ε as replication enzymes. Only DNA polymerase α is associated with priming activity. The heteromeric **primase** is tightly bound to DNA polymerase α and copurifies with it. DNA polymerase β plays no known role in replication but is an important DNA repair enzyme (Table 9.2).

One of the most striking properties of eukaryotic replicative DNA polymerases is their obvious evolutionary relationships to prokaryotic and viral enzymes. Six regions are highly conserved in both primary amino acid sequence and relative position in a wide variety of DNA polymerases, including human DNA polymerase α and the enzymes encoded by several DNA viruses. As discussed in chapters 6 and 7, all template-directed nucleic acid polymerases share several sequence motifs and probably a similar structure, indicating that important features of the mechanisms by which they catalyze synthesis of DNA or RNA from a DNA or RNA template are also common to all these enzymes.

### Identification of Other Replication Proteins via In Vitro Replication Systems

Studies of eukaryotic DNA replication took a quantum leap forward with the development of a cell-free system for synthesis of adenoviral DNA from an exogenous template. The unusual mechanism by which this viral DNA replicates, described below, was reproduced in infected human cell extracts, establishing that faithful replication could be reconstituted in vitro. This breakthrough was soon followed by origin-dependent replication of simian virus 40 DNA in vitro. Because cellular components are largely responsible for simian virus 40 DNA synthesis, this system proved to be the watershed in the investigation of eukaryotic DNA replication, allowing the identification of previously unknown cellular replication proteins and elucidation of the mechanism of DNA synthesis.

### Mechanism of Simian Virus 40 DNA Synthesis

**Origin recognition and unwinding.** The first step in simian virus 40 DNA synthesis is the recognition of the origin by LT, the major early gene product of the virus. In the presence of ATP, LT assembles on its binding sites at the core origin (Fig. 9.3) as two hexamers that encase the entire core origin (Fig. 9.4). Assembly of the hexamers rapidly induces structural transitions in both sequences flanking the LT-binding sites and distortion of the early imperfect palindrome (Fig. 9.4). The next reaction, LT-induced unwinding of the origin, requires cellular replication protein A (Rp-A), which possesses single-stranded-DNA-binding activity and binds specifically to LT (Table 9.3). In concert with replication protein A, the intrinsic $3' \rightarrow 5'$ helicase activity of LT unwinds DNA bidirectionally from the core origin (Fig. 9.4). Formation of the LT complex at the simian virus 40 origin resembles assembly reactions at well-characterized prokaryotic replication origins, such as *Escherichia coli* OriC or the origin of phage λ, in which multimeric protein structures form on AT-rich sequences.

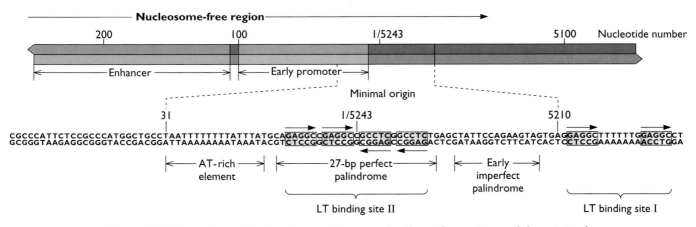

**Figure 9.3  The origin of simian virus 40 DNA replication.** The positions of the minimal origin necessary for simian virus 40 DNA replication in vivo and in vitro and of the enhancer and early promoter (chapter 8) in the viral double-stranded DNA genome are indicated. The sequence of the minimal origin is shown below, with the LT-binding sites (G$^A_C$GGC) in yellow. The AT-rich element and early imperfect palindrome, as well as LT-binding site II, are essential for replication. Other sequences, including the enhancer and Sp1 binding sites of the early promoter, stimulate replication significantly. The transcriptional regulators that bind to the enhancer or promoter might facilitate initiation of replication by various mechanisms. One possibility is that the activation domains of these proteins (described in chapter 8) help recruit essential replication proteins to the origin. This model is consistent with popular ideas about how activators stimulate transcription. Alternatively, the binding of transcriptional activators might induce remodeling of the chromatin structure in the vicinity of the origin. This possibility is particularly attractive in the case of simian virus 40, for, as indicated, the region of the genome containing the origin and transcriptional control regions is nucleosome-free in a significant fraction (about 25%) of minichromosomes in infected cells (but not in virions).

**Leading-strand synthesis.** Binding of DNA polymerase α-primase to both LT and replication protein A at the simian virus 40 origin (Fig. 9.5; Table 9.3) completes assembly of the **presynthesis complex** and sets the stage for the initiation of leading-strand synthesis. The primase synthesizes the RNA primers of the leading strand at each replication fork, while DNA polymerase α extends these to produce short Okazaki fragments (Fig. 9.5). The 3′ OH ends of these small DNA fragments are then bound by cellular replication factor C (Rf-C), proliferating cell nuclear antigen (Pcna), and DNA polymerase δ (Fig. 9.5; Table 9.3). The highly conserved proliferating cell nuclear antigen is the **processivity factor** for DNA polymerase δ. It is required for synthesis of long DNA chains from a single primer and essential for simian virus 40 DNA replication in vitro. This mammalian protein is the functional analog of the β subunit of *E. coli* DNA polymerase III and phage T4 gene 45 protein. These remarkable proteins are "sliding clamps" that track along the DNA template (Fig. 9.6) and serve as movable "platforms" for DNA polymerases. As sliding clamp proteins are closed rings (Fig. 9.6), proliferating cell nuclear antigen cannot load onto DNA molecules

that lack ends, such as the circular SV40 genome or chromosomal domains of genomic DNA. This essential, ATP-dependent step is carried out by the clamp-loading protein replication factor C, which presumably induces transient opening of the proliferating cell nuclear antigen ring. Because replication factor C binds to the 3′ OH end of Okazaki fragments, it places the processivity factor at the replication forks (Fig. 9.5). Binding of these two proteins inhibits DNA polymerase α and induces dissociation of the DNA polymerase α-primase complex. Subsequent binding of DNA polymerase δ completes assembly of a multiprotein structure capable of leading-strand synthesis by continuous copying of the parental template strand.

DNA replication can neither initiate nor proceed without the action of enzymes that unwind the strands of the DNA template. In the simian virus 40 in vitro replication system, LT is the helicase responsible for unwinding of the origin and remains associated with each of the two replication forks (Fig. 9.5), unwinding the template during elongation. LT translocates along DNA in the 3′ → 5′ direction of each parental strand and must therefore be bound to the leading strand in a replication fork.

**Table 9.2** Properties of mammalian nuclear DNA polymerases

| Properties | DNA polymerase | | | |
| --- | --- | --- | --- | --- |
| | Alpha | Delta | Epsilon | Beta |
| Mass (kDa) | | | | |
| Native | >250 | 170 | 256 | 36–38 |
| Catalytic core | 165–180 | 125 | 215 | 36–38 |
| Other subunits | 72, 58, 48 | 48 | 55 | |
| Associated functions | | | | |
| $3' \rightarrow 5'$ Exonuclease | Cryptic | Yes | Yes | No |
| Primase | Yes | No | No | No |
| Characteristics | | | | |
| Responds to proliferating cell nuclear antigen | No | Yes | Yes | No |
| Preferred template | Gapped | Poly(dA) · oligo(dT) | Poly(dA) · oligo(dT) | Gapped |
| Processivity | Low | High[a] | High[a] | Low |
| Fidelity | High | High | High | Low |
| Inhibitors | | | | |
| NaCl (0.15 M) | Strong | Strong | Strong | No |
| Aphidicolin | Strong | Strong | Strong | No |
| Dideoxy-NTPs | No inhibition | Weak | Weak | Strong |
| Replication | Yes | Yes | Yes | No |

[a] In the presence of proliferating cell nuclear antigen.

**Lagging-strand synthesis.** The first primer and Okazaki fragment of the lagging strand are synthesized by the DNA polymerase α-primase complex (Fig. 9.5, step 4). The lagging strand is also synthesized by DNA polymerase δ, at least in the in vitro system, and transfer of the 3' end of the first Okazaki fragment to this enzyme is believed to proceed as on the leading strand. The lagging-strand template is then copied toward the origin of replication. However, synthesis of the lagging strand requires initiation by DNA polymerase α-primase at multiple sites progressively further from the origin (Fig. 9.5). The mechanisms by which leading- and lagging-strand syntheses are coordi-

**Table 9.3** Cellular proteins required for simian virus 40 DNA replication[a]

| Protein | Synonym(s) | Subunits (kDa) | Contact(s) | Functions |
| --- | --- | --- | --- | --- |
| Rp-A | Rf-A, ssB | 70 | Primase, RNA-DNA primers | Binds to single-stranded DNA; origin unwinding in cooperation with LT |
| | | 32, 11 | LT | |
| DNA polymerase α-primase | | 165–180, 72, 58, 48 | LT | Synthesis of RNA primers and Okazaki fragments on leading and lagging strands |
| Rf-C | Activator 1 | 140, 40, 38, 37, 36 | Pcna | ATP-dependent clamp loading; also required for release of Pcna from DNA; largest subunit contacts Pcna and DNA |
| Pcna | | 34 | Rf-C, DNA polymerases δ and ε, Fen-1, DNA ligase 1 | Sliding clamp |
| DNA polymerase δ | | 125, 48 | Pcna | Processive synthesis of leading and lagging strands when bound to Pcna |
| RNase H1 | | Unknown | Unknown | Endonucleotytic degradation of RNA base-paired with DNA and attached to the 5' end of a DNA strand; removal of RNA primers |
| Fen1 | | 44–50 | Pcna | $5' \rightarrow 3'$ exonuclease; removal of RNA primers |
| DNA ligase I | | 125 | Unknown | Sealing daughter DNA fragments |

[a] Rp-A, replication protein A; Rf-A, replication factor A; ssB, single-stranded-DNA-binding protein B; Rf-C, replication factor C; Pcna, proliferating cell nuclear antigen; Fen1, Flap endonuclease 1.

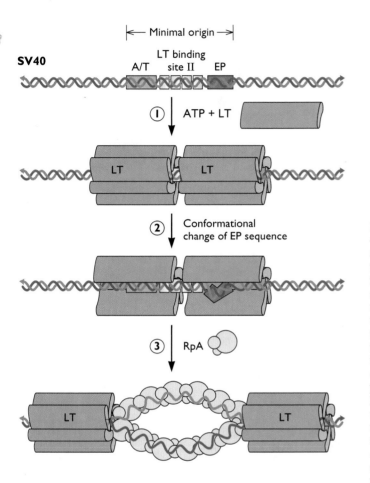

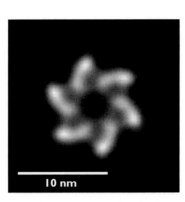

**Figure 9.4 Recognition and unwinding of the simian virus 40 origin.** In the presence of ATP, the viral early protein LT assembles into hexamers. Two hexamers bind to the origin via the pentanucleotide LT-binding sites in the center of the minimal origin (step 1). Each hexamer appears as a propeller-shaped structure with a large central hole, as illustrated in the two-dimensional reconstruction of LT molecules assembled in the presence of a nonhydrolyzable analog of ATP shown in the electron micrograph on the right. (From M. C. San Martin et al., *J. Mol. Biol.* **268:**15–20, 1997, with permission. Courtesy of J. Carazo, Universidad Autónoma de Madrid.) Binding of LT hexamers protects the flanking AT-rich (A/T) and early palindrome (EP) sequences of the minimal origin from DNase I digestion and induces structural changes in them, in particular, distortion of the early palindrome (step 2). However, extensive, stable unwinding of the origin requires the cellular single-stranded-DNA-binding protein replication protein A (Rp-A). This heterotrimeric protein binds to the single strands of the unwound region to stabilize the unwound DNA but appears to fulfill additional roles in simian virus 40 DNA synthesis. Rp-A binds to LT via its smaller subunits (Table 9.3). LT helicase activity in concert with Rp-A progressively unwinds the origin (step 3). For simplicity, the two LT hexamers are shown moving apart, but they probably remain in contact (Box 9.2).

nated are not fully understood. If the replication machinery tracked along an immobile DNA template, the complexes responsible for leading- and lagging-strand syntheses would have to move in opposite directions (Fig. 9.5). Furthermore, lagging-strand synthesis would require repeated assembly of priming complexes at sites that do not contain LT-binding sequences. Therefore, a more attractive alternative is that the DNA template is spooled through an immobile replication complex, the **replisome**, containing all proteins necessary for synthesis of both daughter strands. This mechanism would allow simultaneous copying of the template strands in opposite directions by a single replication machine present at each fork (Fig. 9.7). Replication of chromosomal DNA occurs at fixed sites in the nucleus, as would be predicted by a DNA spooling mechanism. Moreover, spooling of DNA has been observed in the electron microscope during the initial, LT-de-

pendent unwinding of the simian virus 40 origin (Box 9.2). While the replisome model possesses the virtues of elegance and parsimony, it remains to be firmly established.

**Termination and resolution.** The circular SV40 DNA genome possesses no termini. Consequently, its replication requires no special mechanisms to circumvent the "end" problem. Nevertheless, additional cellular proteins are needed for the production of two daughter molecules from the circular template. These essential components of the simian virus 40 replication system are cellular enzymes, topoisomerases I and II, that alter the topology of DNA. These enzymes, which differ in their physical properties, catalytic mechanisms, and functions in the cell (Table 9.4), reverse the winding of one duplex DNA strand around another (supercoiling), regardless of the topology of the su-

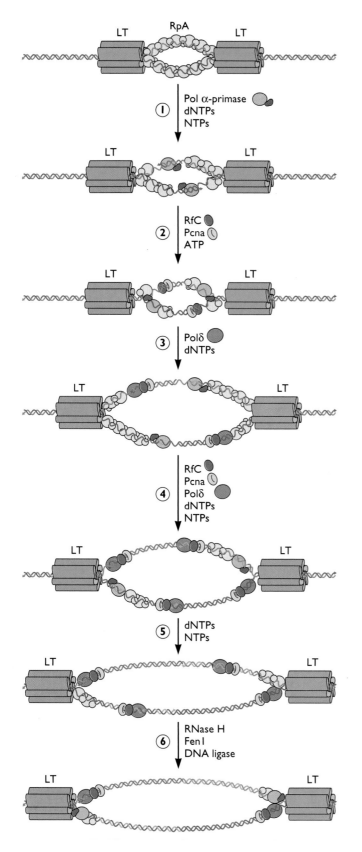

**Figure 9.5 Synthesis of leading and lagging strands.** The DNA polymerase α-primase responsible for the synthesis of Okazaki fragments binds specifically to both the Rp-A and LT proteins assembled at the origin in the presynthesis complex. Once bound, this enzyme complex initiates synthesis of the leading strands with production of RNA primers that are then extended as DNA by the polymerase (step 1). The 3′ OH group of the nascent RNA-DNA fragment (about 30 nucleotides in total length) is then bound by replication factor C (Rf-C) in a reaction that, like assembly of LT at the origin, requires ATP but not its hydrolysis. Rf-C allows ATP-dependent loading of proliferating cell nuclear antigen (Pcna) onto the template (step 2). The conformation of Pcna, a closed ring (Fig. 9.6), must be substantially altered during the loading reaction to allow passage of the replicating DNA into its central channel, but this reaction is not yet understood. DNA polymerase δ then binds to the Pcna/Rp-C complex (step 3). The replication complex thus assembled is competent for continuous and highly processive synthesis of the leading strands (steps 4 and 5). Although DNA polymerase δ is sufficient for leading-strand synthesis during simian virus 40 DNA synthesis in vitro, the second Pcna-responsive enzyme, DNA polymerase ε (Table 9.2), appears to contribute to chromosomal replication in vivo. Lagging-strand synthesis begins with synthesis of the first Okazaki fragment by DNA polymerase α-primase (step 3). While there is evidence indicating that initiation of lagging-strand synthesis takes place after formation of the replication complex on the leading strand, it is not known whether DNA polymerase α-primase transfers directly from the leading to the lagging strands. Transfer of the 3′ OH terminus of the Okazaki fragment to the Rf-C/Pcna/DNA polymerase as in steps 2 and 3 allows synthesis of a lagging-strand segment (step 4). The multiple DNA fragments produced by discontinuous lagging-strand synthesis (step 5) are sealed by removal of the primers by RNase H (an enzyme that specifically degrades RNA hybridized to DNA) and the 5′ → 3′ exonuclease Fen1, repair of the resulting gaps by DNA polymerase δ, and joining of the DNA fragments by DNA ligase I (step 6).

percoils. Because they remove supercoils, topoisomerases are said to relax DNA. In a closed circular DNA molecule like the simian virus 40 genome, the unwinding of duplex DNA at the origin and subsequently at the replication forks is necessarily accompanied by supercoiling of the remainder of the DNA (Fig. 9.8A). If not released, the torsional stress so introduced would act as a brake on movement of the replication forks, eventually bringing them to a complete halt. Both topoisomerases I and II remove such torsional stress during simian virus 40 DNA replication in vitro to allow movement of the replication forks. These enzymes play an analogous role during replication of chromosomal DNA in vivo (at least in yeast). Topoisomerase II is also specifically required for the separation of daughter molecules from late replication intermediates. A cycle of simian virus 40 DNA synthesis

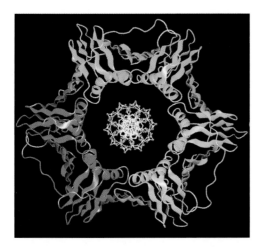

**Figure 9.6 Structure of yeast proliferating cell nuclear antigen.** The protein trimer is represented schematically in ribbon form, with the three subunits colored red, yellow, and orange. Each monomer contains two topologically identical domains. A model of double-stranded B form DNA is shown in the geometric center of the structure to illustrate how the closed ring formed by tight association of the three subunits might encircle DNA. The α-helices from the three subunits line the central channel, forming a surface of positively charged residues favorable for interaction with the negatively charged phosphodiester backbone of DNA. The nonspecificity of such interactions and the orientation of the internal α-helices appear ideal for strong, but nonspecific, contact with the phosphate groups of the backbone of the DNA threaded through the channel, as modeled in the figure. From T. S. R. Krishna et al., *Cell* **79**:1233–1243, 1994, with permission. Courtesy of J. Kuriyan, Rockefeller University.

produces two intertwined (**catenated**) circular DNA molecules that can be separated only by the passage of one through the other mediated by this enzyme (Fig. 9.8B; Table 9.4).

**Replication of chromatin templates.**   The simian virus 40 genome is associated with cellular nucleosomes both in the virion and in infected cell nuclei. It is therefore replicated not as naked DNA but as a minichromosome in which the DNA is wrapped around nucleosomes. This arrangement raises the question of how the replication machinery moves rapidly along a DNA template that is tightly bound to the histones of each nucleosome. A similar problem is encountered during the replication of many viral RNA genomes, when the template RNA is packaged by viral RNA-binding proteins in a ribonucleoprotein. The mechanisms by which replication (or transcription) complexes circumvent the barriers to movement along a DNA template imposed by nucleosomes are not understood in detail. Nevertheless, there is considerable evidence from simple model systems that enzymes that copy DNA templates can displace nucleosomes from their path to a new site on the template.

The organization of the simian virus 40 genome into a minichromosome also implies that viral DNA replication must be coordinated with binding of newly synthesized DNA to cellular nucleosomes. Investigation of the replication of simian virus 40 minichromosomes has identified both passive and active mechanisms for deposition of nucleosomes onto newly replicated DNA. Active deposition is mediated by a previously unrecognized chromatin assembly protein, which coordinates histone production with replication-dependent nucleosome assembly.

**Figure 9.7 A hypothetical simian virus 40 replisome.** The replisome, a replication machine that contains all proteins necessary for both continuous synthesis of the leading strand and discontinuous synthesis of the lagging strand, would assemble at each replication fork. Spooling of a loop of the template DNA strand for discontinuous synthesis would allow the single complex to copy the two strands in opposite directions. Although there is indirect evidence for such a model, formation of this kind of replisome remains to be directly demonstrated.

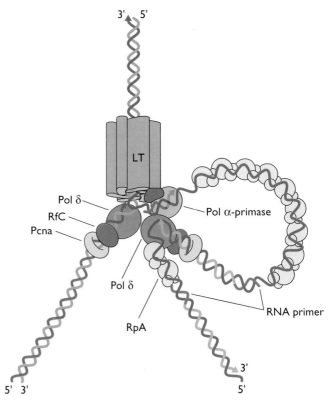

**BOX 9.2**

*Spooling of DNA during unwinding of the simian virus 40 origin*

Electron microscopy of LT-dependent unwinding from the simian virus 40 origin suggests that the two hexamers remain in contact as DNA is unwound. (A) LT bound to the origin, as a characteristic bilobed structure (the double hexamer shown in Fig. 9.4); (B) unwinding intermediates; (C) the intermediate at the bottom right in panel B at higher magnification. This intermediate contains a bilobed LT complex connecting the two replication forks so the single-stranded DNA (ssDNA) is looped out as "rabbit ears." The formation of such structures containing a dimer of the LT hexamer in which each monomer is bound to a replication fork stimulates the helicase activity of LT. This property supports the view that the DNA template is spooled through an immobile replication machine (see text). dsDNA, double-stranded DNA. From R. Wessel et al., *J. Virol.* **66:**804–815, 1992, with permission. Courtesy of H. Stahl, Universität des Saarlandes.

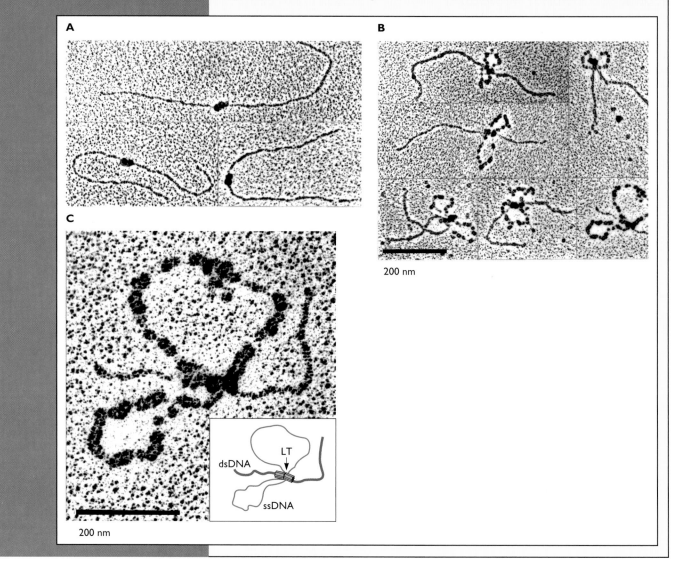

**Summary.** Analysis of simian virus 40 replication in vitro has identified essential cellular replication proteins, led to detailed descriptions of crucial reactions in the very complex process of DNA synthesis, and provided new insights into chromatin assembly. The major question to be addressed in mammalian cell DNA synthesis is how an origin is recognized in reactions analogous to that carried out by LT in the simian virus 40 system. The great interest in mechanisms that control cell cycle progression, including those that ensure that each cellular origin fires exactly

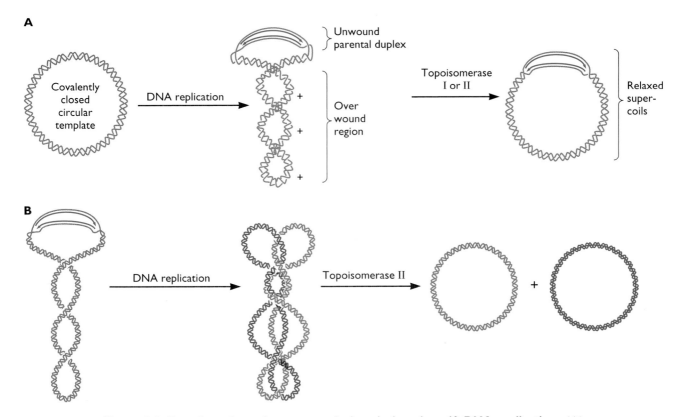

**Figure 9.8 Function of topoisomerases during simian virus 40 DNA replication.** (A) Relief from overwinding. In a closed double-stranded DNA circle like the simian virus 40 genome, unwinding of the template DNA at the origin and two replication forks is necessarily accompanied by overwinding (positive supercoiling) of the DNA ahead of the replication forks (middle). Either topoisomerase I or topoisomerase II can act as a swivel, removing the supercoils to allow continued movement of the replication fork. (B) Decatenation of replication products. The products of a cycle of simian virus 40 DNA replication are two interlocked daughter molecules. Their separation to release daughter molecules identical to the parental genome requires topoisomerase II, which makes a double-stranded break in DNA, passes one double strand over the other to unwind one turn, and reseals the DNA in reactions that require hydrolysis of ATP. During chromosomal replication, this enzyme is likewise essential for separation of daughter chromosomes during mitosis.

once per cell cycle, and the rapid pace of identification of mammalian homologs of yeast origin recognition proteins suggest that the answer to this question will soon be forthcoming. The detailed understanding of the reactions completed by the cellular DNA replication machinery gained through study of simian virus 40 DNA laid the foundation for elucidation of the mechanisms by which other animal viral DNA genomes are replicated and the intricate circuits that regulate DNA synthesis and its initiation.

## Mechanisms of Viral DNA Synthesis

The replication of all viral DNA genomes within infected cells comprises reactions analogous to those necessary for

simian virus 40 DNA synthesis, namely, origin recognition and priming of DNA synthesis, elongation, termination, and often resolution of the replication products. However, the mechanistic problems associated with each of the steps in DNA synthesis are solved by a considerable variety of elegant virus-specific mechanisms.

### Priming and Elongation

Synthesis of viral DNA molecules is initiated by a number of unusual mechanisms in which not only RNA but also DNA and even protein molecules function as primers. Because each of the viral priming strategies has profound consequences for the mechanisms of elongation and com-

**Table 9.4** Cellular DNA topoisomerases

| Topoisomerase I | Topoisomerase II |
|---|---|
| 90- to 135-kDa monomer | 150- to 180-kDa homodimers |
| Relaxes positively and negatively supercoiled DNA in ATP-independent reactions | Relaxes positively and negatively supercoiled DNA in ATP- and $Mg^{2+}$-dependent reactions |
| Catenates and decatenates double-stranded circles, provided one molecule carries a nick or gap | Catenates, decatenates, knots, and unknots double-stranded DNA molecules without nicks or gaps |
| Forms intermediate in relaxation in which one strand of the DNA is cleaved and covalently bound via 3′ phosphate to the protein; changes linking number in steps of one | Cleaves both strands of DNA to form a 4-bp staggered cut; 5′ ends of each strand covalently bound to one subunit of the enzyme; alters linking number in steps of two |
| Cleaves at preferred sequences | Modest sequence specificity |
| Supports replication fork movement in vitro; participates in transcription elongation and recombination; not essential for DNA replication in vivo | Supports replication fork movement; essential in vivo because it is required for separation of daughter chromosomes; decatenates replication products in vitro; participates in recombination; a component of chromosomal scaffolds and nuclear matrices |

plete replication of both parental DNA strands, priming and elongation are considered together.

### Synthesis of Viral RNA Primers

The standard method of priming is synthesis of a short RNA molecule by a specialized primase. As we have seen, cellular DNA polymerase α-primase synthesizes all RNA primers needed for replication of both template strands of papovaviral genomes. A similar mechanism operates at certain origins of some herpesviruses. A characteristic feature of members of this family is the ability to establish **latent infections** in specific cell types. This type of virus-host cell interaction allows the virus to persist in a host for very long periods (chapter 15). From a mechanistic point of view, latent infections of B lymphocytes by the gammaherpesvirus Epstein-Barr virus are among the best characterized. Many cell lines latently infected with this virus have been established from patients with Burkitt's lymphoma, and this state is the usual result of Epstein-Barr virus infection of B cells in culture. Characteristic features of latent Epstein-Barr virus infection include expression of only a small number of viral genes, lack of expression of late genes, the presence of a finite number of episomal viral genomes, and replication from a specialized origin. Because replication from this origin, which is not active in lytically infected cells, is responsible for maintenance of episomal viral genomes in latently infected cells, it is termed the **origin for plasmid maintenance (OriP)**. This origin is specifically recognized by the viral protein Epstein-Barr virus nuclear antigen 1 (EBNA-1) (Table 9.5), but replication of episomal viral DNA is mediated by the cellular machinery described above, including DNA polymerase α-primase.

The integrated proviral genomes of retroviruses are also replicated via RNA primers synthesized by the cellular primase at the origins of the cellular replicon into which the provirus is inserted. Thus, in actively dividing cells, proviral DNA is replicated once per cell cycle by the cellular replication machinery. Such proviral DNA replication makes no contribution to the production of progeny retrovirus particles in a single-cell infectious cycle. However, it does provide an effective means of expanding the population of infected cells, for all descendants of the original infected cell will carry the integrated provirus.

Less common among DNA viruses is synthesis of RNA primers for initiation of DNA synthesis by viral proteins. However, this mechanism is characteristic of herpesviral replication in productively infected cells. For example, during lytic infection the Epstein-Barr virus genome is replicated from two origins distinct from OriP, termed Ori-Lyt. Such lytic replication has been best characterized for herpes simplex virus type 1 when a viral primase synthesizes RNA primers. This primase is a heterotrimer of the products of the viral UL5, UL8, and UL25 genes (Table 9.6).

An inevitable consequence of DNA synthesis from RNA primers by either cellular or viral DNA polymerases is that one of the two parental DNA strands must be copied discontinuously, as described in detail for simian virus 40. When the template is circular, no special mechanism is needed to complete copying of both strands. However, when the template is linear, like the genome of herpesviruses, the ends of the lagging strands cannot be copied by the discontinuous mechanism. In herpesviruses, this problem is circumvented by circularization of the viral genome on entry into the infected cell nucleus to generate "endless" templates for replication. The mechanism of cir-

**Table 9.5** Viral origin recognition proteins[a]

| Virus | Protein(s) | DNA-binding properties | Other activities and functions |
|---|---|---|---|
| Parvovirus<br>  Adeno-associated<br>    virus | Rep 78/68 | Binds to specific sequences in ITR sequence; Rep 78 binds as hexamer | Site- and strand-specific endonuclease; ATPase and ATP-dependent helicase; transcriptional regulator; binding to DNA and endonuclease and ATPase activities stimulated by association with the cellular nonhistone chromosomal protein Hmg1 |
| Papovavirus<br>  Simian virus 40 | LT | Binds cooperatively to origin site II to form double hexamer; distorts origin; binds to two other nearby sites (I, III) in genome; DNA-binding activity regulated by phosphorylation at specific sites | DNA-dependent ATPase and $3' \rightarrow 5'$ helicase; binds to cellular Rp-A and Polα-primase; autoregulates early transcription; activates late transcription; binds cellular Rb protein to induce reentry into, or progression through, the cell cycle; binds cellular p53 protein |
| Bovine papillo-<br>    mavirus type 5 | E1 | Binds origin with low affinity; binds strongly and cooperatively in presence of E2 protein | Binding to E2 essential for viral DNA replication; DNA-dependent ATPase and helicase; helicase active on assembly of a hexamer; binds catalytic subunit Polα-primase |
|  | E2 | Binds as dimer to specific sequences in origin | Regulates transcription by binding to viral enhancers |
| Adenovirus<br>  Human adenovirus<br>    type 2 | DNA poly-<br>    merase–<br>    pre-TP | Binds to origins as complex; binding stimulated by cellular transcriptional activators Nf-1 and Oct-1 | Priming of DNA synthesis via addition of dCMP to pre-TP by the DNA polymerase; continuous synthesis of both strands of viral genome |
| Herpesviruses<br>  Herpes simplex<br>    virus type 1 | UL9 | Binds cooperatively to specific sites at viral origins; distorts DNA to which it binds | ATPase and $3' \rightarrow 5'$ helicase; binds UL29 protein, UL8 subunit of viral primase, and UL42 processivity protein |
| Epstein-Barr virus | EBNA-1 | Binds as dimer to multiple sites in two clusters (FR and DS [Box 9.3]) in viral OriP; may use different DNA-binding motifs to form complexes in which DNA is distorted | Required for maintenance of episomal viral genomes; binding to FR sequences stimulates transcription from specific viral promoters; can bind to cellular Rp-A |

[a]ITR, inverted terminal repetition; Rp-A, replication protein A; Polα-primase, DNA polymerase α-primase complex; pre-TP, preterminal protein.

cularization, which appears to be via direct ligation of the ends of linear viral DNA molecules, has not been worked out in detail. In theory, the formation of a circular template could permit complete copying of both strands of infecting herpesviral genomes, exactly as in the replication of simian virus 40 DNA. However, such a simple replication mechanism explains the properties of neither herpes simplex virus type 1 **replication intermediates**, molecules that contain nascent DNA, nor the products of replication, which are long concatemers containing multiple copies of the genome (see "Resolution and Processing of Viral Replication Products" below).

Regardless of whether cellular or viral proteins are used to make their primers, the replication of DNA viruses that rely on RNA priming is subject to the difficulties intrinsic to this mechanism (e.g., how to copy ends of lagging strands). In contrast, other DNA viruses have evolved alternative priming mechanisms that eliminate the need for discontinuous synthesis.

### Priming via DNA: Specialized Structures in Viral Genomes

Self-priming of viral DNA synthesis via specialized structures in the viral genome is a hallmark of all parvoviruses, the smallest DNA viruses that replicate in animal cells. This virus family includes the dependoviruses, such as adeno-associated viruses, and the autonomous parvoviruses, such as minute virus of mice. The dependoviruses can reproduce only in cells coinfected by a helper adenovirus or herpesvirus, which provides functions needed for efficient expression of parvoviral genes, while replication of the autonomous parvoviruses requires passage of an infected cell through S phase. Replication of all parvoviruses exhibits a number of unusual features, the most striking being self-priming of viral DNA synthesis. This mechanism is exemplified here with adeno-associated virus.

The adeno-associated virus genome is a small molecule (<5 kb) of single-stranded DNA, which carries an **inverted terminal repetition (ITR)**. Genomic DNA is of

**Table 9.6**  Viral proteins that participate in genome replication[a]

| Virus | Protein | Functions and properties |
|---|---|---|
| **Proteins needed for DNA synthesis** | | |
| Adenovirus | | |
|   Human adenovirus type 2 | DNA polymerase | Initiates replication as complex with preterminal protein by covalent linkage of dCMP to the protein primer; completes synthesis of all daughter strands |
| | Preterminal protein | Protein primer for DNA synthesis |
| | Terminal protein | Covalently bound to 5′ termini of the linear genome; facilitates origin unwinding |
| | Single-stranded-DNA-binding protein | Binds displaced single-stranded DNA; confers processivity to viral DNA polymerase |
| Herpesvirus | | |
|   Herpes simplex virus type 1 | UL9 protein | Origin recognition; contains consensus ATP-binding site; can bind to UL42 protein |
| | UL30 protein (DNA polymerase) | Synthesis of viral DNA; associated 3′ → 5′ exonuclease for proofreading; target of several antiherpesviral drugs |
| | UL5, UL8, and UL52 proteins | Form heteromeric complex with primase and helicase activities; UL8 protein binds to the viral DNA polymerase |
| | UL42 protein | Processivity protein for viral DNA polymerase |
| | UL29 protein | Single-stranded-DNA-binding protein essential for viral DNA synthesis; binds DNA cooperatively; binds to the UL42 protein |
| Poxvirus | | |
|   Vaccinia virus | DNA polymerase | Synthesis of vaccinia virus DNA; associated 3′ → 5′ exonuclease activity for proofreading |
| | D5R protein[b] | ? |
| | B1R protein[b] | Serine-theonine kinase |
| | D4R protein[b] | Uracil DNA glycosylase |
| Baculovirus | | |
|   *Autographa californica* nuclear polyhedrosis virus | DNA polymerase[c] | Helicase[c] |
| | LEF-3 | Single-stranded DNA-binding protein |
| | IE-1 | Binds putative origins; transcriptional activator |
| | LEF-1 | Has primase-like motif |
| **Enzymes involved in nucleic acid metabolism** | | |
| Herpesvirus | | |
|   Herpes simplex virus type 1 | Thymidine kinase (UL23 protein, ICP36) | Phosphorylates thymidine and other nucleosides; essential in nondividing cells |
| | Ribonucleotide reductase; $\alpha_2\beta_2$ dimer of UL39 and UL40 proteins | Reduces ribose in ribonucleotides to deoxyribose; essential in nondividing cells |
| | dUTPase (UL50 protein) | Hydrolyzes dUTP to dUMP, preventing incorporation of dUTP into DNA and providing dUMP for conversion to dTMP; not essential for growth in tissue culture |
| | Uracil DNA glycosylase (UL2 protein) | Corrects insertion of dUTP or deamination of C in viral DNA; presumed to be important in DNA repair |
| | Alkaline nuclease (UL12 protein) | Required for efficient virus reproduction; processes newly replicated DNA to remove branched structures |
| Poxvirus | | |
|   Vaccinia virus | Thymidine kinase | Phosphorylates thymidine; required for efficient virus reproduction in animal hosts |
| | Thymidylate kinase | Phosphorylates TMP |
| | Ribonucleotide reductase, dimer | Reduces ribose in ribonucleotides to deoxyribose; essential in nondividing cells |
| | dUTPase | Hydrolyzes dUTP to dUMP (see above) |
| | ATP-dependent DNA ligase | Unknown function; not essential for viral replication in cells in culture |
| | DNase | Unknown function; has nicking-joining activity; present in virion cores |
| | DNA type I topoisomerase | Unknown function; essential; present in virions; structurally and functionally similar to cellular counterpart |

[a]Predicted from coding sequence.

[b]Whether these proteins participate directly in viral DNA synthesis or function indirectly is not known.

[c]Functions inferred from sequence homologies.

both (+) and (−) polarity, for both strands are encapsidated, but in separate virus particles. As illustrated in Fig. 9.9A, palindromic sequences within the central 125 nucleotides of the inverted terminal repetition base pair to form a Y- or T-shaped structure. Formation of this structure at the 3′ end of either strand of viral DNA provides an ideal template-primer for initiation of the first cycle of viral DNA synthesis in infected cells (Fig. 9.9B). Experimental evidence for such self-priming includes the strong dependence of adeno-associated virus DNA synthesis on self-complementary sequences within the inverted terminal repetition as well as the failure to detect RNA primers and viral Okazaki fragments when the virus is replicating within infected cells. Following recognition of the viral DNA primer, the single template strand of an infecting genome can be copied by a continuous mechanism analogous to leading-strand synthesis during replication of double-stranded DNA templates. In subsequent cycles of replication, the same 3′-terminal priming structures form in the duplex replication intermediate produced in the initial round of synthesis (Fig. 9.9B). As such self-priming circumvents the need for both RNA primers and discontinuous DNA synthesis, adeno-associated virus DNA synthesis is always continuous. Consequently, although viral DNA synthesis is carried out by the cellular replication proteins described above, the DNA polymerase α-primase is not required.

On the other hand, a specialized mechanism is required for copying of the sequences that form the priming structure so that replication of parental DNA molecules can be completed. This step is achieved by nicking the parental DNA strand at a specific site, at or near base 125, to liberate a new 3′ OH end for priming of continuous synthesis to the end of the DNA molecule (Fig. 9.9B). Consequently, this part of the 3′ ITR of the adeno-associated virus template DNA is replaced by its complementary sequence during each cycle of viral DNA synthesis (Fig. 9.9B). The nick is introduced by the related viral proteins Rep 78 and Rep 68 (Rep 78/68) (Fig. 9.9C). These proteins are site- and strand-specific endonucleases, which bind to and cut at specific sequences within the inverted terminal repetition. During this **terminal resolution process**, Rep 68 or Rep 78 becomes covalently linked to the cleaved DNA at the site that will become the 5′ terminus of the fully replicated molecules (Fig. 9.9B). Rep 78 and Rep 68 are the only viral gene products necessary for parvoviral DNA synthesis. Following the synthesis of a duplex of the genomic DNA molecule (the replication intermediate [Fig. 9.9B]), formation of the same 3′-terminal priming structure allows continuous synthesis of single-stranded genomes with re-formation of the replication intermediate (Fig. 9.9B, steps 6 and 7). Thus, Rep 78 and Rep 68 can

be considered origin recognition proteins and possess several properties in common with simian virus 40 LT (Table 9.5).

Priming via self-complementary terminal sequences of the viral genome eliminates the need for proteins that make RNA primers or are themselves primers for DNA synthesis (see next section). Such a mechanism would therefore seem especially advantageous for viruses with small genomes, such as the parvoviruses. Nevertheless, a similar self-priming mechanism is believed to initiate replication of the extremely large double-stranded DNA genomes of poxviruses such as vaccinia virus.

### Protein Priming

Judged by the criterion of the simplicity of the reactions necessary to complete a replication cycle, the most effective mechanism of initiation of DNA synthesis is via a protein primer. Nevertheless, this mechanism is rare, restricted to hepadnaviruses and adenoviruses among DNA viruses that infect animal cells. The replication of some viral RNA genomes is also initiated from a protein primer, notably the VPg protein of poliovirus discussed in chapter 6. In a template-dependent reaction, the adenoviral DNA polymerase covalently links the α-phosphoryl group of dCMP to the hydroxyl group of a specific serine residue in the viral preterminal protein (Fig. 9.10). The 3′ OH group of the protein-linked dCMP then primes synthesis of daughter viral DNA strands by this viral enzyme. The nucleotide is added to the preterminal protein only when this protein primer is assembled into preinitiation complexes at the origins of replication. As the origins lie at the ends of the linear genome, each template strand is copied continuously from one end to the other (Fig. 9.10). The parental template strand initially displaced is copied by the same mechanism, following formation of a duplex stem upon annealing of an inverted terminal repeat sequence. This unusual strand displacement mechanism of replication, which was initially deduced from the structures of adenoviral DNA replication intermediates observed in the electron microscope, therefore achieves semiconservative replication even though the two parental strands of viral DNA are not copied at the same replication fork.

## Viral Replication Origins and Their Recognition

As we have seen, origins of replication contain the sites at which viral (or cellular) DNA synthesis begins and can be defined experimentally as the minimal DNA segment necessary for initiation of DNA synthesis in cells or in in vitro reactions. Viral origins of replication support initiation of DNA synthesis by a variety of mechanisms, including some

**A**

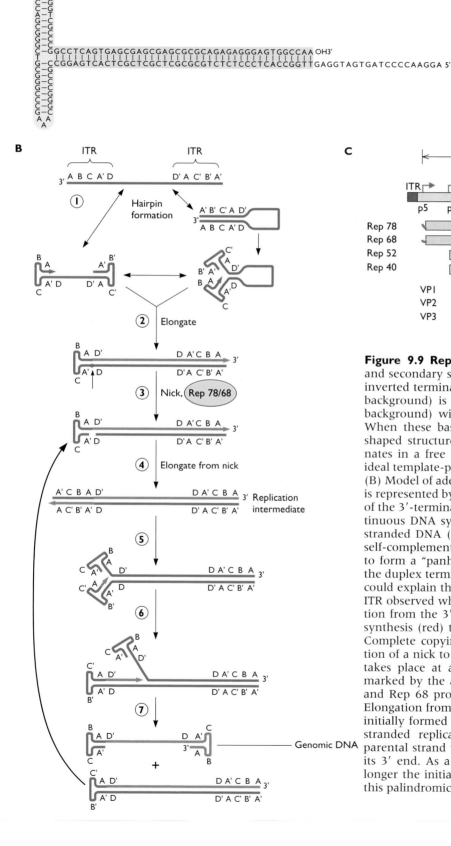

**B**

**C**

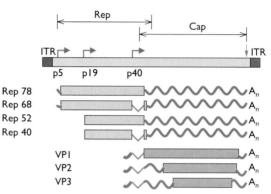

**Figure 9.9 Replication of parvoviral DNA.** (A) Sequence and secondary structure of the adeno-associated virus type 2 inverted terminal repetition (ITR). A central palindrome (tan background) is flanked by a longer palindrome (light blue background) within the central 125 bp of the 145-bp ITR. When these bases pair at the 3′ end of the genome, a T-shaped structure in which the internal duplex stem terminates in a free 3′ OH group is formed. This structure is an ideal template-primer structure for cellular DNA polymerases. (B) Model of adeno-associated virus DNA replication. The ITR is represented by 5′A′B′C′AD′3′ and 5′DA′CBA3′. Formation of the 3′-terminal hairpin provides a primer-template for continuous DNA synthesis to the 5′ end of the parental single-stranded DNA (step 1, left). However, it is possible that the self-complementary terminal repeat sequences first base pair to form a "panhandle," a single-stranded circle stabilized by the duplex terminal sequence (step 1, right). Such a structure could explain the repair of deletions or mutations within one ITR observed when the other is intact. In either case, elongation from the 3′ OH group of the hairpin allows continuous synthesis (red) to the 5′ end of the parental strand (step 2). Complete copying of the parental strand requires introduction of a nick to generate a new 3′ OH (step 3). Such nicking takes place at a specific site (the terminal resolution site) marked by the arrow and is performed by the viral Rep 78 and Rep 68 proteins, which are site-specific endonucleases. Elongation from the nick results in copying of sequences that initially formed the self-priming hairpin to form the double-stranded replication intermediate (step 4). However, the parental strand then contains newly replicated DNA (red) at its 3′ end. As a result, the ITR of the parental strand is no longer the initial sequence but rather its complement. Thus, this palindromic sequence is heterogeneous in populations of

*(continued on next page)*

with no counterpart in cellular DNA synthesis. Nevertheless, the majority are discrete DNA segments that contain sequences to which viral origin recognition proteins bind specifically to seed assembly of multiprotein complexes competent to begin DNA synthesis. When initiation is by self-priming, DNA sequences essential for replication include those needed to form and maintain a specific secondary structure in the template, as well as the sequence at which replication intermediates are cleaved for complete copying of the parental strand (Fig. 9.9A). Even though viral origins may be unconventional (compared to a cellular origin) and differ in number and location in viral genomes, they exhibit a number of common features, as do the proteins that recognize them.

### Number of Origins

In contrast to simian virus 40 and other papovaviruses, the genomes of the larger DNA viruses contain not one but two or three origins. The two identical adenoviral origins at the ends of the linear genome (Fig. 9.10) are essential for the unusual, but highly effective, mechanism of semiconservative adenoviral DNA synthesis, for they are the sites of assembly of complexes containing the viral DNA polymerase and protein primer. Thus, despite protein priming, the adenoviral genome contains discrete origins of replication. The genomes of herpesviruses such as Epstein-Barr virus and herpes simplex virus type 1 contain three origins of replication. As mentioned above, different functions can be ascribed to the different Epstein-Barr virus origins: a single origin (OriP) allows maintenance of episomal genomes in latently infected cells, while two other origins (OriLyt) support replication of the viral genome during lytic infection. The advantage of three origins, two copies of OriS and one copy of OriL (Fig. 9.11), to herpes simplex viruses is less clear, for a full complement of these origins is not necessary for efficient viral

DNA synthesis, at least in cells in culture. The two types of origin possess considerable nucleotide sequence similarity but differ in their organization and can be functionally distinguished. For example, OriL is activated when differentiated neuronal cells are exposed to a glucocorticoid hormone, but OriS is repressed. As glucocorticoids are induced by stress, a condition that induces reactivation of herpes simplex virus type 1 from latency in neurons, it has been suggested that replication from OriL may be particularly important during transition from a latent to a lytic infection. The presence of three origins might contribute to rapid replication of the viral genome in infected cells or be necessary for the mechanisms by which herpes simplex virus DNA is synthesized (see "Synthesis of Viral RNA Primers" above).

### Sequence Features of Viral Replication Origins

Even though the origins of replication of double-stranded DNA viruses are recognized by different proteins and support different mechanisms of initiation, they exhibit a number of common features (Fig. 9.12). The most prominent of these is the presence of AT-rich sequences. In general, AT base pairs contain only two hydrogen bonds between the bases in a duplex DNA molecule, whereas GC pairs interact via three such noncovalent bonds. The AT-rich sequence blocks (Fig. 9.12) are therefore believed to facilitate the unwinding of origins necessary for initiation of viral DNA synthesis from either RNA or protein primers. The best characterized viral origins of DNA replication (e.g., those of simian virus 40 and human adenovirus type 2 genomes) comprise a minimal essential core origin flanked by sequences that are dispensable but nonetheless significantly increase replication efficiency (Fig. 9.12). These stimulatory sequences contain binding sites for cellular transcriptional activators (Fig. 9.12). Yet other viral origins, those of papillomaviruses and OriLyt of Epstein-

**Figure 9.9 (continued)** adeno-associated virus DNA molecules, being present in one of the two orientations listed above. Indeed, such sequence heterogeneity provided an important clue for elucidation of the mechanism of viral DNA synthesis. The newly replicated 3′ end of the replication intermedate can form the same terminal hairpin structure (step 5) to prime a new cycle of DNA synthesis (step 6). One product is a molecule of single-stranded genomic DNA, displaced on copying to the end of the template strand (step 7). The second, which corresponds to the incompletely replicated molecule initially produced, can undergo additional cycles of replication following terminal resolution as in step 4. (C) Organization of coding sequences in the adeno-associated virus genome. The viral genome is depicted at the top, with the locations of the ITR, the promoters for transcription by RNA polymerase II (P5, P19, and P40), and the single site of polyadenylation (vertical red arrow) indicated. The two coding regions with the viral genome are termed Rep (replication) and Cap (capsid) for the functions of the proteins they encode. As shown below, each is expressed as multiple related proteins, which are synthesized from differentially spliced mRNAs transcribed from the different promoters. The two smaller products of the Rep region, Rep 52 and Rep 40, are not required for viral DNA synthesis but appear to facilitate encapsidation of progeny single-stranded genomic DNA molecules during assembly of virus particles.

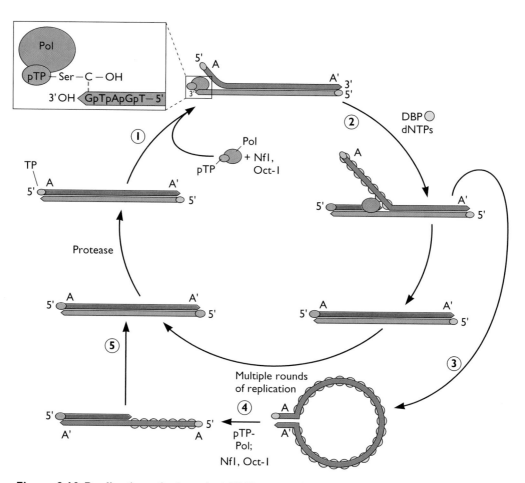

**Figure 9.10 Replication of adenoviral DNA.** Assembly of the viral preterminal protein (pTP) and DNA polymerase (Pol), both early proteins encoded by the E2 transcription unit (appendix, Fig. 15B), into a preinitiation complex at each terminal origin of replication activates covalent linkage of dCMP to a specific serine residue in the preterminal protein by the DNA polymerase (step 1). The cellular transcriptional activators Nf-1 and Oct-1 bind specifically adjacent to the core origin sequence (Fig. 9.12) and facilitate assembly of the preinitiation complex by binding to the viral replication proteins. The free 3′ OH group of preterminal protein-dCMP primes continuous synthesis of viral DNA in the 5′ → 3′ direction by the viral DNA polymerase (step 2). This reaction also requires the viral E2 single-stranded-DNA-binding protein (DBP), which coats the displaced second strand of the template DNA molecule, and a cellular topoisomerase. As the terminal segments of the viral genome comprise an inverted repeat sequence (A and A′), there is an origin at each end, and both parental strands can be replicated by this displacement mechanism. Reannealing of the complementary terminal sequences of the parental strand displaced in single-stranded form during the initial step forms a short duplex stem identical to the terminus of the double-stranded genome (step 3). The origin re-formed in this way directs a new cycle of protein priming and continuous DNA synthesis (steps 4 and 5).

Barr virus, include binding sites for viral proteins that are both transcriptional regulators and essential replication proteins. And all three herpes simplex virus type 1 origins lie between sites at which transcription of viral genes is initiated (Fig. 9.12). A close relationship between origin sequences and those that regulate transcription therefore appears to be a universal feature of viral origins.

## Recognition of Viral Replication Origins

Because they were the first to be identified and have been studied in exquisite detail, polyomaviral LT proteins provide the paradigm for viral origin recognition proteins. We therefore describe the properties of simian virus 40 LT in more detail as a prelude to discussion of other viral proteins with similar functions.

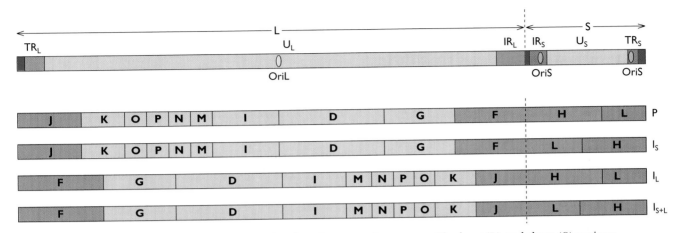

**Figure 9.11 Features of the herpes simplex virus type 1 genome.** The long (L) and short (S) regions of the viral genome that are inverted with respect to one another in the four genome isomers are indicated at the top. Each segment comprises a unique sequence ($U_L$ or $U_S$) flanked by internal and terminal repeated sequences. The locations of the two identical copies of OriS, in repeated sequences, and of the single copy of OriL are indicated. The orientation shown at the top is defined as the prototype (P) genome isomer. The other three isomers differ, with respect to the P form, in the orientation of S ($I_S$), in the orientation of L ($I_L$), or in both S and L ($I_{S+L}$). These differences are illustrated using the *Hin*dIII fragments of herpes simplex virus type 1 DNA. The unusual isomerization of this viral genome was deduced from the presence of fragments that span the terminal or internal inverted repeat sequences at 0.5 and 0.25 M concentrations, respectively, in such *Hin*dIII digests and examination of partially denatured DNA in the electron microscope.

**Figure 9.12 Common features of viral origins of DNA replication.** The single simian virus 40 origin, herpes simplex virus type 1 OriL (Fig. 9.11), and the adenovirus type 2 origin, which is repeated at each end of the viral genome (Fig. 9.10), are illustrated to scale, emphasizing common features. Binding sites for origin recognition proteins and AT-rich sequences are indicated in yellow and green, respectively. Herpes simplex virus type 1 OriL contains two types of binding site for the viral origin recognition protein encoded by the UL9 gene distinguished by the two shades of yellow. This origin, which is a perfectly symmetrical palindrome as indicated by the arrows, is located within the transcriptional control regions of the divergently transcribed UL29 and UL30 genes, which encode the DNA-binding protein ICP8 and the viral DNA polymerase, respectively. The two copies of OriS (Fig. 9.11) are very similar in sequence to OriL but lack the rightmost copy of the UL9 protein-binding site. The small size and relatively simple organization of the adenoviral origin probably reflects the unusual mechanism of initiation. The terminal sequence designated the core origin functions inefficiently in the absence of the adjacent binding site for the transcriptional activator nuclear factor 1 (Nf-1).

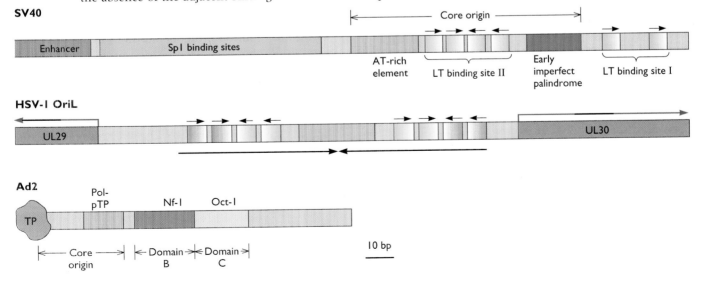

**Properties of simian virus 40 LT.** *Functions and organization.* The LT proteins of polyomaviruses such as simian virus 40 are remarkable proteins that provide essential replication functions and play other important roles in the infectious cycle. As we have seen, simian virus 40 LT is both necessary and sufficient for recognition of the viral origin, where it directs the assembly of cellular proteins into a complex capable of unwinding the origin and completing the synthesis of leading-strand Okazaki fragments. This viral protein also supplies the helicase activity necessary for origin unwinding and movement of the replication fork. LT proteins make a major contribution to the species specificity of polyomaviruses. Although the genomes of simian virus 40 and mouse polyomavirus are closely related in organization and sequence, they replicate only in simian and murine cells, respectively. Such host specificity in viral DNA replication is largely the result of species-specific binding of LT to the largest subunit of DNA polymerase α of the host cell in which the virus will replicate (Table 9.3). Although the precise mechanism remains to be determined, assembly of preinitiation complexes competent for unwinding of the origin does not take place when the LT of one polyomavirus binds to the origin of another.

LT proteins also ensure that the cellular proteins needed for simian virus 40 DNA synthesis are available in the host cell. By binding and sequestering specific cellular proteins, LT perturbs mechanisms that control cell growth and division and can induce infected cells to enter S phase when they would not normally do so (see "Induction of Synthesis of Cellular Replication Proteins by Viral Gene Products" below). In addition, LT carries out a number of transcriptional regulatory functions during the infectious cycle, including autoregulation of its own synthesis and activation of late gene expression.

Sequences of simian virus 40 LT necessary for its numerous activities have been mapped by analysis of the effects of specific alterations in the protein on virus replication in infected cells, DNA synthesis in vitro, or the individual biochemical activities of the protein. The properties of such altered proteins suggest that LT comprises discrete structural and functional domains, such as the minimal domain for specific binding to sites I and II at the viral origin or the discrete sequences by which LT binds to cellular replication proteins (Fig. 9.13). However, the activities of such functional regions defined by genetic and biochemical methods may be influenced by distant sites, as discussed in the next section. The LT protein has resisted the many attempts that have been made to obtain crystals suitable for X-ray diffraction, precluding structural interpretation of its many important functions.

*Regulation of LT synthesis and activity.* The simian virus 40 early enhancer and promoter are active in many cell types (chapter 8). Thus, the viral early gene encoding LT is efficiently transcribed as soon as the viral chromosome enters the nucleus of a host cell. The spliced early mRNA encoding LT is the predominant product of processing of these early transcripts. Although production of LT is not regulated during the early phase of infection, at least until the protein attains concentrations sufficient to repress early gene transcription, its activity is tightly controlled.

Simian virus 40 LT is phosphorylated at multiple Ser and Thr residues, most of which are located within one of two clusters near the N and C termini of the protein (Fig. 9.13). Although the significance of phosphorylation at every site is not known, it is clear that specific modifications regulate the ability of LT to support viral DNA synthesis. For example, phosphorylation of Thr124 is absolutely required for simian virus 40 DNA replication in infected cells and in vitro. This modification specifically stimulates binding of LT to origin site II (Fig. 9.3) and promotes cooperative interactions among LT molecules bound to the core origin and hence assembly of the double-hexamer complex (Fig. 9.4). Moreover, it is essential for unwinding of DNA from the origin by LT. As Thr124 does not lie within the minimal origin-binding domain of LT (Fig. 9.13), such regulation of DNA-binding activity is believed to be the result of conformational change induced by phosphorylation at this site. The best candidate for the protein kinase that activates LT by phosphorylation of Thr124 is cyclin-dependent kinase 2 (Cdk2) associated with cyclin A. The accumulation and activity of this complex are strictly regulated as growing cells traverse the cell cycle, but the mechanisms by which LT ensures that cellular replication proteins are made in simian virus 40-infected cells also result in production of active cyclin A–Cdk2 (see "Induction of Synthesis of Cellular Replication Proteins by Viral Gene Products" below).

**Origin recognition proteins of other viruses.** By definition, other viral origin recognition proteins share with simian virus 40 LT the ability to bind specifically to DNA sequences within the appropriate origin of replication. These proteins also bind to replication proteins (although these may be viral or cellular), and several possess the biochemical activities exhibited by LT (Table 9.5). For example, the UL9 protein, which recruits viral rather than cellular replication proteins, performs functions analogous to those of simian virus 40 LT at the origins of herpes simplex virus type 1 replication. This protein binds cooperatively to specific sites that flank the AT-rich sequences of these origins (Fig. 9.12), which are then distorted, and it possesses an ATP-dependent helicase activity that unwinds

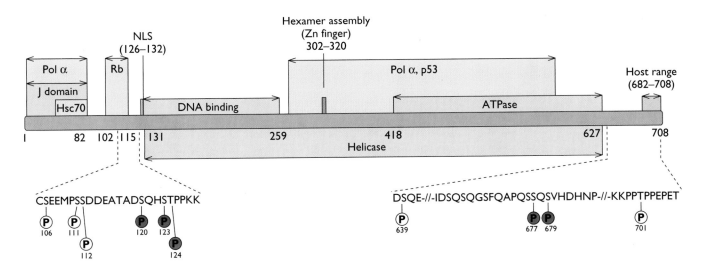

**Figure 9.13 Functional organization of simian virus 40 LT.** The minimal segments of LT required for binding to the DNA polymerase α-primase complex (Polα), to the cellular chaperone Hsc70 (Hsc70), to the cellular retinoblastoma (Rb) and p53 (p53) proteins that negatively regulate cell growth and proliferation, and to sites I and II in the origin of replication (DNA binding), as well as for the helicase and ATPase activities of the protein and hexamer assembly at the origin, are indicated on a linear representation of the 708-amino-acid protein. The nuclear localization signal (NLS) and a C-terminal sequence necessary for virion production but not viral DNA synthesis (Host range) are also shown. The region that binds to Hsc70 lies within an N-terminal segment, amino acids 1 to 82, termed the J domain because it shares sequences and functional properties with the *E. coli* protein DnaJ. The latter protein is a chaperone that assists the folding and assembly of proteins and is required during replication of bacteriophage λ. The chaperone functions of the J domain and Hsc70 are essential for replication and seem likely to assist assembly or rearrangement of the preinitiation complex. Although the three-dimensional structure of LT has not been determined, several observations indicate that it is organized into discrete structural and functional domains. For example, specific protease-resistant fragments that retain DNA binding or ATPase activity have been identified. Furthermore, subsegments of the protein that span the DNA-binding region can be expressed as stable, active proteins in *E. coli*, indicating that this region of LT independently adopts a stable, folded structure. Below are shown the two regions of the protein in which sites of phosphorylation are clustered, indicating modifications that have been shown to inhibit (red) or activate (green) the replication activity of LT.

DNA in the 3′ → 5′ direction (Table 9.5). In several cases, a single viral protein carries out all tasks associated with origin recognition and initial distortion, as well as recruitment of additional proteins essential for replication (Table 9.5). Although simian virus 40 LT is in many respects the best understood of these viral proteins, the only representative for which structural information is available is the Epstein-Barr virus EBNA-1 protein, responsible for recognition of the origin of plasmid maintenance, OriP. The structure of the DNA-binding and dimerization domains of EBNA-1 bound specifically to DNA has provided new insights into mechanisms of distortion of origin DNA (Box 9.3).

Although recognition of viral origins of replication by a single viral protein is common, it is not universal. For example, although the bovine papillomavirus type 1 E1 protein possesses the same activities as simian virus 40 LT

(Table 9.5), to which it is related in sequence, it cannot support papillomaviral DNA replication in infected cells. A second viral protein, the E2 transcriptional regulator, is also essential. It is believed that the E2 protein facilitates stable and specific association of E1 with the viral origin. The two proteins bind to the origin cooperatively and with far greater specificity than either alone. The adenoviral origins of replication are also recognized by a complex of viral proteins, the preterminal protein and viral DNA polymerase. These proteins associate as they are synthesized in the cytoplasm of infected cells and, once within the nucleus, bind specifically to a conserved sequence within the minimal origin of replication (Fig. 9.12). Assembly of the viral proteins into a preinitiation complex at the origin is stimulated by their direct interaction with two cellular transcriptional regulators which bind to sequences adjacent to the minimal origin (Fig. 9.12). These viral and cel-

**BOX 9.3**

*Binding of the Epstein-Barr virus EBNA-1 protein to OriP: a mechanism for origin distortion*

(A) Organization of binding sites for EBNA-1 dimers in OriP shown to scale. Initiation of replication within the dyad symmetry (DS) sequence, which comprises four EBNA-1 binding sites (1 to 4), is indicated by the vertical arrow. The activity of the DS sequence origin is stimulated by the family of repeat (FR) sequence, which is also necessary for maintenance of episomal viral DNA in latently infected B cells. In the absence of the DS origin, DNA synthesis can initiate at EBNA-1 binding sites within the FR se-

quence, suggesting that OriP is functionally complex. Exactly how the FR sequence stimulates replication is not clear, but the ability of EBNA-1 to bind to both DS and FR sequences, which are about 1 kbp apart, looping out the intervening DNA, appears to be important. Within the DS sequence, sites 1 and 2 function as a minimal origin whose activity is critically dependent on the 3-bp spacing of these EBNA-1 binding sites. (B) Three-dimensional structure of the EBNA-1 DNA-binding and dimerization domain dimer

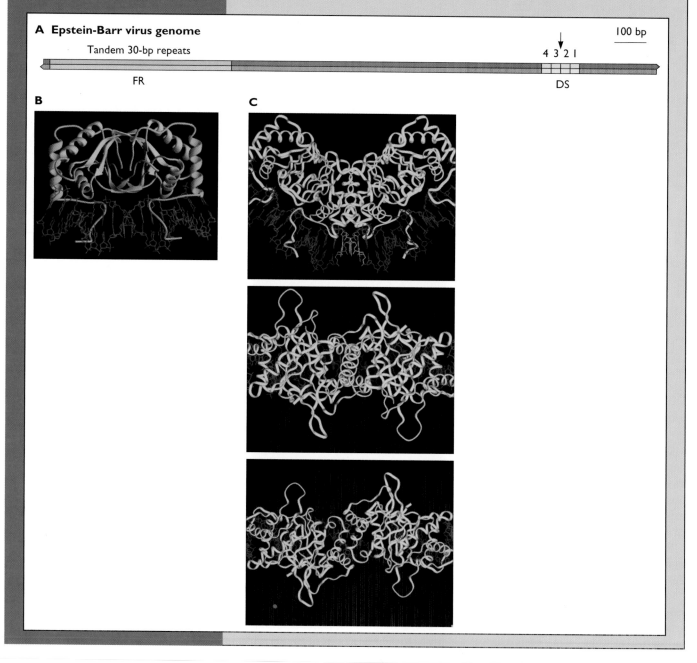

**A Epstein-Barr virus genome**

Tandem 30-bp repeats

100 bp

4 3 2 1

FR          DS

**B**

**C**

**BOX 9.3** *(continued)*

bound to DNA. The core domain of the dimer is shown in blue with the flanking domains in orange. This crystal structure of EBNA-1 bound to a single recognition site yielded a number of interesting surprises. In the first place, the structure of the central domain of the protein is essentially the same as that of the corresponding portion of the bovine papillomavirus E2 protein dimer. As the E2 protein is not the origin recognition protein of this virus (see text), such structural identity was not anticipated. Even more surprising are the different ways in which EBNA-1 and E2 dimers contact DNA despite their structural identity. The EBNA-1 dimer interacts with DNA via its flanking domains (orange) but also possesses two α-helices (white arrows) that are structurally analogous to the DNA recognition helices of the papillomavirus E2 protein. When EBNA-1 is bound to a single site, these helices are too far from the DNA to make contact with it, as

shown. However, it seems likely that when EBNA-1 dimers bind to adjacent sites in a functional origin, the α-helices do indeed make contact with DNA to facilitate distortion (and ultimately unwinding) of the origin as modeled in panel C. (C) Model of the binding of two EBNA-1 dimers to a minimal OriP. The minimal origin comprises EBNA-1 binding sites 1 and 2 (A), separated by 3 bp, to which two EBNA-1 dimers bind cooperatively. The close spacing of these binding sites in the minimal origin precludes binding of two dimers in the same way as in the single dimer-DNA complex shown in panel B: two EBNA-1 dimers collide when modeled on two DNA-binding sites separated by 3 bp (top and middle). The two dimers are blue and yellow, and the 3-bp connection between the DNAs observed in the cocrystal structure (red and purple) is white. The model is shown in the same view as in panel B (top) and viewed from above the DNA (middle).

However, EBNA-1 binding to OriP is known to distort the DNA. Specific distortion of the DNA and twisting of one ENBA-1 dimer relative to the other would relieve the steric clash when dimers bind to the minimal origin (bottom). The two potential DNA-binding motifs of EBNA-1 dimers are postulated to facilitate such distortion, which presumably favors subsequent unwinding of the origin. Additional contacts between the core domain α-helices (white arrows in panel B) and the specifically distorted DNA would stabilize the distorted origin. DNA contact by the α-helices would explain the ability of the core domain of EBNA-1 lacking the flanking regions that contact DNA of a single binding site (B) to bind to DNA. (B and C) From A. Bochkarev et al., *Cell* **84:**791–800, 1996, with permission. Courtesy of A. Bochkarev, University of Oklahoma Health Sciences Center, and L. Frappier, University of Toronto.

---

lular proteins are sufficient to reconstitute initiation of origin-dependent adenoviral DNA synthesis from templates that carry the mature terminal protein covalently linked to their 5′ ends, as in viral chromosomes that enter infected cell nuclei (Fig. 9.10). The terminal protein present on such natural adenoviral DNA templates facilitates unwinding of the duplex termini early in the initiation reaction and may be important for directing viral DNA molecules to the specialized nuclear sites at which replication takes place (see below).

## Viral DNA Synthesis Machinery

In addition to origin recognition proteins, larger viral DNA genomes encode viral DNA polymerases and other essential replication proteins (Table 9.6). The simplest viral DNA replication apparatus, and consequently the best understood, is that of adenoviruses, which comprises the preterminal protein primer and DNA polymerase and only one other protein, a single-stranded-DNA-binding protein. The last protein stimulates initiation and is essential during elongation, when it coats the displaced strands of the template DNA molecule (Fig. 9.10). It binds single-stranded DNA cooperatively and with high affinity, independently of DNA sequence, to stimulate the activity of the viral DNA polymerase as much as 100-fold and induce highly processive DNA synthesis by the viral DNA polymerase.

The most interesting property of the DNA-binding protein is that it supports movement of the replication fork over quite large distances without hydrolysis of ATP. Thus, no helicase activity is needed during adenoviral DNA synthesis. The crystal structure of a large segment of the DNA-binding protein indicates that this protein multimerizes via a C-terminal hook (Fig. 9.14). The formation of long protein chains results in cooperative high-affinity binding to single-stranded DNA and is the driving force for ATP-independent unwinding of the duplex template. Other single-stranded DNA-binding proteins, such as the herpes simplex virus type 1 UL8 protein (Table 9.6) or cellular replication protein A (Table 9.3), may destabilize double-stranded DNA helices by a similar mechanism. Complete copying of adenoviral DNA templates also requires a cellular topoisomerase, such as topoisomerase I (Table 9.4), which is presumed to relieve overwinding of the template upon extensive replication, as during simian virus 40 DNA synthesis.

The simplicity of the adenoviral DNA synthesis machinery stands in stark contrast to the complexity of the cellular replisome, assembled, for example, at simian virus 40 replication forks. This property can be attributed to the very efficient protein priming mechanism, in which single molecules of the preterminal protein-dCMP primer support continuous synthesis of full-length adenoviral daugh-

**A**

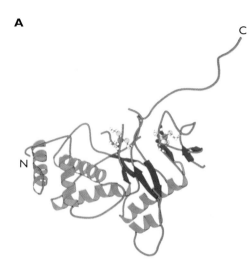

**B**

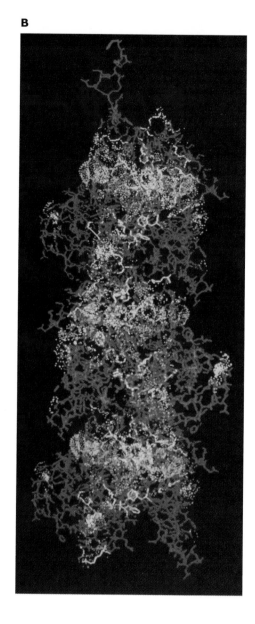

**Figure 9.14 Crystal structure of the adenoviral single-stranded DNA-binding protein.** (A) Ribbon diagram of the C-terminal nucleic acid-binding domain (amino acids 176 to 529) of the human adenovirus type 5 single-stranded-DNA-binding protein, showing the two sites of Zn$^{2+}$ (red atom) coordination. The most prominent feature is the long (≈40-Å) C-terminal extension. This C-terminal hook of one protein molecule invades a cleft between two α-helices in its neighbor in the infinite protein array formed in the crystal. Deletion of the C-terminal 17 amino acids of the DNA-binding protein fragment eliminates cooperative binding of the protein to DNA, indicating that the interaction of one molecule with another via the C-terminal hook is responsible for cooperativity in DNA binding. (B) Likely DNA-binding surface in a chain of three DNA-binding protein molecules. This surface was identified both by the locations of substitutions that inhibit the DNA-binding and replication functions of the protein and as a region of high positive potential (yellow) for interaction with the phosphate backbone of DNA. The regions colored cyan in the model are the ends of the polypeptide chain of each monomer. The organization of the positively charged surface and locations of residues implicated in DNA binding suggests that single-stranded DNA is wound around the DNA-binding protein chain. From P. A. Tucker et al., *EMBO J.* **13:**2994–3002, 1994, with permission. Courtesy of P. C. van der Vliet, Utrecht University.

ter DNA strands (Fig. 9.10). The other DNA viruses that rely on viral DNA polymerases do not enjoy the advantages conferred by protein priming. Consequently, their replication depends on a larger number of accessory replication proteins (Table 9.6). Herpes simplex virus type 1 genes encoding essential replication proteins have been identified by both genetic methods and a DNA-mediated transformation assay for the gene products necessary for plasmid replication directed by a viral origin (Fig. 9.15). Replication from a herpes simplex virus type 1 origin in this assay system requires not only the viral DNA polymerase and origin recognition protein but also five other

viral proteins (Table 9.6). These proteins are functional analogs of essential components of the cellular replication machinery. The products of the UL5, UL8, and UL52 genes form a viral primase, which also functions as a helicase, and the UL42 protein is a processivity factor. Although the herpesviral proteins listed in Table 9.6 provide a large repertoire of replication functions, they do not support viral DNA synthesis in vitro. The additional viral and/or cellular proteins needed to reconstitute herpesviral DNA synthesis have not yet been identified. One may be cellular topoisomerase II, which is essential for replication in infected cells.

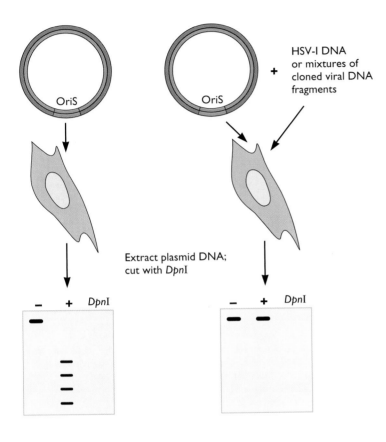

Extract plasmid DNA;
cut with DpnI

**Figure 9.15 DNA-mediated transformation assay for essential herpes simplex virus type 1 replication proteins.** A plasmid carrying a viral DNA fragment spanning OriS is introduced into monkey cells permissive for herpesvirus replication. In the absence of viral proteins, the plasmid DNA is not replicated and retains the methyl groups added to A residues in a specific sequence, which are introduced in *E. coli* by the *dam* methylation system. These sequences include the recognition site for the restriction endonuclease *Dpn*I, which cleaves only such methylated DNA. Thus, the unreplicated plasmid DNA remains sensitive to *Dpn*I cleavage. However, when all viral genes encoding proteins required for OriS-dependent replication are also introduced into the cells (right), the plasmid is replicated. Because the newly replicated DNA is not methylated at *Dpn*I sites, it cannot be cleaved by this enzyme. The sensitivity of the plasmid to *Dpn*I cleavage therefore provides a simple assay for plasmid replication and hence for the identification of viral proteins required for replication from OriS.

## Resolution and Processing of Viral Replication Products

Several of the viral DNA replication mechanisms described in preceding sections yield products that do not correspond to the parental viral genome. For example, replication of simian virus 40 DNA yields two interlocked double-stranded circular DNA molecules that must be separated by cellular topoisomerase II. Such resolution is required whenever circular templates (e.g., episomal Epstein-Barr virus DNA) are replicated as monomers. In other cases, replication yields multimeric "endless" DNA molecules from which linear genomes of fixed length and sequence must be processed for packaging into virions. This situation is exemplified by the herpes simplex virus type 1 genome.

Exactly how head-to-tail concatemers containing multiple copies of the herpes simplex virus type 1 genome are made from the circular viral DNA templates formed in infected cells is not clear. Concatemeric DNA molecules may be synthesized from a circular template by a rolling-circle mechanism in which nicking of one template strand allows the other to be copied continuously multiple times (Fig. 9.16A). Rolling-circle replication is characteristic of

several bacteriophage DNA genomes, such as that of phage λ, and for many years was believed to be responsible for synthesis of herpesviral DNA concatemers in lytically infected cells. However, this mechanism provides no explanation for the nonlinear, branched structure of herpes simplex virus type 1 replication intermediates or the fact that conversion of the genome from one of its four isomers (Fig. 9.11) to another occurs by the time that newly replicated DNA can first be detected in infected cells. Such isomerization is the result of recombination between repeated viral DNA sequences (see "Recombination of Viral Genomes" below). These properties suggest that recombination may be an essential reaction during herpes simplex virus type 1 DNA synthesis. Various mechanisms based on replication of phage or plasmid DNAs can be envisioned. For example, long concatemers could be produced if recombination between repeated sequences (Fig. 9.11) in a replicating, circular DNA template inverted one of the two bidirectional replication forks initially established at an origin. Such a mechanism of recombinational reorientation of the replication forks operates during replication of a yeast plasmid (Fig. 9.16B). As a result of the inversion, the two replication forks move in the same

**A**

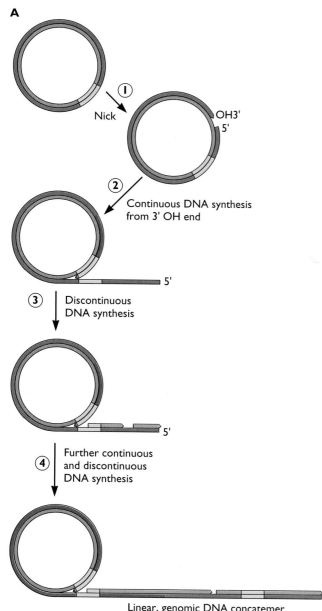

**B**

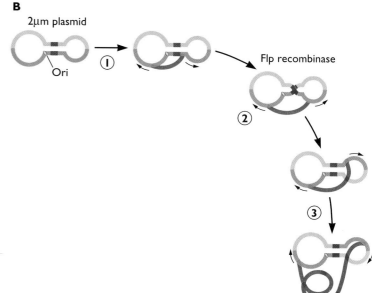

**Figure 9.16 Possible mechanisms of replication of herpesviral DNA.** (A) The rolling circle mechanism of DNA replication. A circular DNA template, like that formed upon ligation of the ends of the linear herpesviral genome, is nicked at a specific site to produce a terminal 3' OH group that can serve as a primer for DNA synthesis (step 1). Continuous copying of the intact template strand from this primer displaces the 5' end of the nicked parental strand (step 2), which is then replicated by discontinuous DNA synthesis from multiple RNA primers (step 3). Multiple cycles of continuous copying of the intact strand of the circular template, followed by discontinuous DNA synthesis on the displaced strand template (step 4) produce double-stranded linear DNA molecules containing multiple copies of the genome (concatemers), as indicated by the repeats of the genomic sequence marked in yellow. Such rolling-circle replication, so called because of the multiple cycles of copying around a circular template, is characteristic of several DNA phages, including φX174 and late replication of phage λ. (B) Model for recombinational reorientation of replication forks illustrated for the yeast 2μm plasmid. Bidirectional DNA replication begins from the single origin of plasmid replication (step 1). The plasmid contains two copies of a 48-bp sequence (purple) recognized by the site-specific recombinase termed the Flp (for flip) protein. This enzyme catalyzes recombination between the repeated sequences, such that the replication forks become oriented in the same direction (step 2). Subsequent DNA synthesis therefore generates long DNA molecules that contain multiple copies of the parental genome (step 3).

direction around the circular DNA template, to produce concatemers. However, until herpes simplex virus type 1 DNA replication is better understood, an entirely novel mechanism cannot be excluded.

Linear herpes simplex virus type 1 DNA molecules with termini identical to those of the infecting genome are lib-

erated from concatameric replication products by cleavage at specific sites within the first of the direct repeats, the *a* repeat, that form the segment present at both ends of the linear viral genome (Fig. 9.11). As described in chapter 13, such cleavage is coupled with encapsidation of viral DNA molecules during virion assembly.

## Mechanisms of Exponential Viral DNA Replication

The details of the mechanisms by which DNA viruses solve the molecular puzzles associated with complete replication of their genomes vary considerably from one virus to another. Nevertheless, each of these strategies generally ensures efficient replication of viral DNA. Production of $10^4$ to $10^5$ viral genomes, or more, per infected cell is not uncommon, as the products of one cycle of replication are recruited as templates for the next. Such **exponential** viral DNA synthesis sets the stage for assembly of a large burst of progeny virions. In this section, we discuss regulatory mechanisms that ensure efficient viral DNA synthesis. These regulatory circuits primarily impinge on the expression of cellular or viral genes encoding the proteins that carry out viral DNA synthesis.

## Induction of Synthesis of Cellular Replication Proteins by Viral Gene Products

With few exceptions, virus reproduction is studied in established cells lines that are permissive for the virus of interest. Such immortal or transformed cell lines grow and divide indefinitely and thus can differ markedly from the cells in which viruses reproduce in nature. For example, highly differentiated cells, such as neurons or the outer cells of an epithelium, do not divide and are permanently in a specialized resting state termed the $G_0$ state. Many other cells in an organism divide only rarely, or only in response to specific stimuli, and therefore spend much of their lives in the $G_0$ state. Such cells do not contain many of the components of the cellular replication machinery and are characterized by generally slow rates of synthesis of RNAs and proteins. The $G_0$ state would not seem to provide a hospitable environment for virus reproduction, which entails the synthesis of large quantities of viral nucleic acids and proteins, often at a rapid rate. Nevertheless, viruses often reproduce successfully within cells in $G_0$. In some cases, such as replication of several herpesviruses in neurons, the virus supplies the replication machinery. Other viruses stimulate resting or slowly growing cells to abnormal activity by disruption of cellular circuits that control cell growth and division. This strategy is characteristic of papova- and adenoviruses. The discovery that these viruses, which rely on quite different strategies for viral DNA synthesis, disrupt the same cellular cell growth control circuits was quite unanticipated and of the greatest importance in elucidating the roles of two critical regulators of cell growth and division, the cellular retinoblastoma (Rb) and p53 proteins.

### Functional Inactivation of the Rb Protein

Loss or mutation of both copies of the cellular retinoblastoma (*rb*) gene is associated with the development of tumors of the retina in children and young adults. Because it is the loss of normal function that leads to tumor formation, the *rb* gene is defined as a **tumor suppressor** gene. As discussed in chapter 16, it is now clear that the Rb protein is an important component of the regulatory program that ensures that cells grow, duplicate their DNA, and divide in an orderly manner. In particular, the Rb protein controls entry into the period of the cell cycle in which DNA is synthesized, the **S phase**, from the preceding $G_1$ phase. Our current appreciation of the critical participation of this protein in the control of cell progression and of the mechanism by which this protein operates stems from the discovery that Rb binds directly to the two adenoviral E1A proteins (see chapter 8). The segments of the E1A proteins necessary for binding to Rb in vitro are also required for these viral proteins to immortalize or transform cells. This correlation suggested that E1A proteins neutralize the growth-inhibitory function of Rb, mimicking genetic inactivation of both *rb* alleles in retinoblastomas and other human tumors. This hypothesis was further strengthened by the similar Rb-binding activities exhibited by segments of papovaviral early proteins that are essential for the ability of these proteins to transform cells.

The generally accepted model for the mechanism of action of the Rb protein (Fig. 9.17) was also initially based on the properties of adenoviral E1A proteins. In uninfected cells prior to entry into S phase, the Rb protein is found in the nucleus bound to the cellular transcriptional activator E2f. This complex, which can bind to specific promoters via the DNA-binding activity of E2f, functions as a repressor rather than an activator of transcription (Fig. 9.17). Binding of adenoviral E1A proteins, simian virus 40 LT, or E7 proteins of highly oncogenic human papillomaviruses to Rb releases E2f from this association and sequesters Rb. Thus, free E2f becomes available to stimulate transcription of cellular genes whose products participate directly or indirectly in DNA synthesis or in control of cell cycle progression (Fig. 9.17; see chapter 16). Because they interact preferentially with hypophosphorylated Rb, the viral proteins override the normal control of E2f activity by cell cycle-dependent phosphorylation of Rb (Fig. 9.17B).

Although the papovaviral LT and E7 and adenoviral E1A proteins all release E2f from inhibitory associations with the cellular Rb protein (Fig. 9.17), the DNA genomes of these viruses replicate by entirely different mechanisms, using largely cellular or mostly viral replication proteins, respectively. Consequently, the benefits conferred by activation of E2f when a papovavirus or an adenovirus infects a cell that is growing slowly or is withdrawn from the cell cycle are virus specific. While the advantages for viral DNA replication can be deduced, it is important to stress that

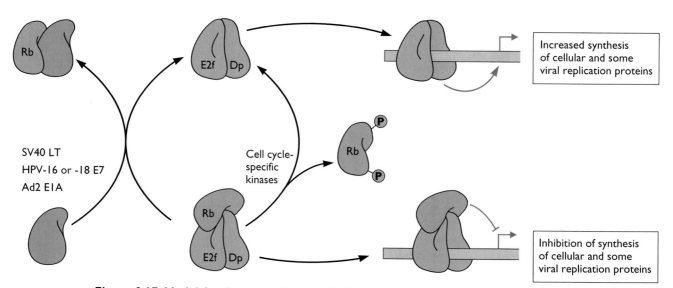

SV40 LT
HPV-16 or -18 E7
Ad2 E1A

Cell cycle-specific kinases

Increased synthesis of cellular and some viral replication proteins

Inhibition of synthesis of cellular and some viral replication proteins

**Figure 9.17 Model for the abrogation of the function of the Rb protein by adenoviral and papovaviral proteins.** The members of the E2f family of transcriptional regulators are heterodimeric proteins, each containing one E2f and one Dp subunit. To date six *e2f* (*e2f1* to *e2f6*) and three *dp* (*dp1* to *dp3*) genes have been identified. As discussed in chapter 16, there is accumulating evidence that these E2f proteins fulfill different functions. On binding to specific promoter sequences, E2f dimers stimulate transcription of cellular genes encoding replication proteins, histones, and proteins that allow passage through the cell cycle (green arrow). Binding of Rb protein does not prevent promoter recognition by E2f. However, Rb protein present in the complex represses transcription (red bar). Phosphorylation of Rb protein at specific sites, a modification that normally takes place to allow cells to enter S phase (see chapter 16), induces its dissociation from E2f and thus activates transcription of cellular genes expressed in S phase. The adenoviral E1A proteins, simian virus 40 LT, and the E7 proteins of certain human papillomaviruses (types 16 and 18) bind to the region of Rb protein that contacts E2f. These proteins can therefore disrupt Rb-E2f complexes and sequester Rb protein to activate E2f-dependent transcription.

essentially all studies of the cell growth control functions of LT, E7, or E1A proteins have been directed toward elucidation of mechanisms of transformation. For example, it has been shown that the smaller adenoviral E1A protein is necessary for efficient viral DNA synthesis when the virus infects quiescent human cells. However, it has not been established that this is because the E1A protein plays a critical role in activation of cellular E2f, as assumed in the model described below.

**E2f and simian virus 40 DNA synthesis.** As papovaviral replication requires the cellular replication machinery, viral DNA synthesis cannot take place unless these host proteins are present in the infected cell. Expression of the cellular genes encoding DNA polymerases, accessory replication proteins, histones, and enzymes that synthesize deoxyribonucleotide substrates for DNA synthesis is tightly controlled during the cell cycle, such that these proteins are made only just before they are needed in S phase. The transcriptional control regions of many of

these genes contain E2f-binding sites, which are thought to be required for their efficient transcription in response to free E2f or repression by an E2f-Rb complex. Expression of simian virus 40 LT (or the papillomavirus E7 protein) in infected cells is therefore believed to activate production of cellular proteins necessary for viral DNA synthesis and thus ensure efficient replication of the viral genome independent of the growth state of the host cell. The ability of simian virus 40 LT both to induce synthesis of components of the cellular replication machinery in the infected cell and to initiate viral DNA synthesis seems likely to coordinate viral replication with the induced entry of the host cell into S phase. Indeed, as discussed previously, LT replication functions are regulated by phosphorylation, probably by the cyclin A–cyclin-dependent kinase 2 complex that accumulates as cells enter S phase.

**E2f and adenoviral DNA synthesis.** Activation of E2f in adenovirus-infected cells is believed to promote viral

DNA synthesis in two ways: stimulation of transcription of cellular genes encoding enzymes that make substrates for DNA synthesis, such as thymidine kinase and dihydrofolate reductase (enzymes that are not supplied by the virus), and activation of expression of all three viral replication proteins. The viral DNA polymerase, preterminal protein primer, and DNA-binding protein are encoded within the E2 transcription unit, whose early promoter contains two binding sites for E2f (Fig. 9.18). In fact, these critical cellular regulators derive their name from the binding sites in this adenoviral promoter, the first to be identified. Binding of E2f is necessary for efficient E2 transcription during the early phase of adenovirus infection. Thus, the conversion of E2f from an Rb-bound repressor of transcription to a free activator by the E1A proteins must

be essential for production of the viral replication proteins during the adenoviral infectious cycle.

The activity of the promoter that directs E2 transcription during the early phase of infection is controlled by the immediate-early E1A proteins by at least three distinct mechanisms (Fig. 9.18). These E1A-dependent regulatory mechanisms presumably operate synergistically, to ensure synthesis of viral mRNAs encoding replication proteins in sufficient quantities to support numerous cycles of viral DNA synthesis. Preferential polyadenylation of the long primary transcripts at the site that produces the mRNA encoding the viral DNA-binding protein allows this protein to be made in the large quantities necessary to coat the single-stranded viral DNA displaced during each replication cycle (Fig. 9.10).

**Figure 9.18 Stimulation of transcription from the adenoviral E2 early promoter by E1A proteins.** The E2E promoter binding sites for the cellular Atf, E2f, and TfIId (TATA sequence) proteins are necessary for E2E transcription in infected cells. The inversion of the two E2f sites and their precise spacing are essential for assembly of an E2f-DNA complex unique to adenovirus-infected cells, in which binding of an E2f heterodimer to each of the two sites is stabilized by binding of two molecules of the E4 ORF6/7 protein. Such binding of the E4 protein greatly promotes cooperative binding of E2f to these sites and increases the lifetime of E2f-DNA complexes. The availability of the cellular E2f and viral E4 ORF6/7 proteins is a result of the action of immediate-early E1A proteins: either the 243R or 289R protein can sequester unphosphorylated Rb to release active E2f from Rb-E2f complexes in infected cells, and the 289R protein stimulates transcription from the E4 promoter. This larger E1A protein can also directly stimulate transcription from the E2E promoter, probably via components of the general initiation protein TfIId, such as Tbp, which binds to the TATA element, and specific members of the Atf family (see chapter 8).

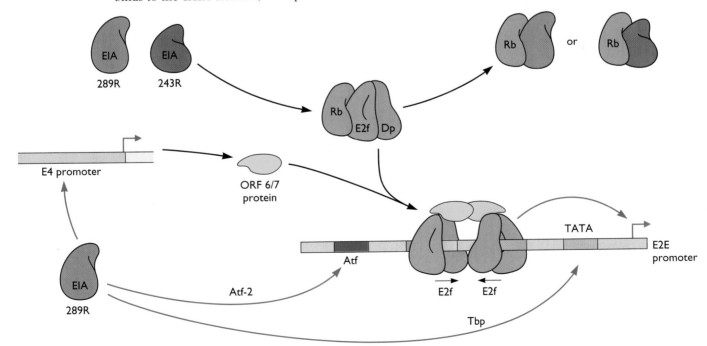

### *Abrogation of the Function of the p53 Protein*

Papova- and adenoviruses encode proteins, made soon after cells are infected, that block the normal functions of a second cellular tumor suppressor, the p53 protein. As discussed in more detail in chapter 16, DNA damage or unscheduled DNA synthesis leads to the accumulation of the p53 protein and either cell cycle arrest or **apoptosis**, programmed cell death. Furthermore, some viral proteins that sequester unphosphorylated Rb protein and its cousins, such as the adenoviral E1A protein, can directly induce p53 protein-dependent apoptosis. It is therefore generally believed that inactivation of p53 protein by papovaviral and adenoviral early proteins provides insurance against deleterious and potentially lethal consequences of the mechanisms that bypass control of the activity of E2f in cells infected by these viruses. At least one protein synthesized after activation of p53 protein, the cyclin-cdk inhibitor p21 (chapter 16), has been reported to bind to and inhibit the activity of the processivity protein proliferating cell nuclear antigen. Consequently, inactivation of p53 may also be of more direct benefit when synthesis of viral DNA depends on cellular replication proteins, as in papovavirus-infected cells.

## Synthesis of Large Sets of Viral Replication Proteins

The DNA genomes of several viruses encode large cohorts of proteins that participate directly or indirectly in viral DNA synthesis. Such viruses, exemplified by herpes simplex virus type 1, rely on two classes of viral replication proteins during lytic infection. The first comprises the viral DNA polymerase and other proteins responsible for viral DNA synthesis discussed previously (Table 9.6). The viral proteins in the second class are enzymes, analogous to host cell enzymes, that catalyze reactions by which deoxyribonucleotides are synthesized (Table 9.6). For example, viral thymidine kinase and ribonucleotide reductase are synthesized in herpesvirus-infected cells. In general, members of this second class of proteins are dispensable for replication in established cells in culture but are essential in animal hosts. Herpes simplex viruses lacking thymidine kinase or ribonucleotide reductase gene cannot replicate in neurons because such terminally differentiated cells are permanently withdrawn from the cell cycle and do not make enzymes that produce substrates for DNA synthesis.

Efficient expression of all herpes simplex virus type 1 replication proteins is primarily the result of the viral transcriptional cascade described in chapter 8. Expression of the genes encoding these viral proteins, which are early (β) genes, is regulated by products of immediate-early genes. These immediate-early proteins operate transcriptionally (e.g., ICP0 and ICP4) or posttranscriptionally (e.g., ICP27) to induce synthesis of viral replication proteins at concentrations sufficient to support exponential replication of viral DNA. Whether the multiple origins of herpes simplex virus type 1 and other herpesviruses increase the rate of replication will not be clear until the viral replication mechanisms have been better characterized.

## Viral DNA Replication Independent of Cellular Proteins

One method guaranteed to ensure replicative success of a DNA virus, regardless of the growth state of the host cell, is the provision of all necessary proteins by the virus itself. On the other hand, this mechanism is genetically "expensive," which may be the reason why it has been adopted only by the viruses with the largest DNA genomes, such as the poxvirus vaccinia virus. This virus, which replicates in the cytoplasm at specialized foci sometimes called **viral factories**, supplies all the proteins needed for viral DNA synthesis. The current catalog of such vaccinia virus proteins includes a DNA polymerase and several other proteins whose roles are not well understood (Table 9.6). The vaccinia virus genome also encodes several other enzymes that would be expected to participate in DNA synthesis or in resolution of replication products, including a type I topoisomerase and a DNase with endonucleolytic activity, as well as several enzymes for synthesis of deoxyribonucleotide precursors (Table 9.6). None of the latter appear to be essential for virus reproduction in actively growing cells in culture. However, several, such as the thymidine kinase, are necessary for efficient virus growth in quiescent cells or animal hosts, where they presumably increase less than optimal dNTP pools. Although much remains to be learned about the mechanism by which the vaccinia virus genome is replicated and the functions of viral replication proteins, it is clear that no cellular proteins are needed, for this and other poxviruses replicate efficiently in enucleated cells.

## Delayed Synthesis of Virion Structural Proteins

During lytic infection by DNA viruses, each cycle of replication increases the number of DNA molecules that can be copied in the subsequent cycle. This increase in the pool of replication templates, a doubling in each cycle of the simpler viruses like simian virus 40 and adenovirus, undoubtedly makes an important contribution to the rapid amplification of genomes that follows initiation of viral DNA synthesis in infected cells. Progeny viral DNA molecules must eventually be encapsidated during the assembly of new virus particles and thus will no longer be available as templates for DNA synthesis. However, assembly and encapsidation of the genome are delayed with respect to initiation of viral DNA synthesis because transcription of late genes encoding virion proteins cannot begin until viral DNA has been replicated within the infected cell. The late, and replication-dependent, synthesis of virion proteins, a common feature of the infectious cycles of DNA viruses, therefore maximizes the number of templates available for viral DNA synthesis.

## Inhibition of Cellular DNA Synthesis

DNA viruses that do not rely heavily on components of the cellular replication machinery often inhibit cellular DNA synthesis, presumably to eliminate competition for substrates from cellular DNA replication. The larger DNA viruses (herpes- and poxviruses) efficiently inhibit synthesis of cellular DNA during productive infection. Adenoviruses also inhibit cellular DNA synthesis when they infect actively growing cells in culture, even though they have evolved elaborate mechanisms to induce quiescent host cells to reenter the cell cycle. Although inhibition of cellular DNA synthesis by these DNA viruses was described in some of the earliest studies of their infectious cycles, very little is known about the mechanisms that shut down this cellular process. Nevertheless, it appears that inhibition is the result of the intervention of virus-specific mechanisms rather than of passive competition between viral and cellular DNA polymerases for the finite pools of deoxynucleotide substrates. For example, infection of actively growing cells by adenovirus or the herpesvirus human cytomegalovirus has been reported to induce cell cycle arrest, although the mechanisms have not been established. Expression of the Epstein-Barr virus Zta protein, a sequence-specific transcriptional regulator and origin-binding protein, also arrests cells at a point in the cell cycle prior to S phase. Thus, cellular DNA synthesis is precluded. Such arrest is the result of increased concentrations of cellular proteins that negatively regulate progression through the cell cycle, such as hypophosphorylated Rb protein. Whether inhibition of cellular DNA synthesis is a primary, or an indirect, consequence of poxvirus infection is not known, although there has been one report of destruction of the cellular genome by the viral DNase (Table 9.6).

## Localization of Replicating Viral DNA to Specialized Intracellular Sites

A common, probably universal, feature of cells infected by viruses with DNA genomes is the presence of virus-specific structures that are the sites of viral DNA synthesis. Vaccinia virus DNA replication takes place in discrete cytoplasmic viral factories containing both the DNA templates and viral replication proteins. The replication of other viral

DNA genomes within infected cell nuclei also takes place in discrete compartments, which can be visualized as distinctive structures specific to virus-infected cells (Fig. 9.19). Such structures, known as **replication centers** or **replication compartments**, have been best characterized in human adenovirus- or herpes simplex virus type 1-infected cells. They contain newly synthesized viral DNA and the viral proteins necessary for viral DNA synthesis (Table 9.6), as well as other viral and cellular proteins. Among the cellular proteins associated with herpes simplex virus type 1 replication centers are several that catalyze DNA synthesis reactions, such as DNA polymerases $\alpha$ and $\gamma$, DNA ligase, and topoisomerase II. Their presence in replication centers might be because they participate in viral DNA synthesis or may be a consequence of the assembly of viral replication centers at sites of initiation of cellular DNA synthesis. The localization of both the templates for viral DNA synthesis and the replication proteins at a limited number of intracellular sites undoubtedly facilitates exponential viral DNA replication. This arrangement increases the local concentrations of proteins that must interact with one another or with viral origin sequences or replication forks, favoring such intermolecular interactions by the law of mass action. In addition, the high local concentrations of replication templates and proteins appear ideally suited for efficient recruitment of the products of one replication cycle as templates for the next.

Much remains to be learned about the composition, assembly, and function of viral replication centers. Nevertheless, there is good reason to believe that these structures also serve as foci for viral gene expression, presumably in part by concentrating templates for transcription with the proteins that carry out or regulate this process. Adenoviral and herpesviral replication centers contain not only replication proteins but also viral and cellular proteins necessary for viral gene expression. The herpes simplex virus type 1 immediate-early ICP4 and ICP27 proteins, transcriptional and posttranscriptional regulators, respectively, of viral gene expression, as well as the host cell's RNA polymerase II, are recruited to these nuclear sites. Similarly, an adenoviral early protein complex necessary for selective export of viral late mRNAs from the nucleus is associated with the periphery of viral replication centers, which contain nascent viral RNA.

Viral replication centers do not assemble at random sites but rather appear to be the result of viral colonization of specialized niches within mammalian cell nuclei. When they enter the nucleus, infecting herpes simplex virus type 1 genomes localize to preexisting nuclear bodies containing a specific cellular protein called the promyelocytic leukemia protein (Pml). Moreover, herpesviral replication compartments are associated with the periphery of these cellular

**A**

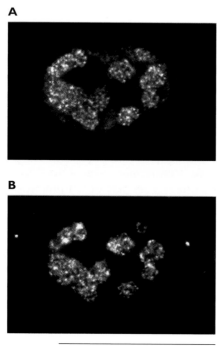

**B**

15 µm

**Figure 9.19 Nuclear replication compartments of herpes simplex virus type I-infected monkey cells.** The locations of the UL29 single-stranded DNA-binding protein (A) and newly replicated DNA (B) were visualized by immunofluorescence with antibodies against UL29 protein and bromodeoxyuridine incorporated into DNA, respectively. Newly synthesized viral DNA and the UL29 protein colocalize to a limited number of globular patches, which contain all seven known viral replication proteins. These viral replication compartments are found at the peripheries of preexisting nuclear domains or structures containing cellular promyelocytic leukemia protein (see text). The infecting viral DNA is found associated with these structures, but there is subsequent reorganization following expression of the viral ICP0 protein. From C. I. Lukonis and S. K. Weller, *J. Virol.* **71:**2390–2399, 1997, with permission. Courtesy of S. Weller, University of Connecticut Health Sciences Center.

structures, which are termed promyelocytic leukemia protein-containing nuclear bodies or nuclear domain 10 structures. The latter term is derived from the average number of the structures present in most cells. Replicating DNA of human adenovirus type 5 and the papovavirus simian virus 40 is also localized to the periphery of nuclear domain 10 structures. The common association of replication centers of different DNA viruses with the same intranuclear structures suggests that these viruses exploit some beneficial feature of the organization of the nuclei of their host cells. Exactly what the advantage might be is not yet clear, for little is known about the molecular functions of such nuclear

structures. Nevertheless, there can be little doubt that further characterization of viral replication centers will shed new light on a poorly understood, indeed controversial, aspect of eukaryotic cell biology, the structural and functional organization of the nucleus.

## Limited Replication of Viral DNA

Exponential replication of viral DNA is the typical pattern when the majority of DNA viruses infect cells in culture. In contrast, several can establish long-term relationships with their hosts and host cells in which the number of genomes produced is limited. This state of affairs stands in stark contrast to the rapid synthesis of geometrically increasing numbers of progeny viral DNA molecules described previously. Various mechanisms responsible for such copy number control are described in this section.

### Replication as Part of the Cellular Genome: Integrated Parvoviral DNA

As mentioned previously, the dependovirus subgroup of the mammalian parvoviruses, the adeno-associated viruses, reproduce only in cells coinfected with a helper adeno- or herpesvirus. Although the latter viruses are widespread in hosts infected by adeno-associated viruses, the chances that a particular host cell will be simultaneously infected by two viruses are very low. The strategy of exploiting other viruses to provide functions for efficient expression of the genetic information of adeno-associated viruses therefore appears to be potentially lethal for individual virus particles. In fact, this is not the case, for adeno-associated virus has an alternative mechanism of survival. When the virus infects a cell that contains no helper virus, its genome becomes integrated into that of the host cell and is replicated as part of a cellular replicon.

Activation of this program for long-term survival of the adeno-associated virus genome depends on expression of its regulatory region (Rep) (Fig. 9.9C) in an infected cell. The two larger proteins encoded by this region, Rep 78/68, are multifunctional (Table 9.5) and control all phases of the viral life cycle. When helper virus proteins, such as adenoviral E1A, E1B, and E4 proteins, allow synthesis of large quantities of Rep 78/68, adeno-associated virus DNA is replicated by the mechanism described previously. In the absence of helper functions, only very small quantities of Rep 78/68 are made, even in healthy, dividing cells. Consequently, there is little viral DNA synthesis, and the genome integrates into that of the host cell. The latter reaction is mediated by Rep 78/68 and depends on specific viral sequences. One of the most unusual features of this integration reaction is that it is also specific with respect to the cellular genome. Unlike other viral DNA molecules that become integrated, adeno-associated virus DNA integrates within a specific chromosomal location on human chromosome 19. Both the viral and cellular sequences required for integration include binding sites for Rep 78/68 and adjacent sequences at which these endonucleases can nick the DNA. This protein can bind simultaneously to both viral and human chromosomal 19 DNA, at least in vitro. The current model of integration therefore proposes that its exquisite specificity is the result of such simultaneous binding to the two DNA molecules by Rep 78/68. A circular adeno-associated virus DNA molecule, as well as replication of both viral and cellular DNA sequences, is also required for integration. The integrated viral genome is preserved within the genome of the host cell and its descendants until a helper virus infects an adeno-associated virus-carrying cell. Following activation of transcription of Rep sequences, the viral genome is excised from that of the host cell, in a Rep 78/68-dependent process, for entry into the productive cycle.

## Regulation of Replication via Different Viral Origins: Epstein-Barr Virus

During herpesviral latent infections, the viral genome is stably maintained at low concentrations, compared to those attained by the end of a lytic cycle, often for long periods. Furthermore, replication of the viral genome can be coordinated with that of the cellular genome. This pattern is characteristic of human B cells latently infected by Epstein-Barr virus.

The Epstein-Barr virus genome is maintained in nuclei of latently infected cells as a stable circular episome, present at 10 to 50 copies per cell. For example, one Burkitt's lymphoma cell line (Raji) has carried about 50 copies per cell of episomal viral DNA for over 30 years in culture. When Epstein-Barr virus infects a B cell, the linear viral genome circularizes by a mechanism that is not well understood. The circular viral DNA is then amplified during S phase of the host cell to the final concentration listed above. Such replication, by the cellular DNA polymerases and accessory proteins described above, requires the single viral protein invariably synthesized in latently infected cells, EBNA-1, and the viral origin for latent viral DNA replication, OriP. In contrast to the lytic replication of this and other viruses, such amplification of the episomal viral genome is limited to a few cycles by an unknown mechanism. Following such limited amplification, the viral DNA is duplicated once per cell cycle, in S phase, such that its concentration is maintained as the lymphocyte host cell divides. The EBNA-1 protein and OriP are sufficient for both such once-per-cell cycle replication of episomal viral genomes and their orderly segregation to daughter cells. The latter process requires the binding of EBNA-1 to its

high-affinity sites in the family of repeat (FR) sequence of OriP (Box 9.3A). The mechanism by which these sequences prevent loss of episomal genomes is not known but may be related to the binding of the segment of the genome containing them to the nuclear matrix in cells containing EBNA-1.

The limited replication of the Epstein-Barr virus genome, once per cell cycle, is quite different from the exponential replication of other viral genomes that also depends on cellular proteins for all but the origin recognition and unwinding steps (e.g., simian virus 40). Thus, while the availability of cellular replication proteins only in late $G_1$ and S accounts for the timing of Epstein-Barr virus replication in latently infected cells, this property cannot explain why such replication is limited. The controlled initiation of replication of episomal viral DNA molecules is analogous to the tight control of initiation of replication from cellular origins, each of which also fires once and only once in each S phase. Such close similarity indicates that specialized mechanisms must prevent rereplication of daughter viral DNA molecules until the cell cycle has been completed and the DNA products segregated to daughter cells. What these mechanisms might be, and in particular whether control of Epstein-Barr virus replication from OriP is exerted by the elaborate circuits that ensure once-per-cycle duplication of chromosomal DNA, is not yet known.

As a latent infection is established, the Epstein-Barr virus genome becomes increasingly methylated at C residues present in CG dinucleotides, although sequences that must function in latently infected cells, such as OriP, generally escape this modification. Such DNA methylation, which is associated with repression of transcription in mammalian cells, contributes to inhibition of expression of viral lytic genes. The genome also becomes associated with cellular nucleosomes and is therefore replicated as a circular minichromosome much like that of simian virus 40. Replication of the Epstein-Barr virus genome once per cell cycle persists unless conditions that induce entry into the viral lytic cycle are encountered. The critical step for this transition is activation of transcription of the viral genes encoding the transcriptional activators Zta and Rta (see chapter 8). These proteins not only induce expression of the viral early genes that encode the viral DNA polymerase and other proteins necessary for replication from one of the two copies of OriLyt but also bind to a sequence of these origins. Indeed, Zta appears to be the OriLyt recognition protein. Consequently, once Zta is made in an Epstein-Barr virus-infected cell, its indirect and direct effects on viral DNA synthesis ensure a switch from OriP-dependent to OriLyt-dependent replication, and progression through the lytic cycle.

## Controlled and Exponential Replication from a Single Origin: the Papillomaviruses

Three different modes of viral DNA replication are associated with papillomavirus infection (Fig. 9.20). Entry of a papillomavirus genome into the nucleus of a host cell initiates a period of amplification of the circular genome, just as during the early stages of latent infection by Epstein-Barr virus. Replication continues until a moderate number of viral genomes (~100) accumulate in the cell. A maintenance replication pattern in which the complement of viral episomes is duplicated on average once per cell cycle is then established. The mechanism responsible for the switch from amplification to maintenance replication is not known. In natural papillomavirus infections, these two types of replication take place in the basal cells of an epithelium. They can also be reproduced in cells in culture transformed by these viruses. The final stage of papillomavirus DNA replication, production of high concentrations of the viral genome for assembly into progeny virions, is restricted to differentiated epithelial cells such as terminally differentiated keratinocytes formed as basal cells move toward the surface of the epithelium (Fig. 9.20). This so-called productive replication is not yet well understood because cell culture systems that support papillomavirus production have been developed only recently. However, the shift from maintenance to productive replication as host cells differentiate appears to entail a switch from a simian virus 40-like to a rolling-circle mechanism of replication.

The single viral origin and the viral E1 and E2 proteins that bind to specific origin sequences are necessary for both the initial amplification of the bovine papillomavirus genome and its maintenance for long periods at a more-or-less constant concentration. However, such maintenance replication does not appear to be the result of a strict once-per-cell-cycle mechanism of replication control. Rather, replication of individual viral episomes appears to occur at random, taking place on average once per cell cycle. This mechanism avoids the need to mark a newly replicated DNA molecule to prevent its further replication until the next cell cycle. Stable maintenance of the viral genome requires an additional sequence, called the

**Figure 9.20 Regulation of papillomaviral DNA replication in epithelial cells.** The outer layers of the skin are shown as depicted in Fig. 4.3. The virus infects proliferating basal epithelial cells, to which it probably gains access after wounding. The double-stranded, circular viral DNA genome is imported into the infected cell nucleus and initially amplified to a concentration of 50 to 100 copies per cell. This concentration of viral DNA episomes is maintained by further limited replication as the basal and parabasal cells of the epithelium divide (maintenance replication). As cells move to the outer layers of the epidermis and differentiate, productive replication of the viral genome to thousands of copies per cell takes place. The mechanism by which differentiation induces the switch from maintenance to productive replication of papillomaviral DNA is not understood.

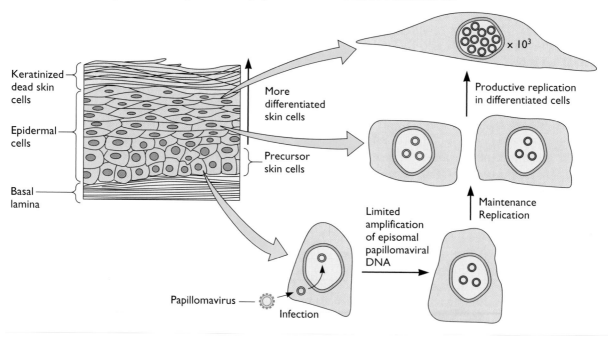

minichromosome maintenance element, which is composed of multiple binding sites for the E2 protein. The mechanism by which the minichromosome maintenance element acts has not been established, but the most attractive possibility is that, like the family of repeat sequences of Epstein-Barr virus OriP, it promotes orderly partitioning of viral genomes during host cell division.

## Origins of Genetic Diversity in DNA Viruses

### Fidelity of Replication by Viral DNA Polymerases

#### Proofreading Mechanisms

Cellular DNA replication is a high-fidelity process with an error rate of only about one mistake in every $10^9$ base pairs synthesized. Such fidelity, which is essential to maintain the integrity of the genome, is based on the accurate pairing of substrate and template deoxyribonucleotide bases prior to synthesis of each phosphodiester bond. Nonstandard base pairs can form quite readily. However, initiation of DNA synthesis requires a primer that is perfectly base paired with the template strand, and DNA synthesis does not begin if the terminal nucleotide or the preceding region of the primer-template is mismatched. In such circumstances, the mismatched base in the primer strand is excised by a $3' \rightarrow 5'$ exonuclease activity present in all replicative DNA polymerases until a perfectly base-paired primer-template is created (Fig. 9.21). Thus, replicative DNA polymerases are self-correcting enzymes, removing errors made in newly synthesized DNA as replication continues (Fig. 9.21). This proofreading mechanism may explain why DNA polymerases do not initiate DNA synthesis de novo.

The cellular DNA polymerases ($\alpha$, $\delta$, and perhaps $\epsilon$) that replicate small viral DNA genomes, such as those of parvo- and papovaviruses, possess such proofreading exonucleases. Such viruses do not inhibit expression of cellular genes during the period in which viral DNA synthesis is made in the infected cell and indeed may induce expression of cellular replication proteins. These viruses can therefore also use the cellular mechanisms of mismatch repair (Fig. 9.22), which correct any errors in the daughter strand of newly replicated DNA missed during proofreading by DNA polymerases. Thus, the replication of the genomes of these small DNAs viruses is expected to be as accurate as that of the genome of their host cells.

#### Proofreading by Viral DNA Polymerases

The question of how accurately viral DNA is replicated by viral DNA polymerases, such as those of adeno-, herpes-, and poxviruses, has received relatively little atten-

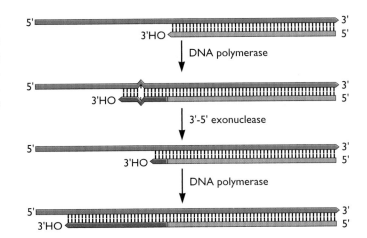

**Figure 9.21 Proofreading during DNA synthesis.** If permanently fixed into the genome, mispaired bases would result in mutation. However, the majority are removed by the proofreading activity of replicative DNA polymerases. A mismatch at the 3′ OH terminus of the primer-template during DNA synthesis by such an enzyme (step 1) activates the $3' \rightarrow 5'$ exonuclease activity of all replicative DNA polymerases, which excises the mismatched region to create a perfect duplex (step 2) for further extension (step 3). In the best characterized case, DNA polymerase I of *E. coli*, the rate of extension from a mismatched nucleotide is much slower than when a correct base pair is formed at the 3′ terminus of the nascent strand. This slow rate of extension allows time for spontaneous unwinding (breathing) of the new duplex region of the DNA and transfer of the 3′ end to the $3' \rightarrow 5'$-exonuclease site for removal of the mismatched nucleotide. Because preferential excision of mismatched nucleotides is the result of differences in the rate at which the polymerase can add the next nucleotide, this mechanism is called **kinetic proofreading**.

tion. However, each of these viral DNA polymerases possesses an intrinsic $3' \rightarrow 5'$ exonuclease which preferentially excises mismatched nucleotides from duplex DNAs in vitro and contains short, conserved sequences related to those that flank the catalytically active exonuclease residues of well-characterized cellular DNA polymerases. Indeed, recent experiments have shown that mutations that impair the exonuclease activity of the herpes simplex virus type 1 DNA polymerase greatly increase the rate of mutation of the viral genome during the infectious cycle. Thus, there can be little doubt that the $3' \rightarrow 5'$ exonucleases of viral DNA polymerases provide proofreading function.

The larger DNA viruses inhibit expression of cellular genes by various mechanisms and can inhibit cellular DNA synthesis. It is therefore possible that the cellular mismatch repair proteins that normally back up proofreading by DNA polymerases are not present in the concentrations necessary for effective surveillance and repair of newly

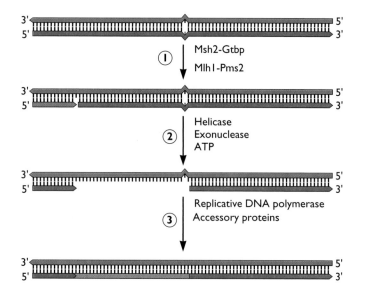

**Figure 9.22 Mismatch repair in newly synthesized DNA.** This activity requires recognition of the newly synthesized strand containing a misincorporated nucleotide and introduction of a nick (step 1), exonucleolytic degradation of the new strand from the nick as it is unwound (step 2), and resynthesis of DNA to repair the gap (step 3). The "long-patch" repair system is responsible for removal of mismatches introduced during DNA synthesis in eukaryotic cells, although not all components have been identified. Those listed, the mammalian Msh2-Gtbp and Mlh1-Pms2 heterodimers, are composed of homologs of the well-characterized *E. coli* mismatch repair proteins MutS and MutL, respectively. This relationship suggests that they participate in steps 1 and 2. The MutS homologs bind to mismatched base pairs in duplex DNA, as well as to mismatches resulting from small insertions or deletions with high specificity, and possesses weak ATPase and DNA helicase activities. The biochemical functions of the mammalian MutL homologs are not well understood, nor is it yet clear how the newly synthesized DNA strand is distinguished. The importance of this repair system is illustrated by the predisposition to tumor development induced by mutations in the human genes encoding MutS or MutL homologs.

synthesized viral DNA in cells infected by these viruses. However, neither the concentrations nor activities of these cellular repair proteins have been examined in cells infected by adeno-, herpes-, or poxviruses. More detailed information about the rates at which viral DNA polymerases introduce errors during DNA synthesis in vitro and the rates of mutation of viral DNA genomes when replicated in infected cells would help establish the role of cellular repair systems in maintaining the integrity of these viral genomes. Similarly, the contributions of viral enzymes that could prevent or repair DNA damage, such as the dUTPase and uracil DNA glycosylases of herpes- and poxviruses (Table 9.6), remain to be established.

## Recombination of Viral Genomes

### General Mechanisms of Recombination

Genetic recombination, the rearrangement of DNA sequences within a genome or between two genomes, is an important reaction in the life cycles of many DNA viruses. Much of our understanding of the mechanisms of recombination is based on studies of bacterial viruses, such as bacteriophage λ. Similar principles apply to recombination of DNA genomes of animal viruses. Two types of recombination are generally recognized: site specific and homologous. In **site-specific recombination**, exchange of DNA takes place at short DNA sequences, specifically recognized by proteins that catalyze recombination. These sequences may be present in only one or both of the DNA sequences that are recombined in this way. Much more common during reproduction of DNA viruses is **homologous recombination**, the exchange of genetic information between any pair of related DNA sequences. Such exchange is generally initiated by introduction of a nick into one DNA strand, which then invades the homologous duplex and base pairs with it (Fig. 9.23). These reactions are facilitated by single-stranded DNA-binding proteins, as well as by specialized recombination proteins that bind to both single- and double-stranded DNA to promote the invasion reaction (Fig. 9.23). The specialized proteins then catalyze branch migration to extend the length of the heteroduplex in which the strands from the two DNA duplexes are paired. The initial invasion reactions are followed by cross-strand exchange, a process that can be achieved in several ways. As illustrated in the generic mechanism shown in Fig. 9.23, isomerization of the structure held together by the exchange of two of the four strands of the recombining DNA molecules (termed a **Holliday junction** or **cross-strand exchange**) followed by cleavage of the crossing strands and ligation results in exchange of DNA between the two original duplexes.

### Viral Genome Recombination

The integration of adeno-associated virus DNA into the cellular genome and its excision when conditions are right are the result of site-specific recombination reactions mediated by the Rep 78/68 viral proteins, which bind to specific sequences in both DNA molecules. In contrast, integration of retroviral DNA (see chapter 7) is site-specific only for the viral DNA.

All viral DNA genomes undergo homologous recombination. These viral genomes do not generally encode specific recombination proteins, although the herpes simplex virus type 1 alkaline nuclease (Table 9.6) may participate in resolution of branched DNA molecules formed by replication. It is therefore believed that general homologous re-

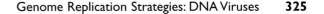

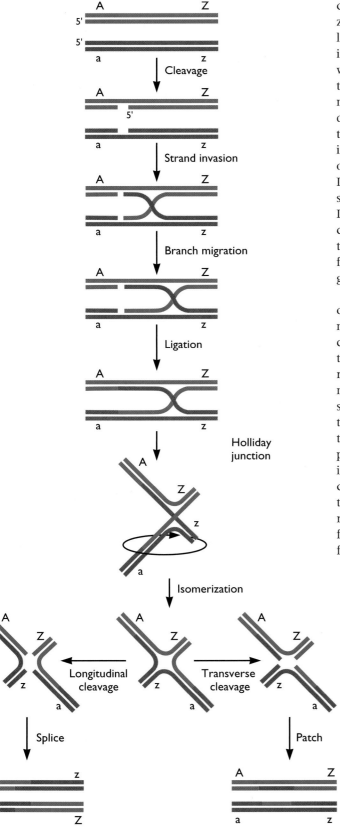

combination of viral genomes is catalyzed by host cell enzymes. Nevertheless, viral recombination is favored by the large numbers of viral DNA molecules present in lytically infected cells and the concentration of replicating DNA within specialized replication centers in the nucleus or cytoplasm. The initial step in recombination, pairing of homologous sequences with one another (Fig. 9.23), depends on random collision and is therefore concentration dependent. Furthermore, the structures of replication intermediates (see below) or the requirement for nicking of viral DNA during replication or packaging, which yields DNA ends that can initiate the invasion process (Fig. 9.23; see also Fig. 7.5), can facilitate recombination among viral DNA molecules. The ease with which viral DNA sequences can recombine is an important factor in the evolution of these viruses. It is also of great benefit to the experimenter, facilitating introduction of specific mutations into the viral genome or construction of viral vectors (see chapter 2).

Although recombination among animal viral DNA sequences has been widely exploited in the laboratory, the mechanisms have not received much attention. One exception is adenovirus recombination. Recombination between markers in two adenovirus genomes, such as restriction endonuclease sites or temperature-sensitive mutations, exhibits many properties typical of this process, such as the dependence of recombination frequency on the distance between the markers. However, several features suggest that recombination is mechanistically coupled with viral DNA synthesis because the initial invasion is by the single strands of viral DNA displaced during replication of the adenoviral genome (Fig. 9.10). In particular, this mechanism accounts for the **polarity** of adenovirus recombination, the decreasing gradient of recombination frequency with distance of the recombining sequences from the ends of the viral genome. Another important ex-

**Figure 9.23 General homologous recombination.** Formation of a Holliday junction (or cross-strand exchange) on alignment of homologous duplexes is initiated by cleavage in one strand of each duplex, which then invades its partner. These reactions and subsequent branch migration are catalyzed by specialized recombination proteins. Homologous recombination can also be initiated by single-stranded DNA by other mechanisms. For example, if a single strand from only one of the two duplexes invaded its partner, DNA synthesis would be necessary to repair the gap left by the invading strand. As illustrated, the sites at which the Holliday junction is cleaved during resolution following its isomerization determine whether short segments of DNA on one strand are exchanged (patch) or whether the duplexes are recombined (splice).

ception is the homologous recombination of DNA sequences of some herpesviruses, including herpes simplex virus type 1, which is responsible for isomerization of the genome.

Populations of viral DNA molecules purified from herpes simplex virus type 1 virions contain four isomers of the genome, defined by the relative orientations of the two unique sequence segments (L and S) with respect to one another (Fig. 9.11). These unique sequences are flanked by several inverted repeats, including the conserved *a* sequence (Fig. 9.11). The viral DNA population isolated from a single plaque contains all four isomers at equimolar concentrations, suggesting that a single virus particle containing just one genome isomer gives rise to all four by recombination between repeated DNA sequences. It was initially believed that the *a* sequences (Fig. 9.11) mediate these recombination reactions by a site-specific mechanism, for example, because insertion of an *a* sequence at an ectopic site induces additional inversions. However, it is now clear that an *a* sequence can act as a "hot spot" for recombination simply because it contains the viral packaging signal. Double-stranded DNA breaks that promote recombination are made within this sequence during cleavage of replication products for packaging of the viral genome during assembly. In fact, the *a* sequences are dispensable for production of all four genome isomers at the normal frequency, and recombination between any of the inverted repeat sequences in the viral genome (Fig. 9.11) promotes inversion of the L and S segments. Such homologous recombination takes place during viral DNA synthesis and requires the viral replication machinery.

Despite some 20 years of study, the function of the unusual isomerization of the genome of herpes simplex virus type 1 and certain other herpesviruses remains enigmatic. The ability of this viral genome to isomerize is not absolutely essential for virus replication in cells in tissue culture, for viruses "frozen" as a single isomer by deletion of internal inverted repeats are viable. On the other hand, the reduced yield of such viruses and the presence of the inverted repeat sequences in all wild-type strains of herpes simplex virus type 1 emphasize the importance of the repeated sequences. It may be that these sequences themselves fulfill some beneficial function (as yet unknown) during herpes simplex virus reproduction. Recombinational isomerization would then be a secondary result of the presence of multiple inverted copies of these sequences in the viral genome. Alternatively, isomerization might be a consequence of an essential role for recombination in replication of the viral genome. The value of the unusual isomerization of the herpes simplex virus type 1 genome may become clearer as understanding of the mechanism of viral DNA synthesis improves.

## Perspectives

Our understanding of mammalian replication proteins and the intricate reactions they carry out during DNA synthesis would still be rudimentary were it not for the simian virus 40 origin of replication. This relatively simple viral DNA sequence, which was initially well characterized by genetic methods, supports origin-dependent replication in vitro when cellular proteins are supplemented with a single viral protein, LT. Many origins have been defined in mammalian cell genomes, and some characterized in considerable detail, but origin-dependent DNA synthesis from them still cannot be reproduced in experimental systems. Thus, the simian virus 40 replication system has been of paramount importance in the identification of cellular replication proteins and elucidation of the numerous reactions needed to complete replication of a duplex DNA molecule.

The mechanism of synthesis of this small viral DNA genome has also provided the conceptual framework with which to appraise the considerable diversity in replication of viral DNA genomes. One parameter that varies considerably is the degree of dependence on the host cell's replication machinery. In contrast to papovaviruses (and parvoviruses), the larger DNA viruses (herpes- and poxviruses) direct the synthesis of not only a complete DNA synthesis system but also accessory enzymes responsible for the production of deoxyribonucleoside triphosphate substrates, while adenoviruses rely on the host cell for the latter but not the former class of proteins. Nevertheless, replication of all viral DNA genomes requires proteins that carry out the reactions first described for simian virus 40 DNA synthesis, notably an origin recognition protein(s), one or more DNA polymerases, proteins that confer processive DNA synthesis, origin-unwinding and helicase proteins, and, usually, proteins that allow priming. Thus, the variation in use of host cell components can be seen as virus-specific solutions to the need for a common set of biochemical activities to complete the complex process of DNA synthesis.

Replication of viral DNA genomes ranges from simple, continuous synthesis of both strands of a linear, double-stranded DNA template (adenovirus) to baroque (and not well understood) mechanisms that produce DNA concatemers (herpesviruses). These various replication strategies represent alternative mechanisms for circumventing the inability of all known DNA-dependent DNA polymerases to initiate DNA synthesis de novo. Some viruses employ the standard mechanism, RNA primers and discontinuous synthesis of the lagging strand, but others have evolved priming mechanisms that allow continuous synthesis of all daughter DNA strands from protein or DNA sequence primers. On the other hand, not even large DNA viruses with genomes that encode all components needed for complete DNA synthesis provide viral DNA polymerases that are primer independent. The fact that viruses have not evolved such an enzyme, despite successful adaptations to synthesis of DNA from an RNA template (retro- and hepadnaviruses) and to novel primers, suggests that the apparently cumbersome and restrictive need for primers for viral and cellular DNA synthesis provides important advantages that we do not yet fully appreciate.

Efficient reproduction of DNA viruses requires the rapid production of very large numbers of progeny viral DNA molecules for assembly of viral particles. Such exponential replication appears to be largely the result of mechanisms that ensure efficient production of the proteins that mediate or support DNA synthesis, be they viral or cellular in origin. However, it is also likely that viral DNA replication at specialized intracellular sites, a common feature of cells infected by these viruses, contributes to efficient viral DNA synthesis. Further exploration of this recently recognized and incompletely understood phenomenon should shed new light on host cell biology, in particular the structural and functional compartmentalization of the nucleus. Also far from well understood are the cues that set the stage for the alternative mode of limited replication characteristic of some DNA viruses, when the number of replication cycles and their timing with respect to the host cell cycle are governed by the nature (alphaherpesviruses), or activity or differentiation state (Epstein-Barr virus and papillomaviruses), of the host cell. Elucidation of the mechanisms that achieve such close integration of viral DNA synthesis with the physiological state of the host cell seems certain to provide important insights into both host cell control mechanisms and the long-term relationships these viruses can establish with their hosts.

# References

## Books
Kornberg, A., and T. Baker. 1992. *DNA Replication*, 2nd ed. W. H. Freeman and Company, New York, N.Y.

## Chapters in Books
Berns, K. I. 1996. Parvoviridae: the viruses and their replication, p. 1017–1042. *In* B. N. Fields, D. M. Knipe, and P. M. Howley (ed.), *Fundamental Virology*. Lippincott-Raven, Philadelphia, Pa.

Weller, S. K. 1995. Herpes simplex virus DNA replication and genome maturation, p. 189–213. *In* G. M. Cooper, R. Greenberg Temin, and B. Sugden (ed.), *The DNA Provirus: Howard Temin's Scientific Legacy*. American Society for Microbiology, Washington, D.C.

## Reviews
Beaud, G. 1995. Vaccinia virus DNA replication: a short review. *Biochimie* **77**:774-779.

Boehmer, P. E., and I. R. Lehman. 1997. Herpes simplex virus DNA replication. *Annu. Rev. Biochem.* **66**:347–384.

Borowiec, J. A., F. B. Dean, P. A. Bullock, and J. Hurwitz. 1990. Binding and unwinding—how T antigen engages the SV40 origin of DNA replication. *Cell* **60**:181–184.

Chow, L. T., and T. R. Broker. 1994. Papillomavirus DNA replication. *Intervirology* **37**:150–158.

Fanning, E. 1994. Control of SV40 DNA replication by protein phosphorylation: a model for cellular DNA replication? *Trends Cell. Biol.* **4**:250–255.

Hamlin, J. L., and P. A. Dijkwel. 1995. On the nature of replication origins in higher eukaryotes. *Curr. Opin. Genet. Dev.* **5**:153–161.

Hay, R. T., A. Freeman, I. Leith, A. Monaghan, and A. Webster. 1995. Molecular interactions during adenovirus DNA replication. *Curr. Top. Microbiol. Immunol.* **199**(Part 2):31–48.

Lindahl, T., P. Karran, and R. D. Wood. 1997. DNA excision repair pathways. *Curr. Opin. Genet. Dev.* **7**:158–169.

Stillman, B. 1996. Cell cycle control of DNA replication. *Science* **274**:1659–1664.

Van der Vliet, P. C. 1995. Adenovirus DNA replication. *Curr. Top. Microbiol. Immunol.* **199**(Pt. 2):1–30.

Waga, S., and B. Stillman. 1998. The DNA replication fork in eukaryotic cells. *Annu. Rev. Biochem.* **67**:721–751.

Weinberg, R. A. 1995. The retinoblastoma protein and cell cycle control. *Cell* **81**:323–330.

## Papers of Special Interest

### Viral Replication Mechanisms and Replication Proteins
Berg, N., and A. Stenlund. 1997. Functional interactions between papillomavirus E1 and E2 proteins. *J. Virol.* **71**:3853–3863.

Bochkarev, A., J. A. Barwell, R. A. Pleutzner, E. Bockhaven, L. Frappier, and E. M. Edwards. 1996. Crystal structure of the DNA-binding domain of the Epstein-Barr virus origin-binding protein, EBNA1, bound to DNA. *Cell* **84**:791–800.

Challberg, M. D. 1986. A method for identifying the viral genes required for herpesvirus DNA replication. *Proc. Natl. Acad. Sci. USA* **83**:9094–9098.

Challberg, M. D., S. V. Desidero, and T. J. Kelly. 1980. Adenovirus DNA replication *in vitro*: characterization of a protein covalently linked to nascent DNA strands. *Proc. Natl. Acad. Sci. USA* **77**:5105–5109.

Coen, D. M., D. P. Aschman, P. T. Gelep, M. J. Retendo, S. K. Weller, and P. Schaffer. 1984. Fine mapping and molecular cloning of mutations in the herpes simplex virus DNA polymerase locus. *J. Virol.* **49**:236–247.

Crate, J. J., T. Tsurumi, L. Zhu, S. K. Weller, P. D. Olivo, M. D. Challberg, E. S. Mocarski, and I. R. Lehman. 1989. Herpes simplex virus 1 helicase-primase: a complex of three herpes-encoded gene products. *Proc. Natl. Acad. Sci. USA* **86**:2186–2189.

Danna, K. J., and D. Nathans. 1972. Bi-directional replication of simian virus 40 DNA. *Proc. Natl. Acad. Sci. USA* **69**:3097–3100.

Dekher, J., N. Kanellopoulos, A. K. Loomstra, J. A. W. M. Van Oosterhout, K. Leonard, P. A. Tucker, and P. C. van der Vliet. 1997. Multimerization of the adenovirus DNA-binding protein is the driving force for ATP-independent DNA unwinding during strand displacement synthesis. *EMBO J.* **16**:1455–1463.

Earl, P. L., E. V. Jones, and B. Moss. 1986. Homology between DNA polymerases of poxviruses, herpesviruses and adenoviruses: nucleotide sequence of the vaccinia virus DNA polymerase gene. *Proc. Natl. Acad. Sci. USA* **83**:3659–3663.

Efstathiou, S., S. Kemp, G. Darby, and A. C. Minson. 1989. The role of herpes simplex virus type 1 thymidine kinase in pathogenesis. *J. Gen. Virol.* **70**:869–879.

Fixman, E. D., G. S. Hayward, and S. D. Hayward. 1995. Replication of Epstein-Barr virus OriLyt: lack of a dedicated virally encoded origin-binding protein and dependence on Zta in co-transfection assays. *J. Virol.* **69**:2998–3006.

Im, D. S., and N. Muzyczka. 1990. The AAV origin-binding protein Rep68 is an ATP-dependent site-specific endonuclease with DNA helicase activity. *Cell* **61**:447–457.

Lechner, R. L., and T. J. Kelly. 1977. The structure of replicating adenovirus 2 DNA molecules. *Cell* **12**:1007–1020.

Li, J. J., and T. J. Kelly. 1984. Simian virus 40 DNA replication *in vitro. Proc. Natl. Acad. Sci. USA* **81**:6973–6977.

Li, J. J., K. W. Peden, R. A. Dixon, and T. J. Kelly. 1986. Functional organization of the simian virus 40 origin of DNA replication. *Mol. Cell. Biol.* **6**:1117–1128.

Ni, T.-H., W. F. McDonald, I. Zolotukhin, T. Melendy, S. Waga, B. Stillman, and N. Muzyczka. 1998. Cellular proteins required for adeno-associated virus DNA replication in the absence of adenovirus coinfection. *J. Virol.* **72**:2777–2787.

Rekosh, D., W. C. Russell, A. J. D. Bellett, and A. J. Robinson. 1977. Identification of a protein linked to the ends of adenovirus. *Cell* **11**:283–295.

Stow, N. D. 1982. Localization of an origin of DNA replication within the TRS/IRS repeated region of the herpes simplex virus type 1 genome. *EMBO J.* **1**:863–867.

Straus, S. E., E. D. Sebring, and J. A. Rose. 1976. Concatemers of alternating plus and minus strands are intermediates in adenovirus-associated virus DNA synthesis. *Proc. Natl. Acad. Sci. USA* **73**:742–746.

Waga, S., and B. Stillman. 1994. Anatomy of a DNA replication fork revealed by reconstitution of SV40 DNA replication *in vitro. Nature* (London) **369**:207–212.

Weller, S. K., A. Spadaro, J. E. Schaffer, A. W. Murray, A. M. Maxam, and P. A. Schaffer. 1985. Cloning, sequencing and functional analysis of OriL, a herpes simplex virus type 1 origin of DNA synthesis. *Mol. Cell. Biol.* **5**:930–942.

## Control of Viral DNA Replication

Adams, A. 1987. Replication of latent Epstein-Barr virus genomes in Raji cells. *J. Virol.* **61**:1743–1746.

Chellappan, S., V. B. Kraus, B. Kroger, M. Munger, P. M. Howley, W. C. Phelps, and J. R. Nevins. 1992. Adenovirus E1A, simian virus 40 tumor antigen and human papillomavirus E7 protein share the capacity to disrupt the interaction between transcription factor E2F and the retinoblastoma gene product. *Proc. Natl. Acad. Sci. USA* **89**:4549–4553.

Doucas, V., A. M. Ishov, A. Romo, H. Juguilon, M. D. Weitzman, R. M. Evans, and G. G. Maul. 1996. Adenovirus replication is coupled with the dynamic properties of the PML nuclear structure. *Genes Dev.* **10**:194–207.

Flores, E. R., and P. F. Lambert. 1997. Evidence for a switch in the mode of human papillomavirus type 11 DNA replication during the viral life cycle. *J. Virol.* **71**:7167–7179.

Gilbert, D. M., and S. N. Cohen. 1987. Bovine papilloma virus plasmids replicate randomly in mouse fibroblasts throughout S phase of the cell cycle. *Cell* **50**:59–68.

Kaelin, W. G., M. E. Ewen, and D. M. Livingston. 1990. Definition of the minimal simian virus 40 large T antigens and adenovirus E1A-binding domain in the retinoblastoma gene product. *Mol. Cell. Biol.* **10**:3761–3769.

Kotin, R. M., M. Siniscalco, R. J. Samulski, X. D. Zhu, L. Hunter, C. A. Laughlin, S. McLaughlin, N. Muzyczka, M. Rocchi, and K. I. Berns. 1990. Site-specific integration by adeno-associated virus. *Proc. Natl. Acad. Sci. USA* **87**:2211–2215.

Lu, M., and T. Shenk. 1996. Human cytomegalovirus infection inhibits cell cycle progression at multiple points, including the transition from $G_1$ to S. *J. Virol.* **70**:8850–8857.

Piirsoo, M., E. Ustav, T. Mandel, A. Stenlund, and M. Ustav. 1996. *Cis* and *trans* requirements for stable episomal maintenance of the BPV-1 replicator. *EMBO J.* **15**:1–11.

Reisman, D., J. L. Yates, and B. Sugden. 1985. A putative origin of replication of plasmids derived from Epstein-Barr virus is composed of two *cis*-acting components. *Mol. Cell. Biol.* **5**:1822–1832.

Spindler, K. R., C. Y. Eng, and A. J. Berk. An adenovirus early region 1A protein is required for maximal viral DNA replication in growth-arrested human cells. *J. Virol.* **53**:742–750.

Weintraub, S. J., C. A. Prater, and D. C. Dean. 1992. Retinoblastoma protein switches the E2F site from positive to negative element. *Nature* (London) **358**:259–261.

Whyte, P., J. J. Buchkovich, J. M. Horowitz, S. H. Friend, M. Raybuck, R. A. Weinberg, and E. Harlow. 1988. Association between an oncogene and an anti-oncogene: the adenovirus E1A proteins bind to the retinoblastoma gene product. *Nature* (London) **334**:124–129.

Zhong, L., and G. S. Hayward. 1997. Assembly of complete, functionally active herpes simplex virus DNA replication compartments and recruitment of associated viral and cellular proteins in transient co-transfection assays. *J. Virol.* **71**:3146–3160.

## Repair and Recombination of Viral DNA

Hayward, G. S., R. J. Jacob, R. C. Wadsworth, and B. Roizman. 1975. Anatomy of herpes simplex virus DNA: evidence for four populations of molecules that differ in the relative orientations of their long and short segments. *Proc. Natl. Acad. Sci. USA* **72**:4243–4247.

Hwang, Y. T., B.-Y. Liu, D. M. Coen, and C. B. C. Hwang. 1997. Effects of mutations in the ExoIII motif of the herpes simplex virus DNA polymerase gene on enzyme activities, viral replication and replication fidelity. *J. Virol.* **71**:7791–7798.

Linden, R. M., E. Winocour, and K. I. Berns. 1996. The recombination signals for adeno-associated virus site-specific integration. *Proc. Natl. Acad. Sci. USA* **93**:7966–7972.

Martin, D. W., and P. C. Weber. 1996. The a sequence is dispensable for isomerization of the herpes simplex virus type 1 genome. *J. Virol.* **70**:8801–8812.

Mocarski, E. S., L. E. Post, and B. Roizman. 1980. Molecular engineering of the herpes simplex virus genome: insertion of a second LS junction causes additional genome inversions. *Cell* **22**:243–245.

Weitzman, M. D., S. R. M. Kyöstiö, R. M. Kotin, and R. A. Owens. 1994. Adeno-associated virus (AAV) rep proteins mediate complex formation between AAV DNA and its integration site in human DNA. *Proc. Natl. Acad. Sci. USA* **91**:5808–5812.

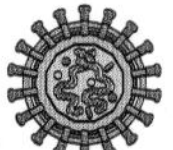

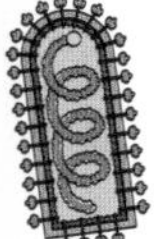

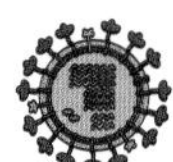

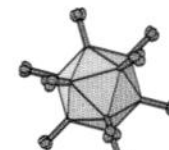

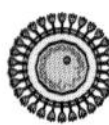

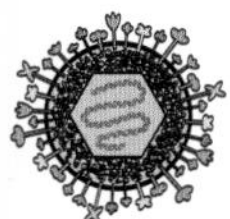

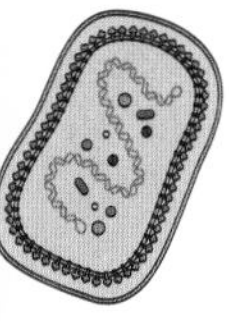

*10*

# Processing of Viral Pre-mRNA

## Introduction

Viral mRNAs are synthesized by either viral enzymes or the cellular transcriptional machinery and may be made in the nucleus or the cytoplasm of the infected cell. Regardless of how and where they are made, all viral mRNAs must be translated by the cellular protein-synthesizing machinery. Consequently, viral mRNAs must conform to the requirements of the cell's translational system. A series of covalent modifications, collectively known as **RNA processing** (Fig. 10.1), endow mRNAs synthesized by viral or cellular RNA polymerases with the molecular features needed for recognition by the protein synthesis machinery and allow translation of the protein-coding sequence by cellular ribosomes. Most RNA-processing reactions were discovered in viral systems, primarily because virus-infected cells provided large quantities of specific mRNAs for analysis.

Two modifications, which are important for efficient translation, made during synthesis of most viral and cellular mRNAs are the addition of $m^7GpppN$ to the 5' end (**capping**) and the addition of multiple A residues to the 3' end (**polyadenylation**) (Fig. 10.1). The enzymes that perform these chemical additions (Box 10.1) may be encoded by viral or cellular genes. When an RNA is produced in the nucleus, another chemical modification, called **splicing**, is possible. During splicing, short blocks of noncontiguous coding sequences (**exons**) are joined precisely to create a complete protein-coding sequence for translation, while the intervening sequences (**introns**) are discarded (Fig. 10.1). Splicing therefore dramatically alters the **precursor mRNA (pre-mRNA)** initially synthesized. As no viral genome encodes even part of the complex machinery needed to catalyze splicing reactions, splicing of viral pre-mRNAs is accomplished entirely by cellular gene products. Some viral pre-mRNAs undergo a different type of internal chemical change in which a single nucleotide is replaced by another or one or more nucleotides are inserted at specific positions (Box 10.1). Such **RNA-editing** reactions introduce nucleotides that are not

**331**

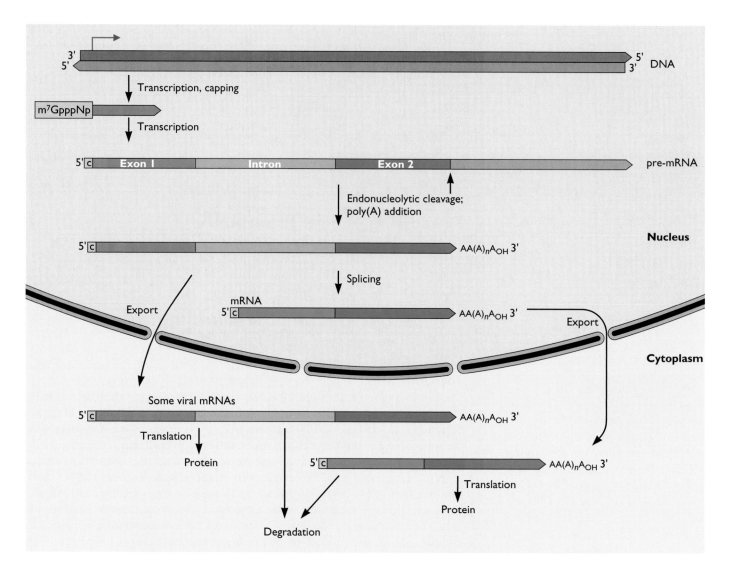

**Figure 10.1 Processing of a viral or cellular mRNA synthesized by RNA polymerase II.** The reactions by which mature mRNA is made from a typical RNA polymerase II transcript are shown. The first such reaction, capping, takes place cotranscriptionally. For clarity, the exons of a hypothetical, partially processed (i.e., polyadenylated but unspliced) pre-mRNA are depicted even though polyadenylation and splicing are often coupled, and in long primary transcripts, splicing of cap-proximal exons occurs before the sequences encoding 3′ exons and polyadenylation signals have been transcribed. Thus, many splicing reactions are cotranscriptional. Most cellular and viral pre-mRNAs synthesized by RNA polymerase II are processed by this pathway. However, some viral mRNAs that are polyadenylated but not spliced, or are incompletely spliced, are exported to the cytoplasm.

encoded in the genome into the mRNA. These altered sequences may change the sequence of the encoded protein.

When a viral pre-mRNA is produced and processed in the nucleus, it must be exported to the cytoplasm for translation (Fig. 10.1). Such export of mature viral and cellular mRNAs is considered to be part of mRNA process-

ing even though the RNA is not known to undergo any chemical change during transport. Viral mRNAs are invariably exported by cellular pathways, but some viruses have evolved mechanisms to allow or facilitate export of their mRNAs. Once within the cytoplasm, an mRNA has a finite lifetime before it is recognized and degraded by ri-

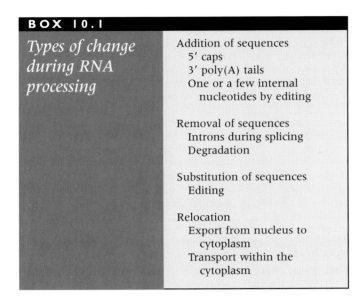

bonucleases. The susceptibilities of individual mRNA species to attack by these destructive enzymes vary greatly and can be altered in virus-infected cells.

The RNA-processing reactions summarized in Fig. 10.1 not only produce functional mRNAs but also provide numerous opportunities for posttranscriptional control of gene expression. In virus-infected cells, regulation of RNA processing can increase the coding capacity of the viral genome, determine when specific viral proteins are made during the infectious cycle, and facilitate selective expression of viral genetic information. In this chapter, we focus on these RNA-processing reactions to illustrate both crucial viral regulatory mechanisms and the seminal contributions of viral systems to the elucidation of essential cellular processes.

## Covalent Modification during Viral Pre-mRNA Processing

### Capping the 5′ Ends of Viral mRNA

The first mRNAs shown to carry the 5′-terminal structure termed the **cap** were those of reovirus and vaccinia virus (Box 10.2). These viral mRNAs are made and processed by virus-encoded enzymes (see below), but subsequent research established that virtually all viral and cellular mRNAs possess the same cap structure, m⁷GpppN. This structure protects mRNAs from 5′ exonucleolytic attack and allows recognition of the 5′-terminal exons of pre-mRNAs made by cellular RNA polymerase II during splicing. In addition, the cap structure is essential for the efficient translation of most viral and cellular mRNAs, for it is recognized by proteins that mediate initiation of translation. The principal exceptions are the uncapped mRNAs

of certain (+) strand viruses, notably picornaviruses and the flavivirus hepatitis C virus, which are translated by the cap-independent mechanism described in chapter 11.

Although most other viral mRNAs carry a 5′-terminal cap structure, there is considerable variation in how this modification is made. Three mechanisms can be distinguished: de novo synthesis by cellular enzymes, synthesis by viral enzymes, and acquisition of preformed 5′ cap structures from cellular pre-mRNA or mRNA (Table 10.1).

### Synthesis of Viral 5′ Cap Structures by Cellular Enzymes

Viral pre-mRNA substrates for the cellular capping enzyme described below are invariably made in the infected cell nucleus by cellular RNA polymerase II (Table 10.1) because the actions of these enzymes are tightly coordinated. The formation of cap structures on the 5′ ends of viral and cellular RNA polymerase II transcripts, the first step in their processing, is a cotranscriptional reaction that takes place when the nascent RNA is only 20 to 30 nucleotides in length. At this early stage, the transcriptional complex pauses and the C-terminal domain of the largest subunit of RNA polymerase II becomes hyperphosphorylated. This modification is the signal for binding of the cellular capping enzyme to this unique domain of RNA polymerase II and capping of the nascent RNA. The intimate relationship between the cellular capping enzyme and RNA polymerase II ensures that all transcripts made by this enzyme are capped at their 5′ ends. It also explains why RNA species synthesized by other cellular RNA polymerases are not modified in this way: these enzymes do not carry an equivalent domain rich in residues that can be phosphorylated.

The 5′ cap structure is constructed by the action of several enzymes (Fig. 10.2). The 5′ ends of nascent RNA polymerase II transcripts carry a triphosphate group from the initiating nucleotide. The terminal phosphate is removed by a 5′ triphosphatase to form the substrate for guanylyltransferase. This enzyme transfers GMP from GTP to the 5′ end of the nascent transcript to form the protective GpppN structure (Fig. 10.2), which is resistant to nucleases that cleave 5′–3′ phosphodiester bonds. In mammalian cells, a single protein, commonly called **capping enzyme**, contains both the 5′ triphosphatase and the guanylyltransferase. Following the action of capping enzyme, the terminal G residue is modified by methylation at specific positions (Fig. 10.2). The cap 1 structure, m⁷GpppNm, is common in viral and mammalian mRNAs. However, the sugar of the second nucleotide can also be methylated, by a cytoplasmic enzyme, to form the cap 2 structure (Fig. 10.2).

## BOX 10.2

### Identification of 5' cap structures and 3' poly(A) sequences on viral mRNAs

The first clues that the termini of mRNAs made in eukaryotic cells possessed special structures came when viral mRNAs that could be obtained in large quantities did not behave as predicted from the known structure of prokaryotic mRNAs.

In the figure, panel A shows 5' cap structures. The 5' end of reoviral (or vaccinia virus) mRNA, in contrast to that of a prokaryotic mRNA, could not be labeled by polynucleotide kinase and [$\gamma$-$^{32}$P]ATP after alkaline phosphatase treatment. This property established that the 5' end did not carry a simple phosphate group but rather was blocked. As indicated below, the structure of the 5' blocking group (termed the cap) was elucidated by labeling the viral mRNA with [$\beta$,$\gamma$-$^{32}$P]GTP, [$^{14}$C]ATP, and [$methyl$-$^{3}$H]-$S$-adenosylmethionine followed by digestion of the

mRNA with nucleases with different specificities. Cleavage with P1 nuclease liberated the expected product, 5'pN-OH3', which contained all the $^{14}$C label as well as a structure that contained all the label incorporated from [$\beta$,$\gamma$-$^{32}$P]GTP and therefore corresponded to the 5' terminus of the mRNA (the only position at which $\beta$ or $\gamma$ phosphates from NTPs are retained in an mRNA). This structure also contained all the $methyl$-$^{3}$H label, indicating that it contained methyl groups. Its digestion with nucleotide pyrophosphatase produced free phosphate with all the $^{32}$P label and two $methyl$-$^{3}$H-labeled products. Because the terminal structure was known to contain G (from incorporation of $^{32}$P from [$\beta$,$\gamma$-$^{32}$P]GTP), these nucleotides were identified as m$^7$Gp and 2'-$O$-methylguanosine monophosphate (i.e., methylated at the 2' position of the sugar

*(continued)*

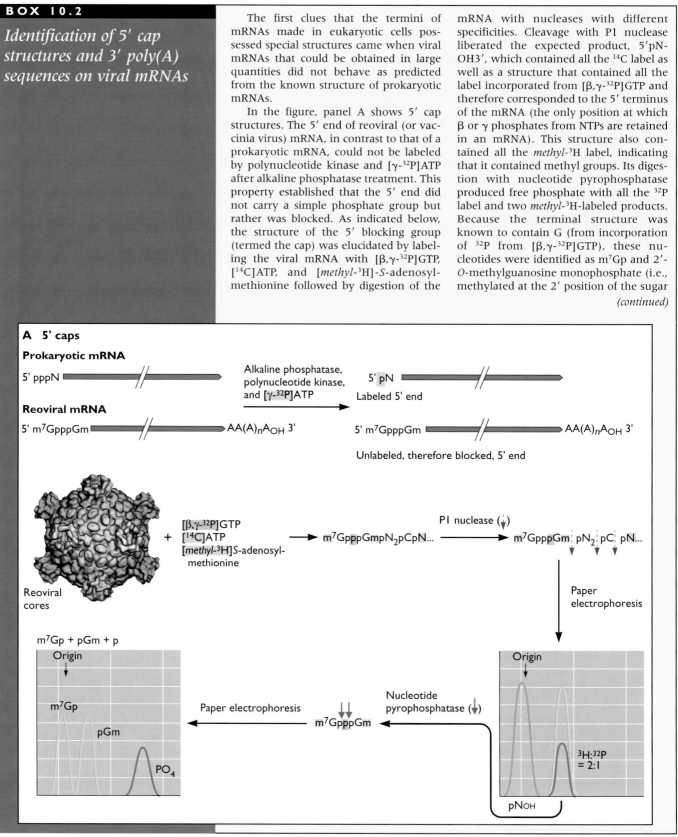

**A  5' caps**

**Prokaryotic mRNA**

5' pppN ▰▰▰ // ▰▰▰▶    Alkaline phosphatase, polynucleotide kinase, and [$\gamma$-$^{32}$P]ATP    5' pN ▰▰▰ // ▰▰▰▶

Labeled 5' end

**Reoviral mRNA**

5' m$^7$GpppGm ▰▰▰ // ▰▰▰ AA(A)$_n$A$_{OH}$ 3'    5' m$^7$GpppGm ▰▰▰ // ▰▰▰ AA(A)$_n$A$_{OH}$ 3'

Unlabeled, therefore blocked, 5' end

Reoviral cores + [$\beta$,$\gamma$-$^{32}$P]GTP  [$^{14}$C]ATP  [$methyl$-$^{3}$H]$S$-adenosylmethionine → m$^7$GpppGmpN$_2$pCpN... → P1 nuclease (↓)  m$^7$GpppGm ¦ pN$_2$ ¦ pC ¦ pN...

Paper electrophoresis

m$^7$Gp + pGm + p

Origin ↓  m$^7$Gp  pGm  PO$_4$

← Paper electrophoresis ← m$^7$GpppGm ← Nucleotide pyrophosphatase (↓)

Origin ↓

$^3$H:$^{32}$P = 2:1

pN$_{OH}$

**BOX 10.2**    *(continued)*

ring) by comigration with these compounds. Panel B shows 3' poly(A) sequences. Polyadenylation of viral mRNAs was first indicated by the observation that an RNA chain resistant to digestion by RNase A, which cleaves after U and C, was produced when vaccinia virus mRNA, but not prokaryotic mRNA, was treated with this enzyme following labeling with [³H]ATP, but not with any other NTP. The presence of a tract of poly(A) was confirmed by the specific binding of vaccinia virus mRNA, but not of prokaryotic mRNA, to poly(U)-Sepharose under conditions that allowed annealing of complementary nucleic acids. The position of the poly(A) sequence in viral mRNA was determined by analysis of the

products of alkaline hydrolysis, when phosphodiester bonds are broken to produce nucleotides with 5' hydroxyl and 3' phosphate (Ap) groups. The liberation of A-OH3' by this treatment indicated that the poly(A) was located at the 3' end of the mRNA.

**Furuichi, Y., M. Morgan, S. Muthukrishnan, and A. J. Shatkin.** 1975. Reovirus messenger RNA contains a methylated, blocked 5'-terminal structure: m⁷G(5')ppp(5')GpCp. *Proc. Natl. Acad. Sci. USA* **72:**362–366.

**Kates, J.** 1970. Transcription of the vaccinia virus genome and occurrence of polyriboadenylic acid sequences in messenger RNA. *Cold Spring Harbor Symp. Quant. Biol.* **35:**743–752.

**Miura, K., K. Watanabe, M. Sugiura, and A. J. Shatkin.** 1974. The 5'-terminal nucleotide sequences of the double-stranded RNA of human reovirus. *Proc. Natl. Acad. Sci. USA* **71:**3979–3983.

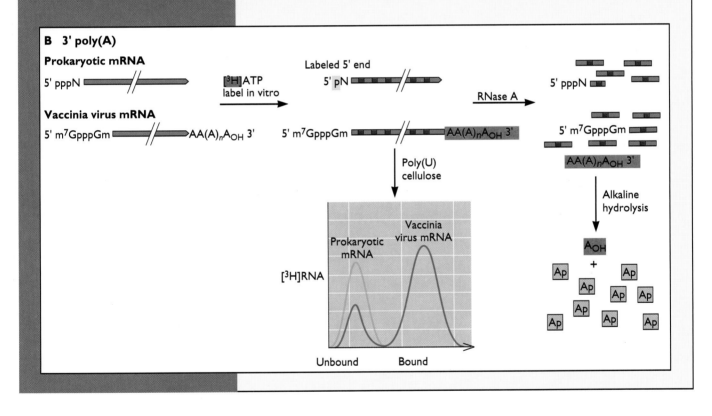

**Table 10.1**    Mechanisms of synthesis of 5'-terminal cap structures of viral mRNAs

| Mechanism | Virus families | Enzyme synthesizing pre-mRNA |
|---|---|---|
| Synthesis by host cell enzymes | *Parvoviridae, Papovaviridae, Adenoviridae, Herpesviridae, Retroviridae, Hepadnaviridae* | Cellular DNA-dependent RNA polymerase II |
| Synthesis by viral enzymes | *Alphaviridae, Rhabdoviridae, Reoviridae* | Viral RNA-dependent RNA polymerase |
|  | *Poxviridae* | Viral DNA-dependent RNA polymerase |
| Acquisition from cellular pre-mRNA or mRNA | *Orthomyxoviridae, Bunyaviridae* | Viral RNA-dependent RNA polymerase |

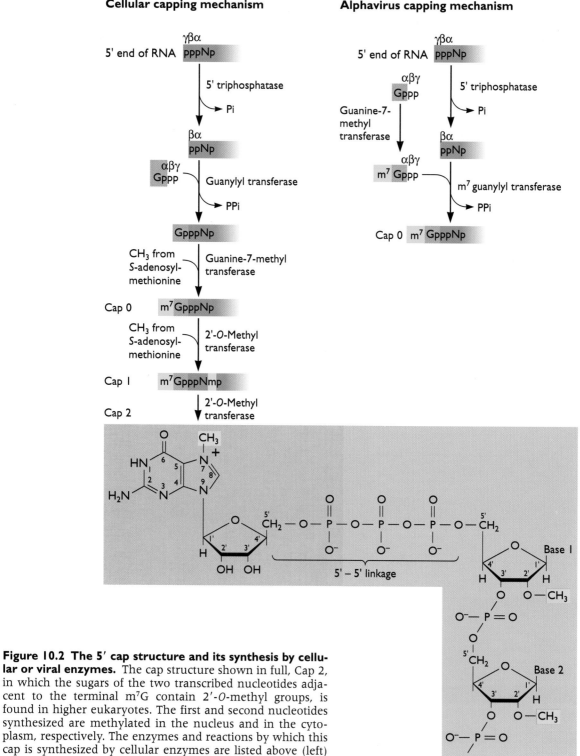

**Cellular capping mechanism**

**Alphavirus capping mechanism**

**Figure 10.2 The 5′ cap structure and its synthesis by cellular or viral enzymes.** The cap structure shown in full, Cap 2, in which the sugars of the two transcribed nucleotides adjacent to the terminal $m^7G$ contain 2′-$O$-methyl groups, is found in higher eukaryotes. The first and second nucleotides synthesized are methylated in the nucleus and in the cytoplasm, respectively. The enzymes and reactions by which this cap is synthesized by cellular enzymes are listed above (left) and compared to the synthesis of the caps of Semliki Forest virus (a togavirus) mRNAs by viral enzymes in the cytoplasm of infected cells (right).

## Synthesis of Viral 5′ Cap Structures by Viral Enzymes

When viral mRNAs are made in the cytoplasm of infected cells, their 5′ cap structures are of necessity synthesized by viral enzymes. These enzymes form cap structures typical of those present on cellular mRNA, although with some variations in the mechanism of cap construction. For example, during synthesis of the caps of vesicular stomatitis virus mRNAs, only the α phosphate group of the initiating nucleotide is retained and GDP is added. The alphaviral capping enzyme methylates the $N^7$ position of GTP prior to transfer of GMP to the 5′ end of viral pre-mRNA, and the resulting cap structure is not modified further (Fig. 10.2). The alphaviral mRNAs therefore carry the cap 0 structure, which is rarely found on mRNAs made in mammalian cells.

Like their cellular counterparts, viral capping enzymes are intimately associated with the viral RNA polymerases responsible for mRNA synthesis. Indeed, in the simplest case, the several enzymatic activities required for synthesis of a 5′ cap structure are supplied by the viral RNA polymerase itself. This strategy is exemplified by the vesicular stomatitis virus L protein, which carries out all reactions necessary for synthesis and capping of viral mRNAs. More complex viruses encode dedicated capping enzymes, such as the λ-2 protein of reovirus particles. This protein, which is the guanylyltransferase and probably also the *N*-methyltransferase, interacts with the viral RNA-dependent RNA polymerase in the core of the virion. One of the first capping enzymes to be analyzed in detail, and the best characterized among those of larger viruses, is the vaccinia virus enzyme. This protein binds directly to the viral RNA polymerase and adds 5′ cap structures cotranscriptionally to nascent viral transcripts approximately 30 nucleotides in length. The capping enzyme of vaccinia virus thus displays striking functional similarities to its host cell counterpart and has provided an important model for the cellular capping machinery and its mechanism of action.

The activities of most viral capping enzymes must be regulated. With the exception of that of vaccinia virus, these enzymes cooperate with viral RNA-dependent RNA polymerases that can synthesize both (−) and (+) strand RNAs. However, only (+) strand RNAs become capped. The mechanisms that coordinate capping activity with viral mRNA synthesis are not well understood. However, in the case of togaviruses, such as Sindbis and Semliki Forest viruses, activation of capping enzymes may be the result of proteolytic processing. As discussed in chapter 6, the viral P1234 polyprotein is responsible for the initial synthesis of (−) strand RNA from the (+) stand viral genome. This polyprotein includes the sequences of the RNA polymerase and the capping enzyme (Nsp2 and Nsp1, respec-

tively), but the latter is inactive. Cleavage of the polyprotein is necessary for synthesis of viral mRNAs (see Fig. 6.12) and thus may also activate the capping enzyme.

## Acquisition of Viral 5′ Cap Structures from Cellular RNAs

Members of two virus families, the *Orthomyxoviridae* and the *Bunyaviridae*, have evolved a unique way to rely on cellular capping enzymes for the production of 5′ cap structures. The 5′ caps of these viral mRNAs are acquired when viral cap-dependent endonucleases cleave cellular transcripts to produce the primers needed for viral mRNA synthesis by viral RNA polymerase, a process called **cap snatching** (see Fig. 6.10B). The 5′-terminal segments and caps of influenza virus mRNAs are obtained from cellular pre-mRNA in the nucleus. On the other hand, bunyaviral mRNA synthesis is primed with 5′-terminal fragments obtained from mature cellular mRNAs in the cytoplasm.

## Synthesis of 3′ Poly(A) Segments of Viral mRNA

Like the 5′ cap structure, a 3′ poly(A) segment was first identified in a viral mRNA (Box 10.2). This 3′-end modification was soon found to be a common feature of mRNAs made in eukaryotic cells. Like the 5′ cap, the 3′ poly(A) sequence stabilizes mRNA and also increases the efficiency of translation. Therefore, it is not surprising that viral mRNAs generally acquire a 3′ poly(A) tail. Those that do not, such as reoviral and arenaviral mRNAs, may survive by virtue of a 3′-terminal stem-loop structure that blocks nucleolytic attack. Such structures are present at the 3′ ends of arenaviral mRNAs as well as the cellular, poly(A)-lacking mRNAs encoding histones.

The addition of 3′ poly(A) segments to viral pre-mRNAs, like capping of their 5′ ends, can be carried out by either cellular or viral enzymes (Table 10.2). However, the majority of viral polyadenylation mechanisms differ markedly from that by which the cellular polyadenylation machinery operates.

## Polyadenylation of Viral Pre-mRNA by Cellular Enzymes

Viral pre-mRNAs synthesized in infected cell nuclei by cellular RNA polymerase II are invariably polyadenylated by cellular enzymes (Table 10.2). Polyadenylation of such nuclear pre-mRNAs occurs at about the same time as their splicing. In fact, as discussed later in this chapter, these two processing reactions are not independent.

Transcription of a viral or cellular gene by RNA polymerase II proceeds beyond the site at which poly(A) will be added. The 3′ end of the mRNA is determined by endonucleolytic cleavage of its pre-mRNA at a specific posi-

**Table 10.2** Mechanisms of addition of poly(A) to viral RNAs

| Mechanism | Enzyme | Viruses |
|---|---|---|
| During mRNA synthesis | | |
| Reiterative copying of short U stretches in the template RNA | Viral | Rhabdoviruses, paramyxoviruses, orthomyxoviruses |
| Copying of a long 5′-terminal U stretch in the template RNA | Viral | Picornaviruses, togaviruses |
| After transcription | | |
| Cleavage of pre-mRNA followed by polyadenylation | Cellular | Papovaviruses, adenoviruses, herpesviruses, retroviruses, hepadnaviruses, hepatitis delta satellite virus |
| Transcription termination followed by polyadenylation | Viral | Poxviruses |

tion. Poly(A) is then added to the new 3′ terminus, while the RNA downstream of the cleavage site is degraded (Fig. 10.3). Cleavage and polyadenylation sites are identified by specific sequences, first characterized in simian virus 40 and adenoviral pre-mRNAs. The 3′ end of mature mRNA is formed 10 to 30 nucleotides downstream of a highly conserved and essential polyadenylation signal, the sequence 5′AAUAAA3′. However, this sequence is not sufficient to specify poly(A) addition. For example, it is found within mRNAs at internal positions that are never used as polyadenylation sites. Sequences at the 3′ side of the cleavage site, notably a GU- or U-rich sequence located 5 to 20 nucleotides downstream, are also required to specify a polyadenylation site, and in many mRNAs (particularly viral mRNAs) additional sequences 5′ to the cleavage site are also important.

The first reaction in polyadenylation is recognition of the 5′AAUAAA3′ sequence by the protein termed cleavage and polyadenylation specificity protein (Fig. 10.3). This interaction is stabilized by other proteins, including the cleavage stimulatory protein, which binds to the downstream U/GU-rich sequence. Poly(A) polymerase is then recruited to the complex prior to cleavage of the pre-mRNA. The enzyme initially adds A residues at a low rate to synthesize a short poly(A) segment of 12 to 20 residues. This structure is then bound by the poly(A)-binding protein PabII, which induces rapid and processive addition of more A residues until a long tail, typically 200 to 250 nucleotides, has been made (Fig. 10.3).

Components of the polyadenylation machinery, like capping enzymes, are associated with RNA polymerase II, and this enzyme is required for efficient polyadenylation in in vitro reactions. Both the cleavage stimulatory protein and the cleavage and polyadenylation specificity protein bind to the C-terminal domain of the largest subunit of RNA polymerase II. Such interactions, which are essential for polyadenylation in vivo, are believed to coordinate synthesis of pre-mRNAs with maturation of the 3′ ends. They may also be required for termination of transcription by RNA polymerase II, for both the 5′AAUAAA3′ and

U/GU-rich polyadenylation signals are necessary for this process, even though the enzyme can continue synthesis for long distances beyond sites of polyadenylation.

### Polyadenylation of Viral Pre-mRNAs by Viral Enzymes

**Formation of 3′ ends by transcription termination.** The 3′ ends of vaccinia virus early mRNAs are formed by termination of transcription by the viral DNA-dependent RNA polymerase (Fig. 10.4), a mechanism with no known counterpart in cellular or other viral mRNA synthesis systems. Unexpectedly, the vaccinia virus capping enzyme was identified as the protein that directs termination of transcription of viral early genes at specific positions. As discussed previously, the capping enzyme binds to the viral RNA polymerase. It is believed to travel with the transcriptional complex until the termination signal 5′UUUUUNU3′ is encountered in the nascent RNA. The termination protein/capping enzyme then induces termination of transcription about 50 nucleotides downstream. The viral poly(A) polymerase adds poly(A) to the 3′ end thus formed, by a two-step process that is remarkably similar to the mechanism of synthesis of poly(A) tails by the cellular machinery (compare Fig. 10.4 and 10.3). The subunit of the viral poly(A) polymerase that induces processive synthesis of full-length poly(A) after the initial production of short chains is also the viral 2′-O-methyltransferase. Thus, all components of the vaccinia virus capping machinery also participate in formation of the 3′ ends of viral mRNAs. This property may be a manifestation of efficient utilization of viral genetic information or may facilitate coordination of the reactions by which viral mRNAs are synthesized.

**Polyadenylation during viral mRNA synthesis.** The synthesis of the poly(A) segments of vaccinia virus early mRNAs resembles the cellular mechanism in several respects. In contrast, the poly(A) sequences of mRNAs made by other viral RNA polymerases are produced by different

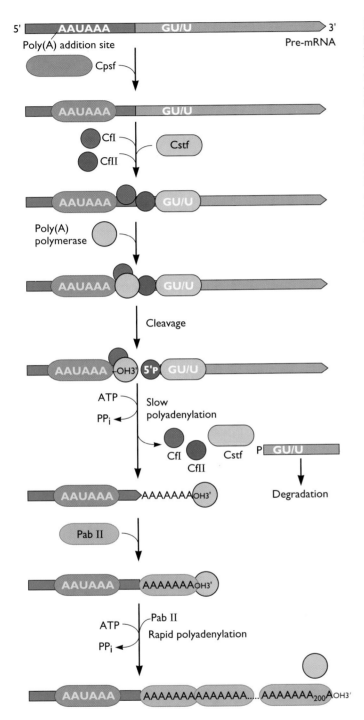

**Figure 10.3 Cleavage and polyadenylation of vertebrate pre-mRNAs.** The cleavage and polyadenylation specificity protein (Cpsf), which contains four subunits, binds to the 5′AAUAAA3′ poly(A) addition signal that lies upstream of the site at which poly(A) will be added. Cleavage stimulatory protein (Cstf) then interacts with the downstream U/GU-rich sequence to stabilize a complex that also contains the two cleavage proteins, CfI and CfII. Binding of poly(A) polymerase is followed by cleavage at the poly(A) addition site when CfI, CfII, Cstf, and the downstream RNA cleavage product are released. The enzyme slowly adds 10 to 15 A residues to the 3′-OH terminus produced by the cleavage reaction. Poly(A)-binding protein II (Pab II) then binds to this short poly(A) sequence and, in conjunction with Cpsf, tethers poly(A) polymerase to the poly(A) sequence. This association facilitates rapid and processive addition of A residues until a poly(A) chain of about 200 residues has been synthesized.

sult of reiterative copying of short stretches of U residues in the (−) strand RNA template, a mechanism discussed in chapter 6.

## Splicing of Viral Pre-mRNA

### Discovery of Splicing

Between 1960 and the mid-1970s, many investigators characterized putative nuclear precursors of eukaryotic mRNAs. These RNAs were found to be larger than the mature mRNAs translated in the cytoplasm and heterogeneous in size. They were therefore named **heterogeneous nuclear RNAs (hnRNAs)**. Such hnRNAs were shown to carry both 5′-terminal cap structures and 3′ poly(A) sequences, leading to the conclusion that both ends of the hnRNA were preserved in the smaller, mature mRNA. Thus, investigators were faced with the dilemma of deducing how smaller mRNAs could be produced from larger hnRNAs while *both* ends of the hnRNA were retained.

The puzzle was solved by two groups of investigators, led by Phillip Sharp and Richard Roberts, who shared the 1993 Nobel Prize in physiology or medicine for this work. Much to everyone's surprise, these investigators showed that adenoviral major late mRNAs were encoded by four separate genomic sequences, three of which are complementary to a short (about 200-nucleotide) 5′-terminal segment present on all major late mRNAs, called the tripartite leader sequence (Box 10.3).

The distribution of adenoviral major late mRNA-coding sequences into four separate blocks in the genome, in conjunction with the large size of major late mRNA precursors and the presence of a single major late promoter in the adenoviral genome, implied that these mRNAs were synthesized by excision of noncoding sequences from primary transcripts with precise joining of coding sequences. The

mechanisms, during synthesis of the mRNA. In the simplest case, exemplified by (+) strand picornaviruses, a poly(U) sequence present at the 5′ end of the (−) strand RNA template is copied directly into a poly(A) sequence of equivalent length at the 3′ end of the (+) strand mRNA. The mRNAs of (−) strand RNA viruses like vesicular stomatis virus and influenza virus are polyadenylated as a re-

**Figure 10.4 The vaccinia virus capping enzyme and 2'-O-methyltransferase process both the 5' and 3' ends of vaccinia virus mRNAs.** After capping the 5' end of nascent viral mRNA chains about 30 nucleotides in length, the capping enzyme remains bound to the nascent RNA chain and to the RNA polymerase as the latter enzyme transcribes the template DNA. The capping enzyme recognizes the 5'UUUUNU3' termination signal in the nascent RNA and induces the polymerase to terminate RNA synthesis about 50 nucleotides downstream. The 2'-O-methyltransferase produces a cap 1 structure. A fraction of the 2'-O-methyltransferase molecules act as an elongation stimulation protein for the viral poly(A) polymerase, analogous to cellular poly(A)-binding protein II. This viral enzyme, like its cellular counterpart (Fig. 10.3), adds poly(A) to the 3' ends of the mRNA in a two-step process.

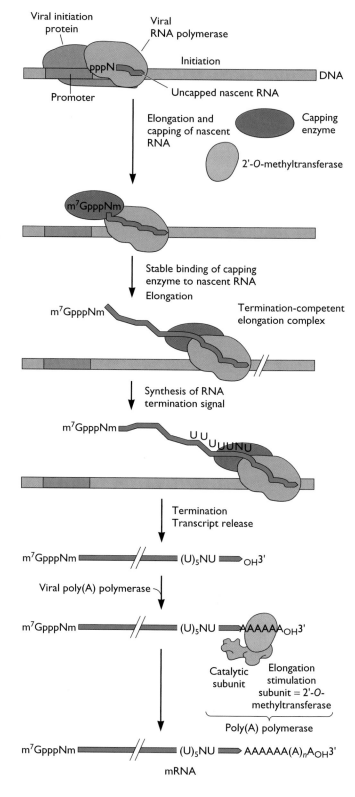

demonstration that nuclear major late transcripts contain the noncoding sequences (introns) confirmed that the mature mRNAs are formed by **splicing** of noncontiguous coding sequences in the pre-mRNA.

This mechanism of mRNA synthesis had great appeal because it could account readily for the puzzling properties of hnRNA described above. Indeed, it was shown within a matter of months that splicing of pre-mRNA is not an obscure virus-specific device. Splicing occurs in all eukaryotic cells, although it is less prevalent in simpler organisms such as the yeast *Saccharomyces cerevisiae*. The great majority of mammalian pre-mRNAs, like the adenoviral major late mRNAs, comprise exons interspersed among introns. In all cases, the capped 5' ends and sequences immediately adjacent to the 3' poly(A) tail are conserved in mature mRNA, as expected from the properties of hnRNA.

The organization of protein-coding sequences into exons separated by introns has profound implications for the evolution of the genes of eukaryotes and their viruses. Introns are generally much longer than exons, and only short sequences at the ends of introns are necessary for accurate splicing (see "Mechanism of Splicing" below). Consequently, introns provide numerous sites at which DNA sequences can be broken and rejoined without loss of coding information. Therefore, they greatly increase the frequency with which random recombination reactions can create new functional genes by rearrangement of exons. Evidence of such "exon shuffling" can be seen in the modular organization of many modern proteins. Such proteins comprise combinations of a finite set of structural and functional domains or motifs, or multiple repeats of a single protein domain, each often encoded by a single exon. In similar fashion, the presence of introns is believed to have facilitated the capture of cellular genetic information by viruses with DNA genomes and by retroviral genomes.

**BOX 10.3**

*Discovery of the spliced structure of adenoviral major late mRNAs*

The demonstration that several different adenoviral major late (ML) mRNAs carry the same capped 11-nucleotide sequence at their 5′ ends (figure panel A) was both surprising and puzzling. Although hybridization studies indicated that these 5′ ends were not encoded adjacent to the main segments of major late mRNAs, it was direct visualization of such mRNAs hybridized to viral DNA that provided the convincing proof that their coding sequences are dispersed in the viral genome. A schematic diagram of one major late mRNA (hexon mRNA) hybridized to a complementary adenoviral single-stranded DNA fragment extending from the left end to a point within the hexon coding sequence is shown in panel B. Three loops of unhybridized DNA (thin lines), designated A, B, and C, bounded or separated by three short segments (1, 2,

and 3) and one long segment (hexon mRNA) of DNA-RNA hybrid (thick lines) were observed. Other adenoviral late mRNAs examined yielded the same sets of hybridized and unhybridized viral DNA sequences at their 5′ ends but differed in the length of loop C and the length and location of the 3′-terminal RNA-DNA hybrid. It was therefore concluded that the major late mRNAs contain a common 5′-terminal sequence (segments 1, 2, and 3) built from sequences encoded at three different sites in the viral genome, termed the tripartite leader sequence. This sequence is joined to the mRNA body, a long sequence complementary to part of the hexon coding sequence in the viral genome in the example shown. (B) Adapted from S. M. Berget et al., *Proc. Natl. Acad. Sci. USA* **74:**3171–3175, 1977, with permission.

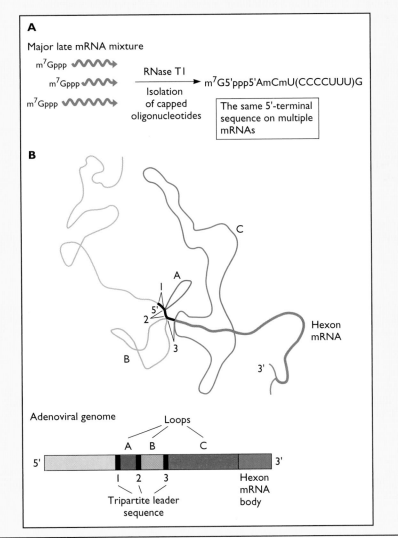

## Mechanism of Splicing

Sequencing of DNA copies of a large number of cellular and viral mRNAs and of the genes that encode them identified short consensus sequences spanning the 5' and 3' **splice sites**, which are joined to each other in mature mRNA (Fig. 10.5). The conserved sequences lie largely within the introns. The dinucleotides GU and AG are found at the 5' and 3' ends, respectively, of almost all introns. Mutation of any one of these four nucleotides eliminates splicing, indicating that all are essential. Elucidation of the mechanism of splicing came with the development of in vitro systems in which model pre-mRNAs (initially of viral origin) are accurately spliced. Numerous components needed for splicing have now been identified by both biochemical studies based on such systems and genetic analyses in organisms such as yeast and the fruit fly *Drosophila melanogaster*.

Pre-mRNA splicing occurs by two transesterification reactions, in which one phosphodiester bond is exchanged for another without the need for an external supply of energy (Fig. 10.6). The first reaction yields two products, the 5' exon and the intron-3' exon in which the 5' end of the intron is linked to a conserved A residue that, in most mRNAs, is located 18 to 40 nucleotides upstream of the 3' splice site (Fig. 10.6). Thus, the intron-3' exon product has the structure of a lariat, in which the conserved intronic A residue serves as the **branch point**. In the second transesterification reaction, the newly formed hydroxyl group at the end of the 5' exon attacks the phosphodiester bond at the 3' splice site, yielding the spliced exon product and the intron lariat (Fig. 10.6).

From a chemical point of view, the splicing of pre-mRNA is a simple process. However, each splicing reaction must be completed with a high degree of accuracy to ensure that coding information is not lost or altered. The chemically active hydroxyl groups (Fig. 10.6) must also be brought into close proximity to the phosphodiester bonds they will attack. Furthermore, some genes contain 50 or more exons that must be spliced in the correct order. It is presumably for such reasons that pre-mRNA splicing requires both many proteins and several RNAs, which associate with the pre-mRNA to form the large structure called the **spliceosome**.

Five small nuclear RNAs (snRNAs) participate in splicing: the U1, U2, U4, U5, and U6 snRNAs. In vertebrate cells, these RNAs vary in length from 100 to 200 nucleotides and are associated with proteins to form **small nuclear ribonucleoproteins (snRNPs)**. The RNA components of the snRNPs recognize splice sites and other sequences in cellular and viral pre-mRNAs. Indeed, they participate in multiple interactions with the pre-mRNA and with each other during the three steps necessary to complete splicing. In the first, assembly of the spliceosome, the U1 and U2 snRNAs base pair with the 5' splice site and the intronic branchpoint, respectively, and the other snRNPs then enter the spliceosome (Fig. 10.7). At this stage, the U4 and U6 snRNAs are extensively hydrogen-bonded to one another and present in a single snRNP. The next step is dramatic rearrangement of the RNA-RNA interactions within the spliceosome. The end result is juxtaposition of the 5' splice site and the branch point for the first transesterification reaction, as the result of interactions of U6 snRNA with the pre-mRNA and with U2 snRNA (Fig. 10.7). Indeed, the current hypothesis is that the U2-U6 RNA complex formed in the spliceosome catalyzes splicing, the final reaction. The U5 snRNA may juxtapose the 5' and 3' exons and thus facilitate catalysis of their joining by the U2-U6 RNA complex.

Although the snRNAs play essential roles in splicing as both guides and catalysts, the spliceosome also contains as

**Figure 10.5 Consensus splicing signals in cellular and viral pre-mRNAs.** Conserved sequences are found at the 5' and 3' splice sites at the junctions of exons (green) and introns (pink) and at the 3' ends of introns. The intronic 5'GU3' and 5'AG3' dinucleotides at the 5' and 3' ends, respectively, of introns and branch point A (highlighted) are present in all but rare mRNAs made in higher eukaryotes.

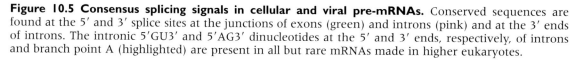
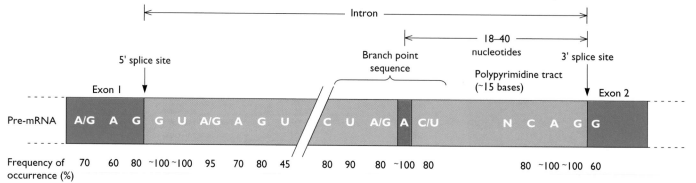

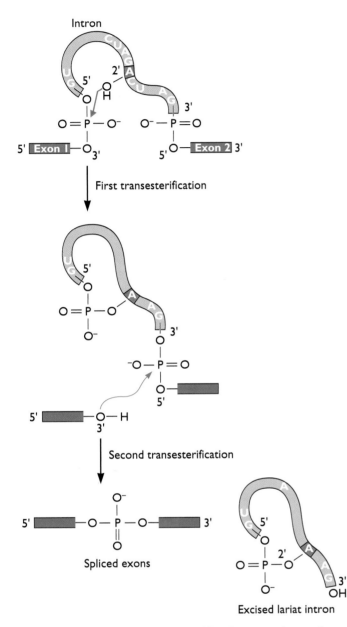

**Figure 10.6 The two transesterification reactions of pre-mRNA splicing.** In the first reaction, the 2′ hydroxyl group of the conserved A residue in the intronic branch point sequence, which is located 18 to 40 nucleotides upstream of the 3′ splice site (Fig. 10.5), makes a nucleophilic attack on the phosphodiester bond at the 5′ side of the GU dinucleotide at the 5′ splice site. In the intron-3′ exon lariat product, this attacking A residue is linked via 3′-5′ and 2′-5′ phosphodiester bonds to two different nucleotides. A second nucleophilic attack by the newly formed 3′ hydroxyl group of the 5′ exon on the phosphodiester bond at the 3′ splice site yields the spliced exons and the intron lariat.

many as 50 to 100 non-snRNP proteins. One class comprises proteins that package the pre-mRNA substrate, termed **heterogeneous nuclear RNP (hnRNP) proteins**. These proteins bind to all pre-mRNAs made by RNA polymerase II, probably as they are synthesized, and play important structural roles. There is some evidence that certain hnRNP proteins may participate more directly in splicing. Many other splicing proteins contain both RNA-binding and protein-protein interaction domains. Such proteins bind to pre-mRNA sequences within or adjacent to exons and act positively or negatively to facilitate splicing or exon recognition. Members of one family of such proteins, called the **SR proteins** because they contain a domain rich in serine (S) and arginine (R), act at early stages of splicing to recruit snRNPs to the spliceosome. For example, the particularly well-characterized SR protein named splicing factor 2 (or alternative splicing factor) facilitates binding of U1 snRNP to 5′ splice sites and stabilizes the prespliceosomal complex (Fig. 10.7). Members of the SR protein family can also recognize exon sequences that stimulate splicing of upstream introns, termed **splicing enhancers**, and can assist juxtaposition of 5′ and 3′ splice sites. Other proteins important for splicing are RNA-dependent helicases, which are thought to catalyze the multiple rearrangements of hydrogen-bonding among different snRNAs and the pre-mRNA substrate (Fig. 10.7). Such helicases are generally ATP dependent, and spliceosome assembly and rearrangement depend on energy supplied by ATP hydrolysis.

Splicing of pre-mRNAs is commonly cotranscriptional, and components of the splicing machinery may associate with RNA polymerase II. The hyperphosphorylated form of the C-terminal domain of the largest subunit of this enzyme can bind to both SR proteins and spliceosomal snRNPs. Peptide mimics of the C-terminal domain or antibodies raised against it inhibit pre-mRNA splicing in vivo or in vitro. As we have seen, association of components of the 5′ capping and 3′ polyadenylation systems with the C-terminal domain of RNA polymerase is necessary for these processing reactions. It is therefore possible that the synthesis of pre-mRNA and all the reactions by which it is processed are coordinated as a result of association of specific proteins needed for each processing reaction with the C-terminal domain of RNA polymerase II. Such a transcription and RNA-processing machine would be analogous to, but much more complex than, that of vaccinia virus described above. This attractive hypothesis is currently receiving much attention. However, it remains to be demonstrated that the C-terminal domain of RNA polymerase II participates in splicing directly, rather than indirectly, for example, as a result of its role in capping, which is required for recognition of first exons during splicing.

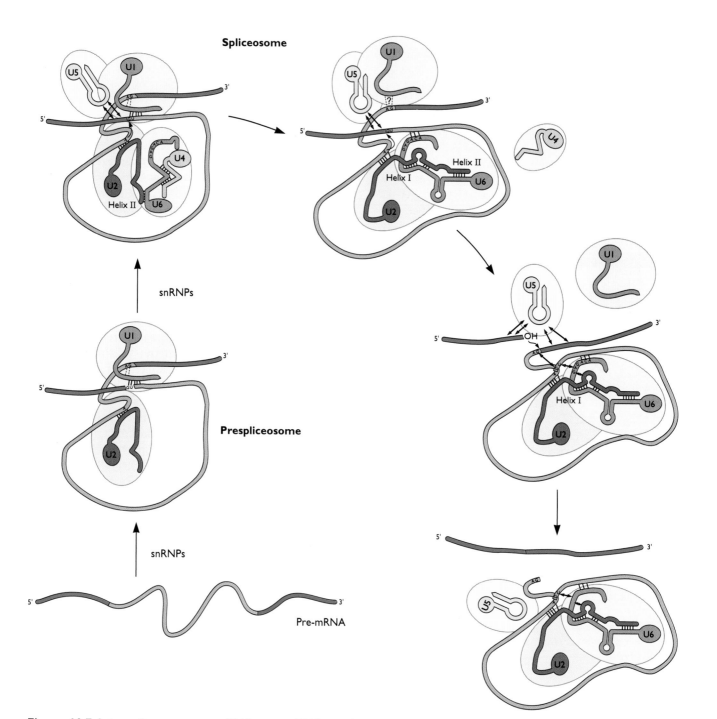

**Figure 10.7 Interactions among snRNAs or snRNAs and pre-mRNA during formation of active spliceosomes.** Base pairs are indicated by dashes and experimentally observed or presumed contacts among the RNA molecules by the two-headed arrows. The U1 and U2 snRNAs initially base pair with the 5' splice site and branch point sequence, respectively, in the pre-mRNA. Usually only three base pairs between U2 snRNA and the pre-mRNA are formed at the branch point sequence. The U2-associated protein facilitates and stabilizes this interaction. The other snRNPs then enter the assembling spliceosome. The U4 and U6 snRNAs, which are present in a single snRNP, are base paired with one another over an extended complementary region. This snRNP binds to the U5 snRNP, and the snRNP complex associates with that containing the pre-mRNA and U1 and U2 snRNAs to form the spliceosome. RNA rearrangements then activate spliceosomes for catalysis of splic-

ing. U4 snRNA dissociates from U6 snRNA, which forms hydrogen bonds with both U2 snRNA and the pre-mRNA. The interaction of U6 snRNA with the 5' splice site displaces U1 snRNA, which plays no further role in splicing. One of the U2 sequences hydrogen-bonded to U6 snRNA (helix I) is adjacent to the U2 snRNA sequence that is base paired with the pre-mRNA branch point region. Thus, the interactions of U2 and U6 snRNAs with each other and with the pre-mRNA juxtapose the branch point and 5' splice site sequences for the first transesterification reaction (Fig. 10.6). The U5 snRNA base pairs to sequences in both the 5' and 3' exons and may align them for the second transesterification reaction. The many proteins that participate in spliceosome assembly and activation or that package the pre-mRNA and snRNAs (hnRNP and snRNP proteins, respectively) (see text) are not shown. Adapted from T. Nilsen, *Cell* **78:**1–4, 1994, with permission.

## Production of Stable Viral Introns

Cells infected by certain herpesviruses provide a remarkable exception to the rule that viral intron-containing pre-mRNAs are spliced just like their cellular counterparts. As discussed in chapter 8, the only abundant viral RNAs detected in neurons latently infected by herpes simplex viruses are the major latency-associated transcripts. These nuclear RNA species not only lack 3' poly(A) sequences but also exhibit the properties of excised introns. The very abundant latency-associated transcripts are not linear molecules and have been reported to contain the branch points characteristic of the intron lariats excised from pre-mRNAs during splicing. In contrast to typical introns removed from other viral and cellular pre-mRNAs, these herpesviral RNAs are very stable and accumulate to high concentrations in the nuclei of latently infected cells. Neither the reason for such unusual stability nor the molecular functions of the latency-associated transcripts, which are believed to be important during reactivation from latency, are yet well understood.

## Alternative Splicing of Viral Pre-mRNA

Many viral and cellular pre-mRNAs contain multiple exons. Splicing of the majority removes all introns and joins all exons in the order in which they are present in the substrate pre-mRNA. Thus, such **constitutive splicing** (Fig. 10.8A) produces a single mature mRNA. However, some cellular and many viral pre-mRNAs yield more than one mRNA as a result of the splicing of different combinations of exons, a process termed **alternative splicing**. Several different types of alternative splicing can be defined, depending on whether the splicing machinery skips an entire exon or recognizes alternative 5' or 3' splice sites (Fig. 10.8B). Alternative splicing generally leads to the synthesis of mRNAs that differ in their protein coding sequences.

The most obvious advantage of this mechanism for viruses that direct production of spliced mRNAs in infected cells is that it can expand the limited coding capacity of viral genomes. For example, the early genes of simian virus 40 and polyomavirus, as well as the adenoviral E1A and E1B genes, each specify two or more proteins as a result of splicing of primary transcripts at alternative 5' or 3' splice sites. Alternative splicing can also be important for temporal regulation of viral gene expression or the control of a crucial balance in the production of spliced and unspliced mRNAs. In many cases, alternative splicing of viral pre-mRNAs is coupled with other RNA-processing reactions or regulated by viral proteins. These more complex phenomena are considered in subsequent sections.

## Examples of Alternative Splicing of Viral Pre-mRNAs

Identification of cellular proteins that control alternative splicing: simian virus 40 early pre-mRNA. The simian virus 40 early pre-mRNA is spliced at one of two suboptimal 5' splice sites to form mRNAs encoding either large T (LT) or small T (sT) antigen (Fig. 10.9). The longer mRNA encodes sT because it includes a translation termination codon that is spliced out of LT mRNA (Fig. 10.9). The two proteins therefore contain a common N-terminal sequence but are unrelated at their C termini. Study of the alternative splicing of simian virus 40 early pre-mRNA provided the first demonstration that specific splicing pro-

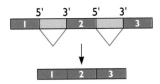

**A Constitutive splicing**

**B Alternative splicing**

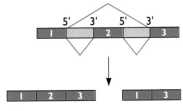

Exon skipping

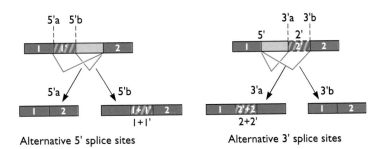

Alternative 5' splice sites

Alternative 3' splice sites

**Figure 10.8 Constitutive and alternative splicing.** In constitutive splicing (A), all exons (green) are joined sequentially and all introns (pink) are excised. Alternative splicing (B) occurs by several mechanisms. In exon skipping, the 3' splice site of exon 2 is sometimes ignored so that this exon is not included in some fraction of the spliced mRNA molecules. Alternatively, one of two 5' splice sites (5'a and 5'b) in exon 1 or one of two 3' splice sites (3'a and 3'b) in exon 2 are recognized. Recognition of different 5' and 3' splice sites produces alternatively spliced simian virus 40 early (Fig. 10.9) and adenoviral major late (Fig. 10.12) mRNAs, respectively.

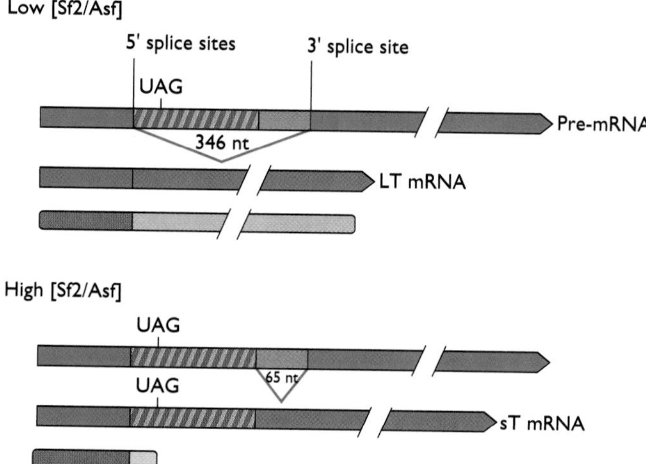

**Figure 10.9 Regulation of alternative splicing of simian virus 40 early pre-mRNA by splicing factor 2.** At low concentrations of splicing factor 2 (Sf2, also known as alternative splicing factor [Asf]), the 5′ splice site for LT mRNA is predominantly recognized in in vitro splicing reactions supplied with simian virus 40 early pre-mRNA. At high concentrations of this cellular protein, the 5′ splice site that produces sT mRNA is activated. Splicing factor 2, an SR protein, can facilitate binding of U1 snRNP to 5′ splice sites and promote interaction of the U1 and U2 snRNPs. High concentrations might therefore improve recognition of the weak sT 5′ splice site or circumvent steric constraints imposed by the very small size of the sT intron.

teins function in both constitutive and alternative splicing. The SR protein splicing factor 2 regulates the recognition of the two 5′ splice sites of this viral pre-mRNA in vitro. As the concentration of this protein is increased above that which supports constitutive splicing, the splicing machinery switches from the LT to the sT 5′ splice site (Fig. 10.9). It has not been demonstrated that splicing factor 2, or any other SR protein, regulates splicing of simian virus 40 pre-mRNA in infected cells. However, these in vitro observations set the stage for many other studies of these proteins, which are important regulators of splicing in both vertebrates and invertebrates.

**Regulation of splicing and virus production by cellular differentiation: papillomaviral late pre-mRNAs.** The late proteins of bovine papillomavirus type 1 are synthesized efficiently only in highly differentiated keratinocytes. The activity of the viral late promoter probably increases upon host cell differentiation, but alterations in splicing of the late pre-mRNA are crucial for production of the mRNA encoding the major capsid protein, the L1 mRNA (Fig. 10.10A). In situ hybridization studies have shown that this mRNA is made only in fully differentiated

cells (Fig. 10.10B). Production of the L1 mRNA requires activation of an alternative 3′ splice site that is not recognized in incompletely differentiated cells. Several *cis*-acting sequences in the viral late pre-mRNA govern the choice of 3′ splice sites (Fig. 10.10C). These sequences are bound by cellular splicing proteins, including SR proteins. It has therefore been proposed that terminal differentiation of keratinocytes is accompanied by changes in the activity or abundance of these cellular proteins such that L1 mRNA is synthesized. Such terminally differentiated keratinocytes are also the only cells in the epithelium that support productive replication of bovine papillomavirus DNA and are the sites of assembly and release of progeny virions. Thus, the major capsid protein is made only at the site of shedding of the virus into the environment. The mechanism that restricts production of the L1 mRNA to the most terminally differentiated cells would therefore be expected to be advantageous to the virus, for it minimizes exposure of the major viral structural protein to the immune system of the host.

**Incomplete splicing essential for virus replication.** The expression of certain coding sequences in the genomes of retroviruses (Gag and Pol) (appendix, Fig. 9B) and orthomyxoviruses (M1 and NS1) (appendix, Fig. 7B) depends on an unusual form of alternative splicing that produces both spliced and unspliced mRNAs within the infected cell. This phenomenon has been well studied in retrovirus-infected cells, in which incomplete splicing of viral transcripts is essential for production of the viral genome.

In cells infected by simple retroviruses, such as avian leukosis virus, a full-length unspliced transcript of proviral DNA serves as the genome and the mRNA for the capsid proteins and viral enzymes, while a singly spliced mRNA specifies the viral envelope protein (Fig. 10.11A). For both mRNAs to be produced, splicing must be incomplete. Retrovirus production depends rather critically on the maintenance of a proper balance between the proportions of unspliced and spliced RNAs. As illustrated in Fig. 10.11A for avian leukosis virus, modest changes in splicing efficiencies cause replication defects. This phenomenon has been used as a genetic tool to select for mutations that affect splicing control. Such mutations arise in different splicing signals at the 3′ splice site and alter the efficiency of either the first or second step in the splicing reaction. Features that maintain the proper splicing balance include a suboptimal 3′ splice site and a splicing enhancer in the adjacent exon. A negative regulatory element located more than 4,000 nucleotides upstream of the 3′ splice site is also important. The current model of the mechanism of action

A

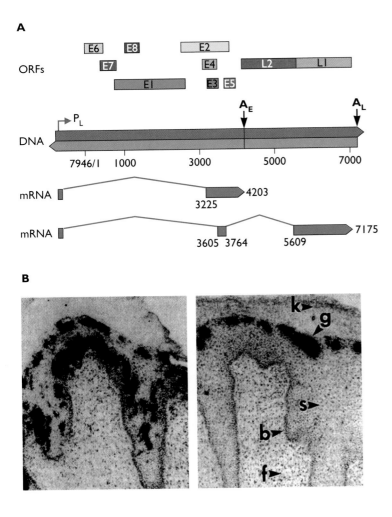

**B**

**Figure 10.10 Control of the production of bovine papillomavirus type 1 late mRNAs by alternative splicing and alternative polyadenylation.** (A) The circular bovine papillomavirus type 1 genome is represented in linear form, with open reading frames (ORFs) shown above. Two of the many mRNAs made from transcripts from the late promoter ($P_L$) are shown to illustrate the changes in recognition of splice sites and of poly(A) addition sites necessary to produce the L1 mRNA. Synthesis of this mRNA depends on recognition of a 3' splice site at position 3605, rather than that at 3225, which is used during the early phase of infection. Polyadenylation of pre-mRNAs must also switch from the early ($A_E$) to the late ($A_L$) polyadenylation sites. (B) In situ hybridization of bovine fibropapillomas to probes that specifically detect mRNAs spliced at the 3225 3' splice site (left) or at the 3605 site (right). The cell layers of the fibropapilloma are indicated in the right panel. Abbreviations: k, keratin horn; g, granular cell layer; s, spinous cell layer; b, basal cell layer; f, fibroma. Note the production of late mRNA spliced at the 3605 3' splice site only in the outermost layer (g) of fully differentiated cells. From S. K. Barksdale and C. C. Baker, *J. Virol.* **69:**6553–6556, 1995, with permission. Courtesy of C. C. Baker, National Institutes of Health. (C) Mechanisms that regulate processing reactions to allow production of L1 mRNA, which are specific to highly differentiated keratinocytes of the granular cell layer. As shown at the top, the sequences that control alternative splicing at the 3225 and 3605 3' splice sites are located between these splice sites. The two splicing enhancers, SE1 and SE2, are recognized by cellular SR proteins. The former and the adjacent sequence that inhibits splicing at the 3605 3' splice site, termed the exonic splicing suppressor (ESS), are believed to facilitate recruitment of U2-associated protein (U2af) and recognition of the branch point sequence upstream of the 3225 3' splice site. Splicing enhancer 2 is located very close to the 3605 3' splice site and may block access to the branch point for splicing at this site until keratinocytes differentiate. However, the mechanisms by which this sequence and the exonic splicing supressor function remain to be determined. Inhibition of polyadenylation at the $A_E$ site by the binding of U1 snRNP to a pseudo-5' splice site located nearby in the primary transcript (see text) is depicted below. Such inhibition may be the result of binding of a U1 snRNP protein to poly(A) polymerase.

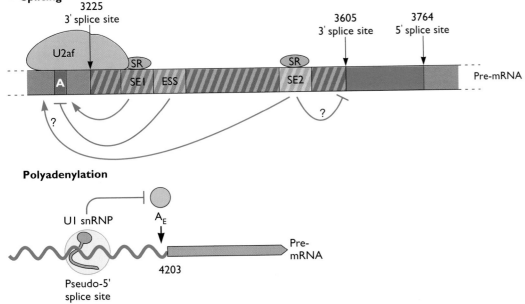

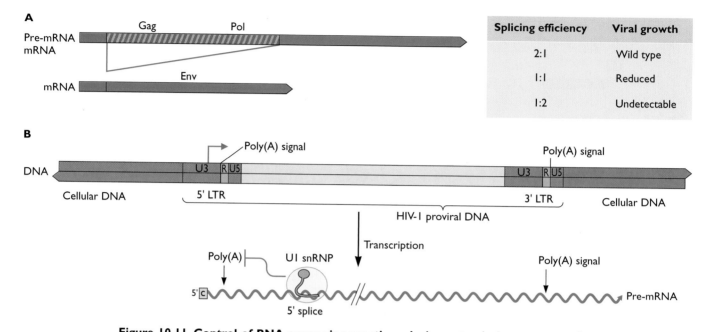

**Figure 10.11 Control of RNA-processing reactions during retroviral gene expression.**
(A) Balanced, incomplete splicing is illustrated for avian leukosis virus mRNAs. A single 3′ splice site is recognized in about one-third of the primary transcripts to produce spliced mRNA encoding the Env protein. The Gag and Pol proteins are synthesized from unspliced transcripts. Even a twofold reduction in the ratio of unspliced to spliced mRNA impairs virus reproduction (right). (B) Suppression of poly(A) site recognition. Polyadenylation signals are present in the long terminal repeats (LTRs) of retroviral proviruses. Thus, primary transcripts of proviral DNA contain identical poly(A) addition signals near both their 5′ and their 3′ ends. In the case of human immunodeficiency virus type 1 (HIV-1), utilization of the 5′ polyadenylation site is inhibited by binding of U1 snRNP to the major 5′ splice site located 195 nucleotides downstream. This snRNP might block access to the 5′ poly(A) addition signal directly, or indirectly via proteins that mediate interaction of the cap structure with U1 snRNP. Alternatively, the U1 snRNP might inhibit the activity of one or more proteins needed for polyadenylation. This possibility is suggested by the ability of a protein specific to the U1 snRNP, the U1A protein, to bind to both poly(A) polymerase and cleavage-polyadenylation specificity protein.

of this sequence dramatically illustrates how the normal functions of components of the cellular splicing machinery can be subverted in virus-infected cells (Box 10.4).

## Interactions between Polyadenylation and Splicing in Viral Gene Expression

Although described separately in previous sections, capping, polyadenylation, and splicing of a pre-mRNA by cellular components are not independent reactions. Rather, one processing reaction governs the efficiency or specificity of another. For example, interaction of the nuclear cap-binding protein with the 5′ end of a pre-mRNA is necessary for both efficient removal of the intron closest to the 5′ end and maximally efficient cleavage at the 3′ poly(A) addition site. Similarly, the presence of a 3′ poly(A) addition signal generally stimulates removal of the intron

closest to it. However, splice sites can either facilitate or suppress recognition of polyadenylation signals. Inhibition of polyadenylation at specific sites is essential for expression of the genes of some retroviruses and can be an important device for the regulation of viral gene expression.

The proviral DNAs of simple and complex retroviruses contain long terminal repeat (LTR) sequences at either end. Each LTR contains the complement of a poly(A) addition signal that in most cases is copied into the pre-mRNA during transcription (Fig. 10.11B). However, only the 5′AAUAAA3′ sequence at the 3′ end of the transcript directs polyadenylation of retroviral pre-mRNA. Various sequences that restrict polyadenylation to this site have been identified in retroviral pre-mRNAs. In the case of that of human immunodeficiency virus type 1, recognition of

## BOX 10.4

*The negative regulator of splicing of retroviral pre-mRNA: a model for competition between major and minor cellular splicing systems*

The negative regulator of splicing (NRS) lies well upstream of the suboptimal 3′ splice site of retroviral mRNA, within the sequence that is removed during splicing. The 5′ segment of this sequence is bound by an SR protein, which facilitates interaction of the 3′ segment with an snRNP that contains a minor snRNA, the U11 snRNA. The U11 snRNP is the functional analog of U1 snRNP (Fig. 10.7) in the cellular system responsible for splicing of a minor class of introns with 5′ and 3′ splice sites distinct from the typical consensus sequences listed in Fig. 10.5. Thus, binding of U11 snRNP to the negative regulatory sequences produces an activated minor class splice site as shown. Such an activated 5′ splice site

can direct splicing only in conjunction with a minor-class 3′ splice site, which is also recognized by specialized snRNPs. It has therefore been proposed that the activated minor-class 5′ splice site formed by the negative regulator of splicing associates nonproductively with the single major-class 3′ splice site of the pre-mRNA. Productive splicing would therefore be reduced as a result of competition between this and the major-class 5′ splice site for the 3′ splice site. nt, nucleotides.

**Gontarek, R. R., M. T. McNally, and K. Beemon.** 1993. Mutation of an RSV intronic element abolishes both U11/U12 snRNP binding and negative regulation of splicing. *Genes Dev.* 7:1926–1936.

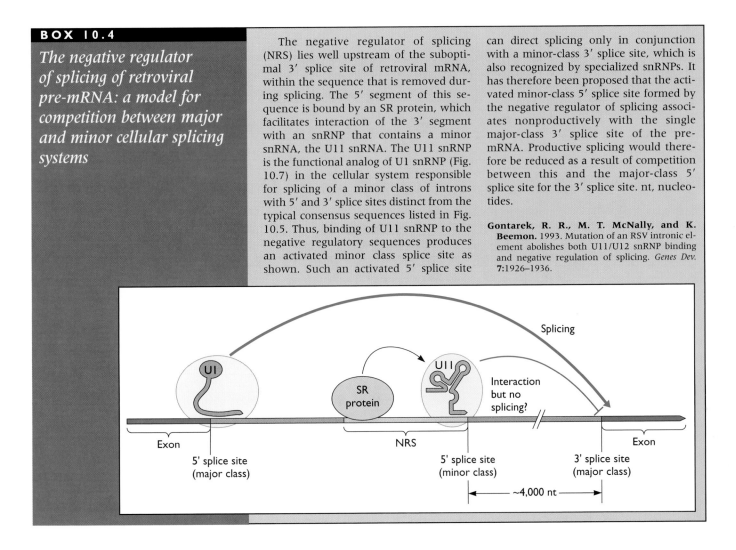

the 5′ poly(A) signal is suppressed by the 5′ splice site lying immediately downstream (Fig. 10.11B). Such inhibition is believed to be mediated by U1 snRNP, which can bind to the 5′ splice site. A related mechanism, binding of U1 snRNP to a pseudo 5′ splice site, appears to suppress recognition of the poly(A) addition site for production of bovine papillomavirus type 1 late mRNAs until the keratinocyte cell is fully differentiated (Fig. 10.10C). In this case, the direct binding to poly(A) polymerase of a protein specific to the U1 snRNP may lead to inhibition of the enzyme.

The production of adenoviral major late mRNAs exemplifies complex alternative splicing and polyadenylation at multiple sites in a pre-mRNA. The major late pre-mRNA contains the sequences of at least 15 different mRNAs (Fig. 10.12). These mRNAs fall into five families (L1 to L5) defined by which of five polyadenylation sites in the pre-mRNA is recognized (Fig. 10.12). Therefore, the frequency with which each site is used must be regulated to allow

production of all major late mRNAs. For example, high-efficiency polyadenylation at the L1 site would prevent efficient synthesis of L2 to L5 mRNAs, a situation clearly detrimental to the virus. In fact, during the late phase of infection each of the five polyadenylation sites directs 3′-end formation with approximately the same efficiency to produce a population of poly(A)-containing major late mRNA precursors in which all mRNA sequences are more or less equally represented. The mechanism responsible for such balanced recognition of multiple poly(A) addition sites is not fully understood. However, it is clear that the activities of cellular polyadenylation proteins are altered as infection proceeds (see "Posttranscriptional Regulation of Viral or Cellular Gene Expression by Viral Proteins" below).

As discussed previously, all major late mRNAs contain the 5′-terminal tripartite leader sequence. The splicing reactions that produce this sequence from three small exons

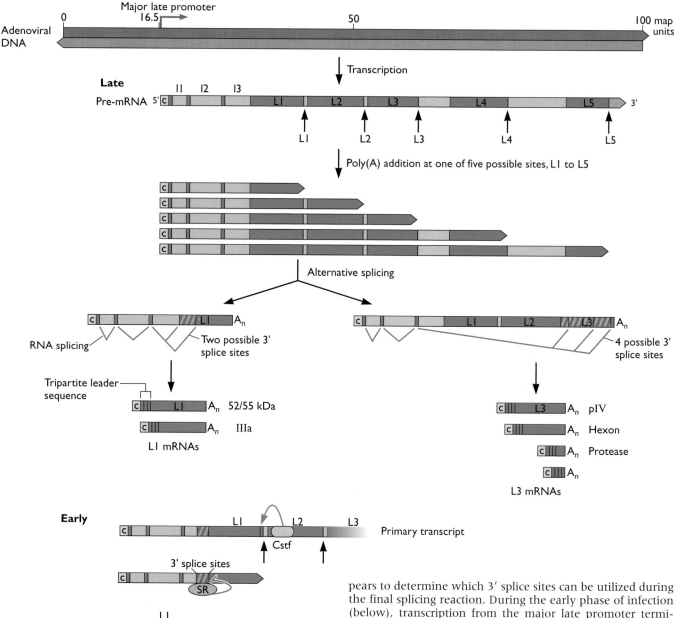

**Figure 10.12 Alternative polyadenylation and splicing of adenoviral major late transcripts.** During the late phase of adenoviral infection (top), major late primary transcripts extend from the major late promoter almost to the right end of the genome. They contain the sequences for at least 15 mRNAs and are polyadenylated at one of five sites, L1 to L5, each of which is used with approximately equal frequency. The tripartite leader sequence, present at the 5' ends of all late mRNAs, is assembled by the splicing of three short exons, l1, l2, and l3. This sequence is then ligated to alternative 3' splice sites. Such joining of the spliced tripartite leader sequence to an mRNA sequence has been reported to take place after polyadenylation of pre-mRNA. Thus, polyadenylation ap- pears to determine which 3' splice sites can be utilized during the final splicing reaction. During the early phase of infection (below), transcription from the major late promoter termi- nates upstream of map unit 70 (see chapter 8). Although the transcripts contain the L1, L2, and L3 poly(A) addition sites, only the L1 site is utilized because its downstream U/GU-rich sequence provides a strong binding site for cleavage stimula- tory protein (Fig. 10.3). The decrease in activity of this protein as infection continues (see text) allows recognition of the L2 to L5 polyadenylation sites, which are less dependent on this protein. During the early phase of infection, only the L1 3' splice site for the 52/55-kDa protein is utilized because bind- ing of SR proteins to the pre-mRNA blocks recognition of the 3' splice site for production of the mRNA for protein IIIa. An E4 protein induces dephosphorylation of these cellular pro- teins by protein phosphatase 2. This modification inhibits binding of the SR proteins to the pre-mRNA and reverses sup- pression of recognition of the 3' splice site for production of mRNA encoding protein IIIa.

can take place before polyadenylation of the primary transcript. The final splicing reaction joins the tripartite leader sequence to one of many mRNA sequences (Fig. 10.12). Because all major late mRNAs include the tripartite leader sequence, each primary transcript yields only a single mRNA even though it contains the sequences for many, and most of its sequence is discarded (Fig. 10.12). It remains something of a mystery why adenoviruses have evolved this bizarre mechanism for synthesis of the majority of their mRNAs. However, one contributing factor may be that it ensures that each major late mRNA molecule carries the 5'-terminal tripartite leader sequence, which is important for efficient translation late in the adenoviral infectious cycle.

## Editing of Viral mRNAs

The term **RNA editing** was first coined in 1980 to describe the presence of uridine residues that are not encoded in the DNA template in a mitochondrial mRNA of trypanosomes. Since this modification was discovered, RNA editing has been identified in many different eukaryotes, as well as in some important viral systems.

Viral mRNAs are edited by either insertion of nucleotides not directly specified in the template during synthesis or alteration of a base in situ. Both mechanisms result in an mRNA sequence that differs from that encoded in the viral genome, altering the sequences and functions of the proteins specified by edited mRNAs. Thus, RNA editing has the potential to make an important contribution to regulation of viral gene expression.

### Editing during Viral Pre-mRNA Synthesis

So far, the mRNAs of two virus families, the *Paramyxoviridae* (e.g., measles and mumps viruses) and the *Filoviridae* (e.g., Ebola virus), are known to be edited. Among paramyxoviruses, this modification occurs only in the mRNAs that encode the RNA-dependent RNA polymerase, the P protein. During mRNA synthesis, the viral RNA polymerase inserts one or two G residues at specific positions in a fraction of the RNA molecules. The genomic RNA template contains a polypyrimidine run at the RNA-editing site. Insertion of G residues is therefore believed to result from a "stuttering" mechanism (Fig. 10.13A) analogous to that by which the viral RNA polymerases synthesize 3' poly(A) tails. The RNA polymerase is postulated to pause near the start of the oligo(C) sequence, with the result that some, but not all, nascent RNA chains slip backwards by one or two nucleotides. Once RNA synthesis resumes, one or two C residues in the template are copied a second time, depending on the specific sequences at the editing sites in the templates (Fig. 10.13A).

Insertional editing of both measles and mumps mRNAs alters the translational reading frame. Thus, this process allows paramyxoviral P genes to encode two distinct proteins (Fig. 10.13B). Similarly, RNA editing determines whether an Ebola virus gene is expressed as a full-length glycoprotein, a type 1 membrane protein that is required for virus entry, or a truncated secreted protein.

### Editing following Viral mRNA Synthesis

Posttranscriptional editing of mRNAs is accomplished by cellular enzymes that deaminate adenine bases in double-stranded RNA regions to form inosine (I). One such double-stranded RNA adenosine deaminase has been implicated in editing of the mRNA of hepatitis delta satellite virus. Two forms of the protein encoded in the genome of this agent, called hepatitis delta antigen, are synthesized in infected cells. Both are necessary for hepatitis delta satellite virus reproduction. The large delta antigen is made when editing at a specific site converts a UAG termination codon in the mRNA to a codon for tryptophan, UGG (Fig. 10.14). As many as 50% of hepatitis delta virus mRNA molecules are modified at this site, but few other sequences in the viral RNA are edited. The mechanisms that restrict action of the cellular double-stranded RNA adenosine deaminase, which exhibits little specificity in vitro, have not been elucidated. The larger form of hepatitis delta antigen can inhibit editing at this site in vivo, a property that is presumably important for production of the essential, smaller form of the protein, which is translated from unedited mRNA molecules (Fig. 10.14).

Double-stranded RNA adenosine deaminases may play a broader role in the biology of RNA viruses. Before the discovery of these enzymes, it was assumed that nucleotide changes in viral RNA genomes arise solely by the introduction of the wrong nucleotide by RNA polymerases that lack proofreading capability, or by RNA recombination. However, the mRNAs and genomes of several viruses exhibit changes that have been attributed to editing. For example, many of the genomic RNAs of defective measles viruses isolated from the brains of patients who died of subacute sclerosing panencephalitis appear to have been edited by cellular double-stranded RNA adenosine deaminases.

## Export of Viral and Cellular RNAs from the Nucleus

Any mRNA made in the nucleus must be transported to the cytoplasm for translation. Other classes of RNA, including small viral RNAs made by RNA polymerase III, must also enter the cytoplasm permanently (e.g., tRNAs, rRNAs) or transiently (snRNAs). The export of viral mRNAs and small RNAs is mediated by the host cell machinery and, in most cases, is indistinguishable from ex-

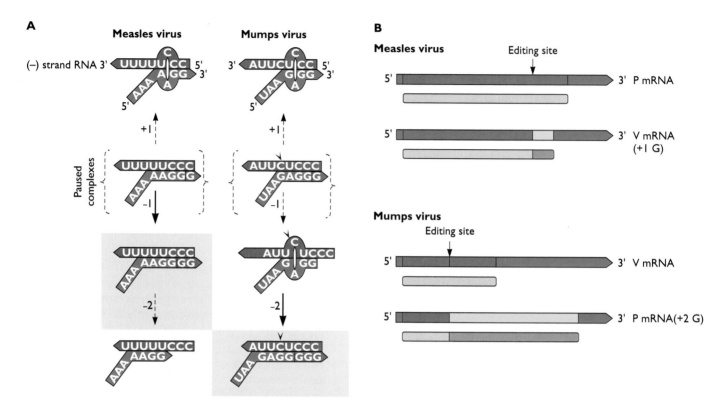

**Figure 10.13 Cotranscriptional editing of paramyxoviral mRNAs.** (A) Proposed mechanism of cotranscriptional editing of measles and mumps mRNAs. The viral polymerase pauses near a junction of $U_n$ and $C_n$ sequences in the template virion RNA after two C residues of the template have been copied into G residues in the nascent mRNA. As a result of this pausing, some fraction of viral RNA polymerase molecules and their attached nascent mRNA chains slip backwards, such that additional nucleotides are incorporated when RNA synthesis resumes. Because of differences in the sequences of the virion RNA templates, the most stable structures (boxed in light green) are formed when the measles virus and mumps virus RNA polymerases slip backward by one ($-1$) and two ($-2$) positions, respectively. Thus, one and two additional G residues, respectively, are incorporated into measles virus and mumps virus mRNAs when RNA synthesis resumes. (B) The P genes of mumps and measles viruses. The unedited measles virus mRNA contains a continuous open reading frame (blue) for the P protein. The addition of one G residue at the editing site changes the translational reading frame to one (yellow) that contains a termination codon. The edited mRNA therefore specifies the V protein, which differs from the P protein in its C-terminal sequence (purple). In contrast, the unedited mumps virus mRNA specifies the V protein. Insertion of two G residues at the editing site changes the reading frame to fuse the V open reading frame with the second, C-terminal open reading frame (yellow) and produce the continuous P protein sequence.

port of analogous cellular RNAs. However, transport to the cytoplasm of mRNA molecules not normally recognized by the cellular export system is a critical step in some viral life cycles. In this section, we describe the cellular export machinery and the mechanisms by which viruses ensure export of unusual substrate mRNAs.

### The Cellular Export Machinery
The substrates for mRNA export are not naked RNA molecules but rather ribonucleoproteins built from the hnRNP proteins. Many of these proteins travel to the cytoplasm

with the mRNA but are then displaced and return to the nucleus. Thus, such hnRNP proteins shuttle between the nucleus and cytoplasm (Box 10.5). One such protein, the abundant hnRNP-A1 protein, contains a short sequence, termed M9, that directs both export and import. This sequence was the first nuclear export signal to be identified. A different kind of nuclear export signal is found in proteins that direct export of certain viral mRNAs and small cellular RNAs (see "Export of Viral mRNA" below). Thus, export of RNA molecules is directed by short sequences present in the *proteins* associated with them.

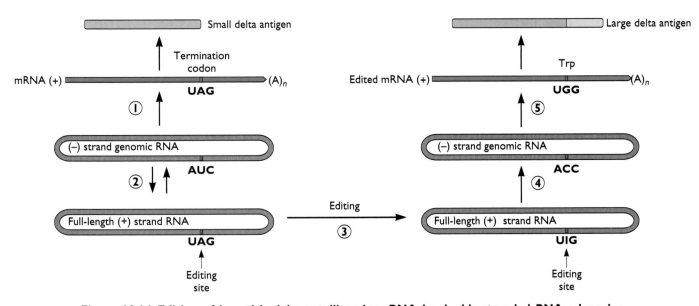

**Figure 10.14 Editing of hepatitis delta satellite virus RNA by double-stranded RNA adenosine deaminase.** The mRNA synthesized from genomic (−) strand RNA specifies the small delta antigen (step 1), which is required for replication of the genome (2). Double-stranded RNA adenosine deaminase acts on the full-length (+) RNA to convert a specific A residue to I (3). Because I base pairs with C, the genome (−) strands copied from edited full-length (+) strands contain a C residue (4). Such edited (−) RNA is therefore copied into (+) mRNA that contains a UGG codon (tryptophan) at the editing site (5) rather than the UAG stop codon. Thus, the mRNA made from edited RNA specifies the large delta antigen, which contains a 19-amino-acid C-terminal extension (orange). This protein inhibits genomic RNA replication and is needed for association of the hepatitis delta satellite viral genome with envelope proteins of its helper virus, hepatitis B virus.

Like proteins entering the nucleus, RNA molecules travel between nuclear and cytoplasmic compartments via the nuclear pore complexes described in chapter 5. Numerous genetic, biochemical, and immunocytochemical studies have demonstrated that specific nucleoporins (the proteins from which nuclear pore complexes are built) participate in nuclear export. Export of RNA molecules also shares several mechanistic features with import of proteins into the nucleus. In both cases, substrates for transport are identified by specific protein signals; some soluble proteins, including the small guanosine nucleotide-binding protein Ran, function in both import and export. Furthermore, RNA export, like protein import, is mediated by receptors that recognize nuclear export signals and direct the proteins and ribonucleoproteins that contain them to and through nuclear pore complexes.

## Export of Viral mRNA

All viral mRNAs made in infected cell nuclei carry the same 5′ and 3′ terminal modifications as cellular mRNAs that are exported. Furthermore, many viral mRNAs, like their cellular counterparts, are produced by splicing of intron-containing precursors. Cellular pre-mRNAs that contain introns and splice sites are ordinarily retained in the nucleus, probably because they remain associated with spliceosomes. Such pre-mRNAs either are spliced to completion and exported or are degraded within the nucleus. However, replication of retroviruses, herpesviruses, and orthomyxoviruses depends on efficient export from the nucleus and translation of mRNAs that are incompletely spliced or not spliced at all. Efforts to address the question of how the normal restrictions on cellular mRNA export are circumvented in cells infected by these viruses have provided important insights into this complex process.

### Export of Intron-Containing mRNAs by the Human Immunodeficiency Virus Type 1 Rev Protein

The human immunodeficiency virus type 1 Rev protein is by far the best understood among the viral proteins that modulate mRNA export from the nucleus. This protein and related proteins of other complex retroviruses promote export of the unspliced and partially spliced viral mRNAs. Rev binds specifically to a sequence lying within *env* in an alternatively spliced intron of human immunodeficiency virus type 1 pre-mRNA, termed the Rev-respon-

**BOX 10.5**

*The interspecies heterokaryon assay for shuttling of proteins between the nucleus and the cytoplasm*

In this assay, human cells expressing the protein of interest, the human immunodeficiency virus type 1 Rev protein in this example, are fused with mouse cells that do not express the protein to form heterokaryons containing multiple nuclei (figure panel C). In this experiment, human (hu) and mouse (mo) nuclei were distinguished by immunofluorescent staining with antibodies specific for the human hnRNP C protein (A). They can also be identified by their distinctive patterns of DNA staining. The location of the protein is then determined by immunofluorescence with antibodies specific for the protein of interest, Rev (B). When a protein can shuttle between the nucleus and the cytoplasm, it is exported from human cell nuclei and reimported into all nuclei present in the heterokaryon. Thus, it is detected in mouse as well as human nuclei (B). Although illustrated here with the Rev protein, this assay was originally developed to examine the shuttling of cellular proteins (nucleolar proteins and the hnRNP-A1 protein) between the nucleus and the cytoplasm. Adapted from B. E. Meyer and M. H. Malim, *Genes Dev.* **8:**1538–1547, 1994, with permission. Courtesy of M. H. Malim, University of Pennsylvania Medical Center.

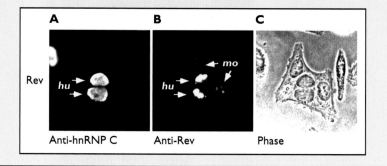

sive element (Fig. 10.15). The Rev-responsive element is some 250 nucleotides in length and contains several stem-loops (Fig. 10.16A). One such structure contains a high-affinity binding site for the arginine-rich RNA-binding domain of Rev (Fig. 10.16B). Comparison of the high-resolution structures of a Rev peptide bound to the high-affinity site (Fig. 10.16C) and free Rev RNA has provided important mechanistic insights into how specific RNA sequences and structural features are recognized. Binding of the protein induces substantial conformational change in the RNA (as occurs in other protein-RNA complexes) to allow extensive interaction of the α-helical RNA-binding domain of Rev with the major groove in the RNA. Export of RNAs that contain the Rev-responsive element depends on the formation of Rev-protein oligomers on this sequence and a nuclear export signal present in Rev.

The C-terminal domain of Rev contains a short (10-amino-acid) nuclear export signal (Fig. 10.16B), which is sufficient to induce export of heterologous proteins when linked to them. This leucine-rich nuclear export signal is recognized by several cellular proteins, including eukaryotic translation initiation protein 5A (eIF-5A) and exportin-1, which is related to the import receptors described in chapter 5 and binds to nucleoporins. The eIF-5A protein interacts with ribosomal protein L5, which participates in

the export of small cellular RNAs from the nucleus. Exportin-1 binds simultaneously to Rev and the GTP-bound form of Ran. These cellular proteins are all required for Rev-dependent RNA export: this process is inhibited by antibodies that recognize eIF-5A or the L5 protein, as well as by the exportin-1-binding drug leptomycin B. This drug also prevents formation of the complex containing Rev, exportin-1, and Ran-GTP. According to the current model, eIF-5A and the L5 protein provide Rev, and the mRNA bound to it, access to exportin-1 and Ran-GTP (Fig. 10.17). Once formed, the Rev–exportin-1–Ran-GTP complex is believed to be translocated to the cytoplasm, mediating the export of human immunodeficiency virus type 1 mRNAs containing the Rev-responsive element (Fig. 10.17). The export complex must disassemble in the cytoplasm, presumably as a result of hydrolysis of GTP bound to Ran to GDP. Rev then shuttles back into the nucleus via a typical nuclear localization signal (Fig. 10.16B).

Perhaps the most interesting aspect of Rev-dependent RNA export is the exit of mRNAs by a pathway that normally does not handle such cargo but rather exports small RNA species of the host cell. The Rev nuclear export signal is similar in sequence to, and can be functionally replaced by, that of the cellular protein TfIIIa. One activity of the latter protein is specific binding to 5S rRNA molecules; it is

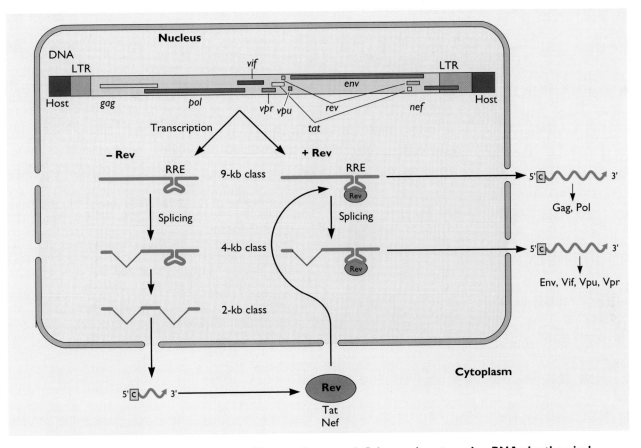

**Figure 10.15 Regulation of export of human immunodeficiency virus type 1 mRNAs by the viral Rev protein.** Before the synthesis of Rev protein in the infected cell, only fully spliced (2-kb class) viral mRNAs are exported to the cytoplasm (left). These mRNAs specify viral regulatory proteins, including Rev. The Rev protein enters the nucleus, where it binds to an RNA structure, the Rev-responsive element (RRE) present in unspliced (9-kb class) and singly spliced (4-kb class) viral mRNAs. This interaction induces export to the cytoplasm of the RRE-containing mRNAs, from which virion structural proteins and enzymes are made (right). Thus, Rev alters the pattern of viral gene expression as the infectious cycle progresses.

required for export of this cellular RNA from the nucleus. When microinjected into *Xenopus* oocytes, peptides containing the Rev nuclear export signal selectively inhibit export of 5S rRNA (and U1 snRNA) but do not impair transport of mRNAs or tRNAs from the nucleus to cytoplasm. Thus, the human immunodeficiency virus type 1 protein circumvents the normal restriction on the export of intron-containing pre-mRNAs from the host cell nucleus by shunting such viral mRNAs to a cellular pathway that handles unspliced RNAs.

### RNA Signals Mediating Export of Intron-Containing Viral mRNAs by Cellular Proteins

The genomes of simple retroviruses do not encode regulatory proteins analogous to Rev, even though unspliced viral mRNAs must reach the cytoplasm. These unspliced

viral mRNAs contain specific sequences that promote export (Table 10.3). Because they must function by means of cellular proteins, such sequences were termed **constitutive transport elements**. The first such sequence was found in the 3' untranslated region of the genome of Mason-Pfizer monkey virus. Deletion of the constitutive transport element prevents the accumulation of unspliced viral mRNA in the cytoplasm. Conversely, this sequence can direct export from the nucleus of other intron-containing RNAs into which it is inserted.

The mechanism by which such constitutive transport elements ensure export of intron-containing mRNAs is not yet known. However, the recently demonstrated ability of even low concentrations of RNA containing the Mason-Pfizer monkey virus constitutive transport element to inhibit export of mature mRNAs when microinjected into

**A**

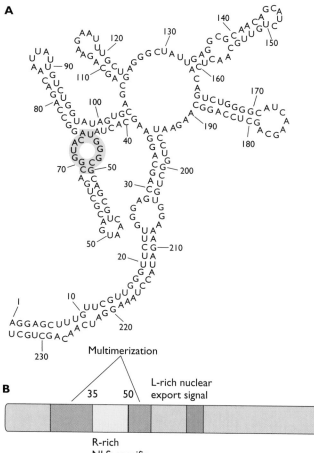

**B**

Multimerization

35    50

L-rich nuclear
export signal

R-rich
NLS; specific
binding to RRE

**C**

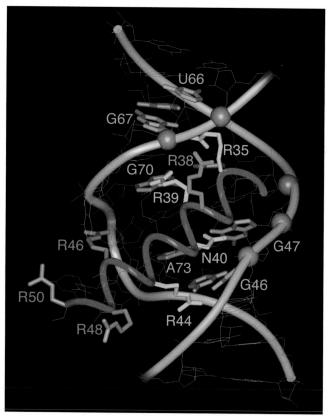

**Figure 10.16 Binding of the Rev protein to the Rev-responsive element.** (A) Predicted secondary structure of the 234-nucleotide Rev-responsive element, with the high-affinity binding site for Rev shaded in yellow. (B) The functional organization of the Rev protein is shown to scale. (C) Structure of a Rev peptide bound to the high-affinity binding site of the Rev-responsive element, determined by nuclear magnetic resonance methods. The Rev peptide corresponding to amino acids 34 to 50 of the full-length protein forms an α-helix (red) that binds to the major groove of the RNA, shown in blue, with phosphates that upon chemical modification impair Rev binding shown as spheres. The bases shown are invariant in RNA molecules selected from a large population for high-affinity binding to Rev in vitro. The major groove of an A-form helix is too narrow to accommodate an α-helix. However, in the Rev-RNA complex, the groove is widened by two purine-purine base pairs and distortion of the RNA backbone. The purine-purine pairs are not observed in the solution structure of the RNA alone, indicating that binding of Rev induces a substantial conformational change. The Rev protein penetrates deeply into the major groove and interacts extensively with the RNA over three to four turns of the α-helix. These interactions include base-specific contacts, for example, via the three arginine and one asparagine residues shown in yellow, and numerous contacts with the phosphate backbone of the RNA, via residues colored orange. From J. L. Battiste et al., *Science* **273:**1547–1551, 1996, with permission. Courtesy of J. R. Williamson, The Scripps Research Institute.

*Xenopus* oocyte nuclei indicates that this retroviral RNA sequence is recognized by components of the mRNA export pathway. Elucidation of the mechanism by which constitutive transport elements promote export promises to yield important information about how mRNAs are normally exported, just as examination of the Rev-dependent pathway operating in human immunodeficiency virus type 1-infected cells has identified signals and proteins important for export of small cellular RNAs.

### The Balance of Incomplete Splicing of Viral Pre-mRNAs with Nuclear Export

A finely tuned balance in the synthesis of fully spliced and unspliced (or incompletely spliced) mRNA is crucial for the successful replication of retroviruses and orthomyxoviruses. This balance is primarily maintained by the relative efficiencies of splicing and export. On the one hand, splicing of viral pre-mRNA must be inefficient to allow export of the essential, intron-containing mRNAs (Fig. 10.11A). Indeed, increasing the efficiency of splicing of human immunodeficiency virus type 1 pre-mRNA, by replacing the natural suboptimal splice sites with efficient ones, leads to complete splicing of all pre-mRNA molecules before Rev can recognize and export the unspliced mRNA to the cytoplasm. On the other hand, when unspliced mRNAs remain in the nucleus (e.g., before Rev is made in infected cells), they are eventually spliced to completion. Thus, efficient export is required to place unspliced mRNA into the cytoplasm for translation or incorporation into virus particles.

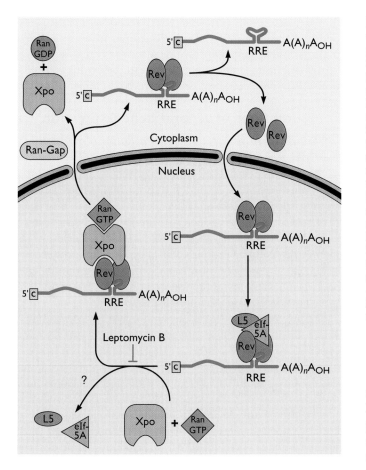

**Figure 10.17 Model of export of human immunodeficiency virus type 1 mRNA containing introns mediated by the Rev protein.** All cellular proteins shown in the nucleus (eIF-5A, ribosomal protein L5, exportin-1 [Xpo], and the GTP-bound form of Ran [Ran-GTP]) have been implicated in Rev-dependent mRNA export (see text). In the presence of Ran-GTP, Rev binds to exportin-1. The binding of exportin-1 to nucleoporins (the proteins of the nuclear pore complex) and the inhibition of both formation of the Rev–exportin-1–Ran-GTP ternary complex and Rev-dependent export by leptomycin B suggest that this ternary complex, with Rev-responsive element mRNA bound to Rev oligomers, is translocated through the nuclear pore complex to the cytoplasm. The eukaryotic initiation factor 5A protein (eIF-5A), which binds to both Rev and ribosomal protein L5, is therefore depicted functioning upstream of the ternary complex to direct Rev to exportin-1. However, the sequence of interactions and the molecular functions of the eIF-5A and the L5 proteins in export remain to be determined. Hydrolysis of GTP bound to Ran to GDP induced by the cytoplasmic Ran GTPase activating protein (Ran-Gap) is presumed to dissociate the export complex, releasing viral mRNA for translation or assembly and Ran for reentry into the nucleus.

### Export of Single-Exon Viral mRNAs

Most of the viral mRNAs made in cells infected by orthomyxo-, herpes-, or hepadnaviruses are not spliced. However, in contrast to the retroviral mRNAs described in previous sections, these viral mRNAs do not arise because of a failure to remove introns. Rather, the viral genes that encode them contain no introns. We therefore designate such mRNAs **single-exon** mRNAs to distinguish them from those that retain introns.

As such viral mRNAs do not contain splice sites, they should not become trapped in the nucleus by a spliceo-

**Table 10.3** Viral RNA sequences that direct export of unspliced mRNAs by cellular proteins

| Virus | Sequence | Properties | Functions |
|---|---|---|---|
| **Retroviruses** | | | |
| Mason-Pfizer monkey virus | Constitutive transport element | 154-nucleotide sequence in 3′ untranslated region; forms a single double-stranded stem with two internal loops; contains degenerate repeat sequences; sequences and structure conserved in RNA of other simian retroviruses | Required for export of unspliced viral mRNA; can substitute for Rev/RRE to promote export of human immunodeficiency virus type 1 intron-containing mRNA; bound by cellular proteins; accesses cellular mRNA export pathway |
| Rous sarcoma virus | Direct repeat (DR) sequence | ∼100-nucleotide sequence flanking Src gene with secondary-structure features similar to constitutive transport elements of simian retroviruses | Required for efficient cytoplasmic accumulation of unspliced mRNA |
| **Hepadnaviruses** | | | |
| Hepatitis B virus | Posttranscriptional regulatory element | 433-nucleotide sequence within viral mRNAs; comprises two subelements, each with low activity, that function cooperatively; partially overlaps a transcriptional enhancer in the viral genome | Required for export to cytoplasm of viral mRNA; can substitute for Rev/RRE to promote export of unspliced human immunodeficiency virus type 1 mRNA; can increase cytoplasmic concentrations of synthetic mRNAs that lack introns |

some retention mechanism. On the other hand, they have few counterparts in uninfected mammalian cells growing under normal conditions, a property that raises the question of how viral single-exon mRNAs are efficiently transported to the cytoplasm. Recent experiments have identified both viral proteins and RNA sequences that promote export of such viral mRNA, analogous to the retroviral Rev protein and constitutive transport elements, respectively, that are needed for export of intron-containing mRNAs.

Most viral mRNAs made in herpes simplex virus-infected cells lack introns. Their efficient export requires the immediate-early protein ICP27. This protein, like the products of all immediate-early genes, is made at the beginning of the infectious cycle. It is a nuclear phosphoprotein that contains a nuclear export signal and shuttles between nucleus and cytoplasm. Viral single-exon mRNAs do not enter the cytoplasm efficiently in cells infected by mutant viruses that do not direct the synthesis of ICP27, but export of viral mRNAs spliced from intron-containing precursors continues normally. Thus, ICP27 specifically facilitates the export of single exon mRNAs. This viral protein binds to such RNAs within infected cells, but the mechanism by which it stimulates export has not yet been identified. It will be of particular interest to learn whether ICP27 operates in the same manner as human immunodeficiency virus type 1 Rev, a possibility suggested by the presence of leucine-rich nuclear export signals in both these viral proteins.

The hepadnaviral mRNAs are made by cellular transcriptional and RNA-processing systems, but all lack introns (appendix, Fig. 11B). The efficient cytoplasmic accumulation of these mRNAs depends on a conserved *cis*-acting sequence, termed the **posttranscriptional regulatory element** (Table 10.3), which can function in the absence of viral proteins. Functionally analogous sequences have recently been identified in one of the uncommon mammalian mRNAs that lack introns (a histone 2A mRNA). It is therefore possible that single-exon hepadnaviral mRNAs leave the nucleus via a cellular pathway that handles a minor fraction of the mRNA traffic in uninfected cells. The majority of influenza virus mRNAs also lack introns. We presently know very little about how these mRNAs, which are synthesized by the viral RNA polymerase, are exported from their nuclear sites of synthesis to the cytoplasm.

## Posttranscriptional Regulation of Viral or Cellular Gene Expression by Viral Proteins

Several viruses that rely on the cellular RNA-processing machinery for production of viral mRNAs encode proteins that regulate one or more of these reactions (Table 10.4). Some of these proteins play crucial roles in temporal regulation of viral gene expression. A second function of viral proteins that modulate RNA-processing reactions is inhibition of the production of cellular mRNAs (Table 10.4).

### Temporal Control of Viral Gene Expression

#### Regulation of Alternative Polyadenylation by Viral Proteins

The frequencies with which alternative poly(A) addition sites within specific viral pre-mRNAs are utilized change during infection by papova-, adeno-, and herpesviruses. As we have seen, bovine papillomavirus type 1 late mRNA polyadenylation is activated by a specific complement of cellular proteins found only in fully differentiated cells of the epidermis. In contrast, viral proteins have been implicated in regulation of polyadenylation in cells infected by the larger DNA viruses.

Despite its name, the adenoviral major late promoter is active during the early phase of infection, prior to the onset of viral DNA synthesis. The major late pre-mRNAs made during this period are polyadenylated predominantly at the L1 mRNA site even though they also contain the L2 and L3 3' processing sites. Such selective recognition of the L1 site depends on the cellular cleavage stimulatory protein, which binds to the U/GU-rich sequence 3' to the cleavage site (Fig. 10.3). As infection continues, the activity of this cellular protein decreases. The recognition of the other four polyadenylation sites present in major late pre-mRNA synthesized during the late phase (Fig. 10.12) is much less dependent on cleavage stimulatory protein. It is therefore believed that these poly(A) addition signals compete more effectively with the L1 site for components of the polyadenylation machinery, as the activity of cleavage-stimulatory protein decreases in adenovirus-infected cells. The mechanism responsible for the change in activity of this protein has not been elucidated, but synthesis of viral proteins is required.

Some herpesviral late pre-mRNAs contain poly(A) addition signals that function poorly in the absence of the ICP27 protein, at least in part because the positions or sequences of the downstream U/GU-rich sequences are not optimal. This viral protein appears to stimulate binding of cleavage stimulatory protein to such suboptimal binding sites in the viral late pre-mRNAs, thus increasing their rate of processing.

#### Regulation of Splicing

**Control of alternative splicing.** Some viral proteins that regulate pre-mRNA splicing in virus-infected cells alter the balance among alternative splicing reactions at

**Table 10.4** Viral proteins that regulate RNA-processing reactions

| Virus | Protein(s) | Properties and functions |
|---|---|---|
| Retrovirus | | |
| Human immunodeficiency virus type 1 | Rev | Mediates export of unspliced and incompletely spliced viral mRNAs; binds to the Rev-responsive element in these mRNA as an oligomer and directs them to the cellular pathway for export of unspliced small RNAs via interactions with cellular eIF-5a and exportin-1 proteins |
| Orthomyxovirus | | |
| Influenza A virus | P proteins | Inhibit splicing of M1 mRNA at the m3 5′ splice site upon binding to 5′ end of the RNA; allow binding of cellular splicing factor 2 to a purine-rich sequence in M1 and activation of the suboptimal M2 mRNA 5′ splice site |
| | NS1 | Can inhibit cleavage and polyadenylation of cellular pre-mRNAs (by binding to cleavage polyadenylation specificity factor), accumulation of cellular mRNAs in the cytoplasm, and splicing (by binding to U6 and U6atac [see Box 10.6] snRNAs) |
| Adenovirus | | |
| Human adenovirus type 2 | E4 ORF4 | Induces dephosphorylation of cellular SR proteins by protein phosphatase 2A; relieves inhibition of L1 pre-mRNA splicing at the IIIa 3′ splice site by phosphorylated SR proteins present early in infection |
| | E1B 55 kDa-E4 ORF6 | The complex of the two proteins inhibits export of fully processed cellular mRNAs to induce selective export of viral late mRNAs; an E4 protein nuclear export signal is activated by binding of the E4 to the E1B protein. |
| Herpesvirus | | |
| Herpes simplex virus type 1 | ICP27 | Stimulates polyadenylation of some viral late mRNAs with suboptimal U/GU-rich sequences at their polyadenylation sites; inhibits splicing of intron-containing cellular and probably viral mRNAs; promotes export of viral single exon mRNAs; contains a leucine-rich nuclear export signal, which may be activated by binding of the protein to such mRNAs |

specific points in the infectious cycle. For example, the ratios of alternatively spliced mRNA products of several adenoviral pre-mRNAs change with the transition into the late phase of infection. In at least one case, a viral early protein is responsible for such regulation. The L1 pre-mRNA, the product of major late transcription and polyadenylation during the early phase of infection, can be spliced at one of two alternative 3′ splice sites (Fig. 10.12). However, only the L1 mRNA that specifies the 52/55-kDa protein is made prior to the onset of viral DNA synthesis. At such early times in the infectious cycle, binding of cellular SR proteins to a negative regulatory sequence located immediately upstream of the 3′ splice site for the L1 IIIa mRNA blocks its recognition (Fig. 10.12). A viral early protein encoded within the E4 transcription unit induces dephosphorylation of the SR proteins, which are highly phosphorylated in uninfected cells. Such dephosphorylated SR proteins lose the ability to repress recognition of the 3′ splice site of the L1 IIIa mRNA.

A cellular SR protein (splicing factor 2) is also an important player in temporal regulation of synthesis of the spliced mRNA that encodes the M2 ion channel protein of influenza A viruses. In this case, the SR protein acts positively, after binding of viral proteins to the splicing sub-strate. Unspliced M1 mRNA, the precursor of spliced M2 mRNA, contains two 5′ splice sites (Fig. 10.18). When the M1 gene is introduced into uninfected cells, only the upstream 5′ splice site, which closely resembles the consensus sequence, is recognized. The product of such splicing, the m3 mRNA, contains a reading frame for a peptide of only nine amino acids (Fig. 10.18). Thus, synthesis of both the M1 and M2 proteins in infected cells depends on inhibition of utilization of the m3 mRNA 5′ splice site. Such inhibition appears to be mediated by cap-dependent binding of the viral RNA polymerase, a complex of the P proteins, to the 5′ end of M1 mRNA. In in vitro splicing systems, this interaction blocks recognition of the m3 mRNA splice site, but recognition of the weak 5′ splice site for production of the M2 mRNA also requires binding of splicing factor 2 to a splicing enhancer in the 3′ exon (Fig. 10.18). However, only about 10% of the M1 mRNA molecules are spliced to form M2 mRNA. The mechanism(s) that limits production of M2 from M1 mRNA in infected cells is not known. Because overproduction of the M2 protein disrupts the cellular secretory pathway and is deleterious to both the host cell and virus reproduction, cells containing relatively low concentrations of this SR protein may be the most suitable hosts for influenza virus repro-

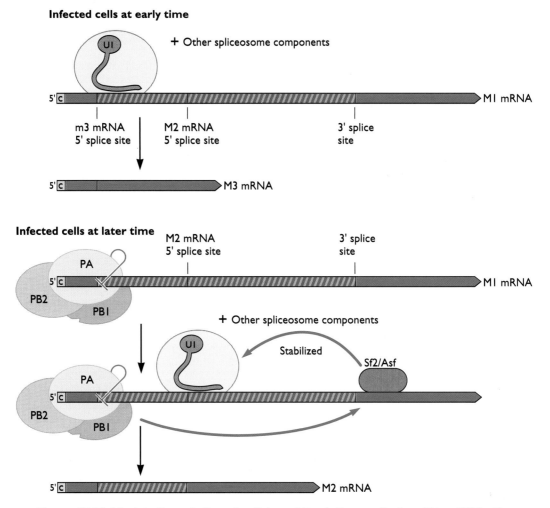

**Figure 10.18 Model of regulation of splicing of the influenza A virus M1 mRNA.** The mRNA synthesized from the M genomic RNA segment encodes the M1 protein but can also be spliced to produce the mRNA specifying the M2 protein. However, at early times of infection (top), before the synthesis of P proteins, the strong 5′ splice site for a third RNA, m3 mRNA, is used exclusively and only m3 mRNA is made. At later times (below), the P proteins bind to a specific sequence near the 5′ end of the influenza A virus mRNAs. In the case of the M1 mRNA, such binding blocks access to the m3 mRNA 5′ splice site, allowing unspliced M1 mRNA to be produced in infected cells. The 5′ splice site for the M2 mRNA is suboptimal, and its recognition depends on binding of the cellular SR protein splicing factor 2 (Sf2/Asf) to a purine-rich splicing enhancer in the 3′ exon of M1 mRNA. Such binding requires interaction of the P proteins with the 5′ end of M1 mRNA, suggesting that the conformation of the RNA is altered upon binding of P proteins.

duction. Indeed, it is possible that the dependence of M2 mRNA synthesis on the cellular SR protein contributes to the tropism of the virus for specific cell types.

### Regulation of mRNA Export

Even though all are encoded within a single proviral transcription unit, the regulatory and virion proteins of the complex retrovirus human immunodeficiency virus

type 1 are made sequentially in infected cells as a result of regulation of mRNA export by the Rev protein. The viral pre-mRNAs initially made in the infected cell are all multiply spliced prior to export from the nucleus to the cytoplasm (Fig. 10.15). These mRNAs encode the viral regulatory proteins Tat and Rev. The Rev protein enters the nucleus and induces export of unspliced and partially spliced viral mRNAs that contain the Rev-responsive element. These mRNAs specify the major structural proteins and enzymes of the virion (Fig. 10.15). Thus, Rev protein regulates a switch in human immunodeficiency virus type 1 gene expression from early production of viral regulatory proteins to a later phase in which virion components are made.

In other cases, viral proteins that modulate mRNA export play subsidiary, but nonetheless crucial, roles in temporal regulation of viral gene expression. For example, the transcriptional program described in chapter 8 results in efficient transcription of herpes simplex virus late genes only following initiation of viral DNA synthesis in infected cells. However, because all but one of the late mRNAs comprises a single exon, their efficient entry into the cytoplasm and the synthesis of viral late proteins requires ICP27. Thus, this posttranscriptional regulator is essential for putting the viral transcriptional program into effect.

## Inhibition of Cellular mRNA Production by Viral Proteins

Because all viral mRNAs are translated by the protein synthesis machinery of the host cell, it is of advantage to viruses to inhibit production of competing cellular mRNAs. The viruses that rely on cellular transcriptional and RNA-processing systems of the nucleus do not have the option of simply inactivating them. Nevertheless, several mechanisms of disrupting cellular gene expression by selective inhibition of cellular RNA processing have evolved.

### Inhibition of Polyadenylation and Splicing by Viral Proteins

The influenza virus NS1 protein can inhibit both polyadenylation and splicing of cellular pre-mRNAs. This viral protein comprises an N-terminal RNA-binding domain (Box 10.6) and a C-terminal segment that is required for inhibition of polyadenylation. Dimers of NS1 bind specifically to several RNAs, including poly(A), and to stem-bulge structures in U6 snRNA and the analogous snRNA of a second splicing system responsible for processing of a minor class of cellular pre-mRNAs that contain alternative, conserved dinucleotides at the 5′ and 3′ ends of their introns. The C-terminal segment of NS1 binds to one of the subunits of cleavage polyadenylation specificity protein, which recognizes the 5′AAUAAA3′ polyadenylation signal in pre-mRNAs (Fig 10.3) both in vitro and in influenza virus-infected cells. These interactions with cellular RNA sequences and cellular proteins inhibit polyadenylation and splicing (by both the major and minor splicing systems) in in vitro systems and in cells that express the viral NS1 gene introduced by DNA-mediated transformation (Fig. 10.19). In the latter circumstance, NS1 also inhibits the accumulation of poly(A)-containing, spliced cellular mRNAs in the cytoplasm, presumably be-

---

**BOX 10.6**

*Unusual RNA-binding properties of the influenza A virus NS1 protein*

The NS1 protein binds specifically to poly(A), stem-bulge structures in U6 snRNA and the equivalent snRNA of the minor cellular splicing system, U6atac snRNA, and double-stranded RNA. Both mutational analyses and RNA competition experiments indicate that these structurally different RNAs share binding sites in the N-terminal RNA-binding domain of NS1. The mechanism of RNA binding of NS1 must therefore be unusual. The high-resolution structure of an N-terminal fragment retaining all RNA-binding activities (and containing the dimerization domain) identified a structural motif not previously seen in proteins that bind to RNA. As shown, this motif comprises a bundle of six α-helices, three contributed by each monomer and shown in cyan and purple, with amino acids believed to stabilize the dimer in yellow. The region rich in basic residues, shown in green, forms a surface well suited for binding to the negatively charged phosphate groups of the RNA molecules to

which NS1 binds. The electrostatic nature of this interaction suggests that, in contrast to the human immunodeficiency virus type 1 Rev protein (Fig 10.16C), the NS1 protein is unlikely to bind to the major groove of a double-stranded segment of the RNAs listed above. From W. Wang et al., *RNA* **5:**195–205, 1999, with permission.

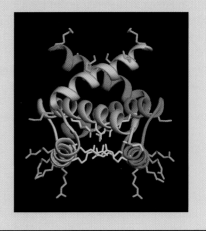

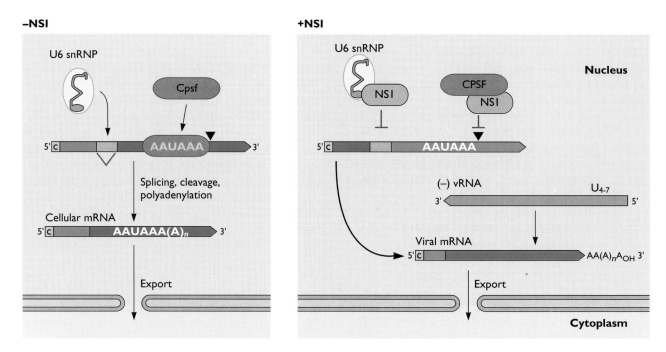

**Figure 10.19 Model of inhibition of cellular pre-mRNA processing by the influenza A virus NS1 protein.** The viral protein binds to a subunit of the cleavage and polyadenylation specificity protein (Cpsf). As indicated (right), this interaction prevents binding of cleavage and polyadenylation specificity protein to the 5'AAUAAA3' polyadenylation signals at the poly(A) addition sites (black arrowhead) of cellular pre-mRNAs and formation of the 3' ends of mature mRNAs. In the absence of 3'-end cleavage, the pre-mRNA probably remains associated with the transcriptional machinery and DNA template and is not available for export to the cytoplasm. Consequently, the intranuclear concentration of capped cellular pre-mRNAs from which the primers for viral mRNA synthesis are acquired would be increased. The viral NS1 protein can also bind to U6 snRNA to inhibit splicing of cellular pre-mRNAs.

cause all steps in polyadenylation, including the initial cleavage reaction (Fig. 10.3) that releases pre-mRNAs from the transcriptional machinery, and splicing are blocked. These properties of NS1 suggest that it could act to increase the intranuclear concentrations and lifetimes of the cellular pre-mRNAs from which the viral cap-dependent endonuclease cleaves the primers for viral mRNA synthesis, thus facilitating the latter process (Fig. 10.19). Such a function of NS1 would also confer the benefit of reducing competition of cellular with viral mRNAs for components of the cellular translational machinery. However, it remains to be demonstrated that NS1 inhibits processing and export of cellular mRNA in influenza virus-infected cells. Indeed, recent experiments with a viable mutant influenza virus lacking all but the N-terminal 10 amino acids of the NS1 coding sequence suggest that a more critical function of NS1 may be to counteract an important host defense mechanism mediated by interferons (see chapters 11 and 14).

The herpes simplex virus immediate-early protein ICP27 also inhibits splicing of pre-mRNAs. This protein induces redistribution of spliceosomal snRNPs from a pattern of dif-

fuse speckles in the nucleus to a limited number of sites with a punctate pattern. The viral protein colocalizes with snRNPs at the latter sites and may bind directly to them. Such abnormal intranuclear distribution of essential splicing components might contribute to the inhibition of splicing, but ICP27 also acts more directly. Extracts prepared from herpes simplex virus type 1-infected cells, but not from cells infected by a mutant virus that cannot express the gene encoding ICP27, are defective for splicing in vitro. Genetic analyses have shown that disruption of cellular RNA processing by ICP27 leads to inhibition of cellular protein synthesis. Moreover, this function is genetically separable from the role of this protein in efficient production of viral late mRNAs. Because herpesviral genes generally lack introns, inhibition of splicing is an effective strategy for the selective inhibition of cellular gene expression.

### Inhibition of Cellular mRNA Export by Viral Proteins

In contrast to the other viruses considered in this section, adenoviruses disrupt cellular gene expression by direct inhibition of export of cellular mRNAs from their nucleus to the cytoplasm. During the late phase of infection, the great majority of newly synthesized mRNAs entering the cytoplasm are viral in origin. However, transcription and processing of cellular pre-mRNAs are unaffected. Rather, cellular mRNAs are not exported and are degraded within the nucleus. Thus, viral mRNAs are selectively exported, a phenomenon that is important for efficient synthesis of viral late proteins. When selective viral mRNA export is prevented by mutations in the viral genome, both the quantities of late proteins made in infected cells and the virus yield are substantially reduced. These same phenotypes are seen in cells infected by a herpes simplex virus mutant expressing an altered ICP27, which is defective only for the inhibition of pre-mRNA splicing. These properties emphasize the importance of posttranscriptional inhibition of cellular mRNA production for efficient virus reproduction.

The preferential export of viral late mRNAs in adenovirus-infected cells requires two viral early proteins, the E1B 55-kDa and the E4 ORF6 proteins. These proteins are found both free and in association with one another in the infected cells. The complex is responsible for the regulation of mRNA export. Like human immunodeficiency virus type 1 Rev and herpesviral ICP27, the E4 protein contains a leucine-rich nuclear export signal and shuttles between the nucleus and cytoplasm. However, the nuclear export signal directs shuttling only when the E4 protein is bound to the E1B protein. In the absence of the E1B protein, the E4 protein is retained in the nucleus by a dominant C-terminal nuclear retention signal. Interaction between the viral proteins is also necessary for entry of the E1B protein into the nucleus because this protein does not contain a nuclear localization signal.

The mechanism by which the E1B-E4 protein assembly induces selective export of viral late mRNAs is not yet clear. Such selectivity is especially puzzling because the viral mRNAs possess all the characteristic features of cellular mRNAs and are made in the same way. One hypothesis is that the viral early protein complex recruits nuclear proteins needed for export of mRNA to the specialized sites within the nucleus at which the adenoviral genome is replicated and transcribed. As a result of such sequestration of essential components of the nuclear export machinery, export of cellular mRNAs would be inhibited.

## Regulation of Turnover of Viral and Cellular mRNAs in the Cytoplasm

Once a viral or cellular mRNA enters the cytoplasm, it is translated by ribosomes and the other components of the protein synthesis machinery. Individual mRNAs may differ in the time and rate at which they are translated and also in such properties as cytoplasmic location and stability. In-

deed, the intrinsic lifetime of an mRNA can be a critical parameter in the regulation of gene expression.

In the cytoplasm of mammalian cells, the lifetimes of specific mRNAs can differ by as much as 100-fold. This property is described in terms of the time required for 50% of the population of the mRNAs to be degraded under conditions in which replenishment of the cytoplasmic pool is blocked, the **half-life** of the mRNA. Many mRNAs are very stable, with half-lives exceeding 12 h. As might be anticipated, such mRNAs encode proteins needed in large quantities throughout the lives of all cells, such as structural proteins (e.g., actin) and ribosomal proteins. At the other extreme are very unstable mRNAs with half-lives of less than 30 min. This class includes mRNAs specifying regulatory proteins whose synthesis is strictly controlled by the external or internal environment of the cell, such as cytokines and proteins that regulate cell growth and division. The short lifetimes of these mRNAs ensure that synthesis of their products can be shut down effectively once they are no longer needed. When synthesis of the mRNA ceases, the preexisting cytoplasmic pool is rapidly cleared as the mRNA is degraded, and the system is reset to the ground state. Specific sequences that signal the rapid turnover of the mRNA in which they reside, such as a 50- to 100-nucleotide AU-rich sequence (Fig. 10.20) within the 3′ untranslated regions, have been identified, as have several mechanisms of mRNA degradation in mammalian cells and yeast.

The stabilities of viral mRNAs have not been examined in much detail, in part because many viral infectious cycles are completed within the normal range of mRNA half-lives. Nevertheless, it is clear that a viral protein that induces RNA degradation plays an important role in selective expression of viral genes in herpes simplex virus-infected cells. Furthermore, regulation of the stability of specific viral or cellular mRNAs has been implicated in the permanent changes in cell growth properties (transformation) induced by some viruses.

## Regulation of mRNA Stability by a Viral Protein

The virion host shutoff protein (Vhs) of herpes simplex virus reduces the stability of mRNAs in infected cells. As its name implies, Vhs is a structural protein of the virion. It is present at low concentrations in the tegument and therefore delivered to infected cells at the start of the infectious cycle, before viral gene expression begins. It remains in the cytoplasm and induces degradation of cellular mRNAs. The Vhs protein may itself be a nuclease that specifically degrades mRNAs, or it may activate a cellular nuclease with such specificity. Alternatively, it might act in concert with cellular proteins that catalyze shortening of poly(A)

**Figure 10.20 Destabilization of cellular mRNAs by 5′AUUUA3′ sequences, repeated in the 3′ untranslated regions.** Such sequences, which are present in short-lived mRNAs encoding cytokines and proteins that regulate cell growth and division, are specifically recognized by one or more proteins. Some proteins stabilize mRNAs upon binding to 5′AUUUA3′ sequences, but binding of others, depicted by the blue oval, appears to induce shortening of the poly(A) tail. This reaction is followed by degradation of the mRNA, in one mechanism via decapping and 5′-3′ exoribonucleolytic attack.

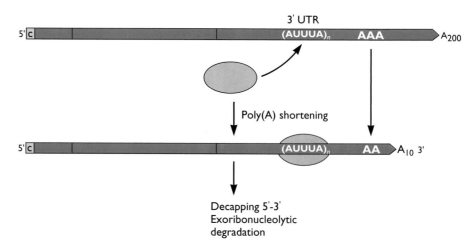

tails or removal of the cap, the initial reactions in some pathways of mRNA degradation. The removal of cellular mRNAs at the start of infection facilitates viral gene expression by reducing or eliminating competition from cellular mRNAs during translation.

The Vhs protein specifically targets mRNAs, but it cannot distinguish those of viral origin from their cellular counterparts. Thus, it induces degradation of both viral and cellular mRNAs. Although more Vhs protein is made in infected cells once the Vhs gene is expressed during the late phase of infection, the protein is sequestered in the teguments of assembling virions by interaction with the viral late VP16 protein. Thus, the activity of Vhs decreases as the infection cycle progresses. This mechanism presumably contributes to the efficient synthesis of viral proteins characteristic of the late phase of infection.

## Regulation of mRNA Stability in Transformation

Stabilization of specific viral mRNAs appears to be important in the development of cervical carcinoma associated with infection by high-risk human papillomaviruses, such as types 16 and 18. As discussed in chapter 16, the E6 and E7 proteins of these viruses induce abnormal growth and division of the cells in which they are made. In benign lesions, the circular human papillomavirus genome is not integrated but persists as an extrachromosomal episome (Fig. 10.21). The E6 and E7 mRNAs synthesized from such templates contain destabilizing AU-rich sequences in their 3′ untranslated regions and possess short lifetimes. This property, as well as repression of transcription from the viral promoter from which the viral pre-mRNAs are transcribed, results in a low steady-state concentration of the E6 and E7 mRNAs. In cervical carcinoma cells, the viral

genome is integrated into the cellular genome. Such reorganization of viral DNA frequently disrupts the sequences encoding the E6 and E7 mRNAs such that their 3′ untranslated regions are copied from cellular DNA sequences (Fig. 10.21). Consequently, the hybrid mRNAs produced from integrated viral DNA templates lack the destabilizing AU-rich sequences and are more stable. This increase in the stability of the viral mRNAs encoding the papillomaviral transforming proteins accounts, at least in part, for their higher concentrations in tumor cells.

The regulation of mRNA stability may also play an important role in the alteration of cell growth by herpesvirus saimiri. This gammaherpesvirus induces lymphomas or leukemias in certain species of New World primates and immortalizes T cells both in vivo and in vitro. Three small viral RNAs produced in large quantities in immortalized cells have been implicated in such transformation. These viral RNAs share several properties with spliceosomal snRNAs, including binding to many of the same proteins. However, they appear to alter cell growth by increasing the stability of cellular mRNAs that contain multiple copies of the sequence 5′AUUUA3′ in their 3′ untranslated regions. Binding of one or more proteins to this sequence in an mRNA leads to shortening of the poly(A) sequence and degradation of the mRNA (Fig. 10.20). Some of the herpesvirus saimiri small RNAs also contain multiple copies of the 5′AUUUA3′ sequence and compete for binding to at least one protein that recognizes this sequence in unstable mRNAs. It has therefore been proposed that the viral small RNAs act as decoys for proteins that initiate degradation of cellular mRNAs marked by the presence of 5′AUUUA3′ sequences. The resulting increase in concentration of these mRNAs would lead to overexpression of cellular proteins that regulate cellular growth and division, many of which are specified by such unstable mRNAs.

**A  Extrachromosomal HPV-16
DNA in benign lesion**

**B  Integrated HPV-16 DNA in cervical cancer**

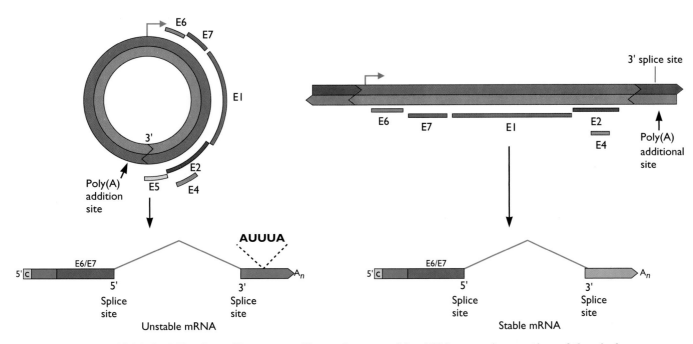

**Figure 10.21 Stabilization of human papillomavirus type 16 mRNAs upon integration of the viral genome into cellular DNA.** (A) In benign lesions, the viral genome is maintained as an extrachromosomal, circular episome. Transcription of such viral DNA and pre-mRNA processing produce various alternatively spliced mRNAs containing the E6 and/or E7 protein-coding sequences, but, as illustrated, all contain destabilizing 5′AUUUA3′ sequences in their 3′ untranslated regions. (B) In cervical carcinoma cells, the viral genome is integrated into cellular DNA (purple) such that the viral genome is disrupted upstream of the E6/E7 mRNA 3′ splice site. The mRNAs encoding these viral proteins are therefore made by using 3′ splice and polyadenylation sites transcribed from adjacent cellular DNA, and they lack the destabilizing sequence.

## Perspectives

Many of the molecular processes required for replication of animal viruses, including such virus-specific reactions as synthesis of genomic and messenger RNAs from an RNA template, were foretold by the properties of the bacteriophages that parasitize prokaryotes. In contrast, the covalent modifications necessary to produce functional mRNAs in eukaryotic cells were without precedent when discovered in viral systems. Study of the processing of viral RNAs has yielded much fundamental information about the mechanisms of capping, polyadenylation, and splicing, as well as the cellular or viral components that complete these reactions. More recently, viral systems have allowed equally important insights into export of mRNA from the nucleus to the cytoplasm. The complexity of the RNA export machinery, the coupling of RNA processing and export, and the lack of in vitro assays for this process have hampered elucidation of export mechanisms. The Rev protein of the complex retrovirus human immunodeficiency

virus type 1 was one of the first shown to contain a discrete nuclear export signal, and subsequent identification and characterization of cellular proteins that bind to Rev led to the first detailed model of RNA export. Perhaps the most important lesson learned from the study of viral mRNA processing is the importance of these reactions in the regulation of gene expression.

Regulation of viral RNA processing can be the result of differences in the concentrations or activities of specific cellular components in different cell types or actively induced by viral gene products. Several mechanisms by which viral proteins or RNAs can regulate or inhibit polyadenylation or splicing reactions, export of mRNA from the nucleus, or mRNA stability have been quite well characterized. However, our understanding of regulation of viral gene expression via RNA-processing reactions is far from complete; the mechanisms of action of several crucial viral regulatory proteins have not been fully elucidated, and many of the specific mechanisms deduced with experimental systems have yet to be confirmed in virus-infected cells. Among the greatest challenges for the future is a full understanding of the recently identified physical and functional couples among the reactions that produce mRNAs in eukaryotic cell nuclei. One view of the intimate relationships among synthesis, capping, polyadenylation, and splicing of pre-mRNAs is that the transcription, RNA-processing, and export machineries are strictly organized within the nucleus to optimize all reactions in the production of a functional mRNA. Viruses that depend on, or perturb, these processes seem likely to be as valuable in addressing such complex issues as many viruses have been in the initial elucidation of RNA-processing reactions and their regulation.

## References

### Chapters in Books

Krug, R. M. 1998. Unique functions of the NS1 protein. *In* R. Nicholson, R. Webster, and A. Hay (ed.), *Textbook of Influenza*. Blackwell Science Ltd., London, United Kingdom.

### Reviews

Banks, J. D., K. L. Beemon, and M. F. Linial. 1997. RNA regulatory elements in the genomes of simple retroviruses. *Semin. Virol.,* 8:194–204.

Colgan, D. F., and J. L. Manley. 1997. Mechanism and regulation of mRNA polyadenylation. *Genes Dev.* 11:2755–2766.

Cullen, B. R. 1998. Retroviruses as model systems for the study of nuclear RNA export pathways. *Virology* 249:203–210.

Emerman, M., and M. H. Malim. 1998. HIV-1 regulatory/accessory genes: keys to unraveling viral and host cell biology. *Science* 280:1880–1884.

Grate, D., and C. Wilson. 1997. Role REVersal: understanding how RRE RNA binds its peptide ligand. *Structure* 5:7–11.

Misteli, T., and D. L. Spector. 1998. The cellular organization of gene expression. *Curr. Opin. Cell Biol.* 10:323–331.

Neugebauer, K. M., and M. B. Roth. 1997. Transcription units as RNA processing units. *Genes Dev.* 11:3279–3285.

Nigg, E. A. 1997. Nucleocytoplasmic transport: signals, mechanisms and regulation. *Nature* (London) 386:779–787.

Nilsen, T. W. 1994. RNA-RNA interactions in the spliceosome: unraveling the ties that bind. *Cell* 78:1–4.

Proudfoot, N. J. 1991. Poly(A) signals. *Cell* 64:671–674.

Ross, J. 1996. Control of messenger RNA stability in higher eukaryotes. *Trends Genet.* 12:171–175.

Shuman, S. 1995. Capping enzyme in eukaryotic mRNA synthesis. *Prog. Nucleic Acid Res. Mol. Biol.* 50:101–129.

Smith, H. C., J. M. Gottand, and M. R. Hanson. 1997. A guide to RNA editing. *RNA* 3:1105–1123.

Staley, J. P., and C. Guthrie. 1998. Mechanical devices of the spliceosome: motors, clocks, springs, and things. *Cell* 92:315–326.

Varani, G. 1997. A cap for all occasions. *Structure* 5:855–858.

Wang, J., and J. L. Manley. 1997. Regulation of pre-mRNA splicing in metazoa. *Curr. Opin. Genet. Dev.* 7:205–211.

Wickens, M., P. Anderson, and R. J. Jackson. 1997. Life and death in the cytoplasm: messages from the 3′ end. *Curr. Opin. Genet. Dev.* 7:220–232.

### Papers of Special Interest

#### Capping of Viral mRNAs

Furuichi, Y., M. Morgan, S. Muthukrishnan, and A. J. Shatkin. 1975. Reovirus messenger RNA contains a methylated, blocked 5′ terminal structure, m⁷G(5′)ppp(5′)GpCp. *Proc. Natl. Acad. Sci. USA* 72:362–366.

Hagler, J., and S. Shuman. 1992. A freeze-frame view of eukaryotic transcription during elongation and capping of nascent mRNA. *Science* 255:983–986.

Lou, Y., X. Mao, L. Deng, P. Cong, and S. Shuman. 1995. The D1 and D12 subunits are both essential for the transcription termination activity of vaccinia virus capping enzyme. *J. Virol.* 69:3852–3856.

Mao, Z. X., and W. K. Joklik. 1991. Isolation and enzymatic characterization of protein lambda 2, the reovirus guanylyl-transferase. *Virology* 185:377–386.

Plotch, S. J., M. Bouloy, I. Ulmanen, and R. M. Krug. 1981. A unique cap (m⁷GpppXm)-dependent influenza virion endonuclease cleaves capped RNAs to generate the primers that initiate viral RNA transcription. *Cell* 23:847–858.

Salditt-Georgieff, M., M. Harpold, S. Chen-Kiang, and J. E. Darnell. 1980. The addition of the 5′ cap structure occurs early in hnRNA synthesis and prematurely terminated molecules are capped. *Cell* 19:69–78.

Wei, C. M., and B. Moss. 1975. Methylated nucleotides block 5′ terminus of vaccinia virus messenger RNA. *Proc. Natl. Acad. Sci. USA* 72:318–322.

#### Polyadenylation of Viral mRNA or Pre-mRNA

Ashe, M. L. P., L. H. Pearson, and N. J. Proudfoot. 1997. The HIV-1 5′ LTR poly(A) site is inactivated by U1 snRNP interaction with the major splice donor site. *EMBO J.* 16:5752–5763.

Fitzgerald, M., and T. Shenk. 1981. The sequence 5′AAUAAA3′ forms part of the recognition site for polyadenylation of late SV40 mRNAs. *Cell* 24:251–260.

Gilmartin, G. M., S.-L. Hung, J. D. DeZazzo, E. S. Fleming, and M. J. Imperiale. 1996. Sequences regulating poly(A) site selection within the adenovirus major late transcription unit influence the interaction of constitutive processing factors with the pre-mRNA. *J. Virol.* 70:1775–1783.

Hwang, L. N., N. Englund, and A. K. Pattnaik. 1998. Polyadenylation of vesicular stomatis virus mRNA dictates efficient transcription termination at the intercistronic gene junctions. *J. Virol.* 72:1805–1813.

Mann, K. P., E. A. Weiss, and J. R. Nevins. 1993. Alternative poly(A) site utilization during adenovirus infection coincides with a decrease in the activity of a poly(A) processing factor. *Mol. Cell Biol.* 13:2411–2419.

McGregor, F., A. Phelan, J. Dunlop, and J. B. Clements. 1996. Regulation of herpes simplex virus poly(A) site usage and the action of immediate-early protein IE63 in the early-late switch. *J. Virol.* 70:1931–1940.

Schneierle, B. S., P. D. Gershon, and B. Moss. 1992. Cap-specific mRNA (nucleoside-$O^2$)-methyltransferase and poly(A) stimulatory activities of vaccinia virus are mediated by a single protein. *Proc. Natl. Acad. Sci. USA* **89**:2897–2901.

Sheldon, R., C. Jurale, and J. Kates. 1972. Detection of polyadenylic acid sequences in viral and cellular eukaryotic RNA. *Proc. Natl. Acad. Sci. USA* **69**:417–421.

## Splicing of Viral Pre-mRNA

Barksdale, S. K., and C. C. Baker. 1995. Differentiation-specific alternative splicing of bovine papillomavirus late mRNAs. *J. Virol.* **69**:6553–6556.

Berget, S. M., C. Moore, and P. A. Sharp. 1977. Spliced segments at the 5′ terminus of adenovirus 2 late mRNA. *Proc. Natl. Acad. Sci. USA* **74**:3171–3175.

Berk, A. J., and P. A. Sharp. 1978. Spliced early mRNAs of simian virus 40. *Proc. Natl. Acad. Sci. USA* **75**:1274–1278.

Bouck, J., X.-D. Fu, A. M. Skalka, and R. A. Katz. 1995. Genetic selection for balanced retroviral splicing: novel regulation involving the second step can be mediated by transitions in the polypyrimidine tract. *Mol. Cell. Biol.* **15**:2663–2671.

Chow, L. T., R. E. Gelinas, T. R. Broker, and R. J. Roberts. 1977. An amazing sequence arrangement at the 5′ ends of adenovirus 2 messenger RNA. *Cell* **12**:1–8.

Ge, H., and J. L. Manley. 1990. A protein factor, ASF, controls cell-specific alternative splicing of SV40 early pre-mRNA in vitro. *Cell* **62**:25–34.

Himmelsprach, M., Y. Cavaloc, K. Chebli, J. Stevinin, and R. Gattoni. 1995. Titration of serine/arginine (SR) splicing factors during adenoviral infection modulates E1A pre-mRNA alternative splicing. *RNA* **1**:794–806.

Kanopka, A., O. Mühlemann, S. Petersen-Mahrt, C. Estmer, C. Öhrmalm, and G. Akusjärvi. 1998. Regulation of adenovirus alternative RNA splicing by dephosphorylation of SR proteins. *Nature* **393**:185–187.

Lamb, R. A., and C. J. Lai. 1980. Sequence of interrupted and uninterrupted mRNAs and cloned DNA coding for the two overlapping non-structural proteins of influenza virus. *Cell* **21**:475–485.

McNally, L. M., and M. T. McNally. 1998. An RNA splicing enhancer-like sequence is a component of a splicing inhibitor element from Rous sarcoma virus. *Mol. Cell. Biol.* **18**:3103–3111.

Wu, T. T., Y. H. Su, T. M. Block, and J. M. Taylor. 1996. Evidence that two latency-associated transcripts of herpes simplex virus type 1 are nonlinear. *J. Virol.* **70**:5962–5967.

Zabolotny, J. M., C. Krummenacher, and N. W. Fraser. 1997. The herpes simplex virus type 1 2.0-kilobase latency-associated transcript is a stable intron which branches at a guanosine. *J. Virol.* **71**:4199–4208.

## Editing of Viral RNA

Lai, M., Y. Liu, and C. E. Samuel. 1998. Adenovirus VA1 RNA antagonizes the RNA-editing activity of the ADAR adenosine deaminase. *Virology* **245**:188–196.

Polson, A. G., B. L. Bass, and J. L. Casey. 1996. RNA editing of hepatitis delta virus antigenome by dsRNA-adenosine deaminase. *Nature* (London) **380**:454–456.

Polson, A. G., H. J. Ley III, B. L. Bass, and J. L. Casey. 1998. Hepatitis delta virus RNA editing is highly specific for amber/W site and is suppressed by hepatitis delta antigen. *Mol. Cell. Biol.* **18**:1919–1926.

Sanchez, A., S. G. Trappier, B. W. Mahy, C. J. Peters, and S. T. Nichol. 1996. The virion glycoproteins of Ebola viruses are encoded in two reading frames and are expressed through transcriptional editing. *Proc. Natl. Acad. Sci. USA* **93**:3602–3607.

Thomas, S. M., R. A. Lamb, and R. G. Paterson. 1988. Two mRNAs that differ by two nontemplated nucleotides encode the amino coterminal proteins P and V of the paramyxovirus SV5. *Cell* **54**:891–902.

## Export of Viral mRNAs from the Nucleus

Battiste, J. L., H. Mao, N.-J. Rao, R. Tan, D. R. Muhandiram, L. E. Kay, A. D. Frankel, and J. R. Williamson. 1996. Alpha helix-RNA major groove recognition in an HIV-1 rev peptide-RRE complex. *Science* **273**:1547–1551.

Ernst, R. K., M. Bray, D. Rekosh, and M.-L. Hammarskjöld. 1997. A structured retroviral RNA element that mediates nucleocytoplasmic export of intron-containing RNA. *Mol. Cell. Biol.* **17**:135–144.

Fischer, U., J. Huber, W. C. Boelens, I. W. Mattaj, and R. Lührmann. 1995. The HIV-1 Rev activation domain is a nuclear export signal that accesses an export pathway used by specific cellular RNAs. *Cell* **82**:475–483.

Huang, Z. M., and T. S. Yen. 1995. Role of the hepatitis B virus posttranscriptional regulatory element in export of intronless transcripts. *Mol. Cell. Biol.* **15**:3864–3869.

Malim, M. H., J. Hauber, S.-Y. Lee, J. V. Maizel, and B. R. Cullen. 1989. The HIV-1 rev trans-activator acts through a structured target sequence to activate nuclear export of unspliced viral mRNA. *Nature* (London) **338**:254–257.

Meyer, B. E., and M. H. Malim. 1994. The HIV-1 Rev trans-activator shuttles between the nucleus and the cytoplasm. *Genes Dev.* **8**:1538–1547.

Pasquinelli, A. E., R. K. Ernst, E. Lund, C. Grimm, M. L. Zapp, D. Rekosh, M.-L. Hammarskjöld, and J. E. Dahlberg. 1997. The constitutive transport element (CTE) of Mason Pfizer monkey virus (MMPV) accesses a cellular mRNA export pathway. *EMBO J.* **16**:7500–7510.

Schatz, O., M. Oft, C. Dascher, M. Schebesta, O. Rosorius, H. Jaksche, M. Dobrovnik, D. Bevec, and J. Hauber. 1998. Interaction of the HIV-1 Rev cofactor with eukaryotic initiation factor 5A and ribosomal protein L5. *Proc. Natl. Acad. Sci. USA* **95**:1607–1612.

## Regulation of RNA Processing by Viral Proteins

Dobbelstein, M., J. Roth, W. T. Kimberley, A. J. Levine, and T. Shenk. 1997. Nuclear export of the E1B 55-kDa and E4 34-kDa adenoviral oncoproteins mediated by a rev-like signal sequence. *EMBO J.* **16**:4276–4284.

Hardy, W. R., and R. M. Sandri-Goldin. 1994. Herpes simplex virus inhibits host cell splicing and regulatory protein ICP27 is required for this effect. *J. Virol.* **68**:7790–7799.

McGregor, F., A. Phelan, J. Dunlop, and J. B. Clements. 1996. Regulation of herpes simplex virus poly(A) site usage and the action of immediate early protein IE63 in the early-late switch. *J. Virol.* **70:**1931–1940.

Nemeroff, M. E., S. M. L. Barabino, Y. Li, W. Keller, and R. Krug. 1998. Influenza virus NS1 protein interacts with the cellular 30 kDa subunit of CPSF and inhibits 3′ end formation of cellular pre-mRNAs. *Mol. Cell* **1:**991–1000.

Phelan, A., J. Dunlop, and J. B. Clements. 1996. Herpes simplex virus type 1 protein IE63 affects the nuclear export of virus intron-containing transcripts. *J. Virol.* **70:**5255–5265.

Pilder, S., M. Moore, J. Logan, and T. Shenk. 1986. The adenovirus E1B 55K transforming polypeptide modulates transport or cytoplasmic stabilization of viral and host cell mRNAs. *Mol. Cell. Biol.* **6:**470–476.

Sandri-Goldin, R. M. 1996. ICP27 mediates HSV RNA export by shuttling through a leucine-rich nuclear export signal and binding viral intronless RNAs through an RGG motif. *Genes Dev.* **12:**868–879.

Shih, S. R., M. Nemeroff, and R. M. Krug. 1995. The choice of alternative 5′ splice sites in influenza virus M1 mRNA is regulated by the viral polymerase complex. *Proc. Natl. Acad. Sci. USA* **92:**6324–6328.

**Stability of Viral and Cellular mRNA**

Jeon, S., and P. F. Lambert. 1995. Integration of human papillomavirus type 16 DNA into the human genome leads to increased stability of E6 and E7 mRNAs: implications for cervical carcinogenesis. *Proc. Natl. Acad. Sci. USA* **92:**1654–1658.

Lam, Q., C. A. Smibert, K. E. Koop, C. Lavery, J. P. Capone, S. P. Weinheimer, and J. R. Smiley. 1996. Herpes simplex virus VP16 rescues viral mRNA from destruction by the virion shutoff function. *EMBO J.* **15:**2575–2581.

Myer, V. E., S. I. Lee, and J. A. Steitz. 1992. Viral small nuclear ribonucleoproteins bind a protein implicated in messenger RNA destabilization. *Proc. Natl. Acad. Sci. USA* **89:**1296–1300.

Zelus, B. D., R. S. Stewart, and J. Ross. 1996. The virion host shutoff protein of herpes simplex virus type 1: messenger ribonucleolytic activity in vitro. *J. Virol.* **70:**2411–2419.

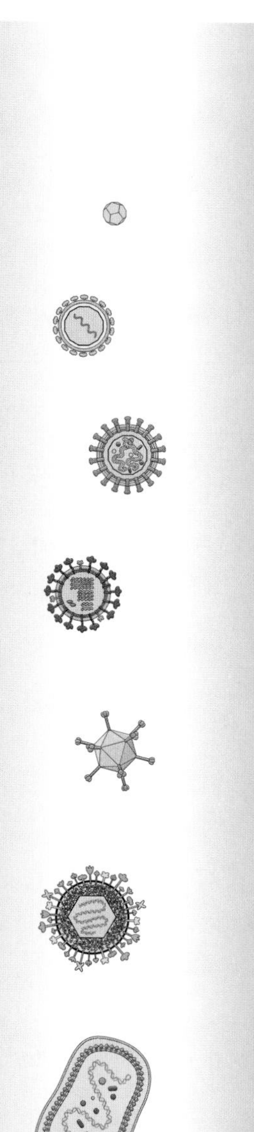

# 11

# Translational Control of Viral Gene Expression

*The difficulty lies, not in the new ideas, but in escaping old ones, which ramify, for those brought up as most of us have been, into every corner of our minds.*

JOHN MAYNARD KEYNES

## Introduction

As obligate intracellular parasites, viruses must use components of the host cell for their growth. Some aspects of viral multiplication depend more on cellular contributions than others. For example, all viruses encode at least one protein needed for viral nucleic acid synthesis. However, animal viruses are completely dependent on the host cell for the translation of their mRNAs because no viral genome encodes any part of the translational machinery. Virus infection often results in modification of the host's translational apparatus so that viral mRNAs can be translated selectively.

Studies of viruses have led to much of our understanding of translation and its regulation. Before the advent of recombinant DNA technology, viruses provided a ready source of large quantities of relatively pure mRNA for in vitro studies of protein synthesis. The 5' cap structure was identified on a viral RNA, and new translation initiation mechanisms, including internal ribosomal entry and ribosome shunting, were discovered during studies of viral mRNAs. Our understanding of how the activity of the multisubunit cap-binding complex, eIF4F, can be regulated originated from the finding that one of its subunits is cleaved in virus-infected cells.

Translation is a universal process in which proteins are produced from mRNA templates read in the 5' → 3' direction, and the growing polypeptide chain is synthesized from the amino to the carboxy terminus. Each amino acid is specified by a genetic code consisting of three bases (a **codon**) in the mRNA. Translation takes place on **ribosomes**, and **transfer RNAs (tRNAs)** are the adapter molecules that link specific amino acids with individual codons in the mRNA. In this chapter, we will first discuss the basic mechanism by which translation occurs in mammalian cells. We will then consider the many ways that viral mRNAs are translated to provide optimal coding capacity in genomes of limited size, followed by a discussion of how translation is regulated in virus-infected cells.

**371**

## Mechanisms of Eukaryotic Protein Synthesis

### General Structure of Eukaryotic mRNA

The general structure of eukaryotic mRNA is shown in Fig. 11.1. Most mRNAs, with the exception of organelle mRNAs and certain viral mRNAs, begin with a 5' 7-methylguanosine (m⁷G) **cap structure**, so called because it literally blocks the 5' end of the mRNA (Fig. 10.2). It is joined to the second nucleotide by a 5'-5' phosphodiester linkage, in contrast to the 5'-3' linkages found in the remainder of the mRNA. The unique cap structure directs pre-mRNAs to processing and transport pathways, regulates mRNA turnover, and is required for efficient translation by the 5'-end-dependent mechanism (see "5'-End-Dependent Initiation" below). Most eukaryotic mRNAs contain **5' untranslated regions**, which may vary in length from 3 to more than 1,000 nucleotides, although most are typically 50 to 70 nucleotides in length. Such 5' untranslated regions often contain secondary structures (e.g., hairpin loops [see Fig. 6.8]) formed by base pairing of the RNA. These double-helical regions

**Figure II.I Structure of eukaryotic and prokaryotic mRNAs.** Both mRNAs contain 5' and 3' untranslated regions (UTR). Eukaryotic mRNAs carry a 5' 7-methylguanosine cap and a 3' poly(A) tail. Most prokaryotic mRNAs are polycistronic: they encode multiple proteins, and each open reading frame is separated from the next one by an untranslated region. In contrast, the vast majority of eukaryotic mRNAs are monocistronic: they encode only one protein. However, some eukaryotic mRNAs are bicistronic, as discussed in the text. Adapted from G. M. Cooper, *The Cell: a Molecular Approach* (ASM Press, Washington, D.C., and Sinauer Associates, Sunderland, Mass., 1997), with permission.

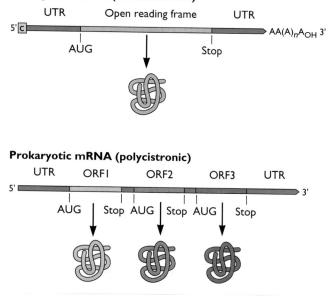

must be unwound to allow passage of ribosomal 40S subunits during translation.

Translation begins at an **AUG initiation codon** and ends at a **termination codon**. The termination codon is followed by a **3' untranslated region**, which can play a role in regulating translation efficiency and mRNA stability. At the very 3' end of the mRNA is a stretch of adenylate residues known as the **poly(A) tail**, which is added to nascent pre-mRNA. The poly(A) tail is necessary for efficient translation and may promote interactions among proteins that bind both ends of the mRNA (see "5'-End-Dependent Initiation" below).

Most prokaryotic mRNAs are said to be **polycistronic**: they encode several polypeptides, and each open reading frame is separated from the next by an untranslated spacer region. Although the vast majority of eukaryotic mRNAs are **monocistronic** in that they encode only a single polypeptide (Fig. 11.1), a small number of eukaryotic mRNAs are functionally polycistronic. In these cases, different strategies have evolved for expressing multiple genes from a single mRNA.

### The Translation Machinery

#### Ribosomes

Mammalian ribosomes, the sites of translation, are composed of two subunits designated according to their sedimentation coefficients: 40S and 60S (Fig. 11.2A). The 40S subunit is made up of an 18S rRNA molecule and 33 proteins, while the 60S subunit contains three RNAs (5S, 5.8S, and 28S rRNAs) and 47 proteins. Actively growing mammalian cells may contain approximately 10 million ribosomes, a reflection of the importance of protein synthesis for metabolism and cell growth.

Ribosomal RNAs (rRNAs) are believed to play a catalytic role in protein synthesis. After removal of 95% of the ribosomal proteins, the 60S ribosomal subunit can still catalyze the formation of peptide bonds. The protein components of ribosomes are now believed to help fold the rRNAs properly, so that they can fulfill their catalytic function, and to position the tRNAs.

#### tRNAs

tRNAs are adapter molecules that align each amino acid with its corresponding codon on the mRNA. Each tRNA is 70 to 80 nucleotides in length and folds into a highly base-paired L-shaped structure (Fig. 11.2B). This shape is believed to be required for a proper fit of tRNAs with ribosomes during translation. The adapter function of tRNAs is carried out by two distinct regions of the molecule. At their 3' ends, all tRNAs carry the sequence 5'-CCA-3', to which amino acids are covalently linked by **aminoacyl-tRNA synthetases**, each of which recognizes

**A**

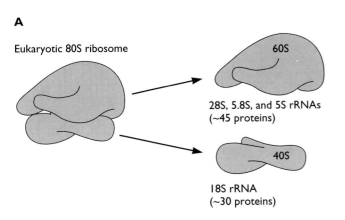

Eukaryotic 80S ribosome

60S

28S, 5.8S, and 5S rRNAs
(~45 proteins)

40S

18S rRNA
(~30 proteins)

**B**

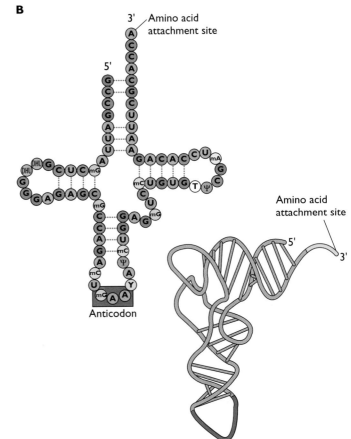

**Figure 11.2 tRNAs and ribosomes.** (A) Model of a eukaryotic ribosome. The 80S ribosome consists of 60S and 40S subunits, which are made of ribosomal proteins and ribosomal RNAs. (B) Structure of tRNA. The model on the left shows how base pairing among the nucleotides of the tRNA results in a cloverleaflike structure. Modified bases include methylguanosine (mG), methylcytosine (mC), dihydrouridine (DHU), ribothymidine (G), a modified purine (Y), and pseudouridine (Ψ). On the right is a folded representation showing the L-shaped structure. In these two models the anticodon loop and amino acid attachment sites are labeled. Adapted from G. M. Cooper, *The Cell: a Molecular Approach* (ASM Press, Washington, D.C., and Sinauer Associates, Sunderland, Mass., 1997), with permission.

a single amino acid and the correct tRNA. At the opposite end of the L is the **anticodon loop**, which base-pairs with the mRNA template. Faithful incorporation of amino acids therefore depends on the specificity of codon-anticodon base pairing, as well as on the correct attachment of amino acids to tRNAs by aminoacyl-tRNA synthetases.

### Initiation Proteins

Many nonribosomal proteins are required for eukaryotic translation (Table 11.1). Some form multisubunit complexes containing as many as eight different polypeptide chains, while others function as single polypeptides. Translation can be separated experimentally into three distinct stages: initiation, elongation, and termination. The proteins involved in each stage are called eukaryotic initiation, elongation, and termination proteins (eIFs, eEFs, and eRFs, respectively, where F stands for factor), to distinguish them from their prokaryotic counterparts. The amino acid sequences of these proteins are conserved from mammals to yeast, indicating that the mechanisms of translation are similar in the two organisms.

### Initiation

The majority of regulatory mechanisms function at initiation because this is the rate-limiting step in the translation of most mRNAs (see "Regulation of Translation during Viral Infection"). Initiation is a complex process in which at least 11 initiation proteins participate and which requires hydrolysis of both ATP and GTP. The end result is formation of a complex containing the mRNA, the ribosome, and the initiator Met-tRNA$_i$ in which the reading frame of the mRNA has been set. The 80S ribosome, which is the predominant species in cells, must be dissociated because it is the 40S subunit that participates in initiation. Three initiation proteins (Table 11.1) promote the dissociation of 80S ribosomes into 60S and 40S subunits. Both eIF1A and eIF3 bind to free 40S subunits to prevent their association with the 60S subunit, while interaction of eIF6 with the larger subunit prevents it from associating with the 40S subunit.

There are two mechanisms by which ribosomes bind to mRNA in eukaryotes. In 5'-end-dependent initiation, by which the majority of mRNAs are translated, the initiation complex binds to the 5' cap structure and moves, or

**Table 11.1**    Mammalian translation proteins[a]

| Name | Subunits | Molecular mass (kDa)[b] | Function |
|---|---|---|---|
| Initiation proteins | | | |
| eIF1 | | 15 | Enhances initiation complex formation; required for location of initiation codon |
| eIF1A | | 17 | Ribosomal dissociation; promotes Met-tRNA$_i$ binding; required for location of initiation codon |
| eIF2 | | 125 | Binds Met-tRNA$_i$ and GTP |
| | α | 36 | Site of phosphorylation on Ser-51 |
| | β | 37 | Binds Met-tRNA$_i$ |
| | γ | 52 | Binds GTP and Met-tRNA$_i$ |
| eIF2B | | 270 | eIF2 recycling |
| | α | 30 | |
| | β | 39 | Binds GTP |
| | γ | 58 | Binds ATP |
| | δ | 66 | Binds ATP |
| | ε | 84 | Phosphorylated by casein kinase II and Gsk-3 |
| eIF2C | | 94 | Stabilizes ternary complex in presence of RNA |
| eIF3 | | 650 | Dissociates ribosomes; promotes Met-tRNA$_i$ and mRNA binding |
| | p35 | 35 | |
| | p36 | 36 | |
| | p40 | 40 | |
| | p44 | 44 | |
| | p47 | 47 | |
| | p66 | 66 | Binds RNA |
| | p115 | 115 | Major phosphorylated subunit |
| | p170 | 170 | |
| eIF4A | | 46 | ATPase, helicase, binds RNA |
| eIF4B | | 80 | Binds RNA, promotes helicase activity |
| eIF4F | | 250 | Binds m⁷G caps, helicase |
| | eIF4E | 25 | m⁷G cap-binding subunit |
| | eIF4A | 46 | ATPase, helicase |
| | eIF4GI/II | 220 | Binds eIF4A, eIF4E, eIF3 |
| eIF5 | | 58 | Promotes GTPase with eIF2 and release of eIFs |
| eIF6 | | 26 | Binds to 60S ribosomes, promotes dissociation |
| Elongation proteins | | | |
| eEF1A | | 51 | GTP-dependent binding of aminoacyl-tRNAs; GTPase |
| eEF1B | | 114 | Guanine nucleotide exchange on eEF1A |
| | α | 30 | GTP exchange activity |
| | β | 36 | GTP exchange activity |
| | γ | 48 | |
| eEF2 | | 95 | Promotes translocation; GTPase |
| Termination proteins | | | |
| eRF1 | | 55 | Recognizes termination codons; promotes peptide hydrolysis |
| eRF3 | | 55.8 | GTPase; stimulates eRF1 activity |

[a]Modified from J. W. B. Hershey et al., *Translational Control*, Cold Spring Harbor Press, Cold Spring Harbor, N.Y., 1996.

[b]Mass determined by sodium dodecyl sulfate-polyacrylamide gel electrophoresis; for many translation proteins, this value differs from the mass calculated from the nucleotide sequence because of posttranslational modification.

"scans," in a 3' direction until the initiating AUG codon is encountered. In internal ribosome entry, the initiation complex binds at, or just upstream of, the initiator codon. Internal ribosome entry was first discovered with picornavirus mRNAs and is now known to be responsible for translation of some cellular mRNAs.

### 5'-End-Dependent Initiation

**How ribosomes assemble at the proper end of mRNA.** The first step in this pathway is recognition of the m⁷G cap by eIF4F, which consists of three proteins, eIF4A, eIF4E, and eIF4G (Fig. 11.3). The cap-binding protein, eIF4E, interacts with the N-terminal domain of eIF4G, while eIF4A binds to its C-terminal domain. Two functional forms of eIF4G, called eIF4GI and eIF4GII, have been found in mammalian cells. Next, eIF1A and eIF3 bind to the 40S ribosomal subunit, which then binds to a ternary complex consisting of eIF2, GTP, and Met-tRNA$_i$. The 43S preinitiation complex thus formed associates with the mRNA through an interaction of eIF3 with the C-terminal domain of eIF4G, forming a 48S preinitiation complex.

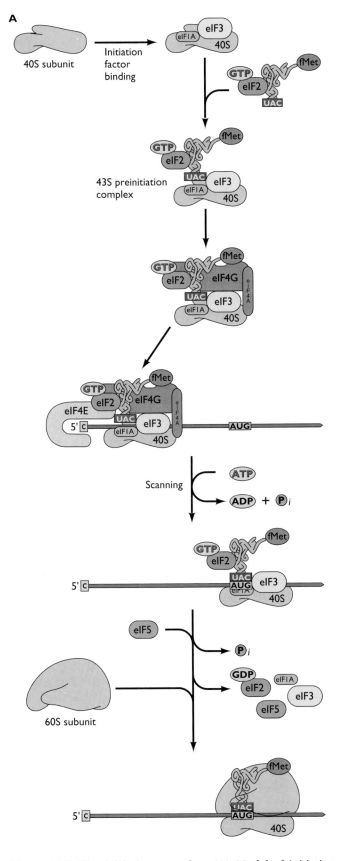

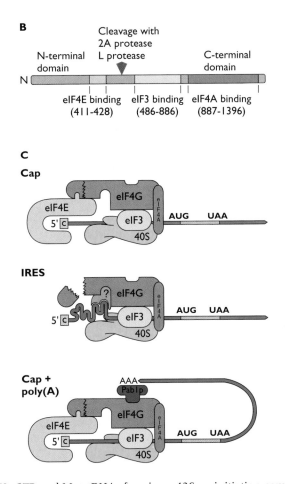

eIF2, GTP, and Met-tRNA$_i$, forming a 43S preinitiation complex. The ribosome then binds eIF4G via eIF3. Alternatively, eIF4G may first join the 43S preinitiation complex and then bind the mRNA via eIF4E bound to the cap. The 40S subunit then scans down the mRNA until the AUG initiation codon is reached. eIF5 triggers GTP hydrolysis, eIF2 bound to GDP is released along with other initiation proteins, and the 60S ribosomal subunit joins the complex. In a variation of this model, the helicases eIF4A/eIF4B scan the mRNA and stop at the initiating AUG, at which point the 40S subunit binds the mRNA. Adapted from G. M. Cooper, *The Cell: a Molecular Approach* (ASM Press, Washington, D.C., and Sinauer Associates, Sunderland, Mass., 1997), with permission. (B) Schematic of eIF4G. Sites for binding of eIF4E, eIF3, and eIF4A are shown, as are the cleavage sites for viral proteases. Adapted from S. J. Morley et al., *RNA* **3:**1085–1104, 1997, with permission. (C) Hypothetical models of initiation complex assembly involving the 5′ cap structure, IRES elements, and poly(A). (Top) In 5′-end-dependent initiation, eIF4F is brought to the mRNA 5′ end by interaction of eIF4E with the cap structure. The N terminus of eIF4G binds eIF4E, and the C terminus binds eIF4A. The 40S ribosomal subunit binds to eIF4G indirectly via eIF3. (Middle) Initiation in the IRES does not depend on the presence of a cap structure but requires the C-terminal fragment of eIF4G to recruit the 40S ribosomal subunit via its interaction with eIF3. Binding of eIF4G to the IRES may occur through an intermediate protein, X, but no such universal IRES-binding protein has been identified. Alternatively, eIF4G may bind directly to the IRES. (Bottom) In yeast, 5′-end-dependent initiation is stimulated by the poly(A)-binding protein Pab1p, which interacts with eIF4G. This interaction may circularize the mRNA and facilitate formation of the initiation complex at the 5′ end. Adapted from M. W. Hentze, *Science* **275:**500–501, 1997, with permission.

**Figure 11.3 The initiation complex.** (A) Model of initiation complex assembly. Initiation proteins eIF3 and eIF1A bind the 40S ribosomal subunit. eIF4F binds the cap via the eIF4E subunit, and the ribosome binds a ternary complex containing eIF4G therefore acts as a "bridge" between the cap structure and the 40S subunit. This important adapter molecule was first discovered as the target of proteolytic cleavage in poliovirus-infected cells that resulted in the abrupt termination of host protein synthesis (see "Regulation of Translation during Viral Infection" below). The 40S ribosomal subunit then moves in a 3' direction on the mRNA in a process called **scanning**. When the 40S subunit reaches the initiating AUG codon, initiation proteins are released, allowing the 60S ribosomal subunit to associate with the 40S subunit, forming the 80S initiation complex.

The 3'-poly(A) sequence of mRNAs may also facilitate recruitment of 40S ribosomal subunits (Box 11.1).

**The role of RNA secondary structure in mRNA translation.** Translation efficiency is reduced by the presence of strong secondary structure in the mRNA 5' untranslated region, particularly when the structure is near the 5' terminus. There are probably two reasons for this effect. The first is that active eIF4E, which is present in cells in limiting quantities, preferentially binds to mRNAs devoid of RNA secondary structure adjacent to the 5' cap. In addition, the presence of secondary structure inhibits ribosome movement.

The eIF4A component of eIF4F possesses ATP-dependent RNA helicase activity which is enhanced by eIF4B (Fig. 11.3). Such activity is believed to unwind any regions of double-stranded RNA (dsRNA) near the 5' end of the mRNA, allowing the 43S preinitiation complex to bind. The helicase may also migrate in a 3' direction, unwinding dsRNA and enabling ribosomes to scan. mRNAs with less secondary structure in the 5' untranslated region have less need for RNA helicase activity during mRNA translation and hence are less dependent on the cap structure through which the helicase is brought to the mRNA. Dependence of translation on the cap structure can be measured experimentally by determining the effect on protein synthesis of cap analogs such as $m^7GDP$ and $m^7GTP$. These compounds competitively inhibit 5'-end-dependent initiation by binding to eIF4E and preventing it from associating with capped mRNAs. For example, the 5' untranslated region of alfalfa mosaic virus RNA segment 4 is largely free of secondary structure, and translation of this mRNA is quite resistant to inhibition by cap analogs.

**Choosing the initiation codon.** For over 90% of mRNAs, translation initiation occurs at the 5'-proximal AUG codon. The efficiency of initiation is influenced by

---

**BOX 11.1**

*The 3'-poly(A) sequence can affect translation initiation at the 5' end*

For years it has been known that the presence of a poly(A) tail can stimulate mRNA translation in vitro. Recent results suggest that the effect of the poly(A) tail on translation may result from interactions between proteins associated with the 5' and 3' ends of the mRNA, promoting 40S subunit recruitment.

In the yeast *Saccharomyces cerevisiae*, the poly(A)-tail-binding protein, Pab1p, is required for efficient mRNA translation. Stimulation of translation in yeast by poly(A), which requires Pab1p, occurs by enhancing the binding of 40S ribosomal subunits to mRNA. Pab1p has been shown to interact with eIF4G N-terminal to the binding site for eIF4E. Mutagenesis of this Pab1p-binding site on eIF4G destroys stimulation of translation by poly(A). These results have led to a model in which Pab1p, bound to the poly(A) tail, interacts with eIF4G bound to the 5' cap, perhaps stabilizing the interaction and assisting in recruiting 40S subunits (Fig. 11.3C). A consequence of these interactions is that the mRNA should be circularized during initiation, although this result has not been demonstrated. Circularization of mRNA might be a mecha-

nism to ensure that only intact mRNAs [containing 5' cap and poly(A)] are translated.

A similar mechanism might not operate in mammalian cells because human poly(A)-tail-binding protein (Pabp) and eIF4G do not interact. However, a mammalian protein that binds to human Pabp has recently been identified. This protein, called Paip-1, is similar to the central portion of eIF4G and can bind to eIF4A. Overexpression of Paip-1 in mammalian cells stimulates translation. It has been suggested that Paip-1 might act as a bridge between Pabp and eIF4A, resulting in mRNA circularization.

**Craig, A. W. B., A. Haghighat, A. T. K. Yu, and N. Sonenberg.** 1998. Interaction of polyadenylate-binding protein with the eIF4G homologue Paip enhances translation. *Nature* (London) **392**:520–523.

**Tarun, S. Z., and A. B. Sachs.** 1996. Association of the yeast poly(A) tail binding protein with translation initiation factor eIF4G. *EMBO J.* **15**:7168–7177.

**Tarun, S. Z., S. E. Wells, J. A. Deardorff, and A. B. Sachs.** 1997. Translation initiation factor eIF4G mediates in vitro poly(A) tail-dependent translation. *Proc. Natl. Acad. Sci. USA* **94**:9046–9051.

the nucleotide sequence surrounding this codon. Studies of the effects of mutating these sequences have shown that the consensus sequence 5'-GCC**A**CC**A**UGG-3' is recognized most efficiently in mammalian cells; the presence of a purine at the −3 position (boldface) is most important for high levels of translation. However, only 5% of eukaryotic mRNAs contain this ideal consensus sequence; most have suboptimal sequences that result in less efficient translation. This finding indicates that not all mRNAs must be translated maximally, only at levels appropriate for the function of the protein product. If a very poor match to this consensus sequence is present, the AUG codon may be passed over by the ribosome and initiation may occur further downstream (see "The Diversity of Viral Translation Strategies"). If the 5'-proximal AUG codon is mutated so that it cannot serve as an initiation codon, translation will start at the next downstream AUG. Insertion of an AUG codon upstream of the initiating codon causes initiation to shift to the more 5'-proximal site.

Although AUG is used to initiate synthesis of most proteins, initiation at ACG, GTG, and CTG codons can also occur in viral and cellular mRNAs, although far less frequently. In all cases, the first amino acid of the protein is a methionine. Precisely how ribosomes recognize non-AUG codons so that methionine is inserted is not known. When non-AUG initiator codons are used, they are always the 5'-proximal start site and in a good sequence context for initiation. Because the efficiency of initiation at these sites is always low, the use of non-AUG initiating codons might be another mechanism for regulating translation efficiency. Not all mRNAs allow initiation from a non-AUG site, suggesting that there may be structural elements in the mRNA that regulate expression from such a codon. For example, RNA secondary structure 3' of the AUG initiation codon might be required for the moving 40S subunit to pause, allowing initiation. Without the secondary structure, the 40S subunit might travel past the AUG initiation codon.

**How does the 40S subunit move on the mRNA?** Translation of most mRNAs begins with the 5'-proximal AUG codon. This observation can be explained by a model in which the 40S ribosomal subunit migrates through the 5' untranslated region until it reaches the AUG initiation codon. Experimental proof for this long-standing hypothesis is still lacking. Hydrolysis of ATP is necessary for translation initiation, but it is not known if this is a requirement only of the ATP-dependent RNA helicase or also of the helicase and of ribosome movement. An alternative hypothesis is that after the 40S subunit binds the mRNA, random movement in both the 5' → 3' and 3' → 5' directions is equally likely. Eventually the 40S subunit reaches the ini-

tiating AUG codon, which gives the appearance of scanning when a single 5' entry site is used.

**Ribosome shunting.** According to the 5'-end-dependent initiation model, stable hairpin structures encountered by scanning 40S ribosomes are melted by helicases. If the AUG initiation codon is buried in the stem of a hairpin, it is still used as an initiation site. However, in some RNAs, ribosomes appear to skip over such hairpin structures in a process that has been called **ribosome shunting** because ribosomes appear to bypass or be shunted across internal sections of the 5' untranslated region. Examples of ribosome shunting are found on the 35S cauliflower mosaic virus RNA and adenovirus late mRNAs. That a particular mRNA is translated by ribosome shunting can be demonstrated by assessing the effect of inserting a very stable stem-loop structure in different parts of the 5' untranslated region (Fig. 11.4). The mechanism by which ribosomes bypass areas of the RNA containing secondary structures remains to be elucidated.

### Internal Ribosome Binding

**The internal ribosome entry site.** For many years it was suspected that the (+) strand genomic RNAs of picornaviruses were translated by an unusual mechanism. Not only did these (+) strand RNAs lack 5'-terminal cap structures, but they were efficiently translated in infected cells despite inhibition of cellular mRNA translation (see "Regulation of eIF4F Activity"). Determination of the nucleotide sequence of poliovirus (+) strand RNA revealed the presence of a 741-nucleotide 5' untranslated region that contains seven AUG codons. These features led to the suggestion that ribosomes might not scan through picornaviral 5' untranslated regions but rather might bind to an internal sequence. In a key experiment, the 5' untranslated region of poliovirus (+) strand RNA was shown to contain a sequence that promotes translation initiation by internal ribosome entry, the internal ribosome entry site (IRES) (Box 11.2).

The IRES of different picornaviruses contains extensive regions of RNA secondary structure, and it is the RNA structure, and less so the nucleotide sequence, that is conserved within type 1 (entero- and rhinoviruses), type 2 (cardio- and aphthoviruses), and hepatitis C virus IRESs (Fig. 11.5C). The only sequence conserved among these three types of IRES is an oligopyrimidine tract located 25 nucleotides upstream of the AUG initiation codon. In a type 1 IRES, this AUG codon is located 50 to 100 nucleotides past the 3' end of the IRES, while it immediately follows the 3' end of a type 2 IRES and is part of the hepatitis C virus IRES. Translation initiation via a type 1 IRES therefore comprises binding of the 40S ribosomal subunit

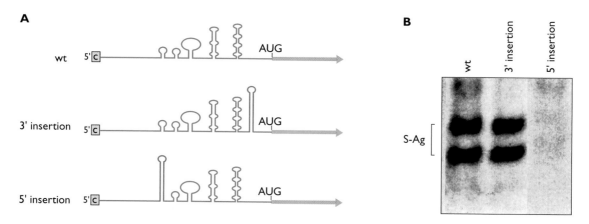

**Figure II.4 Strong secondary structure in the 5′ but not the 3′ end of the adenoviral tripartite leader sequence inhibits initiation.** Plasmids that express mRNAs containing wild-type or mutant forms of the adenovirus tripartite leader sequence (a common 5′ noncoding region of viral late mRNAs that mediates translation by ribosome shunting) linked to the hepatitis B virus S-antigen-coding region were introduced into cells by transformation. Translation was monitored by immunoprecipitation of S antigen from [$^{35}$S]methionine-labeled cell extracts followed by electrophoresis on polyacrylamide gels. (A) Schematic diagram of mRNAs. Strong secondary structure (red) was introduced at the 3′ end or at the 5′ end of the tripartite leader by ligating five copies of a 12-nucleotide *Bam*HI linker at position 202 or 34. (B) Cells were labeled with [$^{35}$S]methionine, and S-antigen synthesis was examined by immunoprecipitation. Two forms of S antigen are synthesized in cells; they have molecular masses of 24 and 27 kDa. Insertion of RNA secondary structure at the 5′ end of the tripartite leader abolished translation, while insertion at the 3′ end had no effect. These results are interpreted to indicate that ribosomes do not scan through the 3′ part of the untranslated region; rather they bypass it by "shunting." The fact that inserting secondary structure at the 5′ end of the leader blocks translation suggests that ribosomes must scan a certain distance before shunting can occur. Adapted from A. Yueh and R. J. Schneider, *Genes Dev.* **10:**1557–1567, 1996, with permission. Courtesy of R. J. Schneider, New York University School of Medicine.

to the IRES, followed by scanning of the subunit to the AUG initiation codon. In contrast, the 40S subunit probably binds at the AUG initiation codon of a type 2 IRES.

An IRES has been identified in other viral mRNAs, including classical swine fever virus (a pestivirus), and in cellular mRNAs such as those encoding the endoplasmic reticulum chaperone Bip, *Drosophila* Antennapedia protein, the fibroblast growth factor 2 protein, and eIF4G. These IRESs have no obvious sequence or structural similarities to each other or to those of picornaviruses. The hepatitis C virus IRES is unusual in that its function depends on RNA sequences that surround and extend beyond the AUG initiation codon. It is believed that the presence of IRESs in cellular mRNAs permits translation when 5′-end-dependent initiation is inhibited by inactivation of eIF4F, such as during heat shock, growth arrest, or mitosis.

The discovery of the IRES makes even more puzzling the rarity of eukaryotic mRNAs that contain several open reading frames reminiscent of prokaryotic mRNAs (Fig. 11.1). In theory, all the open reading frames of a polycistronic mRNA can be translated in a eukaryotic cell as long as each frame is preceded by an IRES. Nevertheless, only one such naturally occurring polycistronic mRNA has

been identified in eukaryotes; so far, all other IRESs have been found only in 5′ untranslated regions.

**How does the IRES function?** Translation initiation mediated by an IRES requires the same set of initiation proteins as translation of mRNAs by 5′-end-dependent initiation. For many years it was believed that the exception was eIF4G, which is cleaved in cells infected with picornaviruses (see "Cleavage of eIF4G" below), inactivating the translation of most cellular mRNAs. It is now known that the C-terminal fragment of eIF4G, which contains binding sites for eIF3 and eIF4A (Fig. 11.3B and C), stimulates translation directed by the IRES. These results have led to a model in which the 40S ribosomal subunit is recruited to the IRES via interaction with eIF3 bound to the C-terminal domain of eIF4G (Fig. 11.3C). Binding of eIF4G to the mRNA does not occur via eIF4E because the N-terminal eIF4E-binding domain is removed by cleavage in picornavirus-infected cells, and in any case, picornavirus mRNAs are not capped. It has been suggested that eIF4G binds to the IRES, either directly to the RNA or via a "linker" protein (Fig. 11.3C). Interestingly, eIF4G mRNA may contain an IRES, suggesting a means for its continued

The hypothesis that poliovirus mRNA is translated by internal ribosome binding was first tested by Nahum Sonenberg, who examined the translation of mRNAs containing two open reading frames (ORFs) separated by the poliovirus 5′ untranslated region (figure panel A). These bicistronic mRNAs directed the synthesis of two proteins, but the second ORF was efficiently translated only if it was preceded by the picornavirus 5′ untranslated region. It was concluded that ribosomes bind within the picornavirus 5′ untranslated region, thereby permitting translation of the second ORF. The region of the 5′ untranslated region that directs internal ribosome entry was called the IRES.

It had long been known that circular mRNAs cannot be translated by 5′-end-dependent initiation. Translation by internal ribosome binding, however, should not require a free 5′ end. To test this hypothesis, Peter Sarnow created circular mRNAs with and without an IRES. The circular mRNA was translated only if an IRES was present (panel B). This experiment formally proved that translation initiation directed by an IRES occurs by internal binding of ribosomes and does not require a free 5′ end.

**Chen, C. Y., and P. Sarnow.** 1995. Initiation of protein synthesis by the eukaryotic translational apparatus on circular RNAs. *Science* **268**:415–417.

**Pelletier, J., and N. Sonenberg.** 1988. Internal initiation of translation of eukaryotic mRNA directed by a sequence derived from poliovirus RNA. *Nature* (London) **334**:320–325.

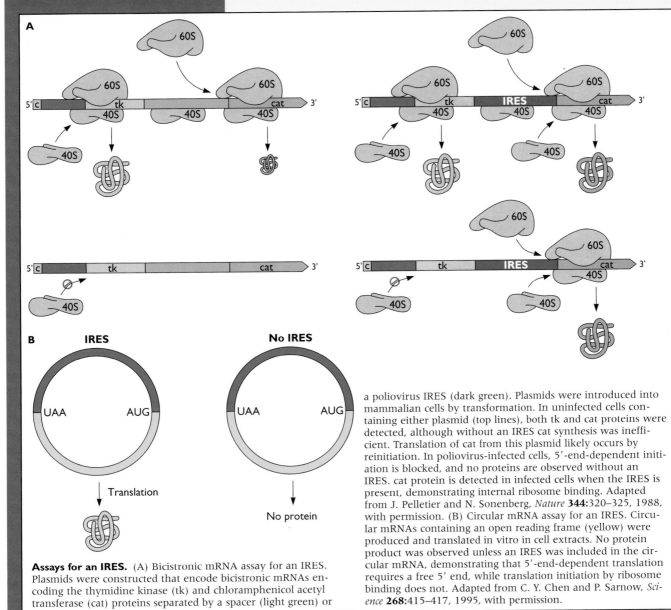

a poliovirus IRES (dark green). Plasmids were introduced into mammalian cells by transformation. In uninfected cells containing either plasmid (top lines), both tk and cat proteins were detected, although without an IRES cat synthesis was inefficient. Translation of cat from this plasmid likely occurs by reinitiation. In poliovirus-infected cells, 5′-end-dependent initiation is blocked, and no proteins are observed without an IRES. cat protein is detected in infected cells when the IRES is present, demonstrating internal ribosome binding. Adapted from J. Pelletier and N. Sonenberg, *Nature* **344**:320–325, 1988, with permission. (B) Circular mRNA assay for an IRES. Circular mRNAs containing an open reading frame (yellow) were produced and translated in vitro in cell extracts. No protein product was observed unless an IRES was included in the circular mRNA, demonstrating that 5′-end-dependent translation requires a free 5′ end, while translation initiation by ribosome binding does not. Adapted from C. Y. Chen and P. Sarnow, *Science* **268**:415–417, 1995, with permission.

**Assays for an IRES.** (A) Bicistronic mRNA assay for an IRES. Plasmids were constructed that encode bicistronic mRNAs encoding the thymidine kinase (tk) and chloramphenicol acetyl transferase (cat) proteins separated by a spacer (light green) or

**A**

**Type 1 IRES**

GNRA sequence

A/C-rich

IV
V
VI

II
III

I

124  162     186  220    234  440     556  561  625
                              448
88

Pyrimidine-rich region

Variable region

743
AUG    Coding region

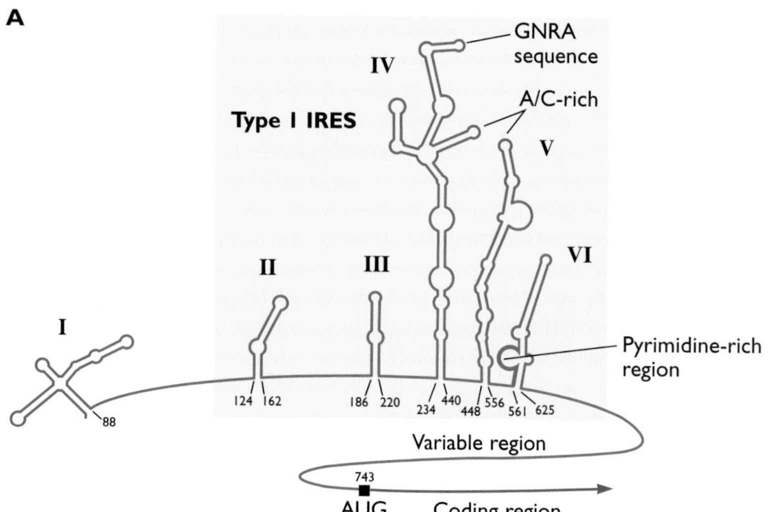

**Figure 11.5 Three types of IRES.** The 5′ untranslated region from poliovirus (A), encephalomyocarditis virus (B), and hepatitis C virus (C) are shown. The IRES is indicated by yellow shading. Predicted secondary and tertiary RNA structures (RNA pseudoknots) are shown. The poliovirus IRES is a type 1 IRES, which is found in the genomes of enteroviruses and rhinoviruses. The type 2 IRES is found in the genomes of aphthoviruses and cardioviruses. Conserved sequence features include the GNRA- and A/C-rich loops and a pyrimidine-rich region. The initiator AUG codon (black box) is part of the IRES of hepatitis C virus. (A and B) Adapted from S. R. Stewart and B. L. Semler, *Semin. Virol.* **8:**242–255, 1997, with permission. (C) Adapted from S. M. Lemon and M. Honda, *Semin. Virol.* **8:**274–288, 1997, with permission.

**B**

GNRA sequence

I    A/C-rich

**Type 2 IRES**

J

A
H
K

D
Pseudoknot?
C    E    F    G
A-rich loop

90
120
B
Pseudoknot?

321  336  347  371  392   448  451   680

L
808

Pyrimidine-rich region

Poly(C) region

AUG
834  Coding region

**C**

IIIb

IIIa
171       228
IIb                    IIIc
85   155         238

144   ~249
IIa                      IIId
70   97              275

5′
5    20
42      131      ~296  IIIe
328       IIIf
383
Coding region
I                   335   351

AUG

IV

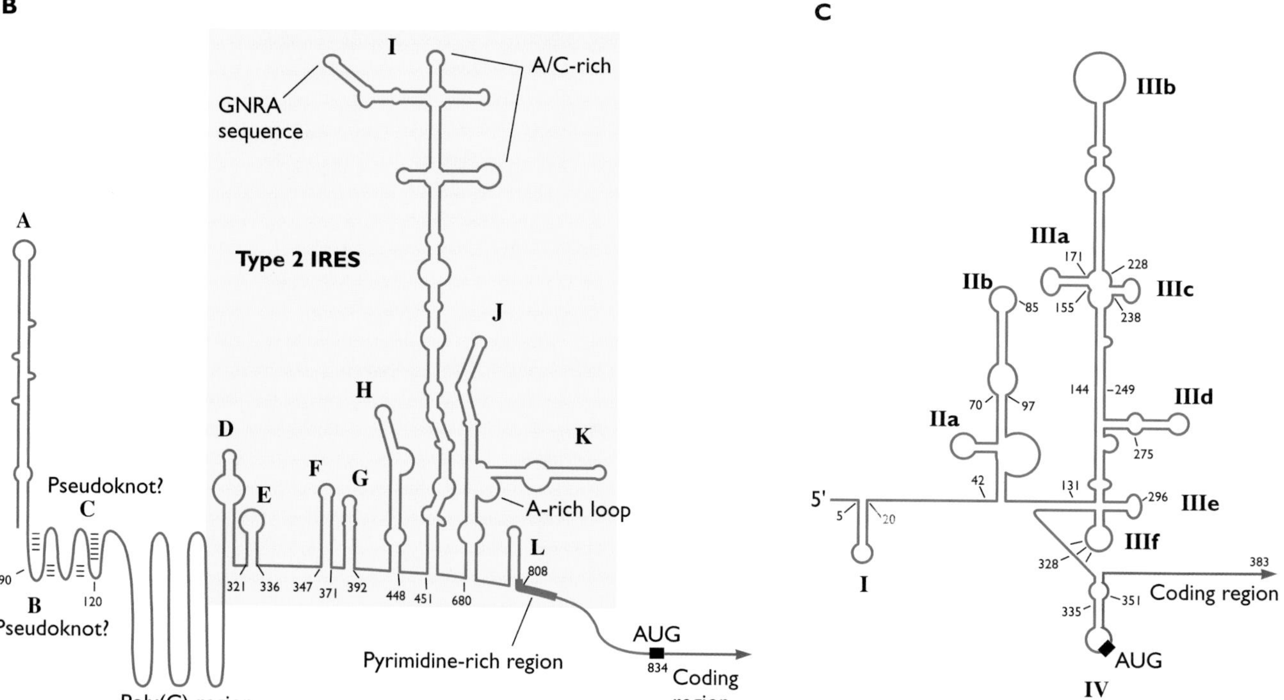

synthesis in picornavirus-infected cells, under conditions where translation initiation by 5′-end scanning is inhibited. It has been reported that IRES function is markedly enhanced in cells in which the poliovirus protease 2A[pro] is expressed. Because 2A[pro] is responsible for cleavage of eIF4G, it may be that such stimulation of translation is due to production of the C-terminal proteolytic fragment of eIF4G.

No IRES functions in vitro in wheat germ extracts, and many function poorly in reticulocyte lysates, in which most capped mRNAs are translated efficiently (Box 11.4). These observations have led to the suggestion that ribosome binding to the IRES requires more than the canonical translation proteins. Candidate proteins have been identified by their ability to bind to the IRES and their capacity to restore IRES function in the reticulocyte lysate.

## Use of the IRES in cloning and expression vectors

The IRES has been used widely in the expression and cloning of foreign genes in eukaryotes. One strategy for the expression of genes in mammalian cells is to produce mRNAs in the cytoplasm by a bacterial DNA-dependent RNA polymerase, such as T7 RNA polymerase. Such mRNAs are poorly translated because they are not capped; inclusion of an IRES in the 5' untranslated region allows them to be translated efficiently.

Another application of the IRES is in the functional cloning of new genes (figure panel A). A DNA library is made by using a cloning vector that produces a bicistronic mRNA encoding both the desired gene and a selectable marker. This library is introduced into cells which are then screened for expression of the desired protein. The use of a selectable marker on the same mRNA increases the efficiency of screening because most transformants also express the encoded gene.

IRESs have also been used in the isolation of mutant mice by homologous recombination in embryonic stem cells. Bicistronic vectors have been designed to produce mRNA encoding the altered protein and the *lacZ* gene, separated by an IRES (panel B). Cells in which this DNA is integrated into the wild-type gene are selected; these cells express a mutant form of the protein and the β-galactosidase enzyme. The latter can be readily assayed in tissue sections by staining with a chromogenic substrate, and since it is encoded on the same mRNA as the targeted gene product, it serves as a marker for mRNA expression.

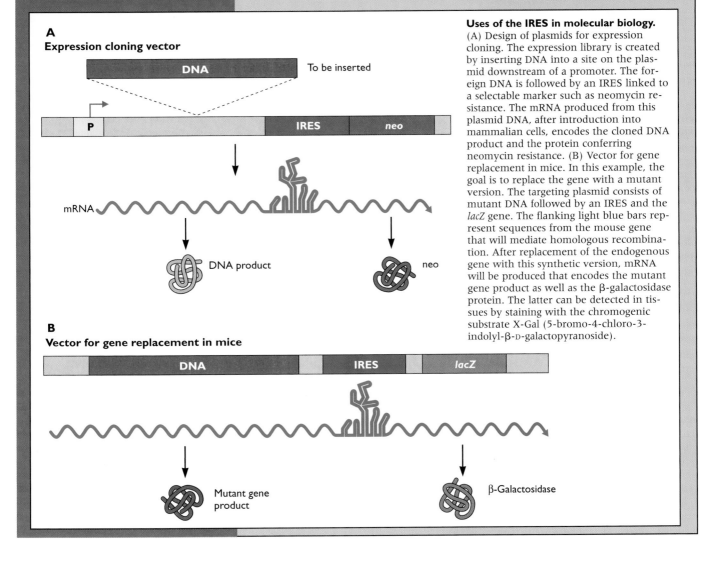

**A**
**Expression cloning vector**

DNA    To be inserted

P    IRES    neo

mRNA

DNA product    neo

**B**
**Vector for gene replacement in mice**

DNA    IRES    *lacZ*

Mutant gene product    β-Galactosidase

**Uses of the IRES in molecular biology.** (A) Design of plasmids for expression cloning. The expression library is created by inserting DNA into a site on the plasmid downstream of a promoter. The foreign DNA is followed by an IRES linked to a selectable marker such as neomycin resistance. The mRNA produced from this plasmid DNA, after introduction into mammalian cells, encodes the cloned DNA product and the protein conferring neomycin resistance. (B) Vector for gene replacement in mice. In this example, the goal is to replace the gene with a mutant version. The targeting plasmid consists of mutant DNA followed by an IRES and the *lacZ* gene. The flanking light blue bars represent sequences from the mouse gene that will mediate homologous recombination. After replacement of the endogenous gene with this synthetic version, mRNA will be produced that encodes the mutant gene product as well as the β-galactosidase protein. The latter can be detected in tissues by staining with the chromogenic substrate X-Gal (5-bromo-4-chloro-3-indolyl-β-D-galactopyranoside).

## Translation in vitro: the reticulocyte lysate and wheat germ extract

Our present understanding of the fundamentals of translation initiation, elongation, and termination, as well as the translation strategies of different viruses, would not be possible without the technique of in vitro translation in cell-free extracts. In this method, cells are lysed and the nuclei are removed by centrifugation. The mRNA is added to the lysate, and the mixture is incubated in the presence of an isotopically labeled amino acid precursor which is incorporated into the translation product.

The ideal extract for in vitro translation has two important properties: high translation efficiency and low levels of protein synthesis in the absence of added mRNA. By the early 1970s, cell-free extracts prepared from Krebs II ascites tumor cells or rabbit reticulocytes were found to translate protein with high efficiency, but the presence of endogenous mRNAs that were also translated complicated the analysis of proteins made from added mRNA. In 1973 Bryan E. Roberts and Bruce M. Patterson reported the development of a cell-free extract from commercial wheat germ that had low background levels of protein synthesis and in which exogenous mRNAs were translated very efficiently. Several years later, Richard J. Jackson and Hugh R. B. Pelham discovered how to eliminate the background in a reticulocyte lysate: by treatment with micrococcal nuclease,

which destroyed the endogenous mRNA. This nuclease requires calcium for its activity, and it was therefore a simple matter of adding EGTA [ethylene glycol-bis(β-aminoethyl ether)-$N,N,N',N'$-tetraacetic acid], a calcium chelator, to the reaction to prevent the degradation of exogenously added mRNA.

Wheat germ extract and reticulocyte lysate are still widely used in translation studies because the cells are abundant and inexpensive and are excellent sources of initiation proteins, leading to high translation efficiencies. Micrococcal nuclease followed by calcium chelation has been successfully used to make mRNA-dependent extracts from many mammalian cell types, although the translation efficiency of such systems does not approach that of wheat germ or reticulocyte lysates. Unfortunately, it has not been possible to consistently prepare translation extracts from normal mammalian tissues, which has hampered the study of tissue-specific translation regulation in virus-infected and uninfected cells.

**Pelham, H. R. B., and R. J. Jackson.** 1970. An efficient mRNA-dependent translation system from reticulocyte lysates. *Eur. J. Biochem.* **67:**247–256.

**Roberts, B. E., and B. M. Patterson.** 1973. Efficient translation of tobacco mosaic virus RNA and rabbit globin 9S RNA in a cell-free system from commercial wheat germ. *Proc. Natl. Acad. Sci. USA* **70:**2330–2334.

---

One host protein identified by this approach is the autoantigen La, which regulates initiation and termination by cellular RNA polymerase III. La protein binds to the 3′ end of the poliovirus IRES. It is present in low concentrations in reticulocyte lysates and when added to them stimulates the function of the poliovirus IRES. However, the amount of protein required is very high and may not be physiologically relevant.

A second IRES-binding protein is polypyrimidine-tract-binding protein, a negative regulator of alternative pre-mRNA splicing. If this protein is removed from a reticulocyte lysate with an RNA affinity column, the activity of the encephalomyocarditis virus IRES is eliminated, but 5′-end-dependent translation is unaffected. The deficiency in IRES function is restored by adding purified protein back to the lysate. However, the depleted lysate still supports the function of the IRES from Theiler's murine encephalomyocarditis virus, another picornavirus, and hepatitis C virus. Furthermore, polypyrimidine-tract-binding

protein does not bind to the IRES of either Bip or Antennapedia mRNAs.

The ribosome-associated poly(rC)-binding proteins interact with multiple sites in the poliovirus IRES. Mutations in the poliovirus 5′ untranslated region that abolish binding of poly(rC)-binding proteins cause decreased translation in vitro. Depletion of poly(rC)-binding proteins from human translation extracts and injection of antibodies against the proteins into *Xenopus* oocytes cause inhibition of translation. These observations indicate that poly(rC)-binding proteins play a role in translation mediated by the poliovirus IRES.

No single protein that is essential for the function of every IRES has been identified. It has been suggested that polypyrimidine-tract-binding protein and La autoantigen act as RNA chaperones, maintaining the IRES in a conformation that allows it to interact directly with eIF4G. IRESs that do not require polypyrimidine-tract-binding protein might fold properly in the absence of an RNA chaperone.

In support of this idea, it has recently been shown that eIF4F can bind directly to the IRES and that eIF4G alone promotes binding of 40S ribosomal subunits to the IRES.

The mechanism of initiation on the IRES of hepatitis C virus (Fig. 11.5C) and classical swine fever virus is dramatically different from initiation mediated by the type 1 or type 2 IRES. The 40S ribosomal subunit can bind di-

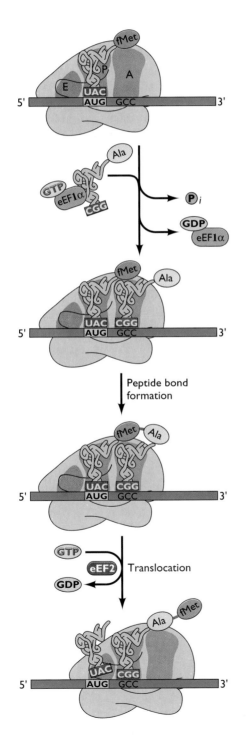

rectly to the IRES in the absence of eIF4A, eIF4B, and eIF4F. Binding is mediated through interaction of the 40S subunit with multiple binding sites on the IRES; no non-canonical translation proteins are required. Addition of only Met-tRNA$_i$, eIF2, and GTP is required to form 48S complexes. Formation of 80S complexes requires only the addition of eIF3. Initiation mediated by these two IRESs therefore differs fundamentally from initiation on all other eukaryotic mRNAs and resembles, in its simple requirements, initiation on prokaryotic mRNAs.

## Elongation and Termination

In the 80S initiation complex, the Met-tRNA$_i$ is bound to the peptidyl (P) site of the ribosome (Fig. 11.6). The **elongation factor** eEF1$\alpha$ bound to GTP brings the next aminoacyl-tRNA to the ribosome, where it enters the aminoacyl (A) site and base pairs with the second codon of the mRNA. Insertion of the tRNA into the A site is accompanied by GTP hydrolysis and release of eEF1$\alpha$. The large ribosomal subunit catalyzes the formation of a peptide bond between the Met and the second amino acid while bound as aminoacyl-tRNAs to the P and A sites, respectively. The ribosome then moves three nucleotides along the mRNA in a step that requires GTP hydrolysis and elongation factor eEF2. This motion brings the peptidyl-tRNA to the P site and the uncharged tRNA to the exit (E) site. The uncharged tRNA is released when a new aminoacyl-tRNA enters the A site. During elongation, the GDP molecule bound to released eEF1$\alpha$ is exchanged for GTP by eEF1$\beta$. mRNAs are usually bound by many ribosomes (**polysomes**), with each ribosome, separated from its neighbors by approximately 100 to 200 nucleotides, synthesizing a polypeptide chain.

The polypeptide chain is elongated until a stop codon (UAA, UAG, or UGA) moves into the A site of the ribosome (Fig. 11.7). The termination codon is usually not recognized by a tRNA but by the **release protein** eRF1, which causes hydrolysis of the bond between the polypep-

**Figure 11.6 Translation elongation.** There are three tRNA-binding sites on the ribosome called peptidyl (P), aminoacyl (A), and exit (E). After the initiating Met-tRNA$_i$ is positioned at the P site, the second aminoacyl-tRNA (alanyl-tRNA is shown) is brought to the A site by eEF1$\alpha$ complexed with GTP. After GTP hydrolysis, eEF1$\alpha$ is released. The peptide bond is then formed, followed by movement of the ribosome three nucleotides along the mRNA, a reaction that requires GTP hydrolysis and eEF2. This moves the peptidyl (Met-Ala) tRNA to the P site, and the uncharged tRNA to the E site. The A site is now empty, ready for another aminoacyl-tRNA. Adapted from G. M. Cooper, *The Cell: a Molecular Approach* (ASM Press, Washington, D.C., and Sinauer Associates, Sunderland, Mass., 1997), with permission.

**Figure 11.7 Translation termination.** When a termination codon is encountered at the A site, it is usually recognized by a release factor (eRF) instead of a tRNA. The peptide chain is then released, followed by dissociation of tRNA and ribosome from the mRNA. Adapted from G. M. Cooper, *The Cell: a Molecular Approach* (ASM Press, Washington, D.C., and Sinauer Associates, Sunderland, Mass., 1997), with permission.

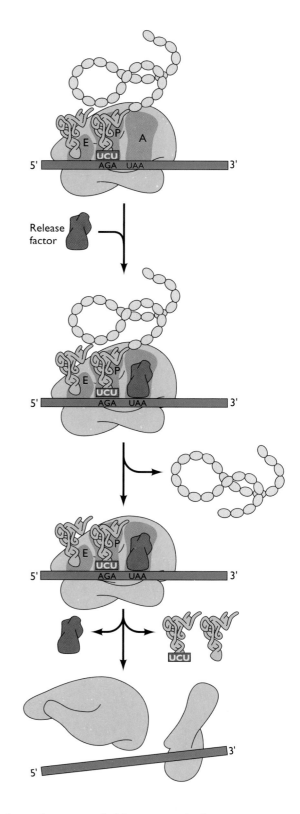

tide and the tRNA at the P site. As a result, the polypeptide is released from the ribosome, and the tRNA falls away from the mRNA. Exactly when ribosomes are released from the mRNA is not clear; this reaction may occur at the termination codon, or ribosomes may continue to transit through the 3' noncoding region to the poly(A) tail. Once ribosomes are released from the mRNA, dissociation into 40S and 60S subunits is favored by the binding of eIF1A, eIF3, and eIF6.

## The Diversity of Viral Translation Strategies

A variety of unusual translation strategies have evolved to expand the coding capacity of viral genomes, which are relatively small, and to allow the synthesis of multiple polypeptides from a single RNA genome (Fig. 11.8). All were discovered in virus-infected cells and subsequently shown to operate also during translation of cellular mRNAs. Nontranslational solutions for maximizing the number of proteins encoded in viral genomes have been discussed in other chapters and include the synthesis of multiple subgenomic mRNAs, mRNA splicing, and RNA editing.

### Polyprotein Synthesis

One strategy allowing the synthesis of multiple polypeptides from a single RNA genome is to synthesize from a single mRNA a polyprotein precursor, which is then proteolytically processed to form functional viral proteins.

Picornaviruses make the most dramatic use of protein processing (Fig. 11.9A): nearly the entire (+) strand genome RNA is translated into a single large polyprotein. Processing of this precursor is carried out by two virus-encoded proteases, 2A$^{pro}$ and 3C$^{pro}$, which cleave between Tyr and Gly and between Gln and Gly, respectively. In both cases, flanking amino acid residues control the extent of cleavage so that not all Tyr-Gly and Gln-Gly pairs in the polyprotein are processed. These two proteases are active in the nascent polypeptide and release themselves from the polyprotein by self-cleavage. Consequently, the polyprotein is not observed in infected cells because it is processed as soon as the proteases' coding sequences have been translated. After the proteases have been released, they cleave the polyprotein in *trans*.

One advantage of this strategy is that gene expression can be controlled by the rate and extent of polyprotein processing. In addition, alternative utilization of cleavage

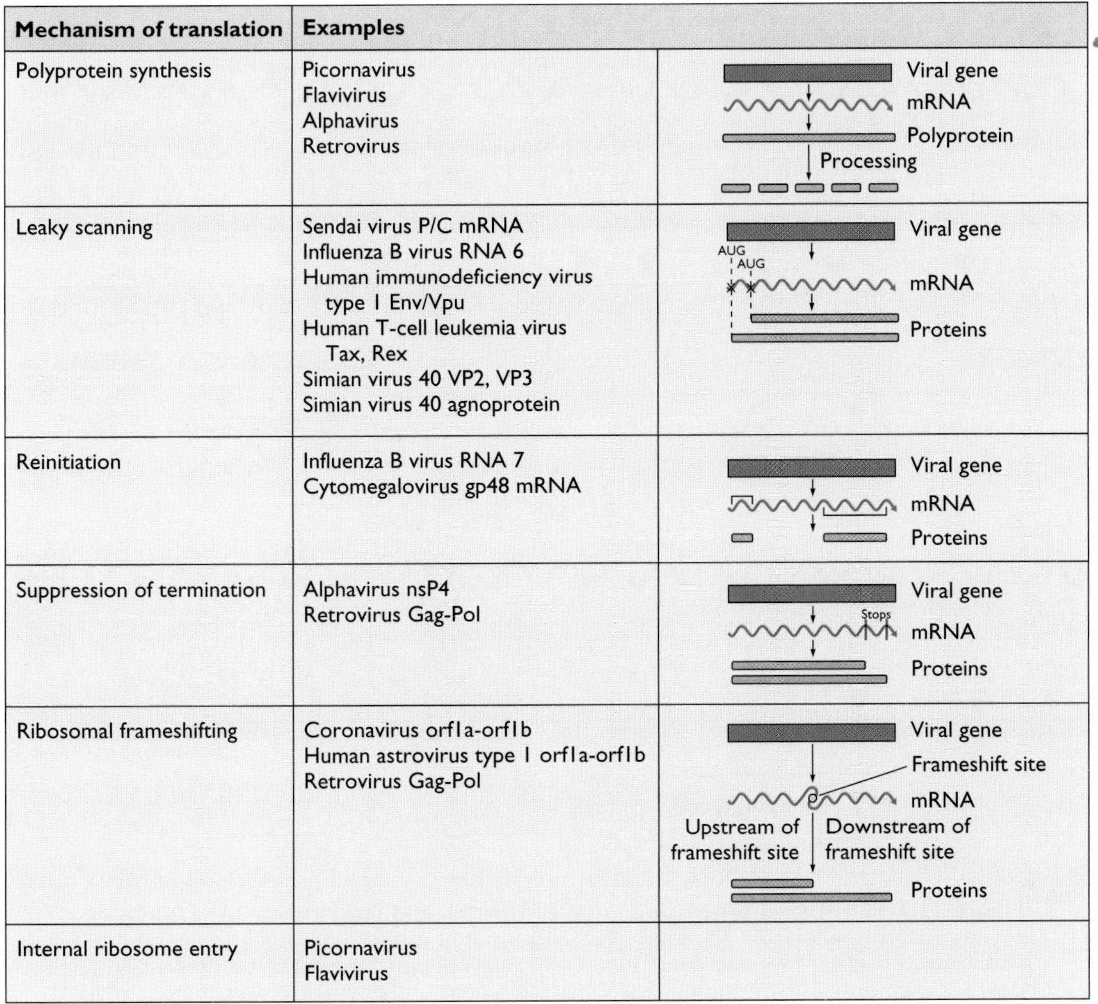

| Mechanism of translation | Examples | |
|---|---|---|
| Polyprotein synthesis | Picornavirus<br>Flavivirus<br>Alphavirus<br>Retrovirus | Viral gene<br>mRNA<br>Polyprotein<br>Processing |
| Leaky scanning | Sendai virus P/C mRNA<br>Influenza B virus RNA 6<br>Human immunodeficiency virus type I Env/Vpu<br>Human T-cell leukemia virus Tax, Rex<br>Simian virus 40 VP2, VP3<br>Simian virus 40 agnoprotein | AUG / AUG<br>Viral gene<br>mRNA<br>Proteins |
| Reinitiation | Influenza B virus RNA 7<br>Cytomegalovirus gp48 mRNA | Viral gene<br>mRNA<br>Proteins |
| Suppression of termination | Alphavirus nsP4<br>Retrovirus Gag-Pol | Viral gene<br>Stops<br>mRNA<br>Proteins |
| Ribosomal frameshifting | Coronavirus orf1a-orf1b<br>Human astrovirus type I orf1a-orf1b<br>Retrovirus Gag-Pol | Viral gene<br>Frameshift site<br>mRNA<br>Upstream of frameshift site \| Downstream of frameshift site<br>Proteins |
| Internal ribosome entry | Picornavirus<br>Flavivirus | |

**Figure 11.8 The diversity of viral translation strategies.** Examples of different mechanisms of translation that occur in virus-infected cells are listed. On the right are schematic diagrams of the strategies used for making multiple proteins from one gene.

sites can produce proteins with different activities. For example, the poliovirus protease 3C$^{pro}$ does not efficiently process the capsid protein precursor P1. Rather the 3C$^{pro}$ precursor, 3CD, is required for processing of P1. By regulating the amount of 3CD produced, the extent of capsid protein processing can be controlled. Because 3CD and 3C$^{pro}$ process Gln-Gly pairs in the remainder of the polyprotein with the same efficiency, an interesting question is why 3CD, which also contains 3D$^{pol}$ protein, is further processed to produce 3C$^{pro}$ (Fig. 11.9A). The answer is that 3CD protein, while proteolytically active, does not possess RNA polymerase activity and therefore some of it must be cleaved to allow RNA replication.

Some viral precursor proteins are processed by cellular proteases. The genome of flaviviruses encodes an open reading frame of more than 10,000 bases (Fig. 11.9B). In infected cells this mRNA is translated into a polyprotein precursor processed by a viral serine protease and by host signal peptidase. The latter enzyme is located in the endoplasmic reticulum, where it removes the signal sequence from proteins translocated into the lumen (further discussed in chapter 12). The viral proteins processed by the host signal peptidase must therefore be inserted into the endoplasmic reticulum.

If the final cleavage products of polyprotein processing were all used, this process would truly be efficient and

**A**

Viral (+) strand genome

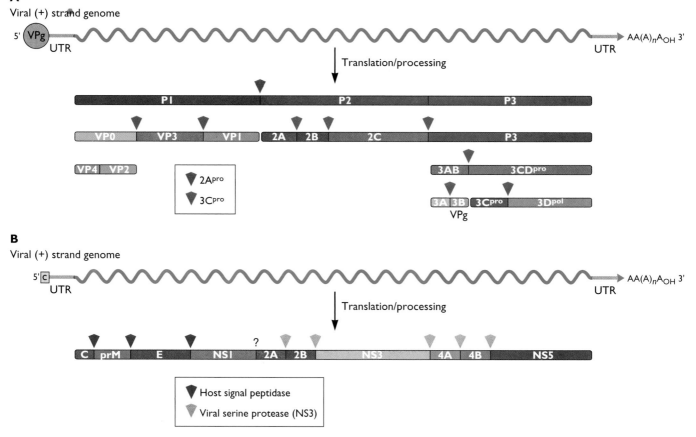

**Figure 11.9 Polyprotein processing in enteroviruses and flaviviruses.** (A) Protein processing map of poliovirus. The viral RNA is translated into a long precursor polyprotein that is processed by two viral proteases, $2A^{pro}$ and $3C^{pro}$, to form viral proteins. Cleavage sites for each protease are shown. (B) Cleavage map of a flavivirus. Processing of the flavivirus precursor polyprotein is carried out either by the host signal peptidase, which occurs in the lumen of the endoplasmic reticulum, or by the viral protease NS3.

widely employed. Unfortunately, polyprotein cleavage is not an ideal expression strategy. For example, some viral proteins (e.g., poliovirus capsid proteins) are produced at levels that are higher than needed. Other mechanisms have therefore evolved to allow translation of all the open reading frames in viral genomes.

## Leaky Scanning

Although the vast majority of eukaryotic mRNAs are monocistronic (Fig. 11.1), some viral mRNAs that encode two proteins in overlapping reading frames have been identified. The P/C gene of Sendai virus is the model for genes that encode mRNAs with such translational flexibility (Fig. 11.10). P protein is translated from an open reading frame beginning with an AUG codon at nucleotide 104. Two additional P proteins with different C-terminal

domains are produced by the paramyxoviral mRNA editing mechanism described in chapter 10. C proteins are produced from a different reading frame, which begins at nucleotide 81, and are completely different from P proteins. No less than four C proteins (called C', C, Y1, and Y2) are produced by translation beginning at four in-frame initiation codons. The first start site is at an unusual ACG codon, and the third, fourth, and fifth are at AUG codons; the result is a nested set of proteins with a common C terminus. mRNA editing has no effect on these proteins because their coding sequence terminates before the editing site.

The first three initiation sites on P/C mRNA are likely to be arranged to permit translation by leaky scanning. The first start site, $ACG^{81/C'}$, is surrounded by a good initiation context but is inefficient because of the unusual start codon. The second initiator, $AUG^{104/P}$, is in a poor initiation

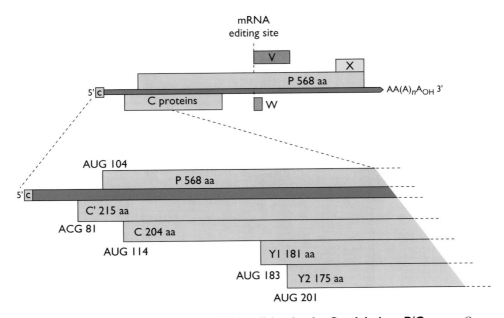

**Figure 11.10 Leaky scanning and mRNA editing in the Sendai virus P/C gene.** Open reading frames are shown as boxes. The P protein open reading frame is shown as a brown horizontal box, and the C protein open reading frame is shown as a blue box. An enlargement of the 5′ end of the mRNA is shown below, indicating the different start sites for four of the C proteins. The site at which G residues are added during mRNA synthesis is shown above; addition of one or two G residues produces the V or W carboxy terminus of P protein. Adapted from J. Curran et al., *Semin. Virol.* **8:**351–357, 1997, with permission.

context (CGC<u>AUG</u>G), while the third, AUG$^{114/C}$, is in a better context (AAG<u>AUG</u>C). Because initiation at the ACG codon is inefficient, some ribosomes can bypass it and initiate at AUG$^{104/P}$; as the context at this AUG is poor, some ribosomes can also find their way to AUG$^{114/C}$, which has a better initiation context. Consistent with this hypothesis, mutagenesis of ACG$^{81/C'}$ to AUG abolishes translation at AUG$^{104/P}$ and AUG$^{114/C}$. When successive initiation codons are used in leaky scanning, they are increasingly efficient as start sites.

The last two C-protein initiation codons, AUG$^{183/Y1}$ and AUG$^{201/Y2}$, are not likely to be translated by leaky scanning because they are in the poorest contexts of all five. Furthermore, mutagenesis of ACG$^{81/C'}$ to AUG has no effect on the synthesis of Y1 and Y2 proteins. Rather, Y1 and Y2 proteins are believed to be translated by ribosome shunting. The initiation complexes enter at the capped 5′ end, bypass the intervening sequences, and are delivered at or near the initiating AUG codons. The precise location of the bypassed sequence on the P/C mRNA has not been defined. Ribosome shunting on the P/C mRNA is dependent on eIF4E and thus differs from shunting on adenovirus late mRNAs, which has a reduced requirement for eIF4E. The 5′ end of the latter is unstructured, which suggests that P/C mRNA might have more RNA secondary struc-

ture in this area, which would increase the dependence of translation on eIF4E.

An interesting question is how the different translation strategies of P/C mRNA are coordinated such that, for example, shunting does not dominate at the expense of translation of upstream AUG codons. The answer to this question is not known, but Y-protein expression relative to that of the other C proteins varies in different cell lines. This result suggests that cellular proteins might regulate ribosome shunting on P/C mRNA, although no such protein has yet been identified.

Translation of overlapping reading frames also occurs in other viral mRNAs (Fig. 11.8). Influenza viruses are classified into three types, A, B, and C; most of the previous discussion in this textbook has concerned work on influenza A virus. The mRNA derived from influenza B virus RNA segment 6 encodes two different proteins, NB and NA, in overlapping reading frames. NB protein translation is initiated at the 5′-proximal AUG codon, while NA protein is initiated from an AUG codon four nucleotides downstream. Although the sequence surrounding the first AUG codon is close to the optimal consensus, both proteins accumulate to approximately equal amounts in infected cells. This observation has been interpreted as support of the model of random ribosome movement in which the

40S subunit moves in both the 5′ → 3′ and 3′ → 5′ directions until the initiating AUG codon is encountered (see "How does the 40S subunit move on the mRNA?" above).

## Reinitiation

Reinitiation is another strategy for producing two proteins from a single mRNA. It is rare in eukaryotes but very common in prokaryotic cellular and viral mRNAs. About 10% of eukaryotic mRNAs contain upstream AUG codons. Many are followed by short open reading frames (upstream open reading frames) that terminate before the downstream open reading frame (Fig. 11.11). The upstream open reading frames may be translated, with reinitiation occurring at the downstream AUG. For example, a 22-amino-acid peptide is synthesized from an open reading frame in the 5′ region of cytomegalovirus gp48 virion glycoprotein mRNA. Other examples are found in retroviruses and in cellular mRNAs such as those encoding *S*-adenosylmethionine decarboxylase and fibroblast growth factor 5. The presence of upstream open reading frames can affect translation of the downstream open reading frame (Box 11.5). The extent of regulation depends on many factors, such as the sequence context of the upstream open reading frame AUG initiation codons, the presence of RNA secondary structure downstream of the upstream open reading frame, and the distance between the upstream open reading frame and the downstream initiating AUG.

Most examples of translation reinitiation are those in which such short upstream open reading frames precede the main open reading frame. Reinitiation involving longer, overlapping reading frames occurs on mRNA of influenza B virus RNA 7, which encodes two proteins, M1 protein and BM2 protein (Fig. 11.11). M1 protein is translated from the 5′-proximal AUG codon, while the BM2 protein AUG initiation codon is part of the termination codon for M1 protein (UAAUG). Translation of the BM2 open reading frame is dependent on the synthesis of M1 protein, as deletion of the M1 AUG codon abrogates BM2 synthesis and is therefore likely to occur by reinitiation. This process is not efficient, as BM2 synthesis is about 25% of M1 synthesis. Translation of the BM2 open reading frame does not occur by ribosomes scanning from the 5′ end of the RNA; this would require bypassing 21 AUG codons and does not occur if the M1 and BM2 open reading frames are fused by mutagenesis. This observation also demonstrates that BM2 synthesis does not occur by internal ribosome entry or ribosome shunting.

## Suppression of Termination

Suppression of termination occurs during translation of many viral mRNAs as a means of generating a second protein with an extended carboxy terminus. In Moloney murine leukemia virus, the *gag* and *pol* genes are encoded in a single mRNA and separated by an amber termination codon, UAG (Fig. 11.12). Infected cells contain a polyprotein precursor called Pr180*gag-pol*. This precursor is synthesized by translational suppression of the amber termination codon for the *gag* gene. The efficiency of suppression is about 4 to 10%. The Gag-Pol precursor is subsequently processed proteolytically to liberate the Gag protein and Pol enzyme. Without this suppression strategy, this essential enzyme would not be produced. In a similar way, translational suppression of a different termination codon, UGA, is required for the synthesis of nsP4 of alphaviruses (Fig. 11.12). In this case, the efficiency of suppression is about 10% of the synthesis of the normally terminated nsP3 polypeptide. Since nsP4 encodes the RNA-dependent RNA polymerase, suppression is essential for viral RNA replication. Translational suppression in eukaryotic mRNAs is extremely rare but has been observed, e.g., in β-globin mRNA.

Translation suppression can be mediated by two types of tRNAs. Suppressor tRNAs that can recognize termination codons and insert a specific amino acid have been detected in eukaryotic cells. One example is a suppressor tRNA that inserts selenocysteine, the 21st amino acid, in place of a UGA codon. Such decoding occurs at 100% efficiency. Normal tRNAs may also misread termination codons, but the efficiency is usually low, e.g., 10%. The misreading of the amber codon in Moloney murine leukemia virus Gag protein for a Gln codon is an example.

The nucleotide sequence 3′ of the termination codon plays an important role in the efficiency of translational suppression. In prokaryotes, the nucleotide next to the termination codon is highly influential. In eukaryotes, the

**Figure 11.11 Reinitiation of translation.** (Top) Some mRNAs contain one or more short upstream open reading frames (uORFs) that may be translated. Expression of the longer, downstream ORF depends on reinitiation. (Bottom) mRNA produced from influenza B virus RNA segment 6 encodes two proteins, M1 and BM2. The initiation AUG codon for BM2 overlaps the termination codon of M1. Synthesis of BM2 occurs by reinitiation.

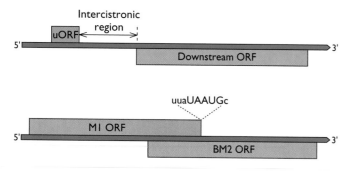

BOX 11.5

*Translational control of yeast GCN4 by upstream open reading frames*

The best characterized example of reinitiation is on *GCN4* mRNA of the yeast *S. cerevisiae*. The *GCN4* gene product activates the transcription of at least 40 different genes that encode biosynthetic enzymes. When nutrients are abundant, four short upstream open reading frames in the 5' untranslated region of this mRNA act in *cis* to prevent ribosomes from initiating at the downstream start site. Under conditions of nutrient deprivation, the upstream open reading frames are unable to prevent ribosomes from translating at the *GCN4* AUG codon. The mechanism of action involves the phosphorylation state of eIF2α (see figure). Under all conditions, all ribosomes trans-late the first upstream open reading frame; half of them continue scanning as 40S ribosomal subunits. When nutrient supplies are adequate, eIF2α is not phosphorylated; ribosomal 40S subunits rapidly rebind to eIF2–GTP–Met-tRNA$_i$ and therefore reinitiate at the second, third, and fourth upstream open reading frames, after which they dissociate and fail to reach the *GCN4* start site. When nutrients are limiting, eIF2α is phosphorylated, reducing the level of eIF2–GTP–Met-tRNA$_i$. As a result, 40S subunits scan the mRNA longer before they bind to a Met-tRNA$_i$, increasing the likelihood that they will reach the *GCN4* start site before initiating translation.

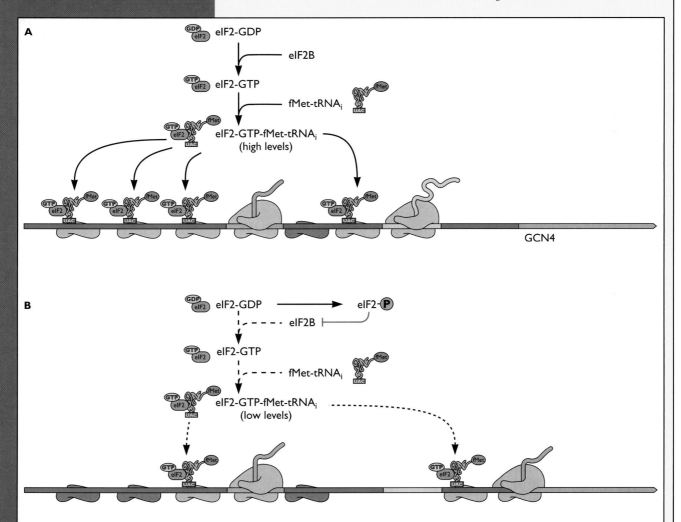

**Model for translational control of yeast *GCN4*.** A schematic of *GCN4* mRNA is shown, with upstream open reading frames 1 (uORF1) (blue) and 4 (uORF4) (orange) and the *GCN4* coding sequence (pink). The gray 40S ribosomal subunits are associated with eIF2, GTP, and Met-tRNA$_i$ and are competent to reinitiate; olive subunits are not associated with this ternary complex and cannot reinitiate. Under nonstarvation conditions (A), high levels of eIF2 are available to form the ternary complex, which reassembles with 40S ribosomes after scanning uORF1, allowing reinitiation at uORF4. Under starvation conditions (B), eIF2α is phosphorylated, leading to limiting amounts of eIF2-GTP. As a result, many 40S subunits cannot initiate at uORF4 and acquire the ternary complex after this open reading frame and initiate translation at *GCN4*. Adapted from A. G. Hinnebusch, Fig. 1, p. 202, *in* J. W. B. Hershey et al., *Translational Control* (Cold Spring Harbor Laboratory Press, Cold Spring Harbor, N.Y., 1996), with permission.

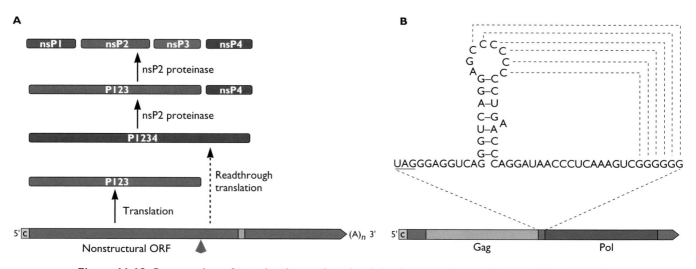

**Figure 11.12 Suppression of termination codons in alphaviruses and retroviruses.** (A) Suppression of termination during the synthesis of alphavirus P123 to produce nsP4, the RNA-dependent RNA polymerase. The termination codon is shown by a red arrowhead. Adapted from J. H. Strauss and E. G. Strauss, *Microbiol. Rev.* **58:**491–562, 1994, with permission. (B) Structure of the termination site between Gag and Pol of Moloney murine leukemia virus. The stop codon that terminates synthesis of Gag is underlined; it is followed by a pseudoknot that is important for suppression of termination.

signals range from very simple to complex. In Sindbis virus, efficient suppression of the UGA codon requires only a single C residue 3' of the termination codon. The effect of this nucleotide on Sindbis virus suppression may be explained in three ways. The 3' nucleotide might influence the recognition of termination codons by release proteins; it might affect the interaction of misreading tRNAs with the codon via stacking energy; or the misreading tRNA and a tRNA recognizing the next codon might interact, affecting the efficiency of elongation and suppression.

In contrast, readthrough of the UAG codon in Moloney murine leukemia virus mRNA requires a purine-rich sequence 3' of the termination codon, as well as a pseudoknot structure further downstream (see chapter 6 for a description of pseudoknots). Because of these differences, it is likely that readthrough can be mediated by different mechanisms. In the case of suppression in Moloney virus, it has been suggested that the pseudoknot causes the ribosome to pause and, together with the eight-nucleotide, purine-rich segment of the suppression signal, allow the suppressor tRNA to compete with release factor at the suppression site.

The RNA sequences and structures required for suppression are not found in most cellular mRNAs. For example, there is a strong bias against cytidine residues at the 3' end of UGA termination codons in cellular mRNAs. This observation explains why suppression of termination is far more prevalent in translation of viral mRNAs than in cellular mRNAs.

## Ribosomal Frameshifting

Ribosomal frameshifting is a process by which, in response to signals in mRNA, ribosomes move into a different reading frame and continue translation in that new frame. It was discovered in cells infected with Rous sarcoma virus and has since been described for many other viruses, including additional retroviruses, eukaryotic (+) strand RNA viruses, and herpes simplex virus. There is only one example of frameshifting in a mammalian gene, that encoding an antizyme protein that binds to ornithine decarboxylase and targets it for degradation by proteasomes.

In retroviruses, the *gag* and *pol* genes may be separated by a stop codon (Fig. 11.12), or they may be in different reading frames, with *pol* overlapping *gag* in the −1 direction (Fig. 11.13A). During translation of Rous sarcoma virus Gag protein, ribosomes frameshift before reaching the Gag stop codon and continue translating Pol, such that a Gag-Pol fusion is produced at about 5% of the level of Gag protein. Studies on the requirements for frameshifting in retroviruses and coronaviruses have identified two essential components: a "slippery" homopolymeric sequence, which is a heptanucleotide stretch with two homopolymeric triplets of the form X-XXY-YYZ (e.g., in Rous sarcoma virus A-AAU-UUA), and an RNA secondary structure, usually a pseudoknot, five to eight nucleotides downstream. These observations led to proposal of the tandem shift model for frameshifting, in which two tRNAs in the zero reading frame (X-XXY-YYZ) slip back by one

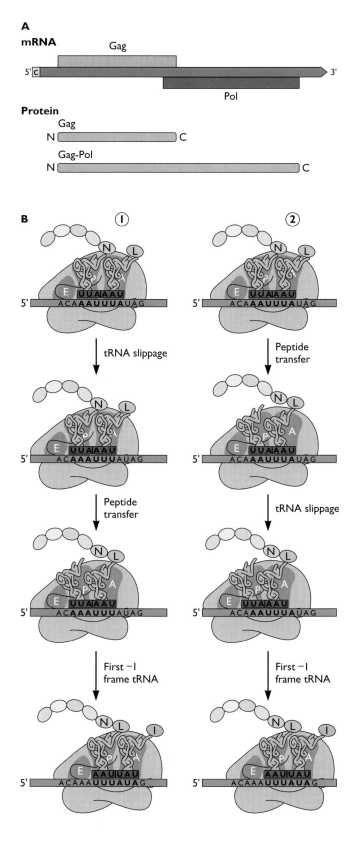

**A**
**mRNA**

**Protein**

**B**

**Figure 11.13 Frameshifting in a retrovirus.** (A) Structure of open reading frames. Rous sarcoma virus mRNA encodes Gag and Pol proteins in reading frames that overlap by −1. Normal translation and termination produce the Gag protein; ribosomal frameshifting to the −1 frame results in the synthesis of a Gag-Pol fusion protein. (B) Tandem shift model for −1 frameshifting. Two different models are illustrated. In model 1, slippage of the two tRNAs occurs after aminoacyl-tRNA enters the A site but before peptide transfer. In model 2, slippage occurs after peptide transfer, when tRNAs occupy the hybrid E-P and P-A sites. In both cases, slippage allows the tRNAs to form only two base pairs with the mRNA. The site shown is that of Rous sarcoma virus. One-letter amino acid codes are used. Adapted from P. J. Farabaugh, *Microbiol. Rev.* **60:**103–134, 1996, with permission.

nucleotide during the frameshift to the −1 phase (XXX-YYY). Each tRNA base pairs with the mRNA in the first two nucleotides of each codon (Fig. 11.13B). The peptidyl-tRNA is transferred to the P site, the −1 frame *pol* codon is decoded, and translation continues to produce the fusion protein. In this model, slippage occurs before peptide transfer, with the peptidyl- and aminoacyl-tRNAs bound to the P and A sites. However, it is possible that the shift occurs after peptide transfer but before translocation of the tRNAs (Fig. 11.13B). These two models cannot be distinguished by mutagenesis or by the sequence of the protein products.

The function of the pseudoknot is not known, but it is believed to cause the ribosome to pause over the slippery sequence, increasing the probability that realignment to the −1 reading frame will occur. Some viruses contain simple stem-loop structures rather than pseudoknots in this position, but the highest levels of frameshifting are associated with the presence of the latter structure. However, when the pseudoknot of a coronavirus is replaced with a stem-loop structure, frameshifting is abolished but ribosomal pausing still occurs. Thus, the pseudoknot might not simply act as a barrier to ribosome movement; perhaps structural features of the pseudoknot are also important to promote frameshifting. It has been suggested that cellular proteins might interact with the pseudoknot to bring about frameshifting, but such proteins have not been identified.

## Regulation of Translation during Viral Infection

Alterations in the cellular translation apparatus are commonplace in virus-infected cells. As part of the antiviral defense, the cell responds to viral infection with measures designed to inhibit protein synthesis and thereby limit virus production. Many viruses encode proteins or nucleic acids that neutralize this response and are targeted to restoring translation and maximizing virus replication. In addition, many viruses encode gene products that modify the host translation apparatus to favor synthesis of viral proteins over those of the cell. As a result, the entire synthetic capability of the cell is turned to the production of new virus particles, which should enhance virus yield and perhaps accelerate replication. This supposition is supported by the growth defects of different viral mutants that cannot inhibit cellular translation. These cellular and viral modifications of the translation apparatus are usually aimed at the initiation stages, which are the rate-limiting steps in translation.

## Inhibition of Translation Initiation as a Cellular Defense against Viral Infection

### Phosphorylation of eIF2α

While higher organisms respond to viral infection with an immune response, there are more rapid first-line defenses, including the induction of interferons, a large family of secreted proteins produced by vertebrates in response to viral infection (discussed in chapter 14). Interferons diffuse to neighboring cells, bind to cell surface receptors, and activate a signal transduction pathway that results in transcription of over 100 cellular genes and the establishment of an "antiviral state." Two interferon-induced genes encode enzymes, protein kinase, RNA-activated (**Pkr**), and **RNase L**, that effectively prevent association of mRNA with polysomes and hence translation. RNase L degrades RNA and will not be considered further, while Pkr phosphorylates eIF2α, inhibiting trans-

lation initiation. Because the block to translation is global, the infected cell may die, but by slowing down viral replication the organism may be spared. Another possibility is that the virus may be catabolized by the cell if it does not replicate.

Pkr is a serine-threonine protein kinase composed of an N-terminal regulatory domain and a C-terminal catalytic domain (Fig. 11.14). Small amounts of an inactive form of Pkr are present in most uninfected mammalian tissues; transcription of its gene is induced 5- to 10-fold by interferon. Pkr is activated by the binding of dsRNA to two dsRNA-binding motifs (Fig. 11.14) at the N terminus of the protein. Such dsRNA is produced in cells infected by either DNA or RNA viruses. It is synthesized during replication of RNA viral genomes and presumably by divergent transcription of double-stranded viral DNA genomes. In addition, it is present in the genomes of some single-stranded RNA viruses. Activation is accompanied by autophosphorylation of Pkr. Low concentrations of dsRNA activate Pkr, but high concentrations are inhibitory, leading to the suggestion that one molecule of Pkr phosphorylates another while bound to the same molecule of dsRNA (Fig. 11.15). Phosphorylation is believed to induce conformational changes in the enzyme, thereby rendering the kinase active without further need for dsRNA. Pkr molecules may also interact via a dimerization region (Fig. 11.14), which could lead to activation of the kinase independently of dsRNA.

Activated Pkr can phosphorylate many substrates, but our discussion will focus on eIF2α, which is modified on serine 51. This initiation protein is part of the complex that also contains GTP and Met-tRNA$_i$ (Fig. 11.3). After GTP hydrolysis, the GDP moiety of eIF2-GDP must be exchanged for GTP to permit the binding of another Met-tRNA$_i$. This exchange is carried out by eIF2B (Fig. 11.16). When the alpha subunit of eIF2 is phosphorylated, eIF2-GDP binds eIF2B with such high affinity that it is effectively trapped; recycling of eIF2 stops, and ternary complexes are depleted. eIF2B is less abundant than eIF2, and phosphorylation of about 10 to 40% of eIF2 (depending on the cell type and the relative levels of eIF2 and eIF2B) results in the complete sequestration of eIF2B, leading to a complete block in protein synthesis. As viral translation is also impaired, the synthesis of new virus particles is diminished.

The function of Pkr in uninfected cells is not well understood. Overexpression of activated Pkr in HeLa cells induces apoptosis, whereas synthesis of catalytically inactive mutant forms produce cells that are tumorigenic. Alteration in the levels of essential nutrients such as amino acids and glucose causes eIF2α phosphorylation, but it is not known whether Pkr participates in these responses.

**Figure 11.14 Schematic structure of dsRNA-activated protein kinase.** A schematic diagram of the protein is shown, with the locations of the regulatory domain, dsRNA-binding sites, and the catalytic domain indicated. Sites where viral gene products interfere with Pkr are shown. These include antagonism of dsRNA binding (adenovirus VA RNA I, Epstein-Barr virus EBERs, influenza virus NS1, vaccinia virus E3L, reovirus σ3, and human immunodeficiency virus type 1 TAR RNA), dimerization (cellular P58$^{IPK}$ and hepatitis C virus NS5A), and substrate interaction (vaccinia virus K3L and human immunodeficiency virus type 1 Tat). Adapted from M. Gale and M. G. Katze, *Pharmacol. Ther.* **78:**29–46, 1998, with permission.

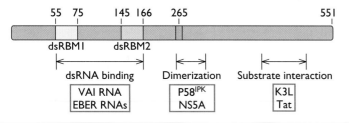

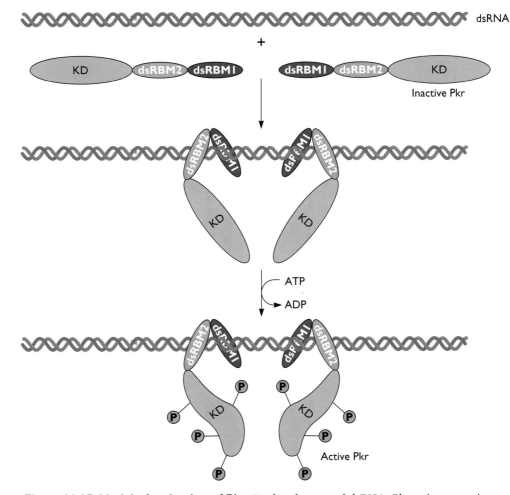

**Figure 11.15 Model of activation of Pkr.** In the absence of dsRNA, Pkr exists as an inactive monomer. When two or more molecules of inactive Pkr bind to one dsRNA molecule, cross-phosphorylation occurs through the physical proximity of the molecules. Phosphorylation is believed to cause a conformational change in the catalytic domain (KD) to allow phosphorylation of other substrates, including eIF2α. Adapted from J. W. B. Hershey et al. (ed.), *Translational Control* (Cold Spring Harbor Laboratory Press, Cold Spring Harbor, N.Y., 1996), with permission.

Eukaryotic cells contain several protein inhibitors of the enzyme, including P58[IPK], which prevents Pkr activation by blocking dimerization (see "Inhibition of Pkr dimerization"). P58[IPK] is usually associated with an inhibitor (I-P58[IPK]) that prevents it from binding to Pkr. These findings indicate that regulation of Pkr activity is important in mammalian cells, although the specific situations where the kinase might be active or inactive remain to be elucidated.

In addition to Pkr, eIF2α may also be phosphorylated by two other protein kinases, heme-regulated protein kinase (in erythroid cells), which is activated under conditions of low iron, hemin, or heat shock, and Gcn2 (in yeast),

which is activated by nitrogen starvation. In all three cases, the resulting inhibition of translation is an attempt to prevent damage to the organism. Phosphorylation of eIF2 is therefore an extremely important global mechanism for reducing rates of protein synthesis.

### Viral Regulation of Pkr

Most viral infections induce activation of Pkr and consequent phosphorylation of eIF2α. Global inhibition of translation is clearly a threat to successful virus replication. At least five different viral mechanisms to block Pkr activation or to stop activated Pkr from inhibiting translation can be distinguished (Table 11.2).

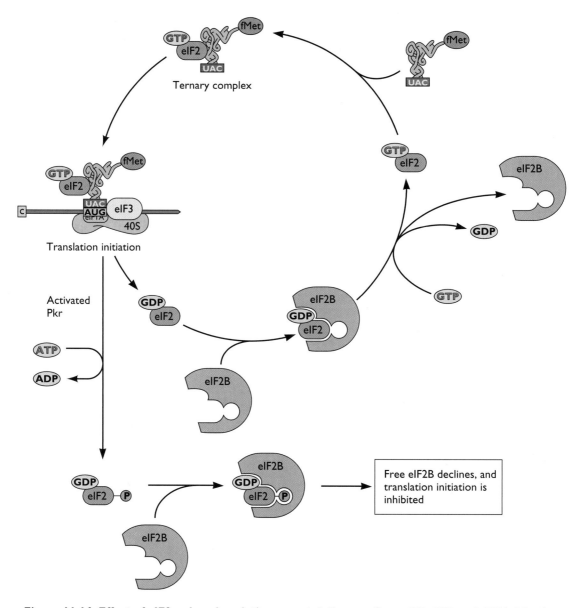

**Figure 11.16 Effect of eIF2α phosphorylation on catalytic recycling.** eIF2-GTP and tRNA-Met$_i$ form the ternary complex required for translation initiation. During initiation, GTP is hydrolyzed to GDP, and in order for initiation to continue, eIF2 must be recharged with GTP. This recycling is accomplished by eIF2B, which exchanges GTP for GDP on eIF2. When eIF2 is phosphorylated on the α subunit, it binds irreversibly to eIF2B, preventing the latter from carrying out its role in recycling GDP. As a result, the concentration of eIF2-GTP declines and translation initiation is inhibited.

**Inhibition of dsRNA binding.** The 166-nucleotide adenovirus VA RNA I, which accumulates to massive concentrations (up to $10^9$ copies per cell) late in infection following transcription of the viral gene by RNA polymerase III, is a potent inhibitor of Pkr. An adenovirus mutant that does not produce VA RNA I is growth impaired. In cells infected with this mutant, eIF2α becomes extensively phos-

phorylated, causing global translation inhibition. VA RNA I binds the dsRNA-binding region of Pkr and blocks activation by dsRNA. It has been suggested that binding of VA RNA I to Pkr blocks interaction with authentic dsRNA and hence prevents activation of the kinase. In another model, translation of viral mRNAs is not impaired because bound VA RNA I inactivates Pkr in the vicinity. However, cellular

**Table 11.2** Targets and mechanisms of viral Pkr inhibitors

| Target(s) | Virus | Inhibitor | Mechanism |
|---|---|---|---|
| dsRBM[a] | Adenovirus | VA RNA I | Inhibition by dsRNA analog |
| | Vaccinia virus | E3L | Sequesters dsRNA |
| | Influenza virus | NS1 | Sequesters dsRNA |
| | Reovirus | σ3 | Sequesters dsRNA |
| | Human immunodeficiency virus type 1 | TAR RNA | Competitive inhibitor of dsRNA binding |
| Dimerization | Influenza virus | P58$^{IPK}$ | Disrupts dimerization |
| | Hepatitis C virus | NS5A | Disrupts dimerization |
| Substrate interaction | Vaccinia virus | K3L | Blocks Pkr-substrate interaction |
| | Human immunodeficiency virus type 1 | Tat | Blocks Pkr-substrate interaction |
| Expression and stability | Poliovirus | Unknown | Degrades Pkr |
| | Human immunodeficiency virus type 1 | Tat | Reduces Pkr expression |
| Regulation of eIF2α | Herpes simplex virus | $_{γ1}$34.5 | Binds phosphatase, may direct to eIF2α |
| | Simian virus 40 | T antigen | Downstream of eIF2α? |

[a]RBM, RNA-binding motif.

mRNAs have no such guardian, and their translation ceases because of local inhibition of eIF2α recycling. Epstein-Barr virus and human immunodeficiency virus type 1 encode small RNAs that inhibit Pkr activation in vitro, but whether they function in a similar manner in infected cells is not clear.

While adenovirus VA RNA I binds Pkr and blocks activation by dsRNA, vaccinia virus encodes a protein that sequesters dsRNA. The vaccinia virus E3L protein contains the same dsRNA-binding region as Pkr; it binds dsRNA and prevents it from activating the kinase. Deletion of the gene encoding the E3L protein renders the virus more sensitive to interferon and results in larger quantities of active Pkr in infected cells. The influenza virus NS1 protein, the reovirus σ3 protein, and the human immunodeficiency virus type 1 TAR RNA also sequester dsRNA.

**Inhibition of Pkr dimerization.** As mentioned previously, Pkr contains a region that can mediate dimerization and therefore activation of the protein in the absence of dsRNA (Fig. 11.14). Influenza virus infection induces the release of P58$^{IPK}$ from its inhibitor, I-P58$^{IPK}$, enabling it to bind to the dimerization domain and block Pkr phosphorylation and activation. NS5A protein of hepatitis C virus is believed to block Pkr activation in the same manner, based on in vitro studies. Hepatitis C virus infections are treated with interferon, and individuals who fail to respond to therapy harbor viruses with mutations in a specific region of NS5A that may mediate its ability to bind Pkr and prevent activation of this enzyme.

**Inhibitors of kinase function.** Several viruses encode proteins that directly inhibit the kinase activity of Pkr, and some do so by acting as pseudosubstrates. For example,

vaccinia virus K3L protein has amino acid homology with the amino terminus of eIF2α but lacks Ser51, which is phosphorylated by Pkr. The protein binds tightly to Pkr within the catalytic cleft and blocks its ability to phosphorylate Pkr. The growth of vaccinia virus mutants lacking the K3L gene is severely impaired by interferon, and human immunodeficiency virus type 1 Tat protein may inhibit Pkr by a similar mechanism. This enzyme is degraded in poliovirus-infected cells by a mechanism that has not been elucidated.

**Inhibitors that function after eIF2α phosphorylation.** Another interesting viral countermeasure is to enlist a phosphatase to reverse the consequences of Pkr activation. In herpes simplex virus-infected cells, Pkr is activated, but eIF2α is not phosphorylated. In cells lacking the viral $_{γ1}$34.5 gene, Pkr is activated and eIF2α becomes phosphorylated, causing global inhibition of protein synthesis. The protein product of the $_{γ1}$34.5 gene interacts with a type 1a protein phosphatase, and it has been suggested that this interaction might target the phosphatase to eIF2α, possibly reversing the effects of activated Pkr and ensuring continued protein synthesis.

## Regulation of eIF4F Activity

The eIF4F complex plays several important roles during 5'-end-dependent initiation, including recognition of the cap, recruitment of the 40S ribosomal subunit, and unwinding of RNA secondary structure. It is not surprising, therefore, that several viruses encode proteins that modify the activity of this protein. The cap-binding subunit eIF4E is frequently a target, probably because its activity can be modulated in at least two ways and because it is present in limiting quantities in cells. The cap-binding complex can also be inactivated by cleavage of eIF4G, a reaction that occurs in cells infected with a variety of picornaviruses.

### Cleavage of eIF4G

Poliovirus infection of cultured mammalian cells results in a dramatic inhibition of cellular protein synthesis. By 2 h after infection, polyribosomes are disrupted, and translation of nearly all cellular mRNAs ceases and is replaced by viral mRNA translation (Fig. 11.17). Translationally competent extracts from infected cells can readily translate poliovirus mRNA, but not capped mRNAs. Studies of these extracts demonstrated that they are defective in only one translation initiation protein, eIF4F. Specifically, in poliovirus-infected cells the eIF4GI/II component of eIF4F is proteolytically cleaved. As the N-terminal domain of eIF4G binds eIF4E, which in turn binds the 5' cap of cellular mRNAs, such cleavage effectively prevents eIF4F from recruiting 40S ribosomal subunits to capped mRNAs (Fig. 11.18). Because poliovirus mRNA is uncapped and is translated by internal ribosome binding, a process that does not require eIF4E, viral mRNA translation is unaffected. However, IRES function appears to require the C-terminal fragment of eIF4G, which, as discussed above (see "The internal ribosome entry site"), is necessary to recruit 40S ribosomal subunits to the IRES. Thus, cleavage of eIF4G not only inhibits cell translation but also is a strategy for stimulating IRES-dependent translation. Although

both eIF4GI and eIF4GII are cleaved in poliovirus- and rhinovirus-infected cells, the kinetics of shutoff of host translation correlates only with cleavage of eIF4GII.

Cleavage of eIF4G is carried out by viral proteases such as 2A$^{pro}$ of poliovirus, rhinovirus, and coxsackievirus, and the L protease of foot-and-mouth disease virus. Purified 2A$^{pro}$ of rhinovirus cleaves eIF4G directly in vitro, although very inefficiently unless eIF4G is bound to eIF4E. Thus, the eIF4F complex appears to be the target of 2A$^{pro}$ cleavage. The binding of eIF4E to eIF4G might induce conformational changes in the protein that make it a more efficient substrate for the protease. Poliovirus 2A$^{pro}$ efficiently cleaves eIF4GI but not eIF4GII, consistent with the differential cleavage of these proteins during viral infection.

### Modulation of eIF4E Activity by Phosphorylation

The activity of eIF4F can be regulated by phosphorylation of its eIF4E subunit, the cap-binding protein. Inhibition of cellular translation during mitosis and heat shock correlates with reduced phosphorylation of eIF4E, and the increased phosphorylation induced by treatment of cultured cells with growth factors or mitogens correlates with enhanced translation. It is presumed that binding of growth factors to cell surface receptors initiates a cascade of reactions that ultimately results in eIF4E phosphoryla-

**Figure 11.17 Inhibition of translation in poliovirus-infected cells.** (A) Rate of protein synthesis in poliovirus-infected and uninfected cells. During poliovirus infection, host cell translation is inhibited by 2 h after infection and is replaced by translation of viral proteins. Adapted from H. Fraenkel-Conrat and R. R. Wagner (ed.), *Comprehensive Virology* (Plenum Press, New York, N.Y., 1984), with permission. (B) Polyacrylamide gel electrophoresis of [$^{35}$S]methionine-labeled proteins at different times after poliovirus infection. In this experiment, host translation is shut off by 5 h postinfection and is replaced by the synthesis of viral proteins, some of which are labeled at the right.

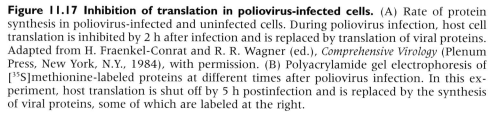
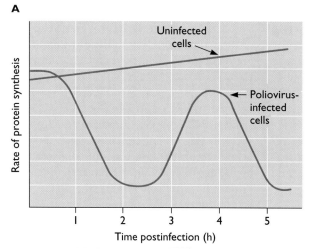
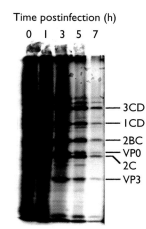

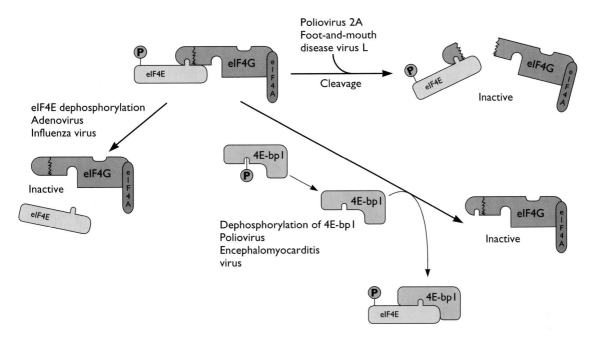

**Figure 11.18 Regulation of eIF4F activity.** The activity of eIF4F can be regulated in at least three ways: phosphorylation of eIF4E, interaction with two eIF4E-binding proteins, and proteolytic cleavage of eIF4G. Dephosphorylation of eIF4E, which occurs in cells infected with adenovirus and influenza virus, causes eIF4E to dissociate from eIF4F. The proteins 4E-bp1 and 4E-bp2 bind eIF4E and block its ability to bind eIF4G. Dephosphorylation of 4E-bp1, which occurs in cells infected with poliovirus and encephalomyocarditis virus, allow it to bind eIF4E. Cleavage of eIF4G occurs in cells infected with poliovirus and foot-and-mouth disease virus. In all cases, the effect is to prevent the interaction of eIF4E and eIF4F, rendering the complex unable to recruit ribosomes to the mRNA via the 5′ cap structure. As a result, translation by 5′-cap-dependent initiation is inhibited.

tion, although the components of this signaling pathway have not been identified. Phosphorylation of eIF4E is carried out by the protein kinase Mnk1, a member of a new class of protein kinases that is activated as a result of mitogen-activated signaling pathways. Whether other protein kinases apart from Mnk1 phosphorylate eIF4E is not known. Protein phosphatases are also likely to contribute to regulation of the phosphorylation state of eIF4E. Phosphorylation of eIF4E leads to increased association with eIF4G and eIF4A. The consequent increase in the concentration of eIF4F facilitates binding to mRNA. It is also likely that phosphorylation of eIF4E moderately increases its affinity for the cap structure.

A decrease in eIF4E phosphorylation appears to be responsible for the shutoff of host translation in cells infected with several viruses. For example, cellular protein synthesis is inhibited at late times in cells infected with adenovirus, a result of virus-induced underphosphorylation of eIF4E. One or more viral late gene products are responsible, although the mechanism of action is not

known. The viral late protein might bind and sequester eIF4E or eIF4E kinase, degrade it, or act as a competitive inhibitor, as we have seen for viral inhibitors of Pkr. Viral infection might also interfere with the signaling cascade that maintains eIF4E phosphorylation. Adenoviral late mRNAs continue to be translated because they possess a reduced requirement for eIF4E. The majority of these viral mRNAs contain the tripartite leader (Fig. 10.12), a common 5′ noncoding region which mediates translation by ribosome shunting. Initiation by this mechanism is less dependent on eIF4F, presumably because the shunting of part of the 5′ untranslated region reduces the requirement for RNA unwinding (helicase) activity associated with cap binding and scanning initiation. Furthermore, adenovirus late mRNAs efficiently recruit the small amounts of phosphorylated eIF4E present late in infection, a feature of mRNAs with little RNA secondary structure near the 5′ cap. Thus, the tripartite leader confers selective translation of viral over host mRNAs under conditions in which eIF4E is underphosphorylated and host translation is inhibited. Adenovirus-induced translation inhibition not only boosts viral late mRNA translation but also enhances cytopathic effects and consequently release of virus from cells.

Influenza virus infection of cells is also accompanied by a shutoff of cellular translation. Infection results in a 60 to 70% reduction in eIF4E phosphorylation, less than that observed in adenovirus-infected cells, and is probably only partially responsible for the selective translation of influenza virus mRNAs. The mechanism by which influenza virus causes eIF4E dephosphorylation is not known. Interestingly, influenza virus mRNAs can be translated during the late phase in adenovirus-infected cells but not in poliovirus-infected cells. Thus, influenza virus mRNAs have a minimal but not a zero requirement for eIF4F.

### Modulation of eIF4E Activity by Two Specific Binding Proteins

Two related low-molecular-weight cell proteins, 4E-bp1 and 4E-bp2, can bind to eIF4E and inhibit translation by 5'-end-dependent scanning but not by internal ribosome entry (Fig. 11.18). The former was found to be identical to a previously described protein, called phosphorylated heat- and acid-stable protein regulated by insulin (Phas-I). This protein was known to be an important phosphorylation substrate in cells treated with insulin and growth factors. Phosphorylation of 4E-bp1 in vitro prevents it from associating with eIF4E. It was subsequently shown that binding of either 4E-bp1 or 4E-bp2 to eIF4E does not prevent it from interacting with the cap but rather inhibits binding to eIF4G. As a result, active eIF4F is not formed. eIF4G and 4E-bp proteins carry a common sequence motif that binds eIF4E. Treatment of cells with hormones and growth factors leads, through signal transduction pathways, to the phosphorylation of 4E-bp1 and its release from eIF4E. mRNAs with extensive secondary structure in the 5' untranslated region, which are translated poorly, are preferentially sensitive to phosphorylation of 4E-bp1. As expected, IRES-mediated translation is not affected. Cellular translation may be independently regulated by the phosphorylation state of eIF4E and/or 4E-bp proteins.

Phosphorylation of 4E-bp1 is inhibited by the drug **rapamycin**. As expected, treatment of cells with this drug inhibits translation initiation by 5'-end-dependent scanning but not by internal ribosome entry. Rapamycin is an immunosuppressant that binds to an immunophilin protein (Fkbp). The latter binds to members of the Fkbp-rapamycin-associated family of protein kinases in humans. Clearly it will be important to identify the reactions that lead to phosphorylation of 4E-bp proteins to establish links between cell surface binding of growth factors and translation initiation.

Infection with several different picornaviruses results in alteration of the phosphorylation state of 4E-bp1 and 4E-bp2 (Fig. 11.18). Host cell translation is inhibited in cells infected with encephalomyocarditis virus, a picornavirus. In contrast to the shutoff that occurs in poliovirus-infected cells, cellular protein synthesis in encephalomyocarditis virus-infected cells occurs late in infection and is not mediated by cleavage of eIF4G. Rather, encephalomyocarditis virus infection induces dephosphorylation of 4E-bp1, which then binds eIF4E to prevent it from forming eIF4F. As a result, translation of cellular mRNAs is inhibited, but because encephalomyocarditis virus mRNA contains an IRES, its translation is unaffected. Dephosphorylation of 4E-bp1 also occurs at late times in cells infected with poliovirus.

### Perspectives

Despite our detailed understanding of how viral proteins regulate eIF4F activity, there are many examples of virus-mediated translation regulation that are poorly understood. For example, the adenovirus nonstructural 100-kDa protein, which is encoded by the L4 region of the major late transcription unit, plays a role in capsid morphogenesis but also binds to viral and cellular mRNAs. A mutation in the 100-kDa protein gene causes reduced translation of adenoviral late mRNAs, but how the protein controls translation is not understood. The translation of certain herpesviral mRNAs also appears to be regulated in infected cells. Transcripts of two cytomegalovirus late genes, ICP27 and ICP36, can be found in infected cells early in infection, yet only small amounts of protein are made. In the late phase in cells infected with herpes simplex virus, mRNA encoding the viral glycoprotein gD accumulates in the cytoplasm, but again only small quantities of the protein are found. How translation of mRNAs encoding ICP27, ICP36, and gD is controlled has not been elucidated. Understanding how viruses regulate translation remains an area of intense research activity. Such studies will not only enhance our understanding of how viruses efficiently infect cells but will contribute to our knowledge of translation during normal and abnormal cell growth.

Our discussion of regulation of translation has been largely restricted to examples relevant to viral infections. For the sake of completeness, it should be noted that there are other mechanisms of translational regulation that have not (yet?) been identified in virus-infected cells. For example, phosphorylation of the ribosomal protein S6 in response to growth-promoting stimuli increases the initiation of translation on mRNAs that encode ribosomal and elongation proteins. The selectivity of such translational stimulation has been correlated with the 5' untranslated regions of the mRNAs. The signal transduction pathways that link growth-promoting stimuli to S6 phosphorylation are beginning to be elucidated. Translation elongation rates can also be modulated by phosphorylation of elongation protein eEF2 in response to growth-promoting stimuli. The activity of this protein is also modified by ADP-ribosylation caused by the toxin produced by the pathogenic bacterium *Clostridium diphtheriae*. Given the enormous complexity of the translation apparatus, it seems likely that many other regulatory points exist and remain to be discovered in virus-infected cells.

# References

## Monographs

Harford, J. B., and D. R. Morris. 1997. *mRNA Metabolism and Post-Transcriptional Gene Regulation.* Wiley-Liss, New York, N.Y.

Hershey, J. W. B., M. B. Mathews, and N. Sonenberg (ed.). 1996. *Translational Control.* Cold Spring Harbor Laboratory Press, Cold Spring Harbor, N.Y.

## Review Articles

Curran, J., P. Latorre, and D. Kolakofsky. 1997. Translational gymnastics on the Sendai virus P/C mRNA. *Semin. Virol.* **8:**351–357.

Farabaugh, P. J. 1996. Programmed translational frameshifting. *Microbiol. Rev.* **60:**103–134.

Gale, M., and M. G. Katze. 1998. Molecular mechanisms of interferon resistance mediated by viral-directed inhibition of Pkr, the interferon-induced protein kinase. *Pharmacol. Ther.* **78:**29–46.

Lemon, S. M., and M. Honda. 1997. Internal ribosome entry sites within the RNA genomes of hepatitis C virus and other flaviviruses. *Semin. Virol.* **8:**274–288.

Morley, S. J., P. S. Curtis, and V. M. Pain. 1997. eIF4G: translation's mystery factor begins to yield its secrets. *RNA* **3:**1085–1104.

Mountford, P. S., and A. G. Smith. 1995. Internal ribosome entry sites and dicistronic RNAs in mammalian transgenesis. *Trends Genet.* **11:**179–184.

Sachs, A. B., P. Sarnow, and M. W. Hentze. 1997. Starting at the beginning, middle and end: translation initiation in eukaryotes. *Cell* **89:**831–838.

Stewart, S. R., and B. L. Semler. 1997. RNA determinants of picornavirus cap-independent translation initiation. *Semin. Virol.* **8:**242–255.

## Papers of Special Interest

### Ribosome Shunting

Yueh, A., and R. J. Schneider. 1996. Selective translation initiation by ribosome jumping in adenovirus infected and heat shocked cells. *Genes Dev.* **10:**1557–1567.

## The IRES

Blyn, L. B., J. S. Towner, B. L. Semler, and E. Ehrenfeld. 1997. Requirement of poly(C) binding protein for translation of poliovirus RNA. *J. Virol.* **71:**6243–6246.

Elroy-Stein, O., T. R. Fuerst, and B. Moss. 1989. Cap-independent translation of mRNA conferred by encephalomyocarditis virus 5′ sequence improves the performance of the vaccinia virus/bacteriophage T7 hybrid expression system. *Proc. Natl. Acad. Sci. USA* **86:**6126–6130.

Jang, S. K., H.-G. Kräusslich, M. J. H. Nicklin, G. M. Duke, A. C. Palmenberg, and E. Wimmer. 1988. A segment of the 5′ nontranslated region of encephalomyocarditis virus RNA directs internal entry of ribosomes during in vitro translation. *J. Virol.* **62:**2636–2643.

Kaminski, A., S. L. Hunt, J. G. Patton, and R. J. Jackson. 1995. Direct evidence that polypyrimidine tract binding protein (PTB) is essential for internal initiation of translation of encephalomyocarditis virus RNA. *RNA* **1:**924–938.

Kolupaeva, V. G., T. V. Pestova, C. U. Hellen, and I. N. Shatsky. 1998. Translation eukaryotic initiation factor 4G recognizes a specific structural element within the internal ribosome entry site of encephalomyocarditis virus RNA. *J. Biol. Chem.* **273:**18599–18604.

Macejak, D. G., and P. Sarnow. 1991. Internal initiation of translation mediated by the 5′ leader of a cellular mRNA. *Nature* (London) **353:**90–94.

Meerovitch, K., Y. V. Svitkin, H. S. Lee, F. Lejbkowicz, D. J. Kenan, E. K. L. Chan, V. I. Agol, J. D. Keene, and N. Sonenberg. 1993. La autoantigen enhances and corrects aberrant translation of poliovirus RNA in reticulocyte lysate. *J. Virol.* **67:**3798–3807.

Pestova, T. V., I. N. Shatsky, S. P. Fletcher, R. J. Jackson, and C. U. Hellen. 1998. A prokaryotic-like mode of cytoplasmic eukaryotic ribosome binding to the initiation codon during internal translation initiation of hepatitis C and classical swine fever virus RNAs. *Genes Dev.* **12:**67–83.

## Leaky Scanning

Williams, M. A., and R. A. Lamb. 1989. Effect of mutations and deletions in a bicistronic mRNA on the synthesis of influenza B virus NB and NA glycoproteins. *J. Virol.* **63:**28–35.

## Reinitiation

Horvath, K. M., M. A. Williams, and R. A. Lamb. 1990. Eukaryotic coupled translation of tandem cistrons: identification of the influenza B virus BM2 polypeptide. *EMBO J.* **9:**2639–2647.

## Translational Suppression

Feng, Y.-X., H. Yuan, A. Rein, and J. G. Levin. 1992. Bipartite signal for read-through suppression in murine leukemia virus mRNA: an eight-nucleotide purine-rich sequence immediately downstream of the *gag* termination codon followed by an RNA pseudoknot. *J. Virol.* **66:**5127–5132.

Li, G., and C. M. Rice. 1993. The signal for translational readthrough of a UGA codon in Sindbis virus RNA involves a single cytidine residue immediately downstream of the termination codon. *J. Virol.* **67:**5062–5067.

Yoshinaka, Y., I. Katoh, T. D. Copeland, and S. Oroszlan. 1985. Murine leukemia virus protease is encoded by the *gag-pol* gene and is synthesized through suppression of an amber termination codon. *Proc. Natl. Acad. Sci. USA* **82:**1618–1622.

### Frameshifting

Chen, X., M. Chamorro, S. I. Lee, L. X. Shen, J. V. Hines, I. Tinoco, and H. E. Varmus. 1995. Structural and functional studies of retroviral RNA pseudoknots involved in ribosomal frameshifting: nucleotides at the junction of the two stems are important for efficient ribosomal frameshifting. *EMBO J.* **14:**842–852.

Hwang, C. B. C., B. Horsburgh, E. Pelosi, S. Roberts, P. Digard, and D. M. Coen. 1994. A net +1 frameshift permits synthesis of thymidine kinase from a drug-resistant herpes simplex virus mutant. *Proc. Natl. Acad. Sci. USA* **91:**5461–5465.

Jacks, T., H. D. Madhani, F. R. Masiarz, and H. E. Varmus. 1988. Signals for ribosomal frameshifting in the Rous sarcoma virus *gag-pol* region. *Cell* **55:**447–458.

### Pkr

He, B., M. Gross, and B. Roizman. 1997. The $_{\gamma_1}$34.5 protein of herpes simplex virus 1 complexes with protein phosphatase 1α to dephosphorylate the α subunit of the eukaryotic initiation translation initiation factor 2 and preclude the shutoff of protein synthesis by double-stranded RNA-activated protein kinase. *Proc. Natl. Acad. Sci. USA* **94:**843–848.

Kitajewski, J., R. J. Schneider, B. Safer, S. M. Munemitsu, C. E. Samuel, B. Thimmappaya, and T. Shenk. 1986. Adenovirus VAI RNA antagonizes the antiviral action of the interferon-induced eIF2 alpha kinase. *Cell* **45:**195–200.

Polyak, S. J., N. Tang, M. Wambach, G. N. Barber, and M. G. Katze. 1996. The p59 cellular inhibitor complexes with the interferon-induced, double-stranded RNA-dependent protein kinase, Pkr, to regulate its autophosphorylation and activity. *J. Biol. Chem.* **271:**1702–1707.

### Control of eIF4F Activity

Feigenblum, D., and R. J. Schneider. 1993. Modification of eukaryotic initiation factor 4F during infection by influenza virus. *J. Virol.* **67:**3027–3035.

Feigenblum, D., and R. J. Schneider. 1996. Cap-binding protein (eukaryotic initiation factor 4E) and 4E-inactivating protein BP-1 independently regulate cap-dependent translation. *Mol. Cell. Biol.* **16:**5450–5457.

Gingras, A.-C., and N. Sonenberg. 1997. Adenovirus infection inactivates the translational inhibitors 4E-bp1 and 4E-bp2. *Virology* **237:**182–186.

Gingras, A.-C., Y. Svitkin, G. J. Belsham, A. Pause, and N. Sonenberg. 1996. Activation of the translational suppressor 4E-bp1 following infection with encephalomyocarditis virus and poliovirus. *Proc. Natl. Acad. Sci. USA* **93:**5578–5583.

Gradi, A., H. Imataka, Y. V. Svitkin, E. Rom, B. Raught, S. Morino, and N. Sonenberg. 1998. A novel functional human eukaryotic translation initiation factor 4G. *Mol. Cell. Biol.* **18:**334–342.

Gradi, A., Y. V. Svitkin, H. Imataka, and N. Sonenberg. 1998. Proteolysis of human eukaryotic translation initiation factor eIF4GII, but not eIF4GI, coincides with the shutoff of host protein synthesis after poliovirus infection. *Proc. Natl. Acad. Sci. USA* **95:**11089–11094.

Haghighat, A., Y. Svitkin, I. Novoa, E. Kuechler, T. Skern, and N. Sonenberg. 1996. The eIF4G-eIF4E complex is the target for direct cleavage by the rhinovirus 2A protease. *J. Virol.* **70:**8444–8450.

Ohlmann, T., M. Rau, V. M. Pain, and S. J. Morley. 1996. The C-terminal domain of eukaryotic protein synthesis initiation factor eIF4G is sufficient to support cap-independent translation in the absence of eIF4E. *EMBO J.* **15:**1371–1382.

Wyckoff, E. E., J. W. B. Hershey, and E. Ehrenfeld. 1990. Eukaryotic initiation factor 3 is required for poliovirus 2A protease-induced cleavage of the p220 component of eukaryotic initiation factor 4F. *Proc. Natl. Acad. Sci. USA* **87:**9529–9533.

Zhang, Y., D. Feigenblum, and R. J. Schneider. 1994. A late adenovirus factor induces eIF4E dephosphorylation and inhibition of cell protein synthesis. *J. Virol.* **68:**7040–7050.

### Other Mechanisms of Translational Regulation

Geballe, A. P., F. S. Leach, and E. S. Mocarski. 1986. Regulation of cytomegalovirus late gene expression: γ genes are controlled by posttranscriptional events. *J. Virol.* **57:**864–874.

Johnson, D. C., and P. G. Spear. 1984. Evidence for translational regulation of herpes simplex virus type 1 gD expression. *J. Virol.* **51:**389–394.

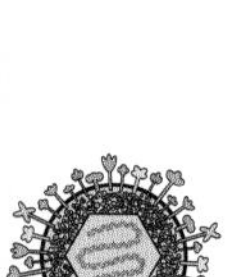

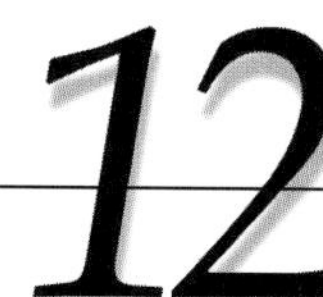

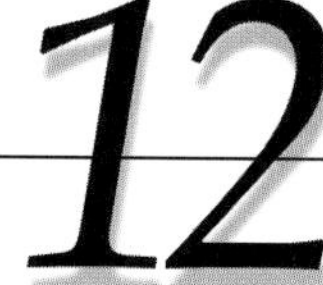

# 12

# Intracellular Transport of Viral Components: Prelude to Assembly

## Introduction

Successful viral reproduction requires the intracellular assembly of progeny virions from their protein, nucleic acid, and, in many cases, membrane components. In preceding chapters, we have considered molecular mechanisms that ensure the production of the macromolecules from which virions are constructed in the host cell. Because of the structural and functional compartmentalization of eukaryotic cells, virion components are generally produced at different intracellular locations and must be brought together for assembly. Thus, intracellular trafficking, or sorting, of viral nucleic acids, proteins, and glycoproteins to the appropriate intracellular sites is an essential prelude to the assembly of all animal viruses.

From our point of view, animal cells are very small, with typical diameters of 10 to 30 $\mu$m. However, in the microscopic world inhabited by viruses, an animal cell is large: the distances over which virion components must be transported within a cell are roughly equivalent to a mile on the macroscopic, human scale. Their transport therefore requires a considerable expenditure of energy, supplied by the host cell. Furthermore, the intracellular trafficking of viral macromolecules must be directed so that individual virion components are delivered to the correct assembly site. Assembly can occur at any one of several intracellular addresses, determined largely by whether the virus is enveloped and by the nature of its nucleic acid genome. All enveloped viruses derive their membranes from one of the host cell's membranes, which are modified by insertion of viral proteins. Assembly of these viruses therefore requires delivery of some viral proteins to the appropriate membrane, as well as transport of internal virion proteins and the nucleic acid genome to the modified membrane. Another common assembly site, used by the majority of viruses with DNA genomes and some that encapsidate RNA, is the cell nucleus. When the viral genome is synthesized within the infected cell nucleus, the minimal requirement for assembly is import of virion structural proteins made in the cytoplasm into the nucleus. This strategy imposes a less complex trafficking problem than assembly of enveloped viruses at membrane sites, but additional mechanisms are required to ensure egress of assembled virions from the cell.

The need for movement of proteins and nucleic acids from one cellular compartment to another, or for insertion of proteins into specific membranes, is not unique to viruses. The majority of cellular proteins are made by translation of mRNAs on cytoplasmic polyribosomes and must then be transported to their sites of operation. Similarly, most cellular RNA species, including mRNAs, rRNAs, and tRNAs, are exported from the nucleus, in which transcription takes place, to the cytoplasm. Eukaryotic cells are therefore constantly engaged in transport of macromolecules among their compartments via intracellular trafficking systems. Viruses make use of the cellular systems that sort macromolecules to each of many possible intracellular destinations, just as they make use of the cellular biosynthetic machinery responsible for transcription, replication, or translation. Indeed, the rapid advances in our understanding of cellular trafficking mechanisms made in recent years can be traced to initial studies with viral membrane or nuclear proteins, such as the vesicular stomatitis virus G protein and simian virus 40 large T antigen, respectively. In the following sections, the cellular transport pathways exploited by viruses are described in the context of the site at which virion assembly takes place.

## Assembly within the Nucleus

Assembly of the capsids or nucleocapsids of the majority of viruses with DNA genomes, including papovaviruses, adenoviruses, and herpesviruses, takes place within infected cell nuclei, the site of viral DNA synthesis. All structural proteins of nonenveloped papovaviruses and adenoviruses are imported into the nucleus following synthesis in the infected cell cytoplasm, allowing complete assembly within this organelle (Fig. 12.1). In contrast, the more complex herpesviruses, which comprise a DNA-containing nucleocapsid surrounded by an additional protein layer (the tegument) and an envelope (appendix, Fig. 17A), complete virion assembly at extranuclear sites (see below). So do some enveloped RNA viruses, such as orthomyxoviruses, whose genomic RNA-containing nucleocapsids assemble in the nucleus as the RNA is replicated. In these cases, a subset of virion proteins must be imported into the nucleus.

As far as we know, all viral structural proteins that must enter the nucleus do so via the normal cellular pathway of nuclear protein import. As discussed in chapter 5, this same pathway is responsible for nuclear import of both viral genomes (or nucleoproteins) and viral, nonstructural proteins that function in the nucleus at the beginning of the infectious cycle. Proteins destined for the nucleus are so labeled by the presence of short, basic nuclear localization signals (Fig. 5.18). These signals are recognized by soluble components of the cellular nuclear import machinery, which dock the protein at the cytoplasmic face of a nuclear pore complex for subsequent transport into the nucleus.

### Import of Viral Proteins for Assembly

All viral proteins that are known to be localized to the nucleus contain specific sequences that target them for import into the nucleus (for examples, see Table 5.1). The nuclear localization signals (NLSs) of viral proteins cannot be distinguished from those of cellular proteins. Indeed, many of the efforts to identify the cellular receptors that dock proteins destined for the nucleus at the nuclear pore complex employed the well-characterized nuclear localization signal of simian virus 40 large T antigen. It is therefore believed that viral proteins destined for the nucleus contain typical nuclear localization signals to ensure their recognition by the cellular import machinery. Nevertheless, such signals have been experimentally verified for only a small fraction of the large number of viral proteins that enter the nucleus. Furthermore, there are a large number of cellular nuclear localization signal receptors, some of which are expressed in tissue-specific fashion. These receptors may recognize different types of viral (and cellular) nuclear localization signals.

The cellular components that mediate import of viral proteins into the nucleus are present at finite concentrations. For example, a typical mammalian cell contains on the order of 3,000 to 4,000 nuclear pore complexes, and the supply of cellular nuclear localization signal receptors and transport proteins is also limited. As very large quantities of viral structural proteins must enter the nucleus prior to assembly, there is potential for competition among viral and cellular proteins for access to receptors, the nuclear pore complex, or transport proteins. Such competition is minimized in cells infected by the larger DNA viruses, such as adenoviruses and herpesviruses; by the time that structural proteins are made during the late phase of infection, cellular protein synthesis is severely inhibited, reducing or eliminating competition for components of the nuclear import machinery. However, the proteins of viruses that do not induce inhibition of cellular protein synthesis, such as those of the papovaviruses, must enter the nucleus despite continual import of cellular proteins. Whether import of viral proteins is favored in such circumstances, for example, by the presence of particularly effective nuclear localization signals, or whether the import machinery can handle the extra load is not known.

Many viral structural proteins that enter the infected cell nucleus form multimeric capsid components. In some

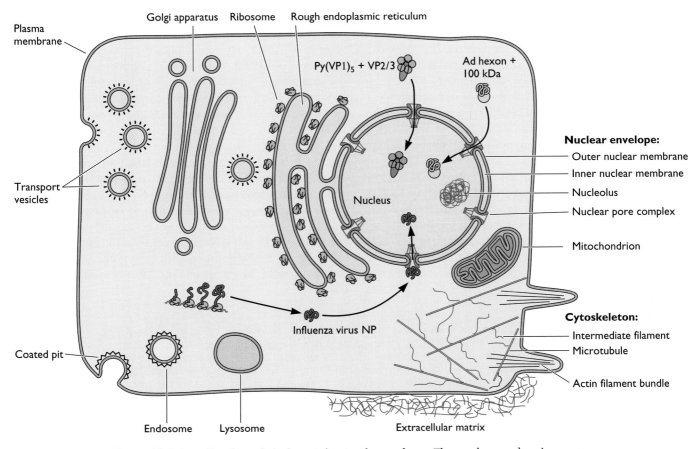

**Figure 12.1 Localization of viral proteins to the nucleus.** The nucleus and major membrane-bound compartments of the cytoplasm, as well as components of the cytoskeleton, are illustrated schematically and not to scale. Transport vesicles are components of the secretory pathway by which proteins travel to the plasma membrane or are secreted from the cell, whereas clathrin-coated vesicles and endosomes mediate uptake of material into the cell. Viral proteins destined for the nucleus are synthesized on cytoplasmic polyribosomes, as illustrated for the influenza virus NP protein. They engage with the cytoplasmic face of the nuclear pore complex and are translocated into the nucleus by the protein import machinery of the host cell. Some viral structural proteins enter the nucleus as preassembled structural units (polyomaviral [Py] VP1 pentamers associated with one molecule of either VP2 or VP3) or in association with a viral chaperone (adenoviral [Ad] hexon monomers bound to the L4 100-kDa protein).

cases, structural units of the virion are assembled in the cytoplasm prior to import into the nucleus. Pentamers of the major capsid protein, VP1, of the papovaviruses simian virus 40 and polyomavirus specifically bind a monomer of either VP2 or VP3, the minor virion proteins, which share C-terminal sequences (appendix, Fig. 13B). Such heteromeric assemblies are the substrates for import into the nucleus. Indeed, efficient nuclear localization of polyomavirus VP2 and VP3 proteins can occur only in cells in which VP1 is also made. Thus, assembly of the heteromeric complex facilitates import of the minor structural proteins even though each contains a nuclear localization

signal. The increased density of nuclear localization signals may allow more effective utilization of essential components of the import pathway, or the nuclear localization signals may be more accessible in the (VP1)$_5$-VP2/VP3 complex.

Despite such potential advantages as increased efficiency of import of viral proteins and import of the structural proteins in the appropriate stoichiometry, import of preassembled capsid components is not universal. For example, adenoviral hexons (trimers of viral protein II), which can be distinguished from monomeric hexon chains by specific monoclonal antibodies, are found only in the nucleus of the infected cell. Association of newly synthesized cytoplasmic protein II with a late, nonstructural protein, the L4 100-kDa protein (appendix, Fig. 15B), is essential both for the entry of hexon subunits into the nucleus and for their assembly into trimers. Thus, the monomeric hexon-L4 protein complex must be the import substrate. The viral L4 100-kDa protein might supply the nuclear localization signal for hexon import or promote nuclear retention by facilitating assembly of hexon trimers within the nucleus.

## Assembly at the Plasma Membrane

Assembly of enveloped viruses frequently takes place at the plasma membrane of infected cells. Before such virions can form, viral integral membrane proteins must be localized to the cellular membrane (Fig. 12.2). The first stages of the pathway by which viral and cellular proteins are delivered to the plasma membrane were identified more than 30 years ago, and the process is now quite well understood. Viruses with envelopes derived from the plasma membrane also contain internal proteins, which may also be membrane associated, and of course nucleic acid genomes. Thus, these internal components must also be sorted to appropriate plasma membrane sites for assembly (Fig. 12.2).

## The Cellular Pathway by Which Viral Membrane Proteins Are Transported to the Plasma Membrane

Viral membrane proteins reach their destinations by the highly conserved, cellular **secretory pathway.** The mechanisms by which proteins enter and travel along this pathway to predetermined sites have been elucidated by morphological, biochemical, and genetic studies, many of

**Figure 12.2 Localization of viral proteins to the plasma membrane.** Viral envelope glycoproteins (red) are cotranslationally translocated into the ER lumen and folded and assembled within that compartment. They travel to and through the Golgi apparatus and from the Golgi apparatus to the plasma membrane by vesicular transport. The internal proteins of the particle (purple) and the genome (olive) are also directed to plasma membrane sites of assembly.

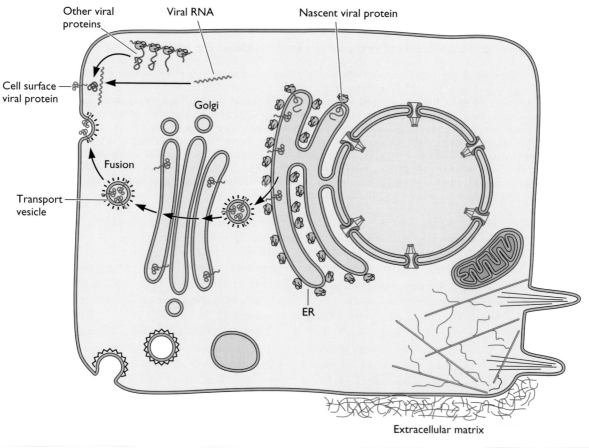

which have employed viral membrane proteins such as the vesicular stomatitis virus G and influenza virus hemagglutinin (HA) proteins. Such viral proteins are characterized by a number of structural features typical of cellular proteins that travel the secretory pathway, including disulfide bonds and covalently linked oligosaccharide chains (Fig. 12.3). Viral envelope proteins generally span the cellular membrane into which they are inserted only once and thus contain a single transmembrane domain. In viral proteins, transmembrane segments (whose properties are described in chapter 4), generally separate large extracellular domains from smaller cytoplasmic domains (Fig. 12.3). The former include the binding sites for cellular receptors, crucial for initiation of the infectious cycle, whereas the latter are important in virus assembly.

As discussed in chapter 3 and summarized in Table 12.1, viral membrane proteins are usually oligomers. Most interactions among the subunits of viral membrane proteins are noncovalent, but some examples of association via covalent, interchain disulfide bonds are known (Table 12.1). Oligomer assembly, like covalent modification, takes place during transit from the cytoplasm to the cell surface, when viral membrane proteins travel through a series of specialized compartments that provide appropriate chemical environments and the enzymatic machinery necessary for their maturation. The first such compartment, the endoplasmic reticulum (ER), is encountered by viral membrane proteins as they are synthesized.

## Translocation of Membrane Proteins into the ER Membrane

All proteins destined for insertion into the plasma membrane, or the membranes of such intracellular organelles as the Golgi apparatus, enter the ER as they are translated (Fig. 12.2). This membranous structure appears as a basketwork of tubules and sacs extending throughout the cytoplasm (Fig. 12.4A). The ER membrane demarcates a geometrically convoluted but continuous internal space, termed the **ER lumen,** from the remainder of the cytoplasm. The ER lumen is a chemically distinctive environment, which is topologically equivalent to the outside of the cell. Proteins that enter the ER during their synthesis are thus separated from cytoplasmic proteins as soon as they are made.

Most proteins, regardless of their final ports of call, are synthesized on polyribosomes in the cytoplasm. However, polyribosomes engaged in synthesis of proteins that will enter the secretory pathway become associated with the cytoplasmic face of the ER membrane soon after translation begins. Those areas of the ER to which polyribosomes are bound form the **rough ER** (Fig. 12.4B). The association of polyribosomes with the ER membrane is directed by a short sequence in the nascent protein termed the **signal peptide.** It is now taken for granted that the primary sequences of proteins include "zip codes" specifying the cellular addresses at which the proteins must reside to fulfill their functions, such as the nuclear localization

**Figure 12.3 Primary sequence features and covalent modifications of the HA protein of the 1968 Hong Kong influenza A virus.** The primary sequence of the HA0 protein, the product of translation of HA mRNA, is depicted in the center, with the orange arrowhead indicating the site of the proteolytic cleavages that generate the HA1 and HA2 chains. Note that an arginine residue is removed during this maturation step, which is necessary to activate the fusion peptide that constitutes the N terminus of the HA2 chain. Residues of the HA1 and HA2 chains are numbered from each N terminus. The fusion peptide, the N-terminal signal sequence which is removed by signal peptidase in the ER, and the C-terminal transmembrane domain are hydrophobic. Disulfide bonds, one of which maintains covalent linkage between the HA1 and HA2 chains following HA0 cleavage, are indicated, as are sites of N-linked glycosylation (oligosaccharides) and palmitoylation (Ac).

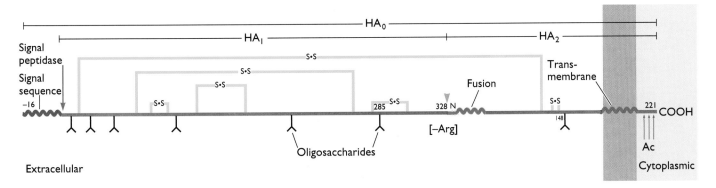

**Table 12.1** Properties of some viral membrane proteins[a]

| Virus | Protein(s) | Oligomeric structure | Efficiency of transport |
|---|---|---|---|
| Alphavirus (Semliki Forest virus) | E1, E2, E3 | (E1-E2-E3)$_3$ | High |
| Rhabdovirus (vesicular stomatitis virus) | G | (G)$_3$ | >90% |
| Orthomyxovirus (influenza A virus) | HA | (HA1-HA2)$_3$ | >90% |
| | NA | (NA-NA)$_2$ | >60% |
| | M2 | (M2)$_4$ | High |
| Retrovirus (Rous sarcoma virus) | Env | (SU-TM)$_3$ | |
| Herpesvirus (herpes simplex virus type 1) | gH, gL | (gHgL)$_?$ | High, but zero when only one protein is synthesized |

[a]The best predicted oligomeric structures of the viral membrane proteins listed are shown in column 3, with hyphens indicating disulfide-bonded subunits. The efficiencies with which these proteins reach their final destinations are listed in the rightmost column. Based on Table 1 of R. W. Doms et al., *Virology* **193**:545–562, 1993.

sequences discussed in the previous section. The signal peptides of proteins that enter the ER lumen were the first such zip codes to be identified and thus established this paradigm about 20 years ago. Signal peptides are commonly found at the N termini of proteins destined for the secretory pathway. They are usually about 20 amino acid residues in length and contain a core of 15 hydrophobic residues. Often, signal peptides are transient structures

**Figure 12.4 The endoplasmic reticulum.** (A) The ER of a mammalian cell in culture. The reticular ER, which extends throughout the cytoplasm, was visualized by fluorescence microscopy of fixed African green monkey kidney epithelial cells stained with the lipophilic fluorescent dye 3,3′-dihexyloxacarbocyanine iodide. This dye also stains mitochondria. The ER membrane accounts for over half of the total membrane of a typical animal cell and possesses a characteristic lipid composition. Courtesy of M. Terasaki, University of Connecticut Health Center. (B) Electron micrograph of the rough ER in rat hepatocytes. Note the many ribosomes associated with the cytoplasmic surface of the membrane. From R. A. Rodewald, Biological Photo Service.

**A**

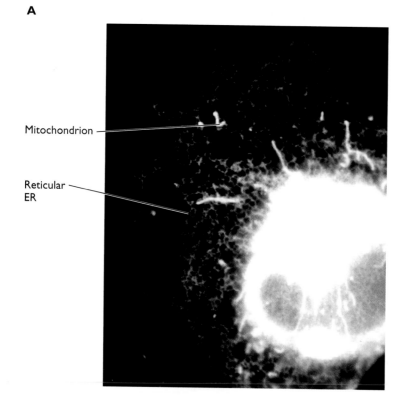

Mitochondrion

Reticular ER

**B**

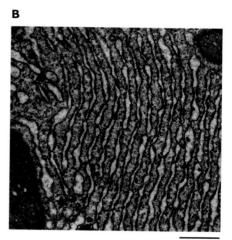

1 μm

that are removed enzymatically during protein translocation into the ER by a protease located in the ER lumen, **signal peptidase.**

Translation of a protein that will enter the ER begins in the normal fashion and continues until the signal peptide emerges from the ribosome (Fig. 12.5). This signal then directs binding of the translation machinery to the ER membrane by means of two components: the signal peptide is recognized by the **signal recognition particle (SRP),** which in turn binds to the cytoplasmic domain of an integral ER membrane protein termed the **SRP receptor** (Fig. 12.5). Binding of the signal recognition particle to the ribosome also temporarily halts translation to allow the polyribosome-nascent protein-signal recognition particle complex time to bind to the ER membrane via the SRP receptor (Fig. 12.5). Both the signal recognition particle and its receptor contain subunits that bind GTP, and efficient targeting requires the presence of this nucleotide. Following the initial docking of the complex at the membrane, the ribosome becomes tightly bound to the membrane and engaged with a protein translocator, the **translocon,** which forms a gated, aqueous channel through the ER

membrane (Fig. 12.5). This interaction is coordinated with release of the signal recognition particle, association of the signal peptide with the translocon, and resumption of translation (Fig. 12.5). Because the ribosome remains bound to the membrane upon release of the signal recognition particle, continued translation facilitates movement of the growing polypeptide chain through the membrane.

**Figure 12.5 Targeting of a nascent protein to the ER membrane.** Translation of an mRNA encoding a protein that will enter the ER lumen proceeds until the signal peptide (purple) emerges from the ribosome. The signal recognition particle (SRP), which contains a small RNA molecule and several proteins, binds to both the signal peptide and to the ribosome to halt or pause translation, upon binding of GTP to one of the protein subunits (step 1). The nascent polypeptide-SRP-ribosome complex then binds to the SRP receptor in the ER membrane (step 2), an interaction that triggers hydrolysis of GTP bound to both the SRP and its receptor, release of the SRP (step 3), and close association of the ribosome with the ER membrane and its multisubunit protein translocator, termed the translocon (step 4). During translocation, the latter structure forms an aqueous pore 40 to 60 Å in diameter, the largest "hole" yet observed in a membrane. The tight association of the ribosome with the membrane blocks the cytoplasmic end of the aqueous channel and maintains the barrier between the cytoplasm and the ER lumen. The hydrolysis of GTP coordinates the release of SRP with binding of the signal peptide to the translocator, and resumption of translation, when the block maintained at the luminal end of the pore by Grp78/Bip before and early in translocation is reversed. The growing polypeptide chain is then transferred through the membrane as its translation continues (step 5). The hydrolysis of ATP by Grp78/Bip within the ER lumen provides the power to effect posttranslational import of certain proteins into the ER. Signal peptidase removes the signal peptide cotranslationally (step 6).

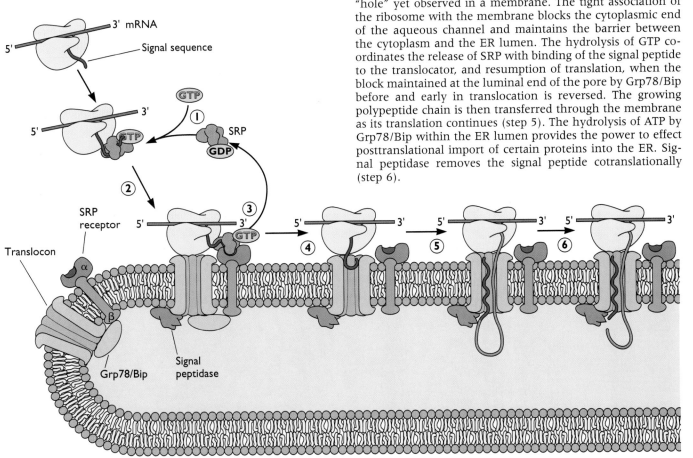

Such coupling of translation and translocation ensures that the polypeptide crosses the membrane as an unfolded chain that can be accommodated by the translocon (Fig. 12.5).

When the protein entering the ER is destined for secretion from the cell, translocation continues until the entire polypeptide chain enters the ER lumen. During translocation, the signal peptide is proteolytically removed by signal peptidase, releasing the soluble protein into the ER. In contrast, translocation of integral membrane proteins, such as viral envelope proteins, halts when a hydrophobic **stop transfer** signal is encountered in the nascent polypeptide chain. This sequence may be the signal peptide it-self or a second, internal hydrophobic sequence. The number, location, and orientation of such sequences within a protein determine the topology with which it is inserted into the ER membrane (Box 12.1). The programming of insertion of proteins into the ER membrane by start transfer and stop transfer signals built into their primary sequences ensures that every molecule of a particular protein adopts the identical topology in the membrane. As this topology is maintained during the many membrane budding and fusion reactions by which proteins reach the cell surface, the way in which a protein is inserted into the ER membrane determines its orientation in the plasma membrane (see, for example, Fig. 12.11).

---

### BOX 12.1

*Establishing the topology of integral membrane proteins that, like the majority of viral envelope proteins, contain a single transmembrane domain*

In some cases (panel A of the figure), the membrane anchor is the signal peptide itself. In the model shown, an internal signal peptide (start transfer signal) binds to the translocon such that the segment of the inserted polypeptide chain C-terminal to the signal peptide is transferred across the membrane. This is because one end of the hydrophobic core of the signal peptide is flanked by a sequence containing positively charged amino acids, which tends to remain on the cytoplasmic side of the ER membrane. When translocation is complete, the pro-tein is discharged into the bilayer. In this case, the C terminus of the polypeptide is located within the ER lumen, and its N terminus is in the cytoplasm (**type II orientation**). A reverse orientation of charge distribution around the hydrophobic core of the start transfer signal peptide results in the opposite orientation of the polypeptide in the ER membrane (i.e., N terminus within the lumen; **type I orientation**). Other proteins with a single transmembrane domain, including the influenza virus HA0 protein, contain not

*(continued)*

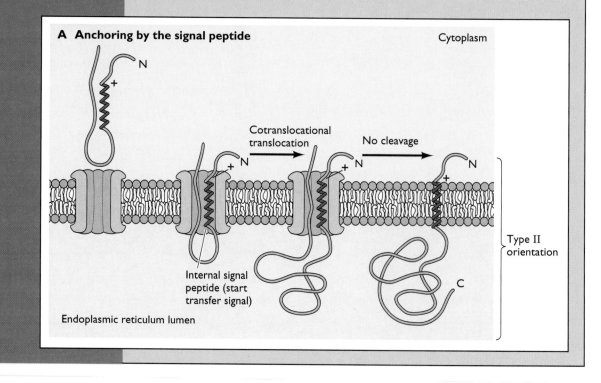

**A Anchoring by the signal peptide**

### Reactions within the ER

The folding and initial posttranslational modification of proteins that enter the secretory pathway take place within the ER, which contains many of the enzymes that catalyze chemical modifications common among these proteins, such as disulfide bond formation and the initial steps in the assembly of oligosaccharide chains (glycosylation).

**Glycosylation.** Viral envelope proteins, like cellular proteins that travel the secretory pathway, are generally modified by the addition of oligosaccharides to either as-paragine (N-linked glycosylation) or serine or threonine (O-linked glycosylation) residues. The presence of oligosaccharides on a protein can be detected by the incorporation of radioactively labeled monosaccharides or, more usually, as changes in the protein's electrophoretic mobility, following exposure of cells to inhibitors of glycosylation or of the protein to enzymes that cleave the oligosaccharide (Fig. 12.6A). The first steps in N-linked glycosylation take place as a polypeptide chain emerges into the ER lumen. Oligosaccharides rich in mannose preassembled on a lipid carrier are added to asparagine residues by oligosaccharyltransferase (Fig. 12.6B). Subse-

---

**BOX 12.1    (continued)**

only an N-terminal signal peptide but also a second hydrophobic sequence (B). The signal peptide initiates cotranslational translocation of the nascent polypeptide through the translocon, as shown in more detail in Fig. 12.5. During translation, the resident ER enzyme signal peptidase cleaves off the signal peptide, generating the N terminus of the mature protein. Translocation continues until the second hydrophobic sequence is encountered. This sequence acts as a stop transfer signal and halts translocation, with lateral discharge of the protein into the ER membrane by a multistep process that is not well understood. This mechanism results in type I orientation, as in the influenza virus HA0 protein. The relative position of the stop transfer peptide within the polypeptide determines the relative sizes of its luminal (extracellular) and cytoplasmic domains. In proteins that contain multiple hydrophobic sequences of this kind, the first one encountered by the signal recognition particle scanning from the N terminus as the polypeptide emerges from the ribosome starts the transfer. Thus, it establishes whether subsequent hydrophobic sequences will stop or start transfer.

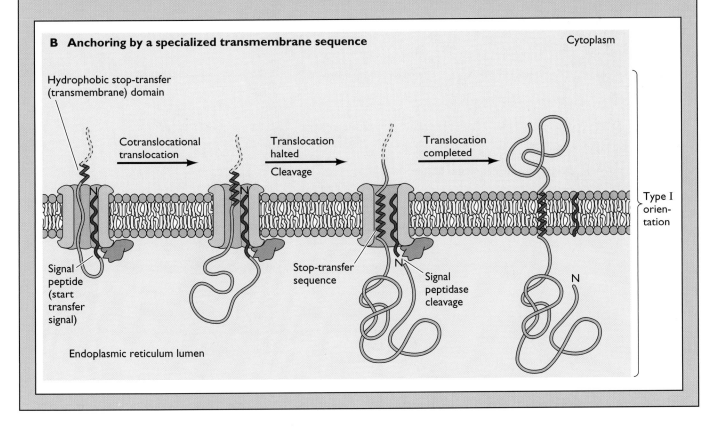

**B  Anchoring by a specialized transmembrane sequence**

**A**

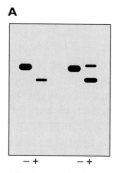

−  +        −  +

Tunica-    Endo H
mycin    *N*-Glycanase

**Figure 12.6 Detection and structure of N-linked oligosaccharides.** (A) Detection of N-linked oligosaccharides. Glycoproteins carrying N-linked oligosaccharides can be distinguished from nonglycosylated proteins by several methods. These include their labeling in cells supplied with radioactive sugars and their specific binding to lectins, proteins that bind to sugar residues. Alternatively, addition to cells of tunicamycin, an inhibitor of the first step in synthesis of the oligosaccharide precursor, prevents N-linked glycosylation, so that the mobility of glycoproteins is altered (left). In vitro treatment of glycoproteins with enzymes (glycosidases) that cleave within the oligosaccharide, such as endoglycosidase H (endo H), or *N*-glycanase can also alter glycoprotein mobility (right). Complex-type N-linked oligosaccharides (B) are not sensitive to the former enzyme. (B) The branched, mannose-rich oligosaccharide added via an *N*-glycosidic bond to asparagine residues of proteins is shown at the left. This common precursor is transferred to N-linked glycosylation sites as proteins are translocated into the ER, from the lipid carrier dolichol upon which the oligosaccharide is built, one sugar unit at a time. While still within the ER, three glucose residues and one mannose residue are trimmed from the core oligosaccharide. Trimming continues, and additional sugars are added as the protein travels through the Golgi apparatus (see Fig. 12.10) to form the three types of mature N-linked oligosaccharide chains shown below the precursor. These chains retain a common core derived from the precursor but differ in the terminal sugar units.

**B**

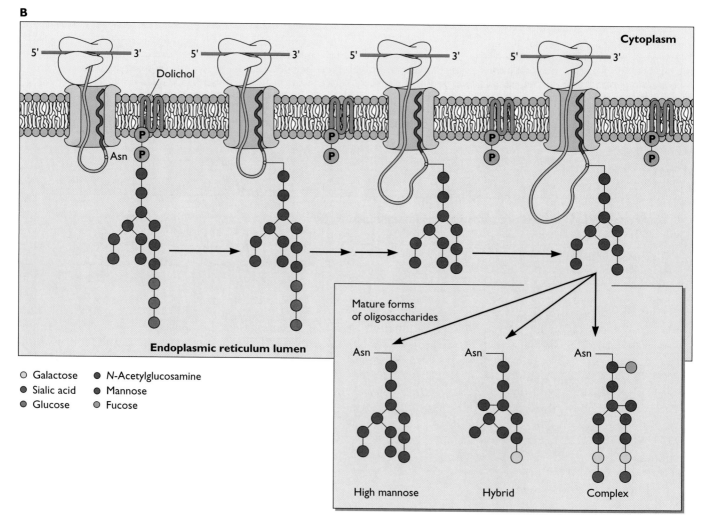

quently, several sugar residues are trimmed from the N-linked core oligosaccharide chains (Fig. 12.6B) in preparation for additional modifications that take place as the protein travels from the ER to the plasma membrane. Sites of N-linked glycosylation are characterized by the sequence NXS/T (where X = any amino acid except proline), but not every potential glycosylation site is modified, and even a single specific site within a protein is not necessarily modified with 100% efficiency. Each glycoprotein therefore exists as a heterogeneous mixture of **glycoforms,** varying in whether a particular site is glycosylated as well as in the composition and structure of the oligosaccharide present at that site. As many viral and cellular proteins contain a large number of potential N-linked glycosylation sites, single proteins can exist in an extremely large number of glycoforms. This property complicates investigation of the physiological function of oligosaccharide chains present on glycoproteins. Nevertheless, this modification has been assigned a wide variety of roles.

As essential components of receptors and ligands, oligosaccharide chains participate in many molecular recognition reactions. These include binding of certain hormones to their cell surface receptors, interactions of cells with one another, and binding of viruses such as influenza A virus, adenovirus, and herpesvirus to their host cells. Glycosylation at specific sites can also determine whether viral envelope proteins are recognized by components of the host's immune defense system. Some sugar units serve as signals, targeting proteins to specific locations, in particular to lysosomes. Glycosylation can also regulate the activity of proteins, for example, of certain hormones. This modification has also been suggested to fulfill more general functions, such as protecting proteins (and viruses) that circulate in body fluids from degradation. Many proteins contain such a large number of glycosylation sites that carbohydrate can contribute as much as 50% of the mass of the mature protein. The hydrophilic oligosaccharides are present on the surface of a protein, where they can form a sugar "shell," masking much of the protein's surface.

Studies of viral glycoproteins have emphasized the importance of glycosylation in protein folding. Elimination (by mutagenesis) of all sites at which the vesicular stomatitis virus G or influenza virus HA proteins are glycosylated blocks the folding of these proteins and their exit from the ER (see "Protein folding" below). The misfolded proteins aggregate and are eventually degraded by the cytoplasmic proteasome, following transport out of the ER. Before a protein folds, its hydrophobic amino acids, which are ultimately buried in the interior, are exposed. Such exposed hydrophobic patches on individual unfolded polypeptide chains tend to interact with one another nonspecifically, leading to aggregation. The hydrophilic oligosaccharide chains are believed to counter this tendency of unfolded polypeptides to form irreversible aggregates and thus to improve their solubility.

**Disulfide bond formation.**    A second chemical modification that is restricted to proteins entering the secretory pathway, and essential for the correct folding of many, is the formation of intramolecular disulfide bonds between pairs of cysteine residues (Fig. 12.3A). These bonds can make important contributions to the stability of a folded protein. However, they rarely form in the reducing environment of the cytoplasm. The ER lumen is more oxidizing and so provides an appropriate chemical environment for disulfide bond formation. This compartment contains high concentrations of protein disulfide isomerase, an enzyme that can catalyze the formation, reshuffling, or even the breakage of disulfide bonds under appropriate redox conditions. As formation of the full and correct complement of disulfide bonds in a protein is often the rate-limiting step in its folding, protein disulfide isomerase is an important catalyst of protein folding in the ER lumen.

**Protein folding.**    A number of other cellular proteins assist the folding of the extracellular domains of viral membrane glycoproteins as they enter the lumen of the ER. In contrast to the enzymes described above, these proteins do not alter the covalent structures of proteins. Rather, their primary function is to facilitate folding by preventing improper associations among unfolded, or incompletely folded, polypeptide chains such as the nonspecific hydrophobic interactions described above. Such **molecular chaperones** play essential roles in the folding of individual polypeptides and in the oligomerization of proteins. The ER chaperones include Grp78/Bip and calnexin. The former, which is a member of the Hsp70 family of stress response proteins, associates transiently with incompletely folded proteins, including the influenza virus HA0, vesicular stomatitis virus G, herpes simplex virus type 1 gB, and human immunodeficiency virus type 1 Env proteins.

Binding of Grp78/Bip, generally at multiple sites in a single protein molecule, is believed to protect against misfolding and aggregation by sequestering sequences prone to nonspecific interaction, such as hydrophobic patches. The binding to and release from unfolded proteins of Grp78/Bip are controlled by the hydrolysis of ATP bound to the chaperone at a site distinct from its peptide-binding site. Thus, multiple cycles of association with and dissociation from Grp78/Bip probably take place as a protein folds. Once the sequences to which the chaperone binds are buried in the interior of the protein, such interactions cease. For example, folded molecules of vesicular stomatitis virus G or influenza virus HA0 proteins that have acquired the full complement of correct disulfide bonds can no longer associate with Grp78/Bip.

Calnexin is an integral membrane protein of the ER that also binds transiently to immature proteins, as does a soluble relative of this protein, calreticulin, present in the ER lumen. In contrast to Grp78/Bip, which recognizes protein sequences directly, calnexin and calreticulin distinguish newly synthesized glycoproteins by binding to immature oligosaccharide chains. For example, the vesicular stomatitis virus G and influenza virus HA proteins bind to calnexin only when trimming of their oligosaccharide chains (Fig. 12.6B) is incomplete. Calnexin and calrecticulin appear to act at later folding stages than Grp78/Bip. Cycles of binding to and release from these chaperones prevent the aggregation of immature glycoproteins, such as the influenza virus HA0 protein. These interactions also prevent degradation of incompletely folded molecules, as well as premature formation of disulfide bonds or oligomerization.

Both types of chaperone perform important quality control functions in addition to assisting folding in the ER lumen. Protein molecules that do not fold correctly are not released from Grp78/Bip or calnexin. For example, a temperature-sensitive vesicular stomatitis virus G protein remains bound to calnexin, and hence to the ER membrane, at the restrictive temperature. These interactions prevent egress of improperly folded proteins from the ER and also target them for degradation. Consequently, nonfunctional proteins are cleared from the secretory pathway at an early step.

**Oligomerization.** Viral membrane proteins are generally oligomers (Table 12.1) that must assemble as their constituent protein chains are folded and covalently modified. Such assembly can begin in the ER, as the surfaces that mediate interactions among protein subunits adopt the correct conformation. For many proteins, these reactions are completed within the ER. For example, influenza virus HA0 protein monomers are restricted to the ER lumen, whereas trimers are found in this and all subsequent compartments of the secretory pathway (see next section). Some cellular and viral proteins cannot leave the ER unless correctly assembled. The heavy and light chains

**Figure 12.7 Compartments in the secretory pathway.** The lumen of each membrane-bound compartment shown is topologically equivalent to the exterior of the cell. Proteins destined for secretion or for the plasma membrane travel from the ER to the cell surface via the Golgi apparatus as indicated by the red arrows. However, proteins can be diverted from this pathway to lysosomes or to secretory granules that carry proteins to the cell surface for regulated release. The return of proteins from the Golgi apparatus to the ER is indicated (purple arrow). The endocytic pathway (blue arrows), discussed in chapter 5, and the secretory pathway intersect in endosomes and the Golgi apparatus.

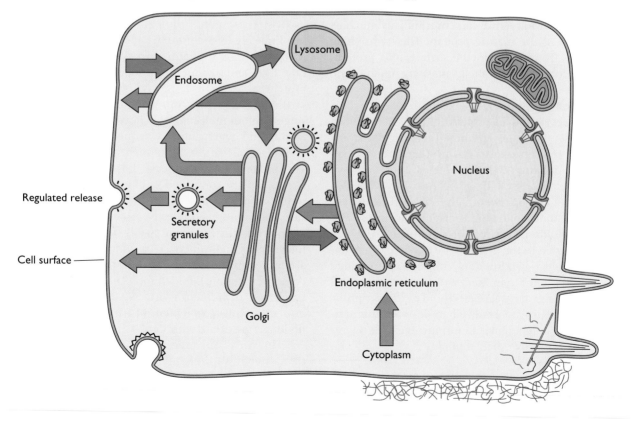

of major histocompatibility complex class I molecules must form heterodimers and bind peptide antigen in order to exit from this compartment. Similarly, the herpes simplex virus type 1 envelope glycoproteins gH and gL must interact with one another for the transport of either from the ER. Indeed, in the absence of gL protein, the gH protein cannot fold correctly. In other cases, oligomer assembly is

completed following exit from the ER: disulfide-linked dimers of the hepatitis B virus surface antigen form higher-order complexes in the next compartment in the pathway, and oligomers of the human immunodeficiency virus type 1 Env protein can be detected only in the Golgi apparatus. At present, we can discern no simple rule describing the relationship of oligomer assembly and trans-

**A**

**B**

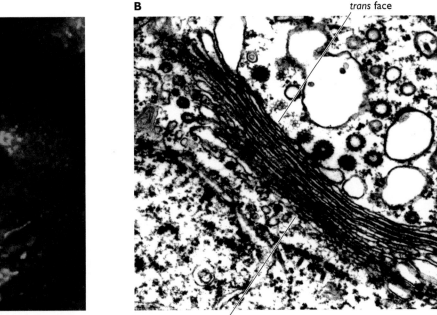

**C**

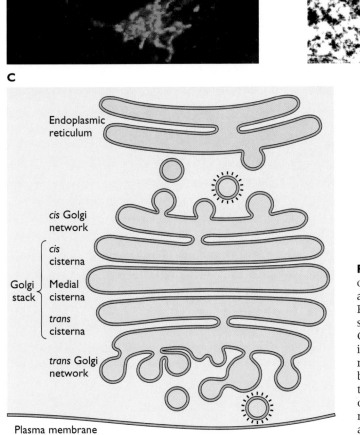

**Figure 12.8 The Golgi apparatus.** (A) Immunofluorescence of pig kidney epithelial cells following staining with antibody against a Golgi protein (p115). Courtesy of A. Brideau, Princeton University. (B) Electron micrograph of negatively stained sections of the alga *Ochromonas danica*. From G. T. Cole, Biological Photo Service. As can be seen, this organelle is usually located near the nucleus. It is composed of flattened membranous sacs, termed cisternae, bordered on either side by more convoluted structures of connected tubules and sacs, the *cis*- and *trans*-Golgi networks (C). Because the cisternae lie on top of one another, they are often called Golgi stacks. The number of cisternae varies considerably from one cell type to another and under different physiological conditions.

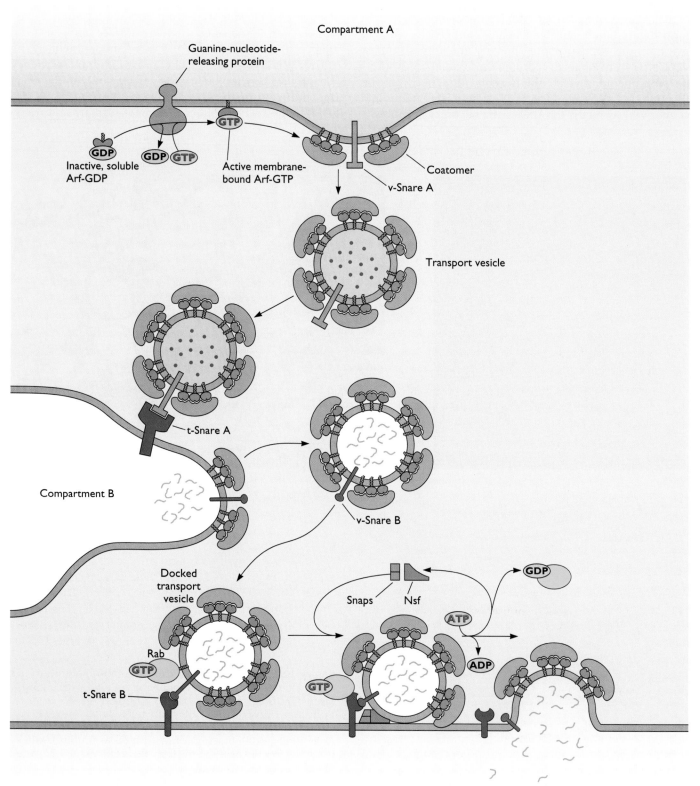

Compartment A

Guanine-nucleotide-
releasing protein

GDP

Inactive, soluble
Arf-GDP

GDP GTP

GTP

Active membrane-
bound Arf-GTP

v-Snare A

Coatomer

Transport vesicle

t-Snare A

Compartment B

v-Snare B

Docked
transport
vesicle

Snaps

Nsf

GDP

ATP

Rab

GTP

ADP

GTP

t-Snare B

Compartment C

◁ **Figure 12.9 Formation and fusion of transport vesicles in the secretory pathway, illustrated with vesicles formed with the Golgi coat complex, COPI.** Coatomer, the major component of COPI coats, contains seven subunits and can be isolated as a preassembled complex from cytoplasmic fractions of mammalian cells. Vesicle formation requires only one other soluble component, the small GTPase, ADP-ribosylation factor (Arf). Active GTP-bound Arf allows recruitment of coatomer, whose binding causes the membrane to bud. Ultimately, the membrane of the coated bud fuses to pinch off and release a coated vesicle. A similar mechanism with different GTP-binding- and coat proteins forms vesicles coated with the COPII coat, which participate in ER-to-Golgi apparatus transport. Vesicles coated with COPI mediate retrograde transport, for example, from the Golgi apparatus to the ER (Fig. 12.7), but their participation in forward (anterograde) transport cannot be excluded. The mechanisms that ensure that each vesicle is loaded with the appropriate cargo are not fully understood. Membrane receptors required for docking of transport vesicles with their target membranes have been proposed to define the specificity with which transport vesicles deliver their cargo. Each target membrane is characterized by a specific receptor, target Snare (t-Snare), which interacts only with a complementary vesicle Snare (v-Snare) present on a transport vesicle, as shown. The identification of various Snares and their essential role in membrane docking and fusion in vitro led to this Snare hypothesis. A requirement for a v-Snare on one membrane and a t-Snare on the second has been directly demonstrated by an elegant combination of genetics and biochemistry in yeast. Docking also requires members of the Rab family of small GTPases, which are specifically localized to different membranes of the cell and appear to regulate the specificity of docking and therefore of membrane fusion. Docking is followed by assembly of a membrane fusion complex via recruitment of two soluble components, *N*-ethylmaleimide-sensitive factor (Nsf) and soluble Nsf attachment proteins (Snaps, α or β, and γ). Nsf, which is an ATPase, and Snaps were originally identified in studies that used in vitro vesicle fusion assays. Whether this complex mediates fusion or whether additional proteins are needed is not known. However, the targeting Snare proteins may represent a minimal membrane fusion machine, for recombinant v- and t-Snare proteins promote docking and at least partial fusion of synthetic lipid vesicles in vitro.

port of membrane proteins from the ER. Nevertheless, oligomerization begins and in some cases may be completed within the ER, where it can be facilitated by the folding catalysts and chaperones characteristic of this compartment.

## *Vesicular Transport to the Cell Surface*

**Mechanism of vesicular transport.** Viral membrane proteins, like their cellular counterparts, travel to the cell surface through a series of membrane-bound compartments and vesicles. The first step in this secretory pathway, illustrated schematically in Fig. 12.7, is transport of the folded protein from the ER to the Golgi apparatus. Within the Golgi apparatus (Fig. 12.8), proteins are sorted according to the addresses specified in their primary sequences or covalent modifications. Proteins travel between compartments via **transport vesicles,** which bud from one compartment and move to the next. Fusion of the membranes of the transport vesicle and the target compartment then releases the cargo of the vesicle into the lumen of that compartment. Consequently, proteins that enter the secretory pathway upon translocation into the ER (and are correctly folded and modified) are never again exposed to the

cytoplasm of the cell. This strategy effectively sequesters proteins that might be detrimental, such as secreted or lysosomal proteases, and avoids exposure of disulfide-bonded proteins to a reducing environment.

Many soluble and membrane proteins participating in vesicular transport have been identified and characterized by biochemical, molecular, and genetic methods. Much of the rapid progress made in this field can be traced to the reconstitution of vesicular transport in vitro. The properties of these proteins suggest that similar mechanisms control the budding and fusion of different types of transport vesicles despite the stringent specificity requirements of vesicular transport. The general mechanism of vesicular transport is understood (Fig. 12.9), but many details of the molecular mechanisms that establish specificity in the loading of cargo or docking of vesicles remain to be elucidated. Budding of transport vesicles from the membranes of compartments of the secretory pathway requires proteins that form the external coats of the vesicles, such as the protein complex, termed COPII, and small GTPases (Fig. 12.9). Transport vesicles then move from compartment to compartment of the secretory pathway, shuttling proteins toward their final destinations. Such movement may be by passive diffusion over short distances, but there is increasing evidence for active transport of vesicles over both short and longer distances. Transport among compartments of the secretory pathway has recently been directly visualized in living cells by time-lapse video microscopy of green fluorescent protein fused to the cytoplasmic tail of the vesicular stomatitis virus G protein. When this protein leaves the ER at a site far from the Golgi apparatus, it enters small clusters of vesicles and sacs that appear to be randomly distributed throughout the cytoplasm. These membrane-bounded structures then move along microtubules to the Golgi apparatus. Transport to specialized membranes of polarized cells (see below) can be mediated by **motor proteins,** which travel along the tracks formed by microtubules driven by energy generated by the hydrolysis of ATP. It has also been proposed that transport vesicles formed within the Golgi apparatus cannot diffuse freely but rather are restricted to processive movement among cisternae by dynamic, fibrous strings that tether the vesicles transiently to one or two cisternae.

The high density of intracellular protein traffic requires considerable specificity during vesicle formation and fusion. For example, vesicles that transport proteins from the ER to the first compartment of the Golgi apparatus, the *cis*-Golgi network (Fig. 12.8), must take up only the appropriate proteins when budding from the ER and must fuse only with the membrane of the *cis*-Golgi network. Analogous specificity constraints apply to each vesicular transport step in the secretory pathway. The mechanisms that ensure specificity of cargo loading during formation of transport vesicles are not fully understood. However, both direct association of cargo proteins with certain of the proteins forming the vesicular coat and indirect interactions via "cargo carrier" proteins have been implicated in specific loading of ER-derived transport vesicles. Specific interactions between distinctive targeting proteins, called **Snares** (defined below), resident in the membranes of both the vesicles and the compartments with which they will fuse are believed to ensure that a particular type of transport vesicle docks only at the appropriate target membrane (Fig. 12.9). The docking reaction also requires small GTPases, termed Rab proteins. Animal cells contain the members of a large family of Rab GTPases, which are thought to regulate many steps in vesicular transport.

Once a vesicle docks at its target membrane, a complex containing proteins necessary to prepare for membrane fusion assembles in a regulated manner and juxtaposes the membranes that will fuse. The components of this complex include *N*-ethylmaleimide-sensitive fusion protein (**Nsf**) and soluble Nsf attachment proteins (**Snaps**), which shuttle between assembled complexes and the cytoplasm (Fig. 12.9). The Nsf protein associates with α-Snap, which in turn binds to Snap receptors (the Snares) in the membranes of the transport vesicle and the compartment with which it will fuse. The final reaction of vesicular transport, membrane fusion, has not been well characterized, and it is not known whether this process requires proteins in addition to those present in the prefusion complex. At present, the best understood examples of membrane fusion are those between enveloped viruses and the membranes of their host cells, such as the HA protein-mediated influenza virus-host cell membrane fusion process described in chapter 5.

**Reactions within the Golgi apparatus.** One of the most important staging posts in the secretory pathway is the Golgi apparatus, which is composed of a series of membrane-bounded compartments and located near the cell nucleus. Proteins enter the Golgi apparatus from the ER via the *cis*-Golgi network, which is composed of connected tubules and sacs (Fig. 12.8). A similar structure, the *trans*-Golgi network, forms the exit face of this organelle. The *cis*- and *trans*-Golgi networks are separated by a variable number of cisternae termed the *cis*, medial, and *trans* compartments (Fig. 12.8). Each of the compartments of the Golgi apparatus, which can comprise multiple cisternae, is the site of specific processing reactions. As illustrated in Fig. 12.10, the N-linked oligosaccharides attached to proteins in the ER are further processed as the proteins move through the Golgi stacks. The enzymes responsible for early reactions in maturation of these oligosaccharides are located in *cis* cisternae, whereas those that carry out later reactions are present only in the medial and *trans* compartments. Such spatial separation ensures that oligosaccharide processing follows a strict sequence as proteins pass through the compartments of the Golgi apparatus. This property has provided an important experimental tool with which to distinguish among and thus localize other reactions to the various Golgi compartments. Synthesis of O-linked oligosaccharides by glycosyltransferases, which add one sugar unit at a time to certain serine or threonine residues of a protein, also takes place in the Golgi apparatus.

## Viral Proteins in the Secretory Pathway

Many of the steps in the secretory pathway described in the previous section have been studied by using viral membrane glycoproteins, which offer several experimental advantages: such viral proteins are frequently synthesized in large quantities, their synthesis is initiated in a controlled fashion following infection, and their transport can be studied by both genetic and biochemical methods of analysis. Viral integral membrane proteins reach the cell surface just as if they were cellular proteins, as summarized in Fig. 12.11 for the influenza virus HA protein.

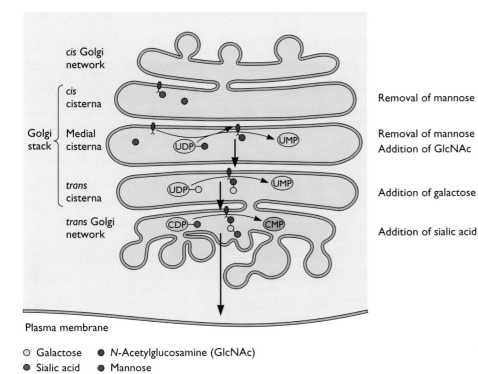

cis Golgi network

cis cisterna

Golgi stack

Medial cisterna

trans cisterna

trans Golgi network

Plasma membrane

○ Galactose    ● N-Acetylglucosamine (GlcNAc)
● Sialic acid    ● Mannose

Removal of mannose

Removal of mannose
Addition of GlcNAc

Addition of galactose

Addition of sialic acid

**Figure 12.10 Compartmentalization of processing of N-linked oligosaccharides.** The reactions by which mature N-linked oligosaccharide chains are produced from the high-mannose core precursor added in the ER (Fig. 12.6B) are shown in the compartment in which they take place. Trimming of terminal glucose and mannose residues of the common core precedes stepwise addition of the sugars found in the mature chains. These additions, which require activated sugar-nucleotide substrates, such as UDP-N-acetylglucosamine, are catalyzed by compartment-restricted transferases.

Along the way, they acquire disulfide bonds and N- and/or O-linked oligosaccharides and assemble into oligomers, again like the cellular proteins that travel this route. A number of viral envelope glycoproteins are also proteolytically processed before they reach the cell surface by cellular enzymes resident in late Golgi compartments.

Retroviral Env glycoproteins are cleaved in the trans-Golgi network to produce two polypeptide chains from the Env polyprotein precursor (Fig. 12.12). The smaller, derived from the C terminus of Env, is a transmembrane protein (TM). The larger, termed the surface protein (SU), associates with TM via noncovalent bonds (Fig. 12.12B), and in many simple retroviruses also disulfide bonds. Similarly, the HA0 protein of certain avian influenza A viruses is cleaved into the HA1 and HA2 chains in the trans-Golgi network (Fig. 12.11). These and other viral membrane proteins (Table 12.2) are processed by members of a family of resident Golgi proteases which cleave after pairs of basic amino acids. The members of this family, which in animal cells includes **furin** and at least five other proteins found in the trans-Golgi network, are serine proteases related to the bacterial enzyme subtilisin. Various furin family members have been shown by genetic and molecular methods to process viral glycoproteins; their normal function is to process cellular polyproteins such as certain hormone precursors.

These proteolytic cleavages are of more than esoteric interest; although not necessary for virus assembly, they can be essential for infectivity. For example, proteolytic processing of envelope proteins of retroviruses and alphaviruses is necessary for infectivity, probably because sequences important in fusion and entry become switched to an active conformation. It is well established that proteolytic processing of the influenza virus HA0 protein liberates the fusion peptide as the N terminus of HA2 (Fig. 12.3A). If HA0 is not cleaved, the fusion peptide is not available to initiate the infectious cycle. As discussed in more detail in chapter 17, virulent strains of avian influenza A virus encode HA0 proteins that can be processed by the ubiquitous furin family proteases such that virus particles carrying fusion-active HA protein are released. It seems likely that the common dependence on furin family proteases (Table 12.2), which act on proteins relatively late in the secretory pathway, helps minimize complications that would arise if viral glycoproteins were initially synthesized with their fusion peptides in an active conformation.

For at least some influenza A viruses, such as fowl plague virus, the ion channel activity of the viral M2 protein also helps to maintain HA in the fusion-incompetent conformation. The M2 protein, which forms a cation-selective ion channel, is found in infected cells at quite high

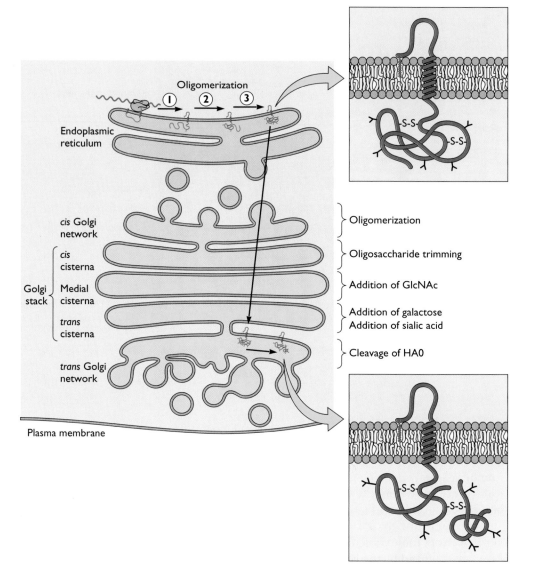

**Figure 12.11 Maturation of influenza virus HA0 protein during transit along the secretory pathway.** The covalent modifications that occur during transit of the influenza virus HA0 protein through the various compartments of the secretory pathway are listed or illustrated. In the ER, these are translocation and signal peptide cleavage (1), disulfide bond formation, and addition of N-linked core oligosaccharide (2), as the protein folds (3). The cytoplasmic domain acquires palmitate (green) while the protein travels to the plasma membrane, but it has not been established when this modification takes place. For simplicity, the protein is depicted as a monomer, although oligomerization also takes place in the ER lumen and the protein travels most of the pathway as a trimer. Note that the protein domain initially introduced into the ER lumen, in this case the N-terminal portion of the protein (type I orientation), corresponds to the extracellular domain of the cell surface protein.

concentrations, localized to the membranes of compartments of the secretory pathway. By increasing the pH of normally acidic compartments, such as those of the *trans*-Golgi network, this protein prevents premature switching of HA to the fusion-active conformation described in chapter 5, following cleavage of HA0 to HA1 and HA2. This activity of M2 appears to be particularly important during assembly and release of fowl plague virus because its HA protein switches to the fusion-competent conformation at a pH higher than that required by HA proteins of human influenza A viruses.

Although the many envelope proteins of viruses that assemble at the plasma membrane travel to the cell surface via the cellular secretory pathway, there is considerable variation in the rate and efficiency of their transport (Table 12.1). A champion is the influenza virus HA protein, which folds and assembles with a half-time of only 7 min, with more than 90% of the newly synthesized HA molecules reaching the cell surface. Many other viral proteins do considerably less well. Parameters determining the rate and efficiency of transport may include the complexity of the protein and the inherent asynchrony of protein folding. With few exceptions (see below), cellular proteins continue to enter and traverse the secretory pathway as enveloped viruses assemble at the plasma membrane. Thus, competition among viral and cellular proteins,

**A**

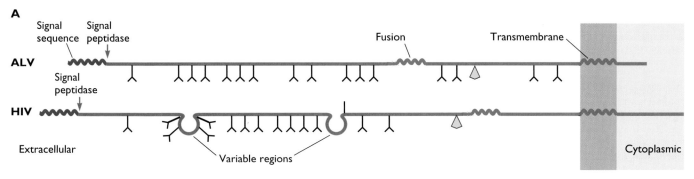

**B**

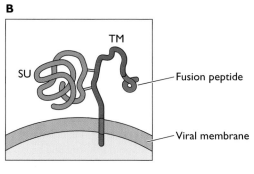

**Figure 12.12 Modification and processing of retroviral Env polyproteins.** Sequence features and modifications of the Env proteins of avian leukosis virus (ALV) and human immunodeficiency virus type 1 (HIV) (A) are depicted as in Fig. 12.3. The variable regions of human immunodeficiency virus type 1 Env differ greatly in sequence among viral isolates. The translocation products shown here are cleaved by the ER enzyme signal peptidase (red arrows) and by furin family proteases in the *trans*-Golgi network (yellow arrowheads). The cleavage products were originally named for their apparent molecular masses. The C-terminal proteins, which contain the transmembrane domains, are now known as TM, and the N-terminal products, which form most of the extracellular domain, are known as SU (surface protein). The interaction between the TM and SU envelope glycoproteins has not been examined by structural methods. Both noncovalent interactions and disulfide bonds (yellow lines) can maintain the association of the SU and TM proteins (B).

which may vary with the nature and physiological state of the host cell, is also likely to affect the transport of viral proteins to the cell surface.

We have focused our discussion of viral envelope proteins that travel the cellular secretory pathway on the well-understood maturation of their extracellular domains during transport to the cell surface. However, the cytoplasmic domains of these proteins are also frequently modified. Many are **acylated** by the covalent linkage of the fatty acid palmitate to cysteine residues in their cytoplasmic domains (Table 12.3). This modification can be necessary for optimally efficient production of progeny virions. For example, inhibition of palmitoylation of the Sindbis virus E2 glycoprotein inhibits virus assembly and budding. It has been suggested that the three large fatty acid chains that are attached to the short C-terminal cytoplasmic domain of this protein anchor this segment to the interior face of the membrane and thus fix the protein in a conformation competent for assembly. However, the functions of such acylation are not well established in other cases. For example, each of three highly conserved cysteine residues present in the short C-terminal cytoplasmic segment of the influenza A virus HA protein can become palmitoylated. The addition of up to three large, hydrophobic fatty acid chains to a molecule of HA would be expected to have a profound effect on its properties and functions, particularly as the three

cysteines are among the best conserved residues in the highly divergent HA protein sequences. However, efforts to identify the consequences of acylation, by elimination of the cysteine residues to which the fatty acid chains are added, have so far yielded contradictory results.

## Sorting of Viral Proteins in Polarized Cells

The fate of viral membrane proteins once they leave the *trans*-Golgi network was not discussed in detail in previous sections. Proteins that are not specifically targeted to an in-

**Table 12.2** Viral envelope glycoprotein precursors processed by furin family proteases

| Virus family | Precursor glycoprotein | Membrane-associated cleavage product(s) |
|---|---|---|
| Alphaviruses | PE2 | E2 |
| Flaviviruses | PrM | M |
| Paramyxoviruses | F0 | F1, F2 |
| Orthomyxoviruses[a] | HA0 | HA1, HA2 |
| Retroviruses | Env | TM, SU |
| Hepadnaviruses | Pre-C | C antigen[b] |

[a]Virulent strains of avian influenza A virus 2.

[b]This cleavage product is largely secreted into the extracellular medium but is also associated with the plasma membrane of infected cells. Cleavage of preC is the only processing reaction listed that has not been implicated in the production of infectious virus particles.

**Table 12.3** Examples of acylated or isoprenylated viral proteins

| Virus | Protein(s) | Lipid | Probable function |
|---|---|---|---|
| Picornavirus | | | |
| Poliovirus | VP0, VP4 | Myristate | Virion assembly; uncoating |
| Togaviruses | | | |
| Sindbis virus | E2 | Palmitate | Efficient budding of virus particles |
| Semliki Forest virus | Nsp1 | Palmitate | Tight association with plasma membrane; movement of Nsp1 (capping enzyme) to filopodia |
| Rhabdovirus | | | |
| Vesicular stomatitis virus | G | Palmitate (cytoplasmic domain) | Unknown |
| Orthomyxovirus | | | |
| Influenza A virus | HA | Palmitate | Unknown |
| | M2 | Palmitate | Unknown |
| Retroviruses | | | |
| Human immunodeficiency virus type 1, murine leukemia virus | Gag, MA | Myristate | Membrane association, assembly, and budding |
| Human immunodeficiency virus type 1 | Env, TM | Palmitate | Unknown |
| Rous sarcoma virus | pp60$^{src}$ | Myristate | Membrane association and transformation (chapter 16) |
| Harvey sarcoma virus | p21$^{ras}$ | Farnesol | Membrane localization |
| Papovaviruses | | | |
| Simian virus 40, polyomavirus | VP2 | Myristate | Virion assembly |
| Hepadnavirus | | | |
| Hepatitis B virus | L (pre-S1) | Myristate | Initiation of infection |
| Hepatitis delta satellite virus | Large delta antigen | Geranylgeranol | Interaction with hepatitis B virus; assembly; inhibition of hepatitis delta satellite virus RNA replication |
| Poxvirus | | | |
| Vaccinia virus | p37 | Palmitate | Targeting p37 to membranes of *trans*-Golgi network; needed for efficient envelopment and release of particles |

tracellular address travel on to the plasma membrane. However, the plasma membrane is not uniform in all animal cells: differentiated cells often devote different parts of their surfaces to specialized functions, and the plasma membranes of such **polarized cells** are divided into correspondingly distinct regions. Many enveloped viruses recognize the cell surface asymmetries of such cells not only during virus entry but also by sorting virion components to a specific plasma membrane region prior to assembly. In this section, we describe the final steps in the sorting of proteins to specialized plasma membrane regions in two types of polarized cell in which animal viruses often replicate, epithelial cells and neurons (Fig. 12.13).

### Epithelial Cells

As discussed in chapter 4, epithelial cells, which cover the external surfaces of vertebrates and line all their internal cavities (such as the respiratory and gastrointestinal tracts), are primary targets of virus infection. The cells of an epithelium are organized into close-knit sheets both by the tight contacts they make with one another and by their interactions with the underlying basal lamina, a thin layer of extracellular matrix (Fig. 4.1). Within the best-characterized epithelia, such as those that line the intestine, each cell is divided into a highly folded **apical** domain exposed to the outside world and a **basolateral** domain (Fig. 4.1 and 12.13). The former performs more

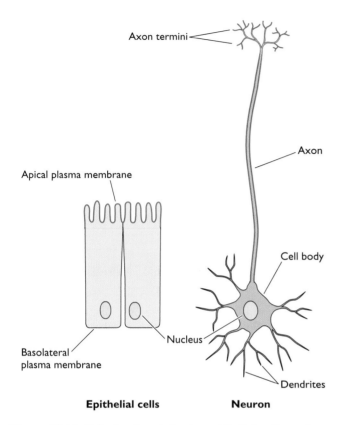

**Figure 12.13 Polarized vertebrate epithelial cells and neurons.** The apical and basolateral membrane domains of an epithelial cell are analogous to the axonal and dendritic (plus cell body) membranes, respectively, of a neuron with respect to the mechanisms used to direct proteins to them.

specialized functions, whereas the latter is associated with cellular housekeeping. These two domains differ in their protein and lipid contents, in part because they are separated by specialized cell-cell junctions (tight junctions) (Fig. 4.1) which prevent free diffusion and mixing of components in the outer leaflet of the lipid bilayer. However, such physical separation does not explain how the polarized distribution of plasma membrane proteins is established and maintained.

Viruses have been important tools in efforts to elucidate the molecular mechanisms responsible for the polarity of typical epithelial cells because certain enveloped viruses bud asymmetrically. For example, in the epithelial cells of all organs studied, influenza A virus buds apically and vesicular stomatitis virus buds basolaterally. The polarity of budding is generally the result of accumulation of their envelope proteins, HA and G, in the apical and basolateral domains, respectively, of the plasma membrane. The most common mechanism for selective localization appears to be signal-dependent sorting of proteins in the *trans*-Golgi network, in which proteins destined for apical or basolateral membrane domains are distinguished and packaged into appropriately targeted transport vesicles. Signals necessary for basolateral targeting comprise short sequences located in the cytoplasmic domains of membrane proteins. Many basolateral targeting signals overlap with those for the localization of proteins to clathrin-coated pits, the first staging post of the endocytic pathway. Indeed, in polarized epithelial cells, certain proteins are transferred through endosomes from one membrane domain to the other, a process termed **transcytosis.**

The signals that direct proteins to the apical membrane are currently a matter of debate, but both N glycosylation and association of proteins with specific lipids may be important. The apical membranes of polarized epithelial cells are enriched in sphingolipids and cholesterol, which form specific microdomains called **sphingolipid-cholesterol rafts**. These structures have been proposed to mediate apical transport of proteins that bind to them. For example, the influenza virus HA protein binds specifically to sphingolipid-cholesterol rafts by its transmembrane domain, but the analogous membrane-spanning segments of viral proteins that are not sorted to apical membranes do not bind.

The specific advantages of polarized assembly and budding of viruses are not fully appreciated. However, as discussed in chapter 17, this process can facilitate virus spread within or among host organisms.

### Neurons

Neurons are probably the most dramatically specialized among the many polarized cells of vertebrates. The axon is typically long and unbranched, whereas the dendrites form an extensive branched network of short projections (Fig. 12.13). Axons are specialized for the transmission of nerve impulses, ultimately via the formation of synaptic vesicles and release of their contents. In contrast, dendrites provide a large surface area for the receipt of signals from other neurons. The nucleus, the rough ER, and the Golgi apparatus are also located in the dendritic region and the cell body of a neuron, which therefore contain the biosynthetic machinery of the cell. Although axonal and dendritic surfaces are not separated by tight junctions, proteins must be distributed asymmetrically in neurons. Several mechanisms contribute to the establishment and maintenance of neuronal polarity. These mechanisms include transport of vesicles in specific directions along the highly organized microtubules of the axon by molecular motors that travel toward one of the two ends (− and +) of a microtubule (**axonal transport**) (Fig. 12.14) and transport of mRNA to dendrites. However, it is believed that sorting and targeting of membrane proteins for delivery to axonal or dendritic surfaces is also essential for po-

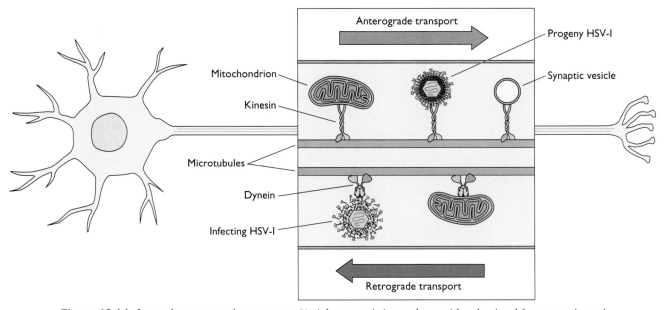

**Figure 12.14 Axonal transport in neurons.** Vesicles containing polypeptides destined for synaptic vesicles (e.g., neurotransmitters) are formed following translocation of the polypeptides into the ER and transport through the Golgi apparatus. Once post-Golgi vesicles are formed and become associated with the molecular motor kinesin, they are transported down the axon on the tracks formed by axonal microtubules. The force generated by kinesin allows transport at rates of 200 to 400 mm/day. Mitochondria are transported to synaptic bulbs in similar fashion. Transport toward the cell body, for example, of degenerating mitochondria, is believed to be mediated by the cytoplasmic motor dynein. Herpes simplex virus type 1 (HSV-1) components can be transported in both directions in infected neurons.

larization. It seems likely that similar mechanisms of vesicle sorting operate in neurons and epithelial cells, for some proteins that travel to the apical or basolateral surfaces of epithelial cells are targeted to the axonal and dendritic membranes, respectively, of neurons.

The directional movement of vesicles and many cellular organelles in neurons is dependent on polarized microtubules and direction-specific motor molecules, which mediate transport both toward the cell body from axons and dendrites and away from the cell body (Fig. 12.14). Viruses that infect neurons depend on these mechanisms during infection, assembly, and egress of particles. An important example is provided by the neurotropic alphaherpesviruses, a group that includes herpes simplex virus type 1 and varicella-zoster virus. Upon infection of sensory neurons, herpesvirus capsids are transported to the nucleus by microtubule-based transport, probably mediated by (−)-end-directed motors such as dynein (Fig. 12.14). The molecular interactions of virion components and these motors are not well understood. After replication and assembly, the enveloped virus is transported away from the neuronal cell body, presumably by distinctive transport mechanisms. A remarkable finding is that the spread of herpesviruses from neuron to neuron occurs at or near sites of synaptic contact, indicating that virus par-

ticles must be targeted to specific areas within neurons for egress. As discussed in chapter 17, this attribute can be exploited to define neuronal connections in a living animal by using the virus as a self-amplifying tracer.

## Inhibition of Transport of Cellular Proteins by Viral Infection

A few viruses encode proteins that interfere with the transport of specific cellular proteins. One such protein is the adenoviral E3 (appendix, Fig. 15B) glycoprotein, gp19, which blocks transport of major histocompatibility complex class I molecules to the cell surface. The E3 gp19 protein binds to these important components of immune defense in the ER and is believed to prevent their exit from this compartment by inhibiting folding and assembly. Two ER transmembrane proteins of the betaherpesvirus human cytomegalovirus also block transport of major histocompatibility complex class I molecules to the cell surface. These proteins, US11 and US2, induce transport of major histocompatibility complex class I molecules from the ER to the cytoplasm for rapid degradation by the cellular proteasome. When this activity was first discovered (for the US11 protein), such **retrograde transport** from the ER to cytoplasm appeared very unusual. However, it is now clear that this is a normal activity of uninfected cells mediated by the translocon that is responsible for cotranslational entry of proteins into the ER.

The human immunodeficiency virus type 1 Vpu protein (appendix, Fig. 9B) also operates within the ER lumen. This transmembrane phosphoprotein stimulates the release of virus particles from infected cells and induces selective degradation of newly synthesized cellular membrane proteins, such as major histocompatibility complex class I proteins and CD4, the major human immunodeficiency virus type 1 receptor. Within the ER, the Env polyprotein binds tightly to CD4. This interaction prevents transit of both proteins to the cell surface, presumably because neither is correctly folded or modified. The Vpu protein binds to CD4 present in Env-CD4 complexes and induces selective degradation of CD4. Consequently, Env is released and can exit from the ER. Like degradation of major histocompatibility complex class I proteins in cells infected by human cytomegalovirus, the Vpu-induced destruction of CD4 depends on the cellular proteasome and its ubiquitin targeting system. Surprisingly, however, CD4 is not transported out of the ER to the cytoplasm, for it can travel along the secretory pathway when proteasome activity is inhibited. This unexpected property suggests that additional mechanisms for the degradation of proteins that enter the ER remain to be elucidated. Such Vpu-dependent degradation of CD4 makes a significant contribution to the decreased cell surface concentrations of this protein that are characteristic of human immunodeficiency virus type 1-infected T cells.

Proteins encoded by certain other viruses exert more drastic effects on the cellular secretory pathway. For example, rotaviruses, which lack a permanent envelope but are transiently membrane enclosed during assembly, encode a protein that disrupts the ER membrane. This protein is believed to allow removal of the transient envelope during virion assembly. Poliovirus and other enteroviruses induce an even more dramatic reorganization of the compartments of the cellular secretory pathway. Virus-induced membranous vesicles, 40 to 50 nm in diameter, appear during the poliovirus infectious cycle and eventually fill the cytoplasm of the infected cell (Fig. 12.15A). These vesicles, whose surface appearance is different from that of transport vesicles, are derived from the membranes of the compartments of the secretory pathway. Indeed, late in infection no Golgi apparatus can be seen in poliovirus-infected cells. Not surprisingly, such dismantling of the cellular secretory pathway results in the inhibition of protein traffic to the cell surface. Because poliovirus particles lack an envelope, the virus has no need for the transport of envelope glycoproteins to the cell surface. However, formation of membranous vesicles in poliovirus-infected cells is essential for replication of the viral RNA genome. As discussed in chapter 6, the replication machinery is membrane-associated, and replication is blocked when vesicle formation is inhibited. These infected-cell-specific vesicles are also the sites of assembly of progeny virions and might participate in the nonlytic release of virus particles described in chapter 13. Furthermore, inhibition of transport of cellular proteins is likely to be advantageous to the virus because cell surface expression of viral antigens bound to major histocompatibility complex class I proteins on the infected cell surface, and consequently detection of the infected cell by immune surveillance systems of the host, is prevented.

Each of the poliovirus 2BC, 2B, and 3A proteins inhibits vesicular transport from the ER when synthesized in mammalian cells in the absence of other viral proteins. The 2B protein can induce disassembly of the Golgi apparatus, whereas the 3A protein appears to preferentially inhibit transport from the ER to the Golgi apparatus. All these viral proteins are found in cytoplasmic membranes of infected cells. The molecular mechanisms by which they divert the normal forward flow of vesicular traffic to the formation of virus-specific vesicles are not yet clear. However, the sensitivity of this process to brefeldin A, a drug that specifically inhibits forward transport from the ER to and through the Golgi stacks, suggests that cellular mechanisms of transport vesicle formation are subverted by the virus (Fig. 12.15B).

## Signal Sequence-Independent Transport of Viral Proteins to the Plasma Membrane

As discussed in chapter 3, many enveloped viruses contain proteins lying between, and making contact with, the inner surface of the membrane of the particle and the capsid or nucleocapsid. In contrast to the integral membrane proteins of enveloped viruses, such internal virion proteins do not enter the secretory pathway but are synthesized in the cytoplasm of an infected cell and directed to the membrane sites at which assembly occurs by membrane-targeting signals. The membrane-targeting signals of some viral proteins are built into their primary sequences and are therefore analogous to the nuclear localization signals or signal sequences described in previous sections. However, in many cases posttranslationally added lipid chains are important components of membrane-targeting signals.

### Lipid Plus Protein Signals

It has been known for many years that cytoplasmic proteins can be modified by the covalent addition of lipid chains (Table 12.3). Best characterized is the addition of the 14-carbon, saturated fatty acid myristate to N-terminal glycine residues or of unsaturated polyisoprenes, such as farnesol ($C_{15}$) or geranylgeranol ($C_{20}$), to a specific C-terminal sequence (Fig. 12.16). Palmitate is also added to

**A**

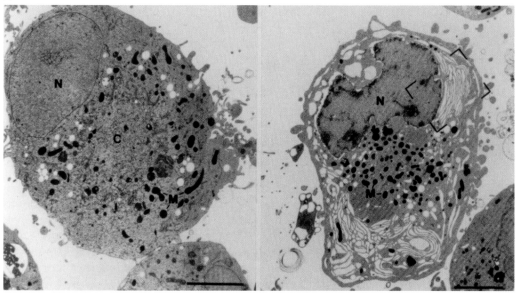

**B**
**Uninfected cells**

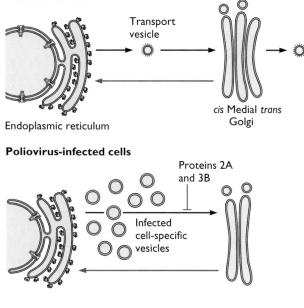

Transport vesicle

Endoplasmic reticulum

*cis* Medial *trans*
Golgi

**Poliovirus-infected cells**

Proteins 2A and 3B

Infected cell-specific vesicles

**Figure 12.15 Inhibition of the cellular secretory pathway in poliovirus-infected cells.** (A) Electron micrographs of Cos-1 monkey cells following introduction of a control plasmid (left) and a plasmid from which the poliovirus 3A protein is expressed (right). Frozen cells were fixed and negatively stained at low temperature prior to sectioning and electron microscopy. Bar = 4 μm. Note the high concentration of membranes of the ER and Golgi apparatus in the center of the cell (C) that does not express the 3A protein (left) and their replacement by dilated membranes with tubular and whorled morphologies when the 3A protein is synthesized (right). N, nucleus. From J. R. Doedens et al., *J. Virol.* **71:**9054-9064, 1997, with permission. Courtesy of K. Kirkegaard, Stanford University School of Medicine. (B) Model for inhibition of vesicular transport from the ER to the cell surface in poliovirus-infected cells, showing a block to vesicular transport by the 2B and 3A proteins between the ER and Golgi apparatus. However, inhibition may occur at other points in the pathway. The block leads to disassembly of the Golgi apparatus and formation of infected-cell-specific vesicles, which are the sites of virus-specific RNA synthesis. The drug brefeldin A, which specifically inhibits anterograde transport from the ER to and through the Golgi apparatus, inhibits poliovirus-induced vesicle proliferation and consequently viral RNA synthesis. This property suggests that the mechanism by which poliovirus-infected cell-specific vesicles are formed is related to the normal mechanism of budding of transport vesicles.

some viral proteins that do not enter the secretory pathway (Table 12.3). The discovery that transforming proteins of oncogenic retroviruses, the Src and Ras proteins, are myristoylated or isoprenylated, respectively, led to a resurgence of interest in these modifications. As discussed in chapter 16, lipid addition is essential for the function of these transforming proteins. In this section, we focus on myristoylation and isoprenylation of viral structural proteins synthesized in the cytoplasm.

Myristoylation of the cytoplasmic Gag proteins of retroviruses and its consequences have been examined in most detail. The internal structural proteins of these viruses, MA (matrix), CA (capsid), and NC (nucleocapsid), are generated by proteolytic cleavage of the Gag polyprotein. The Gag proteins of the majority of retroviruses, but curiously not of avian leukosis virus, are myristoylated at their N-terminal glycines. Mutations that alter the sequence at which murine leukemia virus or human immunodefi-

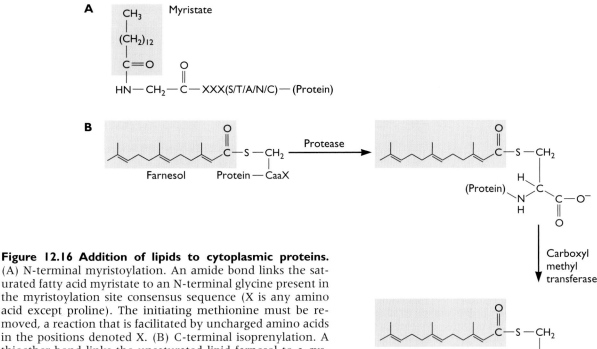

**Figure 12.16 Addition of lipids to cytoplasmic proteins.** (A) N-terminal myristoylation. An amide bond links the saturated fatty acid myristate to an N-terminal glycine present in the myristoylation site consensus sequence (X is any amino acid except proline). The initiating methionine must be removed, a reaction that is facilitated by uncharged amino acids in the positions denoted X. (B) C-terminal isoprenylation. A thioether bond links the unsaturated lipid farnesol to a cysteine in the isoprenylation consensus sequence (a is an aliphatic amino acid). In many proteins, isoprenylation is followed by proteolytic cleavage to expose the C-terminal cysteine, which is then methylated.

ciency virus type 1 Gag proteins are myristoylated prevent interaction of this protein with the cytoplasmic face of the plasma membrane, induce cytoplasmic accumulation of Gag, and inhibit virus assembly and budding. In the case of the human immunodeficiency virus type 1 Gag protein, the myristoylated N-terminal segment and a highly basic sequence located a short distance downstream (Fig. 12.17) form a bipartite signal, described in detail in chapter 3, which allows membrane binding in vitro and virus assembly and budding in vivo.

The hepatitis B virus large surface (L) protein is also myristoylated at its N terminus. However, in contrast to retroviral Gag protein, this hepadnaviral protein is an envelope protein of the virion. Modification of its N terminus must therefore occur while it traverses the secretory pathway. In this case, myristoylation is not necessary for assembly or release of virions but is required for infection of primary hepatocytes. Thus, this modification of the hepatitis B virus L protein appears to be necessary for the initial interaction of the virus with, or its entry into, the host cell.

More surprising is the myristoylation of structural proteins of poliovirus (VP4) and polyomavirus (VP2): although these virions do not contain an envelope, such

modification is necessary for efficient assembly of both viruses. For example, assembly of progeny particles and production of mature (infectious) virons are inhibited when the poliovirus VP0 polyprotein cannot be myristoylated. In mature virions, the myristate chain at the N terminus of VP4, which is processed from VP0, interacts with the VP3 protein on the inside of the capsid (Fig. 3.9B). Thus, the hydrophobic lipid chain must facilitate protein-protein interactions necessary for the production of virions. The fatty acid is also important during entry into cells and the uncoating of poliovirions at the beginning of an infectious cycle.

Among viral structural proteins, only the hepatitis delta satellite virus large delta protein has been found to be isoprenylated. Such isoprenylation is necessary but not sufficient for binding of this protein to the S protein of the helper hepatitis B virus during assembly of hepatitis delta satellite virus. As this hepatitis B virus protein is an integral membrane protein, the isoprenoid tail of large delta protein seems likely to facilitate its interaction with the plasma membrane adjacent to the cytoplasmic domain of S protein in cells infected by both hepatitis B virus and the hepatitis delta virus satellite.

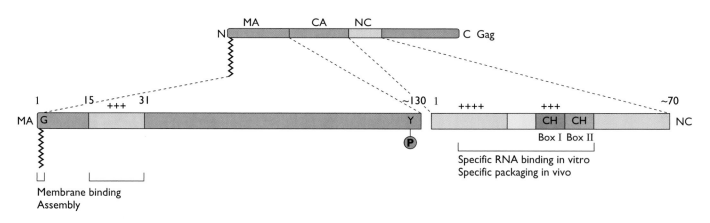

**Figure 12.17 Human immunodeficiency virus type I Gag proteins and their targeting signals.** The locations of the internal structural proteins of human immunodeficiency virus type 1, MA (matrix), CA (capsid), and NC (nucleocapsid), in the Pr55 Gag polyprotein are shown at the top. Sequence features and localization signals (MA) or the RNA-binding domain (NC) are shown below. The lengths of the MA and NC proteins are listed as approximate because of the variation among human immunodeficiency virus type 1 isolates. Specific amino acids are given in the single-letter code. The basic region of MA is not required for membrane binding in simple retroviruses such as avian sarcoma virus. The CH boxes of NC contain three cysteines and one histidine, and each coordinates one atom of $Zn^{2+}$. CH box I is conserved among retroviruses, but CH box II is not.

The addition of lipid chains to virion proteins allows association of these proteins with the cytoplasmic face of the plasma membrane, providing opportunities for interactions with viral membrane glycoproteins that reach the cell surface by the secretory pathway. Therefore, we can now explain (at least in principle) how cytoplasmic structural proteins, such as the Gag and Gag-Pol polyproteins of most retroviruses, associate with the plasma membrane, the site of viral envelope proteins and of virus assembly. It is also clear that the lipid chains added to viral proteins in the cytoplasm can do more than simply provide a "greasy foot" for membrane binding: they can stabilize virions, facilitate virus-host cell interactions, and participate in uncoating reactions.

### Protein Sequence Signals

Although the majority of viral membrane-associated proteins that do not enter the secretory pathway receive lipid chains to facilitate membrane binding, some do not. Among the most important in this latter class are the matrix proteins of several families of (−) strand RNA viruses. Within virions, these proteins lie between the nucleocapsid and the envelope and are essential for correct localization and packaging of RNA genomes. During assembly, matrix proteins, such as M of vesicular stomatitis virus and M1 of influenza A virus, must bind to the plasma membrane of infected cells. These proteins are produced in the cytoplasm but, as far as is known, receive no lipid after

translation. Direct membrane binding in vivo appears to be an intrinsic property of matrix proteins. For example, when the influenza virus M1 protein is expressed in host cells in the absence of any other viral proteins, a significant fraction is tightly associated with cellular membranes. Indeed, both this protein and the vesicular stomatitis virus M protein contain specific sequences that are necessary for their interaction with the plasma membrane in vivo or with lipid vesicles in vitro (Fig. 12.18). This region of the influenza A virus M1 protein contains two hydrophobic sequences (Fig. 12.18A), which might form a hydrophobic surface structure in the folded protein. In addition to an internal hydrophobic segment, membrane association of the vesicular stomatitis virus M protein requires a highly charged, basic N-terminal sequence (Fig. 12.18B). This latter segment might participate directly in membrane binding, like the basic sequence of the human immunodeficiency virus type 1 Gag membrane-targeting signal, or it might stabilize a conformation of the internal sequence favorable for interaction with the membrane.

Assembly of (−) strand RNA viruses requires localization of their matrix proteins to the plasma membrane rather than to the internal membranes of the infected cell. In some cases, such as the vesicular stomatitis virus M protein, specificity for binding to the plasma membrane is an intrinsic property of the protein. However, binding of matrix proteins to the cytoplasmic tails of viral envelope glycoprotein can also be an important determinant of specific

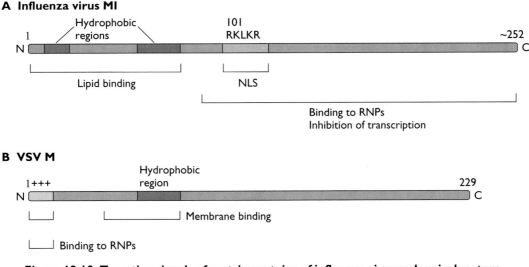

**Figure 12.18 Targeting signals of matrix proteins of influenza virus and vesicular stomatitis virus (VSV).** Sequence features of specific segments of the proteins and the boundaries of targeting and RNP-binding domains are indicated. With the exception of the nuclear localization signal (NLS) of the influenza virus M1 protein (amino acids shown in single-letter code), these domains have not been defined in detail.

localization of these proteins. For example, the cytoplasmic domains of both the NA and HA proteins of influenza A virus stimulate membrane binding by M1 protein, with that of NA protein having the greater effect. Similarly, membrane binding by the matrix protein of Sendai virus (a paramyxovirus) is independently stimulated by the presence of either of the two viral glycoproteins (F or HN) in the membrane.

## Interactions with Internal Cellular Membranes

A variety of viruses acquire their envelopes from internal membranes of the infected cell rather than from the plasma membrane. The majority assemble at the cytoplasmic faces of compartments of the secretory pathway and bud into them, but herpesviruses initially acquire an envelope from the inner nuclear membrane (Table 12.4). Although a single budding reaction is typical, the more complex herpesviruses and poxviruses interact with multiple internal membranes of the cell during assembly and exocytosis. These pathways are described in detail in chapter 13. Even some viruses whose mature forms lack an envelope, such as rotaviruses, can transiently enter the secretory pathway (Table 12.4).

The diversity of the internal membranes with which these viruses associate during assembly and exocytosis (Table 12.4) is the result of variations on a single mechanistic theme: the site of assembly is determined by the in-

tracellular localization of the viral integral membrane protein(s), which traverses the initial stages of the secretory pathway (Fig. 12.19), just as assembly at the plasma membrane is the result of transport of viral membrane proteins to that site. Assembly of viruses at internal membranes therefore requires transport of membrane proteins to, and their retention within, appropriate intracellular compartments.

## Localization of Viral Proteins to Compartments of the Secretory Pathway

The bunyaviruses, a family that includes Uukuniemi virus and Hantaan virus, are among the best-studied viruses that bud into compartments of the secretory pathway, in this case compartments of the Golgi apparatus. Bunyavirus virions contain two integral membrane glycoproteins, designated G1 and G2, which are encoded within a single open reading frame of the M genomic RNA segment. In contrast to the viral envelope proteins discussed previously, the bunyaviral polyprotein containing G1 and G2 is processed cotranslationally by the signal peptidase as the precursor enters the lumen at the ER. However, association of G1 with G2 is required for transport to and retention in Golgi compartments. For example, when expressed independently of one another from vaccinia virus vectors, the G1 protein of Uukuniemi virus localizes to the Golgi complex in normal fashion, but the G2 protein fails to leave the ER. When the two glycoproteins are coex-

**Table 12.4**   Interactions of viruses with internal cellular membranes

| Virus family | Example | Integral membrane protein(s) | Intracellular membrane(s) | Mechanism of envelopment |
|---|---|---|---|---|
| Togaviruses | Rubella virus | E1-E2 | Medial and *trans*-Golgi cisternae | Budding into Golgi cisternae |
| Bunyaviruses | Uukuniemi virus, Hantaan virus | G1, G2 | *cis*- and medial Golgi cisternae | Budding into Golgi cisternae |
| Coronaviruses | Mouse hepatitis virus | M, S | ER, *cis*-Golgi network | Budding into ER and *cis*-Golgi network |
| Rotaviruses | Simian rotavirus | VP7, NS28 | ER | Budding into ER |
| Hepadnaviruses | Hepatitis B virus | L, M, S | ER and other compartments | Budding into ER? |
| Herpesviruses | Herpes simplex virus type 1 | gB | ER, inner and outer nuclear envelope, plasma membrane | Budding at an inner nuclear membrane |
| Poxviruses | Vaccinia virus (immature virus) | A14L, A13L, A17L | *cis*-Golgi network | Engulfment of a particle. |
| | Vaccinia virus (intracellular mature virus) | A56R(HA), F13L, B5R | Late Golgi or post-Golgi compartment | Engulfment to form a structure with four membranes |

pressed, the G2 protein is transported to the Golgi complex. The G1 protein may therefore contain signals necessary for transport along the secretory pathway, or it may be required for correct folding of the G2 protein and its exit from the ER. Both visualization of the intracellular distribution of the G1 and G2 proteins and the structures of their N-linked oligosaccharides indicate that these proteins do not reach the *trans*-Golgi network but accumulate in *cis* or medial compartments. The signals that specify such localization have not been precisely identified but are included within the transmembrane domain and the adjacent segment of the cytoplasmic domain of the G1 protein.

Golgi cisternae are by no means the only compartments of the secretory pathway whose membranes are exploited by viruses. For example, rotaviruses assemble at and bud into the ER but do not retain the envelope they thus acquire, whereas coronaviruses use both the ER and the *cis*-Golgi network. In both cases, it is the localization of viral membrane glycoproteins to specific cellular membranes (Table 12.4) that determines the sites of assembly and budding. As discussed in more detail in chapter 13, the poxvirus vaccinia virus acquires membranes from both the *cis*-Golgi network and a late Golgi or post-Golgi compartment (Table 12.4). This virus does not bud into these cisternae but rather induces engulfment of immature particles by their membranes, a process often called **wrapping.** Thus, unlike enveloped viruses formed by budding, which acquire a typical lipid bilayer, vaccinia virus particles acquire two membranes at each wrapping step. These reactions are not fully understood, but several viral membrane proteins that localize to specific Golgi compartments and are required for the first or second membrane-wrapping reaction have been identified (Table 12.4). As for sim-

pler viruses, it is the localization of these proteins to a specific intracellular membrane that is the first step in acquisition of that membrane by the assembling virus.

## Localization of Viral Proteins to the Nuclear Membrane

The pathways by which herpesviruses are produced in and released from infected cells are unusual, for these are the only enveloped viruses that are known to initiate assembly within, and bud from, the nucleus. The initial association of an assembling herpesvirus with a cellular membrane is therefore budding of the nucleocapsid through the inner membrane of the nuclear envelope. As the outer nuclear membrane is continuous with that of the ER (Fig. 12.19), the nucleocapsid enters the ER lumen. For many years, it was generally believed that the envelope of herpesviruses is acquired from the inner nuclear membrane as a result of this initial budding reaction and thus that all viral envelope glycoproteins must reach this cellular membrane. These glycoproteins can indeed be observed in the nuclear membrane, but they are also found in many other cellular membranes. And, as discussed in chapter 13, there is now good evidence that the envelope of mature virus particles is derived from a cytoplasmic membrane.

The budding of the herpesviral nucleocapsid through the inner membrane of the nuclear envelope takes place at morphologically distinct sites, which appear as thickened, curved regions of the membrane. The herpes simplex virus type 1 gB glycoprotein and its Epstein-Barr virus homolog accumulate at these nuclear membrane sites. To reach this location, gB first enters the ER by the mechanism described above and then acquires oligosaccharide chains.

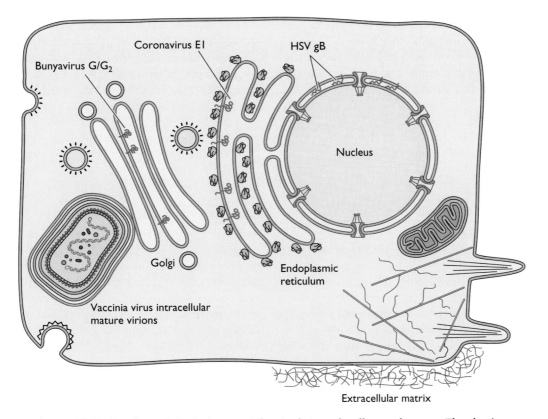

**Figure 12.19 Sorting of viral glycoproteins to internal cell membranes.** The destinations of membrane glycoproteins of viruses that bud into compartments of the secretory pathway (bunyaviruses and coronaviruses) or from the inner nuclear membrane (herpesviruses, e.g., herpes simplex virus [HSV]) and engulfment by cellular membranes during assembly (vaccinia virus) are indicated.

However, the N-linked oligosaccharides lack the features acquired during transit through Golgi compartments (Fig. 12.10), for the nucleus-associated gB protein does not follow the route to the plasma membrane. Rather, it accumulates in the ER and the inner and outer nuclear membranes. The latter localization depends on ER retention, which is determined by a specific protein sequence, a transmembrane segment of 21 amino acids. Lateral diffusion from the ER to the outer nuclear membrane and thence to the inner nuclear membrane via the membranes associated with the nuclear pore (Fig. 5.19) could allow the gB protein to reach the inner nuclear membrane. However, more specific transport reactions cannot yet be excluded.

## Transport of Viral Genomes to Assembly Sites

Like the structural proteins and enzymes of the virion, progeny genomes must be transported to the intracellular site at which assembly takes place. Such transport is fre-quently coupled to mechanisms that prevent participation of the genome in biosynthetic reactions, i.e., replication, transcription, or translation. The packaging of the genomes of most animal viruses with DNA genomes takes place in the infected cell nucleus. In these cases, no transport of progeny DNA genomes from one cellular compartment to another is required to set the stage for assembly. For other DNA viruses, as well as for many with RNA genomes, both synthesis of the nucleic acid genome and assembly of virus particles take place in the cytoplasm of the host cell. However, many of these viruses assemble at membrane sites, necessitating transport of genomic nucleic acid from its site of synthesis to the cytoplasmic face of the appropriate membrane. Both poxviruses and hepadnaviruses acquire membranes from internal compartments of the host cell (Table 12.4), and their DNA genomes appear to reach the appropriate membranes under the direction of protein associated with them early in the assembly process. For example, the hepadnaviral DNA genome is synthesized in the cytoplasm from the pregenomic RNA within cores containing the viral C protein. When infected cells contain a sufficient concentration of the large surface (L) glycoprotein accumulated in the ER membrane, core and L proteins can bind to one another, bringing the genome to the appropriate cell membrane and allowing budding of particles into the ER. Many RNA genomes made in the cytoplasm must be transported to the plasma membrane for assembly of virions. Yet other RNA genomes must travel even further, indeed over the greatest intracellular distances possible: both influenza virus and retroviral genomic RNAs are synthesized within the infected cell nucleus, but progeny virions bud from the plasma membrane.

## Genome RNA Transport from the Nucleus to the Cytoplasm

The transcripts of integrated retroviral proviruses are processed in the infected cell nucleus to generate both unspliced and spliced mRNA species. The unspliced RNA not only serves as the mRNA for synthesis of Gag and Gag-Pol polyproteins but is also the retroviral genome. As discussed in chapter 10, the inefficient export of unspliced mRNAs characteristic of host cells must therefore be circumvented. In the case of complex retroviruses, sequence-specific viral RNA-binding proteins that promote export of unspliced RNA, such as human immunodeficiency virus type 1 Rev protein, overcome this limitation. Simple retroviruses also rely on specific sequences for the export of genomic RNA, but these are recognized by cellular proteins. Although simple and complex retroviruses use different strategies to ensure efficient transport of unspliced genomic RNA to the cytoplasm, they share the mechanism of subsequent transport of the RNA to the plasma membrane described below. The contribution of viral proteins like Rev is thus restricted to the initial stage in transport of viral RNA.

Perhaps the most elaborate requirements for transport of viral RNA species between nuclear and cytoplasmic compartments are found in influenza virus-infected cells, in which both the direction of transport of genomic RNA and the nature of the viral RNA exported from the nucleus vary as a function of the infectious cycle. When the cycle is initiated, viral genomic ribonucleoproteins (RNPs) released into the cytoplasm enter the nucleus under the direction of the nuclear localization signal of the NP protein. As discussed in chapter 10, the mechanisms that ensure export for translation of viral (+) strand mRNAs, the majority of which lack introns, are not fully understood. With the switch to replication, genomic (−) strand RNA segments are synthesized in infected cell nuclei and accumulate as rRNPs containing the NP protein and the three P proteins. These RNPs must be exported to allow virus assembly and completion of the infectious cycle, a reaction

that requires the viral M1 and NS2 proteins. The former is the most abundant protein of the virus particle and following synthesis enters the nucleus by means of a typical nuclear localization signal (Table 5.1). It then binds to genomic RNPs via an RNP-binding domain (Fig. 12.18A), an interaction that both inhibits RNA synthesis and promotes genomic RNP export. The NS2 protein appears to direct genomic RNPs to components of the cellular export machinery. This protein contains an N-terminal nuclear export signal that can substitute functionally for that of the human immunodeficiency virus type 1 Rev protein. The NS2 protein possesses no intrinsic RNA-binding activity but includes a C-terminal M1-binding domain, which is believed to allow recognition of RNPs to which the latter protein is bound. Such an indirect recognition mechanism probably ensures that genomic RNPs are not exported from the nucleus before sufficient quantities of viral mRNA and progeny genomic RNA have been synthesized in the infected cell because the M1 protein is not available in large quantities until the later stages of infection. Furthermore, the RNPs enter the cytoplasm only when they have associated with the protein (M1) that is necessary for guiding them to the plasma membrane.

## RNA Transport from the Cytoplasm to the Plasma Membrane

Like nuclear export, transport of the RNA genomes of enveloped viruses to the appropriate cellular membrane depends on signals present in viral proteins that bind to the RNA that is assembled in RNPs or nucleocapsids. As we have seen, the influenza virus M1 protein contains an N-terminal sequence that allows it to bind to lipids (Fig. 12.18A). This membrane-binding domain of M1 thus allows association of the genomic RNPs with the plasma membrane. Although more limited in its RNA transport functions, the M protein of vesicular stomatitis virus shares several properties with the influenza virus M1 protein.

Replication of vesicular stomatitis virus RNA takes place in the cytoplasm, where newly synthesized genomic RNA molecules assemble with the N, L, and NS proteins to form helical RNPs. Genomic RNA molecules within RNPs can serve as templates for additional cycles of replication or for mRNA synthesis. However, these RNPs must eventually travel to the plasma membrane for association with the G protein and incorporation into virions. Entry into the latter pathway is determined by the viral M protein, which associates via N-terminal sequences with RNPs containing genomic RNA (Fig. 12.18B). This interaction induces formation of a tightly coiled RNA-protein "skeleton," the final structure of the vesicular stomatitis virus virion nucleocapsid. Immunoelectron microscopy suggests that the nucleocapsid is wound around a core built from the M protein. Formation of this structure precludes replication and mRNA synthesis and directs RNPs to the plasma membrane by means of the membrane-binding domains of the M protein (Fig. 12.18B) described above.

The retroviral proteins that mediate membrane association of genomic RNA are similar to the matrix proteins of these (−) strand RNA viruses in several respects. Once within the cytoplasm (see previous section), unspliced retroviral RNA is translated on cellular polyribosomes into the Gag and, at low frequency, Gag-Pol polyproteins. The functions of Gag (Fig. 12.17) include transport of unspliced RNA molecules to membrane assembly sites and packaging of the RNA into assembling capsids. Thus, whether unspliced genomic RNA molecules continue to be translated or are redirected for virion assembly is controlled by the cytoplasmic concentration of Gag polyprotein.

The NC (nucleocapsid) segment of the Gag polyprotein (Fig. 12.17) contains an RNA-binding domain required for specific binding of Gag to the RNA packaging sequences described in chapter 13. The NC region, which functions as an independent protein in mature virions, is very basic and contains at least one copy of a zinc-binding motif comprising an array of cysteine and histidine residues (Fig. 12.18). This "zinc finger" domain makes a major contribution to the specificity with which Gag or NC proteins bind to unspliced retroviral RNA and, in conjunction with basic amino acids located nearby, is responsible for the RNA-packaging activity of Gag polyprotein. However, the N-terminal MA portion of Gag contains fatty acid and protein signals that target the polyprotein to the plasma membrane (Fig. 12.17). Thus, binding of Gag to unspliced retroviral RNA allows delivery of the genome to assembly sites at the plasma membrane, just as binding of matrix proteins to influenza virus or vesicular stomatitis virus genomic RNPs directs these structures to the membrane.

The association of Gag polyprotein molecules with one another appears to be an additional prerequisite for transport to the membrane. For example, several regions in human immunodeficiency virus type 1 Gag, including the spacer between CA and NC and sequences within MA, CA, and NC (Fig. 12.17), are necessary for efficient binding of Gag molecules to one another in a variety of assays. Altered proteins that lack such sequences cannot support assembly of virus particles, and they accumulate at cytoplasmic sites. The rescue of such mutants by wild-type Gag indicates that interaction among Gag molecules is necessary not only for virus assembly but also for transport of the protein and its associated RNA to the plasma membrane.

The mechanisms by which the various viral RNPs containing genomic RNA species, RNA-packaging proteins, and membrane-targeting proteins travel to the plasma membrane are not known. Such transport could be the result of passive, random diffusion and trapping of the RNPs

at membrane sites via binding of M1, M, or Gag protein to the membrane. Alternatively, an active transport process in which the RNPs are carried along microtubules to the membrane by motor proteins might be required.

Although each of the three viruses discussed in this section must solve a unique RNA transport problem, the influenza virus M1, vesicular stomatitis virus M, and retroviral Gag proteins each possess the ability to bind both to RNPs containing genomic RNA and to membranes. In all cases, such interactions commit the genomic RNA to the assembly pathway and prevent its participation in biosynthetic processes. Thus, despite their different sequences and mechanisms of membrane association, these three proteins share the ability to direct genomic RNA to the plasma membrane and to promote interactions among internal and envelope components of virions.

These properties are essential at the end of an infectious cycle, when the primary task is assembly of progeny virions. On the other hand, they would be disastrous for the virus if the interactions could not be reversed before or at the beginning of a new cycle, when the infecting genome must reach nuclear (influenza virus) or cytoplasmic (vesicular stomatitis virus and retroviruses) sites distant from the plasma membrane. In the case of (−) strand viruses, matrix proteins are removed soon after virus entry. Matrix-free RNPs can then enter the nucleus for mRNA synthesis (influenza virus) or begin this process in the cytoplasm (vesicular stomatitis virus). Simple retroviruses exhibit a more elegant mechanism: following virus assembly and budding, Gag (and Gag-Pol) polyproteins are processed by the viral protease to the individual structural proteins listed in Fig. 12.17. Such cleavages place the RNA-binding domain of NC in protein molecules separate from the myristate and protein sequences of MA that mediate membrane binding, so that matrix-free core RNPs can be released into the cell to initiate a new infectious cycle. Membrane association of internal components of human immunodeficiency virus type 1 appears to be controlled by a more subtle mechanism. The N-terminal signal responsible for association of the Gag polyprotein with the plasma membrane (Fig. 12.17) does not support membrane binding by mature MA. A C-terminal α-helix of mature MA sequesters the N-terminal myristate chain. Thus, conformational changes in MA induced when the Gag polyprotein is cleaved appear to release this protein from its membrane association.

## Perspectives

Although they do not produce the viral macromolecules needed to complete an infectious cycle sucessfully, the cellular trafficking systems described in this chapter are just as essential for virus reproduction as the host cell's biosynthetic capabilities. Because all eukaryotic cells are organized into structurally and functionally distinct compartments, assembly of viruses that replicate in animal cells requires transport of at least some viral proteins or of the nucleic acid genome from sites of synthesis to the sites at which progeny virus particles are assembled. The trafficking requirements during the infectious cycle can in fact be quite complex, with transport of viral macromolecules (or structures built from them) over large distances or in opposite directions during different periods, for example, as in cells infected by orthomyxoviruses or herpesviruses. Thus, despite considerable variation in the degree of dependence on cellular systems for the transport and sorting of viral components to appropriate intracellular destinations, all viruses make use of at least one cellular trafficking system prior to assembly of progeny particles.

This property has provided important tools (viral proteins or nucleic acids synthesized in large quantities in infected cells) with which to study these essential cellular processes. Indeed, the fundamental principle of protein sorting, that a protein's final destination is dictated by specific signals within its amino acid sequence and/or covalently attached sugars or lipids, was established in analyses of viral proteins. In some cases, notably entry into the secretory pathway or import into the nucleus, the mechanisms by which such signals are recognized are now understood in considerable detail. Furthermore, the study of viral proteins that enter the secretory pathway has provided much of what we know of the reactions by which proteins are folded and processed within the ER and subsequent compartments, as well as those that clear misfolded proteins from the pathway. It therefore seems certain that viral systems will provide equally important insights into the signals and sorting mechanisms that are presently less well characterized, such as those responsible for the direction of proteins to specialized membrane regions of polarized cells or to the inner nuclear membrane. One of the greatest current challenges in this field is elucidation of the mechanics of the movement of proteins, nucleic acids, nucleoproteins, or transport vesicles from one cellular compartment or site to another. With the exception of axonal transport in neurons, we really know very little about these processes and in many cases cannot even decide between passive diffusion and specific, energy-requiring transport mechanisms such as those mediated by molecular motors. It is to be hoped that continued technical advances, like the use of part of the vesicular stomatitis virus G protein fused to green fluorescent protein to visualize vesicular transport in living cells in real time, will lead to a clearer view of transport mechanics.

Some viruses interfere with the normal transport of specific cellular proteins to the cell surface or completely disrupt the secretory pathway. Such inhibitory consequences of virus infection can facilitate evasion of the host's immune defense systems, for example, because the display of viral antigens bound to major histocompatibility complex class I proteins on the surface of the infected cell is impaired or prevented. However, in most cases, transport of the components from which virus particles are built to assembly sites appears to have little detrimental effect on the host cell despite the large quantities of viral cargo that must be handled. In all cases, the most important consequence of such transport is the formation within infected cells of microenvironments containing high concentrations of viral structural proteins and the nucleic acid genome, ideal niches for the assembly of progeny virions from their multiple parts.

## References

### Book Chapters

Alberts, B., D. Bray, J. Lewis, M. Raff, K. Roberts, and J. D. Watson. 1994. Intracellular compartments and protein sorting, p. 551–598. *In Molecular Biology of the Cell*, 3rd ed. Garland Publishing Inc., New York, N.Y.

Alberts, B., D. Bray, J. Lewis, M. Raff, K. Roberts, and J. D. Watson. 1994. Vesicular traffic in the secretory and endocytic pathways, p. 599–651. *In Molecular Biology of the Cell*, 3rd ed. Garland Publishing Inc., New York, N.Y.

Cooper, G. M. 1997. Protein sorting and transport: the endoplasmic reticulum, Golgi apparatus and lysosomes, p. 347–388. *In The Cell: a Molecular Approach*. ASM Press, Washington, D.C., and Sinauer Associates, Sunderland, Mass.

### Reviews

Doms, R. W., R. A. Lamb, J. K. Rose, and A. Helenius. 1993. Folding and assembly of viral membrane proteins. *Virology* **193:**545–562.

Drubin, D. G., and W. J. Nelson. 1996. Origins of cell polarity. *Cell* **84:**335–344.

Gahmberg, C. G., and M. Tolvanen. 1996. Why mammalian cell surface proteins are glycoproteins. *Trends Biochem. Sci.* **21:**308–311.

Gething, M.-J., and J. Sambrook. 1992. Protein folding in the cell. *Nature* (London) **355:**33–45.

Glick, B. S., and V. Malhotra. 1998. The curious state of the Golgi apparatus. *Cell* **95:**883–889.

Griffiths, G., and P. Rottier. 1993. Cell biology of viruses that assemble along the biosynthetic pathway. *Semin. Cell Biol.* **3:**367–381.

Hammond, C., and A. Helenius. 1995. Quality control in the secretory pathway. *Curr. Opin. Cell Biol.* **7:**523–529.

Hruby, D. E., and C. A. Franke. 1993. Viral acylproteins: greasing the wheels of assembly. *Trends Microbiol.* **1:**20–25.

Lord, J. M. 1996. Go outside and see the proteasome. *Curr. Biol.* **6:**1067–1069.

Mellman, I. 1996. Enigma variations: protein mediators of membrane fusion. *Cell* **82:**869–872.

Rothman, J. E., and T. H. Sollner. 1997. Throttles and dampers: controlling the engine of membrane fusion. *Science* **276:**1212–1213.

Rothman, J. E., and F. T. Wieland. 1996. Protein sorting by transport vesicles. *Science* **272:**227–234.

Schatz, G., and B. Dobberstein. 1996. Common principles of protein translocation across membranes. *Science* **271:**1519–1528.

Schekman, R., and L. Ori. 1996. Coat proteins and vesicle budding. *Science* **271:**1526–1533.

Traub, L. M., and S. Kornfeld. 1997. The trans-Golgi network: a late secretory sorting station. *Curr. Opin. Cell Biol.* **9:**527–533.

## Papers of Special Interest

### Import of Viral Proteins into the Nucleus

Forstova, J., N. Krauzewicz, S. Wallace, A. J. Street, S. M. Dilworth, S. Beard, and B. E. Griffin. 1993. Cooperation of structural proteins during late events in the life cycle of polyomavirus. *J. Virol.* **67:**1405–1413.

Nuesch, J. P., and P. Tattersall. 1993. Nuclear targeting of the parvoviral replicator molecule NS1: evidence for self-association prior to nuclear transport. *Virology* **196:**637–651.

Zhao, L. J., and R. Padmanabhan. 1988. Nuclear transport of adenovirus DNA polymerase is facilitated by interaction with preterminal protein. *Cell* **55:**1005–1015.

### Transport of Viral and Cellular Proteins via the Secretory Pathway

Balch, W. E., W. G. Dunphy, W. A. Braell, and J. E. Rothman. 1984. Reconstitution of the transport of proteins between successive compartments of the Golgi measured by the coupled incorporation of *N*-acetyl glucosamine. *Cell* **39:**405–416.

Chackerjan, B., L. M. Rudensey, and J. Overbaugh. 1997. Specific N-linked and O-linked glycosylation modifications in the envelope V1 domain of simian immunodeficiency virus variants that evolve in the host alter recognition by neutralizing antibodies. *J. Virol.* **71:**7719–7727.

Hallenberger, S., M. Moulard, M. Sordel, H. D. Klenk, and W. Garten. 1997. The role of eukaryotic subtilisin-like endoproteases for the activation of human immunodeficiency virus glycoproteins in natural host cells. *J. Virol.* **71:**1036–1045.

Hammond, C., and A. Helenius. 1994. Folding of VSV G protein: sequential interaction with BiP and calnexin. *Science* **266:**456–458.

Horimoto, T., K. Nakayama, S. P. Smeekens, and Y. Kawaoka. 1994. Preprotein processing endoproteases Pc6 and furin both activate hemagglutin of virulent avian influenza viruses. *J. Virol.* **68:**6074–6078.

Hutchinson, L., H. Browne, V. Wargent, N. Davis-Poynter, S. Primorac, K. Goldsmith, A. C. Minson, and D. C. Johnson. 1992. A novel herpes simplex virus glycoprotein, gL, forms a complex with glycoprotein H (gH) and affects normal folding and surface expression of gH. *J. Virol.* **66:**2240–2250.

Hwang, S. B., and M. M. Lai. 1993. Isoprenylation mediates direct protein-protein interactions between hepatitis large delta antigen and hepatitis B virus surface antigen. *J. Virol.* **67:**7659–7662.

Ivanova, L., and M. J. Schlesinger. 1993. Site-directed mutations in the Sindbis virus E2 glycoprotein identify palmitoylation sites and affect virus budding. *J. Virol.* **67:**2546–2551.

Kreis, T. E., and H. F. Lodish. 1986. Oligomerization is essential for transport of vesicular stomatitis viral glycoprotein to the cell surface. *Cell* **46:**929–937.

Li, Z., A. Pinter, and S. C. Kayman. 1997. The critical N-linked glycan of murine leukemia virus envelope protein promotes both folding of the C-terminal domains of the precursor polypeptide and stability of the postcleavage envelope complex. *J. Virol.* **71:**7012–7019.

Marc, D., M. Girard, and S. van der Werf. 1991. A Gly to Ala substitution in poliovirus capsid protein $VP_0$ blocks its myristoylation and prevents viral assembly. *J. Gen. Virol.* **72:**1151–1157.

Presley, J. F., N. B. Cole, T. A. Schroer, K. Hirschberg, K. J. M. Zaal, and J. Lippincott-Schwartz. 1997. ER-to-Golgi transport visualized in living cells. *Nature* (London) **389:**81–85.

Rothman, J. E., and H. F. Lodish. 1977. Synchronized transmembrane insertion and glycosylation of a nascent membrane protein. *Nature* (London) **269:**775–780.

Scheiffele, P., M. G. Roth, and K. Simons. 1997. Interaction of influenza virus hemagglutinin with sphingolipid-cholesterol membrane domains via its transmembrane domain. *EMBO J.* **18:**5501–5508.

Takeuchi, K., and R. A. Lamb. 1994. Influenza virus M2 protein ion channel activity stabilizes the native form of fowl plague virus hemagglutinin during intracellular transport. *J. Virol.* **65:**911–919.

### Transport of Viral Proteins to Intracellular Membranes

Gilbert, R., K. Ghosh, L. Rasile, and H. P. Ghosh. 1994. Membrane anchoring domain of herpes simplex virus glycoprotein gB is sufficient for nuclear envelope localization. *J. Virol.* **68:**2272–2285.

Hobman, T. C., H. F. Lemon, and K. Jewell. 1997. Characterization of an endoplasmic reticulum retention signal in the rubella virus E1 glycoprotein. *J. Virol.* **71:**7670–7680.

Melin, L., R. Persson, A. Andersson, A. Bergstrom, R. Ronnholm, and R. F. Pettersson. 1995. The membrane glycoprotein G1 of Uukuniemi virus contains a signal for localization to the Golgi complex. *Virus Res.* **36:**49–66.

Salmons, T., A. Kuhn, F. Wylie, S. Schleich, J. R. Rodriguez, D. Rodriguez, M. Esteban, G. Griffiths, and J. K. Locker. 1997. Vaccinia virus membrane proteins p8 and p16 are cotranslationally inserted into the rough endoplasmic reticulum and retained in the intermediate compartment. *J. Virol.* **71:**7404–7420.

### Virus-Induced Inhibition of Transport via the Secretory Pathway

Burgert, H. G., and S. Kvist. 1985. An adenovirus type 2 glycoprotein blocks cell surface expression of human histocompatibility class I antigens. *Cell* **41:**987–997.

Cho, M. W., N. Teterina, D. Egger, K. Bienz, and E. Ehrenfeld. 1994. Membrane rearrangement and vesicle induction by recombinant poliovirus 2C and 2BC proteins in human cells. *Virology* **202:**129–145.

Dales, S., H. J. Eggers, I. Tanm, and G. E. Palade. 1965. Electron microscopic study of the formation of poliovirus. *Virology* **26**:379–389.

Doedens, J. R., T. H. Giddings, Jr., and K. Kirkegaard. 1997. Inhibition of endoplasmic reticulum-to-Golgi traffic by poliovirus protein 3A: genetic and ultrastructural analysis. *J. Virol.* **71**:9054–9064.

Schubert, U., L. C. Antón, I. Bačík, J. H. Cox, S. Bour, J. R. Bennink, M. Orlowski, K. Strebel, and J. W. Yewdell. 1998. CD4 glycoprotein degradation induced by human immunodeficiency virus type 1 Vpu protein requires the function of proteasomes and the ubiquitin-conjugating pathway. *J. Virol.* **72**:2280–2288.

Wiertz, E. J. H., T. R. Jones, L. Sun, M. Bogyo, H. J. Gueze, and H. L. Ploegh. 1996. The human cytomegalovirus US11 gene product dislocates MHC class I heavy chains from the endoplasmic reticulum to the cytosol. *Cell* **84**:709–779.

Willey, R. L., F. Maldarelli, M. A. Martin, and K. Strebel. 1992. Human immunodeficiency virus type 1 Vpu protein induces rapid degradation of CD4. *J. Virol.* **66**:7193–7200.

## Transport of Viral RNA Genomes to Sites of Assembly

Chong, L. D., and J. K. Rose. 1994. Interactions of normal and mutant vesicular stomatitis virus matrix proteins with the plasma membrane and nucleocapsids. *J. Virol.* **68**:441–447.

Enami, M., and K. Enami. 1996. Influenza virus hemagglutinin and neuraminidase stimulate the membrane association of the matrix protein. *J. Virol.* **70**:6653–6657.

Göttlinger, H. C., J. G. Sodroski, and W. A. Haseltine. 1989. Role of capsid precursor processing and myristoylation in morphogenesis and infectivity of human immunodeficiency virus type 1. *Proc. Natl. Acad. Sci. USA* **86**:5781–5785.

Martin, K., and A. Helenius. 1991. Nuclear transport of influenza virus ribonucleoproteins: the viral matrix protein (M1) promotes export and inhibits import. *Cell* **67**:117–130.

O'Neill, R. E., J. Tulon, and P. Palese. 1998. The influenza virus NEP (NS2 protein) mediates the nuclear export of viral ribonucleoproteins. *EMBO J.* **17**:288–296.

Ye, Z. P., R. Pal, W. Fox, and R. R. Wagner. 1987. Functional and antigenic domains of the matrix (M1) protein of influenza A virus. *J. Virol.* **61**:239–246.

Zhou, W., L. J. Parent, J. W. Wills, and M. D. Resh. 1994. Identification of a membrane-binding domain within the amino-terminal region of human immunodeficiency virus type 1 Gag protein which interacts with acidic phospholipids. *J. Virol.* **68**:2556–2569.

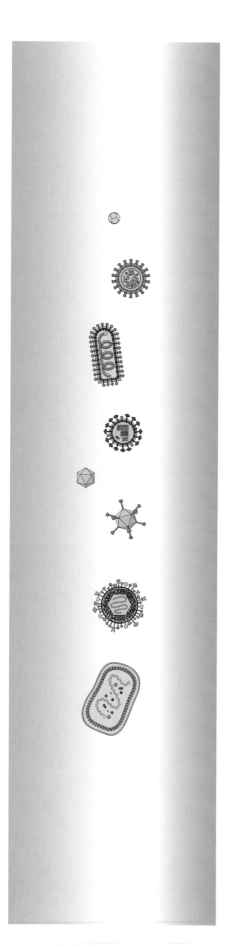

# 13

# Assembly, Exit, and Maturation of Progeny Virions

*The probability of formation of a high complex structure from its elements is increased, or the number of possible ways of doing it diminished, if the structure in question can be broken down in a finite series of successively smaller substrates!*

J. D. BERNAL
*The Origins of Life on Earth (A. I. Oparin, ed.), Pergamon, Oxford, United Kingdom, 1959*

## Introduction

Virus particles exhibit considerable diversity in size, composition, and structural sophistication, ranging from those comprising a single nucleic acid molecule and but one structural protein to complex structures built from many different proteins and other components. Nevertheless, because of their unique mode of replication, all viruses must complete a common set of de novo assembly reactions to ensure reproductive success. These processes include formation of the structural units of the protective protein coat from individual protein molecules, assembly of the coat by interaction among the structural units, incorporation of the nucleic acid genome, and release of newly assembled progeny virions (Fig. 13.1). In many cases, formation of internal virion structures must be coordinated with acquisition of a cellular membrane into which viral proteins have been inserted, or additional maturation steps must be completed to produce infectious particles. Assembly of even the simplest viruses is therefore a remarkable process requiring considerable specificity in, and coordination among, each of multiple reactions. Furthermore, a virus can reproduce successfully only if each of the assembly reactions proceeds with reasonable efficiency and if the overall pathway is irreversible. The diverse mechanisms by which viruses assemble represent powerful solutions to the problems imposed by de novo assembly. Indeed, infectious virus particles are produced in prodigious numbers with great specificity and efficiency.

The structure of a virus particle determines the nature of the reactions by which it is formed (Fig. 13.1), as well as the mechanisms by which it enters a new host cell, reproduces, and promotes pathogenesis within a host animal. For example, the exceptionally stable poliovirus survives passage through the stomach to replicate in the gut. Despite such virus-specific variations in structure and biological properties, all viruses must be metastable structures well suited for protection of the nucleic acid genome in extracellular environments. They must also be built in a way that allows their ready disassembly during

**439**

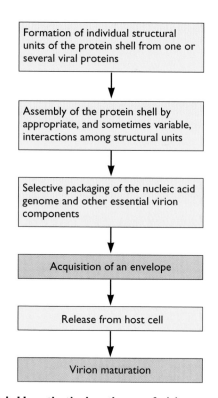

**Figure 13.1 Hypothetical pathway of virion assembly and release.** Reactions common to all viruses are shown in yellow, and those common to many viruses are shown in blue. The structural units that are often the first assembly intermediates are the homo- or hetero-oligomers of viral structural proteins from which virus particles are built (see Table 3.1). The arrows indicate a general sequence that strictly applies to only some viruses (see text). Packaging of the genome can be coordinated with assembly of the capsid or nucleocapsid and, in enveloped viruses, assembly of internal components can be coordinated with acquisition of the envelope.

entry into a new host cell. A number of elegant mechanisms resolve the apparently paradoxical requirements for very stable associations among virion components during assembly and transmission but facile reversal of these interactions when appropriate signals are encountered upon infection of a host cell.

Like synthesis of viral nucleic acids and proteins, production of virus particles depends on host cell components, such as the cellular proteins that catalyze or assist the folding of individual protein molecules or the formation of multimeric proteins. Furthermore, the components from which virions are built are transported to the appropriate assembly site by cellular pathways, for example, via the secretory pathway or nuclear import machinery. Such localization of virion components to a specific intracellular compartment or region undoubtedly facilitates virus production. The concentration of virion proteins at the assembly site must increase the rates of reactions, for exam-

ple, those by which structural units and protein shells are formed. The establishment of specialized microenvironments for assembly also seems likely to restrict the number of interactions in which a particular virion component can engage, thus increasing the specificity and efficiency of assembly reactions.

A successful virus-host interaction, the survival and propagation of a virus in a host population, generally requires dissemination of the virus beyond the cells infected upon initial contact. Progeny virus particles assembled during the later stages of the infectious cycle must therefore escape from the infected cell for transmission to new host cells within the same animal or to new host animals. The majority of viruses leave an infected cell by one of two general mechanisms: they are released into the external environment (upon budding from, or lysis of, the cell), or they are transferred directly into a new host cell.

## Methods of Studying Virus Assembly and Egress

Because they are so complex, mechanisms of virus assembly and release can be understood only with the integration of information obtained by complementary structural, biochemical, and genetic methods of analysis. These methods are introduced briefly in this section.

### Structural Studies of Virus Particles

The mechanisms by which viruses form within, and leave, their host cells are intimately related to their structural properties. Our understanding of these processes therefore improves dramatically whenever the structure of a virus particle is determined. An atomic-level description of the contacts among structural units that maintain the integrity of the virus particle identifies the interactions that mediate assembly and the ways in which these interactions must be regulated. For example, the X-ray crystal structure of the simian virus 40 particle described in chapter 3 not only solved the enigma of how VP1 pentamers could be packed in hexameric arrays but, in so doing, identified three distinct modes of interpentamer contact, mediated by conformationally flexible N- and C-terminal arms of VP1 subunits. Thus, assembly of the simian virus 40 capsid must require specific variations in the way in which pentamers associate, depending on their position in the capsid shell. Such subtle yet sophisticated regulation of association of structural units during assembly was certainly not anticipated and could be revealed only by high-resolution structural information. Such information can also provide important insights into specific features of individual assembly pathways or the mechanisms that ensure that assembly is irreversible, as discussed for picornaviruses in subsequent sections.

The atomic structures of virions of larger naked viruses and enveloped viruses cannot yet be determined by X-ray crystallography. However, the newer methods for structural analysis of such virus particles, for example, cryo-electron microscopy and image analysis, are yielding increasingly detailed information about contacts among the components of these more complex virions and hence the mechanisms by which they assemble.

## Visualization of Assembly and Exit by Electron Microscopy

While high-resolution structural studies of purified virions or virion proteins provide a molecular foundation for understanding virus assembly, they offer no clues about how assembly (or exit) actually proceeds in an infected cell. Electron microscopy can be directly applied to investigation of these processes. Examination of thin sections of cells infected by a wide variety of viruses has provided important information about intracellular sites of assembly, the nature of assembly intermediates, and mechanisms of envelope acquisition and release of particles. This approach can be particularly useful when combined with immunocytochemical methods for identification of individual viral proteins, or the structures they form, via binding of specific antibodies tagged with electron-dense material, usually particles of gold.

## Biochemical and Genetic Analysis of Assembly Intermediates

Although of great value, the static views provided by X-ray crystallography or electron microscopy clearly are not sufficient to describe the dynamic processes of virus assembly and release. An understanding of virus assembly requires identification of the intermediates in the pathway by which individual viral proteins and other virion components are converted to a mature infectious virion.

When extracts are prepared from the appropriate compartment of infected cells under conditions that preserve protein-protein interactions, a variety of viral structures can often be detected by techniques that separate on the basis of mass and conformation (velocity sedimentation in sucrose gradients or gel filtration) or of density (equilibrium centrifugation). These structures range from structural units of the capsid or nucleocapsid (see Table 3.1 for the definition of structural units) to empty capsids and mature virions. Similar methods have identified various subassemblies formed by viral structural proteins in in vitro reactions. Furthermore, such structures can be organized into a logical sequence for assembly, from the least to the most complex. On the other hand, it is often quite difficult to prove that structures identified by these approaches, such as empty capsids, are true intermediates in the pathway.

By definition, the intermediates in any pathway do not accumulate unless the reaction by which they are converted to the next structure is rate limiting. For this reason, assembly intermediates are generally present within infected cells at low concentrations against a high background of the starting material (mono- or oligomeric structural proteins) and the final product (virus particles). This property makes it difficult to establish precursor-product relationships by standard pulse-chase experiments: the large pools of structural proteins initially labeled are converted only slowly and inefficiently into subsequent intermediates in the pathway. Genetic methods of analysis provide a powerful solution to this problem. Mutations that confer temperature-sensitive or other phenotypes and block a specific reaction have been invaluable in the elucidation of assembly pathways. A specific intermediate may accumulate in mutant virus-infected cells and can often be purified and characterized more readily. The judicious use of temperature-sensitive mutants can allow the reactions in a pathway to be ordered with respect to one another, and second-site suppressors of such mutations can identify proteins that interact with one another. Of even greater value is the combination of genetics with biochemistry, an elegant approach pioneered more than 30 years ago during studies of assembly of the complex bacteriophage T4 with the development of in vitro complementation (Box 13.1).

The difficulties inherent in kinetic analyses are compounded by the potential for formation of dead-end products and the unstable nature of some assembly intermediates. Dead-end assembly products are those that form by off-pathway (side) reactions. Because they are not true intermediates, they may accumulate in infected cells and be identified incorrectly as components in the pathway. By definition, such dead-end products differ from true intermediates in some structural property that prevents them from completing assembly. Furthermore, some authentic intermediates may be fragile structures that cannot withstand conditions used to extract mature virus particles from the cell because they lack the complete set of intermolecular interactions that stabilize the virus particle. Less obvious, and recognized more recently, is the conformational instability of some intermediates: such structures do not fall apart during isolation and purification but rather undergo irreversible conformational changes so that the structures studied experimentally do not correspond to any present in the infected cell. This kind of instability is not easy to detect because monoclonal antibodies that distinguish specific structural features (**conformational epitopes**) rather than a simple linear sequence of amino

## Late steps in T4 assembly

As illustrated, the head, tail, and tail fibers of the morphologically complex phage first form separately and are then assembled with one another. The many genes encoding products that participate in T4 morphogenesis are listed for the reaction for which they are required. These gene products and the order in which they act were identified by genetic methods, including identification of second-site suppressors of specific mutations. In general, such second-site mutations are located in a gene encoding a protein that interacts with the protein encoded by the gene carrying the original mutation. The development of in vitro systems in which specific reactions were reconstituted was also of the greatest importance, allowing the development of "in vitro complementation." For example, noninfectious T4 particles lacking tail fibers that accumulate in infected cells when the tail fiber pathway (right part of figure) is blocked by mutation can be converted to infectious phage when mixed in vitro with extracts prepared from cells infected with T4 mutated in the gene encoding the major "head" protein, gene 23. The fact that the phages formed in vitro in this way were infectious established that assembly was accurate. This type of system was used to identify the genes whose products are required for assembly of heads or tails, as well as "scaffolding" proteins essential for assembly but not present in the virus particle. Adapted from W. B. Wood, *Harvey Lect.* **73**:203–223, 1978, and W. B. Wood et al., *Fed. Proc.* **27**:1160–1166, 1968, with permission.

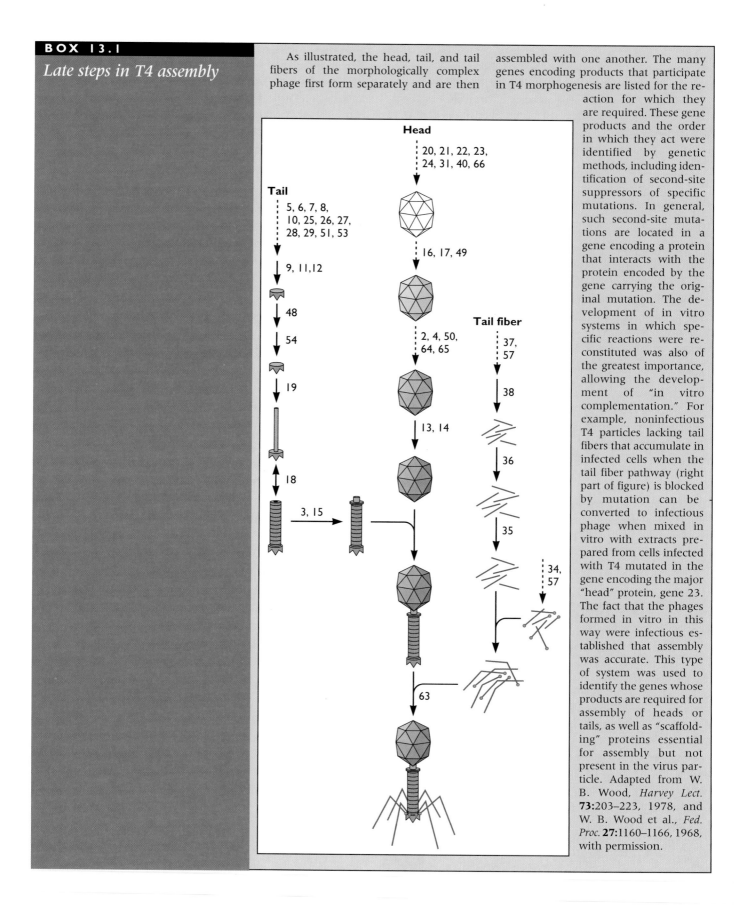

acids are needed. Thus, such conformational change may well escape notice, as was initially the case for poliovirus empty capsids.

## Methods Based on Recombinant DNA Technology

Modern methods of molecular biology and the application of recombinant DNA technology have greatly facilitated the study of virus assembly. Especially valuable is the simplification of this complex process that can be achieved by the expression of an individual viral protein or small sets of proteins in the absence of other viral components (Box 13.2).

## Assembly of Protein Shells

Although virus particles are far simpler in structure than any cell, they are built from multiple components, such as a capsid (or nucleocapsid), a nucleoprotein core containing the genome, and a lipid envelope carrying viral glycoproteins. The first steps in assembly are therefore the formation of the various components of the virion from their component parts. Such intermediates must then associate in ordered fashion, in some cases following transport to the appropriate intracellular site, to complete construction of the virus particle. Application of the techniques described in the previous section has allowed us to sketch the pathways by which many viruses assemble and describe some specific reactions in exquisite detail. In this section, we draw on this large body of information to illustrate strategies by which viruses ensure efficient assembly of protective protein coats for their genomes, the first reaction listed in Fig. 13.1.

## Formation of Structural Units

For some viruses, fabrication of a protein shell is coordinated with binding of structural proteins to the viral genome, as during assembly of the ribonucleoproteins of (−) strand RNA viruses. Consequently, structures built entirely from proteins do not accumulate prior to assembly of the ribonucleoprotein. In other cases, the first assembly reaction is formation of the protein **structural units** from which the protein shell is constructed (Fig. 13.1). Compared to many other steps, this process is relatively simple: individual structural units contain a small number of protein molecules, typically two to six, that must associate appropriately following their synthesis. Nevertheless, several mechanisms have evolved for formation of structural units, and in some cases additional proteins are required to assist the reactions (Fig. 13.2).

### Assembly from Individual Proteins

The structural units of some protein shells, including the VP1 pentamers of the papovavirus simian virus 40, assemble from their individual protein components (Table 13.1; Fig. 13.2A). This straightforward mechanism is analogous to formation of cellular structures containing multiple proteins, such as nucleosomes. In some cases, exemplified by adenovirus pentons, assembly is a two-step process (Fig. 13.2A). In this kind of reaction, the surfaces of individual protein molecules that contact either other molecules of the same protein or another protein (Table 13.1) are formed prior to assembly of the structural unit. This arrangement facilitates specific binding when appropriate protein molecules encounter one another. No energetically costly conformational change is required, and

---

### BOX 13.2

*Assembly of herpes simplex virus type 1 nucleocapsids in a simplified system*

The assembly and egress of herpesviruses from infected cells are complex processes comprising multiple steps (Fig. 13.6 and 13.17). To facilitate analysis of the initial reactions that lead to assembly of the nucleocapsid, viral genes enclding its constituent proteins were cloned in baculovirus vectors. Formation of the nucleocapsid was examined by electron microscopy in insect cells infected with various combinations of the recombinant baculoviruses that directed expression of these herpesviral proteins. Nucleocapsids indistinguishable from those formed in herpes simplex virus type 1-infected cells were observed when the six viral genes encoding VP5 (the major nucleocapsid protein that forms the hexons and pentons), VP19C and VP23 (which form

triplexes linking the VP5 structural units), VP26, VP24, VP21, and VP22a. By omission of individual recombinant baculoviruses, it was shown that VP26 is not necessary for nucleocapsid assembly and that only partial or deformed structures assemble in the absence of VP24, VP21, and VP22a (the protease and scaffolding proteins; see "Viral Scaffolding Proteins: Chaperones for Assembly" below). This approach is also widely used to identify and characterize protein-protein and protein-membrane interactions critical for assembly.

**Tatman, J. D., V. G. Preston, P. Nicholson, R. M. Elliot, and F. J. Rixon.** 1994. Assembly of herpes simplex virus type 1 capsids using a panel of recombinant baculoviruses. *J. Gen. Virol.* **75:**1101–1113.

## A Assembly from individual protein molecules

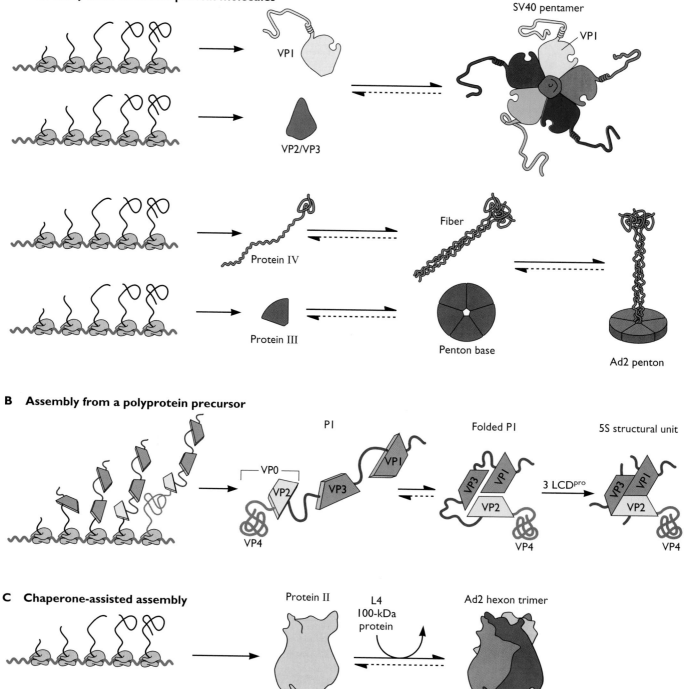

## B Assembly from a polyprotein precursor

## C Chaperone-assisted assembly

subunits that come into contact can simply "interlock" with one another. Indeed, production of these structural units generally can be reconstituted in vitro or in cells that express the component proteins, confirming that all information necessary for accurate assembly is contained within the primary sequence, and hence the folded structure, of the protein subunits. On the other hand, the individual protein subunits must find one another in an intracellular environment in which the concentration of irrelevant (cellular) proteins is as high as that attained in

◁ **Figure 13.2 Mechanisms of assembly of the structural units of virion protein shells.**
(A) Assembly from folded protein monomers, illustrated with simian virus 40 (SV40) VP1 pentamers and adenovirus pentons. The assembly reactions shown are the result of specific interactions among the proteins that form structural units. Some, notably those among VP1 molecules in the simian virus 40 pentamer, have been described at atomic resolution (chapter 3). In other cases, parts of the protein required for assembly have been identified by mutational analysis. For example, formation of the trimeric adenovirus fiber requires both the entire head of each monomer and the region of the shaft immediately adjacent to it. These assembly reactions are driven in a forward direction by the high concentrations of protein subunits synthesized in infected cells, as indicated by the solid arrows. (B) Assembly from a polyprotein precursor, illustrated with the poliovirus polyprotein that contains the four proteins that form the heteromeric structural unit. The latter proteins are synthesized as part of the single polyprotein precursor from which all viral proteins are generated by proteolytic processing. For simplicity, only the P1 capsid protein precursor and its cleavage by the viral 3CD protease following folding and assembly of the proteins of the immature structural unit (VP0, VP3 and VP1) are shown: the covalent connections between VP1 and VP3 and VP3 and VP0 in the P1 precursor, which are flexible regions exaggerated for clarity in the figure, are severed by the protease to form the 5S structural unit. However, VP4 remains covalently linked to VP2 in VP0 until assembly is completed (see text). (C) Assisted assembly. Some structural units are assembled only with the assistance of viral chaperones, such as the adenoviral I4 100-kDa protein, which is required for formation of the hexon trimer.

protein crystals (20 to 40 mg/ml). Such a milieu offers uncountable opportunities for nonspecific binding of viral proteins to unrelated cellular proteins. This property of the intracellular environment may account for the synthesis of viral structural proteins in quantities far in excess of those incorporated in virus particles, a common feature of virus-infected cells. Such high concentrations must facilitate the encounter of viral proteins with one another by random diffusion and would provide a sufficient reservoir to compensate for any loss by nonspecific binding to cellular

components. Another important consequence is that formation of structural units proceeds efficiently (Fig. 13.2A), driving the assembly pathway in the productive direction.

### Assembly from Polyproteins

An alternative strategy for production of structural units of a protective protein coat is their assembly while covalently linked in a polyprotein precursor (Fig. 13.2B). This mechanism circumvents the need for protein subunits to meet by random diffusion, as well as competition from

**Table 13.1**　Mechanisms of assembly of structural units

| Mechanism | Virus | Structural unit |
|---|---|---|
| Association of individual protein molecules | Hepadnavirus (hepatitis B virus) | C (capsid) protein dimers |
| | Papovavirus (simian virus 40) | VP1 pentamer with one molecule of VP2 or VP3 in its central cavity |
| | Adenovirus (adenovirus type 2) | Protein IV trimer (fiber), protein III pentamer (penton base) that forms penton |
| Assisted assembly of protein subunits | Adenovirus (adenovirus type 2) | Hexon trimers of protein II formed with assistance of the L4 100-kDa protein |
| | Herpesvirus (herpes simplex virus type 1) | VP5 pentamers and hexamers, assisted by VP22a |
| From polyprotein precursors | Picornavirus (poliovirus) | Immature 5S structural units, VP0-VP3-VP1 |
| | Togavirus (Sindbis virus) | Capsid (C) protein folds in and cleaves itself from a nascent polyprotein also containing glycoprotein sequences |
| | Retrovirus (avian sarcoma virus) | NC, CA, and MA protein shells assembled via Gag polyprotein |

nonspecific binding reactions. In view of these advantages, it is not surprising that several viruses employ this mechanism (Table 13.1), which is exemplified by assembly of picornaviral capsids. The first intermediate in the poliovirus assembly pathway, which sediments as a 5S particle (Fig. 13.3), is the immature structural unit that contains one copy each of VP0, VP3, and VP1 (Fig. 13.2B). These three proteins are liberated from the capsid protein precursor P1 upon cleavage by the viral 3CD^{pro} protease. However, it is believed that folding of their central β-barrel domains (Fig. 3.8) takes place during synthesis of P1. The poliovirus structural unit can then form by **intramolecular** interactions among the surfaces of these β-barrel domains before the covalent connections that link the proteins in the P1 precursor are severed.

Retroviruses exhibit an elegant and effective variation on the theme of assembly from polyproteins. Mature retrovirus particles contain three protein layers. An inner coat of NC protein, which packages the dimeric RNA genome, is enclosed within the capsid built from the CA protein. The capsid is in turn surrounded by the MA protein, which lies beneath, and is in intimate contact with the inner surface of the viral envelope (appendix, Fig. 9A). These three structural proteins are synthesized as the Gag polyprotein precursor, which contains their sequences in order of the protein layers they form in virus particles, with MA at the N terminus (Fig. 13.4). Retrovirus particles assemble from such Gag polyprotein molecules by a unique mechanism that allows orderly construction of the three protein layers and, as we shall see, coordination of this reaction with encapsidation of the genome and acquisition of the envelope.

### Participation of Cellular and Viral Chaperones

The assembly of viral proteins into structural units is undoubtedly assisted by cellular **chaperones,** specialized proteins that facilitate protein folding by preventing nonspecific, improper associations among exposed "sticky" patches on nascent and newly synthesized proteins. The first chaperone to be identified, the product of the *Escherichia coli groEL* gene, was discovered because it is essential for reproduction of phages T4 and λ. As discussed in chapter 12, the participation of chaperones resident in the endoplasmic reticulum (ER) lumen of animal cells in folding and assembly of oligomeric viral glycoproteins is well established. Cytoplasmic and nuclear chaperones of the host cell seem likely to play equally important roles in the formation of structural units. Several viral structural proteins, including the adenovirus fiber protein IV and the poliovirus P1 precursor, are associated with the cytoplasmic chaperone Hsp70. This protein is believed to facilitate the folding of these proteins and formation of structural units. In some cases, viral chaperones are essential participants in the reactions by which structural units are formed. Production of adenovirus hexon trimers, which form the faces of the icosahedral capsid, depends on such an accessory protein, the viral L4 100-kDa protein (Fig. 13.2C). The latter protein therefore facilitates folding of monomeric hexon subunits or their assembly into trimers, although its biochemical function has not been identified. Similarly, one function of the major herpesviral scaffolding protein described below is to facilitate entry into the nucleus and assembly of the VP5 capsid protein.

### Capsid and Nucleocapsid Assembly

The accumulation of virion structural units within the appropriate compartment of an infected cell sets the stage for the assembly of more complex capsids or nucleocapsids (see Table 3.1 for nomenclature). For reasons discussed previously, the reactions by which these structures are formed are often not understood in detail. Nevertheless, several different mechanisms for their assembly can be distinguished.

#### Intermediates in Assembly

A striking feature of well-characterized pathways of phage assembly (Box 13.1) is the sequential formation of progressively more complex structures: heads, tails, and tail fibers are each assembled in stepwise fashion via defined intermediates. Such an assembly line mechanism appears ideally suited for orderly formation of virus particles, which can be large and morphologically complex. Discrete intermediates also form during assembly of some icosahedral animal viruses. A stepwise assembly mechanism has been well characterized for poliovirus: the 5S structural unit described in the previous section is the immediate precursor of a 14S pentamer, which in turn is efficiently incorporated into virus particles (Fig. 13.3). The pentamer is stabilized by extensive protein-protein contacts and by interactions mediated by the myristate chains present on the five VP0 N termini (see Fig. 3.9C). The contribution of the lipids to pentamer stability is so great that this structure does not form at all when myristoylation of VP0 is prevented by an amino acid substitution in the modification site. The extensive interactions among its components result in molecular interlocking of the five structural units of the pentamer and impart great stability. Thus, formation of the 14S assembly intermediate is irreversible under normal conditions in an infected cell, a property that imposes the appropriate directionality on the entire assembly pathway (Fig. 13.3).

Despite the apparent advantages of a stepwise assembly mechanism, discrete assembly intermediates like the poliovirus pentamers have not been detected for many

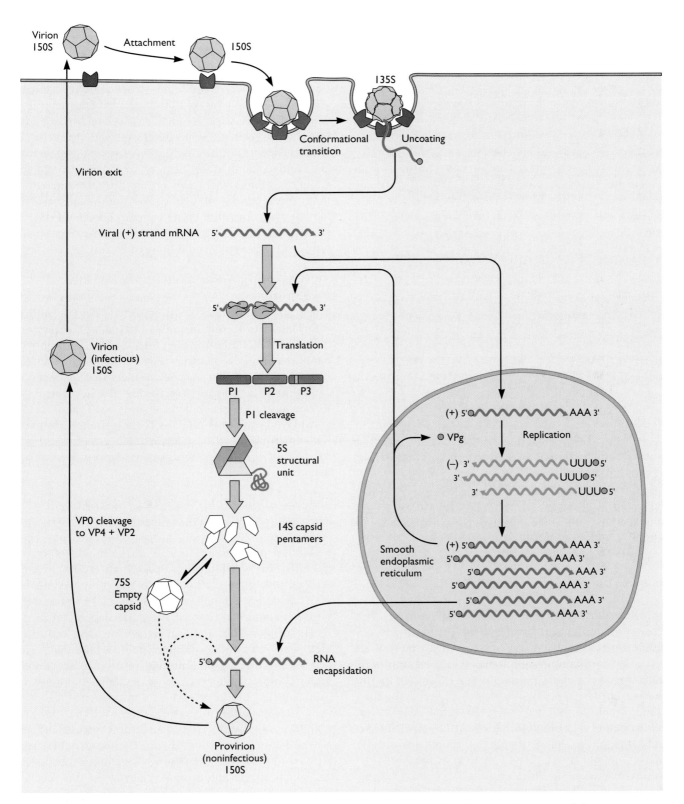

**Figure 13.3 Assembly of poliovirus in the cytoplasm of an infected cell.** Note that most of the assembly reactions are essentially irreversible, either because of proteolytic cleavage (formation of 5S structural units and mature virions) or because of extensive stabilizing interactions in the assembled structure (formation of 14S pentamers and of provirions). Stable, empty capsids, originally considered the precursors of provirions and termed procapsids, do not possess the same conformation as the mature virion, as symbolized by the dashed arrow. They are most likely dead-end products or a storage form of 14S pentamers. The conformational transition upon attachment to the poliovirus receptor, for which the virion is primed by cleavage of VP0 to VP2 and VP4, is also illustrated.

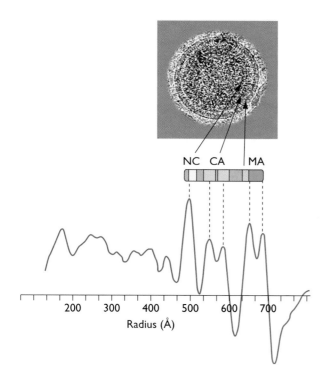

**Figure 13.4 Radial organization of the Gag polyprotein in immature human immunodeficiency virus type I particles.** The model for the arrangement of the Gag polyprotein shown below a cryoelectron micrograph of a viruslike particle assembled from Gag was deduced from radial density measurements of digitized images of the particles. The plot shows density as a function of distance from the particle center in angstroms. Courtesy of T. Wilk, European Molecular Biology Laboratory.

tion and instability. In addition, the formation of assembly intermediates may be rate-limiting, allowing stable structural units to be stockpiled before the final assembly reactions begin. While these possibilities cannot be excluded, it is more likely that assembly of many capsids is a highly cooperative, all-or-none process. Both simple and more complex icosahedral capsids are built by the repetition of interactions among multiple copies of one or a small number of structural units. Consequently, once the first few structural units were associated in the correct manner, assembly of the capsid would proceed rapidly to completion.

## Self-Assembly and Assisted Assembly Reactions

The primary sequences of virion proteins contain sufficient information to specify assembly, including complex reactions like the alternative five- and six-fold packing of VP1 pentamers in the simian virus 40 capsid. When expressed in *E. coli*, VP1 is isolated as pentamers that assemble into capsidlike structures in vitro, as does the hepatitis B virus C protein synthesized in heterologous hosts. Thus, **self-assembly** of structural proteins is the primary mechanism for formation of virus particles. Nevertheless, in some cases other viral components or cellular proteins can assist assembly.

### Viral and Cellular Components That Regulate Self-Assembly

Self-assembly of viral structural proteins may be the mortar for construction of virus particles, but other components of the virion often provide an essential foundation or the blueprint for correct assembly. As we have seen, assembly of the nucleocapsids of (−) strand RNA viruses is both coordinated with and dependent on synthesis of genomic RNA. The RNA must serve as a "template" for productive and repetitive binding of nucleocapsid proteins to one another. In other cases, the viral genome plays a more subtle yet equally important role, ensuring that the interactions among structural units are those necessary for infectivity. For example, poliovirus empty capsids lack internal structural features characteristic of the mature virion (Fig. 3.9) because VP0 is not cleaved to VP4 and VP2. The RNA genome is believed to participate in the autocatalytic cleavage of this precursor, which is essential for the production of infectious particles. Association of structural proteins with a cellular membrane is essential for the assembly of some protein shells, a situation exemplified by many retroviruses: the sequences of MA that specify Gag myristoylation and binding to the cytoplasmic surface of the plasma membrane described in chapter 12 are also required for assembly of the core.

viruses. In some cases, including many retroviruses, the absence of discrete intermediates can be attributed to coordination of assembly of protein shells with binding of the structural proteins to cellular membranes. In others, formation of the protective protein shell is directed by the nucleic acid genome. This mode of assembly is exemplified by the ribonucleoproteins of (−) strand RNA viruses, which assemble as genomic RNA is synthesized. Nucleocapsid formation probably depends on interactions of the protein components (e.g., N and NP in vesicular stomatitis virus and influenza A virus ribonucleoproteins, respectively) with both the nascent RNA and other protein molecules previously bound to the RNA. Because the synthesis of genomic RNA molecules is an all-or-none process, so too is the assembly of ribonucleoproteins.

The general dearth of structures simpler than empty capsids in cells infected by nonenveloped viruses might simply be a consequence of the properties that make such assembly intermediates difficult to detect, notably low concentra-

Binding of structural proteins to the genome or to a cellular membrane might simply raise their local concentration sufficiently to drive self-assembly, might organize the proteins in such a way that their interactions become cooperative, or might induce conformational changes necessary for productive or cooperative association of structural units. These mechanisms, which are not mutually exclusive, have not been distinguished experimentally, but there is evidence for induction of conformational transitions in specific cases. We do not understand adequately the molecular mechanisms by which binding of structural proteins to other virion components directs or regulates particle assembly. However, it is clear that such a requirement offers the important advantage of integration of assembly of protein shells with acquisition of other essential parts of the virion.

Cellular components can also modulate the fidelity with which viral structural proteins bind to one another. For example, the capsidlike structures assembled when simian virus 40 VP1 is expressed in insect or mammalian cells are much more regular in appearance than those formed by bacterially expressed VP1 in vitro. Modification of VP1 (by acetylation and phosphorylation) or the participation of nuclear chaperones must therefore improve the accuracy with which VP1 pentamers associate to form capsids. Similarly, in vitro self-assembly of poliovirus structural proteins is very slow, reduced at least two orders of magnitude compared to the rate observed in infected cells. Furthermore, the empty capsids that form have the altered conformation described previously, unless the reaction is seeded by 14S pentamers isolated from infected cells. This property indicates that the appropriate folding, modification, and/or interaction of the viral proteins that form the pentamer are critical for subsequent assembly reactions to proceed productively. Within infected cells, these crucial reactions are likely to be modulated by cellular chaperones, such as Hsp70, which is associated with the polyprotein during its folding to form 5S structural units. It is clear from such examples that host cells provide a hospitable environment for productive virus assembly, one that is not necessarily well reproduced when viral structural proteins are made or assemble in vitro.

### Viral Scaffolding Proteins: Chaperones for Assembly

Accurate assembly of some large icosahedral protein shells, such as those of adenoviruses (Fig. 13.5) and herpesviruses, is mediated by proteins that are not components of mature virions. Because these proteins participate in reactions by which the capsid or nucleocapsid is constructed but are then removed, they are termed **scaffolding proteins.** Among the best characterized of such proteins is the precurser of the herpes simplex virus type 1 VP22a protein.

This protein is the major component of an interior core present in assembling nucleocapsids (Fig. 3.6A). In the absence of other viral proteins, it forms specific scaffoldlike structures. Such self-association stimulates binding of the scaffolding protein to VP5, the protein that forms the hexameric structural units of the icosahedral nucleocapsid. The VP5 and VP22a proteins form a core to which additional VP5 hexamers and the triplex structures formed by VP19 and VP23 (Table 3.4) are added. The latter are also required for capsid assembly, which occurs by sequential formation of partial domelike structures and the spherical immature nucleocapsid. The interactions of VP5 with the scaffolding protein guide and regulate the intrinsic capacity of VP5 hexamers (and other nucleocapsid proteins) for self-assembly: omission of the scaffolding protein from a simplified assembly system, in which each nucleocapsid protein (Fig. 13.6; Table 3.4) is expressed from a baculovirus vector, leads to the production of unclosed and deformed nucleocapsid shells.

Once nucleocapsids have assembled, scaffolding proteins must be disposed of so that viral genomes can be accommodated (Fig. 13.6). The virion protease (VP24), which is also present in the core of assembling nucleocapsids, is essential for such DNA encapsidation. This enzyme cleaves a short C-terminal sequence from the scaffolding protein. Because this sequence is required for formation of sealed nucleocapsids with a regular structure and binding of the scaffolding protein to VP5, such processing presumably disengages scaffolding from structural proteins once assembly of the nucleocapsid is complete. The protease is also believed to degrade the scaffolding protein so that encapsidation of the genome can begin. Because it is essential for production of virions, the protease is of great interest as a target for development of antiherpesviral drugs.

The adenoviral L1 52/55-kDa proteins, which are differentially phosphorylated forms of a single translation product, exhibit properties expected of scaffolding proteins: they are necessary for assembly of adenovirus particles and are present in all capsid-like structures that can be isolated from infected cells but not in virions (Fig. 13.5). However, their primary function may be in stable encapsidation of the viral DNA genome rather than to provide a framework for assembly of the capsid.

Assembly of simpler protein shells can also depend critically on a viral protein. In addition to its many other functions, the simian virus 40 large T antigen (LT) participates in virion assembly. This protein does not form a scaffold, but an N-terminal domain of LT appears to be essential to organize the capsid shell: mutations within this domain block production of virions and induce accumulation of an

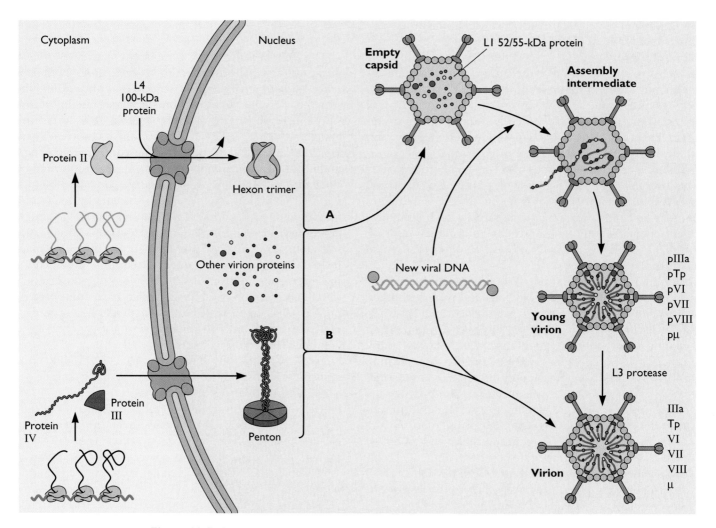

**Figure 13.5 Adenovirus assembly pathways.** Synthesis and assembly of hexons and pentons and their transport into the nucleus set the stage for assembly. The L4 100-kDa protein is required for formation of hexons, but its molecular function is not known. In the pathway originally proposed (A), these structural units and the proteins that stabilize the capsid (see Table 3.5) assemble into empty capsids. The L1 52/55-kDa proteins are present in all assembly intermediates but decrease in concentration as assembly proceeds. They are necessary for the formation of structures that can complete assembly. The DNA is then inserted into this structure via the packaging signal located near the left end of the genome. Premature breakage of DNA in the process of insertion would yield the structure designated "Assembly intermediate," in which an immature capsid is associated with a DNA fragment derived from the left end of the viral genome. Core proteins are encapsidated along with the viral genome to generate noninfectious young virions. Mature virions are produced upon cleavage of the precursor proteins listed to the right of the young virion. In the alternative pathway (B), which is based on the failure of viruses with packing sequence mutations to assemble any capsidlike structures, capsid assembly and encapsidation of the genome are concerted reactions. Empty capsids and the assembly intermediate would then be viewed as dead-end products or artifacts of the methods by which particles are extracted from infected cells. For example, the structure designated "Assembly intermediate" in pathway A could represent unstable structures whose DNA genomes were partially extruded and broken during extraction of intermediates from infected cells.

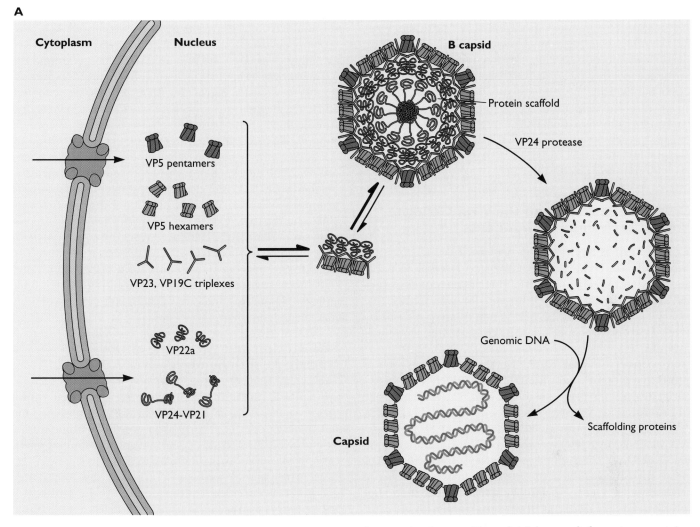

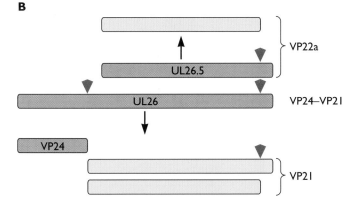

**Figure 13.6 Assembly of herpes simplex virus type 1 nucleocapsids.** (A) Assembly begins as soon as nucleocapsid proteins accumulate to sufficient concentrations in the infected cell nucleus. Intermediates include pentamers and hexamers of the major capsid protein VP5, which form pentons and hexons in the capsid, and triplexes of the minor proteins VP23 and VP19C. Whether structural units assemble prior to transport into the nucleus is not clear. Viral proteins essential for assembly of the nucleocapsid but not present in mature virions, the scaffolding protein (pre-VP22a and VP21) and the viral protease (VP24), must also enter the nucleus. Assembly of nucleocapsids depends on the formation of an internal scaffold around which the protein shell assembles. Further assembly requires the action of herpes simplex virus type 1 protease, VP24, to remove the scaffolding protein, allowing entry of the DNA genome. As discussed in the text, encapsidation is concurrent with cleavage of the concatemeric products of herpesviral DNA replication. (B) Overlapping sequences of herpes simplex virus type 1 scaffolding proteins. The primary translation products of the UL26 and UL26.5 reading frames are shown in purple. The initiating methionine of VP22a protein is within the larger reading frame that encodes the VP24-VP21 polyprotein. Thus, VP21 and VP22a are identical in sequence except that the former contains a unique N-terminal segment. All proteolytic cleavages, at the sites indicated by the red arrowheads, including those that liberate the enzyme itself, are carried out by the VP24 protease.

incomplete structure, which contains both the viral chromatin and VP1, not normally observed during simian virus 40 assembly. This N-terminal segment of LT has recently been shown to possess chaperone activity: it is similar in sequence to a specific domain of cellular chaperones of the DnaJ family and, like these cellular proteins, stimulates the ATPase activity of, and release of unfolded polypeptide substrates from, Hsp70 chaperones. It is believed that the chaperone activity of LT ensures the productive binding of VP1 pentamers to one another and to other components of the virion during assembly.

## Selective Packaging of the Viral Genome and Other Virion Components

### Concerted or Sequential Assembly

Incorporation of the viral genome into assembling particles is often called **packaging.** This process requires specific recognition of genomic RNA or DNA molecules (see below). It is clear that all viral genomes are packaged by one of two general mechanisms, concerted or sequential assembly.

In **concerted assembly,** the structural units of the protective protein shell assemble productively only in association with the genomic nucleic acid. The nucleocapsids of (−) strand RNA viruses form by a concerted mechanism (Fig. 13.7), as do retrovirus particles (Fig. 13.8). In the alternative **sequential assembly** mechanism, the genome is inserted into a preformed protein shell. The formation of herpesviral capsids provides a clear example of this type of assembly (Fig. 13.6). A capsidlike structure that lacks viral DNA but contains internal scaffolding proteins is an early intermediate in the assembly of these complex viruses. Neither mutations that inhibit viral DNA synthesis nor those that prevent DNA packaging block assembly of these capsids. Thus, the DNA genome must enter preformed capsids when proteins that assist capsid assembly described above are lost. In contrast to concerted assembly, encapsidation of the genome in a preformed structure requires specialized mechanisms to pull or push the genome into the capsid, as well as to maintain or open a portal for entry of the genome.

Despite the clear differences between concerted and sequential assembly pathways, it can be quite difficult to decide which mechanism applies to some viruses with icosahedral symmetry. A classic case in point is the poliovirus assembly pathway, which has been studied for more than 20 years. In the first scheme to be proposed, an empty capsid containing 60 copies of the VP0-VP3-VP1 structural unit but lacking the RNA genome was viewed as the precursor of the **provirion** (Fig. 13.3), the RNA-containing, but immature, virion. Such 75S empty capsids,

initially called procapsids, are readily detected in infected cell extracts and also form by self-assembly of 14S pentamers in vitro. In the alternative pathway for poliovirus assembly, pentamers are proposed to condense around the RNA genome (Fig. 13.3). In this mechanism, capsid assembly and packaging of the RNA genome are coupled, and empty capsids are considered dead-end products or a storage form of 14S pentamers. The concerted pathway is supported by several experimental observations, including the ability of radioactively labeled pentamers to form virions without the appearance of a procapsid intermediate. Furthermore, it is now clear that stable empty capsids are produced by irreversible conformational changes that take place during their extraction from infected cells. At present, the weight of the evidence favors the concerted pathway of poliovirus assembly. Nevertheless, it is difficult to exclude the possibility that some form of procapsid, present in infected cells at very low concentrations and difficult to isolate in unaltered form, is a true assembly intermediate in vivo.

### Recognition and Packaging of the Nucleic Acid Genome

The assembly mechanisms described in previous sections are fruitless unless the viral genome is packaged within progeny virions. Despite diversity in size, composition, and morphology, specific incorporation of the genome during assembly of virus particles is achieved by a limited repertoire of mechanisms, which are discussed in the next section. The special problems imposed by segmented genomes are considered subsequently. At the outset, it is important to appreciate that most particles released from an infected cell do not contain a genome. Because synthesis of the viral genome is often rate limiting for assembly, empty particles may be produced by "off-pathway" reactions among structural proteins. Or they may form as a result of failure of genome encapsidation reactions.

#### Mechanisms of Nucleic Acid Recognition and Packaging

During encapsidation, viral nucleic acid genomes must be distinguished from the cellular DNA or RNA molecules present in the compartment in which assembly takes place. This process requires a high degree of discrimination among similar nucleic acid molecules. For example, retroviral genomic RNA constitutes much less than 1% of an infected cell's cytoplasmic RNA population and bears all the hallmarks of cellular mRNAs. Yet it is the RNA packaged in the great majority of retrovirus particles. Such discrimination is the result of specific recognition of sequences or structures unique to the viral genome, termed **packaging signals.** These can be defined experi-

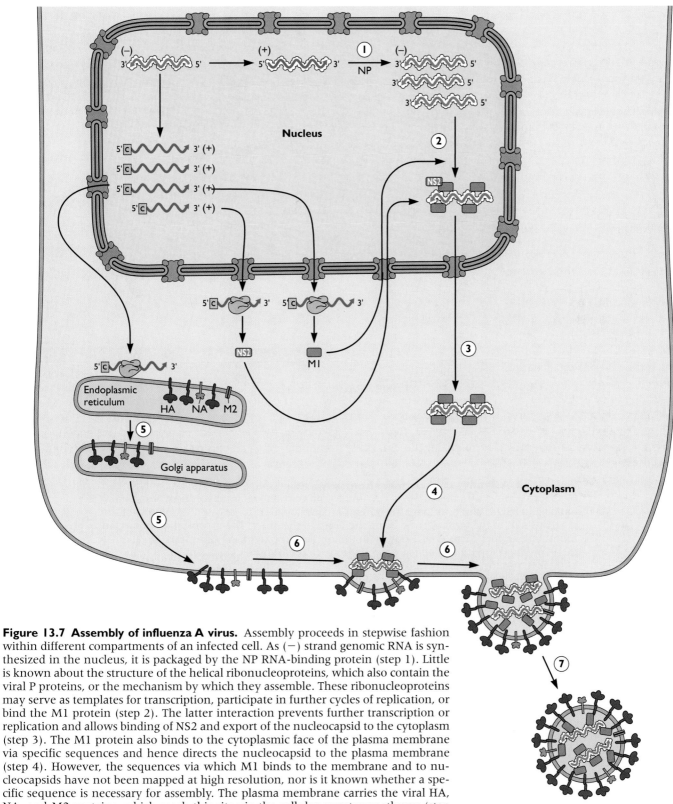

**Figure 13.7 Assembly of influenza A virus.** Assembly proceeds in stepwise fashion within different compartments of an infected cell. As (−) strand genomic RNA is synthesized in the nucleus, it is packaged by the NP RNA-binding protein (step 1). Little is known about the structure of the helical ribonucleoproteins, which also contain the viral P proteins, or the mechanism by which they assemble. These ribonucleoproteins may serve as templates for transcription, participate in further cycles of replication, or bind the M1 protein (step 2). The latter interaction prevents further transcription or replication and allows binding of NS2 and export of the nucleocapsid to the cytoplasm (step 3). The M1 protein also binds to the cytoplasmic face of the plasma membrane via specific sequences and hence directs the nucleocapsid to the plasma membrane (step 4). However, the sequences via which M1 binds to the membrane and to nucleocapsids have not been mapped at high resolution, nor is it known whether a specific sequence is necessary for assembly. The plasma membrane carries the viral HA, NA, and M2 proteins, which reach this site via the cellular secretory pathway (step 5). The M1 protein probably controls budding (step 6). The mechanism by which membranes fuse to release the enveloped particle (step 7) is not known.

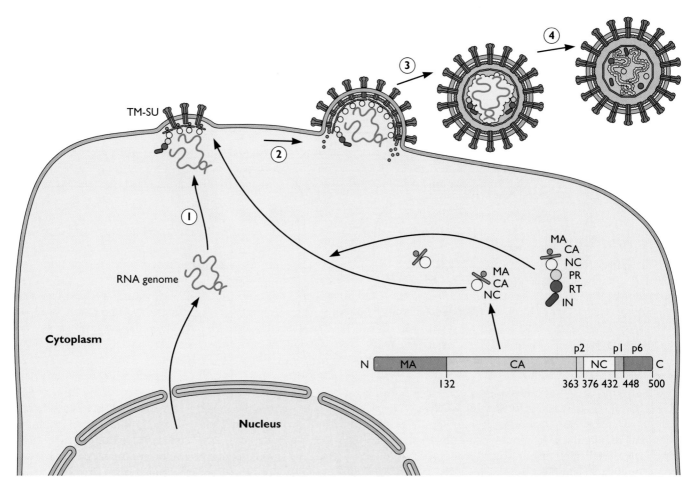

**Figure 13.8 Assembly of a retrovirus from polyprotein precursors.** The Gag polyprotein of all retroviruses contains the MA, CA, and NC proteins linked by spacer peptides that are variable in length and position. The proteins are in the order (from N to C terminus) of the protein shells of the virus particle, from the outer to the inner. The organization of human immunodeficiency virus type 1 Gag is summarized on the right. A minor fraction, about 1 in 10, of Gag translation products carry the retroviral enzymes, denoted by PR, RT, and lN, at their C termini. The association of Gag molecules with the plasma membrane, with one another, and with the RNA genome via binding of NC segments initiates assembly at the inner surface of the plasma membrane (step 1). In some cases, such as human immunodeficiency virus type 1, the MA segment also binds specifically to the internal cytoplasmic domain of the TM-SU glycoprotein, whose synthesis, processing, and transport to the plasma membrane are described in chapter 12. Note that no discrete intermediates are formed, as assembly of the particle continues by incorporation of additional molecules of Gag (step 2). This pathway is typical of many retroviruses, but some (e.g., D-type retroviruses) complete assembly of the core in the interior of the cell prior to its association with the plasma membrane. The dimensions of the assembling particle are determined by interactions among Gag polyproteins. Eventually, fusion of the membrane around the budding particle (step 3) releases the immature noninfectious particle. This fusion process is not well understood. Cleavage of Gag and Gag-Pol polyproteins by the viral protease (PR) produces infectious particles (step 4) with a morphologically distinct core (see Fig. 13.18).

mentally as the genomic sequences whose mutation prevents incorporation of the nucleic acid into the assembling virion or which are sufficient to direct incorporation of foreign nucleic acid. The organization of the packaging signals of several viruses is therefore quite well understood.

**Nucleic acid packaging signals.** *DNA signals.* The products of papova- or adenoviral DNA synthesis are genomic DNA molecules that can be incorporated into assembling virus particles without further modification. The DNA genomes contain discrete packaging signals with several common properties. The signals comprise repeats of short sequences, some of which are also part of viral promoters or enhancers; they are positioned close to an origin of replication, and their ability to direct DNA encapsidation depends on this location. The simian virus 40 DNA packaging signal has been located in the regulatory region of the genome containing the origin of replication, the enhancer, and early and late promoters. Multiple sequences within this region, including the repeated binding sites for the cellular transcriptional regulator Sp1 in the early promoter, contribute to the encapsidation signal. It has been suggested that the internal capsid proteins VP2 and VP3, which can bind to DNA only in a nonspecific manner, recognize this packaging signal indirectly by binding to cellular transcriptional regulators. The encapsidation signal of the adenovirus genome, which is located close to the left inverted repeat sequence and origin, likewise comprises a set of repeated sequences, several of which overlap enhancers that stimulate transcription of viral early genes (Fig. 13.9). Cellular proteins, including a member of the steroid/thyroid hormone receptor family of transcriptional regulators, also bind specifically to the adenovirus DNA-packaging sequence. Recognition of these signals for encapsidation of the genome must prevent viral DNA molecules from participating in transcription or replication reactions and direct encapsidation. However, the molecular mechanisms of packaging are not well understood. Whether the cellular proteins that can bind specifically to the packaging signals are needed for genome encapsidation is not known, nor have the reactions initiated by recognition of these DNA sequences been elucidated.

The replication of herpesviruses produces not genomic DNA molecules but rather concatemers containing many head-to-tail copies of the viral genome. Individual genomes must therefore be liberated from such concatemers for encapsidation. The herpes simplex virus type 1 packaging signals, *pac1* and *pac2*, which lie within the terminal *a* repeats of the genome, are necessary for both recognition of the viral DNA and its cleavage within the adjacent DR repeats (Fig. 13.10A). The proteins that recognize the *pac* sequences must therefore include a specific endonuclease that releases unit-length DNA molecules from concatemers. It is generally believed that cleavage is concomitant with genome encapsidation. In one model (Fig. 13.10B), it is proposed that a protein complex formed on the unique short ($U_S$) *pac* sequence (Fig. 13.10A) is recognized by a capsid protein. Following the first DNA cleavage, a unit-length genome is then reeled into the capsid prior to the second DNA cleavage. This mechanism is analogous to that by which concatemeric DNA products of phage T4 replication are cleaved and packaged by a "terminase" complex, which associates transiently with preformed capsids and a specific portal protein of the capsid. The products of at least seven herpesviral genes are specifically required for cleavage and packaging of the viral genome (but not for DNA replication). The product of one (UL15) exhibits sequence similarities to an ATPase present in the T4 terminase complex, including similarity to an ATP-binding site essential for its function. This protein, as well as two others needed for encapsidation of the herpesviral genome, are associated with immature nucleocapsids and might therefore form the portals through which DNA is inserted. However, much remains to be learned about the biochemical properties and functions of the her-

**Figure 13.9 Packaging signal of human adenovirus type 5.** The location of the repeated sequences (blue arrows) of the packaging signal relative to the left inverted terminal repeat (ITR), the origin of replication (Ori), and the E1A transcription unit is indicated. These repeated sequences are AT rich and functionally redundant. The positions of transcriptional enhancers within this region are also shown. Enhancer 1 specifically stimulates transcription of the immediate-early E1A gene, whereas enhancer II increases the efficiency of transcription of all viral genes.

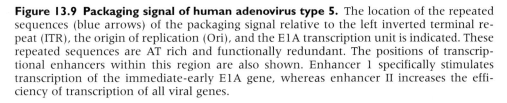

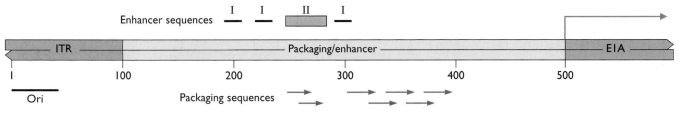

**A**

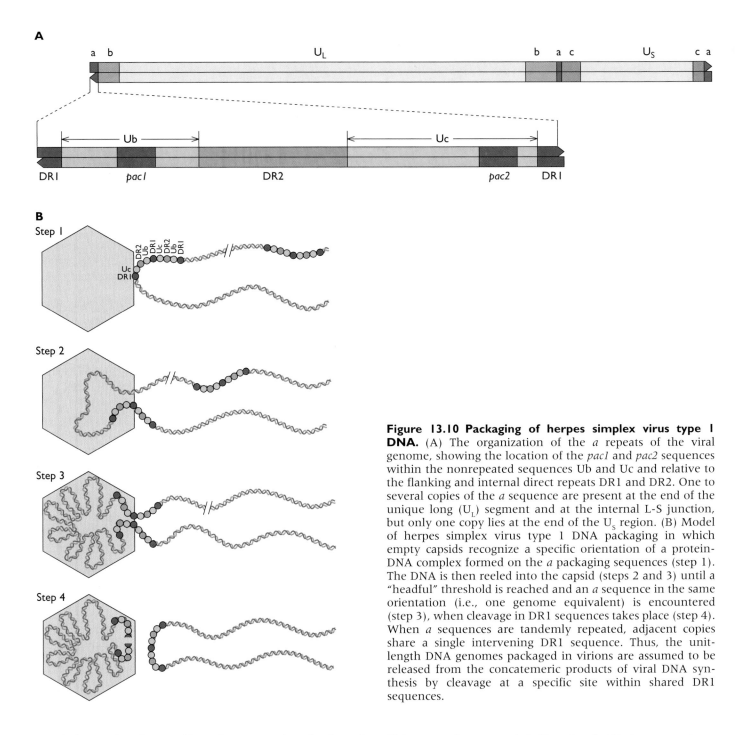

**Figure 13.10 Packaging of herpes simplex virus type 1 DNA.** (A) The organization of the *a* repeats of the viral genome, showing the location of the *pac1* and *pac2* sequences within the nonrepeated sequences Ub and Uc and relative to the flanking and internal direct repeats DR1 and DR2. One to several copies of the *a* sequence are present at the end of the unique long (U$_L$) segment and at the internal L-S junction, but only one copy lies at the end of the U$_S$ region. (B) Model of herpes simplex virus type 1 DNA packaging in which empty capsids recognize a specific orientation of a protein-DNA complex formed on the *a* packaging sequences (step 1). The DNA is then reeled into the capsid (steps 2 and 3) until a "headful" threshold is reached and an *a* sequence in the same orientation (i.e., one genome equivalent) is encountered (step 3), when cleavage in DR1 sequences takes place (step 4). When *a* sequences are tandemly repeated, adjacent copies share a single intervening DR1 sequence. Thus, the unit-length DNA genomes packaged in virions are assumed to be released from the concatemeric products of viral DNA synthesis by cleavage at a specific site within shared DR1 sequences.

pesviral proteins that mediate cleavage and packaging of unit-length genomes.

*RNA signals.* Because it is also an mRNA, the retroviral genome must be distinguished during encapsidation from both cellular mRNA and subgenomic viral mRNA. In addition, two genomic RNA molecules must interact with one another, for the retroviral genome is packaged as a dimer.

This unusual property is believed to facilitate reverse transcription of the genome into proviral DNA at the start of the infectious cycle. In virions, the dimeric genome is in the form of a 70S complex held together by many noncovalent interactions between the RNA molecules. However, most attention has been focused on a short sequence located near the 5' end of each molecule of the genome that allows formation of stable dimers in vitro, termed the

**dimer linkage sequence.** In vitro experiments with human immunodeficiency virus type 1 have provided evidence for base pairing between loop sequences of a specific hairpin within the dimer linkage sequence (Fig. 13.11A) and the formation of an intermolecular four-stranded helical structure (known as a G tetrad or G quartet). The importance of such contacts in mediating dimerization in vivo has yet to be established.

Sequences necessary for packaging of retroviral genomes, termed **Psi (ψ)**, vary considerably in their complexity and location. In some cases, exemplified by Moloney murine leukemia virus, a contiguous ψ sequence of about 350 nucleotides (Fig. 13.11B) is both necessary and sufficient for RNA encapsidation. As this sequence lies downstream of the 5′ splice site, only unspliced genomic RNA molecules are recognized for packaging. The human immunodeficiency virus type 1 genome also contains an RNA-packaging sequence (Fig. 13.11A) that distinguishes the full-length genome from spliced RNA molecules. However, this sequence fails to direct packaging of heterologous RNA species into retrovirus particles, indicating that it is not sufficient. Additional sequences required for human immunodeficiency virus type 1 genomic RNA encapsidation or dimerization reside within the TAR sequence and adjacent poly(A) hairpin (Fig. 13.11A) or at more distant locations. Thus, the genomes of this and some other retroviruses appear to contain packaging signals that are dispersed in the primary sequence but presumably form a distinctive structural feature in folded RNA molecules. The ψ sequences of yet other retroviruses, such as avian leukosis virus, reside only upstream of the 5′ splice site (Fig. 13.11B). Even though both genomic and subgenomic RNA contain this sequence, spliced mRNA molecules are not encapsidated. There must be some additional mechanism to prevent entry of subgenomic avian leukosis virus RNA into assembling particles.

The NC domain of Gag, a basic region that contains at least one copy of a zinc-binding motif (a Cys-His box) (Fig. 13.12A), mediates selective and efficient encapsidation of genomic RNA during retroviral assembly. The central region of NC containing the Cys-His box(es) and adjacent basic sequences binds specifically to RNAs that contain ψ sequences in vitro and is necessary for selective packaging of the genome in infected cells. Furthermore, the interaction of NC with ψ sequences is of sufficient specificity to discriminate the genome of the appropriate retrovirus from those of other retroviral genomes. Because even the simplest retroviral packaging signals comprise long RNA sequences, a complete picture of how NC recognizes ψ sequences is not yet available. However, the structure of human immunodeficiency virus type 1 NC bound to a highly conserved hairpin within the ψ sequence of the

genome (SL3 in Fig. 13.11A) has recently been determined by nuclear magnetic resonance methods. Residues within the RNA-binding region of NC that are highly conserved among human immunodeficiency virus type 1 isolates, or important for specific RNA binding and selective packaging, make numerous specific contacts with the RNA (Fig. 13.12B). The NC protein is inherently flexible. It therefore seems likely that the NC regions of additional Gag molecules bind with analogous specificity (but via distinctive intramolecular interactions) to the other RNA sequences needed for packaging (Fig. 13.11A) to ensure efficient and highly selective encapsidation of genomic RNA in infected cells.

The approximately 2,000 molecules of processed NC present in a mature virion also bind nonspecifically to genomic RNA. Whether specific recognition of the packaging signal by NC-embedded Gag and nonspecific binding take place coordinately or sequentially during assembly is not clear. As discussed in chapter 7, the NC protein may also facilitate dimerization of retroviral genomic RNA.

**Other parameters that govern genome encapsidation.** Specific sequence signals may be required to mark a viral genome for encapsidation, but their presence does not guarantee packaging. The fixed dimensions of the closed icosahedral capsids or nucleocapsids of many viruses impose an upper limit on the size of viral nucleic acid that can be accommodated. Consequently, nucleic acids that are more than 5 to 10% larger than the wild-type genome cannot be encapsidated, even when they contain appropriate packaging signals. This property has important implications for the development of viral vectors. In some cases, the length of the DNA that can be accommodated in the particle (a "headful") is a critical parameter determining encapsidation. This mechanism is exemplified by the coupled cleavage and encapsidation of genomic herpes simplex virus type 1 DNA molecules from the concatemeric products of replication, when both specific sequences and a headful of DNA are recognized. Indeed, the packaging of some viral DNA genomes, such as T4 DNA, depends solely on the latter parameter.

The coupling of encapsidation of a viral genome with its synthesis may also contribute to the specificity with which the viral genome is incorporated into assembling structures. As mentioned previously, such coordination is typical of the assembly of (−) strand RNA viruses (e.g., Fig. 13.7). However, the mechanisms by which nascent genomic RNA molecules are recognized by the structural proteins of viral ribonucleoproteins are not well understood. Coordination of replication and encapsidation may also contribute to the great specificity with which picornaviral genomes are packaged: not only abundant cytoplasmic

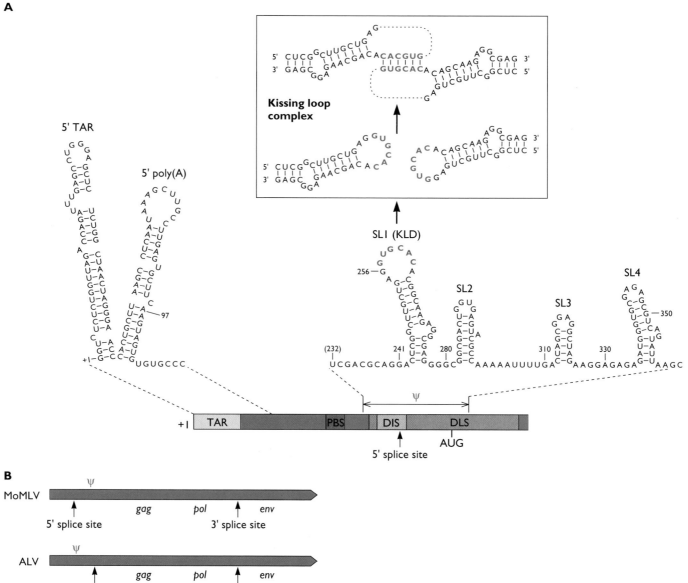

**Figure 13.11 Sequences important in packaging of retroviral genomes.** (A) The 5′ segment of the human immunodeficiency virus type 1 genome is shown to scale at the bottom, indicating the positions of TAR, the tRNA primer-binding site (PBS), the 5′ splice site, a packaging signal designated ψ, the sequence that forms the dimer linkage structure (DLS), and the dimerization initiation site (DIS) within DLS, which can initiate dimerization in vitro. The hairpins formed by TAR and the adjacent sequence containing a poly(A) addition signal, as well as the four hairpins (SL1 to SL4) present in the ψ segment, are shown above. The SL1 hairpin is also the dimer initiation sequence defined above. The loop-loop "kissing" complex proposed to form when two genomic RNA molecules dimerize via the self-complementary sequence shown in red (or the alternative such sequence GCGCGC found in other virus isolates) is depicted at the top. The ψ sequence, which includes intronic sequences, appears to be necessary but not sufficient for encapsidation of genomic RNA. The hairpin containing the poly(A) addition signal near the 5′ end of the RNA and the bottom of the TAR hairpin, as well as more distantly located sequences, have also been reported to be required. (B) Locations within the RNA genomes of sequences necessary for encapsidation of Moloney murine leukemia virus (MoMLV) and avian leukosis virus (ALV) RNAs, designated ψ.

**A**

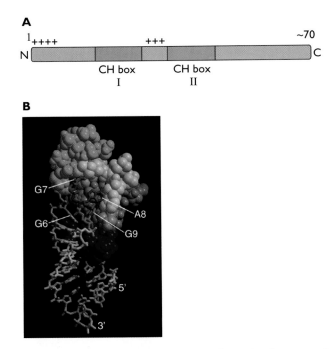

**B**

**Figure 13.12 The retroviral NC RNA packaging protein.** (A) Schematic illustration of the NC protein, showing the locations of the Cys-His (CH) boxes and basic regions necessary for both nonspecific binding to RNA and specific recognition of the packaging signal. (B) Structural model of human immunodeficiency virus type 1 NC protein bound to the SL3 hairpin of the $\psi$ sequence, determined by nuclear magnetic resonance methods, shown as a space-filling models. RNA is shown in gray except for specific G and A residues contacted by the protein, which are shown in pink ($G^7$), violet ($A^8$), and orange ($G^9$). The N- and C-terminal zinc fingers (light blue and green, respectively) and residues within the basic region between them (yellow) make specific contacts with these exposed bases. For example, the N-terminal zinc finger binds to both $G^9$ and $A^8$ via hydrophobic interactions and hydrogen bonds from the protein backbone. The N-terminal segment (royal blue) preceding the first zinc finger includes a helix that binds within the major groove of the double-stranded RNA stem. From R. N. De Guzman et al., *Science* **279**:384–388, 1998, with permission. Courtesy of R. N. De Guzman and M. F. Summers, University of Maryland, Baltimore County.

cellular RNA species (tRNAs, rRNAs, and mRNAs) but also (−) strand poliovirus RNA and poliovirus mRNA lacking VPg are excluded from virions. As discussed in chapter 12, poliovirus infection induces the formation of cytoplasmic vesicles, with which both viral replication and virion assembly are associated. Such sequestration of genomic RNA molecules with the viral proteins that must bind to them clearly could make a major contribution to packaging specificity by reducing competition from cytoplasmic cellular and viral RNAs. There is some evidence that encapsi-

dation of the RNA genome is coupled with RNA replication and that this step in assembly is assisted by the poliovirus 2C protein. This protein is a nucleoside triphosphatase that might facilitate release of RNA from membrane-bound replication complexes.

### Packaging of Segmented Genomes

Segmented genomes pose an intriguing packaging problem. The best-studied example among animal viruses is the influenza A virus genome, which comprises eight molecules of RNA. It has been appreciated for many years that production of an infectious influenza virus particle requires incorporation of at least one copy of each of the eight genomic segments. Nevertheless, random packaging still cannot be distinguished from a selective mechanism for inclusion of a full complement of genomic RNAs.

There is clear precedent for a selective mechanism for incorporation of one copy of each RNA segment. The genome of phage φ6 comprises one copy of each of three double-stranded RNA segments designated S, M, and L. The (+) strand of each segment is packaged prior to synthesis of complementary (−) strands within particles, a mechanism analogous to synthesis of the double-stranded RNA segments of the reovirus genome. The particle-to-PFU ratio of phage φ6 is close to 1, indicating that essentially all particles contain a complete complement of genome segments. Such precise packaging appears to be the result of the serial dependence of packaging of the (+) strand RNA segments: in in vitro reactions, the S segment packages alone, but entry of M RNA requires the presence of S RNA within particles, and packaging of the L segment is dependent on prior entry of both S and M RNAs.

A random packaging mechanism in which any eight RNA segments of the influenza genome were incorporated into virions would generate a maximum of 1 infectious particle for every 400 or so assembled ($8!/8^8$). This ratio might seem impossibly low, but is within the range of ratios of noninfectious to infectious particles found in virus preparations. Furthermore, if packaging of more than eight RNA segments were possible, the proportion of infectious particles would increase significantly. For example, with 12 RNA molecules per virion, 10% of the particles would contain the complete viral genome. Particles containing more than eight RNA segments have been isolated, consistent with random packaging. On the other hand, the specific interference with packaging of wild-type RNA segment 1, but of no other genome segment, by a distinguishable RNA segment 1 argues strongly for a specific mechanism of packaging, at least of segment 1. Until such a mechanism has been identified, the possibility that packaging of influenza virus genomic RNA segments is random cannot be excluded.

## Incorporation of Virion Enzymes and Other Nonstructural Proteins

In many cases, the production of infectious particles requires incorporation into the assembling virion of essential viral enzymes or other proteins that are important in establishing an efficient infectious cycle. Some of these proteins are also structural proteins of the virion, needing no special mechanism to ensure that they enter the assembling virion. For example, the herpes simplex virus type 1 VP16 protein is both a major component of the virion tegument and the activator of transcription of viral immediate-early genes. However, the majority of virion enzymes and other nonstructural proteins must be incorporated by specific mechanisms.

Among the best understood is the simple, yet elegant, mechanism that ensures entry of retroviral enzymes (PR, RT, and IN) into the assembling core of these viruses. In most cases, these enzymes are synthesized as C-terminal extensions of the Gag polyprotein. The organization and complement of these translation products, here designated Gag-Pol, varies among retroviruses, but the important point is that they contain not only Pol but also the sequences specifying Gag-Gag association described below. These are presumed to direct incorporation of Gag-Pol molecules into assembling particles (Fig. 13.8). The low efficiency with which Gag-Pol polyproteins are translated determines their concentrations relative to Gag in the cell and in virions (<1:9). Indeed, a low Gag-Pol/Gag ratio may be necessary for particle assembly: Gag-Pol molecules cannot direct assembly in the absence of Gag, presumably because they are too large to be packed into a protein shell of the size dictated by the interactions among Gag domains. All retrovirus cores also contain the cellular tRNA primer for reverse transcription brought into particles by its base pairing with a specific sequence in the RNA genome and by specific binding to RT.

The enzymes present in other virus particles, such as the RNA-dependent RNA polymerases of (−) strand RNA viruses (see Table 3.8), are synthesized as individual molecules and thus must enter assembling particles by noncovalent binding to the genome or to structural proteins.

## Acquisition of an Envelope

Formation of many types of virus particle requires envelopment of capsids or nucleocapsids by a lipid membrane carrying viral proteins. Most such enveloped viruses assemble by virtue of specific interactions among virion components at a cellular membrane before budding and pinching off of a new virus particle. However, they vary in the intracellular site at which the envelope is acquired and therefore in the relationship of envelopment to release of the virus particle from an infected cell. As discussed in chapter 12, whether particles assemble at the plasma membrane or internal membranes of the host cell is determined by the destination of viral proteins that enter the cellular secretory pathway. Regardless of their sites of assembly, enveloped viruses assemble by one of two mechanisms, distinguished by whether acquisition of the envelope follows assembly of internal structures or whether these processes take place simultaneously.

## Sequential Assembly of Internal Components and Budding from a Cellular Membrane

For most enveloped viruses, the assembly of internal structures of the virion and their interaction with a cellular membrane modified by insertion of viral proteins are spatially and temporally separated. This class of assembly pathways is exemplified by (−) strand RNA viruses such as influenza A virus (Fig. 13.7) and vesicular stomatitis virus. Influenza A virus ribonucleoproteins containing individual genomic RNA segments, NP protein, and the polymerase proteins are assembled in the infected cell nucleus as genomic RNA segments are synthesized. They are then transported to the cytoplasm in the M1- and NS2-dependent reactions described in chapter 12. The viral glycoproteins HA and NA and the M2 membrane protein travel to the plasma membrane via the cellular secretory pathway (Fig. 13.7). The M1 protein, whose concentration appears to be rate limiting for assembly, interacts with both viral nucleocapsids and the inner surface of the plasma membrane to direct assembly of progeny particles at that membrane. Vesicular stomatitis virus assembles in a similar fashion, although no transport of the ribonucleoprotein from nucleus to cytoplasm is required. Thus, the matrix proteins of these (−) strand RNA viruses provide the links among ribonucleoproteins and the modified cellular membrane necessary for assembly and budding. One or more tegument proteins of the more complex herpesviruses (see Fig. 3.13A) must function like these matrix proteins, mediating binding of the preassembled nucleocapsid to the modified cellular membrane from which the virion derives its envelope.

The cellular membranes destined to form the envelopes of virus particles contain viral integral membrane proteins that play essential roles in the attachment of virus particles to, and their entry into, host cells. In simple enveloped togaviruses, direct binding of the cytoplasmic portion of the viral glycoprotein to the single nucleocapsid protein (see Fig. 3.21) is necessary for acquisition of the envelope during budding from the plasma membrane. The crucial role and specificity of these interactions in the final steps in assembly are illustrated by the failure of a chimeric Sindbis virus containing the coding sequence for the E1 glycoprotein of Ross River virus, a second togavirus, to bud efficiently. The PE2 and E1 glycoproteins of the chimeric virus form heterodimers that are correctly processed (by cleavage of PE2 to E2) and transported to the plasma membrane. However, such chimeric glycoproteins exhibit an altered conformation and fail to bind to nucleocapsids at the plasma membrane. Binding of viral glycoproteins to internal components also appears to be important for production of the more complex enveloped viruses. Interactions between the influenza virus M1 protein and the HA or the NA glycoprotein are necessary for formation of virus particles with normal size and morphology (Fig. 13.13).

## Coordination of the Assembly of Internal Structures with the Acquisition of the Envelope

The alternative pathway of acquiring an envelope, in which assembly of internal structures and budding from a cellular membrane are coincident in space and time, is exemplified by many retroviruses. Assembling cores of the majority of retroviruses first appear as crescent-shaped patches at the inner surface of the plasma membrane. These structures extend to form a closed sphere as the plasma membrane wraps around and eventually pinches off the assembling particle (Fig. 13.8). In this unusual mechanism of assembly, the internal structural proteins,

MA, CA, and NC, are not proteolytically processed from the Gag polyprotein until late in the budding process. Formation of the virion is mediated by interaction of Gag polyprotein molecules with one another to form the protein core, with the RNA genome via the NC portion, and with the plasma membrane via the MA segment. Indeed, when expressed alone, Gag directs the budding of viruslike particles (Fig. 13.14), indicating that it contains all the information necessary for assembly and release of retroviral particles.

Segments of Gag that mediate the orderly association of polyprotein molecules with one another and are required for proper assembly have been identified by mutational analyses. These sequences include an essential C-terminal multimerization domain of the CA segment of many simple and complex retroviruses. Deletion of a spacer peptide between the CA and NC domains of Gag of avian sarcoma virus and other retroviruses results in the assembly of misshapen particles, while removal of the p6 segment of human immunodeficiency virus type 1 Gag (Fig. 13.8) leads to defects in the release of virions from the cell surface and the formation of very large particles. Thus, interactions among these segments of Gag polyproteins are important in determining the size and shape of retrovirus virions. Association among Gag polyprotein molecules

**Figure 13.13 Deformed influenza A virus particles.** These particles form when the cytoplasmic (internal) domains of the NA glycoprotein (B) or both NA and HA glycoproteins (C) are truncated. Purified wild-type (A) and mutant virions were negatively stained with phosphotungstic acid and examined by electron microscopy. Bars = 100 μm. The mutant particles are also considerably larger than wild-type virus particles. It is therefore believed that binding of the internal portions of HA, and particularly of NA, to internal virion components (M1 protein and/or the ribonucleoproteins) determines the spherical shape and characteristic size of these virus particles. From H. Jin et al., *EMBO J.* **16:**1236–1247, 1997, with permission. Courtesy of R. A. Lamb, Northwestern University.

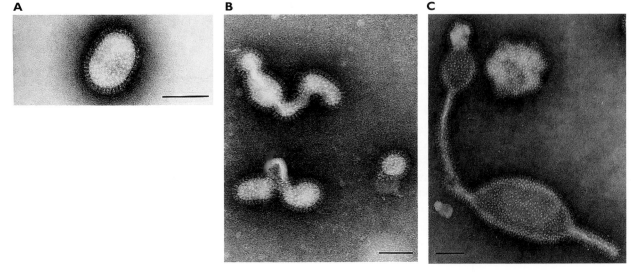

A    B    C

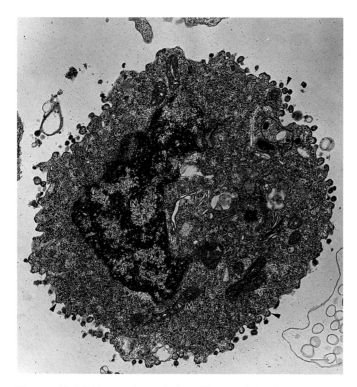

**Figure 13.14 Formation of virus-like particles by a retroviral Gag polyprotein synthesized in the absence of other viral proteins in human T cells.** Shown is an electron micrograph of a thin section of such a cell fixed and stained following introduction of a vector expressing the viral Gag-Pol coding sequence. Prior to electron microscopy, viral particles (arrowhead) were labeled with polyclonal antibodies against the CA protein attached to gold beads. Bar = 1.0 μm. N, nucleus; M, mitochondrion. Courtesy of J. J. Wang, Institute of Biomedical Sciences, Academica Sinica, Taipei, Taiwan, and B. Horton and L. Ratner, Washington University School of Medicine.

must coincide with their binding to the plasma membrane via the N-terminal lipid (myristate) or lipid plus protein signals in the MA segment of the polyprotein described in chapter 12. Particle formation also requires sequences predicted to lie at the interfaces of the MA trimers formed in crystals (see Fig. 3.22). Such binding of Gag molecules to the inner surface of the plasma membrane is essential for productive assembly, probably because Gag adopts an assembly-competent conformation only when its N-terminal MA segment has become firmly anchored in the plasma membrane. This property provides an effective mechanism for ensuring that the assembly of internal protein shells is coordinated with the binding of Gag to the membrane. Conversely, efficient plasma membrane binding of Gag depends on sequences located far from the membrane-binding region of MA, such as a sequence in the N-terminal portion of NC. Because this sequence is not required for production of stable Gag or transport of the polyprotein to

the plasma membrane, it is believed to promote Gag-Gag or Gag-RNA interactions that lead to cooperative and stable binding of Gag molecules to the plasma membrane in the developing bud.

In some cases, the MA segment of Gag also binds to the cytoplasmic tail of the viral envelope glycoprotein, whose synthesis, intracellular processing, and transport to the plasma membrane are described in chapter 12. For example, association of the assembling human immunodeficiency virus type 1 core with the TM-SU glycoprotein requires the N-terminal 100 amino acids of MA. Such Gag-Env interactions ensure specific incorporation of viral glycoproteins into virions. Nonetheless, they do not appear to be universal: glycoprotein-containing virions are produced even when the C-terminal tails of TM of other retroviruses (e.g., avian sarcoma virus) are deleted. Nor can a model based only on Gag-Env interactions account for the ease with which "foreign" viral and cellular glycoproteins are included in the envelopes of all retroviruses. It is thus believed that as a particle assembles at the plasma membrane, Gag-membrane interactions displace cellular membrane proteins connected to internal components of the cell. Such displacement would allow lateral diffusion of viral (and cellular) glycoproteins that are not connected in this way into these regions of the cellular membrane and thus their passive incorporation into assembling virus particles. Although not absolutely required, Gag-Env interactions would facilitate this process.

We have much to learn about the mechanisms governing the final step in retroviral assembly, budding of particles from the membrane. As discussed previously, the dimensions and morphology of the budding structures are set by interactions among Gag polyproteins. The final reaction must be fusion of membrane regions juxtaposed as the particle assembles (Fig. 13.8). Polyprotein-specific regions of Gag are essential for fusion: their deletion or mutation leads to the accumulation of immature particles that remain attached to the host cell by a thin membrane stalk. The segments necessary for membrane fusion during assembly of avian sarcoma virus and human immunodeficiency virus type 1 are located at different positions within Gag and are unrelated in amino acid sequence. Nevertheless, they can substitute for one another. These polyprotein-specific Gag sequences might participate directly in final budding reactions, such as membrane fusion. Alternatively, they might operate earlier in the pathway by which virions are produced, for example, to set the appropriate conformation of Gag or to recruit cellular proteins that mediate fusion.

Efficient release of human immunodeficiency virus type 1 particles also requires the viral Vpu protein. This function of Vpu is distinct from the induction of proteolysis of Cd4 in the ER described in chapter 12. In the absence of Vpu, particles accumulate in intracellular vacuoles or tethered to the infected cell surface. The Vpu protein, which lacks a large extracellular domain, has recently been shown to form an ion channel, a property shared with the influenza A virus M2 protein. Although the ion channel activity of Vpu has been implicated in promoting the release of virus particles, the mechanism has not yet been elucidated.

## Release of Virus Particles

The exit of enveloped viruses from host cells is intimately coupled with the budding process by which the envelope is acquired. The majority of these viruses assemble at and bud from the plasma membrane. Consequently, the final assembly reaction, fusion of the bud membrane, releases the newly formed virion into the extracellular milieu. As discussed in the previous section, the molecular mechanisms (of bud fusion) that put this most effective strategy for release of virus particles into effect are not well understood. When viruses acquire their envelopes from an intracellular membrane, the final step in assembly, budding, is also the first step in egress, which must be followed by transport of the particles to the cell surface. The assembly of enveloped viruses is therefore both mechanistically coupled and coincident with (or at least shortly followed by) their exit from the host cell. Some viruses without envelopes also display specific mechanisms of egress from certain types of host cells. However, such viruses generally rely on destruction (lysis) of the host cell. Large quantities of assembled virions may accumulate within infected cells for hours, or even days, prior to release of progeny on cell lysis. Many enveloped viruses also destroy the cells that support their replication. However, in some cases a nondestructive budding mechanism establishes a long-lasting relationship with the host cell. The progeny of many simple retroviruses are released throughout the lifetime of an infected cell, which is not harmed (but may be permanently altered; see chapter 16). Although the rate of virus release is relatively low, this strategy allows retroviruses to take maximal advantage of permissive host cells that are encountered and consequently minimizes the requirement for transmission to naive cells.

## Release of Nonenveloped Viruses

The most usual fate of host cells permissive for reproduction of nonenveloped viruses is death and destruction (but see chapter 15). In natural infections, the host immune responses to virus-infected cells described in chapter 14 are an important cause of infected-cell destruction. While such host-mediated infected-cell destruction might contribute to virus release, these viruses also destroy their host cells more directly: they are cytopathic in cells in culture. In general, we are remarkably ignorant about the mechanisms by which nonenveloped viruses induce death and lysis of their host cells.

Many viruses, including poliovirus and adenovirus, inhibit expression of cellular genetic information by specific effects on cellular transcription, RNA export from the nucleus, or translation. In the case of adenovirus, whose one-step infectious cycle is a relatively long 1 to 2 days, the shutdown of production of cellular proteins during the late phase of infection makes an important contribution to the eventual destruction of the infected cell. Indeed, this may be the primary function of inhibition of cellular protein synthesis. However, more specific mechanisms have also been implicated in the release of adenovirus and must be essential for exit of poliovirus, which induces host cell lysis more rapidly (within approximately 8 h) than can be explained by inhibition of synthesis of cellular macromolecules. One or more poliovirus capsid proteins have been implicated in lysis, by unknown mechanisms. A number of adenovirus proteins may contribute to virus release by altering cellular structures. Synthesis of the early E1B 19-kDa protein induces both the collapse of some intermediate filaments as a result of cleavage of the protein from which they are built (vimentin) and disruption of the nuclear lamina, a basketlike meshwork of lamin proteins lining the inner surface of the nuclear envelope. Later in the infectious cycle, other intermediate filament components, specific cytokeratins, are cleaved by the L3 protease into polypeptides that can no longer polymerize. Such disruption of intermediate filament networks, which requires inhibition of cellular protein synthesis so that the network cannot be rebuilt, seems likely to damage the structural integrity of the cell and thus facilitate virus release. In addition, a small E3-encoded glycoprotein is necessary for efficient nuclear disruption and lysis of cells in culture. This viral glycoprotein is made in large quantities late in the infectious cycle and accumulates in the nuclear envelope, but its mechanism of action is not yet known.

While cell lysis is the most common means of escape of naked viruses, there is evidence for the release of some viruses without any cytopathic effect. Under certain conditions, polioviruses and other picornaviruses are released under the direction of viral proteins that are incorporated into internal cellular membranes without lysis of the infected cell. Furthermore, when poliovirus replicates in polarized epithelial cells resembling those lining the gastrointestinal tract (a natural site of infection), progeny virions are released exclusively from the apical surface by a nondestructive mechanism. It is unlikely that such directed nonlytic release requires the cellular system for sorting components in polarized cells because poliovirus induces dramatic rearrangement of the cells' internal membranes and inhibition of the secretory pathway. In contrast, the papovavirus simian virus 40, which is also released from permissive cells before induction of cytopathic effects and leaves polarized epithelial cells via their apical surfaces, depends on the vesicular transport pathway of the host cell. It will be of considerable interest to learn how these nonenveloped virions enter the membrane-bound compartments of this pathway.

## Assembly at Internal Membranes: the Problem of Exocytosis

### Compartments of the Secretory Pathway

Several enveloped viruses assemble at the cytoplasmic surfaces of membranes of internal compartments of the secretory pathway under the direction of specifically located viral glycoproteins (Table 12.4). The final step in virion production is budding of the particle into the lumen of one of these compartments, for example, of bunyaviruses into Golgi cisternae. Particles formed in this way must reach the cell surface but lack the cellular proteins that direct transport vesicles to this site; viral particles are topologically "inside out" with respect to transport vesicles, and viral glycoproteins and their transmembrane interactions with internal proteins of the virus displace cellular proteins. It is therefore generally assumed that internal enveloped virus particles must be packaged within cellular transport vesicles for travel along the secretory pathway to the cell surface. Despite almost universal acceptance of this mechanism, we have little direct evidence for this mode of transport.

The budding of virus particles into internal compartments of the secretory pathway is initiated by interactions among the cytoplasmic domains of viral membrane proteins and internal components of the particle. Consequently, this process generally begins as soon as the integral membrane and cytoplasmic viral proteins attain sufficient concentrations in the infected cell. For example, the concentration of viral membrane proteins (surface proteins) determines the fate of hepadnaviral cores, which contain the C protein (Table 13.1), a DNA copy of the pregenomic RNA, and the viral polymerase. Early in infection, the concentration of the large surface protein in ER membranes is too low for efficient envelopment of cores, which therefore enter the nucleus to increase the pool of viral DNA templates for transcription. As the concentration of this protein increases, it interacts with cores and enveloped particles form and bud into the ER for export from the infected cell. The ability of hepadnaviral cores to bind to the viral glycoprotein is also regulated by the nature of the nucleic acid they contain. Nascent cores that

contain the pregenomic RNA do not enter the ER or leave the cell. Synthesis of DNA within the cores must therefore induce some change in their exterior surfaces that allows association with large surface protein molecules within the ER membrane. It is clear that production of (−) strand DNA activates envelopment, but the nature of the signal that regulates acquisition of the hepadnaviral envelope and how it is transmitted remain to be determined.

An intriguing question is why some viruses formed by a budding mechanism assemble at intracellular membranes rather than at the plasma membrane, where budding ensures release of the virion from the host cell. One possible advantage of intracellular budding is that the concentration of viral glycoproteins exposed on the surface of the infected cell is reduced because these proteins accumulate in intracellular membranes. This property would decrease the likelihood that the infected cell would be recognized by host antibodies and other components of the immune system before the maximal number of progeny virions were assembled and released. Alternatively, the simpler cytoplasmic surfaces of the membranes of internal compartments, which are not burdened with cytoskeletal structures and the proteins that attach them to the extracellular matrix, may make for more facile assembly or budding reactions. Or the distinctive lipid composition of internal membranes may confer some special property (as yet unknown) advantageous to these viruses.

Even more unusual is the interaction of the poxvirus vaccinia virus with internal cellular membranes. The DNA and all proteins of these complex particles are synthesized and assembled in the cytoplasm of the infected cell. Although the mechanisms of assembly and release of this virus are only now being worked out, it is clear that several membrane-associated particles form during biogenesis and that infectious particles can leave the cell by several mechanisms. The first structure assembled, the immature virus (Fig. 13.15), contains two membranes acquired from the trans-Golgi network of the infected cell by some poorly defined wrapping mechanism. These membranes become clearly visible when the spherical immature virus matures to the brick-shaped intracellular mature virus (Fig. 13.15). The latter particle is released only upon lysis of the infected cell. However, a variable proportion of the intracellular mature virus population becomes engulfed by the membranes of a second intracellular compartment, probably a post-Golgi or endocytic compartment, to form the intracellular enveloped virus (Fig. 13.15). This particle can be released from the cell as the three-membrane-containing extracellular enveloped virion following transport to the cell surface and fusion of its outer membrane with the plasma membrane (Fig. 13.15). As the intracellular mature virion and extracellular enveloped virus bind to dif-

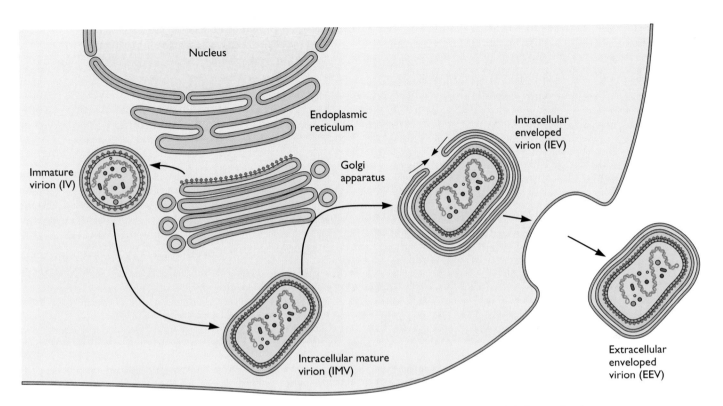

**Figure 13.15 Schematic model of vaccinia virus exocytosis.** The model shows the formation of immature virus (IV) by the acquisition of two cellular membranes from a compartment of the *cis*-Golgi network and its maturation to the infectious intracellular mature virion (IMV). These cellular membranes are modified by the incorporation of viral proteins (e.g., Table 12.4) to form crescent-shaped structures termed viral factories on the periphery of the specific cytoplasmic areas in which viral DNA synthesis takes place. These altered membranes coalesce into spherical structures containing granular material acquired from the viral factories and then mature to form the IMV. Viral proteins that appear to form an internal scaffold (D13L gene product) or are necessary for attachment of membranes to internal material (A14L gene product) have been identified, but this complex assembly process is not well understood. There are at least three routes by which the virus can leave the infected cell: release of the IMV upon cell lysis, engulfment of this particle in a second cellular membrane to form the intracellular enveloped virus (IEV) for transport to the plasma membrane and fusion, or transfer of intracellular enveloped virus to and beyond the cell surface on actin tails (Fig. 13.16).

ferent cell surface receptors, the release of two types of infectious particle may increase the range of cell types that can be infected. However, vaccinia virus has also evolved an amazing alternative mechanism of egress.

In vaccinia virus-infected cells, the intracellular enveloped virion induces a dramatic reorganization of the actin cytoskeleton that allows it to travel to and beyond the cell surface. The number of typical actin stress fibers is significantly decreased in the infected cell because the virus induces the formation of new filamentous actin-containing structures (Fig. 13.16A and B). Each of these, which are termed **actin tails,** is in contact with a single virus particle (Fig. 13.16). Their formation specifically re-

quires viral genes that encode envelope proteins of the intracellular enveloped virion, such as A33R. The tails project from the interior of the cell toward the cell surface and, as the infection progresses, from the cell surface in large microvilli (Fig. 13.16C). Viral particles attached to the tips of actin tails (Fig. 13.16C) are propelled through the cell at an average speed of 2.8 $\mu$m/min, a process that can be directly visualized in remarkable films made by video microscopy. Such movement of the virus, presumably by polymerization of actin at the front end of the tail and its depolymerization at the back end, does not stop at the cell surface: the virus-tipped actin projection extends from and right through the surface of the infected cell into

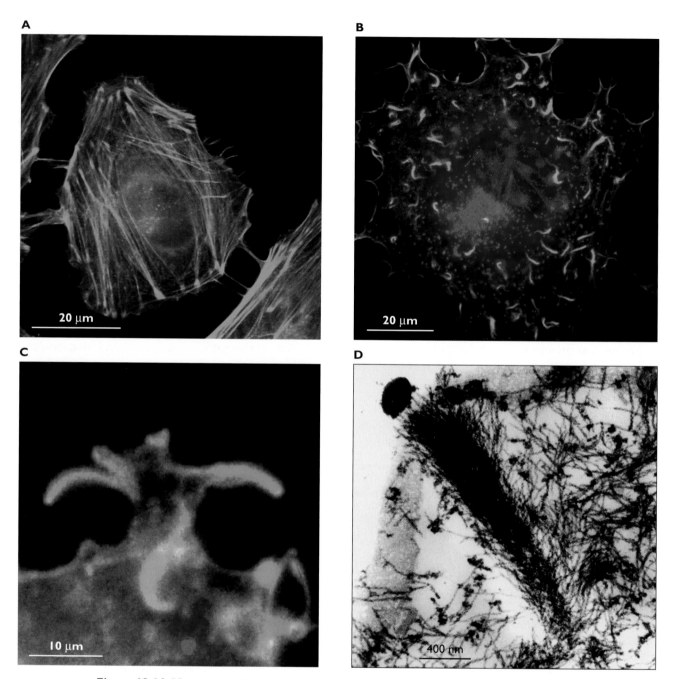

**Figure 13.16 Movement of vaccinia virus on actin tails.** In the immunofluorescence micrographs (A to C), actin is green and vaccinia virus particles are red. The dramatic reorganization of the actin cytoskeleton of an uninfected HeLa cell (A) induced by the virus can be seen (B). (C) A close-up of virus particles projecting from the cell surface projections on actin tails. The coincidence of the tips of the projecting actin and viral particles gives yellow-orange signals (marked by white arrowheads). Thus, the particles are projected from the cell surfaces on the tips of actin tails. When infected cells are plated with uninfected cells, such actin-containing structures to which virus particles are attached can be seen extending from the former into the latter. (D) An electron micrograph of a virus particle attached to an actin tail. Adapted from S. Cudmore et al., *Nature* **378**:636–638, 1995 (C and D); F. Frischknecht et al., *Curr. Biol.* **9**:89–92, 1999 (A); and C. M. Sanderson et al., *J. Gen. Virol.* **79**:1415–1425, 1998 (B), with permission. Courtesy of F. Frischknecht, S. Cudmore, and M. Way, European Molecular Biology Laboratory.

neighboring uninfected cells. Thus, this unusual exploitation of cellular cytoskeletal components allows direct transfer of the viral particles from one cell to another.

### Intranuclear Assembly

The problem of egress is especially acute for the enveloped herpesviruses because the nucleocapsids assemble in the nucleus. Indeed, the pathway by which the virus leaves the cell is a topic of fierce controversy. There is complete agreement that assembled nucleocapsids bud through the inner nuclear membrane. This reaction must be unusual, for this membrane is generally impermeable, with small molecules and macromolecular assemblies moving from the nucleus to the cytoplasm (and vice versa) only via nuclear pore complexes. The controversy concerns the mechanism by which herpesviral nucleocapsids acquire the tegument and envelope prior to exit from the host. Two pathways have been proposed (Fig. 13.17). In the first, the tegument is deposited around the nucleocapsid in the nucleus prior to budding of the particle through the inner nuclear membrane. This nuclear membrane-derived envelope is retained as the particle leaves the nucleus, enters the secretory pathway, and travels to the cell surface (Fig. 13.17A). The second pathway proposes that it is only the assembled nucleocapsid that initially buds through the inner nuclear membrane and, moreover, that the nuclear membrane acquired in this process is lost upon fusion of the particle with the outer nuclear membrane to release the capsid into the cytoplasm (Fig. 13.17B). Addition of the tegument then takes place in the cytoplasm before final envelopment upon budding through a membrane of a compartment of the secretory pathway, probably a Golgi compartment.

Distinguishing between these models has not been as simple as it might seem, in large part because exocytosis is usually studied by electron microscopy: the static images obtained in this way give no information about the direction of reactions. Thus, it has not been possible to establish whether the nonenveloped capsids that accumulate in the cytoplasm under specific experimental conditions represent accumulating intermediates in pathway B or the products of aberrant reactions. The site of assembly of simple viruses is dictated by the transport of viral glycoproteins to, and their accumulation within, specific cytoplasmic membranes. This criterion has not yet proved very helpful for identifying the site at which herpesviruses receive their envelope, in part because each of the mechanisms of exocytosis under consideration requires interactions of the assembling particle with multiple cellular membranes (Fig. 13.17). Despite these difficulties, recent observations are tilting the balance in favor of pathway B. In contrast to most herpesviruses, human cy-

tomegalovirus and human herpesvirus 6 possess dense teguments that are readily visualized in the electron microscope. It can therefore be seen that human herpesvirus 6 capsids present in the perinuclear space between the inner and outer nuclear membranes have no tegument, but tegument-containing structures appear at cytoplasmic invaginations of the outer nuclear membrane, observations consistent with model B (Fig. 13.17). Even more compelling is the fact that the envelope of mature herpes simplex virus type 1 virions possess the lipid composition of Golgi membranes (as predicted by pathway B) rather than that of nuclear membranes.

## Maturation of Progeny Virions

### Proteolytic Processing of Virion Proteins

The products of assembly of several viruses are noninfectious particles, often called immature virions. In all cases, proteolytic cleavage of specific proteins with which the particles are initially built converts them to infectious virions. Such **maturation** cleavages should not be confused with the proteolytic processing of some viral glycoproteins that takes place during their transit to cellular membranes and that can also be necessary for production of infectious particles: the maturation reactions, unlike the latter, are carried out by virus-encoded enzymes and take place in assembled particles late in assembly or following release of immature virions from the host cell. Proteolytic cleavage of virion proteins introduces an irreversible reaction into the assembly pathway, driving it in a forward direction. This modification can also make an important contribution to resolving the contradictory requirements of assembly and virus entry. One consequence of proteolytic processing is the exchange of covalent linkages between specific polypeptide chains for much weaker noncovalent interactions, which can be disrupted if necessary in a subsequent infection. A second is the liberation of a new C terminus and a new N terminus at each cleavage site, and hence opportunities for additional protein-protein contacts. Such changes in chemical bonding among virion proteins clearly facilitate virus entry, for the proteolytic cleavages that introduce them are necessary for infectivity. Thus, viral proteases and the structural consequences of their action are of considerable interest. Moreover, these enzymes are excellent targets for antiviral drugs, a property exemplified by the success of therapeutic agents that inhibit the human immunodeficiency virus type 1 protease.

The alterations in the structure of the virus particle and their functional correlates are best understood for small RNA viruses such as poliovirus. A single cleavage at an N-S dipeptide to liberate VP4 and VP2 from VP0 converts noninfectious provirions to mature infectious

**A**

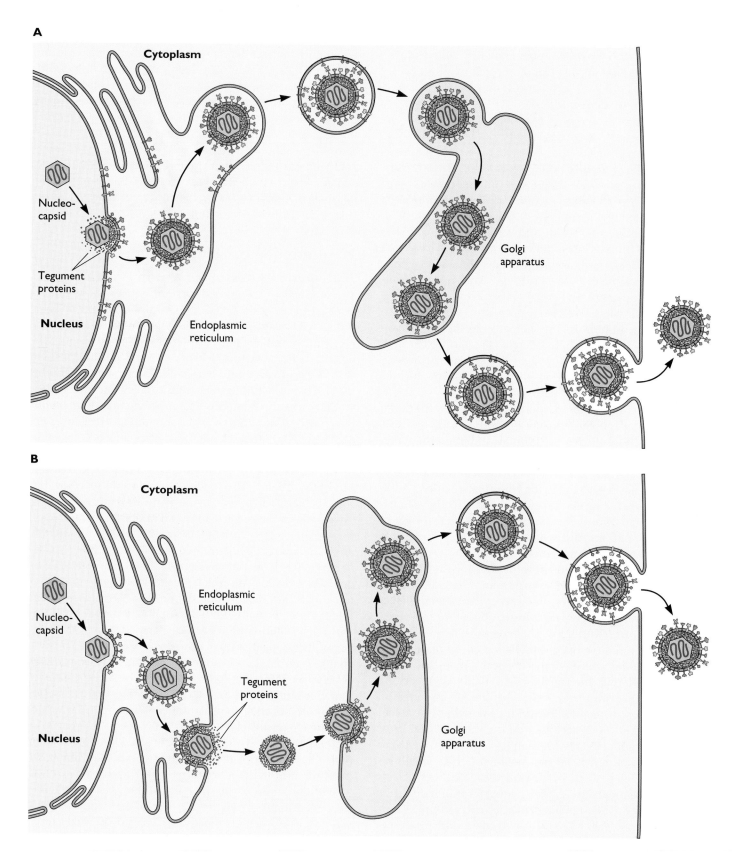

Cytoplasm

Nucleo-
capsid

Tegument
proteins

**Nucleus**

Endoplasmic
reticulum

Golgi
apparatus

**B**

Cytoplasm

Endoplasmic
reticulum

Nucleo-
capsid

Tegument
proteins

**Nucleus**

Golgi
apparatus

◁ **Figure 13.17 Pathways proposed for the exit of herpesvirus from an infected cell.** The major differences between the two models shown are whether the tegument proteins enter the particle in the nucleus (A) or in the cytoplasm (B) and whether the envelope of the mature virus is derived from the inner nuclear membrane (A) or the membrane of a cytoplasmic compartment (B).

---

virions (Fig. 13.3). As the viral proteases are not incorporated into virions, it is believed that VP0 cleavage is catalyzed by a specific feature of the virion itself and that the internal genomic RNA participates in the reaction. The structural changes induced by such maturation cleavage can be described in great detail, for the structure of empty particles in which VP0 is not cleaved (as well as of mature virions [chapter 3]) has been determined at high resolution. Cleavage of VP0 allows the extensive internal structures of the particle, including the network of numerous intra- and interpentamer contacts via segments of VP4 and the N-terminal arms of VP2 and VP1, to be established in its final form. Thus, cleavage of VP0 allows additional protein-protein interactions that make important contributions to the stability of the virion.

Cleavage of VP0 to VP4 and VP2 is also necessary for release of the RNA genome into a new host cell. The conformational transitions that mediate entry of the genome following attachment of the virus to its receptor are not fully understood. However, many mutations that alter receptor binding and entry map to just those regions of the capsid proteins that participate in the structures whose final organization depends on VP0 cleavage. Thus, cleavage of VP0 not only further stabilizes the virion but also "spring-loads" it for the conformational transitions that take place during entry and release of the genome. This unexpected relationship between structures that impart stability to the virion and interactions important during entry suggests that the changes that accompany VP0 cleavage have evolved to maintain an optimal balance between a stable virion and the conformational changes necessary for delivery of the genome into a new host cell.

Less well established are the advantages of delaying VP0 cleavage until capsids containing the genome have been assembled. One reason may be that many of the internal interactions triggered by VP0 cleavage span the boundary between one pentamer and another and therefore cannot be established until assembly of the capsid is complete. In addition, this strategy may be necessary for efficient release of particles: virions assembled from pentamers in which VP0 had been cleaved would be primed for release of the genome should appropriate intracellular conditions, or the poliovirus receptor, be encountered during egress

from the host cell. Clearly, this kind of particle would be poorly suited for efficient exit of progeny genomes and their delivery to new host cells or host organisms.

Upon proteolytic processing of the Gag polyprotein, immature retroviral particles undergo substantial morphological rearrangments. The spherical Gag structure forms a centrally located dense core containing the RNA genome and NC protein encased within a CA protein shell (Fig. 13.18). Although the structural consequences of such processing are not completely understood, it is clear that processing of human immunodeficiency virus type 1 Gag results in conformational change and reorganization of interactions among the products of Gag processing to form the mature particle (Box 13.3). Conformational changes in segments of Gag during assembly may also regulate proteolytic processing. For example, efficient cleavage to liberate human immunodeficiency virus type 1 NC depends on the binding of NC to RNA, at least in vitro.

These processing reactions play an essential part in the mechanisms by which infectious retroviruses are assembled and released. As we have seen, interactions among Gag polyproteins, and of their NC and MA domains with the viral RNA and the plasma membrane, respectively, build and organize the assembling particle. Efficient and orderly assembly also depends on "spacer" peptides that are removed during proteolysis. Furthermore, the membrane-binding signal of MA is exposed when MA is part of Gag but blocked by a C-terminal α-helix of MA in the mature protein. It is therefore very unlikely that particles could be correctly constructed from the processed products of Gag polyproteins. Indeed, mutations that increase the catalytic activity of the retroviral protease inhibit budding and production of infectious particles, indicating that premature processing of the polyproteins is detrimental to assembly. On the other hand, the covalent connection of the virion proteins that is so necessary during assembly is incompatible with release of the virion core following fusion of the viral envelope with the membrane of a new host cell. Such covalent linkage also precludes efficient activity of virion enzymes, which are incorporated as Gag-Pol proteins. The retroviral proteases that sever such connections are therefore absolutely necessary for production of infectious virions even though they are not required for forma-

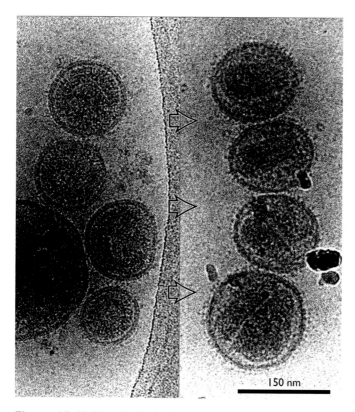

**Figure 13.18 Morphological rearrangement of retroviral particles upon proteolytic processing of the Gag polyprotein.** These two cryoelectron micrographs show the maturation of human immunodeficiency virus type 1 virions. The immature particles (left) contain a Gag polyprotein layer below the viral membrane and its external spikes. Processing of Gag converts such particles to mature virions (right) with elongated internal capsids. Courtesy of T. Wilk, European Molecular Biology Laboratory.

tion of particles. In some retroviruses, including Moloney murine leukemia virus, the protease also removes a short C-terminal segment of the cytoplasmic tail of the TM envelope protein to activate the fusionogenic activity of TM.

Retroviral proteases belong to a large family of enzymes with two aspartic acid residues at the active site (aspartic proteases). The viral and cellular members of this family are similar in sequence, particularly around the active site, and are also similar in three-dimensional structure (see Fig. 19.16). All aspartic proteases contain an active site formed between the two lobes of the protein, each of which contributes a catalytic aspartic acid. In cellular aspartic proteases, the domains that form each lobe are part of a single polypeptide chain. However, the retroviral proteases are homodimers in which each monomer corresponds to a single lobe of their cellular cousins. Consequently, the active site is formed only upon dimerization of two identical subunits. This property undoubt-

edly helps avoid premature activity of the protease within infected cells, in which the low concentration of their polyprotein precursors mitigates against dimerization. Indeed, dimerization of the protease appears to be rate limiting for maturation of virions. Fusion of the protease to the NC domain of Gag also inhibits dimerization. Thus, its synthesis as part of a polyprotein precursor not only allows incorporation of this enzyme into assembling virions but also contributes to regulation of its activity. The Vif protein of human immunodeficiency virus type 1 and other complex retroviruses may also prevent premature activation of the protease. This protein, which is required for efficient production of infectious particles in T lymphocytes and macrophages and acts late in the infectious cycle, inhibits cleavage of Gag and Gag-Pol polyproteins at specific sites by the viral protease in vitro. These properties raise the question of how the protease is activated, a step that requires its cleavage from the polyprotein in which it initially resides. Polyproteins containing the protease (e.g., expressed in bacteria) possess a very low level of activity, sufficient to liberate fully active enzyme at a very low rate in vitro. It is therefore believed that such activity of the polyproteins initially releases protease molecules within the particle. The high local concentrations of protease molecules within the assembling particle would facilitate the initial *trans* cleavages as well as dimerization of protease molecules to form the fully active enzyme.

Like its retroviral counterpart, the adenovirus protease converts noninfectious particles to infectious virions, in this case by cleavage at multiple sites within six virion proteins (Fig. 13.5). Although the adenoviral enzyme does not process polyprotein precursors, the cleavage of so many proteins also appears to alter protein-protein interactions necessary for virion assembly in preparation for early steps in the next infectious cycle. The enzyme is a cysteine protease containing an active site cysteine and two additional cysteines, all highly conserved. One mechanism by which its activity is regulated is by interaction with a small peptide, a product of cleavage of the virion protein pVI, or with pVI itself (Fig. 13.19). The activating pVI sequence, which also contains a cysteine residue, might promote disulfide-thiol exchange reactions to release the active site cysteine from an inactivating disulfide bond, or it might stabilize an active site conformation that is optimal for catalysis.

## Other Maturation Reactions

Newly assembled virus particles appear to undergo few covalent modifications other than the proteolytic processing reactions described in the previous section. One important exception is the removal of terminal sialic acid residues from the complex oligosaccharides added to the envelope

**BOX 13.3**

*Model for refolding of the human immunodeficiency virus type 1 CA protein on proteolytic processing of Gag*

The radial organization of the human immunodeficiency virus type 1 Gag polyprotein (left) is based on cryoelectron micrographs like that in Fig. 13.4. The three-dimensional structures of the processed proteins, MA (green), CA (blue), and NC (violet) bound to the SL3 packaging signal (red), shown on the right are derived from high-resolution structures discussed in this and preceding chapters.

In the X-ray crystal structure of the N-terminal portion of mature CA (right), the charged N terminus is folded back into the protein by a β hairpin formed by amino acids 1 to 13 and forms a buried salt bridge with carboxylate of Asp51. The lack of a charged N terminus prior to cleavage of CA from MA and the steric difficulties of burying the N terminus of CA attached to an MA extension (left) indicate that the β hairpin and buried salt bridge can form only after proteolytic cleavage. Furthermore, the viral protease

recognizes the cleavage site between MA and CA in an extended conformation. As the N terminal β hairpin of mature CA forms a CA-CA interface, it has been proposed that proteolytic cleavage and the consequent refolding of the N terminus of CA facilitate the rearrangements to form the conical core during maturation of virus particles. The inhibition of core assembly and formation of infectious viral particles in vivo by alteration of amino acids in this interface support this model in which refolding of the N-terminal segment of CA upon its proteolytic liberation from Gag drives assembly of the condensed core. Figure courtesy of T. L. Stemmler and W. Sundquist, University of Utah.

von Schwedler, U. K., T. L. Stemmler, V. Y. Klishko, S. Li, K. H. Albertine, D. R. Davis, and W. I. Sundquist. 1998. Proteolytic refolding of the HIV-1 capsid protein amino-terminus to facilitate viral core assembly. *EMBO J.* **17:** 1555–1568.

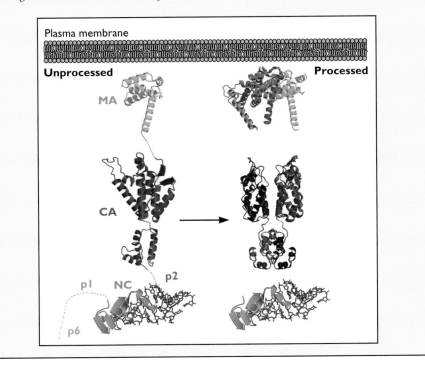

HA and NA glycoproteins of influenza A virus during their transit to the plasma membrane. The influenza A virus receptor is sialic acid, which is specifically recognized by the HA protein. Consequently, newly synthesized virus particles have the potential to aggregate with one another, and with the surface of the host cell, via binding of an HA molecule on one particle to a sialic acid present in an envelope protein of another particle or on cell surface proteins. Such

aggregation is indeed observed when the influenza virus neuraminidase is inactivated, for example, by a temperature-sensitive mutation. The activity of this enzyme, which removes terminal sialic residues from oligosaccharide chains, is essential for effective release of progeny virus particles from the surface of a host cell. Presumably, the neuraminidase eliminates such binding of newly synthesized virions to one another and to cell surface proteins.

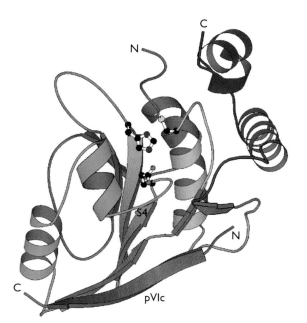

**Figure 13.19 Three-dimensional structure of the human adenovirus type 2 protease bound to the pVI peptide.** This structure, determined by X-ray crystallography of the complex and shown as a ribbon diagram, allowed identification of the active site of the enzyme (at which side chains are shown) by comparison with the secondary and tertiary structures of cellular cysteine protease. The close structural resemblance suggests that the adenoviral protease employs the same catalytic mechanism as papain. The β-strand formed by the activating pVIc peptide (deep red) is covalently linked to the enzyme by a disulfide bond but also makes many noncovalent interactions. The structure suggests that the cofactor might activate the enzyme by stabilizing the conformation of the active site optimal for catalysis or by altering the conformation of the protease active site. Adapted from J. Ding et al., *EMBO J.* **15:**1778–1783, 1995, with permission. Courtesy of W. Mangel, Brookhaven National Laboratory.

## Cell-to-Cell Spread

The raison d'être of all progeny virions is to infect a new host cell in which the infectious cycle can be repeated. Many viruses are released as free particles by the mechanisms described in preceding sections and must travel within the host until they encounter a susceptible cell. The new host cell may be an immediate neighbor of the originally infected cell or a distant cell reached via the circulatory or nervous systems of the host. Virions are designed to withstand such intercellular passage, but they are susceptible to several host defense mechanisms that can destroy virus particles. Localized release of virus particles only at points of contact between an infected cell and its uninfected neighbor(s) can minimize exposure to these host defense mechanisms. Furthermore, some viruses can

spread from one cell to another by mechanisms that circumvent the need for release of progeny virions into the extracellular environment.

In some cases, viruses can be transferred directly from an infected cell to its neighbors, a strategy that avoids exposure to host defense mechanisms targeted against extracellular virus. Herpesviruses provide examples of such cell-to-cell spread, mediated by the gE and gI glycoproteins, which form hetero-oligomers in the membranes of infected cells and in the virus envelope. When transfer of free virus particles from one cell to another is prevented by the addition of neutralizing antibodies to the medium, viruses that lack gE or gI produce no, or only small, plaques, indicating that they cannot spread from cell to cell. In animals, such mutants are also defective in anterograde spread of the virus in neurons, that is, movement of the virus from an axon terminal to an uninfected neuron. Such cell-to-cell spread of herpesviruses is believed to occur at specialized junctions, such as tight junctions of epithelial cells, sites of synaptic contact between individual neurons, and adhesion contacts between fibroblasts in culture. Such junctions between diverse cell types may therefore have more in common than we presently appreciate. Exactly how gE/gI glycoprotein oligomers promote cell-to-cell spread via such junctions is not yet clear. However, the herpesviral gE proteins exhibit significant similarities in amino acid sequence to the *Vibrio cholerae* zona occluda toxin, which induces a signal transduction cascade that alters tight junctions of polarized epithelial cells. Indeed, mutants of human cytomegalovirus with the gene encoding its gE homolog deleted fail to disrupt such tight junctions.

Other viruses deploy more radical mechanisms of transfer. For example, the complex retrovirus human immunodeficiency virus type 1 induces formation of **syncytia,** sheets of multinuclear cells formed by fusion of individual cells. Certain cell types infected by this virus form syncytia when they would not normally do so (Fig. 13.20). This property gives the virus ready access to many cell nuclei, in which proviral DNA integration and transcription take place, without the need to travel from one host cell to another. Perhaps most unusual is the direct transfer of vaccinia virus from an infected cell to its neighbors by the novel actin-containing structures induced by the virus, as described above. In this case, cellular components are reorganized for the specific purpose of direct transfer of vaccinia virus from one cell to another. The importance of this mechanism of spread (at least in cells in culture) is emphasized by the small or tiny plaques made by mutant viruses that cannot induce formation of actin tails.

One alternative strategy by which viruses avoid host defense mechanisms during transmission may well be the production of "decoys," noninfectious particles released in

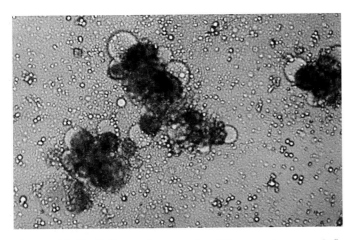

**Figure 13.20 Cell-to-cell spread of human immunodeficiency virus type 1 particles.** Formation of syncytia during infection of peripheral blood mononuclear cells in culture is visualized by phase-contrast microscopy. These giant multinucleated cells result from the fusion of infected cells with uninfected CD4+ cells. From J. A. Levy, *HIV and the Pathogenesis of AIDS,* 2nd ed. (ASM Press, Washington, D.C., 1998), with permission.

sufficiently large quantities to mask virions from immune defense molecules. For example, by far the majority of particles detected in hepatitis B virus-infected humans are empty particles that lack the genome. Another strategy would be to disguise virus particles with normal products of a host cell. Some enveloped viruses, notably retroviruses, retain cellular proteins, such as major histocompability complex class II proteins and the adhesion receptor Icam-1, in their membranes. The latter protein substantially increases the infectivity of human immunodeficiency virus type 1 particles, but the importance of such a "wolf in sheep's clothing" strategy for the spread of a virus from one cell to another in the host has yet to be documented.

## Perspectives

The assembly of even the simplest virus is a complex process in which multiple reactions must be completed in the correct sequence and coordinated in such a way that the overall pathway is irreversible. These requirements for efficient production and release of stable structures must also be balanced with the fabrication of virus particles primed for ready disassembly specifically in response to the appropriate signals encountered at the start of a new infectious cycle. The integration of information collected by the application of structural, biochemical, and genetic methods of analysis has allowed an outline of the dynamic processes of assembly, release, and maturation to be reconstructed for many viruses. Despite the considerable

structural diversity of virions, the repertoire of mechanisms for successful completion of the individual reactions (Fig. 13.1) is limited. Furthermore, we can identify common mechanisms that ensure that assembly proceeds efficiently and irreversibly or that resolve the apparent paradox of great particle stability during assembly and release but facile disassembly at the start of the next infectious cycle. These mechanisms include high concentrations of virion components localized at specific sites within the infected cell and proteolytic cleavage of virion proteins at one or more steps in the production of infectious particles. Indeed, for some smaller viruses, the structural changes that accompany the production of infectious virions from noninfectious precursor particles can be described in atomic detail. Such information has revealed unanticipated relationships between structures that stabilize virions and interactions that prime them for conformational rearrangements during entry. On the other hand, the pathways for production and release of even the least complex virions cannot be fully described, in large part because of the difficulties inherent in the study of assembly and egress in infected cells. The faithful reproduction of even the simplest assembly reaction, formation of structural units competent for accurate assembly from one or a few proteins, in in vitro reactions has also proved more difficult than originally anticipated. This property emphasizes the crucial contributions to virus assembly that can be made by cellular proteins that assist protein folding and oligomerization (chaperones) or covalently modify virion proteins. Historically, assembly reactions have received less attention than mechanisms of viral gene expression or replication of viral genomes. However, the development of new structural methods, coupled with the experimental power and flexibility provided by modern molecular biology, has revitalized investigation of the crucial processes of assembly, release, and maturation of virus particles. This renaissance has been further stimulated by the success of therapeutic agents designed to inhibit virus-specific reactions crucial for the production of infectious particles. Thus, the advances in understanding both specific reactions in pathways of production and release of infectious particles and the biochemical principles that allow the apparently paradoxical requirements of virion production and function to be met illustrated in this chapter seem certain to continue at an accelerated pace.

## References

*Chapters in Books*
Hellen, C. U. T., and E. Wimmer. 1995. Enterovirus structure and assembly, p. 155–170. *In* H. A. Rotbart (ed.), *Human Enterovirus Infections.* American Society for Microbiology, Washington, D.C.

Weldon, R. A., and E. Hunter. 1997. Molecular requirements for retrovirus assembly, p. 381–410. *In* W. Chiu, R. M. Burnett, and R. L. Garcia (ed.), *Structural Biology of Viruses.* Oxford University Press, New York, N.Y.

Wood, W. B., and J. King. 1979. Genetic control of complex bacteriophage assembly, p. 581–633. *In* H. Frankel-Conrat and R. R.Wagner (ed.), *Comprehensive Virology,* vol. 13. Plenum Press, New York, N.Y.

## Reviews

D'Halluin, J. C. 1995. Virus assembly. *Curr. Top. Microbiol. Immunol.* **199:**47–66.

Enquist, L. W., P. J. Husak, B. W. Banfield, and G. A. Smith. 1999. Infection and spread of alphaherpesviruses in the nervous system. *Adv. Virus Res.* **51:**237–347.

Ganem, D. 1991. Assembly of hepadnaviral virions and subviral particles. *Curr. Top. Microbiol. Immunol.* **168:**61–83.

Garot, H., R. Hewson, and D. J. E. Opstelten. 1998. Virus maturation by budding. *Microbiol. Mol. Biol. Rev.* **199:**227–235.

Harrison, S. C. 1995. Virus structures and conformational rearrangments. *Curr. Opin. Struct. Biol.* **15:**157–64.

Rixon, R. J. 1993. Structure and assembly of herpesviruses. *Semin. Virol.* **4:**135–144.

Schlesinger, S., S. Makino, and M. Linial. 1994. *Cis*-acting genetic elements and *trans*-acting proteins involved in the assembly of RNA viruses. *Semin. Virol.* **5:**39–49.

Strauss, J. H., E. G. Strauss, and R. J. Kuhn. 1995. Budding of alphaviruses. *Trends Microbiol.* **3:**346–350.

Weber, J. 1995. The adenovirus endopeptidase and its role in virus infection. *Curr. Top. Microbiol. Immunol.* **199:**227–235.

Wien, M. W., M. Chow, and J. M. Hogle. 1996. Poliovirus: new insights from an old paradigm. *Structure* **4:**763–767.

## Papers of Special Interest

### Assembly of Protein Shells

Ansardi, D. C., D. C. Porter, and C. D. Morrow. 1992. Myristylation of poliovirus capsid precursor P1 is required for assembly of subviral particles. *J. Virol.* **66:**4556–4563.

Basavappa, R., R. Syed, O. Flore, J. P. Icenogle, D. J. Filman, and J. M. Hogle. 1994. Role and mechanism of the maturation cleavage of VP_0 in poliovirus assembly: structure of the empty capsid assembly intermediates at 2.9Å resolution. *Protein Sci.* **3:**1651–1669.

Campbell, S., and V. M. Vogt. 1997. In vitro assembly of virus-like particles with Rous sarcoma virus deletion mutants: identification of the p10 domain as a morphological determinant in the formation of spherical particles. *J. Virol.* **71:**4425–4435.

Desai, P., N. A. DeLuca, J. C. Glorioso, and S. Person. 1993. Mutations in herpes simplex virus type 1 genes encoding VP5 and VP23 abrogate capsid formation and cleavage of replicated DNA. *J. Virol.* **67:**1357–1364.

Edvardsson, B., E. Everitt, E. Jörnvall, L. Prage, and L. Philipson. 1976. Intermediates in adenovirus assembly. *J. Virol.* **19:**533–547.

Falgout, B., and G. Ketner. 1988. Characterization of adenovirus particles made by deletion mutants lacking the fiber gene. *J. Virol.* **62:**622–625.

Ng, S. C., and M. Bina. 1984. Temperature-sensitive BC mutants of SV40: block in virion assembly and accumulation of capsid-chromatin complexes. *J. Virol.* **50:**471–477.

Parks, G. D., and A. C. Palmenberger. 1987. Site-specific mutations at a picornavirus VP3/VP1 cleavage site disrupt *in vitro* processing and assembly of capsid precursors. *J. Virol.* **61:**3680–3687.

Reicin, E. S., A. Ohagen, L. Yin, S. Hoglund, and S. P. Goff. 1996. The role of Gag in human immunodeficiency virus type 1 virion morphogenesis and early steps of the viral life cycle. *J. Virol.* **70:**8645–8652.

Tatman, J. D., V. G. Preston, P. Nicholson, R. M. Elliott, and F. J. Rixon. 1994. Assembly of herpes simplex virus type 1 capsids using a panel of recombinant baculoviruses. *J. Gen. Virol.* **75:**1101–1113.

Yuen, L. K. C., and R. A. Consigli. 1985. Identification and protein analysis of polyomavirus assembly intermediates from infected primary mouse embryo cells. *Virology* **144:**127–136.

### Assembly Chaperones and Scaffolds

Cepko, C. L., and P. A. Sharp. 1982. Assembly of adenovirus major capsid protein is mediated by a non-virion protein. *Cell* **31:**407–415.

Desai, P., S. C. Watkins, and S. Person. 1994. The size and symmetry of B capsids of herpes simplex virus type 1 are determined by the gene products of the UL26 open reading frame. *J. Virol.* **68:**5365–5374.

Dokland, T., R. McKenna, L. L. Hag, B. R. Bowman, N. L. Incardona, B. A. Fane, and M. G. Rossman. 1997. Structure of a viral procapsid with molecular scaffolding. *Nature* (London) **389:**308–313.

Gao, M., L. Matusick-Kumar, W. Hurlburt, S. F. DiTusa, W. W. Newcomb, J. C. Brown, P. C. McCann III, I. Deckma, and R. J. Colonno. 1994. The protease of herpes simplex virus type 1 is essential for functional capsid formation and viral growth. *J. Virol.* **68:**3702–3712.

Gustin, K. E., and M. J. Imperiale. 1998. Encapsidation of viral DNA requires the adenovirus L1 52/55-kilodalton protein. *J. Virol.* **72:**7860–7870.

Hasson, T. B., P. D. Soloway, D. A. Ornelles, W. Doerfler, and T. Shenk. 1989. Adenovirus L1 52- and 55-kilodalton proteins are required for assembly of virions. *J. Virol.* **63:**3612–3621.

Macejak, D. G., and P. Sarnow. 1992. Association of heat shock protein 70 with enterovirus capsid precursor P1 in infected human cells. *J. Virol.* **66:**1520–1527.

Srinivasan, A., A. J. McClellan, J. Vartikar, I. Marks, P. Cantalupo, Y. Li, P. Whyte, K. Rundell, J. L. Brodsky, and J. M. Pipas. 1997. The amino-terminal transforming region of simian virus 40 large T antigen and small t antigens functions as a J domain. *Mol. Cell. Biol.* **17:**4761–4773.

Zhou, Z. H., S. J. Macnab, J. Jakana, R. Scott, W. Chiu, and F. J. Rixon. 1998. Identification of the sites of interaction between the scaffold and outer shell in herpes simplex virus type 1 capsids by difference imaging. *Proc. Natl. Acad. Sci. USA* **95**:2778–2783.

## Packaging the Viral Genome

Ansardi, D. C., and C. D. Marrow. 1993. Poliovirus capsid proteins derived from P1 precursors with glutamine-valine sites have defects in assembly and RNA encapsidation. *J. Virol.* **67**:7284–7297.

Baines, J. D., A. P. W. Poon, J. Rovnak, and B. Roizman. 1994. The herpes simplex virus 1 U<sub>L</sub>15 gene encodes two proteins and is required for cleavage of genomic viral DNA. *J. Virol.* **68**:8118–8124.

Berkowitz, R. D., A. Ohagen, S. Höglund, and S. P. Goff. 1995. Retroviral nucleocapsid domains mediate the specific recognition of genomic viral RNAs by chimeric Gag polyproteins during RNA packaging in vivo. *J. Virol.* **69**:6445–6456.

Berkowitz, R. D., M.-L. Hammarskjöld, C. Melga-Maria, D. Rekosh, and S. P. Goff. 1995. 5′ regions of HIV-1 RNA are not sufficient for encapsidation: implications for the HIV-1 packaging signal. *Virology* **212**:718–723.

Dalyot-Herman, N., O. Ben-nun-Schaul, A. Gordon-Shaag, and A. Oppenheim. 1996. The simian virus 40 packaging signal *ses* is composed of redundant DNA elements which are partly interchangeable. *J. Mol. Biol.* **259**:69–80.

DeGuzman, R. N., Z. R. Wu, C. C. Stalling, L. Pappalardo, P. W. Boner, and M. E. Summer. 1998. Structure of the HIV-1 nucleocapsid protein bound to the SL3 Ψ-RNA recognition element. *Science* **279**:384–388.

Deiss, L. P., and N. Frenkel. 1986. Herpes simplex virus amplicon: cleavage of concatemeric DNA is linked to packaging and involves amplification of the terminally reiterated *a* sequence. *J. Virol.* **57**:933–941.

Duhaut, S. D., and J. W. McCauley. 1996. Defective RNAs inhibit the assembly of influenza virus genome segments in a segment-specific manner. *Virology* **216**:326–337.

Frilander, M., and D. H. Bamford. 1995. *In vitro* packaging of the single-stranded RNA genomic precursors of the segmented double-stranded RNA bacteriophage φ6: the three segments modulate each other's packaging efficiency. *J. Mol. Biol.* **246**:418–428.

Mann, R., R. Mulligan, and D. Baltimore. 1983. Construction of a retrovirus packaging mutant and its use to produce helper-free defective retroviruses. *Cell* **33**:153–159.

Salmon, B., and J. D. Baines. 1998. Herpes simplex virus DNA cleavage and packaging: association of multiple forms of U<sub>L</sub>15-encoded proteins with B capsids requires at least the U<sub>L</sub>6, U<sub>L</sub>17, and U<sub>L</sub>28 genes. *J. Virol.* **72**:3045–3050.

Schmid, S. I., and P. Hearing. 1997. Bipartite structure and functional independence of adenovirus type 5 packaging elements. *J. Virol.* **71**:3375–3384.

## Acquisition of an Envelope

Bruss, V., and D. Ganem. 1991. The role of envelope proteins in hepatitis B virus assembly. *Proc. Natl. Acad. Sci. USA* **88**:1059–1063.

Freed, E. O., and M. A. Martin. 1996. Domains of the human immunodeficiency virus type 1 matrix and gp41 cytoplasmic tail required for envelope incorporation into virions. *J. Virol.* **70**:341–351.

Justice, P. A., W. Sun, Y. Li, Z. Ye, P. R. Grigera, and R. R. Wagner. 1995. Membrane vesiculation function and exocytosis of wild-type and mutant matrix proteins of vesicular stomatis virus. *J. Virol.* **69**:3156–3160.

Lenthoff, R., and J. Summers. 1994. Coordinate regulation of replication and virus assembly by the large envelope protein of an avian hepadnavirus. *J. Virol.* **68**:4565–4571.

Parent, L. J., R. P. Bennett, R. C. Craven, T. D. Nelle, N. K. Krishna, J. B. Bowzard, C. B. Wilson, B. A. Puffer, R. C. Montelaro, and J. W. Wills. 1995. Positionally independent and exchangeable late budding functions of the Rous sarcoma virus and human immunodeficiency virus Gag proteins. *J. Virol.* **69**:5455–5460.

Rodriquez, J. R., C. Risco, J. L. Carrascosa, M. Esteban, and D. Rodriquez. 1998. Vaccinia virus 15-kilodalton (A14L) protein is essential for assembly and attachment of viral crescents to virosomes. *J. Virol.* **72**:1287–1296.

Schell, M. J., L. Buonocore, E. Bonitz, H. P. Ghosh, R. Chernish, and J. K. Rose. 1998. Requirement for nonspecific glycoprotein cytoplasmic domain sequence to drive efficient budding of vesicular stomatis virus. *EMBO J.* **17**:1289–1296.

Vangenderen, I. L., R. Brandimarti, M. R. Torrisi, G. Campadelli, and G. van Meer. 1994. The phospholipid-composition of extracellular herpes-simplex virions differs from that of host-cell nuclei. *Virology* **200**:831–836.

Yao, E., E. G. Strauss, and J. H. Strauss. 1998. Molecular genetic study of the interaction of Sindbis virus E2 with Ross River virus E1 for virus budding. *J. Virol.* **72**:1418–1423.

## Virion Maturation

Bernstein, H., D. Bizub, and A. M. Skalka. 1991. Assembly and processing of avian retroviral Gag polyproteins containing linked protease dimers. *J. Virol.* **65**:6165–6172.

Kohl, N. E., E. A. Emini, W. A. Schleif, J. L. Davis, J. C. Heimbach, R. A. Dixon, E. M. Scolnick, and J. S. Sigal. 1988. Active human immunodeficiency virus protease is required for viral infectivity. *Proc. Natl. Acad. Sci. USA* **85**:4686–4690.

Lee, W.-M., S. S. Monroe, and R. R. Rueckert. 1993. Role of maturation cleavage in infectivity of picornaviruses: activation of an infectosome. *J. Virol.* **67**:2110–2122.

Webster, A., R. T. Hay, and G. Kemp. 1993. The adenovirus protease is activated by a virus-coded disulfide-linked peptide. *Cell* **72**:97–104.

## Release of Virions

Clayson, E. T., L. V. Brando, and R. W. Compans. 1989. Release of simian virus 40 virions from epithelial cells is polarized and occurs without cell lysis. *J. Virol.* **63**:2278–2288.

Cudmore, S., P. Cossart, G. Griffiths, and M. Way. 1995. Actin-based motility of vaccinia virus. *Nature* (London) **378**:636–638.

Maidji, E., S. Tugizov, T. Jones, Z. Zheng, and L. Pereira. 1996. Accessory human cytomegalovirus glycoprotein US9 in the unique short component of the viral genome promotes cell-to-cell transmission of the virus in polarized epithelial cells. *J. Virol.* **70**:8402–8410.

Roper, R. L., E. J. Wolfe, A. Weisberg, and B. Moss. 1998. The envelope protein encoded by the A33R gene is required for formation of actin-containing microvilli and efficient cell-to-cell spread of vaccinia virus. *J. Virol.* **72:**4192–4204.

Tollefson, A. E., A. Scaria, T. W. Hermiston, J. S. Ryerse, L. H. Wold, and W. S. Wold. 1996. The adenovirus death protein (E3-11.6K) is required at very late stages of infection for efficient cell lysis and release of adenovirus from infected cells. *J. Virol.* **70:**2296–2306.

Zhang, Y., and R. J. Schneider. 1995. Adenovirus inhibition of cell translation facilitates release of virus particles and enhances degradation of the cytokeratin network. *J. Virol.* **68:**2544–2555.

Zsak, L., F. Zuckermann, N. Sugg, and T. Ben-Porat. 1992. Glycoprotein gI of pseudorabies virus promotes cell fusion and virus spread via direct cell-to-cell transmission. *J. Virol.* **66:**2316–2325.

# III Pathogenesis and Control

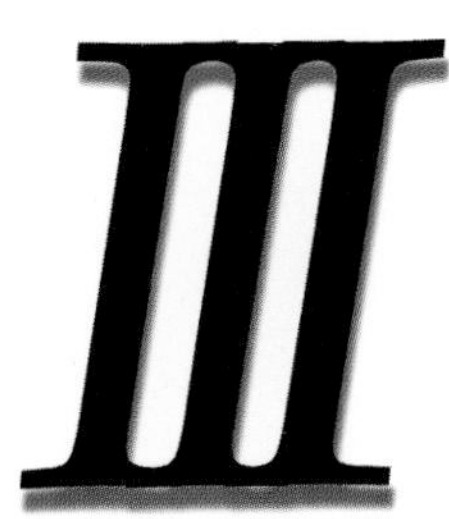

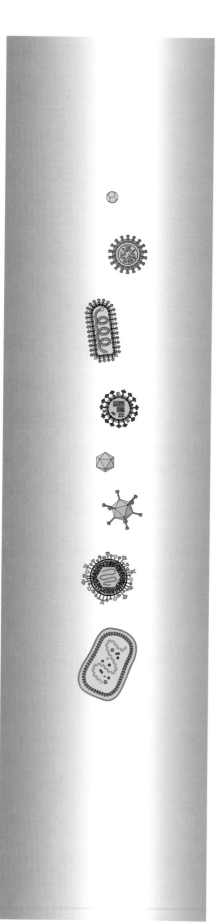

# Virus Offense Meets Host Defense

## The Host Defense against Viral Infections

In previous chapters, we considered viral infection of single cells in culture. However, it is important to keep in mind that cultured cells present a limited view of the propagative power of viruses. These agents have evolved to survive in the harsh world of multicellular, genetically outbred hosts that possess powerful innate and adaptive defense mechanisms for repelling invaders. Life at 37°C in rich tissue culture medium with millions of identical permissive cells is a luxury never encountered in nature. In this chapter, we will begin to consider the multifactorial problem of host-parasite interactions by discussing host defense against viral infections. For many, the labyrinthine complexity of the host defense system is daunting and the details can be overwhelming. However, as we will see, viruses provide both insight and focus.

We live and prosper in a literal cloud of viruses. The numbers of potentially infectious particles that impinge on us daily are astronomical. Seasonal "colds," "flu," childhood rashes, measles, chicken pox, and mumps, as well as AIDS and Ebola fever, all serve notice of our vulnerability. In this chapter, we consider the powerful primary and immune defense systems that ensure our survival.

### Primary Defenses

The primary but often under appreciated host defenses are physical and chemical barriers: the skin, surface coatings such as mucous secretions, tears, acid pH, and surface-cleansing mechanisms. The latter include mucociliary flow, swallowing, and blinking. The skin is the largest "organ" of the body, weighing more than 5 kg in an average adult, and is a strong barrier to infection. Unless broken by cuts, abrasions, or punctures (e.g., insect bites, needle sticks), it is essentially impervious to virus infection because its surface is composed of dead, keratinized cells that are continuously shed. Moreover, many viruses that land on the skin are inactivated by desiccation or by acids or other inhibitors formed by indigenous commensal microorganisms. In contrast, surfaces exposed to the environment but not covered by skin (the oropharyngeal cavity, respiratory tract, conjunctiva, alimentary canal, and urogenital tract) are lined by living cells and therefore depend on other primary defenses for immediate protection. Non-skin-covered surfaces are at risk for virus infection despite the remarkable and continuous action of surface-cleansing mechanisms. For example, in the nasal cavity and lungs, each day a normal individual produces 20 to 200 ml of protective mucus which is swept by cilliary action into the esophagus, where it is swallowed. Nevertheless, essentially all viral infections initiate on the surfaces of epithelial cells that line the apertures of the body exposed to the environment.

## Immune Defenses

When physical and chemical barriers are breached, the powerful **immune system** comes into play. This system has been shaped and selected by the constant activities of invading pathogens. One of its most remarkable attributes is the ability to recognize and destroy virus-infected cells in the context of uninfected cells. How this is accomplished is a matter of considerable research and debate. The paramount importance of the immune system in antiviral defense is amply documented by the devastating viral infections observed in children lacking normal immune function as a result of genetic defects, as well as in patients with AIDS. It should come as no surprise that viruses have provided considerable insight into the functioning of the immune system. Three critical processes in immune defense are **recognition**, **amplification**, and **control**. A viral infection must be recognized early, defenses must be activated quickly, and defensive action must be turned off when the infection has ceased. As we will see, virus-induced pathogenesis can result when any one of these processes is defective.

### *Immune Defense Mechanisms*

The immune response to viral infection consists of an **innate** (nonspecific) and an **adaptive** (specific) defense. The innate response is the "first line of immune defense" because it functions continually in a normal host without any prior exposure to the invading virus. This system is nonspecific in the sense that it responds to essentially any infection. Innate defense systems are encoded by germ line genes that arose early in the evolution of multicellular organisms and remain absolutely essential for human survival. It is now understood that the innate defense response actually dictates the conduct of the adaptive response (Fig. 14.1). How innate defenses are triggered is a matter of intense study. We know that triggering depends on recognition of biochemical characteristics common to or induced by microbial invaders but not by higher life forms. Such characteristics include high concentrations of double-stranded RNA, distinctive cell surface components, and unusual polymers on surfaces.

The innate response to viral infection is composed in part of powerful protein activators, called **interferons (Ifns)**, which induce an antiviral state in infected as well as in neighboring uninfected cells; a complex collection of serum proteins termed **complement**, which when activated destroys virus-infected cells, many virus particles, and other microbes; and cytolytic lymphocytes, called **natural killer cells (NK cells)**, which recognize and destroy virus-infected cells. Macrophages and neutrophils also play important roles in innate defense.

**Figure 14.1 Integration of innate and adaptive immune recognition.** An invading pathogen is first evaluated by molecular interactions with the innate immune system. In this initial and rapid process of molecular recognition and response, molecules of microbial origin are usually detected, resulting in a variety of responses including new cytokine production and release of stimulators of inflammation. These events in turn stimulate cells of the adaptive immune system. Molecules of the invading microbe and the effector molecules of the innate immune system then interact further with the adaptive immune system, often causing clonal expansion of lymphocytes. These lymphocytes produce highly specific effectors such as antibodies and cytotoxic T cells that recognize the invading microbe and the infected cells. This delayed response enables recognition of antigens with a high degree of structural specificity. The innate immune system is the front line not only for immediate deployment but also for reconnaisance and transfer of information to the adaptive immune system. Adapted from D. T. Fearon and R. M. Locksley, *Science* **272**:50–54, 1996, with permission.

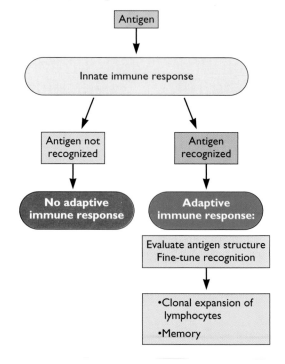

The adaptive defense consists of the **antibody response** and the **lymphocyte-mediated response**, often called the humoral and cell-mediated responses, respectively. This system is called "adaptive" because it not only differentiates infected self from noninfected self but also is tailored to the particular foreign invader such that an appropriate combination of soluble molecules (antibodies and cytokines) and lymphocytes participate in the action. Moreover, the adaptive defense system has "memory"; subsequent infections by the same agent are met almost immediately with a robust and highly specific response that usually stops the infection as soon as it starts. The adaptive immune system receptors that recognize pathogens are created by gene rearrangement processes during the differentiation of white blood cells called lymphocytes. The innate and adaptive responses are interconnected and coordinated by signaling mechanisms that enable them to interact. The cells and cytokines of the adaptive and innate immune response are listed in Table 14.1.

### The Inflammatory Response

Inflammation is of paramount importance in viral infections: not only does it control such infections and contribute to the overt pathogenic effects that follow, but it is essential in *initiating* immune defenses. Inflammation can be induced deliberately by **adjuvants**, such as Freund's adjuvant (killed mycobacterial cells in oil emulsion) or aluminum hydroxide gels, to provide an environment for the induction of a strong immune response to injected viral proteins in vaccines (see chapter 19 for a detailed discussion on adjuvants and vaccines). Cell and tissue damage caused by infection with cytopathic viruses induces a complex sequence of events collectively known as the **inflammatory response** (Fig. 14.2). Noncytopathic viruses do not initiate a strong inflammatory response and, as a consequence, have dramatically different interactions with the host. It is now understood that the early reactions in inflammation determine the type of immune response that will predominate, which in turn can influence the outcome of a viral infection.

While it is common to refer to the inflammatory response as if it were a unique, well-characterized process, its components and the mechanisms of initiation and regulation are only partially understood. For example, inflammation results when cells are damaged or die by unnatural means but not when they die naturally. It is striking that the inflammatory response is never invoked by **programmed cell death (apoptosis)**, a process that must occur during development and tissue remodeling. Cell death that invokes the inflammatory response is often called **necrotic cell death** to distinguish it from apoptosis. The signals that initiate inflammation after necrotic death are only now being characterized.

The nature and extent of the inflammatory response to viral infection depends on the tissue that is infected, as well as on the cytopathic nature of the virus. Many of the cells and proteins that participate in an inflammatory response come from the bloodstream and are directed by chemical signals to the site of viral infection. Some tissues have reduced access to the circulatory system (e.g., the

**Table 14.1**  Important white blood cells that participate in the innate and adaptive defense systems

| Cell | Source/function |
|---|---|
| Lymphocytes | Responsible for specificity of immune responses; recognize and bind to foreign antigens; derived from bone marrow |
| T cells | Differentiate into T-helper cells (Th cells), which secrete cytokines, and cytolytic T cells (CTLs); have T-cell receptors |
| B cells | Produce antibodies |
| NK cells | Natural killer cells; large granular cytolytic cells; do not have T-cell receptors |
| Mononuclear phagocytes | Responsible for phagocytosis and antigen presentation; derived from bone marrow; monocyte lineage; macrophages, Kupffer cells in the liver, alveolar macrophages in the lung |
| Dendritic cells | Responsible for induction of immune response; antigen presentation to T-helper cells; migratory cells found in every tissue except the brain |
| Interdigitating dendritic cells | Bone marrow-derived; present in most organs, lymph nodes, and spleen; Langerhans cells in the skin |
| Follicular dendritic cells | Not bone marrow-derived; present in germinal layers of lymphoid follicles in spleen, lymph nodes, mucosal lymphoid tissue; trap antigens and antigen-antibody complexes for display to B cells |
| Granulocytes | Contain abundant cytoplasmic granules; inflammatory cells |
| Neutrophils | Polymorpholeukocytes; respond to chemotactic signals, phagocytose foreign particles; major leukocyte in the inflammatory response |
| Eosinophils | Function in defense against certain pathogens (like worms) that induce IgE antibody; not critical in antiviral defense; involved in hypersensitivity (allergic) reactions |
| Basophils | Circulating counterparts to tissue mast cells; mediate hypersensitivity caused by IgE antibody response |

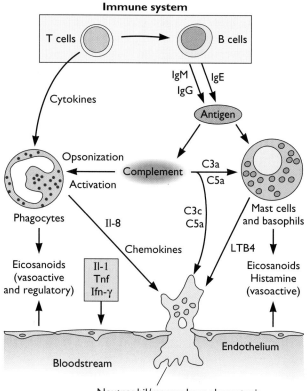

**Immune system**

**Neutrophil/macrophage chemotaxis**

**Figure 14.2 Inflammation provides communication with components of the immune system.** Viral infections at entry sites in the body often result in an inflammatory response. Inflammation reactions can be initiated in several ways: for example, by the classical or alternative pathway of complement activation (Fig. 14.5) or by mast cells that migrate to sites of cell damage where they can be activated by IgE antibody and antigen. C3a, C3c, and C5a are protease digestion products of the complement cascade that stimulate the inflammatory response (Table 14.5). C3a increases vascular permeability and activates mast cells and basophils, C3c stimulates neutrophil release, and C5a increases vascular permeability and chemotaxis of basophils, eosinophils, neutrophils, and monocytes as well as stimulating neutrophils. The cytokines Il-1, Ifn-γ, and Tnf-α act on the local endothelium to enhance leukocyte adhesion and migration. Il-8 and other chemokines promote chemotaxis of lymphocytes and monocytes. Il-1 and Tnf-α bind to receptors on epithelial and mesenchymal cells to cause division and collagen synthesis and stimulate prostaglandin and leukotriene synthesis (eicosanoid compounds). LTB4 is a particularly active leukotriene that is vasoactive and chemotactic. The activities of cells that enter an infected site where inflammation reactions are occurring are controlled by locally produced cytokines, particularly Tnf-α, Il-1, and Ifn-γ. Adapted from D. Male et al., *Advanced Immunology*, 3rd ed. (Mosby, St. Louis, Mo., 1996).

brain and the interior of the eyeball). As a consequence, the kinetics, extent, and outcome of the inflammatory response in such tissues can be markedly different.

The inflammatory response provides an excellent example of the integration of and synergy between the innate and adaptive defense systems. Components of both systems collaborate in the attraction and delivery of monocytes and neutrophils to the site of infection (Fig. 14.2; Table 14.1). These cells play essential roles in eliminating many pathogens and in cleaning up cellular debris. They are directed to the site of infection by the release of several chemotactic agents (chemokines) from damaged cells and from complement components (see Table 14.6) which attract these migratory cells from the circulatory system. Neutrophils are found in abundance in the blood but are normally absent from tissues. They are the earliest phagocytic cells to be recruited to a site of infection; they also produce a variety of cytokines and toxic products that can determine subsequent events. If the infection is not curtailed quickly by the innate response, certain components of the adaptive response are mobilized by the production of cytokines (see Fig. 14.12) and are concentrated at the site of infection.

The four classical signs of inflammation are **redness**, **heat**, **swelling**, and **pain**. These symptoms result from increased blood flow, increased capillary permeability, influx of phagocytic cells, and tissue damage. Increased blood flow occurs when vessels that carry blood away constrict, resulting in engorgement of the capillary network in the area. This produces redness (erythema) and an increase in tissue temperature. Capillary permeability increases, facilitating an efflux of fluid and cells from the engorged capillaries into the surrounding tissue. The fluid that accumulates has a high protein content, in contrast to that of normal fluid found in tissues, and contributes to the swelling. The cells that migrate into the damaged area are largely mononuclear phagocytes (Table 14.1). They are attracted by molecules produced by virus-infected cells, by cytokines produced by local defensive systems, and by subsequent secondary reactions that facilitate adherence of phagocytic cells to capillary walls near sites of damage. Some cytokines and complement components act as chemotactic agents that direct the migration of phagocytes to the region of cell damage (Fig. 14.2; Table 14.2 to 14.4). Neutrophils and monocytes are essential to the process: they both remove cell debris, but monocytes are also important in healing reactions after the infection is cleared.

The molecules that stimulate or control the inflammatory response are only partially characterized. Some are produced by the invading virus and are often called **virulence factors**. They include the soluble cytokine receptors expressed by poxviruses, alphaherpesvirus complement component-binding proteins (e.g., glycoprotein C), and various cytokines and growth factors whose expression is

**Table 14.2**    Some cytokines that function in the immune response to viral infection

| Cytokine | Source | Target/action |
|---|---|---|
| Ifn-γ | T cell/NK cell | Activates macrophages; promotes adhesion of Th cells to vascular endothelium; inhibits Il-6 effects; induces antiviral state |
| Ifn-α/β | Many cell types | Induces antiviral state; inhibits cell proliferation; stimulates growth and cytolytic function of NK cells; increases expression of major histocompatibility complex (MHC) class I and decreases expression of MHC class II molecules |
| Tnf-α | T cell/macrophage | Activates neutrophils; induces inflammatory response and fever and initiates catabolism of muscle and fat (cachexia); induces adhesion molecules on vascular endothelial cells; potentiates lysis of some virus-infected cells |
| Tgf-β | T cell/macrophage | Negative regulator of immune responses; inhibits T-cell proliferation; inhibits macrophage activation; induces IgA |
| gIp10 | T cell, monocyte keratinocytes, endothelial cells, fibroblasts | Chemoattractant for monocytes |
| Tca-3 | T cell, mast cell | Chemoattractant for monocytes |
| Gm-csf | T cell, macrophage fibroblast, endothelial cell | Induces myelomonocytic cell growth and differentiation |
| Mcp-1 | Macrophage, monocyte, T cell, endothelium, smooth muscle, keratinocytes, fibroblast | Induces monocyte chemotaxis |
| Mif | T cell | Inhibitor of macrophage migration |
| Mip-1α | Macrophage, T cell, B cell, neutrophil, Langerhans cell | Chemoattractant for T and B cells, monocytes, and eosinophils |
| Mip-1β | T cell, macrophage, B cell | Chemoattractant for T cells and monocytes |
| Rantes | T cell, platelets | Chemoattractant for T cells, monocytes, and eosinophils |
| Il-1 | Macrophage, T cell, B cell, epithelial cells | Costimulator of T cells; initiates inflammatory response; affects brain to produce fever; causes metabolic wasting (cachexia); induces acute-phase protein synthesis in the liver |
| Il-2 | T cell | Induces proliferation of T cells; stimulates growth and cytolytic function of NK cells; induces antibody synthesis in B cells |
| Il-3 | Th cell | Multilineage colony-stimulating factor; stimulates growth and differentiation of precursor cells in bone marrow; shares many activities with Il-4 |
| Il-4 | Mast cell, bone marrow | Induces B-cell proliferation and differentiation; required for production of IgE; inhibits macrophage function; blocks effects of Ifn-γ on macrophages; stimulates growth and differentiation of Th2 cells; stimulates expression of adhesion molecules on vascular endothelium |
| Il-5 | Th2 cell | Induces B-cell differentiation with Il-2 and Il-4; increases expression of eosinophil killing functions |
| Il-6 | Macrophage, T cells, vascular endothelial cells, fibroblasts | Made in response to Il-1 and Tnf; stimulates B-cell growth; costimulator of T cells; stimulates hepatocytes to synthesize acute-phase proteins |
| Il-7 | Marrow stromal cells | Acts to stimulate differentiation of B lymphocytes |
| Il-10 | T and B cells | Promotes growth and differentiation of B cells; inhibits macrophage function |
| Il-12 | Macrophage, monocytes | Potent stimulator of NK cell growth and killing activity; promotes differentiation of Th cells to Th1 subset; stimulates differentiation of immature CD8$^+$ T cells to functionally active CTLs |

induced by beta- and gammaherpesviruses. Others are released by antigen-activated lymphocytes (e.g., Ifn-γ, tumor necrosis factor alpha [Tnf-α], interleukin-5 [Il-5], Il-10, and Il-12) (Tables 14.2 and 14.3; see Table 14.5) or by action of the complement system (Table 14.4). The cytokines produced by T-helper (Th) lymphocytes are major cofactors controlling inflammation. Indeed, as discussed below, the Th1 response is often called the **proinflammatory response**.

The transcriptional activator protein Nf-κb is central in the activation of inflammation (see also chapter 8). When the cell binds inducers of Nf-κb such as the inflammatory cytokine Tnf-α or Il-1, transcription of many target genes ensues. This simple control circuit is extraordinary in that it integrates a large number of diverse inducing signals, such as ultraviolet light and cytokines, that culminate in activation of Nf-κb to initiate the complex inflammatory response.

**Table 14.3**  Chemokines and their receptors[a]

| Receptor[b] | Chemokine ligand(s) |
|---|---|
| Ccr1 | Mip-1α, Rantes, Mcp-3, Hcc-1, Mpif-1 |
| Ccr2 | Mcp-1, -3, and -4 |
| Ccr3 | Eotaxin, Mcp-2, -3, -4, and -5, Rantes, Mpif-2 |
| Ccr4 | Tarc |
| Ccr5 | Rantes, Mip-1α, -β |
| Ccr6 | Larc/Mip-3α |
| Ccr7 | Elc, Mip-3β |
| Ccr8 | I309 |
| Cxcr1 | Il-8, Gcp-2 |
| Cxcr2 | Il-8, Groα, -β, and -γ, Nap-2, Ena-78, Gcp-2 |
| Cxcr3 | Ip-10, Mig, I-Tac |
| Cxcr4 | Sdf-1 |
| Cxcr5 | Bca-1/Blc |

[a]Data from C. R. Mackay, *Curr. Biol.* **7:**R384–R386, 1997.

[b]The four families of chemokine receptors are distinguished by the pattern of cysteine residues near the amino terminus and are abbreviated as CXC, CC, C, and CX3C. The CXC family has an amino acid between two cysteines; the CC family has none; the C family has only one cysteine; and the CX3C family has three amino acids between two cysteines. Subfamilies of these four major groups also exist.

## The Innate Immune Response

The innate immune response is crucial in antiviral defense because it can be activated so quickly and can begin functioning within hours of a viral infection. Such rapid action contrasts with the activation of the adaptive response, which is orders of magnitude slower than the replication cycles of some viruses; it takes days to weeks to amplify an effective response of antibodies and activated lymphocytes specifically tailored for the infecting virus. While the rapidity of the innate response is important, it must also be transient because, as we shall see, continued activity of this defense system is damaging to the host.

Not surprisingly, there is considerable interplay between the innate and adaptive immune defense systems; indeed the innate system is sometimes said to "jump-start" the adaptive response (Fig. 14.2). Considerable evidence indicates that the innate response provides essential information to the adaptive immune system about the nature of the potential hazard confronting the host. Below, we discuss four important players in the innate response: cytokines, complement, collectins, and NK cells.

## Cytokines

**Cytokines** are regulatory proteins that mediate essential processes in intercellular communication during an antiviral defense. Indeed, their presence is one of the first indications that the host has been infected. Although many cells are capable of synthesizing cytokines, they do so only under appropriate conditions. Moreover, cytokines tend to

**Table 14.4**  Biological activities of proteins and peptides released during complement activation

| Substance | Biological activity |
|---|---|
| C5b, C6, C7, C8, and C9 | Lytic membrane attack complex |
| C3a | Peptide mediator of inflammation, smooth-muscle contraction; vascular permeability increase; degranulation of mast cells, eosinophils, and basophils; histamine release; platelet aggregation |
| C3b | Opsonization of particles and solubilization of immune complexes; facilitation of phagocytosis |
| C3c | Neutrophil release from bone marrow; lyses leukocytes |
| C3dg | Molecular adjuvant; profoundly influences adaptive response |
| C4a | Smooth-muscle contraction; vascular permeability increase |
| C4b | Opsonin for phagocytosis; processing and clearance of antibody-antigen immune complexes |
| C5a | Peptide mediator of inflammation, smooth-muscle contraction; vascular permeability increase; degranulation of mast cells, basophils, and eosinophils; histamine release; platelet aggregation; chemotaxis of basophils, eosinophils, neutrophils, and monocytes; hydrolytic enzyme release from neutrophils |
| Bb | Inhibition of migration and induction of monocyte and macrophage spreading |
| C1q | Opsonin for phagocytosis, clearance of apoptotic cells, processing and clearance of antibody-antigen immune complexes |

act locally, near the cells that make them, rather than at a distance like endocrine hormones.

Cytokines participate in almost every phase of the host response to viral infection, including control of inflammation, induction of an antiviral state in cells, and regulation of the adaptive immune response. They constitute a major means of communication between the innate and adaptive arms of the immune system.

While cytokines work locally in antiviral defense, they also invoke global responses that are both essential and familiar to all who experience viral infections. For example, they act directly on the nervous system to produce many of the characteristic behaviors and responses exhibited by infected individuals, including somnolence, lethargy, myalgias, appetite suppression, and nausea. After an inflammatory viral infection, cytokines stimulate the liver to produce characteristic **acute-phase proteins**, such as C-reactive protein, fibrinogen, complement proteins, α1-antitrypsin, and α1-acid glycoprotein, many of which are required to deal with tissue damage and for clearing the infection. Members of the **colony-stimulating factor (Csf)** class of cytokines are produced in the bone marrow

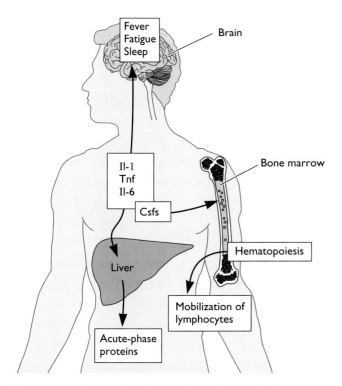

**Figure 14.3 Systemic effects of cytokines in inflammation.**
A localized viral infection often produces global effects, including fever and lethargy, lymphocyte mobilization (swollen glands), and production of new proteins in the blood. The inflammatory cytokines Il-1, Il-6, and Tnf all act on the brain (particularly the hypothalamus) to produce a variety of effects, including these typical responses to viral infection. These cytokines also act in the liver to cause the release of iron, zinc, and acute-phase proteins including mannose-binding protein, fibrinogen, C-reactive protein, and serum amyloid protein. Acute-phase proteins have innate immune defense capabilities: e.g., C-reactive protein binds phosphorylcholine on microbial surfaces and activates complement. The colony-stimulating factors (Csfs) activated by an inflammatory response have long-range effects in the bone marrow on hematopoiesis and lymphocyte mobilization. Adapted from A. S. Hamblin, *Cytokines and Cytokine Receptors* (IRL Press, Oxford, United Kingdom, 1993), with permission.

after inflammatory viral infection and control the growth and maturation of lymphocytes and other cells essential in antiviral defense. Figure 14.3 depicts the diverse and far-reaching actions of three cytokines required to elicit the inflammatory response, Il-1, Tnf, and Il-6.

Cytokines exert their activities by binding to specific receptors and activating gene expression in the cell bearing these receptors. Cytokines are potent agents capable of inducing a response at nanogram-per-milliliter concentrations. More than 35 cytokines are known, and the list is growing. Some of the more important in antiviral defense are listed in Table 14.2. The terms "lymphokines" and "monokines" were originally used to denote secreted proteins produced by activated lymphocytes or monocytes,

**Table 14.5**  The interferons: antiviral cytokines

| Interferon[a] | Producer cells | Inducers |
|---|---|---|
| Ifn-α | Leukocytes | Virus infection, dsRNA |
| Ifn-β | Fibroblasts, epithelial cells | Virus infection, dsRNA |
| Ifn-γ | T cells, NK cells | Antigens, mitogens, Il-2, Il-12 |

[a]The molecular mass of each Ifn is approximately 20 kDa. Human Ifn-α is encoded by more than 24 closely related, intronless genes. Human Ifn-β is encoded by a single, intronless gene that has about 30 to 45% homology to Ifn-α genes. Ifn-γ is encoded by a single-copy gene with three introns. There is little homology with Ifn-α and Ifn-β. Ifn-α and Ifn-β are sometimes called type I interferons. Ifn-γ is often called type II or immune interferon.

respectively. Similarly, the term "interleukin" identified proteins that communicated signals between different populations of white blood cells and other nonhematopoietic cells. The three types of interferons (Table 14.5) are the most important cytokines in the innate response against viral infection.

## Interferons

In 1957 A. Isaacs and J. Lindenmann found that chicken cells exposed to inactivated influenza virus contained a substance that interfered with the infection of other chick cells by live influenza virus. They called the inhibitor interferon. Most cells produce Ifn when exposed to viruses. Indeed, Ifns are produced by mammals, birds, reptiles, and fish. As originally noted by Isaacs and Lindenmann, virus replication is not required to induce Ifn production. In fact, a variety of agents and treatments, including natural double-stranded RNA, synthetic double-stranded RNA (e.g., polyinosine/polycytosine), endotoxin, and even inhibitors that block transcription (actinomycin D) or translation (cycloheximide), promote the synthesis of Ifn. There are three classes of interferons (Table 14.5). In following sections, we use the abbreviation Ifn to mean both Ifn-α and Ifn-β. We will refer specifically to Ifn-γ, which is induced only when certain lymphocytes are stimulated to replicate and divide after binding a foreign antigen. In contrast, Ifn-α and Ifn-β are induced directly by viral infection of almost any cell type.

### Ifn Is Produced by Infected Cells

Virus-infected cells produce Ifn, but the signals that trigger its synthesis are not completely understood. It is commonly thought that the important signal is accumulation of double-stranded RNA (dsRNA), an inevitable product of RNA and DNA virus replication. Consistent with this idea, synthetic dsRNA molecules are known to be powerful Ifn inducers. Single-stranded RNA molecules, including viral mRNA, invariably contain stable double-stranded structures formed by base pairing between complementary sequences. In cells infected by DNA viruses, dsRNA can also form from RNA transcripts of complementary strands of DNA. The hypothesis is that Ifn synthesis is stimulated by the high concentration of viral dsRNA in infected cells compared to the low and normal levels of dsRNA in uninfected cells.

Other signals of viral infection can lead to Ifn production. Structural proteins of some viruses stimulate Ifn synthesis by as yet uncharacterized mechanisms. In other cases, virus-induced degradation of specific cellular transcription regulatory proteins (e.g., IκBα) activates Ifn gene transcription. Infection by some viruses directly activates transcriptional regulatory proteins specific for interferon-activated genes. No matter what the mechanism, induction of Ifn in infected cells is rapid, transient, and relatively nonspecific; it occurs within hours of viral infection and declines rapidly in less than 10 h.

The quantity of Ifn that is released from cells infected by a given virus is astonishingly variable, for reasons that are not always understood. In the case of vesicular stomatitis virus infection, the Ifn concentration can vary over a 10,000-fold range depending on the serotype and genotype of the infecting virus. The Indiana serotype either stimulates production of Ifn or does not, depending on the isolate, while the New Jersey serotype is invariably an effective inducer of Ifn synthesis. As discussed later, many viral proteins regulate the amount of Ifn produced, as well as its action.

### Ifn Affects Only Cells with Specific Cell Surface Ifn Receptors

Although much remains to be learned about the mechanism(s) leading to transcription of the appropriate Ifn genes after viral infection, expression of the genes subsequently induced by Ifn has been studied extensively. It is known that induction occurs at the transcriptional level, generally does not require new protein synthesis, and occurs only on binding of Ifn to specific receptors.

There are two different cell surface Ifn receptors: one for both Ifn-α and Ifn-β and a separate one for Ifn-γ. All three Ifn proteins bind to their receptors with high affinity (equilibrium dissociation constant [$K_D$] of about $1 \times 10^{-10}$ M). A simplified outline of signal transduction induced by Ifn binding to its receptor is shown in Fig. 14.4. (See also Box 14.1.)

### Ifn Action Produces an Antiviral State

Shortly after viral infection, newly produced Ifn released from infected cells can be found circulating in the body, but its concentration is highest at the site of infection. Such local action of Ifn is important in reducing or eliminating the infection. A cell that has bound and re-

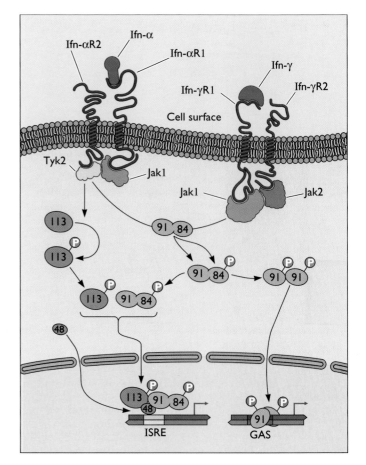

**Figure 14.4 Overlapping signal transduction pathways for Ifn-α/β and Ifn-γ.** The receptors for Ifn-α/β and Ifn-γ are different but share components of the Jak/Stat signal transduction pathway. Binding of either Ifn-α/β or Ifn-γ to the appropriate receptor leads to tyrosine phosphorylation of tyrosine kinases (Tyk2 and Jak1 for Ifn-α/β and Jak1 and Jak2 for Ifn-γ) as well as the receptor itself. This in turn is followed by tyrosine phosphorylation of the Stat proteins, p113, p91, and p84 (also known as Isgf-3α proteins) for Ifn-α/β and p91/84 for Ifn-γ. The phosphorylated Stat proteins form a collection of dimers that enter the nucleus with another cellular protein called p48 (also called Isgf-3γ) to form defined complexes that can bind to specific transcriptional control sequences of Ifn-α/β- or Ifn-γ-inducible genes (called interferon-stimulated response elements [ISREs] for Ifn-α/β-inducible genes and gamma-activated site [GAS] elements for Ifn-γ-inducible genes).

sponded to Ifn is said to be in an **antiviral state**—it is primed to stop replication of a variety of viruses should they enter that cell. Indeed, the replication of some viruses like vesicular stomatitis virus is so sensitive to Ifn that this property is used to titrate the biological activity of Ifn. On the other hand, some viruses are relatively resistant to Ifn action. Ironically, influenza virus is rather resistant to the

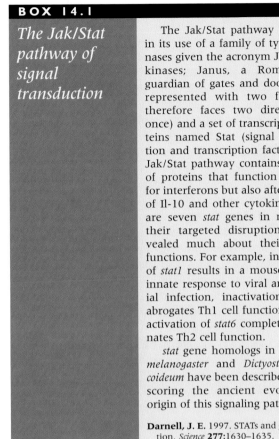

**BOX 14.1**

*The Jak/Stat pathway of signal transduction*

The Jak/Stat pathway is unique in its use of a family of tyrosine kinases given the acronym Jak (Janus kinases; Janus, a Roman god, guardian of gates and doorways, is represented with two faces and therefore faces two directions at once) and a set of transcription proteins named Stat (signal transduction and transcription factors). This Jak/Stat pathway contains a family of proteins that function not only for interferons but also after binding of Il-10 and other cytokines. There are seven *stat* genes in mice, and their targeted disruption has revealed much about their various functions. For example, inactivation of *stat1* results in a mouse with no innate response to viral and bacterial infection, inactivation of *stat4* abrogates Th1 cell function, and inactivation of *stat6* completely eliminates Th2 cell function.

*stat* gene homologs in *Drosophila melanogaster* and *Dictyostelium discoideum* have been described, underscoring the ancient evolutionary origin of this signaling pathway.

**Darnell, J. E.** 1997. STATs and gene regulation. *Science* **277:**1630–1635.

effects of Ifn even though historically it was instrumental in the discovery of this cytokine.

What is the antiviral state? Ifn can induce the transcription of more than 100 genes. Many of the products of these genes possess potent, broad-spectrum antiviral activities, but the relevant molecular mechanisms of only a few are understood. In general, prolonged activity of Ifn-induced proteins leads to cell death by apoptosis, a response that certainly limits spread of virus from one cell to another. Ifn not only causes the death of infected cells but also ensures that uninfected cells in the vicinity of an infection are induced to kill themselves should they become infected. Such a local cauterizing response that prevents spread of virus has led some to characterize Ifn as a "firebreak" to infection.

In addition to its antiviral activities, Ifn has marked effects on uninfected cells and functions as an important cytokine in the body. It blocks cell proliferation, increases the ability of NK cells to lyse infected cells, reduces amino acid biosynthesis, and alters transcription of major histocompatibility complex (MHC) genes in uninfected cells (see

"Antigen Presentation and Activation of Immune Cells" below). Because Ifn induces the expression of many deleterious gene products and is potentially lethal to any cell that has specific receptors (most cells in our bodies), the production of large amounts of Ifn in an infected individual has dramatic physiological consequences, notably the appearance of such common symptoms as fever, chills, nausea, and malaise. The ability of diverse viruses to induce Ifn production is one reason why most viral infections give rise to these common "flulike" symptoms. Thus, while Ifn is a powerful weapon against viruses, it is a blunderbuss rather than a "magic bullet." Therapeutic use of Ifn has been effective for treatment of some persistent infections, particularly those caused by hepatitis B virus.

### Ifn-Induced Antiviral Responses

In the case of Ifn-sensitive viruses (e.g., picorna-, orthomyxo-, retro-, pox-, herpes-, adeno-, papilloma-, and hepadnaviruses), it has been known for years that both viral and cellular protein synthesis stops abruptly in Ifn-treated infected cells but not in uninfected cells. This lethal defense is carried out by at least two cellular proteins, dsRNA-activated protein kinase (Pkr) (described in chapter 11) and ribonuclease L (RNase L). For both proteins, establishment and activation of the antiviral state is a two-step process whereby Ifn first promotes the increased expression and accumulation of an inactive protein that subsequently can become activated when the cell is infected.

All cells contain low concentrations of inactive Pkr, a serine/threonine kinase that not only has antiviral properties but also has antiproliferative and antitumor functions. The signal transduction cascade initiated by Ifn binding to its receptor leads to a dramatic increase in the concentration of inactive Pkr. If the cell is infected, Pkr proenzyme can be activated by viral dsRNA. Activated Pkr then phosphorylates the alpha subunit of the eIF2 translation initiation factor (eIF2$\alpha$) (see chapter 11), effectively rendering it incapable of contributing to new protein synthesis in the cell. Other Pkr substrates, such as the Nf-$\kappa$b inhibitor I-$\kappa$b and the human immunodeficiency virus type 1 protein Tat, have recently been described, but their contributions to antiviral defense are not well characterized.

RNase L is a nuclease capable of degrading most cellular and viral RNA species. Its concentration increases 10- to 1,000-fold after Ifn treatment, but the enzyme remains inactive unless another enzyme is produced. This enzyme, 2′-5′-oligo(A) synthetase, produces 2′,5′ oligomers of adenylic acid but only when activated by dsRNA. These unusual poly(A) oligomers then activate RNase L, which in turn begins to degrade all host and viral mRNA in the cell. It is known from the properties of mouse mutants de-

**Table 14.6**  How viruses block the action of interferons

| Virus | Viral gene product | Mechanism of inhibition |
|---|---|---|
| Adenovirus | E1A protein | Blocks signal transduction |
| | VA RNA | Blocks activation of Pkr |
| Cowpox virus | crmA protein | Serpin, protease inhibitor, blocks activation of Il-1β |
| Epstein-Barr virus | EBNA-2 protein | Blocks signal transduction |
| | EBER RNA | Blocks activation of Pkr |
| | BCRF1 (viral Il-10) | Il-10 homolog (inhibits Ifn-γ, Il-1, Il-2, Tnf synthesis) |
| Hepatitis B virus | Terminal protein | Blocks signal transduction |
| Human immunodeficiency virus | Tat protein | Reduces activity of Pkr |
| | TAR RNA | Blocks activation of Pkr |
| Herpes simplex virus | Unknown | Blocks activation of RNase L |
| Influenza virus | NS1 | Binds dsRNA, blocking Pkr activation |
| | Unknown | Activates cellular inhibitor of Pkr called p58 |
| Myxomavirus | M-T7 protein | Soluble Ifn-γ receptor (decoy) |
| | T2 protein | Soluble Tnf receptor (decoy) |
| Poliovirus | Unknown | Activates cellular inhibitor of Pkr |
| Reovirus | Sigma 3 protein | Binds dsRNA and blocks activation of Pkr |
| Vaccinia virus | E3L protein | Binds dsRNA and blocks activation of Pkr |
| | K3L protein | Pkr pseudosubstrate |
| | B15R protein | Soluble Il-1 receptor (decoy) |
| | A18R protein | Regulates dsRNA production |

fective in RNase L that this enzyme participates not only in Ifn-mediated antiviral defense but also in apoptosis. As a death effector, RNase L contributes to the antiviral activity of Ifn by eliminating infected cells as well as by destroying viral RNA. The in vivo substrates by which RNase L stimulates apoptosis remain to be discovered.

An Ifn-γ-inducible protein with important antiviral activities is nitric oxide synthase. This enzyme exists as several isoforms, each of which has a distinctive tissue distribution. Nitric oxide synthase produces nitric oxide during the conversion of arginine to citrulene. The nitric oxide so produced has a variety of effects including inhibition of poxvirus and herpesvirus replication. Another Ifn-inducible protein is promyelocytic leukemia protein (Pml). This protein is specifically associated with nuclear bodies (the ND10 structures described in chapter 9) and, when overexpressed, confers resistance to infection by vesicular stomatitis virus and influenza A virus but not by encephalomyocarditis virus. The molecular mechanisms of such selective inhibition are not known. Even less is known about the other Ifn-inducible proteins with antiviral effects. In contrast to Pkr and RNase L, these proteins do not block translation, yet they can inhibit a given virus at different steps in the viral life cycle. For example, unknown Ifn-induced proteins block penetration and uncoating of simian virus 40 and some retroviruses. Others block the maturation, assembly, and release of vesicular stomatitis virus, herpes simplex virus, and some retroviruses.

Unlike the broad-spectrum antiviral effects of activated Pkr and RNase L, at least one Ifn-induced mouse protein and two human homologs appear to be directed against specific viruses. Mouse strains that have an Ifn-inducible gene called *mx1* are completely resistant to influenza virus infection. The Mx1 protein is part of a small family of Ifn-inducible GTPases with potent antiviral activity against various (−) strand RNA viruses. After Ifn induction, this protein accumulates in the nucleus and inhibits the unique "cap-snatching" mechanism used by influenza virus. It is likely that the Mx1 protein interferes with the function of one of the viral polymerase subunits called PB2. If this viral protein is overproduced, the antiviral effect of Mx1 protein is blocked. The significance of the *mx1* gene in the biology of influenza virus or of mice is not at all clear, as influenza virus does not circulate among wild mice. Moreover, in these animals, about one-quarter of the population lacks a functional *mx1* gene with no clear positive or negative advantages.

There are two human homologs of mouse *mx1* called *mxA* and *mxB*. Expression of these two genes is also induced by Ifn, but unlike the murine protein, the human proteins reside in the cytoplasm. MxA, but not MxB, blocks replication of influenza virus. Interestingly, in contrast to murine Mx1, which inhibits only influenza virus, the human MxA protein also inhibits replication of vesicular stomatitis virus, measles virus, human parainfluenza virus type 3, and some bunyaviruses. It is of some interest to understand how MxA inhibits the replication of several different RNA viruses.

### Viral Anti-Ifn Strategies
Clearly, Ifn is a broad-spectrum, highly effective antiviral agent. Because of the multifaceted nature of the Ifn defense system, one might predict that it would be difficult

for a virus to evolve measures to counter it. This is wishful thinking, as viruses have numerous mechanisms for confounding Ifn action (Table 14.6; Box 14.2).

One obvious inference from the various Ifn countermeasures encoded by diverse viruses is that Ifn is an essential defense component for the host. For many viruses, this argument has merit; however, viruses without any known Ifn defense do exist, and they obviously prosper. They apparently avoid Ifn effects in vivo but cannot do so in vitro. For example, vesicular stomatitis virus has no known anti-Ifn defense: Ifn treatment of infected tissue culture cells results in at least a 100-fold reduction in vesicular stomatitis virus protein synthesis and a 50,000-fold reduction in infectious virus yield. Nevertheless, in nature, vesicular stomatitis virus is a highly successful animal pathogen.

## Complement

### A Complex Collection of Interrelated Serum Defense Proteins That Can Also Amplify Immune Reactions

The complement system in the blood is a major defense and a clearance component of innate defense and the adaptive immune response. It was identified in 1890 as a heat-labile factor in serum that lysed bacteria in the presence of antibody (it "complemented" antibody). As part of the innate defense system, complement is activated when a key component, an enzyme called C3 convertase, binds to the surface of the infectious agent, presumably through recognition of unique patterns of molecules on microbial surfaces. This system is called the **alternative pathway** (Fig. 14.5B). Such binding initiates a cascade of protein cleavage reactions that results in destruction of the invader. The alternative pathway is thought to play an important role in innate defense against enveloped viruses.

Complement also functions as part of the adaptive response in the **classical pathway** (Fig. 14.5A). A protein cleavage cascade similar to that activated by the alternative pathway is initiated when a complement protein called C1 binds to antigen-antibody complexes. When this occurs, complement becomes a major effector of the antibody response.

### The Complement Cascade

Complement is not a single protein but rather at least 18 distinct serum proteins and an almost equal number of membrane proteins that act sequentially in a cascade (Fig. 14.5). This cascade amplifies the consequences of a specific antibody-antigen or of an antigen-complement protein interaction to ever-increasing local production of inflammation mediators and cell-killing complexes. Indeed, complement proteins cooperate in an elegant pathway of sequential protease cleavage reactions leading to the assembly of a multiprotein **membrane attack complex** on the surface of the infected cell (Fig. 14.5). The membrane attack complex kills cells literally by disrupting the lipid bilayer, permitting the entry of water and small ions but not

---

**BOX 14.2**

*Viruses versus interferon*

Influenza virus and vaccinia virus counterattack Pkr and RNase L action by producing proteins that bind and sequester dsRNA. After infection by herpes simplex virus, Ifn-treated cells accumulate complex mixtures of 2',5'-oligoadenylates that either fail to activate or competitively inhibit RNase L activation by authentic 2',5'-oligo(A), such that the digestion of RNA does not occur.

Influenza virus and herpes simplex virus infections induce synthesis of cellular proteins that compromise the function of the Pkr enzyme. One such host protein, called p58, is normally bound to the heat shock protein hsp40 but is released after influenza virus infection by an as yet uncharacterized mechanism. Free p58 then binds Pkr, inhibiting its activity.

Adenovirus not only produces VA-RNA I, which blocks activation of Pkr, but also encodes E1A proteins that reduce activation of interferon response genes by inhibiting the DNA-binding activity of specific transcriptional regulators (e.g., Isgf3) (Fig. 14.4).

Probably the most remarkable anti-Ifn defense is elaborated by poxviruses. Vaccinia virus not only encodes two proteins that block Pkr activation and inhibit activation of RNase L but also expresses soluble Ifn-binding proteins that bind all three types of Ifn, effectively removing all species of active Ifn from the area of infection. In addition, vaccinia virus contains a gene called A18R whose protein product controls the production of viral dsRNA. The A18R protein is required for efficient virus replication and has homology to excision-repair DNA helicase enzymes. Conditional mutations in the A18R gene result in increased dsRNA synthesis from aberrant transcription of opposite strands of viral DNA.

**Griffin, D. E.** 1994. Cytokines in viral infections. *Semin. Virol.* **5:**403–463.

**A Classical complement pathway**

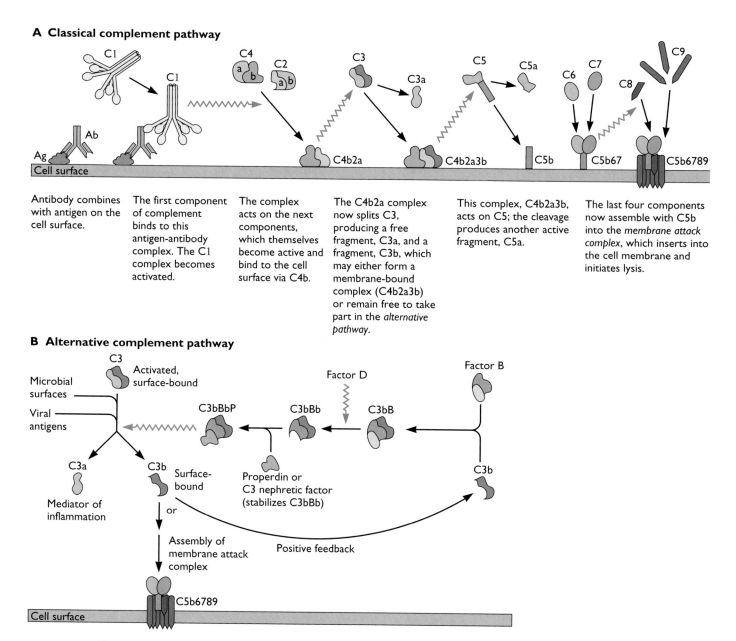

Antibody combines with antigen on the cell surface.

The first component of complement binds to this antigen-antibody complex. The C1 complex becomes activated.

The complex acts on the next components, which themselves become active and bind to the cell surface via C4b.

The C4b2a complex now splits C3, producing a free fragment, C3a, and a fragment, C3b, which may either form a membrane-bound complex (C4b2a3b) or remain free to take part in the *alternative pathway.*

This complex, C4b2a3b, acts on C5; the cleavage produces another active fragment, C5a.

The last four components now assemble with C5b into the *membrane attack complex*, which inserts into the cell membrane and initiates lysis.

**B Alternative complement pathway**

**Figure 14.5 The classical and alternative pathways for activation of complement.** (A) The classical pathway is activated when C1 binds to antigen-antibody complexes. (B) The alternative pathway is activated when C3b binds directly to viral surface antigens. Both pathways converge in the formation of an enzyme called C3 convertase which cleaves complement component C3 into its active peptides. The C4b2a complex (classical pathway) and the C3bBbP complex (alternative pathway) are active C3 convertase enzymes. C3 convertase activity produces the membrane attack complex, which forms holes in membranes of cells and viruses, and is instrumental in the formation of C3a, a potent mediator of inflammation.

the exit of proteins. As a result, the cells burst by osmotic lysis. The complex can lyse a broad spectrum of microorganisms, viruses, red blood cells, and nucleated cells.

More than 90% of complement components in the plasma are produced in the liver. However, complement components are made elsewhere as well. For example, C1 is synthesized mainly in the gut epithelium. Monocytes, macrophages, lymphocytes, fibroblasts, endothelial cells, and cells lining kidney glomeruli or synovial cavities all produce most proteins of the complement system. One recent finding is that astrocytes in the brain can synthesize the full panoply of the complement system when stimulated by inflammatory cytokines.

The complement cascade has four major biological functions, namely, **cytolysis, activation of inflammation, opsonization**, and **solubilization of immune complexes**. Cytolysis and lysis of foreign cells and enveloped viruses occur when specific activated complement components (C6, C7, C8, and C9) polymerize on a foreign cell or enveloped virus, forming pores or holes that disrupt the lipid bilayer and compromise its function. The cell or virus lyses by the osmotic effects described above. Inflammation is stimulated by several peptide products of proteolysis during establishment of the complement cascade (Fig. 14.2; Table 14.4). These peptides (C3a, C4a, and C5a) bind to vascular epithelial cells and various classes of lymphocytes to stimulate inflammation and to enhance responses to foreign antigens. Complement proteins (typically C3b) can bind to virus particles so that phagocytic cells carrying appropriate receptors can then engulf the coated viruses and destroy them; this process is called **opsonization**. Finally, noncytopathic virus infections commonly result in pathological accumulations of antigen-antibody complexes in lymphoid organs and kidneys. Complement proteins can solubilize these complexes by binding both antigen and antibody and thus facilitate their clearance from the circulatory system.

One important consequence of activating the complement cascade is the initiation of a brisk local defense. Complement components aid in recruitment of monocytes and neutrophils (Fig. 14.2) to the site of infection, as well as stimulating their activities and increasing vascular permeability. Thus, the antiviral effects of complement are both direct and indirect; infected cells and enveloped viruses can be killed by the membrane attack complex, while the local inflammatory response stimulated by complement components can limit viral infection.

Complement contributes to antiviral defense against enveloped viruses, as the protective membranes can be lysed by activated complement. This system also plays a modest role in inactivating nonenveloped viruses. When complement is activated by antibody binding to the capsids of such viruses, large aggregates that reduce the number of infectious particles are formed. This mechanism may be effective in slowing virus spread. For example, with as few as two antibody molecules bound per virion, polyomavirus can be neutralized by aggregate formation on binding of C3b.

Not all antiviral antibodies are capable of activating complement; notably, many antibodies specific for viral envelope proteins cannot "fix" complement. The basis for this inability to activate complement is not completely understood but most likely represents an evolutionary adaptation of these viruses to escape complement-mediated lysis. In addition, the antibody isotype stimulated by viral infection may or may not be able to interact with complement. For example, antibodies produced during the Th1 response (see below) are predominantly the immunoglobulin G2a (IgG2a) isotype and actively stimulate the complement cascade. In contrast, antibodies produced during the Th2 response typically are IgG1 (mice), IgG4 (humans), and IgE, which bind neither to complement nor to macrophage Fc receptors (see Table 14.8).

A main line of complement defense in humans and higher primates is triggered by the antigen **galactose α(1,3)-galactose** (also called αGal) found as a terminal sugar on glycosylated cell surface proteins. Humans and higher primates do *not* produce this antigen because they lack the enzyme galactosyltransferase. Because of constant exposure to bacteria expressing αGal in the gut, human serum contains high levels of antibodies specific for this antigen; indeed, more than 2% of the IgM and IgG populations is directed against this sugar. It is this antibody that triggers the complement cascade and subsequent lysis of foreign cells and enveloped viruses bearing αGal antigens. The anti-αGal antibody/complement reaction is probably the primary reason why humans and higher primates are not infected by enveloped viruses of other animals despite the ability of many of these viruses to infect human cells efficiently in culture. Such viruses susceptible to human complement/anti-αGal inactivation include arenaviruses, paramyxoviruses, togaviruses, herpesviruses, retroviruses, and rhabdoviruses. Evidence supporting this assertion is that when such viruses are grown in nonhuman cells, they are sensitive to inactivation by fresh human serum, but when they are grown in humans cells, they are resistant.

### Regulation of the Complement Cascade

Complement is a double-edged sword with a capacity to harm as well as to heal. Any amplified host antiviral defense must be regulated with precision. A primary requirement is that the cascade be initiated only in response to infection; it cannot be initiated in a random fashion. Such tight control is accomplished in several ways. Some regulation is intrinsic to the complement proteins themselves. Many are large and cannot leave blood vessels to attack infected tissues unless there is local tissue damage that exposes cells directly to blood. Consequently, many minor infections do not activate a substantial complement response. Many cascade intermediates are short-lived, with millisecond half-lives, and therefore do not exist long enough to diffuse far from the site of infection. Further control is maintained by complement-inhibitory proteins present in the serum and by integral membrane proteins found on the surface of many cells (e.g., the complement receptor type 1 protein [Cr1], decay accelerating factor [CD55], protectin [CD59], and membrane cofactor protein [CD46]). These proteins, which are unique to vertebrates, limit the cascade by binding complement components such as C3b and C4b (Fig. 14.5). Some enveloped viruses do not carry CD55 or Cr1 on their surfaces and therefore are susceptible to the action of complement. Others, such as human immunodeficiency virus type 1 and the extracellular form of vaccinia virus, incorporate CD46, CD55, and CD59 in their envelopes and are protected from complement-mediated lysis.

Interestingly, several viruses use complement control proteins as receptors for their own attachment. For example, measles virus binds CD46, whereas the picornaviruses coxsackie B virus and echovirus bind to CD55. Other viral genomes encode complement-binding proteins that disrupt the complement cascade. For example, alphaherpesvirus glycoprotein C binds the C3b component in a host-specific fashion, and several poxvirus proteins bind C3b and C4. While the latter interactions can be demonstrated to be important for the viruses to avoid complement inactivation, the significance of the use of complement components as viral receptors remains an intriguing mystery.

## Collectins

Mammalian serum contains soluble collagenous lectins (glycoproteins) called **collectins**, such as the mannose-binding protein and conglutinin in plasma and pulmonary surfactant proteins A and D. As lectins, these proteins bind to polysaccharides on a wide variety of microbes and, as part of the innate defense system, act as opsonins or activate complement (e.g., mannose-binding protein). Collectins bind the glycoproteins of a number of enveloped viruses, including human immunodeficiency virus type 1, herpes simplex virus, and influenza virus. They display antiviral activity in the laboratory, but their in vivo role has not been well studied. Virologists have long pondered the significance and function of glycosylation of viral membrane proteins. The existence of collectins suggests that the sugar moieties on viral proteins may represent previously unrecognized interfaces of host defense and virus offense.

## NK Cells

NK cells represent a distinct, abundant lymphocyte population (Table 14.1) patrolling the blood and lymphoid tissues. They are large, granular lymphocytes, distinguished from others by the absence of antigen receptors found on B and T cells. NK cells can constitute as much as 30% of the circulating lymphocyte population and act during the early phase of many viral infections by spontaneously killing virus-infected cells. We know this because NK cell depletion and reconstitution experiments in mice have shown that such cells are vital in controlling a variety of viral infections. In humans, NK cells are particularly important in controlling primary infection by many herpesviruses, as patients with NK cell deficiencies suffer from severe infections with varicella-zoster virus, human cytomegalovirus, and herpes simplex virus.

NK cells are in the immediate front line of defense. Their numbers expand quickly after viral infection and then decline as the acquired immune response is established. NK cells not only eliminate infected cells by direct cell lysis but also influence subsequent immune responses by secreting potent cytokine mediators of inflammation such as Ifn-$\gamma$ and Tnf-$\alpha$ on recognition of an infected cell. In turn, NK cells can be stimulated to divide and produce more cytokines by interferons and Il-2 produced by other cells responding to infection.

In contrast to B and T cells, NK cells do not appear to have the classical antigen-specific receptors that can distinguish infected self from uninfected self (see "General Features" below). NK cells are easily activated on binding to a target cell, particularly endothelial cells, but are prevented from killing a healthy cell by a variety of inhibitory interactions. NK cells carry many surface receptors, some of which facilitate binding to cells with reduced concentrations of surface molecules such as **major histocompatibility complex class I** proteins that are important in the adaptive response (see "Antigen Presentation and Activation of Immune Cells"). NK cells can also recognize and kill cells with increased concentrations of cell surface proteins such as lymphocyte function-associated antigen 3 (Lfa-3). Other NK cell receptors are lectins that bind specific carbohydrates.

It is of some importance to understand how the NK cell can distinguish a virus-infected cell so quickly after infection. One idea is that at least two receptor-binding interactions are required: one to activate the NK cell and the other to block this activation. The activating signal might be a new carbohydrate ligand on the target cell that binds a receptor on the NK cell. For example, virus-infected cells may present new glycoproteins on their surface that interact with these activating receptors. As a consequence, a signal transduction cascade is initiated and the NK cell is stimulated to kill the target cell. However, a negative regulatory signal is produced by an MHC class I-specific receptor on the NK cell. When the target cell expresses MHC class I that can be bound by the NK cell receptor, an inhibitor of the activating signal transduction cascade is produced. Some virus-infected cells may be killed by NK cells, in part because many viruses actively reduce cell surface expression of MHC class I molecules, as discussed in chapter 15. Thus, normal cells expressing MHC class I are not killed by NK cells even if they have novel ligands on their surface. Much work is in progress to test this idea.

Recently, it has been found that NK cells can recognize and spare target cells expressing HLA-E (a unique MHC class I molecule that binds peptides derived from the signal sequences of other MHC class I molecules). Such recognition is mediated by the NK cell CD94-Nkg2 cell surface receptor. Interestingly, infection by human cytomegalovirus induces the expression of HLA-E, suggesting one mechanism by which this virus evades destruction by NK cells.

## The Adaptive Immune Response

### General Features

The adaptive response comprises two complex actions, the **humoral response** (actions of serum and lymph proteins) and the **cell-mediated response** (actions of specific helper and effector cells) (Fig. 14.6 and 14.7; Table 14.7). As we discuss the cells and actions that characterize each response, it is important to understand that *both* are essential in antiviral defense and function in concert but that the relative contribution of each in any given infection varies with the nature of the infecting virus as well as with the host. While innate responses tend to engage the initial infection, adaptive responses are essential for eradicating the infected cells (**clearing the infection**) and for providing long-term, specific defense. In general, then, antibodies bind to virus in the bloodstream and at mucosal surfaces, decreasing the number of cells that could have been infected, whereas T cells recognize and kill infected cells.

A definitive characteristic of the adaptive response to viral infection is the ability of the system to distinguish infected self from uninfected self. This feat is accomplished in part by molecular recognition mediated by two antigen recognition receptors, the membrane-bound antibody on B cells and the T-cell receptor on T cells, *and* two kinds of protein that display fragments of internal cellular proteins on the cell surface. MHC class I proteins display protein fragments on the surface of almost all cells, whereas MHC class II proteins are found only on the surface of macrophages, dendritic cells, and B cells. While B- and T-cell receptors both bind foreign antigens, they do so in markedly different ways. The B-cell receptor is an antibody and binds discrete **epitopes** (contiguous sequences or unique conformations) in regions in *intact proteins*. In contrast, the T-cell receptor binds short, linear peptides derived from *proteolytically processed proteins*. The act of binding an epitope or peptide by either receptor has profound effects on the cell bearing that receptor: it responds by producing cytokines, by replicating and dividing, by killing the cell that bears the foreign protein or peptide, or by synthesizing antibodies. The entire response initiated by the binding of foreign peptides or epitopes comprises the adaptive immune response.

As in all defense mechanisms, an uncontrolled or inappropriate adaptive response can be damaging to the host. Therefore, precise initiation and rapid cessation of the response are as important characteristics as its amplification. Regulation is achieved largely by multiple interactions between the surfaces of lymphocytes and infected cells, the short life spans of the activated cells, and the short half-lives of the chemical mediators. When this regulation is disrupted actively or passively by large-scale viral infection, or when the cells of the immune system themselves are infected by viruses, severe consequences ensue. As discussed in chapters 15, 17, and 18, many pathological effects associated with viral infection are manifestations of inappropriate reactions of the host immune defense system. On the other hand, some viruses have evolved mechanisms that bypass these host defenses and can therefore establish long-lived persistent or latent infections in the face of an active, adaptive immune system (see chapter 15).

Unlike the innate response, the adaptive response is tailored specifically to the invading organism or substance and requires more time (days to weeks) after exposure to become fully active. However, once a specific response has been established and the viral infection is subdued, the individual is immune to subsequent infection by the same invader (Fig. 14.8). Such memory of previous infections is one of the most powerful features of the adaptive immune response. Memory enables the individual to mount a rapid, robust, highly specific response to a subsequent invasion months or years after the initial infection. Unlike the primary response, which takes many days to reach its optimum, a subsequent encounter with an invader in an immune individual engenders a ferocious response that occurs within hours of the infection. The innate defenses (cytokines, NK cells, and complement) are also stimulated during this secondary response to infection. The cellular and molecular mechanisms for maintenance of memory are the focus of considerable research and debate, but a subset of lymphocytes called **memory cells** is maintained after each encounter with a foreign antigen. These cells survive for years in the body and are ready to respond immediately to any subsequent encounter by rapid proliferation and efficient production of their defensive products. Because such a secondary response is usually stronger than the primary response, childhood infection protects adults and vaccination can last for years.

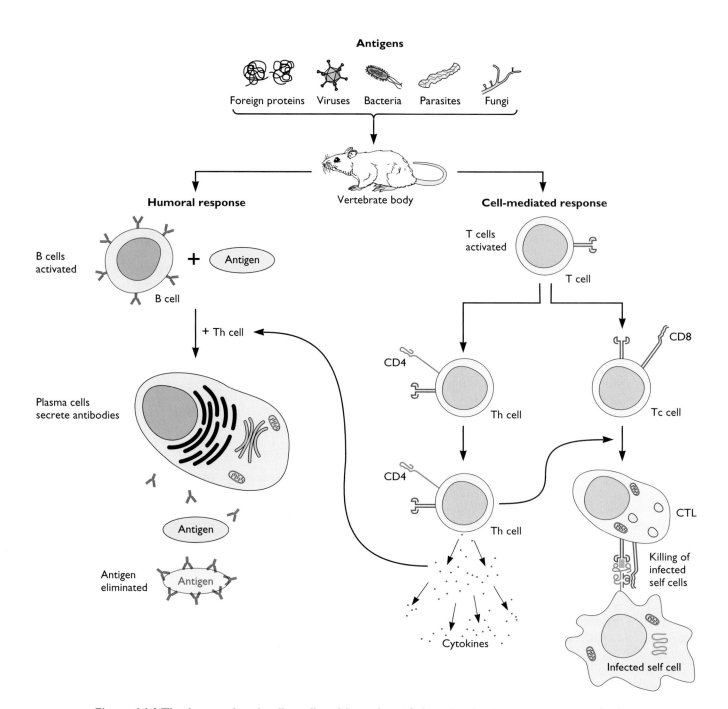

**Antigens**

Foreign proteins  Viruses  Bacteria  Parasites  Fungi

Vertebrate body

**Humoral response**

B cells activated

B cell

+ Antigen

+ Th cell

Plasma cells secrete antibodies

Antigen

Antigen eliminated

Antigen

**Cell-mediated response**

T cells activated

T cell

CD4

Th cell

CD8

Tc cell

CD4

Th cell

Cytokines

CTL

Killing of infected self cells

Infected self cell

**Figure 14.6 The humoral and cell-mediated branches of the adaptive immune system.** The humoral branch comprises lymphocytes of the B-cell lineage. Antibodies are the important effector molecules produced by this response. The process begins with the interaction of a specific receptor on precursor B lymphocytes with antigens. Binding of antigen promotes differentiation into antibody-secreting cells (plasma cells). The cell-mediated branch comprises lymphocytes of the T-cell lineage. Cytotoxic T cells (Tc cells) and T-helper cells (Th cells) are the key effectors of this response. The process is initiated when T lymphocytes recognize antigens on the surface of self cells. The antigens on self cells can be recognized only by a receptor on the surface of T cells (the T-cell receptor) when they are bound to the MHC family of membrane proteins. Two subpopulations of T cells are illustrated. The Th cells recognize antigens bound to MHC class II molecules and produce powerful cytokines that "help" other lymphocytes (both B and T cells) by promoting or inhibiting cell division and gene activity. Once activated by Th cells, Tc cells differentiate into CTLs that can kill virus-infected cells.

**A**

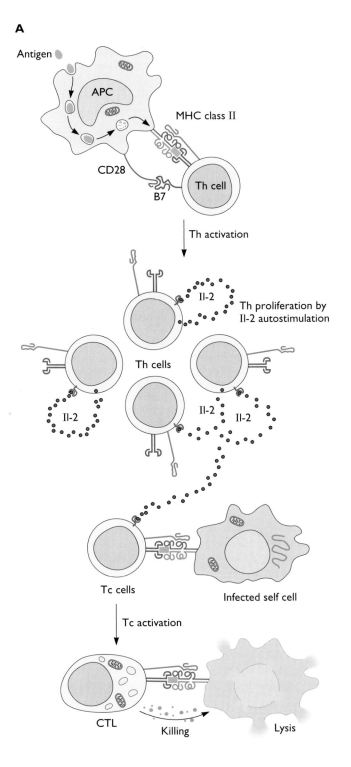

**B**

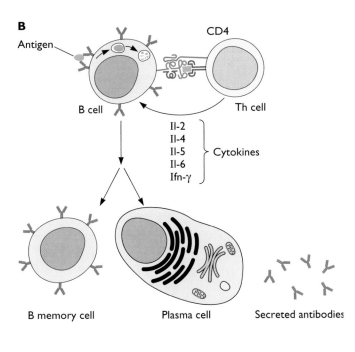

**Figure 14.7 The production of both humoral and cellular adaptive responses requires activation and proliferation of Th cells.** (A) The cellular adaptive response begins with the interaction of a T-cell receptor with an MHC class II-peptide complex on the surface of an antigen-presenting cell (APC). This interaction, together with binding of coreceptors such as B7 (on the Th cell) and CD28 (on the APC), activates a signal transduction pathway leading to new gene expression in the Th cell. One common result is the secretion of Il-2, which binds to the Il-2 receptor on the Th cell in an autocrine loop, stimulating Th cell proliferation. Il-2 also stimulates Tc cells to become CTLs. When the T-cell receptor on CTLs engages an MHC class I-viral peptide complex on the surface of a self cell, the CTL actively injects lethal enzymes that kill the virus-infected cell. (B) The humoral response begins when antigen binds to several receptors on the surface of a B cell (called receptor cross-linking). Some of the antigen enters the cell by endocytosis, is processed into peptides complexed with MHC class II molecules, and is transported to the surface of the B cell. The Th cell recognizes the antigen-MHC class II complex and secretes various cytokines that activate the B cell to proliferate and differentiate into antibody-secreting plasma cells and memory cells.

One idea of how memory is maintained is that follicular dendritic cells in lymph nodes bind antibody-antigen complexes and keep such complexes on their surfaces for long periods. When the circulating antibody concentration drops, as happens after a viral infection is cleared, antigen is released gradually from the immune complexes. This slow release of antigen provides continual stimulation of the immune response.

**Table 14.7** The major humoral and cell-mediated immune responses to viral infections

| Response | Effector | Activity |
|---|---|---|
| Cell mediated | Secretion of Ifn-γ by Th and cytotoxic T lymphocytes | Induces antiviral state |
| | Cytotoxic T lymphocytes (CTLs) | Destroy virus-infected cells |
| | NK cells and macrophages | Destroy virus-infected cells directly or by antibody-dependent cell-mediated cytotoxicity (ADCC) |
| Humoral | Antibody (primarily secretory IgA) | Inhibits virion/host attachment |
| | Antibody (primarily IgG) | Inhibits fusion of enveloped viruses with host membrane |
| | IgG and IgM antibody | Enhances phagocytosis (opsonization) after binding to virions |
| | IgM antibody | Agglutinates virions |
| | Activation of complement (IgG and IgM) | Lysis of enveloped viruses; opsonization by C3b-antibody complex |

## Cells of the Adaptive Immune System

The adaptive response depends on two important cell groups: lymphocytes and antigen-presenting cells (Table 14.1). **Lymphocytes** are white blood cells produced by hematopoiesis in the bone marrow. They are migratory cells, leaving the bone marrow to circulate in the blood and lymphatic systems and often settling in various lymphoid organs throughout the body (Fig. 14.9). The small, bean-shaped organs called lymph nodes are important centers for initiation of the adaptive immune response. **Antigen-presenting cells**, including the dendritic Langerhans cells in the skin, Kupffer cells in the liver, dendritic cells in lymph nodes and mucosa, macrophages, and B cells, are found in or adjacent to a variety of lymphoid tissues. B cells and dendritic cells are particularly important for the presentation of viral antigens in the adaptive response.

Two major classes of lymphocytes are recognized: B cells and T cells. T cells (lymphocytes that mature in the thymus) and particular cytokines are the primary components of the cell-mediated response. Plasma cells, whose immediate precursors are B cells (lymphocytes derived from the bone marrow, the mammalian equivalent of the avian bursa), produce **antibodies** that bind to foreign molecules and define the antibody response (also known as the humoral response) (Figs. 14.6 and 14.7).

### The Mucosal and Cutaneous Arms of the Immune System

While we often speak of the "immune system" as if it were one centralized machine, there is considerable decentralization of the immune cells and tissues. As most viral infections occur at mucosal surfaces, the **mucosal immune system** is often the first adaptive defense to be

**Figure 14.8 The specificity, self-limitation, and memory of the adaptive immune response.** This general profile of a typical adaptive antibody response demonstrates the relative concentration of serum antibodies (*y* axis) after time (weeks) of exposure to antigen A or a mixture of antigens A and B (*x* axis). The antibodies that recognize antigen A are given by the red line, and the antibodies that recognize antigen B are given by the blue line. The primary response to antigen A takes about 3 to 4 weeks to reach a maximum. When the animal is injected with a mixture of both antigens A and B at 7 weeks, the secondary response to antigen A is more rapid and more robust than the primary response. However, the primary response to antigen B takes about 3 to 4 weeks to reach a maximum and is less robust than the secondary reponse to antigen A. This response demonstrates immunological memory. Antibody levels (also termed titers) decline with time after each immunization. This property is called self-limitation or resolution. From A. K. Abbas et al., *Cellular and Molecular Immunology* (The W. B. Saunders Co., Philadelphia, Pa., 1994), with permission.

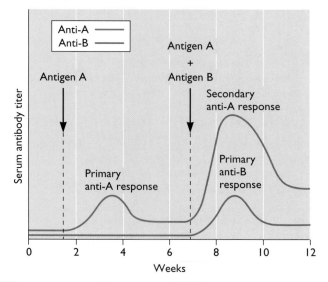

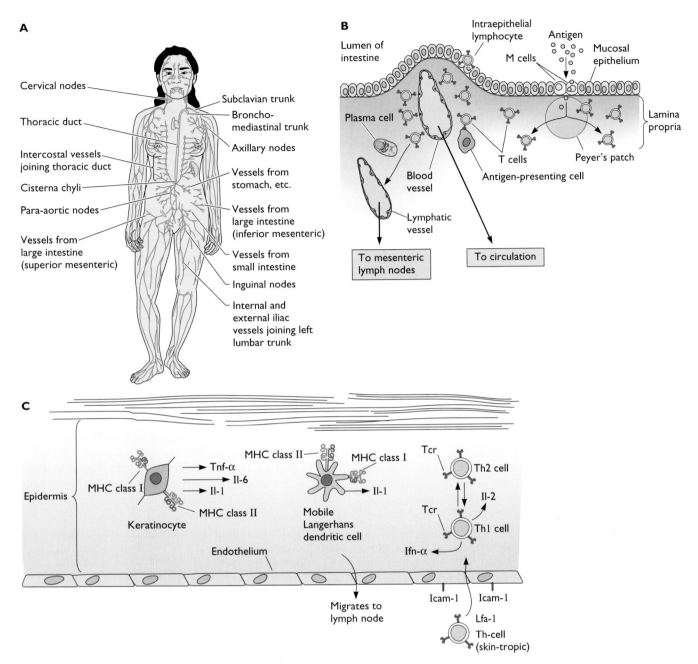

**Figure 14.9 Components of the human lymphatic and mucosal immune systems.** (A) The primary lymphatic system is illustrated in green. Many clusters of lymphatic tissue, called nodes, are found throughout the lymphatic system. Lymph nodes provide the interactive environment where mobile dendritic cells patrolling the local area exchange information with lymphocytes in the circulation. (B) Cellular components of the mucosal immune system in the gut (mucosa-associated lymphoid tissue). The lumen of the small intestine is at the top of the figure. The mucosal epithelial cells are shown with their basal surface oriented toward the dark purple line marking the start of the lamina propria. A cross section of a lymphatic vessel and a capillary are shown, illustrating their juxtaposition to cells of the mucosal immune system. M cells (yellow) have large intraepithelial pockets filled with B and CD4 T lymphocytes, macrophages, and dendritic cells. M cells and intraepithelial lymphocytes are important in the transfer of antigen from the intestinal lumen to the lymphoid tissue in Peyer's patches, where an immune response can be initiated (see also Fig. 17.3). (C) The cutaneous immune system (skin-associated lymphoid tissue) comprises three cell types: keratinocytes, Langerhans cells, and T cells. Keratinocytes actively secrete various cytokines including Tnf-α, Il-1, and Il-6 and have phagocytic activity. They also express both MHC class I and MHC class II proteins and can present antigens to T and B cells if stimulated by Ifn-γ. Langerhans cells are migratory dendritic cells and are the major antigen-presenting cells in the epidermis. They express both MHC class I and MHC class II proteins. Special skin-tropic T cells can cross the endothelium to enter the epidermis where they can mature into Th1 or Th2 cells depending on the antigen and cytokine millieu. Activated T cells express cytokines, including Ifn-γ, that activate MHC expression from keratinocytes and Langerhans cells. Tcr, T-cell receptor. Adapted from A. K. Abbas et al., *Cellular and Molecular Immunology* (The W. B. Saunders Co., Philadelphia, Pa., 1994), with permission.

**497**

engaged. The lymphoid tissues below the mucosa of the gastrointestinal and respiratory tracts (often called mucosa-associated lymphoid tissue) (Fig. 14.9B) are vital in antiviral defense. These clusters of lymphoid cells include the dome-shaped collection of cells called **Peyer's patches** in the lamina propria of the small intestine, the **tonsils** in the pharynx, the **submucosal follicles** of the upper airways, and the **appendix**. A specialized epithelial cell of such mucosal surfaces is the **M cell** (microfold or membranous epithelial cell), which samples and delivers antigens to the underlying lymphoid tissue by transcytosis of antigens via endocytic vesicles from the apical or luminal surface to the basal surface. M cells have invaginations of their membranes (pockets) that harbor B and CD4 T lymphocytes, macrophages, and dendritic cells. The secreted antibody called immunoglobulin A (IgA) (important in antiviral defense at mucosal surfaces; see below) is produced by B cells accumulating at adhesion sites in these pockets. The passage of antigens and viruses through M cells puts them in intimate contact with immune cells and represents an essential step for the development of mucosal immune responses, as well as the pathogenesis of many viral infections.

The outermost skin layer comprises the epidermis, a highly organized stratified squamous epithelium containing keratinocytes at different stages of differentiation. The epidermis also harbors small numbers of melanocytes, Langerhans cells, Merkel cells, and lymphocytes. Merkel cells, are modified epithelial cells that cap the terminal branches of mechanosensory nerve endings and may provide sites of entry to the peripheral nervous system for some viruses. Lymphocytes and Langerhans cells (specific dendritic antigen-presenting cells) comprise the **cutaneous immune system** (also called skin-associated lymphoid tissue) (Fig. 14.9C). These cells are important in the initial response and resolution of viral infections of the skin. In particular, Langerhans cells function as the sentinels or outposts of early warning and reaction. They are the predominant scavenger antigen-presenting cells of the epidermis. These mobile cells are capable of sampling antigens and migrating to regional lymph nodes to transfer information to T cells and also are able to activate B lymphocytes directly. A class of T cells in the circulation has tropism for the skin and, after binding to the endothelium, can enter the epidermis to interact with Langerhans cells and keratinocytes.

Viruses can interact with lymphoid cells associated with mucosal and cutaneous immune systems at the primary site of infection. Such events can govern the outcome of the host-virus interaction and often establish the characteristic pattern of infection for a given virus. For example, human immunodeficiency virus is thought to gain access to T cells after primary infection by accumulating in and around follicular dendritic cells in lymphoid tissue. As T cells interact with these antigen-presenting dendritic cells, they are infected by human immunodeficiency virus virions lying in ambush. The M cells in the mucosal epithelium have been implicated in the spread from the pharynx and the gut to the lymphoid system of a variety of viruses including poliovirus, enteric adenovirus, human immunodeficiency virus type 1, and reovirus. The antigen-presenting Langerhans cells of the cutaneous immune system have been suggested to be sites of persistent or latent infection for a number of viruses, including herpes simplex virus.

### The Immune System and the Brain

In the past, the central nervous system was considered to be an "immunoprivileged" organ physically isolated from the immune system. This view was further expanded by some to mean that the brain and spinal cord were unable to mount any immune defenses. In fact, the central nervous system has the capacity to initiate and regulate immune responses by allowing restricted entry of immune cells circulating in the periphery, as well as by providing an environment conducive to the activation of immune effector cells by resident glial cells. Activated T cells can enter and travel through the brain, performing immune surveillance. At least two glial cell types, astrocytes and microglia, can transfer information to CD4$^+$ T cells. Astrocytes, the most numerous cell type in the central nervous system, respond to a variety of cytokines produced by cells of the immune system.

### Acquired Immunity Results from the Action of Lymphocytes That Express Distinct Antigen Receptors

The antigen receptors are formed through somatic gene rearrangement, and the specificities are built originally without regard to any one target. Subsequently, specific cells bind a particular antigen and are killed if they recognize self-antigens or stimulated to divide and prosper if they recognize a foreign invader or infected cell. Lymphocytes are responsible for specificity, memory, and self/nonself discrimination (Fig. 14.6).

As noted earlier, each T or B cell carries on its surface a specific receptor that binds and responds to a particular peptide or epitope (Fig. 14.10). Antiviral defense is possible because T and B cells bearing these specific receptors have survived the normal selective process that eliminates cells that respond to self peptides and epitopes. It is thought that the cells emerging from such a selection enter the circulation and pass through the lymphatic system or are retained in various tissues in the body. They are not

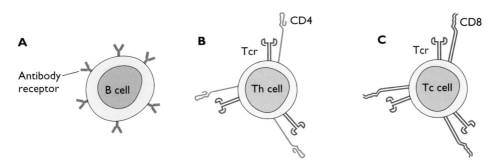

**Figure 14.10 The antigen receptors on the surface of B and T cells.** (A) B cells have about 100,000 molecules of a single membrane-bound receptor antibody per cell. Every receptor antibody on a given B cell has identical bivalent specificity for one antigen epitope. (B and C) T lymphocytes have about 100,000 T-cell receptors (Tcrs) per cell, each with identical specificity. (B) T cells bearing the surface membrane protein CD4 always recognize peptide antigens bound to MHC class II proteins and generally function as helper T cells (Th). (C) T cells bearing the surface membrane protein CD8 always recognize peptide antigens bound to MHC class I proteins and generally function as cytotoxic T cells (Tc).

completely differentiated or "armed" to produce their ultimate immune effector response (e.g., antibody production or cell killing) but are able to react to foreign (infected self) signals with remarkable diversity and specificity. On such binding, they rapidly acquire their specific immune effector activity.

The initial encounter with any foreign epitope, whether in lymphoid tissues or elsewhere in the body, involves only a few cells. For example, the frequency of lymphocytes that recognize a virus on first exposure is as few as 1 in 10,000 to 1 in 100,000. So few cells certainly are not sufficient for protection against a virus that is infecting and spreading in the host. What makes the adaptive response so powerful is that the initial encounter can be amplified substantially during the ensuing 1 or 2 weeks. The number of virus-specific lymphocytes increases more than 1,000-fold because the original encounter stimulates previously uncommitted precursor lymphocytes to differentiate, replicate, divide, and produce antibodies or cytokines. The antigen-stimulated cell is said to be "activated." It then undergoes numerous rounds of cell division such that each daughter cell has the same specific immune reactivity as the original parent.

## B Cells

B cells are produced in the bone marrow. As they mature, each expresses a unique antigen receptor, a membrane-bound antibody (Fig. 14.6, 14.7, and 14.10). When an antigen binds specifically to the membrane-bound antibody, a signal transduction cascade is initiated in the B cell. As a consequence, new gene products are made and the cell begins to divide rapidly. The sister cells produced by each division differentiate into memory B cells and ef-

fector cells called **plasma cells**. As their name implies, **memory B cells** or their clonal progeny are long-lived and continue to express the parental membrane-bound antibody receptor. In contrast, plasma cells live only a few days and no longer make membrane-bound antibody but instead synthesize the same antibody in a secreted form. A single plasma cell can secrete more than 2,000 antibody molecules per second (Table 14.8).

## T Cells

T-cell precursors are also produced in the bone marrow, but in contrast to a B-cell precursor, a T-cell precursor must migrate to the thymus gland to mature (Fig. 14.6, 14.7, and 14.10). The maturation process comprises two types of selection: positive selection for T cells that can bind appropriate surface molecules (see below) and negative selection that efficiently kills T cells recognizing cells expressing self peptides on their surface. As a result, only 1 to 2% of all immature T cells entering the thymus emerge as mature cells able to defend the host against viral infections.

As noted above, T cells express a membrane-bound receptor, the T-cell receptor, which is capable of binding antigens. The T-cell receptor is a disulfide-linked heterodimer composed of either alpha and beta ($\alpha/\beta$ complex) or gamma and delta ($\gamma/\delta$ complex) protein chains. Cells with the $\gamma/\delta$ T-cell receptor represent a minor population of T cells whose specificity and function are unknown, although recent work indicates that they may clear some viral infections at certain mucosal surfaces. The peptide-binding site of the T-cell receptor and the epitope-binding site of the B-cell receptor are very similar structures formed by the folding of three regions in the

**Table 14.8**    The five classes of immunoglobulins

| Property | IgA | IgD | IgE | IgG | IgM |
|---|---|---|---|---|---|
| Function | Mucosal, secretory | Surface of B cell | Allergy, anaphylaxis; epithelial surfaces | Major systemic immunity; memory responses | Major systemic immunity; primary response; agglutination |
| Subclasses | 2 | 1 | 1 | 4 | 1 |
| Light chain | κ, λ | κ, λ | κ, λ | κ, λ | κ, λ |
| Heavy chain | α | δ | ε | γ | μ |
| Concentration in serum (mg/ml) | 3.5 | 0.03 | 0.00005 | 13 | 1.5 |
| Half-life (days) | 6 | 2.8 | 2 | 25 | 5 |
| Complement activation[a] | | | | | |
| Classical | − | − | − | + | ++ |
| Alternative | −/+ | − | + | − | − |

[a] −, no activation; −/+, sporadic weak activation; +, measurable but low activation; ++, moderate activation.

amino-terminal domains of the proteins that participate in antigen recognition (the so-called hypervariable regions). However, unlike the B-cell receptor, which can recognize the epitope as part of an intact folded protein, the T-cell receptor can recognize only a peptide fragment produced by proteolysis. Furthermore, for recognition by the T-cell receptor, the peptide must be bound to MHC cell surface proteins (see below).

When the T-cell receptor on a T cell engages an MHC molecule carrying the appropriate antigenic peptide, a signal transduction cascade that results in new gene expression is initiated. As a result, the stimulated T cell is capable of differentiating to form memory T cells and various effector T cells. The memory T cells have a long life span in the body, whereas the various effector T cells (e.g., helper cells and cytotoxic cells; see below) have considerably shorter life spans.

### T-Helper Cells and Cytotoxic T Lymphocytes

In general, lymphocytes can be distinguished by the presence of specific surface antigens called **cluster of differentiation (CD)** markers (e.g., CD3, CD4, CD8). Over 130 individual CD markers are known and are invaluable in identifying lymphocytes of a particular lineage or differentiation stage. Two well-known subpopulations of T cells are identified by the presence of either CD4 or CD8 surface proteins, which are coreceptors for MHC class II or MHC class I, respectively (Fig. 14.11). CD4$^+$ T cells are generally **T-helper (Th) cells,** and CD8$^+$ T cells are **cytotoxic T lymphocytes (CTLs).** Th cells synthesize cytokines and growth factors that stimulate (help) specific classes of lymphocytes. Cytotoxic T lymphocytes recognize foreign peptides on MHC class I molecules at the cell surface and, when productively engaged, actively destroy the cell presenting the peptides. Mature CTLs play important roles in eliminating virus-infected cells from the body. Production

of these specialized cells is a precisely orchestrated process. T cells are not fully differentiated when they emerge from the thymus. But, as they bind antigen and encounter specific cytokines, they become activated, express new gene products, and differentiate to attain their mature potential. Such cells are said to be activated.

### Th1 and Th2 Cells

Immature Th cells can differentiate into two classes, **Th1** and **Th2** (Fig. 14.12). These two cell types can be distinguished by the cytokines they produce (Table 14.2). Th1 and Th2 cells have distinct functions and respond to antigenic stimulation with a transient burst of characteristic cytokine production that influences the subsequent activation and proliferation of selected immune cells.

Th1 cells promote the cell-mediated proinflammatory response by causing maturation of immature CTLs. They accomplish this in part by producing Il-2 and Ifn-γ. In addition, Th1 cells arm the antigen-presenting cell so that it can communicate with CTLs. We know from in vitro experiments that if Il-12 is present at the time of antigen recognition, immature Th cells will differentiate into Th1 cells. Il-12 also stimulates NK and Th1 cells to secrete Ifn-γ, a cytokine important in increasing the activity of inflammatory cells such as macrophages. Th1 cells are important for controlling most but not all viral infections.

In the presence of Il-4, immature Th cells differentiate into Th2 cells. Th2 cells promote the antibody response by causing maturation of immature B cells and resting macrophages. They also reduce the inflammatory response by producing Il-4, Il-6, and Il-10, but not Il-2 or Ifn-γ. Th2 cells are more active after invasion by extracellular bacteria or multicellular parasites.

B cells produce specific antibody isotypes when stimulated by Th1 cells. One predominant isotype is IgG2a, which actively stimulates the complement cascade and

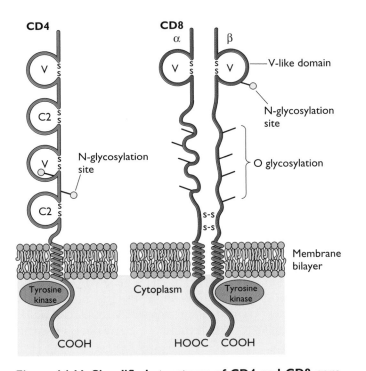

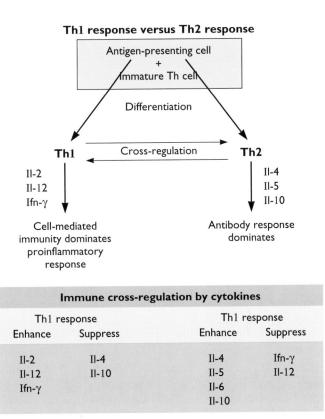

**Figure 14.11 Simplified structures of CD4 and CD8 coreceptor molecules.** These two molecules associate with the T-cell receptor on the surface of T cells. The CD4 molecule is a glycosylated type 1 membrane protein and exists as a monomer in membranes of T cells. It has four characteristic immunoglobulinlike domains labeled here as V and C2. The V domains are similar to the variable domain of immunoglobulin in the tertiary structure. The first two domains fold into a rigid rodlike structure linked to the other two domains by a flexible arm and are thought to be the domains for binding MHC class II proteins. The cytoplasmic domain forms a complex with specific tyrosine kinases, endowing CD4 with signal transduction properties. The CD8 molecule is a type I membrane protein with both N and O glycosylation. It is a heterodimer of an α- and a β-chain associated by disulfide bonds that interacts with MHC class I proteins. The two polypeptides are quite similar in sequence, each having an immunoglobulinlike variable domain (V) thought to exist in an extended conformation. Tyrosine kinases also associate with the CD8 cytoplasmic domain and participate in signal transduction events.

**Figure 14.12 Th cells: the Th1 versus the Th2 response.** Immature Th cells differentiate into two general subtypes called Th1 and Th2, defined functionally according to the cytokines they secrete. Th1 cells produce cytokines that promote the inflammatory response and activity of cytotoxic T cells, and Th2 cells synthesize cytokines that stimulate the antibody response. The cytokines made by one class of helper tend to suppress those of the other class.

binds to Fc receptors of macrophages. In contrast, B cells produce the IgG1 (mice), IgG4 (humans), and IgE isotypes when stimulated by Th2 cells. Such antibodies do not bind complement, nor do they bind to macrophage Fc receptors.

In general, Th1 and Th2 responses have a yin-yang relationship; as one increases, the other decreases (Fig. 14.12). Thus, while Ifn-γ turns up the Th1 response, it also inhibits the synthesis of Il-4 and Il-5 by Th2 cells, effectively dampening the latter response.

An obvious but unanswered question is how particular infectious agents such as viruses induce the secretion of Il-

12, a cytokine that promotes the Th1 response. One idea gaining credence is that CpG sequences, rare in vertebrate DNA but common in the DNA of certain pathogens, are immunostimulatory and thus effect Il-12 induction (see chapter 19). Other signals are bound to exist but remain unidentified. It is likely that many viral proteins influence the Th1/Th2 balance. For example, Epstein-Barr virus and equine herpesvirus type 2 express proteins homologous to Il-10, a regulatory cytokine produced by Th2 cells. These viruses infect B cells in their life cycles and stimulate B-cell development through their Il-10 homolog. In doing so, they foil the Th1 antiviral defense and promote survival of the cells important for their long-term survival.

It is an oversimplification to think of a "pure" Th1 or Th2 response. For most viral infections, a given Th response represents a spectrum of some Th1 and some Th2 cells, and consequently a mixture of cytokines, depending on the infectious agent and the nature of the infection. Establishment of the proper spectrum of the Th cells therefore is an important early event in host defense; an inappropriate response has far-reaching consequences. How this fundamental decision is made is under much scrutiny in laboratories around the world.

## Antigen Presentation and Activation of Immune Cells

T cells distinguish infected cells from uninfected cells by specific receptor-peptide interactions mediated by their T-cell receptors and two types of membrane proteins, the major histocompatibility complexes (MHC class I and MHC class II) on the target cell (Fig. 14.13). MHC class I proteins are found on the surface of nearly all nucleated cells (except neurons, cells of the placenta, and sperm cells at certain stages of differentiation). They are composed of two chains, an α-chain (sometimes called the heavy chain) and β2-microglobulin (the light chain). Lymphocytes possess the highest concentration of MHC class I protein, with about $5 \times 10^5$ molecules per cell. In contrast, fibroblasts, muscle cells, and liver hepatocytes possess much lower quantities, sometimes 100 or fewer per cell. There are three MHC class I loci in humans (A, B, and C) and two in mice (K and D). Because there are many allelic forms of these genes in outbred populations, MHC class I genes are

said to be **polymorphic**. For example, at the human MHC class I locus HLA-B, more than 149 alleles with pairwise differences ranging from 1 to 49 amino acids have been identified. When cells bind cytokines such as Ifn and Ifn-γ, transcription of MHC class I α-chains, β2-microglobulin, and the linked genes for processing peptides (proteasome and peptide transporter genes) is markedly increased.

Expression of MHC class II proteins occurs only in specialized cells, sometimes called **professional antigen-presenting cells**, such as dendritic cells, macrophages, and B cells. Other cell types, including fibroblasts, pancreatic β cells, endothelial cells, and astrocytes, can express MHC class II molecules on exposure to Ifn-γ. Human genomes contain three MHC class II loci (DR, DP, and DQ), and mouse genomes contain two (IA and IE). MHC class II molecules contain two chains called α and β. The human DR cluster contains an extra β-chain gene whose product can pair with the DR α-chain. Thus, the three sets of human genes give rise to four types of MHC class II molecules. As for MHC class I, there are many alleles of MHC class II genes.

The defining property of MHC proteins is that they specialize in binding short peptides and presenting them for recognition by lymphocytes with T-cell receptors on their surfaces. Peptides are bound to both MHC proteins *inside the cell* and are then transported to the cell surface, where they are displayed or presented to T cells. Both classes of MHC protein have a **peptide-binding cleft** that is sufficiently flexible to enable the binding of many peptides (Fig. 14.13 and 14.14). Because the specificity of the peptide-binding site is determined by the allelic differences in the MHC genes inherited by an individual, the ability of MHC molecules to bind and display peptides on the cell surface varies from individual to individual.

In addition to MHC class 1 and MHC class II proteins, mammals express an additional class of MHC proteins called CD1. Humans carry five CD1 genes (CD1a through CD1e), and mice carry two (mCD1.1 and mCD1.2, both related to human CDd). Like conventional MHC proteins, CD1 proteins are antigen-presenting molecules, but the antigens they present are not peptides but rather are lipids and glycolipids. CD1 proteins play important roles in immune regulation by interacting with a specialized subclass of CD4+ T cells called NK1. Mouse CD1.1 is expressed mainly by professional antigen-presenting cells. Moreover, when NK1 T cells engage CD1, they secrete cytokines such as Il-4 that are known to control the differentiation of T lymphocytes. It is thought that CD1 lipid/glycolipid antigen presentation to the NK1 subset of CD4+ T cells is an important part of the preadaptive immune response. Many microbial invaders with distinctive membranes can be recognized by this recently discovered antigen-present-

**Figure 14.13 The MHC class I and MHC class II molecules.** (A) The MHC class I molecule is a heterodimer of a membrane-spanning type I glycoprotein called the α-chain (43 kDa) which is noncovalently associated with β2-microglobulin (12 kDa) that does not span the membrane. The α-chain folds into three domains, 1, 2, and 3. Domains 2 and 3 fold together to form the groove where peptide binds, and domain 1 folds into an immunoglobulinlike structure. (B) The MHC class II molecule is a heterodimer of two membrane-spanning type I glycoproteins called the α-chain (34 kDa) and β-chain (29 kDa). Each chain folds into two domains, 1 and 2, and together the α- and β-chains fold into a structure similar to that of MHC class I. The two amino-terminal domains from α- and β-chains form the groove where peptide binds. Unlike the closed MHC class I peptide groove, the MHC class II peptide-binding groove is open at both ends. The second domain of each chain folds into an immunoglobulinlike structure.

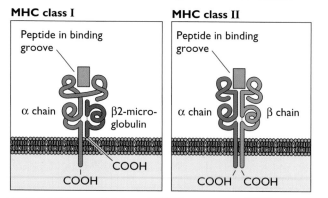

**MHC class I**

Peptide in binding groove

α chain    β2-micro-globulin

COOH
COOH

**MHC class II**

Peptide in binding groove

α chain    β chain

COOH   COOH

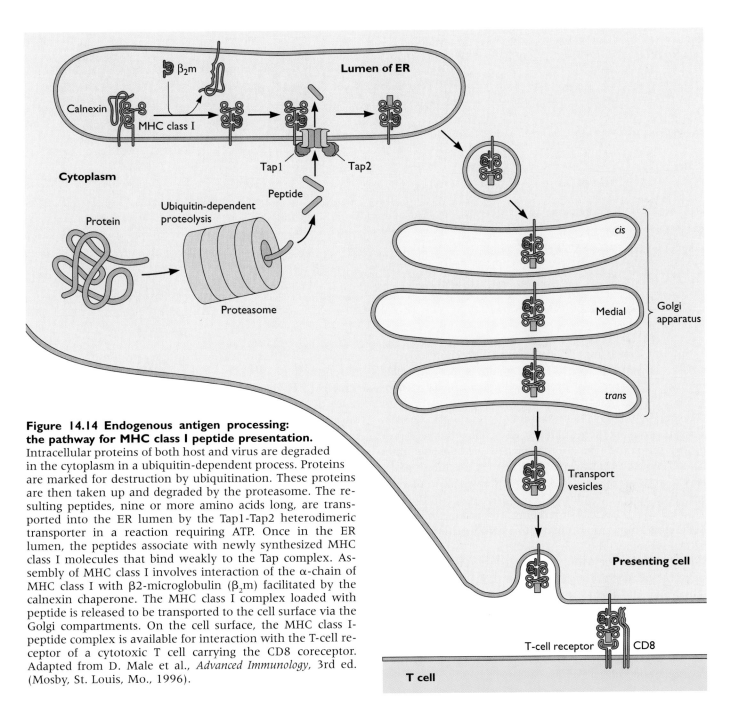

**Figure 14.14 Endogenous antigen processing: the pathway for MHC class I peptide presentation.** Intracellular proteins of both host and virus are degraded in the cytoplasm in a ubiquitin-dependent process. Proteins are marked for destruction by ubiquitination. These proteins are then taken up and degraded by the proteasome. The resulting peptides, nine or more amino acids long, are transported into the ER lumen by the Tap1-Tap2 heterodimeric transporter in a reaction requiring ATP. Once in the ER lumen, the peptides associate with newly synthesized MHC class I molecules that bind weakly to the Tap complex. Assembly of MHC class I involves interaction of the α-chain of MHC class I with β2-microglobulin (β$_2$m) facilitated by the calnexin chaperone. The MHC class I complex loaded with peptide is released to be transported to the cell surface via the Golgi compartments. On the cell surface, the MHC class I-peptide complex is available for interaction with the T-cell receptor of a cytotoxic T cell carrying the CD8 coreceptor. Adapted from D. Male et al., *Advanced Immunology*, 3rd ed. (Mosby, St. Louis, Mo., 1996).

ing system. Nevertheless, the role of the CD1 system in antiviral defense remains to be established.

## Recognition of Infected Cells: the MHC Class I System

The immune system must not only *distinguish* virus-infected cells from uninfected cells but also *destroy* the former and *ignore* the latter. As viruses can infect many cell types, the recognition/destruction system must operate on all cells with high fidelity. As discussed below, recognition and selective killing are accomplished by CTLs that constantly patrol the body and scan the surface of essentially every cell. Infected cells are identified in part because they display small viral peptides on their surfaces via the **endogenous pathway of antigen presentation** (Fig. 14.14).

**BOX 14.3**

*Viruses provide Nobel Prize–winning insight: MHC restriction*

In 1974 at the John Curtin School of Medical Research in Canberra, Australia, Rolf Zinkernagel and Peter Doherty performed a classic experiment that provided insight into how CTLs can recognize virus-infected cells. Initially, they teamed up to determine the mechanism of the lethal brain destruction observed when mice are infected with lymphocytic choriomeningitis virus, an arenavirus that does not directly kill the cells it infects. They anticipated that the brain damage was due to CTLs responding to replication of the noncytopathic virus in the brain.

When they infected mice of a selected MHC type with lymphocytic choriomeningitis virus and then isolated T cells capable of recognizing virus-infected cells, these T cells lysed virus-infected target cells in vitro *only* when the target cells and the T cells were of identical MHC haplotype. Uninfected target cells were not lysed even when they shared identical MHC alleles. This requirement for MHC matching was called **MHC restriction.**

Their prize-winning insight was that a CTL must recognize two determinants on a virus-infected cell; one specific for the virus (a short peptide derived from the virus) and one specific for the MHC of the host. CTLs recognize foreign antigens only when the T-cell receptor engages the foreign peptides bound to MHC class I proteins on the surface of target cells.

**Zinkernagel, R. M., and P. C. Doherty.** 1974. Restriction of in vitro T-cell mediated cytotoxicity in lymphocytic choriomeningitis within a syngeneic or semiallogenetic system. *Nature* (London) **248:**701–702.

In uninfected and infected cells, a fraction of most newly synthesized proteins is broken down in a controlled manner by a large host protease complex called the **proteasome** (Fig. 14.14). The targeted protein is marked for digestion by the covalent attachment of multiple copies of a small protein called ubiquitin and, following ATP-promoted unfolding, the protein is degraded in the inner chamber of the proteasome. The peptide products are released and transported into the endoplasmic reticulum (ER) by a specific peptide transporter system and then assembled with MHC class I proteins. This interaction allows MHC class I molecules to fold into a conformation that can be transported to the Golgi apparatus and then to the cell surface. Hence, the MHC class I pathway displays an "inside-out" picture of the cell to the T cell. Patrolling T cells move over the surface of potential target cells, engaging MHC class I peptide complexes by their T-cell receptors. Normal cells (self) are ignored, but when a viral peptide-MHC class I complex is bound by the T-cell receptor, a series of reactions is triggered such that the CTL becomes activated and kills the infected cell (see below). It is noteworthy that the T-cell receptor has a low affinity, 1 $\mu$M or less, for its peptide-MHC class I ligand.

### Professional Antigen-Presenting Cells Use MHC Class II Proteins

Both antibody and CTL responses are carefully controlled by cytokines produced by Th cells. Precise regulation is achieved in part by a specific antigen recognition process distinctly different from the MHC class I antigen presentation system described above. The Th cell is activated only when the peptide antigen is presented on the surfaces of the professional antigen-presenting cells defined previously. Antigen-presenting cells are mobile (Table 14.1; Fig. 14.7) and are characterized by the presence of MHC class II glycoproteins on their surfaces. Viral antigens are delivered to these cells from the outside as antigen-antibody complexes, as soluble proteins, as virus particles, or as infected cells in the circulation. In addition to external delivery, many viral membrane proteins can be internalized by endocytosis from the cell surface to intracellular compartments containing MHC class II proteins. The antigen-presenting cells then migrate to the regional lymph node and present their information to the T and B cells waiting for instructions.

External viral proteins and particles or newly synthesized viral membrane proteins on the plasma membrane are internalized and degraded, and peptides are subsequently displayed on the membranes of these antigen-presenting cells by a process called **exogenous antigen presentation** (Fig. 14.15). Such presentation need not require active viral infection of the presenting cell. Rather these cells internalize virus particles and proteins by phagocytosis or endocytosis. They then process the viral proteins in lysosomelike compartments to produce small peptides from the viral proteins. The peptides then bind to the MHC class II molecule, and MHC class II-peptide complexes are transported to the cell surface, where they are available to interact with appropriate T cells. Interaction of the T-cell receptor on the Th cell with the MHC class II-peptide complex on antigen-presenting cells induces a series of membrane changes in the Th cell, leading to its activation (Fig. 14.7). In many instances, full activation requires the interactions of other surface proteins and co-

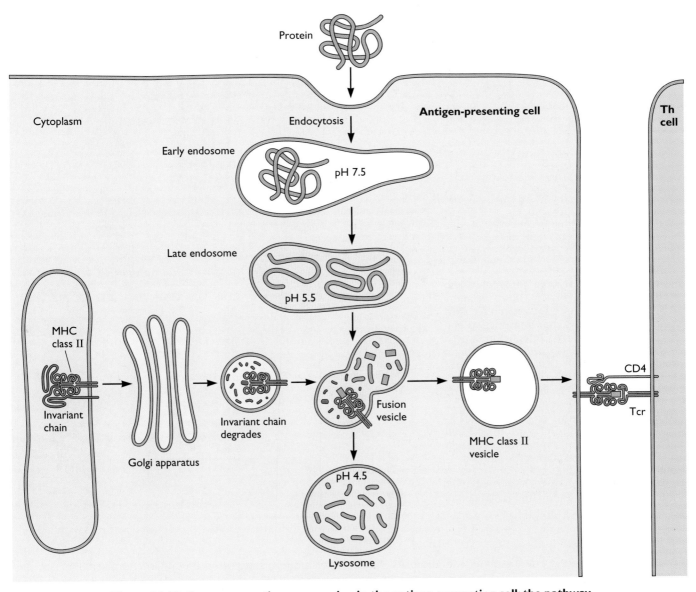

**Figure 14.15 Exogenous antigen processing in the antigen-presenting cell; the pathway for MHC class II peptide presentation.** (Left) Peptides in the ER lumen of the antigen-presenting cell are prevented from binding to the MHC class II peptide groove by association of a polypeptide called the invariant chain with MHC class II molecules. The complex is transported through the Golgi compartments to a post-Golgi vesicle, where the invariant chain is removed from the MHC class II heterodimer by proteolysis. This reaction activates the MHC class II complex to accept peptides. The peptides are not derived from endogenous proteins but rather from extracellular proteins that enter the antigen-presenting cell. In some antigen-presenting cells, the protein enters by endocytosis (top). In the case of B cells, antigens are captured by the surface antibodies. In both cases, the antigens are internalized to early endosomes with neutral luminal pH. Endosome vesicles traveling to the lysosome via this pathway are characterized by a lowering of pH as they "mature" into late endosomes. The lower pH activates vesicle proteases which degrade the exogenous protein into peptides. The internalized endosomes with their peptides fuse at some point with the vesicles containing activated MHC class II. The newly formed peptide-MHC class II complexes then become competent for transport to the cell surface. On the cell surface, the MHC class II-peptide complex is available for interaction with the T-cell receptor (Tcr) of a Th cell carrying the CD4 coreceptor. Adapted from D. Male et al., *Advanced Immunology*, 3rd ed. (Mosby, St. Louis, Mo., 1996).

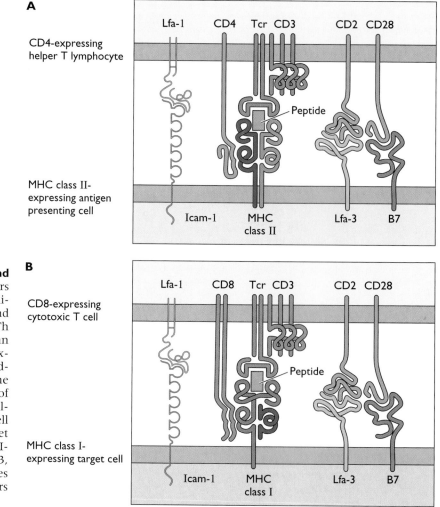

**A**

CD4-expressing helper T lymphocyte

MHC class II-expressing antigen presenting cell

**Figure 14.16 T-cell surface molecules and ligands.** The interactions of these receptors and ligands are important for antigen recognition and initiation of signal transduction and other T-cell responses. (A) Interaction of a Th cell expressing the CD4 coreceptor with an antigen-presenting cell (APC). The APC exhibits an MHC class II-peptide complex in addition to Icam-1, Lfa-3, and B7 membrane proteins. These complexes all are capable of binding cognate receptors on the Th cell as illustrated. (B) Interaction of a cytotoxic T cell expressing the CD8 coreceptor with its target cell. The target cell exhibits an MHC class I-peptide complex in addition to Icam-I, Lfa-3, and B7 membrane proteins. These complexes all can be recognized and bound by receptors on the cytotoxic T cell as illustrated.

**B**

CD8-expressing cytotoxic T cell

MHC class I-expressing target cell

stimulators (Fig. 14.16). For example, the stimulus that enables the antigen-presenting cell to activate a CTL is thought to include the binding of CD40 ligand (CD40L) expressed on the antigen-presenting cell to CD40 receptor present on T cells. Another necessary interaction is the binding of costimulator B7 protein on the antigen-presenting cell to CD28 on the T cell.

Th cells activated in this fashion produce Il-2, as well as a high-affinity receptor for this cytokine. The secreted Il-2 binds to the new receptor to induce autostimulation of the Th cell, leading to proliferation. Such **clonal expansion** of specific Th1 or Th2 cells then promotes the activation of CTLs and B lymphocytes, which go on to develop the cell-mediated response and antibody response, respectively (Fig. 14.6).

### The Cell-Mediated Adaptive Response

While antibodies can neutralize free virus particles in the circulation and can exert antiviral effects in certain labora-

tory models, they often have modest or no effect in natural infections caused by viruses that spread by cell-to-cell contact (e.g., neurotropic viruses like alphaherpesviruses) or by viruses that infect circulating immune cells (e.g., lentiviruses and paramyxoviruses). In these cases, to recover from a viral infection, it is essential to eliminate virus-infected cells without damage to uninfected cells. This is usually accomplished by the cell-mediated response (Fig. 14.6 and 14.7). In contrast to humoral or antibody responses, in which a specific secreted antibody binds to a virus or the surface of an infected cell, an important function of the cell-mediated response requires binding of a CTL to an infected cell.

### *Cytotoxic T Lymphocytes*

CTLs are superbly equipped to kill virus-infected cells, and once they complete one killing, they can detach and kill again. Most CTLs express the CD8 surface protein as a coreceptor and recognize foreign peptides bound to MHC class I proteins on the cell surface. CTLs are not completely differentiated when they leave the thymus. Accordingly, they can recognize antigen by their T-cell receptor, but they cannot lyse target cells. For realization of their full potential, CTLs require at least three additional reactions, namely, interaction of a T-cell receptor with a foreign antigen presented by MHC class I molecules, binding additional surface proteins on the potential CTL to those on the infected cell, and binding of specific cytokines produced by Th1 cells in the vicinity. Signaling through the T-cell receptor when it engages the peptide antigen-MHC complex also requires clustering (aggregation) of a number of T-cell receptors. Such signaling is stimulated strongly by clustering of the CD8 coreceptors as well. Only after these reactions have taken place does the CTL become capable of lysing an infected cell.

CTLs kill by two mechanisms: first, exocytosis of the contents of cytoplasmic granules from the CTL and transfer to the target cell, and second, induction of apoptosis. These killing systems are formed during the differentiation process. The maturing CTL fills with cytoplasmic granules that contain macromolecules required for target cell lysis, such as **perforin**, a membrane pore-forming protein, and **granzymes**, members of a family of serine proteases. Granules are released by CTLs when in direct membrane contact with the target cell and are taken up by the target cell via receptor-mediated endocytosis. Perforin, as its name implies, makes holes in the endosomal membrane, allowing the release of granzymes, whose actions result in apoptosis of the infected cell. Activated CTLs also can induce apoptotic cell death via binding of the Fas ligand on their surfaces to the Fas receptor (CD95) on target cells. Many activated CTLs also secrete Ifn-$\gamma$, which, as we have seen, is a potent inducer both of the antiviral state in neighboring cells and MHC class I and II glycoprotein expression. Activated CTLs also secrete powerful cytokines, such as Il-16, and chemoattractant cytokines (chemokines) such as Rantes (regulated on activation, normal T-cell expressed and secreted) protein. Their release by virus-specific CTLs following recognition of an infected target cell may assist in coordination of the antiviral response. Some chemokine ligands and their receptors are listed in Table 14.3. In some cases, molecules secreted by CTLs reduce viral gene expression in infected cells without lysing them. This mechanism is thought to be important in controlling human immunodeficiency virus infections and has been proposed to facilitate establishment of herpes simplex virus latent infections in peripheral sensory ganglia.

The primary interaction of MHC-peptide with the T-cell receptor is often called signal 1. A second signal produced by binding of accessory molecules on the CTL and appropriate ligands on the infected cell is also required for efficient activation (Table 14.9) (Fig. 14.16). In the absence of such second signals, interaction of MHC-peptide with T-cell receptor can induce apoptosis in the T cell, or **anergy**, a nonresponsive state. One required second interaction is that of CD28 protein on the CTL with a surface protein called B7 on the target cell. Other examples of accessory interactions include those that facilitate adhesion of T cell and target cell to one another, such as binding of CD2 or leukocyte function antigen 1 (Lfa-1) on the CTL with Lfa-3 or Icam-1, respectively, on the infected cell. Viral infection can affect the abundance of these accessory molecules on the surfaces of infected cells and thus alter CTL recognition.

Typically, for a cytopathic viral infection, CTL activity appears within 3 to 5 days after infection, peaks in about a week, and declines thereafter (see Fig. 14.20). The magnitude of the CTL response depends on variables such as virus concentration, route of infection, and age of the host. The type of viral infection is also an important parameter. For example, some noncytopathic viruses evoke a delayed or reduced CTL response. Some viruses are able to confound CTLs by disguising or reducing antigen presentation by MHC class I molecules.

The critical role of CTLs in antiviral defense is demonstrated by adoptive transfer experiments in which virus-specific CTLs from an infected animal can be shown to confer protection on unimmunized recipients. However, despite their importance, CTLs can also cause direct harm (immunopathology) by large-scale cell killing. This is often the case with noncytopathic virus infections, when cells can be infected yet still function. For example, the liver damage caused by hepatitis viruses is actually due to CTL killing of persistently infected liver cells. Other examples of immunopathology are discussed in chapter 17.

Interestingly, recent studies suggest that astrocytes, the most abundant glial cells in the brain, express the cytolytic protein perforin. Perforin was once believed to be found exclusively in mature CTLs. It is thought that although the brain has reduced contact with the peripheral immune system, astrocytes may use CTL-like killing mechanisms (perforin and Fas-Fas ligand) to destroy pathogens in the

**Table 14.9** Representative ligand-receptor interactions important in CTL activation

| Molecules on the T cell | Ligands on the antigen-presenting cell |
| --- | --- |
| CD4, T-cell receptor | MHC class II |
| CD8, T-cell receptor | MHC class I |
| Il-1 receptor | Il-1 (soluble) |
| CD28/Ctla4 | CD80 (B7.1), CD86 (B7.2) |
| Lfa-1 | CD54 (Icam-1), CD102 (Icam-2) |
| CD2 | Lfa-3 |

brain. These CTL-like activities may contribute to the loss of glia and neurons in many neuropathologies.

### T-Cell-Mediated, Delayed-Type Hypersensitivity

Activated Th1 cells and macrophages that have accumulated at a site of viral infection respond by synthesis of cytokines, including Il-2 and Ifn-γ. These cytokines act locally to increase capillary permeability, a reaction that is responsible in part for a characteristic local reaction referred to as **delayed-type hypersensitivity (DTH)**. This response is typically measured in a previously infected mouse by an ear-swelling assay. Viral antigen is injected intradermally on the dorsal pinna of a mouse ear, while the other ear is injected similarly with control antigen. Twenty-four hours later, the antigen-injected ear becomes inflamed and swells as a result of DTH, but the control-injected ear does not. It is known from the following adoptive transfer experiment that DTH is due to T cells. If nonimmune animals are injected with pure activated T cells from a virus-infected animal and the animal is then given an intradermal inoculation of viral antigen, a lesion rich in macrophages and lymphocytes is seen 24 h later at the site of injection. This reaction is responsible for many virus-promoted rashes and lesions with fluid-filled vesicles.

### Antibody-Dependent Cell-Mediated Cytotoxicity: Specific Killing by Nonspecific Cells

Macrophages and some NK cells carry on their surfaces antibody receptors (called **Fc receptors**) that bind to the carboxy-terminal, more conserved region of an antibody molecule. If an antiviral antibody is bound in this manner, the antigen-binding site at its amino-terminal region is still free to bind viral antigen on the surface of the infected cell. In this way, the antiviral antibody can bring the NK cell or macrophage to the infected cell for elimination. This process is called **antibody-dependent cell-mediated cytotoxicity (ADCC)**. The antibody provides the specificity, but the killing cell is nonspecific. While ADCC is well documented in vitro, its importance in controlling in vivo viral infections is uncertain.

## The Antibody Response

### Specific Antibodies Are Made by Activated B Cells Called Plasma Cells

Once the B cell emerges from the bone marrow into the circulation and travels to lymph and lymphoid organs, it must be activated by binding its specific antigen. Antigen must bind to several antibody receptors on the B cell simultaneously, clustering (aggregating) the receptors in the

---

**BOX 14.4**

*Transgenic mouse models prove useful in defining an antiviral role of CTLs*

The use of mice deficient in immune defense genes required for CTL killing (specific granzymes, perforin, or Fas), B-cell function (various immunoglobulin regulatory genes), and Ifn requirements (selected STAT signaling genes) has provided a wealth of information on the roles of these gene products in immune defense. The only requirement is that the virus of interest must be able to infect the transgenic mouse.

For some noncytopathic viruses such as lymphocytic choriomeningitis virus that can cause a lethal and persistent infection of mice, perforin-mediated killing, but not granzyme-mediated killing or Fas-mediated killing, is an important component of host CTL defense. These studies also demonstrate that the perforin activity of lymphocytic choriomeningitis virus-specific CTLs is responsible not only for the clearance of virus but also for the onset of pathology associated with disease. In contrast, granzyme-deficient mice recover completely from viral infection. One provocative finding of these studies was that when lymphocytic choriomeningitis virus was not cleared in perforin-defective mice, the resulting persistent infection led to a lethal over-

production of cytokines, including Tnf-α and Ifn-γ.

Transgenic models also demonstrate an unexpectedly limited role for the killing functions of CTLs in defense against certain cytopathic viruses. Vaccinia virus and vesicular stomatitis virus pathogenesis was not affected by lack of CTL killing systems, but rather recovery from primary infection depended largely on production of neutralizing IgG antibodies. In fact, cytokines secreted by CD4 T-helper cells and CD8 T cells are required for early immune defense but not for recovery from infection by many different cytopathic viruses (including vesicular stomatitis virus, herpes simplex virus, influenza virus, and Semliki Forest virus).

One way to think about this conclusion is that recognition and lysis of infected cells by CTLs may be too slow in acute cytopathic infections to be effective in primary defense. For such infections, the rapid innate defenses and early cytokine responses of lymphocytes are the key front-line systems.

**Yeung, R. S. M., J. Penninger, and T. W. Mak.** 1994. T cell development and function in gene knockout mice. *Curr. Opin. Immunol.* **6:**298–307.

membrane. Such receptor clustering activates a signal transduction cascade through Src family tyrosine kinases which associate with the cytoplasmic domains of the clustered receptors. A B-cell coreceptor, such as CD19, CD21, or CD81, is thought to enhance signaling by recruiting tyrosine kinases to the clustered coreceptor and antigen receptors. When antigen-binding and coreceptor interactions occur in the presence of appropriate cytokines made by Th2 cells and growth factors made by macrophages, the signal transduction cascade causes the immature B cell to produce a variety of new gene products and to differentiate into antibody-producing plasma cells and memory B cells (Fig. 14.7) (Tables 14.2 and 14.7).

The B cell itself is an antigen-presenting cell using the MHC class II system and exogenous antigen processing (Fig. 14.15). Antigen bound by the membrane-anchored antibody molecule on B cells is internalized by the endosomal pathway.

When the T-cell receptor of Th2 cells recognizes antigen bound to MHC class II proteins on the B-cell surface, the Th2 cell produces a locally high concentration of stimulatory cytokines as well as CD40 ligand (a protein homolo-gous to Tnf). The engagement of CD40 ligand with CD40 (a Tnf receptor family member) on the surface of B cells facilitates the local exchange of cytokines that further stimulates proliferation of the activated B cell and promotes its differentiation, described in a previous section. B-cell-activating cytokines produced by the Th2 cell are Il-2, Il-4, Il-6, and Ifn-γ. The mature plasma cells make prodigious amounts of antibody: the rate of synthesis for IgG can be as high as 30 mg per kg of body weight per day (Table 14.8).

### Antibodies

Antibodies (immunoglobulins) have the common structural features illustrated in Fig. 14.17. Five classes of immunoglobulin, IgA, IgG, IgD, IgE, and IgM, are defined by their distinctive heavy chains, α, γ, δ, ε, and μ, respectively. Their properties are summarized in Table 14.8. IgG, IgA, and IgM are commonly produced after viral infection. During plasma cell differentiation, "switching" of the constant-region antibody genes occurs by recombination and is regulated in part by specific cytokines. Consquently, the mature cell now produces a specific type of antibody.

---

## BOX 14.5

### Identifying virus-specific T cells

Since the discovery of T cells and their unique antigen receptor, scientists have tried with little success to identify individual T cells based on their unique antigen recognition properties. This failure has been attributed to the low affinity and fast off rates of the MHC-peptide complex and T-cell receptor. Without this information, it was difficult, if not impossible, to quantitate antigen-specific T-cell responses. T cells are routinely identified by function, such as their ability to kill cells, but not by their peptide specificity.

A new advance that is changing our ideas about T cells is the use of an artificial **MHC tetramer** as an antigen-specific, T-cell-staining reagent. The extracellular domains of MHC class I proteins are produced in *Escherichia coli*. These engineered MHC class I molecules have a unique C-terminal 13-amino-acid sequence that enables them to be biotinylated by an enzyme called BirA. The truncated MHC class I proteins are folded in vitro with a synthetic peptide that will be recognized by a specific T cell. Tetrameric complexes are purified after biotinylation and reaction with a streptavidin derivative. Isolated T-cell pools from virus-infected animals are reacted with this reagent, and individual T cells that bind the biotinylated MHC class I-peptide com-plex are detected by a variety of immuno-histochemical techniques. Cell-sorting techniques can be used, and stained cells are viable.

With this new technology, several groups have found that there is a massive CD8+ T-cell expansion after acute primary infections by viruses such as lymphocytic choriomeningitis virus (LCMV) and Epstein-Barr virus. For example, more than 50% of CD8+ T cells from the spleen of an LCMV-infected mouse were specific for a single LCMV peptide. The response reached a maximum 8 days after infection, but up to 10% of tetramer-staining cells were still detectable after a year.

The new technology raises a number of questions. Why is there such a large primary expansion of particular CD8 T cells? Is it necessary or simply an overreaction? How is this rapid expansion regulated? Is this response typical of all primary viral infections? What is the relationship of T-cell function (e.g., cell killing) to peptide specificity? Do antigen-specific cells reflect the antiviral potential of the expanded population? For now, this assay has shed unexpected light and raised important questions concerning CD8+ T-cell responses to viruses.

**Doherty, P. C.** 1998. The numbers game for virus-specific CD8+ T cells. *Science* **280**:227.

During the **primary antibody response**, which follows initial contact with antigen or viral infection, the antibodies produced by activated B cells follow a characteristic course over time. IgM appears first, followed by IgA on mucosal surfaces or IgG in the serum. IgG is the major antibody of the response and is remarkably stable in the serum, with a half-life of 7 to 21 days. Specific IgG molecules remain detectable in the serum for years, reflecting the presence of memory B cells. A subsequent challenge with the same antigen or viral infection promotes a rapid antibody response, the **secondary response** (Fig. 14.8).

### Virus Neutralization by Antibodies

While the cell-mediated response clearly plays an important role in eliminating virus-infected cells and clearing such an infection, the antibody response also contributes to containing many viral infections. Any virus that infects mucosal surfaces is likely to encounter secretory IgA antibodies. Similarly, any virus that spreads in the blood (a viremia, or hematogenous spread) (chapter 17) is likely to encounter circulating IgG and IgM antibody molecules. We know that antibodies are an important defense against many viruses (e.g., rabies virus, vesicular stomatitis virus, and enteroviruses) from experiments with immunodeficient-animal models and through passive immunization, in which animals are protected by injection with virus-specific antiserum or purified monoclonal antibodies. Additional confirmation that antibodies affect the outcome of a virus infection comes from studies on eastern and western equine encephalitis viruses (alphaviruses in the family

**Figure 14.17 The structure of an antibody molecule.** This is a schematic representation of an IgG molecule delineating the chain and domain structures. The light and heavy chains are indicated in colors, as are the disulfide bonds that hold them together (yellow bars). The variable regions of the heavy ($V_H$) and light ($V_L$) chains, as well as the constant regions of the heavy ($C_H$) and light ($C_L$) chains, are indicated on the left part of the molecule. The hypervariable regions and invariable regions of the antigen-binding domain (Fab) are emphasized. The constant region (Fc) harbors many important biological functions, including complement binding (activation of the classical pathway) and binding to Fc receptors found on macrophages and other cells. Clusters of papain protease cleavage sites are indicated, as this enzyme is used to define the Fab and Fc domains.

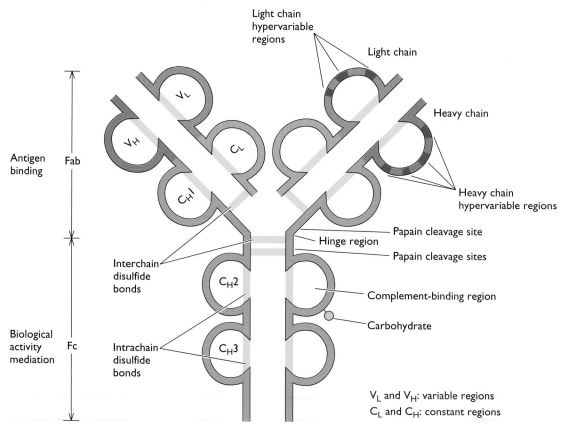

*Togaviridae*). Here, the speed with which virus-specific antibody appears after infection correlates well with the ability of patients to recover from encephalitis; those with low antibody titers are more likely to die.

The type of antibody produced can significantly influence the outcome of a viral infection. For example, while poliovirus infection stimulates strong IgM and IgG responses in the blood, it is the mucosal IgA that is vital in defense against this virus. IgA can neutralize poliovirus directly in the gut, the site of primary infection. The live attenuated Sabin poliovirus vaccine is effective because it elicits a strong mucosal IgA antibody response.

IgA is synthesized by local plasma cells that underlie the mucosal epithelium. This antibody is secreted as dimers of two conventional immunoglobulin subunits. These IgA dimers bind the polymeric immunoglobulin receptor on the basolateral surface of the epithelial cells lining the mucosal membrane (Fig. 14.18). This complex is then endocytosed and transcytosed across the cell to the apical surface. Protease cleavage of the receptor (the so-called secretory component) releases dimeric IgA into mucosal secretions, where it is free to interact with incoming virus particles.

Antibody neutralization of virus particles is usually demonstrated in vitro when binding of virus-specific antibodies to a virus leads to a reduction in the number of infectious particles. Enveloped viruses often can be destroyed in a test tube through membrane disruption by antiviral antibodies and serum complement. On the other hand, in the absence of complement, it is not always clear how an antibody can neutralize a virus. Clearly, antibodies can bind to a virion and block it from productive interaction with receptors (Fig. 14.19). Antibodies can also cause virus particles to aggregate and thus reduce the concentration of infectious particles. The combination of antibodies and complement can activate the classical complement pathway to lyse enveloped viruses. Nonneutralizing antibodies to viruses are also prevalent after infection; they bind specifically to virus particles but do not appear to interfere with viral infectivity.

In general, *many antibody molecules can bind to a single virus particle to effect neutralization* (Fig. 14.19). Much of what we know about antibody neutralization in vitro comes from the isolation and characterization of "antibody escape" mutants or monoclonal antibody-resistant (MAR) mutants. These viruses are selected by their ability to grow in the presence of neutralizing antibody. Analysis of the mutant viruses allows a precise molecular definition, not only of antibody binding sites but also of regions involved in fusion or uncoating of the virus particle.

While it is straightforward to study neutralization mechanisms in vitro, it is more difficult to do so in an infected host, and seemingly enigmatic results that do not correlate with those from in vitro neutralization experiments have been obtained. For example, it is often possible to protect mice from lethal viral infection in passive-immunity experiments in which nonneutralizing antiviral antibodies are administered. An obvious inference from these experiments is that the role of antibodies in antiviral defense is broader than simply interfering with virus binding to cells.

Antibodies can provoke other remarkable responses in virus-infected cells. For example, antibodies to the surface proteins of many enveloped viruses (e.g., alphaviruses, paramyxoviruses, arenaviruses) can clear infectious virus from persistently infected cells by a noncytolytic process (complement independent). In this way, they act synergistically with Ifn. Virus-specific antibodies bound to the surface of infected cells can inhibit virion budding at the plasma membrane and can also affect expression of viral membrane proteins. Antiviral antibodies injected into the peritoneal cavities of mice can block the neurotropic transmission of reovirus, poliovirus, and herpesvirus to the brain, even when the virus is injected at sites far removed from the peritoneum. The molecular bases for these intriguing effects remain elusive.

Recently, an interesting activity of IgA antibodies, neutralization of virus infectivity *inside* the infected mucosal epithelial cell, was observed (Fig. 14.18). Because IgA must pass through the mucosal epithelial cell en route to secretion from the apical surface, it is available during transit for interaction with viruses that may be replicating within the cell. The antigen-binding domain of intracellular IgA lies in the lumen of the endoplasmic reticulum, the Golgi compartment, and any transport vesicles of the secretory system. It is therefore available to bind to the external domain of any type I viral membrane protein that has the cognate epitope of that IgA molecule. Such interactions have been demonstrated with Sendai virus and influenza virus in tissue culture infections. Two-color immunofluorescence analysis established that IgA antibodies colocalized with viral antigen only when the IgA could bind the particular virus envelope protein. These studies suggested that intracellular neutralization of virus by IgA antibody molecules is possible. If this assertion is correct, mechanisms of clearing viral infection of mucosal surfaces need not be limited to the lymphoid cells of the adaptive immune system.

## Perspectives

The regular occurrence of viral disease tells us that despite substantial host defenses, viral offensive countermeasures can be remarkably effective. Given the complexity of the interactions, it is useful to outline a hypothetical general

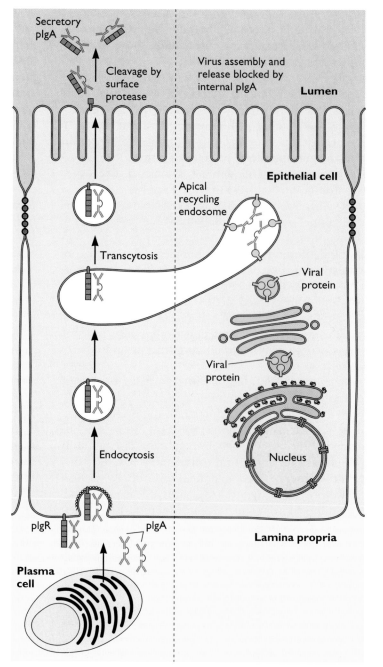

**Figure 14.18 Secretory antibody, IgA, plays a key role in antiviral defense at mucosal surfaces.** (Bottom) Antibody-producing B cells (plasma cells) in the lamina propria of a mucous membrane specifically secrete the IgA antibody (also called polymeric IgA or pIgA). pIgA is a dimer, joined at its Fc ends. For IgA to be effective in defense, it must be moved to the surface of the epithelial cells that line body cavities. This process is called transcytosis and begins by the binding of pIgA to its receptor (pIgR) on the basolateral surface of an epithelial cell. The IgA-receptor complex is taken into the epithelial cell by endocytosis and then delivered to vesicles that transport the complex to the apical surface. Once on the apical surface, pIgA is released by cleavage of the receptor by a surface protease which releases IgA antibody into the lumen. A portion of the receptor remains bound to the dimeric IgA antibody (the so-called secretory component). The dimeric IgA antibody can bind antigens at many steps along this pathway. If antigen is bound before the IgA enters the epithelial cell, antigen-pIgA-pIgR complexes can be transported through the cell to the lumen. If virus is in the lumen, free IgA can bind and block infection at the apical surface. Interestingly, virus infecting an epithelial cell potentially can be bound by *internal* IgA if virus components intersect with the IgA in the lumen of vesicles during transcytosis. This is likely to occur for enveloped viruses, as their membrane proteins are processed in many of the same compartments as those used during transcytosis. Adapted from M. E. Lamm, *Annu. Rev. Microbiol.* **51:**311–340, 1997, with permission.

host defense and viral offense for a typical viral infection. The following description is highly generalized to facilitate discussion. Many variations are possible, depending on such parameters as the site of infection, the mechanism of cell-to-cell spread of the virus, the virulence and cytopathic nature of the virus, physiology of the host, host genetic variations, and environmental conditions (chapter 15) (Fig. 14.20).

To initiate the primary infection, a virus must broach the host's physical barriers and enter a permissive cell at one of the many apertures of the body. Virus replication begins while the cell often counters by synthesizing Ifn or by initiating apoptosis. The virus responds by expressing anti-Ifn or antiapoptosis gene products. Progeny virus particles are released from the infected cells to infect other cells in which similar reactions occur. If the newly infected

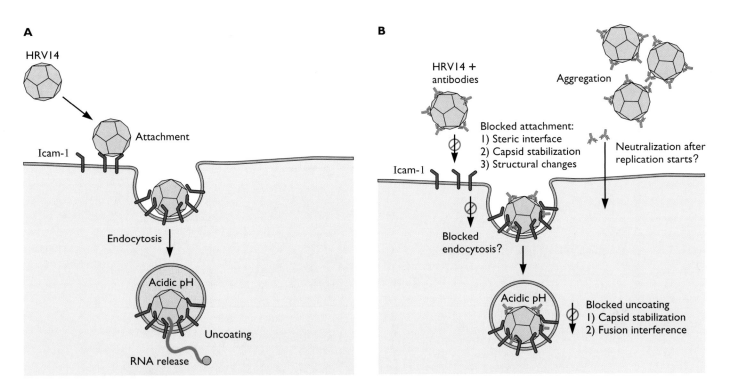

**Figure 14.19 Human rhinovirus 14 (HRV14) infection and potential interactions with antibodies.** (A) The normal route of HRV14 infection. The virus attaches to the Icam-I receptor and enters by endocytosis. As the internal pH of the endosome decreases, the particle uncoats and releases its RNA genome into the cytoplasm. (B) Possible mechanisms of neutralization of HRV14 by antibodies. With well-characterized monoclonal antibodies, at least three modes of antibody neutralization have been proposed and are illustrated: (1) binding of antibody molecules to virus causes steric interference with virus receptor binding; (2) antibodies bound to virus fix the capsid in a stable conformation so pH-dependent uncoating is not possible; (3) antibody molecules binding to the capsid can alter the capsid structure, affecting other steps in entry and uncoating. In addition, antibody can aggregate virus particles (upper right), affecting concentration of virus particles and facilitating their destruction by phagocytes. Antibody itself may be taken up by endocytosis and interact with virus in the cell after infection starts. Adapted from T. J. Smith et al., *Semin. Virol.* **6**:233–242, 1995, with permission.

cells have already bound Ifn produced by nearby infected cells, protein synthesis is inhibited when the virus makes dsRNA. Soon thereafter, NK cells can recognize and kill the infected cells because of new surface antigens and a low or aberrant display of MHC class I proteins on the cell surface. The Ifn produced by infected cells stimulates the NK cells to intensify their activity. In some cases, serum complement can be activated through the alternative pathway and the resulting complement cascade can destroy enveloped viruses and infected cells. Some viruses counter complement by expressing proteins that bind complement components. The innate defenses generally bring most viral infections to an uneventful close by limiting the extent of viral replication before the adaptive response is required.

If viral replication outpaces the innate defense, increased Ifn expression elicits a more global host response, and flulike symptoms (fever, anorexia, fatigue, aches) are noted by the infected individual. As viral replication continues, viral antigens reach the lymph nodes or spleen via blood or lymph or encounter lymphoid tissue in the tonsils or Peyer's patches in the gut. Viral particles, processed viral antigens on migratory macrophages, and monocytes interact with antigen-presenting cells (dendritic cells), which then migrate to regional lymph nodes to establish sites of information exchange with the cells of the adaptive response. Dendritic cells may engulf infected cells dying of apoptosis and present antigens that stimulate CTLs. T-cell recirculation is shut down because of the massive recruitment of lymphocytes into lymphoid tissue. The swelling of

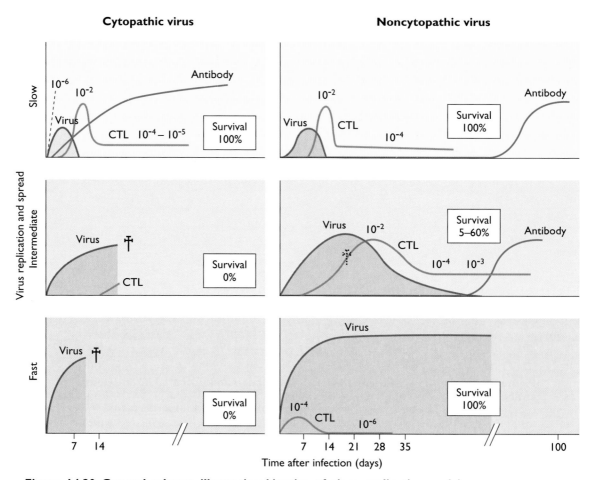

**Figure 14.20 General schema illustrating kinetics of virus replication and immune responses.** Three infections by two types of viruses are illustrated. The measurement of virus replication and spread on the $y$ axes with respect to time after infection on the $x$ axes is illustrated. The three infections are characterized as slow, localized spread (top), intermediate spread (middle), and fast, disseminated spread (bottom). A cytopathic virus (left) is compared to a noncytopathic virus (right). The percentage of animals that survive each infection is indicated in the box. The relative virus titers are depicted by the red lines, and the shaded areas under these curves indicate the period of disease. The immune responses as measured by CD8$^+$ T cells (CTL; green lines) or neutralizing antibody titers (blue lines) are superimposed on the graphs. The frequencies of CTL precursors in the spleen are noted at $10^{-2}$ to $10^{-6}$ values, where $10^{-2}$ indicates that there was 1 CTL in 100 spleen cells. The death of animals is indicated by a cross. The point to be noted is that the kinetics of virus replication and the kinetics of the defensive response both affect the outcome. For example, in this hypothetical scheme only the virus spread kinetics were changed from slow to fast, but the overall interaction between virus and immune response can also vary similarly if immune responses shift from fast and disseminated to slow and localized. Adapted from R. M. Zinkernagel, *Science* **271**:173–178, 1996, with permission.

lymph nodes so often characteristic of infection is a sign of this action.

Within days, Th cells and CTLs appear; these are the first signs of activation of the adaptive immune response. Th cells produce cytokines that begin to direct the amplification of the adaptive response. The antibody response quickly follows with the synthesis first of IgM and then of

other isotypes, including IgG, IgA, and IgE. The prevalence of these isotypes is governed by the route of infection and the pattern of cytokines produced by the Th cells. As the immune response is amplified, CTLs kill infected cells and antibodies bind to virus particles and infected cells. The specific antibody-virus complexes can be recognized by macrophages and NK cells to induce ADCC and can also activate the classical complement pathway. Both processes lead to the specific killing of infected cells and enveloped viruses by nonspecific cells.

A strong inflammatory response often occurs as infected cells die and innate and adaptive responses develop. Both systems release cytokines, chemotactic factors, and vasodilators at the site of infection. Soluble factors, cell death, and invading cells, including macrophages, neutrophils, eosinophils, and various complement components, all contribute to produce the swelling, redness, heat, and pain characteristic of the inflammatory response. Many viruses express proteins that abrogate the action of specific cytokines and modulate the inflammatory response and the subsequent activation of immune cells.

If virus spreads from the primary site of infection (e.g., by viremia or by cell-to-cell spread), second and third rounds of infection can occur in other organs. As T cells are activated, they can cause DTH (usually demonstrated by a characteristic rash or lesion) at the sites of later replication. Immunopathology, particularly after infection by noncytopathic viruses, is not uncommon. It is caused by the cytokines, antibodies, and cells mobilized by the viral infection and, in some infections, can contribute to the severity of the disease.

Finally, the combination of innate and adaptive responses clears the infection, and the host is immune to further infection because of the presence of memory T and B cells and long-lived antibodies. The high concentrations of immune lymphocytes drop dramatically as these cells die and the system returns to normal. Some viruses can avoid the adaptive response completely or in part by infecting organs or tissues that have no or poor immune responses, by avoiding detection because of a high mutation rate and generation of new variants, or by spreading within cells so that they are never exposed to effector cells or antibodies. These processes are discussed in chapter 15.

# References

*Textbooks*

Janeway, C. A., and P. Travers. 1997. *Immunobiology: the Immune System in Health and Disease*, 3rd ed. Garland Publishing, Inc., New York, N.Y.

Male, D., M. Cooke, J. Owen, J. Trowsdale, and B. Champion. 1996. *Advanced Immunology*, 3rd ed. The C. V. Mosby Co., St. Louis, Mo.

*Reviews*

Bendelac, A., M. N. Rivera, S.-H. Park, and J. H. Roark. 1997. Mouse CD1-specific NK1 T cells: development, specificity and function. *Annu. Rev. Immunol.* **15:**535–562.

Biron, C. A. 1997. Activation and function of natural killer cell responses during viral infections. *Curr. Opin. Immunol.* **9:**24–34.

Brutkiewicz, R. R., and R. M. Welsh. 1995. Major histocompatibility complex class I antigens and the control of viral infections by natural killer cells. *J. Virol.* **69:**3967–3971.

Darnell, J. E. 1997. STATs and gene regulation. *Science* **277:**1630–1635.

Doherty, P. C. 1993. Inflammation in virus infections. *Semin. Virol.* **4:**117–122.

Fearon, D. T., and R. M. Locksley. 1996. The instructive role of innate immunity in the acquired immune response. *Science* **272:**50–54.

Griffin, D. E. 1994. Cytokines in viral infections. *Semin. Virol.* **5:**403–463.

Lachman, P. J., and A. Davies. 1997. Complement and immunity to viruses. *Immunol. Rev.* **159:**69–77.

Lamm, M. E. 1997. Interaction of antigens and antibodies at mucosal surfaces. *Annu. Rev. Microbiol.* **51:**311–340.

Marrack, P., and J. Kappler. 1994. Subversion of the immune system by pathogens. *Cell* **76:**323–332.

McMichael, A. J., and C. R. M. Bangham. 1996. Presentation of viral antigens to cytotoxic T cells. *Semin. Virol.* **7:**1–92.

McMichael, A. J., and C. A. O'Callaghan. 1998. A new look at T-cells. *J. Exp. Med.* **187:**1367–1371.

Morgan, B. P., and P. Gasque. 1996. Expression of complement in the brain: role in health and disease. *Immunol. Today* **17:**461–466.

Pitha, P. M. 1995. Interferon and interferon inducers. *Semin. Virol.* **6:**141–213.

Ramshaw, I. A., A. J. Ramsay, G. Karupiah, M. S. Rolph, S. Mahalingam, and J. C. Ruby. 1997. Cytokines and immunity to viral infections. *Immunol. Rev.* **159:**119–135.

Schluger, N. W., and S. N. Rom. 1997. Early responses to infection: chemokines as mediators of inflammation. *Curr. Opin. Immunol.* **9:**504–508.

Sen, G. C., and P. Lengyel. 1992. The interferon system: a bird's eye view of its biochemistry. *J. Biol. Chem.* **267:**5017–5020.

Sha, W. C. 1998. Regulation of immune response by NK-kB/Rel transcription factors. *J. Exp. Med.* **187:**143–146.

Silverman, R. H., and N. M. Cirino. 1997. RNA decay by the interferon-regulated 2-5A system as a host defense against viruses, p. 295–309. *In* J. B. Harford and D. R. Morris (ed.), *mRNA Metabolism and Post-Transcriptional Gene Regulation*. Wiley-Liss Inc., New York, N.Y.

Smith, T. J. 1995. Antibody recognition of viruses. *Semin. Virol.* **6:**215–267.

Smyth, M. J., and J. A. Trapani. 1998. The relative role of lymphocyte granule exocytosis versus death receptor-mediated cytotoxicity in viral pathophysiology. *J. Virol.* **72:**1–9.

Sprecher, E., and Y. Becker. 1993. Role of Langerhans cells and other dendritic cells in viral diseases. *Arch. Virol.* **132:**1–28.

Timonen, T., and T. S. Helander. 1997. Natural killer cell-target interactions. *Curr. Opin. Cell Biol.* **9:**667–673.

Weiss, R. A. 1998. Transgenic pigs and virus adaptation. *Nature* (London) **391:**327–328.

Zinkernagel, R. M. 1996. Immunology taught by viruses. *Science* **271:**173–178.

*Selected Papers*

Bachmann, M. F., U. Kalinke, A. Althage, G. Freer, C. Burkhart, H. P. Roost, M. Aguet, H. Hengartner, and R. M. Zinkernagel. 1997. The role of antibody concentration and avidity in antiviral protection. *Science* **276:**2024–2027.

Callan, M. F. C., L. Tan, N. Annels, G. S. Ogg, J. D. K. Wilson, C. A. O'Callaghan, N. Steven, A. J. McMichael, and A. B. Rickinson. 1998. Direct visualization of antigen-specific CD8+ T cells during the primary immune response to Epstein-Barr virus in vivo. *J. Exp. Med.* **187:**1395–1402.

Isaacs, A., and J. Lindenmann. 1957. Virus interference. I. The interferon. II. Some properties of interferon. *Proc. R. Soc. Lond. B* **147:**258–267, 268–273.

Landis, H., A. Simon-Jödicke, A. Klöti, C. Di Paolo, J.-J. Schnorr, S. Schneider-Schaulies, H. P. Hefti, and J. Pavlovic. 1998. Human MxA protein confers resistance to Semliki Forest virus and inhibits the amplification of a Semliki Forest virus-based replicon in the absence of viral structural proteins. *J. Virol.* **72:**1516–1522.

Marcus, P. I., and M. J. Sekellick. 1976. Cell killing by viruses. III. The interferon system and inhibition of cell killing by vesicular stomatitis virus. *Virology* **69:**378–393.

Marcus, P. I., L. L. Rodriguez, and M. J. Sekellick. 1998. Interferon induction as a quasispecies marker of vesicular stomatitis virus populations. *J. Virol.* **72:**542–549.

Matzinger, P., and E. J. Fuchs. 1996. Beyond "self" and "nonself": immunity is a conversation, not a War. *J. NIH Res.* **8:**35–39.

Rother, R. P., and S. P. Squinto. 1996. The α-galactosyl epitope: a sugar coating that makes viruses and cells unpalatable. *Cell* **86:**185–188.

Samuel, C. E. 1991. Antiviral actions of interferon-regulated cellular proteins and their surprisingly selective antiviral activities. *Virology* **183:**1–11.

Vanderplasschen, A., E. Mathew, M. Hollinshead, R. Sim, and G. Smith. 1998. Extracellular enveloped vaccinia virus is resistant to complement because of incorporation of host complement control proteins into its envelope. *Proc. Natl. Acad. Sci. USA* **95:**7544–7549.

Zinkernagel, R. M., and P. C. Doherty. 1974. Restriction of in vitro T cell-mediated cytotoxicity in lymphocytic choriomeningitis within a syngeneic or semiallogeneic system. *Nature* (London) **248:**701–702.

# 15

# Patterns of Infection: a Delicate Balance

## Introduction

### Infection Strategies

Viral infections of individuals in populations differ from viral infections of tissue culture cells in the laboratory. In the former, initiation of the infection, and its eventual outcome, rests upon complex variables such as host defenses and the environment. Despite such complexity and the plethora of viruses and hosts, common patterns of infection appear. In general, natural infections can be rapid and self-limiting (**acute infections**) or long-term (**persistent infections**). Variations and combinations of these two modes abound. While we can provide detailed descriptions of individual patterns of infection, we are in the early days of understanding the molecular mechanisms required to initiate or maintain any specific one.

### Life Cycles and Patterns of Viral Infections

A cursory examination of the animal viruses that grow in cultured cells identifies many distinctive life cycles with common features. Some viruses rapidly kill the cell while producing a burst of new infectious particles (**cytopathic viruses**). Others infect cells and actively produce infectious particles without causing immediate host cell death (**noncytopathic viruses**). Alternatively, some viruses infect but neither kill the cell nor produce any virus progeny. These apparently diverse life cycles defined in vitro comprise the two primary patterns of infection in the host: acute and persistent infections (Fig. 15.1). Variations on these archetypes occur repeatedly. For example, a **latent infection** is an extreme version of a persistent infection. Similarly, **slow, abortive**, and **transforming infections** are more complicated variants of a persistent infection.

Before we discuss patterns of infection, we will review the parameters important in initiating and establishing any viral infection.

**519**

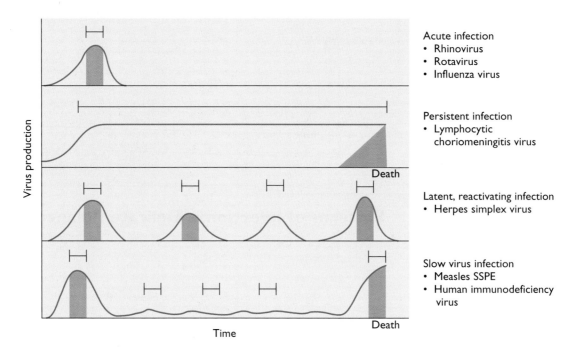

**Figure 15.1 General patterns of infection.** Relative virus production is plotted as a function of time after infection. The time when symptoms appear is indicated by the red shaded area, and the period in which infectious virus is released (available to infect other hosts) is indicated by the bracket. The top panel is the typical profile of an acute infection, in which virus is produced, symptoms appear, and virus is cleared within 7 to 10 days after infection. The second panel is the typical profile of a persistent infection, in which virus production continues for the life of the host. Symptoms may or may not appear just before death, depending on the virus. Infectious virus is usually produced throughout the infection. The bottom two panels are variations of the persistent infection. The third panel depicts a latent infection, in which an initial acute infection is followed by a quiescent phase and repeated bouts of reactivation. Reactivation may or may not be accompanied by symptoms but generally results in the production of infectious virus. The fourth panel depicts a slow virus infection, in which a period of years intervenes between a typical primary acute infection and the usually fatal appearance of symptoms. Depending on the virus, the production of infectious virus during the long period between primary infection and fatal outcome may be continuous (e.g., human immunodeficiency virus) or absent (e.g., measles virus subacute sclerosing panencephalitis [SSPE]). The brackets indicating infectious virus release are placed arbitrarily to indicate this phenomenon. Adapted from F. J. Fenner et al., *Veterinary Virology* (Academic Press, Inc., Orlando, Fla., 1993), with permission.

## Initiating an Infection

### Basic Requirements

Three requirements must be met to ensure successful infection in an individual host: sufficient virus must be available to initiate infection, the cells at the site of infection must be susceptible and permissive for the virus, and the local host antiviral defense systems must be absent or at least initially ineffective.

The first requirement erects a substantial barrier to any infection and forms a significant weak link in the transmission of infection from host to host. Free virus particles face both a harsh environment and rapid dilution that can reduce their concentration. Viruses spread by contaminated water and sewage must be stable in the presence of osmotic shock, pH changes, and sunlight and must not adsorb irreversibly to debris. Aerosol-dispersed viruses must stay wet and highly concentrated to infect the next host. Such viruses do best in populations in which individuals are in close contact. Viruses that are spread by biting insects, contact of mucosal surfaces, or other means of direct contact, including contaminated needles, have little environmental exposure.

Even if one virus particle survives the passage from one host to another, infection may fail simply because the concentration is not sufficient. In principle, a single virus particle should be able to initiate an infection, but host physical and immune defenses, coupled with the complexity of the infection process itself, demand the participation of many particles. How many particles are required to initiate and maintain an infection? Unfortunately, there is no simple answer to this question because the success of an infection depends on the particular virus, the site of infection, and the physiology and age of the host. However, some basic facts help to guide us.

Statistical analysis of infections in cultured cells demonstrates that on average a single virus particle can initiate an infection, but that many perfectly competent virions fail to complete this process. Such failure can be explained in part by the complexity of the infectious cycle: there are many distinct reactions, and the probability of a virus particle completing any one is not 100%. For example, virus particles face many potentially nonproductive interactions with debris and extracellular material during their initial encounter with the cell surface. Even if a virus attaches successfully to a permissive cell, it may be delivered to a digestive lysosome upon entry. It is not that any of the steps in infection are inherently inefficient, but rather that many of these false starts or inappropriate interactions are irreversible, aborting infection by the virion.

In addition, populations of viruses often contain particles that are not capable of completing an infectious cycle. For example, defective particles can arise from mistakes during virus replication or from interaction with inhibitory compounds in the environment. In the laboratory, a quantitative measure of the proportion of infectious viruses is the particle to plaque-forming unit (PFU) ratio. As described in chapter 2, the number of physical particles in a given preparation is counted, usually with an electron microscope, and compared to the number of infectious units or PFU per unit volume. This ratio is a useful indicator of the quality of a virus preparation, as it should be constant for a given virus prepared by identical or comparable procedures.

## Tropism

All patterns of infection are dominated by the property of viral **tropism**. Tropism is a predilection of viruses to infect certain tissues and not others. For example, an **enterotropic** virus replicates in the gut, whereas a **neurotropic** virus replicates in cells of the nervous system. Some viruses are **pantropic**, infecting and replicating in many cell types and tissues. Tropism may be determined by the distribution of receptors for entry (**susceptibility**), it may be the result of a requirement of the virus for differentially expressed intracellular gene products to complete the infection (**permissivity**), or it may indicate that the virus is physically prevented from interacting with tissues that otherwise could support virus growth. In most cases, tropism is determined by a combination of two or more of these parameters.

Tropism affects the pattern of infection, pathogenesis, and long-term virus survival. The role of tropism in pathogenesis will be discussed in greater detail in chapter 17. Consider the example of herpes simplex virus, an alphaherpesvirus. This human herpesvirus is often said to be neurotropic because of its noteworthy ability to infect and reactivate from the nervous system. But in fact, herpes simplex virus is pantropic and replicates in many cells and tissues in the host. By infecting neurons, it establishes a stable latent infection (see "Two Viruses That Produce Latent Infections" below), but because it is pantropic it spreads to other tissues and host cells. One serious consequence is that if the virus escapes host defenses at the site of infection, it will spread widely, causing disseminated disease, as can occur when herpes simplex virus infects babies and immunocompromised adults. Herpes simplex virus neurotropism leads to yet another serious result of infection. On rare occasions, this virus can enter the central nervous system and cause an encephalitis which is often fatal.

Similar alterations of tropism in a single virus infection can be observed for many RNA viruses. Consider poliovirus, a picornavirus with *enteric* tropism. The poliovirus receptor is found on the surface of almost every cell of the body, but in the common natural infection, poliovirus replicates in the gut; it rarely spreads to other tissues. It is efficiently spread from host to host by fecal-oral transfer. As an enteric virus, poliovirus is not usually a major health problem. However, *neurotropic* strains of poliovirus that spread from the gut to motor centers of the central nervous system arise. Infections by such viruses result in paralysis and often death, a considerably more serious aspect of poliovirus infection. At least two questions arise. What are the molecular changes in the virus that alter its tropism and virulence? Is there a selective advantage of the change in tropism for virus survival? We have some data with regard to the former question (see chapter 17 and Fig. 19.6B) but can only speculate about the latter.

## Successful Infections Must Evade Host Defenses

The sites of virus entry in a host are described in chapter 17. To initiate an infection at these sites, viruses must counter the host defenses by an active or passive mechanism or by a combination of both. This is a simple but often overlooked fact; most of the information encoded in viral genomes never has a chance to be expressed because most infections are blocked before anything happens. Moreover, the physical defenses exemplified by skin and mucous surfaces, in combination with the innate defenses described in chapters 14 and 17, may block or limit infection before the acquired immune system is activated.

Some of these defenses may be overcome passively by an overwhelming inoculum of virus falling on a mucosal surface. Single droplets found in the aerosol produced by sneezing can contain as many as 100 million rhinovirus particles; a similarly high number of hepatitis B virus particles can be found in 1 µl of blood from a hepatitis patient. At these concentrations, it may be impossible for physical and innate defenses to block every infecting virus particle. Free passage of virus through the primary physical barriers of skin and mucous layers made possible by a cut, abrasion, or needle stick may also allow passive evasion of defenses. A more egregious breech of both primary and secondary defenses may occur during organ transplantation, which places viruses in direct contact with potentially susceptible cells in immunosuppressed patients. Some viruses (e.g., herpesviruses, papovaviruses, and rabies virus) adopt a passive stance simply by infecting organs or cells not exposed to antibodies or cytotoxic lymphocytes.

Many viruses have evolved active mechanisms for bypassing or disarming host defenses (Table 15.1). For example, some viruses express proteins that block the general suicide program activated in most virus-infected cells (apoptosis), while the high mutation rates of most RNA viruses result in the production of altered proteins that allow the virus to evade immune defenses.

### Apoptosis (Programmed Cell Death)

It is axiomatic that viruses require an intact and functioning cell during the initiation of infection because premature demise of the cell would effectively block virus

**Table 15.1** Active evasion of immune defense

| Process | Virus examples |
| --- | --- |
| Exchange of genetic information by reassortment of genomic segments to replace entire surface proteins | Influenza virus, rotavirus |
| Blocking of specific immune defenses | Adenovirus, herpesviruses, poxviruses, human immuno-deficiency virus |
| Initiation of a noncytopathic infection to delay or avoid a robust immune response | Arenaviruses, paramyxo-viruses, polyomavirus, papillomaviruses, parvoviruses |

replication and spread. Therefore, it should not be surprising that cell death is a frequent antiviral defense. When the biochemical alterations initiated by viral infection are detected, a process of self-destruction called **apoptosis** is initiated (Fig. 15.2). This pathway was first discovered in 1842 by Carl Vogt, and the term "apoptosis" was coined in 1972. This process is controlled by a variety of interacting signals that monitor the orderly processes of growth regulation, cell cycle progression, and metabolism in metazoan cells. Survival signals from the cell's environment and internal signals reporting on cell integrity normally keep the apoptosis response in check. When these signals are perturbed, cell death invariably ensues (Fig. 15.3). Apoptosis can be activated by a large variety of both external and internal stimuli. As an example of the former, cytotoxic T cells may kill their target by apoptosis initiated by Fas

**Figure 15.2 Apoptosis, the process of programmed cell death.** Apoptosis can be recognized by several distinct changes in cell structure. A normal cell is shown at the left. When programmed cell death is initiated, as indicated by the second cell, the first event visible is the compaction and segregation of chromatin into sharply delineated masses that accumulate at the nuclear envelope (dark blue shading around periphery of nucleus). The cytoplasm also condenses, and the outline of the cell and nuclear membranes changes, often dramatically. The process can be rapid, so that within minutes the nucleus fragments and the cell surface convolutes, giving rise to the characteristic "blebs" and stalked protuberances illustrated. These blebs then separate from the dying cell and are called apoptotic bodies. Macrophages (the cell at the right) engulf and destroy these apoptotic bodies. Adapted from J. A. Levy, *HIV and the Pathogenesis of AIDS*, 2nd ed. (ASM Press, Washington, D.C., 1998), with permission.

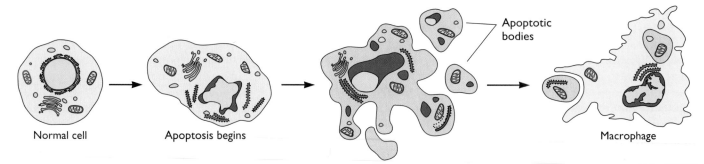

Normal cell    Apoptosis begins    Apoptotic bodies    Macrophage

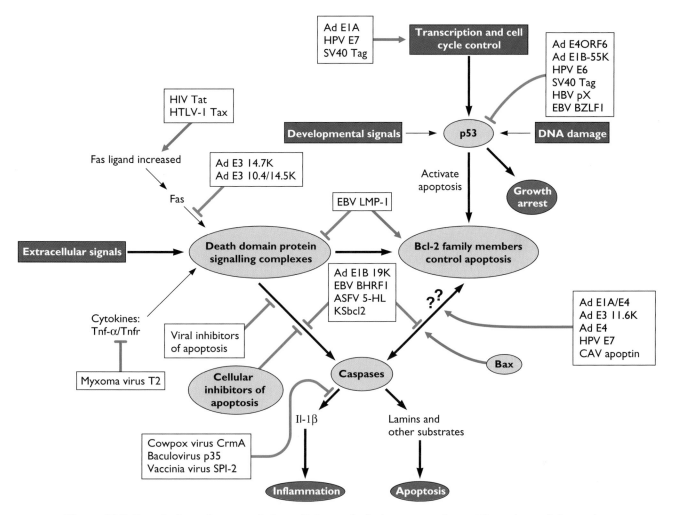

**Figure 15.3 Regulation of apoptosis by cellular and viral gene products.** Flow chart of the major processes and selected signals in the pathway mediating cell growth and programmed death (tan). A variety of signals from the cell cycle, developmental pathways, and DNA damage activate p53 (dark red boxes). Activated p53 participates either in growth arrest or apoptosis (blue). Apoptosis can be abrogated by a variety of cellular regulatory proteins illustrated here by Bcl-2 family members. Expression of these gene products is affected by a variety of extracellular signals (purple box) such as cytokines like tumor necrosis factor (Tnf) and the interaction of cell surface ligands and receptors, including Fas ligand and Fas receptor. If Bcl-2-like proteins do not block the apoptotic pathway, then cysteine proteases (caspases) are activated and act in casade to carry out the final stages of apoptosis. One such caspase family member is caspase 1, the interleukin-1β-converting enzyme (ICE) protease. This enzyme activates interleukin-1β, which in turn facilitates inflammation (blue). Other caspases cleave a variety of substrates, including lamins and pronucleases. As a result of caspase action, the cell architecture is altered and cellular DNA is degraded. Viral proteins known to affect the pathway of apoptosis are indicated in the white boxes. Green arrows indicate induction, stimulation, or activation; red bars indicate inhibition. Abbreviations: HTLV-1, human T-cell leukemia virus type 1; SIV, simian immunodeficiency virus; HIV, human immunodeficiency virus; CAV, chicken anemia virus; HPV, human papillomavirus; Ad, adenovirus; SV40, simian virus 40; ASFV, African swine fever virus; HBV, hepatitis B virus; EBV, Epstein-Barr virus; KS, Kaposi's sarcoma herpesvirus (human herpesvirus 8); Tnfr, Tnf receptor; Tag, large T antigen.

ligand on the T cell binding to Fas receptor on the target cell. Similarly, apoptosis is initiated when the cytokine tumor necrosis factor (Tnf) binds to its receptor on a virus-infected cell. Common intracellular initiators include DNA

damage and ribonucleotide depletion. In these situations, the cell cycle regulatory protein p53 is activated (see chapter 16) and apoptosis ensues. How activated p53 stimulates this process is still unresolved. Regardless of the nature of the initiation signal, the result is activation of common effectors, the caspases. Caspases are members of a family of cysteine proteases that specifically cleave proteins after asparagine residues. They carry out limited proteolysis of many cellular substrates in a protease cascade, not unlike blood clotting or the complement cascade. The principle is similar: a modest initial signal can be amplified significantly, culminating in an all-or-none result. In apoptosis, after caspases are activated (no matter how diverse the initial activating signal), the end results are always the same: cell and organelle dismantling, vesicle and membrane bleb formation, and DNA cleavage to nucleosome-sized fragments (Fig. 15.4). As one might expect, apoptosis is a tightly regulated process, and the cell encodes several suppressors including members of the Bcl-2 family. Despite the complexity, it is best to think of apoptosis as a default pathway held in check by the continuous action of a variety of regulatory molecules.

While apoptosis can be a powerful antiviral defense, it is also a normal host process essential for orderly development of many organisms. Indeed, cell-encoded inducers of apoptosis (e.g., Bax and Bad [Fig. 15.3; Table 15.2]) must be synthesized for the maintenance of important aspects of normal cell physiology and homeostasis, such as the regulation of cell numbers in development and proper functioning of the immune system. Apoptosis also has an important role in presenting antigens to cytotoxic T cells. When influenza virus-infected cells undergo apoptosis, the cellular debris containing viral antigens is ingested by dendritic cells which present the antigen to T cells. If apoptosis is blocked in the infected cells, dendritic cells do not pick up antigen and T cells are not activated.

Like all cellular antiviral defenses, apoptosis can be a double-edged sword. Viruses have evolved to survive despite apoptosis (Box 15.1). Some viruses are able to bypass it by synthesizing proteins that interfere with the program at a number of distinct steps (see below; Fig. 15.3). Others actually incorporate apoptosis as part of their life cycles. Most DNA viruses induce apoptosis upon infection because they require all or part of the host's transcription, translation, and replication machinery (Fig. 15.3). In many infections, the target cell is quiescent and hence unable to provide the enzymes and other proteins needed by the infecting virus. Consequently, to replicate in these cells, the virus must actively engage the host's cell cycle machinery and induce the cell to leave the resting state. However, cell cycle checkpoint proteins then respond to these unscheduled events by inducing apoptosis (Fig. 15.3). Typically, activation of the cell cycle is accomplished by viral early proteins (e.g., adenovirus E1A proteins or simian virus 40 [SV40] large T antigen). The genomes of many viruses that activate apoptosis as part of their life cycle carry additional genes whose products block this potentially lethal process long enough for the virus to replicate its genome and produce infectious particles.

**Viral proteins that inhibit apoptosis.** Viral mutants unable to inhibit apoptosis were detected originally because the host DNA of mutant infected cells was unstable, the cells lysed prematurely, and, as a consequence, viral yields were reduced, resulting in small plaques. The mutant gene products are analogs of host Bcl-2, caspase inhibitors, and other proteins that prevent or delay apoptotic death of infected cells (Fig. 15.3; Table 15.1).

Some viral proteins that inhibit apoptosis have rather remarkable properties. For example, the adenovirus E1B 19K protein inhibits apoptosis by at least two distinct mechanisms. In one, the protein binds to the apoptosis-promoting Bax protein to prevent caspase activation. In the other, the protein interferes with the function of adaptor molecules that interact with and activate caspases. Some herpes- and poxviruses encode other types of apoptosis inhibitors. **Death receptors** are cell surface proteins that transmit apoptosis signals on binding **death ligands**. These receptors are part of the tumor necrosis factor receptor gene superfamily. A characteristic of such receptors is a short amino acid sequence, called the **death domain**, which defines a surface of the cytoplasmic portion of the receptor that engages the cell's apoptotic machinery when the appropriate ligand is bound. Cellular adaptor proteins that bind this domain in this way include Fas-associated death domain protein. This complex of Fas-associated death domain protein and tumor necrosis factor or Fas receptor then activates caspase 8 to initiate the caspase cascade. Computer searches of viral DNA sequences revealed several viral proteins with sequences similar to the death domain of cellular death receptors. The viruses expressing such proteins are found in two DNA virus groups, the gammaherpesviruses (e.g., human herpesvirus 8, equine herpesvirus 2, and bovine herpesvirus 4) and the poxviruses (e.g., molluscum contagiosum virus). When these viral proteins were expressed in cells, they blocked Fas ligand- and Tnf-induced apoptosis. Work is now in progress to determine the role of these novel apoptosis inhibitors in natural infections.

RNA virus infections are also modulated by apoptosis. In the case of Sindbis virus, apoptosis directs the pattern of infection. In some vertebrate tissue culture cell lines, Sindbis virus infection is acute and cytopathic because apoptosis is induced. However, the virus establishes a persistent infection of postmitotic neurons in culture because the cell death pathway is not activated. A corollary of the in vitro experiment can be observed in animals. When virus is injected into an adult mouse brain, it establishes a persistent

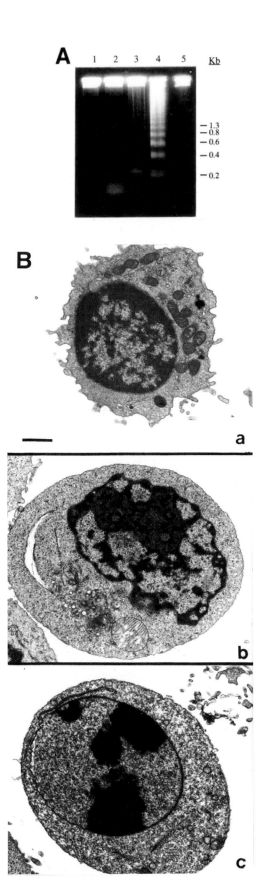

**Figure 15.4 Apoptosis in a CD4⁺ T cell from peripheral blood.** (A) Agarose gel electrophoresis of total T-cell DNA after treatment of cells with various antibodies (lanes 2, 3, and 5) or gamma irradiation (lane 4). Lane 1 contains DNA from untreated cells. The typical ladder pattern of DNA digestion associated with apoptosis is observed after gamma irradiation. (B) Ultrastructural morphology of a normal CD4⁺ T cell (a) and a cell in early stages of apoptosis (b) and terminal stages of apoptosis (c). Bar = 1 μm. Reprinted from J. A. Levy, *HIV and the Pathogenesis of AIDS*, 2nd ed. (ASM Press, Washington, D.C., 1998), with permission.

noncytopathic infection, but in neonatal mouse brains, the virus is cytopathic and lethal. Sindbis virus induces lethal apoptosis in neonatal mouse brains because cellular inhibitors of apoptosis are not produced in developing neurons as they are in adult neurons.

### Antiviral Cytokine Defense

Viral infection invariably stimulates cells to make interferons and cytokines. These molecules are parts of the host primary innate defense system. To establish an infection, some viruses, particularly cytopathic viruses, produce cytokine antagonists to counteract this most potent early host defense. For example, poxvirus genomes encode a variety of anti-immune defense proteins including soluble receptors that bind many host cytokines so that the cytokines cannot reach their proper receptors on cells (Fig. 15.5; see also Table 14.1). Remarkably, soluble gamma interferon (Ifn-γ) receptors are produced by at least 17 orthopoxviruses; obviously, this cytokine has an important role in thwarting poxvirus infection. These **decoy cytokine receptors** are examples of the so-called **unselfish defense**, as their presence affects the entire population of infecting viruses and microbes, not just the virus that directs their synthesis.

Tumor necrosis factor is a multifunctional cytokine produced primarily by activated monocytes and macrophages. It can induce an antiviral response when it binds to receptors on virus-infected cells. Such cells are especially sensitive to tumor necrosis factor, most likely because viral infection blocks or alters host protein synthesis. Within seconds, the combination of infection and binding of tumor necrosis factor to its receptor initiates a signal transduction cascade that activates caspases and thus apoptosis. Many adenoviruses can counteract the lytic effect of tumor necrosis factor with several small proteins encoded in the E3 region of the genome (Fig. 15.6; Table 15.3).

### Antigenic Variation

Antigenic variation, which results when viruses interact with the immune system, is an important mechanism of virus evolution and survival (Box 15.2). Invariably, muta-

**Table 15.2** Selected viral and cellular regulators of apoptosis

| Gene product(s) | Function |
|---|---|
| **Inducers** | |
| Viral | |
| Viral infection (general) | DNA damage, host shutoff |
| Adenovirus E1A, E4, E3, SV40 large T, human papillomavirus E7 | Cell cycle control, proliferation, transformation, transcription |
| HIV, simian immunodeficiency virus Tat | Transcription |
| Parvovirus B19 NSP | Nonstructural protein, replication, transcription |
| Chicken anemia virus VP3 | Structural protein |
| Cellular | |
| c-myc protein | Proliferation, transformation, transcription |
| Tumor necrosis factor | Antiviral and antitumor activities |
| Fas antigen | Negative selection of lymphocytes |
| p53 protein | Cell cycle control, tumor suppressor, transcription |
| **Inhibitors** | |
| Cellular | |
| Bcl-2 | Blocks apoptosis in hematopoietic cells in response to growth factor withdrawal, CD3, irradiation, glucocorticoids; maintains B-cell memory; role in T-cell maturation |
| BclXL | Like Bcl-2, but functions in neurons |
| Viral | |
| Adenovirus E1B 19K | Blocks apoptosis by tumor necrosis factor, Fas, p53; transformation |
| Adenovirus E1B 55K | Inactivates p53; transformation |
| Adenovirus E3 14.7K | Blocks apoptosis by tumor necrosis factor |
| Epstein-Barr virus latent membrane protein 1 | Increases expression of Bcl-2; latency, transformation |
| Ksbcl-2 | Human herpesvirus 8 Bcl-2 homolog |
| Herpesvirus saimiri ORF16 | Bcl-2 homolog |
| Human cytomegalovirus IE1/IE2 | Immediate-early proteins 1 and 2; block apoptosis induced by tumor necrosis factor alpha and adenovirus E1A but not that induced by UV irradiation |
| Baculovirus p35 | Inhibits apoptosis in vertebrate and insect cells; inhibits interleukin-1β-converting enzyme (ICE-like family) protease |
| Baculovirus IAP protein | Inhibits apoptosis by a different mechanism than p35 |
| Poxvirus CrmA | Serpin; serine protease inhibitor; ICE protease inhibitor |
| Hepatitis B virus pX | Blocks p53-mediated apoptosis |

tions accumulate as viruses replicate, and, depending on the selective forces to which they are exposed, some mutants propagate while others are eliminated. In an immunocompetent host, viral antigenic variation comes about by two distinct processes called antigenic drift and antigenic shift. **Antigenic drift** is the appearance of virus with a slightly altered surface protein (antigen) structure following passage in the natural host. Mutants expressing

**BOX 15.1**

*The many ways in which viruses perturb apoptotic pathways*

**At the cell surface**
Apoptosis-inducing cytokines/receptors (e.g., receptor mimicry)
Membrane disruption (e.g., virion adsorption/engagement of receptors)

**In the cytoplasm**
Metabolic inhibitor (e.g., arrest of host translation)
Cytoskeletal modification (e.g., disruption of actin microfilaments)
Signal transduction blocker (e.g., death domain proteins)

**In the nucleus**
Host DNA degradation
Altered gene expression (e.g., increased expression of heat shock genes)

**Miller, L. K., and E. White (ed.).** 1998. Apoptosis in viral infections. *Semin. Virol.* **8**:443–523.

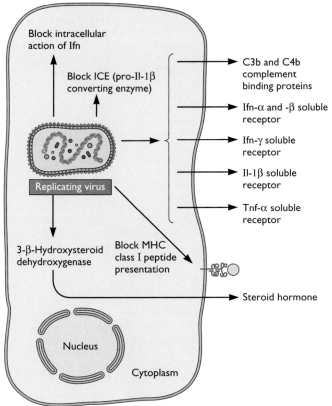

**Poxvirus-infected cell**

Block intracellular action of Ifn

Block ICE (pro-Il-1β converting enzyme)

Replicating virus

C3b and C4b complement binding proteins

Ifn-α and -β soluble receptor

Ifn-γ soluble receptor

Il-1β soluble receptor

Tnf-α soluble receptor

3-β-Hydroxysteroid dehydroxygenase

Block MHC class I peptide presentation

Steroid hormone

Nucleus

Cytoplasm

**Figure 15.5 Immune evasion mechanisms encoded by poxviruses.** After infection of a cell by a poxvirus, a variety of viral proteins that influence host defenses are produced. Several examples of proteins that block different arms of the host response to infection (e.g., complement activation, cytokine function, antigen presentation to cytotoxic T lymphocytes, inflammation) or mimic host cytokines or cytokine receptors are represented. Modified from A. Alcami and G. L. Smith, *Immunol. Today* **16:**474–478, 1995, with permission.

altered surface proteins often escape the host immune response and are propagated in the population. In contrast, **antigenic shift** is a major change in the surface protein of a virus as completely new surface proteins are acquired by the virus. This process occurs when viruses with segmented genomes exchange segments after coinfection. The new reassortant viruses display dramatic changes in surface proteins that facilitate escape from immune surveillance (Box 15.3).

As surface proteins are often absolutely essential for the attachment and entry of viruses into cells, their function cannot be compromised during antigenic variation. Some viruses, like human immunodeficiency virus and influenza virus, are more tolerant of antigenic variation than others and are said to be **plastic** or to have **structural plasticity**. In contrast, nonenveloped RNA viruses, like poliovirus, can tolerate only modest antigenic drift in their capsid proteins. Capsid proteins, unlike envelope proteins, participate in multiple protein-protein and protein-nucleic acid interactions required to package the genome and to assemble or disassemble the icosahedral capsid. Mutations that reduce antibody binding during antigenic drift are likely to affect these processes and thus are often detrimental to virus reproduction.

**Antigenic drift in human immunodeficiency virus infections.** Virus-infected individuals produce virus for years before they develop the symptoms of AIDS. In the early stages of infection, more than $10^9$ new virus particles are produced in an infected patient each day, and these viruses are confronted continually by the immune system. However, the virus is not completely eliminated, as witnessed by the continuing replication of virus for years. Until the final stages of AIDS, the immune system continually selects viruses that can grow from the pool of newly replicated virus. At these rates of replication and selection, progeny viruses are many generations removed from the

**Table 15.3**   Functions of adenovirus E3 proteins[a]

| Host response | Protein | Intracellular location(s) | Function or mechanism |
|---|---|---|---|
| CTL | gp19K | ER membrane | Binds to MHC class I antigens; blocks their transport from ER to cell surface; prevents CTL recognition |
| Tnf | 14.7K | Nucleus and cytoplasm | Prevents Tnf cytolysis; inhibits Tnf-induced inflammatory response by blocking activation of phospholipase A2 |
|  | RIDα (10.4K), RIDβ (14.5K) | Plasma membrane | Prevents Tnf cytolysis by a different mechanism than 14.7K |
| Egf/Fas | RIDα (10.4K), RIDβ (14.5K) | Plasma membrane | Stimulates internalization and degradation of Egf receptor; inhibits Fas agonist-induced apoptosis |
| Cell lysis | ADP (11.6K) | Nuclear membrane | Promotes release of virus from lysed cells |

[a]See also Fig. 15.6. Abbreviations: CTL, cytotoxic T lymphocyte; MHC, major histocompatibility complex; ER, endoplasmic reticulum; Tnf, tumor necrosis factor; Egf, epidermal growth factor.

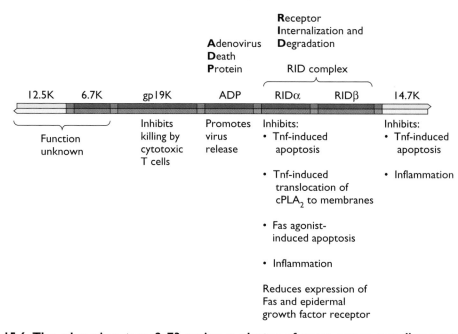

**Figure 15.6 The adenovirus type 2 E3 region, a cluster of seven genes encoding proteins that mediate host defense evasion.** The proteins were named initially according to their apparent molecular masses (e.g., 14.7K for 14.7 kDa). Recently, some of the proteins have been given names that reflect their known functions. RID is an acronym for "receptor internalization and degradation." The RID protein complex (previously called E3-10.4K/14.5K) is composed of RIDα (10.4 kDa) and RIDβ (14.5 kDa). RID proteins have multiple functions, as indicated in the figure. These functions may or may not represent the same molecular mechanism. ADP is an abbreviation for "adenovirus death protein." Even though the ADP gene is in the E3 cluster, it is expressed as a late gene by alternative splicing from the major late promoter. ADP promotes cell lysis and virus release after virus replication is complete. Integral membrane proteins are indicated by reddish shading. gp19K is a glycoprotein that reduces MHC class I protein expression and inhibits killing by cytotoxic T cells. The 14.7K protein inhibits Tnf-induced apoptosis. The functions of the 12.5K and 6.7K proteins have not been identified. Adapted from W. S. M. Wold and A. E. Tollefson, *Semin. Virol.* **8:**515–523, 1998, with permission.

original infecting virus and soon comprise a variety of mutants selected in part through the process of antigenic drift. We will discuss the pathogenesis of human immunodeficiency virus more fully in chapter 18.

Such antigenic drift also is believed to facilitate escape from immune surveillance by a second mechanism, a response to the high variation of individual viral peptides presented on the surfaces of infected cells by major histocompatibility complex (MHC) class I molecules. Some cytotoxic T cells specific for a given viral peptide recognize the cognate mutant peptide on infected cells, but because the peptides are similar but not identical, the cytotoxic T cell is inactivated, or "anergized." Instead of being destroyed, the infected cell actually inactivates a cytotoxic T cell that could have destroyed other cells containing different virus mutants. This mechanism of inactivating T cells by antigen "mimics" or "decoys" is likely to function for any virus exhibiting high rates of antigenic drift.

## Interference with Expression and Function of MHC Proteins

The cytotoxic-T-cell response is one of the most powerful adaptive host defenses against viral infection. This response depends in part on the ability of host T cells to detect viral antigens on the surfaces of infected cells and to kill these cells. The recognition of infected cells requires the presentation of viral peptides by MHC class I proteins. The pathway by which peptide antigens are produced and presented on the infected cell surface is discussed in chapter 14 (Fig. 14.14). Obviously, any viral strategy that stops viral peptides from appearing on MHC class I molecules on the surface of cells provides an important selective advantage.

Peptide presentation by MHC class I can be reduced by lowering expression of the MHC genes directly, by interfering with the production of peptides by the proteasome, or by interfering with subsequent assembly and transport

**BOX 15.2**

*Influenza virus provides the classic paradigm of antigenic shift and drift*

Influenza, which has been recognized as a human disease for centuries, is notable because although it causes a typical acute infection and elicits a strong immune response, it continues to occur regularly in the population. Influenza can occur in many nations almost simultaneously (pandemics) with serious consequences.

After more than 65 years of work, we now have some reasonable ideas why this virus causes epidemics of "flu" annually and pandemics at frequent intervals. Influenza virus not only is a human virus but has a complex relationship that includes replication in several hosts (see chapter 20). For example, influenza A virus infects not only humans but also pigs, birds, horses, mink, and some aquatic mammals such as seals and whales.

Epidemiologists can classify a given influenza A virus by its antigenic composition, usually on the basis of serologic reactions of the two envelope proteins hemagglutinin (HA or H) and neuraminidase (NA or N). A common nomenclature simplification is to refer to combinations of HA and NA as HxNy (where $1 \leq x \leq 14$ and $1 \leq y \leq 9$), e.g., H1N1 or H5N2. At least 14 subtypes of viral hemagglutinin are known in viruses that infect birds. Three of these subtypes are present in viruses that can infect humans, and at least two can infect pigs, horses, and aquatic mammals.

More than nine NA subtypes are known, and, like HA subtypes, viruses with these subtypes have a characteristic host range. Thus, if antisera that recognize each of these HA and NA subtype proteins are available, it is possible to trace a virus population through multiple hosts and its movement around the world.

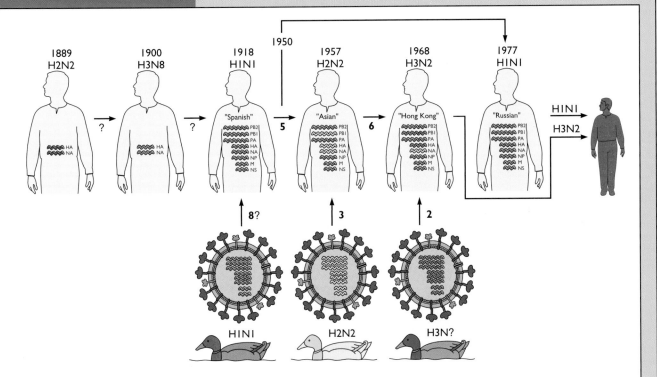

**Postulated evolution of human influenza A viruses from 1889 to 1977.** The figure depicts the appearance and transmission of distinct serotypes of influenza A virus in humans. The bottom part shows the nature of the avian influenza viruses that reassort with human viruses. The color of the genome segments represents a particular viral genotype. Segments of the predominant influenza virus genome and its gene products are indicated in each human silhouette for each year. The number next to the arrow indicates how many segments of the viral genome are known to have been transmitted. The earliest serology data we have are from 1889 and suggest that H2N2 was the predominant class in humans. In 1900 the predominant serotype was H3N8. No data exist for the other influenza virus genes present at these times, and these segments are not illustrated. Phylogenetic evidence is consistent with the appearance by 1918 of an influenza virus with eight distinct segments. This virus is thought to be of avian origin and is characterized by the H1N1 serotype (red). It was also found in pigs and was carried from North America to Europe by U.S. soldiers. It gained notoriety as the cause of the catastrophic Spanish influenza pandemic of 1918. In 1957 the Asian pandemic was caused by a virus that acquired three genes (PB1, H2, and N2) from avian viruses infecting wild ducks (yellow) and retained five other genes from the circulating human strain with H2N2 (red). As the Asian strains appeared, the H1N1 strains disappeared from the human population. In 1968, the Hong Kong pandemic virus acquired two new genes (PB1 and H3) from the wild duck reservoir (blue) and kept six genes that were circulating in human viruses (red and yellow). The pandemic virus had a characteristic H3N2 serotype. After the appearance of this virus, the Asian H2N2 strains could no longer be detected in humans. In 1977 the Russian H1N1 strain that had circulated in humans in the 1950s reappeared and infected young adults and children. One theory was that this virus escaped from a laboratory. It has continued to cocirculate with H3N2 influenza viruses in the human population. Adapted from R. G. Webster and Y. Kawaoka, *Semin. Virol.* **5:**103–111, 1994, with permission.

## BOX 15.3

*Antigenic shift, not drift, was the driving force for the five pandemics of human influenza during the past century (1890, 1900, 1918, 1957, and 1968)*

Viruses causing these worldwide infections can be typed into HA and NA subtypes (see also Box 15.2). Each pandemic is characterized by a new combination of HA and NA:

The 1918 flu was H1N1.
The 1957 "Asian flu" was H2N2.
The 1968 "Hong Kong flu" was H1N2.
The 1977 "swine flu" was H1N1.

These dramatic shifts of H and N serotypes result from the exchange of genome segments by mammalian and avian influenza viruses. In general, influenza A viruses that grow well in birds are not efficient at infecting humans, and vice versa. This assertion is currently under scrutiny, as evidence of direct infection of humans by an avian influenza virus (H5N1) was documented in 1997. This H5N1 combination of antigens has never been observed previously in human infections despite its occurrence in virulent avian viruses that have caused major domestic bird epidemics. Since no humans have immunity to the H5N1 viruses, the appearance of those viruses in humans was of major concern. It now appears that these viruses are incapable of efficient spread in human populations.

Virologists have demonstrated that certain combinations of H and N are better selected in avian hosts than in humans. An important observation was that *both* avian and human viruses replicate well in certain species such as pigs, no matter what the H/N composition is. Indeed, the lining of the throats of pigs contains receptors for both human and avian influenza viruses, providing an environment in which both can flourish. Thus, the porcine host is a good nonselective host for mixed infection of avian and human viruses in which reassortment of H and N segments can occur, creating new viruses that can reinfect the human population.

At first glance, one might think that this combination of human, bird, and pig

*(continued)*

**Genetic reassortment between avian and human influenza A viruses in swine.** Studies of Italian pigs provide evidence for reassortment between avian and human influenza viruses. The figure shows how the avian H1N1 viruses in European pigs reassorted with H3N2 human viruses. The color of the segments of the influenza genome indicates the origin: blue segments are from avian viruses found in swine in 1979, and red segments are from human viruses found in swine in 1968. The host of origin of the influenza virus genes was determined by partial sequencing and phylogenetic analysis. These studies support the hypothesis that pigs can serve as an intermediate host in the emergence of new pandemic influenza viruses. Adapted from R. G. Webster and Y. Kawaoka, *Semin. Virol.* **5**:103–111, 1994, with permission.

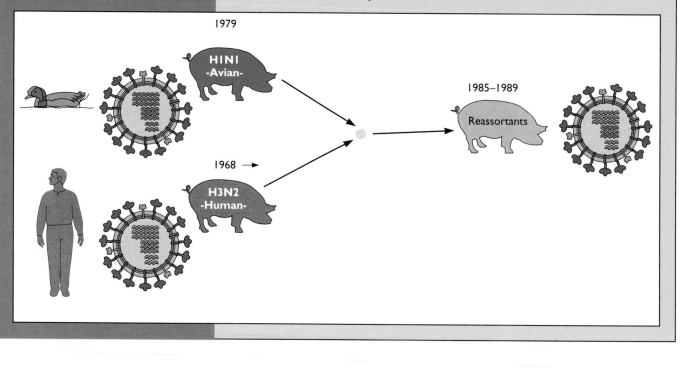

infections must be extremely rare. However, the dense human populations in Southeast Asia that come in daily contact with domesticated pigs, ducks, and domestic fowl remind us that these interactions are likely to be frequent. Indeed, epidemiologists can show that the 1957 and 1968 pandemic influenza A virus strains originated in the People's Republic of China and that the human H and N

serotypes are circulating in wildfowl populations. In the United States, turkeys as well as swine may serve as intermediate hosts for mixing of avian and mammalian influenza viruses, but direct transfer from turkeys to humans has not yet been demonstrated.

Southeast Asia should not be the only focus of attention, as intense production of domestic swine and turkeys in Europe

and the United States, coupled with the major migratory paths of wild ducks and geese, is likely to make these regions centers of interspecies transfer as well.

Ito, T., J. N. S. S. Couceiro, S. Kelm, L. G. Baum, S. Krauss, M. R. Castrucci, I. Donatelli, H. Kida, J. C. Paulson, R. G. Webster, and Y. Kawaoka. 1998. Molecular basis for the generation in pigs of influenza A viruses with pandemic potential. *J. Virol.* **72:**7367–7373.

of the MHC-peptide complex to the cell surface. Several viruses express proteins that block MHC class I function at various points in the pathway (Fig. 15.7; Table 15.4). The MHC class I pathway is of obvious importance in immunology, but the existence of many of these MHC-processing or regulatory steps was not known until viral inhibitors were characterized.

Human cytomegalovirus deserves special mention because it has evolved multiple strategies to prevent MHC class I molecules from exposing viral antigens on the cell surface. The human cytomegalovirus US6 protein inhibits peptide translocation into the endoplasmic reticulum lumen, although by a different mechanism than the herpes simplex virus ICP47 protein (Fig. 15.7; Table 15.4). The human cytomegalovirus US3 protein detains MHC class I proteins in the endoplasmic reticulum by direct interaction, while the US11 and US2 proteins eject MHC class I molecules from the endoplasmic reticulum into the cytoplasm, where they are degraded by cellular proteases. Why human cytomegalovirus encodes so many proteins to block antigen presentation remains an open question (Box 15.4). One possibility is that multiple gene products act additively or synergistically to effect the block of MHC class I function. Other ideas are that different proteins are required to block MHC class I function in different cell types or that the concentrations of viral proteins vary in different cell types.

Human cytomegalovirus and Epstein-Barr virus have evolved another strategy for avoiding cytotoxic-T-cell destruction. The first indication of this phenomenon was the observation that Epstein-Barr virus-infected individuals do not produce cytotoxic T cells capable of recognizing the viral protein Epstein-Barr virus nuclear antigen 1 (EBNA-1). This phosphoprotein is found in the nuclei of latently infected cells and is the only protein regularly detected in all malignancies associated with Epstein-Barr virus. T cells specific for other Epstein-Barr virus proteins are made in abundance, indicating that Epstein-Barr virus nuclear

antigen 1 must possess some special features. Indeed, this protein contains a remarkable amino acid sequence consisting of arginine-glycine motifs surrounding an internal glycine-alanine repeat. The sequence inhibits the host proteasome so that relevant Epstein-Barr virus nuclear antigen 1 peptides are not produced at all. This inhibitory sequence is nonspecific, for it can be fused to other proteins to inhibit their processing and subsequent presentation of peptide antigens normally produced by them.

Human cytomegalovirus uses a different twist of the same strategy to block proteasome processing of its major immediate-early protein. Upon infection, the pp65 kinase, a viral tegument protein encoded by the UL83 gene, phosphorylates the newly synthesized major immediate-early protein so that it cannot be processed into antigenic peptides. As a result, peptides of immediate-early protein are not presented on the surface by MHC class I molecules, and the infected cell escapes early destruction by cytotoxic T cells.

## Many Other Variables Govern the Result of Infection

In addition to the parameters described in the preceding sections, other complex variables can determine the course and result of infections. Of these, the age and immune status of the host, host population density, host interactions, and environmental conditions are predominant (see chapter 17 for more details).

## Acute Infections

### Definition and Requirements

An acute infection is one of the best understood of all viral infection patterns because it is characteristic of many viruses that grow well in animals and in cultured cells. By the term "acute," we mean rapid production of infectious virus followed by rapid resolution and clearing of the infection by the host. Acute infections are the typical, ex-

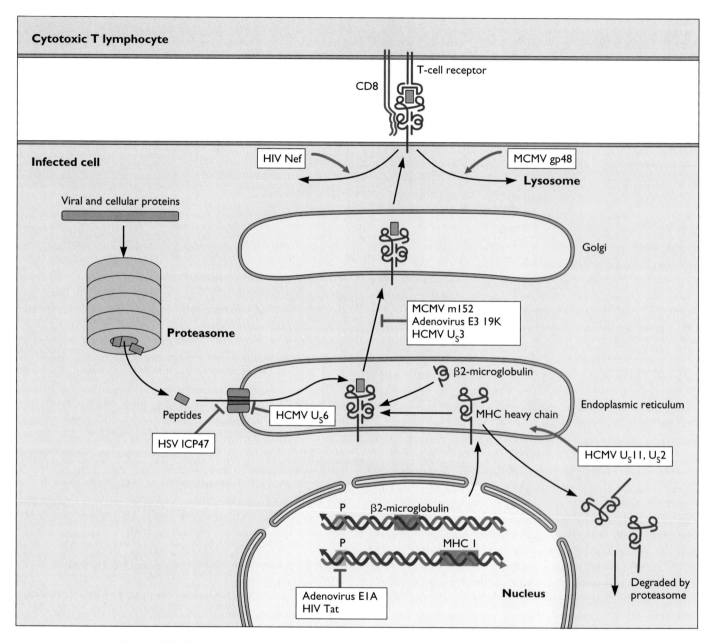

**Figure 15.7 Viral proteins block cell surface antigen presentation by the MHC class I system.** In almost every cell, a fraction of newly synthesized proteins translated in the cytoplasm is targeted to the proteasome, where the molecules are digested into peptide fragments (orange). These peptides are transported to the lumen of the endoplasmic reticulum (lavender) by the Tap transporters in the endoplasmic reticulum membrane (pink channel). Peptides then bind to a cleft in the newly synthesized MHC class I protein complex (blue) consisting of an MHC class I heavy chain and a β2-microglobulin chain. The complete peptide-MHC class I complex moves into the Golgi apparatus (light blue) and then to the cell surface, where it can be recognized by the T-cell receptors on the surfaces of CD8$^+$ T cells that are interacting with the cell. Specific viral gene products block (red bars) this process at almost every step along the pathway. Green arrows indicate stimulation. In the nucleus, transcription of MHC class I genes can be blocked by E1A or Tat at the promoter (P), as indicated. HSV, herpes simplex virus; HCMV, human cytomegalovirus; MCMV, mouse cytomegalovirus; HIV, human immunodeficiency virus.

**Table 15.4**  Viral regulation of MHC class I antigen expression[a]

| Virus | Observed effect or postulated mechanism |
|---|---|
| **RNA** | |
| Mouse hepatitis virus | Decrease in transcription of specific MHC class I genes |
| Respiratory syncytial virus | Decrease in MHC class I gene transcription |
| HIV-1 | Tat protein-induced reduction in MHC class I gene transcription |
| HIV-1 | Vpu interferes with an early step in synthesis of MHC class I proteins |
| **DNA** | |
| Adenovirus | E3 19K retains MHC class I in ER; reduced transcription of MHC class I genes induced by E1A proteins |
| Epstein-Barr virus | EBNA-4 may block generation of antigenic peptides or their transport from the cytoplasm to the ER; allele-specific decrease in MHC class I appearance on cell surface |
| Human cytomegalovirus | $U_s3$ retains MHC class I molecules in the ER; $U_s6$ inhibits peptide translocation by Tap (ER luminal domain); $U_s11$ and $U_s2$ dislocate MHC class I molecule from the ER lumen to the cytoplasm; $U_L83$ blocks IE-1 peptide presentation |
| Herpes simplex virus | ICP47 binds to Tap transporter and blocks import of peptides into the ER |
| Vaccinia virus | Lower abundance of MHC class I on cell surface induced by unknown mechanisms |

[a]Abbreviations: HIV-1, human immunodeficiency virus type 1; ER, endoplasmic reticulum; Tap, transporter of antigenic peptide; EBNA-4, Epstein-Barr virus nuclear antigen 4.

pected course for agents like influenza virus and rhinovirus (Fig. 15.8). These infections are relatively brief, and in a healthy host, virus particles and virus-infected cells are completely eliminated (cleared) by the immune system within days. Nevertheless, an acute infection is an effective survival strategy for a virus, as some progeny are invariably available for infection of other hosts before the infection is resolved (Box 15.5).

### Inapparent Acute Infections

It is important to distinguish an **inapparent acute infection** from an unsuccessful infection. Inapparent infections are successful acute infections that produce no symptoms or disease. Sufficient virus is produced to maintain the virus population, but the amount is below the threshold required to produce symptoms in the host. The usual way an inapparent infection is detected is by a rise in antiviral antibody concentrations in an otherwise healthy individual. In a healthy host, most acute infections

are inapparent because they are quickly confronted and cleared by the host immune system. Well-adapted pathogens often follow this infection pattern, as demonstrated by poliovirus, in which more than 90% of infections are inapparent.

### Acute Infections Present Common Public Health Problems

An acute infection is most frequently associated with serious epidemics of disease affecting millions of individuals every year (e.g., polio, influenza, and measles). The nature of an acute infection presents difficult problems for physicians, epidemiologists, drug companies, and public health officials. The main problem is that by the time people feel ill or mount a detectable immune response, most acute infections are essentially complete and the virus has spread to the next host. Such infections can be difficult to diagnose retrospectively or to control in large populations or in crowded environments (e.g., day care centers, military

---

**BOX 15.4**

*Human cytomegalovirus versus the host MHC class I system*

Human cytomegalovirus expresses at least five gene products that act directly to reduce the function of MHC class I molecules. Surprisingly, both human and mouse cytomegaloviruses produce a protein that resembles host MHC class I proteins. These proteins form a complex with β2-microglobulin and contain a peptide-binding groove.

One hypothesis is that these proteins act as *decoys* to distract natural killer cells. When such MHC class I decoy proteins are displayed on the infected cell's surface,

natural killer cells may ignore them despite the reduced host MHC class I expression. Evidence supporting and negating this hypothesis has been published, and it remains to be seen why these viruses have their own MHC-like proteins.

An inescapable conclusion is that MHC protein must present an important host-virus interface in cytomegalovirus infections.

Ploegh, H. L. 1998. Viral strategies of immune evasion. *Science* **280**:248–253.

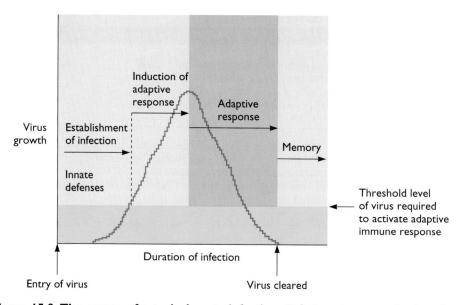

**Figure 15.8 The course of a typical acute infection.** Relative virus growth plotted as a function of time after infection. The concentration of infectious virus increases with time as indicated by the jagged red line. During the establishment of infection, only the innate defenses are at work. If the virus reaches a certain threshold level characteristic of the virus and host (purple), the adaptive responses initiate. After 4 to 5 days, effector cells and molecules of the adaptive response begin to clear the infection by removing virus particles and infected cells (green). When virus is cleared and infected cells are eliminated, the adaptive response ceases. Antibodies, residual effector cells, and "memory" cells provide lasting protection should the host be reinfected at a later date. Redrawn from C. A. Janeway, Jr., and P. Travers, *Immunobiology: the Immune System in Health and Disease* (Current Biology Ltd. and Garland Publishing, New York, N.Y., 1996), with permission.

camps, college dormitories, nursing homes, schools, and offices). Effective antiviral drug therapy requires treatment early in the infection, often before symptoms are manifested, because by the time the patient feels ill the viral infection has been resolved. Antiviral drugs can be given in anticipation of an infection, but this strategy demands that the drugs be safe and free of side effects. Moreover, as we discuss in chapter 19, our arsenal of antiviral drugs is very small, and drugs effective for many common acute viral diseases simply do not exist.

### Defense against Acute Infections

The typical immediate host response that limits most acute infections is the innate response: synthesis of interferons and lytic attack by natural killer cells. The antiviral defense strategy of killing a few cells as a "firebreak" is quite common and very effective. In a naive host, the adaptive immune response (antibody and activated cytotoxic T cells) does not influence virus growth for several days, but it is essential for final clearance of virus and infected cells from the infected host, as well as for providing memory for de-

fense against subsequent infections. In immunocompromised individuals, acute infections can be disastrous, primarily because the infection does not remain localized to the primary site of infection. This property highlights containment of an acute infection as a central role of the immune system in the delicate balance between virus offense and host defense. As we discuss in chapter 17, the protective immune response that follows acute infection by some viruses, such as dengue virus, actually makes subsequent infection by dengue virus of a different serotype much more severe.

### Multiple Acute Infections in a Single Host

An initial acute infection by some viruses may be followed by a second or third round of infection in the same animal. In these cases, virus spreads from the primary site of infection to other tissues, where a second acute infection can occur. We will discuss mechanisms of spread in more detail in chapter 17. In **hematogenous spread**, virus particles enter the bloodstream from the primary site via the lym-

An influenza virus infection begins in the upper respiratory tract by inhalation of droplets from a sneeze or cough by an infected individual. Virus replicates in ciliated columnar epithelial cells of the respiratory epithelium, releasing progeny virus that spreads to nearby cells. Infectious virus can be isolated for 1 to 7 days, with the peak of released virus occurring on the fourth or fifth day after infection. About 48 h after the initial infection, symptoms appear abruptly such that infected individuals can almost pinpoint the hour that they noticed they had the flu. Symptoms last for about 3 days and then begin to abate.

The infection typically resolves within a week through action of the innate and acquired immune systems, but it may take several weeks before the individual feels completely well because of the lingering effects of the host defensive responses.

phatic system or subepithelial blood vessels serving that site. Viruses can also infect mobile cells that enter the blood (e.g., lymphocytes) and migrate to new tissues. The first appearance of virus particles in the blood is called a **primary viremia**. Once in the circulation, the virus or virus-infected lymphocytes can initiate another acute infection of internal organs, including the liver, spleen, lungs, or heart. Progeny virus resulting from this second acute infection can also enter the bloodstream, producing a **secondary viremia** with a potential for even more serious problems for the host. Secondary viremias are usually characterized by virus titers in the blood much higher than those of primary viremias. Classic examples of viruses that establish acute primary and secondary viremias are mousepox, smallpox, and measles viruses. Virus can also spread from the primary site of infection to the peripheral and central nervous systems by infecting neurons in close proximity to the primary infection (**neurological spread**). This pattern of multiple acute infections, viremias, and spread to the nervous system is characteristic of varicella-zoster virus, an alphaherpesvirus that causes the familiar childhood disease chickenpox (Fig. 15.9).

## Pathogenic Effects of an Acute Infection

While many symptoms of an acute infection (e.g., fever, malaise, aches, nausea) are actually due to a vigorous host immune response, an acute viral infection of internal organs, the gut, the respiratory system, or the nervous system can cause considerable damage because infected cells are killed by infection and by the immune system. If a sufficient number of cells are infected, severe problems may result. Symptoms such as diarrhea, poor lung or liver function, and breakdown of capillary beds or the blood-brain barrier can arise from direct viral damage to important epithelial and endothelial surfaces of the body. Local cell damage caused by active replication, by viral proteins, or by local necrotic reactions also can have serious consequences, including secondary infections by opportunistic bacteria. As discussed in chapter 17, several viral membrane proteins possess intrinsic activities that may be toxic, such as the human immunodeficiency virus gp120 envelope protein, which damages neurons. In the case of enteric rotaviruses, a single nonstructural viral protein (NSP4) alone may be sufficient to cause pathological fluid loss in gut epithelial cells.

## Persistent Infections

### Definition and Requirements

Unlike an acute infection, a persistent infection is not cleared quickly and virus particles or viral products continue to be produced for long periods. Infectious virus may be produced continuously or intermittently for months or years (Fig. 15.1). In some instances, viral genomes remain long after viral proteins can no longer be detected. Distinctions have been made between persistent infections that are eventually cleared (**chronic infections**) and those that last the life of the host (latent infections or slow infections; see next sections).

Many viruses can establish a persistent infection (Table 15.5). They not only shut down any lethal or cytolytic activities they encode or initiate but also must avoid host antiviral defenses for extended periods. For example, some arenaviruses, like lymphocytic choriomeningitis virus, are inherently noncytopathic in their natural hosts and maintain a persistent infection if the host cannot clear the infected cells. Other viruses, like the herpesvirus Epstein-Barr virus, use alternative transcription and replication programs to maintain the viral genome in some cell types with no production of virus particles. In any case, the evasion of host immune defenses is paramount in establishing a persistent infection.

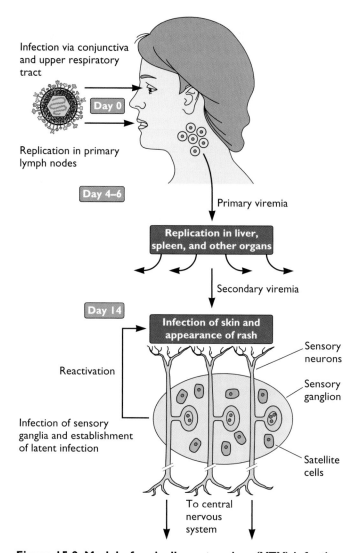

Infection via conjunctiva and upper respiratory tract

Day 0

Replication in primary lymph nodes

Day 4–6

Primary viremia

**Replication in liver, spleen, and other organs**

Secondary viremia

Day 14

**Infection of skin and appearance of rash**

Reactivation

Infection of sensory ganglia and establishment of latent infection

Sensory neurons

Sensory ganglion

Satellite cells

To central nervous system

**Figure 15.9 Model of varicella-zoster virus (VZV) infection and spread.** The virus infects the conjunctiva or mucosa of the upper respiratory tract and then moves to infect regional lymph nodes. After 4 to 6 days from the initial infection, infected T cells enter the bloodstream, causing a primary viremia. These infected cells subsequently invade the liver, spleen, and other organs, causing a second round of infection. Virus and virus-infected cells are then released into the bloodstream in a secondary viremia, subsequently infecting the skin. This third round of infection gives rise, after 2 weeks from the initial infection, to the characteristic vesicular rash of chicken pox. Following acute replication in the skin, VZV then infects sensory ganglia of the peripheral nervous system, where it establishes a new pattern of infection called latency. The nervous system is not subject to vigorous immune surveillance and is an excellent site for a virus to hide. Later in life, probably as the specific immune response to VZV wanes, the latent virus occasionally reactivates and initiates another acute infectious cycle. This causes the characteristic recurrent disease called shingles, often accompanied by a painful condition called postherpetic neuralgia.

## Infections of Tissues with Reduced Immune Surveillance

Cells and organs of the body differ in the degree of their immune defense. Tissues with surfaces exposed to the environment (e.g., skin, glands, bile ducts, and kidney tubules) do not employ active immune surveillance, presumably because they are exposed to foreign matter on a routine basis and possess other mechanisms to contend with this situation. If a virus can infect such tissue, it may establish a persistent infection. Other organs, such as the eye, avoid damaging lymphocyte invasion by expression of Fas ligand. When invading T cells recognize and bind Fas ligand on cells of the eye, the T cells die by apoptosis. Certain compartments of the body, including the central nervous system, vitreous humor of the eye, and areas of lymphoid drainage, are thought to be devoid of effectors of humoral defense such as the complement system. The eye and brain may be isolated from routine immune surveillance because they are highly susceptible to damage that might result from the fluid accumulation, swelling, and ionic imbalances produced by an inflammatory response. Moreover, because most neurons do not regenerate, immune defense by cell death is obviously detrimental. Tight junctions between the epithelial cells that line brain capillaries and ventricles, called the **blood-brain barrier**, limit entry of some molecules and cells. Lymphocytes are able to enter the central nervous system but are not retained unless they encounter properly presented foreign antigens. However, neurons (as well as muscle and other cells) do not express MHC proteins readily and in normal circumstances are not "seen" by cytotoxic T cells. As a consequence, the brain is a favored site for the establishment of a persistent infection by a variety of viruses.

Herpesviruses, papovaviruses, and some complex retroviruses are prime examples of infectious agents that establish persistent infections, in part by infecting tissues with reduced immune surveillance. By replicating on luminal surfaces of glands and ducts with poor immune surveillance (kidney, salivary, and mammary glands), human cytomegalovirus will be shed almost continually in secretions. Possibly the most extreme example of immune avoidance is represented by papillomaviruses that cause skin warts. Productive replication of these infectious particles occurs only in the outer, terminally differentiated skin layer where an immune response is impossible. Dry skin is continually flaking off, ensuring efficient spread of virus. One can verify this assertion by running a finger along a clean surface in the most hygienic hospital and noticing a white film, which is 70 to 80% keratin from human skin. Molecular biologists often discover this abundance of dried skin in the laboratory when examining silver-stained

**Table 15.5**  Persistent viral infections of humans

| Virus | Site of persistence | Consequence(s) |
|---|---|---|
| Rubella virus | CNS[a] | Progressive rubella panencephalitis |
| Hepatitis C virus | Liver | Cirrhosis, hepatocellular carcinoma |
| Measles virus | CNS | Subacute sclerosing panencephalitis, measles inclusion body encephalitis |
| Human immunodeficiency virus | CD4+ T cells, macrophages, microglia | AIDS |
| Human T-cell leukemia virus types 1 and 2 | T cells | Leukemia |
| Hepatitis B virus | Liver, lymphocytes | Cirrhosis, hepatocellular carcinoma |
| Hepatitis D virus | Liver | Pathological synergy with hepatitis B virus |
| Polyomavirus JC | Kidneys, CNS | Progressive multifocal leukoencephalopathy |
| Polyomavirus BK | Kidneys | Hemorrhagic cystitis |
| Papillomavirus | Skin, epithelial cells | Papillomas, carcinomas |
| Adenovirus | Adenoids, tonsils, lymphocytes | None known |
| Varicella-zoster virus | Sensory ganglia | Zoster (shingles), postherpetic neuralgia |
| Herpes simplex virus types 1 and 2 | Sensory and autonomic ganglia | Cold sores, genital herpes |
| Epstein-Barr virus | B cells, nasopharyngeal epithelia | Lymphoma, carcinoma |
| Human cytomegalovirus | Kidneys, salivary glands, lymphocytes, macrophages, stromal cells | Pneumonia, retinitis |

[a]CNS, central nervous system.

protein gels; the major band is often contaminating keratin.

## Direct Infection of the Immune System Itself

Some viruses infect cells at the very heart of the host antiviral defense system, lymphocytes and macrophages. Such cells not only play direct roles in the immune defense but also migrate to the extremes of the body, providing easy transport of virus to new areas and hosts. Infection of such cells is likely to reduce immune function directly, and as a result virus should be able to persist in the host. The effectiveness of this strategy is demonstrated by the surprisingly large number of viruses known to infect lymphocytes and monocytes (Table 15.6).

Human immunodeficiency virus provides a powerful reminder of how effective infection of the immune system can be (see also chapter 18). The virus infects not only CD4+ T-helper cells but also monocytes/macrophages that can transport virus to the brain and other organs. Professional antigen-presenting cells, such as the dendritic cells in the spleen discussed in chapter 14, are also infected. An untreated infected individual continues to produce prodigious quantities of virus for years. Disease is characterized by persistent immune activation. One line of research suggests that very early reactions in virus attachment to susceptible T cells play a major role in this phenomenon. When the virus attaches to susceptible T-helper cells in tissue culture by binding of envelope gp120 protein to the

**Table 15.6**  Viruses that infect lymphocytes and monocytes

(+) strand RNA viruses
  Poliovirus
  Rubella virus
  Caliciviruses

(−) strand RNA viruses
  Lymphocytic choriomeningitis virus
  Measles virus
  Mumps virus
  Respiratory syncytial virus
  Influenza A virus
  Vesicular stomatitis virus
  Parainfluenza virus

Retroviruses
  Murine leukemia virus
  Feline leukemia virus
  Human T-cell leukemia virus types 1 and 2
  Human immunodeficiency virus
  Endogenous C-type virus

Single-stranded DNA viruses
  Porcine parvovirus
  Minute virus of mice

Double-stranded DNA viruses
  Hepatitis B virus
  Papovaviruses
  Group C adenoviruses
  Herpes simplex virus
  Varicella-zoster virus
  Epstein-Barr virus
  Human herpesvirus 6
  Human herpesvirus 7
  Human cytomegalovirus
  Leporipoxviruses

CD4 and Ccr5 chemokine receptors, a signal transduction cascade is activated. As a result, the infected T cells express putative chemoattractant molecules that attract more uninfected T cells (new host cells) to the site of infection In addition, the virus replicates efficiently in these activated cells. Thus, as a result of engagment of the receptor and coreceptor, *both* virus replication and cell-cell spread are accelerated.

## Two Viruses That Cause Persistent Infections

### Lymphocytic Choriomeningitis Virus

Lymphocytic choriomeningitis virus, a member of the family *Arenaviridae*, was the first virus associated with aseptic meningitis in humans. Perhaps more importantly, its study has illuminated fundamental principles of immunology and viral pathogenesis, particularly those that underlie persistent infection and cytotoxic-T-cell recognition and killing. Lymphocytic choriomeningitis virus is not cytopathic and also infects both mice and humans. It was noted early on that the virus spreads from rodents (the natural host) to humans, in which it can cause severe neurological and developmental damage. Infected rodents normally produce large quantities of infectious virus, which is excreted in feces and urine throughout their lives without any apparent detrimental effect. These mice are called **carriers** because of such lifelong production of virus. The carrier state occurs in part because of two phenomena: the virus is not cytopathic; and, if mice are infected congenitally or immediately after birth, the virus is not recognized as "foreign" and thus cannot be cleared by the murine immune system. However, if the virus is injected into the brains of healthy adult mice in the laboratory, the mice die of acute immunopathologic encephalitis. This disease is similar to that contracted by humans who develop aseptic meningitis after lymphocytic choriomeningitis virus infection.

From results with the laboratory mouse model, we now understand that cytotoxic T cells are required both for clearing lymphocytic choriomeningitis virus and for the lethal response to intracerebral infections. If adult mice are depleted of such cells, direct injection of virus into the brain is no longer fatal. Instead, the mice express infectious virus throughout their lifetimes, precisely as seen in persistent infections of neonates. When lymphocytic choriomeningitis virus-responsive, cytotoxic T cells are added back to neonates with a persistent infection, all virus is cleared after several weeks. In the case of neonatal persistent infection, virus is not in the brain, so no immunopathologic encephalitis is promoted by the activated T cells. How the neonatal infection effectively "silences" the effector system from clearing virus is currently under investigation. Recent experiments have implicated an ac-

tive process of clonal deletion of T lymphocytes capable of recognizing lymphocytic choriomeningitis virus antigens.

### Measles Virus

Measles virus provides a provocative example of the delicate balance that determines the pattern of infection. Many important questions remain unanswered about how this human pathogen switches from an acute to a persistent infection and does so despite an active immune response.

Measles virus, a member of the family *Paramyxoviridae*, is a common human pathogen with no known animal reservoir. The genome organization and replication strategy are similar to those of the rhabdovirus vesicular stomatitis virus. Measles is one of the most contagious human viruses, with about 40 million infections occurring worldwide each year, resulting in 1 to 2 million deaths. Normally, it causes an acute infection only once in a lifetime because a single infection routinely protects the individual. Measles virus is a highly adapted human pathogen, persisting only in populations sufficient to produce a large number of new hosts (children). Population geneticists calculate that communities of 200,000 to 500,000 individuals are required in order to maintain measles virus.

The receptor for measles virus is the human complement regulatory cofactor protein CD46. Despite a rather broad distribution of its receptor, measles virus shows tropism for the respiratory tract, the site of primary infection. In an acute infection, the disease course runs about 2 weeks—so-called uncomplicated measles (Fig. 15.10). An acute infection causes cough, fever, and conjunctivitis and, as already noted, confers lifelong immunity. The familiar rash is due to a hypersensitivity reaction.

On closer examination, the hallmarks of acute measles infection are sinister, and as we now appreciate, presage the subsequent potentially serious problem of persistent infection. Such hallmarks include the following features: measles virus kills cells by cell-cell plasma membrane fusion, not by shutting off host macromolecular DNA, RNA, and protein synthesis; the virus can replicate in a variety of tissues, including cells of the nervous system and the immune system; the majority of acutely infected individuals exhibit an abnormal electroencephalograph indicating subtle neurological damage from uncomplicated measles; and finally, an initial acute infection results in both virus dissemination via viremia and *general immunosuppression.*

Immunosuppression is responsible for many deaths from secondary infections; it may also be necessary to establish a persistent infection. The immunosuppressive effect is not long-lived, and with proper care the vast majority of patients recover with no further problems. The molecular basis for immunosuppression is not yet fully

**A**

**Pleomorphic particles**
100–300 nm

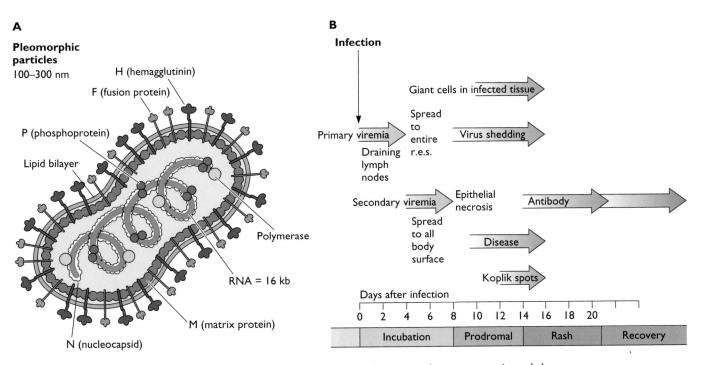

**Figure 15.10 Infection by measles virus.** (A) Diagrammatic representation of the structure of the pleomorphic measles virion. (B) Course of clinical measles infection and events occurring in the spread of the virus within the body are illustrated. Four clinically defined temporal stages occur as infection proceeds and are illustrated at the bottom. As virus spreads by primary and secondary viremia from the lymph node to the entire reticuloendothelial system (r.e.s.) and finally to all body surfaces, characteristic symptoms and clinical findings appear. The timing of typical reactions that correspond to the clinical stages is shown by the colored arrows. The telltale spots on the inside of the cheek (Koplik's spots) and the skin lesions of measles consist of pinhead-sized papules on a reddened, raised area. They are typical of immunopathology in response to measles virus antigen. Redrawn from A. J. Zuckerman et al., *Principles and Practice of Clinical Virology*, 3rd ed. (John Wiley & Sons, Inc., New York, N.Y., 1994), with permission.

understood, but several mechanisms have been suggested (see also chapter 17). We know that the virus infects T and B cells, as well as macrophages, arresting them in the late $G_1$ phase of the cell cycle. As a result, the infected cells cannot perform their normal functions. Uninfected lymphocytes can also be suppressed, as demonstrated by cocultivation with virus-infected cells. Suppression of this type requires cell-cell contact and expression of both the viral hemagglutinin (HA) protein and the fusion protein. When the viral HA binds to CD46 receptor protein, expression of interleukin-12 (Il-12) is suppressed. As a result, the Th1 immune response is not activated efficiently.

On rare occasions, measles virus genomes and antigens may persist for years in a single individual. The mechanisms responsible are only now being characterized. Paradoxically, these studies indicate that antibodies produced to check the acute infection play a central role in driving the virus into persistent infection. While the mechanism for this unexpected effect in humans is not understood, some insight has come from studying the effect of antibodies during infection of tissue culture cells. Measles virus-specific antibodies reduce expression of viral membrane proteins on the cell surface by an unknown mechanism. Because one of these proteins, viral fusion protein, is responsible for the cell-cell fusion that causes cell death, exposure to antibodies effectively blocks cell killing and thereby allows viral persistence.

Measles virus can enter the brain by infecting lymphocytes that traverse the body during the viremia following primary infection. Such a secondary infection has a number of consequences. One is **acute postinfectious encephalitis**, which occurs in about 1 in 3,000 infections. The other is a rare, but delayed and often lethal, brain infection called **subacute sclerosing panencephalitis**.

This disease results from a slow infection, a unique variation of a persistent infection (see "Slow Infections" below and Fig. 15.1). Six to eight years after young adults and children contract measles, about one in a million develop subacute sclerosing panencephalitis. It appears that this disease results when intracellular host proteins in cells of the central nervous system interfere with acute infection by inhibiting viral gene expression, especially synthesis of envelope proteins. Available evidence indicates that components essential for the assembly and budding of mature infectious virus particles are absent in brains of afflicted patients. One line of thinking focuses on aberrant expression of the matrix (M) protein as an important participant in this fatal infection. The matrix proteins of (−) strand RNA viruses are essential for assembly of particles, and the M protein is often poorly expressed in persistent paramyxovirus infections. Much remains to be discovered, and important questions remain. Does immunosuppression during an acute infection facilitate infection in the brain? Are defects in M protein expression and particle assembly necessary and sufficient to cause disease, or are they effects of other selection processes in the brain? Do the defects in viral gene expression require long exposure to the host, or are they caused by host cell proteins that reduce viral gene expression? Transgenic mice expressing the human measles virus receptor CD46 are now available and should allow these important questions to be addressed in a rigorous and controlled fashion.

## Latent Infections

### An Extreme Variation of the Persistent Infection

Latent infections can be characterized by four general properties: a nonreplicating cell is infected or the viral genome is replicated in conjunction with host DNA replication so that the cell cycle is not interrupted; immune detection of the cell harboring the latent genome is reduced or eliminated; expression of productive cycle viral genes is absent or inefficient; and the viral genome itself persists intact so that at some later time a productive acute infection can be initiated to ensure spread of its progeny to a new host (Fig. 15.1). The latent genome can be maintained as a nonreplicating chromosome in a nondividing neuron (herpes simplex virus), become an autonomous, self-replicating chromosome in a dividing cell (Epstein-Barr virus), or be integrated into a host chromosome (adeno-associated virus).

Such "long-term parking" of a viral genome is remarkable for its stability, which requires a delicate balance among the regulators of viral and cellular gene expression. Generally, only a restricted set of viral gene products are made during a latent infection. Viral proteins required for productive replication may not be produced at all, a pattern exemplified by herpes simplex virus. This virus establishes a latent infection in nondividing neurons, and only characteristic **latency-associated transcript (LAT)** RNAs can be detected in the nuclei. In contrast, several viral proteins may be required to maintain the latent infection, as is the case for Epstein-Barr virus. This virus establishes latency in B lymphocytes when a specific transcription program produces at least seven viral proteins needed to replicate and maintain the viral genome as the latently infected cells divide.

If latency is to have any value as a survival strategy, the latent virus must have a mechanism for reactivation so that it can spread to other hosts. Reactivation usually follows trauma, stress, or other insults, conditions that may mark the host as a poor place to continue the latent infection. In the case of herpes simplex virus, reactivation can also provide a mechanism for reinfection and establishment of a latent infection in more neurons in the same individual. This is the so-called **round-trip strategy**: virus reactivates from neurons and replicates at mucosal surfaces, and the progeny enter more neuronal termini to repopulate ganglia. The latent infection is remarkable for its simplicity and effectiveness as a survival strategy.

## Two Viruses That Produce Latent Infections

### Herpes Simplex Virus

Over 40 million people in the United States (about 20% of the population) harbor latent herpes simplex virus in their peripheral nervous system and will experience reactivation of their own private virus sometime in their lifetime. Many millions more carry latent herpes simplex virus in their nervous system but never report reactivated infections. Herpes simplex virus is a highly efficient pathogen as demonstrated by its widespread prevalence in humans, its only known host. No animal reservoirs are known, although several laboratory animals, including rats, mice, guinea pigs, and rabbits, can be infected. Such success results from the high efficiency of both the productive and latent infections in humans. Herpes simplex virus is unique in establishing latent infections predominately in terminally differentiated, nondividing neurons. These cells are excluded from some forms of immune surveillance. They neither replicate their DNA nor divide, and so once established, a viral genome need not replicate to persist for the life of the neuron. Finally, sensory neurons are highly connected through synapses and thus serve as excellent conduits for transport of virus particles to and from mucosal surfaces and other segments of the nervous system. The ability of the virus to establish a latent infection is an effective survival mechanism, as neither vaccines nor antivirals can attack the virus in its latent form. Once infected with these viruses, the host is infected for life—latency is absolute persistence.

Many details of the molecular aspects of herpes simplex virus latency await discovery and explanation, but the general pathway and critical issues are well established. Virus infects neurons of sensory and autonomic ganglia following primary rounds of replication in cells of mucosal or epidermal surfaces (Fig. 15.11). The general outline of putative regulatory steps necessary for the establishment, maintenance, and reactivation of a viral infection after primary infection is shown in Fig. 15.12. A typical primary infection of a mouse, showing the time course of production of infectious virus and establishment of a latent infection, is seen in Fig. 15.13. One or two weeks after primary infection of the ganglion, infectious virus can no longer be isolated—the operational definition of an established latent infection. If the animal survives the primary infection, establishment of the latent infection is inevitable. The time frame for this process varies depending on the animal species, the concentration and genotype of the infecting virus, and the site of primary infection. It is not clear how neurons in the ganglia survive the primary infection. We do not yet know how the viral genes whose products are necessary to complete the productive infection are turned off as latent infections are being established. Plausible hypotheses are that terminally differentiated, nondividing neurons may lack proteins required for viral gene transcription or may contain repressors of transcription not found in "permissive" cells. Indeed, there is evidence for both mechanisms. Viral gene products may also serve to repress viral gene expression. The latent genome persists as a nonintegrated circular DNA molecule associated with nucleosomes and is probably tethered to a specific site in the nucleus (Box 15.6). Some latently infected neurons express the latency-associated transcripts mentioned above. They are discussed in detail in chapter 8.

Nonneuronal cells and the immune system play important roles in establishing the pattern of herpes simplex virus infection. For example, only 10% of the cells in a typical sensory ganglion are neurons; the remaining 90% are nonneuronal satellite cells and Schwann cells associated with a fibrocollagenous matrix. These cells are in intimate contact with ganglionic neurons. Some of the nonneuronal cells are infected during initial invasion of the ganglion. The peripheral nervous system is accessible to antibodies, complement, cytokines, and lymphocytes of the innate and adaptive immune system. In murine models, the immune response to productive infection in the ganglion actively influences the outcome of primary infection of neurons. For example, cytotoxic T cells that recog-

nize viral antigens and passive immunization with antibodies to virus facilitate efficient establishment of latency. How this is accomplished without death of the infected neurons remains controversial.

Establishment of a reactivatable latent infection in sensory ganglia that service mucosal surfaces is a particularly effective means of ensuring virus transmission because mucosal contact is widespread among affectionate humans. Virus can be spread both by infection at mucosal surfaces or by cuts and abrasions in the skin. However, a person must be actively producing infectious virus to transfer herpes simplex virus to another person. Local spread of reactivated virus is rapidly curtailed because the host is immunized during the primary infection. The ability of herpes simplex virus to spread among mucosal epithelial cells after reactivation in an immune host may be facilitated by action of the viral protein ICP47, which blocks MHC class I presentation of viral antigens to the T cells that initiate the cellular immune response. Such activity may provide the virus with sufficient time for a few rounds of replication before elimination by activated cytotoxic T lymphocytes.

Some individuals with latent herpes simplex virus experience reactivation *every 2 to 3 weeks*, while others have only rare or no episodes of reactivation. The signaling mechanisms that reactivate the latent infection are not well characterized, although sunburn, stress, nerve damage, depletion of nerve growth factor, steroids, heavy metals, and trauma (including dental surgery) all promote reactivation. We can imagine that such diverse exogenous signals converge to activate specific cellular proteins needed for transcription of herpes simplex virus immediate-early genes and thus activate the productive transcriptional program. Alternatively, reactivation may not require action of viral transcriptional regulators but may result from activation of cellular analogs of these immediate-early proteins capable of direct activation of viral early genes (Fig. 15.12). Reactivation may be an all-or-none process requiring but a single reaction, such as inactivation of a repressor or production of an activator, to "flip the switch" that triggers the cascade of gene expression of the lytic pathway. Glucocorticoids are excellent examples of such activators, as they stimulate transcription rapidly and efficiently while inducing an immunosuppressive response. These properties explain the observation that clinical administration of glucocorticoids frequently results in reactivation of latent herpesvirus. An interesting hypothesis is that spontaneous reactivation results from single reactions in individual neurons, each of which leads to a small burst of herpes simplex virus transcription. When the stimulus is strong enough, such sporadic transcription ultimately passes a threshold, resulting in replication and reactivation. This idea is consis-

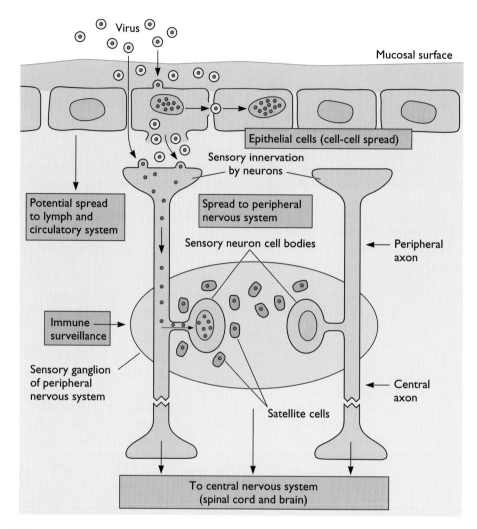

**Figure 15.11 Herpes simplex virus primary infection of a sensory ganglion.** Virus replication occurs at the site of infection (primary infection), usually mucosal surfaces, and the infection may or may not be apparent. Host innate defenses normally limit the spread of virus at this stage. Virus may infect local immune effector cells, including dendritic cells and infiltrating natural killer cells. Virus also spreads locally between epithelial cells and has access to lymph and the circulation as well as to neurons that service the mucosal surfaces. Virus replication in epithelial cells may not be an absolute requirement for infection of neurons because replication-defective herpes simplex virus mutants can establish a latent infection in model systems. During the primary infection, virus is taken up at nerve endings of the local sensory and autonomic nerves innervating the area of infection and is transported to the neuronal cell bodies in the peripheral ganglia by transport systems operating within the neurons. It is likely that the virus loses its envelope on entry into these neurons and the capsid is transported to the neuronal cell body, where it delivers the viral DNA to the nucleus. In some neurons, virus enters the productive cycle forming many infectious virus particles that then spread to other neurons in synaptic contact with the infected neuron (transsynaptic spread). Such spread could be reflected in local spread among cells in the ganglion, more distant spread to other ganglia that are in synaptic contact, or spread to the central nervous system (CNS; the spinal cord or the brain). Spread to the CNS from the peripheral nervous system is frequently fatal, but fortunately it is a rare consequence of primary herpes simplex virus infection. Primary infection is normally constrained to the ganglion. Viral proteins can also be detected in the nonneuronal satellite cells of the ganglion during the acute infection. Sensory and autonomic ganglia are in close contact with the bloodstream and do not have the so-called blood-brain barrier found in the CNS. The ganglion can become inflamed and be transiently visited by various lymphoid cells that leave the circulatory system in response to the infection. Infection of the ganglion is usually resolved within 7 to 14 days after primary infection, virus particles are cleared, and a latent infection of some neurons in the ganglion is established (see also Fig. 15.13).

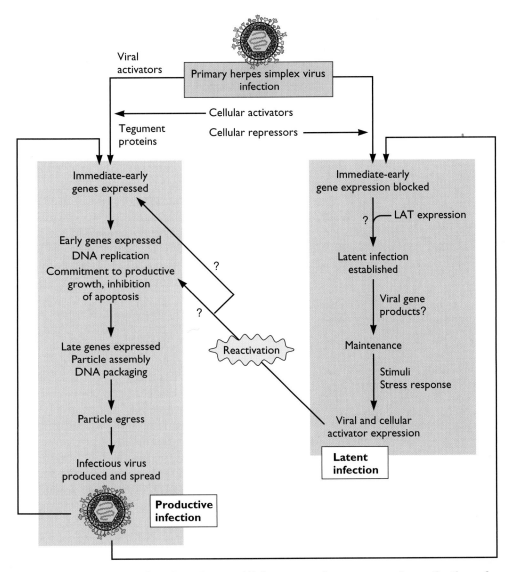

**Figure 15.12 General flowchart for establishment, maintenance, and reactivation of a latent infection by herpes simplex virus.** The green box at the top indicates the primary infection by virus particles at mucosal surfaces. The productive infection is shown by the pathway on the left, and the latent infection is indicated by the pathway on the right. The question marks indicate our lack of knowledge concerning expression and function of viral proteins at the indicated steps. Infectious particles produced by the productive pathway may infect other cells and enter either the productive or latent pathway as indicated. Reactivation is indicated by the diagonal arrow from the latent infection to the start of the productive infection. The question marks note the current controversy as to whether reactivation requires "going back to go" (immediate-early gene expression) or expression of early genes required for in DNA replication. Adapted from M. A. Garcia-Blanco and B. R. Cullen, *Science* **254:**815–820, 1991, with permission.

tent with the low levels of immediate-early and early transcripts that can be detected in ganglia harboring a latent viral infection. Currently, we know only that reactivation of the latent genome requires resumption of transcription and initiation of viral DNA replication.

## Epstein-Barr Virus

Epstein-Barr virus, which infects only humans, is exceptionally efficient at establishing latent infections in B lymphocytes. About 90% of the world's population carries Epstein-Barr virus as a latent virus. The virus has two major

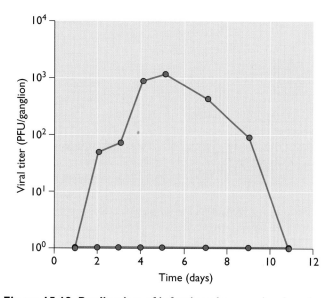

**Figure 15.13 Replication of infectious herpes simplex virus type I in mouse trigeminal ganglia during acute infection.** Mice were anesthesized and infected by a standard strain of herpes simplex virus by dropping approximately $10^5$ plaque-forming units of virus onto the cornea of one eye that had been lightly scratched with a sterile needle. After a few minutes, the liquid was blotted and the animal was allowed to recover. At selected time points, animals were euthanized and the trigeminal ganglia were dissected quickly and frozen. Each point on the graph (red line) represents the geometric mean titer in plaque-forming units from eight individual ganglia from two different experiments tested at the indicated time after infection. Uninfected animal controls are indicated by the blue line.

target tissues in vivo, B lymphocytes and epithelia. People carrying latent virus in their B cells not only maintain cytotoxic T cells directed against proteins made by the virus during the latent infection but also shed small quantities of infectious virus. How latency is maintained in the face of an active immune response is an important question.

The primary site of Epstein-Barr virus infection is the oropharyngeal cavity. The virus probably replicates in differentiating epithelial cells. Children and teenagers are commonly afflicted, usually following close oral contact (hence the name "kissing disease"). This acute infection requires expression of most of the genes contained in the viral genome. Spread of virus to B cells in an individual with a normal immune system can induce substantial immune and cytokine responses, resulting in a disease called **infectious mononucleosis**. Epstein-Barr virus establishes a latent infection in B cells by an active process that requires the expression of a specific subset of viral genes. An important property of one class of virus-infected B cells is their ability to proliferate indefinitely (i.e., they are "immortalized"). The latent virus genome is maintained as a circular episome that replicates via a program distinct from the one used during productive replication, which was discussed in chapter 9. B cells latently infected by Epstein-Barr virus contain a set of nuclear proteins, termed Epstein-Barr virus nuclear antigens, as well as two membrane proteins that are important in altering the properties of the host cells (Table 15.7).

There are at least two distinct phenotypes of viral latency distinguished by the viral gene products made in an infected B cell. First, a B-cell growth-promoting program is

---

**BOX 15.6**

*Neurons harboring latent herpes simplex virus often contain hundreds of viral genomes*

The number of neurons in a ganglion that will ultimately harbor latent genomes depends on the host, the virus, the concentration of infecting virus, and the conditions at the time of infection. It is possible to infect as few as 1% to as many as 50% of the neurons in a ganglion. In the mouse trigeminal ganglion (about 20,000 neurons per ganglion), this means that less than 200 or more than 10,000 neurons may carry latent genomes. In controlled experiments in mice, the number of latently infected neurons increases as the titer of infecting virus increases.

Interestingly, many infected neurons contain multiple copies of the latent viral genome varying from fewer than 10 to more than 1,000; a small number have more than 10,000 copies. This variation in copy number has been enigmatic. Does

it reflect multiple infections of a single neuron or is it the result of replication in a stimulated permissive neuron after infection by one particle? If it is the latter, how does the neuron recover from what should be an irreversible commitment to the productive cycle?

Recent experiments indicate that viruses that cannot replicate or whose replication is blocked by antiviral drugs exhibit a significant reduction in the number of latently infected neurons with multiple genomes. Thus, it is likely that a single neuron can be infected by multiple viruses, each of which participates in the latent infection.

**Sawtell, N. M.** 1997. Comprehensive quantification of herpes simplex virus latency at the single-cell level. *J. Virol.* **71:**5423–5431.

active during acute infection, and these infected B cells express all the known latency-associated genes (sometimes called stage 3 latency) (Table 15.7). The multiple viral proteins produced are required to establish the latent infection and to avoid elimination of infected cells by the immune system. A second phenotype is found in which only EBNA-1 protein is expressed (sometimes called stage 1 latency). Cells of this phenotype are found in Burkitt's lymphoma but have been difficult to detect in virus-infected individuals. Recent evidence indicates that major sites of viral persistence in the peripheral blood are *resting* B cells. Such nonreplicating cells express an mRNA that specifies latent membrane protein 2A but not Epstein-Barr virus nuclear antigen 1. These cells do not express the B7 coactivator receptor on their surfaces (see chapter 14) and thus are not killed by cytotoxic T cells.

A remarkable feature of Epstein-Barr virus persistence is the equilibrium established between active immune elimination of infected cells and viral persistence. Although normal humans infected with virus maintain cytotoxic T cells directed against many of the viral proteins synthesized by latently infected B cells, these cells are not eliminated. This is because some proteins, like latent membrane protein 1, inhibit apoptosis or immune recognition of the latently infected cells. Moreover, peptides of Epstein-Barr virus nuclear antigen 1 are not presented to T cells as discussed above. When the equilibrium between latently infected B-cell proliferation and the immune response is altered (e.g., after immunosuppression of Epstein-Barr virus-positive patients following bone marrow transplantation), the virus-immortalized B cells can form lymphomas (see chapter 16).

The signals that reactivate latent Epstein-Barr virus infection in vivo are not well understood, but considerable information has been obtained by studies of activation of latent B-cell infections in cultured cells. In these circumstances, activation can be achieved by stimulating certain signal transduction cascades or by providing an essential virus transcriptional activator, Zta (Z or zebra protein [see below]). These studies are all consistent with the hypothesis that the latent state is not maintained by a repressor of productive infection.

Many signal transduction pathways efficiently reactivate Epstein-Barr virus from the latent state. They can be activated by diverse interactions, including clustering of the B-cell antigen receptor CD21 induced by anti-immunoglobulin antibodies (activation of tyrosine kinases), binding of phorbol esters (stimulation of protein kinase C), and introduction of calcium ionophores. Therefore, it is surprising that latent virus infection of B cells is so stable in vivo. We now know that virus-encoded latent membrane protein 2A makes an important contribution to maintaining the latent infection. This protein inhibits tyrosine kinase signal transduction pathways. It is the first example of a viral protein that blocks reactivation of a virus. As virus reactivates efficiently in vivo, a second signal transduction pathway that bypasses the latent membrane protein 2A block must exist, but thus far such a pathway has not been found.

B cells harboring latent Epstein-Barr virus genomes can also be reactivated by superinfection with a virus that expresses a virus immediate-early gene product called Zta. While the role of viral superinfection in virus reactivation in vivo is not well understood, this phenomenon has pro-

**Table 15.7**  Epstein-Barr virus proteins required for establishment and maintenance of latent infection

| Epstein-Barr virus protein[a] | Function(s) |
|---|---|
| EBNA-1 | Maintains replication of the latent Epstein-Barr virus genome during the S phase of the cell cycle. It is a sequence-specific DNA-binding protein and binds to a unique origin of replication called oriP that is distinct from the origin used in the productive replication cycle. |
| EBNA-2 | A transcription factor that coordinates Epstein-Barr virus and cell gene expression in the latent infection by activating the promoters for the LMP-1 gene and cellular genes like CD23 (low-affinity immunoglobulin E Fc receptor), CD21 (the Epstein-Barr virus receptor, CD23 ligand, and receptor for complement protein C3d) |
| EBNA-LP | Required for cyclin D2 induction in primary B cells in cooperation with EBNA-2 |
| EBNA-3A and EBNA-3C | Play important roles early in establishment of the latent infection |
| LMP-1 | An integral membrane protein required to protect the latently infected B cell from the immune response. LMP-1 stimulates the expression of several surface adhesion molecules in B cells, a calcium-dependent protein kinase, and the apoptosis inhibitor Bcl-2. |
| LMP-2 | An integral membrane protein required to block activation of the src family signal transduction cascade; an inhibitor of reactivation from latency. Two spliced forms exist: LMP-2A and LMP-2B. LMP-2B lacks a receptor binding domain and may act to modulate LMP-2A. |

[a]Abbreviations: EBNA, Epstein-Barr virus nuclear antigen; LMP, latent membrane protein.

vided significant insight into the action of Zta. The pattern of gene expression after exposure to this protein is remarkably similar to that seen following reactivation by phorbol esters. Indeed, as described in chapter 8, Zta functions in part by increasing transcription of virus lytic genes. This protein also represses the latency-associated promoters and is responsible for recognition of the virus lytic origin of replication.

In contrast to the nonpathogenic latent state of herpes simplex virus, the latent state of Epstein-Barr virus is implicated in several important diseases, including infectious mononucleosis and at least two kinds of tumor, Burkitt's lymphoma and nasopharyngeal carcinoma.

## Slow Infections

### Sigurdsson's Legacy: Icelandic Sheep and Fatal Degenerative Diseases

Many fatal brain diseases characterized by ataxia or dementia stem from another extreme variation of persistent infection, a pattern called slow infection (Fig. 15.1). It may be years from the time of initial contact of the infectious agent with the host until the appearance of recognizable symptoms. Elucidating the molecular mechanisms responsible for an infectious disease process of such long duration is a formidable challenge. Experimental analysis of these unusual diseases began in the 1930s when a flock of Karakul sheep was imported from Germany to Iceland, where they infected the native sheep, causing a disease called maedi/visna. Thanks to the many years of careful work by Bjorn Sigurdsson and colleagues, we now know that the maedi/visna syndrome is caused by a lentivirus quite similar to human immunodeficiency virus. The striking feature that Sigurdsson discovered is the slow progression to disease after primary infection—often more than 10 years. He developed a framework of experimentation for studying the slow, relentless, usually progressive and fatal brain infections, including those now proposed to be caused by prions.

## Slow Viruses and "Unconventional Agents"

Many slow infections are caused by well-known viruses, including lentiviruses, flaviviruses, rubiviruses, rhabdoviruses, and paramyxoviruses. Slow viral diseases include subacute sclerosing panencephalitis caused by measles virus (see "Measles Virus" above), tick-borne encephalitis caused by Russian spring-summer encephalitis virus, progressive rubella panencephalitis, human T-cell lymphotropic virus type 1-mediated tropical spastic paraparesis, and dementia produced after human immunodeficiency virus infection. While the causative viruses have been identified, we have much to learn about the molecular mechanisms that promote the slow but relentless progression to disease.

The identity of infectious agents mediating one group of slow diseases, called **transmissible spongiform encephalopathies (TSEs)**, remains controversial. These diseases are fatal neurodegenerative disorders afflicting humans and other mammals (Box 15.7). Some of the controversy stems from the varied interpretation of experiments concerning the physical nature of the infectious agent. Some investigators contend that TSEs are caused by viruses or viruslike particles, whereas others propose that they are caused by infectious proteins called prions.

### Human TSE

Initially, the human TSEs were not thought to be transmissible, but studies of the human disease called kuru, a fatal encephalopathy found in the Fore people of New Guinea, by Carleton Gajdusek and colleagues proved otherwise. Kuru spread among women and children by ritual cannibalism of the brains of deceased relatives. When cannibalism stopped, so did kuru. Insightful comparison of the pathology of scrapie-infected (see below) and kuru-infected brains by William Hadlow led to experiments demonstrating the transmission of kuru disease from humans to chimpanzees and other primates. Thus, since 1957, the TSE diseases of animals and humans have been

---

**BOX 15.7**

*Transmissible Spongiform Encephalopathies (TSEs)*

| TSE diseases of animals | TSE diseases of humans |
|---|---|
| Scrapie in sheep and goats | Kuru |
| Transmissible mink encephalopathy (TME) | Creutzfeldt-Jakob disease (CJD) |
| Chronic wasting disease (CWD) (deer, elk) | Fatal familial insomnia (FFI) |
| Bovine spongiform encephalopathy (BSE) ("mad cow disease") | Gerstmann-Sträussler syndrome (GSS) |
| Feline spongiform encephalopathy (FSE) (domestic and great cats) | Prusiner, S. B. (ed.). 1996. Prion diseases. *Semin. Virol.* 7:157–223. |
| Exotic ungulate encephalopathy (EUE) (nyala and greater kudu) | |

considered to have common features. In humans, TSE diseases fall into three classes, infectious, familial, and sporadic, distinguished by how the disease is initially acquired (Box 15.8).

### Hallmarks of TSE Pathogenesis

For most TSEs, the presence of an infectious agent can be detected definitively only by injection of organ homogenates into susceptible recipient species. Clinical signs of infection commonly include cerebellar ataxia (defective motion or gait) and dementia, with death occurring after months or years. The infectious agent first accumulates in the lymphoreticular and secretory organs and then spreads to the nervous system. In model systems, spread of the disease from site of inoculation to other organs and the brain appears to require B cells. The disease agent then appears to invade the peripheral nervous system and spreads from there to the spinal cord and brain. Once in the central nervous system, the characteristic pathology includes severe astrocytosis, vacuolization (hence the term "spongiform"), and loss of neurons. Occasionally, dense fibrils or aggregates (sometimes called plaques) can be detected in brain tissue at autopsy. There is little indication of inflammatory, antibody, or cellular immune responses. The time course, degree, and site of cytopathology within the central nervous system are dependent on the particular TSE agent and the genetic makeup of the host.

### Identification of the Scrapie Agent

One of the best studied TSE diseases is **scrapie**, so-called because infected sheep tend to scrape their bodies on fences so much that they rub themselves raw. A second characteristic symptom, tremors caused by skin rubbing over the flanks, also led to the French name for the disease, *tremblant du mouton*. Soon after, motor disturbances appear as a wavering gait, staring eyes, and paralysis of the hindquarters. There is no fever, but infected sheep lose weight and die, usually within 4 to 6 weeks of the first appearance of symptoms. Scrapie has been recognized as a disease of European sheep for more than 200 years. It is endemic in some countries, like Great Britain, where it affects 0.5 to 1% of the sheep population per year.

Sheep farmers discovered that animals from diseased herds could pass the affliction to a scrapie-free herd, suggesting that an infectious agent was involved. In 1936, infectivity from extracts of scrapie-affected sheep brains was shown to pass through filters with pores small enough to retain all but viruses. In the 1970s, ultracentrifugation studies indicated that the agent was heterodisperse, and it could not be banded in density gradients. Even to this day, purification of the infectious agent to homogeneity has not been achieved.

### Biological Assay of the Scrapie Agent

The only assay for biological activity relies on animal infection. The endpoint assay for infectivity (the highest dilution capable of infecting 50% of the animals, also called the 50% infective dose [$ID_{50}$]) is notoriously difficult, requiring 2 to 8 months under the best of conditions (i.e., with laboratory-adapted agents) or years (i.e., assaying primary isolates). Nevertheless, endpoint dilution is a highly sensitive assay. However, accuracy is a problem, and typical errors range from 5- to 10-fold. Thus, it is impossible to distinguish 10 and 100% infectivity with confidence. In addition, as deduced from sedimentation analyses, the infectious agent forms variable aggregates that confound accurate measurement of infectivity.

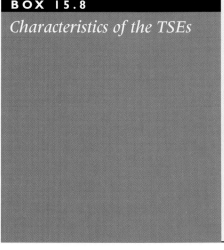

**BOX 15.8**

*Characteristics of the TSEs*

An **infectious TSE** is exemplified by kuru and iatrogenic spread of disease to healthy individuals by transplantation of infected corneas, the use of purified hormones, or transfusion with blood from patients with Creutzfeldt-Jakob disease (CJD). Recently, the epidemic spread of bovine spongiform encephalopathy (BSE, or "mad cow disease") among British cattle may have resulted from the practice of feeding processed animal by-products to cattle as a protein supplement. There is continued concern that consumption of BSE-infected beef will transmit bovine TSE to humans.

**Sporadic CJD** is a disease affecting 1 million to 2 million people worldwide, usually late in life (from age 50 to 70). As the name indicates, the disease appears with no warning or epidemiological indications. Kuru may have originally been established in the small population of Fore people in New Guinea by eating the brain of an individual with sporadic CJD.

**Familial TSE** is associated with an autosomal dominant mutation in the PrP gene (see below). Familial CJD, for example, in contrast to sporadic CJD, is an inherited disease.

The important point is that diseases of all three classes usually can be transmitted experimentally or naturally to primates by inoculation or ingestion of infected tissue.

### Physical Nature of the Scrapie Agent

A major point of contention is the physical nature of the infectious agent. Studies as early as 1966 showed that scrapie infectivity was considerably more resistant than most viruses to ultraviolet (UV) and ionizing radiation. For example, the scrapie agent is 200-fold more resistant to UV irradiation than polyomavirus and 40-fold more resistant than a mouse retrovirus. Other TSE agents exhibit similar UV resistance. The scrapie agent is also more resistant to chemicals, such as 3.7% formaldehyde, and autoclaving routinely used to inactivate viruses. It is possible to reduce infectivity by 90 to 95% after several hours of such treatments, but some residual infectivity remains. The inability to inactivate all infectivity has been used to infer novel properties for the agent. However, at this level of analysis, the resistant residual fraction provides no information about the physical properties of the majority fraction that was sensitive. On the basis of relative resistance to UV irradiation, some investigators have argued that TSE agents are viruses well shielded from irradiation or especially efficient in nucleic acid repair. Others have claimed that TSE agents have little or no nucleic acid at all. Suffice it to say that TSE agents are not typical infectious agents.

### Prions and the PrP Gene

The unconventional physical attributes and slow infection pattern of TSE agents have prompted many to argue that these agents are not viruses at all. For example, in 1967 the mathematician J. S. Griffith made three suggestions as to how scrapie may be mediated by a host protein, not by a nucleic acid-carrying virus. His thoughts were the first of the "protein-only" hypotheses to explain TSE.

In 1981 an important experimental observation was made by P. Mertz et al., who described scrapie-associated fibrils in infected brains. This work, and previous studies indicating that the scrapie agent could be concentrated by centrifugation, led to the development by Stanley B. Prusiner and colleagues of an improved bioassay, as well as a fractionation procedure that allowed the isolation of an unusual protein from scrapie-infected tissue. This protein is insoluble and relatively resistant to proteases. Sequence analysis led to the cloning of a gene called PrP, which is highly conserved in the genomes of many animals including humans. The PrP gene is now known to be essential for the pathogenesis of common TSEs.

The PrP gene encodes a 35-kDa membrane-associated glycoprotein expressed widely in the brain. The protein can adopt several topological forms in the endoplasmic reticulum. It can be anchored in the membrane with a type I or a type II orientation via a transmembrane domain, it can be anchored to the membrane via a phosphatidylinositol linkage, or it can be secreted into the lumen. The three-dimensional structure of a portion of the mouse PrP protein has been determined; the amino-terminal half of the protein forms a random coil, while the remaining protein consists of three α-helices and a short β-pleated sheet. Current evidence indicates that at least one form of PrP protein binds copper and is sequestered in caveolae, unique detergent-resistant membrane vesicles enriched in cholesterol. Despite such basic information, the function of PrP protein and the role of the various topological forms remain unknown. Mice lacking the PrP gene develop normally and have few obvious defects that can be directly attributed to lack of PrP. At least 18 specific mutations in the human PrP gene are associated with familial TSE diseases. Furthermore, specific PrP mutations appear be associated with susceptibility to different strains of TSE (see below).

Prusiner named the scrapie infectious agent a **prion** (an anagram of "proin," from "proteinaceous infectious particle") and proposed that an altered form of the PrP protein caused the fatal encephalopathy characteristic of scrapie disease. Occasionally, the term "prion" is used by some investigators as a synonym for the infectious agent and for PrP protein. Adriano Aguzzi and Charles Weissman provide a useful definition: a prion is the agent of TSE with unconventional properties. The term does not have structural implications other than that a protein is an essential component.

Prusiner's protein-only hypothesis holds that the essential pathogenic component is an altered conformation of the host-encoded PrP protein, called PrPsc ("PrP-scrapie"; also called PrPres for "protease-resistant form"). Furthermore, in the simplest case, PrPsc is proposed to have the property of converting normal PrP protein into more copies of pathogenic PrPsc. An important finding in this regard is that mice lacking both copies of their PrP gene are resistant to infection. In recognition of his work on this problem, Prusiner was awarded the Nobel Prize in physiology or medicine in 1997.

### Strains of Scrapie Prion

As a result of many serial infections with infected sheep brain homogenates of different strains of mice and hamsters, investigators have derived distinct strains of scrapie prion distinguished by length of incubation time before the appearance of symptoms, brain pathologies, relative abundance of various glycoforms of PrP protein, and electrophoretic profiles of protease-resistant PrPsc. Some strains also have a different host range. For example, mouse-adapted scrapie prions cannot propagate in hamsters, but hamster-adapted scrapie prions can propagate in mice. Sue Priola and Bruce Chesebro discovered that a single amino acid substitution in the hamster protein enables it to be efficiently converted by mouse PrPsc into hamster PrPsc. Thus, the barrier to interspecies transmission is in the sequence of the PrP protein. Bovine spongiform encephalopathy prions have an unusually broad host range, infecting a number of meat-eating animals, including domestic and wild cats, and humans. A striking finding is that different scrapie strains can be propagated in the same inbred line of mice yet maintain their original phenotypes. Stable inheritance suggests to some investigators that a nucleic acid must be an essential component and has been used as support for the existence of a virus in TSE disease. The protein-only hypothesis explains the existence of strain variation by postulating that each strain represents a unique conformation of PrPsc. Each of these distinctive pathogenic conformations is then postulated to convert the normal PrP protein into a conformational image of itself.

### TSE Research Directions

This is an exciting time in TSE research, as many groups are formulating testable predictions of hypotheses that TSEs are caused by viruses or by infectious proteins. The relationship of the protease-resistant form of PrP and PrP-enriched fibrils to the disease process remains to be discovered. Several neurodegenerative diseases, including Alzheimer's, Parkinson's, and Huntington's diseases, are associated with deposits of insoluble aggregates of cellular proteins in the brain. These proteins, like PrP, have poorly structured native conformations which can be destabilized further by genetic mutations to adopt β-sheet structures. Perhaps aberrant protein folding is a common feature of these diseases. Even so, we have little understanding of how altered or aggregated PrP participates in the fatal pathogenesis of TSE. Similarly, we do not understand the molecular nature of the species barrier that exists for transmission of some prions. We do not yet have a firm understanding of the role of the immune system in TSE disease. For example, normal mice do not develop an immune response against the normal or pathological form of PrP. However, mice lacking the PrP gene develop an immune response to PrP. In fact, most good monoclonal antibodies against PrP are made in mice lacking the PrP gene. The normal tolerance to PrP may be important for the persistence of PrPsc and its pathological effects.

While most investigators find the protein-only hypothesis compelling, a few maintain that the PrP protein is not the only participant in the TSE story and that an unrecognized virus is responsible for infection. In their view, the PrP protein would be a cofactor, a receptor, or a protein that determines susceptibility to infection. The experiment that would certainly resolve the basic controversy would be to produce the putative pathogenic conformation of PrP in vitro and demonstrate that it causes a specific TSE.

The existence of TSE as a human genetic disease and the presence of prions in the human and animal food supply are causes for concern, for we have little under-standing of either the molecular nature of the infectious agent or the processes of pathogenesis. At a minimum, effort must be focused on basic research toward protection of the food supply and finding targets for therapeutic intervention.

## Other Patterns of Viral Infections

### Abortive Infections

An abortive infection is one in which virus infects a susceptible cell or host but does not complete replication, usually because an essential viral or cellular gene is not expressed. When the host cell is defective for a necessary viral cofactor, it is said to be a nonpermissive cell, as opposed to a resistant cell, which lacks specific viral receptors on the cell surface.

Clearly, an abortive infection is nonproductive with respect to making more infectious virus. Even so, it is not necessarily benign for the infected host. Viral interactions at the cell surface and subsequent uncoating can initiate membrane damage, disrupt endosomes, or activate signaling pathways that cause apoptosis and interferon production. For some viruses, an abortive infection may proceed far enough for viral early proteins to be made so that the infected cell is recognized by cytotoxic T cells. Such an infection would then induce an interferon, as well as an inflammatory response which may be damaging to the host if sufficient cells are so infected. Recently, it was discovered that the human immunodeficiency virus structural protein Vpr can damage cells when it is associated with virions containing noninfectious genomes. Vpr is required for the infection of nondividing cells such as macrophages, and it can induce cell cycle arrest in the $G_2$ phase. When these noninfectious viruses bind to T cells, the cells arrest in the $G_2$ phase. It has been suggested that most of the particles in an infected individual are noninfectious, but because of the presence of virion Vpr, these particles participate in immune suppression due to loss of T-cell function.

With the advent of modern viral genetics, virologists can construct defective viruses which, in the absence of a complementing gene product, initiate an abortive infection. One popular idea is to use such "synthetic" defective viruses as vectors to deliver genes for gene therapy or vaccine production. To be effective, cytopathic genes of a prospective viral vector must be eliminated. Many of these nonlethal viral vectors are missing essential genes and are designed to express only the therapeutic cloned gene after infection. Care must be taken to ensure that the infection is truly noncytotoxic. Cytotoxicity and inflammatory host responses are of particular concern if the therapeutic gene is to be delivered to a substantial number of cells, a process that requires administration of many virus particles.

## Transforming Infections

A transforming infection is a special class of persistent infection. A cell infected by certain DNA viruses or retroviruses may exhibit altered growth properties and begin to proliferate faster than uninfected cells. Often this change is accompanied by integration of viral genetic information. Virus particles may no longer be produced, but some or all of their genetic material generally persists. We characterize this pattern of persistent infection as **transforming** because of the change in cell behavior. It is also considered oncogenic because transformed cells can cause cancer in animals. This important infection pattern will be discussed in detail in chapter 16.

## Perspectives

The patterns of infection confront us with several questions that challenge our basic understanding of virology. A particular infection pattern can be a defining characteristic of a virus family, yet why this should be is not always obvious. Why has one particular pattern been selected over another? As we discuss in chapter 20, virus evolution occurs when new virus populations emerge from selection pressures. What are the selective advantages or disadvantages of a given infection pattern? Is an acute infection any better as a survival strategy than a persistent infection? How do we rationalize the characteristic pattern of infection in terms of **pathogenesis** (the spectrum of events leading to disease)?

One hypothesis is that successful viruses establish an infection pattern that results in **benign symbiosis** with their hosts. In this case, a successful virus neither helps nor harms the host. Lewis Thomas sagely wrote that pathogenesis is an aberration of symbiosis, an overstepping of boundaries. Another hypothesis is that a successful virus need not have a static or benign relationship with its host. Rather, virus and host populations are in constant flux, and the successful relationship is better described as an approach to equilibrium (benign symbiosis may never be attained). In this case, viruses engage in a contentious relationship with the host population, meeting defensive measures with countermeasures. The relationship may harm individuals in the host population in the short run to achieve long-term survival of the virus population. In this model, pathogenesis may be a necessary survival feature of the virus and would be selected during evolution of the relationship. We will continue the discussion of viral pathogenesis in more detail in chapter 17.

## References

### Books

Mims, C. A., N. Dimmock, A. Nash, and A. Stephen. 1995. *Mims' Pathogenesis of Infectious Disease*, 4th ed. Academic Press, Orlando, Fla.

Richman, D. D., R. J. Whitley, and F. G. Hayden (ed.). 1997. *Clinical Virology*. Churchill Livingstone, Inc., New York, N.Y.

Roizman, B., R. Whitley, and C. Lopez (ed.). 1993. *The Human Herpesviruses*. Raven Press, New York, N.Y.

### Reviews

#### Evasion of Host Defenses

Anderson, P. 1997. Kinase cascades regulating entry into apoptosis. *Microbiol. Mol. Biol. Rev.* **61:**33–46.

Franco, A., C. Ferrari, A. Sette, and F. V. Chisari. 1995. Viral mutations, TCR antagonism and escape from the immune response. *Curr. Opin. Immunol.* **7:**524–531.

Griffin, D. E., and J. M. Hardwick. 1997. Regulators of apoptosis on the road to persistent alphavirus infection. *Annu. Rev. Microbiol.* **51:**565–592.

Hengel, H., and U. H. Koszinowski. 1997. Interference with antigen processing by viruses. *Curr. Opin. Immunol.* **9:**470–476.

Krajcsi, P., and W. S. M. Wold. 1998. Viral proteins that regulate cellular signalling. *J. Gen. Virol.* **79:**1323–1335.

Miller, L. K., and E. White (ed.). 1998. Apoptosis in virus infection. *Semin. Virol.* **8:**443–523.

Multiple authors. 1998. Apoptosis. *Science* **281:**1301–1326.

Neipel, F., J.-C. Albrecht, and B. Fleckenstein. 1997. Cell-homologous genes in the Kaposi's sarcoma-associated rhabdinovirus human herpesvirus 8: determinants of its pathogenicity? *J. Virol.* **71:**4187–4192.

O'Brien, V. 1998. Viruses and apoptosis. *J. Gen. Virol.* **79:**1833–1845.

Oldstone, M. B. A. 1996. Virus-lymphoid cell interactions. *Proc. Natl. Acad. Sci. USA* **93:**12756–12758.

Ploegh, H. L. 1998. Viral strategies of immune evasion. *Science* **280:**248–253.

Smith, G. L. (ed.). 1998. Immunomodulation by virus infection. *Semin. Virol.* **8:**359–442.

Spriggs, M. K. 1996. One step ahead of the game: viral immunomodulatory molecules. *Annu. Rev. Immunol.* **14:**101–130.

Teodoro, J. G., and P. E. Branton. 1997. Regulation of apoptosis by viral gene products. *J. Virol.* **71:**1739–1746.

Ter Meulen, V. (ed.). 1994. Pathogenesis of persistent virus infections. *Semin. Virol.* **5:**259–324.

Weinberg, R. A. 1997. The cat and mouse games that genes, viruses and cells play. *Cell* **88:**573–575.

#### Slow Viruses, Scrapie, Prions, and PrP

Aguzzi, A., and C. Weissman. 1997. Prion research: the next frontiers. *Nature* (London) **389:**795–798.

Bloom, M. E., H. Kanno, S. Mori, and J. B. Wolfinbarger. 1994. Aleutian mink disease: puzzles and paradigms. *Infect. Agents Dis.* **3:**279–301.

Borrow, P., and M. B. A. Oldstone. 1995. Measles virus—mononuclear cell interactions. *Curr. Top. Microbiol. Immunol.* **191:**85–100.

Chesebro, B., and B. Caughey. 1997. Transmissible spongiform encephalopathies (prion protein diseases), p. 1285–1304. *In* D. D. Richman, R. J. Whitley, and F. G. Hayden (ed.), *Clinical Virology*. Churchill Livingstone, Inc., New York, N.Y.

Cohen, F. E., and S. B. Prusiner. 1998. Pathological conformations of prion proteins. *Annu. Rev. Biochem.* **67**:793–819.

Farquhar, C. F., R. A. Somerville, and M. E. Bruce. 1998. Straining the prion hypothesis. *Nature* (London) **391**:345–346.

Nowak, M., et al. 1998. Prion infection dynamics. *Integr. Biol.* **1**:3–15.

Prusiner, S. B., M. R. Scott, S. J. DeArmond, and F. E. Cohen. 1998. Prion protein biology. *Cell* **93**:337–348.

Prusiner, S. B. 1998. Prions. *Proc. Natl. Acad. Sci. USA* **95**:13363–13383.

### Herpes Latency

Block, T. M., and J. M. Hill. 1997. The latency associated transcripts (LAT) of herpes simplex virus: still no end in site. *J. Neurovirol.* **3**:313–321.

Longnecker, R., and C. L. Miller. 1996. Regulation of Epstein-Barr virus latency by latent membrane protein 2. *Trends Microbiol.* **4**:38–42.

Specke, S. H., T. Chatila, and E. Flemington. 1997. Reactivation of Epstein-Barr virus: regulation and function of the BZLF1 gene. *Trends Microbiol.* **5**:399–405.

Wagner, E. K. (ed.). 1994. Herpesvirus latency. *Semin. Virol.* **5**:198–258.

## Selected Papers

### Evasion of Host Defenses

Bertin, J., R. C. Armstrong, S. Ottilie, D. A. Martin, Y. Wang, S. Banks, G.-H. Wang, T. G. Senkevich, E. S. Alnemri, B. Moss, M. J. Lenardo, K. J. Tomaselli, and J. I. Cohen. 1997. Death effector domain-containing herpesvirus and poxvirus proteins inhibit both Fas- and TNFR1-induced apoptosis. *Proc. Natl. Acad. Sci. USA* **94**:1172–1176.

Griffith, T. S., T. Brunner, S. M. Fletcher, D. R. Green, and T. A. Ferguson. 1995. Fas ligand-induced apoptosis as a mechanism of immune privilege. *Science* **270**:1189–1192.

Ito, T., J. N. S. S. Couceiro, S. Kelm, L. G. Baum, S. Krauss, M. R. Castrucci, I. Donatelli, H. Kida, J. C. Paulson, R. G. Webster, and Y. Kawaoka. 1998. Molecular basis for the generation in pigs of influenza A viruses with pandemic potential. *J. Virol.* **72**:7367–7373.

Karp, C. L., M. Wysocka, L. M. Wahl, J. M. Ahearn, P. J. Duomo, B. Sherry, G. Trinchieri, and D. E. Griffin. 1996. Mechanism of suppression of cell-mediated immunity by measles virus. *Science* **272**:228–231.

Levitskaya, J., M. Coram, V. Levitsky, S. Imreh, P. M. Steigerwald-Mullen, G. Klein, M. G. Kurilla, and M. G. Masucci. 1995. Inhibition of antigen processing by the internal repeat region of the Epstein-Barr virus nuclear antigen-1. *Nature* (London) **375**:685–688.

Nowak, M. A., S. Bonhoeffer, A. M. Hill, R. Hoehme, H. C. Thomas, and H. McDade. 1996. Viral dynamics in hepatitis B virus infection. *Proc. Natl. Acad. Sci. USA* **93**:4398–4402.

Reyburn, H. T., O. Mandelboim, M. Vales-Gomez, D. M. Davis, L. Pazmany, and J. L. Strominger. 1997. The class I MHC homologue of human cytomegalovirus inhibits attack by natural killer cells. *Nature* (London) **386**:514–517.

Shimeld, C., J. L. Whiteland, N. A. Williams, D. L. Easty, and T. J. Hill. 1997. Cytokine production in the nervous sys-

tem of mice during acute and latent infection with herpes simplex virus type 1. *J. Gen. Virol.* **78**:3317–3325.

Shisler, J. L., T. G. Senkevich, M. J. Berry, and B. Moss. 1998. Ultraviolet-induced cell death blocked by a selenoprotein from a human dermatotropic poxvirus. *Science* **279**:102–105.

Thome, M., P. Schneider, K. Hofmann, H. Fickenscher, E. Meinl, F. Neipel, C. Mattmann, K. Burns, J.-L. Bodmer, M. Schroter, C. Scaffidi, P. H. Krammer, M. E. Peter, and J. Tschop. 1997. Viral FLICE-inhibitory proteins (FLIPs) prevent apoptosis induced by death receptors. *Nature* (London) **386**:517–521.

Wiertz, E. J. H. J., T. R. Jones, L. Sun, M. Bogyo, H. J. Geuize, and H. L. Ploegh. 1996. The human cytomegalovirus US11 gene product dislocates MHC class I heavy chains from the endoplasmic reticulum to the cytosol. *Cell* **64**:769–779.

York, I. A., C. Roop, D. W. Andrews, S. R. Riddell, F. L. Graham, and D. C. Johnson. 1994. A cytosolic herpes simplex virus protein inhibits antigen presentation to CD8+ T lymphocytes. *Cell* **77**:525–535.

### Herpes Latency

Kramer, M. F., and D. Coen. 1995. Quantification of transcripts from the ICP4 and thymidine kinase genes in mouse ganglia latently infected with herpes simplex virus. *J. Virol.* **69**:1389–1399.

Miyashita, E. M., B. Yang, G. J. Babcock, and D. A. Thorley-Lawson. 1997. Identification of the site of Epstein-Barr virus persistence in vivo as a resting B cell. *J. Virol.* **71**:4882–4891.

Sawtell, N. 1998. The probability of in vivo reactivation of herpes simplex virus type 1 increases with the number of latently infected neurons in ganglia. *J. Virol.* **72**:6888–6892.

Speck, P., and A. Simmons. 1998. Precipitous clearance of herpes simplex virus antigens from the peripheral nervous systems of experimentally infected C57Bl/10 mice. *J. Gen. Virol.* **79**:561–564.

Zabolotny, J. M, C. Krummenacher, and N. W. Fraser. 1997. The herpes simplex virus type 1 2.0-kilobase latency-associated transcript is a stable intron which branches at a guanosine. *J. Virol.* **71**:4199–4208.

### Slow Viruses, Scrapie, Prions, and Prp

Griffith, J. S. 1967. Self-replication and scrapie. *Nature* (London) **215**:1043–1044.

Hadlow, W. J. 1959. Scrapie and kuru. *Lancet* **ii**:289.

Hegde, R. S., J. A. Mastrianni, M. R. Scott, K. A. DeFea, P. Tremblay, M. Torchia, S. J. DeArmond, S. B. Prusiner, and V. R. Lingappa. 1998. A transmembrane form of the prion protein in neurodegenerative disease. *Science* **279**:827–834.

Merz, P. A., R. A. Sommerville, H. M. Wisniewski, and K. Iqbal. 1981. Abnormal fibrils from scrapie-infected brain. *Acta Neuropathol.* (Berlin) **54**:63–74.

Prusiner, S. B. 1982. Novel proteinaceous infectious particles cause scrapie. *Science* **216**:136–144.

Sigurdsson, B. 1954. Rida, a chronic encephalitis of sheep with general remarks on infections which develop slowly and some of their special characteristics. *Br. Vet. J.* **110**:341–354.

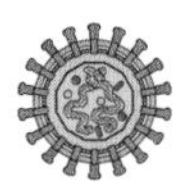

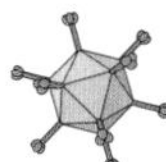

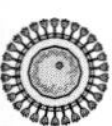

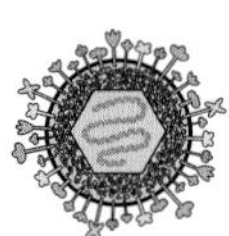

# 16

# Viral Transformation and Oncogenesis

## Introduction

Cancer is a leading cause of death in developed countries: about 500,000 individuals succumb each year in the United States alone. Efforts to understand and control this deadly disease have therefore long been high priorties of public health institutions. Our general understanding of the mechanisms of **oncogenesis**, the development of cancer, as well as of normal cell growth, has improved enormously in the past 35 years. Such progress can in large part be traced to efforts to elucidate how members of several virus families cause cancer in animals. In fact, as we discuss in this chapter, study of such oncogenic viruses has led to a detailed understanding of the molecular basis of the development of cancer.

It is now clear that cancer (defined in Box 16.1) is a genetic disease: oncogenesis comprises the processes that result in the growth of successive populations of cells in which mutations have accumulated (Box 16.2). These mutations affect various steps in the pathways that regulate cellular communication, growth, and division and lead to uncontrolled growth, increasing cellular disorganization and, ultimately, cancer. One or more mutations may be inherited (Box 16.2), or they may arise as a consequence of exposure to environmental carcinogens or infectious agents, including viruses. It is estimated that viruses are a contributing factor in approximately 20% of all human cancers. For some, such as liver and cervical cancer, they are the major cause.

Understanding the development of cancer in animals ultimately depends on knowledge of how individual cells behave *within* the animals. As described in chapter 17, analysis of viral pathogenesis must encompass a consideration of the organism as a whole, especially the body's immune defenses. However, understanding how members of several virus families cause cancer in animals began with studies of cultured cells in the laboratory. In particular, early investigators noticed that the growth properties and morphologies of some cultured cells could be changed upon infection with certain viruses. We therefore

**BOX 16.1**

*Some cancer terms*

| | |
|---|---|
| Benign | An adjective used to describe a growth that does not infiltrate into surrounding tissues; opposite of malignant |
| Cancer | A malignant tumor; a growth that is not encapsulated and that infiltrates into surrounding tissues, replacing normal with abnormal cells; it is spread by the lymphatic vessels to other parts of the body; death is caused by destruction of organs to a degree incompatible with their function, by extreme debility and anemia, or by hemorrhage |
| Carcinogenesis | The complex multistage process by which a cancer develops |
| Carcinoma | A cancer of epithelial tissue |
| Endothelioma | Overproduction of erythrocytes |
| Fibroblast | A cell derived from connective tissue |
| Fibropapilloma | A solid tumor of cells of the connective tissue |
| Hepatocellular carcinoma | A cancer of liver epithelial cells |
| Leukemia | A cancer of white blood cells |
| Lymphoma | A cancer of lymphoid tissue |
| Malignant | An adjective applied to any disease of a progressive and fatal nature; opposite of benign |
| Neoplasm | An abnormal new growth, i.e., a cancer |
| Oncogenic | Causing a tumor |
| Retinoblastoma | A cancer of cells of the retina |
| Sarcoma | A cancer of fibroblasts |
| Tumor | A swelling, caused by abnormal growth of tissue, not resulting from inflammation; may be benign or malignant |

consider such cells **transformed**. The advantages of these cell culture systems are many. The molecular virologist can focus attention on particular cell types, can manipulate their behavior in a controlled manner, and can easily distinguish effects specific to the virus. In many cases, cells transformed by viruses in culture can form tumors when implanted in animals. But this does not always occur, and it is important to realize that transformed cultures are *not* tumors. The major benefit of cell culture systems is that they allow researchers to study the molecular events that establish an oncogenic *potential* in virus-infected cells.

## Properties of Transformed Cells

### Cellular Transformation

The multiplication of cells in the body is a strictly regulated process. In a young animal, total cell multiplication exceeds cell death as the animal grows to maturity. In an adult, the processes of cell multiplication and death are carefully balanced. For some cells, high rates of multiplication are required to maintain this balance. For example, human intestinal cells and white blood cells have half-lives of only a few days and need to be replaced rapidly.

On the other hand, red blood cells live for over 100 days, and healthy neuronal cells rarely die. Occasionally, this carefully regulated process breaks down, and a particular cell begins to grow and divide even though the body has sufficient numbers of its type; such a cell behaves as if it were immortal. Acquisition of this property, **immortality**, is generally acknowledged to be an early step in oncogenesis. An immortalized cell may acquire one or more additional genetic changes to give rise to a clone of cells that is able to expand, ultimately forming a mass called a **tumor**. Some tumors are **benign**; they cease to grow in the body after reaching a certain size and generally do no great harm. Other tumor cells grow and divide indefinitely to form **malignant** tumors (see Box 16.1 for definition) which damage and impair the normal function of organs and tissues. Occasionally, some cells in a tumor acquire additional mutations (Box 16.2) that confer the ability to escape the boundary of the mass, to invade surrounding tissue, and to be disseminated to other parts of the body, where the cells take up residence. There they continue to grow and divide, giving rise to secondary tumors called **metastases**. Such cells cause the most serious and life-threatening disease.

## BOX 16.2

*Genetic alterations associated with the development of colon carcinoma*

The clinical stages in the development of human colon cancer, which is the fourth most common cancer in the United States, are particularly well defined. Furthermore, as shown, several genes that are frequently mutated to allow progression from one stage to the next have been identified. The early adenomas or polyps that initially form are benign lesions. Their conversion to malignant metastatic carcinomas correlates with the acquisition of additional mutations in the *p53* and *dcc* (deleted in colon carcinoma) genes. Inherited mutations in the genes listed can greatly increase the risk that an individual will develop colon carcinoma. For example, patients with familial adenomatous polyposis can inherit defects in the *apc* (adenomatous polyposis coli) gene that result in the development of hundreds of adenomatous polyps. The large increase in the number of these benign lesions increases the chance that some will progress to malignant carcinomas. In contrast, patients with hereditary nonpolyposis colorectal cancer develop polyps at the same rate as the general population. However, polyps develop to carcinomas more frequently because defects in genes encoding proteins that correct mismatched bases in DNA lead to a higher mutation rate. Consequently, the likelihood that an individual polyp will develop into a malignant lesion increases from 5% to 70%.

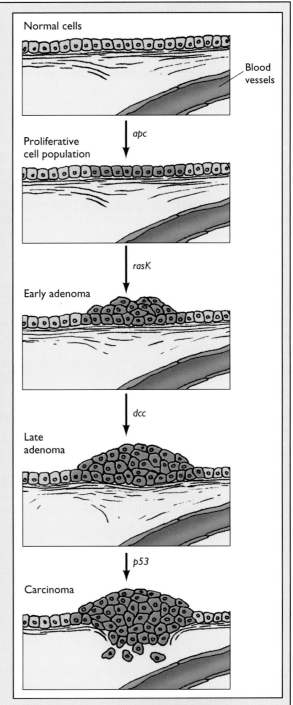

Many studies of the molecular biology of oncogenic animal viruses employed primary cultures of normal cells, for example, rat or mouse embryo fibroblasts, obtained by the methods described in chapter 2. The advantage of these cells is that all the molecular changes necessary for the virus to convert them to the oncogenic state can be studied. On the other hand, cells of such primary cultures, like their normal counterparts in the body, have a finite capacity to grow and divide in culture. Cells from some animal species, such as rodents, undergo a spontaneous transformation when maintained in culture. Immortalized cells appear in these cultures after a "crisis" period in which the great majority of the cells die (Fig. 16.1). As these cells are otherwise normal and do not induce tumors when introduced into animals, they can be used to identify viral gene products needed for steps in transformation subsequent to immortalization. For reasons that are not fully understood, human and simian cells rarely undergo spontaneous transformation to immortality when passaged in culture. In fact, established lines of human cells generally can be derived only from tumors or following exposure of primary cells to chemical carcinogens or oncogenic RNA or DNA viruses (or their transforming genes). The realization that, regardless of how they were obtained, such transformed cells share a number of common properties provided a major impetus for the investigation of viral transformation.

## Properties That Distinguish Transformed and Normal Cells

The definitive characteristic of transformed cells is their lack of response to the signals or conditions that normally control growth, DNA replication, and division. This property is illustrated by the list of growth parameters and behaviors provided in Table 16.1. As described above, transformed cells are **immortal**: they can grow and divide indefinitely, provided that they are diluted regularly into fresh medium. Production of telomerase, an enzyme that maintains telomeric DNA at the ends of chromosomes, which is not normally made in most postembryonic human cells, has recently been implicated in immortalization. In addition, transformed cells typically grow in serum concentrations much lower than those required by normal

**Figure 16.1 Stages in the establishment of a cell culture.** (A) Human cells. When an initial explant is made (e.g., from foreskin), some cells die and others (mainly fibroblasts) start to grow; overall, the growth rate increases. If the surviving cells are diluted regularly, the cell strain grows at a constant rate for about 50 cell generations, after which growth begins to slow. Eventually, all the cells die. The generation time of human cells varies from 24 to 72 h. (B) Mouse or other rodent cells. In a culture prepared from mouse embryo cells, cell death initially occurs, and then healthy growing cells emerge. As these cells are maintained in culture, they soon begin to lose growth potential and most cells die (the culture goes into crisis). Very rare cells do not die but continue growth and division until their progeny overgrow the culture. These cells constitute a cell line, which will grow indefinitely if it is appropriately diluted and fed with nutrients: the cells are immortal.

**A  Human cells**

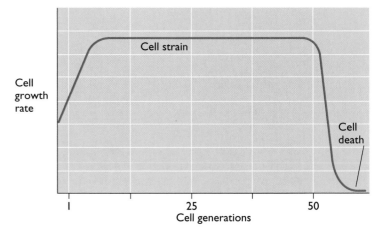

**B  Mouse cells**

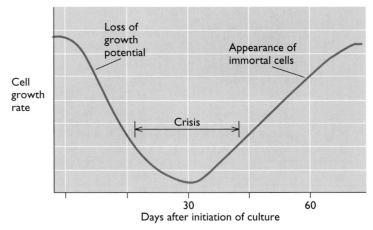

**Table 16.1** Growth parameters and behavior of transformed cells

Immortal (can grow indefinitely)

Reduced requirement for serum growth factors

Loss of capacity for growth arrest upon nutrient deprivation

High saturation densities

Loss of contact inhibition (can grow over one another or normal cells)

Anchorage independence (can grow in soft agar)

Altered morphology (appear rounded and refractile)

Tumorigenicity

---

cells because they have a reduced requirement for the growth factors present in serum. Some transformed cells actually produce their own growth factors and the cognate receptors, providing themselves **autocrine growth stimulation**. Normal cells cease to grow and enter a quiescent state (called $G_0$, described below) when essential nutrient concentrations drop below a threshold level. Transformed cells are deficient in this capacity, and some may even kill themselves by trying to continue to grow in an inadequate environment.

Transformed cells grow to high densities. This characteristic is manifested by the cells piling up over or under each other. They also grow on top of untransformed cells,

forming visibly identifiable clumps called **foci** (Fig. 16.2). Transformed cells behave in this manner because they have lost **contact inhibition**, a response by which normal cells cease growth and movement when they sense the presence of their neighbors. Unlike normal cells, many transformed cells have also lost the need for a surface on which to adhere, and we describe them as being **anchorage independent**. Some anchorage-independent cells can also form isolated colonies in semisolid media (e.g., 0.6% agar). This property correlates well with the ability to form tumors in animals and is often used as an experimental surrogate for malignancy. Finally, transformed cells *look* different from normal cells; they are more rounded, with fewer processes, and as a result appear more refractile when observed under a microscope (Fig. 16.2).

There are other ways in which transformed cells can be distinguished from their normal counterparts. These properties include metabolic differences and characteristic changes in the cell surface and cytoskeletal components. However, the list in Table 16.1 comprises the standard criteria used to judge whether cells have been transformed by viruses.

## Control of Cell Growth and Division

### Sensing the Environment

Normal cells possess elaborate pathways that receive and process growth-stimulatory or growth-inhibitory sig-

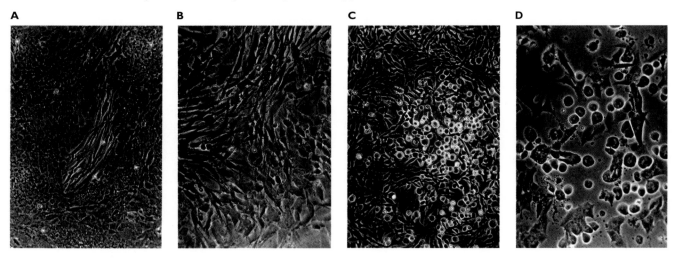

**Figure 16.2 Foci formed by avian cells transformed with two strains of the Rous sarcoma virus.** Differences in morphology are due to genetic differences in the *src* oncogene transduced. (A) A focus of infected cells with fusiform morphology shown on a background of flattened, contact-inhibited, uninfected cells. (B) Higher magnification of a fusiform focus showing lack of contact inhibition among the transformed cells. (C) A focus of highly refractile infected cells with rounded morphology and reduced adherence. (D) Higher magnification of rounded infected cells showing tightly adherent normal cells in the background. Courtesy of P. Vogt, The Scripps Research Institute.

A          B          C          D

nals transmitted by other cells in the tissue or organism. Much of what we know about these pathways comes from the study of the oncogenic retroviruses and the cellular genes transduced or activated by them. Signaling often begins with the secretion of a growth factor by a specific type of cell. The growth factor may enter the circulatory sys-

tem, as in the case of many hormones, or may simply diffuse through the spaces around cells inside a tissue. Growth factors bind to the external portion of specific receptor molecules on the surface of the same or other types of cells in the tissue (described in chapter 4). Alternatively, signaling can be initiated by binding of a receptor on one

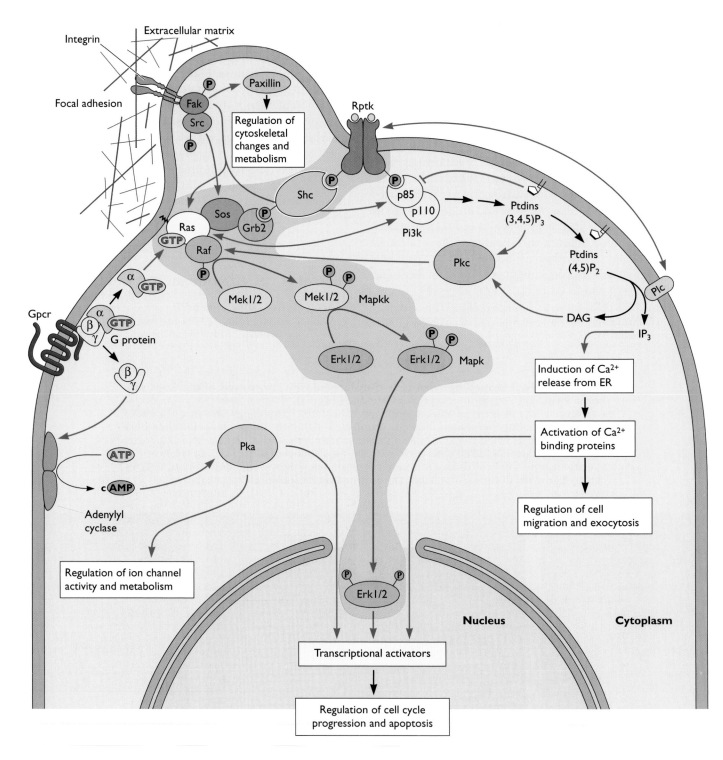

cell to a specific protein (or proteins) present on the surface of another cell or to components of the extracellular matrix. The binding of the ligand triggers a change, often via aggregation of receptor molecules, that is transmitted to the cytoplasmic portion of the receptor. In the case illustrated in Fig. 16.3, the cytoplasmic domain of the receptor possesses protein tyrosine kinase activity, and interaction with the growth factor ligand triggers autophosphorylation. This modification sets off a **signal transduction cascade**, a chain of sequential physical interactions among, and biochemical modifications of, membrane-bound and cytoplasmic proteins (Fig. 16.3). Membrane-associated guanine nucleotide-binding proteins with GTPase activity (**G proteins**) regulate the activities of some of the membrane-bound components of signaling pathways. A common biochemical modification of cytoplasmic protein components of a signaling cascade is phosphorylation of serine and/or threonine residues

(e.g., via the mitogen-activated protein [Map] kinases). As illustrated in Fig. 16.3, small molecules generated by some membrane-bound proteins in the cascade, such as cyclic nucleotides, and lipids act as diffusible **second messengers** in the signal relay. Changes in ion flux across the plasma membrane, or in membranes of the endoplasmic reticulum, may also contribute to signal transmission.

Relay of the signal can terminate at cytoplasmic sites to alter metabolism or cell morphology and adhesion. However, many signaling cascades culminate in the modification of transcriptional activators or repressors and thus alter the expression of specific cellular genes. The products of these genes either allow the cell to progress through another growth cycle or cause the cell to stop growing, to differentiate, or to die, whichever is appropriate to the situation. Errors in the signaling pathways that regulate cell cycle progression can lead to cancer.

◁ **Figure 16.3 Some signal transduction pathways and their connections.** The pathway extending from the cell surface to the nucleus (pink background) is the mitogen-activated protein (Map) kinase cascade, which is highly conserved among eukaryotes. Signal transduction is initiated by binding of ligand to the extracellular domain of a receptor protein tyrosine kinase (Rptk), for example, the receptor for epidermal growth factor or the receptor for platelet-derived growth factor. Binding of ligand induces receptor dimerization and autophosphorylation of tyrosine residues. Adapter proteins like Shc and the Grb2 component of the Grb2-Sos complex are recruited to the membrane by binding to phosphotyrosine-containing sequences in the cytoplasmic domains of activated receptor protein tyrosine kinases or to a substrate phosphorylated by the activated receptor, along with Ras. Juxtaposition of Sos and Ras at the membrane activates exchange of GDP for GTP bound to Ras by Sos, the guanine nucleotide exchange protein for Ras. The GTP-bound form of Ras binds to members of the Raf family, thus bringing these serine/threonine protein kinases to the plasma membrane. Raf is then activated by an unknown mechanism and becomes autophosphorylated. Such activation of Rafs initiates the Map kinase cascade. The pathway shown contains Mek1 and Mek2 (Mek1/2) (Map kinase kinases [Mapkk]) and Erk1 and Erk2 (Erk1/2) (Map kinases). Activated Erk1/Erk2 can enter the nucleus, where they phosphorylate and activate transcriptional regulators. These kinases can also regulate transcription indirectly, by activation of other protein serine kinases. As shown, signals also feed into this pathway from other types of receptors. These include seven transmembrane domain receptors, which transmit signals by means of heterotrimeric G proteins, termed G protein coupled receptors (Gpcr), or integrin receptors, which bind to components of the extracellular matrix at focal adhesions (sites at which the cell is in close contact with the extracellular matrix). The latter interaction induces binding of focal adhesion kinase (Fak) to the cytoplasmic domains of integrins and autophosphorylation of Fak on tyrosine residues. Activated Fak then binds Src and other members of the Src family of protein tyrosine kinases, an interaction that may lead to further phosphorylation of Fak, activation of the Src kinase, and/or binding and phosphorylation of other proteins that recognize sequences containing phosphotyrosine. The Fak-Src complex not only activates the Mapk pathway but also induces alterations in the cytoskeleton, via paxillin, and activation of phosphoinositol 3-kinase (Pi3k). As shown, this enzyme is also activated upon binding of its p85 subunit to activated receptor protein tyrosine kinases. The enzyme synthesizes phosphatidylinositol-3-phosphate [Ptdins(3)P], which is a substrate for synthesis of other phosphoinositides, some of whose autoregulatory effects and connections to the Mapk pathway are depicted. The lipid products of activation of phospholipase C (Plc) act as second messengers. For example, inositol 1,4,5-triphosphate (IP3) stimulates release of $Ca^{2+}$ from storage within the endoplasmic reticulum (ER) following binding to a receptor (Ip3r), in the ER membrane, while diacylglycerol (DAG) activates members of the protein kinase C (Pkc) family. Cyclic AMP (cAMP) produced upon activation of adenylyl cyclase by heterotrimeric G proteins is also a second messenger which activates protein kinase A (Pka). Physical associations between proteins are shown directly or indicated by a blue arrow, with green arrows and the red bar indicating positive and negative effects, respectively, of the interaction.

## *Phases of the Cell Cycle*

The capacity of cells to grow and divide is regulated by a cellular timer. The timer comprises an assembly of proteins that integrate stimulatory and inhibitory signals received by, or produced within, the cell. Eukaryotic cells do not attempt to divide until all their chromosomes have been duplicated and are precisely organized for separation into daughter cells. Nor do they initiate DNA synthesis and chromosome duplication until the previous cell division is complete or unless the extra- and intracellular environments are appropriate. Thus, the cellular timer controls a tightly ordered **cell cycle** comprising intervals, or **phases**, devoted to specific processes.

The duration of each of the separate phases in the cell cycle shown in Fig. 16.4 is typical of those of many mammalian cells growing actively in culture. However, there is considerable variation in the length of the cell cycle,

largely reflecting differences in the **gap phases** ($G_1$ and $G_2$). For example, early embryonic cells of animals such as the fruit fly *Drosophila* and the toad *Xenopus laevis* dispense with $G_1$ and $G_2$, do not increase in mass, and move immediately from the **DNA synthesis phase (S)**, to **mitosis (M)**, and again from M to S. Thus, they possess extremely short cycles of 10 to 60 min. At the other extreme are cells that have ceased growth and division and have withdrawn from the cell cycle. The variability in duration of this specialized resting state, termed $G_0$, accounts for the large differences in the rates at which cells in multicellular organisms proliferate. As discussed in chapter 9, many viruses replicate successfully in cells that spend all or most of their lives in $G_0$, a state that has been likened to cell cycle sleep. In many cases, synthesis of viral proteins in such resting or slowly cycling cells induces them to reenter the cell cycle and grow and divide more rapidly. In

**Figure 16.4 The phases of a eukaryotic cell cycle.** Most obvious morphologically, and hence the first to be identified, is mitosis, or M phase, the process of nuclear division that precedes cell division. During this period, the nuclear envelope breaks down. Duplicated chromosomes become condensed and aligned on the mitotic spindle and are segregated to opposite poles of the cell, where nuclei reform upon chromosome decondensation (top). The end of M phase is marked by **cytokinesis**, the process in which the cell divides in two. Despite this remarkable reorganization and redistribution of cellular components, M phase occupies only a short period within the cell cycle. During the long interval from one mitosis to the next, **interphase**, cells grow continuously. Interphase was divided into three parts with the recognition that DNA synthesis takes place only during a specific period, the synthetic or S phase, beginning about the middle of interphase. The other two periods, which appeared as gaps between defined processes, are designated the $G_1$ and $G_2$ (for gap) phases.

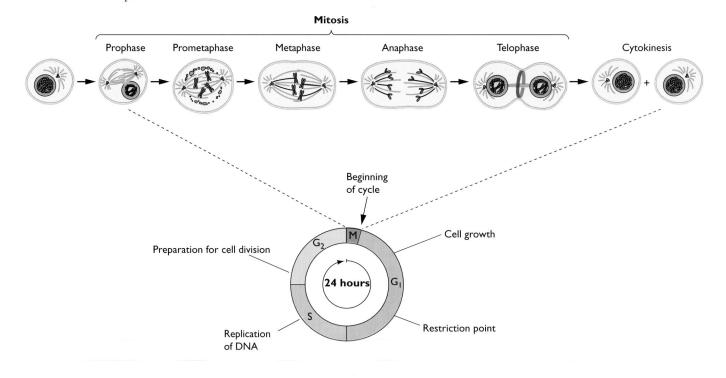

order to describe the mechanisms by which these viral proteins transform cells and induce such abnormal activity, we will first introduce the molecular mechanism that control passage through the cell cycle.

### The Cell Cycle Engine

The orderly progression of eukaryotic cells through periods of growth, chromosome duplication, and nuclear and cell division is driven by intricate regulatory circuits whose elucidation must be considered a tour de force of contemporary biology. The first experimental hint that cells contain proteins that control transitions from one phase of the cell cycle to another came more than 25 years ago when nuclei of slime mold (*Physarum polycephalum*) cells in early $G_2$ were found to enter mitosis immediately following fu-

sion with cells in late $G_2$ or M. This crucial observation led to the conclusion that the latter cells must contain a **mitosis-promoting factor**. Subsequently, similar experiments with mammalian cells in culture identified an analogous S-phase-promoting factor. The convergence of many observations from the biochemical characterization of amphibian mitosis-promoting factor, genetic studies of yeast, and investigation of the growth and division of mammalian cells in culture eventually led to the identification of the highly conserved cell cycle engine components (Fig. 16.5A). Mitosis-promoting factor purified from amphibian oocytes proved to be an unusual protein kinase: its catalytic subunit is activated by the binding of an unstable regulatory subunit. Furthermore, the concentration of the regulatory subunit was found to oscillate

**Figure 16.5 The mammalian cyclin-Cdk cell cycle engine.** (A) The phases of the cell cycle are denoted on the inner circle. The progressive accumulation of specific cyclin-Cdk complexes and their abrupt disappearance from the cell are represented by the broadening arrows, with the arrowheads marking the time of disappearance. The cyclins are related in sequence to one another, and they share such properties as activation of cyclin-dependent kinases and controlled destruction by the proteasome. In early amphibian oocytes, the first system in which oscillation of a cyclin (cyclin B) was observed, such regulation of cyclin concentrations is posttranscriptional: the cyclins are made at more or less constant rates throughout the cell cycle, but their proteolysis is initiated abruptly at specific points in the cycle. In mammalian cells, proteolysis is also very important in resetting the concentrations of individual cyclins at specific points in the cycle, but production of cyclin mRNAs is also regulated. For example, the concentrations of mRNAs encoding cyclins E, A, and B, like those of the proteins, oscillate during the cell cycle. Regulation of nuclear export, which is inhibited by cyclin B phosphorylation at the $G_2$/M transition, makes an important contribution to nuclear accumulation of the B-type cyclins specifically in the M phase. The cyclin-dependent kinases are also related in sequence and activity. (B) The production, accumulation, and activities of both cyclins and Cdk are regulated by numerous mechanisms. Activating and inhibitory reactions are indicated by green arrows and red bars, respectively. For example, activation of Cdks can require not only binding to the appropriate cyclin but also phosphorylation at specific sites and removal of phosphate groups at other sites. The activities of the kinases are also controlled by association with members of two families of cyclin-dependent kinase-inhibitory proteins, which control the activities of only $G_1$ Cdks (Ink4 proteins) or all cyclin-cyclin dependent kinase complexes (Cip/Kip proteins). Both types of inhibitor play crucial roles in cell cycle control. For example, the high concentration of p27$^{Kip1}$ characteristic of quiescent cells falls as they enter $G_1$, and inhibition of synthesis of this protein prevents cells from becoming quiescent.

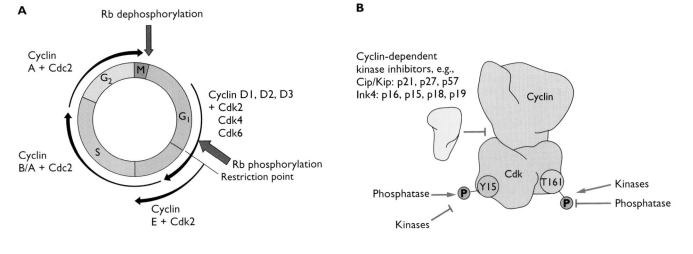

reproducibly during each and every cell cycle. The regulatory subunit was therefore given the descriptive name **cyclin**, and the associated protein kinase was termed **cyclin-dependent kinase (Cdk)**. Similar proteins were implicated in cell cycle control in yeast, and it soon became clear that all eukaryotic cells contain multiple cyclins and Cdks, which operate in specific combinations to control progression through the cell cycle.

Figure 16.5A lists the various mammalian cyclin-Cdk complexes and shows the phases in the cell cycle in which they accumulate. The patterns of accumulation illustrate a critical feature of the cell cycle: individual cyclin-Cdk complexes, the active protein kinases, accumulate in successive waves. The concentration of each increases gradually during a specific period in the cycle but decreases abruptly as the cyclin subunit is degraded. This orderly activation and deactivation of specific kinases governs passage through the cell cycle. For example, cyclin E synthesis is rate-limiting for the transition from $G_1$ to S phase in mammalian cells, and the cyclin E-Cdk2 complex accumulates during late $G_1$ (Fig. 16.5A). Soon after cells have entered S phase, cyclin E and the complex it forms with Cdk2 rapidly disappear from the cell; its task is completed until a new cycle begins.

While the oscillating waves of active Cdk accumulation and destruction are thought of as the gears that advance the cell cycle timer, it is important not to interpret this metaphor too literally. The orderly and reproducible sequence of DNA replication, chromosome segregation, and cell division is not determined solely by the oscillating concentrations of individual cyclin-Cdk complexes. Rather, the cyclin-Cdk cycle serves as a device for integrating numerous signals from the exterior and interior of the cell into appropriate responses, and the regulatory circuits that feed into and from the cycle are both many and complex (Fig. 16.5B). These regulatory signals ensure that the cell increases in mass and divides only when the environment is hospitable or, in multicellular organisms, when the timing is correct. Thus, many signal transduction pathways that convey information about the local environment or the global state of the organism converge on the cyclin-Cdk cycle. In addition, surveillance mechanisms monitor such internal parameters as DNA damage, problems with DNA replication, and proper assembly and function of the mitotic spindle, to protect cells against potentially disastrous consequences of continuing a cell division cycle that cannot be completed correctly. As we shall see, it is primarily these signaling and surveillance mechanisms that are compromised during transformation by oncogenic viruses.

## Oncogenic Viruses

Oncogenesis is the result of the accumulation of discrete genetic changes that alter the expression or function of proteins that play critical roles in the control of cell growth and division. Oncogenic viruses cause cancer by inducing changes that affect these processes. Indeed, it was study of the mechanisms of viral transformation and oncogenesis that laid the foundation for our current understanding of cancer, for example, with the identification of both oncogenes and tumor suppressor genes (Fig. 16.6). Specific members of a number of different virus families have been implicated in naturally occurring or experimentally induced cancers in animals (Table 16.2). It has been estimated that approximately 20% of all cases of human cancer are associated with one of five viruses as the direct causative agent or as a cofactor: hepatitis C virus, human T-cell leukemia virus type 1, human papillomaviruses, Epstein-Barr virus, and hepatitis B virus. In this section, we introduce oncogenic viruses and general features of their transforming interactions with host cells.

### Discovery of Oncogenic Viruses

Oncogenic viruses were discovered almost 100 years ago when Vilhelm Ellerman and Olaf Bang (1908) first showed that avian leukemia could be transmitted by filtered extracts (i.e., viruses) of leukemic cells or serum from infected birds. Because leukemia was not recognized as cancer in those days, the significance of this discovery was not generally appreciated. Shortly thereafter (in 1911),

**Figure 16.6 A genetic paradigm for cancer.** The division cycle for all eukaryotic cells has an intrinsic pace that can be modulated both positively and negatively by different sets of gene products. Cancer arises from a combination of dominant gain-of-function mutations in proto-oncogenes and recessive loss-of-function mutations in tumor suppressor genes. The function of either type of gene product can be affected by oncogenic viruses.

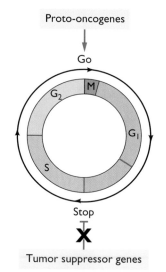

**Table 16.2** Oncogenic viruses and cancer

| Families | Associated cancers |
|---|---|
| **RNA viruses** | |
| *Flaviviridae* | |
| Hepatitis C virus | Hepatocellular carcinoma |
| *Retroviridae* | Hematopoietic cancers, sarcomas, and carcinomas |
| **DNA viruses** | |
| *Hepadnaviridae* | Hepatocellular carcinoma |
| *Papovaviridae* | |
| Papillomaviruses | Papillomas and carcinomas |
| Polyomaviruses | Various solid tumors |
| *Adenoviridae* | Various solid tumors |
| *Herpesviridae* | Lymphomas, carcinomas, and sarcomas |
| *Poxviridae* | Myxomas and fibromas |

Peyton Rous demonstrated that solid tumors could be produced in chickens by using cell-free extracts from a transplantable sarcoma that had appeared spontaneously. Despite the viral etiology of this disease, the cancer viruses of chickens were thought to be oddities until similar murine malignancies, as well as mammary tumors, were found to be induced by viruses. These oncogenic viruses all proved to be members of the retrovirus family.

Early researchers classified the oncogenic retroviruses into two groups depending on the rapidity with which they caused neoplasia (Table 16.3). The first group comprises rapidly transforming, **transducing** oncogenic retroviruses. These are all highly carcinogenic agents that cause malignancies in nearly 100% of infected animals in a matter of days. They were later discovered to have the ability to transform susceptible animal cells in culture. The second class, **nontransducing** oncogenic retroviruses, includes less-carcinogenic agents. Not all animals infected with these viruses develop tumors, which appear only weeks or months after infection. In the late 1980s, a third type of oncogenic retrovirus, a **long-latency** virus for which tumorigenesis is very rare and occurs months or

even years after infection, was identified in humans. Each group of oncogenic retroviruses induces tumors by a distinct mechanism. As their name implies, transducing retroviruses cause cancer because their genomes contain transduced cellular genes that become **oncogenes** (genes whose products cause transformation or tumorigenesis) when expressed in the viral context. The virally transduced versions of cellular genes are called **v-oncogenes**, and their normal cellular counterparts are called **c-oncogenes** or **proto-oncogenes**. In the case of nontransducing retroviruses, transcription of proto-oncogenes is inappropriately activated as a consequence of the nearby integration of a provirus in the host cell genome. In either situation, the oncogene products play no role in the reproductive cycle of the retroviruses themselves. Nevertheless, the study of v-oncogenes and proto-oncogenes that are affected by retroviruses has been of great importance in advancing our understanding of the origins of cancer.

The first oncogenic DNA virus to be discovered was the papillomavirus that causes warts (papillomas) in cottontail rabbits, isolated by Richard Shope in 1933. Important properties of the virus and its DNA genome were identified with remarkable speed given the techniques available at the time. But the lack of cell culture systems for papillomaviruses initially precluded use of these viruses as experimental models for oncogenesis. Other viruses (in particular polyomaviruses, like simian virus 40, and human adenoviruses) proved much more tractable and soon dominated early studies of transformation and tumorigenesis by DNA viruses. Although neither simian virus 40 nor adenoviruses are associated with oncogenesis in their natural hosts, it was shown soon after their discovery that these viruses can induce tumors in rodents and transform cultured mammalian cells. In contrast to the v-oncogenes of retroviruses, the transforming genes of these DNA viruses do play important roles in virus reproduction. The proteins encoded by some contribute to oncogenesis or transformation by altering the activities of

**Table 16.3** Oncogenic retroviruses

| Property or characteristic | Transducing viruses | Nontransducing viruses | Nontransducing, long-latency viruses |
|---|---|---|---|
| Efficiency of tumor formation | High ($\cong$100% of infected animals) | High to intermediate | Very low (<5%) |
| Tumor latency | Short (days) | Intermediate (weeks, months) | Long (months, years) |
| Infecting viral genome | Virus-cell recombinant; normally replication-defective | Intact; replication-competent | Intact; replication-competent |
| Oncogenic element | Cell-derived oncogene carried in viral genome | Cellular oncogene activated in situ by a provirus | Virus-encoded regulatory protein controlling transcription |
| Mechanism | Oncogenic transduction | *cis*-acting provirus | *trans*-acting protein? |
| Ability to transform cells in culture | Yes | No | No |

products of the cellular progenitors of certain retroviral v-oncogenes. This important discovery, initially made in studies of the early middle T protein (mT) of mouse polyomavirus in the early 1980s, provided the first indication that retroviruses and DNA viruses can transform cells by related mechanisms. Investigation of biochemical properties of proteins encoded in other transforming genes of these DNA viruses led to equally important insights, in particular the characterization of cellular proteins that can stop cell growth and division, the products of **tumor suppressor genes**.

Since the discovery of Epstein-Barr virus in cells derived from Burkitt's lymphoma in 1966, it has been appreciated that herpesviruses can be associated with the development of tumors in humans and other animals. Members of this family can immortalize or induce typical transformed phenotypes in appropriate cells in culture. Poxviruses can also induce cell proliferation, which may be prolonged or rapidly followed by cell death, depending on the virus. Indeed, some members of this family, such as Shope fibroma virus, cause tumors of the skin. However, poxviruses do not transform cells in culture, in part because they are highly cytotoxic. The large sizes of the herpes- and poxviral genomes initially presented a major impediment to elucidation of the molecular mechanisms by which these viruses can alter the proliferative properties of infected cells. However, we now know the molecular functions of many of the relevant viral proteins, and it appears that these viruses generally alter cell growth and proliferation by mechanisms related to those responsible for transformation by the smaller DNA viruses or retroviruses.

Oncogenic viruses associated with human disease continue to be isolated with some regularity. The most recently discovered (in 1994) is a previously unknown member of the family *Herpesviridae*, human herpesvirus 8, which was isolated from tumor cells from Kaposi's sarcoma, a cancer common in AIDS patients. Although this virus was only recently identified, it is already clear that the genome of human herpesvirus 8, like those of transducing retroviruses, contains homologs of cellular proto-oncogenes. Perhaps an even greater surprise was the realization that RNA viruses other than retroviruses can be associated with cancer: hepatitis C virus, a (+) strand RNA virus belonging to the family *Flaviviridae* identified in 1989, is associated with a high risk for hepatacellular carcinoma in the 60 to 70% of infected humans who become persistently infected. Humans are the only known natural hosts for this virus, and a tissue culture system that can support its proliferation is not yet available. As hepatitis C virus is estimated to be the cause of about 30% of all acute liver disease in the United States, studies of cloned viral genes are being actively pursued in many academic and industrial laboratories.

Although members of different families (Table 16.2), the oncogenic viruses share several general features with respect to their ability to induce cellular transformation. Transformation is a single-hit process (defined in chapter 2) in the sense that infection of a susceptible cell with a single virus particle is sufficient to cause transformation. All or part of the viral genome is usually integrated into the host cell DNA, resulting in an irreversible genetic change in the cell. With few exceptions, cellular transformation is accompanied by the continuous expression of specific viral sequences. On the other hand, while specific viral genes are present and expressed, transformed cells need not and (except in the case of some retroviruses) do not produce infectious virus particles. Most importantly, viral transforming proteins alter cell growth and proliferation by a limited repertoire of molecular mechanisms.

## Viral Genetic Information in Transformed Cells

### State of Viral DNA

Cells transformed by oncogenic viruses generally retain viral DNA in their nuclei. These DNA sequences correspond to all or part of the proviral DNA made in retrovirus-infected cells or of the infecting DNA genome. In most cases, viral DNA sequences are maintained as a permanent addition to the cellular genome following their integration (Box 16.3).

As discussed in chapter 7, integration of proviral DNA by the viral integrase is an essential step in the retroviral life cycle. Such proviral integration can occur at essentially any site in cellular DNA but preserves a fixed order of viral genes and control sequences (see Fig. 7.14). When the provirus carries a v-oncogene, where it is integrated into the cellular genome is of no importance provided that the viral transcription unit is not silenced by integration into transcriptionally inert regions of cellular chromosomes. In contrast, integration of proviral DNA within specific regions of the cellular genome is a hallmark of, and essential for, the induction of tumors by nontransducing retroviruses. The proviral sequences present in cells of tumors induced by members of this class are found in the same chromosomal location, and the tumors are **monoclonal**, having arisen from a single transformed cell. Although infection is initiated with a nondefective retrovirus, the proviruses in the resulting tumor cells have usually lost some or most of the proviral sequences. Nevertheless, at least one long terminal repeat (LTR) containing the transcriptional control region is always present. Thus, viral transcription signals, but not protein-coding sequences, are required for transformation by nontransducing retroviruses. The significance of these properties became apparent when it was discovered that, in several tumors,

*Detection and characterization of integrated viral DNA sequences in transformed cells*

Integration of viral into cellular DNA in transformed cells was initially demonstrated by the cosedimentation of viral DNA with high-molecular-mass cellular DNA under strongly denaturing conditions. The development of restriction endonucleases as molecular tools and of the DNA blotting assay allowed direct proof of such integration and characterization of integrated viral DNA, as illustrated for a circular double-stranded viral DNA genome.

Cleavage of genomic DNA isolated from transformed cells with enzyme A (no sites in the viral genome) yields one high-molecular-mass fragment of lower mobility than free viral DNA for each site of integrated viral DNA. The number of viral integrants can therefore be counted. When used with appropriate standards, the number of copies of viral sequences at each integration site can also be estimated. Restriction enzyme B (one site in the viral genome) generates two fragments differing in mobility from the free, linear viral DNA. The results obtained with enzymes A and B demonstrate that viral DNA must be covalently attached to cellular DNA. Enzyme C, with two cleavage sites in the viral genome, produces three fragments from transformed cell DNA. The comigration of one with the smaller product of digestion of free viral DNA indicates that this segment of the viral genome is intact in the integrated viral DNA. Extension of this approach to restriction endonucleases that cut the viral genome at multiple sites allows identification of which genomic sequences are preserved following recombination, and mapping of the sites of integration within the viral genome.

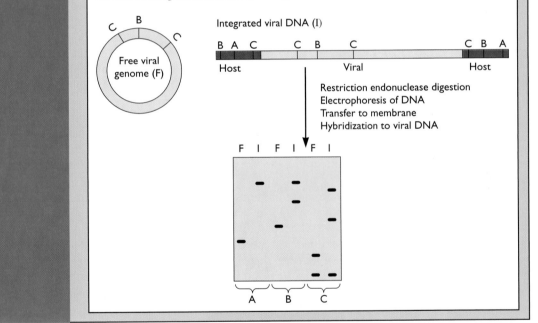

proviruses are integrated in the vicinity of some of the same cellular oncogenes captured by transducing retroviruses. The study of nontransducing oncogenic retroviruses has also led to the identification of some additional proto-oncogenes, as illustrated for two murine retroviruses in Table 16.4. As integration of retroviral DNA into the host genome is essentially random, the long latency for tumor induction by the nontransducing viruses can be explained in part by the limited probability that integration will occur in the vicinity of an oncogene.

Integration of genomic DNA sequences is not a prerequisite for successful reproduction of any oncogenic DNA viruses. Indeed productive replication of these viruses and transformation of host cells are mutually exclusive processes. Nevertheless, integration is the rule in polyomavirus- or adenovirus-transformed cells, which contain specific subsets of viral genes inserted into the cellular genome. Such integration is the result of rare recombination reactions (catalyzed by cellular enzymes) between generally unrelated cellular and viral DNA sequences. Integration can therefore occur at many sites in the cellular genomes and, in contrast to proviral DNA integration, does not maintain specific viral DNA sequences at the junctions with cellular DNA. The low probability that viral

**Table 16.4** Some previously unknown loci targeted in tumors induced by two nontransducing simple retroviruses[a]

| Virus | Neoplasia | Cellular gene | Normal function |
|-------|-----------|---------------|-----------------|
| Mouse mammary tumor virus | Mammary tumors | Int-1 (wnt-1) | Secreted glycoprotein important in pattern formation in early embryonal development |
| | | Int-2 (hst) | Secreted protein that may act as a growth factor |
| | | Int-3 | Protein presumed to function in development (notch family) |
| | | Int-H/Int-5 | Converts androgens to estrogens |
| Moloney murine leukemia virus | Pre-B lymphoma | Ahi-1 | Adaptor protein |
| | T-cell lymphoma | cyclin D2/Vin-1 | G₁ cyclin |
| | | Lck | Nonreceptor protein tyrosine kinase |
| | | Notch-1/Mis-6 | Transmembrane receptor presumed to function in development |
| | | Pim-1 | Serine/threonine kinase |
| | | Tpl-2 | Serine/threonine kinase |
| | B-cell lymphoma | Pim-2 | Serine/threonine kinase |

[a]Adapted from J. M. Coffin et al. (ed.), *Retroviruses*, p. 428–484 (Cold Spring Harbor Laboratory Press, Plainview, N.Y., 1997), with permission.

DNA will become integrated into the cellular genome and the fact that only a fraction of these recombination reactions will maintain the integrity of viral transforming genes are major factors contributing to the low efficiencies of transformation by these viruses: for simian virus 40, one focus-forming unit (see chapter 2) corresponds to $10^2$ to $10^4$ plaque-forming units, while the ratio is even lower for human adenoviruses.

The great majority of cells transformed by polyoma- or adenoviruses retain only partial copies of the viral genome. The genomic sequences integrated can vary considerably among independent lines of cells transformed by the same virus, but a common, minimal set of genes is always integrated. This property provided one of the first ways in which the genes of DNA viruses needed to maintain the transformed state were identified. The number of copies of viral DNA that can be integrated and the number of integration sites range from 1 copy to 10 to 20 copies per cell of specific viral DNA sequences at a single site to multiple copies at several different sites. However, there is little indication that such variation is associated with phenotypic differences among the cells transformed by these viruses.

A second mechanism by which viral DNA can persist in transformed cells is as a stable extrachromosomal episome. Such episomal viral genomes are a characteristic feature of B cells immortalized by Epstein-Barr virus and can also be found in cells transformed by papillomaviruses. The viral episomes are maintained at concentrations of tens to hundreds of copies per cell by both replication of the viral genome in concert with cellular DNA synthesis and orderly segregation of viral DNA to daughter cells (see chapter 9). Because of these requirements, permanent alteration of cell growth properties depends on the viral proteins necessary for the survival of viral episomes, in addition to the viral gene products that directly modulate cell growth and proliferation (Table 16.5).

## Identification and Properties of Viral Transforming Genes

Transforming genes of oncogenic viruses have been identified by classical genetic methods, characterization of

**Table 16.5** Epstein-Barr virus gene products implicated in growth transformation of B lymphocytes[a]

| Gene product(s) | Activities |
|-----------------|------------|
| EBNA-1 | Replication, and orderly segregation to progeny cells, of the episomal viral genome via binding to specific viral DNA sequences (chapter 9) |
| EBNA-2 | Activation of transcription of specific viral (e.g., LMP-1) and cellular (CD23, Fgr, cyclin D2) genes via binding to a cellular sequence-specific DNA-binding protein (Cbf-1) involved in notch signaling; sequences needed for transformation are those required for transcriptional activation |
| EBNA-3A, -3B, -3C | Related nuclear proteins; 3A and 3C required for transformation of primary B lymphocytes; may increase expression of viral (LMP-1) and cellular genes; 3A and 3C bind to Cbf-1; 3C can bind to Rb |
| LMP-1 | Alone alters the growth and proliferation of human B-lymphocyte cell lines, transforms established rodent cells, and induces B-cell lymphomas in transgenic mice; acts as a constitutively active receptor to alter cellular gene expression (see text) |

[a]Many of these proteins are also required in order to establish and maintain Epstein-Barr virus latent infection (chapter 15).

the viral genes present and expressed in transformed cell lines, and analysis of the transforming activity of viral DNA fragments directly introduced into cells (Box 16.4). For example, analysis of transformation by temperature-sensitive mutants of mouse polyomavirus established as early as 1965 that the viral early transcription unit is necessary and sufficient to initiate and maintain transformation. This conclusion was reinforced by the finding that only this viral transcription unit was invariably present and expressed in cells transformed by the virus. Of even greater value were mutants of retroviruses, in particular two mutants of Rous sarcoma virus isolated in the early 1970s. One mutant, which had suffered a spontaneous deletion of approximately 20% of the viral genome, could no longer transform the cells it infected, but it could still replicate. The second mutant was temperature-sensitive for transformation, but the virus could replicate at both temperatures. Thus, results with both mutants showed unequivocally that transformation and viral replication are distinct processes. More importantly, the deletion mutant allowed the preparation of the first nucleic acid probe specific for a v-oncogene. Complementary (−) strand DNA was prepared by reverse transcription of the Rous sarcoma virus RNA genome and then hybridized to (+) strand genomic RNA of the mutant. The nonhybridizing DNA, separated from the double-stranded hybrids, comprised the viral oncogenic (*src*) information. This *src*-specific probe was found to hybridize to cellular DNA, providing the first conclusive evidence that v-oncogenes are of cellular and not viral origin. This finding, for which J. Michael Bishop

and Harold Varmus received the 1989 Nobel Prize in physiology or medicine, had far-reaching significance because it immediately suggested that such cellular genes might become oncogenes by means *other* than viral transduction.

The acquisition of cellular oncogenes turned out to be the definitive characteristic of transducing retroviruses, as illustrated in Fig. 16.7. With the exception of Rous sarcoma virus, these viruses are all replication defective, having lost all or most of the viral coding sequences during oncogene capture. Such defective transducing viruses can, however, be propagated in mixed infections with replication-competent "helper" viruses, which provide all the necessary virion proteins. Such mixed viral stocks have been the source of many transducing oncogenic viruses. The majority carry a single v-oncogene in their genomes, but some include more than one oncogene (e.g., *erbA* and *erbB* in avian erythroblastosis virus ES4). In such cases, one is sufficient for transformation; the second enhances oncogenicity.

In many transducing retroviruses, the viral and cellular protein-coding sequences are fused (Fig. 16.7). The presence of viral sequences can enhance the efficiency of oncogene translation, stabilize the protein, or determine its location in the cell. With some v-oncogenes (e.g., *myc* and *mos*), unregulated expression or overexpression of the cellular sequence from the viral promoter is sufficient to cause transformation. However, in most cases, the captured oncogenes have undergone additional changes that contribute to their transforming potential. These alterations, which include nucleotide changes, truncations at one or both ends,

---

*Multiple lines of evidence identified the transforming proteins of the papovavirus simian virus 40*

Early gene products are necessary and sufficient to initiate transformation.

1. Viruses carrying temperature-sensitive mutations in the early transcription unit (tsA mutants), but no other region of the genome, fail to transform at a nonpermissive temperature.
2. Simian virus 40 DNA fragments containing only the early transcription unit transform cells in culture; DNA fragments containing other regions of the genome exhibit no activity.

Early gene products are necessary to maintain expression of the transformed phenotype.

1. Many lines of cells transformed by simian virus 40 tsA mutants at a permissive temperature revert to a normal phenotype when shifted to a nonpermissive temperature.

2. Integration of viral DNA sequences disrupts the late region of the viral genome but not the early transcription unit, and early gene products are synthesized in all transformed cell lines.

Both LT and sT contribute to transformation.

1. Simian virus 40 mutants carrying deletions of sequences expressed only in sT fail to transform rat cells to anchorage-independent growth.
2. Introduction and expression of LT cDNA are sufficient for induction of transformation, but expression of sT can stimulate transformation (especially at low LT concentrations), is necessary for expression of specific phenotypes in specific cells, and is required for transformation of resting cells.

## Avian transducing retroviruses

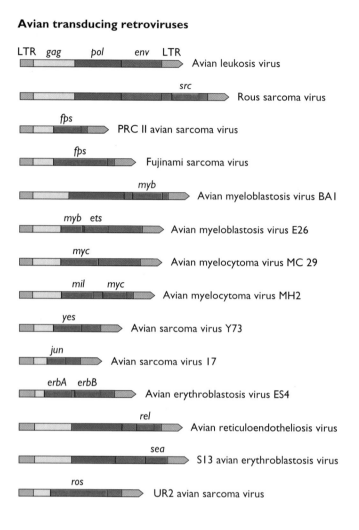

## Mammalian transducing retroviruses

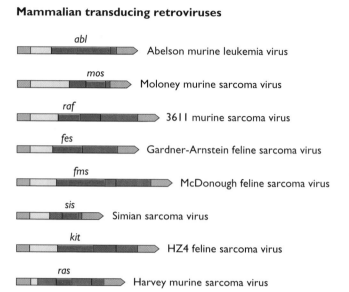

**Figure 16.7 Genome maps of avian and mammalian transducing retroviruses.** Avian leukosis virus (e.g., Rous-associated virus) is a prototypical retrovirus. Its genome contains the three major coding regions *gag* (pink), *pol* (blue), and *env* (brown) and regulatory sequences that constitute the long terminal repeat (LTR) (lavender) of the provirus. In Rous sarcoma virus, the oncogene *src* is added to the complete viral genome. In all other avian and mammalian transducing retroviruses, some of the viral coding information is replaced by cell-derived oncogene sequences (red). Consequently, such transducing viruses are defective in replication. In some cases, additional cellular DNA sequences (green) were also captured in the viral genome. Adapted from T. Benjamin and P. Vogt, p. 317–367, *in* B. N. Fields et al. (ed.), *Fields Virology*, 2nd ed., (Raven Press, New York, N.Y., 1990), with permission.

or other rearrangements, increase or decrease activity or the normal catalytic function of the gene product.

DNA-mediated transformation allows assessment of the effects of individual viral genes (or DNA sequences encoding a single viral protein) on cell growth and proliferation. This method has been especially valuable in identification of transforming proteins encoded by DNA viruses. Transformation of primary cells by these viruses requires the products of two or more viral genes, as illustrated for papova- and adenoviruses in Table 16.6. The majority of these genes exhibit some ability to alter the properties of the cells in which they are expressed in the absence of other viral proteins. However, some are required only for the induction of specific transformed phenotypes or only under certain conditions (e.g., simian virus 40 small T antigen [sT], bovine papillomavirus type 1 E7), and several exhibit no activity on their own (Table 16.6). A classic example of the latter phenomenon is provided by the adenoviral E1B gene: this gene, together with the E1A gene, was initially shown to be essential for transformation of rodent cells in culture, but it possesses no intrinsic ability to induce any transformed phenotype. This apparent paradox has been resolved with elucidation of the molecular functions of these gene products: E1A proteins induce apoptosis, but E1B proteins suppress the pathways that lead to programmed cell death and therefore allow cells synthesizing E1A proteins to survive and express transformed phenotypes. A requirement for multiple viral genes is also the rule for transformation by the larger DNA viruses (e.g., see Table 16.5).

For some viruses, a complete protein-coding sequence is not necessary for oncogenesis or transformation. This situation is exemplified by nontransducing retroviruses: as we have seen, the minimal proviral sequences associated with oncogenesis are those that comprise the transcriptional control sequences of the long terminal repeat (LTR).

**Table 16.6**  Transforming gene products of papova- and adenoviruses

| Virus | Gene product | Activities |
|---|---|---|
| **Papovaviruses** | | |
| Polyomavirus | LT | Immortalizes primary cells; required to induce but not to maintain transformation of primary cells |
| | mT | Transforms established cell lines; required both to induce and maintain transformation of primary cells |
| | sT | See simian virus 40 |
| Simian virus 40 | LT | Immortalizes primary cells; required to induce and maintain transformation of primary and established cells; in some recipient cells, an N-terminal segment (amino acids 1 to 121) is sufficient, but in most cells, most LT sequences are required |
| | sT | Required under many conditions, depending on LT concentration, genetic background of recipient cells, and transformation assay; necessary for transformation of resting cells |
| Human papillo-mavirus types 16 and 18 | E6 | Required for efficient immortalization of primary human fibroblasts and keratinocytes |
| | E7 | Transforms established rodent cells; cooperates with E6 to transform primary rodent cells; required for efficient immortalization of primary human fibroblasts or keratinocytes |
| | E5 | Induces some transformed phenotypes in established cells; can increase the proliferative capacity of human keratinocytes |
| Bovine papillo-mavirus type 1 | E5 | Transforms bovine and established rodent fibroblasts in culture; the major viral protein synthesized in transformed cells |
| | E6 | Required for expression of typical transformed phenotypes in primary mouse cells; can transform established cells |
| | E7 | Required for expression of full transformed phenotype in primary mouse cells; no activity alone but can cooperate with the E6 protein |
| **Adenoviruses** | | |
| Human adeno-virus type 2 | E1A: 243R and 289R | Essential for transformation of established and primary rodent cells; cooperates with E1B products to transform primary cells; not sufficient for establishment of transformed cell lines |
| | E1B: 55 and 19 kDa | Necessary for E1A-dependent transformation of primary and established cells; when expressed alone, cannot alter cell growth or proliferation |
| | E4: ORF6 | Cooperates with E1A to transform primary cells; converts nontumorigenic E1A + E1B-transformed cells to cells tumorigenic in nude mice |
| | E4: ORF1 | Transforms established cells; the adenovirus type 9 protein is necessary for induction of estrogen-dependent mammary tumors by this virus |

## The Origin and Nature of Viral Transforming Genes

Two classes of viral oncogenes can be distinguished on the basis of their similarity to cellular sequences. The oncogenes of transducing retroviruses and certain herpesviruses (e.g., human herpesvirus 8) are so closely related to cellular genes that it is clear that they were captured relatively recently from the genomes of cells in which these viruses replicate. Such acquisition of cellular genetic information must be the result of recombination between viral and cellular nucleic acids, a process that has been documented for transducing retroviruses. Retrovirions contain some cellular RNAs, and rare recombination reactions during reverse transcription can give rise to transducing retroviruses. The limiting factor appears to be the frequency with which cellular mRNA molecules are encapsidated into virions. Two mechanisms that can increase the likelihood of such encapsidation, and thus increase the frequency of gene capture, have been described (Fig. 16.8). Both mechanisms depend on integration of a provirus in or near a cellular gene and final recombination reaction(s) between largely nonhomologous sequences. Model systems have provided experimental support for these more efficient mechanisms for producing transducing retroviruses, but it is still unclear exactly how these viruses arise in nature.

Many of the cellular proto-oncogenes from which v-oncogenes are derived have been highly conserved throughout evolution; many vertebrate examples have homologs in yeast. Thus, such genes must fulfill functions that are indispensable for a wide variety of eukaryotic cells. Furthermore, as single copies of v-oncogenes are suf-

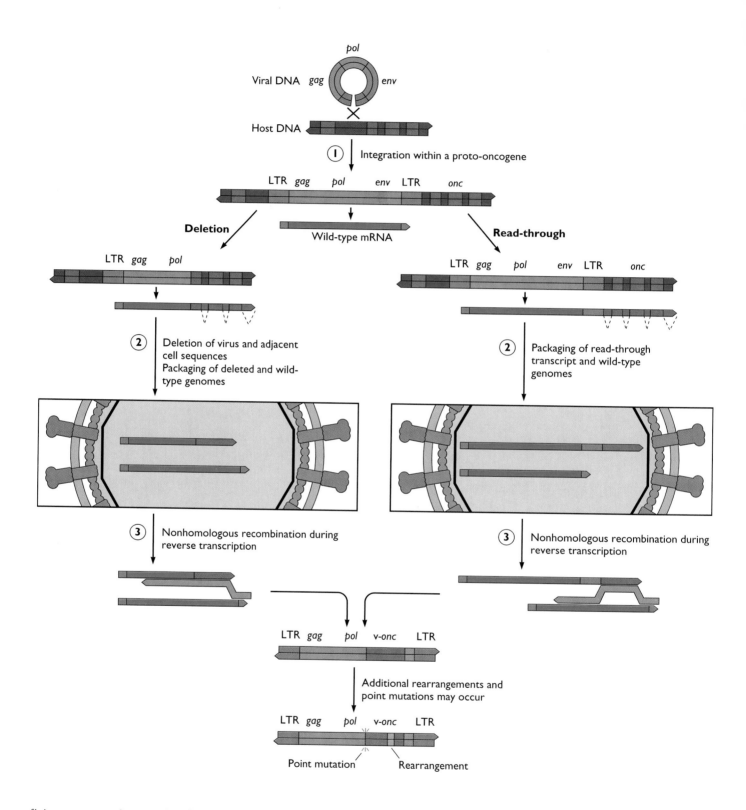

ficient to transform infected cells, their functions must override those of the resident, cognate proto-oncogenes. Accordingly, v-oncogenes function as **dominant transforming** genes.

Members of the second class of viral oncogenes are not obviously related to cellular genes. However, the products of these genes may contain short amino acid sequences also present in cellular proteins (for example, see Fig.

◁ **Figure 16.8 Possible mechanisms for oncogene capture by retroviruses.** The first step in each of two mechanisms shown is integration of a provirus in or near a cellular gene. The **deletion mechanism** (left) requires removal of the right end of the provirus, thereby linking cellular sequences to the strong viral transcriptional control region in the left long terminal repeat (LTR). Thus, the first recombination step in this mechanism takes place at the DNA level. It leads to expression of a chimeric RNA in which viral sequences from the left end of the provirus are joined to cellular sequences. Should the chimeric RNA include the viral packaging signal, it could be incorporated efficiently into virions as a heterozygote with a wild-type genome produced from another provirus in the same cell. A second recombination reaction, during reverse transcription (as described in chapter 7), is then required to add right-end viral sequences to the recombinant. At a minimum, these right-end sequences must include signals for subsequent integration of the recombinant viral DNA into the genome of the newly infected host cell, from which the transduced gene is then expressed. The **read-through mechanism** (right) does not require a chromosomal deletion. It was suggested by the finding that viral transcription does not always terminate within the right LTR, with the result that cellular sequences downstream of an integrated provirus are occasionally incorporated into viral transcripts. Such chimeric transcripts can then be incorporated into virions together with the normal viral transcript. The cellular sequences can then be captured by recombination during reverse transcription, although a double crossover is required. In either case, important additional mutations and rearrangements probably occur during subsequent virus replication. Adapted from J. M. Coffin et al. (ed.), *Retroviruses* (Cold Spring Harbor Laboratory Press, Plainview, N.Y., 1997), with permission.

16.19). The precise origins of such oncogenes, like those of viruses themselves, therefore remain shrouded in mystery but must have been nucleic acids of more complex organisms (see chapter 20).

## Functions of Viral Transforming Proteins

Many approaches have been employed to determine the specific functions of viral oncogenes. In some cases, the sequence of a viral transforming gene can immediately suggest the function of its protein product. For example, the genomes of certain herpes- and poxviruses include coding sequences that are closely related to cellular genes encoding growth factors, cytokines, and their receptors. Or the protein may contain predicted amino acid motifs characteristic of specific biochemical activities, such as tyrosine kinase or sequence-specific DNA binding. With some retroviral v-oncogene products, it has been possible to identify important biochemical properties, such as enzymatic activity, binding to a hormone or growth factor, or sequence-specific binding to nucleic acids. Table 16.7 includes a list of these v-oncogenes and the proteins they express, organized according to the known or deduced functions of their normal cellular homologs.

The watershed in elucidation of the mechanisms by which other viral proteins change cellular growth control so dramatically came with mutational analyses that correlated their transforming activity with binding to specific cellular proteins. For example, two of the regions of adenoviral E1A proteins necessary for transformation mediate binding to the cellular retinoblastoma tumor suppressor protein, Rb. The similar relationship between Rb binding and the transforming activities of simian virus 40 large T antigen (LT) and the E7 protein of oncogenic human papillomaviruses rapidly established the general importance of interaction with Rb in transformation.

Viral transforming proteins exhibit great diversity in all their properties, from primary amino acid sequence to biochemical activity. They also differ in the number and nature of the cellular pathways they alter. Transduced viral oncogenes, such as those of many retroviruses, generally act on a limited number of cellular targets. In contrast, the transforming activities of the products of other viral oncogenes have been correlated with effects on multiple, different cellular pathways and proteins (Table 16.8). Despite such variation, the best-characterized mechanisms of transformation by viral proteins fall into one of two classes, permanent activation of cellular signal transduction cascades or disruption of the circuits that regulate cell cycle progression.

## Viral Transformation by Activation of Cellular Signal Transduction Pathways

The products of transforming genes of both RNA and DNA viruses can alter cellular signal transduction cascades. The consequence is invariably permanent activation of pathways that promote cell growth and proliferation. However, as discussed in subsequent sections, these viral proteins can intervene at various points in the cascades that transmit signals from the exterior of the cell to nuclear components, and they operate in several different ways.

**Table 16.7** Functional classes of oncogenes transduced by retrovirus[a]

| Transduced oncogene | Viral oncoprotein[b] | Function of cellular homolog |
|---|---|---|
| **Growth factor** | | |
| *sis* | $p28^{env\text{-}sis}$ | Platelet-derived growth factor (Pdgf) |
| **Tyrosine kinase growth factor receptors** | | |
| *erbB* | $gp65^{erbB}$ | Epithelial growth factor (Egf) receptor |
| *fms* | $gp180^{gag\text{-}fms}$ | Colony-stimulating factor 1 (Csf-1) receptor |
| *sea* | $gp160^{env\text{-}sea}$ | Receptor; ligand unknown |
| *kit* | $gp80^{gag\text{-}kit}$ | Hematopoietic receptor; product of the mouse W locus |
| *ros* | $p68^{gag\text{-}ros}$ | Receptor, ligand unknown |
| *mpl* | $p31^{env\text{-}mpl}$ | Member of the hematopoietin receptor family |
| *eyk* | $gp37^{eyk}$ | Receptor, ligand unknown |
| **Hormone receptor** | | |
| *erbA* | $p75^{gag\text{-}erbA}$ | Thyroid hormone receptor |
| **G proteins** | | |
| H-*ras* | $p21^{ras}$ | GTPase |
| K-*ras* | $p21^{ras}$ | GTPase |
| **Adapter protein** | | |
| *crk* | $p47^{gag\text{-}crk}$ | Signal transduction |
| **Nonreceptor tyrosine kinases** | | |
| *src* | $pp60^{src}$ | Signal transduction |
| *abl* | $p460^{gag\text{-}abl}$ | Signal transduction |
| *fps*[c] | $p130^{gag\text{-}fps}$, $p105^{gag\text{-}fps}$ | Signal transduction |
| *fes*[c] | $p85^{gag\text{-}fes}$ | Signal transduction |
| *fgr* | $p70^{gag\text{-}actin\text{-}fgr}$ | Signal transduction |
| *yes* | $p90^{gag\text{-}yes}$, $p80^{gag\text{-}yes}$ | Signal transduction |
| **Serine/threonine kinases** | | |
| *mos* | $p37^{env\text{-}mos}$ | Required for germ cell maturation |
| *raf*[d] | $p75^{gag\text{-}raf}$ | Signal transduction |
| *mil*[d] | $p100^{gag\text{-}mil}$ | Signal transduction |
| *akt* | $p86^{gag\text{-}akt}$ | Signal transduction |
| **Nuclear proteins** | | |
| *jun* | $p65^{gag\text{-}jun}$ | Transcriptional regulator (Ap-1 complex) |
| *fos* | $p55^{fos}$ | Transcriptional regulator (Ap-1 complex) |
| *myc* | $p100^{gag\text{-}myc}$, $p90^{gag\text{-}myc}$, $p200^{gag\text{-}pol\text{-}myc}$, $p59^{gag\text{-}myc}$ | Unknown; possibly transcriptional regulator |
| *myb* | $p45^{myb}$, $p135^{gag\text{-}myb\text{-}ets}$ | Transcriptional regulator |
| *ets* | $p135^{gag\text{-}myb\text{-}ets}$ | Transcriptional regulator |
| *rel* | $p64^{rel}$ | Transcriptional regulator |
| *maf* | $p100^{gag\text{-}maf}$ | Transcriptional regulator |
| *ski* | $p110^{gag\text{-}ski\text{-}pol}$ | Transcriptional regulator |
| *qin* | $p90^{gag\text{-}qin}$ | Transcriptional regulator of the forkhead/Hnk-3 family |

[a]Adapted from J. Nevins and P. Vogt, p. 301–343, in B. N. Fields et al. (ed.), *Fields Virology*, 3rd ed. (Lippincott-Raven Publishers, Philadelphia, Pa., 1996), with permission.

[b]Designations for viral proteins: p, protein; gp, glycoprotein; pp, phosphoprotein. The last is not applied consistently but is used mainly in conjunction with the *src* product. The numbers give the estimated molecular mass in kilodaltons, and the superscript lists the genes from which the coding information is derived in the 5′ → 3′ direction. The listing of more than one protein for an oncogene signifies its inclusion in independent virus isolates.

[c]*fps* and *fes* are the same oncogene derived from the avian and feline genomes, respectively.

[d]*raf* and *mil* are the same oncogene derived from the murine and avian genomes, respectively.

## Viral Mimics of Cellular Signaling Molecules

### The Transduced Cellular Genes of Acutely Transforming Retroviruses

**The v-*src* paradigm.** The protein product of v-*src* was the first retroviral transforming protein to be identified, when serum from rabbits bearing tumors induced by Rous sarcoma virus was shown to immunoprecipitate a 60-kDa phosphoprotein (pp60$^{v\text{-}src}$) (Table 16.7). The v-Src protein was soon found to possess protein tyrosine kinase activity, a property that provided the first clue that specific phosphorylation of cellular proteins is critical to oncogenesis. The discovery of this protein tyrosine kinase led to the identification of a large number of other proteins with

**Table 16.8**  Viral transforming proteins can bind to multiple cellular proteins

| Viral protein | Cellular proteins that bind to sequences needed for transforming activity | Function of cellular protein |
|---|---|---|
| Human papillomavirus type 16 or 18 E6 | p53 | Nuclear transcriptional regulator that induces $G_1$ arrest or apoptosis in response to DNA damage or stress; E6 binding stimulates ubiquitination and degradation |
| | Paxillin | Cytoplasmic protein that participates in signal transduction from focal adhesions to the cytoskeleton; E6 binding implicated in transformation |
| | Interferon regulatory factor 3 (Irf3) | Transcriptional regulator activated in response to interferon or viral infection; E6 binding inhibits induction of Irf3 mRNA |
| Adenoviral E1A proteins | Rb family members (Rb, p107, and p130) | Nuclear phosphoproteins that regulate cell cycle progression, in part by binding E2f family members and repressing transcription; E1A proteins disrupt such Rb-E2f complexes and sequester Rb |
| | p300/Cbp | Nuclear coactivators required for stimulation of transcription by many sequence-specific activators, including p53; E1A protein binding inhibits their activity |
| | pCaf | Histone acetyltransferase that acetylates nucleosomal histones and can stimulate transcription; inhibited by E1A binding via CR1 residues |
| | p27$^{Kip1}$ | Nuclear inhibitor of $G_1$/S phase cyclin-Cdks; E1A binding inhibits cell cycle arrest induced by the protein |
| | Ubc9 | Ubiquitin-conjugating enzyme |

similar enzymatic activity and important roles in cellular signaling. The Src protein contains a tyrosine kinase domain and two domains mediating protein-protein interactions (Fig. 16.9). The first, called the Src homology region 2 (SH2) domain, binds to phosphotyrosine-containing sequences, whereas the SH3 domain has affinity for proline-rich sequences. Both domains are found in other proteins that participate in intracellular signal transduction pathways. One transforming protein, Crk, is made up of only an SH2 and an SH3 domain; it functions as an **adapter**, bringing other proteins in the signal transduction pathway together (Table 16.7). A fourth Src domain (SH4) includes the N-terminal myristoylation signal that directs Src to the plasma membrane. All four domains are required for transforming activity.

The Src kinase phosphorylates itself at specific tyrosine residues; these modifications regulate its enzymatic activity. Phosphorylation of Y416 in the kinase domain activates the enzyme, whereas phosphorylation of Y527 in the C-terminal segment inhibits activity. The recently determined crystal structures of cellular Src and another member of the Src family emphasize the importance of the SH2 and SH3 domains in such regulation. As illustrated in Fig. 16.9, the phosphorylated C-terminal segment of the cellular Src protein is bound by the SH2 domain in the crystal structure. This interaction brings a polyproline helix located between the SH2 and kinase domains into proxim-

ity to SH3. Binding of SH3 to the helix deforms the kinase domain, accounting for the inactivity of the Y527-phosphorylated form of the protein. Exchange of these intramolecular associations for binding of the SH2 or SH3 domain to phosphotyrosines or polyproline sequences, respectively, in other proteins initiates conformational changes that activate the kinase. This autoregulatory mechanism explains earlier findings that transduction and overproduction of the normal Src protein do not lead to cellular transformation and that the oncogenic activity of v-*src* requires loss or mutation of the Y527 codon. Thus, the v-Src protein is constitutively active.

Since its kinase activity was first discovered, numerous proteins that can be phosphorylated by v-Src have been identified. Most are probably not relevant to transformation. Two likely exceptions are an SH2-domain-containing protein (Shc), which is an adapter in the pathway leading to activation of Ras, and focal adhesion kinase (Fak), which phosphorylates cytoskeletal proteins (Fig. 16.3). Phosphorylation of the latter may account for the altered shape and poor adhesion of v-*src*-transformed cells.

**Other transduced oncogenes.**  The transduced oncogenes of retroviruses are homologs of cellular genes that encode many components of signal transduction cascades, from the external signal molecules (e.g., v-Sis) and their receptors (v-ErbB and v-Kit) to the nuclear proteins at the

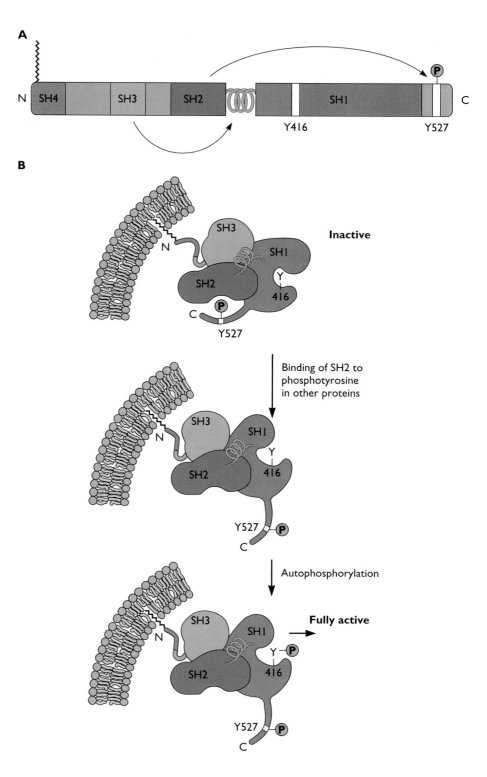

**Figure 16.9 Organization and regulation of the c-Src tyrosine kinase.** (A) The functional domains of the protein. The SH4 domain contains the site for addition of the myristate chain that serves to anchor the protein to the cell membrane. The SH2 and SH3 domains mediate protein-protein interactions by binding to phosphotyrosine-containing and proline-rich sequences, respectively. Arrows represent intramolecular interactions observed in the repressed-state crystal structures of Src. (B) The interactions and their reversal. When Y527 is phosphorylated, the C-terminal region of c-Src in which this residue lies is bound to the SH2 domain. This interaction brings a helix located between the SH3 and SH1 domains into contact with the SH3 domain, as illustrated at the top. Such intramolecular associations maintain the kinase domain (SH1) in an inactive conformation. A conformational change that activates the kinase can be induced as shown, as well as by binding of the SH3 domain to proline-rich sequences in other proteins and probably by dephosphorylation of Y527 (see Fig. 16.15). Once released from the autoinhibited state in this way, Y416 in the kinase domain is autophosphorylated, a modification that stabilizes the active conformation of the SH1 domain. The v-Src protein is not subject to such autoinhibition because the sequence encoding the C-terminal regulatory region of c-Src was deleted during transduction of the cellular gene.

end of the relay (v-Fos and v-Myc) (Fig. 16.3; Table 16.7). Thus, it seems likely that any positively acting protein in such a cascade has the potential to act as an oncoprotein. The oncogenic potential of such transduced oncogenes is realized by two nonexclusive mechanisms, genetic alter-

ations that lead to constitutive protein activity and inappropriate production or overproduction of the protein. The former mechanism applies to most of the retroviral oncogenes listed in Table 16.7. For example, like other small guanine nucleotide-binding proteins, Ras normally cycles between active GTP-bound and inactive GDP-bound conformations. Such cycling is under the control of GTPase-activating proteins and guanine nucleotide exchange proteins. The latter (e.g., Sos) (Fig. 16.3) stimulate the release of GDP once bound GTP has been hydrolyzed. However, v-Ras proteins fail to hydrolyze GTP efficiently and therefore persist in the active GTP-bound conformation that relays signals to downstream pathways such as the Map kinase cascade (Fig. 16.3). Such constitutive activity is the result of mutations that lead to substitution of specific, single amino acids in the protein (at residue 12, 13, or 61) and render the protein refractory to the GTPase-activating protein. Analogous mutations are common in certain human tumors, such as colorectal cancers (Box 16.2), and were the first discrete genetic changes in a proto-oncogene linked to neoplastic disease in humans.

Less commonly, over- or misexpression of the transduced oncogene is sufficient to disrupt normal cell behavior. This type of mechanism is best characterized for *myc*. In normal cells, the expression of this gene is tightly regulated such that the Myc protein is made only during a short $G_1$ period in the cell cycle and is not synthesized when cells withdraw from the cycle or differentiate. The production of even small quantities of Myc-containing proteins specified by retroviruses, such as avian myelocytoma virus MH2, at an inappropriate time results in cellular transformation.

## Other Viral Homologs of Cellular Genes

The genomes of some larger DNA viruses also contain coding sequences that are clearly related to cellular genes encoding signal transduction molecules. Human herpesvirus 8, a gammaherpesvirus related to Epstein-Barr virus, has been strongly implicated as an infectious cofactor for Kaposi's sarcoma, a malignancy common in AIDS patients. Its structural proteins and viral enzymes are closely related to those of other herpesviruses. The genome also contains several homologs of cellular genes that encode signaling proteins, many of which are clustered in a single region (Fig. 16.10). Among the best characterized of these genes is that specifying a guanosine nucleotide-binding protein-coupled receptor (v-*gpcr*) that is most closely related to a cellular receptor for CXC chemokines. Members of this family of receptors bind chemokines released at sites of inflammation to activate signal transduction. In contrast, v-Gpcr is fully active in the absence of any ligand and can trigger specific cellular signal transduction pathways. When introduced into mouse NIH 3T3 cells in culture, the v-*gpcr* gene induces their morphological transformation and secretion of vascular endothelial growth factor. A characteristic feature of Kaposi's sarcoma is extensive **angiogenesis**, the proliferation of new blood vessels, which is essential for tumor progression. Angiogenesis appears to be the result of the production of large quantities of vascular endothelial

**Figure 16.10 Some genes of human herpesvirus 8 and herpesvirus saimiri that are homologous to cellular genes.** The two viral genomes are shown in orientations that align genes conserved among herpesviruses, the core gene blocks shown at the top. These conserved genes encode proteins needed for virus reproduction and assembly. These two herpesviral genomes, but not that of closely related Epstein-Barr virus, carry homologs of cellular genes (arrowheads) interspersed among the core gene blocks. Those shown in purple are related to cellular chemokines (vIl-6, vIl-17, and macrophage inflammatory factor [v-Mip] 1α or 1β), chemokine receptors (v-Gpcr), or other signaling molecules (interferon-responsive protein [v-Irf] and an *n-cam* family transmembrane protein involved in intercellular signaling [vOx2]). Viral genes shown in red are related to cellular genes whose products regulate cell growth or apoptosis (*cyclin D* [v-cyclin] and *bcl-2* [v-Bcl-2]).

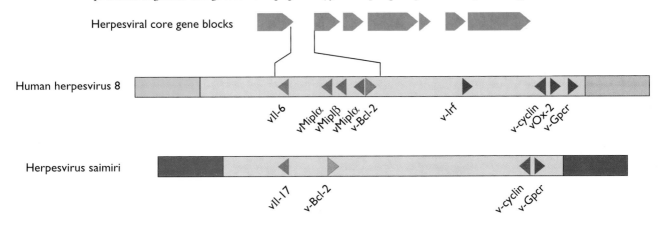

growth factor by spindle cells that proliferate in the tumor. These properties suggest that the v-*gpcr* gene plays a critical role in human herpesvirus 8 oncogenesis in human hosts.

The functions of v-*gpcr* or any other homologs of cellular genes encoding signal transduction proteins (Fig. 16.10) in human herpesvirus 8-infected human cells have not yet been examined, for de novo infection of cells in culture by this virus was reported only very recently. In these experiments, infection of primary human endothelial cells induced immortalization and spindle-shaped morphology. However, only a small fraction (1 to 6%) of the cells contained the virus, even through all proliferated far beyond their normal life span. The propagation and survival of uninfected cells appear to be the result of a paracrine mechanism. These cells increase production of vascular epithelial growth factor in response to an as yet unidentified protein released from the infected cells. It therefore appears that human herpesvirus 8 infection leads to immortalization by a previously unknown mechanism. However, retention and expression of the viral genome is necessary for the appearance of other transformed phenotypes, notably anchorage-independent growth.

Viral homologs of cellular receptors and the ligands that bind to them have also been identified in the genomes of poxviruses. For example, vaccinia virus growth factor is a functional analog of the host's epidermal growth factor and can bind to the appropriate receptor to activate signal transduction pathways that result in anchorage-independent growth. Closely related genes are present in the genomes of other poxviruses, including variola virus and Shope fibroma virus. Mutant vaccinia viruses with both copies of the growth factor gene deleted fail to induce proliferation of ectodermal and endodermal cells of the chick chorioallantoic membrane, reproduce less efficiently, and are less pathogenic than their wild-type parents in infected animals. There can therefore be little doubt that the poxviral homologs of cellular epidermal growth factor are important in the hyperplasia or development of tumors associated with infection by members of this family.

### Virus-Specific Signal Transduction Molecules

Some viruses alter the growth and proliferation of infected cells by the action of viral signal transduction proteins with no known cellular counterparts. The best-understood example of this mechanism is provided by Epstein-Barr virus latent membrane protein 1 (LMP-1), one of several viral gene products implicated in immortalization of human B lymphocytes (Table 16.5). This viral protein can also inhibit differentiation of epithelial cell lines in culture and induce typical transformed phenotypes in established

lines of rodent fibroblasts. The LMP-1 protein, which is an integral protein of the plasma membrane, functions as a constitutively active receptor. In the absence of any ligand, LMP-1 oligomerizes to form patches in the cellular membrane and activates the cellular transcriptional regulator Nf-κb. The LMP-1 protein binds to the same intracellular proteins as the active ligand-bound form of the tumor necrosis factor receptor (Fig. 16.11). The C-terminal segment of LMP-1 to which these proteins bind is sufficient for both immortalization of B cells and activation of Nf-κb when localized to the plasma membrane. It is therefore believed that LMP-1 activates the kinase cascade that normally releases this transcriptional regulator from association with cytoplasmic inhibitors (Fig. 16.11).

When LMP-1 is made in the absence of other Epstein-Barr virus proteins, its constitutive signaling leads to many of the alterations in both cellular properties and cellular gene expression typically observed when primary B cells are infected with, and immortalized by, the virus. These changes include increased expression of certain cell adhesion molecules and consequently increased cell adhesion and clumping. In infected cells, the EBNA-2 protein, which is a transcriptional activator, ensures efficient production of the LMP-1 protein. The EBNA-2 protein also stimulates transcription of several cellular genes encoding proteins that might be expected to influence cell growth, such as the Fgr tyrosine kinase (Table 16.7).

### Alteration of the Expression or Activity of Cellular Signal Transduction Proteins

#### Insertional Activation by Nontransducing Simple Retroviruses

Most tumors induced by nontransducing retroviruses arise as a result of increased transcription of cellular genes located in the vicinity of integrated proviruses. This mechanism of oncogenesis is known as **insertional activation**. As in the case of the transducing retroviruses, Rous sarcoma-derived avian viruses played a seminal role in delineating the mechanisms of insertional activation. The original stocks of viruses isolated by Rous included replication-competent leukosis viruses called Rous-associated viruses 1 and 2. These viruses do not carry an oncogene, but in young chickens they induce B-cell tumors that originate in the bursa of Fabricius, the major lymphoid organ of these birds. A provirus was found integrated in the vicinity of the cellular *myc* gene in each of these tumors. Although the exact integration site varies from tumor to tumor, many lie in the intron between exon 1 (a noncoding exon) and exon 2 (Fig. 16.12). However, in some tumors proviruses are located upstream or downstream of the cellular *myc* gene. In this avian system, inappropriate

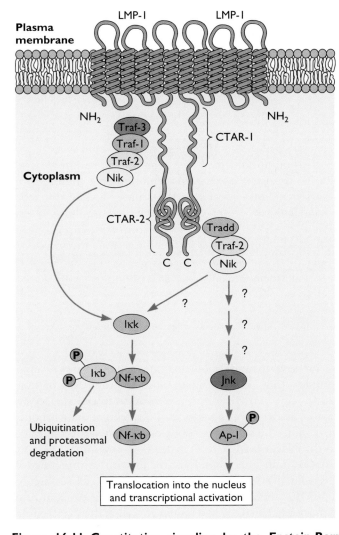

**Figure 16.11 Constitutive signaling by the Epstein-Barr virus LMP-1 protein.** The LMP-1 protein, which possesses six membrane-spanning segments but no large extracellular domain, oligomerizes in the absence of ligand, a property represented by the LMP-1 dimer depicted. The long, cytoplasmic, C-terminal domain of the viral protein contains two segments implicated in activation of signaling, designated C-terminal activation regions (CTAR) 1 and 2. As shown, multiple members of the tumor necrosis factor receptor-associated protein family (Trafs) bind to CTAR-1. Binding of Traf-2 leads to activation of Nf-κb, through the signaling relay that includes Nf-κb-inducing kinase (Nik) and Iκb kinase (Iκk), and induces release of Nf-κb from association with the cytoplasmic inhibitors. The same pathway is activated in uninfected cells by binding of tumor necrosis factor to its receptor. The CTAR-2 domain of LMP-1 is responsible for activation of Ap-1 via the Jun N-terminal kinase (Jnk) pathway. The first reaction appears to be indirect association of this region of LMP-1 with Traf-2 via a second cellular protein (Tradd), a reaction that may lead to activation of both Nf-κb and Ap-1, as shown. The latter effect is not essential for the ability of LMP-1 to alter the proliferation of B cells, but it may play an augmenting role.

synthesis of the cellular Myc protein is associated with lymphomagenesis; no changes in the protein are required. Analysis of the sites of proviral DNA integration and the gene products formed in these tumors provided the first evidence for two types of insertional activation, **promoter insertion** and **enhancer insertion** (Fig. 16.13).

The first, promoter insertion, results in production of a chimeric RNA in which sequences transcribed from the proviral LTR are linked to cellular proto-oncogene sequences. If transcription originates from the left-end LTR, some viral coding sequences may be included. However, transcription from the right-end LTR seems to be more common, and in these cases the proviral left-end LTR has usually been deleted. Such deletion is probably important because transcriptional read-through from the left-end promoter reduces transcription from the right-end LTR. Proviral integration often occurs within the cellular proto-oncogene, truncating cellular coding sequences and eliminating noncoding domains that may include negative regulatory sequences. Some chimeric transcripts formed by promoter insertion are analogous to the intermediates required for oncogene capture by the transducing retroviruses (compare Fig. 16.13 and 16.8). Indeed, it has been possible to isolate newly generated oncogene-transducing retroviruses from tumors arising as a result of promoter insertion. In the second type of insertional activation, enhancer insertion, viral and cellular transcripts are not fused. Instead, activation of the cellular gene is mediated by the strong viral enhancers, which increase transcription from the cellular promoters (Fig. 16.13). Because enhancer activity is independent of orientation and can be exerted over long distances, the provirus need not be oriented in the same direction as the proto-oncogene and may even lie downstream of it.

### Alteration of the Activities of Cellular Signal Transduction Molecules

**Cell surface receptors.** Many signal transduction cascades are initiated by binding of external growth factors to the extracellular portions of cell surface receptor tyrosine kinases. Such ligand binding induces oligomerization of receptor molecules and autophosphorylation of specific tyrosine residues in their intracellular domains. Activated receptors are internalized rapidly (within 10 to 15 min) by endocytosis. Following acidification of the endosomes, the ligand is released and all but a small fraction of the receptor molecules are usually degraded. As a result, the initial signal is short-lived. The E5 proteins of bovine and human papillomaviruses interfere with the normal mechanisms that control activation of this class of receptors.

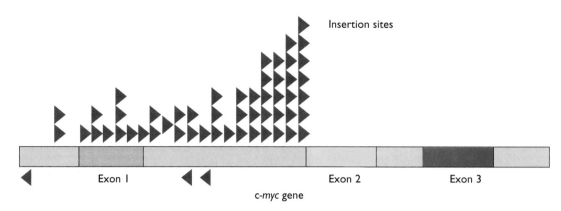

Insertion sites

Exon 1    Exon 2    Exon 3

c-*myc* gene

**Figure 16.12 Insertional activation of c-*myc* by avian leukosis viruses.** In avian cells derived from avian leukosis virus-induced B cell lymphomas, individual proviral integration sites are clustered as shown (arrowheads) within noncoding exon 1 and intron 1 of the *myc* gene. Most integrated proviruses are oriented in the direction of *myc* transcription (arrowheads pointing to the right). Adapted from J. Nevins and P. Vogt, p. 301–343, *in* B. N. Fields et al. (ed.), *Fields Virology* (Lippincott-Raven Publishers, Philadelphia, Pa., 1996), with permission.

The bovine papillomavirus type 1 E5 protein, a hydrophobic protein of only 44 amino acids, is the major transforming protein of this virus (Table 16.6). Transformation of fibroblasts by this protein is the result of its binding to, and activation of, the platelet-derived growth factor receptor. Because the E5 protein is a dimer, it is believed that it induces ligand-independent dimerization of the receptor and hence activation of its tyrosine kinase. This mechanism is likely to be important in the oncogenicity of the virus in its natural hosts, in which it induces fibropapillomas (Box 16.1).

Other papillomaviruses, including many human serotypes, cause epithelial tumors by transforming cells that lack the platelet-derived growth factor receptor. In these cells, the primary target appears to be the epidermal growth factor receptor. Expression of the human papillomavirus type 16 E5 protein in cells containing this receptor stimulates and prolongs activation of signaling molecules in downstream pathways, such as the Map kinase cascade (Fig. 16.3). The viral protein binds to the enzyme responsible for the acidification of endosomes, vacuolar ATPase, and significantly delays the decrease in endosomal pH that normally takes place within 20 min of receptor internalization (Fig. 16.14). One consequence is that a larger fraction of the internalized receptor survives passage through internal compartments and is recycled to the cell surface. Because dissociation of the ligand from the receptor requires a drop in pH, such inhibition of acidification may account for the stimulation of signaling observed in E5-expressing cells. However, the E5 protein may also modulate the activity of the epidermal growth factor receptor more directly by binding to and stabilizing the active ligand-receptor complex (Fig. 16.14). Integration of proviruses of nontransducing retroviruses can also activate cell surface receptors as a result of alteration of the cellular gene product. In certain chicken lines, Rous-associated virus-1 induces erythroblastosis (Box 16.1) instead of lymphoma. These tumors contain intact nondefective proviruses integrated in the cellular *erbB* gene, which encodes the cell surface receptor for the epithelial growth factor. The proviral integrations are clustered in a region encoding the extracellular portion of this receptor, and read-through transcription (Fig. 16.13) produces chimeric RNAs. The proteins translated from these RNAs are truncated growth factor receptors missing the ligand-binding domain. Consequently, they produce a constitutive mitogenic signal. A similar truncation is seen in the protein produced from the v-*erbB* gene captured by transducing viruses.

**Src family tyrosine kinases.** Members of both the *Papovaviridae* and the *Herpesviridae* encode proteins that permanently activate cellular signal transduction pathways as a result of binding to Src family tyrosine kinases (Table 16.9). This mechanism was first encountered in studies of the mouse polyomavirus middle T (mT) protein, a viral early gene product with no counterpart in the genome of simian virus 40 (Fig. 16.15A). This protein can fully transform established rodent cell lines (Table 16.7) and induce endotheliomas (Box 16.1) when overexpressed in transgenic animals.

Middle T antigen is localized to the plasma membrane of transformed or infected cells, where it associates with

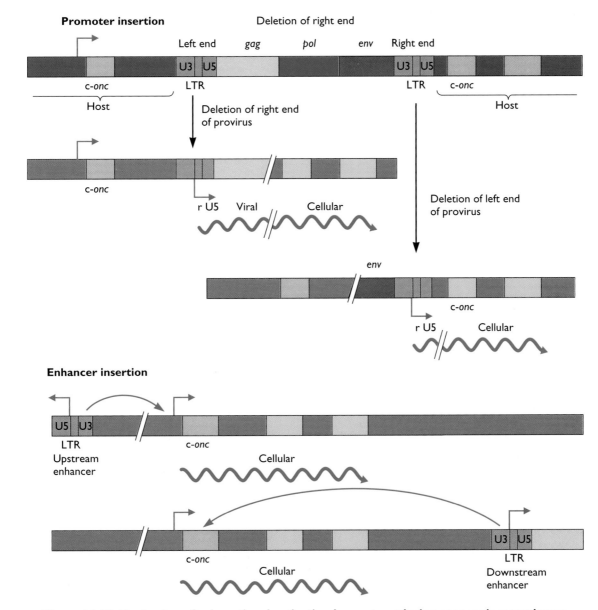

**Figure 16.13 Mechanisms for insertional activation by nontransducing oncogenic retroviruses.**

Src or the related tyrosine kinases Yes and Fyn. No increase in Src concentration is observed in mT-transformed cells, but the catalytic activity of Src bound to mT is increased by an order of magnitude. This change in tyrosine kinase activity is the result of inhibition of the autoregulatory mechanism, in which binding of the SH2 domain of Src to phosphorylated Y527 inhibits its kinase activity. Middle T antigen appears to be able to sequester the Y527-containing segment of Src and induce its dephosphorylation (see below) to stabilize the active conformation of the kinase. However, it is unlikely that mT can disrupt preformed interaction of these Src domains, for only about 10% of the Src molecules, and a similar fraction of mT molecules, are bound to each other in transformed or infected cells. Such nonstoichiometric association suggests that mT can bind to Src molecules only during their transient activation by normal cellular mechanisms. While it remains a mystery why so little of the mT protein is associated with cellular tyrosine kinases, the permanent alteration of cell proliferation induced via such a small fraction of these cellular enzymes is remarkable.

The mT protein also binds to cellular protein phosphatase 2A via an N-terminal sequence that is also present in sT (Fig. 16.15A). Binding of mT to this cellular enzyme

**Figure 16.14 The human papillomavirus E5 protein increases the cell surface epidermal growth factor receptor concentration.** The E5 protein of human papillomavirus type 16 binds to one subunit of the cellular vacuolar ATPase, the enzyme responsible for acidification of endosomes. As indicated by the bar, this interaction inhibits endosome acidification. The dissociation of epidermal growth factor (Egf) from the receptor and degradation of the receptor (dashed arrows) are thus slowed substantially. Such prolonged survival of the active Egf receptor may contribute to the stimulation of signaling induced by the viral E5 protein. The fraction of internalized Egf receptor that is not degraded but rather recycled to the cell surface increases significantly (green arrow). The E5 protein may also increase the concentration of activated receptors on the cell surface more directly; E5 proteins of some serotypes have been reported to bind to and stabilize ligand-bound Egf receptors on the cell surface, as indicated. Modified from M. A. Nead and D. J. McCance, p. 225–251, in D. J. McCance (ed.), *Human Tumor Viruses* (ASM Press, Washington, D.C., 1998), with permission.

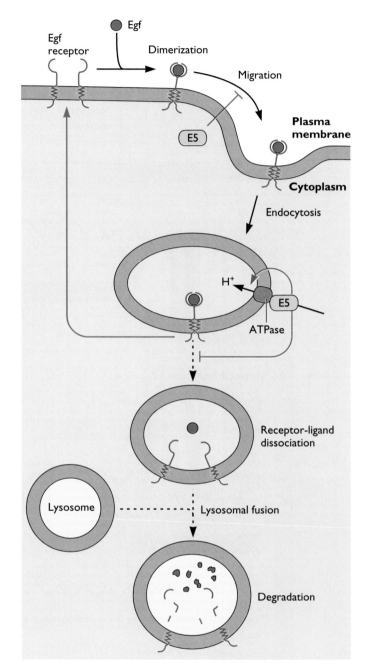

appears to be necessary for association of the viral protein with Src. This requirement would ensure that the phosphatase would be brought into close physical association with Src and accounts for the fact that Y527 is dephosphorylated in mT-bound Src molecules. As we have seen in the context of the v-Src protein, the absence of phosphotyrosine at this position results in activation of the kinase.

It was initially surprising that mT-transformed cells do not contain generally elevated levels of phosphotyrosine, despite activation of Src family kinases. It is now clear that mT itself is a critical substrate of this cellular enzyme: phosphorylation of specific mT tyrosine residues by activated Src allows a number of cellular proteins that contain phosphotyrosine-binding domains to bind to mT (Fig. 16.15B). These proteins include the adapter Shc and phosphoinositol 3-kinase, an enzyme responsible for production of lipid second messengers (Fig. 16.3). In all cases, mutations that disrupt binding of the cellular protein to mT impair the transforming activity of the viral protein. When bound to mT, such signaling proteins are tyrosine-phosphorylated by the activated Src kinase to trigger signal transduction, as illustrated for the activation of Ras and the Map kinase pathway in Fig. 16.15B. Thus, mT both bypasses the normal mechanism by which the kinase activity of Src is regulated and also serves as a virus-specific adapter, bringing cellular signal transduction proteins together when they would not normally be associated.

The X protein of hepatitis B virus, which may contribute to the development of hepatocellular carcinoma, also activates Src and some other members of this family of tyrosine kinases. Production of the X protein results in prolonged activation of these enzymes, activation of Ras, and activation of the Map kinase pathway (Fig. 16.3).

Whether the viral X protein, like polyomaviral mT, binds directly to Src to trigger this signal transduction cascade remains to be seen.

**Protein phosphatase 2A.** In preceding sections, we have illustrated transformation by viral gene products as a result of permanent or prolonged activation of signal transduction pathways that control cell growth and proliferation. In normal cells, such signaling is a transient process, for the molecular components are reset to the ground state

**Table 16.9** Viral proteins that bind to Src family tyrosine kinases

| Viral protein | Cellular tyrosine kinase(s) bound | Consequences of interaction |
|---|---|---|
| Mouse polyomavirus mT | Src, Fyn, Yes | Activation of mT-bound kinase; tyrosine phosphorylation of mT and adapter proteins bound to it; transformation of established cell lines |
| Herpesvirus saimiri 484 tyrosine kinase-interacting protein (Tip) | Lck | Activation of the cellular kinase and signal transduction; transformation of human and simian T lymphocytes to inter leukin-2-independent growth; Tip required for viral transformation of primary T cells in culture and oncogenic disease in vivo |
| Epstein-Barr virus LMP-2 | Fyn, Lym | Inhibition of signal transduction via the cellular kinase; may prevent B-cell activation and the transition from latent to productive infection |

once they have transmitted the signal. Inhibition of the reactions that shut down signaling can therefore also contribute to transformation, a mechanism exemplified by the small T protein of polyomaviruses such as simian virus 40.

Small T protein is not necessary for transformation of many types of cell but can stimulate transformation by simian virus 40 LT and is required for transformation of resting cells. It also stimulates such cells to proliferate. In

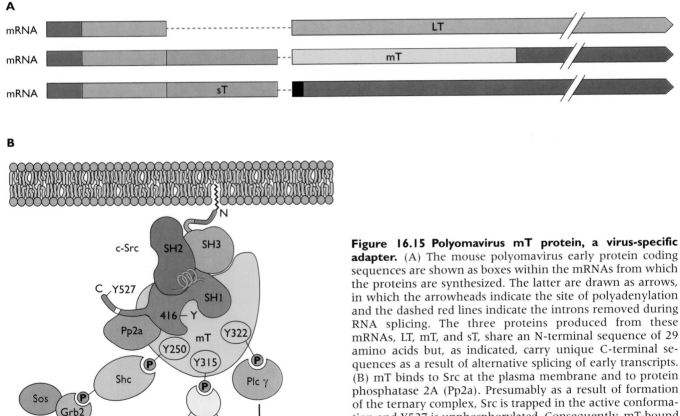

**Figure 16.15 Polyomavirus mT protein, a virus-specific adapter.** (A) The mouse polyomavirus early protein coding sequences are shown as boxes within the mRNAs from which the proteins are synthesized. The latter are drawn as arrows, in which the arrowheads indicate the site of polyadenylation and the dashed red lines indicate the introns removed during RNA splicing. The three proteins produced from these mRNAs, LT, mT, and sT, share an N-terminal sequence of 29 amino acids but, as indicated, carry unique C-terminal sequences as a result of alternative splicing of early transcripts. (B) mT binds to Src at the plasma membrane and to protein phosphatase 2A (Pp2a). Presumably as a result of formation of the ternary complex, Src is trapped in the active conformation and Y527 is unphosphorylated. Consequently, mT-bound Src is catalytically active and phosphorylates specific tyrosines in mT. These are then bound by cellular proteins that contain phosphotyrosine-binding motifs, such as Shc, phospholipase C-γ (Plc γ), and phosphoinositol 3-kinase (Pi3k). These proteins can then be tyrosine phosphorylated by Src and activated. The lipids produced upon activation of Plc γ and Pi3k act as second messengers, relaying signals to various pathways (Fig 16.3).

both infected and transformed cells, this viral protein binds to protein phosphatase 2A, a widespread, abundant serine/threonine protein phosphatase. This enzyme is a heterotrimer composed of a two-subunit core enzyme bound to a regulatory subunit. Small T antigen binds to the core enzyme at sites that include those contacted by the regulatory subunit and partially inhibits enzyme activity. One important consequence of this interaction is in vivo stimulation of the Map kinases, which are normally inactivated by dephosphorylation of threonine or tyrosine residues (Fig. 16.16). Consequently, sT increases the activity of sequence-specific transcriptional activators such as activator protein 1 (Fig. 16.16). The viral protein also prevents dephosphorylation and inactivation of the nuclear cyclic AMP-responsive element-binding protein by protein phosphatase 2A. The increased activities of these transcriptional stimulators when protein phosphatase 2A is inhibited by sT result in stimulation of transcription of a $G_1$ cyclin, cyclin D1 (Fig. 16.16), and account for the ability of the viral protein to circumvent the need for growth factors

**Figure 16.16 Inhibition of protein phosphatase 2A by simian virus 40 sT.** Inhibition of the activity of protein phosphatase 2A (Pp2a) by sT results in activation of cellular transcriptional regulators, both via the Map kinase pathway and by inhibition of dephosphorylation of activated Creb within the nucleus. Production of sT within cells stimulates cyclin D1 transcription, as shown.

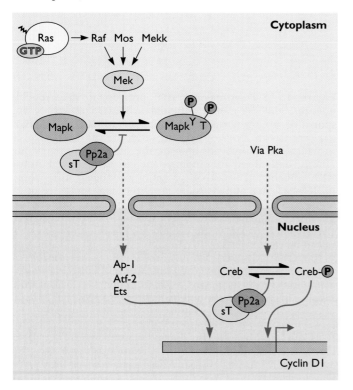

or other mitogens during transformation by simian virus 40.

## Viral Transformation via Cell Cycle Control Pathways

One end point of many cellular signal transduction pathways is the transcription of genes encoding proteins that control entry into or progression through the cell cycle. Consequently, permanent activation of such pathways by viral proteins, by any of the mechanisms described in the previous section, can result in an increased rate of cell growth and division or in proliferation of cells normally in the resting state. Other viral proteins intervene directly in the intricate circuits by which cell cycle progression is mediated and regulated.

### Abrogation of Restriction Point Control Exerted by the Rb Protein

#### The Restriction Point in Mammalian Cells

In mammalian cells, passage through $G_1$ into S and reentry into the cell cycle from $G_0$ depend on extracellular mitogens. Late in $G_1$, cells responding to such external cues become committed to enter S and to divide and complete the cell cycle; during this period, they are refractory to extracellular signals that regulate growth. Cells that have entered this state are said to have passed the $G_1$ **restriction point** (Fig. 16.5). Normal cells respond to mitogenic signals by mobilization of the $G_1$ Cdk complexes containing D-type cyclins (Fig. 16.5). Expression of genes encoding one or more of these cyclins is induced by such signals via Ras and Map kinases. The activation of Cdks by assembly with a cyclin D protein also depends on mitogenic stimulation. When such stimulation is continuous, Cdk activity appears at mid-$G_1$ and increases to a maximum near the $G_1$-to-S phase transition (Fig. 16.5). Such activity must be maintained until the restriction point has been passed but then becomes dispensable. This property implies that the kinase activity of the cyclin D-dependent Cdks is necessary for exit from $G_1$. The best-characterized substrate of the $G_1$ Cdks is the Rb protein. Many lines of evidence indicate that cyclin D-dependent Cdks initiate the transition through the restriction point solely by phosphorylation of Rb. For example, inhibition of cyclin D expression or function prevents entry into S phase in Rb-containing cells, but this cyclin is not required in Rb-negative cells. As discussed in chapter 9, the Rb protein controls the activity of members of a family of sequence-specific transcriptional regulators known collectively as E2f.

Hypophosphorylated Rb present at the beginning of $G_1$ binds to specific members of the E2f family. Because Rb is a transcriptional repressor, these Rb-E2f complexes inhibit transcription of E2f-responsive genes (Fig. 16.17A). The Rb protein is phosphorylated at numerous sites by the $G_1$ cyclin-Cdk complexes. Such phosphorylated Rb can no longer bind to E2f, which is thus free to activate transcription from E2f-responsive promoters (Fig. 16.17A). This set includes those of the genes encoding the kinase Cdk2, the cyclins that activate this kinase (Fig. 16.5), and E2f proteins themselves. Thus, the initial release of E2fs from association with Rb triggers a positive feedback loop that augments both Rb phosphorylation and activation of E2fs, resulting in a rapid increase in the concentrations of active E2fs and cyclin E-Cdk2 complexes (Fig. 16.17A). In this way, cell cycle progression becomes independent of the mitogens necessary for the production of active cyclin D-Cdk complexes. The regulatory circuits described so far might suggest that cells containing no Rb protein would proceed into S phase in the absence of mitogenic signals. However, this does not take place. Furthermore, cyclin E is required for entry into S phase in Rb-negative cells. Thus, molecular reactions other than Rb phosphorylation must contribute to passage through $G_1$ and the restriction point.

The positive feedback loop for activation of cyclin E-dependent kinases and E2fs late in $G_1$ is subject to several checks and balances imposed by negative regulatory proteins (Fig. 16.5B). These inhibitory proteins must therefore be inactivated or destroyed to allow progression into S phase. The synthesis of at least one member of the Ink4 family of proteins, which bind specifically to and inhibit cyclin D-dependent kinases, is also induced in response to free E2f. It is therefore believed that the accumulation of this inhibitor establishes a feedback loop that blocks the activity of the cyclin D-Cdks and hence the ability of cells to respond to external signals (Fig. 16.17A).

Although E2f proteins are the best-characterized targets of Rb, the latter protein can also bind to numerous other proteins that mediate or stimulate transcription, as well as to regulatory proteins such as the Abl tyrosine kinase (Table 16.7). These interactions can lead to activation or repression of transcription and, at least in some cases, have been implicated in inhibition of cell cycle progression and cell growth by Rb. The Rb protein has also recently been shown to regulate DNA synthesis directly, at least in part by binding to a protein component of the complex that ensures that the genome is replicated only once per cell cycle.

### *Inhibition of Rb Function by Viral Proteins*

The products of oncogenes of several DNA viruses bypass the sophisticated circuits that impose restriction point control, and hence the dependence of passage into S phase on environmental cues. These proteins, the adenoviral E1A proteins, simian virus 40 LT, and the E7 proteins of oncogenic human papillomavirus (types 16 and 18), are mitogenic, inducing DNA synthesis and cell proliferation. Such mitogenic activity requires regions of the viral proteins that are necessary for their binding to hypophosphorylated Rb. All three viral proteins bind to the two noncontiguous regions by which Rb associates with E2f family members (regions A and B in Fig. 16.17B). Thus, the adeno- and papovaviral proteins that sequester the inhibitory form of Rb can disrupt Rb-E2f complexes to induce transcription of E2f-dependent genes and inappropriate entry of cells into S phase (Fig. 16.17A).

Disruption of E2f-Rb complexes by the viral proteins appears to be an active process rather than the result of passive competition for binding to Rb. Induction of cell cycle progression by simian virus 40 LT requires not only the Rb-binding site but also the N-terminal J domain located nearby (Fig. 16.18). As the J domain functions as a molecular chaperone (see chapter 9), it has been suggested that LT actively dismantles the Rb-E2f complex once bound to it (Fig. 16.18). The CR1 sequence of the adenoviral E1A proteins (Fig. 16.19) fulfills a similar function. Exactly how these viral proteins disrupt the association of Rb with E2f remains to be determined, but conformational alteration of Rb seems likely to be important.

Unless bound to Rb (or to the related proteins discussed in the next section), members of the E2f family such as E2f-1 and E2f-4 are unstable proteins that are degraded by the proteasome following ubiquitination. Such instability is dictated by a C-terminal sequence that lies close to the binding site for Rb and appears to be shielded in Rb-E2f complexes. Recent experiments indicate that the adenoviral E1A proteins not only induce accumulation of free E2f but also stabilize these proteins. The mechanism(s) responsible for such stabilization have not yet been elucidated.

### **Inhibition of Negative Regulation by Rb-Related Proteins**

The Rb protein is the founding member of a small family of related gene products, which also contains the proteins called p107 and p130. The latter proteins were discovered by virtue of their binding to adenoviral E1A proteins (Fig. 16.19), and both simian virus 40 LT and human papillomavirus type 16 and 18 E7 proteins also associate with them. The strong similarity of p107 and p130 to Rb is most pronounced in the A and B sequences needed for binding of Rb to both E2f and the viral transforming proteins (Fig. 16.17B). Indeed, the residues by which Rb contacts the common Rb-binding motif of the viral proteins (LXCXE) are invariant among the other family members. Binding of

**Figure 16.17 Regulation of E2f activity by the Rb protein.** (A) Either phosphorylation of Rb by cyclin D- and cyclin E-dependent Cdks or binding of adenoviral E1A or papovaviral LT or E7 proteins disrupts binding of Rb to E2fs, which comprise heterodimers of an E2f protein and a Dp protein, such as Dp-1. In uninfected cells, Rb is phosphorylated at many sites by both cyclin D-Cdk4/6 and cyclin E-Cdk2. The latter cyclin, which appears in mid- to late $G_1$ (Fig. 16.5), is required for entry into S phase and, when expressed ectopically, can drive quiescent cells to reenter the cell cycle. Its modification of Rb depends on the prior action of cyclin D-Cdk4/6. As discussed in the text, it appears that the viral proteins actively disrupt Rb-E2f complexes, rather than simply competing for binding to regions of Rb that contact E2fs. At least in some cases, the viral proteins also alter Rb metabolism. The human papillomavirus type 16 and 18 E7 proteins induce proteasome-mediated degradation of Rb in transformed epithelial cells and can bind directly to one subunit of this multicatalytic complex. The E2f-Rb complex represses transcription when bound to E2f recognition sites in promoters of E2f-responsive genes. Rb is the transcriptional repressor and this function requires the binding of histone deacetylases (Hdac). Binding to Hdac by the Rb family members p107 and p130 is also necessary for their repression of E2f-dependent transcription. Free E2f-Dp-1 heterodimers activate transcription from such promoters, including those of the genes encoding cyclins E and A, Cdk2, and E2f proteins themselves, to establish a positive autoregulatory loop. Induction of transcription of the gene coding for the Ink4 family proteins would lead to inhibition of cyclin D-dependent Cdks and thus block the ability of cells to respond to external mitogenic signals, a characteristic

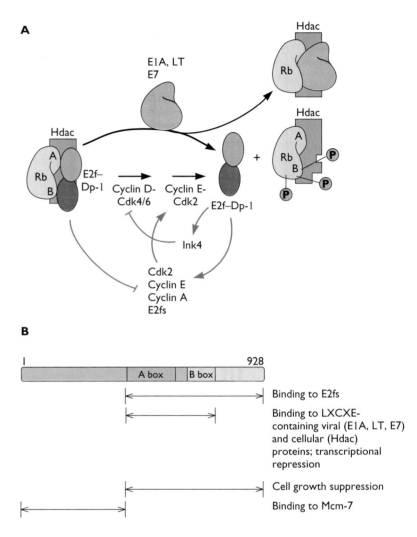

property of cells that have passed the $G_1$ restriction point. (B) Functional domains of the human Rb protein are shown to scale. The A- and B-box regions form the so-called pocket domain, which is necessary for binding of Rb to both E2fs and the viral proteins described in the text. This segment is also sufficient to repress transcription when fused to a heterologous DNA-binding domain, and it is required for binding to Hdac. Like the papova- and adenoviral Rb-binding proteins, Hdac contains the motif LXCXE within the region that binds to Rb. Additional C-terminal sequences of Rb are required for transcriptional repression and binding to E2fs, as well as inhibition of cell growth. The N-terminal segment of the protein, which is also important for suppression of cell proliferation, binds to human Mcm-7, a component of a chromatin-bound complex required for DNA replication and control of initiation of DNA synthesis.

the viral proteins to p107 and p130 can make important contributions to transforming activities. For example, the LXCXE sequence of simian virus 40 LT is required for transformation of fibroblasts derived from Rb knockout mice. Furthermore, the J domain of LT is also necessary and induces hypophosphorylation of p107 and p130, concomitant with their increased turnover.

The Rb, p107, and p130 proteins bind preferentially to different members of the E2f family during different phases of the cell cycle. Hypophosphorylated Rb binds pri-

marily to E2f-1, -2, or -3 during the $G_1$ phase. In contrast, p107 is largely associated with E2f-4 during the $G_1$, S, and $G_2$ phases. Such complexes also contain substoichiometric concentrations of cyclin E-Cdk2 and cyclin A-Cdk2. These properties, and the targeting of p107 by transforming proteins of the smaller DNA viruses, indicate that inhibition or more subtle regulation of the activity of E2f-4 must be important for orderly progression through the cell cycle. Furthermore, E2f-4 must play a distinctive regulatory role; in contrast to E2f-1, -2, or -3, E2f-4 is only a weak inducer

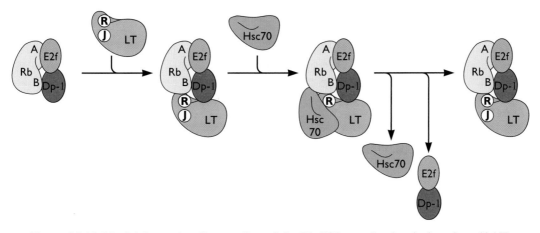

**Figure 16.18 Model for active dismantling of the Rb-E2f complex by simian virus 40 LT.** The LT protein binds to the Rb A- and B-box domains via the sequence that contains the LXCXE motif (amino acids 101 to 118), designated R. The adjacent, N-terminal J domain (amino acids 1 to 82) is not necessary for binding of LT to Rb but is required for induction of cell cycle progression. It has been proposed that the J domain recruits the cellular chaperone Hsc70 to the complex. The chaperone then acts to release E2f–Dp-1 heterodimers from their association with Rb, by a mechanism that is not yet understood but that is believed to depend on ATP-dependent conformational change.

of entry into S phase when overexpressed in quiescent mammalian cells. Until this role is better understood, the consequences of binding of the p107 to these transcriptional activators, or the disruption of this interaction by viral transforming proteins, remain obscure.

On the other hand, binding of p130 to E2f-4 and E2f-5 appears to be critical for maintaining cells in the quiescent state. As discussed in chapter 9, such complexes predominate in mammalian cells in $G_0$. Their disruption by adeno- or papovaviral transforming proteins is believed to allow such cells to reenter the cycle, in part because these E2f family members stimulate transcription of genes encoding both the E2f proteins (E2f-1 and -2) and the cyclin-dependent kinase (Cdk2) needed for entry into S phase.

**Figure 16.19 Organization of the larger adenoviral E1A protein.** Regions of the protein are shown to scale. Those designated CR1 to CR3 are conserved in the E1A proteins of human adenoviruses. The CR3 region, most of which is absent from the smaller E1A protein because of alternative splicing, is not necessary for transformation. The locations of the Rb-binding motif and of the regions required for binding to the other cellular proteins discussed in the text are indicated.

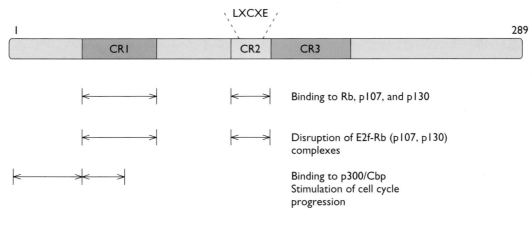

## Production of Virus-Specific Cyclins

Human herpesvirus 8 and its close simian relative herpesvirus saimiri encode functional cyclins. The cyclin gene of the former virus, designated v-cyclin, has 31% identity and 58% similarity to the human gene encoding cyclin D2, and its product binds predominantly to Cdk6. Like its cellular counterpart, v-cyclin activates this protein kinase. However, the viral cyclin and the active enzyme it forms with cellular Cdk6 are resistant to even high concentrations of the proteins that inhibit cellular cyclin-Cdks (Fig. 16.5). Normally, synthesis of these inhibitory proteins, such as p21$^{cip1}$ or p16$^{Ink4a}$ (Fig. 16.5), inhibits cell cycle progression (see "Activation and Functions of the Cellular p53 Protein" below for an example). However, neither Cip nor Ink4 family members bind well to the v-cyclin. As a result, synthesis of the viral cyclin can overcome the G$_1$ arrest imposed when either type of inhibitory protein is made in human cells and can induce cell cycle progression in quiescent cells. Thus, these herpesviruses have evolved a very simple, effective mechanism to ensure that their host cells progress through the cell cycle. The specific advantages conferred by production of the viral cyclins during the infectious cycle have not been identified. However, it would be surprising if they do not contribute to the oncogenicity of these herpesviruses in their natural hosts.

## Inhibition of p53 Functions

As described in chapters 14 and 15, the genomes of many viruses encode proteins that protect infected cells against an antiviral defense of last resort, programmed cell death (apoptosis). This program is activated by damage to the cellular genome or unscheduled DNA synthesis. Consequently, viral transforming proteins that induce cells to enter S phase when they would not normally do so also promote the apoptotic response. This potentially fatal side effect of unscheduled passage through the cell cycle is counteracted by specific viral proteins to allow viral DNA replication or, under appropriate circumstances, survival of transformed cells.

### Activation and Functions of the Cellular p53 Protein

Transformation by several DNA viruses requires inactivation of a second cellular tumor suppressor gene product, the p53 protein. This protein, first identified by virtue of its binding to simian virus 40 LT, is a critical component of regulatory circuits that determine the response of cells to damage to their genomes, as well as to low concentrations of nucleic acid precursors or hypoxia. The importance of this protein in the appropriate response to such damage or stress is emphasized by the fact that it is the most frequently mutated gene in human tumors. When all is well, the intracellular concentration of p53 is very low, because the protein is targeted for proteasomal degradation by binding of the Mdm-2 protein to an N-terminal sequence (Fig. 16.20). However, DNA damage, such as double-strand breaks produced by γ-irradiation or the accumulation of DNA repair intermediates following UV irradiation, leads to the stabilization of p53 and a substantial increase in its concentration. The rate of translation of the protein may also increase. Various proteins that appear to be important for activation of p53 have been identified. These include a human protein mutated in ataxia telangiectasia, a genetic disease with a broad spectrum of defects, including hypersensitivity to X-ray and ionizing radiation. This protein, designated Atm, recognizes potentially genotoxic DNA alterations. Cells lacking the Atm protein do not accumulate the p53 protein or arrest at the G$_1$/S boundary in response to DNA damage. However, the signaling pathway(s) that results in the stabilization of p53 has not been fully elucidated. Recent experiments suggest that one component may be the Abl tyrosine kinase (Table 16.7). The p53 protein can itself bind to double-stranded DNA ends, as well as to sites of excision-repair damage. This property has led to the hypothesis that the p53 protein becomes localized at sites of DNA damage, with the proteins that detect and respond to such injury. Thus, p53 could be activated rapidly whenever potentially dangerous injury to the genome was detected.

The p53 protein is a sequence-specific transcriptional regulator, containing an N-terminal activation and a central DNA-binding domain (Fig. 16.20). Its ability to stimulate transcription of p53-responsive genes is tightly regulated. The p53 protein must form a tetramer via its oligomerization domain (Fig. 16.20) before it can bind specifically to double-stranded DNA. Sequence-specific DNA-binding activity is also regulated by the C-terminal basic domain, which is responsible for the binding of p53 to DNA ends described above. Various modifications within this short C-terminal sequence, including phosphorylation of specific residues or acetylation of five conserved lysines by the coactivator p300 (see "Alteration of p53 Activity via the p300/Cbp Proteins" below), stimulate the specific DNA-binding activity of p53 and its ability to activate transcription. The DNA-binding domain itself (Fig. 16.20) is frequently altered as a result of the mutations that inactivate p53 in human tumors. The biochemical functions of the N-terminal activation domain can also be regulated. For example, binding of the Mdm-2 protein to this segment inhibits p53-dependent transcription. The many mechanisms by which the accumulation or activity of p53 can be regulated provide the means to integrate the multiple signals that are monitored to ensure that this potent protein alters cell physiology only under extreme conditions.

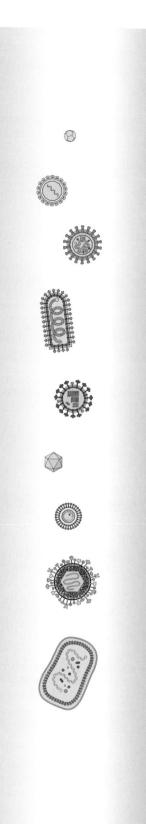

# 17

Rous, P. 1910. A transmissible avian neoplasm: sarcoma of the common fowl. *J. Exp. Med.* **12:**696–705.

Sicheri, F., I. Moarefi, and J. Kuriyan. 1997. Crystal structure of the Src family tyrosine kinase Hck. *Nature* **385:**602–609.

Stehelin, D., H. E. Varmus, J. M. Bishop, and P. K. Vogt. 1976. DNA related to the transforming gene(s) of avian sarcoma viruses is present in normal avian DNA. *Nature* **260:**170–173.

Xu, W., S. C. Harrison, and M. J. Eck. 1997. Three-dimensional structure of the tyrosine kinase c-Src. *Nature* **385:**595–602.

## Transformation by Papovaviruses

Bolen, J. R., C. J. Thiele, M. A. Israel, W. Yonemoto, L. A. Lipsich, and J. S. Brugge. 1984. Enhancement of cellular *src* gene product-associated tyrosyl kinase activity following polyoma virus infection and transformation. *Cell* **38:**367–377.

DeCaprio, J. A., J. W. Ludlow, J. Figge, J. Y. Shew, C. M. Huang, W. H. Lee, E. Marsilio, E. Paucha, and D. M. Livingston. 1988. SV40 large tumor antigen forms a specific complex with the product of the retinoblastoma susceptibility gene. *Cell* **54:**275–283.

Dyson, N., P. M. Howley, and E. Harlow. 1988. The human papillomavirus E7 oncoprotein is able to bind to the retinoblastoma gene product. *Science* **243:**934–937.

Lane, D. P., and L. V. Crawford. 1979. T antigen is bound to a host protein in SV40-transformed cells. *Nature* **278:**261–263.

Linzer, D. I., and A. J. Levine. 1979. Characterization of a 54K dalton cellular SV40 tumor antigen present in SV40-transformed cells and uninfected embryonal carcinoma cells. *Cell* **17:**43–52.

Scheffner, M., B. A. Werness, J. M. Huibregtse, A. J. Levine, and P. M. Howley. 1990. The E6 oncoprotein encoded by human papillomavirus types 16 and 18 promotes the degradation of p53. *Cell* **63:**1129–1136.

Sontag, E., S. Fedorov, C. Kamibayashi, D. Robbins, M. Cobb, and M. Mumby. 1993. The interaction of SV40 small tumor antigen with protein phosphatase 2A stimulates the map kinase pathway and induces cell proliferation. *Cell* **75:**887–897.

Symonds, H., L. Krall, L. Remington, M. Saenz-Robles, S. Lowe, T. Jacks, and T. VanDyke. 1994. p53-dependent apoptosis suppresses tumor growth and progression in vivo. *Cell* **78:**703–711.

Talmage, D. A., R. Freund, A. T. Young, J. Daki, C. J. Dawe, and T. L. Benjamin. 1989. Phosphorylation of middle T by pp60$^{c-src}$: a switch for binding of phosphatidylinositol 3-kinase and optimal tumorigenesis. *Cell* **54:**55–65.

Zalvide, J., H. Stubdal, and J. DeCaprio. 1998. The J domain of simian virus 40 large T antigen is required to functionally inactivate Rb family proteins. *Mol. Cell. Biol.* **18:**1408–1415.

## Transformation by Adenoviruses

Egan, C., S. T. Bayley, and P. E. Branton. 1989. Binding of the Rb1 protein to E1A products is required for adenovirus transformation. *Oncogene* **4:**383–388.

Martin, M. E. D., and A. J. Berk. 1998. Adenovirus E1B 55K represses p53 activation in vitro. *J. Virol.* **72:**3146–3154.

Moore, M., N. Horikoshi, and T. Shenk. 1996. Oncogenic potential of the adenovirus Orf6 protein. *Proc. Natl. Acad. Sci. USA* **93:**11295–11301.

Rao, L., M. Debbas, D. Sabbatini, D. Hockenberry, S. Korsmeyer, and E. White. 1992. The adenovirus E1A proteins induce apoptosis which is inhibited by the E1B 19K and Bcl-2 proteins. *Proc. Natl. Acad. Sci. USA* **89:**7742–7746.

Somasundaram, K., and W. S. El-Deiry. 1997. Inhibition of p53-mediated transactivation and cell-cycle arrest by E1A through its p300/CBP-interacting region. *Oncogene* **14:**1047–1057.

Van der Eb, A. J., C. Mulder, F. L. Graham, and A. Houweling. 1977. Transformation with specific fragments of adenovirus DNAs. I. isolation of specific fragments with transforming activity of adenovirus 2 and 5. *Gene* **2:**133–146.

Whyte, P., N. M. Williamson, and E. Harlow. 1989. Cellular targets for transformation by the adenovirus E1A proteins. *Cell* **56:**67–75.

Yang, X.-J., V. V. Ogryzko, J. Nishikawa, B. H. Howard, and Y. Nakatani. 1996. A p300/CBP-associated factor that competes with the adenoviral oncoprotein E1A. *Nature* **382:**319–324.

## Transformation by Herpesviruses

Bias, C., B. Santomasso, O. Coso, L. Arvanitakis, E. Raaka, J. S. Gutkind, A. S. Asch, E. Cesarman, M. C. Gerhengorn, and E. A. Mesri. 1998. G-protein-coupled receptor of Kaposi's sarcoma-associated herpesvirus is a viral oncogene and angiogenesis activator. *Nature* **391:**86–89.

Flore, O., S. Rapil, S. Ely, J. J. O'Leary, E. M. Hyjek, and E. Cesarman. 1998. Transformation of primary human endothelial cells by Kaposi's sarcoma-associated herpesvirus. *Nature* **394:**588–592.

Gires, O., U. Zimber-Strobl, R. Gonnella, M. Ueffing, G. Marschall, R. Zeidler, D. Pick, and W. Hammerschmidt. 1997. Latent membrane protein 1 of Epstein-Barr virus mimics a constitutively active receptor molecule. *EMBO J.* **16:**8131–8140.

Kaye, K., K. Izumi, and E. Kieff. 1993. Epstein-Barr virus latent membrane protein 1 is essential for B lymphocyte transformation. *Proc. Natl. Acad. Sci. USA* **90:**9150–9154.

Kulwichi, W., R. H. Edwards, E. M. Davenport, J. F. Baskara, V. Godfrey, and N. Raab-Traub. 1998. Expression of the Epstein-Barr virus latent membrane protein 1 induces B cell lymphomas in transgenic mice. *Proc. Natl. Acad. Sci. USA* **95:**11963–11968.

Swanton, C., D. J. Mann, B. Fleckenstein, F. Neipel, G. Peters, and W. Jones. 1997. Herpesviral cyclin/cdk6 complexes evade inhibition by CDK inhibitor proteins. *Nature* **390:**184–187.

Sylla, B. D., S. C. Hung, D. M. Davidson, E. Hatzivassiliou, N. L. Malinin, D. Wallach, T. D. Gilmore, E. Kieff, and G. Mosialos. 1998. Epstein-Barr virus-transformation protein latent infection membrane protein 1 activates transcription factor Nf-κB inducing kinase and the IκB kinases Iκκα and Iκκβ. *Proc. Natl. Acad. Sci. USA* **95:**10106–10111.

# Viral Pathogenesis

## Introduction

There is no easy answer to the seemingly straightforward question, How does a viral infection cause disease in its host? This chapter will provide a framework for dissection of this question by defining and discussing the basic principles of **viral pathogenesis**, the entire process by which viruses cause disease. The term **virulence** (synonymous with **pathogenicity**) describes the capacity of a virus to cause disease. Viral pathogenesis is a sum of the effects on the host due to virus replication and the immune response.

Interest in viral pathogenesis stems in large part from the desire to treat or eliminate viral diseases that affect humans. However, progress in understanding the molecular basis of viral pathogenesis has come not from experiments with human subjects but from studies of animal models. In some cases, nonhuman hosts can be infected with the same viruses that infect humans, but close relatives of human viruses must often be used. While the knowledge obtained from animal models is essential for understanding how viruses cause disease in humans, the results of such studies must be interpreted with caution. What is true for a mouse may not be for a human. For example, although transgenic mice that express the human poliovirus receptor develop paralytic disease after inoculation with poliovirus, they cannot be infected by ingestion of virus, the principal route of infection in humans.

Because animals are complex organisms, the study of viral pathogenesis encompasses a broad range of issues, such as how viruses enter and spread in the host, why particular tissues are sites of replication, how some viruses cause direct cellular damage, how the immune system causes indirect damage to infected and uninfected tissue, and how viruses spread to other hosts. These questions are the topics of this chapter.

## Viral Dissemination in the Host

### Viral Entry

To infect its host, a virus must first infect cells at one or more of the body surfaces. Alternatively, a virus may be introduced by parenteral inoculation through a needle, a wound, or a bite of an insect or verte-brate. Common sites of entry include the mucosal linings of the respiratory, alimentary, and urogenital tracts, the outer surface of the eye (conjunctival membranes or cornea), and the skin (Fig. 17.1).

### Respiratory Tract

Probably the most common route of viral entry is through the respiratory tract. There are about 300 million terminal sacs called **alveoli** in a human lung that function in gaseous exchange between inspired air and the blood. To accomplish this function, each sac is in close contact with capillary and lymphatic vessels. The combined absorptive area of the human lung is almost 140 m². Humans have a resting ventilation rate of 6 liters of air per min, which introduces large numbers of foreign particles and

**Figure 17.1 Sites of virus entry into the host.** A representation of the human host is shown with sites of virus entry and shedding indicated. The body is covered with skin, which has a relatively impermeable (dead) outer layer, but it cannot cover the entire body. Layers of living cells must be present to absorb food, exchange gases, and release urine and sexual products. These layers offer easier pathways for the entry of viruses than the skin does. Viruses can also be introduced through the skin by a scratch or injury, a vector bite, or inoculation with a needle.

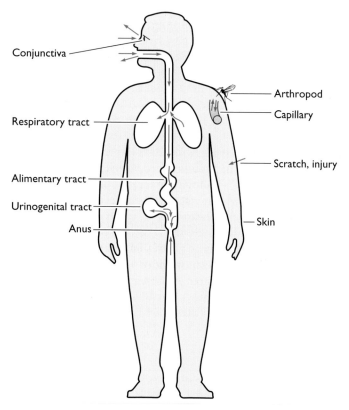

aerosolized droplets into the lungs each day. Many of these particles and droplets contain viruses. Consequently, there are numerous host defense mechanisms available to block respiratory tract infection. Mechanical barriers play a significant role in antiviral defense. For example, the tract is lined with a mucociliary blanket consisting of ciliated cells, mucus-secreting goblet cells, and subepithelial mucus-secreting glands (Fig. 17.2). Foreign particles deposited in the nasal cavity or upper respiratory tract are trapped in mucus, carried to the back of the throat, and swallowed. In the lower respiratory tract, particles trapped in mucus are brought up from the lungs to the throat by cilliary action. The lowest portions of the tract, the alveoli, lack cilia or mucus, and macrophages lining the alveoli are responsible for ingesting and destroying particles. Other cellular and humoral immune responses also intervene.

Many viruses (e.g., rhinoviruses, influenza viruses, coronaviruses, arenaviruses, adenoviruses, herpesviruses, and poxviruses) enter the respiratory tract in the form of aerosolized droplets generated by an infected individual by coughing or sneezing (Table 17.1). Infection can also spread through contact with saliva from an infected individual. Larger virus-containing droplets are deposited in the nose, while smaller droplets find their way into the airways or the alveoli. To infect the respiratory tract successfully, viruses must not be swept away by mucus, neutralized by immunoglobulin A (IgA), or destroyed by alveolar macrophages. Virus replication may be restricted to the epithelium of the respiratory tract (e.g., rhinoviruses) or may spread to other tissues (e.g., measles virus).

### Alimentary Tract

The alimentary tract, the tube that connects the oral cavity to the anus, is different from the respiratory tract in that it is always in motion. Eating, drinking, and some social activities routinely place viruses in the alimentary tract. It is designed to mix, digest, and absorb food, so the contents are always moving, providing a good opportunity for viruses to encounter a susceptible cell and to interact with the circulatory, lymphatic, and immune systems. The alimentary tract is a common route of infection and dispersal for a variety of viruses (Table 17.1). At first glance, the alimentary tract appears to be a hostile environment. The stomach is acidic, the intestine is alkaline, digestive enzymes and bile detergents abound, mucus lines the epithelium, and the luminal surfaces of intestines contain antibodies and phagocytic cells.

Viruses that have evolved to infect by the intestinal route must, at a minimum, be resistant to extremes of pH, proteases, and bile detergents. Indeed, viruses that lack these features are destroyed when exposed to the alimentary tract and must use other sites to establish an infection.

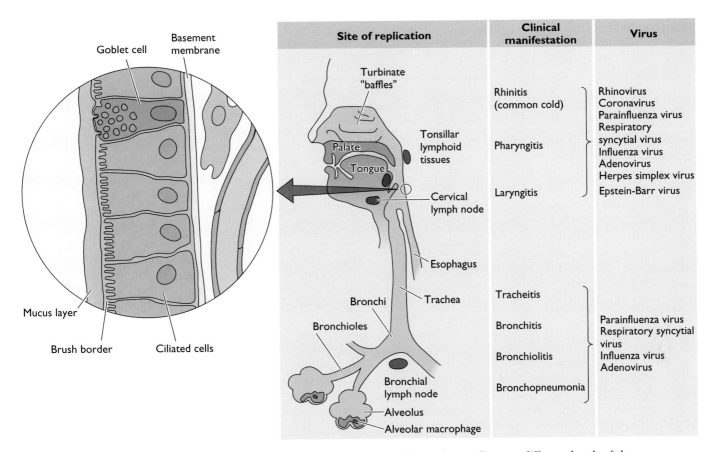

**Figure 17.2 Sites of virus entry in the respiratory tract.** Viruses that replicate at different levels of the respiratory tract are indicated on the right along with the associated clinical syndromes. A detailed view of the respiratory epithelium is shown. A layer of mucus, produced by goblet cells, is a formidable barrier to virus attachment to cells. Viruses that pass through this layer may multiply in the ciliated cells or pass between them, reaching another physical barrier, the basement membrane. Beyond this membrane are tissue fluids from which viruses may be taken into lymphatic capillaries and reach the blood. Local macrophages patrol the tissue fluids in search of foreign particles. Adapted from C. A. Mims et al., *Mims' Pathogenesis of Infectious Disease* (Academic Press, Orlando, Fla., 1995), with permission.

The family *Picornaviridae* comprises both acid-labile viruses (e.g., rhinoviruses) and acid-resistant viruses (e.g., poliovirus) (Box 17.1). Rhinoviruses are respiratory pathogens and cannot infect the upper intestine. They have evolved to spread in a population by means of inhaleable aerosols. In contrast, poliovirus can survive ingestion and establish an infection of the upper intestine. This virus has evolved to spread by the fecal-oral route. Some viruses have evolved to *require* the hostile environment of the alimentary tract to facilitate infection. For example, reovirus particles in the intestinal lumen are converted by host proteases into infectious subviral particles, the forms that subsequently infect intestinal cells. As might be expected, most enveloped viruses do not initiate infection in the alimentary tract because viral envelopes are susceptible to dissociation by detergents such as bile

salts. Enteric coronaviruses are notable exceptions, but it is not known why these enveloped viruses can withstand the harsh conditions in the alimentary tract.

Nearly all of the intestinal surface is covered with columnar villous epithelial cells whose apical surfaces are densely packed with microvilli (Fig. 17.3). This brush border, together with a surface coat of glycoproteins and glycolipids and the overlying mucous layer, is permeable to electrolytes and nutrients but presents a formidable barrier to microorganisms. Nevertheless, viruses such as enteric adenoviruses and Norwalk calicivirus replicate extensively in the intestinal epithelial cells. The mechanisms by which they bypass the physical barriers and enter susceptible cells are not well understood. Scattered throughout the intestinal mucosa are lymphoid follicles which are covered on the luminal side with a specialized follicle-associated

**Table 17.1**    Different routes of virus entry into the host

| Location | Viruses |
| --- | --- |
| **Respiratory tract** | |
| Localized upper tract | Rhinovirus, coxsackievirus, coronavirus, arenaviruses, hantavirus, parainfluenza virus types 1–4, respiratory syncytial virus, influenza A and B viruses, human adenovirus types 1–7, 14, and 21 |
| Localized lower tract | Respiratory syncytial virus, parainfluenza virus types 1–3, influenza A and B viruses, human adenovirus types 1–7, 14, and 21, infectious bovine rhinotracheitis (herpesvirus) |
| Entry via respiratory tract followed by systemic spread | Foot-and-mouth disease virus, rubella virus, arenaviruses, hantavirus, mumps virus, measles virus, varicella-zoster virus, poxviruses |
| **Alimentary tract** | |
| Systemic | Enterovirus, reovirus, adenovirus |
| Localized | Coronavirus, rotavirus |
| **Urogenital tract** | |
| Systemic | Human immunodeficiency virus type 1, hepatitis B virus, herpes simplex virus |
| Localized | Papillomavirus |
| **Eyes** | |
| Systemic | Enterovirus 70, herpes simplex virus |
| Skin | Togavirus, flavivirus, hepatitis C and D viruses, rhabdovirus, bunyavirus, reovirus, human immunodeficiency virus, hepatitis B virus, papillomavirus (localized), cytomegalovirus |

epithelium consisting mainly of columnar absorptive cells and M cells (membranous epithelial cells). The M-cell cytoplasm is very thin, resulting in a membranelike bridge that separates the lumen from the subepithelial space. As discussed in chapter 14, M cells ingest and deliver antigens to the underlying lymphoid tissue by **transcytosis**. In this process, material taken up on the luminal side of the M cell traverses the cytoplasm virtually intact and is delivered to the underlying basal membranes and extracellular space (Fig. 17.3). It is thought that M-cell transcytosis provides the mechanism by which some enteric viruses gain entry to deeper tissues of the host from the intestinal lumen. After crossing the mucosal epithelium in this manner, a virus could enter lymphatic vessels and capillaries of the circulatory system, facilitating spread within the host. One well-studied example of virus transcytosis is provided by reovirus. After attaching to the M cell, reovirus particles are transported to cells underlying the lymphoid follicle, where they replicate and spread to other tissues. Rather than spread by transcytosis across the M cell, some viruses actively replicate in the M cell and do not spread to underlying tissues. For example, human rotavirus and the coronavirus transmissible gastroenteritis virus destroy M cells, resulting in mucosal inflammation and diarrhea.

It is possible for a virus to enter the body through the lower gastrointestinal tract without passing through the upper tract and its defensive barriers. For example, human immunodeficiency virus can be introduced into the body

**BOX 17.1**

*Why is rhinovirus but not poliovirus sensitive to low pH?*

Rhinoviruses are sensitive to low pH, which causes irreversible disassembly of the viral capsid. In contrast, poliovirus is not inactivated by low pH; the changes in the capsid that occur under acidic conditions are fully reversible. Although high-resolution crystallographic structures of rhinovirus and poliovirus have been solved, they do not provide information about the structural basis for the difference in pH sensitivity. Acid-resistant mutants of rhinovirus contain amino acid changes near the fivefold axes of symmetry. Rhinovirus acid-resistant mutants behave similarly to poliovirus in that low pH leads to reversible conformational changes in the capsid. The mutations responsible for acid resistance probably result in stabilization of the particle so that it does not dissociate at low pH. Such mutations are not selected during natural infections because they are not necessary for the virus to replicate in the respiratory tract.

Giranda, V. L., B. A. Heinz, M. A. Oliveira, I. Minor, K. H. Kim, P. R. Kolatkar, M. G. Rossmann, and R. R. Rueckert. 1992. Acid-induced structural changes in human rhinovirus 14: possible role in uncoating. *Proc. Natl. Acad. Sci. USA* **89**:10213–10217.

Skern, T., H. Torgersen, H. Auer, E. Kuechler, and D. Blaas. 1991. Human rhinovirus mutants resistant to low pH. *Virology* **183**:757–763.

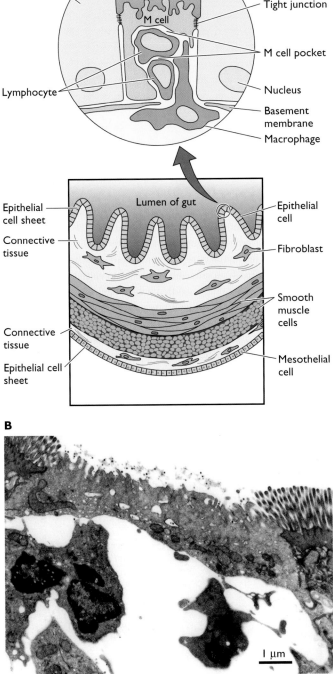

**A**

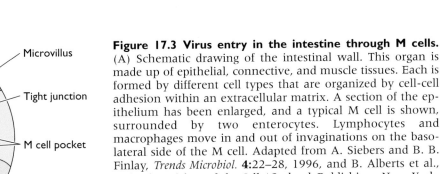

**B**

**Figure 17.3 Virus entry in the intestine through M cells.** (A) Schematic drawing of the intestinal wall. This organ is made up of epithelial, connective, and muscle tissues. Each is formed by different cell types that are organized by cell-cell adhesion within an extracellular matrix. A section of the epithelium has been enlarged, and a typical M cell is shown, surrounded by two enterocytes. Lymphocytes and macrophages move in and out of invaginations on the basolateral side of the M cell. Adapted from A. Siebers and B. B. Finlay, *Trends Microbiol.* **4:**22–28, 1996, and B. Alberts et al., *Molecular Biology of the Cell* (Garland Publishing, New York, N.Y., 1994), with permission. (B) Reovirus attached to, and within vesicles of, an M cell. An electron micrograph of the gut epithelium shows reovirus (small black dots) attached to the surface of an M cell and also within intracellular vesicles. Reprinted from J. L. Wolf and W. A. Bye, *Annu. Rev. Med.* **35:**95–112, 1984, with permission. Photo courtesy of R. Finberg, Harvard Medical School.

## Urogenital Tract

Many viruses enter the urogenital tract as a result of sexual activities of the host (Table 17.1). Such infections are common for human immunodeficiency virus type 1, herpes simplex viruses, and certain human papillomaviruses. The urogenital tract is well protected by physical barriers, including mucus and low pH (in the case of the vagina). Normal sexual activity can result in minute tears or abrasions in the vaginal epithelium or the urethra, allowing viruses to enter. Some viruses infect the epithelium and produce local lesions (e.g., human papillomaviruses, which cause genital warts). Other viruses gain access to cells in the underlying tissues and infect cells of the immune system (e.g., human immunodeficiency virus), or sensory and autonomic neurons (in the case of herpes simplex viruses). These two viruses invariably spread from the urogenital tract to establish lifelong persistent or latent infections, respectively.

## Eyes

The epithelium covering the exposed part of the **sclera** (the outer fibrocollagenous coat of the globe of the eye) and the inner surface of the eyelid (**conjunctiva**) is the route of entry for several viruses. Every few seconds the eyelid passes over the sclera, bathing it in secretions that wash away foreign particles. There is usually little opportunity for viral infection of the eye unless it is injured by abrasion. Direct inoculation into the eye may occur during ophthalmologic procedures or from environmental contamination (e.g., improperly sanitized swimming pools and hot tubs). In most cases, replication is localized and results in inflammation of the conjunctiva, a condition called **conjunctivitis**. Systemic spread of the virus from the eye is rare, although it does occur (e.g., paralytic illness after

by anal intercourse. As M cells abound in the lower colon, these cells are likely to provide a portal of entry for this virus into susceptible lymphocytes in the underlying lymphoid follicles. Once in the follicle, the virus can infect migratory lymphoid cells and spread throughout the body.

enterovirus 70 conjunctivitis). Herpesviruses, in particular herpes simplex virus type 1, can also infect the cornea, mainly at the site of a scratch or other injury. Such an infection may lead to immune destruction of the cornea and blindness. Inevitably, herpes simplex virus infection of the cornea is followed by spread of the virus to sensory neurons that innervate the cornea and then to neuronal cell bodies in the sensory ganglion (e.g., the trigeminal ganglia), where a latent infection is established.

### Skin

The skin protects the body yet provides sensory contact with the environment. The external surface of skin, or **epidermis**, is composed of several layers, including a basal germinal layer (stratum malpighii), a granular layer of dying cells (stratum granulosum), and an outer layer of dead, keratinized cells (stratum corneum) (Fig. 17.4). The skin of most animals is an effective barrier against viral infections, as the dead outer layer cannot support viral growth. Entry through this organ occurs primarily when its integrity is breached by breaks or punctures. Examples of viruses that can gain entry in this manner are human papillomaviruses and certain animal poxviruses (e.g., myxoma virus) that are transmitted mechanically by insect vectors such as the arthropod depicted in Fig. 17.1 (Table 17.1). Replication is usually limited to the site of entry because the epidermis is devoid of blood or lymphatic vessels that could provide pathways for further spread. The epidermis is supported by the highly vascularized **dermis**. Other viruses (e.g., bunya-, reo-, toga-, and flaviviruses) can gain entry to the dermis through the bites of arthropod vectors such as mosquitoes, mites, ticks, and sandflies. Even deeper inoculation, into the tissue and muscle below the dermis, can occur by hypodermic needle punctures, body piercing or tattooing, or sexual contact where body fluids are mingled through skin abrasions or ulcerations. Typical viruses that invade by these mechanisms are human immunodeficiency virus, hepatitis B virus, human cytomegalovirus, and Epstein-Barr virus. Animal bites introduce rabies virus into tissue and muscle rich with nerve endings by which the virus can invade the peripheral nervous system. In contrast to the strictly localized replication of viruses in the epidermis, viruses that initiate infection in dermal or subdermal tissues can reach nearby blood vessels, lymphatic tissues, and cells of the nervous system. As a consequence, they may spread to other sites in the body.

## Viral Spread

Following replication at the site of entry, virus particles can remain localized or can spread to other tissues. Local replication in the respiratory tract is characteristic of influenza virus, parainfluenza virus, rhinovirus, and respiratory coronavirus; replication of rotaviruses and enteric corona- and adenoviruses is restricted to the alimentary tract, and replication of some papillomaviruses is confined to the skin. Local spread of the infection in the epithelium occurs when newly released virus infects adjacent cells. These infections are usually contained by the physical constraints of the tissue and brought under control by the innate immune defenses discussed in chapters 14 and 15. In general, an infection that spreads beyond the primary site of infection is said to be **disseminated**. If many organs become infected, the infection is described as **systemic**. For an infection to spread beyond the primary site, physical and immune barriers must be breached. After crossing the epithelium, virus particles reach the **basement membrane** (Fig. 17.3). Its integrity may be compromised by epithelial cell destruction and inflammation. Below the basement membrane are subepithelial tissues, where the

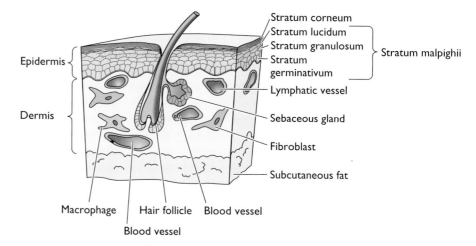

**Figure 17.4 Schematic diagram of the skin.** The epidermis consists of a layer of dead, keratinized cells (stratum corneum) over the stratum malpighii. The latter may have two layers of cells with increasing amounts of keratin granules (stratum granulosum and stratum lucidum) and a basal layer of dividing epidermal cells (stratum germinativum). Below this is the basement membrane. In the dermis are blood vessels, lymphatic vessels, fibroblasts, and macrophages. A hair follicle and a sebaceous gland are shown. Adapted from F. Fenner et al., *The Biology of Animal Viruses* (Academic Press, New York, N.Y., 1974), with permission.

virus encounters tissue fluids, the lymphatic system, and phagocytes. All three play significant roles in clearing foreign particles but also may disseminate infectious virus from the primary site of infection.

One important mechanism for avoiding local host defenses and facilitating spread within the body is the **directional release** of virus particles from polarized cells at the mucosal surface. Virions can be released from the apical surface, from the basolateral surface, or from both (Fig. 17.5). The consequences of directional release are striking. After replication, virus released from the apical surface is back where it started, outside the host. Such directional release facilitates the dispersal of many newly replicated enteric viruses in the feces (e.g., poliovirus). In contrast, virus particles released from the basolateral surfaces of polarized epithelial cells have been moved away from the defenses of the luminal surface. Directional release is therefore a major determinant of the infection pattern. In general, viruses released at apical membranes establish a localized or limited infection. In this case, local lateral spread from cell to cell occurs in the infected epithelium, but virus particles rarely invade the underlying lymphatic

and circulatory vessels. On the other hand, release of viruses at the basal membrane provides access to the underlying tissues and may facilitate systemic spread.

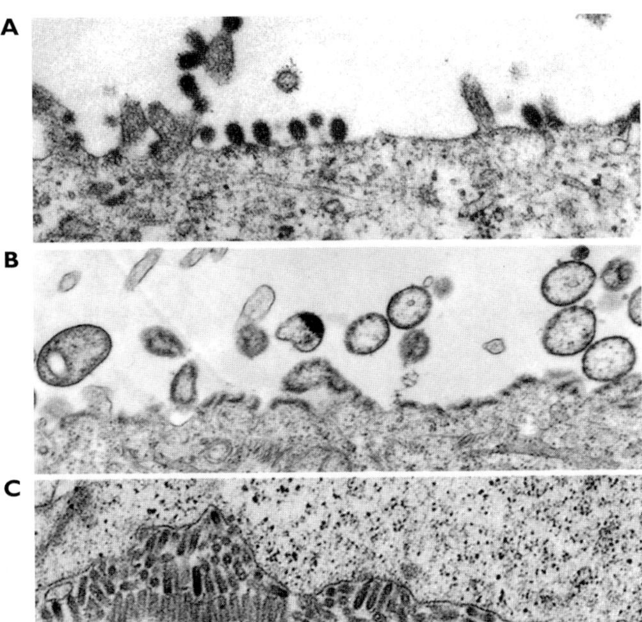

Directional release of virus from a polarized cell can be the major determinant of whether virus replication is localized or disseminated. This assertion is underscored by studies of Sendai virus, which is normally released from the apical surfaces of polarized epithelial cells. Consequently, this virus establishes only a localized infection of the respiratory tract. A mutant virus, called F1-R, is released at *both* apical and basal surfaces. In striking contrast to that by wild-type virus, infection by this mutant virus is disseminated, with replication in a wide variety of tissues.

### Hematogenous Spread

Viruses that escape from local defenses to produce a disseminated infection often do so by entering the bloodstream (**hematogenous spread**) (Table 17.2). Viruses may enter the blood directly through capillaries, by replicating in endothelial cells, or through inoculation by a vector bite. Once in the blood, a virus has access to almost every tissue in the host (Fig. 17.6). Hematogenous spread begins when newly replicated particles produced at the entry site are released into the extracellular fluids, which can be taken up by the local lymphatic vascular system

**Figure 17.5 Polarized release of viruses from cultured epithelial cells visualized by electron microscopy.** (A) Influenza virus released by budding from the apical surface of MDCK cells. (B) Budding of measles virus on the apical surface of Caco-2 cells. (C) Release of vesicular stomatitis virus at the basolateral surface of MDCK cells. Magnification, ×24,000. Reprinted from D. M. Blau and R. W. Compans, *Semin. Virol.* 7:245–253, 1996, with permission. Courtesy of D. M. Blau and R. W. Compans, Emory University School of Medicine.

**Table 17.2** Cell-associated and plasma viremia[a]

| Cell type | Virus | Duration of viremia |
|---|---|---|
| Monocyte/macrophage | Dengue virus | Acute |
| | Rubella virus | Persistent |
| | Measles virus | Acute |
| | Lymphocytic chorio-meningitis virus | Persistent |
| | Human immunodeficiency virus | Persistent |
| | Cytomegalovirus | Acute |
| | Mousepox virus | Acute |
| B lymphocytes | Murine leukemia virus | Persistent |
| | Epstein-Barr virus | Persistent |
| T lymphocytes | Human immunodeficiency virus | Persistent |
| | Human T-cell leukemia virus type 1 | Persistent |
| | Human herpesviruses 6 and 7 | Acute |
| Erythrocytes | Colorado tick fever virus | Acute |
| Neutrophils | Influenza virus | Acute |
| Free in plasma | Poliovirus | Acute |
| | Yellow fever virus | Acute |
| | Hepatitis B virus | Persistent |

[a] The associations may not play a role in pathogenesis.

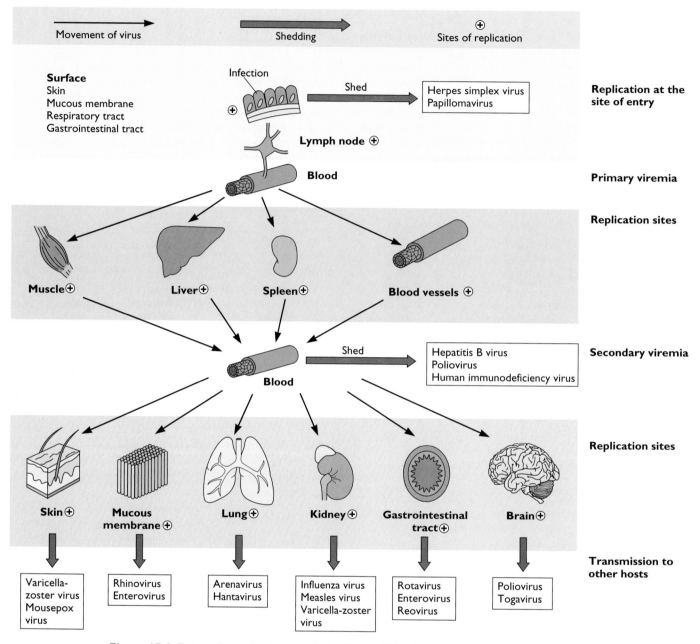

**Figure 17.6 Entry, dissemination, and shedding of blood-borne viruses.** Shown are the target organs for selected viruses that enter at epithelial surfaces and spread via the blood. The sites of virus shedding (red arrows), which may lead to transmission to other hosts, are shown. Adapted from N. Nathanson (ed.), *Viral Pathogenesis* (Lippincott-Raven Publishers, Philadelphia, Pa., 1997), with permission.

(Fig. 17.7). Lymphatic capillaries are considerably more permeable than circulatory system capillaries, facilitating virus entry. As the lymphatic vessels ultimately join with the venous system, virus particles in lymph have free access to the bloodstream. In the lymphatic system virions pass through lymph nodes, where they encounter lym-

phocytes and other migratory cells of the immune system. Viral pathogenesis resulting from the direct infection of immune system cells (e.g., human immunodeficiency virus, measles virus) is initiated in this fashion. Some viruses replicate in the infected lymphoid cells, and progeny are released into the blood plasma. The infected lym-

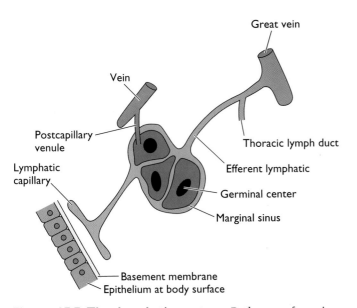

**Figure 17.7 The lymphatic system.** Pathways for virus spread from the epithelium to the blood via the lymphatic system are shown. Lymphocytes flow from the blood into the lymph node through postcapillary venules. Adapted from C. A. Mims et al., *Mims' Pathogenesis of Infectious Disease* (Academic Press, Orlando, Fla., 1995), with permission.

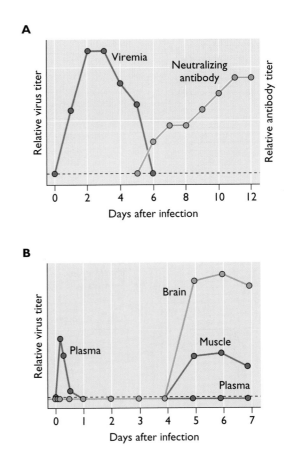

**Figure 17.8 Characteristics of active and passive viremia.** (A) Active viremia. Poliovirus was inoculated intramuscularly into monkeys, and the presence of virus and neutralizing antibody in the blood was determined at different times after infection. Viremia ceases just as serum antibody appears. (B) Passive viremia. La Crosse virus, a bunyavirus, was injected into weanling mice, and virus titers in plasma, brain, and muscle were determined at different times afterward. No virus can be detected in the blood after 1 day; however, beginning at 5 days after infection virus can be found in the brain and muscle. The latter is progeny virus produced by multiplication of the inoculum in the target organs. Adapted from N. Nathanson (ed.), *Viral Pathogenesis* (Lippincott-Raven Publishers, Philadelphia, Pa., 1997), with permission.

phoid cell may also migrate away from the local lymph node to distant parts of the circulatory system. There may be little virus replication while the cell is in the bloodstream. However, when the infected cell receives chemotactic signals that direct it to enter other tissues, new virus particles may be produced in the activated cell.

The term **viremia** describes the presence of infectious virus particles in the blood. These virions may be free in the blood or contained within infected cells such as lymphocytes. **Active viremia** is produced by virus replication, while **passive viremia** results when virus particles are introduced into the blood without viral replication at the site of entry (e.g., injection of a virus suspension into a vein results in passive viremia) (Fig. 17.8). Progeny virions released into the blood after initial replication at the site of entry constitute **primary viremia**. The concentration of virus particles during primary viremia is usually low. However, the subsequent disseminated infections that result from this primary viremia are often extensive, resulting in the release of considerably more virus particles. Such delayed appearance of a high concentration of infectious virus in the blood is termed **secondary viremia**. The two phases of viremia were first described in the classic studies of mousepox by Frank Fenner (Fig. 17.9).

The concentration of virus particles in blood is determined by the rates of their synthesis in permissive tissues and by how quickly they are released into, and removed

from, the blood. Circulating virus particles are removed by phagocytic cells of the reticuloendothelial system in the liver, lung, spleen, and lymph nodes. When serum antibodies appear, virions in the blood may bind these antibodies and be "neutralized" as described in chapter 14. Formation of a complex of antibodies and virus particles facilitates uptake by Fc receptors expressed by macrophages lining the circulatory vessels. Virus-antibody complexes can accumulate in significant quantities in the kidney, spleen, and liver, where they produce pathological effects. The time a virus spends in the blood usually varies from 1 to 60 min,

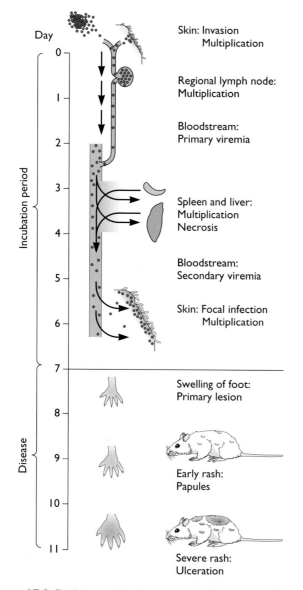

**Day**

0 — Skin: Invasion
Multiplication

1 — Regional lymph node:
Multiplication

2 — Bloodstream:
Primary viremia

3 —

Spleen and liver:
Multiplication
Necrosis

4 —

5 — Bloodstream:
Secondary viremia

6 — Skin: Focal infection
Multiplication

7 —

8 — Swelling of foot:
Primary lesion

9 —

Early rash:
Papules

10 —

11 — Severe rash:
Ulceration

Incubation period

Disease

**Figure 17.9 Pathogenesis of mousepox.** Sequence of events in the pathogenesis of mousepox after inoculation of virus into the footpad. The figure is based on the classic studies of Frank Fenner, which were the first to demonstrate how disseminated viral infections develop from local multiplication to primary and secondary viremia to the target organ(s). In the case of mousepox, after local multiplication in the foot, the host response leads to swelling at the site of inoculation, and after viremia, the host response to replication in the skin results in a rash. Adapted from F. Fenner et al., *The Biology of Animal Viruses* (Academic Press, New York, N.Y., 1974), with permission.

depending on such parameters as the physiology of the host (e.g., age, health) and the size of the virus (larger particles are cleared more rapidly than small particles).

While viremias are usually of short duration, some viral infections are noteworthy for the long-lasting presence of infectious virus in the blood. Hosts infected with hepatitis

B and C viruses or lymphocytic choriomeningitis virus may have viremias that last for years. One source of prolonged viremia is viral replication in macrophages (Table 17.2). Such infected macrophages often cannot perform normally, leading to delays in virus clearance and exacerbation of disease symptoms.

In a pregnant female, viremia may result in infection of the developing fetus. This can occur by direct infection of the placental tissues and subsequent invasion of the fetal tissue, or infected circulating cells such as monocytes may enter the fetal bloodstream directly. Many viral infections of the mother lead to infection of the fetus, especially if infection occurs early in pregnancy. The risk of fetal infection in infants whose mothers have been infected with rubella virus during the first trimester is approximately 80%. Similarly, intrauterine transmission of human cytomegalovirus occurs in approximately 40% of pregnant women with primary infection. While human immunodeficiency virus can certainly be transmitted from mother to infant, the incidence of intrauterine infection is not yet established. There is ample evidence that infection of the baby can occur during delivery or breast-feeding.

Viremias obviously are of diagnostic value and can be used to monitor the course of infection, but they also present practical problems. Infections can be spread inadvertently in the population when pooled blood from thousands of individuals is used directly for therapeutic purposes (i.e., transfusions) or as a source of therapeutic proteins (e.g., gamma globulin or blood-clotting factors). We have learned from unfortunate experience that hepatitis and AIDS can be spread by contaminated blood and blood products. Obviously, sensitive detection methods and stringent purification protocols are required to protect those who use and dispense these products. Frequently it may be difficult or technically impossible to quantify infectious virus in the blood, which is certainly the case for hepatitis B virus. Accordingly, the presence of characteristic viral proteins provides surrogate markers of the viremia. Immunoblotting, hemagglutination, and complement fixation are techniques frequently used to detect viral proteins in blood.

### Neural Spread

Many viruses spread from the primary site of infection by entering local nerve endings (Table 17.3). For certain viruses (e.g., rabies virus and alphaherpesviruses), neuronal spread is the definitive characteristic of their pathogenesis. For other viruses (e.g., poliovirus and reovirus), invasion of the nervous system is a less frequent diversion from their normal replication and hematogenous spread. Some viruses (e.g., mumps virus, human immunodeficiency virus, and measles virus) may replicate in the brain,

**Table 17.3**   Steps required to establish an infection of neurons

1. *Enter* the neuron at the axon, sensory terminal, or cell body, depending on the site of infection.
2. *Transport* the virus particle or subviral particle (e.g., nucleocapsid) toward the cell body of the neuron, where replication occurs.
3. *Replicate* the genome.
4. Assemble virus particles that *egress* from the infected neuron in a directional manner (Fig. 17.11).

but spread is by the hematogenous route and not through neurons. The molecular mechanisms that dictate spread by neural or hematogenous pathways are not well understood for most viruses. Clues may be found through the study of different strains of the same virus that differ in their ability to spread by these two routes (Table 17.4).

Neurons are polarized cells whose functionally and structurally distinct processes (axons and dendrites) can be separated by enormous distances. For example, in adult humans the axon terminals of motor neurons that control stomach muscles can be 50 cm away from the cell bodies and dendrites in the brain stem. We currently have a limited understanding of how viral particles move in and among cells of the nervous system. It is likely that virions enter neurons by the same mechanisms used to enter other cells. Because protein synthesis does not occur in the extended processes of neuronal cells, virus particles must be transported over relatively long distances to the site of viral replication. All evidence indicates that viruses are transported in the infected neuron by cellular systems, but the direction of spread may be facilitated by viral proteins. Those who study virus spread in the nervous system often use the words "retrograde" and "anterograde" to describe direction. Unfortunately, confusion arises because the terms can be used to describe directional movement of virus inside a cell as well as spread of virus between synaptically connected neurons. Spread from the primary neu-

ron to the second-order neuron in the direction of the nerve impulse is said to be anterograde spread (Fig. 17.10). Spread in the opposite direction is said to be retrograde spread (Fig. 17.10). Inside a neuron, spread is defined by microtubule polarity. Transport on microtubules that goes from (−) to (+) ends is said to be anterograde, while transport on microtubules that goes from (+) to (−) ends is said to be retrograde.

Directionality of movement within a neuron is most likely mediated by microtubules and their attendant motor proteins, including kinesin and dynein. For example, drugs such as colchicine that disrupt microtubules efficiently block the spread of many neurotropic viruses from the site of peripheral inoculation to the central nervous system. All microtubules in the axon are oriented with their plus ends toward the synapse and their minus ends toward the cell body. From experiments with cultured cells, dynein has been implicated in the movement of viral

**Table 17.4**   Viral spread to the central nervous system

| Pathway | Viruses |
| --- | --- |
| Neural | Poliovirus, yellow fever virus, mouse hepatitis virus, Venezuelan encephalitis virus, rabies virus, reovirus (type 3 only; type 1 is spread by viremia), herpes simplex virus types 1 and 2, pseudorabies virus |
| Olfactory | Poliovirus (experimental), herpes simplex virus, coronavirus |
| Hematogenous | Poliovirus, coxsackievirus, arenavirus, mumps virus, measles virus, herpes simplex virus, cytomegalovirus |

**Figure 17.10 Anterograde and retrograde spread of virus in nerves.** (A) Retrograde spread of infection. Virus invades at axon terminals and spreads to the cell body, where replication ensues. Newly replicated virus particles spread to a neuron at sites of synaptic contact. Particles enter the axon terminal of the second neuron to initiate a second cycle of replication and spread. (B) Anterograde spread of infection. Virus invades at dendrites or cell bodies and replicates. Virus particles then spread to axon terminals, where virions cross synaptic contacts to invade dendrites or cell bodies of the second neuron to initiate a second cycle of replication and spread.

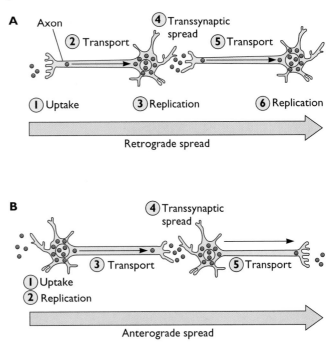

nucleocapsids in the retrograde direction. It is likely that kinesinlike motors facilitate anterograde movement of virus particles within a neuron, but the actual molecular motors that move viruses in vivo in neurons have not been identified. Furthermore, the precise intracellular form of any virus that spreads in the nervous system has not been established. Both mature virions and nucleocapsids have been observed in axons and dendrites of animals infected by rabies virus and alphaherpesviruses.

While viruses that infect the nervous system are often said to be **neurotropic** (Box 17.2), they are generally capable of infecting a variety of cell types. Viral replication usually occurs first in nonneuronal cells, with virions subsequently spreading into afferent (e.g., sensory) or efferent (e.g., motor) nerve fibers innervating the infected tissue (Fig. 17.11). Concurrent with the primary infection in nonneuronal cells at the epithelial surface, viruses that invade neurons must complete four general processes to establish the neuronal infection (Table 17.3). Virus particles may be taken up at axon terminals at neuromuscular junctions, at sensory terminals, or at cell bodies of neurons (e.g., the olfactory neurons in the nasopharyngeal cavity) (Fig. 17.12). Under some circumstances, virions can enter neurons directly with no prior replication in nonneuronal cells. However, this is probably an infrequent event, as nerve endings are rarely exposed directly to the environment. Unless virus particles are injected directly into the brain or spinal cord, enter olfactory neurons, or infect retinal ganglion cells, the first neurons to be infected are constituents of the peripheral nervous system. These neurons represent the first cells in circuits connecting the innervated peripheral tissue with the spinal cord and brain (the central nervous system). Once in the peripheral nervous system, alphaherpesviruses, some rhabdoviruses (e.g., rabies virus), and some coronaviruses can spread among neurons connected by synapses. Infection of nonneuronal support cells and satellite cells in ganglia may also occur. If two neurons are in synaptic contact, then virus can spread from the first infected neuron (defined as the first-order neuron) to a second neuron (the second-order neuron) by **transsynaptic spread**. Virus spread by this mode can continue through chains of connected neurons of the peripheral nervous system to the spinal cord and brain, with devastating results (Box 17.3).

An important component of neuronal infection is movement and release of infectious particles in polarized neurons. Directional release of infectious virus from neurons affects the outcome of infection, as does the directional release of virions from polarized epithelial cells discussed earlier. For example, after primary infection, alphaherpesviruses become latent in peripheral neurons that innervate the site of infection (Fig. 17.13). Reactivation from the latent state results in viral replication in the primary neuron and subsequent transport of progeny viruses from the neuron cell body *back* to the innervated peripheral tissue. Following the reactivation of herpes simplex virus from latency in peripheral neurons, virus can spread from the peripheral to the central nervous system, or it can spread back to the peripheral site serviced by that particular group of neurons in the ganglion (Fig. 17.13). The direction taken is the difference between a minor local infection (a cold sore) or a lethal viral encephalitis. Luckily, spread back to the peripheral site is by far more common. The molecular mechanisms that promote such two-way traffic are under investigation.

### Tissue Invasion

Once virions enter the blood and are dispersed from the primary site, any subsequent replication requires invasion of new cells and tissues. There are three main types of

---

**BOX 17.2**

*Infection of the nervous system: definitions and distinctions*

A **neurotropic** virus can infect neural cells; infection may occur by neural or hematogenous spread initiating from a peripheral site. A **neuroinvasive** virus can enter the central nervous system (spinal cord and brain) after infection of a peripheral site. A **neurovirulent** virus can cause disease of nervous tissue; this is manifested by neurological symptoms and often death.

Examples:

Herpes simplex virus has low neuroinvasiveness of the central nervous system but high neurovirulence. This virus always enters the peripheral nervous system but rarely enters the central nervous system. When it does, the consequences are almost always severe, if not fatal.

Mumps virus has high neuroinvasiveness but low neurovirulence. Most infections lead to invasion of the central nervous system, but neurological disease is mild.

Rabies virus has high neuroinvasiveness and high neurovirulence. This virus readily infects the peripheral nervous system and spreads to the central nervous system with 100% lethality unless antiviral therapy is administered shortly after infection.

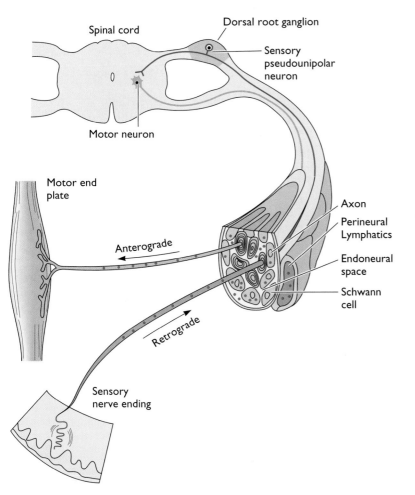

**Figure 17.11 Possible pathways for the spread of viruses in nerves.** Viruses may enter the sensory or motor neuron endings. They may be transported within axons, in which case viruses taken up at sensory endings reach dorsal root ganglion cells, while those taken up at motor endings reach motor neurons. Viruses may also travel in the endoneural space, in perineural lymphatics, or in infecting Schwann cells. Directional transport of virus inside the sensory neuron is defined as anterograde [movement from the (−) to the (+) ends of microtubules] or retrograde [movement from the (+) to the (−) ends of microtubules]. Adapted from R. T. Johnson, *Viral Infections of the Nervous System* (Raven Press, New York, N.Y., 1982), with permission.

blood vessel-tissue junctions that provide routes of tissue invasion (Fig. 17.14). In some tissues the endothelial cells are continuous with a dense basement membrane. At other sites the endothelium contains gaps, and at still others there may be sinusoids in which macrophages form part of the blood-tissue junction.

### *The Liver, Spleen, Bone Marrow, and Adrenal Glands*

The liver, spleen, bone marrow, and adrenal glands are characterized by the presence of sinusoids lined with macrophages. Such macrophages, known as the **reticuloendothelial system**, function to filter the blood and remove foreign particles. They often provide a portal of entry into various tissues. For example, viruses that infect the liver, the major filtering and detoxifying organ of the body, usually enter from the blood. The presence of virus particles in the blood invariably results in the infection of **Kupffer cells**, the macrophages that line the liver sinusoids (Fig. 17.15). Virions may be transcytosed across

Kupffer and endothelial cells without replication to reach the underlying hepatic cells. Alternatively, viruses may multiply in Kupffer and endothelial cells and then infect underlying hepatocytes. Either mechanism may result in inflammation and necrosis of liver tissue, a condition termed **hepatitis**.

### *The Renal Glomerulus, Pancreas, Ileum, and Colon*

Some tissues lack sinusoids, and therefore macrophages are not present to facilitate the invasion of viruses (Fig. 17.14). To enter such tissues, viruses must first adhere to the endothelial cells, usually in capillaries or venules, where the blood flow is slowest and the walls are thinnest. Endothelial cells are not highly phagocytic, and therefore adhesion of virus particles to these cells requires the presence of cellular receptors. To increase the chances of adhesion, the virus must be present in a high concentration and circulate for a long enough period. Clearly there is a race between adhesion and removal of the virus by the

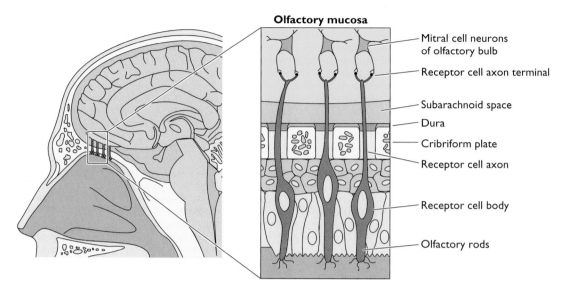

**Figure 17.12 Pathways for virus entry into the central nervous system by olfactory routes.** Olfactory neurons are unusual in that their cell bodies are present in the olfactory epithelia and their axon terminals are in synaptic contact with olfactory bulb neurons. The olfactory nerve fiber passes through the skull via an opening called the arachnoid. Adapted from R. T. Johnson, *Viral Infections of the Nervous System* (Raven Press, New York, N.Y., 1982), with permission.

reticuloendothelial system. Once blood-borne viruses have adhered to the vessel wall, they can readily invade the renal glomerulus, pancreas, ileum, or colon because the endothelial cells that make up the capillaries are fenestrated (with "windows" between cells; loosely joined together), permitting the virus or virus-infected cells to cross into the underlying tissues (e.g., poliovirus). Some viruses (e.g., herpes simplex, yellow fever, and measles viruses) cross the endothelium while being carried by infected monocytes or lymphocytes in a process called **diapedesis**.

### Central Nervous System, Connective Tissue, and Skeletal and Cardiac Muscle

In the central nervous system, connective tissue, and skeletal and cardiac muscle, capillary endothelial cells are generally not fenestrated and are backed by a dense basement membrane (Fig. 17.14 and 17.16). The complexity of this membrane differs according to the organ and is less well developed in the fetus. In the central nervous system the basement membrane is the basis for the blood-brain barrier. However, it is not so much a barrier as a selective permeability system. Much study has been devoted to determining how viruses enter the brain from the blood. Routes of virus invasion into the central nervous system are summarized in Fig. 17.17.

In several well-defined parts of the brain, the capillary epithelium is fenestrated and the basement membrane is sparse. These highly vascularized sites include the **choroid plexus**, a tissue found primarily around the lateral, third, and fourth ventricles. This tissue produces more than 70% of the cerebrospinal fluid that bathes the spinal cord and ventricles of the brain. Some viruses (e.g., mumps virus and some togaviruses) pass through the capillary endothelium and enter the stroma of the choroid plexus, where they may cross the epithelium into the cerebrospinal fluid either by transcytosis or replication and directed release. Once in the cerebrospinal fluid, these viruses infect the ependymal cells lining the ventricles and invade the underlying brain tissue. Other viruses may directly infect, or be transported across, the capillary endothelium (e.g., picornaviruses, togaviruses, bunyaviruses, and parvoviruses). Some viruses cross the endothelium while being carried by infected monocytes or lymphocytes (e.g., human immunodeficiency virus, measles virus, and mumps virus). Increased local permeability of the capillary endothelium caused, for example, by vasogenic amines may also permit virus entry into the brain and spinal cord.

Coxsackieviruses, members of the *Picornaviridae*, multiply in skeletal and cardiac muscle, and rubella virus, a togavirus, infects connective tissue. The mechanisms by which these viruses enter muscles from the blood has not been adequately studied. It is known that certain coxsackieviruses can replicate in cultured endothelial cells. These viruses may therefore be released into muscle cells after multiplying within cells of the capillary walls.

BOX 17.3

## Tracing neuronal connections in the nervous system with viruses

The identification and characterization of synaptically linked multineuronal pathways in the brain are important in understanding the functional organization of neuronal circuits. Conventional tracing methodologies have relied on the use of markers such as wheat germ agglutinin-horseradish peroxidase or fluorochrome dyes (Fluoro-Gold and fast blue) to delineate anterograde (the direction of the nerve impulse) and retrograde (the direction opposite that of the nerve impulse) pathways. The main limitations of these tracers are their limited specificity and sensitivity. During experimental manipulations, it is difficult to restrict the diffusion of certain conventional neuronal circuit tracers to a particular cell group or nucleus, and so uptake of the tracer occurs in neighboring neurons or adjacent nonconnected axon fibers, producing false-positive labeling of a circuit. Second, neurons located one or more synapses away from the injection site receive progressively less label because the tracer is diluted at each stage of transneuronal transfer. The ideal tracer should be specific only for those neurons connected within a particular circuit and sensitive enough to label all neurons (first-, second-, third-order, etc.) in multineuronal pathways.

Several neurotropic viruses, including two alphaherpesviruses (herpes simplex virus and pseudorabies virus) and one rhabdovirus (rabies virus), have demonstrated considerable promise as self-amplifying tracers of synaptically connected neurons. Under proper conditions, second- and third-order neurons show the same labeling intensity as first-order neurons. Moreover, the specific pattern of infected neurons observed in tracing studies is consistent with transsynaptic passage of virus rather than lytic spread through the extracellular space.

These three viruses can infect all main categories of primary sensory neurons, motor neurons, autonomic (sympathetic and parasympathetic) neurons, and central nervous system sites in a variety of mammalian species. Herpes simplex virus and pseudorabies virus have been used extensively to define circuits that modulate the output of both the sympathetic and parasympathetic components of the autonomic nervous system; sensory (afferent) neurons of the olfactory, optic, and trigeminal system involving the ophthalmic, maxillary, and mandibular nerve branches; and motor neurons of the hypoglossal, phrenic, ulnar, and median nerves.

The detection of viruses typically involves immunohistochemical localization of viral antigens by light microscopy. More recently, reporter genes have been introduced into alphaherpesvirus genomes for more direct localization of viral infection. These include the *lacZ* gene from *Escherichia coli* and the green fluorescent protein gene from *Aequorea victoria*. Dual-labeling experiments following the inoculation of viruses that express different markers have identified several brain regions coregulating multiple peripheral target organ functions.

**Enquist, L. W., P. J. Husak, B. W. Banfield, and G. A. Smith.** 1999. Infection and spread of alphaherpesviruses in the nervous system. *Adv. Virus Res.* **51:**237–347.

## Tropism

Most viruses do not infect all the cells of a host but are restricted to specific cell types of certain organs. As discussed in chapter 15, the spectrum of tissues infected by a virus in its animal host is known as its **tropism**. The tropism of some viruses is strictly limited, whereas others are **pantropic**, capable of infecting many organs.

### Cell Receptors for Viruses

A cell may be susceptible to infection if the viral receptor(s) is present and functional. If the proper form of viral receptor is not expressed, the tissue cannot become infected. The site of expression of the receptor might also be a determinant of cellular susceptibility to infection. If the cellular receptor is expressed only on the basal cell membrane of polarized epithelial cells, a virus cannot replicate unless it first traverses the epithelium. In some cases, virus particles bound to antibodies can be taken up by Fc receptors on nonsusceptible cells (see, for example, "Immunopathological Lesions Associated with B Cells").

As cellular receptors for viruses have been identified and studied, it has become evident that they can be determinants of tropism. For example, with few exceptions, the distribution of CD4, the receptor for human immunodeficiency virus type 1, mimics closely the tissue tropism of this virus. Similarly, the tropism of Epstein-Barr virus can be explained by the distribution of its receptor, the complement protein CR2, on B cells. On the other hand, re-

Normal maintenance cycle      Pathways leading to serious disease

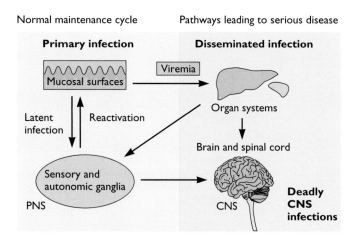

**Figure 17.13 Outline of the spread of alphaherpesvirus infection and its relationship to disease.** Abbreviations: PNS, peripheral nervous system; CNS, central nervous system.

ceptor expression may not be sufficient for infection. For example, Pvr, the receptor for poliovirus, can be found in many human tissues, including those that are not sites of poliovirus replication. Similarly, the receptor for influenza A virus is the ubiquitous complex sugar sialic acid, yet infection in humans is restricted to the respiratory tract. Despite the simplicity of the receptor hypothesis, the distribution of cellular receptors in host tissues generally is more widespread than the observed tropism of the virus. Clearly, receptors are necessary for infection but are not sufficient to explain viral tropism.

**Figure 17.14 Blood-tissue junction in a capillary, venule, and sinusoid.** (Left) Continuous endothelium and basement membrane found in the central nervous system, connective tissue, skeletal and cardiac muscle, skin, and lung. (Center) Fenestrated endothelium found in the choroid plexus, villi of the intestine, renal glomerulus, pancreas, and endocrine glands. (Right) Sinusoid, lined with macrophages of the reticuloendothelial system, as found in adrenal gland, liver, spleen, and bone marrow. Adapted from C. A. Mims et al., *Mims' Pathogenesis of Infectious Disease* (Academic Press, Orlando, Fla., 1995), with permission.

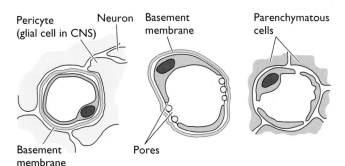

## Cellular Proteins That Regulate Viral Transcription

Sequences in viral genomes that control transcription of viral genes such as enhancer regions may be determinants of viral tropism. For example, in the brain, JC papovavirus replicates only in oligodendrocytes (Box 17.4). The results of in vitro and in vivo experiments indicate that the JC virus enhancer is active only in this cell type. Similar results have been obtained with the liver-specific enhancer of hepatitis B virus, the keratinocyte-specific enhancer of human papillomavirus type 11, and the enhancers in the long terminal repeat of certain avian leukosis virus strains.

## Cell Proteases

Many viruses require cellular proteases to cleave viral proteins to form the mature infectious virus particle. A cellular protease cleaves the influenza virus HA precursor into two subunits so that fusion of the viral envelope and cell membranes can proceed. In mammals, the replication

**Figure 17.15 How viruses enter the liver.** Two layers of hepatocytes are shown, with the sinusoid at the center lined with Kupffer cells. Endothelial cells are not shown. Virus may pass through the sinusoid without entering; enter the Kupffer cell and be destroyed; pass through the Kupffer cell, multiply in hepatocytes, and perhaps be excreted in the bile; multiply in the Kupffer cell; or multiply in the Kupffer cell, multiply in hepatocytes, and perhaps be excreted in the bile. Adapted from C. A. Mims et al., *Mims' Pathogenesis of Infectious Disease* (Academic Press, Orlando, Fla., 1995), with permission.

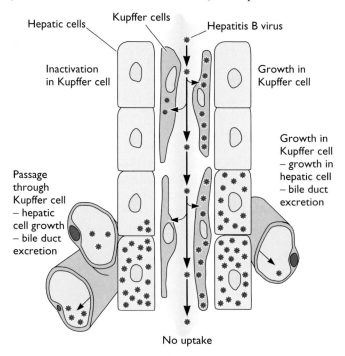

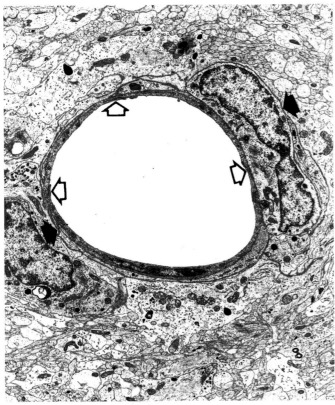

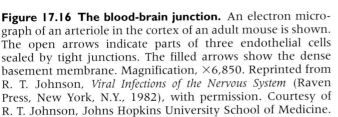

**Figure 17.16 The blood-brain junction.** An electron micrograph of an arteriole in the cortex of an adult mouse is shown. The open arrows indicate parts of three endothelial cells sealed by tight junctions. The filled arrows show the dense basement membrane. Magnification, ×6,850. Reprinted from R. T. Johnson, *Viral Infections of the Nervous System* (Raven Press, New York, N.Y., 1982), with permission. Courtesy of R. T. Johnson, Johns Hopkins University School of Medicine.

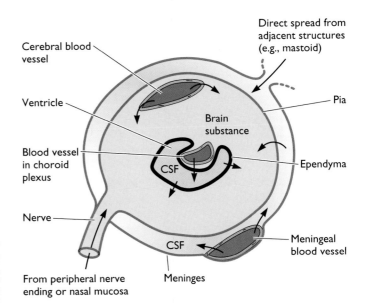

**Figure 17.17 How viruses gain access to the central nervous system.** A summary of the mechanisms by which viruses can enter the brain is shown. Adapted from C. A. Mims et al., *Mims' Pathogenesis of Infectious Disease* (Academic Press, Orlando, Fla., 1995), with permission.

of influenza virus is restricted to the epithelial cells of the upper and lower respiratory tract. The tropism of this virus is thought to be influenced by the limited expression of the protease that processes HA. This serine protease, called tryptase Clara, is secreted by nonciliated Clara cells of the bronchial and bronchiolar epithelia (Fig. 17.18). The purified enzyme can cleave HA and activate HA in virions in vitro. It is also clear that alteration of the HA cleavage site so that it can be recognized by other cellular proteases dramatically influences the tropism of the virus and its subsequent pathogenicity. For example, some highly virulent avian influenza virus strains contain an insertion of multiple basic amino acids at the cleavage site of HA. This new sequence permits cleavage by ubiquitously expressed intracellular proteases such as furins. As a result, these mutant viruses are released in an active configuration and are able to infect many organs of the bird, including the

spleen, liver, lung, kidney, and brain. Naturally occurring mutants of this type cause high mortality in poultry farms. Avian influenza viruses isolated from 16 people in Hong Kong contained similar amino acid substitutions at the HA cleavage site. Indeed, many of these individuals had gastrointestinal, hepatic, and renal as well as respiratory symptoms. A virus with such an HA site alteration had not been previously identified in humans. Isolation of such a virus led to fears that an influenza pandemic was imminent. This concern was based on the knowledge that a pantropic influenza virus strain isolated over 50 years ago possessed a previously unknown mechanism for HA cleavage by ubiquitous proteases (see Box 17.5).

### The Site of Entry Often Establishes the Pathway of Spread

For viruses that spread by neural pathways, the innervation at the site of inoculation determines the neuronal circuits that become infected. Herpes simplex virus damages circumscribed areas of the brain or spinal cord only if the infection site contains axon terminals or dendrites of neurons connected to the particular neurons in the spinal cord or brain. After peripheral inoculation, poliovirus never reaches certain areas of the spinal cord and brain. However, replication at these sites occurs if the virus is placed directly into the tissues. The conclusion is that *accessibility* of susceptible cells, not a lack of susceptibility or permissiveness, can also determine the pathway of infection.

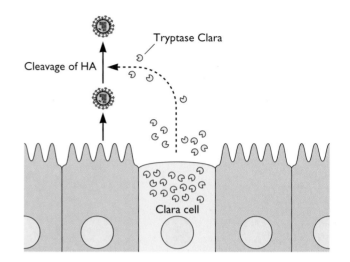

**Figure 17.18 Cleavage of influenza virus HA0 by tryptase Clara.** Influenza viruses are produced by respiratory epithelial cells in humans. These viruses contain the uncleaved form of HA (HA0) and are noninfectious. Clara cells secrete the protease tryptase Clara, which cleaves the HA0 of extracellular virus, thereby rendering it infectious. Adapted from M. Tashiro and R. Rott, *Semin. Virol.* **7:**237–243, 1996, with permission.

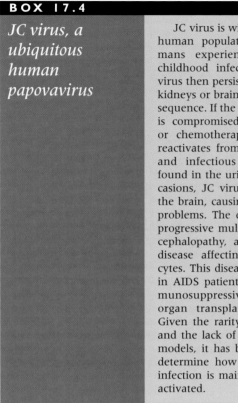

## BOX 17.4

### JC virus, a ubiquitous human papovavirus

JC virus is widespread in the human population. Most humans experience inapparent childhood infections, but the virus then persists for life in the kidneys or brain with little consequence. If the immune system is compromised by pregnancy or chemotherapy, virus often reactivates from kidney tissues and infectious virus can be found in the urine. On rare occasions, JC virus reactivates in the brain, causing more serious problems. The disease is called progressive multifocal leukoencephalopathy, a demyelinating disease affecting oligodendrocytes. This disease is often seen in AIDS patients and after immunosuppressive therapy for organ transplant procedures. Given the rarity of the disease and the lack of suitable animal models, it has been difficult to determine how the permissive infection is maintained and reactivated.

The pathway of spread markedly influences the disease process, but the pathway taken by a virus is not always predictable. As we have discussed, the brain can be infected either by hematogenous or neural spread. When the neurotropic NWS strain of influenza is inoculated intraperitoneally into mice, it is disseminated to the brain by hematogenous spread and replicates in the meninges, choroid plexus, and ependymal cells lining the brain and spinal cord. However, when the virus inoculum is placed in the nose, there is no viremia. Instead, virus enters the brain by neural spread. It first replicates in olfactory epithelia and then enters the sensory neurons that richly endow the nasopharygeal cavity (Fig. 17.12). Virions then can be found in the trigeminal ganglia, neurons in the brain stem that are in contact with trigeminal neurons, and the olfactory bulb in the brain. The experiment with the NWS influenza virus strain makes a significant point: the *site of inoculation* can determine the pathway of spread of a particular virus.

### Virus Shedding and Transmission

The release of infectious viruses from an infected host is called **shedding**. Shedding is usually an absolute requirement for virus survival. The exceptions are direct transmission of viral genomes in the germ lines of their hosts and infections transmitted by cannibalism, such as the prion diseases kuru in humans and mad cow disease in cattle. Processes that promote shedding maintain the virus in a susceptible population. During localized infections, shedding occurs from the primary site of replication at one of the body openings. For viruses that cause disseminated infections, virus release can occur at many sites. As discussed in chapter 15, effective transmission of virions from one host to another depends directly on the concentration of released viruses and the mechanisms by which the virions are introduced into the next host. The shedding of small quantities of virus particles may be irrelevant to transmission, while the shedding of high concentrations of infectious virus may facilitate transmission even when minute quantities of tissue or fluid are involved. For example, the concentration of hepatitis B virus in blood can be so high that as little as a few microliters can be sufficient to initiate an infection. The ability of a virus to survive in the environment also influences transmission.

### Respiratory Secretions

Most viruses that replicate in the respiratory tract are shed into respiratory secretions. Transmission depends on the production of airborne particles, or aerosols, that contain viruses. Aerosols are produced during speaking,

BOX 17.5

*A mechanism for expanding the tropism of influenza virus is revealed by analyzing infections that occurred in 1940*

Until the isolation of the H5N1 virus from 16 persons in Hong Kong, the HA cleavage site mutation that permits cleavage by intracellular furin proteases had not been found in humans. However, the WSN/33 strain of influenza virus, produced in 1940 by passage of a human isolate in mouse brain, is pantropic in mice. Unlike most human influenza virus strains, WSN/33 can replicate in cell culture in the absence of added trypsin because its HA can be cleaved by serum plasmin. It was subsequently found that the NA of WSN/33 is necessary for HA cleavage in cell culture by serum plasmin. This NA contains an in-frame deletion that removes a glycosylation site at residue 46. Recently it was found that the mutant NA can bind plasminogen, sequestering it on the cell surface, where it can be converted to the active form, plasmin (see figure). Plasmin then cleaves HA into HA1 and HA2. Thus, a mutation in NA, not in HA, allows cleavage of HA by a ubiquitous cellular protease. This property may in part explain the pantropic nature of WSN/33.

WSN/33 can replicate in the mouse brain, causing encephalitis. Interestingly, replication in the brain requires the presence of the M and NS genes in addition to the NA from WSN/33; recombinant viruses containing only NA from WSN/33 did not produce encephalitis. Thus, the virulence of WSN/33 is polygenic and cannot be explained solely by the expanded tropism of this strain.

**Goto, H., and Y. Kawaoka.** 1998. A novel mechanism for the acquisition of virulence by a human influenza A virus. *Proc. Natl. Acad. Sci. USA* **95:**10224–10228.

**Taubenberger, J. K.** 1998. Influenza virus hemagglutinin cleavage into HA1 and HA2: no laughing matter. *Proc. Natl. Acad. Sci. USA* **95:**9713–9715.

**Proposed mechanism for activation of plasminogen and cleavage of HA.** (A) Plasminogen binds to NA, which has a lysine at the carboxy terminus. A cellular protein converts plasminogen to the active form, plasmin. Plasmin then cleaves HA0 into HA1 and HA2. (B) When NA does not contain a lysine at the carboxy terminus, plasminogen cannot interact with NA and is not activated to plasmin. Therefore, HA is not cleaved. Adapted from H. Goto and Y. Kawaoka, *Proc. Natl. Acad. Sci. USA* **95:**10224–10228, 1998, with permission.

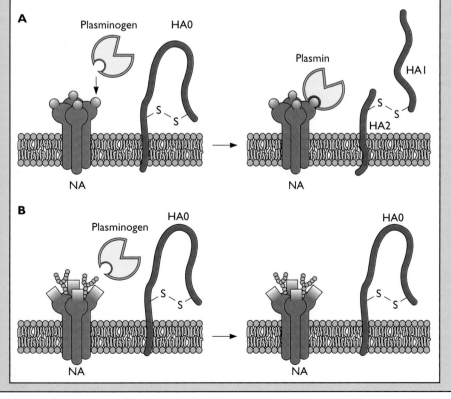

singing, and normal breathing, while coughing produces even more forceful expulsion. Shedding from the nasal cavity requires sneezing and is much more effective if infection induces the production of nasal secretions. Up to 20,000 droplets are produced by a sneeze (in contrast to several hundred produced by coughing), and all may contain rhinovirus if the individual has a common cold. The largest droplets fall to the ground within a few meters, while the smallest droplets evaporate quickly. Many viruses are inactivated by drying (e.g., measles virus, influenza virus, and rhinovirus), and therefore close proximity is required for transmission. The remaining droplets travel a distance related to their size. Of these, the smallest droplets (1 to 4 $\mu$m in diameter) may remain suspended indefinitely because air is continually in motion. Such particles may reach the lower respiratory tract. Nasal secretions also frequently contaminate hands or handkerchiefs. The infection may be transmitted when these objects contact another person's fingers and that person in turn touches his or her nose or conjunctiva. In today's crowded society, the physical proximity of people selects for those viruses that spread most efficiently by this route.

### Saliva

Some viruses that replicate in the lungs, nasal mucosa, or salivary glands are shed into the oral cavity. Transmission may occur through aerosols, as discussed above, or by contaminated fingers, kissing, or spitting. Animals that lick, nibble, and groom may also transmit infections in saliva. Human cytomegalovirus, human herpesviruses 6 and 7, mumps virus, and some retroviruses are known to be transmitted by this route.

### Feces

Enteric and hepatic viruses are shed in the feces and are generally more resistant to inactivation by environmental conditions than viruses released from other sites. An important exception is hepatitis B virus, which is shed in bile into the intestine but is inactivated as a consequence and not transmitted by feces. Instead, hepatitis B virus is transmitted through the blood. Viruses shed in feces usually survive dilution in water as well as drying. Inefficient or no sewage treatment, contaminated irrigation systems, and the use of animal manures are prime sources of fecal contamination of food, water supplies, and living areas. Any one of these conditions provides an efficient mode for continued reentry of these viruses into the alimentary canal of their hosts. Two hundred years ago, such contamination was inevitable in most of the world. Disposal of human feces in the streets was a common practice. With modern sanitation, the flow of feces into human mouths has been largely interrupted in developed countries but still is common throughout the rest of the world. Even so, the closing of beaches because of high coliform counts and the contamination of clam and oyster beds by sewage outflows provide regular reminders of our continuous exposure to enteric and hepatic viruses.

### Blood

Viremias are a common feature of many viral infections, and viremic blood is a prime source for virus transmission. For example, arthropods acquire viruses when they bite viremic hosts. Although hepatitis and AIDS are not spread this way, they can be transmitted by virus-laden blood during transfusions and injections. Infections may be transmitted from viremic blood during coitus or childbirth, and eating raw meat may place viremic blood in contact with the alimentary and respiratory tracts. Health care workers, dentists, and emergency rescue workers are exposed routinely to viremic blood. Indeed, for many of the fatal hemorrhagic fevers caused by viruses (e.g., members of the *Bunyaviridae* and *Filoviridae*), their only mode of transmission to humans is by blood and body fluids. Consequently, health care workers often are the first people to die in an outbreak of such viruses.

### Urine, Semen, Milk, and Skin Lesions

Urine is a common contaminant of food and water supplies and may contain virus particles. The presence of virus particles in the urine is called **viruria**. Hantaviruses and arenaviruses that infect rodents cause persistent viruria. Consequently, humans may be infected by exposure to dust that contains dried urine of infected rodents. A variety of morbilliviruses and picornaviruses replicate in tubular epithelial cells in the kidney and are shed in urine. Lymphocytic choriomeningitis virus replicates in the transitional epithelium of the bladder and is shed directly into the urine. However, for most viruses, viruria is not important for transmission of infection.

Some retroviruses, herpesviruses, and hepatitis B virus are shed in semen and transmitted during coitus. Herpesviruses and papillomaviruses that infect the genital mucosa are shed from lesions and transmitted by genital secretions.

Mouse mammary tumor virus is spread primarily by milk, as are some tick-borne encephalitis viruses. Mumps virus and cytomegalovirus are shed in human milk but probably not often transmitted by this route. Many viruses replicate in the skin, and some can be transmitted from skin lesions, including poxviruses, herpes simplex virus, varicella-zoster virus, papillomaviruses, and Ebola virus.

## Viral Virulence

Once infected, the host may develop a wide range of disease symptoms, as discussed in chapter 14. Virulence refers to the capacity of a virus to cause disease in an infected host. It is a quantitative statement of the degree or extent of pathogenesis. In general, a **virulent virus** causes significant disease, whereas an **avirulent** or **attenuated virus** causes no or reduced disease, respectively.

From the earliest days of experimental virology, it was recognized that viral strains often differ in virulence. Virologists believed that study of these attenuated strains would answer the question of how viruses cause disease. This belief has stood the test of time; indeed, the study of attenuated viruses is still a common strategy. We have learned how to alter viral virulence by direct and indirect methods and have produced viruses of such limited virulence that they can be used as live vaccines (see chapter 19). Today, the methods of recombinant DNA technology permit a more systematic and unbiased analysis by introducing defined mutations into viral genomes so that virulence genes can be identified. The goal of these studies is to understand the mechanisms by which viral and cellular gene products control viral virulence.

## Measuring Viral Virulence

Virulence can be quantified in a number of ways. One approach is to determine the concentration of virus that causes death or disease in 50% of the infected animals. This parameter is called the 50% lethal dose ($LD_{50}$) when lethality is measured, the 50% paralytic dose ($PD_{50}$) when paralysis is measured, and the 50% infectious dose ($ID_{50}$) when infection is measured. Other measurements of virulence include mean time to death or appearance of symptoms and measurement of fever or weight loss. Virus-induced tissue damage can be measured directly by examining histological sections or the blood. For example, the safety of live attenuated poliovirus vaccine is determined by assessing the extent of pathological lesions in the central nervous system in experimentally inoculated monkeys. The reduction in blood concentration of $CD4^+$ lymphocytes caused by human immunodeficiency virus type 1 infection is another example. Indirect measures of virulence include assays for liver enzymes (alanine or aspartate aminotransferase) that are released into the blood as a consequence of virus-induced liver damage.

It is important to recognize that the virulence of a single virus strain may vary dramatically depending on the dose and the route of infection, as well as on the species, age, gender, and susceptibility of the host. The effect of inoculation route on virulence is illustrated in Table 17.5. Clearly, virulence is a relative property. Thus, when the levels of virulence of two very similar viruses are compared the assays must be conducted in an identical fashion. Quantitative terms such as $LD_{50}$ cannot be used to compare virulence among different viruses.

## Genetic Determinants of Virulence

A major goal of animal virology is to identify viral and host genes that control virulence. Once such genes have been identified, defined mutations can be made, gene products can be purified, and hypotheses can be tested. For example, if the molecular mechanisms by which viruses cause disease are known, drugs that block these processes may be synthesized. It may be possible to test the hypothesis that virulence affects the long-term survival of viruses in nature. In addition, the role of virulence genes in establishing the characteristic patterns of viral infections can be determined.

### Alteration of Viral Virulence

To identify viral virulence genes, it is necessary to compare viruses that differ only in their degree of virulence. Before the era of modern virology, several approaches were used to attain this goal. Occasionally, natural isolates of some avirulent viruses could be identified. For example, although wild-type strains of poliovirus type 2 readily cause paralysis after intracerebral inoculation into monkeys, a type 2 isolate from the feces of healthy children was shown to be completely avirulent after inoculation by the same route. Serial passage of virulent viruses either in animal hosts or in cell culture yielded viruses with reduced virulence. In animals, repeated passage by a particular route of inoculation led to the selection of viruses with reduced virulence when inoculated by different routes. An example is the reduction of the parenteral virulence of yellow fever virus in monkeys after repeated passage in mice by intracerebral inoculation. With the development of cell

**Table 17.5** Effect of route of inoculation on viral virulence

| Virus | Plaque-forming units per $LD_{50}$ | | | |
|---|---|---|---|---|
| | Intracerebral infection in suckling mice | Subcutaneous infection in suckling mice | Intracerebral infection in adult mice | Subcutaneous infection in adult mice |
| Wild-type La Crosse virus | ~1 | ~1 | ~1 | ~10 |
| Attenuated La Crosse virus mutant | ~1 | >$10^5$ | >$10^6$ | >$10^7$ |

cultures for use in virology, serial passage of viruses in such cells became a common method for isolating attenuated viruses.

Although these approaches were useful, they were laborious. To facilitate the isolation of attenuated viruses, viral genomes often were mutagenized as described in chapter 2, and the resulting viruses were assayed for virulence in animals. However, controlling the degree of mutagenesis was difficult, and multiple mutations were often introduced. Until the advent of recombinant DNA technology, the ability to introduce and localize mutations in a specific gene was limited. More recently, rapid sequencing of entire viral genomes, polymerase chain reaction (PCR) amplification of selected genomic segments, and site-directed mutagenesis have become routine procedures in the quest to identify and understand viral virulence genes.

### Viral Virulence Genes

Despite modern technological advances, the identification and analysis of virulence genes in a systematic way have not been straightforward. Part of the problem is that there are few simple tissue culture assays for virulence. For example, many of the pathogenic effects promoted by viruses are a result of action of the innate and adaptive immune system, and it is not possible to reproduce the complex actions of the immune system in a tissue culture dish. Another problem confronting investigators taking a reductionist approach is the simple fact that it is not clear what to look for. Consequently, most studies begin with the premise that a virus harboring a defective virulence gene would cause reduced or no disease in an animal host. Such a strategy is not without merit, but there is an element of circular reasoning that can confound the analysis. There is an adage in genetics that says you always get what you select, but you may not get what you want. As discussed earlier, the first virus-encoded genes that affected virulence were identified in attenuated viruses obtained by empirical efforts (see chapter 19 for a discussion of the isolation of live vaccine viruses). Such attenuated viruses were often found to harbor multiple mutations. Often the contributions of the individual mutations to the attenuation phenotype were difficult to ascertain. However, by reversion of point mutations, repairing of deletions, and crossing of mutants and wild-type viruses many relevant defects were identified. Some mutations reduced, eliminated, or augmented protein function, while others affected the binding of transcription, translation, or replication proteins.

The viral genes affecting virulence fall into four general classes (and some may fall into more than one): those that affect the ability of the virus to replicate, defeat the host's defense mechanisms, enable the virus to spread in the host

and among hosts, and produce gene products that are directly toxic. As might be expected, mutations in these genes often have minimal or no effect on replication in cell culture and, as a consequence, often are called "nonessential genes," an exceedingly misleading appellation. Virulence genes require careful definition, as exemplified by the first general class listed above (ability of the virus to replicate). Any defect that impairs virus reproduction or propagation often results in reduced virulence. In many cases, this observation is not particularly insightful or useful. The difficulty in distinguishing an indirect effect due to inefficient replication from an effect directly relevant to disease has plagued the study of viral pathogenesis for years.

A significant drawback to studies of virulence phenotypes of viruses that infect humans is that relevant animal models of disease are not always readily available. Nevertheless, considerable progress has been made in recent years. In the following sections, we will discuss examples of viral virulence genes that can be placed in one of the four general classes. It should be understood, however, that the vast majority of such genes have not been studied sufficiently to place them in one of these categories.

**Genes that affect the ability of the virus to replicate.** Gene that affect viral replication and virulence can be placed in one of two broad classes (Fig. 17.19). Classification is accomplished by analyzing mutants defective in the gene of interest. Viral mutants of the first class exhibit reduced or no replication in the animal host and in many cultured cell types. It is likely that mutations resulting in weakly replicating virus can be identified in any viral gene and that any of these mutants is attenuated when compared to the parental virus. Reduced virulence results from failure to produce sufficient numbers of viruses to cause disease.

The second class of mutants are those exhibiting impaired virulence in animals but no replication defects in cells in culture (except perhaps in cell types representative of the tissue in which disease develops). Such host range mutants should provide valuable insight into the basis of viral virulence because they identify genes specifically required for disease. Host range mutations in a wide variety of viral genes that encode proteins that participate in many of the steps in viral replication have been described. We will discuss three gene families found in alphaherpesviruses as specific examples of this second class of mutants.

A primary requirement for the replication of DNA viruses is access to large pools of deoxyribonucleoside triphosphates. This need poses a significant problem for viruses that replicate in terminally differentiated nonreplicating cells such as neurons. Many DNA viruses with small

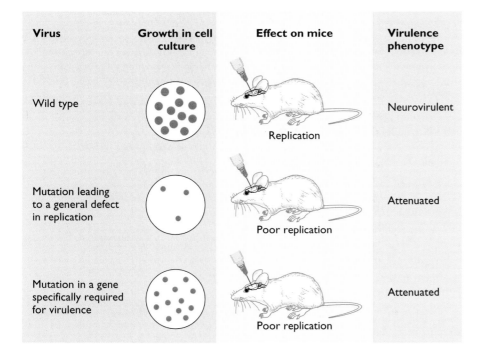

| Virus | Growth in cell culture | Effect on mice | Virulence phenotype |
|---|---|---|---|
| Wild type | | Replication | Neurovirulent |
| Mutation leading to a general defect in replication | | Poor replication | Attenuated |
| Mutation in a gene specifically required for virulence | | Poor replication | Attenuated |

**Figure 17.19 Different types of virulence genes.** Examples of virulence genes that affect viral growth, using intracerebral neurovirulence in adult mice as an example. Wild-type viruses grow well in cell culture; after intracerebral inoculation into the brains of mice, they replicate and are virulent. Mutants with broad defects in replication do not grow well in cultured cells or in the mouse brain and are attenuated. Mutants with a defect in a gene specifically required for virulence replicate well in certain cultured cells, but not in the mouse brain, and are attenuated. Adapted from N. Nathanson (ed.), *Viral Pathogenesis* (Lippincott-Raven Publishers, Philadelphia, Pa., 1997), with permission.

genomes circumvent this problem by encoding proteins that alter the cell cycle such that the cellular DNA synthesis machinery is activated. An additional solution, exemplified by alphaherpesviruses, is to encode enzymes that function in nucleotide metabolism and DNA synthesis, such as thymidine kinase and ribonucleotide reductase. Mutations in these genes often reduce the neurovirulence of herpes simplex virus because the mutants cannot replicate in neurons or in any cell unable to complement the deficiency.

Several attenuated herpesvirus strains harbor mutations in the viral DNA polymerase gene that alter virus replication in neurons but not in other cell types. Such mutants are attenuated after direct inoculation into the brains of mice yet replicate well in the periphery after inoculation into the external ear. Furthermore, transfer of the DNA polymerase gene from a neuroinvasive herpes simplex virus type 1 strain to a nonneuroinvasive herpes simplex virus type 2 strain yielded a herpes simplex virus type 2 recombinant virus that readily spread to the brain after eye infection in mice. In another example, point mutations that affect the helicase activity of the helicase-primase complex result in a virus that cannot replicate in neurons and is attenuated. These observations imply that neuron-specific proteins cooperate with the viral DNA replication machinery to promote DNA synthesis. This requirement must not be as stringent in dividing cells or in cells of nonneuronal origin. Such attenuating mutations are found in other virus families as well. For example, a missense mutation in the RNA polymerase of La Crosse virus, a bunyavirus, resulted in markedly reduced neurovirulence after intracerebral inoculation into adult mice.

Deletion of the herpes simplex virus gene encoding ICP34.5 protein results in a mutant virus so dramatically attenuated that it is difficult to determine an $LD_{50}$ even when it is injected directly into the brain. Such mutants can replicate in some, but not all, cell types in culture. Notably they are unable to grow in postmitotic neurons. The molecular basis for the cell type specificity of ICP34.5 mutants has yet to be determined. It is not yet clear if their lack of virulence is related to the interaction of ICP34.5 protein with components of the interferon-activated Pkr pathway, as described in chapter 11.

**Noncoding sequences that affect the ability of the virus to replicate.** The attenuated strains used for the live Sabin poliovirus vaccine are examples of viruses with attenuating mutations that are not in protein-coding sequences. Each of the three serotypes present in the vaccine contains a mutation in the 5' noncoding region of the viral RNA that impairs the ability of poliovirus to replicate in the brain after intracerebral inoculation. They also reduce the translation of viral mRNA in cultured cells of neuronal origin but not in other cell types. An interesting finding is that attenuated viruses bearing these mutations apparently do not replicate efficiently at the primary site of

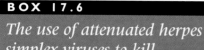

**BOX 17.6**

*The use of attenuated herpes simplex viruses to kill human brain tumors*

Malignant glioma, a common brain tumor, is almost universally fatal despite advances in surgery, radiation, and chemotherapy. Patients rarely survive longer than a year after diagnosis. Several groups have proposed the use of cell-specific replication mutants of herpes simplex virus to kill glioma cells in situ. One such virus under study carries a deletion of the ICP34.5 gene and the large subunit of ribonucleotide reductase. These mutant viruses replicate well in dividing cells such as glioma cells, but not in nondividing cells such as neurons. The theory is that attenuated virus injected into the glioma will replicate and kill the dividing tumor cells but will not replicate or spread in the nondividing neurons.

This idea works in principle: studies of mice have indicated that direct injection of this virus into human gliomas transplanted into mice result in clearing of the tumor. The virus is attenuated and safe: injection of 1 billion virus particles into the brain of *Aotus nancymai* (a monkey highly sensitive to herpes simplex virus brain infections) had no pathogenic effect on the animal. This level of attenuation is remarkable because the virus replicates to reasonable titers in nonneural monkey and human cells.

Several human trials involving only a few individuals are in progress to test safety and dosage. Low concentrations of virus have been injected directly into the brain tumors of patients. If no safety problems are identified after these initial experiments, higher concentrations of virus will be used until a therapeutic effect is attained.

**Andreansky, S., L. Soroceanu, E. R. Flotte, J. Chou, J. M. Markert, G. Y. Gillespie, B. Roizman, and R. J. Whitley.** 1997. Evaluation of genetically engineered herpes simplex viruses as oncolytic agents for human malignant brain tumors. *Cancer Res.* **57:**1502–1509.

**Mineta, T., S. D. Rabkin, T. Yazaki, W. D. Hunter, and R. L. Martuza.** 1995. Attenuated multi-mutated herpes simplex virus-1 for the treatment of malignant gliomas. *Nat. Med.* **1:**938–943.

infection in the gut. Consequently, much less virus is available for hematogenous or neural spread to the brain. Mutations in the 5′ noncoding regions of other picornaviruses also affect virulence in animal models. For example, deletions within the long poly(C) tract within the 5′ noncoding region of mengovirus reduce virulence in mice without affecting viral replication in cell culture.

**Genes that modify the host's defense mechanisms.** The study of viral virulence genes has identified a diverse array of viral proteins that sabotage the body's innate and adaptive defenses. These viral proteins are called **virokines** (secreted proteins that mimic cytokines, growth factors, or similar extracellular immune regulators) or **viroceptors** (homologs of host receptors or cell surface immune molecules). Mutations in genes encoding either class of protein affect virulence, but these genes are not required for growth in cell culture. Most virokines and viroceptors have been discovered in the genomes of large DNA viruses (see Fig. 15.5). These mechanisms have been discussed in detail in chapters 14 and 15.

As discussed in chapter 15, many viruses induce cellular apoptosis, a response that contributes significantly to viral pathogenesis. For example, the pattern of infection of Sindbis virus changes from persistent to acute depending on whether the infected cell can mount an apoptotic response. Interestingly, after viral infection, apoptosis can be either indirect (uninfected cells die) or direct (infected cells die), and both responses influence subsequent pathogenesis. African swine fever, a highly contagious disease of pigs caused by an iridovirus (a double-stranded DNA virus transmitted by insects), is an example of a disease caused by indirect apoptosis. The severe lesions and hemorrhages are striking, but intense destruction of lymphoid tissue is characteristic of the disease. This destruction is the result of apoptosis of uninfected lymphocytes induced by cytokines and apoptotic mediators released from virus-infected macrophages. Direct pathogenic effects of virus-induced apoptosis have been suggested in a model of herpes simplex virus-mediated fulminant hepatitis in which massive hepatocyte death occurs within 24 h of virus injection into the bloodstream. In this disease model, virions in the blood are removed rapidly by cells of the reticuloendothelial system, primarily those found in the liver. If the phagocytic activity of macrophages is impaired, virus replication ensues in the liver followed by massive apoptotic death throughout the organ.

**Genes that enable the virus to spread in the host.** The mutation of some viral genes disrupts the ability of the virus to spread from peripheral sites of inoculation to the organ in which disease is produced. For example, after intramuscular inoculation in mice, reovirus type 1 spreads to the central nervous system through the blood, while type

3 spreads by neural routes. Studies of viral recombinants between types 1 and 3 indicate that the gene encoding the viral outer capsid protein s1, which recognizes the cell receptor, determines the route of spread. Only one plaque-forming unit of the bunyavirus La Crosse virus causes lethal encephalitis after intracerebral injection into mice, but subcutaneous inoculation of over $10^7$ plaque-forming units causes no disease because no viremia is produced (Table 17.5). Poliovirus causes paralysis in only 1 in 150 human infections, perhaps because it does not produce a viremia sufficient to reach the central nervous system.

Viral membrane proteins have been implicated in neuroinvasiveness. For example, the change of a single amino acid in the gD glycoprotein of herpes simplex virus type 1 blocks the ability of the virus to spread to the central nervous system via nerves after footpad inoculation. Similarly, studies of neuroinvasive and nonneuroinvasive strains of bunyaviruses indicate that the G1 glycoprotein is an important determinant of entry into the central nervous system from the periphery. Although it is tempting to speculate that these viral glycoproteins, which participate in entry of the virus into cells, facilitate direct entry into nerve termini, the mechanisms by which they control neuroinvasiveness are unknown. They may also influence the ability of the host's humoral immune system to clear virus from the primary site of infection prior to entry into the peripheral nervous system.

**Toxic viral proteins.** Some viral gene products cause cell injury directly, and mutation of these genes reduces viral virulence. Evidence of their intrinsic activity is usually obtained by adding purified proteins to cultured cells or by expressing the proteins from expression plasmids or viral vectors. The most convincing example of a viral protein with intrinsic toxicity relevant to the viral disease is the NSP4 protein of rotaviruses, which cause gastroenteritis and diarrhea. NSP4 is a nonstructural glycoprotein that participates in the formation of a transient envelope as the viruses bud into the endoplasmic reticulum. When this protein is expressed in insect cells, it causes an increase in intracellular calcium concentration. When fed to young mice, NSP4 causes diarrhea by potentiating chloride secretion. Thus, NSP4 acts as a viral enterotoxin and triggers a signal transduction pathway in the intestinal mucosa.

The SU and TM glycoproteins of human immunodeficiency virus are toxic to cultured cells. Unfortunately the contribution of this toxicity to pathogenesis in humans remains untested: these glycoproteins cannot be deleted without destroying viral infectivity, and there is no animal model for AIDS. TM causes death of cultured cells most likely through alterations in membrane permeability. The addition of SU to cells results in a high influx of calcium.

The human immunodeficiency virus proteins Tat and Nef also have been suggested to be toxic when added to cells (chapter 18).

### Cellular Virulence Genes

Disruption of cellular genes that encode components of the host's immune response—proteins required for T- and B-cell function and cytokines—may have enormous effects on the outcome of viral infection. In some cases viral disease becomes more severe, while in others disease severity is lessened. Such cellular genes can therefore be considered virulence determinants. A consideration of immune response genes can be found in chapter 14.

The host proteins required for viral translation, genome replication, and mRNA synthesis are prime candidates for virulence determinants. Tissue-specific differences in such proteins could in principle influence virus replication. A common approach in studying viral pathogenesis is to construct mice with specific gene disruptions. However, as one might expect, it is unlikely that mice lacking proteins required for translation, mRNA synthesis, and DNA replication can be isolated. As described in chapter 6, replication and translation of RNA viral genomes may require host proteins that serve unrelated functions in the uninfected cell. It may be possible to derive mice lacking such proteins to study viral virulence. Few studies have been done in this largely uncharted aspect of animal virology, an area that may be one of the most interesting yet to be explored in viral pathogenesis.

## Injury Induced by Viruses

The clinical symptoms of viral disease in the host (e.g., fever, tissue damage, aches, pains, nausea) result primarily from the host response to infection. This response is initiated by cell injury caused by virus replication. Cell injury can result from the direct effects of virus replication on the cell, from the consequences of the host's innate and adaptive immune responses, or from both. In this section we summarize the mechanisms of direct and indirect injury of cells by viruses.

## Direct Effects of Primary Infection by Cytolytic Viruses

Infection of cultured cells by many viruses results in visible changes in the cells collectively called cytopathic effect, a topic of chapter 2. Direct alteration of the cell by viral infection can clearly account for some of the damage observed during infections in an animal host. For example, poliovirus induces cytopathic effects in cultured neuronal cell lines; virus-induced lysis of neurons in the central nervous system could account for the paralytic symptoms characteristic of poliomyelitis.

Because virus-induced cytopathic effects are clearly relevant to viral pathogenesis, much effort has been devoted to understanding how viruses alter cell structure and metabolism. One major process producing visible cell damage is apoptosis after viral infection. As described in chapter 15, many viruses encode genes that both induce and block apoptosis, facilitating completion of the infectious cycle and the production of new progeny virions. Apoptosis can be a major component of virulence.

Viral infection may also result in the cessation of essential host processes such as translation, DNA and RNA synthesis, and vesicular transport (Table 17.6). A general consequence of these events may be the increased permeability of cell membranes. One possible consequence would be that lysosomal contents would diffuse into the cytoplasm, resulting in autolytic digestion of the cell. Indeed, the 2B protein of coxsackieviruses enhances membrane permeability and may cause such enzyme leakage from infected cells.

## Pathogenic Effects of Noncytolytic Viruses

Most disease caused by noncytolytic viruses is a consequence of the immune response, as described in the next section. An example of a different mechanism is infection of mice by a strain of lymphocytic choriomeningitis virus. The inoculation of newborn mice produces a persistent infection in which the virus replicates in cells of the pituitary gland that produce growth hormone. The virus does not destroy these cells, but replication leads to a twofold reduction in the synthesis of growth hormone. As a result, the mice fail to grow and perish within a month. Two viral proteins determine the ability to cause growth hormone deficiency: a glycoprotein and a nucleoprotein. The former is presumably required for the virus to enter the cells of the pituitary gland. Nucleoprotein decreases the level of a transcription protein specifically required for synthesis of growth hormone mRNA. Thus, viral infection does not result in cell death but interferes with the synthesis of a critical hormone needed for survival of the host.

## Immunopathology: Indirect Effects of Viral Infection

Disease does not result simply because viruses kill cells. Indeed, most virus-induced pathology is caused by the host's immune system (Table 17.7). This damage is called **immunopathology**, and it may be the price paid by the host to eliminate a viral infection. For noncytocidal viruses it is likely that the immune response is the sole cause of disease. In fact, most viral infections with an immune pathological component are persistent and cause little cell killing. Most immune pathology induced by viruses is caused by T cells, but there are also examples of disease caused by antibodies.

### Immunopathological Lesions

**Lesions associated with CD8$^+$ T cells.** The best-characterized example of CD8$^+$-T-cell-mediated immunopathology occurs during the infection of mice with lymphocytic choriomeningitis virus. This virus is a noncytocidal one that causes tissue damage only in immunocompetent animals. Experiments using adoptive transfer of T-cell subtypes, depletion of cells, and gene knockout

**Table 17.6** Examples of virus-induced alteration of the host

| Virus | Protein(s) | Effect | Target |
|---|---|---|---|
| Poliovirus | 2A$^{pro}$ | Inhibition of 5'-end-dependent translation | eIF-4G |
| | 2B, 3A | Inhibition of endoplasmic reticulum-to-Golgi protein traffic | Unknown |
| | 2C, 2BC | Membranous vesicle accumulation | Unknown |
| | Unknown | Alteration of Map4 by unknown mechanism | MAP4 |
| | 3C | Inhibition of transcription | Tbp, TfIIc complex |
| Sindbis virus | Unknown | Increased plasma membrane permeability | Na,K-ATPase? |
| Paramyxovirus | F | Fusion of cell membranes, syncytium formation | Plasma membrane |
| Adenovirus | E1B-55K, E4-34K | Blocking of accumulation of host mRNA in cytoplasm | Cellular protein involved in mRNA transport? |
| | Unknown | Inhibition of 5'-end-dependent translation | eIF-4E (dephosphorylation) |
| Herpesviruses | *vhs* gene product (a ribonuclease) | Disassembly of polysomes | Host mRNA |
| Herpes simplex virus | ICP27 | Inhibition of transport and processing of host RNA | Unknown |
| Many viruses | Unknown | Depolymerization of cytoskeleton | Actin filaments, microtubules, intermediate filaments |

**Table 17.7**  Examples of virus-induced immunopathology

| Proposed mechanism | Virus(es) |
|---|---|
| CD8$^+$ T cell-mediated | Coxsackievirus B, lymphocytic chorio-meningitis virus, Sin Nombre virus, human immunodeficiency virus, hepatitis B virus |
| CD4$^+$ T cell-mediated | |
| Th1 | Theiler's virus, mouse coronavirus, Semliki Forest virus, measles virus, visna virus, herpes simplex virus |
| Th2 | Respiratory syncytial virus |
| Antibody-mediated | Dengue virus, feline infectious peritonitis virus |

and transgenic mice have clearly shown that tissue damage during infection requires CD8$^+$ T cells. The mechanism by which these cells cause this damage is not clear. It has been suggested that tissue damage is a result of the cytotoxicity of CD8$^+$ T cells. This hypothesis is supported by the finding that knockout mice lacking perforin (chapter 14) develop less severe disease after infection with virus. CD8$^+$ T cells may also release proteins that recruit inflammatory cells to the site of infection, which in turn elaborate proinflammatory cytokines such as interleukin-1 (Il-1), Il-6, and tumor necrosis factors alpha and beta (Tnf-$\alpha$ and -$\beta$).

Liver damage caused by hepatitis B virus also appears to depend on a CD8$^+$ T-cell component. Expression of the hepatitis B virus envelope proteins in transgenic mice has no effect on the animals. When the mice are injected with CD8$^+$, hepatitis B virus-specific cytotoxic T cells, liver lesions resembling those observed in acute human viral hepatitis develop. Cytotoxic lymphocytes first attach to hepatocytes and induce apoptotic cell death. Next, cytokines released by these lymphocytes recruit neutrophils and monocytes, which cause even more extensive cell damage. Death from hepatitis can be prevented by the administration of antibody to gamma interferon or by depletion of macrophages. Thus, CD8$^+$ T cells are required for immune pathology, but tissue damage is largely mediated by nonspecific cytokines and cells recruited to the site of infection. Cytotoxic T lymphocytes must also have access to infected cells to promote damage. When hepatitis B virus proteins are expressed in other tissues, such as kidney or brain, disease does not result when cytotoxic T lymphocytes are injected into the blood. However, disease will occur if the lymphocytes are inoculated directly into the infected organ. Thus, availability of viral antigens to patrolling cytotoxic T lymphocytes is an important determinant of immunopathology.

Myocarditis (inflammation of the heart) caused by coxsackievirus B infection of mice also requires the presence of CD8$^+$ T cells. In particular, perforin, the major cytolytic protein of cytotoxic T lymphocytes, is a major determinant of myocarditis. Mice lacking the perforin gene develop a mild form of heart disease yet are still able to clear the infection. The disease is probably also exacerbated by the action of cytokines such as Tnf-$\alpha$ and Il-1$\beta$. Chemokines, small chemotactic cytokines, also contribute to disease by controlling directional migration of lymphoctyes into infected tissues. For example, mice lacking the chemokine macrophage inflammatory protein 1$\alpha$ (Mip-1$\alpha$) do not develop myocarditis. Inflammation of heart muscle following infection by this virus is an example of the combined action of immune-mediated tissue damage and virus-induced cytopathology. The recently recognized acute and often fatal respiratory disease caused by hantaviruses is characterized by prominent infiltration of CD8$^+$ T cells into the lung.

**Lesions associated with CD4$^+$ T cells.**  CD4$^+$ T lymphocytes elaborate far more cytokines than CD8$^+$ T cells do and recruit and activate many nonspecific effector cells. Such inflammatory reactions are usually called **delayed-type hypersensitivity responses**. Most of the recruited cells are neutrophils and mononuclear cells, which are protective and cause tissue damage. Immunopathology is the result of release of proteolytic enzymes, reactive radicals such as peroxide and nitric oxide (see below), and cytokines such as Tnf-$\alpha$. For noncytopathic persisting viruses, the CD4$^+$-mediated inflammatory reaction is largely immunopathological. Against cytopathic viruses, this response is protective, although there may be cases where immune pathology occurs. As discussed in chapter 14, CD4$^+$ T cells are divided into two classes called Th1 and Th2 depending on which cytokines and chemokines they produce. Th1 cells promote the **cell-mediated proinflammatory response** by causing maturation of immature cytotoxic T lymphocytes, and Th2 cells promote the **antibody response** by causing maturation of immature B cells and resting macrophages.

*CD4$^+$ Th1 T cells.*  There appears to be a crucial role for CD4$^+$ Th1 T cells in central nervous system disease of rodents infected with Theiler's murine encephalomyelitis virus (a picornavirus), mouse coronavirus, or Semliki Forest virus (a togavirus). These viruses persistently infect the central nervous system and cause inflammatory reactions leading to demyelination. In Theiler's virus-infected mice, CD4$^+$ Th1 cells are believed to cause inflammatory responses by the activation of monocytic phagocytes that mediate demyelination of neurons. It is not known how demyelination occurs, but perhaps the activated macrophages release superoxide and nitric oxide radicals in addition to Tnf-α and Il-6. Tnf-β elaborated by Th1 cells may also destroy oligodendrocytes, which are the source of myelin. It is not clear whether demyelination observed in humans after infection with measles, vaccinia, varicella-zoster, rubella, mumps, influenza, or Epstein-Barr virus is a product of similar mechanisms. That similar pathology is caused by so many different viruses tells us that an underlying immune-mediated mechanism is at work.

Herpes stromal keratitis, one of the most common causes of vision impairment in developed countries of the world, is an example of an infection by a cytolytic virus in which the pathology is mainly immune-mediated. In humans, the virus causes lesions on the corneal epithelium, and repeated infections result in opacity and reduced vision. Studies of a mouse model for this disease have demonstrated the importance of CD4+ Th1 T cells. Mice lacking T cells do not develop ocular disease following infection. Lymphocytes isolated from the corneas of mice with active disease are CD4+ T cells that produce cytokines typical of the Th1 subset. Interestingly, viral replication occurs in the corneal epithelium, while inflammation is restricted to the stroma. Moreover, the expression of viral genes and proteins has ceased by the time that CD4+ T cells attack the stroma. These and other observations have led to the belief that herpes simplex virus infection may induce the exposure of a corneal autoantigen which then perpetuates the inflammatory reaction. Herpes stromal keratitis may therefore be an autoimmune disease (see below) in which the virus acts as a "hit-and-run" agent.

*CD4+ Th2 cells.* Disease caused by respiratory syncytial virus appears to be mediated by cytokines produced by CD4+ Th2 cells. This virus, which does not cause extensive cytopathic effect, is an important cause of lower respiratory tract disease in infants and the elderly. In immunosuppressed mice, lesions of the respiratory tract are minor but become severe after transfer of virus antigen-specific T cells, particularly CD4+ Th2 cells. The mechanisms that mediate Th2-mediated tissue damage in the lung are not understood. Lesions in the respiratory tract contain many eosinophils, which may be responsible for pathology. One likely possibility is that these eosinophils are recruited by products of CD4+ Th2 T cells.

The previous discussion focused on the idea that an inappropriate Th1 or Th2 response can have pathogenic effects. Recent experiments indicate that it may be possible to alter the balance of Th1 or Th2 CD4+ T-cell responses by vaccination. This concept is being exploited not only to produce safe vaccines but also to facilitate analysis of immunopathogenic responses. For example, respiratory syncytial virus causes respiratory infections that can be prevented by vaccination in young children. Children who received a live virus vaccine were protected and suffered

no ill effects. However, when children were vaccinated with a formalin-inactivated whole-virus vaccine, they were not protected from subsequent infection. Instead, on infection, they developed an atypically severe disease characterized by increased infiltration of eosinophils into the lungs. The atypical augmented disease experienced in children after immunization with inactivated virus appears to be caused by CD4+ Th2 cells. Current vaccination efforts are based on preparations that induce only the Th1 response, which does not cause immunopathological disease.

### Immunopathological Lesions Associated with B Cells

When extensive virus replication occurs at sites inaccessible to the immune system or continues in the presence of an inadequate immune response, extensive formation of virus-antibody complexes occurs. Such complexes are not efficiently cleared by the reticuloendothelial system and circulate in the blood. They become deposited in the smallest capillaries and cause lesions that are exacerbated when the complement system is activated (Fig. 17.20). Deposition of these immune complexes in blood vessels, kidney, and brain may result in vasculitis, glomerulonephritis, and mental confusion. Such an immune complex disease was first described in mice infected with lymphocytic choriomeningitis virus. Although immune complexes have been demonstrated in humans, viral antigen has been found in the complexes only in hepatitis B virus infections.

Antibodies may also cause an enhancement of viral infection. This mechanism probably accounts for the pathogenesis of dengue hemorrhagic fever. This disease is transmitted by mosquitoes and is endemic in the Caribbean, Central and South America, Africa, and Southeast Asia, where billions of people are at risk. The primary infection is usually asymptomatic but may result in an acute febrile illness with severe headache, back and limb pain, and a rash. It is normally self-limiting, and patients recover in 7 to 10 days. There are four serotypes of the virus, and although patients produce antibodies to the infecting serotype, these are not protective against any of the other three serotypes. After infection by another serotype of dengue virus, nonprotective antibodies bind the virus and *facilitate* uptake into Fc-receptor-expressing peripheral blood monocytes that normally are not susceptible to infection. Consequently, the infected monocytes produce proinflammatory cytokines, which in turn stimulate T cells to produce more cytokines. This vicious cycle results in high concentrations of cytokines and other chemical mediators that are believed to trigger the plasma leakage and hemorrhage characteristic of dengue hemorrhagic fever. There may be so much internal bleeding that the often fatal dengue shock syndrome results. Dengue hemorrhagic

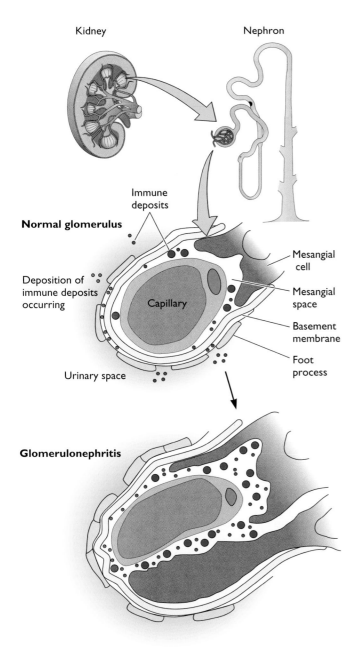

**Figure 17.20 Deposition of immune complexes in the kidney leading to glomerulonephritis.** (Top) Normal glomerulus and its location in the nephron and kidney. Red dots are immune complexes. The smaller complexes pass to the urine, and the larger ones are retained at the basement membrane. (Bottom) Glomerulonephritis. Complexes have been deposited in the mesangial space and around the endothelial cell. The function of the mesangial cell is to remove complexes from the kidney. In glomerulonephritis, the mesangial cells enlarge into the subepithelial space. This results in constriction of the glomerular capillary, and foot processes of the endothelial cells fuse. The basement membrane becomes leaky, filtering is blocked, and glomerular function becomes impaired, resulting in failure to produce urine. Adapted from C. A. Mims et al., *Mims' Pathogenesis of Infectious Disease* (Academic Press, Orlando, Fla., 1995), with permission.

fever occurs in approximately 1 in 14,000 primary infections. However, after secondary infection by a second dengue virus serotype, the incidence of hemorrhagic fever increases dramatically to 1 in 90 infections and the shock syndrome is seen in as many as 1 in 50 infections.

### Autoimmune Diseases

Human autoimmune diseases result when the immune response is directed against host tissues (often described as "breaking immune tolerance"). Viral infection can trigger potent autoimmune responses in laboratory animals, but it has been difficult to find evidence that viral infections contribute to human autoimmune disease. One model for the role of viruses in autoimmune disease posits that replication at an anatomically sequestered site leads to the release of self-antigens against which immune responses are then generated. A modified version of this hypothesis proposes that viral infection leads to exposure of cellular determinants normally hidden from the immune system and thus not recognized as self-antigens. It has been suggested that the unmasking of such antigens might be caused by cytokines or even virus-antibody complexes that modulate the activity of proteases in antigen-presenting cells to liberate cryptic peptides. Another possibility is that elaboration of cytokines from cells in the area of a viral infection might render host membrane proteins autoreactive by altering their surface expression.

A popular hypothesis for virus-induced autoimmunity states that viruses share antigenic determinants with host tissues. This idea is often called **molecular mimicry**. Infection leads to the elaboration of immune responses against such shared determinants, resulting in an autoreactive response. Although many peptide sequences are shared between viral and host proteins, direct evidence for this popular theory has been difficult to obtain. A set of experiments with herpes simplex virus has provided new impetus, however. As discussed earlier, herpes simplex virus infection of the cornea produces stromal keratitis, a disease in which the inflammatory response that follows infection results in T-cell-mediated destruction of corneal tissue, culminating in blindness. The destructive T cells apparently recognize a host antigen on corneal tissue. A herpes simplex virion protein called UL6 contains seven of eight consecutive amino acids identical or similar to a stretch of amino acids in mouse IgG2 antibody, and this sequence is also thought to be present on a corneal protein. Significantly, virus mutants that lack the cross-reactive UL6 epitope are unable to induce the autoimmune disease in mice.

Despite recent progress with the infectious stromal keratitis model, it remains difficult to study the events that trigger human autoimmune diseases because of the long lag period that often occurs before the onset of clinical symptoms. To circumvent this problem, transgenic mouse models in which the products of foreign genes are expressed as self antigens have been established. This approach permits a precise study of how the autoreactive immune response is initiated and what cell types and soluble mediators are involved (Box 17.7).

### Superantigens "Short-Circuit" the Immune System

Some viral proteins are extremely powerful T-cell mitogens known as **superantigens**. These proteins bind to major histocompatibility complex class II molecules on antigen-presenting cells and interact with the Vβ chain of the T-cell receptor. Approximately 2 to 20% of all T cells express the particular Vβ chain that binds the superantigen. Thus, instead of being presented to the T-cell receptor by the major histocompatibility complex class II molecule in the form of peptides, superantigen binding to the Vβ chain short-circuits the interaction. As a result, rather than activation of a small, specific subset of T cells (e.g., only 0.001 to 0.01% of T cells usually respond to a given antigen), all subsets of T cells expressing the Vβ chain to which the superantigen binds are activated and proliferate. Superantigens clearly interfere with a coordinated immune response and divert the host's defenses.

All known superantigens are microbial products. Many are produced by viruses, such as mouse mammary tumor virus, rabies virus, Epstein-Barr virus, and cytomegalo-virus, and are thought to play essential roles in viral pathogenesis. For example, mouse mammary tumor virus, which is transmitted from mother to offspring via milk, replicates poorly in most somatic tissues. In the gut of newly infected animals, the virus infects B cells, in which the superantigen is produced and presented to T cells expressing the appropriate T-cell receptor Vβ chain. The large number of activated T cells produce growth factors and other molecules that stimulate B cells, allowing efficient viral replication. The B cells then carry the virus to the mammary gland. The viral superantigen is therefore required to stimulate the immune system to permit migration of the virus from the gut to the mammary gland. Mouse mammary tumor viral superantigen, which is unrelated to the other viral proteins, is encoded in the U3 region of the 3' long terminal repeat. Deletion of this gene leads to inefficient replication and spread in the mouse and minimal transmission to offspring.

### Mechanisms Associated with Free Radicals

Two free radicals, superoxide ($O_2^-$) and nitric oxide (NO), are produced during the inflammatory response and may play important roles in virus-induced pathology. Superoxide is produced by the enzyme xanthine oxidase present in phagocytes (Fig. 17.21). The production of $O_2^-$ is significantly increased in the lungs of mice infected with influenza virus or cytomegalovirus. Inhibition of xanthine

---

**BOX 17.7**

*Transgenic mouse models of virus-induced autoimmune disease*

Transgenic mice that express proteins of lymphocytic choriomeningitis virus in β cells of the pancreas or oligodendrocytes of the brain develop insulin-dependent diabetes mellitus and central nervous system demyelinating disease, respectively. The viral transgene, present in the mouse throughout development, is a self-antigen. Infection with virus leads to an immune response in which the self-antigen is recognized, leading to disease. Such transgenic models have the advantage that the properties that lead to breaking of self-tolerance against the viral protein can be carefully dissected.

Among transgenic mice that express lymphocytic choriomeningitis virus proteins in the pancreas, viral infection causes diabetes in more than 90% of the animals. CD8+ lymphocytes are essential for this immunopathology. Such virus-specific lymphocytes are present in the peripheries of uninfected mice, as determined from in vitro studies. The mechanisms that maintain unresponsiveness to self-antigens and prevent autoimmunity can be elucidated with such transgenic models. Whether tolerance to self-antigens is broken depends on the number of autoreactive T cells and their affinity for peptides presented by the class II major histocompatibility complex, the number of memory cells induced by infection, and the cytokine milieu.

These transgenic mouse models also can be used to explore ways to prevent autoimmune disease. For example, virus-induced diabetes is prevented when mice are first immunized with virus-specific peptides to make antiself T cells tolerant. Immunodominant peptides can also be administered to mice to modulate cytotoxic T-lymphocyte responses that mediate autoimmune disease. Such experiments provide guidelines for potential therapy of human autoimmune diseases.

**von Herrath, M. G., and M. B. A. Oldstone.** 1996. Virus-induced autoimmune disease. *Curr. Opin. Immunol.* **8:**878–885.

oxidase by allopurinol protects mice from virus-induced lethality. However, $O_2^-$ is not toxic for some cells and viruses, and its effects might be the result of formation of peroxynitrite ($ONOO^-$) through interaction with nitric oxide.

Nitric oxide is produced in abundance in virus-infected tissues during inflammation as part of the innate immune response. This gas has been shown to inhibit the replication of many viruses in cultured cells and in animal models. It probably acts intracellularly to inhibit virus replication, but the molecular sites of action are not well understood. Nitric oxide is produced by three different interferon-inducible isoforms of nitric oxide synthase. While low concentrations of NO have a protective effect, high concentrations or prolonged production have the potential to contribute to tissue damage, as shown by treating infected animals with inhibitors of nitric oxide synthase. Although NO is relatively inert, it rapidly reacts with $O_2^-$ to form peroxynitrite ($ONOO^-$), which is much more reactive than either radical and may be responsible for cytotoxic effects on cells (Fig. 17.21). In the central nervous system, NO is produced by activated astrocytes and microglia and may be directly toxic to oligodendrocytes and neurons.

## Host Susceptibility to Viral Disease

When viruses are introduced into populations of humans or animals, many different outcomes are possible. Some hosts may be highly resistant to infection, some may become infected, and still others may fall in the spectrum between the two extremes. Those who become infected may develop disease ranging from asymptomatic to fatal. Thus, susceptibility to infection and disease vary independently. Understanding the basis for such variation is important because it may suggest methods for preventing viral disease. Such studies have revealed that both host genes and nongenetic factors control how populations respond to viral infections.

### Genetic Determinants of Susceptibility

The genes that control the humoral and cellular immune responses are determinants of susceptibility to viral infections. The class I and class II major histocompatibility complex proteins, for which there are many genes, control the presentation of foreign peptides to T cells. The ability of these proteins to interact with peptides derived from viral proteins determines in part how efficiently the infection is cleared. Diversity in major histocompatibility protein genes make it more likely that there will be a suitable molecule to present peptides for a given infectious agent. Populations from isolated islands have less polymorphism in these genes, which may account for their greater susceptibility to infection.

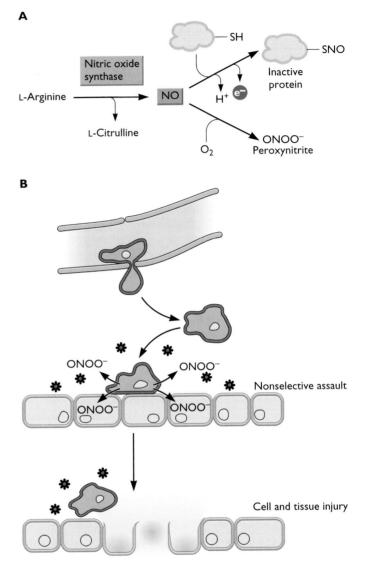

**Figure 17.21 Consequences of nitric oxide production.** (A) Formation of NO and $ONOO^-$. Nitric oxide, which is produced from L-arginine by nitric oxide synthase, reacts with proteins, inactivating them, or with $O_2^-$, forming peroxynitrite, $ONOO^-$. Adapted from C. S. Reiss and T. Komatsu, *J. Viral.* **72:**4547–4551, 1998, with permission. (B) Biological effects of NO/$O_2^-$/$ONOO^-$. Phagocytes move into the area of active virus replication and produce NO/$O_2^-$/$ONOO^-$, which destroys cells, causing tissue injury but perhaps limiting virus yields. Adapted from T. Akaike et al., *Proc. Soc. Exp. Biol. Med.* **217:**64–73, 1998, with permission.

Genes other than those that control the immune response are likely to govern susceptibility to viral infections. Several examples have been studied in animal systems. The best characterized is the *mx* gene of mice, which confers resistance to infection with influenza A virus. The *mx* gene encodes a GTPase that inhibits influenza virus replication as discussed in chapter 14. Resistance of mice to flavivirus disease has been mapped to the *flv* locus. Although these viruses replicate in resistant mice, the virus titers are 1,000 to 10,000-fold lower than in susceptible animals, and the infection is cleared before disease symptoms develop. The product(s) of the *flv* locus is not yet known but appears to inhibit viral RNA replication. The *fv-1* locus confers resistance of mice to infection with Moloney murine leukemia virus. This gene has been shown to encode the Gag protein of a defective endogenous provirus. It is believed that the *fv-1* gene product in some way blocks viral infection by interfering with receptor function.

*mx* and *flv* genes were identified in animal models but have not been shown to play roles in human infections. The best examples of human genes influencing virus susceptibility are those encoding the chemokine receptors that serve as cofactors for entry of human immunodeficiency virus type 1 into cells, as discussed in chapter 18.

## Nongenetic Determinants of Susceptibility

The age of the host plays an important role in determining the outcome of viral infections. Some infections become milder with advancing age. The increased susceptibility of infants and young children is likely due to the immaturity of their immune responses. Sindbis virus causes lethal encephalitis in newborn but not in older mice. This virus kills cells by inducing apoptosis, which can be blocked by the oncogene *bcl-2*. The resistance of older mice may be due to the expression of higher levels of this protein. Although young animals may have an immature immune response and become infected more frequently, they also have greater freedom from immunopathology. Thus, intracerebral inoculation of lymphocytic choriomeningitis virus in adult mice is lethal because of the T-cell response, while infant mice survive because of their weaker cell-mediated response. In infants, infections with enteric coronaviruses are severe because the alimentary canal is not fully active. In addition, the pH tends toward neutrality and digestive enzymes are not available.

Other physiological and physical differences may explain age-dependent variation in susceptibility. As animals age, their alveoli become less elastic, the respiratory muscles weaken, and the cough reflex is diminished. These changes may explain in part why elderly people have increased susceptibility to respiratory infections. Respiratory syncytial virus causes severe lower respiratory tract infections in infants but only mild upper tract infections in adults. It is not known whether the difference is due to host defenses or to variations in the susceptibility of cells to viral infection.

Some viral infections are less severe in children than in adults. An example is poliomyelitis, which is more frequently asymptomatic in childhood. Paralysis is more commonly observed when infection occurs later in life. Epidemics of paralytic poliomyelitis were unheard of until the early 1900s, when the development of modern sanitation delayed infection until later in life. The basis for this difference is not known.

Males are more susceptible to viral infections than females, but the difference is very slight and the reasons are not understood. Hormonal differences, which affect the immune system, may be in part responsible. Pregnant women are more susceptible to infectious disease than nonpregnant women, probably for similar reasons. Hepatitis A, B, and E are more lethal, and paralytic poliomyelitis is more common, in pregnant women than in others.

Malnutrition increases susceptibility to infection because the physical barriers to infection, as well as the immune response, are compromised. An example is the increased susceptibility to measles of children with protein deficiency. For this reason the disease is 300 times more lethal in developing countries than in Europe and North America. When children are malnourished, Koplik spots become massive ulcers, the skin rash is much worse, and lethality may approach 10 to 50%. Such severe measles infections are observed in children in tropical Africa and in aboriginal children in Australia.

Many other parameters influence susceptibility to infection. Corticosteroid hormones have large effects on infection because they are essential for the body's response to the stress of infection. These hormones have an anti-inflammatory effect which is believed to limit tissue damage in infections. Cigarette smoking is believed to increase susceptibility to respiratory infections, but the data are conflicting. Increased susceptibility to infection is correlated with a poor mental state such as occurs in stressful life situations (e.g., a death in the family, injury, or the loss of a job). Our mothers told us that exposure to changes in temperature increases susceptibility to infection. Fortunately, studies with rhinovirus have failed to reveal any relationship between exposure to low temperature and increased susceptibility to the common cold.

## Perspectives

Most of this chapter has been concerned with descriptions: how viruses enter the host, how they spread, and how they cause disease. Our intention is to stress the importance of recognizing and understanding the consequences of viral infections. Without this understanding, it is impossible to define the underlying molecular mechanisms, the quest of the molecular virologist. While the field of viral pathogenesis is almost 100 years old, the viral and host genes that control this process are only now being identified. Even though many genes that affect pathogenesis have been identified, this information rarely provides more than a glimpse of the molecular mechanisms responsible. For example, knowing that a gene controls the spread of a virus by the blood or by the nerves does not answer the question of how the gene product determines which route will be taken. Although it is difficult to obtain such mechanistic knowledge, many investigators believe that viral pathogenesis is the most exciting field in virology today simply because more fundamental questions remain to be answered than in any other area. The challenge is to build on our understanding of viral replication and host defenses embodied in the previous 16 chapters of this book to provide a comprehensive molecular description of how viruses cause disease. One obvious consequence of such knowledge will be new approaches for preventing, treating, and curing viral disease.

# References

## Monographs

Becker, Y., and G. Darai (ed.). 1994. *Frontiers of Virology*, vol. 3. *Pathogenicity of Human Herpesviruses due to Specific Pathogenicity Genes.* Springer-Verlag, Berlin, Germany.

Johnson, R. T. 1982. *Viral Infections of the Nervous System.* Raven Press, New York, N.Y.

Mims, C. A., N. Dimmock, A. Nash, and A. Stephen. 1995. *Mims Pathogenesis of Infectious Disease*, 4th ed. Academic Press, Orlando, Fla.

Nathanson, N. (ed.). 1997. *Viral Pathogenesis.* Lippincott-Raven Publishers, Philadelphia, Pa.

Notkins, A. L., and M. B. A. Oldstone (ed.). 1984. *Concepts in Viral Pathogenesis.* Springer-Verlag, New York, N.Y.

Richman, D. D., R. J. Whitley, and F. G. Hayden (ed.) 1997. *Clinical Virology.* Churchill Livingstone, Inc., New York, N.Y.

Roizman, B., R. J. Whitley, and C. Lopez (ed). 1993. *The Human Herpesviruses.* Raven Press, New York, N.Y.

Scheld, W. M., R. J. Whitley, and D. T. Durack (ed.). 1991. *Infections of the Central Nervous System.* Raven Press, New York, N.Y.

Wright, P. F. 1996. *Seminars in Virology*, vol. 7. *Viral Pathogenesis.* The W. B. Saunders Co., Philadelphia, Pa.

## Review Articles

Bielefeldt-Ohmann, H. 1997. Pathogenesis of dengue virus diseases: missing pieces of the jigsaw. *Trends Microbiol.* **5**:409–413.

Biondi, M., and L. Zannino. 1997. Psychological stress neuroimmunomodulation and susceptibility to infectious diseases in animals and man: a review. *Psychother. Psychosom.* **66**:3–26.

Enquist, L. W., P. J. Husak, B. W. Banfield, and G. A. Smith. 1999. Infection and spread of alphaherpesviruses in the nervous system. *Adv. Virus Res.* **51**:237–347.

Graham, B. S. 1996. Immunological determinants of disease caused by respiratory syncytial virus. *Trends Microbiol.* **4**:290–294.

Kennedy, P. G. E. 1992. Molecular studies of viral pathogenesis in the central nervous system; the Linacre lecture 1991. *J. R. Coll. Physicians Lond.* **26**:204–214.

Lane, T. E., and M. J. Buchmeier. 1997. Murine coronavirus infection: a paradigm for virus-induced demyelinating disease. *Trends Microbiol.* **5**:9–14.

Oldstone, M. B. A. 1996. Principles of viral pathogenesis. *Cell* **87**:799–801.

Reiss, C. S., and T. Komatsu. 1998. Does nitric oxide play a critical role in viral infections? *J. Virol.* **72**:4547–4551.

Rouse, B. T. 1996. Virus-induced immunopathology. *Adv. Virus Res.* **47**:353–376.

Simas, J. P., and S. Efstathiou. 1998. Murine gammaherpesvirus 68: a model for the study of gammaherpesvirus pathogenesis. *Trends Microbiol.* **7**:276–282.

Streilein, J. W., M. R. Dana, and B. R. Ksander. 1997. Immunity causing blindness: five different paths to herpes stromal keratitis. *Immunol. Today* **18**:443–449.

Takada, A., and Y. Kawaoka. 1998. Pathogenesis of Ebola virus infection: recent insights. *Trends Microbiol.* **6**:258–259.

Tracy, S., N. M. Chapman, J. Romero, and A. I. Ramsingh. 1996. Genetics of coxsackievirus B cardiovirulence and inflammatory heart muscle disease. *Trends Microbiol.* **4**:175–179.

## Papers of Special Interest

Chou, J., E. R. Kern, R. J. Whitley, and B. Roizman. 1990. Mapping of herpes simplex virus-1 neurovirulence to $\gamma_1 34.5$, a gene nonessential for growth in culture. *Science* **250**:1262–1266.

Dong, Y., C. Q.-Y. Zeng, J. M. Ball, M. K. Estes, and A. P. Morris. 1997. The rotavirus entertoxin NSP4 mobilizes intracellular calcium in human intestinal cells by stimulating phospholipase C-mediated inositol 1,4,5-trisphosphate production. *Proc. Natl. Acad. Sci. USA* **94**:3960–3965.

Dropulic, L. K., J. M. Hardwick, and D. E. Griffin. 1997. A single amino acid change in E2 glycoprotein of Sindbis virus confers neurovirulence by altering an early step of virus replication. *J. Virol.* **71**:6100–6105.

Evans, C. F., M. S. Horwitz, M. V. Hobbs, and M. B. A. Oldstone. 1996. Viral infection of transgenic mice expressing a viral protein in oligodendrocytes leads to chronic central nervous system autoimmune disease. *J. Exp. Med.* **184**:2371–2384.

Gebhard, J. R., C. M. Perry, S. Harkins, T. Lane, I. Mena, V. C. Asensio, I. L. Campbell, and J. L. Whitton. 1998. Coxsackievirus B3-induced myocarditis: perforin exacerbates disease, but plays no detectable role in virus clearance. *Am. J. Pathol.* **153**:417–428.

Haller, B. L., M. L. Barkon, G. P. Vogler, and H. W. Virgin IV. 1995. Genetic mapping of reovirus virulence and organ tropism in severe combined immunodeficient mice: organ-specific virulence genes. *J. Virol.* **69**:357–364.

Irie, H., H. Koyama, H. Kubo, A. Fukuda, K. Aita, T. Koike, A. Yoshimura, T. Yoshida, J. Shiga, and T. Hill. 1998. Herpes simplex virus hepatitis in macrophage-depleted mice: the role of massive, apoptotic cell death in pathogenesis. *J. Gen. Virol.* **79**:1225–1231.

Levine, B., J. E. Goldman, H. H. Jian, D. E. Griffin, and J. M. Hardwick. 1996. *Bcl-2* protects mice against fatal alphavirus encephalitis. *Proc. Natl. Acad. Sci. USA* **93**:4810–4815.

Morimoto, K., D. C. Hooper, H. Carbaugh, Z. F. Fu, H. Koprowski, and B. Dietzschold. 1998. Rabies virus quasispecies: implications for pathogenesis. *Proc. Natl. Acad. Sci. USA* **95**:3152–3156.

Oura, C. A. L., P. P. Powell, and R. M. E. Parkhouse. 1998. African swine fever: a disease characterized by apoptosis. *J. Gen. Virol.* **79**:1427–1438.

Posavad, C. M., D. M. Koelle, M. F. Shaughnessy, and L. Corey. 1997. Severe genital herpes infections in HIV-infected individuals with impaired herpes simplex virus-specific CD8[+] cytotoxic T lymphocyte responses. *Proc. Natl. Acad. Sci. USA* **94**:10289–10294.

Price, G. E., H. Smith, and C. Sweet. 1997. Differential induction of cytotoxicity and apoptosis by influenza virus strains of differing virulence. *J. Gen. Virol.* **78**:2821–2829.

Ramsingh, A. I., W. T. Lee, D. N. Collins, and L. E. Armstrong. 1997. Differential recruitment of B and T cells in coxsackievirus B4-induced pancreatitis is influenced by a capsid protein. *J. Virol.* **71**:8690–8697.

Ren, R., and V. Racaniello. 1992. Human poliovirus receptor gene expression and poliovirus tissue tropism in transgenic mice. *J. Virol.* **66**:296–304.

Tripp, R. A., A. M. Hamilton-Easton, R. D. Cardin, P. Nguyen, F. G. Behm, D. L. Woodland, P. C. Doherty, and M. A. Blackman. 1997. Pathogenesis of an infectious mononucleosis-like disease induced by a murine γ-herpesvirus: role for a viral superantigen? *J. Exp. Med.* **185**:1641–1650.

Tufariello, J., S. Cho, and M. S. Horwitz. 1994. The adenovirus E3 14.7-kilodalton protein which inhibits cytolysis by tumor necrosis factor increases the virulence of vaccinia virus in a murine pneumonia model. *J. Virol.* **68**:453–462.

Tyler, K. L., H. W. Virgin IV, R. Bassel-Duby, and B. N. Fields. 1989. Antibody inhibits defined stages in the pathogenesis of reovirus serotype-3 infection of the central nervous system. *J. Exp. Med.* **170**:887–900.

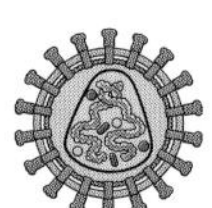

# 18

# Multiple Facets of Human Immunodeficiency Virus Pathogenicity

*Nature is not human-hearted.*

LAO TZU

*Tao Te Ching*

## Introduction

### Worldwide Scope of the Problem

Acquired immunodeficiency syndrome (AIDS) is the name given to end-stage disease caused by infection with human immunodeficiency virus (HIV). By almost any criteria, HIV qualifies as one of the world's deadliest scourges. First recognized as a clinical entity in 1981, by 1992 AIDS had become the major cause of death in individuals 25 to 44 years of age in the United States. The current worldwide statistics are even more staggering, with the developing countries of Africa and parts of Asia being especially hard-hit (Table 18.1). As of December 1998, 13.9 million people in the world had already died of AIDS. An end-of-year report from the United Nations' AIDS program estimated the number of new HIV infections in 1998 to be 5.8 million, bringing the total number of infected people worldwide to 33.4 million. This number corresponds to approximately 1 in every 100 adults, ages 15 to 49, in the world's population. In certain parts of sub-Saharan Africa, 25 to 30% of the adult population had become infected with the virus, and 2 million people in the region had died of AIDS. It is estimated that by 2000 one-third of the children under 15 years of age in these areas will have lost one or both parents to AIDS. In these places a whole generation of human beings will have succumbed to this fatal disease.

At its present rate, HIV is spreading faster in the world than any known chronic infectious agent in the last half-century. The recent clinical emergence is likely the consequence of a number of political, economic, and societal changes, including the breakdown of national borders, economic distress with the migration of large populations, and the ease and frequency of travel throughout the world.

HIV has become the most intensely studied of any infectious agent. Research on the virus has contributed to our understanding of AIDS and related veterinary diseases and provided new insights into retrovirology, cellular biology, and the functioning of the immune system. This chapter describes many

**Table 18.1**    Numbers of people infected with HIV[a]

| Region | No. of infected persons |
|---|---|
| Sub-Saharan Africa | 22,500,000 |
| South and Southeast Asia | 6,700,000 |
| Latin America | 1,400,000 |
| North America | 890,000 |
| Western Europe | 500,000 |

[a] Data from the Joint United Nations Programme on HIV/AIDS as of December 1998.

of the facets of HIV-induced pathogenesis and what has been learned through its analysis. The complexities involved will illustrate the enormous scope of the challenges faced by biomedical researchers and physicians in their efforts to control this virus, which strikes at the very heart of the body's defense system.

## HIV Is a Lentivirus

### Discovery and Characterization

The first clue to the etiology of AIDS came in 1983 upon isolation of a retrovirus from the lymph node of a patient with lymphadenopathy by Françoise Barré-Sinoussi, in the laboratory of Jean-Claude Chermann, collaborating with a group headed by Luc Montagnier at the Pasteur Institute in Paris. Although not fully appreciated initially, the significance of this finding became apparent in the following year with the isolation of a cytopathic, T-cell-trophic retrovirus from combined blood cells of AIDS patients by Robert Gallo and coworkers at the U.S. National Institutes of Health and of a similar retrovirus from blood cells of a single AIDS patient by Jay Levy and coworkers at the University of California, San Francisco. Although the Gallo isolate was later shown to be a contaminant of virus received from the Pasteur Institute, the Levy virus and subsequent isolates in the Gallo laboratory were unique. As commonly happens, each laboratory gave its isolate a different name: LAV, lymphadenopathy-associated virus, by the Montagnier group; HTLV-III, human T-cell lymphotropic virus type III, by the Gallo group; and ARV, AIDS-associated retrovirus, by the Levy group. Electron microscopic examination revealed that these viruses were morphologically similar to a known group of retroviruses, lentiviruses, and further characterization confirmed this relationship. In 1986, the International Committee on Taxonomy of Viruses recommended the current name, human immunodeficiency virus (HIV).

Lentiviruses comprise a separate genus of the family *Retroviridae* (Table 18.2). Of the lentiviruses, equine infectious anemia virus was one of the first viruses to be identified in nature. Discovered in 1904, this virus causes episodic autoimmune hemolytic anemia in horses. Lentiviruses of sheep (visna/maedi virus) and goats (caprine arthritis-encephalitis virus) have also been known for many years. Like the equine lentivirus, these viruses are associated with long incubation periods, a characteristic that has earned them the name **slow viruses**. The discovery of HIV led to a search for additional lentiviruses and their subsequent isolation from cats (feline leukemia virus) and a variety of nonhuman primates (simian immunodeficiency virus). In 1986, François Clavel, Montagnier, and their coworkers discovered a distinct type of HIV that is prevalent in certain regions of West Africa. It was called HIV type 2 (HIV-2) to distinguish it from the original type (HIV-1). Individuals infected with HIV-2 also develop AIDS, but with a longer clinical latency period and lower morbidity. For example, some people progress to AIDS in 9 to 10 years, as with HIV-1; others are long-term nonprogressors.

Many independent isolates of both HIV-1 and HIV-2 have been characterized over the last decade. Nucleotide sequence comparisons allow us to distinguish two major groups among HIV-1 isolates: **group M** includes most HIV-1 isolates, and **group O** represents what appear to be relatively rare "outliers" (Box 18.1). There are at least seven

**Table 18.2**    Lentiviruses[a]

| Virus | Host infected | Primary cell type infected | Clinical disorder |
|---|---|---|---|
| Equine infectious anemia virus | Horse | Macrophages | Cyclic infection in the first year, autoimmune hemolytic anemia, sometimes encephalopathy |
| Visna/maedi virus | Sheep | Macrophages | Encephalopathy/pneumonitis |
| Caprine arthritis-encephalitis virus | Goat | Macrophages | Immune deficiency, arthritis, encephalopathy |
| Bovine immunodeficiency virus | Cattle | Macrophages | Lymphadenopathy, lymphocytosis, central nervous system disease (?) |
| Feline immunodeficiency virus | Cat | T lymphocytes | Immune deficiency |
| Simian immunodeficiency virus | Primate | T lymphocytes | Immune deficiency and encephalopathy |
| Human immunodeficiency virus | Human | T lymphocytes | Immune deficiency and encephalopathy |

[a] Adapted from Table 1.1 of J. A. Levy, *HIV and the Pathogenesis of AIDS*, 2nd ed. (ASM Press, Washington, D.C., 1998), with permission.

## BOX 18.1

### *The earliest record of HIV-1 infection*

The earliest record of HIV-1 infection comes from a serum sample obtained early in 1959 from a Bantu male in the city now known as Kinshasa, in the Democratic Republic of Congo. Phylogenetic analyses verify that this is the oldest known HIV-1 infection and place the viral sequence (ZR59) near the ancestral node of clades B and D. These studies suggest that these two subtypes, and perhaps all of group M, evolved from a single transspecies transmission into the African population in the 1940s or early 1950s. Its rapid evolution, represented by the starburst pattern of subtypes in the figure, seems to have occurred near the end of or just after World War II.

**Phylogenetic analyses of the ZR59 sequence.** A weighted-parsimony tree shows that the virus ZR59 is closely related to the common ancestor of modern HIV-1 subtypes B, D, and F and that virus YBF30 defines a new group, N-, that lies phylogenetically between groups M and O. Adapted from Fig. 2 of T. Zhu et al., *Nature* **391:**594–597, 1998, and F. Simon et al., *Nat. Med.* **4:**1032–1037, 1998.

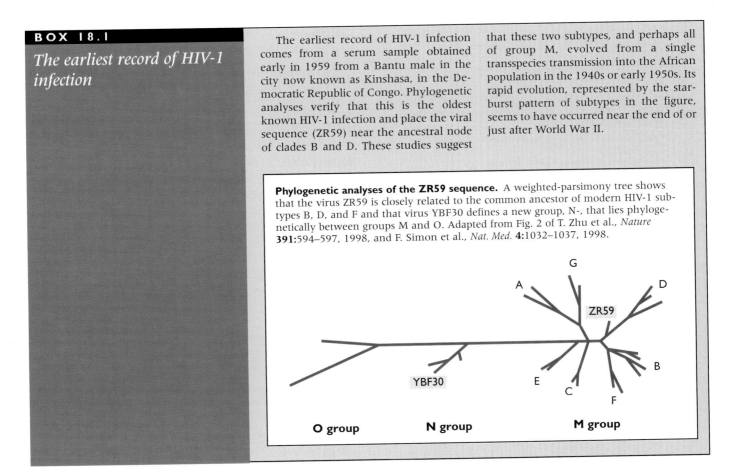

subtypes in group M (called **clades**), each of which is prevalent in a different geographic area. For example, clade B is the most common subtype in North America and Europe. Five clades of HIV-2 have also been identified on the basis of similar criteria. A new **group N** was proposed for HIV-1 in 1998 based on a virus isolate, YBF30, obtained from an AIDS patient in Cameroon. The nucleotide sequence of this virus is more closely related to group M than to group O. Identification of related strains in Cameroon supports a three-pronged radiating of HIV-1 groups.

Figure 18.1 shows the phylogenetic relationships among the lentiviruses, based on sequences of their *pol* genes. The African monkey isolates are endemic to each of the species from which they were obtained and do not appear to cause disease in their native hosts. However, a fatal AIDS-like disease is caused by infection of Asian macaques with virus originating from the African sooty mangabey ($SIV_{SMM}$). This phenomenon and the observation that several isolates of HIV-2 are nearly indistinguishable in nucleotide sequence from $SIV_{SMM}$ support the hypothesis that HIVs emerged via interspecies transmission between nonhuman primates and humans. In this chapter we will use the abbreviation HIV to describe properties shared by all the HIVs, and will specify the type when referring to one or the other.

As summarized in Table 18.2, all lentiviruses cause immune deficiencies and disorders of the hematopoietic and central nervous systems, and sometimes arthritis and autoimmunity. Lentiviruses also possess relatively large genomes with more genes than most retroviruses (Fig. 18.2). In addition to the three structural polyproteins Gag, Pol, and Env, encoded by all oncogenic retroviruses, lentiviral genomes encode a number of additional **auxiliary** proteins. Two HIV auxiliary proteins (Table 18.3), Tat and Rev, preform **regulatory** functions that are essential for viral replication. The remaining four, Nef, Vif, Vpr, and Vpu, are not essential for virus replication in most immortalized T-cell lines and hence are known as **accessory** proteins. However, accessory proteins do modulate virus replication, and they are essential for efficient virus production in vivo.

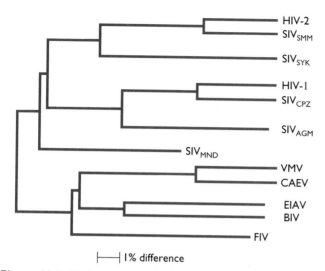

**Figure 18.1 Phylogenetic relationships among lentiviruses.** Representative lentiviruses compared by using *pol* gene nucleotide sequences for establishing phylogenetic relationships. Five groups of primate lentiviruses are shown: HIV-1, HIV-2, simian immunodeficiency virus from the sooty mangabey monkey (SIV$_{SMM}$), Sykes monkey (SIV$_{SYK}$), chimpanzee (SIV$_{CPZ}$), African green monkey (SIV$_{AGM}$), and mandrill (SIV$_{MND}$). Nonprimate lentiviruses are visna/maedi virus (VMV), caprine arthritis-encephalitis virus (CAEV), equine infectious anemia virus (EIAV), bovine immunodeficiency virus (BIV), and feline immunodeficiency virus (FIV). The scale indicates the percentage difference in nucleotide sequences in the *pol* gene. The branching order of the primate lentiviruses is controversial. Adapted from Fig. 2 of P. A. Luciw, p. 1881–1952, *in* B. N. Fields et al. (ed.), *Fields Virology*, 3rd ed. (Lippincott-Raven, Philadelphia, Pa., 1996), with permission.

## Distinctive Features of the HIV Replication Cycle and the Roles of Auxiliary Proteins

### *cis*-Acting Regulatory Sequences and the Virus Regulatory Proteins Tat and Rev

**The long terminal repeat.** As in all retroviruses, expression of integrated HIV DNA is regulated by sequences in the trancriptional control region of the viral long terminal repeat (LTR) which are recognized by the host cell's transcriptional machinery. The HIV-1 LTR functions as a promoter in a variety of cell types in vitro, but its basal level is very low. As described in chapter 8 (Fig. 8.12), the LTR of HIV includes an enhancer sequence that binds a number of cell-type-specific transcription activators, for example, the T-cell protein Nf-κb (see Fig. 18.4). The release of Nf-κb from its cytoplasmic inhibitor in activated T cells may explain the T-cell activation requirement for HIV replication.

**TAR and Tat.** Just downstream of the site of initiation of transcription in the HIV LTR is a unique viral regulatory sequence, TAR (Fig. 18.3). As described in chapter 8 (Fig. 8.13), TAR RNA forms a stable, bulged stem-loop structure that binds the 14-kDa viral regulatory protein Tat (Table 18.3). In the absence of Tat, viral transcription is poorly processive and usually terminates prematurely. The principal role of Tat is believed to be the facilitation of efficient elongation.

The Tat protein can be cytotoxic to cells in culture and is neurotoxic when inoculated intracerebrally into mice. The protein can also cause increased expression of adhesion molecules on endothelial cells, depolarization of cells, and degeneration of cell membranes, by mechanisms still unknown. It has been reported that transgenic expression of Tat in mice causes a disease that resembles Kaposi's sar-

**Figure 18.2 Genome organization of HIV-1 and HIV-2.** Vertical positions denote each of the three different reading frames that encode viral proteins. The long terminal repeats (LTRs) contain sequences necessary for transcriptional initiation and termination, reverse transcription, integration, and binding the viral transactivator Tat. The position of the RRE is indicated in the HIV-1 *env* gene. Adapted from Fig. 1 of M. Emerman, *Curr. Biol.* **6:**1096–1103, 1996, with permission.

**A HIV-1**

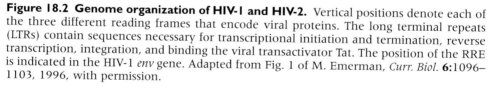

**B HIV-2**

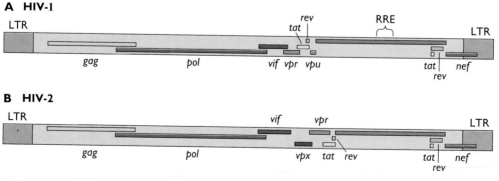

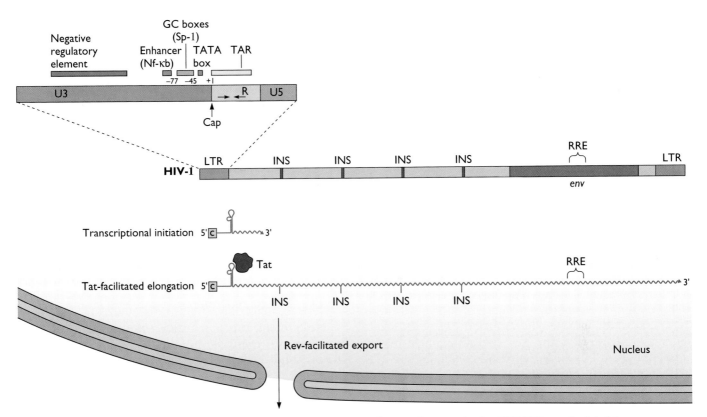

**Figure 18.3 Mechanisms of *tat* activation.** Some regulatory elements in the HIV LTR are depicted in the expanded section at the top. The numbers refer to positions relative to the initiation of transcription (cap site). The opposing arrows in R represent a palindromic sequence that folds into a stem-loop structure in the transcribed mRNA to which Tat binds (center). Tat is required for efficient elongation during HIV-1 RNA synthesis. The position of the RRE in the *env* transcript and the presence of *cis*-acting repressive sequences, also known as instability elements (INS), are also illustrated.

**Table 18.3**  HIV auxiliary proteins[a]

| Protein[b] | Size (kDa) | Function | Location |
|---|---|---|---|
| **Regulatory** | | | |
| Tat | 14 | Stimulation of transcription; binds TAR to facilitate initiation and elongation of viral transcription | Primarily in cell nucleus |
| Rev | 19 | Regulation of viral mRNA production; binds RRE and facilitates nuclear export of unspliced or singly spliced RNAs | Primarily in cell nucleus |
| **Accessory** | | | |
| Nef | 27 | Pleiotropic, can increase or decrease virus replication; reduces expression of MHC class I and the CD4 receptor; affects T-cell activation; enhances virion infectivity | Cell cytoplasm, plasma membrane |
| Vif | 23 | Increases virus infectivity; affects virion assembly and/or viral DNA synthesis | Cell cytoplasm |
| Vpr | 15 | Causes $G_2$ arrest; facilitates nuclear entry of preintegration complex | Virion |
| Vpu[c] | 16 | Affects virus release; disrupts Env-CD4 complexes; CD4 degradation | Integral cell membrane protein |
| Vpx[d] | 15 | Nuclear entry of preintegration complexes | Virion |

[a] Adapted from Table 1.5 of J. A. Levy, *HIV and the Pathogenesis of AIDS*, 2nd ed. (ASM Press, Washington, D.C., 1998), with permission.
[b] See Fig. 18.2 for location of the viral genes on the HIV genome.
[c] Present only with HIV-1. Expression appears to be regulated by Vpr.
[d] Encoded only by HIV-2. May be a duplication of Vpr.

coma and, although the human disease is almost certainly caused by a herpesvirus (human herpesvirus 8 [HHV-8]), Tat may serve as a growth factor for this malignancy in AIDS patients.

**Multiple splice sites and the role of Rev.** Unlike those of the simpler oncogenic retroviruses, the full-length HIV transcript contains numerous 5′ and 3′ splice sites. The regulatory proteins Tat and Rev and the accessory protein Nef are synthesized early in infection from multiply spliced mRNAs (appendix, Fig. 9B). As Tat then stimulates transcription, these mRNAs are found in abundance at this early time. However, the accumulation of Rev protein brings about a change in the pattern of mRNAs, leading to a temporal change in viral gene expression.

The Rev protein (Table 18.3) is an RNA-binding protein that recognizes a specific sequence within a structural element in *env* called the **Rev-responsive element (RRE)** (Fig. 18.3). As discussed in chapter 10 (Fig. 10.15 to 10.17), Rev activates the nuclear export of any RRE-containing RNA. Thus, as the Rev concentration increases, unspliced or singly spliced transcripts containing the RRE are exported from the nucleus. In this way, Rev facilitates synthesis of the viral structural proteins and enzymes and ensures the availability of full-length genomic RNA to be incorporated into new virus particles. The accessory proteins Vif, Vpr, and Vpu (for HIV-1) or Vpx (for HIV-2) are also expressed later in infection from singly spliced mRNAs whose export to the cytoplasm is Rev dependent (appendix, Fig. 9B).

The dependence of HIV gene expression on Rev is due in part to *cis*-acting repressive sequences, also called **instability elements**, present in the unspliced or singly spliced transcripts. These sequences, some of which are characterized by a high A/U content, have been mapped to several locations within the regions encoding Gag and Pol. Mutations in these sequences have been shown to increase the stability, nuclear export, and translatability of the transcripts in the absence of Rev. The response to their presence appears to be cell dependent, but the mechanism(s) by which these elements act, and exactly how Rev counteracts their effects, is not understood. This phenomenon does, however, provide an explanation for the failure of early attempts to express individual HIV-1 structural proteins and enzymes in primate or human cells from mRNAs that did not also encode Rev.

## The Accessory Proteins

**Nef protein.** As noted above, Nef (initially, for negative factor) is translated from multiply spliced early transcripts. The 5′ end of the Nef transcript includes two initiation codons, and as both are utilized, two forms of Nef are produced in infected cells. The apparent size of these proteins can vary from 27 to 37 kDa because of cellular differences in posttranslational modification. Nef is myristylated posttranslationally at its N terminus (Fig. 18.4 and 18.5) and thereby anchored to the inner surface of the plasma membrane, probably in a complex with a cellular serine kinase. There are numerous potential threonine and

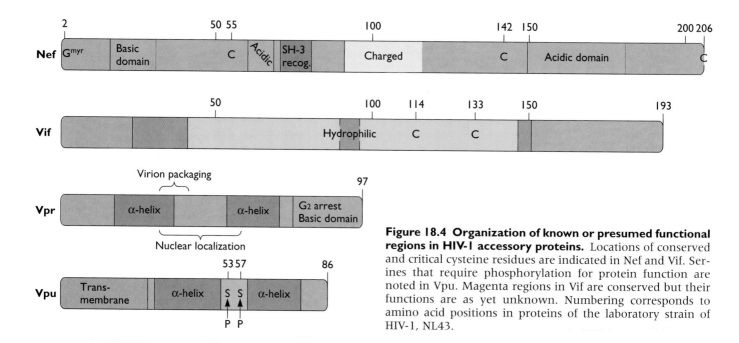

**Figure 18.4 Organization of known or presumed functional regions in HIV-1 accessory proteins.** Locations of conserved and critical cysteine residues are indicated in Nef and Vif. Serines that require phosphorylation for protein function are noted in Vpu. Magenta regions in Vif are conserved but their functions are as yet unknown. Numbering corresponds to amino acid positions in proteins of the laboratory strain of HIV-1, NL43.

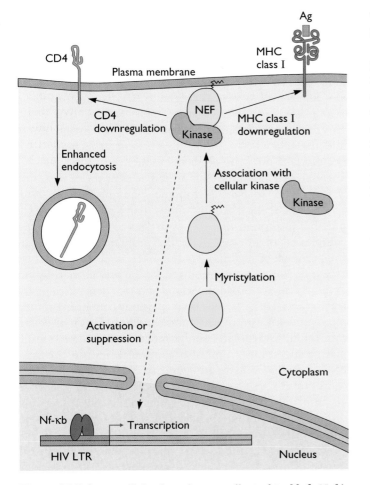

**Figure 18.5 Intracellular functions attributed to Nef.** Nef is myristylated posttranslationally; the jagged protrusion represents myristic acid covalently linked to the glycine residue at position 2 in Nef. Myristylation enables Nef to attach to cell membranes and interact with a cellular serine kinase and perhaps other cellular proteins. The cytoplasmic domain of CD4 contains the target sequences for Nef-mediated downregulation. Nef reduces cell surface expression of CD4 by enhancing endocytosis; major histocompatibility complex class I expression is also reduced. By an unknown mechanism, Nef increases the activity of the cellular transcriptional activator Nf-κb and perhaps other cellular transcription proteins, thereby augmenting expression directed by the viral LTR. Nef has also been reported to suppress viral gene expression.

serine phosphorylation sites in Nef, and the protein is reported to be phosphorylated, but the significance of such a modification is as yet unknown. The Nef protein also includes an SH3 recognition domain ($P^{69}xxPxxPxxP^{78}$), which might facilitate interaction with components of intracellular signaling pathways.

Most laboratory strains of HIV-1 that have been adapted to grow well in T-cell lines contain deletions or mutations in the *nef* gene. Restoration of *nef* reduces the efficiency of

virus replication in these cells, hence the name "negative factor." Multiple functions have been attributed to Nef (Table 18.3; Fig. 18.5), and it is now clear that Nef exerts pleiotropic effects on infected cells. The functions reported for the Nef protein vary for different strains of the virus and with different cell types. Nef has been shown to decrease cell surface expression of CD4 as a result of interaction with the cytoplasmic portion of this integral membrane protein and induction of its endocytosis (and lysosomal degradation). Nef also decreases cell surface expression of major histocompatibility complex (MHC) class I molecules in infected primary lymphocytes. As a strong cytotoxic lymphocyte response against viral infection requires recognition of viral epitopes presented by major histocompatibility complex class I molecules, this inhibitory activity of Nef may allow infected cells to evade lysis by cytotoxic T cells and could be a major factor contributing to HIV-1 pathogenesis.

Although the initial tissue culture experiments suggested a negative effect on virus production, subsequent animal experiments indicated that Nef augments HIV pathogenesis significantly. Rhesus macaques inoculated with a Nef-defective mutant of SIV had low virus titers in their blood during early stages of infection, and the later appearance of high titers of virus was associated with reversion of the mutation. More importantly, adult macaques inoculated with a virus strain containing a deletion of *nef* did not progress to clinical disease and were in fact immune to subsequent challenge with wild-type virus. The observation that *nef* had been deleted in HIV-1 isolates from some individuals who remained asymptomatic for long periods and from transfusion recipients who did not develop AIDS also suggests that this viral protein can contribute significantly to pathogenesis. Initial hopes that intentional deletion of *nef* might facilitate the development of a vaccine strain for humans were dashed when it was discovered that newborns of the female macaques who had been immunized with the *nef*-minus strain of SIV developed an AIDS-like disease.

**Vif protein.** Vif stands for viral infectivity factor. This protein (Table 18.3; Fig. 18.4) accumulates in the cytoplasm and at the plasma membrane of infected cells. Early studies showed that mutant viruses lacking the *vif* gene were approximately 1,000 times less infectious than the wild type in certain CD4⁺ T-cell lines, although cell-to-cell transmission was only slightly lower than normal. Thus, virions produced in the absence of Vif are defective in some way.

Production of Vif from an independent expression vector in susceptible host cells does not compensate for its absence in the cell that produces virions. Rather, it appears that Vif is needed at the time of virus assembly or maturation in the producing cells. Virions produced from *vif*-defective genomes contain the normal complement of progeny RNA; they are able to enter susceptible cells and to initiate reverse transcription, but full-length double-stranded viral DNA is not produced. Thus, the defect is manifested at a very early stage in replication, perhaps indicating a problem with uncoating or in nucleocapsid organization. Clearly, many questions remain concerning the role of Vif in viral infection. The conservation of *vif* homologs in all lentiviruses save equine infectious anemia virus indicates that this protein is likely to contribute to lentivirus propagation in natural host populations.

**Vpr protein.** The viral protein R, Vpr (Table 18.3; Fig. 18.4), derives its name from the early observation that it affects the rapidity with which the virus replicates in and destroys T-cell lines. Most T-cell-adapted strains of HIV-1 carry mutations in *vpr*. The Vpr protein is encoded in an open reading frame lying between *vif* and *tat* in the genomes of primate lentiviruses (Fig. 18.2). The SIV and HIV-2 genomes include a second, related gene, *vpx*, which appears to have arisen as a duplication of *vpr*. The other lentiviruses do not contain sequences related to *vpr* but do include small open reading frames that could perform similar functions.

Vpr and Vpx are assembled into virions. Vpr incorporation is dependent on specific interactions with a proline-rich domain (p6) at the C terminus of the Gag polyprotein. As a consequence of these interactions, Vpr is associated with the nucleocapsid of the virion and is present at concentrations close to those of the viral structural proteins. This property is consistent with the observation that Vpr function is required at some early stage in the virus replication cycle.

Two different functions have been recognized for HIV-1 Vpr. The protein causes a $G_2$ cell cycle arrest, and it promotes entry of viral nucleic acids into the nucleus. In HIV-2 these functions are segregated; the $G_2$ arrest is encoded in *vpr*, whereas the nuclear entry function is encoded in *vpx*. The biological advantage of preventing infected cells from entering mitosis is not clear, but the fact that viral LTR promoters are more active in the $G_2$ phase of the cell cycle may lead to enhanced virus production. The ability to promote nuclear entry is of obvious importance, as such entry allows the virus to integrate its DNA and complete a replication cycle in fully differentiated, nondividing macrophages. Vpr has been shown to bind to nuclear pore proteins; as noted in chapter 5, these interactions are believed to facilitate docking of the HIV-1 preintegration complex at the nuclear pore complex in preparation for

import. Studies with SIV-infected macaques indicate that deletion of *vpr* attenuates but does not eliminate viral pathogenicity. However, deletion of both *vpr* and *vpx* does reduce virus replication in these animals. Thus, it seems likely that HIV-1 Vpr, which combines the functions of the two SIV gene products, is crucial to HIV pathogenesis.

**Vpu protein.** This small protein is unique to HIV-1 and the related SIV$_{CPZ}$ (Fig. 18.1), hence the name viral protein U (Vpu). The predicted sequence of Vpu includes a stretch of 27 hydrophobic amino acids at the N terminus that is likely to be a membrane-spanning domain (Fig. 18.4). Biochemical studies show that Vpu is an integral membrane protein that self-associates to form oligomeric complexes. In infected cells, the protein accumulates in the perinuclear region.

Expression of Vpu is required for the proper maturation and targeting of progeny virions and for their efficient release (Table 18.3). In its absence, virions containing multiple cores are produced, and budding is targeted to intracellular vacuolar compartments rather than to the plasma membrane. Vpu also appears to reduce the syncytium-mediated cytopathogenicity of HIV-1, perhaps because the efficient release of virions prevents the excessive accumulation of virion Env proteins at the cell surface that promotes formation of syncytia. A second function of Vpu is the degradation of CD4. Vpu traps this receptor in the endoplasmic reticulum via specific interactions with its cytoplasmic domain, preventing transport of CD4 to the cell surface and triggering its destruction by the proteosome. These two activities of Vpu are distinct, as the facilitation of virion release is independent of Env or CD4 expression. It is interesting to note that expression of Vpu also enhances the release of virions produced by other, simple retroviruses, such as Moloney murine leukemia virus.

The Vpu protein has structural and biochemical features similar to those of the influenza virus M2 protein. As noted in chapter 5, M2 is an ion channel protein that modulates the pH in the Golgi compartment, thereby protecting the newly formed hemagglutinin protein from changing conformation prematurely in the secretory pathway. Thus, a putative ion channel activity of Vpu may be required for proper virion maturation and assembly.

## Cellular Targets

As discussed in chapters 4 and 5, virus attachment and entry into host cells are dependent on the interaction between viral surface proteins and receptors on the surface of susceptible cells. The major receptor for the HIV envelope proteins, SU and TM, is the cell surface CD4 molecule, whose normal function is to bind to major histocompatibility complex class II molecules on antigen-presenting cells in conjunction with the T-cell receptor. The HIV envelope proteins must also interact with a coreceptor molecule to trigger fusion of the viral and cellular membranes and gain entry into the cytoplasm. The ability to bind to specific coreceptor molecules is a critical determinant of the cell tropism of different HIV-1 strains. For example, use of the α-chemokine receptor Cxcr4 is a definitive feature of strains that infect T-cell lines. Infection with these strains also causes T cells to fuse, forming syncytia. Use of the β-chemokine receptor Ccr5 is characteristic of non-syncytium-inducing M (monocyte/macrophage)-tropic strains. However, some M-tropic strains can also infect T cells. The physiological importance of these two chemokine receptors is demonstrated by the finding that people who carry a mutation in the *ccr5* gene and produce a defective receptor protein are resistant to HIV-1 infection, as are individuals who carry a mutation in the gene for a ligand of Cxcr4 (Tables 15.2 and 15.3). The latter mutation is hypothesized to lead to increased availability of the ligand, which then blocks virus entry by binding to the coreceptor. This idea is consistent with earlier studies showing that chemokines that bind to these receptors can inhibit the infectivity of specific strains of HIV in cell culture (Fig. 18.6). Cells of the hematopoietic lineage that bear CD4 and one or more of these chemokine receptors are the main targets of HIV infection, and they produce the highest titers of progeny virions.

Several additional coreceptors for HIV and simian immunodeficiency virus have been identified in tissue culture experiments in various laboratories, but it is not yet known what roles such molecules might play in pathogenesis. It is hypothesized that these additional coreceptors may allow the virus to enter a broader range of cells than first appreciated. Some of the coreceptors are found on cells of the thymus gland and the brain, and they could play a role in infection in infancy or of cells in the nervous system. It has also been hypothesized that binding to these additional coreceptors may trigger signals that affect virus replication in target cells or that harm nontarget cells, producing a "bystander" effect.

The entry of HIV may also occur through direct cell-cell contact or by binding to cell surface galactosyl ceramide or to Fc receptors when complexed with antibodies. The virus has been found in small amounts in several tissues of the body, and it can infect many different types of human cells in culture. As discussed below, some of these susceptibilities are likely to be relevant to HIV-1 pathogenesis in humans.

**Figure 18.6 Coreceptors for macrophage/monocyte- and T-cell-tropic strains of HIV-1.** Cxcr4 is the major coreceptor for T-cell-tropic strains; entry of such strains is inhibited by the receptor's natural ligand, Sdf-1. Ccr5 is the major coreceptor for macrophage/monocyte (M)-tropic strains, and their entry is inhibited by the receptor's natural ligands, Rantes, Mip-1α, and Mip-1β. Primary T cells and monocytes express both coreceptors; primary T cells are susceptible to both strains, but monocytes can be infected only by M-tropic strains for reasons that are not yet clear. Adapted from Fig. 3 of A. S. Fauci, *Nature* **384:** 529–533, 1996, with permission.

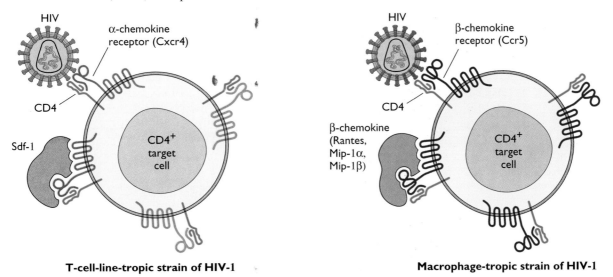

**T-cell-line-tropic strain of HIV-1**          **Macrophage-tropic strain of HIV-1**

## Routes of Transmission

### Sources of Virus Infection

Even before HIV-1 was identified, epidemiologists had established the most likely routes of its transmission to be via sexual contact, via blood, and from mother to child. As might be anticipated, the efficiency of transmission is influenced greatly by the concentration of the virus in the body fluid to which an individual is exposed. Table 18.4 provides estimates for the percentage of infected cells and the concentration of HIV-1 in different body fluids. The highest values are observed in peripheral blood monocytes, in blood plasma, and in cerebrospinal fluid, but semen and female genital secretions also appear to be important sources of virus. Other routes of transmission have been shown to be relatively unimportant or nonexistent, at least for HIV-1; among these are casual, nonsexual contact, exposure to saliva or urine from infected individuals, and exposure to blood-sucking insects.

Fortunately, HIV-1 infectivity is reduced rapidly upon air drying (by 90 to 99% within 24 h). The virus is inactivated by heating (56 to 60°C for 30 min), exposure to standard germicides (such as 10% bleach or 70% alcohol), or exposure to pH extremes (e.g., <6 or >10 for 10 min). This information and results from epidemiology studies have been used to establish safety regulations to prevent transmission in the public sector and in the health care setting.

### Modes of Transmission

Modes of HIV-1 transmission vary in different geographic locations (Fig. 18.7). In the United States, the major mode is via homosexual contact, whereas heterosexual transmission is most common in the rest of the world. In both cases, the likelihood of infection is related to the number of partners. The presence of venereal diseases also increases the probability of transmission, presumably because infected inflammatory cells may be present in both seminal and vaginal fluids. Genital ulceration and conse-

**Table 18.4** Isolation of infectious HIV-1 from body fluids[a]

| Fluid | Virus isolation[b] | Estimated quantity of virus[c] |
|---|---|---|
| Cell-free fluid | | |
|   Plasma | 33/33 | 1–5,000[d] |
|   Tears | 2/5 | <1 |
|   Ear secretions | 1/8 | 5–10 |
|   Saliva | 3/55 | <1 |
|   Sweat | 0/2 | ND[e] |
|   Feces | 0/2 | ND |
|   Urine | 1/5 | <1 |
|   Vaginal-cervical | 5/16 | <1 |
|   Semen | 5/15 | 10–50 |
|   Milk | 1/5 | <1 |
|   Cerebrospinal fluid | 21/40 | 10–10,000 |
| Infected cells | | |
|   PBMC[f] | 89/92 | 0.001–1%[d] |
|   Saliva | 4/11 | <0.01% |
|   Bronchial fluid | 3/24 | ND |
|   Vaginal-cervical fluid | 7/16 | ND |
|   Semen | 11/28 | 0.01–5% |

[a]Adapted from Table 2.1 of J. A. Levy (ed.), *HIV and the Pathogenesis of AIDS*, 2nd ed. (ASM Press, Washington, D.C., 1998), with permission.

[b]Number of samples positive/number analyzed.

[c]For cell-free fluid, units are infectious particles per milliliter; for infected cells, units are percentages of total cells capable of releasing virus. Results are from studies in the laboratory of J. A. Levy. ND, not determined.

[d]High levels associated with acute infection and advanced disease (~5 × 10⁶ PBMCs per ml in blood).

[e]ND, none detected.

[f]PBMC, peripheral blood mononuclear cells.

**Figure 18.7 Modes of transmission of HIV in the United States and other parts of the world.** From January 1998 data published by the Joint United Nations Programme on HIV/AIDS and the U.S. Centers for Disease Control and Prevention.

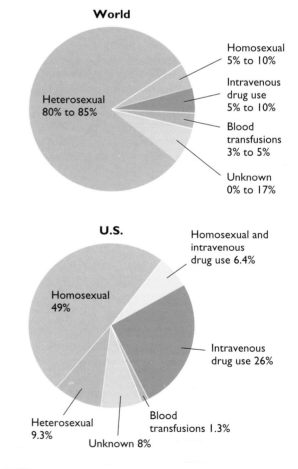

quent direct exposure to blood cells also increase the likelihood of transmission. In both types of sexual contact, the recipient partner is the one most at risk.

Intravenous drug use is the next most common route of transmission, owing to the common practice of sharing contaminated needles and other drug paraphernalia. Here again, the probability of transmission is a function of frequency of exposure and the prevalence of HIV-1 in the area. Of course, heterosexual partners of drug users are also at increased risk.

Until 1985, when routine HIV antibody testing of donated blood was established in the United States and other Western industrialized countries, individuals who received blood transfusions or certain blood products, such as clotting factors VIII and IX, were at high risk of becoming infected. Transfusion of a single unit (500 ml) of blood from an HIV-1-infected individual nearly always leads to infection of the recipient. Appropriate heat treatment of clotting factor preparations has now virtually eliminated the possibility of transmission from this source, but this safeguard is small comfort to the many hemophiliacs who contracted the disease before this route was understood. Other blood products, such as pooled immunoglobulin, albumin, and hepatitis B vaccine, seem not to have been implicated in HIV-1 transmission.

Transmission of HIV from mother to child can occur across the placenta or, most frequently, at the time of delivery as a consequence of exposure to a contaminated genital tract. The virus can also be transmitted via infected cells in the mother's milk during breastfeeding. Rates of transmission from an infected mother to a child range from as low as 11% to as high as 60%, depending on the severity of infection in the mother and the prevalence of breastfeeding. Antiviral drug therapy administered during pregnancy can reduce the frequency of transmission to a newborn. Even a single treatment with a nonnucleoside reverse transcriptase inhibitor early in labor can reduce the incidence significantly. Unfortunately, because of the cost, this is not often an option in underdeveloped countries, where the risk is greatest. Current efforts are focused on discouraging breastfeeding by infected mothers. As of 1997, an estimated 830,000 children had been infected worldwide.

## Mechanics of Spread

Except in cases of direct parenteral inoculation or blood transfusion, HIV enters the body through mucosal surfaces, as do most viruses (discussed in chapter 17). In the case of sexual transmission, the initial target cells in the rectum or genital tract have not been clearly identified. The most likely sources of transmission are virus-infected cells, as they can be present in much larger numbers than free infectious virus in vaginal or seminal fluids. Cell culture studies show that HIV-1 can be transferred directly to CD4⁻ epithelial cells via cell-cell contact (Fig. 18.8). Whether such transfer is relevant to the situation in vivo is unknown. Results from analyses of tissue biopsies of bowel mucosae and cervical and uterine epithelia suggest that cells in these layers can be infected in the absence of any injury. This infection might occur by interaction of the virus with galactosyl ceramide receptors on the surface of these epithelial cells or by attachment of virus-antibody complexes present in these fluids to Fc receptors on the mucosal cells.

**Figure 18.8 HIV-1-infected HUT 78 cell (top) cocultivated for 1 h with an ME1809 cervix-derived epithelial cell (bottom).** Virus can be seen at the cell-cell interface. It is noteworthy that virus is produced by the T-cell line only at the point of contact with the epithelial cells. A role of cytokines in this localized induction of virus production should be considered. Magnification, × 8,700. Reprinted with permission from Fig. 2.6 of J. A. Levy, *HIV and the Pathogenesis of AIDS*, 2nd ed. (ASM Press, Washington, D.C., 1998).

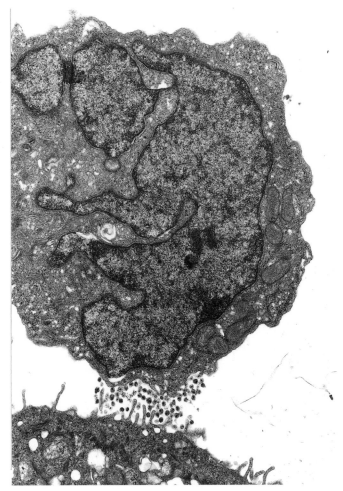

Other routes (most likely the main routes) of initial infection include acquisition of virus by cells of the mucosal and cutaneous immune system described in chapter 14; these include M cells present in the bowel epithelium and the dendritic, antigen-presenting CD4$^+$ Langerhans cells in the vaginal and cervical epithelia. Activated T cells, coincidental with genital infections that cause sores, are also likely targets. Although the insertive partner is at relatively low risk for infection, transmission to the male can occur through cells in the lining of the urethral canal of the penis, presumably from infected macrophages or Langerhans cells in the cervix or the intestinal mucosa of the infected partner. Uncircumcised males have a twofold-increased risk of infection, suggesting that the mucosal lining of the foreskin may be susceptible to HIV infection. Both male and female hormones appear to facilitate HIV transmission by stimulating cell-cell contact (prostaglandins) and erosion of the vaginal lining (progestin).

After replicating in the initial target cells, free virus or virus-infected cells enter draining lymph nodes or the circulatory system, where the next major targets of HIV are encountered, namely, susceptible cells that bear the CD4 receptor. As nonactivated peripheral blood mononuclear cells are not very permissive for infection and few activated CD4$^+$ lymphocytes are circulating in the blood at any given time, the first CD4$^+$ cells in the blood to be infected are probably macrophages. The macrophages are in a differentiated state, permissive for viral replication, and can pass progeny virus to activated lymphocytes in the lymph nodes. From this point the infection runs its protracted but usually inevitable course.

## The Course of Infection

### Patterns of Virus Appearance and Immune Cell Indicators of Infection

Pathological conditions associated with different phases in HIV-1 infection are summarized in Table 18.5.

### The Acute Phase

In the first few days after infection, the virus is produced in large quantities by the activated lymphocytes in lymph nodes, sometimes causing the nodes to swell (lymphadenopathy) or generating "flulike" symptoms. Virus released into the blood can be detected by infectivity with appropriate cell cultures or by screening directly for viral RNA or proteins (Fig. 18.9). Up to $5 \times 10^3$ infectious particles or $10^7$ viral RNA molecules per ml of plasma can be found during this stage. Generally this viremia is greatly curtailed within a few weeks, most likely via the cell-mediated immune response to HIV-1, as the number of CD8$^+$ cytotoxic lymphocytes increases before neutralizing anti-

**Table 18.5** Pathological conditions associated with HIV-1 infection[a]

Acute phase
  Mononucleosis-like syndrome: fever, malaise, pharyngitis, lymphadenopathy, headache, arthralgias, diarrhea, maculopapular rash, meningoencephalitis

Asymptomatic phase
  Often none, but patients may present sporadically with one or more of the following symptoms: fatigue, mild weight loss, generalized lymphadenopathy, thrush, oral hairy leukoplakia, shingles

Symptomatic phase and AIDS
  200–500 CD4$^+$ T cells per ml: generalized lymphadenopathy, oral lesions (thrush, hairy leukoplakia, aphthous ulcers), shingles, thrombocytopenia, molluscum contagiosum, basal cell carcinomas of the skin, headache, condyloma acuminata, reactivation of latent *Mycobacterium tuberculosis*
  Less than 200 CD4$^+$ T cells per ml
    Protozoal infections: *Pneumocystis carinii, Toxoplasma gondii, Isospora belli*, cryptosporidia, microsporidia
    Bacterial infections: *Mycobacterium avium-M. intracellulare, Treponema pallidum*
    Fungal infections: *Candida albicans, Cryptococcus neoformans, Histoplasma capsulatum*
    Viral infections and malignancies: human cytomegalovirus, recurrent bouts of oral or genital HSV, lymphoma (mostly EBV, some HHV-8), Kaposi's sarcoma (HHV-8), anogenital carcinoma (HPV)
    Neurological symptoms: aseptic meningitis; myelopathies, such as vacuolar myelopathy; pure sensory ataxia; paresthesia/dysesthesia; peripheral neuropathies, such as acute demyelinating polyneuropathy, mononeuritis multiplex, and distal symmetric polyneuropathy; myopathy; AIDS dementia complex (ADC)

[a] Adapted from Table 1 of A. S. Fauci and R. C. Desrosiers, p. 587–635, *in* J. M. Coffin et al. (ed.), *The Retroviruses* (Cold Spring Harbor Laboratory Press, Plainview, N.Y., 1997), with permission. HSV, herpes simplex virus; EBV, Epstein-Barr virus; HHV-8, human herpesvirus 8; HPV, human papillomavirus.

bodies can be detected. The CD4$^+$ cell count decreases during this phase but then returns to near normal levels.

During the period of acute infection, the virus population is relatively homogeneous. In most cases, it appears that the predominant virus was a minor variant in the population present in the blood of the contaminating individual. These early-stage isolates are generally macrophage-tropic, but the reason for their apparent selective transmission is unknown.

### The Asymptomatic Phase

By 3 to 4 months after infection, viremia is usually reduced to low levels, with small bursts of virus appearing from time to time. It is known that the degree of viremia at this stage of infection, the so-called virologic set point, is a direct predictor of how fast the disease will progress in a

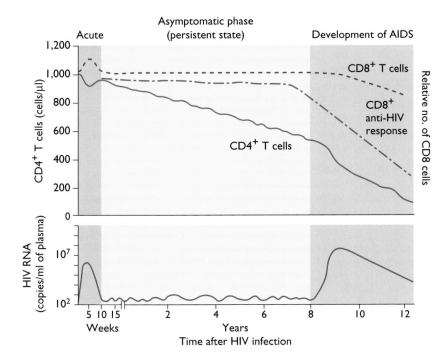

**Figure 18.9 Schematic diagram of events occurring after HIV-1 infection.** Adapted from Fig. 13.1 of J. A. Levy, *HIV and the Pathogenesis of AIDS*, 2nd ed. (ASM Press, Washington, D.C., 1998), with permission.

particular individual: the higher the viremia, the faster the progression. During this time, CD4$^+$ cell numbers decrease at a steady rate estimated to be approximately 60 cells/$\mu$l/year. The cause of this decline remains a mystery, although direct cytopathogenicity by the virus and/or apoptosis due to inappropriate cytokine production may be responsible. In this protracted asymptomatic period, which can last for years, the CD8$^+$ T-cell level remains slightly elevated, but virus replication continues at a low rate, mainly in the lymph nodes. In lymphoid tissues a relatively large, stable pool of virions can be detected bound to the surface of follicular dendritic cells. A small number of infected T cells are also observed. During this persistent state, known also as **clinical latency**, only 1 in 300 to 400 infected cells in the lymph nodes may actually release virus. It is believed that, as in acute infection, virus propagation is suppressed at this stage by the action of antiviral CD8$^+$ cytotoxic lymphocytes. The number of these specific lymphocytes decreases toward the end of this stage. During the asymptomatic phase, the virus population becomes more heterogeneous, probably because of continual selection for specific mutations as a result of immunological pressures.

### The Symptomatic Phase and AIDS

The end stage of disease, when the infected individual develops symptoms of AIDS, is characterized by a CD4$^+$ cell count below 200/ml and increased quantities of virus. The total CD8$^+$ cytotoxic lymphocyte count also decreases, probably owing to the precipitous drop in the number of

those specific for HIV. In the lymph nodes, virus replication increases with concomitant destruction of lymphoid cells and of the normal architecture of lymphoid tissue. The cause of this lymph node degeneration is not clear; it may be due directly to virus replication or an indirect result of chronic immune stimulation.

In this last stage, the virus population again becomes relatively homogeneous and, generally, T-cell-tropic. Properties associated with increased virulence predominate, including an expanded cellular host range, ability to cause formation of syncytia, rapid replication kinetics, and CD4$^+$ T-cell cytopathogenicity. Late-emerging virus also appears to be less sensitive to neutralizing antibodies and more responsive to antibodies that enhance infectivity. In some cases, strains that have enhanced neurotropism or increased pathogenicity for other organ systems emerge. Where analyzed, these changes can be traced to specific mutations, for example, in the viral envelope gene or in a regulatory gene (e.g., *tat*).

### Variability of Response to Infection

Studies of large cohorts of HIV-1-infected adults show that approximately 10% progress to AIDS within the first 2 to 3 years of infection. Over a period of 10 years, approximately 80% of infected adults will show evidence of disease progression, and of these 50% will have developed AIDS. Of the remainder, 10 to 17% will be AIDS free for over 20 years; a very small percentage of these individuals will be completely free of any symptoms, with no evidence of progression to disease. What are the parameters that contribute to this variability?

One parameter is the degree to which an individual's immune system may be stimulated by infection with other pathogens. HIV-1 replicates most efficiently in activated T cells, and it is known that virus quantities increase when the immune system of an infected individual is activated by opportunistic infections with other microorganisms. Such activation can explain the fact that HIV-1 disease is generally more aggressive in sub-Saharan Africa, where chronic infection by parasites and other pathogens is frequent. As might be expected for an outbred population, variations in an individual's genetic makeup can modulate immune response to HIV infection and affect survival. As already noted, differences in chemokines or chemokine receptors and, probably, in any one of several components of the immune system can have an impact on the course of the disease.

Clearly, variation in the virus itself can also affect the course of the disease. As noted above, some long-term survivors of HIV-1 infection harbor viruses with deletions in the *nef* gene. Others appear to be infected with otherwise attenuated strains that produce low titers in tissue culture cells and have restricted cell tropism. Low virus titers and the presence of viral neutralizing and not enhancing antibodies (see Fig. 18.10) are other characteristics of these infections.

It can be hoped that further study of the rare, fortunate individuals who are nonprogressors or long-term survivors of HIV-1 infection will lead to a better understanding of viral pathogenesis and suggest new strategies for effective prevention or therapy.

## Origins of Cellular Immune Dysfunction

The defining feature of HIV disease is impaired immune cell function. This defect is eventually devastating because these defenses are vital in the body's battle against this virus as well as other pathogens. At first the immune system appears to keep the HIV infection in check. However, the virus is not eliminated, and the infection that persists in the asymptomatic stage leads to increasing dysfunction among immune cells. In the end, most AIDS patients actually succumb to opportunistic infections with microorganisms that are little threat to individuals with healthy immune systems. The impairment in immune cell function typical of HIV-1 disease can result from direct attack of the virus on particular cell types, or it can come about indirectly as a response of uninfected cells to viral gene products or specific proteins made in infected or stimulated cells (Table 18.6).

### CD4$^+$ T Lymphocytes

The major reservoir of HIV in the peripheral blood is the CD4$^+$ T cell. Even before the profound depletion of CD4$^+$

T cells that is a signal of end-stage disease, abnormalities in a variety of CD4$^+$ T-cell functions can be detected. These include decreased ability to form colonies when grown in tissue culture, decreased expression of interleukin-2 (Il-2) and the Il-2 receptor, and reduced proliferative response to various antigens. The precise causes of these impairments are unknown. Some loss of function can be explained by direct killing of cells by the virus. In addition, as surface expression of CD4 is reduced by some viral proteins (SU, Vpu, and Nef), major histocompatibility complex class II-CD4 interactions become less frequent, leading to a decrease in antigen-specific responses. Trapping of infected CD4$^+$ lymphocytes in lymphoid organs may account for some of the decline. Destruction of infected CD4$^+$ lymphocytes in the circulation by anti-HIV-1 CD8$^+$ cytotoxic lymphocytes also reduces their total number.

Noninfectious virions and viral proteins shed from infected cells can bind to or enter uninfected CD4$^+$ lymphocytes, triggering inappropriate responses. For example, HIV-1 envelope proteins and purified Tat protein have been reported to suppress proliferative responses of T cells to antigens and mitogens. It is possible that SU bound to CD4 interferes sterically with its interaction with major histocompatibility complex class II molecules. Such binding may also disturb the normal signaling that must take place during antigen-specific responses. Exposure to HIV-1 SU protein also inhibits production of the lymphokine Il-2 and Il-2 receptor expression by uninfected T cells. Finally, changes in cytokine production by HIV-1-infected macrophages can trigger apoptosis in uninfected CD4$^+$ T cells.

### CD8$^+$ T Lymphocytes

As mentioned previously, the number of CD8$^+$ T cells is abnormally high following the acute phase and decreases precipitously during the end stage of the disease. The early increase may be the result of an imbalance brought about as the immune system attempts to achieve homeostasis of CD8$^+$ and CD4$^+$ cells at the same time that CD4$^+$ cells are being destroyed. The reduced numbers of anti-HIV-1 CD8$^+$ cytotoxic lymphocytes at late times can be understood in part by the direct infection and killing of CD8$^+$ progenitor cells. Additionally, and most importantly, because CD8$^+$ T-cell proliferation and function depend on inductive signals from CD4$^+$ T cells (Fig. 14.6 and 14.7), the decline of the CD4$^+$ population also contributes to CD8$^+$ T-cell dysfunction.

### Monocytes and Macrophages

HIV-1-infected macrophages can be detected readily in tissues throughout the body of an infected individual. However, as only a small proportion of monocytes/macrophages in the blood are infected with the virus, it

**Table 18.6**   Immune cell dysfunction associated with the onset of AIDS

| Affected cell type | Dysfunction | Known and/or postulated causes | |
|---|---|---|---|
| | | **Direct** | **Indirect** |
| CD4+ T cells | Colony formation ↓ Proliferative response to antigen ↓ Expression of Il-2 and Il-2 receptor Total number in circulation ↓ | Direct killing by HIV-1 CD4 downregulated in infected cells (by SU, Vpu, Nef) Trapping of infected cells in lymphoid organs Destruction of infected cells by anti-HIV-1 CD8+ cytotoxic lymphocytes | SU component of noninfectious virions binds to cell surface CD4, suppresses immune response, and causes inappropriate intracellular signaling |
| CD8+ T cells | Abnormally high numbers following acute phase Loss of anti-HIV cytotoxic lymphocyte activity Loss in numbers toward end stage | Infection and killing of progenitor CD4+/CD8+ and CD8+ immature thymocytes CD8+ immature thymocytes | Attempt of the immune system to establish homeostasis results in a higher production of CD4+ *and* CD8+ cells as CD4+ cells are depleted Loss of Il-2 production as CD4+ cell pool is depleted |
| Monocytes (dendritic cells)/ macrophages | Defects in chemotaxis Monocyte-dependent T-cell proliferation ↓ Antigen-presenting cell activity ↓ Fc receptor function ↓ Complement C3 receptor-mediated clearance/ oxidative burst ↓ Decrease in numbers | Only few circulatory cells are infected, but changes in cytokine production by such cells (e.g., Tnf-α ↓; Il-1 ↓) can cause apoptosis of CD4+ cells and other abnormalities | Exposure to noninfectious virions (SU, TM), downregulates chemotactic ligand receptors and causes abnormal secretion of certain chemokines Decreased expression of costimulatory molecules (e.g., B7) Cells killed by eliciting cytotoxic response in cytotoxic lymphocytes |
| B cells | Abnormal proliferation Hypergammaglobulinemia Poor response to additional antigen signals Production of autoantibodies | | Exposure to noninfectious virions (TM) causes polyclonal activation Loss of CD4+ T-cell helper function Molecular mimicry of host proteins by viral proteins |
| Natural killer cells | Cytotoxicity function ↓ | | Loss of Il-2 as CD4+ cell pool is depleted |

seems likely that the functional impairment seen in this population of cells is due to indirect effects. Monocyte/macrophage abnormalities include defects in chemotaxis, inability to promote T-cell proliferation, and defects in Fc receptor function and complement-mediated clearance. In vitro experiments suggest that some of these effects are caused by exposure to the viral envelope proteins.

## B Cells

HIV-1-infected individuals produce abnormally large quantities of immunoglobulin G (IgG), IgA, and IgD. Such production is indicative of B-cell dysfunction that may result from increased proliferation of cells of the lymph nodes. Infected individuals also show poor responses to primary and secondary immunization. During end-stage disease, there may also be a decrease in the total number of B cells. Some of the B-cell loss may be attributable to the decline in CD4+ T-cell helper function (see Fig. 14.6). B cells isolated from infected individuals proliferate in culture without stimulation and are also defective in their response to specific antigens or mitogens. Binding of viral

proteins (e.g., TM) has been shown to induce polyclonal B-cell activation, a property that might explain such spontaneous proliferation. Infection by Epstein-Barr virus and human cytomegalovirus, common in AIDS patients, may also contribute to this abnormality.

## Natural Killer Cells

Impairment of natural killer cell function is observed throughout the course of HIV-1 infection, becoming more severe during end-stage disease. As natural killer cell cytotoxicity depends on Il-2, these abnormalities may be a consequence of impaired CD4+ T-cell function and the reduced production of this cytokine. This reduction in natural killer cell function cripples the innate immune response to infection by other microorganisms.

## Autoimmunity

Because of the imbalance of the immune system, T-cell dysfunction, and abnormal B-cell activation described above, it is not surprising that immune disorders, such as a breakdown in the system's ability to distinguish self from nonself, accompany HIV-1 infection. In early studies, antibodies against platelets, T cells, and peripheral nerves were detected in AIDS patients. Subsequently, autoantibodies to a large number of normal cellular proteins have been found in infected individuals (Table 18.7). The specific reason for the appearance of such antibodies is not clear, but their production might be stimulated in part by cellular proteins on the surface of HIV-1 particles or by viral proteins, regions of which may resemble cellular proteins (called molecular mimicry) (Table 18.8).

## Immune Response to HIV

### Humoral Responses

Antibodies to HIV-1 can be detected shortly after acute infection, sometimes as early as a few days after exposure to the virus but generally within 1 to 3 months. These antibodies are secreted into the blood and on mucosal surfaces of the body and can be detected in genital and other body fluids. This phenomenon has been exploited in the design of home kits for detecting anti-HIV antibodies in the blood or urine. Among the various isotypes, IgG1 antibodies are known to play a dominant role at all stages of infection, giving rise to an antibody-dependent cellular cytotoxicity response (known as ADCC), complement-dependent cytotoxicity, and neutralizing and blocking responses as described in chapters 14 and 15. Levels of other classes of

antibody (IgM, IgA, IgG2, IgG4, and IgD) may vary at different clinical times, but there is no known correlation between the isotype and the clinical stage of disease.

Neutralizing antibodies are likely to play a role in limiting viral replication during the early, asymptomatic stage of infection. However, the titers of these antibodies are generally very low. This property is consistent with recent structural analyses of the HIV-1 envelope protein which indicate that neutralizing epitopes are likely to be hidden from the immune system. The low titer may favor selection of resistant mutants. Indeed, many individuals produce antibodies that neutralize earlier virus isolates but not those obtained at the time of plasma collection, suggesting effective immune "escape" by the virus. Some studies show loss of neutralizing antibodies with progression to AIDS, but the clinical relevance of such antibodies during the later stages of infection remains obscure.

Neutralizing antibodies generally bind to specific sites on the viral envelope complex, SU-TM (see Fig. 4.12). Variable region 3 (V3) in HIV-1 SU is one of the initial targets. Anti-V3 antibodies appear to block coreceptor interactions that occur after the virus attaches to the major CD4 receptor. Because of the high sequence variation within the V3 loop (hence "variable"), neutralizing antibodies to this region are usually strain-specific. Consequently, de-

**Table 18.7**   Autoantibodies detected in HIV infection[a]

| Antibodies to: | Associated clinical condition |
|---|---|
| Lymphocytes | Loss of CD4+, CD8+, and B lymphocytes |
| Platelets | Thrombocytopenia |
| Neutrophils | Neutropenia |
| Erythrocytes | Anemia |
| Nerves (myelin) | Peripheral neuropathy |
| Sperm, seminal plasma | Aspermia |
| Cellular components (Golgi complex, centriole, vimentin) | Immune disorder |
| Nuclear protein (antinuclear antibody) | Autoimmune symptoms |
| Lupus anticoagulant (phospholipid, cardiolipid) | Neurological disease (?), thrombosis (?) |
| Myelin basic protein | Dementia, demyelination |
| Collagen | Arthritis (?) |
| CD4 | CD4+ cell loss |
| Hla | Lymphocyte depletion |
| Hydrocortisone | Addison's-like disease |
| Thymic hormone | Immune disorder |
| Thyroglobulin | Thyroid disease |

[a]Adapted from Table 10.3 of J. A. Levy, *HIV and the Pathogenesis of AIDS*, 2nd ed. (ASM Press, Washington, D.C., 1998), with permission.

**Table 18.8**   Regions of HIV that resemble normal cellular proteins[a]

| Normal cellular protein | HIV protein(s) | Relation |
|---|---|---|
| Hla (major histocompatibility complex) | SU | Sequence |
|  | TM | Serology |
|  | MA | Sequence |
| Il-2 | TM | Sequence |
| Il-2 receptor | Nef | Sequence |
| Thymosin | MA | Serology |
| Vasoactive intestinal polypeptide | SU (peptide T) | Sequence, serology |
| Neuroleukin (phosphohexose) | SU | Sequence, serology |
| Immunoglobulin | SU | Sequence |
| Neurotoxin | Nef, Tat, TM | Sequence |
| Protein kinase | Nef | Sequence |
| Epithelial cells | MA | Serology |
| Astrocytes | MA, TM | Serology |
| Platelet glycoprotein | SU | Serology |
| Platelets | CA | Serology |
| Brain | SU (V3 loop) | Serology |
| Fas | SU | Sequence |
| Nuclear export sequences | Rev | Sequence |

[a]Adapted from Table 10.5 of J. A. Levy, *HIV and the Pathogenesis of AIDS*, 2nd ed. (ASM Press, Washington, D.C., 1998), with permission.

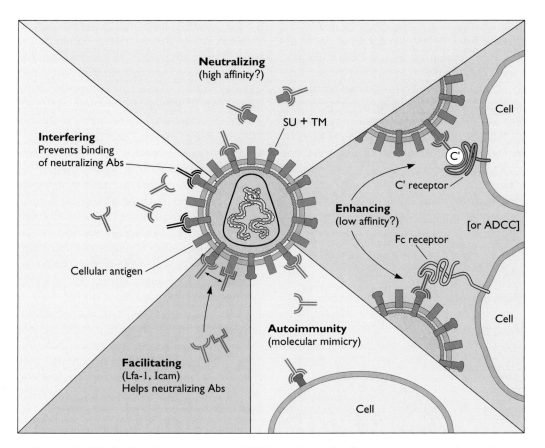

**Figure 18. 10 Antibody responses to HIV infection.** The figure summarizes the various responses described in the text. "Cellular antigen" refers to cellular membrane proteins that are incorporated in the virus envelope. C', complement; ADCC, antibody-dependent cellular cytotoxicity response.

spite its relatively strong antigenicity, V3 is not a good target for the development of vaccines or broadly specific antiviral drugs. Members of another class of neutralizing antibody block the binding of SU to the CD4 receptor. These antibodies bind to numerous sites on SU and are usually more broadly reactive than anti-V3 antibodies. Other conserved and variable regions on both SU and TM can be targets for neutralizing antibodies. Broad neutralizing activity against carbohydrate-containing regions of the viral envelope protein has also been detected, but it is still not clear whether such activity represents true anticarbohydrate reactions or conformational epitopes on SU. Even less well understood is the relative importance of antibodies to other molecules on the surface of virions, such as the adhesion molecules Lfa-1 and Icam. Antibodies to these cell surface adhesion molecules may inhibit syncytium formation, because the cells are less likely to form clumps. In some instances, these antibodies seem to enhance neutral-

ization of HIV-1 by antiviral antibodies (Fig. 18.10, bottom left).

In addition to neutralizing antibodies, an antibody-dependent cellular cytotoxicity response and interfering antibodies and infectivity-enhancing antibodies can be detected in HIV-1-infected individuals. In one antibody-dependent cellular cytotoxicity reaction, the complement system destroys cells that are coated with antibodies bound to viral proteins expressed or associated with the cell surface. Such cells may also be killed by NK cells or by monocytes/macrophages that bear Fc receptors. Antibody-dependent cellular cytotoxicity certainly has the potential to protect against infection, but its actual role is far from clear.

Some antibodies, called **interfering antibodies**, can bind to virions or infected cells and block interaction with neutralizing antibodies (Fig. 18.10, left). Others, called **enhancing antibodies**, can actually facilitate infection by allowing virions coated with them to enter susceptible cells (Fig. 18.10, right). In complement-mediated antibody enhancement, the complement receptors Cr1, Cr2, and Cr3 appear to play a critical role in attaching virion-antibody complexes to susceptible cells. In Fc-mediated enhancement, attachment is via Fc receptors that are abundant not only on monocytes/macrophages and natural killer cells but also on other human cell types. As HIV-1 has been shown to replicate in cells that lack CD4 but express an Fc receptor, binding to the CD4 receptor is probably not required for Fc-mediated enhancement.

It is noteworthy that the same cellular receptors (for complement and Fc) are implicated in infection enhancement and the antibody-dependent cellular cytotoxicity response. In the case of enhancement, the receptors allow *antibody-coated virions* to enter susceptible cells bearing such receptors. In the case of antibody-dependent cellular cytotoxicity, such receptors on cytotoxic lymphocytes, natural killer cells, or monocytes/macrophages mediate the recognition and killing of *antibody-coated infected cells.*

Both neutralizing and enhancing antibodies recognize epitopes on SU and TM. Consequently it has been difficult to identify the features that specify either response. Indeed, polyclonal antibodies against SU and the SU V3 loop have been shown to possess *both* neutralizing and enhancing activities, and it has not been possible to decide how either effect might predominate. One idea, still untested, is that the relative affinities of the antibodies may be critical. According to this hypothesis, high-affinity antibodies neutralize the virus by binding tightly to SU, causing it to become detached from the virion; low-affinity antibodies bind to SU but not tightly enough to cause its detachment. Conformational changes in SU that might occur as a consequence of such low-affinity binding would then facilitate viral entry. This idea and the distinctions between other classes of anti-HIV-1 antibodies are summarized in Fig. 18.10.

The clinical importance of antibody-dependent enhancement is uncertain, but the fact that circulating, infectious virus-antibody complexes have been described suggests that this phenomenon may contribute to HIV-1 pathogenesis. In addition, infectivity-enhancing antibodies have been demonstrated in individuals who progress to disease. Results from studies of other lentiviral infections (equine infectious anemia virus, caprine arthritis-encephalitis virus, and simian immunodeficiency virus), as well as infections with other viruses (dengue viruses, coronaviruses, and others), show a correlation between increased symptoms of disease and increased quantities of enhancing antibodies. These observations and the finding that enhancing antibodies have been found in individuals who were vaccinated with bacterially expressed Env from a different HIV-1 strain certainly complicate strategies for the development of an effective vaccine against HIV-1. As it is increasingly clear that antibodies do not play a major role in controlling HIV infections, attention has been focused on the role of cellular responses. Understanding HIV disease and developing an effective vaccine may depend on greater knowledge of the cellular immune response to HIV.

## The Cellular Immune Response

Antigen-specific cellular immune responses include activities of cytotoxic T cells and T-helper cell activities that stimulate both B cells and the cytotoxic lymphocytes. The great majority of cytotoxic lymphocytes are CD8$^+$, and their general role in limiting or suppressing viral replication is discussed in chapters 14 and 15. Such lymphocytes can also block virus replication by production of a cell antiviral factor (Caf), a protein that is distinct from all known cytokines. This protein appears to suppress viral gene transcription.

CD8$^+$ cytotoxic lymphocytes programmed to recognize virtually all HIV-1 proteins have been detected in infected individuals. As noted previously, there is a direct correlation among a good cytotoxic lymphocyte response, a low virus load, and slower disease progression. Furthermore, a broadly reactive response appears to correlate with a less fulminant course of disease. End-stage disease is characterized by a rapid drop in anti-HIV-1 CD8$^+$ cytotoxic lymphocyte activity. Animal studies with severe combined immunodeficient (SCID) mice that have been reconstituted with human lymphoid cells show that adoptive transfer of human anti-HIV-1 cytotoxic lymphocytes provides some protection against subsequent challenge with HIV. These results all demonstrate a significant role for such lymphocytes in fighting HIV-1 disease. Why, then, do these cytotoxic lymphocytes eventually fail to control infection? While the answer to this most pressing question is not yet known, it seems quite likely that most of the failure is due to reduced cytokine expression from infected and dysfunctional CD4$^+$ T cells (Table 18.6). However, loss of recognition of peptides by the cytotoxic lymphocyte receptor due to virus mutation, or simply inadequate response of these cells in the face of the high rates of virus production in AIDS patients, may also contribute to this failure.

## Summary: the Critical Balance

HIV-1 replication is controlled by what may be thought of as a finely balanced scale that can be tipped in either direction by a number of stimulatory or inhibitory host proteins. Among these, various cytokines can have important but opposing effects. Immune responses and the production of specific chemokines can inhibit virus replication, whereas immune cell activation and certain antibodies can be stimulatory (Fig. 18.11). The challenge of modern med-

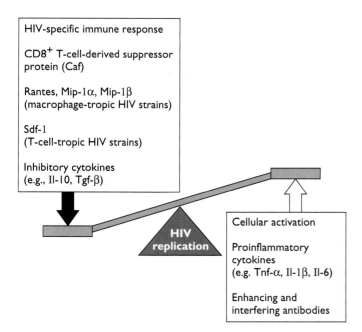

**Figure 18.11 Control of HIV-1 replication and the progress of HIV-induced disease by the balance of host proteins.** Cellular activation and proinflammatory cytokines drive viral replication. These are counterbalanced by inhibitory proteins, including both the immune system and cytokines such as Il-10 and transforming growth factor β (Tgf-β), as well as coreceptor ligands such as the β-chemokines and the α-chemokine Sdf-1. Other inhibitory proteins are secreted by CD8+ T cells (e.g., cell antiviral protein [Caf]). Adapted from Fig. 4 of A. S. Fauci, *Nature* **384:**529–533, 1996, with permission.

icine is to find practical interventions that will tip the balance in favor of patients' survival.

## Dynamics of HIV-1 Replication in AIDS Patients

The availability of potent drugs that block HIV-1 replication by inhibiting the activity of the viral enzymes reverse transcriptase (RT) and protease (PR) has made it possible to measure the dynamics of virus production in vivo. Clinical studies performed with patients near end stage, whose CD4+ T cell counts are in decline (500 or fewer per ml), have revealed the magnitude of the battle between the virus and the immune system in HIV-1 disease. Within the first 2 weeks of treatment with a combination of these drugs, an exponential decline in viral RNA in the plasma is observed; this is followed by a second rate of decline. The initial drop represents clearance of free virus and loss of virus-producing CD4+ lymphocytes from the blood. The most important contributor to the second drop is presumed to be the loss of long-lived infected cells, such as tissue macrophages and dendritic cells, with a minor contribution from the clearance of latently infected nonactivated lymphocytes (Fig. 18.12).

At steady state, in the absence of drugs the rate of virus production must equal the rate of virus clearance. Thus, mathematical analyses of the data from these clinical studies can provide estimates for the rates of HIV-1 production in the blood and other compartments of the body, as well as for the rate of loss of virus and virus-infected cells. The results calculated for HIV-1 production are nothing less than astonishing. The minimal rate estimated for release into the blood is on the order of $1 \times 10^{10}$ virions per day. This minimal number computes to approximately 240 HIV-1 infectious cycles each year. Continuous high-replication capacity is undoubtedly the principal engine that drives viral pathogenesis at this stage. Because of the high mutation rate of HIV-1, on average every possible mutation at every position in the genome is predicted to occur numerous times each day. This enormous variation and the continuous onslaught of infection must present a prodigious challenge to the immune system.

As illustrated in Fig. 18.12, 93 to 99% of the viremia is sustained by productively infected CD4+ lymphocytes with average half-lives of only ca. 1.1 days. A smaller percentage, approximately 1 to 7%, comes from long-lived cells in other compartments, with half-lives from 8.5 up to 145 days. If de novo synthesis of virus can be blocked completely by drug treatment, it could take approximately 3 to 5 years of treatment to eliminate virus from these long-lived compartments. Sadly, this can be considered only a minimal estimate. To achieve complete eradication might take even longer, as some virus may still lurk in even longer-lived, undetected "sanctuary" compartments or at sites that are not accessible to some drugs, such as the brain. As all potent HIV-1 inhibitors can produce undesirable side effects, it is still not clear how patients will tolerate such long-term treatment.

The other dramatic change that takes place in patients treated with a combination drug regimen is a resurgence of the CD4+ lymphocyte count in the blood. From the initial rates of recovery, it has been calculated that during an ongoing infection, as many as $4 \times 10^7$ of these cells are replaced in the blood each day. Lymphocytes in the peripheral blood comprise a relatively small fraction (ca. 1/50) of the total in the body, and lymphocyte trafficking, homing, and recirculation are complex processes. Thus, it is still uncertain whether such CD4+ lymphocyte replacement following drug treatment represents newly generated cells or a redistribution from other compartments. If one assumes that the increase in the circulation is proportional to the total, then as many as $2 \times 10^9$ new CD4+ cells are produced each day. This estimate is controversial because it seems to exceed significantly the normal proliferative capacity of CD4+ T cells. However, recent studies suggest that HIV-1-infected individuals who are treated with potent drug regimens do produce new CD4+ T cells at rates higher

**Blood compartment**                    **Other compartments**

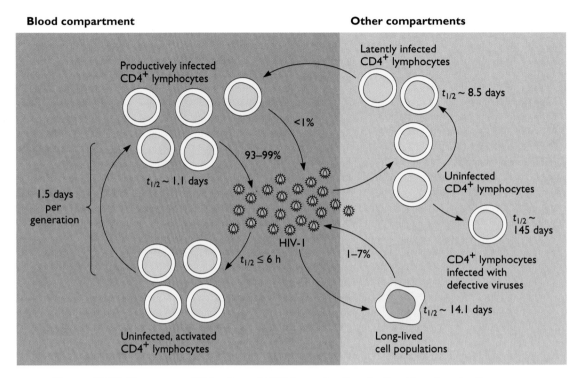

**Figure 18.12 Summary of kinetics of HIV-1 production in the blood and other compartments in vivo.** The percentages indicate the relative quantities of virus in blood plasma calculated to be produced by the various cell populations illustrated. The average time in days for 50% of the cells in each population to be destroyed or eliminated is indicated as the half-life ($t_{1/2}$). The average time in hours (h) for 50% of the virus to be eliminated from the plasma ($t_{1/2}$) is also shown. Adapted from Fig. 1 of D. D. Ho, *J. Clin. Investig.* **99:**2565–2567, 1997, with permission.

than normal. Perhaps both increased rate of synthesis and redistribution contribute to the observed resurgence in CD4[+] cells in the periphery.

Long-term studies of treated patients should help to clarify some of these issues. From recent analyses it is unlikely that a simple inverse correlation exists between virus load and number of CD4[+] cells. For example, in certain patients the virus reappears after 6 months, reaching levels near those recorded before treatment, even though CD4[+] cell counts remain high. Whether this situation can be explained by a genetic change in the virus population or some sort of reactivation of the immune system remains to be determined.

## Effects of HIV on Different Tissues and Organ Systems

### Lymphoid Organs
Lymphoid organs of the body include the lymph nodes, spleen, and gut-associated lymphoid tissue. The majority of the body's lymphocyte pool resides in such organs. Because

these lymphocytes may differ from those in the peripheral blood in significant ways, owing to the nature of cytokines in the immediate environment and the way in which the lymphocytes are exposed to virus, virus replication in the lymphoid organs may differ from that in the periphery.

The function of these organs is to filter invading pathogens, such as HIV, and to present them to immunocompetent cells. The majority of virus in lymph nodes is trapped within the germinal centers that comprise networks of follicular dendritic cells with long interdigitating processes which surround lymphocytes. These follicular dendritic cells also trap antibodies and complement and present antigens to B cells. The process of cellular activation that takes place in these organs could make resident and recruited lymphocytes permissive for HIV-1 replication. Lymph nodes appear to contain a much larger percentage of virus-infected cells than the peripheral blood.

Early in infection the germinal centers in lymph nodes appear to remain intact, although there is some hyperplasia due to the proliferation of activated immune cells (see Fig. 8.13). Early in the asymptomatic stage, infected indi-

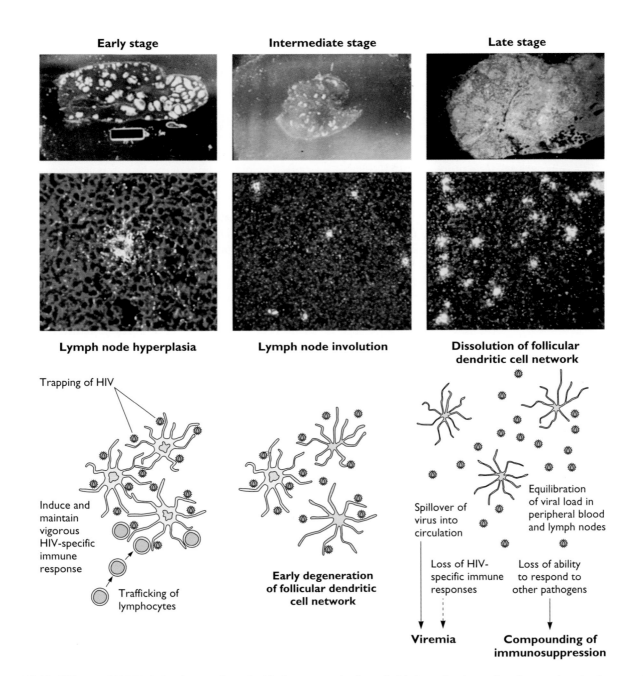

**Figure 18.13 Effects of HIV-1 infection on lymphoid tissue.**
Top panels indicate the changes in lymph node germinal centers as determined by selective staining (above) and by the location of virus replication in lymph node tissue (below). HIV-1 replication was detected by polymerase chain reaction procedures. The examples illustrate conditions in the early, intermediate, and late stages of HIV infection. The bottom panels illustrate events that take place in lymph node germinal centers during various stages of HIV-1 disease. In early-stage HIV disease (left), the virus is trapped on the follicular dendritic cell network of the germinal centers of lymphoid tissue. CD4+ T cells that migrate into the area to provide help to B cells in the germinal center that are responding to the virus are susceptible to infection by these virions. At this stage of disease, the follicular dendritic cell network is relatively intact. Over time,

the lymphoid tissue begins to involute and early degeneration of the network is observed (middle). The trapping ability of these follicular dendritic cells is then compromised. In advanced-stage disease (right), there is dissolution of the network; virus can no longer be effectively trapped and spills over readily into the circulation. At this stage, there is loss of HIV-1-specific immune responses as well as loss of the ability to respond to other pathogens. Profound immunosuppression ensues. (Top) Reprinted from Color Plate 11 of J. A. Levy, *HIV and the Pathogenesis of AIDS*, 2nd ed. (ASM Press, Washington, D.C., 1998), with permission. (Bottom) Adapted from Fig. 10 of A. S. Fauci and R. C. Desrosiers, p. 587–635, *in* J. M. Coffin et al. (ed.), *The Retroviruses* (Cold Spring Harbor Laboratory Press, Cold Spring Harbor, N.Y., 1998), with permission.

viduals often have palpable lymphadenopathy at two or more sites as a result of follicular dendritic cell hyperplasia and capillary endothelial cell proliferation. Later, during an intermediate stage of disease (CD4$^+$ counts of 200 to 500 per ml), the nodes begin to deteriorate, there is evidence of cell death, and the trapping efficiency of the follicular dendritic cells declines. At a more advanced stage of disease (below 200 CD4$^+$ cells per ml), the architecture of the lymphoid tissue is almost completely destroyed and the follicular dendritic cells disappear. At this point, there is also a significant increase of HIV-1 in the peripheral blood. This increase may be due in part to an inability of the lymphoid system to limit the amount of virus or infected cells released into the blood. Finally, because the lymphoid tissue is no longer able to mediate an immune response, there is increased incidence of opportunistic infections.

## The Nervous System

Approximately one-third of AIDS patients are diagnosed with neurological disorders at some time during the course of their disease. HIV-1 enters the nervous system early after infection, either as free virions in plasma or via infected monocytes or lymphocytes from the bloodstream. The most prevalent disorder associated with infection of the brain is **subacute encephalitis**, also called **AIDS dementia complex**; nearly two-thirds of all HIV-1-infected

individuals ultimately develop AIDS dementia. The disease progresses slowly over a period of up to 1 year, but mean survival time from the onset of severe symptoms is less than 6 months. Several AIDS-associated opportunistic infections can also produce neurological damage, and neurological abnormalities in infected individuals can be caused by lymphomas of the central nervous system (Table 18.9).

In general, brain-derived isolates of HIV-1 are macrophage-tropic and distinct from blood-derived strains from the same individual. Viral genome sequencing suggests that the brain isolates represent a distinct lineage selected for neurotropism. The precise mechanisms by which these isolates cause neuronal abnormalities and damage are not known. In animal lentivirus infections central nervous system disease usually results from direct infection of glial cells, astrocytes, and possibly brain macrophages called microglia. Infection of astrocytes and microglia also appears to be important in HIV-1 pathogenesis of the central nervous system. The virus can also be detected in brain capillary endothelial cells and has been shown to infect such cells in culture. One plausible hypothesis for how infection of the central nervous system is initiated (Fig. 18.14A) is that virus replication in the endothelial cells and astrocytes causes damage to the blood-brain barrier, allowing access of activated infected CD4$^+$ lymphocytes, macrophages, or free virus from the blood. Alternatively,

**Table 18.9** Major neurological disorders associated with HIV-1 infection[a]

| Disorder | Prevalence (%) | Clinical features[b] |
|---|---|---|
| Opportunistic infections | | |
| *Toxoplasma gondii* | 10 | Focal seizures, altered consciousness, CT ring-enhancing lesions |
| *Cryptococcus neoformans* | 4–8 | Headache, confusion, seizures, CSF cryptococcal antigen |
| Progressive multifocal leukoencephalopathy | 1–4 | Limb weakness, gait abnormalities, visual loss, altered mental state, CT-magnetic resonance imaging white matter lesions |
| Lymphoma | 5–10 | Diplopia, weakness, encephalopathy, CT contrast-enhancing lesions |
| HIV-1-related | | |
| Subacute encephalitis | 90 | Cognitive deficits, memory loss, psychomotor slowing, pyramidal tract signs, ataxia, weakness, depression, organic psychosis, incontinence, myoclonus, seizures |
| Peripheral neuropathies | 10–50 | |
| Chronic distal symmetric polyneuropathy | | Painful dyesthesias, numbness, paresthesia, weakness, autonomic dysfunction |
| Chronic inflammatory myelinating polyneuropathy | | Weakness, sensory deficits, mononeuropathy multiplex, cranial nerve palsies, hyporeflexia or areflexia, CSF pleocytosis |
| Vacuolar myelopathy | 11–22 | Gait ataxia, progressive spastic paraparesis, posterior column deficits, incontinence |
| Aseptic meningitis | 5–10 | Headache, fever, meningeal signs, cranial nerve palsies, CSF pleocytosis |

[a] Adapted from Table 4 of M. S. Hirsch and J. Curran, p. 1953–1995, *in* B. N. Fields et al. (ed.), *Fields Virology*, 3rd ed. (Lippincott-Raven, Philadelphia, Pa., 1996).
[b] CSF, cerebrospinal fluid; CT, computed tomography.

**A**

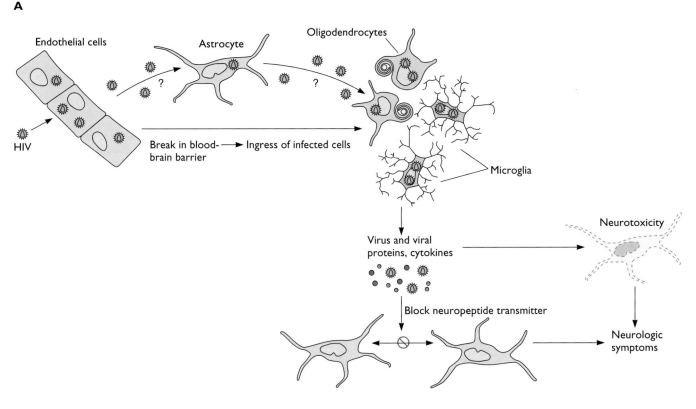

**Figure 18.14 HIV-1 neuropathogenesis.** (A) It is proposed that the virus infects capillary endothelial cells of the brain and passes into the brain via the basolateral surface of these cells. Infection of both these cells, and perhaps astrocytes, could lead to a breakdown in the blood-brain barrier and ingress of infected T cells and macrophages. The ultimate result is an infection of other brain cells (e.g., oligodendrocytes and microglia) by HIV-1. Production by these cells of HIV-1 and viral proteins ensues, as does the release of various cytokines and other cellular products. These products could lead to an interruption of cell-to-cell transmission through a blockage of neurotropic factors. Direct infection of cells, as well as high levels of viral proteins (e.g., SU, TM, Tat, and Nef) and cytokines could lead to direct neurotoxicity through detrimental effects on the cell membrane. (B) The role of infected microglia/macrophages is illustrated. After exposure to HIV or to viral envelope proteins, microglia/macrophages may be induced to produce cytokines (transforming growth factor β [Tgf-β], Inf-α, Il-1, Il-6, and Il-10) and chemokines (Mip-1α, and Mip-1β) that can be toxic to cells of the central nervous system. Endothelin-1, a monocyte-derived protein whose production can be increased on microglia/macrophage activation, has potent vasoconstricting activity in the brain. Arachidonic acid metabolites and platelet-derived activating factor (Paf) produced by activated microglia/macrophages have also been implicated in toxicity to cells of the central nervous system. High levels of quinolinic acid also produced by such activated cells have been found in the spinal fluid and brain tissue of HIV-1-infected patients and are implicated in central nervous system diseases of infected children. Elevated amounts of a monocyte-typical isoform of nitric oxide syn-

**B**

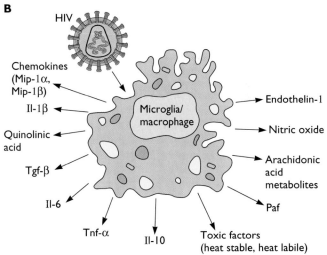

thase have been detected in the central nervous systems of patients with dementia. Production of nitric oxide may explain the death of neuronal cells that come into direct contact with infected or activated microglia/macrophages. Other low-molecular-weight toxic products (heat stable and heat labile) produced by such activated cells have been detected in high levels in the blood of HIV-1-infected patients with dementia. Adapted from p. 216 and Fig. 9.11 of J. A. Levy, *HIV and the Pathogenesis of AIDS*, 2nd ed. (ASM Press, Washington, D.C., 1998), with permission.

virus could pass through the endothelial layer to infect the adjacent astrocytes. Because astrocytes trigger production of cytokines that help maintain the blood-brain barrier, as well as cytokines that are harmful to the brain, they could play a significant role in this pathology. Neurotropic viral mutants that replicate well in microglia might later trigger production of more virus as well as harmful cytokines and other neurotoxic products (Fig. 18.14B). As HIV-1 does not appear to infect neurons, direct replication is unlikely to explain their loss. It is more probable that the release of toxic cellular products and viral proteins (e.g., SU, TM, Nef, Rev, and Tat) from infected macrophages or microglial cells are responsible for the damage to neurons. There is ample evidence from cell culture experiments that these viral proteins could contribute to neuropathogenesis. In addition, transgenic mice expressing HIV-1 *env* under the control of neuronal promoters show abnormalities in astrocytes, dendrites, and neurons similar to those seen in HIV-1-infected individuals. Finally, a role for HIV-1 proteins in neuropathogenesis is further supported by the observation that 3'-azido-3-deoxythymidine (zidovudine, or AZT) can reduce the symptoms of AIDS dementia significantly in some patients, at least for a period.

### The Gastrointestinal System

The more advanced stages of HIV disease are often associated with damage to the gastrointestinal system. Diarrhea and chronic malabsorption, with consequent malnourishment and weight loss, are frequently observed. In Africa, this has been called "slim disease." In some cases, the disorders are associated with opportunistic infections with other microbial agents, including human cytomegalovirus and herpes simplex viruses. It is possible that the combined effects of infection with human cytomegalovirus and HIV-1 can cause damage in the bowel and elsewhere. Nevertheless, in cases where no opportunistic agent can be identified, HIV-1 itself may be the primary cause of gastrointestinal pathogenesis. As noted previously, rectal tissue may be a route of entry of HIV-1, and macrophage-tropic strains can be isolated from bowel tissue. As in the central nervous system, the tissue macrophage appears to be a major target cell of HIV-1 replication in the gastrointestinal system. HIV-1-inflicted damage is thought to result from virus replication, toxic effects of viral proteins, and/or indirect destructive effects of certain cytokines.

### Other Organ Systems

Dysfunction of the hematopoietic system has been discussed above. In addition, HIV has been found in the lungs of patients with pneumonia, in the hearts of some with cardiomyopathy, in the kidneys of some with renal injury, and in the joint fluid of patients with arthritis. The role of the virus in these pathologies is not clearly understood. Opportunistic infection of the lungs is common, with *Pneumocystis carinii* being the most prevalent pathogen. Other microorganisms may also cause pulmonary infections, most notably *Mycobacterium tuberculosis*, *Mycobacterium avium*, fungi, and human cytomegalovirus. The known effects of SU on membrane permeability or other toxic effects of viral proteins might explain some of the lung damage or the electrophysiological abnormalities associated with heart disease. Direct infection of endothelial cells or other cells in the kidney has been proposed as a potential cause of tubular destruction in this organ. In addition, deposition of antigen-antibody complexes could also account for some kidney damage. The virus also has been identified in the adrenal glands of infected individuals. The significance of this observation is unclear, as decreased adrenal function could also result from cytokines produced by infected hematopoietic cells.

## HIV and Cancer

It is not surprising that the immune dysfunction resulting from HIV-1 infection leads to an increased incidence of neoplastic malignancies. Contributory factors might include the absence of proper immune surveillance directed against other (oncogenic) viruses or transformed cells. Also, high levels of cytokine production associated with HIV infection might induce inappropriate proliferation of transformed cells, activate oncogenic-virus replication, and promote the generation of blood vessels (angiogenesis) in developing tumors. Indeed, cancers that develop in HIV-infected individuals are generally more aggressive than those in uninfected individuals. These malignancies can develop in a number of tissues and organs, but certain types, namely, Kaposi's sarcoma and B-cell lymphoma, are especially prevalent in AIDS patients. The significance of such prevalence is not clear. However, it may be important that in these cases the neoplastic cells are derived from the immune system. The endothelial cells thought to give rise to Kaposi's sarcoma can act as accessory cells in lymphocyte activation; they also express Fc and complement C3 receptors. One reasonable hypothesis is that proliferation of endothelial cells, B cells, and the epithelial cells that give rise to carcinomas may be promoted by cytokines produced by immune cells in HIV-1-infected individuals. As discussed in this section, many malignancies that arise in HIV-1-infected individuals are associated with infection by oncogenic viruses.

### Kaposi's Sarcoma

This disease was first reported by the Hungarian physician Moritz Kaposi in 1872. It is a multifocal cancer that differs from more common tumors in that the lesions contain many cell types; the dominant type is a called a **spindle** cell, thought to be of endothelial origin. The tumors contain infiltrating inflammatory cells and many newly formed blood vessels. Kaposi's sarcoma was typically found in older men from the Mediterranean rim region and eastern Europe. In these areas, and in immunocompetent individuals, Kaposi's sarcoma normally appears in a nonaggressive (classical) form confined to the skin and extremities and is rarely lethal. The classical form as well as a more aggressive, sometimes lethal, form is found in sub-Saharan Africa where there are more immunocompromised individuals. In HIV-1-infected men, Kaposi's sarcoma appears in the more aggressive form, affecting both mucocutaneous and visceral areas. This disease occurs in about 20% of HIV-1-infected homosexual men (Fig. 18.15) but in only about 2% of HIV-1-infected women, transfusion-infected recipients, and blood product-infected hemophiliacs in the United States. This distribution suggests a specific type of sexual transmission.

**Figure 18.15 Kaposi's sarcoma in a young man infected with HIV-1.** Note the distribution of the lesions, suggesting lymphatic involvement. Photo courtesy of P. Volberding. Reprinted with permission from Color Plate 13 of J. A. Levy, *HIV and the Pathogenesis of AIDS*, 2nd ed. (ASM Press, Washington, D.C., 1998).

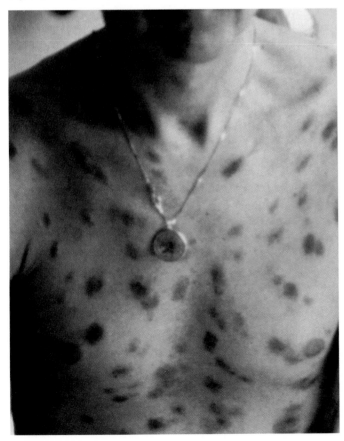

Spindle cell cultures established from Kaposi's sarcoma tumors are not fully transformed, but they do produce a variety of proinflammatory and angiogenic proteins. It is thought that these products are responsible for recruiting the other cell types in these tumors. Spindle cells from AIDS patients are also not infected with HIV-1, and the disease occurs in individuals who are not infected with the virus. Thus, HIV cannot be the sole factor in the development of Kaposi's sarcoma. Epidemiological studies suggested that another sexually transmitted virus was the inducing agent. Subsequently, a new member of the gammaherpesviruses, a lymphotropic subgroup that includes Epstein-Barr virus, was found to be associated with Kaposi's sarcoma (see Fig. 16.10). This new virus, called **human herpesvirus 8 (HHV-8)**, can infect spindle cells, and in situ hybridization studies show RNA transcripts from this virus in the vast majority of Kaposi's sarcoma lesions, irrespective of the presence or absence of HIV-1. Human herpesvirus 8 can also infect B cells and has been linked to certain AIDS-associated B-cell lymphomas (Fig. 18.16).

The evidence that infection with human herpesvirus 8 is necessary for the development of Kaposi's sarcoma is quite strong. Antibodies to this virus can be found in up to 6% of the general population but in 25% of the population in classic areas of endemicity, indicating widespread exposure. Nevertheless, very few people develop this disease. Thus, other parameters are likely to be important in its etiology. The immune deficiency associated with HIV-1 is certainly one explanation for its prevalence in AIDS patients, and expression of viral proteins, such as Tat, may be another. The exact role of human herpesvirus 8 in Kaposi's sarcoma remains to be elucidated. The viral genome includes a number of open reading frames with sequence homology to genes encoding cellular proteins known to be important in growth control, cell signaling, and immunoregulation (see Fig. 16.10). Certain viral proteins (e.g., v-Mip-1 and -11) have angiogenic properties significant for tumors of endothelioblasts. It seems likely that many of these viral proteins play a role in the pathogenesis of human herpesvirus 8.

It is ironic that one of the first reports of AIDS in 1981 described seven patients with Kaposi's sarcoma. While HIV-1 was identified soon afterward, the discovery of human herpesvirus 8 did not occur until many years later even though both viruses were present and responsible for the pathology in these first patients.

## B-Cell Lymphomas

B-cell lymphomas are 60 to 100 times more common in AIDS patients than in the general population. The incidence is especially high among patients whose survival has

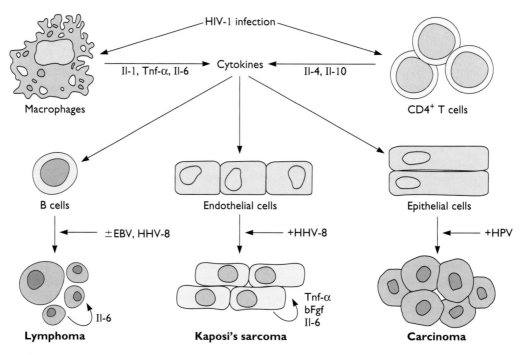

**Figure 18.16 Induction of cancers in HIV-1 infection.** Virus infection of macrophages, CD4$^+$ T cells, or other cells could lead to the production of cytokines that enhance the proliferation of certain other cells, such as B cells, endothelial cells, and epithelial cells. The enhanced replication of these cells, either through cytokine production or through subsequent viral infection, could lead to eventual development of the malignancies noted. In some cases, such as B-cell lymphomas and Kaposi's sarcoma, ongoing cytokine production by the tumor cells maintains the malignant state. Adapted from Fig. 12.11 of J. A. Levy, *HIV and the Pathogenesis of AIDS*, 2nd ed. (ASM Press, Washington, D.C., 1998), with permission.

been prolonged by anti-HIV-1 drugs. Tumors can be found in many locations, including lymph nodes, the intestine, the central nervous system, and the liver. B-cell lymphomas in the peritoneal or other body cavities, called primary effusion lymphomas or body cavity-based lymphomas, are almost always associated with human herpesvirus 8. Both polyclonal and monoclonal B-cell lymphomas are found in the central nervous system, with monoclonal types being most frequent. Epstein-Barr virus is found in all AIDS-associated primary lymphomas in the brain. Why these two B-cell lymphoma-inducing herpesviruses show such site specificity remains unknown. On the other hand, approximately 60% of the tumors outside the brain show no evidence of infection with either herpesvirus, indicating that B-cell transformation in HIV-1-infected individuals does not require infection with such viruses.

Lymphomas may arise because of the destruction of germinal centers in the lymphatic system. Lysis of antigen-presenting follicular dendritic cells would render B cells less sensitive to normal apoptotic processes, allowing them to live longer and to replicate. Epstein-Barr virus latent mem-

brane protein 1 (LMP-1) (see Table 16.5 and Fig. 16.11) also inhibits apoptosis. Proliferation of B cells as a result of production of cytokines by macrophages or CD4$^+$ T cells (Fig. 18.16), or even some viral proteins (e.g., SU or TM), may also play a role in this process. Excess proliferation, by whatever mechanism, could lead to the chromosomal changes required for cell transformation and malignancy.

## Anogenital Carcinomas

These malignancies are two to three times more frequent in HIV-1-infected individuals than in the general population and are linked to human papillomavirus infections that are typically spread through sexual contact. These papillomaviruses, particularly the high-risk human serotypes 16 and 18, have been found to be associated with both cervical and anal carcinomas. Such cancers often arise in areas of squamous metaplasia near the glandular epithelium and reach more advanced stages in immunosuppressed individuals. The proposed mechanism of transformation is mutation in the p53 tumor suppressor gene in cases where a human papillomavirus is not involved, or binding and inactivation of the tumor suppressor proteins p53 and Rb by the viral proteins E6 and E7, as described in chapter 16.

# Prospects for Treatment and Prevention

## Antiviral Drugs and Therapies

This topic will be discussed in a broader context in the following chapter. Some of the more clinically relevant details are summarized here. The first drug to be licensed in the United States for the treatment of AIDS was the nucleoside analog inhibitor of the viral reverse transcriptase, AZT. AZT is converted to the triphosphate form after being taken up into cells. On the basis of laboratory experiments, it appears to act both as a competitor of normal nucleoside triphosphates and as a chain terminator when incorporated into viral DNA by reverse transcriptase. However, its exact mechanism of action in the body is not known. Effective treatment of HIV-infected patients with AZT was limited by the rapid appearance of resistant mutants. However, as noted earlier, this drug can be used alone as prophylactic treatment for pregnant woman to reduce infection of newborns. Indeed, a brief course of administration has proven both effective and economical for this purpose in developing countries, where the availability of drugs is a major problem. AZT has also been used after needle stick injuries and is a drug of choice after other types of acute exposure to the virus. When used in this fashion, the drug can block infection in approximately 79% of the exposed individuals.

### Reverse Transcriptase and Protease Inhibitors

Nowadays, a number of other inhibitors of HIV-1 reverse transcriptase and protease are available (Table 18.10). For maximum effectiveness, these drugs are given

**Table 18.10**  Anti-HIV drugs[a]

| Names of drug[b] | Studied most in combination with: | How given | Side effects |
|---|---|---|---|
| **Reverse transcriptase inhibitors** | | | |
| Retrovir (zidovudine, AZT) | ddI, ddC, 3TC, saquinavir, ritonavir, indinavir, nevirapine | Two or three times daily | Anemia, granulocytopenia |
| Videx (didanosine, ddI) | AZT, d4T, indinavir, nevirapine | Twice daily, 1 h before or 2 h after eating | Pancreatitis, peripheral neuropathy |
| Hivid (zalcitabine, ddC) | AZT, saquinavir | Three times daily 1 h before or 2 h after eating | Peripheral neuropathy, pancreatitis |
| Zerit (stavudine, d4T) | ddI, 3TC, nelfinavir | Twice daily | Peripheral neuropathy |
| Epivir (lamivudine, 3TC) | AZT, d4T; AZT plus 3TC is well studied with all protease inhibitors | Twice daily; children who have had pancreatitis should use AZT plus 3TC only if they cannot take other anti-HIV drugs | Adults: mild headache, nausea, fatigue |
| Viramune (nevirapine) | AZT, AZT plus ddI | Once daily for first 2 weeks, then twice daily, with or without food | Rash, usually transient |
| Rescriptor (delavirdine) | AZT, ddI, AZT plus ddI | Three times daily | Rash, usually transient |
| **Protease inhibitors** | | | |
| Invirase (saquinavir) | AZT, ddC, AZT plus ddC, ritonavir | Three times daily | Mostly mild; nausea, diarrhea |
| Norvir (ritonavir) | AZT plus 3TC, AZT plus ddC, saquinavir | Twice daily with meals if possible; advised to start with 300 mg twice daily, then build up to full dose, 600 mg twice daily, within 2 weeks | Nausea, numbness around mouth, diarrhea common, especially in first weeks of therapy |
| Crixivan (indinavir) | AZT, AZT plus 3TC, AZT plus ddI | Three times daily 1 h before or 2 h after eating or with a light low-fat meal | Painful kidney stones (high water intake lowers the chance of this side effect) |
| Viracept (nelfinavir) | d4T, AZT plus 3TC | Three times daily with food | Diarrhea, usually transient |

[a]Adapted from Table 14.2 of J. A. Levy, *HIV and the Pathogenesis of AIDS*, 2nd ed. (ASM Press, Washington, D.C., 1998), with permission.

[b]The first drug name in each group, spelled with a capital letter, is the brand name (i.e., the official name approved by the U.S. Food and Drug Administration). The second name, spelled without a capital letter, is the generic name (i.e., the one used during later studies of the drug). Nucleosides are usually referred to by abbreviations of the chemical names of the drugs (i.e., AZT, ddI, ddC, d4T, and 3TC). For therapy, AZT, d4T, and nevirapine can pass the blood-brain barrier; ddI, ddC, and 3TC are most active in quiescent infected cells.

in combinations, as has proven advantageous in anti-cancer therapy. AZT and lamivudine (3TC) are often combined, and, as will be seen in the next chapter, other drug combinations are also used. The most dramatic decrease in viral RNA levels in AIDS patients has been seen with triple-drug therapy in which two anti-reverse transcriptase and one antiprotease drug are combined. Most reverse transcriptase inhibitor drugs are nucleoside analogs. Non-nucleoside inhibitors (nevirapine and delavirdine) bind to a hydrophobic pocket in the enzyme near the active site (see Fig. 19.28 and 19.31). These drugs are more strain specific; for example, nevirapine inhibits HIV-1 but not HIV-2, whose reverse transcriptase lacks a feature critical to drug binding. At the moment, all protease inhibitors are peptide mimics (peptidomimetics) that resemble a region of a substrate protein that fits into the enzyme's binding pocket (see Fig. 19.29 and 19.30). As is often the case in science and medicine, the development of inhibitors for both enzymes benefited partly from serendipity. AZT had been developed initially as a potential anticancer drug, and the earliest protease inhibitors were derivatives of renin inhibitors being developed as antihypertensive drugs. In addition, the firm base of both structural and biochemical knowledge available from research with these enzymes was a great advantage in developing this first group of drugs and is still being exploited to design the next generation of improved drugs. As can be seen in Table 18.10, all the available anti-HIV drugs must be taken frequently and have significant side effects. Thus, noncompliance is a large factor in treatment failure. Hopefully, newly developed drugs will help to minimize this problem.

### Other Approaches for Antiviral Agents

Many steps in the HIV-1 replication cycle offer opportunities for possible intervention (Table 18.11); a number have been or are currently being exploited. One obvious candidate is the third retroviral enzyme integrase. Structural data and biochemical information have become available for this protein, and a search for effective antiintegrase drugs is ongoing. Similar information for the viral proteins MA, SU, and TM has become available, opening up potential new avenues for drug design. Analogs of myristic acid, which is needed for membrane association of Gag (via MA) and Nef, can reduce viral replication, although possible deleterious effects on cellular growth have not yet been fully evaluated. Drugs that disrupt the $Zn^{2+}$ finger domain of NC may also have utility. Finally, some of the more actively pursued approaches have been directed toward the destruction of viral RNA by RNA enzymes (ribozymes) and antisense RNAs. One roadblock to their use is the necessity for devising ways to allow these RNA molecules to enter target cells.

**Table 18.11**　Potential approaches to therapies for HIV-1 infection[a]

Anti-HIV virions
　Soluble CD4
　Neutralizing antibodies (polyclonal, monoclonal)[b]
Virus entry
　CD4 analogs
　Chemokine analogs
　Galactosyl ceramide analogs
　Neutralizing antibodies[b]
　Saliva glycoprotein
　Sulfated polysaccharides
　Peptide T
Early steps prior to integration
　Anti-reverse transcriptase, e.g., nucleoside analogs (e.g., AZT, ddI, ddC, d4T), nonnucleoside inhibitors (e.g., TIBO,[c] nevirapine, foscarnet)
　Antinucleotides (e.g., MPA,[d] hydroxyurea)
　Anti-integrase[e]
　Interferons
　Hypericin
　Anti-RNA (e.g., hammerhead RNA)
Postintegration
　Anti-Tat (Ro 24-7429)
　Antiprotease[e]
　Antisense nucleotides
　Hypericin
　Ribozymes
　Cell antiviral protein
　Cytokines (immune modulation)
Virus budding
　Glycosylation inhibitors (castanospermine)
　Interferons
Anti-infected cells
　Trichosanthin (compound Q)
　Antiviral antibodies, antibody-dependent cellular cytotoxicity
　rCD4[f] toxin
　Anti-gp120 toxin
　UV-A plus psoralen
　CD8+ cells

[a] Adapted from Table 14.1 of J. A. Levy, *HIV and the Pathogenesis of AIDS*, 2nd ed. (ASM Press, Washington, D.C., 1998), with permission.
[b] By either passive immunotherapy or active immunization.
[c] TIBO, thiobenzimidazolone agents.
[d] MPA, mycophenolic acid.
[e] Retroviral enzymes.
[f] rCD4, recombinant CD4.

### Immune System-Based Therapies

Because the progression of HIV-1 disease is influenced significantly by particular cytokines, manipulation of their production is an obvious approach to therapy. For example, the administration of Il-2 alone or in combination with tumor necrosis factor alpha (Tnf-α) has produced encouraging results in that CD4+ cell counts were maintained or increased. However, as Il-2 alone activates CD4+ cells, this treatment leads to an increase in virus replication. Furthermore, the nature of these new CD4$^+$ T cells needs to be evaluated. Other cytokines, as well as drugs or antibodies that affect the levels of cytokines, have also been tested, with variable results. The administration of certain chemokines or compounds that affect the HIV coreceptor function of chemokine receptors is another approach currently under consideration. It is clear that the balance of the beneficial and detrimental effects of these agents needs to be considered carefully. Unfortunately, it is not possible to test organ system toxicity for such agents in the laboratory. Possible ways of restoring the function of the immune system in patients who are receiving antiviral therapy may also be devised, but it is not clear how well new lymphocytes can be produced in adults, as their organ of origin, the thymus, degenerates with age. Finally, investigations are under way to determine if CD8$^+$ T cells active against HIV-1 can be grown in large quantities in culture to be administered as a cell-based therapy. Despite the potential problems, it is possible that a combined approach of antiviral drugs and immune system-based therapy will ultimately be the best way to control HIV-1 disease.

### Future Challenges

All future treatment regimens must cope with the ability of the virus to hide out in inaccessible tissues or in quiescent cells. For example, it will be necessary to develop drugs that can penetrate the blood-brain barrier and enter the central nervous system to block replication of virus at this site. As latently infected cells can have long half-lives,

one suggested strategy is to treat patients with agents (e.g., cytokines or growth factors) that activate the cells to produce virus that can then be attacked by antiviral drugs. It has even been suggested that "treatment holidays," which allow virus reappearance, might stimulate the recovered immune system to mount a protective response. All of these ideas and others remain to be evaluated.

### Vaccine Development

The most potent defense against viral infections is a vaccine, a topic that is discussed in more detail in the next chapter. For HIV-1, vaccine development presents unique challenges (Box 18.2), not the least of which is the fact that the virus attacks the immune system. Other confounding issues are the large genetic diversity and high mutation rates of the virus, the fact that a vaccine must attack not only virions but also infected cells, and the danger of producing an inappropriate immune response that enhances infection. In addition, adequate animal models of HIV that are faithful to the human disease are lacking. The simian immunodeficiency virus-macaque model has been useful, and it is possible that the discovery and cloning of HIV-1 coreceptors might make it possible to design other, transgenic models. Ultimately, however, any protective vaccine must be tested in humans. Here there are many societal and ethical issues to be considered. At the present time, the best counseling must be to adjust behavior so as to avoid infection.

Despite the problems there is hope that an effective vaccine can be developed. Some encouragement can be

---

**BOX 18.2**

*The correlates of protective immunity against HIV-1 have not been well defined*

Vaccine development requires a deeper understanding of the specific immune responses essential for protection against HIV-1.

Is the humoral response essential?

Must the antibodies be neutralizing or should they be able to activate killer cells (ADCC)?

Is mucosal immunity in regions such as the rectum and vagina essential?

Are T cells essential, and, if so, which kind?

The answer to many of these questions will dictate not only the nature of the vaccine but also the manner and site of its administration.

**Possible requirements for protective immunity against HIV-1 infection$^a$**

Cytotoxic lymphocyte activity (CD4$^+$ and CD8$^+$ cells)

CD8$^+$ cell noncytotoxic antiviral response

Natural killer cell activity

Antibody-dependent cellular cytotoxicity (ADCC)

Neutralizing antibodies

$^a$Adapted from Table 15.2 of J. A. Levy, *HIV and the Pathogenesis of AIDS*, 2nd ed. (ASM Press, Washington, D.C., 1998), with permission. The presence of these responses at mucosal sites is likely to be important. If attenuated virus is used, sufficient replication of the virus must occur to induce protective immune responses.

drawn from the cross-protection against HIV-1 obtained in HIV-2-infected individuals and from other lentiviral systems, such as feline immunodeficiency virus infection of cats and equine infectious anemia virus infection of horses in which inoculation with killed, cell-free viruses has been shown to induce protection against high-dose virus challenges. Some individuals may have better innate defenses, more efficient complement, natural killer cells, and interferon responses that abrogate any primary infection. Others who have specific genetic traits that confer protection are individuals with mutations in specific chemokines or chemokine coreceptors. In addition, as indicated by a study of a group of Kenyan and Gambian prostitutes who remained uninfected despite continuous exposure to HIV-1, particular major histocompatibility complex class I alleles are associated with protection. Such protection may signify that peptides presented by these major histocompatibility complex molecules specify strongly immunogenic regions of HIV-1 proteins. The identification of such peptides might present a fruitful approach for vaccine development.

In some instances in which prior infection with HIV appears to offer protection, the infecting virus may be defective (i.e., attenuated) or the dose of infecting virus may be such that the immune system is activated and the infection is cleared. Long-term protection against simian immunodeficiency virus in adult macaques was achieved by immunization with a live attenuated virus carrying a deletion of the Nef coding sequences. Unfortunately, as discussed above, immunized female animals were found to bear infected infants that succumbed to the attenuated virus. Whether a live attenuated vaccine will ever be acceptable for humans is unclear. However, it is also important to appreciate that an AIDS vaccine need not be 100% effective to be useful. With such a deadly disease, even partial protection that might spare 20 to 40% of potential victims would extend millions of lives and reduce transmission significantly.

Research thus far suggests that the safest and most effective vaccination regimen is immunization with a live vector that expresses HIV-1 proteins, or with a DNA vaccine, followed by a booster with a viral protein or a protein-expressing vector. There are several AIDS Vaccine Evaluation Units in the United States that can carry out initial trials. However, such studies are limited by the relatively small number of potential participants in this country. To determine efficacy properly, vaccines will have to be tested in large, high-incidence populations. Such trials will require international collaboration, extensive testing in developing countries where such populations exist, and acceptance of the fact that the vaccines will have to be directed against strains that are not necessarily important to the developed world.

## Perspectives

*Pneumocystis* Pneumonia—Los Angeles. In the period October 1980–May 1981, 5 young men, all active homosexuals, were treated for biopsy-confirmed *Pneumocystis carinii* pneumonia at three different hospitals in Los Angeles, California. Two of the patients died. . . .

M. S. Gottlieb et al.
(Centers for Disease Control),
*Morbid. Mortal. Weekly Rep.* **30**:250–252, 1981

Thus began the first warning, soon echoed in large urban centers throughout the United States and Europe where physicians were being confronted by a puzzling and ominous new disease that was killing young homosexual men. In a deceptively low-key editorial note with this report it was observed that *"Pneumocystis* pneumonia is almost exclusively limited to severely immunosuppressed patients," that the occurrence of the disease "in these 5 previously healthy individuals is unusual," and that "the fact that these patients were all homosexuals suggests an association between some aspect of a homosexual lifestyle or disease acquired through sexual contacts. . . ." Soon the disease was to ravage this group and also the hemophiliac community, whose lives depended on blood products. This state of distress lead to unprecedented social activism that demanded significant public investment for AIDS research and put patient advocates on scientific review panels for the first time.

Initial progress in AIDS research was impressive. The etiological agent of the disease was identified within two years, and screening assays to safeguard the blood supply were developed shortly thereafter. As a broad base of knowledge about retroviruses already existed, it seemed that treatment and a vaccine should soon be available. Unfortunately, nearly 20 years later, AIDS is still an emerging epidemic whose death toll rises every year, with about 16,000 people newly infected each day. As is the case for most sexually transmitted diseases, efforts at prevention are fraught with difficulty and complicated by societal factors. Although a godsend to some, the treatments that have been developed to date are expensive, difficult to tolerate, and unavailable to most of the world's victims. And an effective vaccine is not yet on the horizon. Why has the initial optimism been so frustrated?

A major difficulty in combating HIV infection is the fact that the virus attacks the immune system, destroying the delicate balance needed for the system to function properly. The pathology is complex, and the physician is forced to battle not only HIV but also many other infectious agents, some rarely seen in immunocompetent patients. Other problems include the enormous replicative capacity of the virus and the lack of fidelity of reverse transcriptase that produces vast populations of mutants. Thus, drug and vaccine developers face an ever-changing target, with no good animal models to help them. Furthermore, HIV is much more complicated than was first imagined, with several auxiliary genes and structural proteins that are themselves toxic to certain cell types. Perhaps, in the end, this very complexity will provide a key to effective and affordable treatment. There are numerous steps in the virus replication cycle, and several additional viral and host proteins may be good targets for such intervention. Although we already know a great deal about the biology of HIV, to exploit these targets, uncover others, and learn what is needed to develop an effective vaccine, much more knowledge will have to be gained.

What is it that we expected from our shamans, millennia ago, and still require from the contemporary masters of the profession? To *do* something, that's what.

Lewis Thomas, *The Fragile Species*

# References

## Books
Levy, J. A. 1998. *HIV and the Pathogenesis of AIDS*, 2nd ed. ASM Press, Washington, D.C.

## Reviews
Bloom, B. R. 1996. A perspective on AIDS vaccines. *Science* **272**:1888–1890.

Boshoff, C., and R. Weiss. 1998. Kaposi's sarcoma-associated herpesvirus. *Adv. Cancer Res.* **75**:57–86.

Desrosiers, R. C., and A. S. Fauci. 1995. Retroviral pathogenesis, part II, p. 587–635. *In* H. Varmus, S. Hughes, and J. Coffin (ed.), *Retroviruses*. Cold Spring Harbor Laboratory Press, Cold Spring Harbor, N.Y.

Emerman, M., and M. H. Malim. 1998. HIV-1 regulatory/accessory genes: keys to unraveling viral and host cell biology. *Science* **280**:1880–1884.

Fauci, A. S. 1996. Host factors and the pathogenesis of HIV-induced disease. *Nature* **384**:529–533.

Haynes, B. F., G. Pantaleo, and A. S. Fauci. 1996. Toward an understanding of the correlates of protective immunity to HIV infection. *Science* **271**:324–328.

Hirsch, M. S., and J. Curran. 1996. Human immunodeficiency viruses, p. 1953–1975. *In* B. N. Fields, D. M. Knipe, P. M. Howley, R. M. Chanock, J. L. Melnick, T. P. Monath, B. Roizman, and S. E. Straus (ed.), *Fields Virology*, 3rd ed. Lippincott-Raven Publishers, Philadelphia, Pa.

Ho, D. D. 1997. Perspectives series: host/pathogen interactions. *J. Clin. Investig.* **99**:2565–2567.

Luciw, P. A. 1996. Human immunodeficiency viruses and their replication, p. 1881–1952. *In* B. N. Fields, D. M. Knipe, P. M. Howley, R. M. Chanock, J. L. Melnick, T. P. Monath, B. Roizman, and S. E. Straus (ed.), *Fields Virology*, 3rd ed. Lippincott-Raven Publishers, Philadelphia, Pa.

Subbramanian, R. A., and E. A. Cohen. 1994. Molecular biology of the human immunodeficiency virus accessory proteins. *J. Virol.* **68**:6831–6835.

## Articles of Special Interest
Barré-Sinoussi, F., J.-C. Chermann, F. Rey, M. T. Nugeyre, S. Chamaret, J. Gruest, C. Dauguet, C. Axler-Blin, F. Vézinet-Brun, C. Rouzioux, W. Rozenbaum, and L. Montagnier. 1983. Isolation of a T-lymphotropic retrovirus from a patient at risk for acquired immune deficiency syndrome (AIDS). *Science* **220**:868–871.

Berger, E. A., R. W. Doms, E.-M. Fenyö, B. T. M. Korber, D. R. Littman, J. P. Moore, Q. J. Sattentau, H. Schuitemaker, J. Sodroski, and R. A. Weiss. 1998. A new classification for HIV-1. *Nature* **391**:240.

Collins, K. L., B. K. Chen, S. A. Kalams, B. D. Walker, and D. Baltimore. 1998. HIV-1 Nef protein protects infected primary cells against killing by cytotoxic T lymphocytes. *Nature* **391**:397–401.

Emerman, M. 1996. HIV-1, Vpr and the cell cycle. *Curr. Biol.* **6**:1096–1103.

Popov, S., M. Rexach, G. Zybarth, N. Reiling, M.-A. Lee, L. Ratner, C. M. Lane, M. S. Moore, G. Blobel, and M. Bukrinsky. 1998. Viral protein R regulates nuclear import of the HIV-1 pre-integration complex. *EMBO J.* **17**:909–917.

Schneider, R., M. Campbell, G. Nasioulas, B. K. Felber, and G. N. Pavlakis. 1997. Inactivation of the human immunodeficiency virus type 1 inhibitory elements allows rev-independent expression of Gag and Gag/protease and particle formation. *J. Virol.* **71**:4892–4903.

Simon, F., P. Mauclère, P. Roques, I. Loussert-Ajaka, M. C. Müller-Trutwin, S. Saragosti, M. C. Georges-Courbot, F. Barré-Sinoussi, and F. Brun-Vézinet. 1998. Identification of a new human immunodeficiency virus type 1 distinct from group M and group O. *Nat. Med.* **4**:1032–1037.

Vodicka, M. A., D. M. Koepp, P. A. Silver, and M. Emerman. 1998. HIV-1 Vpr interacts with the nuclear transport pathway to promote macrophage infection. *Genes Dev.* **12**:175–185.

Winkler, C., W. Modi, M. W. Smith, G. W. Nelson, X. Wu, M. Carrington, M. Dean, T. Honjo, K. Tashiro, D. Yabe, S. Buchbinder, E. Vittinghoff, J. J. Goedert, T. R. O'Brien, L. P. Jacobson, R. Detels, S. Donfield, A. Willoughby, E. Gomperts, D. Vlahov, J. Phair, ALIVE Study, Hemophilia Growth and Development Study (HGDS), Multicenter AIDS Cohort Study (MACS), Multicenter Hemophilia Cohort Study (MHCS), San Francisco City Cohort (SFCC), and S. J. O'Brien. 1998. Genetic restriction of AIDS pathogenesis by an SDF-1 chemokine gene variant. *Science* **279**:389–393.

Zhu, T., B. T. Korber, A. J. Nahmias, E. Hooper, P. M. Sharp, and D. D. Ho. 1998. An African HIV-1 sequence from 1959 and implications for the origin of the epidemic. *Nature* **391**:594–597.

## Web Site
http://www.us.unaids.org

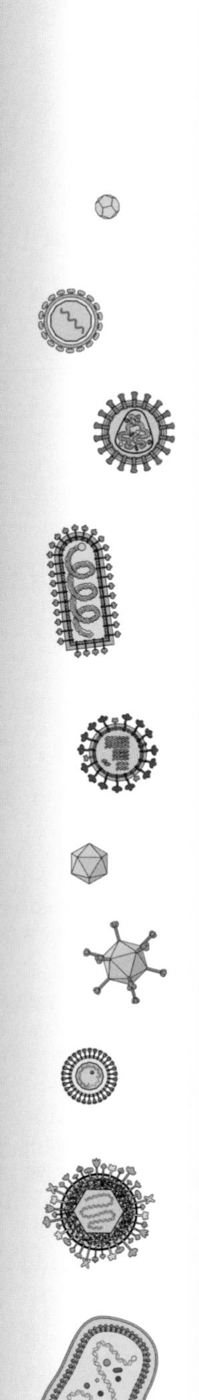

# 19

# Prevention and Control of Viral Diseases

## Introduction

In this chapter we discuss the two main efforts of antiviral defense: prevention of viral infections by vaccines and treatment of viral disease with drugs. When applied properly, both approaches can be remarkably effective. However, the proviso "when applied properly" warns of the complexity of host-virus interactions and the problems of controlling disease. While we will discuss vaccines and antiviral drugs as distinct arms of antiviral defense, there are two common threads that connect them. First, any agent or method that blocks virus replication or reduces viral pathogenesis *imposes selection* for viral mutants capable of bypassing the blocking agent. Second, as viruses are obligate intracellular parasites, any intervention in the host-virus relationship carries *inherent risks* for the host.

Good public health, proper nutrition, and effective personal hygiene policies make major contributions to the prevention of viral infections. If the sources and avenues of viral spread are eliminated and if natural defenses are maintained, many viral diseases can be reduced dramatically without resorting to vaccines or drugs. Such social measures include clean drinking water, adequate sewage disposal, insect control, good medical practice, and an uncontaminated blood supply. A highly effective, yet often unappreciated, antiviral defense is proper nutrition because the malnourished have reduced innate and acquired immune defenses. These fundamental policies must be bolstered by public awareness, understanding of potential hazards and risks, appreciation of the importance of early detection, and reporting of the incidence of infections. However, in a complex world even these measures can be ineffective, and more directed methods of preventing and treating viral infections must be employed. We begin our discussion with vaccination, the most effective antiviral intervention, which mobilizes the host's immune system to prevent viral infections. While this discussion focuses on common human viruses and vaccines, the same principles also apply to nonhuman viruses.

**663**

## Vaccines: the Proven Best Defense against Viruses

### Smallpox: a Historical Perspective

Smallpox has been called the most destructive disease in history; smallpox virus has killed, crippled, or disfigured nearly *one-tenth* of all humankind. In the 20th century alone, more than 300 million people succumbed to this disease. Nevertheless, it was the first to be eliminated by directed human intervention. It is fascinating to follow the convoluted path of discovery that made it possible to orchestrate the demise of such a deadly agent. Several thousand years ago, the Chinese knew and wrote about diseases that occur only once in a lifetime. The practical aspects of this observation were recognized by Chinese and Indian physicians of the 11th century who injected pus from smallpox lesions into healthy individuals in hope of inducing mild disease that would provide lifelong protection. Despite such early insight, little progress was made for centuries. In fact, the ancient Chinese would have recognized the practice called **variolation**, in which dried crusts of smallpox lesions were rubbed into cuts in the skin of healthy people. This dangerous practice was introduced into Europe in the beginning of the 18th century. Unfortunately, many people contracted smallpox from variolation, and the practice not only never became popular but actually was prevented by law in many countries.

The horror of smallpox is now vanishing from our collective consciousness, but its legacy for vaccines is legendary. Edward Jenner (1749–1823), a naturalist and country doctor whose previous claim to fame was a seminal paper titled "Observations on the Natural History of the Cuckoo," seems an unlikely candidate to conceive of and establish the means by which smallpox was eventually eradicated. Jenner's insight was that milkmaids might be spared the facial disfigurations of smallpox because they had cowpox infections on their hands. He put this idea to the test on 14 May 1796, when he injected pus from a cowpox lesion on the finger of milkmaid Sarah Nelmes under the skin of James Phipps, a healthy young boy. Phipps developed a fever and a lesion typical of cowpox at the site of infection. Jenner then deliberately infected Phipps with smallpox 2 weeks later, and the boy survived this potentially lethal challenge. Needless to say, such an experiment would not be possible today.

Despite this amazing result, Jenner's paper describing these experiments was rejected for publication by the Royal Society, and so he ultimately resorted to publishing his work privately. While it is Jenner's name that has been remembered, his friend, William Woodville, a prominent physician, was responsible for the first large-scale clinical trial that confirmed Jenner's observations. Thanks to Woodville's careful clinical research, the idea of harnessing the body's natural defenses as a public health measure was placed on firm ground.

The next practical vaccine for a viral disease did not appear until almost 100 years later when Louis Pasteur prepared a rabies vaccine and introduced the term **vaccination** (from *vacca*, Latin for "cow") in honor of Jenner's pioneering work. Even with Pasteur's success, other antiviral vaccines were slow to follow, largely because viruses were difficult to identify, propagate, and study. Indeed, the next antiviral vaccines (against yellow fever and influenza viruses) did not appear until the mid-1930s.

### Large-Scale Vaccination Programs Can Be Dramatically Effective

As illustrated by the eradication of smallpox, a vaccine represents the most effective antiviral defense. More success is expected: the World Health Organization (WHO) has targeted poliovirus for eradication by 2003 and measles virus for eradication shortly thereafter. Significant progress with poliovirus has already been made (Fig. 19.1). The number

**Figure 19.1 Profiles of successful vaccination campaigns.** The consequences of poliovirus (top) and measles virus (bottom) infection have been reduced or eliminated in the United States after massive vaccination programs. Subacute sclerosing panencephalitis (SSPE) is a rare but fatal brain disease that can occur 10 to 15 years after primary measles virus infection. Blocking primary measles virus infection by vaccination also eliminates SSPE. Adapted from C. A. Janeway, Jr., and P. Travers, *Immunobiology: the Immune System in Health and Disease* (Current Biology Limited, Garland Publishing Inc., New York, N.Y., 1996), with permission.

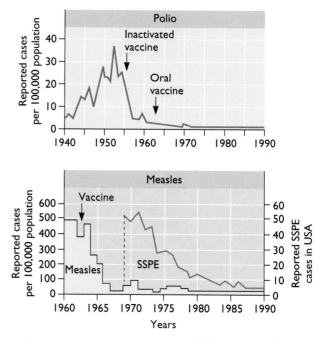

of cases reported worldwide dropped from about 350,000 to less than 35,000 by 1997. The wild virus has been eliminated from the Western Hemisphere with the last case reported in Peru in 1991. Pockets of poliovirus exist in the Eastern Hemisphere, but intense immunization programs by the World Health Organization are expected to wipe them out (Box 19.1). As an example of the magni-

tude of the effort, the World Health Organization once immunized 127 million Indian children in more than 650,000 villages in a single day. It is encouraging that measles (one of the world's five major child killers) and a rare but lethal brain infection, subacute sclerosing panencephalitis, also caused by measles virus, are rapidly disappearing in the United States because of effective vaccines

---

**BOX 19.1**

*The poliovirus eradication effort: should vaccine eradication be next?*

The worldwide effort to eradicate polio is likely to reach its goal by 2003. The World Health Organization, in cooperation with the Centers for Disease Control and Prevention, has indicated that vaccination will cease shortly thereafter, followed by the destruction of all poliovirus stocks. However, several aspects of these crucial final steps require reconsideration. Because the WHO is relying on oral poliovirus vaccine and because the Sabin oral strains readily mutate back to virulent forms, potentially pathogenic viruses are still being released into the aquifers. Immunocompromised individuals who have received live oral vaccine may also shed virus for years in the absence of clinical symptoms. We do not know how long such viruses, some of which may cause vaccine-associated poliomyelitis, will persist in the postvaccine era and pose a threat to unvaccinated individuals.

Before vaccination can be stopped, it will be necessary to destroy most existing viral stocks and restrict access to the remainder to prevent both accidental and

deliberate release into the environment. In the case of smallpox, virus stocks were located at only a few institutions before eradication, and so control of the inventory was possible. There is no central record of poliovirus stocks, which are distributed among hundreds, or possibly thousands of sites. Without an accurate inventory, it is unlikely that all virus stocks can be found and destroyed.

Even if total virus destruction could be accomplished, the small size of the poliovirus genome (7.5 kb), whose sequence is known and whose complementary DNA is infectious, would make it possible for a terrorist to synthesize a new stock of virus de novo. In the postvaccine world, the susceptible population would increase each year, and these potential sources of reintroduction would soon constitute a major threat to public health.

**Dove, A. W., and V. R. Racaniello.** 1997. The polio eradication effort: Should vaccine eradication be next? *Science* **277**:779–780.

**Hull, H. F., and R. B. Aylward.** 1997. Ending polio immunization. *Science* **277**:780.

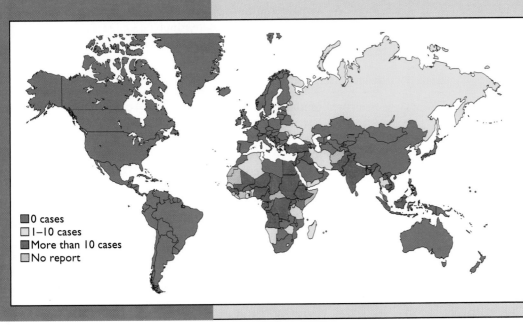

**Globally reported incidence of indigenous poliomyelitis in 1996.** The map depicts the distribution of 3,755 cases. The WHO hopes to wipe out the lingering pockets of poliovirus by 2003.

☐ 0 cases
☐ 1–10 cases
☐ More than 10 cases
☐ No report

(Fig. 19.1). Unlike the polio vaccine, which can be given orally, the current measles vaccine is injected. This requirement imposes a number of logistic and practical problems that will make worldwide eradication of measles virus more difficult.

The concept of eradicating a virus from the planet deserves careful contemplation. Given the remarkable survival strategies of viruses, the hubris of declaring a virus "eradicated" is obvious. What makes eradication conceivable? A number of parameters are important, but two are absolutely essential: the virus must have no alternative hosts, and infection (or vaccination) must produce lifelong immunity. By definition, a vaccine that renders the human population immune to subsequent infection by a virus that can grow only in humans effectively eliminates that virus. In contrast, a virus with several alternative hosts in which to propagate cannot be eliminated by vaccination of only humans. As we learned from the successful eradication of smallpox virus, a variety of other factors, summarized in Table 19.1, are also important in eradication programs (see also Box 19.2).

## Preparation of Vaccines

### The Fundamental Challenge

Given the remarkable success of the smallpox and polio vaccines, it might seem reasonable to prepare effective vaccines for all viral diseases. Unfortunately, despite considerable progress in research, we cannot predict with confidence the efficacy or side effects of vaccines. For most

**Table 19.1**  Features of smallpox that enabled its eradication

Virology and disease aspects
  No secondary hosts; it is a human-only virus
  Long incubation period
  Infectious only after incubation period
  Low communicability
  No persistent infection
  Subclinical infections are not a source of spread
  Easily diagnosed
Immunology
  Infection confers long-term immunity
  One stable serotype
  Effective vaccine available
  Vaccine is stable and cheap
Social political aspects
  Severe disease with high morbidity and mortality
  Considerable savings to developed, nonendemic countries
  Eradication from developed countries demonstrated its feasibility
  Few cultural or social barriers to case tracing and control

---

**BOX 19.2**

*Now that wild smallpox virus has been eliminated, should laboratory stocks of smallpox virus be destroyed?*

The important issues are the following:

Should we destroy biodiversity and gene pools that are not well understood?

Is wild smallpox virus simply a bad collection of genes that should be destroyed?

Should we sequence the genome and save it in a computer and then destroy the virus and all its DNA?

How do we ensure that all stocks have been destroyed?

**Fenner, F.** 1996. History of smallpox, p. 25–37. *In* H. Koprowski and M. B. A. Oldstone (ed.), *Microbe Hunters, Then and Now*. Medi-Ed Press, Bloomington, Ill.

---

viruses, we lack sufficiently detailed knowledge of the important mechanisms of immune protection. In fact, only when a vaccine is effective or when it fails can we learn what actually does or does not constitute a protective response. We know that cytotoxic T cells, antibodies, or both, are required for good antiviral defense against some viruses, but the regulatory network required to reproduce an effective antiviral response to a given virus is understood only in outline. The fundamental challenge now is to capitalize on the discoveries in molecular virology to make more effective and safe vaccines. Nevertheless, those working on vaccine discovery research and development remain optimistic because of our remarkable history of success and our increasingly detailed knowledge of the molecular nature of viruses and viral disease.

### Vaccine Basics

Many successful commercial vaccines simply comprise attenuated or killed viruses (Fig. 19.2); their preparation is based on principles that would be understood by Pasteur (Table 19.2; Box 19.3). Vaccines can be classified broadly in three general groups: **attenuated viruses**, **inactivated** (killed) **viruses**, and **purified viral components** (subunit vaccines).

#### Vaccines Stimulate Immune Memory of Viruses

Vaccines work primarily because the immune system can remember the identity of a specific virus years after the initial encounter, a phenomenon called **immune memory** (Box 19.4). While the molecular aspects of the establishment and maintenance of memory continue to be active topics of research and debate in immunology, the

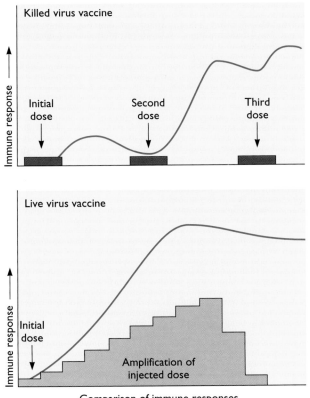

Killed virus vaccine

Initial dose    Second dose    Third dose

Live virus vaccine

Initial dose

Amplification of injected dose

Comparison of immune responses to live and killed viruses

**Figure 19.2 Comparison of the predicted immune responses to live and killed viruses used in vaccine protocols.** (Top) Immune responses plotted against time after injection of a killed virus vaccine (red curve). Three doses of killed virus were administered at the indicated positions. (Bottom) Results after injection of a live attenuated virus vaccine. A single dose was administered at the start of the experiment. The filled histogram (purple) under the curve displays the titer of infectious attenuated virus. Redrawn from C. A. Mims et al., *Mims' Pathogenesis of Infectious Disease*, 4th ed. (Academic Press, Orlando, Fla., 1995), with permission.

resounding practical success of immunization in stimulating long-lived immune memory stands as one of humankind's greatest achievements.

In general, we define immune memory as the presence of specifically dedicated T and B lymphocytes that remain after an infection has waned and maintain a heightened ability to respond to and proliferate on subsequent infection (Fig. 19.3). Antiviral vaccines establish immunity and memory without the pathogenic events typical of the initial encounter with a virulent virus. Ideally, an effective vaccine is one that induces and maintains significant concentrations of specific antibodies (products of B cells) in serum and at points of viral entry (e.g., mucosal surfaces). At the same time, T cells responsible for specific cellular immunity must be maintained in a precursor state, ready

to make their lethal products (e.g., granzymes and perforins) but not activated until challenged by subsequent infection.

### Immunization Can Be Active or Passive

An **active immunization** employs a modified form of the virus, or material originating from the virus, that induces an immunologically mediated resistance to disease. In **passive immunization**, the *products* of the immune response (e.g., antibodies or stimulated immune cells) obtained from an appropriate donor(s) are introduced directly into the patient. This is a preemptive response, usually given when a virus epidemic is suspected. In 1997, consumption of contaminated fruit led to a widespread outbreak of hepatitis A infections in the United States. Pooled human antibodies (also called immune globulin) were given to individuals who might have eaten the fruit in an attempt to block the spread of infection and reduce disease. Pooled human antibodies carry the collective experience of many individual infections and provide instant protection. When stimulated immune cells (e.g., T cells) are used, the process is called **adoptive transfer**. Passive immunization is expected to produce short-term effects whose duration depends on the biological half-life of the antibodies or immune cells. Mothers passively immunize their babies through colostrum (antibody-rich first milk) or by transfer of maternal antibody to the fetus via the placenta, providing a protective umbrella against a number of pathogens (Fig. 19.4). This protective effect can be detrimental if active immunization of infants is attempted too early, as maternal antibody may block the vaccine from stimulating immunity in the infant.

### Vaccines Must Be Safe, Efficacious, and Practical

The major prerequisites for an effective vaccine are presented in Table 19.3. The overriding requirement for any vaccine is safety: the vaccine cannot cause undue harm. It is imperative that vaccines containing inactivated virus or only viral proteins be completely free of infectious virus or contaminating viral nucleic acid. Live attenuated vaccines must not revert to virulence. Contamination of vaccines with adventitious agents introduced during production must be avoided. Some of these potential problems can be caused by human error, and some are built into the biology of the host-virus relationship. Another safety issue is immunopathology induced in response to the vaccine. An overreaction by the immune system can be as harmful as some viral infections. Almost every aspect of an immune response has the capacity both to heal and to harm. Untoward side effects have been the death knell of many otherwise effective vaccines. For some viruses such as dengue

**Table 19.2** Viral vaccines licensed in the United States[a]

| Disease or virus | Type of vaccine | Indications for use | Schedule |
|---|---|---|---|
| Hepatitis A | Inactivated whole virus | Travellers, other high-risk groups | 0, 1, and 6 mo |
| Polio (inactivated) | Inactivated whole viruses of types 1, 2, and 3 | Universal vaccination of infants | 2, 4, and 12–18 mo of age; then 4–6 yr |
| Polio (live) | Live attenuated oral mixture of types 1, 2, and 3 | Universal vaccination | 2, 4, and 6–18 mo of age |
| Yellow fever | Live attenuated | Travel in areas where infection is common | One dose every 10 yr |
| Rubella | Live attenuated | Universal vaccination of infants | Same as measles, given as MMR[b] |
| Japanese encephalitis | Inactivated whole virus | Travelers or inhabitants of high-risk areas in Asia | 0, 7, and 30 days |
| Influenza | Inactivated, viral subunits | Elderly and other high-risk groups | Two-dose primary series, then one seasonal dose |
| Measles | Live attenuated | Universal vaccination of infants | 12 mo of age; 2nd dose, 6–12 yr of age |
| Mumps | Live attenuated | Universal vaccination of infants | Same as measles, given as MMR |
| Rabies | Inactivated whole virus | Exposure to rabies, actual or prospective | 0, 3, 7, 14, and 28 days postexposure |
| Hepatitis B | Yeast-produced recombinant surface protein | Universal in children, exposure to blood, sexual promiscuity | 0, 1, 6, and 12 mo |
| Adenovirus | Live attenuated oral | Military recruits | One dose |
| Varicella | Live attenuated | Universal vaccination of infants | 12–18 mo of age |
| Smallpox | Live vaccinia virus | Certain laboratory workers | One dose |

[a]Adapted from D. D. Richman et al., *Clinical Virology* (Churchill Livingstone, New York, N.Y., 1997), with permission.
[b]MMR, measles-mumps-rubella.

virus and human immunodeficiency virus (HIV), some classes of antiviral antibodies (called enhancing antibodies) provide an alternative mechanism for entry into cells; they bind to virus, and the virus-antibody complex is taken up by macrophages or other cells with Fc receptors. In this case, a vaccine can actually potentiate the disease. Safety is paramount because vaccines are given to millions of people, many of whom may never be infected by the virus targeted by the vaccine. Moreover, as immunizations are often given to infants and expectant mothers, even a low probability of a side effect is unacceptable.

The next requirement is that the vaccine induce protective immunity in the population as a whole. Viruses must have access to a critical number of susceptible hosts to perpetuate their species. Consequently, not every individual in the population must be immunized to stop virus spread,

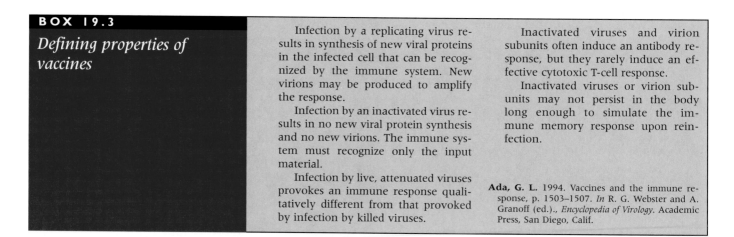

**BOX 19.3**

*Defining properties of vaccines*

Infection by a replicating virus results in synthesis of new viral proteins in the infected cell that can be recognized by the immune system. New virions may be produced to amplify the response.

Infection by an inactivated virus results in no new viral protein synthesis and no new virions. The immune system must recognize only the input material.

Infection by live, attenuated viruses provokes an immune response qualitatively different from that provoked by infection by killed viruses.

Inactivated viruses and virion subunits often induce an antibody response, but they rarely induce an effective cytotoxic T-cell response.

Inactivated viruses or virion subunits may not persist in the body long enough to simulate the immune memory response upon reinfection.

Ada, G. L. 1994. Vaccines and the immune response, p. 1503–1507. *In* R. G. Webster and A. Granoff (ed.)., *Encyclopedia of Virology*. Academic Press, San Diego, Calif.

**BOX 19.4**

*A natural experiment demonstrating immune memory*

A striking example of immune memory comes from a natural experiment in the 18th century on the Faroe Islands. In 1781 a devastating measles outbreak drastically reduced the island's population. For the next 65 years, the island remained measles-free and the surviving population flourished. In 1846 measles struck the island again, infecting over 75% of the population with similar devastating results. An astute Danish physician noted that none of the aged people who survived the 1781 measles epidemic were infected. However, their age-matched peers who had not been infected earlier were ravaged by measles.

This natural experiment illuminates two important points: immune memory lasts a long time, and memory is maintained during this time without reexposure to virus.

**Ahmed, R., and D. Gray.** 1996. Immunological memory and protective immunity: understanding their relation. *Science* **272**:54–60.

but enough people must become immune to impede virus transmission in the population. Once the virus load drops below the threshold required to sustain the infection in the population, the virus is eliminated. This has been called **herd immunity**. The threshold is agent and population specific but generally corresponds to 80 to 95% of the population having vaccine-induced immunity. This is a very large number, and achieving such high levels of immunity by vaccination is a daunting task. Moreover, if the virus remains in other populations or in alternative hosts, reinfec-

tion is always possible. Thus, epidemics can occur if the immunity of a population falls below a critical level because an available vaccine is not used. As a consequence, public complacency or reluctance to be immunized is dangerous to any vaccine program. Closed populations (e.g., military training camps) can be immunized easily, but extended populations spread over large areas of the country present serious logistic and social problems.

**Figure 19.3 Antibody and effector T cells are the basis of protective immunity.** The relative concentration of antibody and T cells is shown as a function of time after first (primary) infection. Antibody levels and activated T cells decline after the primary viral infection is cleared (purple). A second, inapparent infection is noted (tan) and recognized by the rapid increase in specific antibodies and T cells. This infection is "inapparent" because the patient shows no symptoms of the infection. The rapid immune response clears this infection with minimal effects. Reinfections at later times (years later), even if mild or inapparent, are marked by rapid and robust immune response because of "memory." Adapted from C. A. Janeway, Jr., and P. Travers, *Immunobiology: the Immune System in Health and Disease* (Current Biology Limited, Garland Publishing Inc., New York, N.Y., 1996), with permission.

**Figure 19.4 Passive transfer of antibody from mother to infant.** The fraction of the adult concentration of various antibody classes is plotted as a function of time, from conception to adult. Newborn babies have high levels of circulating IgG antibodies derived from the mother during gestation (passively transferred maternal IgG), enabling the baby to benefit from the broad immune experience of the mother. This passive protection falls to low levels at about 6 months of age as the baby's own immune response takes over. Thus, total antibody concentrations are low from about 6 months to 1 year after birth, which may lead to susceptibility to disease. Premature infants are particularly at risk to infections because the level of maternal IgG is lower and their immune system is less well developed. The time course of production of various isoforms of antibody (IgG, IgM, and IgA) produced by the baby are indicated. Adapted from C. A. Janeway, Jr., and P. Travers. *Immunobiology: the Immune System in Health and Disease* (Current Biology Limited, Garland Publishing Inc., New York, N.Y., 1996), with permission.

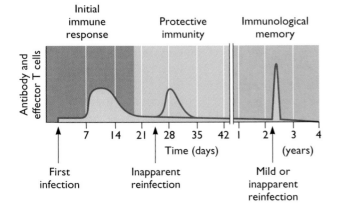

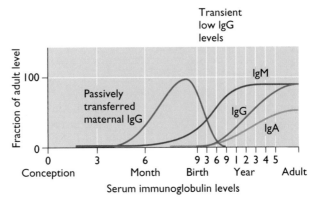

**Table 19.3**   Requirements of an effective vaccine

| | |
|---|---|
| Safety | Vaccine must not cause disease. Side effects must be minimal. |
| Induction of protective immune response | Vaccinated individual must be protected from illness due to pathogen. Proper innate, cellular, and humoral responses must be evoked by vaccine. |
| Practical issues | Cost per dose, biological stability, ease of administration, public acceptance |

The protection provided by a vaccine must also be long-term, lasting many years. Some vaccines cannot provide lifelong immunity after a single administration. For these vaccines, **booster shots** (inoculations given after the initial vaccination) to stimulate waning immunity are effective but may be impractical in large populations. Protective immunity also requires that the proper immune response be mounted at the site of potential viral infection in the host. Primary infection by some viruses, such as polio virus, can be blocked only when a robust humoral response is evoked by vaccination. For other viruses, such as many herpesviruses, a potent cellular immune response is required for protection against disease. Thus, a vaccine must be tailored to fit its viral target.

Outbred populations always have varied responses to vaccination. Some individuals exhibit a robust response, while others may not respond as well (a "poor take"). While many factors are responsible for this variability, the age and health of the recipient are major contributors. For example, the yearly "flu" vaccine is far more effective in young adults than in the elderly. Weak immune responses to vaccination present several problems. Obviously, protection against subsequent infection may be inadequate, but another concern is that on such subsequent infection, viruses will be able to replicate in the presence of the weak immune effectors. Mutants that can escape the host's immunity can then be selected and may be able to spread even in the immunized population. These mutants are particularly worrisome for vaccines directed against RNA viruses with high spontaneous mutation rates. Indeed, vaccine "escape" mutants are well documented. Commercial developers of vaccines spend considerable time and effort to ensure that their vaccines produce a protective response in the target population.

Important practical requirements for an effective vaccine include stability, ease of administration, and low cost. If a vaccine can be stored at room temperature rather than refrigerated or frozen, it can be used more effectively in the field. In 1996 the WHO stated that unless cost falls to less than $1 per dose, much of the world will not be able to afford a vaccine. However, the research and develop-

ment expense for a standard vaccine is in the range of $50 million to $350 million. Covering the liability of the vaccine producer if there should be problems is a substantial cost. Liability expenses can be astronomical in a litigious society and have forced many companies to forego vaccine development completely. Thus, there is an inherent conflict between providing a good return on investment to vaccine developers and supplying vaccines to people and government agencies with limited ability to bear the cost.

## Three Types of Vaccines

### Live Attenuated Virus Vaccines

Virulent viruses can be made less virulent (**attenuated**) by growing them in cells other than those of the normal host or by propagating them at nonphysiological temperatures (Fig. 19.5). Mutants able to propagate better than the wild-type virus under these selective conditions arise during viral replication. When such mutants are isolated, purified, and subsequently tested for pathogenicity in appropriate models, some may be less pathogenic than their parent. These mutants are good vaccine candidates because they no longer grow well in the natural host but replicate enough to stimulate an immune response. Temperature-sensitive and cold-adapted mutants are often less pathogenic than the parental virus because of reduced capacity for replication and spread in the warm-blooded host. Cold-adapted influenza viruses with mutations in almost every gene have been used as vaccines in the former Soviet Union. For viruses with segmented genomes (e.g., arena-, orthomyxo-, bunya-, and reoviruses), attenuated viruses can be produced by mixed infections with pathogenic and nonpathogenic viruses that give rise to reassortants (viruses with mixed genomes due to segment exchange).

Live oral poliovirus vaccine comprises three attenuated strains selected for their reduced neurovirulence. Types 1 and 3 vaccine strains were isolated by the passage of virulent viruses in different cells and tissues until variants with reduced neurovirulence in experimental animals were obtained (see Fig. 19.6A for a description of how type 3 was derived). The type 2 component was derived from a naturally occurring attenuated isolate. The mutations responsible for the attenuation phenotype of all three serotypes are shown in Fig. 19.6B. Curiously, each of the three serotypes contains a different mutation in the 5' noncoding region that may affect mRNA translation. Mutations in capsid proteins are believed to influence viral assembly.

The live measles virus vaccine currently in use was derived from the original wild virus isolated in 1954 and called the Edmonston strain. Attenuated viruses were isolated following serial passages through human kidney, human amnion, sheep kidney, and chick embryo cells

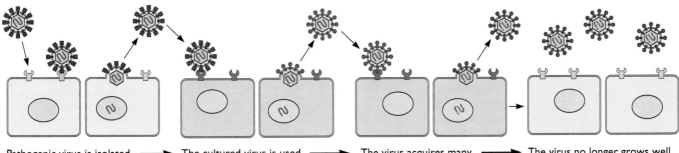

Pathogenic virus is isolated $\longrightarrow$ The cultured virus is used $\longrightarrow$ The virus acquires many $\longrightarrow$ The virus no longer grows well
from a patient and grown    to infect monkey cells    mutations that allow it to    in human cells and may be a
in human cultured cells                                 grow well in monkey cells    candidate for a vaccine

**Figure 19.5 Viruses specific for humans may become attenuated by passage in nonhuman cell lines.** The four panels show the process of producing an attenuated human virus by repeated transfers in culture cells. The first panel depicts isolation of the virus with cultured human cells (orange). The second panel shows passage of the new virus in monkey cells (purple). During the first few passages in nonhuman cells, virus yields may be low. Viruses that grow better can be selected by repeated passage as shown in the third panel. These viruses usually have several mutations, facilitating growth in nonhuman cells. The last panel shows one outcome in which the monkey cell-adapted virus now no longer grows well in human cells. This virus may also be attenuated (have reduced ability to cause disease) after human infection. Such a virus may be a candidate for a live vaccine if it will produce immunity but not disease. Adapted from C. A. Janeway, Jr., and P. Travers, *Immunobiology: the Immune System in Health and Disease* (Current Biology Limited, Garland Publishing Inc., New York, N.Y., 1996), with permission.

(Fig. 19.6D). Even though this approach was empirical, the viruses that emerged differed from wild-type virus in a number of properties. Most obvious were their temperature sensitivity and the fact that they caused markedly less disease in primates. Subsequently, the vaccine strain was demonstrated to have a reduced capacity to induce expression of an adhesion molecule that facilitates the penetration of infected monocytes into tissues. The vaccine strain also lacked the ability to reduce expression of the host complement protein CD46 (the viral receptor), a property that leads to complement-mediated lysis of infected cells. The relationship of these phenotypes to the molecular mechanisms of attenuation is under active study.

The results presented in the previous two paragraphs should give one pause. It seems ironic in this age of modern biology that the mutations in these vaccine viruses were made not by site-directed mutagenesis of genes known to be required for viral virulence but rather by selection of viruses that could replicate in various cell types. However, through this seemingly undirected or blind approach, we are actually learning about the molecular basis of measles virus pathogenicity. As a result, more defined and hypothesis-based studies on vaccine virus construction should be possible.

Successful vaccines comprising live attenuated viruses are effective for three reasons: the viruses replicate to

some extent in the host, they have a reduced capacity to spread from the site of replication, and they cause mild or inapparent disease. Live attenuated enteric viruses are administered by injection (e.g., measles-mumps-rubella vaccine) or by mouth (e.g., poliovirus, rotavirus, and adenovirus vaccines). One virtue of the oral delivery method for enteric viruses is that it mimics the natural route of infection and corresponding immune response. Another is that it bypasses the traditional need for hypodermic needles required for intramuscular or intradermal delivery. The highly effective Sabin poliovirus vaccine is administered as drops to be swallowed, and enteric adenovirus vaccines are administered as virus-impregnated tablets. Live attenuated respiratory viruses can also be delivered by nasal spray, simulating the natural route of infection and immune response, but this technology has not yet been approved in the United States.

Attenuated virus vaccines have inherent problems. For example, despite a reduced capacity of attenuated viruses to spread in the vaccinee, some virus can be shed and can then infect unvaccinated individuals. In some rare instances after administration of live attenuated poliovirus vaccine, the vaccine virus has been isolated from unvaccinated family members. When new selections are imposed after reintroduction of an attenuated strain in the natural host, virulent virus may arise as the attenuating mutations revert to wild type or as compensating mutations accumu-

**A** Derivation of Sabin type 3 attenuated poliovirus

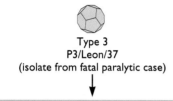

Type 3
P3/Leon/37
(isolate from fatal paralytic case)

Passages

21 passages in vivo (intracerebrally in monkeys)
8 passages in vitro (monkey testicle cultures)
39 passages in vitro (monkey kidney cultures)
3 plaque purifications (monkey kidney cultures)
3 passages in vitro (preparative, monkey kidney cultures)

P3/Leon 12a1b KP3/56 Sabin vaccine strain

**B** Determinants of attenuation in the Sabin vaccine strains

| Virus | Mutation (location/nucleotide position) |
|---|---|
| P1/Sabin | 5'-UTR (480) <br> VP1 (1106) <br> VP1 (1134) <br> VP3 (3225) <br> VP4 (4065) |
| P2/Sabin | 5'-UTR (481) <br> VP1 (1143) |
| P3/Sabin | 5'-UTR (472) <br> VP3 (3091) |

**C** Reversion of P3/Sabin

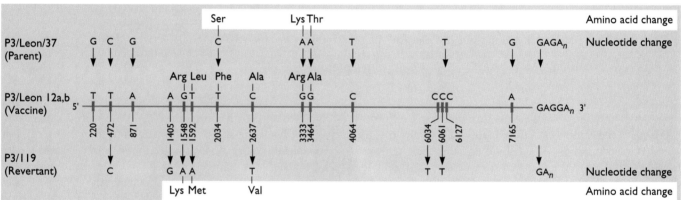

**D** Derivation of attenuated measles virus

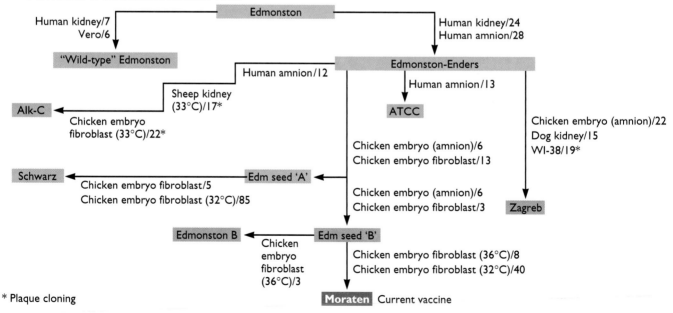

* Plaque cloning

◁ **Figure 19.6 Live attenuated oral poliovirus vaccines and measles vaccines.** (A) Derivation of the type 3 component of oral poliovirus vaccine. The virulent P3/Leon strain isolated from a fatal case of paralytic poliomyelitis was passaged serially in monkeys, monkey testicle cell cultures, and monkey kidney cell cultures. At various intervals, viruses were cloned by limiting dilution, and the virulence was determined in monkeys. An attenuated strain was selected to be the type 3 component of the live attenuated vaccine. (B) Determinants of attenuation in the Sabin vaccine strains. The mutations responsible for the reduced neurovirulence of each serotype of the live poliovirus vaccine are shown. Mutations in the 5' untranslated region (5'-UTR) are given as nucleotide numbers. Mutations in the capsid proteins VP1, VP3, and VP4 are given as a four-digit number in which the first digit is the capsid protein and the next three digits identify the amino acid within the capsid protein. (C) Reversion of P3/Sabin vaccine strain. Differences in the nucleotide sequences of the virulent P3/Leon strain and the attenuated P3/Sabin vaccine strain are shown above the cartoon of the parental viral RNA. A virus (P3/119) isolated from a case of vaccine-associated poliomyelitis is shown below. Comparison of the sequences demonstrates that P3/119 is a true revertant of P3/Sabin. (D) Passage histories of live attenuated measles virus vaccines derived from John Enders original isolate of Edmonston virus (top, blue). The current vaccine is called the Moraten strain (bottom, red). The cells used in passaging the virus to select attenuated mutants were as follows: human kidney, human amnion, Vero (African green monkey fibroblasts), chick embryo intraamniotic cavity passage, chick embryo fibroblast, dog kidney, WI-38 (human diploid fibroblasts), and sheep kidney. The temperature during growth is given in parentheses, and the number of passages of virus in the particular cell or cell line follows the slash. An asterisk indicates that a single plaque was picked for further propagation. Adapted from D. D. Richman et al., *Clinical Virology* (Churchill Livingstone, New York, N.Y., 1997), with permission.

late during replication. In the United States, the rare reported cases of polio in recent years (about 1 per 790,000 vaccinated individuals) were most likely caused by virulent revertants of the live attenuated poliovirus vaccine (Fig. 19.6C).

Given the high rate of mutation associated with RNA virus replication, a reversion to virulence should not surprise us. Such reversion is one of the main obstacles to developing effective live attenuated vaccines and is formally equivalent to the emergence of drug-resistant mutants (see "Resistance to Antiviral Drugs" below). While virulent revertants are a serious problem for vaccines, considerable insight into viral pathogenesis and the immune response can be obtained by determining the changes responsible for increased virulence.

Because our knowledge about virulence is so limited, it is difficult to predict how a live attenuated virus will behave in individuals and in the population. As a result of the attenuating mutations, unexpected diversions from the natural infection and expected host response may occur. The attenuated virus may be eliminated from the vaccinated individual before it can induce a protective response, it may infect new niches in the host with unpredictable effects, or it may initiate atypical infections (e.g., slow or chronic infections) that can trigger immunopathological responses such as **Guillain-Barré syndrome**. This syndrome may follow viral illness or vaccination and is characterized by rapidly progressing symmetric weakness of the extremities. It is the most frequent cause of acute generalized paralysis.

Ensuring purity and sterility of the product is a problem inherent in the production of biological reagents on a large scale. Attenuated viruses grown in tissue culture cells can be contaminated with viruses resident in these cells. For example, in the 1950s, early lots of poliovirus vaccine were grown in monkey cells that harbored a then unknown simian vacuolating virus, simian virus 40. It is estimated that from 10 million to 30 million persons received one or more doses of live simian virus 40 in contaminated vaccines. Many even developed antibody to simian virus 40 virion proteins. Luckily, there have been no documented complications arising from this unfortunate event.

Alternatives to the classical empirical approach to attenuation based on modern virology and recombinant DNA technology can now be applied. Viral genomes can be isolated and sequenced, and their genetic information can be expressed and tested in other biological systems. Viral genes required for pathogenesis in model systems can be identified in systematic fashion. Thus, viruses can be attenuated *deliberately*. For example, deletion mutations with exceedingly low probabilities of reversion can be created (Fig. 19.7). In another approach that relies on genome segment reassortment of influenza viruses and reoviruses, genes contributing to virulence are replaced with those from related but nonpathogenic viruses to cre-

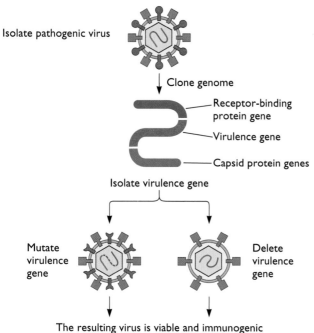

Isolate pathogenic virus

Clone genome

Receptor-binding protein gene

Virulence gene

Capsid protein genes

Isolate virulence gene

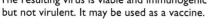

Mutate virulence gene

Delete virulence gene

The resulting virus is viable and immunogenic but not virulent. It may be used as a vaccine.

**Figure 19.7 Construction of attenuated viruses by using recombinant DNA technology.** Once the genome of a pathogenic virus is cloned in a suitable system, deletions, insertions, and point mutations can be introduced by standard recombinant DNA techniques. If the cloned genome is infectious or if mutations on plasmids can be transferred to infectious virus, it is possible to systematically mutate viral genes to find those required for producing disease. The virulence gene can then be isolated and mutated, and attenuated viruses can be produced. Such viruses can be tested for their ability to be effective live attenuated vaccines. The mutations in such attenuated viruses may be point mutations (e.g., temperature-sensitive mutations) or deletions. Multiple point mutations or deletions are preferred to reduce or eliminate the probability of reversion to virulence. Adapted from C. A. Janeway, Jr., and P. Travers, *Immunobiology: the Immune System in Health and Disease* (Current Biology Limited, Garland Publishing Inc., New York, N.Y., 1996), with permission.

ate attenuated strains. No matter which technology is applied for attenuation, the genetic engineer and the classical virologist must satisfy the same fundamental requirements: construction of a virus with low pathogenic potential that retains the ability to induce a long-lived protective immune response.

### Inactivated or "Killed" Virus Vaccines

The inactivated poliovirus vaccine, the commonly used influenza virus vaccines, the hepatitis A vaccine, and the rabies virus vaccine are examples of effective killed virus vaccines used in humans (Table 19.2). In addition, a large number of successful veterinary products comprise inacti-

vated virus vaccines. With this approach, the natural pathogenic virus is produced in quantity and inactivated by chemical or physical procedures designed to eliminate the infectivity of the virus without loss of antigenicity (i.e., the ability to induce the desired immune response). Common techniques for accomplishing the latter goal include treatment with formalin or β-propriolactone or extraction of enveloped virus particles with nonionic detergents like Triton X-100. Such vaccines are relatively uncomplicated and inexpensive to produce.

Theoretically, killed virus vaccines are very safe, but accidents can and do happen. In the 1950s, a manufacturer of Salk polio vaccine allegedly did not inactivate it completely. As a result, more than 100 children were infected with virulent poliovirus and developed disease. Incomplete inactivation of live virus and contamination of vaccine stocks with potentially infectious viral nucleic acids have been singled out as major problems with inactivated virus vaccines.

The flu vaccine is currently the most important measure for reducing influenza virus-induced morbidity and mortality. In the United States alone, seasonal influenza virus infections cause over 20,000 deaths every year and cost at least $12 billion in health care. Every year, millions of citizens seeking to avoid infection receive their flu shot, which contains several inactivated influenza viruses that have been predicted to reach the United States that year. The magnitude of this undertaking is noteworthy. For example, 75 million doses of inactivated influenza virus vaccine were manufactured in 1995. It is an amazing fact that most of the chicken eggs produced in the United States are used in the production of influenza virus vaccines. Typically, these vaccines are formalin-inactivated whole virus, detergent-disrupted or chemically disrupted virus (often called a "split" virus preparation), or purified viral membrane proteins. The viruses that are mass-produced in embryonated chicken eggs can be actual wild-type virus adapted by serial passage or reassortant viruses constructed to express the appropriate hemagglutinin (HA) or neuraminidase (NA) protein from the virulent virus.

Currently, a typical influenza vaccine dose is standardized to comprise 15 μg of each viral HA protein but may contain other viral structural proteins as well. Recent vaccines have contained antigens from two influenza A virus subtypes and antigens from one influenza B virus. The efficacy of inactivated influenza virus vaccines varies considerably and is reportedly about 60 to 90% in healthy children and adults under the age of 65 after exposure to the homologous virus (the same virus that is in the vaccine). Inactivated vaccines are all highly immunogenic in young adults but less so in the elderly, immunosuppressed individuals, and people with chronic illnesses. Protection against illness correlates with the concentration of serum antibodies produced after vaccination that react with viral HA and NA proteins. Immunization may also stimulate limited mucosal antibody production and cytotoxic T-cell activities, but these responses vary widely.

As discussed, influenza viruses continually change their envelope proteins by antigenic shift and drift during their sojourn in various animal hosts around the world. Consequently, protection one year does not guarantee protection the next. To deal with this ever-changing agent, vaccine manufacturers must reformulate influenza virus vaccine *every year* so that it contains antigens from the predicted next generation of viruses. The particular strains are chosen by the Food and Drug Administration in conjunction with World Health Organization-designated laboratories that monitor influenza infections. Timing is critical, as the final decision for the virus composition in the vaccine must be made within the first few months to allow sufficient lead time for production of the vaccine. Any delay or error in the process, from prediction to manufacture, has far-reaching consequences given the millions of people who are vaccinated and expect safe protection. Even if the vaccine contains the appropriate viral antigens and is made promptly and safely, inactivated influenza virus vaccines have the potential to cause side effects in some individuals who are allergic to the eggs in which the vaccine strains are grown. As an example of other problems, the 1997 H5N1 avian virus that infected humans was extraordinarily cytopathic to chicken embryos, making it difficult to propagate for vaccine purposes. Reassortants had to be constructed by placing the new H5N1 segments in viruses with less cytopathic genomes. The risk, of course, is that such reassortants will not provide the proper immune protection against the original strain.

### Adjuvants

Killed viruses often do not induce the same immune response as live attenuated viruses unless mixed with a substance that stimulates early processes in immune recognition, particularly the inflammatory response. Such immunostimulatory substances are called **adjuvants**. These agents help to produce a more robust acquired immune response that requires less antigen. Adjuvants work in at least three different ways: by presentation of antigen as particles, by localization of antigen to the site of inoculation, and by direct stimulation of the immune response.

Different chemicals or preparations can be used as adjuvants. They vary from complex mixtures of killed mycobacteria and mineral oil (complete Freund's adjuvant) to lipid vesicles to mixtures of aluminum salts (Table 19.4). The discovery of specific adjuvants has been largely empirical, and it is not always clear how adjuvants potentiate the immune response. Some adjuvants like alum (microparticulate aluminum hydroxide gel) are widely used for human vaccines such as hepatitis A and B vaccines. Other adjuvants, such as complete Freund's adjuvant, are mixed with viral antigens to form particles or emulsions that stimulate the inflammatory response and allow release of antigen locally over long periods (the depot effect). Freund's adjuvant is often used in the laboratory but is not licensed for use in humans because it causes tissue damage and other undesirable effects. When pro-

**Table 19.4** Vaccine delivery systems and adjuvants

| | |
|---|---|
| Aluminum salt | Aluminum hydroxide or phosphate (alum). Forms precipitates with soluble antigen, making the complexes more immunogenic; antigen "depot" at site of injection; complement activation. The only adjuvant approved for use in humans. |
| Emulsions | Freund's complete adjuvant: antigen suspended in water/mineral oil emulsion with killed *Mycobacterium tuberculosis* bacteria or muramyl di- or tripeptide to stimulate strong T-cell responses. Freund's incomplete adjuvant: antigen suspended in water/mineral oil emulsion. |
| Microspheres | Antigen encapsulated in polymers of lactic and glycolic acids. They are biodegradable and cause slow release of antigen. |
| ISCOMs | Immune-stimulating complexes composed of glycosides in an adjuvant called QuilA (a purified saponin from the plant *Quillaja saponaria*), cholesterol, phospholipids, and antigens. Form spheres 30 to 40 nm in diameter that incorporate antigen. |
| Liposomes | Antigens encapsulated in lipid vesicles; vesicles provide targeting to antigen-presenting cells on injection. Ligands can be inserted into liposome membranes to increase specificity of vesicle targeting. |
| Nucleic acid vaccines | Genes encoding antigens expressed from strong promoters are introduced directly into muscle or skin using physical methods or liposomes, leading to intracellular protein expression and presentation of antigen to the immune system. |
| Engineered viruses | Genes encoding foreign antigens are introduced into a viral genome (the vector) such that the new protein is expressed following infection. Common viral vectors are retroviruses, vaccinia virus, adenovirus, and baculovirus. Many other viruses can also be modified to express foreign proteins. |

[a]Adapted from C. A. Mims et al., *Mims' Pathogenesis of Infectious Disease*, 4th ed. (Academic Press, Orlando, Fla., 1995), with permission.

duced in large quantities, some viral structural proteins (e.g., hepatitis B virus capsid protein) spontaneously form empty capsids that act as natural adjuvants essential for stimulation of a protective immune response. New particle-forming adjuvants like liposomes and immune-stimulating complexes (ISCOMs), micelles 30 to 40 nm in diameter with viral proteins exposed on the surface (Table 19.4; Fig. 19.8), are currently being evaluated in human trials.

Powerful adjuvant effects have been obtained by mixing purified viral proteins with immunomodulators. In fact, the mycobacterial component of complete Freund's adjuvant is such a potent immunomodulator that it can cause toxic effects (e.g., local tissue necrosis) and cannot be used with human vaccines. Two of the active immunomodulatory components in Freund's adjuvant have been identified as muramyl dipeptide and lipid A. Less toxic muramyl dipeptide and lipid A derivatives, along with saponins and so-called block copolymers (linear polymers of clustered hydrophobic and hydrophilic monomers) are promising immunomodulating adjuvants.

Some new adjuvants enable exogenous proteins to enter the major histocompatibility complex (MHC) class I

pathway normally reserved for proteins synthesized de novo (Fig. 19.8). In addition, adjuvants of the saponin group, ISCOMs, and muramyl tripeptide-phosphatidylethanolamine also are reported to induce macrophages and dendritic cells to produce interleukin 12 (Il-12), which stimulates cellular immune responses. As a result, when mixed with purified viral proteins, these adjuvants have the potential to stimulate a Th1 response. Vaccine developers are seeking to uncover the mechanism involved. Other proteins, including inactivated toxins such as cholera toxin and *Escherichia coli* toxins, are being tested for their ability to induce mucosal immune responses.

Vaccine researchers speak of "tailoring" a vaccine by using different combinations of adjuvant and vaccine to produce a protective immune response. The principle is promising, but much work remains to be done to demonstrate the efficacy and safety of this approach.

### Subunit Vaccines

It is possible to make a vaccine consisting of only a subset of viral proteins, as demonstrated by successful hepatitis B and influenza vaccines. Vaccines formulated with

**Figure 19.8 ISCOMs as peptide delivery vehicles.** ISCOMs are lipid micelles that can fuse with the plasma membranes of cells. The figure shows how an ISCOM can be used to introduce defined peptides into cells to activate cytotoxic CD8+ T-cell responses. (A) When ISCOMs are formed with peptides in the solution, the resulting micelles have peptides trapped inside. (B) These trapped peptides can be delivered to the cytoplasm of antigen-presenting cells by fusion of the ISCOM with the cell membrane. (C) Some internalized peptide can be transported to the endoplasmic reticulum lumen by the TAP transporter system, where it can bind newly synthesized MHC class I proteins. (D) The peptide-MHC class I complex is then transported to the cell surface, where the peptide-MHC class I complex may be recognized by CD8+ T cells. Adapted from C. A. Janeway, Jr., and P. Travers, *Immunobiology: the Immune System in Health and Disease* (Current Biology Limited, Garland Publishing Inc., New York, N.Y., 1996), with permission.

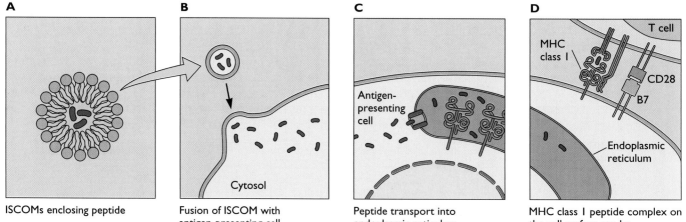

A — ISCOMs enclosing peptide

B — Fusion of ISCOM with antigen-presenting cell / Cytosol

C — Peptide transport into endoplasmic reticulum / Antigen-presenting cell

D — MHC class I peptide complex on the cell surface can be recognized by CD8 T cells / T cell / MHC class I / CD28 / B7 / Endoplasmic reticulum purified components of viruses rather than the whole virus are called **subunit vaccines**. We can deduce what viral proteins to include in a vaccine by determining what is recognized by the antibodies and cytotoxic T cells found in a person or animal recovering from viral disease. In many instances, the viral structural proteins, particularly proteins found on the virion surface, are major targets of the immune response. However, in other cases, nonstructural proteins are major targets.

Recombinant DNA methods now allow cloning of appropriate viral genes into nonpathogenic viruses, bacteria, yeast, insect cells, or plant cells (the latter potentially yielding edible vaccines) to produce the immunogenic protein. As only a portion of the viral genome is used in these systems, there can be no contamination of the resulting vaccine with the original virus, solving a major problem inherent in inactivated whole-virus vaccines. Viral proteins can be produced inexpensively in quantity by engineered organisms under conditions that simplify purification and quality control. For example, problems with egg allergies after vaccination can be eliminated completely when influenza virus HA and NA proteins are produced in *E. coli* or yeast.

Even with adjuvants, many candidate subunit vaccines fail because they do not produce an immune response sufficient to protect against viral challenge. *Protection against infection* is the "gold standard" of any vaccine. To achieve this standard, many factors, including the nature, dose, and virulence of the challenging virus, as well as the route of immunization and the age and health of the host, come into play. The immune repertoire usually evoked by a live virus infection may be only partially represented in a response to a subunit vaccine. In particular, appearance of mucosal antibodies (e.g., immunoglobulin A [IgA], the class that can be highly effective against the initial infection at the portal of entry) is rarely stimulated by purified protein antigens.

Hepatitis B subunit vaccine produced by recombinant DNA technology, which induces protective, long-lived immunity, has been an encouraging commercial success. This vaccine contains a single viral structural protein (hepatitis B virus surface antigen [HBsAg]) that assembles spontaneously into viruslike particles. Whether produced in yeast, *E. coli*, or mammalian cell lines, HBsAg is capable of forming such empty particles that are stable during purification. This property appears to be important, as purified monomeric capsid protein does not induce an effective, protective immune response (see adjuvant effect of particles in "Adjuvants"). In mice, as little as 0.025 μg of HBsAg particles can elicit antibody production without the presence of an adjuvant. Typically, in the human vaccine 10 to 20 μg of capsid protein in the form of empty capsids is

given in three doses over a 6-month period with the result that more than 95% of recipients develop antibody against the surface antigen. This response is sufficient to reduce the infection rate to about 3%, compared to 25% for placebo controls. The capsid proteins of other viruses often self-assemble into viruslike structures, a phenomenon that may be useful in the development of effective vaccines.

Synthetic peptides about 20 amino acids in length can induce specific antibody responses when chemically coupled to certain protein carriers that can be taken up, degraded, and presented by MHC class II proteins. A typical experiment begins with the identification of an "immunogenic" viral protein by demonstrating that the purified protein can induce the desired immune response in an animal. Next, the viral gene encoding this protein is cloned and sequenced. Short peptides corresponding to sections in the sequence are then synthesized chemically and tested for their ability to promote an immune response. Many experiments have demonstrated that peptides exposed on the surfaces of proteins (e.g., exposed loops) are often the targets of natural antiviral antibodies and are the best peptide immunogens. Such peptides can be identified by using computer algorithms that determine hydrophilicity (interaction with aqueous solvents) or from high-resolution structures of virions or virion proteins. Sometimes the antibodies so produced can bind to native viral proteins or are able to neutralize the virus in vitro. In principle, synthetic peptides should be the basis for an extremely safe, well-defined vaccine. To date, however, peptide vaccines remain to be exploited in commercially successful preparations mainly because synthetic peptides are expensive to make in sufficient quantity, and the antibody response they elicit is often weak and short-lived. As discussed earlier, a weak immune response can be far worse than no response at all because viral mutants that can escape the response may be selected. Given the simplicity of the antipeptide response (usually a single epitope is represented by the peptide), selection of immune escape mutants is highly probable.

Perhaps the most extensive work on peptide vaccines has been directed against foot-and-mouth disease virus, a picornavirus. Economically, it is the most important disease of cattle and other farm animals throughout the world. Several countries have mounted a considerable effort in research and field work to evaluate foot-and-mouth disease virus peptide vaccines, but results have not been encouraging. The antipeptide response is not always predictable, partial protection is common, and virus mutants with amino acid substitutions at the major antigenic site represented in the peptide vaccines can be isolated. Nevertheless, some investigators remain optimistic about the

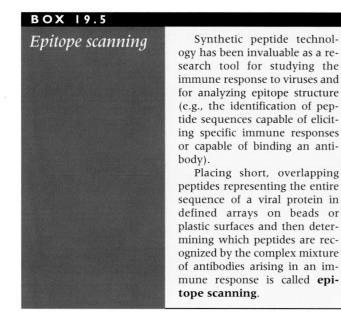

*Epitope scanning*

Synthetic peptide technology has been invaluable as a research tool for studying the immune response to viruses and for analyzing epitope structure (e.g., the identification of peptide sequences capable of eliciting specific immune responses or capable of binding an antibody).

Placing short, overlapping peptides representing the entire sequence of a viral protein in defined arrays on beads or plastic surfaces and then determining which peptides are recognized by the complex mixture of antibodies arising in an immune response is called **epitope scanning**.

future of peptide-based vaccines. Problems may be overcome by including multiple synthetic peptides (Box 19.5) recognized by both B and T cells, which should promote a more potent, multifaceted immune response, or by using ISCOMs or liposomes to deliver peptides to the MHC class I processing pathway (Fig. 19.8).

## New Vaccine Technology

### Viral Vectors

A nonpathogenic virus can be used to immunize a host against a pathogenic virus. Such a virus (often called a **vector**) can be constructed to present the pathogenic viral proteins to the immune system. Thus, the pathogen's proteins are delivered to the immune system in the context of the nonpathogenic infection promoted by the vector virus. This approach merges subunit vaccine and live attenuated virus technologies. In theory, such a system provides all the "benefits" of a viral infection with respect to the immune response with none of the pathogenesis associated with the virulent virus. The vector infection initiates a local inflammatory response and the host mobilizes all arms of its immune defense to this site, where the vaccine protein is presented and recognized and its memory is implanted in the immune system. In reality, many vectors can produce pathogenic side effects, especially in children, the elderly, and immunocompromised individuals.

Vaccinia virus provided the foundation of the vaccine program for eradicating smallpox. It was injected into millions of people and has proved to be a safe, reliable infectious agent. It continues to serve as a unique viral vector for production of foreign proteins in the cytoplasm of

mammalian cells, both as a research tool and as a live vaccine vector. Highly attenuated viruses have been isolated and studied in detail. Construction of recombinant viruses is straightforward. A wide variety of systems are available for the construction of vaccinia virus recombinants that cannot replicate in mammalian cells but allow the efficient expression of cloned gene products that retain their immunogenicity. Attenuated vaccinia virus vectors can accommodate more than 25 kb of new genetic information. A simplified outline of how to construct a recombinant vaccinia virus that expresses influenza virus HA protein is presented in Fig. 19.9. Vaccinia virus recombinants can also be used to dissect the immune response to a given protein from a pathogenic virus. This application is illustrated in Fig. 19.10. Other poxviruses, including raccoonpox, canarypox, and fowlpox viruses, offer additional possibilities because they are able to infect, but not replicate in, humans.

The successful use of an oral rabies vaccine for wild animals in Europe and the United States demonstrates that recombinant vaccinia virus vaccines have considerable potential. Live vaccinia virus recombinants that express the major envelope protein of rabies virus are formulated in edible pellets that are spread in the wild. The pellets are designed to attract the particular animal to be immunized (e.g., foxes or raccoons). The ingested vaccinia virus recombinant replicates and directs synthesis of the rabies virus envelope protein, and a protective immune response is induced.

Two general problems arise when live viral vectors are considered as vaccines. First, in most cases, the host is immunized against the viral vector as well as the vaccine antigen. As a result, subsequent uses of the particular viral vector may result in a weak response, no response, or an immunopathological response. Second, introducing live viral vectors, even though they are attenuated, into the population at large may have long-term effects. For example, immunocompromised individuals may be infected, with adverse consequences.

### DNA Vaccines

In 1992, a variation of the subunit vaccine approach was introduced. In this case, the vaccine was not a protein but rather naked DNA consisting simply of a DNA plasmid encoding a viral gene that can be expressed inside a cell (Fig. 19.11). In the simplest case, the plasmid encodes only the immunogenic viral protein under control of a strong eukaryotic promoter (the human cytomegalovirus immediate-early promoter is commonly used) and a selectable marker to facilitate maintenance of the plasmid in bacteria. The plasmid DNA produced in bacteria can be prepared free of contaminating protein and has no capacity to

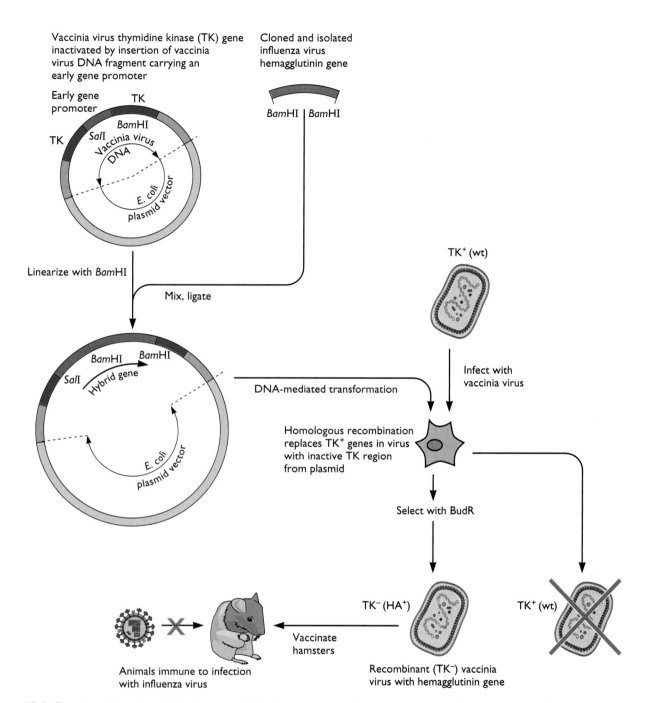

**Figure 19.9 Construction of an infectious vaccinia virus capable of expressing the influenza virus hemagglutinin gene.** An *E. coli* plasmid vector containing a segment of vaccinia virus DNA encoding the thymidine kinase (TK) gene (dark blue), as well as some contiguous viral sequences (yellow), is constructed. Next, a small segment of vaccinia virus DNA (a *Sal*I-*Bam*HI restriction fragment [light blue]) containing an early gene promoter is inserted into the TK gene, disrupting the gene. A *Bam*HI fragment of a cDNA encoding the influenza virus HA gene (red) is inserted into the *Bam*HI site immediately following the vaccinia virus early promoter. The HA gene can now be transcribed from the vaccinia virus early promoter by vaccinia virus RNA polymerase. This hybrid gene is transferred to a vaccinia virus genome by homologous recombination as follows. Plasmid DNA is introduced into a permissive cell line that has been infected by vaccinia virus with

an intact TK gene. A few percent of the progeny virus will have replaced the functional viral TK gene with the inactive hybrid TK gene containing the influenza virus HA coding sequences. These TK⁻ viruses can be selected from the mixture of virus progeny by treatment with bromodeoxyuridine (BUdR). BUdR, which is incorporated into DNA only when phosphorylated, is lethal to wild-type nonrecombinant viruses with an intact TK gene. The BUdR-resistant viruses are expected to express the influenza virus HA gene after infection. Animals infected with these recombinant viruses respond by making both antibodies and cytotoxic T cells specific for the influenza HA protein. Such animals are protected from challenge by virulent influenza virus. Adapted from N. J. Dimmock and S. B. Primrose, *Introduction to Modern Virology*, 4th ed. (Blackwell Science, Cambridge, Mass., 1994), with permission.

**679**

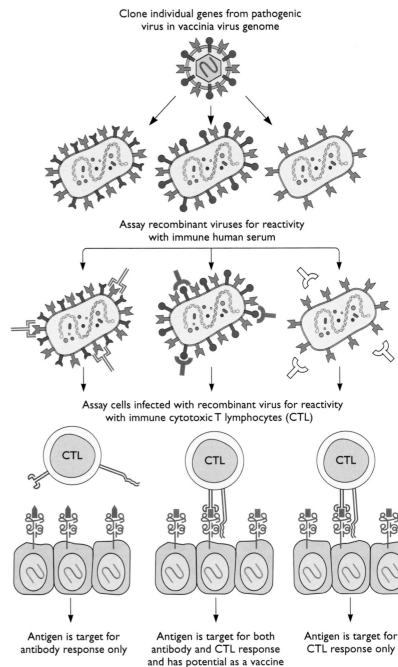

Clone individual genes from pathogenic
virus in vaccinia virus genome

Assay recombinant viruses for reactivity
with immune human serum

Assay cells infected with recombinant virus for reactivity
with immune cytotoxic T lymphocytes (CTL)

Antigen is target for
antibody response only

Antigen is target for both
antibody and CTL response
and has potential as a vaccine

Antigen is target for
CTL response only

**Figure 19.10 Use of recombinant vaccinia viruses to identify and analyze T- and B-cell epitopes from other viral pathogens.** (Top) Individual genes from a viral pathogen suspected of playing key roles in stimulating host immune defense are cloned and expressed from several recombinant vaccinia viruses by methods described in Fig. 19.9. The recombinant viruses are used to infect cells, and the resulting viruses or infected cells are tested with antiviral human antibodies or T cells (second and third panels). As each recombinant virus expresses a single gene from the viral pathogen, it is possible to determine if a particular viral protein contains a B-cell epitope (binds antibodies), a T-cell epitope (recognized by CTLs), or both. This is accomplished by reacting antibodies and T cells from immune individuals with cells infected with each recombinant virus. Subsequent site-directed mutational analysis of the viral genes enables precise localization of these epitopes on the viral protein. Adapted from C. A. Janeway, Jr., and P. Travers, *Immunobiology: the Immune System in Health and Disease* (Current Biology Limited, Garland Publishing Inc., 1996), with permission.

replicate in the vaccinated host. Remarkably, unlike the requirements for standard protein subunit vaccines, no adjuvants or special formulations are necessary to stimulate an immune response. All that is needed is that the viral protein encoded by the DNA vaccine be recognized by the immune system. An immune response is produced by injection into muscle or skin tissue of an aqueous solution containing a few micrograms of the plasmid DNA encoding an immunogenic protein from the pathogen. Another

effective method of delivery uses a "gene gun" that literally shoots DNA-coated microspheres through the skin to introduce the plasmid into dermal tissue. Once in the dermal or muscle tissue, the DNA resides in intracellular spaces until some of it enters cells by an unknown mechanism. Because the viral protein is produced de novo inside a cell, a fraction of it is presented on the surface of the producing cells by MHC class I molecules, and the cell is recognized by T cells. Both the humoral response and

Functional components of DNA vaccine plasmid

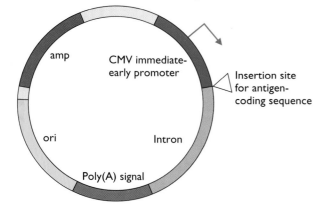

**Figure 19.11 Functional components of a DNA vaccine expression vector.** The system is based on a double-stranded plasmid, typically from *E. coli*, carrying its own origin of replication (ori) and a selectable marker (amp [ampicillin resistance]). A key component is a strong eukaryotic promoter element (here it is the major immediate-early promoter from human cytomegalovirus [CMV]). The antigen-encoding gene is placed downstream of this strong promoter at an engineered insertion site. Expression of the antigen may be enhanced by including an intron sequence (to facilitate transport of mRNA from the nucleus) and a 3' untranslated sequence that provides a polyadenylation signal as well as stability to the antigen-encoding mRNA.

cytotoxic T cells have been reported to be stimulated by DNA vaccination, and most importantly, the vaccine protects against challenge in some animal models (Table 19.5).

The striking property of DNA vaccination is that a relatively low dose of immunizing protein seems sufficient to produce long-lasting immune responses: the quantity of encoded protein produced is predicted to be extremely small, perhaps in the nanogram range, as determined by sensitive reporter proteins such as luciferase. The type of immune response appears to be dictated by the method of inoculation. After DNA immunization by injection of an aqueous DNA solution into muscle, free DNA rapidly leaves the site of injection and travels to the spleen, where the immune response occurs. Animals vaccinated in this way usually produce high levels of gamma interferon (Ifn-γ) and Il-2, low levels of Il-4 or Il-10, and cytotoxic T cells that recognize the viral protein encoded by the DNA vaccine. These patterns are typical of the Th1 response, which leads to a strong defense against many viruses. In contrast, after DNA immunization by gene gun, the DNA is taken up by skin cells and keratinocytes produce the viral antigen. Langerhans cells in the skin acquire the antigen and move to the draining lymph node, where the immune response is induced. In this case, the response produced

**Table 19.5**   DNA vaccines in animal models[a]

| Pathogen | Proteins | Induction of[b]: | | |
|---|---|---|---|---|
| | | **AB** | **CTL** | **Protection** |
| Hepatitis C virus | Nucleocapsid | + | + | + |
| Influenza virus | NP, HA, M1 | + | + | + |
| Rabies virus | Glycoprotein, NP | + | + | + |
| Lymphocytic choriomeningitis virus | NP | + | + | + |
| Human immuno-deficiency virus | Env, Gag, Rev | + | + | ND |
| Human T-cell leukemia virus type 1 | Env | + | ND | ND |
| Hepatitis B virus | Surface, core antigens | + | + | + |
| Bovine herpesvirus | Glycoprotein | + | ND | + |
| Herpes simplex virus | gB, gD, ICP27 | + | + | + |

[a]Adapted from J. B. Ulmer et al., *ASM News* **62**:476–479, 1996.

[b]ND, not detected; AB, antibody; CTL, cytotoxic T lymhocytes. "Protection" refers to survival after a lethal virus challenge.

often is more typical of a Th2 response, with high levels of Il-4 and the production of antibodies.

An interesting, but not well understood, observation is that DNA may contain short immunostimulatory sequences with adjuvant effects. The plasmid vector, which contains DNA of bacterial origin, has intrinsic immunostimulatory activity in mammalian cells. DNA motifs containing unmethylated CpG dinucleotides flanked by two 5' purine nucleotides and two 3' pyrimidines induce synthesis of immunomodulating cytokines such as interleukin-6 (Il-6), gamma interferon (Ifn-γ), Il-12, and tumor necrosis factor alpha (Tnf-α). For example, human monocytes transformed with plasmids containing the sequences 5'-GACGTC-3' and 5'-AGCGCT-3' are often stimulated to produce cytokines. It is thought that the strong adjuvant effects of complete Freund's adjuvant may be due at least in part to the mycobacterial DNA. Interestingly, the DNA of many microbes and invertebrates, including bacteria, yeast, insects, nematodes, and mollusks, has adjuvant effects, whereas the DNA of many vertebrates, including frogs and fish, is nonstimulatory. It is not clear how unmethylated CpG motifs stimulate polyclonal activation of B cells and antigen-presenting cells, but these sequences are in effect DNA-based adjuvants: they stimulate synthesis of powerful cytokines that also enhance cytotoxic lymphocyte and Th1 immune responses after DNA vaccination.

Research and development in this area have only begun, and so we cannot be sure if DNA vaccine technology has a commercial future. Most of the initial work has been done in animals, and there is little experience with DNA vaccines in humans. From a basic-research point of view, there is also much to be investigated. For example, the best immune responses occur when DNA is injected into striated muscle or skin (intradermal injection). Some DNA vaccines also work when introduced as a liquid into the noses of rodents, but it is not clear what cells take up DNA or even how DNA enters cells by this route. In particular, the types of immune responses induced by viral DNA molecules administered by different routes remain to be determined. Nevertheless, considerable data demonstrate that the response induced by DNA vaccines is protective in animals (Table 19.5). DNA vaccines directed toward influenza virus, hepatitis B virus, and herpes simplex virus are being tested in humans, for whom the safety of these vaccines is under close scrutiny. Some possible dangers might be integration of plasmid DNA leading to insertional mutagenesis, induction of autoimmune responses such as anti-DNA antibodies, and induction of immune tolerance.

An exciting possibility is a **genomic vaccine**, which comprises a library of *all* the pathogen's genes prepared in multiple DNA plasmid vectors. The entire plasmid mixture would be injected into an animal. Such a vaccine should be safe (no live organisms are used) and may be efficacious because every gene product of the pathogen can be presented to the immune system. Moreover, from a basic research perspective, this approach has much to offer: because each plasmid expresses a unique protein from a constitutive promoter, it should be straightforward to determine the gene or genes necessary and sufficient to produce a protective immune response.

## Vaccine Delivery Technology

Improvement of administration or delivery is an important goal of vaccine developers. At the present time vaccines can be administered by a variety of methods including the traditional hypodermic needle injection and the "air gun" injection of liquid vaccines under high pressure through the skin. As already noted, in some cases oral delivery can be effective in stimulating appearance of antibodies (IgA) at mucosal surfaces of the gut and in inducing a more systemic response. Specially constructed edible plants that express viral proteins represent an attractive new approach for oral delivery. Transgenic plants can be produced, or plant viruses expressing viral proteins can be used to infect food plants. Early experiments are promising: when such a plant is eaten, antibodies to the viral structural protein can be demonstrated in the animal's serum. Another direction of oral vaccine research, **formulation technology**, involves exploitation of the tactics of enteric viruses.

These viruses have evolved protective mechanisms that facilitate their survival during passage through the stomach, despite its low pH and digestive enzymes. To overcome the sensitivity of common vaccine viral vectors to passage through the gut, they can be encapsulated in biopolymers. These molecules can be formulated to resist the low pH of the stomach but dissolve in the more hospitable regions of the upper gut. Such coating technology has been used for oral delivery of adenovirus vaccines and the vaccinia virus-based wildlife rabies vaccine discussed previously. Other formulation methods under consideration include new emulsions and artificial particles.

## Immunotherapy

Patients already infected with viruses that cause persistent infections or that reactivate from latency in the face of an immune response present special problems for vaccine development. Common viruses that fall into this category are listed in Table 19.6. **Immunotherapy** provides the already infected host with antiviral cytokines, other immunoregulatory agents, antibodies, or lymphocytes in order to stimulate and bolster the immune response and reduce viral pathogenesis. As noted in chapter 18, a variety of endogenous cytokines are known to regulate HIV replication in CD4$^+$ cells, each of which is a potential imunotherapy candidate for this virus infection. Immunotherapy can be administered by introduction of the purified compounds or of a gene encoding the immunotherapeutic molecule (a gene therapy approach). A live attenuated virus or a DNA vaccine can be modified to express cytokines that stimulate the appropriate immune response to the protein encoded by the DNA vaccine. It is possible to isolate lymphocytes from patients, infect these

**Table 19.6**   Human viruses that cause persistent infections

| Virus | Disease or consequence |
|---|---|
| Enteroviruses | Chronic recurrent meningoencephalitis |
| Hepatitis C virus | Chronic hepatitis, hepatocellular carcinoma |
| Measles virus | Subacute sclerosing panencephalitis |
| Human immuno-deficiency virus | AIDS |
| Human T-cell leukemia virus | Leukemia |
| Hepatitis B virus | Chronic hepatitis, hepatocellular carcinoma |
| Parvovirus B19 | Agranulocytosis, granulocytic aplasia |
| Papillomavirus | Cervical carcinoma, warts, papillomatosis |
| Cytomegalovirus | Retinitis, pneumonia, encephalitis |
| Epstein-Barr virus | Lymphoproliferative disorder |
| Herpes simplex virus types 1 and 2 | Recurrent mucocutaneous infections, encephalitis |
| Varicella-zoster virus | Recurrent infections (shingles or zoster) |

cells with a defective virus vector (e.g., a retrovirus) carrying an activated immunoregulatory molecule, and then introduce the transduced cells back into the patient. If these cells survive and express the immunotherapeutic gene, the patient's immune response may be bolstered. If stem cells are transduced, a long-term effect may be achieved, as these cells will replicate and propagate the transgene. Of course, the consequences of long-term expression of immune modulators, which are not known, must be investigated before this principle is put into practice.

Immunotherapy can be effective. Historically, it was pioneered by using interferons to treat a variety of diseases including those caused by persistent viruses. For example, Ifn-α-2b is one drug approved in the United States for treatment of chronic hepatitis B and C. Its effect on chronic hepatitis B virus infection is remarkable: as many as 50% of the patients treated experience an apparent "cure." This result illustrates the potential of immunotherapy. However, treatment of hepatitis C virus infection with Ifn-α-2b has been much less successful, for reasons that are not clear. For treatment of hepatitis B and C virus infections, Ifn is delivered by injection, three times a week for 6 months, at a cost of over $1,000 per course of treatment. A combination therapy with the antiviral drug ribavirin [see "Ribavirin (Virazole)" below] and Ifn-α-2b has also been tested with some success and is now approved for use. Other limitations of Ifn therapy (and probably cytokine therapy in general) are that the biological activity of Ifn is not sustained for a prolonged time, side effects are significant, and patients over the age of 50 tend to be less responsive.

Immunomodulating agents, including antiviral interferons, cytokines that stimulate the Th1 response (e.g., Il-2), and certain chemokines, are being studied individually and in combination for their ability to reduce virus load and complications of persistent infections caused by papillomavirus and HIV. Cytokines that stimulate natural killer (NK) cells (e.g., Il-12 and Ifn-γ) may have promise as well. An alternative to using cytokines or immunoregulatory molecules to stimulate the immune system is the introduction of more antibodies or activated lymphocytes into an infected individual. A fundamental problem with any immunotherapy approach is that immune stimulation can lead to immunopathology or provide new avenues for virus spread in the host.

## Antiviral Drugs: Small Molecules That Block Virus Replication

### Paradox: So Much Knowledge, So Few Antivirals

Vaccines have been very successful in preventing some viral diseases; however, they provide modest to no thera-

peutic effect in individuals that are already infected. Indeed, for some viral diseases like dengue fever, an immune response to one serotype can have adverse effects if the individual is subsequently infected by a second serotype. Consequently, the second arm of antiviral defense has been the development and use of antiviral drugs capable of preventing an infection or stopping it once started. However, despite almost 50 years of research, our arsenal of antiviral drugs remains small. Because we know so much about some viruses, this dearth of antiviral drugs is perhaps unexpected. There are many reasons, but a primary one is that compounds interfering with virus growth often have adverse effects on the host. Another reason is that antiviral compounds must be virtually 100% efficient in blocking virus growth; as we will see, any virus that escapes is potentially dangerous. Achieving such high potency is remarkably difficult. Additionally, many medically important viruses are difficult or impossible to grow in the laboratory (e.g., hepatitis B virus, hepatitis C virus, and papillomaviruses); some have no available animal model of human disease (e.g., measles virus and hepatitis C virus). As a result, the testing and development of safe, efficacious compounds are difficult. Finally, as noted in chapter 15, many acute infections are of short duration, and by the time the individual feels ill, the virus is no longer replicating and indeed may already have been cleared from the host. Antiviral drugs for these viruses must be given early in the infection or prophylactically to populations at risk. It takes time not only to identify the specific virus (no broad-spectrum antiviral agents are currently available) but also to obtain and dispense the prescription antiviral drug. Under these circumstances, it is not surprising that the drug cannot be given or is not administered in time to be useful. The lack of rapid diagnostic reagents alone has hampered the development and marketing of antiviral drugs for many acute viral diseases despite the existence of effective therapies.

### Historical Perspective

The first antiviral drugs were derived from sulfonamide antibiotics in the early 1950s with the synthesis of thiosemicarbazones active against poxviruses. In the 1960s and 1970s, spurred on by success in the treatment of bacterial infections with antibiotics, drug companies launched huge **blind-screening** programs to find chemicals with antiviral activity. Despite much effort, there was relatively little success. One notable exception was amantadine (Symmetrel), approved in the late 1960s for treatment of influenza A virus infections. These antiviral discovery programs were called blind screening because no effort was made to focus discovery on a particular virus or a virus-specific mechanism (now called **mechanism-based**

**screening).** Instead, random chemicals and natural product mixtures were tested for their ability to block replication of a variety of viruses in cell culture systems. "Hits," compounds or mixtures that blocked in vitro viral replication, were then purified, and fractions were tested in various tissue culture and animal models for safety and efficacy. Promising molecules, called "leads," were modified systematically by medicinal chemists to reduce toxicity, increase solubility and bioavailability, or improve biological half-life. As a consequence, hundreds if not thousands of molecules were made and screened before a specific antiviral compound was tested in humans. Moreover, the mechanism by which these compounds inhibited the virus was often unknown. For example, in the case of amantadine, the mechanism of action was not deduced until the early 1990s.

## Discovering Antiviral Compounds

### Blind Screening Is No Longer Attractive

With the advent of modern molecular virology and recombinant DNA technology, the random, blind-screening procedures described above have been all but discarded. Viral genes essential for virus growth can be cloned and expressed in genetically tractable organisms, and their products can be purified and analyzed in molecular and atomic detail. The life cycles of many viruses are known, revealing many common targets for intervention (Table 19.7). Because of recombinant DNA technology, inhibitors of these processes can be found, even for viruses that cannot be propagated in tissue culture.

### The New Lexicon of Antiviral Discovery

Mechanism-based screens, cell-based assays, high-throughput screens, rational drug design, bioinformatics,

and combinatorial chemistry are now the jargon of modern antiviral discovery. A general outline of the pathway in modern antiviral drug discovery is shown in Fig. 19.12. This discovery process does not find drugs (compounds approved and licensed for use in humans), but rather it detects hits and supplies directions for developing them as leads. This procedure is used because the new methods cannot tell if a compound will get to the right place in the body at the appropriate concentration (bioavailability) or if it will persist in the body long enough to be effective (pharmacokinetics). They also fail to address questions of toxicity and specificity.

### Viral Genetics and Antiviral Drug Discovery

As viral genomes are small, it is possible to determine the complete sequence of essentially any viral genome in short order. With modern genetic methods, many viruses can be manipulated to test the **proof of principle** for an antiviral compound (Table 19.8). For example, if the hypothesis is that inhibition of a particular viral enzyme or

**Table 19.7** Possible targets for antiviral chemotherapy

| Target | Prototype drug |
| --- | --- |
| Attachment of virion to cell receptor | Receptor analogs, WINs[a] |
| Uncoating | Amantadine |
| Primary viral RNA synthesis in RNA viruses | RNA-dependent RNA polymerase inhibitors |
| Reverse transcription | AZT (zidovudine) |
| Regulation of RNA synthesis | HIV Tat inhibitors |
| Processing of RNA transcripts | Ribavirin |
| Translation of viral mRNA | Interferons |
| Protease processing/maturation | Protease inhibitors |
| Replication of DNA viruses | Acyclovir |
| Replication of RNA viruses | RNA-dependent RNA polymerase, helicase inhibitors |

[a]WINs, compounds from Sterling/Winthrop Inc.; initials often indicate the company name.

**Table 19.8** Key points for drug hunters seeking commercially viable antiviral drugs

**Virology issues**

Must obtain proof of principle as soon as possible.
- What is the molecular mechanism? Does the drug work in tissue culture and in animal models as predicted?

Compound should block viral spread early to limit cytopathology and host cytokine/inflammatory response.
- Drug must block replication completely and must not make an infection "persistent" by slow replication and spread.

Resistance to the antiviral drug must be manageable.
- Resistance mutations arise when any virus is permitted to replicate.
- Viruses with resistance to other drugs must not be resistant to the new drug.
- The virulence of resistant mutants must be understood.
- Noncompliance by patient may select for drug resistance or may encourage persistent infection; multiple dosing, stringent dietary requirements, or unpleasant taste or side effects affect compliance.

**Business issues**

Compound should be safe with no side effects.

Compound should be inexpensive to manufacture.

Compound should be easy to formulate and deliver.
- A pill to be swallowed is much preferred over injection.

Compound must satisfy an unmet medical need.
- That is, it must be better than any competitive drug or, better yet, have no competition.

Ultimately, a profit should be possible.
- The market should be large enough to allow a profit to be made.

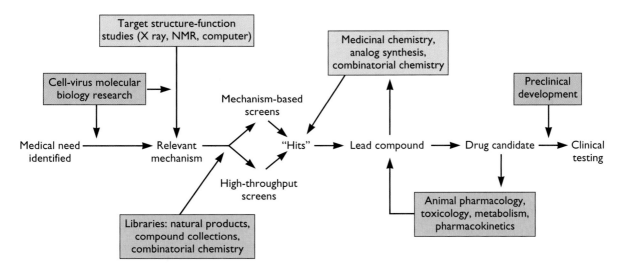

**Figure 19.12 Path of drug discovery.** The flow of information and action followed by modern drug discovery programs that ultimately yield compounds that can be tested clinically for efficacy. The process begins with recognition of a disease or medical need that can be attacked with molecular biology research. Cloning, sequencing, protein purification, and gene manipulation are carried out such that defined assays with purified components can be established to identify the gene products responsible for the disease or medical need. Once identified, the molecular mechanisms can then be studied in detail through structure-function experiments. Collections of small molecules from various sources can be tested to determine if they inhibit or activate the desired mechanism using various screening strategies. This is an iterative process because once a prospective compound is found (a hit), it can be modified and retested in the simple assays. After formulation of a hit to make it more soluble or tractable for further analysis, animal studies must be completed to determine various pharmacological parameters including toxicology and metabolism. Only then is a lead compound submitted for clinical testing. NMR, nuclear magnetic resonance.

process blocks virus replication and disease, then construction of a mutant virus in which the gene of interest is inactivated or deleted provides a rigorous test. If the mutant is unable to grow at all or cannot cause disease in model systems, the investigator has increased confidence that an inhibitor will have antiviral activity.

## Screening for Antiviral Compounds

### Mechanism-Based Screens

As the name "mechanism-based screen" implies, the specific virus-encoded enzyme or molecular interaction to be inhibited is known (or predicted), and compounds that affect this particular mechanism are sought. Such enzymes (proteases, nucleases, polymerases, helicases, kinases), transcriptional activators, cell surface receptors, and ion channels are popular targets of mechanism-based screens because the disappearance of substrate or appearance of product can be assayed directly or indirectly. Often this screening is done with purified protein in vitro with microtiter dishes or other formats that facilitate the assay of many samples. Two examples of mechanism-based screens

designed to identify inhibitors of a viral protease required for assembly are given in Fig. 19.13.

### Cell-Based Screens

Rather than depend on purified components, cell-based assays engineer essential elements of the specific mechanism to be inhibited (e.g., a viral enzyme plus its substrate) into a bacterium, yeast, vertebrate, or invertebrate cell such that inhibitors can be identified by their effects on the cell after application. Two examples of cell-based screens are illustrated in Fig. 19.14. Changes in cell growth, fluorescence, pH, or ion content can be measured manually or by sensitive electronics when the particular reaction being tested is stimulated or inhibited. Such cell-based assays can provide information not only on inhibition of the target reaction but also about cytotoxicity and specificity if suitable controls are included.

### High-Throughput Screens

High-throughput screens are mechanism-based or cell-based screens that allow very large numbers of compounds to be tested in an automated fashion. It is not unusual for

**A**

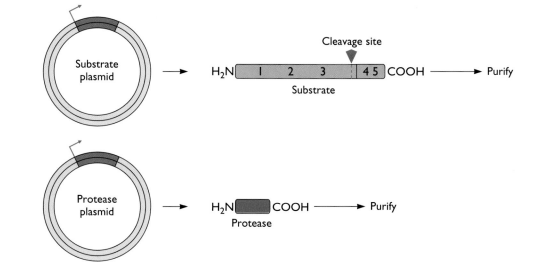

Assay:
1) Mix a constant amount of substrate with a "high" and "low" concentration of protease.
2) Controls: protease and substrate alone.
3) Incubate for set periods under appropriate conditions with and without inhibitor.
4) Sample; separate proteins by electrophoretic mobility.

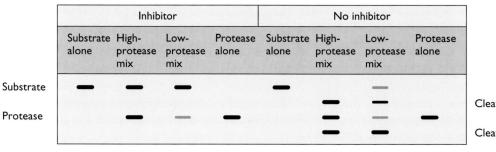

**B**

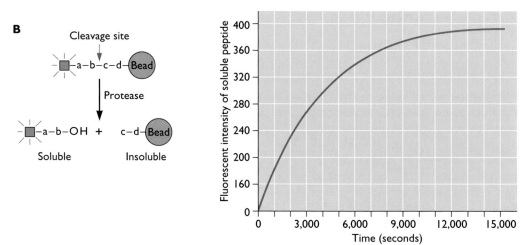

◁ **Figure 19.13 Assays for viral proteases.** (A) A mechanism-based assay for activity of a viral protease. The plasmids used to produce the protease and the protease substrate are indicated by circles at the top of the figure. The red arrows indicate the promoter and direction of transcription. The resulting proteins are indicated on the right. The amino terminus ($NH_2$) and carboxy terminus (COOH) of each protein are indicated. The substrate protein has a defined cleavage site for the protease (red arrowhead). The protease is purified as an active enzyme from *E. coli* extracts. The assay is performed as indicated. A specific inhibitor of the protease activity is added to one set of reactions. A diagram of a typical electropherogram is shown at the bottom of the figure; the four samples on the left were from reactions containing the inhibitor, and the four samples on the right were not exposed to the inhibitor. The position expected for the uncleaved substrate and the protease are shown on the left side of the image, and the specific cleavage products are noted on the right side. Experiments of this type can be used to titrate kinetic parameters of enzyme activity and inhibition. (B) Fluorogenic, mechanism-based screen for a viral protease. The substrate is a short peptide encoding the protease cleavage site. A fluorogenic molecule is covalently joined to the amino terminus of the peptide, and the entire complex is attached via its carboxy terminus to a polystyrene bead. When the peptide/bead suspension is exposed to active protease, the peptide is cleaved such that the fluorogenic N terminus is released into the soluble fraction, which can be quickly and cleanly separated from the insoluble beads containing the nonfluorogenic product as well as the fluorogenic unreacted substrate. Protease activity is assayed by the appearance of soluble fluorescent peptide as a function of time.

pharmaceutical companies to process more than 50,000 compounds per day, a rate inconceivable for early antiviral drug hunters. Compounds to be screened are usually arrayed in multiwell, interchangeable plastic dishes with a few microliters or less of compound solution per well. Robots then apply these compounds to other plastic dishes containing the cell-free or cell-based screen, and after incubation the signal created by the reporter gene or other output is read and recorded. Data are stored in computers for retrieval and analysis. Not only numerical data but also images of cells or reactions can be captured, stored, and analyzed.

### Sources of Compounds Used in Screening

Many pharmaceutical and chemical companies maintain large **libraries** of chemical compounds. Usually, a sample of every compound synthesized by the company for any project is archived, its history stored in a database. Chemical libraries of half a million or more distinct compounds are not unusual for a large company. In addition, similar libraries of natural products collected from all over the world, including "broths" from microbial fermentations, extracts of plants and marine animals, and perfusions of soils containing diverse mixtures of unknown compounds, can be searched for components that may have antiviral activities. Another type of chemical library is now available from application of **combinatorial chemistry**, a technology that provides drug hunters access to unprecedented numbers of small synthetic molecules for screening (Fig. 19.15). Before implementation of this technology, a medicinal chemist could reliably make only about 50 compounds a year. Combinatorial chemistry produces all possible combinations of a basic set of modu-

lar components, often on uniquely tagged microbeads or other chemical supports, such that active compounds in the mixtures can be traced, purified, and identified with relative ease. Making thousands of unique compounds in days is now routine with this technology.

## Designer Antivirals and Computer-Based Searching

### Structure-Based Drug Design

Structure-based design depends on having the atomic structure of the target molecule, usually provided by X-ray crystallography. Computer programs, known and predicted mechanisms of enzyme action, fundamental chemistry, and personal insight all aid an investigator in the design of ligands that will bind at the active site of an enzyme, or at another critical point, and inhibit its activity. Currently, the atomic structures of more than 6,000 macromolecules, including important viral proteins, have been determined. An example of how structure-based design has been used in the discovery of new influenza virus NA inhibitors is detailed below (see "Virus Spread: Structure-Based Design of Influenza Virus Neuraminidase" below). HIV protease inhibitors are the most successful antiviral agents that have been designed and are also being improved by structure-based analyses (Fig. 19.16) (see "Examples of Drugs Approved by the Food and Drug Administration for Treatment of HIV Infection" below).

### Genome Sequencing Provides New Information for Antiviral Drug Discovery

We now can anticipate having the entire sequence not only of any viral genome but also of the host genome.

**A**

Active tetracycline efflux protein; insertion of protease site has no effect

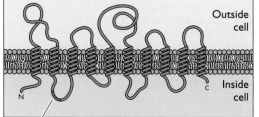

Outside cell

**Tetracycline-resistant bacteria**

Inside cell

Engineered HIV protease site

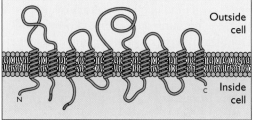

Outside cell

Inside cell

HIV protease

Coexpression of HIV protease leads to inactivation of the tetracycline protein

Inactive tetracycline efflux protein

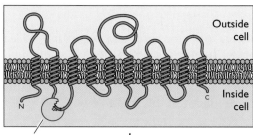

Outside cell

**Tetracycline-sensitive bacteria**

Inside cell

Active tetracycline efflux protein

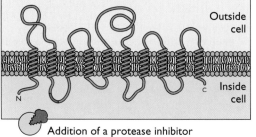

Outside cell

**Tetracycline-resistant bacteria**

Inside cell

Addition of a protease inhibitor blocks cleavage, leaving an active tetracycline efflux protein

**B**

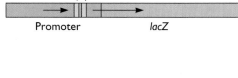

Transcription regulatory protein binding sites

Promoter            *lacZ*

A transcription regulatory protein, by virtue of its DNA binding ability, can block transcription in *E. coli,* leading to loss of reporter gene activity.

Promoter            *lacZ*

A small molecule that directly blocks DNA binding of the protein or blocks a protein-protein interaction required for the cooperative binding can be identified by the restoration of reporter gene activity.

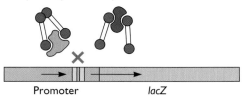

Promoter            *lacZ*

◁ **Figure 19.14 Two cell-based screens for a viral protease or a transcription regulatory protein.** (A) Cell-based assay for HIV protease, using tetracycline resistance of genetically engineered bacteria. Tetracycline resistance in *E. coli* is mediated by the tetracyline efflux protein, which pumps the antibiotic out of the cell. The idea is to insert an HIV protease cleavage site into the tetracycline efflux protein such that the hybrid protein remains active and still confers tetracycline resistance (first panel). When the HIV protease is coexpressed in the same cell, the hybrid tetracycline efflux protein is cleaved and no longer functions (second panel). As a result, the cell becomes tetracycline sensitive (third panel). When inhibitors of HIV protease are added to cells expressing the protease, cleavage may be blocked, leaving an active tetracycline efflux protein (fourth panel). Inhibitor activity is scored readily by the appearance of tetracycline-resistant bacteria. Inhibitors must be able to cross the cell membrane and not be toxic to the cell. To facilitate uptake of small molecules that might inhibit the protease, a variety of *E. coli* strains that have reduced permeability barriers are available. It is important to include many controls and secondary assays to identify false-positive and false-negative results. This assay is described in more detail in R. H. Grafstrom et al., *Adv. Exp. Med. Biol.* **312:**25–40, 1992. (B) Cell-based assay for inhibitors of a viral transcription regulatory protein. As in panel A, this assay requires construction of a specific expression cassette to be inserted into the test organism. This is accomplished by having the cassette as part of a plasmid or integrated into the test organism's genome. The system illustrated uses a *lacZ* reporter gene with a viral transcription protein binding site between the *lacZ* promoter and the *lacZ* open reading frame. When this specific viral protein is coexpressed in the cell harboring this hybrid *lacZ* gene, it binds to its cognate site and represses transcription. The cell does not produce β-galactosidase, and this phenotype can be scored by a variety of colony color assays on indicator media. A small molecule that blocks binding of the viral transcription protein to the DNA or protein-protein interaction can be identified through restoration of *lacZ* transcription and the appearance of active β-galactosidase. It is prudent to include a variety of controls and secondary assays to identify false-positive and false-negative results. Courtesy of R. Menzel, Small Molecule Therapeutics, Inc., Princeton, N.J.

---

Such sequences will provide a wealth of information on new targets for antiviral drug discovery. High-density arrays of DNA fragments on a solid surface can be used to assess the expression of thousands of genes in a single experiment. At the time of this writing, more than 10,000 unique human gene sequences can be put on such a "DNA chip" (the size of a microscope slide). Such global analysis of complicated processes in host-parasite interactions is unprecedented. Thus, with this technology, it will be possible to identify human genes whose transcription is activated or repressed in response to viral infection in various tissues. It is likely that new human antiviral gene products and new defensive pathways will be identified in the not too distant future.

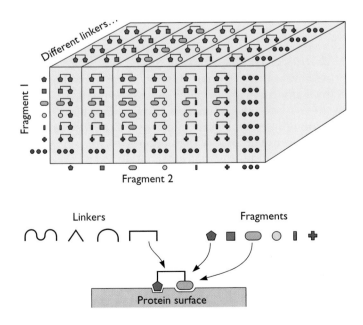

**Figure 19.15 Combinatorial chemistry and the building block approach to chemical libraries.** Small organic molecules predicted to bind to different pockets on the surfaces of proteins can be grouped into subsets of distinctive chemical structures, and these molecules can be synthesized. With automated procedures, these chemical entities (fragments) can be joined together by various chemical linkers to produce a large but defined library of small compounds. For example, if assembled pairwise with 10 linkers, a collection of 10,000 small molecules yields a library of 1 billion new combinations. These defined chemical libraries allow a detailed exploration of the binding surfaces of complex proteins and are predicted to provide new antiviral compounds. To take advantage of these chemical libraries, technology is being developed to enable medicinal chemists to reconstruct the specific combination of linkers and fragments that give rise to any useful compound found in the library. Adapted from P. J. Hajduk et al., *Science* **278:**497–499, 1997, with permission.

**A**   **B**

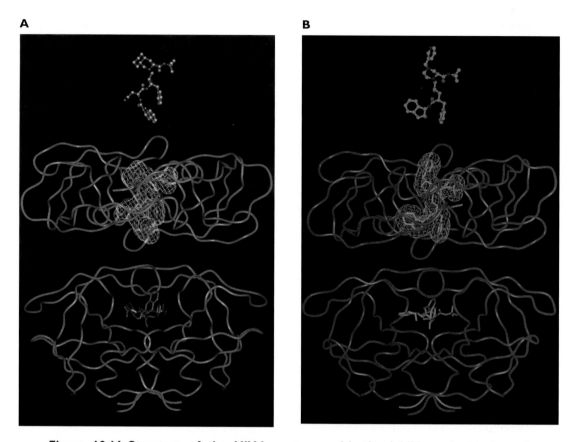

**Figure 19.16 Structure of the HIV-1 protease with the inhibitors indinavir and saquinavir.** Structures of HIV-1 protease with bound AIDS drugs saquinavir (Roche) and indinavir (Merck), as well as the structures of these drugs by themselves. (A) Tracing of the main chain of HIV-1 protease in two views rotated by 90° (middle and bottom), with the ball-and-stick model of saquinavir (top), saquinavir covered by a net representing the atomic surface (middle), and stick representation of saquinavir alone (bottom). (B) Similar images of indinavir. The images were prepared by J. Vondrasek and are reproduced with permission from the HIV protease database (http://www-fbsc.ncifcrf.gov/HIVdb; J. Vondrasek et al., *Nat. Struct. Biol.* **4:**8, 1997).

## The Difference between "R" and "D"

### Antiviral Drugs Are Expensive To Discover, Develop, and Bring to the Market

It is not unusual for the cost of bringing an antiviral drug to market to exceed $100 to $500 million. Even with modern methods, it is common for thousands of leads to yield but one promising candidate for further development (Fig. 19.17). **Research and lead identification**, the "R" of "R&D," represent only the beginning of the process of producing a drug. The "D" of "R&D" is **development**, comprising all the steps involved in taking an antiviral lead compound through safety testing, scale-up of synthesis, formulation, pharmacokinetic studies, and clinical trials. With rare exceptions, it takes 5 to 10 years *after* the initial lead is found to get a drug to the market, even with the latest technology.

### Antiviral Drugs Must Be Safe

As in vaccine development, safety is the overriding concern of any company developing an antiviral drug. Toxicity to test cells and animals is the first indication that a compound may not be safe. More promising leads are discarded because of toxicity than because of any other problem, confounding the molecular virologist whose prime concern is finding a compound that blocks viral replication. In the simplest case, an antiviral compound must be more toxic to the virus than to the cell (or to the host). This parameter is quantified as the **cytotoxic index** (for cells) or the **therapeutic index** (for hosts). These indices are defined as the dose that inhibits virus divided by the dose that is toxic to cells or host. The smaller the index the better; indices of 1/1,000 or smaller are preferred. Toxicity and efficacy measurements can be difficult, as initial lead compounds often are not very soluble in

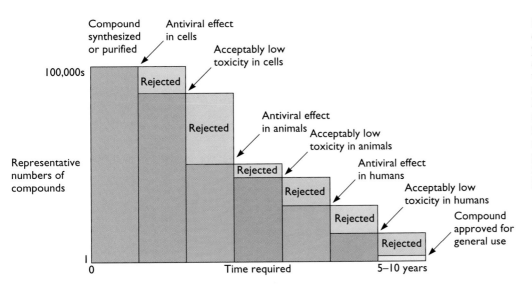

aqueous solutions or may have only modest antiviral activities.

The pharmacology and toxicology of a drug must also be thoroughly investigated in a variety of experimental animals, including primates, because no human trial can be initiated without detailed toxicology studies in diverse animal species. Compounds that may be used in long-term treatment of persistent infections must be evaluated for chronic toxicity, allergic effects, mutagenicity, teratogenicity, and carcinogenicity. Many drugs will be given to children, to pregnant women, and to individuals with reduced immune responses, all situations that present special problems. *Safety overrides efficacy* in most cases. On the other hand, when there are no other effective treatments, as in the early days of the HIV epidemic, even drugs that caused some undesirable side effects can be licensed for human use (e.g., zidovudine).

### Drug Formulation and Delivery

The science of formulation and delivery is an essential part of any antiviral discovery program. After introduction into a person, a drug must reach the proper place, at an effective concentration, and for an appropriate time to affect virus replication. A compound that cannot enter the bloodstream after ingestion is not likely to be effective. Simple entry into the bloodstream may not be sufficient, as many compounds bind to albumin or other proteins in the blood and thus are rendered ineffective. Other compounds may be metabolized as they pass through the liver and therefore cleared rapidly from the body. Such problems are not easily predicted and often are discovered empirically. Insoluble compounds or chemicals that are not able to enter the bloodstream after ingestion (poor bioavailability) can be modified by the addition of new

side chains that may improve absorption from the gut (see Fig. 19.21 for such a modification of acyclovir). In addition, new delivery vehicles like liposomes, minipumps, or slow-release capsules may improve bioavailability. Other desirable features, including stability (shelf life), cost-effective synthesis in quantity (literally tons of precursor materials are needed to produce commercial quantities of an antiviral drug), and ability to be formulated in tablets for oral ingestion, must be optimized. Every time a lead compound fails to cross one of these hurdles, the research director must decide whether to rework the failed compound or to stop the project.

### Examples of Commercially Available Antiviral Drugs for Viruses other than HIV

#### Acyclovir

Acyclovir is an example of a specific nontoxic drug that is highly efficacious against herpes simplex virus (genital and oral herpes) and, to some extent, varicella-zoster virus (chickenpox and shingles). It was discovered in 1974, but it was not until the mid-1980s that its full potential as an antiherpes drug was realized. It can be formulated to be administered by injection, as a topical cream, or as an ingestible capsule. Acyclovir is a nucleoside analog similar to guanosine but contains an acyclic sugar group (Fig. 19.18 and 19.19). It is a **prodrug**—a precursor of the active antiviral compound (Fig. 19.20). Activation of the drug requires three kinase activities to be present in the cell to convert acyclovir to a triphosphate derivative, the actual antiviral drug. The first kinase activity which converts acyclovir to the monophosphate is not found in an uninfected cell. Thus, when acyclovir enters a normal cell, it has no effect on host DNA replication because it cannot be phos-

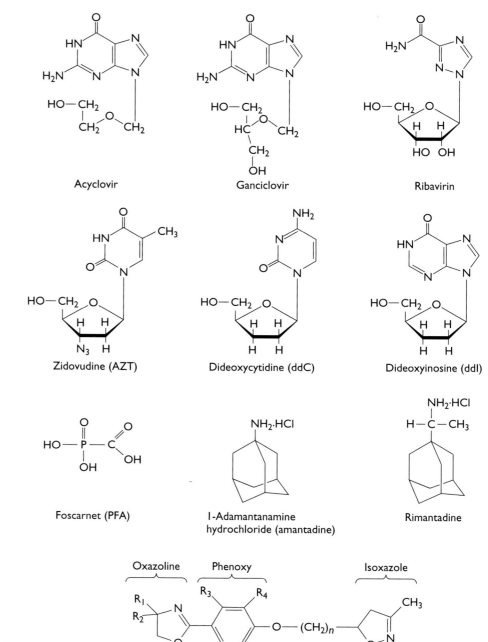

**Figure 19.18 Chemical structures of effective antiviral compounds in use today.**

phorylated and cannot be incorporated into DNA. Herpes simplex virus and varicella-zoster virus, however, encode a kinase that can phosphorylate the prodrug to produce the monophosphate. Cellular enzymes complete synthesis of the triphosphate, which is then incorporated into DNA by the viral DNA polymerase. As acyclovir lacks the 3'-OH group of the sugar ring, the growing DNA chain terminates, and further DNA replication is blocked. While this drug is highly specific for cells infected by herpes simplex virus and varicella-zoster virus, the specificity comes primarily from the virus-encoded kinase. This enzyme, thymidine kinase, phosphorylates thymidine to thymidine monophosphate and also phosphorylates a wide range of other substrates. This property enables thymidine kinase to phosphorylate synthetic molecules like acyclovir. Indeed, if herpes simplex virus thymidine kinase is produced

**Nucleoside**
**(DNA component)**

**Nucleoside analog**
**(DNA chain terminator)**

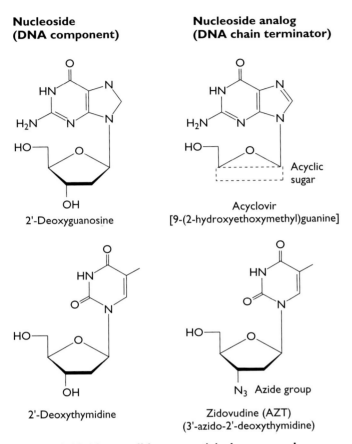

2'-Deoxyguanosine

Acyclovir
[9-(2-hydroxyethoxymethyl)guanine]

2'-Deoxythymidine

Zidovudine (AZT)
(3'-azido-2'-deoxythymidine)

**Figure 19.19 Many well-known antiviral compounds are nucleoside and nucleotide analogs.**

in an uninfected cell and acyclovir is added, the cell will die because its DNA replication will also be blocked by the chain-terminating phosphorylated acyclovir base analog. This phenomenon is the basis of several strategies for selective cell killing in gene therapy and of embryonic stem cells during construction of transgenic mice.

Because herpesviruses encode essentially all their own DNA replication enzymes, DNA replication has been a fruitful target for the discovery of antiherpes drugs (Table 19.9). Many variations of chain-terminating nucleoside analogs have been developed. Chemical modifications of antiherpes nucleoside analogs can lead to dramatic increases in efficacy or bioavailability (Fig. 19.21). As a result, the latest anti-herpes simplex virus drugs, valacyclovir (Valtrex) and famciclovir (Famvir), are much more effective and more widely used than acyclovir.

### Ganciclovir (Cytovene)

Ganciclovir (Fig. 19.18) is a derivative of acyclovir that was developed to treat human cytomegalovirus infections. However, in contrast to acyclovir, initial formulations of ganciclovir were given intravenously and were quite toxic.

Consequently, ganciclovir was used only for life-threatening human cytomegalovirus infections, including pneumonia in AIDS patients and transplant recipients, retinitis, and colitis. Recently, ganciclovir has been approved in an oral formulation that is effective for prophylaxis and long-term use for human cytomegalovirus infections in immunocompromised patients. The oral formulation appears to be much less toxic than the original form delivered intravenously, and thus effective concentrations of drug can be delivered orally.

### Foscarnet (Foscavir, Trisodium Phosphonoformate)

Foscarnet is the only nonnucleoside DNA replication inhibitor of herpesviruses (Fig. 19.18). It is a noncompetitive inhibitor of the pyrophosphate-binding site of herpesvirus DNA polymerase. It also inhibits hepatitis B virus polymerase and the reverse transcriptase of HIV. As it accumulates in bone and causes kidney toxicity, foscarnet is recommended only for life-threatening infections for which other antiviral drugs are no longer effective.

### Ribavirin (Virazole)

Ribavirin, a nucleoside analog (Fig. 19.18), was synthesized in 1972 and was purported to have broad-spectrum antiviral activity against many DNA and RNA viruses. However, it is relatively toxic, and its development and indications for use have been controversial. In fact, this drug is not licensed for use in many countries. Despite its long history, its mechanism of antiviral action is not clear. Ribavirin monophosphate is a competitive inhibitor of cellular inosine monophosphate dehydrogenase, and such inhibition leads to reduced GTP pools in the cell. It also inhibits initiation and elongation by virus-specific RNA polymerase and interferes with 5' capping of mRNA. Even with its toxicity, ribavirin is used as an aerosol in treatment of infants suffering from respiratory syncytial virus infection. Ribavirin also has been used to treat Lassa fever virus and hantavirus infections.

### Amantadine (Symmetrel)

The three-ringed symmetric amine known as amantadine (Fig. 19.18) was developed by Du Pont chemists more than 30 years ago. It was the first highly specific, potent antiviral drug effective against any virus. Amantadine specifically inhibits influenza A virus. The target of the drug was shown only recently to be M2 protein, a tetrameric transmembrane ion channel that transports protons. The specificity of amantadine for influenza A and not influenza B virus can now be understood, as influenza B viruses do not express M2 protein. Influenza A virus mutants resistant to amantadine arise after therapy (Fig.

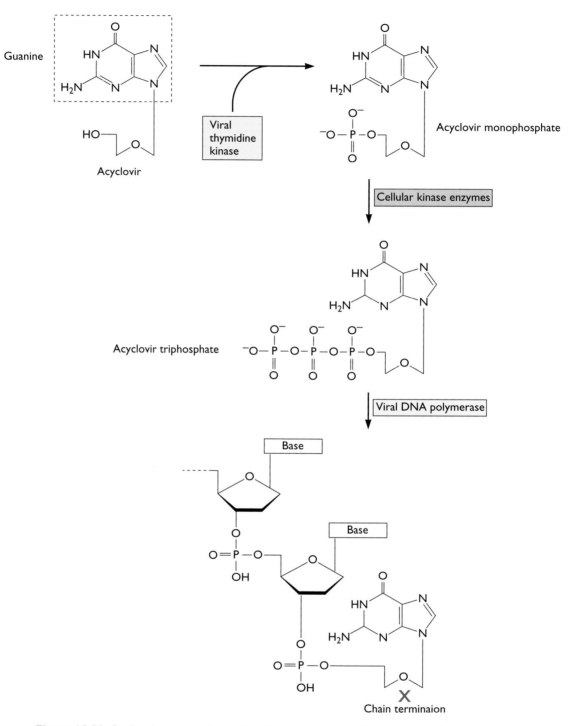

**Figure 19.20 Activation of acyclovir.** Acyclovir is a prodrug that must be phosphorylated in the infected cell before it can inhibit viral DNA replication. The viral thymidine kinase, but not the cellular kinase, adds one phosphate to the 5′-hydroxyl group of acylovir. The monophosphate compound is a substrate for cellular kinases that convert the monophosphate compound to acyclovir triphosphate. The triphosphate compound is recognized by the viral DNA polymerase and incorporated into viral DNA. As acyclovir has no 3′-hydroxyl group, the growing DNA chain is terminated and viral replication ceases. Adapted from D. R. Harper, *Molecular Virology* (Bios Scientific Publishers Ltd., Oxford, United Kingdom, 1994), with permission.

**Table 19.9**  Some commercially available antiviral drugs

| Drug | Chemical name | Virus[a] | Effect |
|---|---|---|---|
| Acyclovir | 9-(2-Hydroxyethoxymethyl) guanine | HSV, VZV | Chain terminator; inhibits viral DNA replication |
| Amantadine | Tricyclo[3,3,1$^{3,7}$]dec-1-ylamine HCl | Influenza A virus | Ion channel blocker; inhibits M2 function |
| Famciclovir | 9-(4-Acetoxy-3-acetoxy methylbut-1-yl guanine | VZV | Chain terminator; inhibits viral DNA replication |
| Foscarnet | Trisodium phosphonoformate hexahydrate | HSV and other herpesviruses | Noncompetitive inhibitor of viral DNA polymerase (pyrophosphate-binding site) |
| Ganciclovir | 9-(1,3-Dihydroxy-2-propoxymethyl) guanine | HCMV, HSV | Competitive inhibitor of viral DNA polymerase |
| Idoxuridine | 5-Iodo-2'-deoxyuridine | HSV | Disrupts transcription and translation after incorporation into viral DNA |
| Ribavirin | 1-β-D-Ribofuranoxyl-1,2,4 triazole 3-carboxamide | RSV | Unclear mechanism |
| Trifluorothymidine | | HSV | Like idoxuridine |
| Vidarabine | 9-β-D-Arabinofuranoxyl adenine monohydrate | HSV, VZV | DNA synthesis inhibitor, mechanism unknown |

[a]Abbreviations: HSV, herpes simplex virus; VZV, varicella-zoster virus; HCMV, human cytomegalovirus; RSV, respiratory syncytial virus.

**Famciclovir:**
**9-(4-acetoxy-3-acetoxymethylbut-1-yl)guanine (Famvir)**

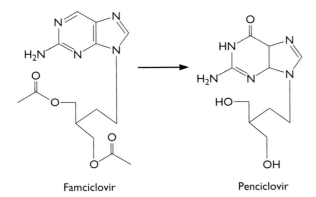

Famciclovir          Penciclovir

**Valacyclovir:**
**L-valyl ester of acyclovir (Valtrex)**

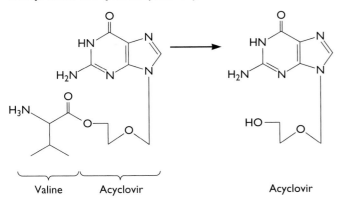

Valine    Acyclovir          Acyclovir

19.22). These mutants all have amino acid changes in the M2 transmembrane sequences predicted to form the ion channel, but it is unknown if the drug binds directly in the channel or if a mutation that alters the channel affects binding of the drug to another site. In any case, amantadine blocks the channel so that protons cannot enter the virus particle, effectively preventing the uncoating of influenza A virus.

Amantadine must be administered in the first 24 to 48 h of infection and be given at high doses for at least 10 days. Thus, it is most effective when given to susceptible patients in anticipation of influenza virus infection. However, side effects are common, particularly those affecting the central nervous system. On follow-up of these side effects, amantadine was found to be useful for relieving symptoms of Parkinson's disease in some patients. Today,

**Figure 19.21 Derivatives of acyclovir with improved oral bioavailability.** Acyclovir is not taken up efficiently after oral ingestion. Two new compounds, famciclovir and valacyclovir, have as much as fivefold-higher oral bioavailability than acyclovir (serum levels relative to the dose of drug given orally). The addition of a new side group to acylocvir allows increased passage of drug from the digestive tract to the circulation. These new acyclovir derivatives are **proprodrugs**, which are converted to penciclovir or acyclovir inside the cell by cellular enzymes that cleave off the side chain. Penciclovir and acyclovir are then activated by the viral thymidine kinase to inhibit viral DNA replication. Adapted from D. R. Harper, *Molecular Virology* (Bios Scientific Publishers, Ltd., Oxford, United Kingdom, 1994), with permission.

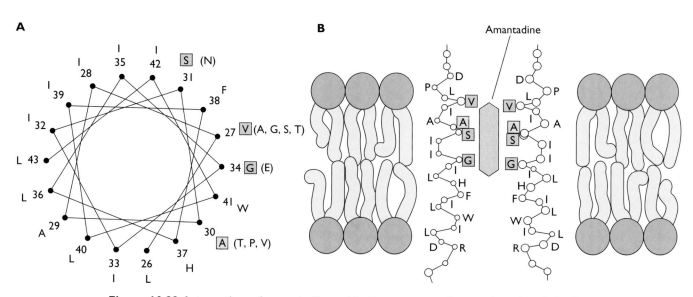

**Figure 19.22 Interaction of amantadine with the transmembrane domain of the influenza A virus M2 ion channel.** Amantadine is thought to exert its antiviral effect by blocking the virus-encoded M2 ion channel activity in infected cells. This inference is supported by genetic data and models of M2 protein, but the points in the virus life cycle critical for inhibition are still controversial. M2 protein is a tetramer with an aqueous pore in the middle of the four subunits. (A) Helical wheel representation of the transmembrane domain of M2 protein. The sequence Asp24-Arg45 of A/chicken/Germany/27(H7N7) is given in single-letter code. Polar amino acids (Ser, Thr, Asn) and charged amino acids (His, Glu) are found on the right side of the helical wheel while nonpolar amino acids cluster on the left. The four locations of single amino acid changes in different amantadine-resistant mutants are boxed, and the substitutions most commonly observed are shown in parentheses. (B) Diagram of the putative interaction of amantadine with two diagonally located α-helices of the M2 tetramer in a lipid bilayer. The polar faces of the helices form the interior of the proposed ion channel. Adapted from A. J. Hay, *Semin. Virol.* **3:**21–30, 1992.

more amantadine is sold for central nervous system disease than for antiviral treatment. Rimantidine (Fig. 19.18), a methylated derivative, cannot cross the blood-brain barrier and therefore has fewer central nervous system side effects.

## New Targets for Antiviral Drugs

### Entry Inhibitors

The first step of virus replication has long been an attractive target for antiviral discovery, as virus-receptor interactions offer the promise of high specificity. Early enthusiasm for entry inhibitors came from experiments with monoclonal antibodies that inhibited virus attachment or entry in cultured cells. Specific receptor-virus interactions can be identified if one can prove that an antibody inhibits virus entry by direct means rather than by steric hindrance or other indirect effects. Passive immunization with these antibodies often can protect animals from virus challenge, suggesting that entry inhibitors may have possibilities as antiviral drugs. As it is usually impractical to develop antibodies as antiviral agents, these

molecules have been most valuable in identifying interacting proteins and entry mechanisms.

The antibody-antigen binding site of such entry-specific antibodies provides a starting point for screening chemical libraries or for design of small molecules that block virus entry. It is important that such compounds interfere with virus-host interaction but not with the normal function of the cellular receptor. Obviously, an important assumption in this strategy is that no other routes of entry for the virus exist. For many viruses, like herpesviruses and HIV, alternative receptors are available and blocking a single virus-receptor interaction may not be totally effective.

Membrane fusion, the ubiquitous process by which enveloped viruses enter cells, is also a target for chemotherapeutic intervention. To identify inhibitors of influenza virus fusion, a computer program was first used to predict which small molecules might bind into a pocket of the HA molecule and possibly prevent low-pH-induced conformational change. From the list of molecules produced in this way, several benzoquinone- and hydroquinone-containing compounds were found to inhibit HA-mediated membrane fusion at low pH. One of these, called BMY-27709, inhibits influenza virus replication and has been shown to block fusion by preventing the low-pH-induced conformational changes in the HA required for fusion. It is believed that the compound binds within the membrane-proximal region of the HA molecule and may interact with the amino-terminal region of HA2, which contains the fusion peptide. Although it is not know how binding of the drug prevents fusion, it has been suggested that it acts a a "molecular glue" to prevent movement of the fusion peptide.

### Proteases

Many viruses require a specific viral protease to cleave polyproteins to functional units or to release structural components during ordered particle assembly (maturational proteases). As discussed in chapter 13, all herpesviruses encode a serine protease that is absolutely required for formation of nucleocapsids. Many features of these enzymes and their substrates are conserved among the members of the family *Herpesviridae*. The structure of the human cytomegalovirus protease is shown in Fig. 19.23. Enthusiasm for this enzyme as a target is based on the unusual serine protease fold and mechanism of catalysis of the herpesvirus enzyme, properties that bode well for the design of specific inhibitors. As homologs of this

enzyme appear to be encoded by all herpesviruses, it holds promise as a target for broad-spectrum antiherpes compounds and may provide a welcome alternative or adjunct to nucleoside analog drugs.

Hepatitis C virus infects four times more people than HIV in the United States. Currently there are 4 million Americans, 9 million Europeans, and 170 million people worldwide infected with hepatitis C virus. As a result, an intensive antiviral drug discovery effort is in progress. A viral protease called NS3 is a major target. The virus encodes at least two polyprotein-processing proteases (Fig. 19.24). NS3 protein is an attractive antiviral target because its activity is thought to be required for virus replication. At this time, hepatitis C virus cannot be propagated in cultured cells, and so the effect of inhibiting NS3 protease on virus replication cannot be tested directly. However, all available evidence from related flaviviruses that encode similar enzymes suggests that such protease inhibitors will be effective antiviral compounds. The crystal structure of NS3 protease has been solved by several companies, and major efforts in structure-based design, as well as mechanism-based screening, are under way. Given the large number of people infected with hepatitis C virus and the lack of useful therapy, there is considerable impetus to find effective antiviral agents.

**Figure 19.23 Human cytomegalovirus protease.** (A) The β-strands are shown as arrowed ribbons in green, α-helices are orange, and the connecting loops are indicated in purple. The active site residues (Ser132, His63, and His157) are also indicated. The secondary-structure elements are labeled starting from the N terminus, the β-strands numerically (β1 to β8), and the α-helices alphabetically (αA to αG). The open end of the β-barrel is toward the top corner. (B) Topological drawing of the backbone fold of human cytomegalovirus protease. Strand β3 is broken at the crossover position, and the breakpoints are represented by black squares. The secondary-structure elements are labeled. The loops are numbered from the N terminus (L1 to L6). The locations of Ser132, His63, and His157 are also shown. Adapted from L. Tong et al., *Nature* **383**:272–275, 1996, with permission.

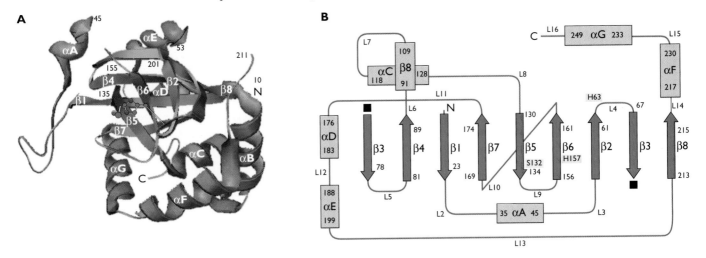

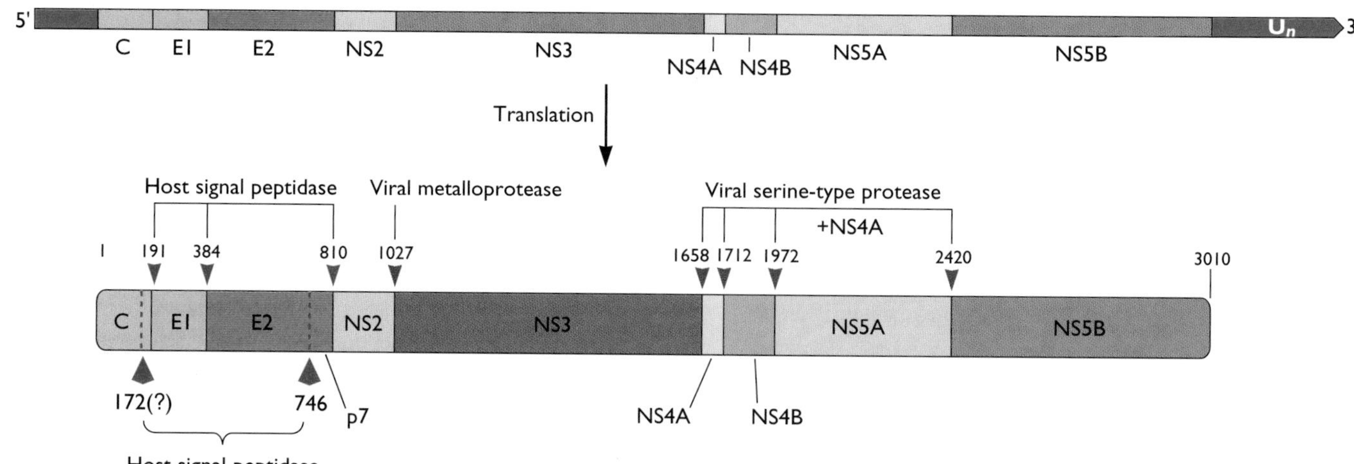

**Figure 19.24 Genome structure of hepatitis C virus.** Hepatitis C virus is a human flavivirus with a (+) strand RNA genome. The viral proteins are encoded in one large open reading frame that is translated into a polyprotein. The polyprotein is processed by cellular and viral proteases to release the viral proteins. The numbers above the polyprotein indicate the amino acid of the N terminus of each hepatitis C virus protein, except the last number, 3010, which indicates the amino acid number of the carboxy terminus of the polyprotein. The proteases and their cleavage sites are indicated by arrows. Signal peptidase is the host enzyme in the endoplasmic reticulum that cleaves signal peptides from membrane proteins. The dotted lines represent cleavage sites that are inefficiently processed or not well studied. The viral metalloprotease that cleaves at position 1027 is viral protease 1, whose coding sequence is located between the carboxy terminus of NS2 and the amino terminus of NS3. The viral serine-type protease is the NS3 protein complexed with NS4A that cleaves at positions 1658, 1712, 1972, and 2420. The hepatitis C virus NS3 protease is essential for the assembly of virus particles and is a target for antiviral drug discovery. Adapted from D. D. Richman et al., *Clinical Virology* (Churchill Livingstone, New York, N.Y., 1997), with permission.

## Virus-Specific Nucleic Acid Synthesis and Processing

Virus replication enzymes are primary targets for antiviral drugs. From genetic analyses, we know that DNA polymerase accessory proteins that promote processivity, origin-binding proteins that direct polymerases to viral origins of replication, and virus-encoded primase-helicases required to make primers are all essential for virus replication and thus represent attractive new targets for antiviral drugs. The RNA-dependent RNA polymerases of any RNA virus appear to be unique to the virus world—there are no known cellular counterparts in animal cells. The only exception is hepatitis delta virus, which relies on the host RNA polymerase II enzyme. These polymerases and their varied virus-specific activities, including synthesis of primers and cap snatching and recognition of RNA secondary structure in viral genomes, mark them as targets for drug discovery. The unique helicases encoded by many RNA viruses are already in many companies high-throughput screens, but no lead compounds have progressed to the market as of this writing.

Newly replicated herpesviral DNA is found in large multigenome-length polymers (concatemers) which are cleaved into monomeric units for packaging into virions. These unusual processes, which are essential for virus replication, are carried out by specific viral enzymes and represent new targets for drug discovery. For example, 5-bromo-5,6-dichloro-1-β-D-ribofuranosyl benzimadazole binds to the human cytomegalovirus UL89 gene product, which is thought to be part of the "terminase" complex responsible for cleaving and packaging replicated concatemeric DNA. Members of the UL89 gene family are highly conserved among all herpesviruses and have homology with a known ATP-dependent endonuclease encoded in bacteriophage T4. Thus, this class of compounds may be the basis for discovery of broad-based inhibitors of herpesvirus replication.

## Regulatory Proteins

Many viruses control their replication cycles through the action of unique regulatory proteins. These proteins often exert their action by binding to specific viral nucleic acid sequences or to other viral proteins. Many of these proteins are known to be essential for virus growth and are therefore prime targets for antiviral screens.

### Virus Spread: Structure-Based Design of Influenza Virus Neuraminidase

Influenza virus NA protein (Fig. 19.25) cleaves terminal sialic acid residues from glycoproteins, glycolipids, and oligosaccharides. It plays an important role in the spread of infection from cell to cell because in cleaving sialic acid residues it releases virus bound to the surface of infected cells, may facilitate viral diffusion through respiratory tract mucus, and can activate latent transforming growth factor β by removing sialic acid from the inactive protein. Because activated transforming growth factor β can induce apoptosis, NA may thus influence the host response to virus infection. Because of these properties, NA is a particularly intriguing target for drug hunters (Box 19.6).

One of the most promising features of one NA inhibitor was that when it was given to infected animals by in-

tranasal spray, it blocked the virus shedding and fever that normally accompany infection. Several such compounds are currently being evaluated in human clinical trials.

An important consideration for any influenza virus therapy is the emergence of drug-resistant variants and their subsequent role in influenza virus disease in the human population. Given the high rate of mutation during influenza virus replication, the appearance of resistant mutants seems to be inevitable. Indeed, they have been detected in tissue culture experiments. These drug-resistant mutants should provide insight into the molecular biology of the enzyme as well as its role in viral pathogenesis.

### Antiviral Gene Therapy and Transdominant Inhibitors

In 1988 a trio of scientists implanted a mutant herpes simplex virus gene in a cell, and that cell line subsequently proved to be resistant to further infection. Later, when the same mutant gene was inserted into the genome of mice,

**Figure 19.25  Structure of influenza A virus NA.** (A) A MOLSCRIPT diagram of influenza A NA with α-sialic acid bound in the active site of the enzyme. The molecule is viewed down the fourfold axis of the tetramer. The molecule is an N2 subtype from A/Tokyo/3/67. The C terminus is on the outside surface near a subunit interface. The six β-sheets of the propeller fold are indicated in colors. (B) The packing of the N2 NA tetramer viewed from above and down the symmetry axis. From J. N. Varghese, p. 459–486, *in* P. Verrapandian (ed.), *Structure-Based Drug Design* (Marcel Dekker, New York, N.Y., 1997), with permission. Courtesy of J. Varghese. Information on MOLSCRIPT may be found in P. J. Kraulis, *J. Appl. Crystallogr.* **D50:**869–873, 1991.

**A**

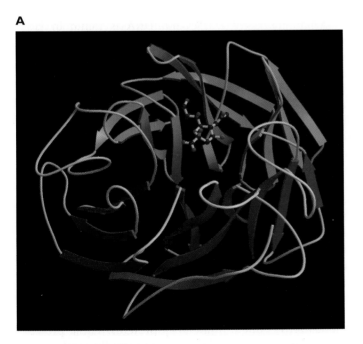

**B**

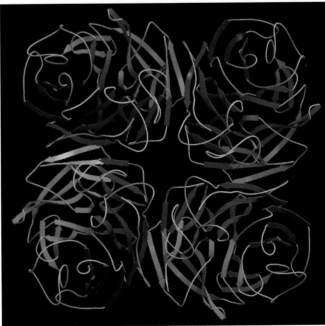

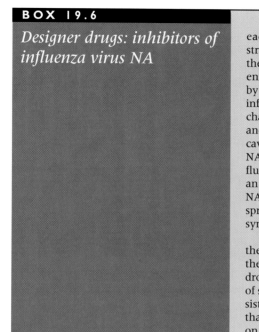

**BOX 19.6**

*Designer drugs: inhibitors of influenza virus NA*

NA is a tetramer of identical subunits, each of which consists of six four-stranded antiparallel sheets arranged like the blades of a propeller (Fig. 19.25). The enzyme active site is a deep cavity lined by identical amino acids in all strains of influenza A and B viruses that have been characterized. Because of such invariance, compounds designed to fit in this cavity would be expected to inhibit the NA activity of all A and B strains of influenza virus, a highly desirable feature in an influenza antiviral drug. Moreover, as NA inhibitors are predicted to block virus spread, they may be effective in reducing symptoms after the infection has begun.

The sialic acid-NA complex fits into the active site cleft in such a fashion that there is an empty pocket near the hydroxyl at the 4' position on the sugar ring of sialic acid. On the basis of computer-assisted analysis, investigators predicted that replacement of this hydroxyl group on sialic acid with either an amino group or guanidinyl group would fill the empty pocket and would increase the binding affinity by establishing new contacts with one or more neighboring glutamic acid residues.

These predictions were verified by direct structural analysis of the new compounds bound to NA. Significantly, these two compounds were inhibitors of both influenza A and B viruses in tissue culture. They did not inhibit other nonviral NAs, an important requirement for safety and lack of potential side effects.

**Varghese, J. N., V. C. Epa, and P. M. Colman.** 1995. Three dimensional structure of the complex of 4-guanidino-Neu5Ac2en and influenza virus neuraminidase. *Protein Sci.* **4:**1081–1087.

**von Itzstein, M., W. Y. Wu, G. B. Kok, M. S. Pegg, J. C. Dyason, B. Jin, T. Van Phan, M. L. Smythe, H. F. White, S. W. Oliver, P. M. Coleman, J. N. Varghese, D. M. Ryan, J. M. Woods, R. C. Bethell, V. J. Hotham, J. M. Cameron, and C. R. Penn.** 1993. Rational design of potent sialidase-based inhibitors of influenza virus replication. *Nature* (London) **363:**418–423.

the mice became resistant to infection by herpes simplex virus. The gene was VP16, which encodes a viral tegument protein with powerful transcriptional activation properties. The mutation in the VP16 gene conferred a "dominant negative" phenotype: the altered protein interfered with the function of the normal VP16 protein. Dominant negative mutations are not new in genetics, but the idea that such an altered protein could be used to protect an animal against virus infection was novel.

The term **intracellular immunization** was coined in an editorial response to the experiments based on the VP16 dominant negative mutation described above. The original term described the general process of making a cell or organism resistant to HIV infection by inserting an interfering gene into the cell. The principle is simple: any cell type easily renewed from precursors could be protected from virus infection if its precursors could be engineered to contain a gene product that interferes with virus replication. For viruses like HIV that replicate in lymphocytes, the implications were obvious, and considerable creative effort has since been expended on building genes encoding products that interfere with virus replication and pathogenesis.

Inhibition of replication by a viral gene product conferring a dominant negative phenotype can be obtained when the wild-type protein functions as a multimer rather than as a monomer. Mixed multimers containing a normal and a mutant protein may be inactive, depending on the nature of the mutation.

Many natural and engineered gene products that should block intracellular virus growth when present in an infected cell can be imagined, and some of them may yield effective antiviral therapies (Table 19.10). For example, RNA-binding proteins, DNA-binding proteins, ribozymes, and antisense oligonucleotides offer interesting possibilities (Box 19.7). Some clever ideas are based on hybrid proteins composed of a target domain and a "killer" domain. The target domain brings the killer domain to the virus-infected cell. For example, nucleases fused to viral capsid proteins block virus propagation if the viral genome is degraded during virion assembly. Toxins have been fused to antibodies that bind to viral proteins on the surface of infected cells. When the hybrid protein is taken up by endocytosis, it kills the infected cell.

The effectiveness of dominant negative mutations and killer or "suicide" molecules has been validated in tissue culture systems for many viruses; indeed, the current strategies (now classified under the rubric "gene therapy") are too numerous to mention. While this approach often captures the fancy of molecular virologists, its practical utility in the treatment of disease remains uncertain. Such compounds or approaches provide interesting leads, but they face the same hurdles as more conventional antiviral lead compounds (Fig. 19.17). The basic requirements of

**Table 19.10** Molecules and gene products capable of inhibiting HIV-1 replication

| Molecule or product | Putative mechanism |
| --- | --- |
| Antisense RNA | Blocks translation of target HIV gene |
| Ribozymes | Cleave HIV RNA |
| Dominant negative mutants of *tat, rev, gag* | Inhibit normal function |
| RNA decoys (TAR, RRE) | Compete for essential viral regulatory proteins |
| Intracellular antibodies | Bind to HIV proteins intracellularly |
| Capsid-nuclease hybrid proteins | Destroys viral genome during packaging |
| Inducible toxins or killer proteins | HIV long terminal repeat fused to toxin; turned on upon HIV infection and kills infected cells (e.g., herpes simplex virus TK, diphtheria toxin, interferon) |
| | HIV infection-inducible host shutoff functions (e.g., herpes simplex virus *vhs* gene; expression results in destruction of all polysomes in HIV-infected cell) |

safety and efficacy cannot be predicted with confidence until human trials are conducted.

## Resistance to Antiviral Drugs

Because viruses replicate so efficiently with moderate to high mutation frequencies, resistance to any antiviral drug must be anticipated when one administers compounds that inhibit virus growth. This property is a special concern during the extended therapy required for chronic infections. Indeed, viral mutants resistant to *every* antiviral drug produced to date have been detected clinically, a discon-

certing fact because our arsenal of antiviral drugs is small (Table 19.9). Drug-resistant viruses are dangerous for patients, who cannot be treated for the resistant infection, and are potentially hazardous for the population who might be exposed to them. Nevertheless, a genetic analysis of resistance can provide powerful insight into the antiviral mechanism and may identify new strategies to reduce or circumvent the problem.

Consider acyclovir-resistant mutants of herpes simplex virus, which arise spontaneously during virus replication and are selected after exposure to the drug. These mutants are unable to phosphorylate the prodrug or are unable to incorporate the phosphorylated drug into DNA. The majority of mutations that confer resistance are in the viral TK gene and inactivate kinase function. We can deduce that TK is essential for the action of the drug. As TK is not required for herpes simplex virus to grow in tissue culture, such mutants can be identified with ease. A subset of mutations leading to acyclovir resistance can be found in the viral DNA polymerase gene. The altered polymerases possess reduced ability to incorporate phosphorylated drug into DNA. Similar patterns of resistance have been reported when other nucleoside analog inhibitors are used against varicella-zoster virus.

The appearance of TK⁻ mutants after acyclovir therapy is of concern in the clinic. Although TK-defective viruses have a markedly reduced ability to invade and replicate in the central nervous system in animal models, they can be devastating in AIDS patients: they are still able to cause disseminated disease and are often resistant to other nucleoside analogs that share the same antiviral mechanism. Foscarnet can be used in some instances to treat these infections. Acyclovir-resistant DNA polymerase mutants present a special problem because they can also be resistant to foscarnet, leaving essentially no recourse.

---

**BOX 19.7**

*RevM10, a dominant negative inhibitor of HIV replication*

The HIV mutant RevM10 contains a mutation in the effector domain of Rev. The altered RevM10 protein forms mixed oligomers with the normal Rev protein which are inactive in facilitating the nuclear export of unspliced viral pre-mRNAs. To determine the ability of RevM10 protein to protect human cells against HIV-1 infection, the RevM10 mutant was integrated into the genome of CD4⁺ human T lymphocytes via a recombinant retrovirus. A majority, though not all, of the T-cell lines cloned from the mixed cell population were resistant to HIV-1 replication.

On the basis of these results, RevM10 protein is currently in phase I clinical tri-

als for the treatment of AIDS. A major obstacle to the use of this type of treatment is that the inhibitor is a protein, and biologically active proteins are notoriously difficult to deliver effectively to patients; the protein cannot be administered directly but must be delivered with a viral vector containing the gene. Such technology is not well developed at this time and is difficult, costly, and of unproved effectiveness.

**Malim, M. H., S. Bohnlein, J. Hauber, and B. R. Cullen.** 1989. Functional dissection of the HIV-1 Rev trans-activator—derivation of a transdominant repressor of Rev function. *Cell* **58**:205–214.

## HIV and AIDS

The recognition that HIV was associated with AIDS in the early 1980s provided the impetus for unprecedented focus on antiviral drug discovery. As noted in chapter 18, the first effective drugs acted on the viral reverse transcriptase. The strategy was based on well-known abilities of nucleoside analogs to block replication of herpesviruses and tumor cells. Scientists quickly deciphered many distinctive steps of the virus life cycle, identifying numerous additional targets (Fig. 19.26). Today, not only new, effective nucleoside inhibitors (Fig. 19.27) but also nonnucleoside inhibitors of reverse transcriptase (Fig. 19.28) and potent protease inhibitors (Fig. 19.29 and 19.30) are available in the clinic. Scientists are also finding more compounds directed to specific steps in the virus life cycle, including inhibitors of virus-cell interaction (attachment, fusion, and entry), integration of viral DNA, and nucleocapsid assembly (see also chapter 18).

### Examples of Drugs Approved by the Food and Drug Administration for Treatment of HIV Infection

#### Nucleoside Analogs

Five nucleoside analogs that are effective inhibitors of reverse transcriptase have been approved for the treatment of AIDS (Fig. 19.27). Azidodeoxythymidine (AZT), or zidovudine, was the first drug to be licensed. This drug was initially discovered during screens for antitumor cell compounds rather than for antiviral agents. Unlike acyclovir, which is phosphorylated by a viral kinase, AZT is phosphorylated to the active form by cellular enzymes.

**Figure 19.26 Important steps in the replication of HIV.** Five steps in the life cycle of HIV are highlighted as potential targets for antiviral compounds. Known or hypothetical inhibitory compounds are listed under each step. Adapted from D. D. Richman et al., *Clinical Virology* (Churchill Livingstone, New York, N.Y., 1997), with permission.

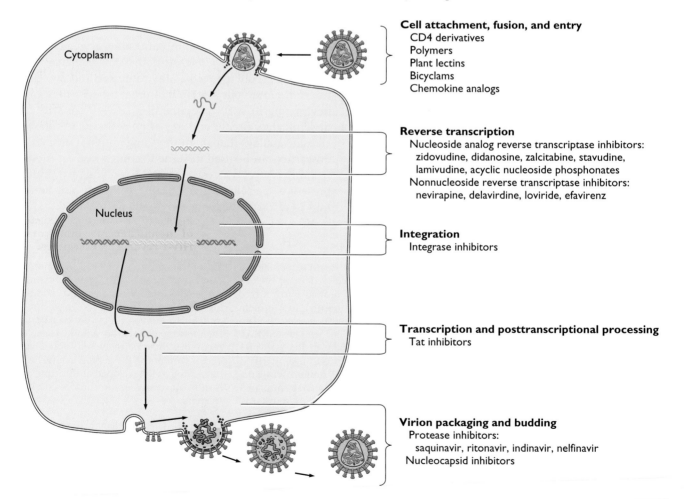

**Cell attachment, fusion, and entry**
CD4 derivatives
Polymers
Plant lectins
Bicyclams
Chemokine analogs

**Reverse transcription**
Nucleoside analog reverse transcriptase inhibitors:
zidovudine, didanosine, zalcitabine, stavudine,
lamivudine, acyclic nucleoside phosphonates
Nonnucleoside reverse transcriptase inhibitors:
nevirapine, delavirdine, loviride, efavirenz

**Integration**
Integrase inhibitors

**Transcription and posttranscriptional processing**
Tat inhibitors

**Virion packaging and budding**
Protease inhibitors:
saquinavir, ritonavir, indinavir, nelfinavir
Nucleocapsid inhibitors

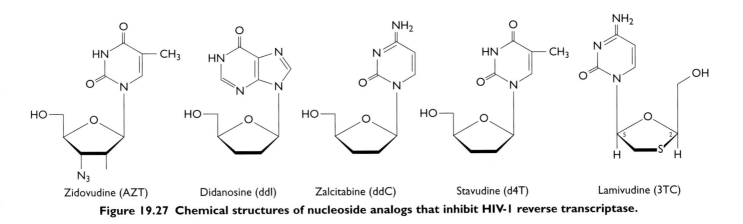

Zidovudine (AZT)    Didanosine (ddl)    Zalcitabine (ddC)    Stavudine (d4T)    Lamivudine (3TC)

**Figure 19.27 Chemical structures of nucleoside analogs that inhibit HIV-1 reverse transcriptase.**

Like acyclovir, it acts as a chain terminator if incorporated into DNA because the 3′ OH of adenosine is replaced by an azido (N3) group. Although phosphorylated AZT is not a good substrate for most cellular polymerases, it is incorporated into the DNA copy of the viral genome by the viral reverse transcriptase. The relative selectivity of this drug depends on the fact that reverse transcription takes place in the cytoplasm, where the drug appears first and in the highest concentration. Because AZT monophosphate competes with thymidine monophosphate for the formation of nucleoside triphosphate, its presence causes depletion of the intracellular pool of TTP. Therefore, AZT is much less selective than acyclovir and has substantial side effects. AZT can be given orally and is absorbed quickly but is also degraded rapidly by liver glucuronidation enzymes so that its effective half-life in a patient is only about 1 h. Patients must take the drug two or three times daily to maintain an effective antiviral concentration. If the illness is severe, AZT can be administered intravenously. As we will discuss below, a short half-life is problematic because if an effective concentration of any antiviral drug is not maintained, resistant mutants will be selected. At best, AZT provides modest and transient effects in infected adults with a concomitant increase in CD4+ T-cell counts. Its most effective use is in prophylactic treatment in the case of accidental needle sticks and in pregnant women, in whom treatment can reduce considerably the probability of delivering an HIV-infected baby.

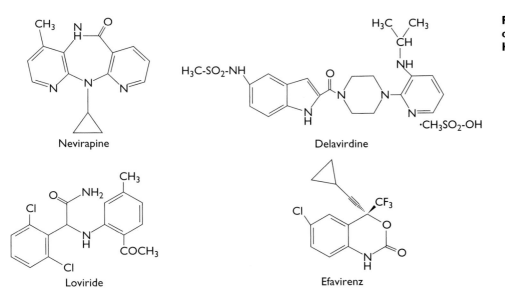

Nevirapine

Delavirdine

Loviride

Efavirenz

**Figure 19.28 Chemical structures of nonnucleoside inhibitors of HIV-1 reverse transcriptase.**

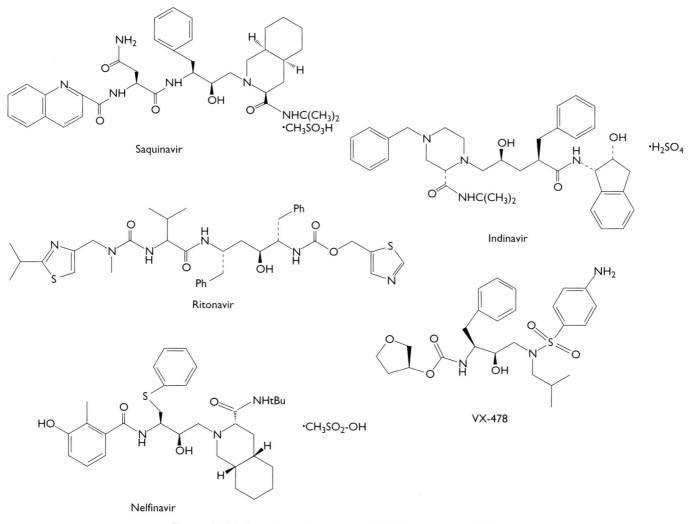

**Figure 19.29 Chemical structures of HIV-1 protease inhibitors.**

AZT toxicity is of great concern when the drug must be given for long periods. AZT treatment can damage bone marrow, resulting in the reduction of neutrophils, a problem whose treatment requires transfusions of red blood cells. In addition, muscle wasting, nausea, and severe headaches are but a few of AZT's other side effects that must be endured. Despite this, the drug has been used extensively because until recently there simply was no alternative.

Considerable effort has been devoted to discovering alternatives to AZT. Particular emphasis has been placed on finding reverse transcriptase inhibitors that are effective against AZT-resistant virus. Several new nucleoside analogs that have therapeutic value are now available, including didanosine (ddI), zalcitabine (ddC), stavudine (d4T), and lamivudine (3TC) (Fig. 19.27).

## Nonnucleoside Inhibitors of Reverse Transcriptase

Nonnucleoside inhibitors of viral reverse transcriptase do not bind at the nucleotide-binding site of the enzyme (Fig. 19.28 and 19.31). Examples of these compounds are nevirapine and the tetrahydroimidazobenzodiazepinone (TIBO) class of compounds. Initially they offered the hope of complementing nucleoside analog inhibitors, but the rapid emergence of resistant mutants dampened enthusiasm. Although they are effective inhibitors of reverse transcriptase, a substitution in any of seven residues that line their binding site on the enzyme confers resistance (Fig. 19.31). Because of the development of resistant mutants, nonnucleoside inhibitors cannot be used by themselves (**monotherapy**). Rather, their value lies in **combination therapy** with other drugs (see below). Efavirenz (Sus-

**A  Natural substrate of the HIV-1 protease**

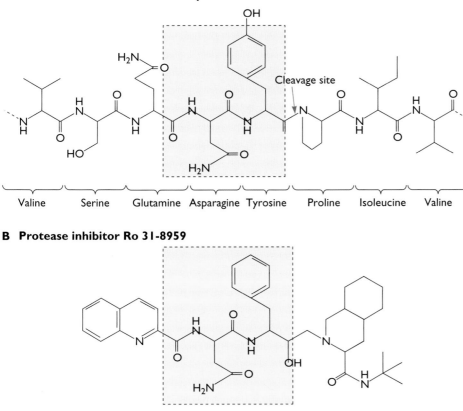

| Valine | Serine | Glutamine | Asparagine | Tyrosine | Proline | Isoleucine | Valine |

**B  Protease inhibitor Ro 31-8959**

**Figure 19.30 Comparison of one natural cleavage site for HIV-1 protease with a peptidomimetic protease inhibitor.** (A) The chemical structure of eight amino acids comprising one of the cleavage sites in the Gag-Pol polyprotein. The cleavage site between the tyrosine and proline is indicated by an arrow. (B) The chemical structure of an inhibitory Roche peptide mimic (Ro 31-8959). The dotted box indicates the region of similarity. Adapted from D. R. Harper, *Molecular Virology* (Bios Scientific Publishers Ltd., Oxford, United Kingdom, 1994), with permission.

tiva), the most recently approved nonnucleoside inhibitor, is so potent that a single tablet each day is effective.

### Protease Inhibitors

HIV protease, which is encoded in the *pol* gene, is essential for production of mature infectious virus particles. The enzyme cleaves itself from the Gag-Pol precursor polyprotein and then cleaves at seven additional sites in Gag-Pol to yield six *gag*-encoded proteins (MA, CA, p2, NC, p1, and p6) and three enzymes (protease itself, reverse transcriptase, and integrase). Active protease has been produced in high yields in many recombinant organisms and has even been synthesized chemically. It was the first HIV enzyme to be crystallized and studied at the atomic level (Fig. 19.16). The active enzyme is a small (only 99 amino acids) aspartyl-protease (similar to renin and pepsin) that functions as a dimer.

The seven cleavage sites in HIV Gag-Pol are similar but not identical. In the early stages of research to find protease inhibitors, it was essential to understand how the enzyme bound to and cleaved these sites. In pursuing this goal, an important discovery was made: the enzyme can recognize and cleave small peptide substrates in vitro. It

was subsequently determined that the active site is large enough to accommodate seven amino acids. The parameters of specific peptide binding and protease activity were determined by screening synthetic peptides containing variations of the seven cleavage sites for their ability to be recognized and cleaved by the enzyme. The first inhibitor leads were **peptide mimics** (peptidomimetics) modeled after inhibitors of other aspartyl proteases such as human renin, an enzyme involved in hypertension (Fig. 19.30). Subsequent screens for mechanism-based inhibitors and structure-based inhibitors designed de novo have yielded several distinct, powerful peptidomimetic inhibitors of the viral protease (Fig. 19.16 and 19.29).

### Other Viral Targets

While reverse transcriptase and protease are proven targets for antiviral therapy, at least 12 other virus-specific proteins offer promising targets for drug hunters (see also Table 18.14). Integrase protein (IN) is particularly attractive because biochemical and structural data are available. The absence of IN blocks insertion of viral DNA into the host genome and hence the ability of the virus both to persist in the cell and to produce progeny genomes.

**A**

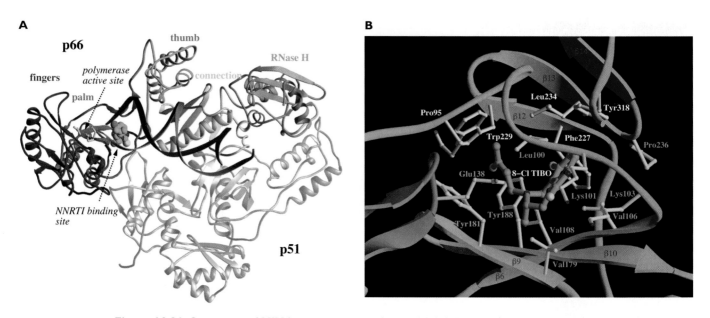

**B**

**Figure 19.31 Structure of HIV-1 reverse transcriptase highlighting the polymerase active site and the nonnucleoside reverse transcriptase inhibitor binding site.** (A) Structure of the reverse transcriptase p66-p51 heterodimer complexed with a double-stranded DNA template-primer, showing the relative locations of the polymerase active site and the site for binding nonnucleoside reverse transcriptase inhibitors (NNRTIs). Data from A. Jacobo-Molina et al., *Proc. Natl. Acad. Sci. USA* **90:**6320–6324, 1993. (B) Structure of 8-CL TIBO (a prototype NNRTI) bound to reverse transcriptase. Amino acid side chains corresponding to sites of drug resistance mutations are shown in orange; residues with cyan side chains are sites at which NNRTI resistance mutations have not been detected. Data from J. Ding et al., *Nat. Struct. Biol.* **2:**407–415, 1995. Courtesy of K. Das and E. Arnold, Center for Advanced Biotechnology and Medicine, Rutgers University.

The main neutralizing domain, identified by antibodies that block viral attachment, is the third variable domain (the so-called V3 loop) of the SU protein (Fig. 4.17). The V3 loop is required for virus-cell fusion. Compounds that interfere with V3 loop activity include specific antibodies and polysulfated or polyanionic compounds such as dextran sulfate, curdlan sulfate, and suramin. These compounds were identified early in the search for antiviral agents but were found to be ineffective because of intolerable side effects such as anticoagulant activity. Although considerable effort was expended in the development of inhibitors of the SU-CD4 receptor interaction, including a "soluble CD4" that theoretically would act as a competitive inhibitor of infection, no effective antiviral agents have been found with this strategy. The reason for this lack of success can be attributed in part to the high concentration of SU on the virus, as well as to the alternative mechanisms for spread in an infected individual.

Nevertheless, such studies have provided much insight into how the virus enters cells and have focused attention on other targets in the process. For example, it was curious that viruses resistant to neutralizing antibodies have clustered mutations in the V3 loop, yet virus-cell fusion is not affected. The implication was that CD4-V3 interactions were not involved in virus entry. As described in chapter 5, we now understand that for the entry of virus, the target cells must express not only CD4 but also any one of several seven-transmembrane-segment chemokine receptors such as Ccr5 on macrophages and Cxcr4 on T cells. The V3 loop of SU appears to interact with the chemokine receptor, not CD4. Chemokine receptors are attractive targets because individuals who are homozygous for mutations in one such receptor (Ccr5) are resistant to virus infection and apparently have no other problems. Recent studies have shown that infected individuals carrying a mutation in the Ccr2 chemokine receptor experience a delay of from 2 to 4 years in the progression to AIDS. Thus, development of a safe chemokine receptor antagonist that delays the onset of AIDS for a long period, if not indefinitely, may be feasible.

**BOX 19.8**

*Highly specific designed inhibitors may have unexpected activities*

The discovery and development of structure-based inhibitors of HIV protease has been pronounced a triumph of rational drug design. Structural biology and molecular virology came together to provide the protease inhibitors that anchor today's highly active antiretroviral therapy. However, patients receiving protease inhibitors often respond in unexpected ways, and the extent of immunological recovery with treatment is being debated. One idea was that protease inhibitors affected the immune response in addition to blocking virus replication. Recently, it was discovered that the protease inhibitor ritonavir also inhibits the chymotrypsin-like activity of the proteasome. As a result, the protease inhibitor blocks the formation and subsequent presentation of peptides to cytotoxic lymphocytes by MHC class I proteins.

The challenge now is to determine if this secondary activity helps or hinders AIDS therapy. As discussed in chapter 17, cytotoxic lymphocytes not only kill virus-infected cells but are also responsible for significant immunopathology in persistent infections. Perhaps ritonavir blocks this immunopathology. On the other hand, reduction in immunosurveillance by cytotoxic T cells potentiates persistent infections. In this case, the secondary activity of ritonavir may signal long-term problems. We now know that ritonavir and saquinavir inhibit proteasome activity, while indinavir and nelfinavir do not. Therefore, it will be of some interest to follow the antiviral effects and lymphocyte functions over time in patients under treatment with these different protease inhibitors. As noted by the investigators who found this surprising activity, the modulation of antigen presentation by protease inhibitors can be exploited for treatment of autoimmune disease, chronic immunopathologies, and diseases of transplantation.

Andre, P., M. Groettrup, P. Klenerman, R. DeGiuli, B. L. Booth, Jr., V. Cerundolo, M. Bonneville, F. Jotereau, R. M. Zinkernagel, and V. Lotteau. 1998. An inhibitor of HIV-1 protease modulates proteasome activity, antigen presentation, and T-cell responses. *Proc. Natl. Acad. Sci. USA* **95**:13120–13124.

## The Problem of Drug Resistance and the Use of Combination Therapy To Combat Infection

### HIV Accumulates Base Changes in Its Genome at a High Frequency

Viral mutants that grow with impunity in the presence of AZT appeared almost immediately after the drug was approved for general use. Resistance results from amino acid substitutions corresponding to single base pair changes at one of at least four sites in the reverse transcriptase gene. Reverse transcriptase enzymes bearing these substitutions no longer bind phosphorylated AZT, but they retain enzymatic activity. Mutants resistant to other nucleoside analogs as well as to protease inhibitors also arise with disheartening frequency. Resistance to protease inhibitors is found only after acquisition of several distinct amino acid substitutions, but mutant viruses resistant to these drugs have been isolated from patients. Thus, drug-resistant mutants threatened to undermine the entire antiviral effort until combination treatment became feasible in 1995.

### Indicators of Disease

To understand how antiviral drugs affect viral pathogenesis, sensitive measures of virus replication and cellular immune responses needed to be developed. As noted in chapter 18, two indicators of virus-induced illness are the CD4[+] T-cell count and the virus load. More than 500 CD4[+] lymphocytes per $\mu l$ is considered a normal value; fewer than 200 is the definition of AIDS. Viral load is estimated from the number of copies of viral RNA per milliliter of serum or plasma. The lowest detectable level with current methodology is about 400, and values greater than $10^6$ have been recorded. As discussed in detail in chapter 18, the late stage of infection is characterized by high levels of virus replication (often more than $10^9$ virions produced per day) and a high turnover of virus (50% turnover in less than a day). In view of the high mutation rate and the immense numbers of virions being produced daily in a single patient, the rapid appearance of drug-resistant HIV is not surprising.

### The Reality of Virus Biology Complicates Analysis of Drug Resistance

One of the most insidious aspects of RNA virus replication is that mutations arise at high frequency. For HIV, on average every new genome can be expected to carry at least one new mutation, and a single patient will harbor many new viral genomes in various tissues (Fig. 19.32). However, we cannot determine if, and to what extent, these mutations will become established in the virus population residing in an infected patient. Thus, it is difficult if

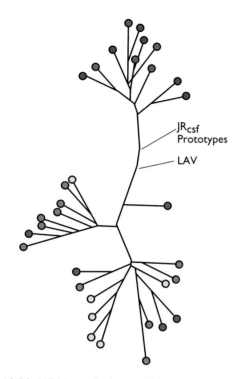

JR<sub>csf</sub>
Prototypes

LAV

**Figure 19.32 HIV-1 evolution within a single patient.** An unrooted phylogenetic tree showing relationships among 33 HIV-1 variants coexisting within a single infected patient. Two prototype viruses, LAV and Jrcsf, are shown in the center of the tree. Colors denote tissue sources for virus isolation: blue, lymph nodes; red, peripheral blood; yellow, spleen; green, lung; purple, spinal cord; brown, dorsal root ganglion. From M. Ait-Khaled et al., *AIDS* **9**:675–683, 1995, with permission.

not impossible to calculate the precise mutation frequency for any viral gene in any one infected individual. The different microniches for replication and the genetic diversity that exists in a single virus-infected individual provide the fuel for an explosion of more fit viruses after any contact with a selective force (Fig. 19.33). Even viruses with detrimental mutations are maintained in the population if their defects can be complemented by coinfecting viruses and they are integrated in nonactivated cells. Drug resistance does not have to be absolute because any increase in fitness, no matter how subtle, can result in large changes in the virus population. The reality is that we do not know the principal factors controlling virus load in an infected patient. Future drug and vaccine trials providing quantitative data on immune responses will go far to solve this problem.

### Antiviral Therapy Has the Potential To Promote or Prevent the Emergence of Resistant Viruses

An obvious but often ignored fact is that mutations can appear *only when the virus replicates its genome.* For retro-

viruses, the process of genome replication comprises making DNA copies by reverse transcriptase and RNA genomes by the host RNA polymerase II. Replication and associated recombination create the mutations that provide diversity for evolution in the population. As demonstrated by Salvador Luria and Max Delbrück more than 50 years ago, mutations arise spontaneously in the absence of selection (see Fig. 19.33 for an example). However, if viral replication is blocked, no drug-resistant mutants can arise. Thus, if an individual harboring a small virus population with no relevant preexisting mutations is given enough drug to block all viral replication, the virus will be held in check (Fig. 19.34). In contrast, if the same antiviral drug is given after the virus population becomes large, or if the drug is given in insufficient quantity to block replication, viruses harboring mutations conferring resistance will survive and will continue to replicate and evolve. By blocking viral replication with potent antiviral drug therapy when the virus population includes a few drug-resistant mutants, the infection may still be cleared by the host's immune system before resistant viruses take over. If resistance to an antiviral drug requires multiple mutations (e.g., resistance to protease inhibitors), the chance that all mutations preexist in a single genome is lower than if only a single mutation is required. But if replication is allowed in the presence of the inhibitor, resistant mutants will accumulate.

### The Use of Two or More Antiviral Drugs To Combat the Resistance Problem

The use of two or more treatments simultaneously is well known in tuberculosis therapy and cancer treatment. Combining two mechanistically different treatments often leads to more effective killing of the bacterium or tumor, thereby circumventing the appearance of bacteria or tumor cell variants resistant to one treatment or the other. This principle also applies to antiviral therapy. In theory, if resistance to one drug occurs once in every $10^3$ virions, and resistance to a second occurs once in every $10^4$ virions, then the likelihood that a genome carrying both mutations will arise is the product of the two probabilities, or once in every $10^7$ virions. However, until physicians were confronted with the specter of drug-resistant virus, the combination therapy approach for viral diseases was never investigated clinically. Current experience has established that combination therapy for viral diseases has considerable merit. One result is that many individual drug companies are developing multiple agents for treatment of the same viral infections.

### Cross-Resistance to Similar Inhibitors

HIV mutants resistant to different nucleoside analogs often carry different amino acid substitutions in the re-

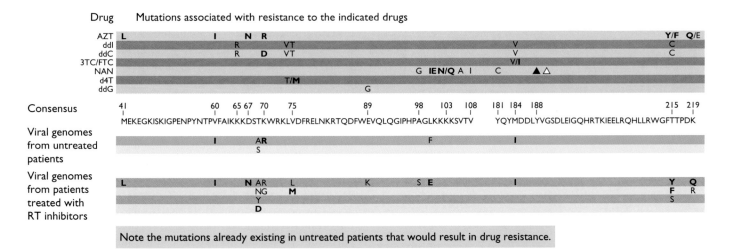

**Figure 19.33 Unselected mutations specifying resistance to HIV-1 reverse transcriptase inhibitors.** In this study, investigators reported the nucleotide sequences of the *pol* gene from 60 HIV-1 genomes obtained directly from primary lymphocytes from infected individuals who were treated or not treated with inhibitors of reverse transcriptase. Representative data are illustrated in this figure. The single-letter amino acid code is used. The reverse transcriptase inhibitors are indicated at the top left; the chemical structures of AZT, ddI, ddC, 3TC/FTC, and d4T are given in Fig. 19.27. NAN is a nonnucleoside inhibitor, and ddG is 2′,3′-dideoxyguanosine. The consensus amino acid sequence in the middle is that of HIV-1 group A reverse transcriptase. Amino acid replacements known to be involved in resistance to reverse transcriptase inhibitors are listed above the consensus sequence. A slash separating two amino acids denotes their association with resistance to the same inhibitor. The solid triangle corresponding to position 188 for NAN indicates that resistance has been associated with amino acids C, H, and L; the open triangle at position 190 indicates that resistance has been associated with amino acids A, E, and S. The green box below the consensus sequence indicates the amino acid changes found in individual viruses obtained from patients not exposed to reverse transcriptase inhibitors (untreated). Note the presence of amino acid changes that would result in resistance. Examples of viruses selected with antiviral drugs are indicated in the four groups in the blue boxes. Mutations related to drug resistance found in the absence of selection are the expected consequence of the statistical distribution of mutations along the *pol* gene. The presence of such important amino acid replacements in virus populations reminds us that viral quasispecies are reservoirs of diversity available for selection. Adapted from I. Najera et al., *J. Virol.* **69**:23–31, 1995, with permission.

verse transcriptase. Furthermore, a mutation conferring resistance to one reverse transcriptase inhibitor may suppress resistance to another (Table 19.11). Consequently, combinations of nucleoside analogs were clinically tested with the expectation that double-resistance mutants would be rare, perhaps inviable, or at least severely crippled. While initially promising, many combinations failed miserably, with mutants resistant to both drugs arising after less than a year of therapy. Furthermore, alterations in reverse transcriptase that conferred resistance to different nucleoside analogs appeared simultaneously. The barrier to resistance to many pairwise combinations of nucleoside and nonnucleoside inhibitors was higher than that for any single drug, but not high enough.

Experience with protease substrate analog inhibitors has been similar; resistance to two inhibitors emerges almost as fast as resistance to either one alone. As current protease inhibitors are all peptide mimics that bind to the substrate-binding pocket of the enzyme (see Fig. 19.16 to see the binding of two different protease inhibitors at the active site of the enzyme), a change in binding pocket residues can affect the binding of more than one inhibitor.

Treatment of a patient with one antiviral drug at a time clearly is of limited clinical value because of the possibility of selecting resistant mutants. Furthermore, the use of a single drug must be contemplated carefully because this same drug may be used for combination therapy in the future.

## Combination Therapy

### Problems and Promise

Although it is promising both in theory and in preliminary experiments, combination therapy does not always

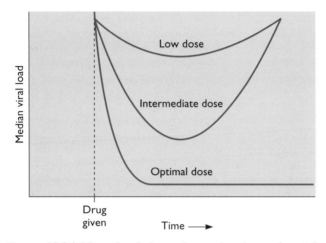

**Figure 19.34 Virus load depends on the dose of antiviral drug.** If virus replication is allowed in the presence of an antiviral drug, mutants resistant to that drug will be selected. This assertion is illustrated by plotting median virus load in relative units on the y axis as a function of time after exposure to a drug on the x axis. An antiviral drug is administered to a patient at a given time (drug given), and the concentration of virus in the blood or tissue sample is determined at various intervals thereafter. In the top curve (low dose), the concentration of antiviral drug is insufficient to block virus replication, and virus load is reduced transiently, if at all. Viruses that replicate may be enriched for resistant mutants. In the middle curve (intermediate dose), the concentration of antiviral drug appears to be successful in lowering viral load initially, indicating that some replication was blocked. In this example, virus replication was not blocked completely, and resistant viruses overwhelmed the patient. In the bottom curve (optimal dose), the concentration of the antiviral drug is such that *all* virus replication at the time of administration is blocked. As a consequence, no drug-resistant mutants can arise, and virus load drops dramatically. Redrawn from J. H. Condra and E. A. Emini, *Sci. Med.* **4**(1):14–23, 1997, with permission.

clear an infection. Resurgence of virus may be due not only to the appearance of drug-resistant mutants but also to a concomitant increase in the number of susceptible $CD4^+$ cells. Moreover, this type of therapy is very demanding for physician and patient and is not as easy as it may first appear. Dozens of pills must be swallowed every day without fail. If other infections are being treated, as they almost always are in AIDS patients, then more than 50 pills a day may be required. Other problems arise because simple storage and keeping track of different medications is a daunting task for an ill patient. To compound the problems, every drug has side effects, and some are severe (see also Table 18.13). For example, the gastrointestinal problems that accompany many protease inhibitors are particularly stressful for patients. Some side effects, such as the wasting of limbs and face with fat accumulation in the gut (lipodystrophy), may appear only after months of continuous use of current antiprotease drugs. Because of these problems, some patients simply do not take their medication. The most insidious failure lies in wait when the patient begins to feel better and stops taking the pills. Virus replication resumes when the inhibitors are removed. Replication means mutation, and in such cases combination therapy may be ineffective if ever reinstated. Finally, combination therapy is very expensive, currently costing more than $12,000 per year for some regimens. In addition to the costs of medication, tests for virus load and CD4 counts must be performed regularly to monitor therapy. Such tests cost in excess of $100 each. At these prices, combination therapy is clearly not accessible to everyone.

Even in the face of these seemingly formidable issues, there is considerable optimism that combination therapy, particularly for long-term control of virus infection, will be

**Table 19.11** Representative resistance mutation interactions[a]

| Agent | Resistance substitution(s) | Resistance interaction |
|---|---|---|
| Zidovudine | T215F in reverse transcriptase[b] | Didanosine resistance mutation (L74V) restores zidovudine susceptibility. |
| | | Lamivudine resistance mutation (M184V) restores zidovudine susceptibility. |
| | | Nevirapine and loviride resistance mutation (Y181C) restores zidovudine susceptibility. |
| | | Foscarnet resistance mutation (W88G) restores zidovudine susceptibility. |
| Delavirdine | P236L in reverse transcriptase | Increased susceptibility to nevirapine and R82913 (TIBO) |
| VX-478 | M46I, I47V, I50V in protease | In the presence of saquinavir, VX-478 susceptibility is restored while saquinavir resistance develops. Novel mutations include G48V, I50V, I84L. |
| | | In the presence of indinavir, VX-479 resistance persists while indinavir resistance develops. Novel mutations include V32I, A71V. |
| Foscarnet | E89K, L92I, S56A, Q161L, H208Y in reverse transcriptase | Increased susceptibility to zidovudine, nevirapine, and R82150 (TIBO) |

[a]Adapted from D. D. Richman et al., *Clinical Virology* (Churchill Livingstone, New York, N.Y., 1997.
[b]T215F indicates that the threonine at position 215 is changed to a phenylalanine.

effective. In one study, approximately 85% of patients treated with AZT, 3TC, and indinavir continued to exhibit significant reductions in the concentrations of circulating virus in the plasma, without apparent selection of resistant virus, for at least 68 weeks of therapy. As many as 80% of patients whose viral loads have dropped to below the limit of assay detection have sustained this response, with the longest duration being more than 160 weeks. In another study, the viral load in 45 patients went down for about 6 months but then rebounded. Yet after 18 months, their CD4 counts remained higher than they were before treatment began. Some investigators speculate that when treatment lowers the viral load, even for a few months, the immune system has a chance to revive and to renew the battle. Many believe that the clinical success of combination therapy is truly remarkable and represents one of the high points in the battle against AIDS.

### Lessons Learned

Despite the complexities of virus replication, the difficulties in estimating mutation frequencies, and the conceptual problems of understanding viral "fitness" in an infected individual, some conclusions that are applicable to many viral infections are possible. For example, it is prudent to avoid therapies in which resistance can be achieved by a single mutation in the virus. If drug resistance can be acquired by a single mutation, the probability that this mutation preexists in any given infected individual is high. As a corollary, therapies in which the virus must accumulate multiple mutations before becoming resistant should be employed, as the probability that a resistant mutant preexists in a patient never treated with the drug is lower. Combinations of drugs will be most effective if the patient has not been treated previously with either drug. Unfortunately, with a limited number of potent antiviral drugs available, this is often an unattainable luxury. Finally, the hope of combination therapy is that as the cells of the immune system in which the virus replicates ultimately die and are replaced, the virus concentration will be reduced to a point where the infection cannot be sustained. Even if a cure is wishful thinking, combination therapies with drugs of the future may convert HIV infection to a chronic treatable disease rather than a death sentence.

## The Quest for an AIDS Vaccine

In 1984, several years after HIV-1 was identified, the U.S. government predicted that an AIDS vaccine would be available within 3 years. Although research has continued for more than 14 years with unprecedented intensity, we still do not know how to produce a vaccine that will protect against AIDS. Such ill-founded optimism has a prece-

dent in the history of vaccination. Poliovirus was isolated in 1908, and 3 years later Simon Flexner of the Rockefeller Institute announced that a vaccine would be prepared in 6 months. Fifty years of research on poliovirus was necessary to provide the knowledge of pathogenesis and immunity that allowed the development of effective poliomyelitis vaccines.

Preliminary experiments on HIV with standard vaccine approaches (Table 19.12) have provided useful information but no clear indications of how to induce an effective immune defense in humans. Some of the problems to be surmounted involve the antigenic drift of the virus, its infection of the immune system itself, and the fact that it establishes a persistent infection that can produce billions of virus particles per day for years. An effective vaccine therefore must protect not only against transmission of the genetically diverse population of virus particles that arise after every round of replication but also against the diverse quasispecies that now exist worldwide. Furthermore, such a vaccine must protect against virus-infected cells that can transmit the infection from person to person.

These problems are daunting, but there are others as well. The infection is maintained in the face of a vigorous immune response. For some scientists, this is the crux of the challenge: how do we induce an immune response *superior* to the normal response that apparently is not good enough to clear the infection? Furthermore, testing of potential vaccines is a special problem, as animal models for virus replication and pathogenesis are not available. Current model systems that use simian immunodeficiency virus or HIV infection of primates are expensive and do not reproduce all features of the human disease. There are complicated social, ethical, and political issues that must be faced if vaccines are to be tested in humans. For example, can vaccine trials be conducted so that appropriate placebo controls can be evaluated? Vaccine developers face costly litigation if unanticipated reactions occur. On a larger scale, significant geopolitical issues arise when decisions on vaccine development are made by Western lead-

**Table 19.12** Possible HIV vaccine approaches

Inactivated virus

Attenuated virus

Subunit vaccine

   Envelope glycoproteins

   Gag proteins

   Viral vector expressing selected immunogenic HIV proteins or derivatives

   Viral particles (Gag-Pol cores)

   Synthetic peptides from immunogenic HIV proteins

DNA vaccine expressing immunogenic HIV proteins

ers for vaccines that must be used in developing countries. Resolution of these problems will not be easy.

A number of approaches—killed viruses, subunit vaccines based on single viral proteins, and passive immunization—have already been tested with no clear success. In particular, subunit vaccines, although capable of inducing strong antibody responses, were markedly inefficient in eliciting a cytotoxic-T-cell response. However, recombinant poxvirus vaccines (both vaccinia virus and canarypox virus vectors) have proved to be far more potent in activating cytotoxic T cells. A promising area of research is use of a live attenuated poxvirus vaccine followed by a subunit or DNA vaccine to provide strong cellular and humoral responses. Live attenuated vaccines, modeled after the successful live poliovirus vaccines, present difficult scientific and ethical problems: injecting thousands of healthy uninfected volunteers with a living, albeit attenuated, retrovirus to induce protective immunity against a wild virus may introduce unacceptable risks.

## Perspectives

Viral infection is a complex process making intervention a challenging enterprise. The built-in survival mechanisms of both virus and host provide both problems and opportunities in our quest to control viral disease. While vaccines and antiviral drugs can be effective, our primary defenses remain good personal health, proactive public health measures, and a public that understands how to avoid exposure to and spread of viral disease. The goal for vaccine research is to understand the molecular mechanisms of antiviral immune defense so that vaccines can be tailored to a particular virus. We must find ways to deliver viral antigen into both the MHC class I and class II pathways in order to induce a well-tolerated long-lasting immunity without significant vaccine-associated risk to the patient.

We have a limited arsenal of antiviral drugs and an urgent need for more. The paucity of drugs that inhibit RNA viruses is particularly troublesome. In addition, we are unable to diagnose acute infections accurately and in a timely fashion. Persistent viral infections present special problems for any antiviral intervention, as these infections may require the combined application of vaccines, immunotherapy, and antiviral drugs. For example, it should be possible to reduce virus load by antiviral drugs and then clear the remaining infection with bolstered immune responses. However, viruses that are resistant to immune defenses, as well as to antiviral drugs, will be selected. This certainty is a specter that continues to haunt antiviral research and public health. As the search for effective therapy to treat HIV infection demonstrates, we still have much to learn.

## References

### Books
Blair, E., G. Darby, G. Gough, E. Littler, D. Rowlands, and M. Tisdale. 1998. *Antiviral Therapy*. Bios Scientific Publishers, Oxford, United Kingdom.

Janeway, C. A., Jr., and P. Travers. 1996. *Immunobiology: the Immune System in Health and Disease*. Current Biology Limited, Garland Publishing Inc., New York, N.Y.

Mims, C., N. Dimmoch, A. Nash, and J. Stephan. 1995. *Mims Pathogenesis of Infectious Disease*, 4th ed. Academic Press, Orlando, Fla.

Richman, D. D., R. J. Whitley, and F. G. Hayden. 1997. *Clinical Virology*. Churchill Livingstone, New York, N.Y.

Spilker, B., and P. Cuatrecasas. 1990. *Inside the Drug Industry*. Prous Science Publishers, Barcelona, Spain.

Veerapandian, P. (ed.). 1997. *Structure-Based Drug Design: Diseases, Targets, Techniques and Developments*, vol. 1. Marcel Dekker, Inc., New York, N.Y.

### Historical Papers
Jenner, E. 1788. Observations on the natural history of the cuckoo. *Philos. Trans. R. Soc.* **78**:219–237.

Jenner, E. 1798. *An Inquiry into the Causes and Effects of the Variolae Vaccinae, a Disease Discovered in Some of the Western Counties of England, especially Gloucestershire, and Known by the Name of Cow Pox*. Sampson Low, London, England.

Metchnikoff, E. 1905. *Immunity in the Infectious Diseases*, 1st ed. Macmillan Press, New York, N.Y.

Woodville, W. 1796. *The History of the Inoculation of the Smallpox in Great Britain; Comprehending a Review of all the Publications on the Subject: with an Experimental Inquiry into the Relative Advantages of Every Measure Which Has Been Deemed Necessary in the Process of Inoculation*. James Phillips, London, England.

### Reviews and Perspectives
Ada, G. 1993. Vaccination in third world countries. *Curr. Opin. Immunol.* **5**:683–686.

Ada, G. L. 1994. Vaccines and immune response, p. 1503–1507. *In* R. G. Webster and A. Granoff (ed.), *Encyclopedia of Virology*. Academic Press, San Diego, Calif.

Ahmed, R. (ed.). 1996. Immunity to viruses. *Semin. Virol.* **7**:93–155.

Ahmed, R., and D. Gray. 1996. Immunological memory and protective immunity: understanding their relation. *Science* **272**:54–60.

Audibert, F. M., and L. D. Lise. 1993. Adjuvants: current status, clinical perspectives and future prospects. *Immunol. Today* **14**:281–284.

Babe, L. M., and C. S. Craik. 1997. Viral proteases: evolution of diverse structural motifs to optimize function. *Cell* **91**:427–430.

Balter, M. 1997. HIV survives drug onslaught by hiding out in T cells. *Science* **278**:1227.

Baltimore, D. 1988. Intracellular immunization. *Nature* (London) **335**:395–396.

Berger, E., B. Moss, and I. Pastan. 1998. Reconsidering toxins to eliminate HIV infection: you gotta have HAART. *Proc. Natl. Acad. Sci. USA* **95**:11511–11513.

Coffin, J. M. 1995. HIV population dynamics in vivo: implications for genetic variation, pathogenesis and therapy. *Science* **267**:483–489.

Cohen, J. 1997. Exploiting the HIV-chemokine nexus. *Science* **275**:1261–1264.

Condra, J. H., and E. A. Emini. 1997. Preventing HIV-1 drug resistance. *Sci. Med.* **4**(1):14–23.

DesJardin, J. A., and D. R. Snydman. 1998. Antiviral immunotherapy. A review of current status. *Biodrugs* **9**:487–507.

Elion, G. B. 1986. History, mechanisms of action, spectrum and selectivity of nucleoside analogs, p. 118–137. *In* J. Mills and L. Corey (ed.), *Antiviral Chemotherapy: New Directions for Clinical Application and Research.* Elsevier Science Publishing Co., New York, N.Y.

Ellis, R. W. 1996. The new generation of recombinant viral subunit vaccines. *Curr. Opin. Biotechnol.* **7**:646–652.

Evans, J. S., K. P. Lock, B. A. Levine, J. N. Champness, M. R. Sanderson, W. C. Summers, P. J. McLeish, and A. Buchan. 1998. Herpesviral thymidine kinases: laxity and resistance by design. *J. Gen. Virol.* **79**:2083–2092.

Field, A. K. 1994. Antivirals, p. 42–49. *In* R. G. Webster and A. Granoff (ed.), *Encyclopedia of Virology.* Academic Press, San Diego, Calif.

Fields, B. N. 1994. AIDS: time to turn to basic science. *Nature* (London) **369**:95–96.

Hajduk, P. J., R. P. Meadows, and S. W. Fesik. 1997. Discovering high-affinity ligands for proteins. *Science* **278**:497–499.

Hassett, D. E., and J. L. Whitton. 1996. DNA immunization. *Trends Microbiol.* **4**:307–311.

*Intelligent Drug Design.* 1996. *Nature* (London), supplement to vol. 384.

Marsden, H. (ed.). 1992. Antiviral therapies. *Semin. Virol.* **3**:1–75.

Moss, B. 1996. Genetically engineered poxviruses for recombinant gene expression, vaccination and safety. *Proc. Natl. Acad. Sci. USA* **93**:11341–11348.

Nathanson, N., and O. Kew (ed.). 1995. Molecular epidemiology of viruses. *Semin. Virol.* **6**:357–418.

Nolan, G. P. 1997. Harnessing viral devices as pharmaceuticals: fighting HIV-1's fire with fire. *Cell* **90**:821–824.

Norkin, L. C. 1995. Virus receptors: implications for pathogenesis and design of antiviral agents. *Clin. Microbiol. Rev.* **8**:293–315.

Ridky, T., and J. Leis. 1995. Development of drug resistance to HIV-1 protease inhibitors. *J. Biol. Chem.* **270**:29621–29623.

Stewart, P. L., and G. R. Nemerow. 1997. Recent structural solutions for antibody neutralization of viruses. *Trends Microbiol.* **5**:229–233.

*Therapeutic Horizons.* 1998. *Nature* (London) **392**(Suppl.):1–35.

*Vaccines: Frontiers in Medicine.* 1994. *Science* **265**:1371–1404.

*Vaccines.* 1998. *Nat. Med.,* supplement to vol. 4, no. 5.

Wade, R. C. 1997. "Flu" and structure-based drug design. *Structure* **5**:1139–1145.

*Selected Papers*

**Antiviral Drug Discovery**

Bonhoeffer, S., R. M. May, G. M. Shaw, and M. A. Nowak. 1997. Virus dynamics and drug therapy. *Proc. Natl. Acad. Sci. USA* **94**:6971–6976.

DeLucca, G. V., S. Erickson-Viitanen, and P. Y. S. Lam. 1997. Cyclic HIV protease inhibitors capable of displacing the active site structural water molecule. *Drug Discov. Today* **2**:6–18.

Kim, J. L., K. A. Morgenstern, C. Lin, T. Fox, M. D. Dwyer, J. A. Landro, S. P. Chambers, W. Markland, C. A. Lepre, E. T. O'Malley, S. L. Harbeson, C. M. Rice, M. A. Murcko, P. R. Caron, and J. A. Thompson. 1996. Crystal structure of the hepatitis C virus NS3 protease domain complexed with a synthetic NS4A cofactor peptide. *Cell* **87**:343–355.

Perelson, A. S., P. Essunger, Y. Cao, M. Vesanen, A. Hurley, K. Saksela, M. Markowitz, and D. D. Ho. 1997. Decay characteristics of HIV-1 infected compartments during combination therapy. *Nature* (London) **387**:188–191.

Underwood, M. R., R. J. Harvey, S. C. Stanat, M. L. Hemphill, T. Miller, J. C. Drach, L. B. Townsend, and K. K. Biron. 1998. Inhibition of human cytomegalovirus DNA maturation by a benzimidazole ribonucleoside is mediated through the UL89 gene product. *J. Virol.* **72**:717–725.

Varghese, J. N., V. C. Epa, and P. M. Colman. 1995. Three dimensional structure of the complex of 4-guanidino-Neu5Ac2en and influenza virus neuraminidase. *Protein Sci.* **4**:1081–1087.

von Itzstein, M., W. Y. Wu, G. B. Kok, M. S. Pegg, J. C. Dyason, B. Jin, T. Van Phan, M. L. Smythe, H. F. White, S. W. Oliver, P. M. Colman, J. N. Varghese, D. M. Ryan, J. M. Woods, R. C. Bethell, V. J. Hotham, J. M. Cameron, and C. R. Penn. 1993. Rational design of potent sialidase-based inhibitors of influenza virus replication. *Nature* (London) **363**:418–423.

**Vaccine Discovery**

Bolker, B. M., and B. T. Grenfell. 1996. Impact of vaccination on the spatial correlation and persistence of measles dynamics. *Proc. Natl. Acad. Sci. USA* **93**:12648–12653.

Borrow, P., H. Lewicki, X. Wei, M. S. Horwitz, N. Peffer, H. Meyers, J. A. Nelson, J. E. Gairin, B. H. Hahn, M. B. Oldstone, and G. M. Shaw. 1997. Antiviral pressure exerted by HIV-1-specific cytotoxic T lymphocytes (CTLs) during primary infection demonstrated by rapid selection of CTL escape virus. *Nat. Med.* **3**:205–211.

Brochier, B., M. P. Kieny, F. Costy, P. Coppens, B. Baudilin, J. P. Lecocq, B. Languet, G. Chappuis, P. Desmettres, K. Afiademanyo, R. Lilbois, and P. P. Pastoret. 1991. Large scale eradication of rabies using recombinant vaccinia-rabies vaccine. *Nature* (London) **354**:520–522.

Chu, R. S., O. S. Targoni, A. M. Krieg, P. V. Lehmann, and C. V. Harding. 1997. CpG oligodeoxynucleotides act as adjuvants that switch on T helper 1 (Th1) immunity. *J. Exp. Med.* **186**:1623–1631.

Gans, H. A., A. M. Arvin, J. Galinus, L. Logan, R. Detlovitz, and Y. Maldonado. 1998. Deficiency of the humoral immune response to measles vaccine in infants immunized at age 6 months. *JAMA* **280**:527–532.

Kuklin, N., M. Daheshia, K. Karem, E. Manickan, and B. T. Rouse. 1997. Induction of mucosal immunity against herpes simplex virus by plasmid DNA immunization. *J. Virol.* **71:** 3138–3145.

Modelska, A., B. Dietzschold, and V. Yusibov. 1998. Immunization against rabies with plant-derived antigen. *Proc. Natl. Acad. Sci.* **95:**2481–2485.

Mortara, L., F. Letourneur, H. Gras-Masse, A. Venet, J.-G. Guillet, and I. Bourgault-Villada. 1998. Selection of virus variants and emergence of virus escape mutants after immunization with an epitope vaccine. *J. Virol.* **72:**1403–1410.

O'Neal, C. M., S. E. Crawford, M. K. Estes, and M. E. Conner. 1997. Rotavirus virus-like particles administered mucosally induce protective immunity. *J. Virol.* **71:**8707–8717.

Rodriguez, F., J. Zhang, and J. L. Whitton. 1997. DNA immunization: ubiquitination of a viral protein enhances cytotoxic T-lymphocyte induction and antiviral protection but abrogates antibody induction. *J. Virol.* **71:**8497–8503.

Taboga, O., C. Tami, E. Carrillo, J. I. Nunez, A. Rodriguez, J. C. Saiz, E. Blanco, M. L. Valero, X. Roig, J. A. Camarero, D. Andreu, M. G. Mateu, E. Giralt, E. Domingo, F. Sobrino, and E. L. Palma. 1997. A large-scale evaluation of peptide vaccines against foot-and-mouth disease: lack of solid protection in cattle and isolation of escape mutants. *J. Virol.* **71:**2606–2614.

Xu, L., A. Sanchez, Z. Yang, S. R. Zaki, E. G. Nabel, S. T. Nichol, and G. J. Nabel. 1998. Immunization for Ebola virus infection. *Nat. Med.* **4:**37–42.

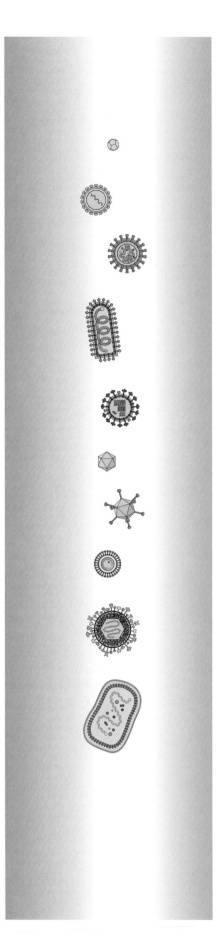

# 20

# Virus Evolution and the Emergence of New Viruses

Anything produced by evolution is bound to be a bit of a mess.

SYDNEY BRENNER

Around here, it takes all the running you can do just to stay in the same place.

LEWIS CARROLL
The Red Queen to Alice in Alice in Wonderland

## Virus Evolution

### Virus Evolution Is Contemporary

The word "evolution" conjures up images of fossils, dusty rocks, and ancestral phylogenetic trees covering eons. Modern virology shatters this staid image and reminds us that evolution not only is contemporary (we can see it happening in real time) but also carries substantial implications for the future. Indeed, as host populations grow and adapt, so must viruses that infect them. It also works the other way: viruses can be significant selective forces in the evolution of host populations. If a host population cannot adapt to a lethal virus, it may be exterminated. This is the basic nature of a host-parasite relationship, magnified because viruses display enormous genetic diversity. The public is made aware of virus evolution by frequent, often sensationalized reports in the press of new viral diseases, the human immunodeficiency virus pandemic, and our regular bouts with influenza and common cold viruses. Scientists are challenged regularly to devise new antiviral drugs, vaccines, diagnostics, and treatments for these "new, emerging diseases." In truth, we are only beginning to develop an understanding of the selective forces that drive virus evolution and the corresponding implications for our own survival.

### How Do Viruses Evolve?

By definition, to exist, successful viruses must always be one step ahead of their host. Because of their genetic simplicity, viruses appear poorly equipped for survival. Moreover, the harsh realities of maintaining a submicroscopic infectious agent in a large, often dispersed host population and in ever-changing environments appear to present overwhelming barriers to survival. Obviously this line of thinking is flawed, as viruses may be the most successful life forms on the planet. As will be discussed in this chapter, the primary reason for their remarkable success is that despite a minimal set of genes, viral populations display spectacular **diversity.** It is such diversity, manifested in the large collection of genomic permutations that are present in any viral population at any given time, that provides constant new opportunities for survival. The sources of this diversity are mutation, recombination, and selection. The constant change of a viral population in the face of selection pressures is the definition of **virus evolution.**

Positive and negative selection for particular preexisting mutants can occur at any step in a viral life cycle. The selective forces working on viruses are imposed not only by the environment but also by the constraints of the viral genome and life cycle. The requirement to spread within the cells of an infected host and among individuals in the population exposes the virus to a variety of host antiviral defenses. The population density, the social behavior, and the health of potential hosts clearly represent but a few of the powerful selective forces determining the survival of the virus.

Notice that we defined virus evolution in terms of a **virus population.** Viruses are capable of rapid, often explosive, evolution because their large populations comprise diverse arrays of mutants that are produced in short replication cycles. For some RNA viruses, thousands of progeny are produced after a single cycle of replication, and almost every new genome differs from every other. This property has important ramifications. For example, it is misleading and confusing to think of an "average" virus. Virologists must consider entire virus populations as well as the individuals that comprise it. As Stephen J. Gould put it, the median is *not* the message when it comes to evolution. As we will see, the diversity of the population provides surprising avenues for survival, yet every individual is a potential winner. Occasionally even the most rare genotype in a current population may be the most common after a single selective event. It is worth keeping this simple fact in mind: the population of viruses under study today has descended from survivors of past selections.

The trajectory of evolution has long been a subject of debate. Scientists and philosophers have considered many questions, such as the following. Is there a predictable direction for evolution, and if so, what is the path? Does change occur in a series of small steps leading to higher and higher degrees of fitness? Does selection impose a landscape of distinct peaks and valleys representing various fitness niches in which single mutations can take a fit organism back into a valley to start over, perhaps on a new course? Virology provides a productive area for research and insight into these questions. A primary lesson is that we must avoid judging outcomes as good or bad. While making such judgments is a uniquely human activity, anthropomorphic assessments and the comfort of analogies must be avoided when virus evolution is considered. Fur-

thermore, from the first principle that there is no goal but survival, we can deduce that evolution does not move a virus from "simple" to "complex" or along some trajectory aimed at "perfection." Rather, change is effected by removing the ill-adapted, not on building something better. At the most basic level of virus evolution, much can be learned by distinguishing viruses that simply survive a selection event (survival of the survivors) from those that survive because they are more fit (survival of the fittest).

### Viruses Produce Large Numbers of Progeny

Infection of a cell by a single virus results in the production of enormous numbers of progeny, providing a rich resource for selection. For example, poliovirus can produce about 10,000 virus particles per cell, and, in theory, after three or four cycles of replication at this rate could produce enough virus to infect every cell in the adult human body. This does not happen for a variety of reasons, including a vigorous host defense and a tropism for certain types of cells. Nevertheless, high rates of replication over short periods and the resulting accumulation of large quantities of infectious particles are hallmarks of many common acute and persistent viral infections. This feature is illustrated in Table 20.1 for human immunodeficiency virus and hepatitis B virus, two important human viruses. Total viral release in a single infected person can range from more than $10^9$ particles per day for human immunodeficiency virus, up to a staggering $10^{11}$ particles per day for hepatitis B virus. In the case of human immunodeficiency virus, the time from release of a virus particle to infection and lysis of another target cell is estimated to be 2.6 days during the later stages of infection. What is more striking is the 50 to 90% turnover rate (virus replication minus elimination by host defenses) in plasma. Such measurements reveal the prodigious, relentless production of new viruses in the face of a vigorous host defense. These are fertile grounds for selection and evolution.

**Table 20.1** In vivo dynamics of hepatitis B virus and human immunodeficiency virus

| Virus location and characteristic | Hepatitis B virus | Human immunodeficiency virus[a] |
|---|---|---|
| Virus in plasma | | |
| Half-life | 24 h | 6 h |
| Daily turnover | 50% | 90% |
| Total production in blood | >$10^{11}$ | >$10^9$ |
| Virus in infected cell | | |
| Half-life | 10–100 days | 2 days |
| Daily turnover | 1–7% | 30% |

[a] As measured in the symptomatic phase.

## Replicating Viruses Produce Large Numbers of Mutant Genomes

Evolution is possible only when mutations occur in a population. Viruses provide vivid proof of this assertion: when viruses replicate, mutations invariably accumulate in their progeny genomes at substantial rates.

**RNA viruses.** Most RNA virus genomes, or genomes that are replicated through an RNA intermediate (e.g., retroviruses and hepadnaviruses), are copied with considerably less fidelity than DNA virus genomes (see chapters 6 and 7). The average error frequencies reported for RNA genomes are about one misincorporation in $10^4$ or $10^5$ nucleotides polymerized, more than one million times greater than the rate for a host genome. Given a typical RNA viral genome of 10 kb, a mutation frequency of 1 in $10^5$ nucleotides polymerized results in about one mutation per 10 genomes; a 1 in $10^4$ error rate means that there will be an average of one mutation in every replicated genome.

**DNA viruses.** Unlike RNA viruses, DNA viruses tend to have low error rates during genome replication because of error-correcting DNA polymerases. Nevertheless, large numbers of progeny provide DNA viruses with ample diversity for evolution, even at this relatively low mutation rate. Furthermore, the mutation rate tends to be somewhat higher than that observed for the host genome during cellular DNA replication. Such increased genetic instability is best seen by propagating DNA viruses in vitro. For example, the 48-kb, double-stranded DNA genome of bacteriophage lambda can be replicated either by host enzymes (when the genome is integrated into the host genome as a prophage) or by viral enzymes during lytic propagation. When replicated by the host, the number of errors per lambda genome is 20 to 100 times lower than when the same genome is copied during lytic replication.

As discussed below (see "DNA Virus Relationships Deduced by Nucleic Acid Sequence Analysis"), DNA viruses that establish latent infections do not replicate their genomes extensively. As a result, diversity is reduced, but in turn, many selective forces that would normally operate on replicating viruses are avoided. For example, alphaherpesviruses establish permanent latent infections in nondividing neurons and may complete as few as two or three productive cycles of replication during the life of their host (which may be years). As one might predict, these viral genomes exhibit low sequence variation, even among isolates from diverse geographic areas. Because of the low sequence variation and the long-lived latent infection, herpesviral genomes change at about the same rate as those of their hosts.

**Error rates are difficult to quantify.** While we are certain that viruses have high mutation rates, estimates of the numbers must be viewed with caution. Absolute error rates (measured as the number of misincorporations per nucleotide polymerized) for any nucleic acid polymerase are difficult, if not impossible, to determine. A variety of technical issues must be considered, including sampling problems and potential artifacts of experimental design. Estimates of error rate can vary substantially for a given virus depending on the experimental method by which they are assessed. For example, polymerase chain reaction (PCR) technology is commonly used to sample viral genomes, but the polymerase used in the amplification may itself introduce copying errors that must be factored into the analysis. Reporter genes (e.g., the *lacZ* gene, which encodes β-galactosidase) whose expression can be monitored easily are often cloned in a viral genome so that errors in the reporter gene can be scored by inspection. The error rate for the viral genome is then extrapolated from that determined for the reporter gene. While this method is relatively simple, it can yield misleading results because errors of incorporation are not uniformly distributed as each genome is copied and are often dependent on the particular sequence analyzed. For example, the reverse transcriptase from human immunodeficiency virus is inaccurate when measured in vitro, having an average error rate per nucleotide incorporated of 1 in 1,700. However, certain positions in the human immunodeficiency virus genome can be "hot spots" for mutation where the error rate can be as high as 1 per 70 polymerized nucleotides.

**RNA viruses and the quasispecies concept.** With error frequencies approaching one error in every genome, it is inappropriate to envision an RNA virus with a single genome. Rather these viruses exist as dynamic distributions of nonidentical but related replicons often called **quasispecies,** a concept developed by Manfred Eigen. Thus, the wild-type virus is not an individual genome but rather an ensemble of closely related genomes. While the members of all outbred species have genomic polymorphisms (by definition), the viral quasispecies principle emphasizes the extreme variation commonly exhibited by viruses and its often counterintuitive consequences. The term serves to remind us that the classical definition of a species (an interbreeding population of individuals) may be inadequate for viruses. For example, most viral infections are not initiated by a single virion but rather by a population of particles. Selective pressures acting on an infected cell or an infected organism affect the population of viruses produced by that cell or organism. As we will discuss below, even this expanded definition of a virus has limitations. For example, reaching the steady-state equilibrium population of a given viral quasispecies requires vast numbers of particles. Such an equilibrium cannot be attained in small populations typically produced in infections in nature or in the laboratory; indeed when small populations of viruses replicate, extreme fluctuations in genotype and phenotype are possible.

For a given RNA virus, the genome sequences cluster around a consensus or average sequence, but *virtually every genome can be different from every other.* Thus, while a rare genome with a particular mutation may survive a selection event, this mutation can be found in genomes carrying other mutations. As a result, the product of selection after replication is most certainly not a population of identical genome sequences but rather a new, diverse population of viruses that simply share the selected mutation. The ramifications of this phenomenon in virus evolution and pathogenesis are only now being appreciated.

**Increased mutation rates are selected during virus evolution.** Previously, we made the assertion that creation of diversity was the primary reason for virus survival. This conclusion follows in part from the isolation of certain mutations located in viral polymerases that reduce the frequency of incorporation errors during in vitro growth. In the case of herpes simplex virus, infection by particular mutant viruses resistant to DNA polymerase inhibitors produces 10- to 100-fold fewer spontaneous mutations in the viral thymidine kinase gene than does infection by the wild-type virus. Such mutants are called **antimutators.** Similarly, for human immunodeficiency virus, certain spontaneous mutants that are resistant to the reverse transcriptase inhibitor lamivudine exhibit a 3.2-fold reduction in error frequency. As the wild-type viruses have maintained high mutation rates, we can infer that lower rates are neither advantageous nor selected in nature. Moreover, these antimutator virus mutants can often be shown to have a selective disadvantage compared to the parent virus. In the case of human immunodeficiency virus, the seemingly modest 3.2-fold increase in fidelity was associated with a significant growth disadvantage in infected individuals.

The high mutation rate of RNA viruses has survived countless bouts of selection; the capacity to produce prodigious numbers of mutations is a powerful advantage. Yet, there must be a point at which selection and survival balance genetic fidelity and mutation rate. Intuitively, mutation rates higher than one error in 1,000 nucleotides incorporated must begin to challenge the very existence of the viral genome, as precious genetic information can be irreversibly lost. This limit is called the **error threshold,** a mathematical parameter that measures the complexity of the information to be maintained in the genome and the positive selective features of the resultant quasispecies.

RNA viruses tend to evolve close to their unique error threshold, while DNA viruses have evolved to exist far below it. We can infer these properties from experiments with mutagens. If one treats a cell culture infected with an RNA virus such as vesicular stomatitis virus or poliovirus with a base analog such as 5-azacytidine, the error frequency per genome increases only two- to threefold at best. In contrast, a similar experiment done with a DNA virus such as herpes simplex virus or simian virus 40 reveals an increase in single-site mutations of several orders of magnitude. One conclusion is that because RNA viruses cannot sustain any significant increase in mutation rate, they must be prospering at the precarious edge of their error threshold (Box 20.1).

The error threshold concept is more complicated than it might first appear. Many parameters contribute, including a complex property called **fitness**—the replicative adaptability of an organism to its environment. In the laboratory setting, fitness may be determined by comparison of growth rates or virus yields, but it is difficult to measure under more natural conditions such as infection of complex organisms that live in large, interacting populations. Another essential component equally difficult to measure is the stability or predictability of the *environment* as it affects viral propagation. Host population dynamics and seasonal variation are but two examples of the many complicated environmental variables that exist. Finally, given the diversity in any one viral population, determining the fitness of one population versus that of another involves the mathematics of population genetics, a subject beyond the scope of this text.

Another critical parameter affecting virus evolution is the **genetic bottleneck**—extreme selective pressures on small populations that result in loss of diversity, accumulation of nonselected mutations, or both. A simple experiment illustrating this idea is easily produced in the laboratory. A single RNA virus plaque on tissue culture cells is picked and expanded. Next, a single plaque is picked from the expanded stock, and the process is repeated over and over. Thus, the bottleneck arises by restricting further viral replication to the progeny found in a single plaque, which contains a few thousand progeny viruses derived from a single founder virus. The perhaps surprising result is that after about 20 or 30 cycles of single-plaque amplification, many resulting virus populations are barely able to grow. They are markedly less fit than the original population. The environment is constant, and the only apparent selection is that imposed by the ability of the population of viruses from a single plaque to replicate. Why does fitness plummet?

The answer lies in a phenomenon dubbed **Muller's ratchet:** small, asexual populations decline in fitness over time if the mutation rate is high. Replicating RNA viruses

## BOX 20.1

*Pushing human immuno-deficiency virus over the "error threshold": the Achilles' heel of the viral quasispecies?*

At the end stage of AIDS, individuals infected with human immunodeficiency virus produce more than $10^{10}$ virions per day, and, on average, each genome contains one mutation. Microniches in various tissues yield diverse viral populations in every infected individual. Interestingly, most of the virions appear to be nonviable, suggesting that the virus exists at its error threshold and cannot tolerate many additional mutations. What would happen if one could intentionally push the virus past its error threshold with a mutagenic nucleoside analog? Is lethal mutagenesis of human immunodeficiency virus a practical approach to treating RNA virus infections? Lawrence Loeb and colleagues selected antiviral nucleoside analogs that lacked toxicity for human cells but were incorporated into the DNA of human immunodeficiency virus genomes by the viral reverse transcriptase. These are not chain-terminating compounds like zidovudine; rather, they promote base misincorporations when copied. One compound, 5-hydroxydeoxycytidine, was promising in that it was not toxic to human cells but was efficiently incorporated into replicating viral genomes.

The experiments comprised serial passage of virus in human cells in the presence of the nucleoside analog. In seven of nine experiments, there was a precipitous loss of viral replication after 9 to 24 serial passages in the presence of the compound. Sequence analysis showed that the progeny genomes accumulated G-to-A substitutions. Loss of virus replication was not observed in 28 control infections in which the virus was serially passaged without the analog.

The concept of lethal mutagenesis is intriguing. Emergence of resistant virus may be less likely because nucleotide substitutions that alter base pairing affect interactions between the DNA template and the incoming deoxynucleoside triphosphate; they may not associate directly with the enzyme. Furthermore, mutant enzymes that are resistant to such mutagenic analogs are also likely to exhibit simultaneous reduced affinity for the complementary nucleoside triphosphates and thus may be less fit. Only a small fraction of accumulated mutations resulting from incorporation of the mutagenic nucleoside will lead to resistance. On the other hand, a significant proportion of mutations in other parts of the viral genome is likely to result in reduced fitness. In contrast to the situation with chain-terminating nucleosides like zidovudine there is a reduced likelihood of an immediate selective growth advantage with resistance to a mutagenic analog.

More studies must be done to validate the concept of pushing viruses over their error threshold. Certainly, the safety of using mutagenic nucleoside analogs in uninfected as well as infected individuals must be examined closely.

**Coffin, J. M.** 1995. HIV population dynamics in vivo: implications for genetic variation, pathogenesis and therapy. *Science* **267**:483–489.

**Loeb, L. A., J. M. Essigmann, F. Kazazi, J. Zhang, K. D. Rose, and J. I. Mullins.** 1999. Lethal mutagenesis of HIV with mutagenic nucleoside analogs. *Proc. Natl. Acad. Sci. USA* **96**:1492–1497.

---

produce many mutations; indeed, they survive close to their error threshold. By restricting population growth to serial single founders (the bottleneck) under otherwise nonselective conditions, mutations that exceed the threshold accumulates during replication and thus fitness decreases.

The ratchet metaphor should be clear: each of the new mutations works like a ratchet, "clicking" relentlessly as they accumulate at every replication cycle. Each mutation has the potential to erode the fitness of subsequent populations. Simple studies such as the serial plaque transfer experiment indicate that Muller's ratchet can be avoided if a more diverse viral population is replicated by serial passage (e.g., opening the bottleneck by increasing the number of clonal pools). One such study showed that pools of virus from 30 individual plaques were required in serial

transfer to maintain the culture's original fitness. This observation can be explained as follows: more diversity in the replicating population facilitates construction of a mutation-free genome by recombination or reassortment, thus removing or compensating for mutations that affect growth adversely. Even if such a recombinant is a rare species, it has a powerful selective advantage in this experimental paradigm. Indeed, its progeny will ultimately take over the population. The message is simple but powerful: diversity of a viral population is important for the survival of individual members; remove diversity, and the population suffers.

While the particular bottleneck of single-plaque passage is obviously artificial, infection by a limited virus population and subsequent amplification are often found in nature. Examples include the small droplets of suspended virus during aerosol transmission, the activation of a latent virus from a limited population of cells, or the small volume of inoculum introduced in infection by insect bites. An important question is, How do viruses that spread in nature by these routes escape Muller's ratchet?

### Viruses Exchange Information

Viruses can exchange genetic information by **recombination** or **reassortment** of genome segments (see chapters 6, 7, and 9). Recombination creates new combinations of mutations that may be essential for survival under selective pressures. As discussed above, this process allows the construction of viable genomes from debilitated ones and avoids Muller's ratchet. Recombination occurs when the polymerase changes templates **(copy choice)** during replication or when nucleic acid segments are broken and rejoined. The former mechanism is common among RNA viruses, whereas the latter is more typical of double-stranded DNA virus recombination. Another mechanism for exchange of genetic information is reassortment among genomic segments when cells are coinfected with segmented RNA viruses. It is an important source of variation, as exemplified by orthomyxoviruses and reoviruses.

Insertion of nonviral nucleic acid into a viral genome (sequence-independent recombination) is well documented in virology. Indeed, viruses that have incorporated cellular genetic information into their genomes (transducing viruses) not only are a mainstay of viral genetics in the laboratory but also play an essential role in viral evolution. Incorporation of cellular sequences can lead to defective genomes or to more pathogenic viruses, as exemplified by the appearance of a cytopathic virus in an otherwise nonpathogenic infection by the pestivirus bovine viral diarrhea virus (see chapter 6). The acquisition of activated oncogenes from the cell genome is the hallmark of acutely transforming retroviruses like Rous sarcoma virus (see chapter 16).

Information can be exchanged in a variety of unexpected ways during viral infections. For example, a host can be infected or coinfected by many different viruses during its lifetime. In fact, serial and concurrent infections are commonplace. Both have a profound effect on virus evolution. In the simplest case, propagation of a virus quasispecies in an infected individual allows coinfection of individual cells, phenotypic mixing, and genetic complementation. As a result, recessive mutations are not immediately eliminated despite the haploid nature of most viral genomes. Of course, such coinfection also provides an opportunity for physical exchange of genetic information.

Viruses capable of affecting the host's immune response have a marked effect on concurrent or subsequent infec-

tions of the same host by very different viruses. Indeed, human immunodeficiency virus and measles virus provide dramatic examples of viruses capable of producing a global reduction in host defense, facilitating subsequent infections by diverse pathogens. This is exemplified by AIDS initiated by human immunodeficiency virus infection: most AIDS patients suffer and die from a variety of infections normally contained by the immune system. The poxviruses produce a number of secreted proteins that block the host immune defenses, providing many coinfecting unrelated viruses an opportunity to grow relatively unimpeded (Fig. 15.5). Conversely, the ability of one virus to *activate* host defenses can impair subsequent infection by a different virus. The existence of these multifactorial interactions of diverse viruses reinforces the idea that selection acts not only on one population of viruses but also on interacting populations of diverse viruses.

### Two General Pathways for Virus Evolution

As viruses are absolutely dependent on their hosts, viral evolution tends to take one of two general pathways. In one pathway, viruses coevolve with their hosts so that both share a common fate; as the host prospers, so does the virus. However, given no other host, the virus faces a bottleneck because it can be eliminated with antiviral measures or by loss of the host. In the other pathway, viruses occupy broader niches and have multiple host species. Then, when one host species is compromised, the virus can replicate in another. In general, as discussed below, DNA viruses tend to follow the first pathway while many RNA viruses follow the second. Both pathways provide excellent strategies for virus survival, and both suggest important avenues for interruption of the host-virus relationship.

## The Origin of Viruses

We presented a brief introduction to the origin of viruses in chapter 1. The primordial origin of viruses is a matter of conjecture and debate centering around three nonexclusive ideas (Table 20.2). The **regressive theory** suggests that viruses are derived from intracellular parasites that have lost all but the most essential genes required for replication and maintenance. The **cellular origin theory** proposes that viruses arose from cellular components that gained the ability to replicate autonomously in the host cell. The third theory postulates that **viruses coevolved with cells** from the origin of life itself. As there is no fossil record and there are few viral stocks more than 80 years old, we are not in a good position to test the three hypotheses for the primordial origins of viruses. Nevertheless, we can offer some insight into viral evolution by examining contemporary viruses.

**Table 20. 2** Three general theories of the origin of viruses

Regressive evolution: viruses degenerated from previously independent life forms, lost many functions that other organisms possess, and have retained only what is needed for their parasitic lifestyle

Cellular origins: viruses are subcellular functional assemblies of macromolecules that have gained the capacity to move from cell to cell

Independent entities: viruses evolved, on a course parallel to that of cellular organisms, from the self-replicating molecules believed to have existed in the primitive prebiotic RNA world

## DNA Virus Relationships Deduced by Nucleic Acid Sequence Analysis

Nucleic acid sequence analysis has identified many relationships among diverse viral genomes, thus providing considerable insight into the origin of viruses. The herpesviruses have been studied extensively by gene sequencing, and enough data have been accumulated for a detailed molecular phylogenetic analysis. The three main subfamilies of the family *Herpesviridae* (*Alphaherpesvirinae*, *Betaherpesvirinae*, and *Gammaherpesvirinae*) are easily distinguished by this method even though the original separation of these families was based on general, often arbitrary biological properties (Fig. 20.1). By assuming a constant "molecular clock" (the relatively constant incorporation of errors over time by DNA replication systems) and by comparing the sequence of a gene shared by most herpesviruses (e.g., uracil-DNA glycosylase) with cellular homologs of this gene, it has been possible to put herpesviral genome evolution on an evolutionary time scale previously established for their host lineages. The fascinating result is that for most herpesviruses, points of divergence coincide with well-established points of divergence of the vertebrate host. This property can be seen in the overlapping, branching pattern of the host and virus evolutionary trees. The conclusion is that an early herpesvirus infected an ancient host progenitor, and subsequent viruses developed by cospeciation with their hosts. By knowing the times of divergence of the hosts on the basis of a more classical study of fossils and assuming a constant molecular clock, we can show that the three major groups of herpesviruses arose approximately 180 million to 220 million years ago, indicating that these three subfamilies of viruses were in existence before mammals spread over the planet 60 million to 80 million years ago.

Coevolution with a host is a property relevant not only to herpesviruses but also to small DNA viruses (parvoviruses and papovaviruses). For these viruses, the evidence for coevolution comes not from comparison of host and viral genes but rather from finding close association of a given viral DNA sequence with a particular host group.

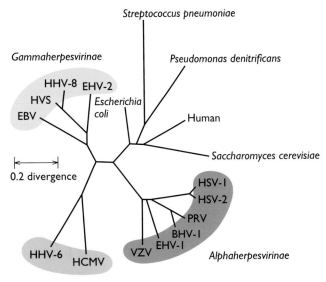

**Figure 20.1 Herpesvirus family relationships deduced by amino acid sequence comparison of the common enzyme uracil DNA glycosylase.** The amino acid sequences were deduced from the DNA sequences of cloned genes from a variety of organisms. The figure is a neighboring-joining distance tree (from alignment with gap positions removed). It illustrates clusters of related sequences and the degree of their relatedness. The length of lines joining sequences illustrates the calculated divergence from a putative precursor. The three main subfamilies of the family *Herpesviridae* (*Alphaherpesvirinae*, *Betaherpesvirinae*, and *Gammaherpesvirinae*) are easily distinguished by this method even though the original separation of these families was based on general biological properties. Virus abbreviations: BHV-1, bovine herpesvirus 1; EBV, Epstein-Barr virus; EHV-1 and -2, equine herpesviruses 1 and 2; HCMV, human cytomegalovirus; HHV-6 and -8, human herpesviruses 6 and 8; HSV-1 and -2, herpes simplex virus types 1 and 2; HVS, herpesvirus saimiri; PRV, pseudorabies virus; VZV, varicella-zoster virus. Data from D. J. McGeoch et al., *J. Mol. Biol.* **247:**443–458, 1995, with permission.

The linkage of host to virus was particularly striking when human papillomavirus types 16 and 18 were compared: the distribution of distinct viral genomes is congruent with the racial and geographical distribution of the human population. Another example of the same phenomenon is obtained with JC virus, a human polyomavirus associated with a rare, fatal brain infection of oligodendrocytes. This virus exists as five or more genotypes identified in the United States, Africa, and parts of Europe and Asia. Recent PCR analyses of these subtypes indicate that JC virus not only coevolved with humans but did so within specific human subgroups. Probably the most striking finding was that the JC virus of a particular human group provides a convenient marker for human migrations from Asia to the Americas in both prehistoric and modern times.

How can viral evolution be linked to specific human populations in a manner akin to vertical transmission of a host gene? We can begin to appreciate this perhaps counterintuitive phenomenon from the unusual biology of human papillomaviruses. These viruses establish a persistent infection in the skin of adults, replicating the genomes only as basal keratinocyte skin cells differentiate and finally producing viral particles only as cells reach terminal differentiation. When mothers infect newborns with high efficiency, because of close contact or of reactivation of persistent virus during pregnancy or birth, the virus appears to spread vertically in preference to the more standard horizontal spread between hosts. This mode of transmission appears to be the predominant mechanism for papillomavirus and stands in contrast to that of most acutely infecting viruses which are spread by aerosols, contaminated water, and food.

### RNA Virus Relationships Deduced by Genome Sequence Analysis

The relationships among RNA viruses can also be deduced from sequence analyses, but some difficulties arise because these viruses accumulate mutations with high frequency during their replication. Moreover, their genomes are often smaller and carry few if any "nonessential" genes that might be useful for comparative studies like those conducted with DNA viruses. And, unlike large DNA viruses, RNA viruses have few if any genes in common with their hosts that can be used to track and correlate their molecular evolution. Nevertheless, the nucleotide sequences of many (+) and (−) strand RNA viruses reveal clear "superfamilies" defined by **homologous gene arrangements** and **common coding strategies.** These similarities suggest a common ancestry or that such viruses evolved common solutions to similar problems (Fig. 20.2 and 20.3).

If one examines the sequences of many (−) strand RNA genomes, the first obvious common feature is the limited number of encoded proteins (as few as 4 and not more than 13). Next, one observes that these proteins can be arranged in three homologous blocks of function: core proteins that interact with the RNA genome, glycoproteins that are in the envelope and are required for attachment and penetration of viral particles, and a polymerase required for replication and mRNA synthesis (Fig. 20.2). Comparative studies also indicate that (−) strand viruses all depend on host proteases to cleave their polyproteins [in contrast to the (+) strand viruses, most of which use both virus- and host-encoded proteases].

The (+) strand RNA viruses (excluding the retroviruses) are the largest and most diverse subdivision of viruses: the genome sequences of more than 100 viruses representing

at least 30 distinct groups are available for analysis. The number of distinct proteins encoded by (+) strand RNA viruses ranges from 3 to more than 12, and these proteins can be divided into three groups by function: those required for RNA replication, encapsidation, and accessory functions. This comparison has resulted in the identification of three virus "supergroups" (Fig. 20.3A). A unifying feature is that, for (+) strand RNA viruses, the RNA polymerase gene appears to be the most highly conserved, implying that it arose once in the evolution of these viruses. As each of the supergroups contains members that infect a broad variety of animals and plants, the primordial RNA polymerase gene might have been provided by an ancestor present before the separation of kingdoms. Alternatively, the ancestral (+) strand virus could have radiated horizontally among plants, animals, and bacteria.

It is instructive to consider the postulated evolution of supergroup 3, which contains the alphaviruses. In doing so, we must make certain assumptions about the rate of RNA genome divergence and the constancy of virus evolution in diverse hosts and environments. These assumptions cannot be tested directly by experiment, but for what follows they are reasonable. Given this caveat, consider the assertion that the 26 distinct viruses in the alphavirus family arose by divergence from a common ancestor that existed *about 4,000 years ago* as shown in Fig. 20.3B. The dendrogram represents the relationship of viral genome sequences based on similarity. The illustration also provides a rough estimate of the time scale for divergence of the current virus families deduced from assumptions discussed in the legend. One conclusion of this analysis was that *recombination* established a new viral lineage, the western equine encephalitis viruses. The modern western equine encephalitis virus lineage can be explained by a recombination event occurring approximately 1,600 years ago between a member of the eastern equine encephalitis virus lineage and a member of the Sindbis virus group. It is interesting that in similar sequence comparisons made with flavivirus genomes, the postulated precursor of modern lineages arose 5,000 years ago, a date not much different from that asserted for the origin of the 4,000-year-old precursor of alphaviruses. One implication is that these two arboviral families (arthropod-borne viruses) encountered a genetic bottleneck at about the same time. It may be no coincidence that this time coincides with the last ice age, a climatic event that would have had a devastating effect on many insect hosts and would thereby have interrupted the arbovirus-insect-vertebrate host transmission cycle.

### The Protovirus Theory

The origin of retroviruses may center on the unique enzyme reverse transcriptase and is embodied in the **pro-**

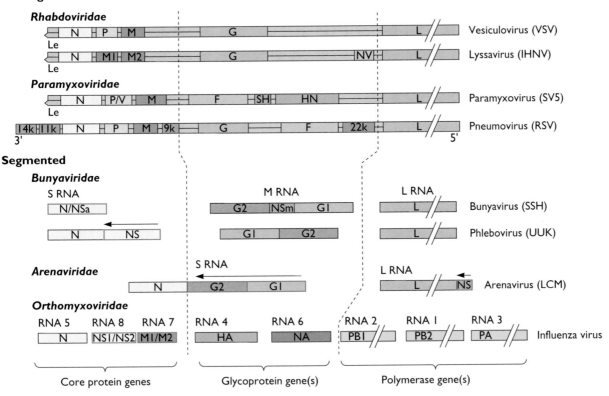

**Figure 20.2 Genome organization among (−) strand RNA viruses.** Gene maps of the *Rhabdoviridae, Paramyxoviridae, Bunyaviridae,* and *Arenaviridae* are aligned to maximize the similarity of gene products. The individual gene segments of the *Orthomyxoviridae* are arranged according to functional similarity to the two other groups of segmented viruses. Virus abbreviations: VSV, vesicular stomatitis virus; IHNV, infectious hematopoietic necrosis virus; RSV, respiratory syncytial virus; SV5, simian virus 5; SSH, snowshoe hare virus; UUK, Uukuniemi virus; LCM, lymphocytic choriomeningitis virus. Le is a nontranslated leader sequence. Gene product abbreviations: N, nucleoprotein; P, phosphoprotein; M (M1, M2), matrix proteins; G (G1, G2), membrane glycoproteins; F, fusion glycoprotein; HN, hemagglutinin/neuraminidase glycoprotein; L, replicase; NA, neuraminidase glycoprotein; HA, hemagglutinin glycoprotein; NS (NV, SH, NSs, and NSm), nonstructural proteins; PB1, PB2, and PA, components of the influenza virus replicase. Within a given genome, the genes are approximately to scale. Blue-outlined genes are those in which multiple proteins are produced from different open reading frames. Red-outlined genes are expressed by the ambisense strategy. Figure derived from J. H. Strauss and E. G. Strauss, *Microbiol. Rev.* **58:**491–562, 1994, with permission.

**tovirus theory** elaborated by Howard Temin, who shared the Nobel Prize for the discovery of reverse transcriptase. In this theory, a cellular reverse transcriptase-like enzyme copied segments of cellular RNA into DNA that were then inserted in the genome to form "retroelements." These in turn acquired more complex sequences, including integrase and RNase H domains, regulatory sequences, and structural genes encoding capsid and envelope proteins (Fig. 20.4). This theory predicts that remnants of this

process might still exist in the genomes of mammals and other species. Indeed, many of the predicted intermediates are found in abundance, including pseudogenes, retroposons, retrotransposons, and a variety of endogenous retroviruses (see Box 7.2). In humans these endogenous retroviruses (human endogenous retroviruses [HERVs]) are surprisingly abundant, comprising several percent of the human genome, and may represent footprints of ancient germ cell infections.

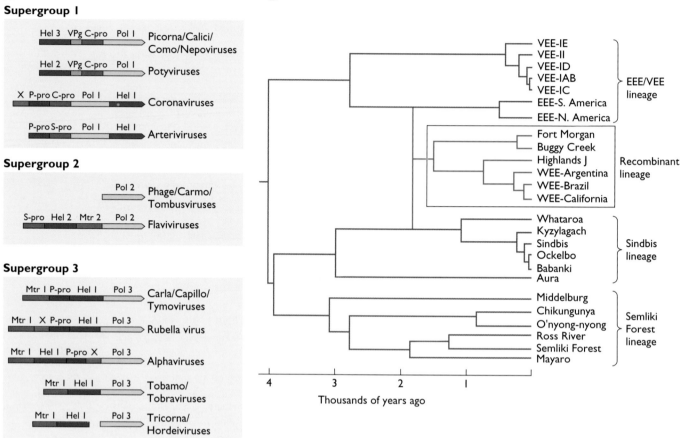

**Figure 20.3 RNA virus genomes and evolution.** (A) Organization of (+) strand RNA genomes. The genomes of (+) strand RNA viruses comprise several genes for replicative functions that have been mixed and matched in selected combinations. These functions include a helicase (purple), a genome-linked protein (orange), a chymotrypsinlike protease (red), a polymerase (yellow), a papainlike protease (brown), a methyltransferase (dark blue), and a region of unknown function, X (green). Some of these gene functions are encoded in more than one sequence (such as Pol 1, Pol 2, and Pol 3) as defined by amino acid sequence similarity. In this figure, the genes are not shown to scale and the structural proteins have been omitted for clarity. Derived from J. H. Strauss and E. G. Strauss, *Microbiol. Rev.* **58:**491–562, 1994, with permission. (B) Dendrogram depicting a proposed phylogenetic tree of modern alphaviruses based on nucleic acid sequence comparisons. The time scale in thousands of years is given at the bottom, and the four modern lineages and representative members are shown on the right. New World viruses are at the top (eastern and Venezuelan equine encephalitis [EEE/VEE] virus lineage), and Old World viruses are at the bottom (Sindbis and Semliki Forest virus lineages). The boxed lineage in red contains the western equine encephalitis (WEE) virus subgroup. This subgroup arose by a recombination event that occurred about 1,600 years ago between a member of the eastern equine encephalitis virus lineage and the Sindbis virus lineage. This estimate was deduced from the rate of divergence of WEE virus lines (about $3 \times 10^{-4}$ substitutions per nucleotide per year) and comparison of WEE virus strains isolated at different times in physically separated regions of the world. Redrawn from J. H. Strauss and E. G. Strauss, *Semin. Virol.* **8:**85–94, 1997, with permission.

## Contemporary Virus Evolution

Even with sophisticated genome sequence analysis and powerful computers, it remains difficult if not impossible to answer the question of the origin of viruses because we lack physical evidence from ancient times. However, as discussed below, it may be possible to define modern precursors of new viruses by studying emerging viruses. In the final analysis, it is probably safe to assume that every living thing is infected with viruses. Studying the viral strategies underlying the emergence of new contemporary

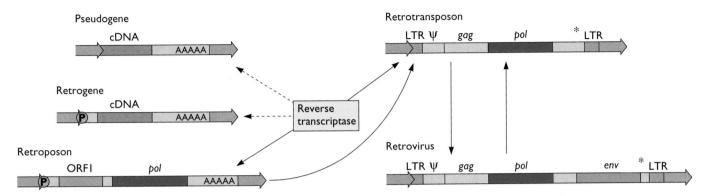

**Figure 20.4 Putative evolution of retroelements results from the action of a primordial reverse transcriptase.** A speculative scheme shows the evolution of various retroelements resulting from the action of a primordial reverse transcriptase enzyme. Dashed arrows depict the emergence of genetic elements by reverse transcription; the complementary DNA copies are shown integrated into the host genome (gray). Solid arrows show the acquisition of new elements by recombination. cDNA, double-stranded DNA copied by reverse transcriptase; ORF, open reading frame; *gag,* capsid protein gene; *pol,* polymerase gene coding for reverse transcriptase, RNase H, integrase, and other enzymes; *env,* envelope protein gene; LTR, long terminal repeat; P, promoter; AAAAA, poly(A) tail; Ψ, packaging signal; *, polypurine tract. Direct repeats of cellular DNA are indicated in gray. Adapted from R. J. Loewer et al., *Proc. Natl. Acad. Sci. USA* **93:**5177–5184, 1996, with permission.

viruses seems likely to prove more helpful in divining the present and future evolution of viruses than explaining evolution in their distant past.

What can we say about modern virus evolution? If we concede that viruses are not likely to arise de novo, then modern and future viruses must arise from progenitors that already exist. Hence, even human immunodeficiency virus, which has arisen within the past few decades, has many relatives in nature and probably descended from one, or a combination of them (Fig. 20.5). Therefore, the sources of "new" viruses are strictly limited to three possibilities: a mutant virus already existing in an infected host can be selected, a virus can enter a naive population from an infected population of the same species, or a virus can enter a naive population from an entirely different infected species. These simple possibilities notwithstanding, the interactions of host and virus required to achieve a successful conclusion are remarkably complex.

## Evolution Is Constrained by the Fundamental Properties of Viruses

We can recognize a herpesvirus, adenovirus, retrovirus, or influenza virus genome by sequence analysis no matter how much replication, mutation, and selection have occurred. This fact is emphasized by sequence analysis of human immunodeficiency virus strains isolated from patients around the world. As much as 10% of the total viral genome can vary from isolate to isolate, yet despite this, human immunodeficiency virus isolates fall into consistent subgroups called **clades.** Figure 20.6 shows some selected sequence relationships of the human immunodeficiency virus type 1 and type 2 genomes (based on alignments of their *env* or *gag* genes) for a variety of isolates from around the world. These viruses have replicated in widely dispersed geographical locations and have very different histories, yet each sequence is clearly recognizable as human immunodeficiency virus. Each clade differs from the others in amino acid sequence by at least 20% for Env proteins and 15% for Gag proteins. Differences within a clade can be as much as 8 to 10%, emphasizing the rather arbitrary delineation.

The important message is clear: viral populations frequently maintain quite stable "master" or consensus sequences despite opportunities for extreme variation. Diversity exists and indeed is necessary for virus survival, but the consensus sequence remains despite many years of selection and growth. How is stability maintained in the face of mutation, recombination, and selection? What keeps a viral genome from evolving into something totally new or slipping into oblivion? The answer is that despite amazing diversity, all viruses share fundamental characteristics that define and constrain them. Those that can function within the constraints survive. Comprehending the evolution of viruses requires an understanding of these shared properties and their interaction.

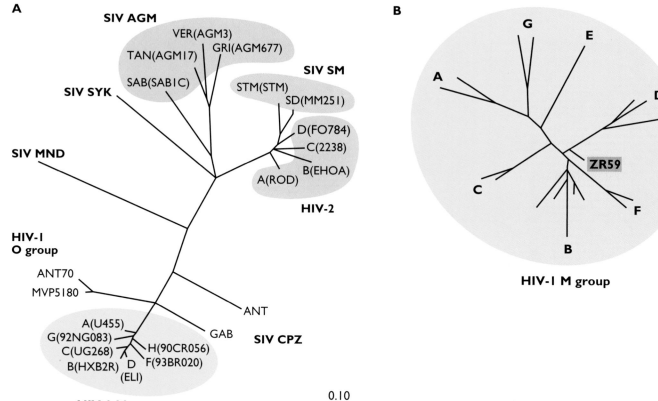

**A**

SIV AGM

VER(AGM3)
TAN(AGM17)     GRI(AGM677)
SAB(SAB1C)

SIV SYK

SIV SM

STM(STM)
SD(MM251)

D(FO784)
C(2238)
B(EHOA)
A(ROD)

HIV-2

SIV MND

HIV-1
O group

ANT70
MVP5180

ANT

A(U455)
G(92NG083)                GAB          SIV CPZ
C(UG268)     H(90CR056)
B(HXB2R)   D   F(93BR020)
(ELI)

HIV-1 M group

0.10

**B**

G
E
A
D
ZR59
C
F
B

HIV-1 M group

**Figure 20.5 Human immunodeficiency virus (HIV) and simian immunodeficiency virus (SIV) phylogeny.** (A) Phylogenetic relationships among HIV and SIV isolates. The genetic distance between any two viral isolates is found by following the shortest path relating them in the tree. Units are in terms of nucleotide substitutions according to the indicated scale. HIV type 1 (HIV-1) is divided into two groups, major (M) and outlier (O). The HIV-1 M group is responsible for the global AIDS epidemic. HIV-1 M is further subdivided into several subtypes (shown here as A through G), and representatives of subtypes A, B, C, D, F, G, and H are illustrated (the actual isolate names are given in parentheses). Subtype B is the dominant form in America and Europe, while all the other subtypes are found in Africa. In Asia, subtypes B, C, and E predominate. HIV-1 O strains are found mainly in western Africa. Interestingly, at least two monkey viruses, chimpanzee (SIV CPZ) and mandrill (SIV MND) viruses, are more closely related to HIV-1 than HIV-2 is. HIV-2 has also been suggested to be divided into subtypes (shown here as subtypes A, B, C, and D) and is closely related to the sooty mangabey virus group (SIV SM). The different African green monkeys are each apparently infected by their own specific SIV variants (SIV AGM SAB, TAN, VER, and GRI). The relationships among the Sykes' monkey virus SIV SYK, SIV AGM, and the SIV STM–HIV-2 group, as well as among HIV-1 M and O and SIV CPZ, must be regarded as uncertain at this time (indicated by zero-length branches). This tree was constructed from nu-cleotide sequences of the complete *gag* gene (676 sites after the removal of insertions and deletions) by the maximum likelihood method. Branch lengths were optimized by using an evolutionary model that includes the special substitution pattern of these viruses (J. Felsenstein, PHYLIP: Phylogeny Inference Package, version 3.52 [University of Washington, Seattle, 1993]; Z. Yang, PAML: Phylogenetic Analysis by Maximum Likelihood, version 1.1 [Institute of Molecular Evolutionary Genetics, Pennsylvania State University, University Park, 1995]). (B) Relationships of the subtypes of the M group of HIV-1 (see also Box 18.1). Representative sequences from each of the more common subtypes display the starlike shape of the M group. A sequence from the earliest known HIV-1 sample, ZR59, is included in the tree. This sample tells us that HIV-1 originated at least as far back as 1959 and that subtypes B, D, and F probably diverged shortly before that. This tree was calculated with the maximum likelihood method and was based on three short fragments from the *env* gene (468 sites after removing insertions and deletions). The current HIV database contains more than 16,000 entries of sequences. The trees were calculated and kindly provided by Thomas Leitner and Bette Korber, Theoretical Biology and Biophysics, Los Alamos National Laboratory, Los Alamos, N.Mex. [B. Korber et al., *Human Retroviruses and AIDS 1997: a Compilation and Analysis of Nucleic Acid and Amino Acid Sequences* (Los Alamos National Laboratory, Los Alamos, N.Mex., 1998)]. See also http://hiv-web.lanl.gov.

## Fundamental Properties of Viruses Constrain Their Evolution

The very characteristics that enable us to define and classify viruses are the primary barriers to major genetic change. Extreme changes in the virus master plan obviously do not survive selection. What are the constraints that define and confine virus evolution? We know that all viruses contain their nucleic acid genomes within particles assembled following symmetry principles. The nature of the particle may constrain the amount of nucleic acid sequence that can be packaged. For example, helical nucleocapsids have no theoretical size limit, while icosahedral capsids have a defined internal space that fixes the amount of packaged nucleic acid. Once an icosahedral capsid is encoded, genome size is essentially fixed. Another constraint is that the genomes of all viruses are nucleic acid based. Thus, solutions to replicating and decoding have common mechanisms and are limited. All viruses have an infectious cycle in which every step requires direct and indirect interactions with host cell machinery. Any change in a viral component without a compensating change in the interacting host component may have a direct effect on the virus. Similarly, inappropriate expression, concentration, or location of a viral component is likely to be detrimental to the virus. Finally, all viruses establish a relationship with their host (or hosts) ranging from benign to lethal to ensure their propagation and maintenance in the host population. To do so, viruses must elaborate mechanisms to overcome a broad spectrum of host defenses, including physical barriers to viral access and the vertebrate immune system. A mutant that is too efficient in bypassing host defenses will kill its host and suffer the same fate as one that does not replicate efficiently enough: it will be eliminated.

## Possibilities of Variation within Viral Genomes Can Be Astronomical

Even in the seemingly constrained context of a given virus, the number of all possible viable mutants is astronomical if not inconceivable. It is so large in fact that all the possibilities can never be tested in nature. Sequence comparisons of several RNA virus genomes have demonstrated that well over half of all nucleotides can accommodate mutations. On average, this means that for a 10-kb viral RNA genome there are more than $4^{5,000}$ sequence permutations that define all possible mutants. If one considers deletions, recombination, and reassortment, the numbers become even larger. Considering that there are roughly $4^{135}$ atoms in the visible universe, this is a large number indeed. Thus, even with the increased rates of replication and mutation inherent in viruses, we can be sure that only a minuscule fraction of all possible viral genomes has arisen since life began: even the most efficient virus cannot make all the possible permutations for a run through the gauntlet of selection in nature. The conclusion is inescapable: virus evolution is relentless; new mutants will always arise, and the possibilities literally are unimaginable (Box 20.2).

## Emerging Viruses

We define an **emerging virus** as one that either has newly appeared in the population or is rapidly expanding its range of hosts with a corresponding increase in detectable disease. While the term became part of the popular lexicon in the 1990s (usually with dire implications), emerging viruses have long been recognized as an important manifestation of virus evolution. Some examples of emerging viruses are given in Table 20.3. Each represents a story of contemporary virus evolution that by definition requires an understanding of host-virus interactions on a large scale. First, we will focus on the spectrum of host-virus interactions and their role in the maintenance of a virus in nature, and then we will consider how apparently new viruses can appear in a given host population.

### The Spectrum of Host-Virus Interactions

Viruses obviously require hosts for their survival, but it is important to consider the spectrum of possible interactions and how these interactions contribute to the apparent emergence of viruses in populations (Fig. 20.6). We define four hypothetical interactions: **stable, evolving, dead-end,** and **resistant.** These definitions are arbitrary and are not meant to imply some permanent division. They are snapshots of the extremes of dynamic host-virus interactions designed to emphasize essential concepts. In Fig. 20.7, arrows suggest the hypothetical flow of virus from one situation to another and thus the continuity of viral interactions in nature. In addition, it is important to understand that these definitions apply to *host and virus popu-*

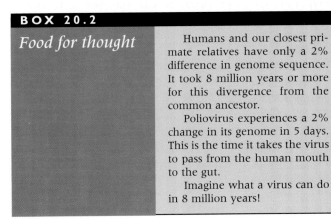

**BOX 20.2**

*Food for thought*

Humans and our closest primate relatives have only a 2% difference in genome sequence. It took 8 million years or more for this divergence from the common ancestor.

Poliovirus experiences a 2% change in its genome in 5 days. This is the time it takes the virus to pass from the human mouth to the gut.

Imagine what a virus can do in 8 million years!

**Table 20.3** Examples of emerging viruses[a]

| Virus | Family | Emergence factor(s) |
|---|---|---|
| Influenza virus | *Orthomyxoviridae* | Integrated pig/duck agriculture, mobile population |
| Dengue virus | *Flaviviridae* | Urban population density, open water storage favors mosquito breeding (e.g., millions of used tires) |
| Sin Nombre virus | *Bunyaviridae* | Natural increase in deer mice and subsequent human-rodent contact |
| Rift Valley fever virus | *Bunyaviridae* | Dams, irrigation |
| Hantaan virus | *Bunyaviridae* | Agriculture techniques favor human contact with rodents |
| Machupo virus | *Arenaviridae* | Agriculture techniques favor human contact with rodents |
| Junin virus | *Arenaviridae* | Agriculture techniques favor human contact with rodents |
| Ebola virus | *Filoviridae* | Human contact with unknown natural host in Africa, importation of monkeys in Europe and United States |
| Marburg virus | *Filoviridae* | Unknown; importation of monkeys in Europe |
| Human immunodeficiency virus | *Retroviridae* | Transfusions and blood products, sexual transmission, needle transfer during drug abuse |
| Human T-cell leukemia virus | *Retroviridae* | Transfusions and blood products, contaminated needles, social factors |
| Canine parvovirus | *Parvoviridae* | Spontaneous mutation in feline panleukopenia virus resulted in new tropism and pathogenesis |
| Norwalk/gastroenteritis viruses | *Caliciviridae* and others | New methods for detection, infectious diarrhea |

[a]From S. Morse (ed.), *Emerging Viruses* (Oxford University Press, Oxford, United Kingdom, 1993), with permission.

*lations* (e.g., large numbers of individuals) as opposed to a single individual.

### Stable Interactions

We define a **stable host-virus interaction** as one in which both members survive and multiply. As a result, the virus is maintained in the ecosystem. Such interactions are essential for the continued existence of the virus and may influence host survival as well. Arguably, this state is optimal for a host-parasite relationship, but the relationship need be neither benign nor maintained in an outbred population. Infected individuals can become ill, recover, develop immunity, or even die, yet in the long run both populations survive. While often described as an equilibrium, this interaction is dynamic and fragile. Viruses may evolve to be more or less virulent if such a change enables them to be maintained in the population, while hosts may evolve to attenuate the more debilitating effects of their viruses.

A stable host-virus interaction may involve only a single species. For example, humans are thought to be the sole natural host for a variety of viruses, including measles virus, herpes simplex virus, and smallpox virus, whereas certain monkey species are the natural hosts for simian virus 40, monkeypox virus, and simian immunodeficiency virus. In contrast, some viruses, including influenza virus and arboviruses, are capable of propagating in a variety of species. In these instances, the host that maintains the virus may not be apparent or obvious. For example, influenza A virus infects humans, wild birds, and pigs, but

birds may be its natural hosts. This conclusion is discussed in "Human Disease Resulting from Genetic Reassortment of Viruses in Animals" below, but it follows from the finding that avian influenza virus genomes appear to be less variable (mostly neutral mutations appear) than those genomes circulating in humans.

**Figure 20.6 The general interactions of hosts and viruses.** Four hypothetical host-virus interactions are indicated in the boxes. These are artificial constructs that serve to demonstrate the dynamics of virus interactions with the host population. The emergence of new viruses is one possible outcome. The stable interaction maintains the virus in the ecosystem. The evolving interaction describes the passage of a virus from "experienced" populations to naive populations in the same host species. The dead-end interaction represents one-way passage of a virus to different species. The host usually dies, or if it survives, the virus is not transmitted efficiently to the new host species. The resistant host interaction represents situations in which the host completely blocks infection. The arrows indicate possible flow of a virus from one situation to another or the possible transformation of one interaction into another. Mutation and selection, the environment, age of hosts, social behavior, floods, and wars all have a major influence on the establishment and maintenance of these interactions.

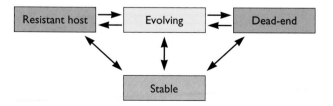

Evolution is an exercise played out over time: what is successful today may be suicidal at another time. For example, some DNA viruses establish latent infections that may be stable over millennia (e.g., herpesviruses). However, once a virus becomes completely dependent on one and only one host, it has entered a potential bottleneck that may constrain its further evolution. As a result, if humans become extinct, it is likely that human herpesviruses, including the herpes simplex viruses, varicella-zoster virus, Epstein-Barr virus, and human cytomegaloviruses, will face the same fate. Acutely infecting viruses with unique hosts face the same bottleneck. Eradication of smallpox virus was possible because humans are the only hosts and worldwide immunization was achieved. Thus, the stable host-parasite interaction represents a fragile, dynamic state sensitive to many variables that we are only beginning to understand.

As the arrows in Fig. 20.7 indicate, there is no reason to believe a priori that once a stable relationship is established, it is permanent. What we observe today is not necessarily the optimal state for any virus-host relationship. Because unpredictable selective agents are always present, other interactions between the host and virus are in constant play. Some of these result in new, often unexpected situations.

### The Evolving Host-Virus Interaction

The notion of an **evolving host-virus interaction** emphasizes that stability is unlikely to be established instantly. We highlight this interaction to illuminate the dynamic consequences of a virus spreading among populations of the same or perhaps closely related species. The hallmarks of this interaction are *instability* and *unpredictability*. These properties are to be expected, as selective forces are applied to both host and virus. For example, the virus may cause severe disease or death in some infections while the host may survive in others, or some subpopulation of hosts may experience high infection rates while others are unaffected. Such an interaction is typified by the introduction of smallpox and measles to natives of the Americas by Old World colonists and slave traders. Europeans previously had experienced the same horror when these diseases spread to Europe from Asia. A virus with reduced virulence or a new, more resistant host population may be selected to give rise to a more stable relationship. Interestingly, in the case of human immunodeficiency virus, relatively resistant individuals who harbor mutations in chemokine or chemokine receptor genes have been identified. It has been suggested that these mutations were selected after ancient encounters of humans with a virus like human immunodeficiency virus. As the recent appearance of human immunodeficiency virus could not have been anticipated in evolution, the selective advantage that has maintained these mutations remains to be discovered. Other opportunities to enter the evolving host-virus interaction may arise if the virus in a stable host-virus interaction acquires a new property that increases its virulence or spread (as exemplified by neurotropic poliovirus) or if the host population suffers a far-reaching catastrophe that reduces resistance (e.g., famine or mass population changes during wars).

### The Dead-End Interaction

A commonly encountered host-virus interaction is described rather vividly as a **dead-end interaction.** It has much in common with the evolving host-virus interaction, as both represent departures from a stable relationship, with often lethal consequences. A dead-end interaction is a frequent outcome of cross-species infection (but not of intraspecies infections, as noted in reference to the evolving host-virus interaction). Importantly, the dead-end interaction has captured the public's interest and provided inspiration for the more sensational implications of the term "emerging virus."

In a dead-end interaction, the host is often killed so quickly that there is little or no subsequent transmission of the virus to others. This outcome may be seen as a Pyrrhic victory for the virus; it is overly successful in one individual yet fails to establish itself in the population. Classical examples of dead-end interactions are presented by the many arboviruses (carried by arthropods like ticks and mosquitoes) that cycle in the wild between insects and a vertebrate host in a stable relationship. Occasionally the infected insect bites a new species (e.g., humans) and transmits the virus (Fig. 20.7). As a consequence, the virus can inflict severe pathogenic effects in the infected human but is not transmitted further; i.e., it reaches a dead end.

As the human is not part of the natural, stable host-virus relationship, dead-end infections have little if any effect on the evolution of the virus and its natural host. To a first approximation, dead-end hosts are transparent participants. However, such logic may be too simplistic; infection by a less virulent mutant or infection of a more resistant individual may be the first step in establishing a new host-virus interaction.

Many animal models of disease might be considered examples of dead-end interactions. For example, herpes simplex virus is a human virus, but when it is introduced into mice, rabbits, or guinea pigs in the laboratory, these animals become infected and show pathogenic effects that mimic some aspects of the human disease. However, these animals play no known role in the transmission or survival of the virus in nature and thus are experimental models of the dead-end interaction.

Human diseases like yellow fever and dengue fever provide excellent examples of the dead-end interaction in na-

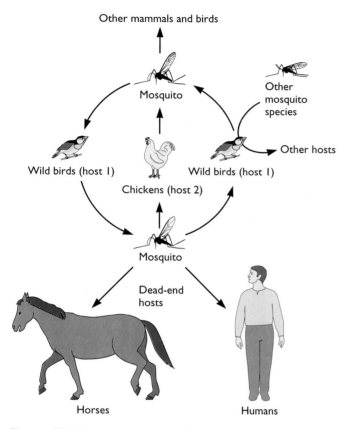

**Figure 20.7 A complex host-virus relationship.** This arbovirus is in equilibrium with host 1 (wild birds) and host 2 (domestic chickens) and is spread by a mosquito vector. The interaction of hosts and vector propagates the virus while sustaining the hosts. The virus occasionally infects a third host (horses or humans) not important in maintenance of the equilibrium of the virus with hosts 1 and 2. Horses and humans are dead-end hosts and play little to no role in the spread or sustenance of the virus. Occasionally an infected host transmits virus to a new species of insect vector and, as a consequence, new hosts can be infected.

ture. The viruses causing these diseases are endemic in the tropics and maintain an equilibrium with their natural host and insect vectors. When humans begin to develop a tropical area by clearing forests and building roads, dams, canals, and towns, they are at risk for being bitten by mosquitoes and other insects. The natural host-virus interaction that existed before the intrusion is now changed, and humans may experience new viral infections spread by insect vectors. A similar situation exists for some flaviviruses transmitted by ticks, which in turn are carried by rodents. Humans are often dead-end hosts and experience European tick-borne encephalitis (Fig. 20.8).

Rodent vectors can play critical roles in the introduction of new viruses into populations in areas where these animals abound. Most hemorrhagic disease viruses, including

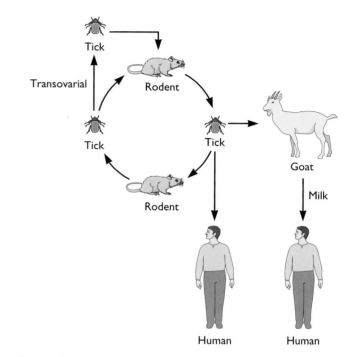

**Figure 20.8 Infectious cycle of the central European tick-borne flavivirus.** Congenital transmission in the tick maintains the virus in the tick population as they feed on rodents. The virus is transmitted from the tick to a variety of animals, including cows, goats, and humans. Humans can also be infected by drinking milk from an infected goat.

Lassa, Junin, and Sin Nombre viruses, are endemic in rodents, their natural hosts. The viruses establish a persistent infection, and the rodents show few if any ill effects. Substantial numbers of viral particles are excreted in urine, saliva, and feces to maintain the virus in the rodent population. However, such rodent viruses can cause local, often lethal, outbreaks in humans (dead-end hosts). Humans become infected only because they happen to come in contact with rodent excretions containing infectious viral particles.

The high lethality of the Marburg and Ebola filoviruses in humans is typical of a dead-end host infection. These are *the* emerging viruses in the eyes of the layperson, thanks to the popular press. Filoviruses are (−) strand RNA viruses causing severe hemorrhagic manifestations in infected humans. Disease onset is sudden, with 25 to 90% fatality rates reported. Virus spreads through the blood and replicates in many organs, causing focal necrosis of the liver, kidneys, lymphatic organs, ovaries, and testes. Capillary leakage, shock, and acute respiratory disorders are observed in fatal cases. Patients usually die of intractable shock without evidence of an effective immune response. Even when recovery is under way, survivors do not have detectable neutralizing antibodies. The virus clearly overwhelms a particular human host but apparently does not spread widely; human infections tend to cluster in local areas. Such viruses can be transmitted to other humans but only by close personal contact with infected blood and tissue. The molecular mechanisms of filovirus pathogenicity in humans are only now being established. For example, Marburg virus causes destruction of endothelial cells, which may be responsible for the hemorrhagic disease and shock syndrome typical of this infection. Not all natural Ebola virus strains are human pathogens: the Reston strain appears to infect certain monkeys but does not cause disease in humans. The molecular basis of this species specificity is of some interest and may result from differential expression of a host protease that cleaves the viral envelope protein. Although the identities of the natural hosts for many filoviruses remain elusive, these viruses must have established a stable interaction with an animal host and infect humans only inadvertently. Some investigators have argued that the host is responsible for the apparent "geographical containment" of Ebola virus in central Africa.

### The Resistant Host

It is instructive to include the **resistant host** when discussing viral evolution. All living things are exposed continuously to viruses of all types, yet the vast majority of these interactions are uneventful. This may be because the host cells are not susceptible, not permissive, or both. A more likely possibility is that the primary physical and innate defenses are so strong that potential invaders are diverted or destroyed on contact. In other examples, organisms actually may become infected, but the virus is cleared rapidly with no ill effect on the host or without activation of the host's acquired immune system. This outcome is in contrast to an inapparent infection, in which an immune response is mounted but the individual exhibits no signs of disease.

We include the resistant host in our analysis because the jump between inability and ability to replicate virus need not be insurmountable. Indeed, xenotranplantation (the use of animal organs in humans) offers thought-provoking examples. Not only does transplantation bypass physical and innate defenses by surgery, but also the acquired immune system is suppressed by drugs. Thus, the virus passengers in a xenograft are given the opportunity to explore the once-resistant host from a new and relatively defense-free perspective. As many of these viruses have the capacity to replicate in human cells or have close human-adapted relatives, the xenotransplantation patient represents a nidus of new viral diversity. This opportunity to experience viral evolution in real time deserves careful contemplation.

### The Need for Experimental Analysis of Host-Virus Interactions

The definitions of different host-virus interactions listed above are useful for discussion, but experimental studies are essential in testing hypotheses. Unfortunately, such experiments are complicated by a variety of issues, including safety, cost, concerns for animal rights, and uncertainty of the relevance of laboratory models to natural disease and virus spread. Infections in nature occur in outbred populations with considerable diversity, features that are difficult to model in the laboratory. An example of how molecular techniques can be used to study the evolution of retroviral pathogenesis in an animal is shown in Box 20.3. Many important questions about the evolution of host-virus interactions can be posed, such as the following. What is the relationship between viral virulence (ability to cause disease) and its rate of transmission in the population? How important is "species jumping" in the establishment of new viruses in populations? Are quasispecies maintained during selection in a population? How important to survival is the ability of a virus to cause disease in its host?

We understand in principle that a virus survives in nature only because of serial infections (a chain of transmission) among individuals. Nevertheless, quantitative measures of these interactions, determination of the molecular mechanisms responsible for them, and development of model systems that can predict them are sorely needed. Mathematical models are being developed to describe patterns of disease transmission in complex populations, to determine the critical population size necessary to support the continuous transmission of viruses with differing incubation periods, to describe the dynamics of persistent viral infections, and for the calculation of variables in age-dependent infection and spread. The value of these models remains to be seen simply because it is difficult to perform a controlled experiment in nature to test hypotheses.

### Accidental Natural Infections

Our understanding of the dynamics of a viral infection in a large outbred human population is rudimentary at best. What we do know is based predominantly on a limited number of accidental "experiments." Two classic examples are provided by hepatitis B virus and poliovirus infections. During World War II, large doses of infectious hepatitis B virus were accidentally introduced into approximately 45,000 soldiers when they were injected with a contaminated yellow fever vaccine. Surprisingly, only 900 soldiers (2%) came down with clinical hepatitis, and fewer than 36 of them developed severe disease. Similarly, in 1955, 120,000 grade school-aged children were vacci-

**BOX 20.3**

*Studies of evolution of pathogenesis using molecular clones of simian immunodeficiency virus*

In 1998 a collaboration of several research teams provided a detailed analysis of the molecular and population changes required for pathogenesis of simian immunodeficiency virus in macaques. Simian immunodeficiency virus infection of macaques is used as a model for lentivirus pathogenesis, transmission, tissue tropism, antiviral therapy, and virus evolution. cDNA copies of simian immunodeficiency virus RNA were prepared and were found to be infectious but markedly attenuated: virus derived from these clones failed to induce pathogenesis in macaques within a year. Scientists then tracked the natural history of infection of two such cloned attenuated strains as viruses derived from them gained virulence during repeated passages in macaques.

Detailed sequence comparisons of the original attenuated molecular clones and the newly evolved virulent strains yielded two important findings. First, evolution of *individual variants* with increased pathogenicity occurred rather than selection of a more diverse quasispecies. This property was demonstrated by cloning and testing individual viral genomes from the pathogenic quasispecies. Second, *multiple and dispersed mutations* were required in each new pathogenic simian immunodeficiency virus clone, including elongation of TM and Nef coding sequences; changes in RNA splice donor and acceptor sites, the TATA sequence, and Sp1 binding sites; multiple changes in the V2 region of SU, including a consensus neutralization epitope; and five new N-linked glycosylation sites in SU.

The conclusion is that for these strains of simian immunodeficiency virus evolution of virulence is a multistep, multigenic process understandable only in hindsight. We clearly have much to learn about the selective forces involved in virus evolution.

**Edmonson, P., M. Murphey-Corb, L. N. Martin, C. Delahunty, J. Heeney, H. Kornfeld, P. R. Donahue, G. H. Learn, L. Hood, and J. I. Mullins.** 1998. Evolution of a simian immunodeficiency virus pathogen. *J. Virol.* **72:**405–414.

nated with an improperly inactivated poliovirus vaccine. About half of them had preexisting antibody to poliovirus thanks to inapparent infections by wild virus, but of the remainder about 10 to 25% were infected by the vaccine, as determined by the appearance of new antibody. More than 60 cases of paralytic poliomyelitis were documented among these infected children, and the remainder escaped disease. These two experiments tell us that even when a large number of individuals are infected with a virulent virus, the results vary widely; we have only a rudimentary understanding of why this is so (Box 20.4).

### Population Density and Health Are Important Factors

Two predominant factors influencing infection in a given population are the **population density** and the **health status** of the individuals in that population. High population density (not necessarily overall size) is an important parameter. Close personal contact either by direct methods (e.g., aerosols or sexual contact) or by indirect action (e.g., water or sewage) is also required. The **critical population size** of interacting hosts varies considerably for a given virus and depends on such parameters as duration of immunity and amount of virus produced and shed. As discussed in chapter 15, at least half a million people in a more or less confined urban setting are required to ensure a large enough annual supply of susceptible hosts to maintain measles virus in a human population. When this large population of interacting hosts is not available to the virus, it dies out. One can reduce the effective population by splitting the half-million hosts into small groups physically separated from social discourse (exemplified by quarantine or the use of sanitariums to isolate infected patients) (Fig. 20.9B) or by immunizing the group such that the large majority are immune and cannot propagate the virus. If these or similar actions are not taken, our ever-expanding and increasingly interactive human population will provide viruses such as measles virus with ample opportunity to maintain themselves.

The **clinical status** of any potential host population is also important. For example, babies and the elderly commonly are more susceptible to a given virus than the general population. As a consequence, prevention of infection in these groups tends to reduce the overall infection rate in a population. Similarly, individuals in poor health due to malnutrition are more susceptible to disease than those who are well fed.

Obviously, the health and genetic variation in any given outbred host population contributes in part to the unpredictable nature of viral infection. Perhaps not so obvious are the other modulating variables. There are distinct seasonal variations in types of viral infections affecting a population. Respiratory infections caused by adenoviruses and rhinoviruses often occur in the spring, respiratory infec-

**BOX 20.4**

*Classical experiment in virus evolution: deliberate release of rabbitpox virus in Australia*

In 1859, the European rabbit was introduced into Australia for hunting purposes and, lacking natural predators, reproduced to plague proportions in a short time. The rabbitpox virus myxoma virus was released in Australia in the 1950s in an attempt to rid the continent of rabbits. The natural hosts of myxoma virus are the cottontail rabbit, the brush rabbit of California, and the tropical forest rabbit of Central and South America. The virus is spread by mosquitoes, and infected rabbits develop superficial warts on their ears. However, European rabbits are a different species and are killed rapidly by myxoma virus. In fact, the infection is 90 to 99% fatal!

In the first year, the released virus was amazingly efficient in killing rabbits, with a 99.8% mortality rate. However, by the second year the mortality dropped dramatically to 25%. In subsequent years, the rate of killing was lower than the reproductive rate of the rabbits, and hope for 100% eradication was dashed. Careful epidemiological analysis of this artificial epidemic provided important information about the evolution of viruses and hosts.

The virus spread rapidly during spring and summer, when mosquitoes are abundant, but slowly in winter, as expected. Given the large numbers of rabbits and viruses and the almost 100% lethal nature of the infection, advantageous mutations present in both the virus and host populations were quickly selected. Within 3 years, less virulent viruses appeared, as did rabbits that survived the infection. The host-virus interaction observed was that predicted for an evolving host coming to an equilibrium with the pathogen. A balance is struck whereby some infected rabbits die, but they die more slowly, and many rabbits survive.

What was learned? To paraphrase the Rolling Stones: you always get what you select, but you often don't get what you want. Probably the most obvious lesson was that the original idea to eliminate rabbits with a lethal virus was flawed. Powerful selective forces that could not be controlled or anticipated were at work. In this experiment, the virus evolved to kill fewer rabbits and to extend the life of lethally infected rabbits so that virus could overwinter and spread with the spring mosquitoes. Moreover, the rabbits evolved to become more resistant or tolerant of the invader. Obviously, evolution works on both host and parasite.

Surprisingly, more experiments in the biological control of rabbits with a lethal rabbit calicivirus are under way in Australia.

---

tions caused by coronaviruses and respiratory syncytial viruses tend to occur in the winter, and most arbovirus infections are experienced in the summer. Many anecdotal explanations are given for these basic observations, but solid experimental data are not always available.

### Known Viruses Can Expand Their Niche because of Changing Host Populations and Environment

#### A Disease of Modern Sanitation

Host populations change with time, and each change can have dramatic effects on virus evolution. Poliovirus and poliomyelitis, the disease it causes in humans, provide an instructive example. This is an ancient disease, postulated by some to be present over 4,000 years ago (Fig. 1.1B). For centuries, the host-virus relationship was stable, and infection was endemic in the human population at a low but continuous level. Poliomyelitis epidemics were unheard of, but occasional bouts of disease were obvious in scattered areas. This state of affairs changed dra-

matically in the first half of the 20th century, when vast annual outbreaks of poliomyelitis appeared in Europe, North America, and Australia. The virus did not change or evolve substantially, yet poliovirus obviously emerged as a significant disease-causing agent (Fig. 20.9A). How can this be explained?

The answer is that humans changed their lifestyle on an unprecedented scale. Poliomyelitis is caused by an enteric virus spread by oral-fecal contact. As a consequence, endemic disease was characteristic of life in rural communities which generally had poor sanitation and small populations. Because virus circulated freely in the local populations, most children were infected at an early age and developed antibody to at least one of the serotypes. Maternal antibody, which protected newborns, was also prevalent, as most mothers had experienced a poliovirus infection at least once. An important consideration is that most infected children do not develop paralysis, the most visible symptom of poliomyelitis. Paralysis is a more frequent result when older individuals are infected. Even the most virulent poliovirus strains cause 100 to 200 subclini-

**A**

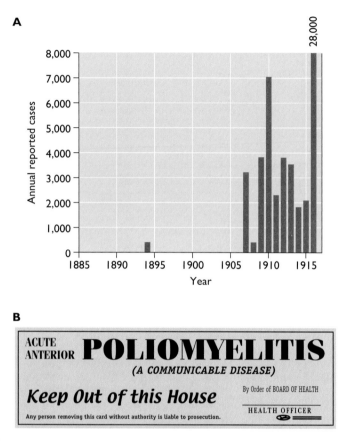

**B**

ACUTE ANTERIOR **POLIOMYELITIS**
*(A COMMUNICABLE DISEASE)*

**Keep Out of this House**    By Order of BOARD OF HEALTH

Any person removing this card without authority is liable to prosecution.    HEALTH OFFICER

**Figure 20.9 Poliovirus in the early 20th century.** (A) The emergence of paralytic poliomyelitis in the United States, 1885 to 1915. From N. Nathanson, *ASM News* **63**:83–88, 1997, with permission. (B) Board of Health quarantine notice, San Francisco, Calif., circa 1910.

cal infections (inapparent infections) for every case of poliomyelitis (see chapter 15). Thus, these inapparent infections in children provided a form of natural vaccination. As a variety of paralytic diseases were not uncommon in rural populations in general, the few individuals who developed poliomyelitis were not seen as out of the ordinary. No one noticed endemic poliovirus.

However, during the 19th and 20th centuries, industrialization and urbanization changed the pattern of poliovirus transmission. Dense populations and increased travel provided new opportunities for rapid transmission of virus. Improved sanitation broke the normal pattern of virus transmission and effectively stopped natural vaccination. As a result, children tended to encounter the virus for the first time at a later age, without the protection of maternal antibodies, and thus were at far greater risk for developing paralytic disease. Consequently, epidemic poliovirus infections emerged time and time again in communities across Europe, North America, and Australia.

Until vaccines became available, quarantine was the only public health defense (Fig. 20.9B).

Widespread use of inexpensive, effective poliovirus vaccines has since controlled the epidemic in industrial countries. As the virus has no host other than humans, it is possible to eradicate it by vaccinating sufficient people so that spread of the virus is impossible. Accordingly, the World Health Organization is targeting the eradication of poliovirus by 2003 with a massive worldwide vaccination program. For the eradication program to succeed, lessons learned from history must not be forgotten, and we must not drop our guard. Today, many parents do not have their children vaccinated because of the cost or from fear of side effects. Complacency and the misguided idea that viral diseases are no longer a problem are also prevalent. Poverty, social problems, and economic conditions often conspire to prevent vaccination of children in inner cities and in poorer countries. While the poliovirus eradication program appears to be well on target, the virus is almost guaranteed to emerge again if vaccination is not effectively maintained until the virus is eliminated, as smallpox virus was.

### Diseases of Exploration and Colonization

Many examples of explosive epidemic spread are due to viruses finding susceptible hosts in the larger population that have never been exposed (the evolving host-virus interaction). To this subpopulation, the viruses appear to "come out of the blue," but they are not new viruses. Geographic isolation of human populations is a case in point. Consider the classic lethal epidemics of measles and smallpox caused by two diverse but well-known viral scourges of human populations throughout history. These viruses require a high population density to be maintained. Hence, they tend to exist in large dense populations but not in isolated small subpopulations. They are also uniquely human viruses with no other hosts. History records that smallpox reached Europe from the Far East in A.D. 710 and attained epidemic proportions in Europe in the 18th century as the population grew and became concentrated. The effects on society are hard to imagine today, but as an example, at least five reigning monarchs died of smallpox. The survivors were immune to new infections and may also have been less susceptible to infection to begin with, although evidence for this is weak at best.

Smallpox virus continued its spread around the world when European colonists and slave traders moved to the Americas and Australia. The role that this virus played in changing the balance of human populations in the New World is striking for its speed and vicious effect. Some say that viruses were responsible for the elimination of a myriad of native languages and the nearly exclusive use of

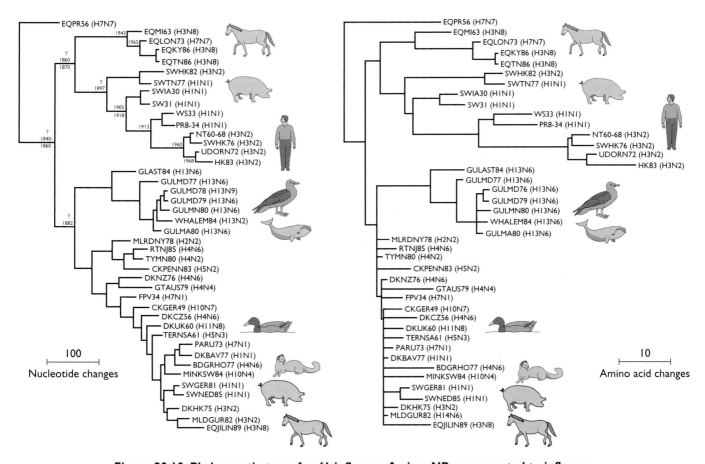

**Figure 20.10 Phylogenetic trees for 41 influenza A virus NP genes rooted to influenza B virus NP (B/Lee/40).** The nucleotide tree is shown on the left. The horizontal distance is proportional to the minimum number of nucleotide differences to join nodes and NP sequences. Vertical lines are for spacing branches and labels and have no other meaning. Blue animal symbols denote the five host-specific lineages. Pink animal symbols denote viruses in the lineage that have been transmitted to other hosts. The amino acid tree is shown on the right. Adapted from R. G. Webster and Y. Kawaoka, *Semin. Virol.* **5**:103–111, 1994, with permission.

Spanish and Portuguese in South America. While this may seem like hyperbole, consider this history. The first recorded outbreak of smallpox in the Americas occurred among African slaves on the island of Hispaniola in 1518, and the virus rapidly spread through the Caribbean islands. Within 2 years, this toehold of smallpox in the New World enabled the conquest of the Aztecs by European colonists. When Cortez first visited the Yucatan Peninsula in 1518 and began his conquest, his soldiers infected no one with smallpox virus. However, in 1520, an African slave infected with smallpox accompanied an expedition from the governor of Cuba whose mission was to contain Cortez. Within 2 years, 3.5 million Aztecs were dead, more than could be accounted for by the bullets and swords of the small band of conquistadors. Smallpox spread like

wildfire in the native population (which unfortunately was highly interactive and of sufficient density for efficient virus replication), reaching as far as the Incas in Peru *before* Pizarro made his initial invasion in 1533. As is true in most smallpox epidemics, some Aztecs and Incas survived, but those who did were then decimated by measles virus, also brought in by Cortez's and Pizarro's men. Conquest occurred by a one-two virological punch rather than by military prowess. Slave traders (who were most likely immune to infection) were populating Brazil with their infected human cargo at approximately the same time, with the same horrible result. The devastation of indigenous peoples by these viruses was documented in the colonization of North America and continued into the 20th century as contaminated explorers infected isolated groups of

Alaskan Inuit and other native populations in New Guinea, Africa, South America, and Australia.

### Human Disease Resulting from Genetic Reassortment of Viruses in Animals

"Old" diseases can be caused by "new" viruses. Influenza, a disease whose symptoms have remained unchanged for centuries, is caused by a constantly changing virus (Fig. 20.10). Indeed, influenza is the paradigm for a situation in which continued evolution of the virus in several host species is essential for its maintenance. Several documented pandemics of influenza occurred between 1957 and 1986, and each was associated with more than 10,000 deaths in the United States alone. The catastrophic pandemic in 1918 claimed over 25 million lives and affected more than 200 million people worldwide. Massive troop movements and the disruption of significant segments of the population as a consequence of World War I played important roles in the pandemic. The unprecedented virulence of the 1918 influenza virus strain is still unexplained and remains of concern. PCR analysis of tissue from victims that succumbed during that epidemic has yet to reveal any clues (Box 20.5).

The life cycle of influenza virus, while comparatively well understood at the molecular level, is remarkable for its complexity in nature. A given human influenza virus can evolve rapidly by antigenic shift (genetic reassortment between a human and a nonhuman virus in a nonhuman host) and antigenic drift (accumulation of mutations that facilitate evasion of the host immune response). New influenza viruses constantly emerge from migratory aquatic bird populations to infect humans, swine, horses, domestic poultry, and sea mammals. In birds, influenza virus replicates in the gastrointestinal tract and is excreted in quantity, a most efficient virus distribution system. The widespread dispersal of virus in water, the facile changing of hosts, and the ease of genetic reassortment form an engine for creation of new pathogenic strains. One hypothesis is that pigs are an important intermediate host between birds and humans, providing the "mixing vessel" for genetic exchange between avian and human viruses. Passage through pigs may also provide the selective pressures that enrich for new viruses that will emerge and infect new hosts.

Humans are not the only hosts who suffer from this ever-changing virus. When pigs are infected experimentally with an avirulent influenza virus mutant, a virulent strain can arise within a few days and cause swine disease. Similarly, a single mutation (changing a threonine to lysine in the hemagglutinin gene) in a relatively attenuated strain caused the 1983 influenza virus epidemic that devastated commercial chicken production in Pennsylvania. The direct transfer of a virulent avian virus to humans, first documented in 1997, reminds us how quickly host-virus interactions can change.

Even in this myriad of hosts and genetic exchanges, there is some remarkable stability that is critical to the survival of influenza virus. One surprising finding has been that avian influenza viruses have not changed much in more than 60 years, in contrast to human and other nonavian viruses. Avian viruses have mutation and reassortment rates as high as those of human and swine viruses, but viruses with neutral mutations are selected. Experts note that influenza virus appears to be in "evolutionary stasis" in birds. The avian host apparently provides the stable reservoir for influenza virus gene sequences that emerge as recombinants for reintroduction into the human population.

**BOX 20.5**

*The hunt for the deadly 1918 influenza virus in the Norwegian permafrost*

In October 1918, seven coal miners died of influenza on the island of Spitsbergen in the high Arctic. Johan Bjerk, William Richardsen, Ole Kristoffersen, Magnus Gabrielsen, Tormod Albrigtsen, Hans Hansen, and Kristian Hansen were fishermen and farmers, 18 to 29 years old, who came to this desolate island to earn extra money in the coal mines. They contracted influenza on the boat trip from Norway and died soon after. Their graves are only 3 feet deep but are just above the Spitsbergen permafrost, which has encased them in a tomb of ice. The bodies were not embalmed and have been undisturbed for more than 80 years. While live virus is unlikely to be found, the hope is that the permafrost has preserved tissues and blood and slowed the degradation of proteins and nucleic acids. Studies of tissues of the dead miners may help determine the genetic makeup and the protein structure of the 1918 influenza virus. In August 1998 scientists began the detailed analysis by taking needle biopsies of the lungs, viscera, and heart blood of the frozen bodies of six of the seven men. Extreme care is being taken to avoid possible infection should any virus remain viable.

## Human Disease Resulting from Changing Climate and Animal Populations

A **zoonosis** is a disease or infection that is naturally transmitted between vertebrate animals and humans. Unanticipated viral zoonoses are often classic cases of emerging viruses that reflect our changing environment.

In 1993 a small but alarming epidemic of a highly lethal infectious disease occurred in the Four Corners area of New Mexico. Individuals who were in excellent health developed flu-like symptoms which were quickly followed by a variety of pulmonary disorders, including massive accumulation of fluid in the lungs, and death. Quick action by local health officials and a prompt response by the Centers for Disease Control and Prevention in Atlanta showed that these patients had low-level cross-reacting antibodies to hantaviruses, viruses in the family *Bunyaviridae* previously associated with renal diseases in Europe and Asia. Members of this virus family were well known to scientists, who associated them with viral hemorrhagic fever during the Korean War. Hantaviruses commonly infect rodents and are endemic in these populations around the world. Using powerful PCR technology, scientists from the Centers for Disease Control and Prevention found that the patients were indeed infected with a new hantavirus. Concerted efforts with field biologists showed that this new virus was found in a common New Mexican rodent called the deer mouse (*Peromyscus maniculatus*). Humans are dead-end hosts for this virus. Hantavirus pulmonary syndrome has invariably been associated with the presence of this virus in New Mexico as well as a few other isolated incidents around North America. The virus, which was given the name Sin Nombre virus (no-name virus), is an example of an emerging virus, endemic in rodents, that causes severe problems when it crosses the species barrier and infects humans.

The reason why humans were infected with Sin Nombre virus is being debated, but one popular idea is that a dramatic increase in the deer mouse population was an important factor. In 1992 and 1993, higher than normal rainfall resulted in a bumper crop of piñon nuts, a favorite deer mouse and local human food. Mouse populations rose in response, and contacts with humans increased as well. Hantavirus infection is asymptomatic in mice, but the virus is excreted in large amounts in mouse urine and droppings, where it is quite stable. Human contact with contaminated blankets, dust from floors, or dust from granaries or other food storage areas provided ample opportunities for infection. Hantavirus syndrome is rare because humans are not the natural host and apparently are not efficient vehicles for virus spread. However, this infection shows that casual intrusion by humans into a natural host-

virus relationship can result in a frequently fatal, but thankfully rare, disease. The hantavirus zoonotic infection serves as a warning that potential human diseases lurk in the wild if we inadvertently intrude on another host-parasite relationship.

Caliciviruses represent an underappreciated class of zoonotic viruses. These (+) strand RNA viruses are among the most common causes of virus-induced vomiting and diarrhea in humans. Other common calicivirus-promoted symptoms include skin blistering, pneumonia, abortion, encephalitis, myocarditis, myositis, hepatitis, and coagulation/hemorrhage. Calicivirus origins are obscure, mainly because they are difficult to study and diagnose. In fact, the most common way to identify them is by their unique structure in the electron microscope—hardly a facile diagnostic tool. Yet, as more is learned about them, it appears that the ocean and its marine animals are a major reservoir of these interesting viruses. It is clear that humans often become infected via water and food sources. Four genera are recognized in the family *Caliciviridae*: Sapporo-like virus, Norwalk-like viruses, marine caliciviruses, and rabbit hemorrhagic disease virus. The first two are known to cause human disease, and all have resisted attempts at cultivation. Only the marine caliciviruses can be propagated in vitro and are known to infect terrestrial hosts, including humans. The host range of these marine viruses is astounding. One virus serotype, SMSV-5, can infect five genera of seals, cattle, three genera of whales, donkeys, foxes, opaleye fish, horses, domestic swine, primates, and humans. The degree of exposure of terrestrial hosts to marine caliciviruses may be substantial, especially in confined areas. Infected whales excrete more than $10^{13}$ caliciviruses daily, and the viruses remain viable more than 2 weeks in cold seawater. Despite our ignorance of this agent, human interactions with marine animals and contaminated water and sewage guarantee more outbreaks of calicivirus disease.

## Known Viruses Can Expand Their Host Range or Pathogenicity by Mutation or Recombination

Given the large numbers of viruses and ample opportunities for infection, we should expect to witness new host-virus interactions as they are established from time to time. Because the new host population will have had no experience with the virus, the infection will have unpredictable consequences (e.g., the dead-end or the evolving host-virus scenario). This invasion may be heralded by only an immune response and no obvious disease. Infection may result in mild disease, the symptoms of which may or may not be unusual, and occasionally, astute health care workers may notice unusual symptoms in local population centers. However, sometimes the invasion can be dramatic, as in the global epidemics caused by canine parvovirus and human immunodeficiency virus.

### Cat-to-Dog Host Range Change by Two Mutations

In 1978 canine parvovirus, a member of the family *Parvoviridae*, was first identified simultaneously in several countries around the world. It apparently evolved from feline panleukopenia virus after two mutations made it pathogenic and highly transmissible in the dog population. This new virus also found hosts in cats, mink, and raccoons. As canine parvovirus appeared less than 30 years ago, it has been possible to analyze dog and cat tissue collected in Europe in the early 1970s to search for the progenitor canine parvovirus. Some fascinating studies have tracked and identified the new virus mutants causing the epidemic.

The ancestor of canine parvovirus began infecting dogs in Europe during the early 1970s, as deduced by the appearance of antibodies in dogs in Greece, The Netherlands, and Belgium in 1974, 1976, and 1977, respectively. In 1978 virus-positive dog sera were collected in Japan, Australia, New Zealand, and the United States, suggesting that canine parvovirus had spread around the world in less than 6 months. The stability of the new virus, its efficient fecal-oral transmission, and the universal susceptibility and behavior of the world's dog population were important factors in the emergence of this new virus. We now understand at the molecular level the genetic changes that created a new pathogen from a well-known virus. Only two mutations in a capsid gene were necessary to change the tropism of the virus from cats to dogs. We know this because introduction of codons 93 and 323 from the VP2 structural gene into the corresponding sites in the VP2 gene of feline panleukopenia virus enabled this virus to replicate in canine cells in culture as well as in dogs. Conversely, changing these same codons in canine parvovirus to the corresponding feline panleukopenia virus codons eliminated the tropism for canine cells. Furthermore, parvovirus replication is accompanied by mutation: substitution rates in the capsid protein gene occur at about one nucleotide change per 10,000 bases. This rate is several orders of magnitude higher than the rates for other DNA viruses. Therefore, canine parvovirus continues to change as it replicates in its new host. The original antigenic variant, called type 2, was only a transient player in virus evolution, as it has been replaced by new strains called types 2a and 2b. The evolutionary challenges and solutions presented by this emerging virus can be studied, and the an-

swers will go far to help us understand virus-host adaptation and host range shifts.

### Host Range Jump from Monkey to Human by Rapid, Multiple Changes

The origin of HIV was an enigma until a remarkable study was published in 1998 (see also Box 18.1). A human serum sample taken in 1959 was found to be positive for human immunodeficiency virus type 1. This observation is noteworthy because the human immunodeficiency virus epidemic came to our attention in the early 1980s. The patient from whom the sample was taken had been infected over 20 years before! Careful PCR amplification of portions of the viral genome (called ZR59) yielded sufficient DNA for sequencing and analysis. When the ZR59 sequences were compared to known human immunodeficiency virus type 1 sequences, the analysis revealed significant homology with all the major clades of human immunodeficiency virus. Most significantly, ZR59 fell near the *center* of the M group (Fig. 20.5). It was clear that ZR59 was only a few nucleotides away from the putative progenitor of all known human immunodeficiency virus strains. Given the rapid rate of nucleotide change during human immunodeficiency virus replication, the founder virus is postulated to have emerged in the human population only 10 to 15 years before, no earlier than the late 1940s or early 1950s. Because of the sequence homology to simian viruses (Fig. 20.5A), it seems likely that the founder virus had its origins in a simian lentivirus from subhuman primates. This precursor of contemporary human immunodeficiency virus acquired new mutations, providing it tropism for humans. A chance encounter of this mutant virus and a human established human immunodeficiency virus on the planet.

Shortly after the recognition of human immunodeficiency virus type 1 in the 1980s, a second related subtype, called human immunodeficiency virus type 2, was identified in Portugal and subsequently has been found in many countries (Fig. 20.5A; see also Fig. 18.1). Type 2 is not as widely distributed as type 1, being more commonly found in Africa. Interestingly, human immunodeficiency virus type 2 membrane proteins have homology to those of simian immunodeficiency virus, a simian lentivirus. In fact, humans infected with human immunodeficiency virus type 2 have antibodies that cross-react with simian immunodeficiency virus proteins. Moreover, simian immunodeficiency virus can cause an AIDS-like disease in rhesus macaques or can be nonpathogenic in some species, like African green monkeys. This observation has led some scientists to speculate that human immunodeficiency virus type 2 also was derived from simian immunodeficiency virus mutants that crossed species into humans.

### *Pathogenicity Change by Recombination*

Retroviruses also provide instructive examples of the acquisition of entirely new genes by recombination with the host genome, producing viruses with unexpected and frequently lethal pathogenic potential. In general, a retrovirus can acquire a cellular oncogene and become an acutely oncogenic virus. As an example, feline leukemia virus is a well-known, highly conserved retrovirus infecting cats throughout the world. The virus causes lymphosarcomas and a range of degenerative diseases, including anemia and thymic atrophy. However, the sarcomas are caused not by feline leukemia virus but rather by a new virus resulting from recombination of feline leukemia virus with host cell genes that yielded defective viruses that are acutely oncogenic. These new viruses are defective because their envelope genes have been replaced by cellular oncogenes. As a consequence of unregulated expression of the oncogene from the integrated defective virus, tumors arise rapidly, often with lethal consequences. Not one, but several, host genes, including c-*abl*,c-*fes*, c-*fgr*, c-*fms*, c-*kit*, and c-*sis* have been acquired by different feline sarcoma viruses (see chapter 16). Similarly, other oncogenes are acquired and expressed by the recombinant viruses implicated in lymphosarcomas.

## Some Emergent Viruses Are Truly Novel

Classically, viruses were identified by the diseases they caused and, more recently, by growing them in culture to characterize their components. It is not difficult to imagine the existence of viruses that have never been described or grown in tissue culture. Many of these "unculturable" viruses cause serious, widespread disease. For example, rotaviruses infect every child under the age of 4 years in the United States, causing severe diarrhea, yet these viruses were not recognized until the 1970s simply because the technology for finding them did not exist and the viruses could not be propagated in the laboratory. Associating a disease or syndrome with a particular infectious agent used to be impossible if the agent could not be cultured. Fortunately, with powerful recombinant DNA, hybridization, and nucleic acid amplification technology, we are now able to detect and characterize such hitherto unknown viruses with comparative ease.

### *Discovery of New Hepatitis Viruses in the Blood Supply by Recombinant DNA Technology*

One of the first examples of this quantum leap in diagnostic capacity occurred in the recognition of hepatitis C virus, a unique member of the family *Flaviviridae*. With the development of specific diagnostic tests for hepatitis A and B viruses in the 1970s, it became clear that most cases of hepatitis that occur after blood transfusion are caused by other agents. Even after more than 10 years of research, the identity of these so-called non-A, non-B (NANB) agents remained elusive. In the late 1980s, Chiron Corporation, a California biotechnology company, isolated a cDNA clone derived from a chimpanzee with NANB hepatitis. The cDNAs were cloned in a bacteriophage vector, enabling the efficient expression of cDNA-encoded polypeptides in *Escherichia coli*. Clones were screened with serum from a patient with chronic NANB hepatitis. Screening about a million such clones resulted in the identification of two positive samples. Chiron scientists proved that the cDNAs were overlapping and were derived not from host DNA but from a (+) strand RNA molecule of about 10,000 nucleotides present in the infected animal. The sequence analysis of these cDNA clones indicated high homology to known flavivirus genomes. This hepatitis C virus-specific cDNA enabled the design and implementation of diagnostic reagents that effectively eliminated the virus from the U.S. blood supply, reducing the incidence of transfusion-derived NANB hepatitis significantly. Hepatitis C virus cannot be grown in culture, and progress in testing effective antiviral agents or vaccines has been slow. Nevertheless, its discovery provides an excellent example of the identification of a previously unknown human pathogen by cloning technology. Interestingly, hepatitis C virus was not the only new hepatitis virus hidden in the human blood supply; the same technology permitted the discovery of several more hepatitis viruses.

### *Discovery through PCR*

A powerful technique for finding previously unknown nucleic acid sequences (i.e., viruses) is PCR, which allows the amplification of minute quantities of any nucleic acid, RNA or DNA, to concentrations that can be analyzed by standard recombinant DNA technology. One useful PCR-based technology, **representational difference analysis (RDA)**, compares DNA sequences present in diseased and nondiseased tissues. RDA is done by amplifying DNA, providing a "representation" of DNA from diseased and from normal tissue. The PCR-amplified DNA representation is then ligated to a unique primer and hybridized to an excess of the amplified representation of DNA from normal tissue. This competition reaction ensures that only the unique sequences found in the diseased tissue are amplified preferentially from the primer. This technique was used to identify a new herpesvirus called human herpesvirus 8 present in Kaposi's sarcomas, a common malignancy in AIDS patients.

## A Paradigm Shift in Diagnostic Virology

Recombinant DNA and PCR technology not only are expanding our ability to define and understand the etiologies of complex diseases but also are causing an upheaval in diagnostic virology. New nucleic acid sequences can be associated with diseases and characterized in the absence of standard virology techniques. The time-honored Koch's postulates, which require that the infectious agent be isolated from a diseased patient, grown, and used to cause the same disease, are bypassed by this new technology. Because of its sensitivity and the potential for contamination by adventitious viruses, care must be taken with any PCR-based methodology. Without proper controls, these techniques have the potential to associate a particular virus incorrectly with a disease, thus confounding the etiology. An excellent example of this was the spurious association of multiple sclerosis with infection by human T-cell leukemia virus type 1. Similarly, a variety of studies have been published in which PCR identified human herpesviruses in the brains of patients who had died of Alzheimer's disease, implying a causal link between the disease and the virus. These conclusions are called into question by the finding of the same viruses in the tissues of similarly aged patients who died of other causes.

Clearly, the ability to detect new viruses by any technique has the potential to be a double-edged sword. While early detection and precise identification of viruses is important in public health and must be pursued with vigor, problems can arise if proper experimental controls are not applied or if no treatment or follow-up is available for the putative disease-causing agent. We do not have a robust armamentarium of safe vaccines or antiviral drugs against even the common viruses. Therefore, developing treatments for previously unknown viruses discovered by these powerful techniques will be challenging.

## Perceptions and Possibilities

### Infectious Agents and Public Perceptions

One reason why emerging viruses have had such an impact on the public consciousness is that less than 30 years ago, many were ready to close the book on infectious diseases. The public perception was that wonder drugs and vaccines had microbes "under control." Obviously, in a few short years, this optimistic view has changed dramatically. Regular announcements of new and destructive viruses and bacteria appear with increasing frequency (Table 20.3). The discovery of human immunodeficiency virus and its effects at every level of society have attracted worldwide attention, while exotic viruses like Ebola virus capture front-page headlines. Movies and books bring viruses to the public consciousness more effectively than

ever before. Consequently, it is important to separate fact from fiction and hyperbole. The widespread interest in emerging viruses reminds us that a little knowledge can be a dangerous thing: scientists, as well as the public, can be misled about host-virus interactions by semantics.

### Virus Names Can Be Misleading

It is common to name a virus by the host from which it was isolated. By calling the virus that causes AIDS *human immunodeficiency virus*, we give short shrift to its nonhuman origins. Canine parvovirus indeed causes disease in puppies, but it clearly is a feline virus that recently changed hosts. Similarly, canine distemper virus is not confined to dogs but can cause disease in lions, seals, and dolphins. Well-known viruses can cause new diseases when they change hosts. Much is implied and more is ignored about the host-virus interaction when the virus is given a host-specific name.

### Pathogenic versus Nonpathogenic Viruses

It is not unusual to think that disease-causing viruses are important while nonpathogens are uninteresting and irrelevant. As we have seen, a well-behaved virus in one host may be devastating when it enters a different species. Another type of confusion arises from human-centered thinking. How often do we think that viruses causing human diseases are more important than those that infect mammals, birds, fish, or other hosts? Naming viruses as "human" gives them artificial importance. Similarly, by changing the environment, thinking only of human needs, we are blind to the multiple levels of interaction in host-virus relationships. Indeed, viruses have a broad smorgasbord of hosts, and humans often represent but one stop in the evolution of a virus.

### Human Economic Interests Have Significant Effects

In a society centered on money, it is common to direct attention to viruses that have economic consequences as opposed to those that do not. Moreover, companies and governments may be tempted to concentrate resources on viruses that affect First World economies rather than those that affect Third World countries simply because they will be rewarded monetarily for their efforts. As a corollary, they may be tempted to focus research on antiviral drugs and therapies that offer greater economic rewards as opposed to those aimed at smaller markets and needs. Succumbing to this temptation can have important consequences, as some of the most fertile grounds for new viruses are developing countries. Because of the mobility of society and our expanding populations, economic impulses must be more prudently balanced by a larger world view.

## What Next?

### *What Will Be the Next Pandemic Virus?*

Despite our knowledge that there are many viruses potentially capable of causing problems, scientists are often asked to predict which *one* lethal virus has the potential to bring humankind to its knees. On reflection, the sobering reality is that some of the most dangerous viruses are not the popularized highly lethal filoviruses such as Ebola virus, the hemorrhagic disease viruses such as Lassa virus, or an as yet undiscovered virus lurking in the wild. Rather, the most dangerous viruses are the well-adapted but highly evolving viruses already in the human population. Influenza virus fits the description perfectly. Its yearly visits show no signs of diminishing, genes promoting pandemic spread and virulence are already circulating in the virus population, and the world is ever more prone to its dissemination. Indeed, a pandemic of influenza virus on the scale of the 1918 outbreak is thought by many to be the most likely next emerging disease to affect humans on an unprecedented scale (Table 20.4). We were reminded of this in 1997 with reports from Hong Kong that an avian strain of influenza virus (H5N1) apparently had spread directly to humans. The HA5 allele of viral hemagglutinin had never been reported in a human-adapted virus and thus no humans have antibodies directed toward this protein. Obviously, once such a virus enters the human population, it has the potential for pandemic spread. Scientists from around the world are now following the trajectory of the H5N1 combination as it reassorts among the influenza virus population and interacts with the many hosts available around the world.

In this context, it is important to reflect on the implications of the eradication of human viruses by vaccination, as exemplified by the successful campaign against smallpox virus. The human population is rapidly losing immunity to this virus, as would be predicted given that the natural virus has been eliminated except for stocks of virus maintained in research laboratories. Some fear that the emerging global susceptibility to this devastating virus could be exploited in developing a terrorizing bioweapon by unscrupulous individuals. While this possibility is horrible to contemplate, we must recognize the danger and respond to it.

### *Many Emerging Viruses Illuminate Immediate Problems and Issues*

Obviously, it would be a mistake to concentrate on influenza virus or smallpox virus to the exclusion of others because many emerging and reemerging infectious diseases pose immediate and urgent problems for the world. The human immunodeficiency virus pandemic has shown the ease with which a new virus can enter the human population. Human immunodeficiency virus and the secondary infections that accompany it have exposed the fragility of the world's health care system. Xenotransplantation (transplanting animal organs into humans) bypasses primary physical defenses and exposes humans directly to animal viruses, often under conditions of immunosuppression. The known outbreaks of new, exotic viruses and the possible cross-species interactions promoted by new technologies challenge the standard methods of diagnosis, epidemiology, and treatment.

### *Humans Are Constantly Providing New Ways To Meet Viruses*

To understand and control emerging viruses, humans must recognize the scale and highly interactive nature of the problem. The examples of poliovirus, measles virus, and smallpox virus remind us that viruses can suddenly cause illness and death on a catastrophic scale following a change in human behavior. Current human technological advances and changing social behaviors continue to influence the spread of viruses. Some human-introduced environmental changes that are significant to virology are listed in Table 20.5. Most did not exist 50 years ago, and each one brings humans and viruses into new situations. Probably the most critical fact is that the human population is as large as it ever has been, and there are few signs that our growth is slowing down. As a consequence, humans are interacting among themselves and with the environment on scales unprecedented in history.

Many changes listed in Table 20.5 have major effects on the transmission of viruses by vectors such as insects and rodents. Any one change that puts humans in the path of a natural host-virus interaction provides opportunities for cross-species infection. Population movements, the transport of livestock and birds, irrigation systems, and deforestation provide not only new contacts with mosquito and

**Table 20.4** The 1918–1919 influenza pandemic: one of history's most deadly events[a]

| Event | Deaths (millions) |
|---|---|
| Influenza pandemic (1918–1919) | 20–40 |
| Black Death (1348–1350) | 20–25 |
| World War II (1937–1945) | 15.9 |
| AIDS pandemic (through 1997) | 11.7 |
| World War I (1914–1918, military deaths) | 9.2 |

[a] Data from *The New York Times,* 21 August 1998.

**Table 20.5** Large-scale changes in virus ecology introduced by humans

Air travel

Dams and water impoundments

Irrigation

Rerouting of wildlife migration patterns

Wildlife parks

Air conditioning

Blood transfusion

Xenotransplants

Long-distance transport of millions of head of livestock and birds

Moral and societal changes with regard to sex and drugs of abuse

Deforestation of massive proportions

Millions of used tires

Uncontrolled urbanization

Day care centers

tick vectors but also mechanisms for transport of infected hosts to new geographic areas (Box 20.6).

Humans can provide novel habitats for viruses, as demonstrated by used tires, an unexpected venue for movement of viruses and their hosts. Several species of tropical mosquitoes (e.g., *Aedes* species) prefer to breed in small pockets of water that accumulate in tree trunks and flowers in the tropics. The used tire has provided a perfect mimic of this breeding ground and, as a consequence, the millions of used tires (almost all carrying a little puddle of water inside) accumulating around the world provide a mobile habitat for mosquitoes and their viruses. In the United States, such used tires are shipped all around the country for recycling, transmitting mosquito larvae to new environments literally overnight. Thus, the mosquito host for dengue and yellow fever viruses may be provided the chance to establish a new range in New Jersey, thanks to the shipment of used tires for recycling from the South. It remains to be seen how the new insect vector will impinge on human health in the Northeast, but mosquito-borne viral diseases such as dengue fever clearly are of major concern.

Puerto Rico had five dengue fever epidemics in the first 75 years of this century but has had six epidemics in the past 11 years with major economic costs. Simultaneously, Puerto Rico, Brazil, Nicaragua, and Cuba experienced their first dengue fever epidemics in 50 years. Brazil reported 180,000 cases in 1996. It is suspected that this virus is emerging because the mosquito host has new avenues for spread and many governments have poor, or no, vector control programs.

Another contemporary example of humans moving viruses to new hosts comes from transport of livestock.

African swine fever virus, a member of the family *Iridoviridae,* causes the most serious viral disease threatening the swine industries of developing and industrialized countries and is the only known DNA virus that can be designated an arbovirus. It is spread by soft ticks of the genus *Ornithodoros.* African swine fever virus was spread from Africa to Portugal in 1957, to Spain in 1960, and to the Caribbean and South America in the 1960s and 1970s by long-distance transport of livestock and their resident arthropods (ticks) which carried the virus.

The construction of dams and irrigation systems can change host-virus interactions on a large scale. The creation of vast areas of standing water has introduced new sources of viruses, hosts, and vectors. The 1987 outbreak of Rift Valley fever in Mauritania along the Senegal River was associated with the new Diama Dam, which created conditions for mosquito propagation. Not only do dams and water impoundments affect insects, but they also alter population patterns of waterfowl and other animals and the viruses they carry. New hosts and new viruses are brought together as a consequence.

In industrialized countries, the increasing need for day care centers on a large scale has led to new opportunities for viral transmission and other types of day care-associated illness. In the United States, more than 12 million children are in day care centers for several hours a day, and the vast majority are under 3 years of age. As most new parents can testify, respiratory and enteric infections are most common, and infections spread easily to other children, day care workers, and the family at home.

### What Can We Do about Emerging Viruses?

We have discussed how the modernization of society has led to virus spread and evolution. We cannot turn back the clock, but we can use common sense and learn from our experiences. The requirements for control of the emergence of new viruses range from the obvious to the idealistic. Implementation of these ideas will take money and integration of the public and private sectors at a level almost unprecedented today. The most obvious requirement is worldwide implementation of old-fashioned public health measures, including vector management (mosquito, tick, and rodent control), water and sewage management, and, perhaps most important, better nutrition so that natural human defenses are up to the task. Recognizing when human activities are likely to disrupt endemic disease patterns would be a major step forward. Viruses do not respect the boundaries between the agencies that manage human and animal care and those that deal with public health. Therefore, an interesting idea is the promotion of more communication among these agencies to reflect more accurately the interaction be-

*Yellow fever virus: humans change the pattern and pay the price*

In the jungles, yellow fever virus is maintained by a monkey-mosquito cycle involving Old World monkeys. Neither the monkey nor the mosquito is the worse for wear. Various mosquito species serve as vectors, including *Aedes* species in Africa and *Haemagogus* species in the Americas. If humans blunder into mosquito-infected areas, they stand a chance of being bitten by the infected insects and contracting the disease. Yellow fever came to the New World with the colonists and slave trade, where it wreaked havoc among the indigenous human populations.

Humans were not the only species to suffer from the invasion. New World monkeys also died when infected with yellow fever virus, indicating that they were as unequipped to handle this new virus as were the indigenous humans.

Yellow fever virus and its mosquito vector spread by ship to the burgeoning populations of the U.S. South and East Coast with ease. New Orleans; Charleston, S.C.; Philadelphia; New York; and Boston were hit hard. In 1800 Thomas Jefferson lamented, "Yellow Fever will discourage the growth of great cities in our nation." In a striking example of an emerging disease, 15% of the population of Philadelphia died of yellow fever in 1793. Neighboring New Jersey and Maryland attempted to bar panicked Philadelphians from entering their states. It was not just an American problem, as the residents of Cadiz, Spain, found in 1730. Ships returning from South America carried yellow fever that killed 2,300 people. The role of mosquitoes in the transmission of yellow fever was first glimpsed by Carlos Finlay in Cuba in 1880 and was established firmly by Walter Reed in 1900 when the disease was a problem during construction of the Panama Canal in the Central American jungle. A vaccine was developed in 1937 to contain the disease. Even so, more than 10,000 people die of yellow fever every year in South America.

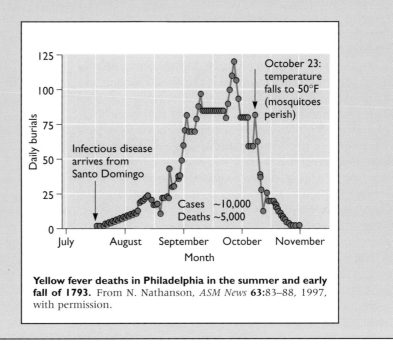

**Yellow fever deaths in Philadelphia in the summer and early fall of 1793.** From N. Nathanson, *ASM News* **63**:83–88, 1997, with permission.

tween hosts and parasites in the evolution of disease. Because of the potential for rapid spread of viruses via air travel and urban development, a system of global surveillance and early warning is required for primary care physicians and health care workers. When a new (or old) virus disease is suspected, the virus must be identified and characterized and the information must be shared widely. Active or passive vaccines should be made available, although the infrastructure for accomplishing anything more than local action is almost nonexistent and solutions are expensive to implement. To do this, methods have to be developed for rapid diagnosis in the field, and primary health care workers must be "wired into" databases. Certainly, baseline data on endemic disease and vector prevalence need to be compiled and included in this "early warning" database. Finally, responsible public and professional education is essential at all levels, as is more scientific research on host-parasite interactions. Inherent in this last proposal is the need for the scientific community to develop better mechanisms for sharing and developing the fruits of scientific research and making their work known, appreciated, and applied by the general public.

## Perspectives

The relationships of viruses and their hosts are in constant flux. The perspective of an evolutionary biologist is required in order to understand why this is so. At present, the interplay of environment and genes, as well as the interactions of virus and host populations, can barely be articulated, much less studied in the laboratory. For viruses, the rapid reproduction of huge numbers of progeny, their tolerance of amazing population fluctuations, and their capacity to produce enormous genetic diversity provide the adaptive palette that ensures their survival. As Joshua Lederberg once said, "In the battle by attrition, humans have a real problem competing with microorganisms. Here we are, here are the bugs, they're looking for food, we're their meal in one sense or another, how do we compete?" One obvious reason for our success in the competition is that our immune defense system is capable of recognizing and destroying invading viruses. A less obvious reason is that the current viral hosts on the planet represent progeny of survivors of past interactions with viruses. Survivors provide a vast gene pool reflecting this experience. Indeed, heterozygosity at the many major histocompatibility complex alleles is known to provide a strong selective advantage in human populations in the battle against microbes. Viruses have far-reaching effects on their hosts ranging from shaping of the immune system in survivors to eliminating entire host populations. Clearly, survival of host and virus is poised on a delicate balance: once this balance is disrupted, the host, the virus, or both will be changed or even eliminated. The metaphor of the balance may be too simple, as it implies that an equilibrium between virus and host can be attained. Rather, reaching an equilibrium may be impossible given changing host and viral populations and the drastic modifications of the ecosystem that accompany the current human population explosion. Perhaps the most sobering fact for humans is that despite our brains and ability to adapt, viruses are still able to infect and even kill us. In some senses, they are and probably always will be on top of the evolutionary ladder. The challenge is to keep them at bay and to strive to be at least one step ahead.

## References

### Books

Biddle, W. 1995. *A Field Guide to Germs.* Henry Holt and Company, New York, N.Y.

Bjornsson, J., R. I. Carp, A. Love, and H. M. Wisniewski (ed.). 1994. *Annals of the New York Academy of Sciences,* vol. 724. *Slow Infections of the Central Nervous System.* New York Academy of Sciences, New York, N.Y.

Diamond, J. 1997. *Guns, Germs and Steel: the Fates of Human Societies.* W. W. Norton and Company, New York, N.Y.

Ewald, P. W. 1994. *Evolution of Infectious Disease.* Oxford University Press, Oxford, United Kingdom.

Gould, S. J. 1996. *Full House.* Three Rivers Press, Pittsburgh, Pa.

Harvey, P. H., A. J. Leigh Brown, J. Maynard Smith, and S. Nee (ed.). 1996. *New Uses for New Phylogenies.* Oxford University Press, Oxford, United Kingdom.

Koprowski, H., and M. B. A. Oldstone (ed.) 1996. *Microbe Hunters Then and Now.* Medi-Ed Press, Bloomington, Ill.

Morse, S. S. (ed.). 1993. *Emerging Viruses.* Oxford University Press, Oxford, United Kingdom.

Scheld, W. M., D. Armstrong, and J. M. Hughes (ed.). 1998. *Emerging Infections 1.* ASM Press, Washington, D.C.

Scheld, W. M., W. A. Craig, and J. M. Hughes (ed.). 1999. *Emerging Infections 2.* ASM Press, Washington, D.C.

### Reviews

**Virus Evolution**

Becker, Y. 1996. A short introduction to the origin and molecular evolution of viruses. *Virus Genes* **11:**73–77.

Best, S. P., R. LeTissier, and J. P. Stoye. 1997. Endogenous retroviruses and the evolution of resistance to retroviral infection. *Trends Microbiol.* **5:**313–318.

Domingo, E., and J. J. Holland. 1997. RNA virus mutations and fitness for survival. *Annu. Rev. Microbiol.* **51:**151–178.

Ebert, D. 1998. Experimental evolution of parasites. *Science* **282:**1432–1435.

Koblet, H. 1990. The "merry-go-round": alphaviruses between vertebrate and invertebrate cells. *Adv. Virus Res.* **38:**343–402.

Koonin, E. V. 1992. Evolution of viral genomes. *Semin. Virol.* **3:**311–417.

Loewer, R., J. Loewer, and R. Kurth. 1996. The viruses in all of us: characteristics and biological significance of human endogenous retrovirus sequences. *Proc. Natl. Acad. Sci. USA* **93:**5177–5184.

Shadan, F. F., and L. P. Villarreal. 1996. The evolution of small DNA viruses of eukaryotes: past and present considerations. *Virus Genes* **11:**239–257.

Smith, D. B., J. McAllister, C. Casino, and P. Simmonds. 1997. Virus 'quasispecies': making a mountain out of a molehill? *J. Gen. Virol.* **78:**1511–1519.

Strauss, J. H., and E. G. Strauss. 1994. The alphaviruses: gene expression, replication, and evolution. *Microbiol. Rev.* **58:** 491–562.

Sturtevant, A. H. 1937. Essays on evolution. I. On the effects of selection on mutation rate. *Q. Rev. Biol.* **12:**464–467.

Van Regenmortel, M. H. V., D. H. L. Bishop, C. M. Fauquet, M. A. Mayo, J. Manilkoff, and C. H. Calisher. 1997. Guidelines to the demarcation of virus species. *Arch. Virol.* **142:**1505–1518.

Villareal, L. P. 1997. On viruses, sex, and motherhood. *J. Virol.* **71:**859–865.

## Emerging Viruses

Jackson, A. C. 1995. Acute viral infections. *Curr. Opin. Neurol.* **8:**70–174.

Murphy. F. A., and N. Nathanson. 1994. New and emerging virus diseases. *Semin. Virol.* **5:**85–87.

Nathanson, N. 1997. The emergence of infectious diseases: societal causes and consequences. *ASM News* **63:**83–88.

Strauss, J. H., and E. G. Strauss. 1997. Recombination in alphaviruses. *Semin. Virol.* **8:**85–94.

Webster R. G., W. J. Bean, O. T. Gorman, T. M. Chambers, and Y. Kawaoka. 1992. Evolution and ecology of influenza A viruses. *Microbiol. Rev.* **56:**152–179.

Weiss, R. A. 1998. Transgenic pigs and virus adaptation. *Nature* (London) **391:**327–328.

## *Selected Papers*

### Virus Evolution

Agostini, H. T., R. Yanagihara, V. Davis, C. F. Ryschkewitsch, and G. L. Stoner. 1997. Asian genotypes of JC virus in Native Americans and in a Pacific island population: markers of viral evolution and human migration. *Proc. Natl. Acad. Sci. USA* **94:**14542–14546.

Baric, R. S., B Yount, L. Hensley, S. A. Peel, and W. Chen. 1997. Episodic evolution mediates interspecies transfer of a murine coronavirus. *J. Virol.* **71:**1946–1955.

Bonhoeffer, S., and M. A. Nowak. 1994. Intra-host versus inter-host selection: viral strategies of immune function impairment. *Proc. Natl. Acad. Sci. USA* **91:**8062–8066.

Chao, L. 1990. Fitness of RNA virus decreased by Muller's ratchet. *Nature* (London) **348:**454–455.

Chao, L. 1997. Evolution of sex and the molecular clock in RNA viruses. *Gene* **205:**301–308.

Clarke, D. K., E. A. Duarte, A. Moya, S. F. Elena, E. Domingo, and J. J. Holland. 1993. Genetic bottlenecks and population passages cause profound fitness differences in RNA viruses. *J. Virol.* **67:**222–228.

Clarke, D. K., E. A. Duarte, S. Elena, A. Moya, E. Domingo, and J. J. Holland. 1994. The Red Queen reigns in the kingdom of RNA viruses. *Proc. Natl. Acad. Sci. USA* **91:**4821–4824.

Drake, J. W., and E. F. Allen. 1968. Antimutagenic DNA polymerases of bacteriophage T4. *Cold Spring Harbor Symp.* **33:**339–344.

Eigen, M. 1971. Self organization of matter and the evolution of biological macromolecules. *Naturwissenschaften* **58:**465–523.

Eigen, M. 1996. On the nature of viral quasispecies. *Trends Microbiol.* **4:**212–214.

Guidotti, L. G., P. Borrow, M. V. Hobbs, B. Matzke, I. Gresser, M. B. A. Oldstone, and F. V. Chisari. 1996. Viral cross talk: intracellular inactivation of the hepatitis B virus during an unrelated viral infection of the liver. *Proc. Natl. Acad. Sci. USA* **93:**4589–4594.

Hall, J. D., D. M. Coen, B. L. Fisher, M. Weisslitz, S. Randall, R. E. Almy, P. T. Gelep, and P. A. Schaffer. 1984. Generation of genetic diversity in herpes simplex virus: an antimutator phenotype maps to the DNA polymerase locus. *Virology* **132:**26–37.

Herniou, E., J. Martin, and M. Tristem. 1998. Retroviral diversity and distribution in vertebrates. *J. Virol.* **72:**5955–5966.

Mayr, E. 1997. The objects of selection. *Proc. Natl. Acad. Sci. USA* **94:**2091–2094.

McGeoch, D. J., S. Cook, A. Dolan, F. E. Jamieson, and E. A. R. Telford. 1995. Molecular phylogeny and evolutionary timescale for the family of mammalian herpesviruses. *J. Mol. Biol.* **247:**443–458.

Muller, H. J. 1964. The relation of recombination to mutational advance. *Mutat. Res.* **1:**2–9.

Nichol, S. T. K., J. E. Rowe, and W. M. Fitch. 1993. Punctuated equilibrium and positive Darwinian evolution in vesicular stomatitis virus. *Proc. Natl. Acad. Sci. USA* **90:**10424–10428.

Oude Essink, B. B., N. K. T. Back, and B. Berkhout. 1997. Increased polymerase fidelity of the 3TC-resistant variants of HIV-1 reverse transcriptase. *Nucleic Acids Res.* **25:**3212–3217.

Preston, B. D., B. J. Poiesz, and L. A. Loeb. 1988. Fidelity of HIV-1 reverse transcriptase. *Science* **242:**1168–1171.

Saiz, J.-C., and E. Domingo. 1996. Virulence as a positive trait in viral persistence. *J. Virol.* **70:**6410–6413.

Schrag, S. J., P. A. Rota, and W. Bellini. 1999. Spontaneous mutation rate of measles virus: direct estimation based on mutations conferring monoclonal antibody resistance. *J. Virol.* **73:**51–54.

Spiegelman, S., N. R. Pace, D. R. Mills, R. Levisohn, T. S. Eikhom, M. M. Taylor, R. L. Peterson, and D. H. L. Bishop. The mechanism of RNA replication. *Cold Spring Harbor Symp.* **33:**101–124.

Tufariello, J., M. S. Cho, and M. S. Horwitz. 1994. Adenovirus E3 14.7-kilodalton protein, an antagonist of tumor necrosis factor cytolysis, increases the virulence of vaccinia virus in severe combined immunodeficient mice. *Proc. Natl. Acad. Sci. USA* **91:**10987–10991.

Webster, R. G., W. G. Laver, G. M. Air, and G. C. Schild. 1982. Molecular mechanisms of variation in influenza viruses. *Nature* (London) **296:**115–121.

Zhu, T., B. T. Korber, A. J. Nahmias, E. Hooper, P. M. Sharp, and D. D. Ho. 1998. An African HIV-1 sequence from 1959 and implications for the origin of the epidemic. *Nature* (London) **391:**594–597.

## Emerging Viruses

Henige, D. 1986. When did smallpox reach the New World (and why does it matter)?, p. 11–26. *In* P. E. Lovejoy (ed.), *Africans in Bondage*. University of Wisconsin Press, Madison.

Jin, L., D. W. G. Brown, M. E. B. Ramsay, P. A. Rota, and W. J. Bellini. 1997. The diversity of measles virus in the United Kingdom, 1992–1995. *J. Gen. Virol.* **78:**1287–1294.

Leitner, T., S. Kumar, and J. Albert. 1997. Tempo and mode of nucleotide substitutions in *gag* and *env* gene fragments in human immunodeficiency virus type 1 populations with a known transmission history. *J. Virol.* **71:**4761–4770.

Morimoto, K., M. Patel, S. Corisdeo, D. C. Hooper, Z. F. Fu, C. E. Rupprecht, H. Koprowski, and B. Dietzschold. 1996. Characterization of a unique variant of bat rabies virus responsible for newly emerging human cases in North America. *Proc. Natl. Acad. Sci. USA* **93:**5653–5658.

Murray, K., P. Selleck, P. Hooper, A. Hyatt, A. Gould, L. Gleeson, H. Westbury, L. Hiley, L. Selvey, B. Rodwell, and P. Ketterer. 1995. A morbillivirus that caused fatal disease in horses and humans. *Science* **268:**94–97.

Parrish, C. R. 1997. How canine parvovirus suddenly shifted host range. *ASM News* **63:**307–311.

Reid, A. H., T. G. Fanning, J. V. Hultin, and J. K. Taubenberger. 1999. Origin and evaluation of the 1918 "Spanish" influenza virus hemagglutinin gene. *Proc. Natl. Acad. Sci. USA* **96:**1651–1656.

Sharp, G. B., Y. Kawaoka, D. J. Jones, W. J. Bean, S. P. Pryor, V. Hinshaw, and R. G. Webster. 1997. Coinfection of wild ducks by influenza A viruses: distribution patterns and biological significance. *J. Virol.* **71:**6128–6135.

Smith, A. W., D. E. Skilling, N. Cherry, J. H. Mead, and D. O. Matson. 1998. Calicivirus emergence from ocean reservoirs: zoonotic and interspecies movements. *Emerg. Virus Infect.* **4:**13–20.

Subbarao, K., A. Klimov, J. Katz, H. Regnery, W. Lim, H. Hall, M. Perdue, D. Swayne, C. Bender, J. Huang, M. Hemphill, T. Rowe, M. Shaw, X. Xu, K. Fukuda, and N. Cox. 1998. Characterization of an avian influenza A (H5N1) virus isolated from a child with a fatal respiratory illness. *Science* **279:**393–396.

Taubenberger, J. K., A. H. Reid, A. E. Krafft, K. E. Bijwaard, and T. G. Fanning. 1997. Initial genetic characterization of the 1918 "Spanish" influenza virus. *Science* **275:**1793–1795.

# APPENDIX
# Structure, Genomic Organization, and Infectious Cycles of Selected Animal Viruses Discussed in This Book

# Picornaviruses

**Figure 1 Structure and genomic organization of the picornavirus poliovirus. (A) The virion.** (Left) Electron micrograph of negatively stained poliovirus. Magnification, ×270,000. Courtesy of N. Cheng and D. M. Belnap, National Institutes of Health. (Right) Diagram of the polioviral virion, showing the names and locations of component proteins and genomic RNA. The capsid consists of 60 structural units (each made up of a single copy of VP1, VP2, VP3, and VP4, colored blue yellow, red, and green, respectively) arranged in 12 pentamers. One of the icosahedral faces has been removed to illustrate the locations of VP4 and the viral RNA. **(B) Genome organization.** Polioviral RNA is shown with the VPg protein covalently attached to the 5' end. The genome is of (+) polarity and encodes a polyprotein precursor. The polyprotein is cleaved during translation by two virus-encoded proteases, 2A$^{pro}$ and 3C$^{pro}$, to produce structural and nonstructural proteins. The P1 protein is cleaved into the virion capsid proteins, while the P2 and P3 proteins are cleaved to form the proteases and the proteins that participate in viral RNA synthesis.

**Figure 2 Single-cell reproductive cycle of poliovirus.** The virus binds to a cellular receptor **(step 1)**; the mechanism and site of uncoating of the RNA genome are unknown **(2)**. The VPg protein, depicted as a small, orange circle at the 5' end of the virion RNA, is removed, and the resulting RNA associates with ribosomes **(3)**. Translation is initiated at an internal site 741 nucleotides from the 5' end of the viral mRNA, and a polyprotein precursor is synthesized **(4)**. The polyprotein is cleaved during and after its synthesis to yield the individual viral proteins **(5)**. Only the initial cleavages are shown here. P1 contains the viral structural proteins. P2 and P3 are cleaved to produce the proteases and the proteins involved in viral RNA synthesis, with the orange bar corresponding to VPg. The proteins that participate in viral RNA synthesis are transported to membrane vesicles **(6)**. RNA synthesis occurs on the surfaces of these infected-cell-specific membrane vesicles. The (+) strand RNA is transported to these membrane vesicles **(7)**, where it is copied into (−) strands carrying VPg at their 3' ends **(8)**. These (−) strands serve as templates for the synthesis of (+) strand genomic RNAs **(9)**. Some of the newly synthesized (+) strand RNA molecules enter the translational system after the removal of VPg **(10)**. Structural proteins formed by partial cleavage of the P1 precursor **(11)** associate with (+) strand RNA molecules that retain VPg to form progeny virions **(12)**, which are released from the cell by lysis **(13)**.

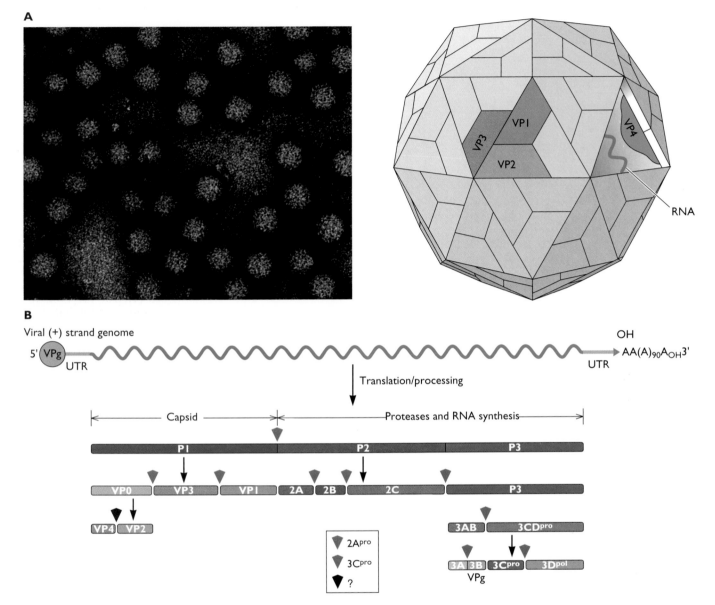

**Figure 1 Structure and genomic organization of the picornavirus poliovirus.**

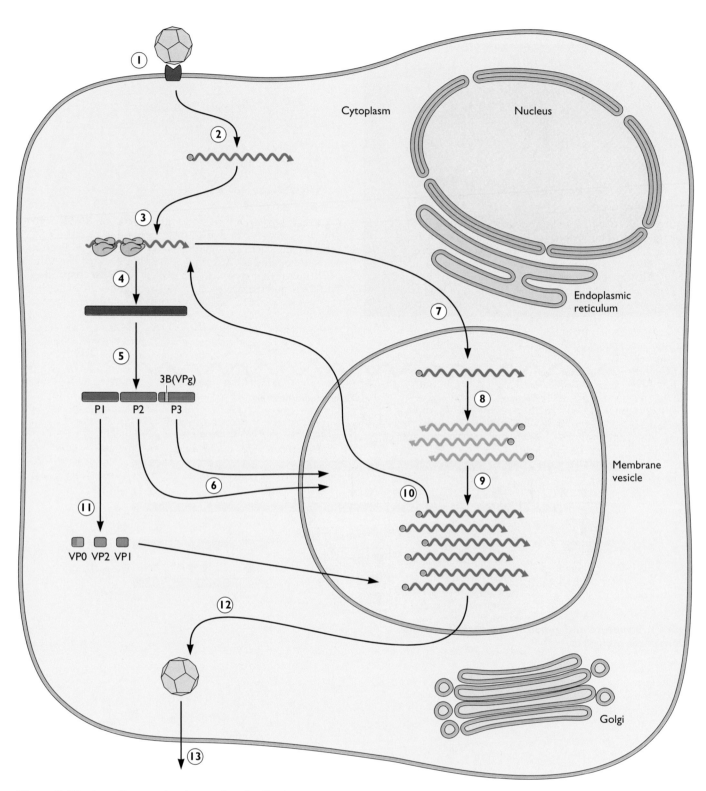

**Figure 2 Single-cell reproductive cycle of poliovirus.**

# Togaviruses

**Figure 3 Structure and genomic organization of to-gaviruses. (A) The virion.** (Left) Cryoelectron micrograph of Ross River virus. Magnification, ×110,000. Courtesy of N. Olson, Purdue University. (Right) Three-dimensional image reconstruction of Sindbis virus (courtesy of B. V. V. Prasad, Baylor College of Medicine), showing the intact virus (yellow, with glycoprotein spikes visible) and the nucleocapsid only (blue). Also shown is a cross section of the particle (green), illustrating the relationship among the spike glycoproteins (S), the lipid membrane (M), the capsid (C), and the viral RNA genome (RNA). **(B) Genome organization.** The first two-thirds of togavirus genomic RNA, which is of (+) polarity and carries a 5′ cap, is translated to produce the polyprotein P1234, the precursor of the RNA polymerase. The proteins encoded in the terminal one-third are produced from a subgenomic mRNA that is copied from a full-length (−) strand RNA intermediate. The subgenomic mRNA encodes the structural proteins.

**Figure 4 Single-cell reproductive cycle of Sindbis virus.** The virus binds to a cellular receptor and enters the cell via receptor-mediated endocytosis **(step 1)**. Upon acidification of the vesicle, viral RNA is uncoated **(2)** and translated to form the proteins that comprise the RNA polymerase, nsp1234 **(3)**. The RNA polymerase copies (+) strands into full-length (−) and (+) strands **(4)** and catalyzes synthesis of the subgenomic mRNA **(5)**. This mRNA is translated by free cytoplasmic ribosomes to produce the capsid protein **(6)**; proteolytic cleavage to liberate the capsid protein exposes a hydrophobic sequence of PE2 that induces the ribosomes to associate with the endoplasmic reticulum **(7)**. As a result, the PE2, 6K, and E1 proteins enter the secretory pathway. The membrane topology of these proteins is shown in detail, with cleavage sites for cellular proteases indicated by red arrows. The glycoproteins are transported to the cell surface **(8 and 9)**. The capsid protein and (+) strand genomic RNA assemble to form capsids **(10)** that migrate to the plasma membrane and associate with viral glycoproteins **(11)**. The capsid acquires an envelope by budding at this site **(12)**, and mature virions are released **(13)**.

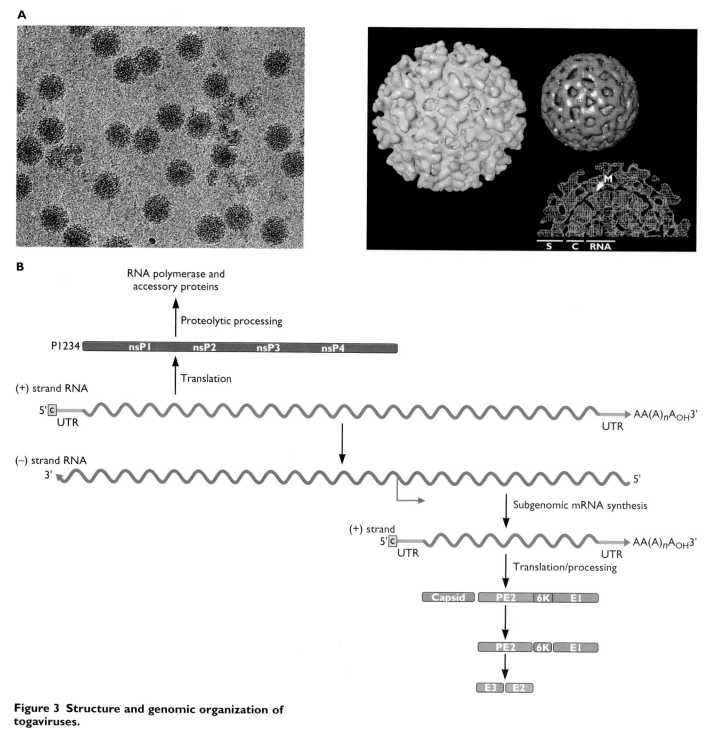

**Figure 3  Structure and genomic organization of togaviruses.**

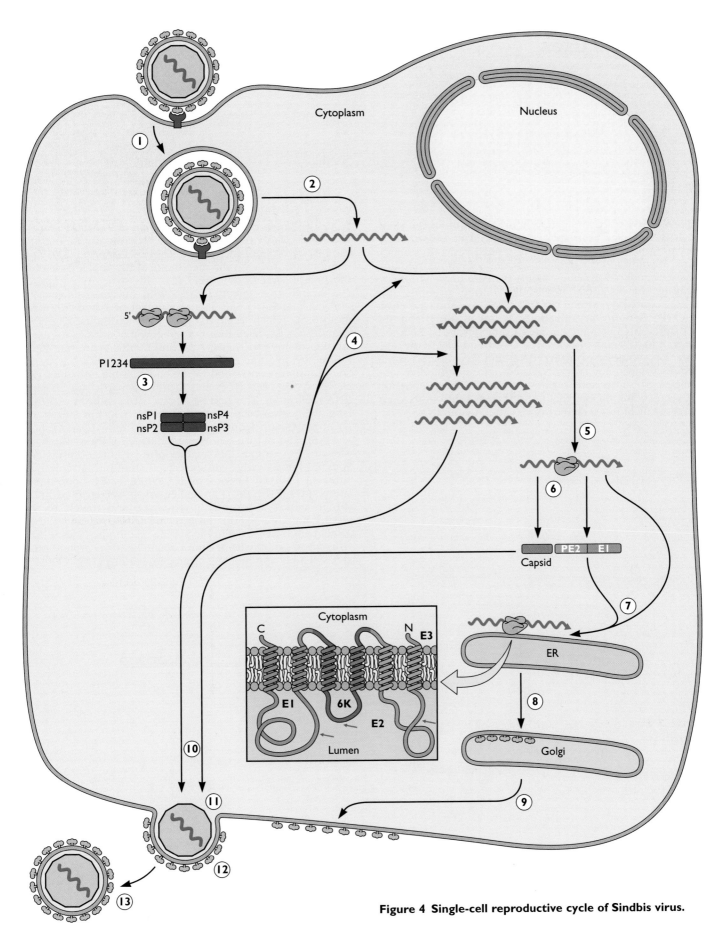

**Figure 4 Single-cell reproductive cycle of Sindbis virus.**

# Rhabdoviruses

**Figure 5 Structure and genomic organization of the rhabdovirus vesicular stomatitis virus. (A) The virion.** (Left) Electron micrograph of negatively stained vesicular stomatis virus. Magnification, ×129,000. Courtesy of J. Rose, Yale University School of Medicine. (Right) Diagram of the virion, indicating the location of the five virion proteins, the nonsegmented (−) strand genomic RNA, and the viral envelope. **(B) Genome organization.** Starting at the 3' end, the (−) strand RNA genome encodes a small leader (l) RNA and the N, P, M, G, and L proteins. The (−) strand RNA is the template for synthesis of leader RNA and five monocistronic mRNAs (capped and polyadenylated) encoding the five viral proteins. The intergenic regions, which are short sequences between coding regions that are not copied into mRNA, are indicated by brackets above the (−) strand viral RNA.

**Figure 6 Single-cell reproductive cycle of vesicular stomatitis virus.** The virus binds to a cellular receptor and enters the cell via receptor-mediated endocytosis **(step 1)**. The viral membrane fuses with the membrane of the vesicle, releasing the helical viral nucleocapsid **(2)**. This structure comprises (−) strand RNA (green) coated with nucleocapsid protein molecules (yellow) and a small number of L and P protein molecules, which catalyze viral RNA synthesis. The (−) strand RNA is copied into five subgenomic mRNAs by the L and P proteins **(3)**. The N, P, M, and L mRNAs are translated by free cytoplasmic ribosomes **(4)**, while G mRNA is translated by ribosomes bound to the endoplasmic reticulum **(5)**. Newly synthesized N, P, and L proteins participate in viral RNA replication. This process begins with synthesis of a full-length (+) strand copy of genomic RNA, which is also in the form of a ribonucleoprotein containing the N, L, and P proteins **(6)**. This in turn serves as a template for the synthesis of progeny (−) strand RNA in the form of nucleocapsids **(7)**. Some of these newly synthesized (−) strand RNA molecules enter the pathway for viral mRNA synthesis **(8)**. Upon translation of G mRNA, the G protein enters the secretory pathway **(9)**, in which it becomes glycosylated and travels to the plasma membrane **(10)**. Progeny nucleocapsids and the M protein are transported to the plasma membrane **(11 and 12)**, where association with regions containing the G protein initiates assembly and budding of progeny virions **(13)**.

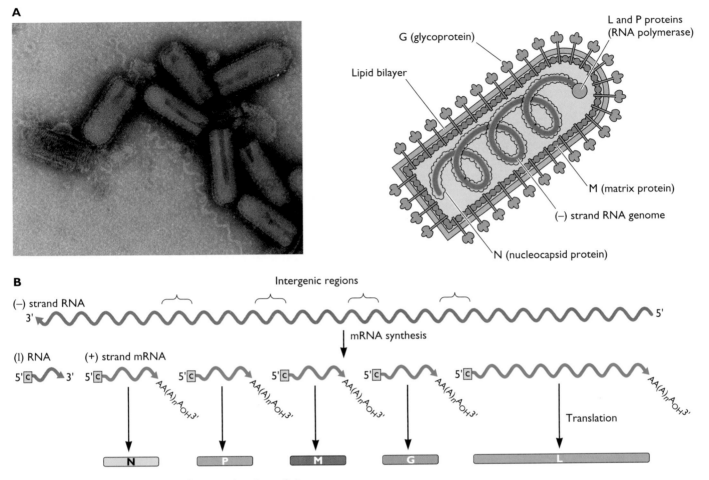

**A**

**B**

Intergenic regions

(–) strand RNA

3'                                        5'

mRNA synthesis

(l) RNA      (+) strand mRNA

5' C   3'    5' C        5' C        5' C        5' C        5' C

AA(A)$_n$A-OH 3'    AA(A)$_n$A-OH 3'    AA(A)$_n$A-OH 3'    AA(A)$_n$A-OH 3'    AA(A)$_n$A-OH 3'

Translation

| N | P | M | G | L |

**Figure 5  Structure and genomic organization of the rhabdovirus vesicular stomatitis virus.**

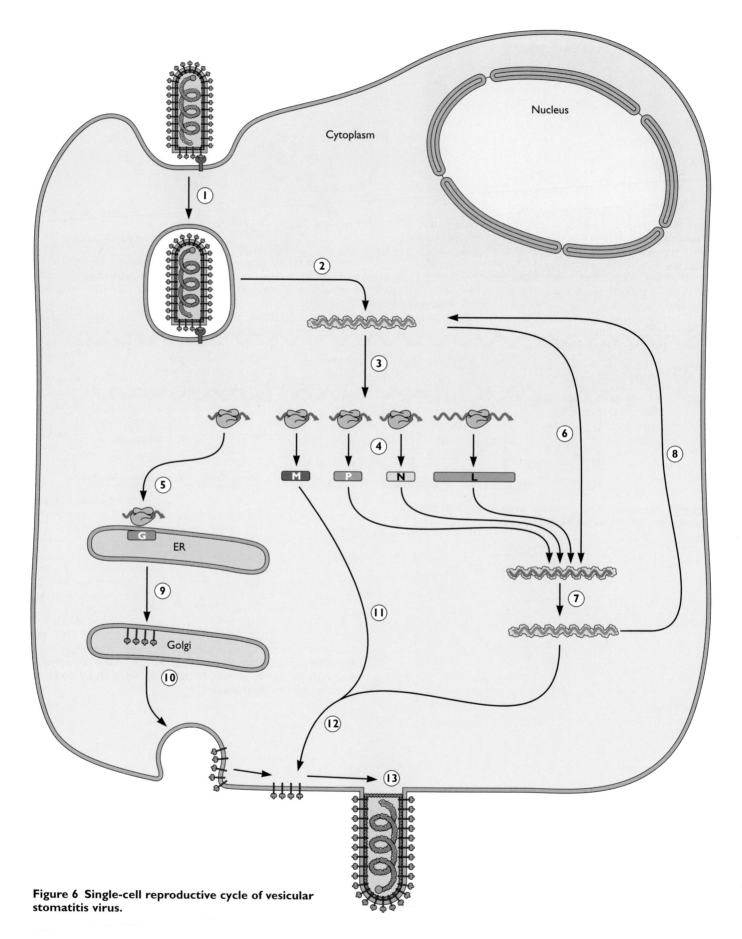

**Figure 6 Single-cell reproductive cycle of vesicular stomatitis virus.**

# Orthomyxoviruses

**Figure 7 Structure and genomic organization of the orthomyxovirus influenza A virus. (A) The virion.** (Left) Electron micrograph of negatively stained influenza A virus particles. Magnification, ×300,000. Courtesy of P. Palese, Mt. Sinai School of Medicine. (Right) Diagram of the influenza A virion, indicating the location of the nine virion proteins, the viral envelope, and the eight genomic RNA segments. **(B) Genome organization.** The (−) strand RNA genome comprises eight segments, each of which encodes at least one viral protein as shown. The two smallest genomic RNA segments, 7 and 8, each code for two proteins, for some fraction of their (+) RNA copies are spliced by host cell enzymes. The NS (nonstructural) proteins are so named because the larger protein, NS1, is not incorporated into virus particles. Blue boxes containing the letter C denote the capped host mRNA fragments acquired by viral enzymes during mRNA synthesis.

**Figure 8 Single-cell reproductive cycle of influenza A virus.** The virus binds to a sialic acid-containing cellular surface protein or lipid and enters the cell via receptor-mediated endocytosis **(step 1)**. Upon acidification of the vesicle, the viral membrane fuses with the membrane of the vesicle, releasing the viral nucleocapsids into the cytoplasm **(2)**. The viral nucleocapsids containing (−) strand genomic RNA, multiple copies of the NP protein, and the P proteins are transported into the nucleus **(3)**. The (−) strand RNA is copied by virion RNA polymerase into viral mRNA, using the capped 5′ ends of host pre-mRNAs and mRNAs as primers to initiate synthesis **(4)**. The mRNAs are transported to the cytoplasm **(5)**, following splicing in the case of the mRNAs encoding NS2 and M2 **(6)**. The mRNAs specifying the viral membrane proteins (HA, NA, M2) are translated by ribosomes bound to the endoplasmic reticulum (ER) **(7)**. These proteins enter the host cell's secretory pathway, in which they undergo glycosylation. All other mRNAs are translated by ribosomes in the cytoplasm **(8 and 9)**. The PA, PB1, PB2, and NP proteins are imported into the nucleus **(10)**, where they catalyze the synthesis of full-length (+) strand RNAs **(11)** and then (−) strand virion RNAs **(12)**, both of which are synthesized in the form of nucleocapsids. Some of the newly synthesized (−) strand RNAs enter the pathway for mRNA synthesis **(13)**. The M1 protein and the NS1 protein are transported into the nucleus **(10)**. Binding of the M1 protein to newly synthesized (−) strand RNAs shuts down viral mRNA synthesis and, in conjunction with the NS2 protein, induces export of progeny nucleocapsids to the cytoplasm **(14)**. The HA and NA proteins are transported to the cell surface **(15)** and become incorporated into the plasma membrane **(16)**. The virion nucleocapsids associated with the M1 protein **(17)**, and the NS2 protein **(18)**, are transported to the cell surface and associate with regions of the plasma membrane containing the HA and NA proteins. Assembly of virions is completed at this location by budding from the plasma membrane **(19)**.

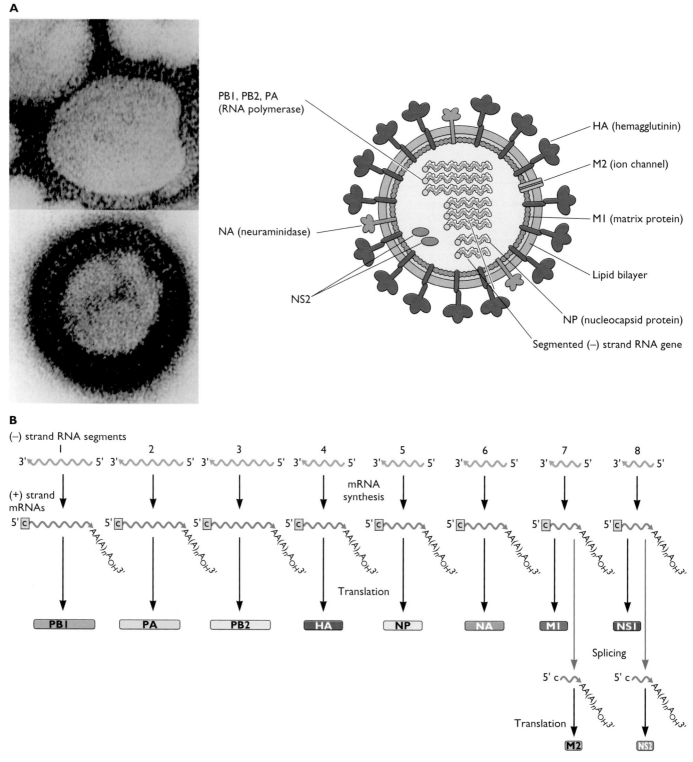

**A**

PB1, PB2, PA
(RNA polymerase)

NA (neuraminidase)

NS2

HA (hemagglutinin)

M2 (ion channel)

M1 (matrix protein)

Lipid bilayer

NP (nucleocapsid protein)

Segmented (–) strand RNA gene

**B**

(–) strand RNA segments

1  2  3  4  5  6  7  8

3' ⌇⌇⌇ 5'  3' ⌇⌇⌇ 5'  3' ⌇⌇⌇ 5'  3' ⌇⌇⌇ 5'  3' ⌇⌇ 5'  3' ⌇⌇⌇ 5'  3' ⌇⌇ 5'  3' ⌇⌇ 5'

(+) strand
mRNAs

mRNA
synthesis

5'[c]⌇⌇AA(A)nAOH-3'   5'[c]⌇⌇AA(A)nAOH-3'   5'[c]⌇⌇AA(A)nAOH-3'   5'[c]⌇⌇AA(A)nAOH-3'   5'[c]⌇⌇AA(A)nAOH-3'   5'[c]⌇⌇AA(A)nAOH-3'   5'[c]⌇⌇AA(A)nAOH-3'   5'[c]⌇⌇AA(A)nAOH-3'

Translation

PB1   PA   PB2   HA   NP   NA   M1   NS1

Splicing

5' c⌇⌇AA(A)nAOH-3'   5' c⌇⌇AA(A)nAOH-3'

Translation

M2   NS2

**Figure 7 Structure and genomic organization of the orthomyxovirus influenza A virus.**

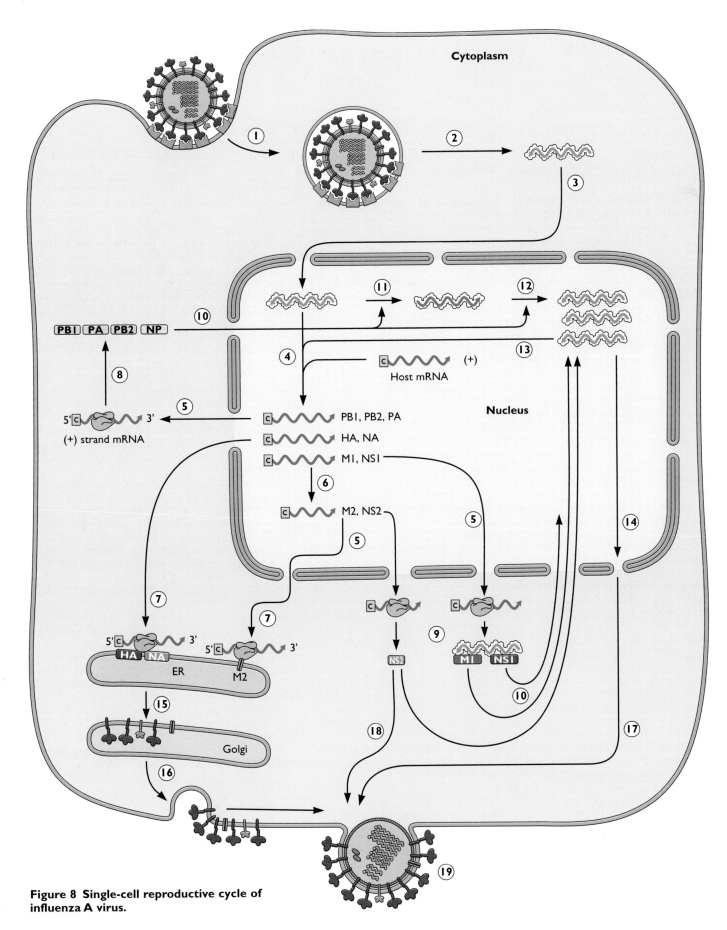

**Figure 8 Single-cell reproductive cycle of influenza A virus.**

# Retroviruses

**Figure 9 Structure and genomic organization of retroviruses. (A) The virion.** (Left) Electron micrograph of negatively stained Rous sarcoma virus, a simple retrovirus. Magnification, ×160,000. Courtesy of R. Katz and T. Gales, Fox Chase Cancer Center. The method of preparation for electron microscopy (staining of thin sections) does not allow visualization of the envelope protein projections. (Right) Diagram of avian leukosis virus, indicating the names and locations of the component proteins, genomic RNA, and envelope. **(B) Genomic organization.** (Left) A simple retrovirus. (Top) Genetic map of the avian leukosis virus (ALV) provirus. Colored boxes delineate open reading frames which, as indicated, are overlapping. (Bottom) The map of genomic RNA shows the genes common to all replication-competent retroviruses, *gag*, *pol*, and *env*. The ends of the RNA include short terminal repeats. Near the termini of the genomic RNA, sections containing *cis*-acting sequences that are required in replication are also shown: U5 (unique to the 5' end) and U3 (unique to the 3' end). In the integrated proviral DNA, the U5 and U3 sections are duplicated and comprise long terminal repeats (LTRs) at the borders between cellular and viral DNA. Initiation of DNA transcription and polyadenylation of pre-mRNA occur within the LTRs to produce progeny viral genomes and mRNAs. The viral mRNAs are translated to produce the indicated polyprotein precursors, which are eventually processed proteolytically to form the mature viral proteins. (Right) A complex retrovirus. (Top) Genetic map of the human immunodeficiency virus type 1 provirus. Genes are encoded in all three reading frames as indicated by the overlaps. (Bottom) Human immunodeficiency virus type 1 transcripts, which fall into one of three classes. The first type is an unspliced transcript of 9.1 kb, identical in function to that of the simple retrovirus avian leukosis virus. The second type consists of five singly spliced mRNAs (average length, 4.3 kb) that result from splicing from a 5' splice site upstream of *gag* to any one of a number of 3' splice sites near the center of the genome. Among these transcripts is one specifying the Env polyprotein precursor, as illustrated for the simple retrovirus avian leukosis virus. The others specify the human immunodeficiency virus type 1 accessory proteins Vif, Vpr, and Vpu. Both unspliced RNA and singly spliced mRNAs require the Rev protein for transport out of the nucleus. This transport is mediated by the attachment of Rev to the Rev-responsive element (RRE) in the *env* gene. The third type of mRNAs, multiply spliced molecules, average 1.8 kb in length; all lack the region of *env* that includes the RRE. This group includes a complex class of ~16 mRNAs derived by exhaustive splicing from 5' and 3' splice sites throughout the genome. These multiply spliced products are the first mRNAs to accumulate after infection; they specify the regulatory proteins Tat and Rev and the accessory protein Nef. As Rev protein accumulates, more and more of the 9.1- and 4.3-kb mRNAs are exported from the nucleus, allowing for the production of virion proteins and formation of progeny virus.

**Figure 10 Single-cell reproductive cycle of a simple retrovirus.** The virus attaches by binding of SU and TM to specific receptors on the surface of the cell **(step 1)**. The identities of receptors (which are normal components of the cell surface) are known for several retroviruses. The viral core is deposited into the cytoplasm **(2)** following viral protein-assisted fusion of the virion and cell envelopes. The viral RNA genome is reverse transcribed by the virion RT **(3)** within a fast-sedimenting subviral particle. The product is a linear double-stranded viral DNA whose ends are shown juxtaposed in preparation for integration. The viral preintegration complex, which includes viral DNA and IN, enters the nucleus **(4)**, where it gains access to the host chromatin. Integrative recombination **(5)**, catalyzed by IN, results in site-specific insertion of the viral DNA ends into numerous (essentially random) sites in the host genome. Integrated viral DNA is called the **provirus**. Transcription of proviral DNA by host cell RNA polymerase II **(6)** produces full-length RNA transcripts which are used for multiple purposes. Some full-length RNA molecules are exported from the nucleus and serve as mRNA **(7)**, which is translated by cytoplasmic ribosomes to form the viral Gag and Gag-Pol polyprotein precursors **(8)**. Some full-length RNA molecules become encapsidated as progeny viral genomes **(9)** following transport into the cytoplasm. Other full-length RNA molecules are spliced within the nucleus **(10)** to form mRNA for the Env polyprotein precursor proteins. Env mRNA is translated by ribosomes bound to the endoplasmic reticulum (ER) **(11)**. The Env proteins are transported through the Golgi apparatus **(12)**, where they are glycosylated and cleaved to form the mature SU + TM complex. These mature envelope proteins are then delivered to the surface of the infected cell **(13)**. Virion components (viral RNA, Gag and Gag-Pol precursors, and SU + TM) assemble at budding sites **(14)** with the help of *cis*-acting signals encoded in each. Type C retroviruses assemble at the inner face of the plasma membrane, as illustrated. Other types (A, B, and D) assemble on internal cellular membranes. The nascent virions bud from the surface of the cell **(15)**. Maturation (and infectivity) requires the action of the virus-encoded protease (PR), which is itself a component of the core precursor polyprotein. During or shortly after budding, PR cleaves at specific sites within the Gag and Gag-Pol precursors **(16)** to produce the mature virion proteins. This process causes a characteristic condensation of the virion cores.

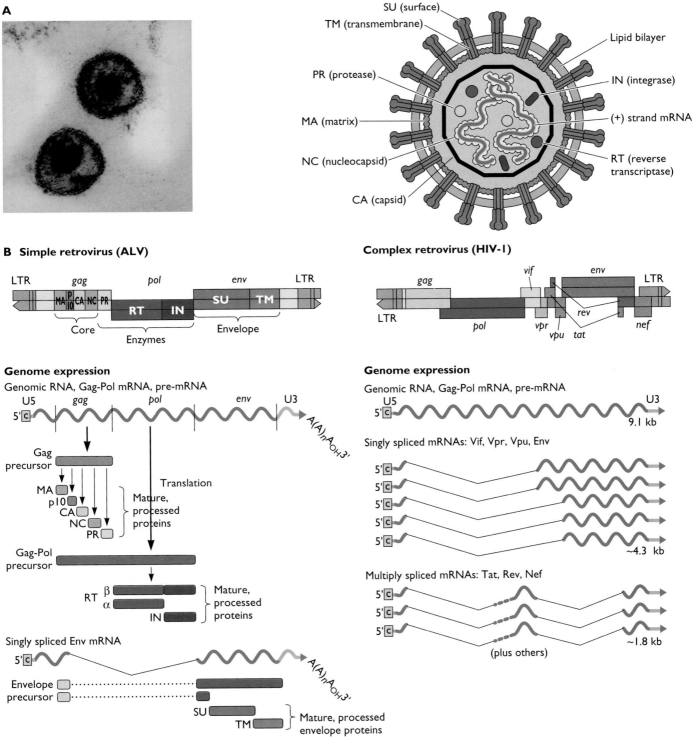

**Figure 9** Structure and genomic organization of retroviruses.

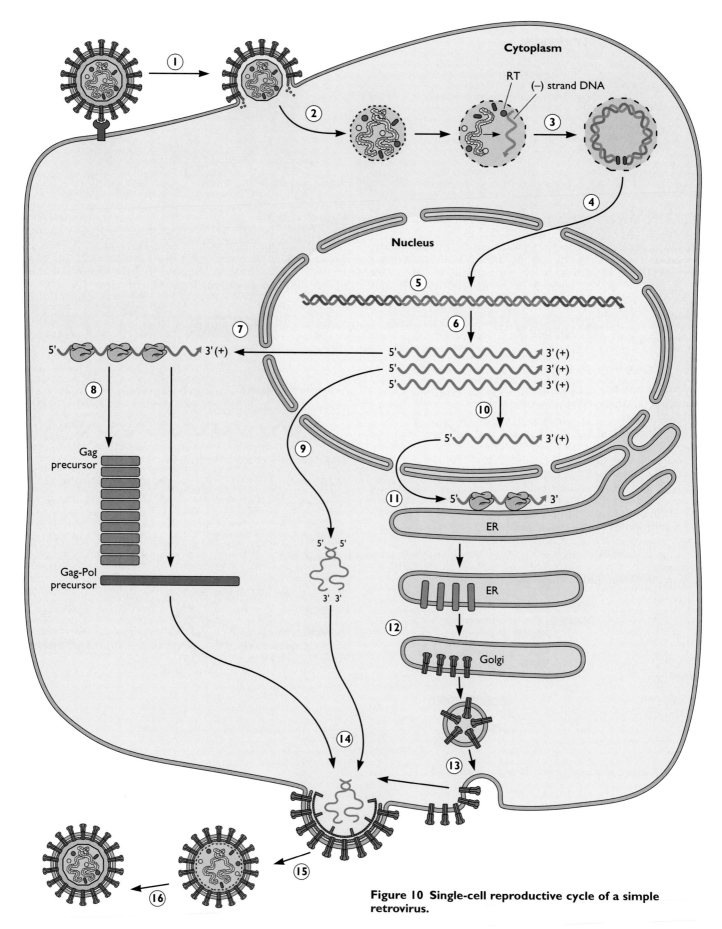

**Figure 10  Single-cell reproductive cycle of a simple retrovirus.**

# Hepadnaviruses

**Figure 11 Structure and genome organization of hepadnaviruses. (A) The virion.** (Left) Electron micrograph of negatively stained woodchuck hepatitis virus, a mammalian hepadnavirus related to human hepatitis B virus. Magnification, ×320,000. Courtesy of W. Mason and T. Gales, Fox Chase Cancer Center. (Right) Diagram of hepatitis B virus and hepatitis B virus-related particles that accumulate in the serum of infected individuals. **(B) Genome organization.** A viral protein (blue arc) is bound to the 5′ end of the (−) DNA strand, which includes a short terminal redundancy signified by the small overlap in the diagram. This strand contains a nick, whereas the (+) strand of the genomic DNA contains a large gap. Nucleotides are numbered from the single *Eco*RI site shown. The (−) strand contains 3,227 bases of unique sequence. Synthesis of subgenomic viral RNA transcripts is initiated at the several sites indicated, but all are polyadenylated at the site marked "t." The pregenome transcript is longer than genome length and begins at DR1 and ends at the polyadenylation site (t). The positions of the open reading frames are indicated in the outer circle. The hepadnaviral genome packs a lot of information into its DNA by encoding genes in overlapping reading frames and by using multiple translational start sites. Every nucleotide in the hepadnaviral DNA contributes to the encoding of at least one protein, and every *cis*-acting sequence is also part of one or more open reading frames. Although this represents remarkable efficiency for the virus, it is a nightmare for the molecular biologist trying to manipulate the viral genome.

**Figure 12 Single-cell reproductive cycle of hepatitis B virus.** The virion attaches to a susceptible hepatocyte **(step 1)** through recognition of a cell surface receptor(s) that has yet to be identified. The mechanism of virus uptake **(2)** is also unknown, and repair of the gapped (+) DNA strand is accomplished **(3)** by as yet unidentified enzymes. The DNA is translocated to the nucleus **(4)**, where it is found in a covalently closed circular form called CCC DNA. The (−) strand of such CCC DNA is the template for transcription by cellular RNA polymerase II **(5)** of a longer-than-genome-length RNA called the pregenome and shorter, subgenomic transcripts (Fig. 11B), all of which serve as mRNAs. Viral mRNAs are transported from the nucleus **(6)**. The shorter viral mRNAs are translated by ribosomes bound to the endoplasmic reticulum (ER) **(7)**, and the proteins destined to become surface antigens (HBsAg) in the viral envelope enter the secretory pathway. The pregenome RNA is translated at low efficiency **(8)** to produce a 90-kDa polymerase protein, P, which possesses reverse transcriptase activity. This protein then binds to a specific site at the 3′ end of its own transcript, where viral DNA synthesis is eventually initiated. The pregenomic RNA also serves as mRNA for the capsid protein **(9)**. Concurrently with capsid formation, the RNA-P protein complex is packaged **(10)** and reverse transcription begins with synthesis of (−) strand DNA **(11)**. At early times after infection, the DNA is recirculated to the nucleus **(12)**, where the process is repeated, resulting in the eventual accumulation of 10 to 30 molecules of CCC DNA and a concomitant increase in viral mRNA concentrations. At later times, and perhaps as a consequence of the accumulation of sufficient viral protein, mature nucleocapsids are formed **(13)**. These structures acquire envelopes by budding into the ER, where viral morphogenesis is completed **(14)**. Progeny enveloped virions are released from the cell by exocytosis **(15)**.

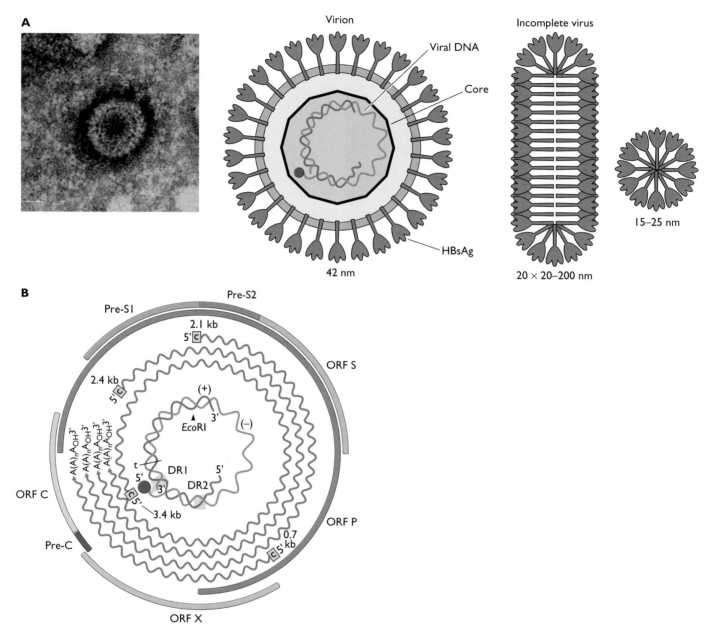

**Figure 11 Structure and genome organization of hepadnaviruses.**

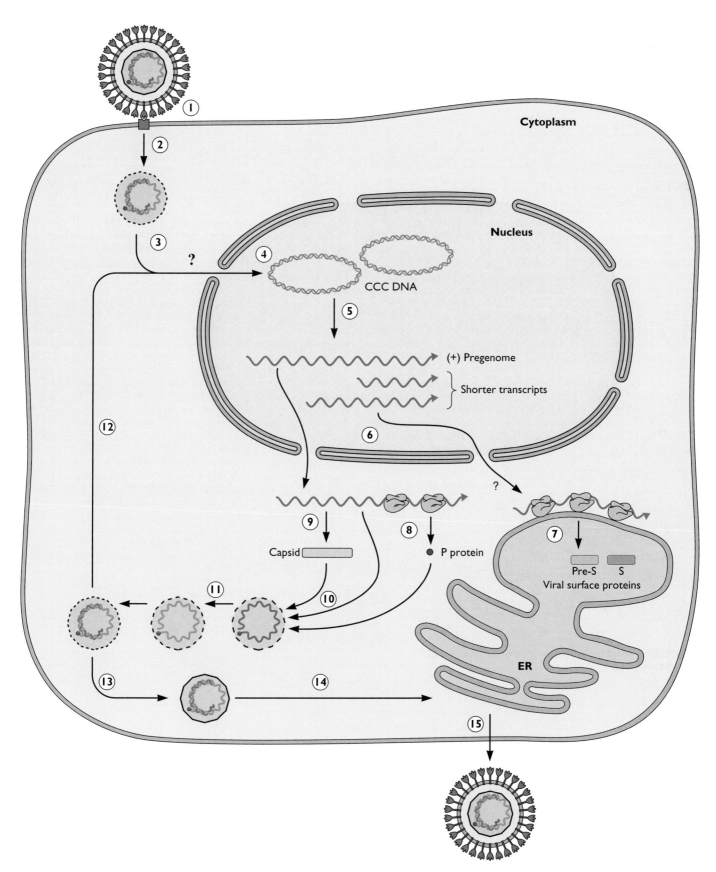

**Figure 12 Single-cell reproductive cycle of hepatitis B virus.**

# Papovaviruses

**Figure 13 Structure and genome organization of the papovavirus simian virus 40. (A) The virion.** (Left) Electron micrograph of negatively stained simian virus 40 virions. Magnification, ×625,000. From F. A. Andered et al., *Virology* **32:**511–523, 1967, with permission. (Right) Diagram of the virion, showing the names and locations of virion proteins and the organization of the 5,243-bp, circular, double-stranded DNA genome into approximately 25 nucleosomes by the cellular histones H2A, H2B, H3, and H4 (the core histones). One molecule of either VP2 or VP3, which possess a common C-terminal sequence (B), is believed to be associated with each VP1 pentamer. **(B) Genome organization.** Locations of the origin of viral DNA synthesis (Ori) and of the early and late mRNA sequences encoding the large and small T antigens (LT and sT) and the virion structural proteins VP1, VP2, and VP3 are indicated. The late mRNA species generally contain additional open reading frames in their 5'-terminal exons, such as that encoding leader protein 1 (LP1).

**Figure 14 Single-cell reproductive cycle of simian virus 40.** The virus attaches to permissive monkey cells upon binding of VP1 to a receptor on the surface of the cell. Polyomavirus, but not simian virus 40, employs cell surface sialic acid residues for attachment. The virion is deposited in the cytoplasm **(step 1)** following entry. It is transported to the nucleus and uncoated **(2)** to release the viral genome packaged by cellular nucleosomes. The **early** transcription unit is transcribed by host cell RNA polymerase II **(3)**. After alternative splicing and export to the cytoplasm **(4)**, the early mRNAs are translated by free cytoplasmic ribosomes into the early proteins LT and sT **(5)**. The former is imported into the nucleus **(6)**, where it binds to the simian virus 40 origin of replication to initiate viral DNA synthesis **(7)**. Apart from LT, all components needed for viral DNA replication are provided by the host cell. As they are synthesized, daughter viral DNA molecules are packaged by cellular nucleosomes to form the viral nucleoproteins often called minichromosomes. LT also stimulates **late** transcription from replicated viral DNA templates **(8)**. Processed late mRNAs are exported to the cytoplasm **(9)**, in which the virion structural proteins VP1, VP2, and VP3 are synthesized from them **(10)**. These structural proteins are imported into the nucleus **(11)** and assemble around viral minichromosomes to form progeny virions **(12)**. Virions are released by an unknown mechanism **(13)**.

**A**

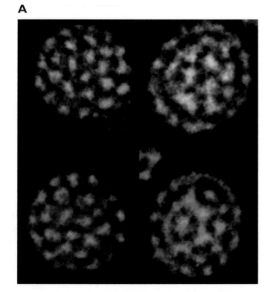
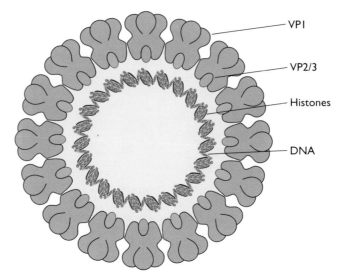

VP1

VP2/3

Histones

DNA

**B**

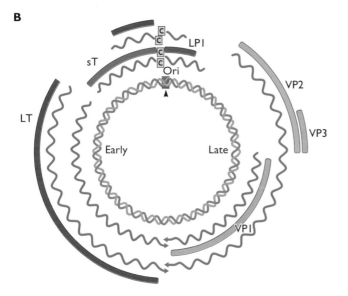

Figure 13 Structure and genome organization of the
papovavirus simian virus 40.

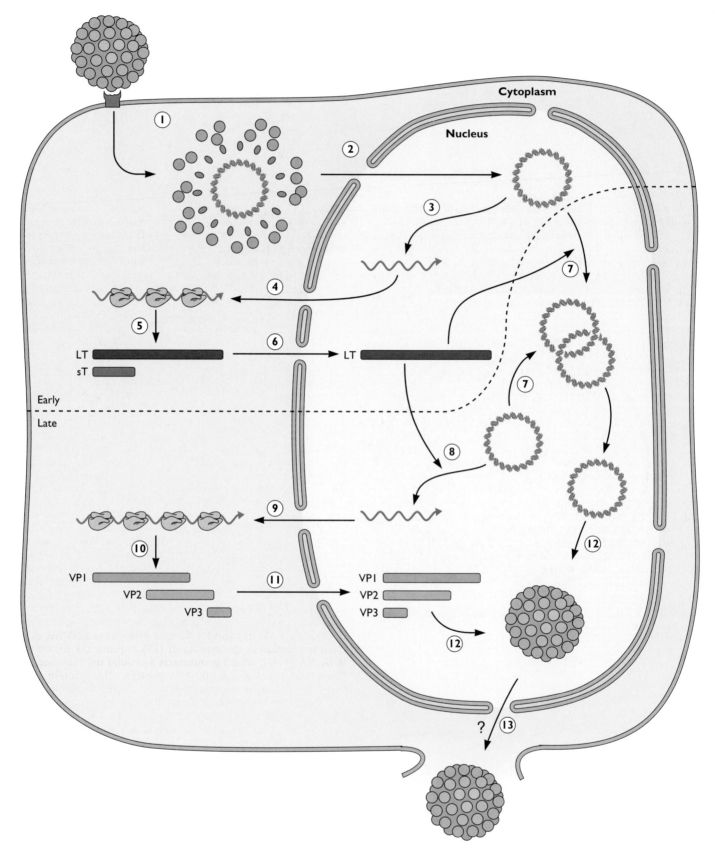

**Figure 14  Single-cell reproductive cycle of simian virus 40.**

# Adenoviruses

**Figure 15 Structure and genome organization of the adenovirus human adenovirus type 2. (A) The virion.** (Left) Electron micrographs of negatively stained adenovirus type 2 particles showing the triangular faces of the icosahedral capsid (top; magnification, ×200,000) and the fibers (bottom; magnification, ×425,000). Courtesy of M. Vayda, University of Maine. (Right) Diagram of the virion illustrating the major structural units of the capsid, hexons and pentons, additional proteins (e.g., VI and IX) that stabilize the capsid, and the viral core proteins. The latter are associated with the linear double-stranded DNA genome of nearly 36 kbp, which carries covalently linked terminal protein at its 5′ ends. **(B) Genome organization.** The origins of replication (Ori) at each end of the genome, the terminal protein (TP) covalently attached to each end of the genome, and the eight RNA polymerase II (green arrows) and three RNA polymerase III (tan arrows) transcription units are shown. Arrows indicate the direction of transcription, with the arrowheads at the sites of polyadenylation of RNA polymerase II primary transcripts. The three small major late (ML) exons designated l1, l2, and l3 are spliced to form the 202-nucleotide tripartite leader common to all mRNA species processed from major late primary transcripts. These mRNAs, which form the five families (L1, L2, L3, L4, and L5) defined by their common 3′ polyadenylation sites, encode all but one (polypeptide IX) of the virion structural proteins, as shown for several of the capsid proteins. The coding sequences for proteins that fulfill related functions, such as the viral replication proteins (E2 transcription unit) and the core proteins (L2 segment of ML transcription unit), are often organized together in the viral genome.

**Figure 16 Single-cell reproductive cycle of human adenovirus type 2.** The virus attaches to a permissive human cell via interaction between the fiber and the adenoviral receptor on the cell surface. The virus enters the cell by endocytosis **(step 1)**, a step that depends on the interaction of a second virion protein, penton base, with a cellular integrin protein. Within the cell, uncoating takes place and the viral genome, associated with core protein VII, is imported into the nucleus **(2)**. The host cell RNA polymerase II transcription system transcribes the **immediate-early** E1A gene **(3)**. Following alternative splicing and export of E1A mRNAs to the cytoplasm **(4)**, E1A proteins are synthesized by the cellular translation machinery **(5)**. These proteins, which are extensively modified by phosphorylation, are imported into the nucleus **(6)**, where they regulate transcription of both cellular and viral genes and the proliferation state of the host cell. The larger E1A protein stimulates transcription of the viral **early** genes by cellular RNA polymerase II **(7)**. Transcription of the VA genes by host cell RNA polymerase III also begins during the early phase of infection. The early mRNA species are processed, exported to the cytoplasm **(8)**, and translated into the early proteins **(9)**. These early proteins include the viral replication proteins, which are imported into the nucleus **(10)** and cooperate with a limited number of cellular proteins in viral DNA synthesis **(11)**. Replicated viral DNA molecules can serve as templates for further rounds of replication **(12)** or for transcription of **late** genes **(13)**. Some late promoters are activated simply by viral DNA replication, but maximally efficient transcription of the major late transcription unit requires the late IVa$_2$ protein and a second, unidentified, infected-cell-specific protein. Processed late mRNA species are selectively exported from the nucleus as a result of the action of early E1B 55-kDa and E4 34-kDa proteins **(14)**. Their efficient translation in the cytoplasm **(15)** requires the major VA RNA, VA RNA-I, which counteracts a cellular defense mechanism, and the late L4 100-kDa protein. This protein also serves as a chaperone for assembly of trimeric hexons as they and the other structural proteins are imported into the nucleus **(16)**. Within the nucleus, capsids are assembled from these proteins and progeny viral genomes to form noninfectious immature virions **(17)**. Such virions contain the precursors of the mature forms of several proteins. Mature infectious virions are formed **(18)** when these precursor proteins are cleaved by the viral protease, encoded within the L3 region, and assembled into the virion core. Progeny virions are released **(19)**, usually upon destruction of the host cell.

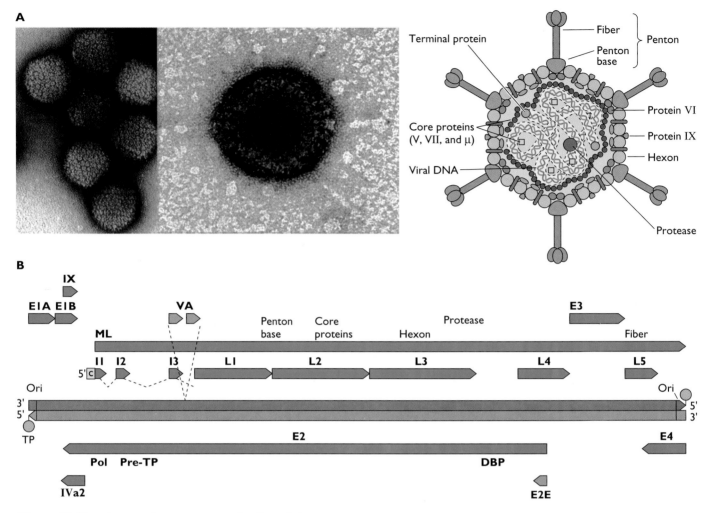

**A**

**B**

IX

E1A E1B

VA

Penton
base

Core
proteins

Protease

E3

ML

Hexon

Fiber

Terminal protein

Fiber ⟩ Penton

Penton
base

Protein VI

Core proteins
(V, VII, and μ)

Protein IX

Viral DNA

Hexon

Protease

I1 I2

I3

L1

L2

L3

L4

L5

5' C

Ori

Ori

3'

5'

5'

3'

TP

E2

E4

Pol   Pre-TP

DBP

IVa2

E2E

**Figure 15 Structure and genome organization of the
adenovirus human adenovirus type 2.**

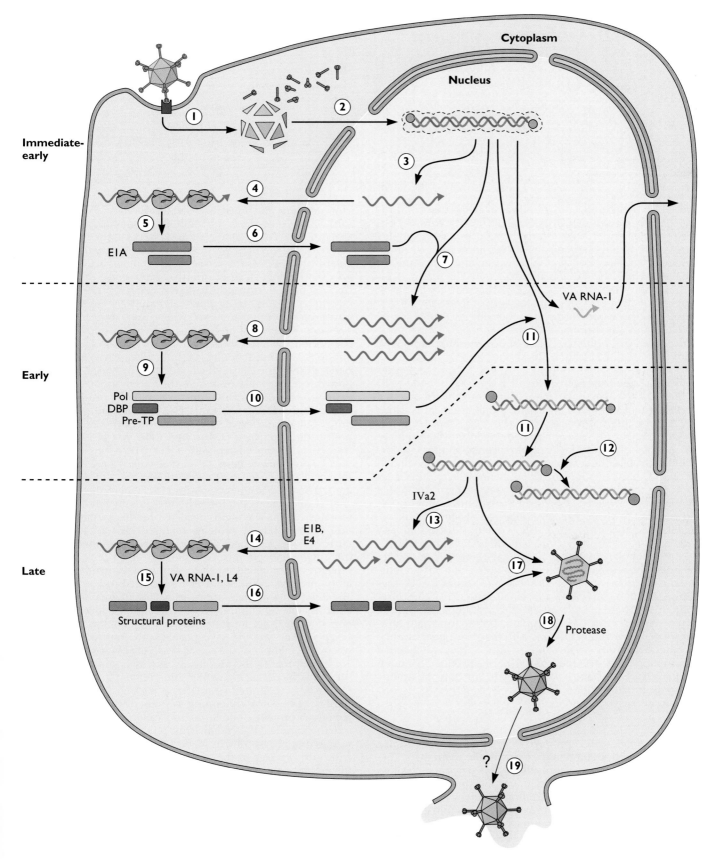

**Figure 16 Single-cell reproductive cycle of human adenovirus type 2.**

# Herpesviruses

**Figure 17 Structure and genome organization of herpesvirus herpes simplex virus type 1. (A) The virion.** (Left) Electron micrograph of a negatively stained herpes simplex virus type 1 virion. Staining and dehydration have distorted the virion, enabling the stain to penetrate and reveal the internal structure. The virion is approximately 200 nm in diameter. Magnification, ×80,000. (Right) Diagram of the virion showing the lipid envelope studded with at least 10 viral glycoproteins, the tegument layer comprising at least 15 viral proteins, and the icosahedral capsid containing the linear double-stranded genome. **(B) Genome organization.** The genome is organized into a 126-kb long region and a 26-kb short region of double-stranded DNA. The unique sequences of the long and short regions are called $U_L$ (108 kb) and $U_S$ (13 kb), respectively. The long region is bracketed by an inverted repeat sequence of about 9 kb. The terminal repeat is called $TR_L$, and the internal repeat is called $IR_L$. The short region is similarly bracketed by a different inverted repeat sequence ($TR_S$ and $IR_S$) of about 7 kb. A short repeating sequence of 200 to 500 bp, called the *a* sequence, is found at the ends and at the $IR_L/IR_S$ junction (not indicated in the figure). The *a* sequence is required for cleavage of concatemeric replicated DNA and for packaging of the genome within the capsid. The herpes simplex virus type 1 genome can "isomerize" or recombine via the inverted repeat sequences such that any given population within an infected cell, or in virions, consists of four equimolar isomers in which the $U_L$ and $U_S$ sequences are inverted with respect to each other. The viral genome contains three origins of replication, one within the $U_L$ segment (OriL) and two within a repeated sequence of the $U_S$ region (OriS). There are at least 84 open reading frames in the herpes simplex virus type 1 genome. The approximate locations of genes encoding proteins with different functions during the infectious cycle and of the latency-associated transcription unit are indicated. Genes encoding common functions have the same color shading to illustrate the finding that gene functions tend to be dispersed, rather than clustered, in herpesviral genomes.

**Figure 18 Single-cell reproductive cycle of herpes simplex virus type 1.** Virions bind to the extracellular matrix (e.g., heparan sulfate or chondroitan sulfate proteoglycans) through specific viral membrane protein interactions (primarily gB and gC) **(step 1)**. Another viral membrane protein (gD) interacts with a second cellular receptor **(2)**. This contact initiates fusion of viral and plasma membranes mediated by viral membrane glycoproteins (gD, gB, and gH/gL) **(3)**. After membrane fusion, the virus is uncoated, liberating tegument proteins and the nucleocapsid into the cytoplasm **(4)**. Viral nucleocapsids attach to microtubules and are transported to the nucleus **(5a)**. Certain tegument proteins, like VP16, activate transcription of the

viral genome and must also be transported to the nucleus **(5b)**. Others, like virion host shutoff factor (Vhs), act in the cytoplasm **(5c)**. The viral nucleocapsid docks at the nuclear pore **(6)**, releasing viral DNA into the nucleus, where the DNA circularizes **(7)**. VP16, the virion transcriptional activator, interacts with host transcription components to stimulate transcription of **immediate-early** genes by host cell RNA polymerase II **(8)**. Immediate-early mRNAs are spliced and transported to the cytoplasm, where they are translated **(9)**. The immediate-early proteins (α proteins) are imported into the nucleus, where they activate transcription of **early** genes and regulate transcription of immediate-early genes **(10)**. Early gene transcripts, which are rarely spliced, are transported to the cytoplasm, where they are translated **(11)**. The early proteins (β proteins) function primarily in DNA replication and production of substrates for DNA synthesis. Some β proteins are transported to the nucleus **(12a)**, and some function in the cytoplasm **(12b)**. Viral DNA synthesis is initiated from viral origins of replication **(13)**. DNA replication produces long concatemeric DNA molecules, the templates for **late** gene expression **(14)**. Late mRNAs are rarely spliced and are transported to the cytoplasm **(15)**, where they are translated. Late proteins (γ proteins) are primarily virion structural proteins and those proteins needed for virus assembly and particle egress. Some γ proteins required for formation of the virus envelope are made on, and inserted into, membranes of the rough endoplasmic reticulum **(16a)**. These membrane proteins are modified by glycosylation. Precursor viral membrane proteins are thought to be localized to both the outer and inner nuclear membranes, as well as to membranes of the endoplasmic reticulum **(16b)**. These precursors are transported to the Golgi apparatus for further modification and processing **(16c)**. Mature glycoproteins are transported to the plasma membrane of the infected cell **(16d)**. Some γ proteins are transported to the nucleus for assembly of the nucleocapsid and DNA packaging, while some remain in the cytoplasm **(17)**. Newly replicated viral DNA is packaged into preformed capsids **(18)**. DNA-containing capsids, together with some tegument proteins, bud from the inner nuclear membrane into the lumen of the endoplasmic reticulum, acquiring an envelope thought to contain precursors of viral membrane proteins **(19)**. Passage of the enveloped virus from the lumen of the endoplasmic reticulum to the cell surface is not well understood, but two possible mechanisms are shown in pathways 19a to 19d (de-envelopment pathway) and pathways 20a to 20e (luminal pathway). In the de-envelopment pathway, enveloped virus fuses with the endoplasmic reticulum membrane from within **(19b)**, releasing a capsid in the vicinity of a late Golgi compartment that contains mature viral membrane proteins **(19c)**. Tegument proteins added in the nucleus remain with the capsid, and others are added from the cytoplasm at this step. Capsids bud into the late Golgi com-

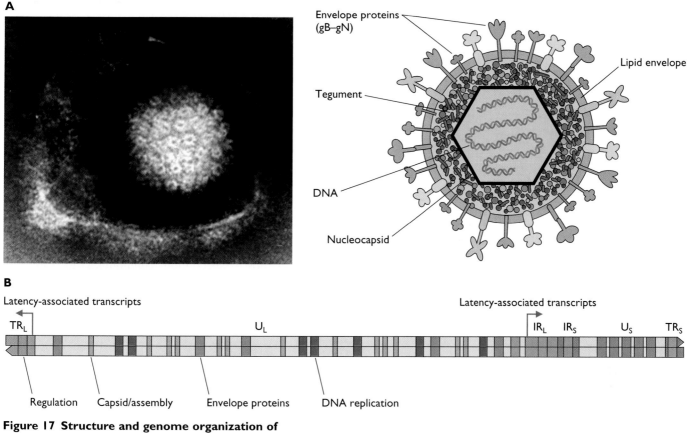

**A**

Envelope proteins (gB–gN)

Tegument

Lipid envelope

DNA

Nucleocapsid

**B**

Latency-associated transcripts

Latency-associated transcripts

TR$_L$  U$_L$  IR$_L$  IR$_S$  U$_S$  TR$_S$

Regulation  Capsid/assembly  Envelope proteins  DNA replication

**Figure 17  Structure and genome organization of herpesvirus herpes simplex virus type 1.**

partment, acquiring a membrane envelope containing mature viral envelope proteins **(19d)**. Enveloped virus is then transported to the plasma membrane for release by exocytosis **(21 and 22)**. In this model, the envelope of the mature virus is derived from Golgi membranes. In the luminal model **(20a)**, enveloped virus in the lumen of the endoplasmic reticulum is engulfed in a transport vesicle **(20b)** and delivered to Golgi compartments **(20c and 20d)**. The precursor viral envelope proteins are modified and processed by Golgi enzymes while part of the virion. Enveloped virus is transported to the plasma membrane for release by exocytosis **(21 and 22)**. In the luminal model, the membrane of mature virions originates from the inner nuclear membrane. In both models, exocytotic vesicles fuse with the plasma membrane **(22)** to release fully mature enveloped virus into the extracellular space **(23)**.

**Figure 19  Herpes simplex virus latent infection.** Latent infection occurs primarily in neurons found in sensory and autonomic ganglia. In the simplest model, the initiation of latent infection occurs as in **steps 1** to **7** of the productive infection. It is unclear if tegument proteins are transported into the nucleus **(8?)**. How the normal transcription cascade is blocked remains a matter of research and debate. However, when the latent state is established, transcription is severely restricted such that a single pre-mRNA is produced from the latency-associated transcript (LAT) promoter **(9)**. Low-level or sporadic transcription of immediate-early and early genes can occur but is not sufficient to initiate a productive infection. The latency-associated transcript initially synthesized is spliced, and a stable intron in the form of a lariat, called the 2-kb LAT, is produced in the nucleus **(10)**. The spliced LAT mRNA is transported to the cytoplasm, where several small open reading frames (e.g., ORF-O and ORF-P) may be translated into proteins **(11)**. The function of LAT RNAs and the expression of open reading frames as proteins are not well understood. After months to years, changes in neuronal physiology induced by trauma, hormonal changes, or other stressful conditions confer permissivity on neurons for productive infection. The genome is then transcribed more efficiently and replicated, and new virus is produced as shown in Fig. 18.

**775**

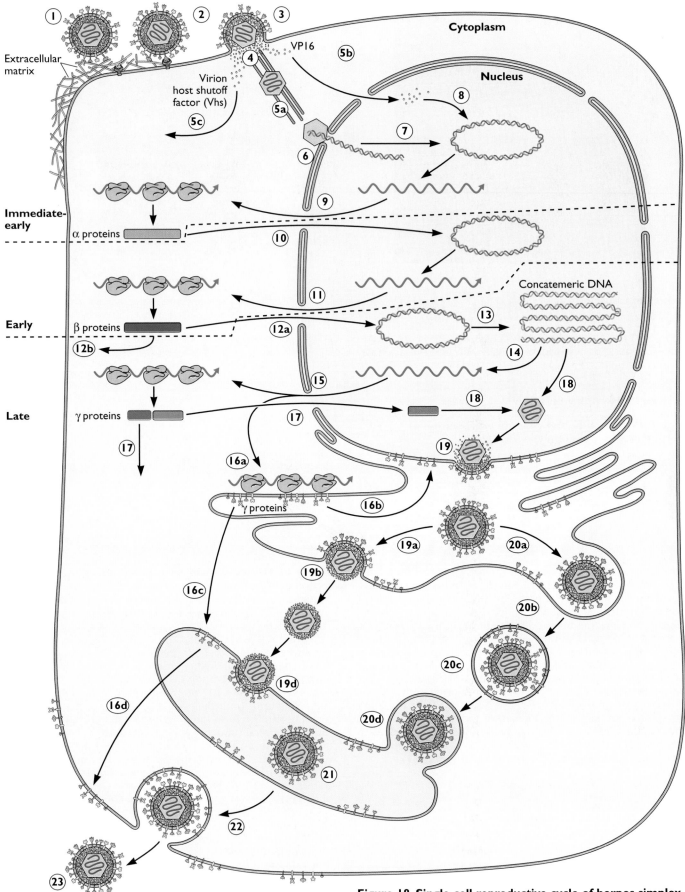

**Figure 18 Single-cell reproductive cycle of herpes simplex virus type 1.**

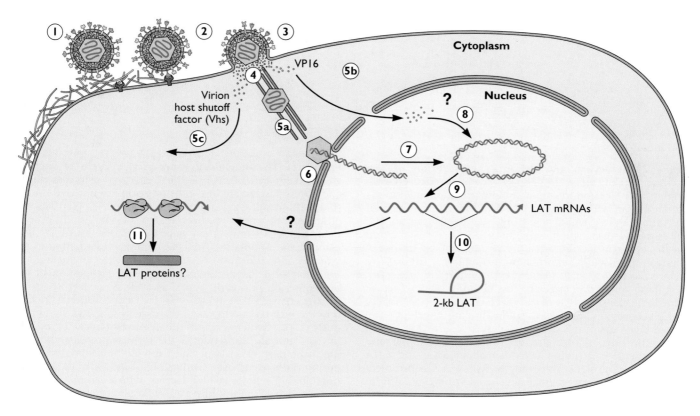

**Figure 19  Herpes simplex virus latent infection.**

# Poxviruses

**Figure 20 Structure and genome organization of the poxvirus vaccinia virus. (A) The intracellular mature virion (IMV).** (Left) Electron micrograph of negatively stained intracellular mature virion. Magnification, ×195,000. Courtesy of M. Adrian and J. Dubochet, Université de Lausanne. The well-resolved surface tubules are marked T. (Right) Structural features of the intracellular mature virion, deduced from cryoelectron microscopy under different conditions. The particle comprises a 300-Å-thick surface layer bounded by the outer membrane and core envelope membranes. The structure termed the palisade is a dense layer of spikes, presumably viral glycoproteins, projecting from the core envelope. The surface layer surrounds a homogeneous core within which resides the large double-stranded viral DNA genome and virion enzymes, including DNA-dependent RNA and DNA polymerases, and RNA-processing enzymes. A second infectious particle, the extracellular enveloped virion, contains an additional membrane. Modified from J. Dubochet et al., *J. Virol.* **68**:1935–1941, 1994, with permission. **(B) Structural features and organization of the genome.** The >191-kbp genome of vaccinia virus, depicted at the top, is a double-stranded DNA molecule whose ends are covalently connected by hairpin loops of 101 nucleotides. The sequences that form the hairpins are AT-rich and lie within the 12-kbp inverted terminal repetition (ITR). As indicated in the expansion of the left end of the genome, the ITR contains multiple repeated sequences. Determination of the complete sequence of the Copenhagen strain of vaccinia virus has identified some 185 unique protein-coding sequences and provided a detailed genetic map. Features of the organization of viral genes are illustrated by the coding sequences of *Hin*dIII fragment D, shown below. These reading frames are depicted by the arrows indicating their orientation in the genome and named according to the *Hin*dIII fragment in which the first ATG of the frame lies, the order of their ATGs within the fragment, and their orientation (left [L] or right [R]). The identities of the encoded proteins, where known, are also listed. As this segment of the vaccinia virus genetic map illustrates, coding sequences are densely packed in the genome and may be present in either strand. Sequences encoding proteins that fulfill different functions (e.g., structural proteins, membrane proteins, and virion or infected-cell-specific enzymes) are interspersed throughout the genome, as are genes expressed during different phases of the infectious cycle. For example, the mRNA capping enzyme, an RNA polymerase subunit, and uracil DNA glycosylase are products of early genes, whereas the genes encoding structural proteins are expressed only during the late phase of infection.

**Figure 21 The single-cell reproductive cycle of vaccinia virus.** The mechanisms by which vaccinia virus attaches to and enters susceptible host cells **(step 1)** are not well understood. The result is release of the core into the cytoplasm, indicating that entry requires fusion of viral with cellular membranes. All subsequent steps in the infectious cycle take place in the cytoplasm. The core contains, in addition to the viral genome, the viral DNA-dependent RNA polymerase and initiation proteins necessary for specific recognition of the promoters of viral early genes, as well as several RNA-processing enzymes that modify viral transcripts. Upon its release into the host cell's cytoplasm, the core synthesizes viral **early mRNAs (2)**, which exhibit the features typical of cellular mRNAs and are translated by the cellular protein-synthesizing machinery **(3)**. About half the viral genes are expressed during the early phase of infection. Some early proteins are homologs of cellular growth factors, which can induce proliferation of neighboring host cells following their secretion, or proteins that counteract host immune defense mechanisms **(4)**. Synthesis of early proteins also induces a second uncoating reaction in which a nucleoprotein complex containing the genome is released from the core **(5)**. Such core disassembly leads to inhibition of viral early gene expression, probably because early transcription is shut down. Other early proteins mediate replication of the viral DNA genome **(6)**. Newly synthesized viral DNA molecules can serve as templates for additional cycles of genome replication **(7)** and are the templates for transcription of viral **intermediate** genes **(8)**. Activation of transcription of intermediate genes also requires viral initiation proteins, the products of early genes, that confer specificity for intermediate promoters on the viral RNA polymerase, and a cellular protein (Vitf2), which relocates from the infected cell nucleus to the cytoplasm. Proteins made upon translation of intermediate mRNAs **(9)** include those necessary for transcription of **late** genes **(10)**. The latter genes encode the proteins from which virions are built and the virion enzymes and other essential proteins, such as the early initiation proteins, that must be incorporated into virus particles during assembly. As these late proteins are synthesized by the cellular translation machinery **(11)**, assembly of progeny virus particles begins, probably in association with internal membranes of the infected cell **(12)**. The initial assembly reactions result in formation of the immature virion **(13)**, a spherical particle that is believed to possess a double membrane acquired upon wrapping of the membranes of cellular compartments of the *cis*-Golgi network about the assembling particle. This virus particle then matures into the brick-shaped intracellular mature virion (IMV) **(14)**, which is released only upon cell lysis **(15)**. However, this particle can acquire a second, double membrane from a *trans*-Golgi or early endosomal compartment to form the intracellular enveloped virion (IEV) **(16)**. Fusion of the latter particles with the plasma membrane releases enveloped virions from the infected cell **(17)**. However, IEVs can also be transferred directly from one host cell to a neighboring cell via reorganization of the cell's actin cytoskeleton induced by the virus **(18)**.

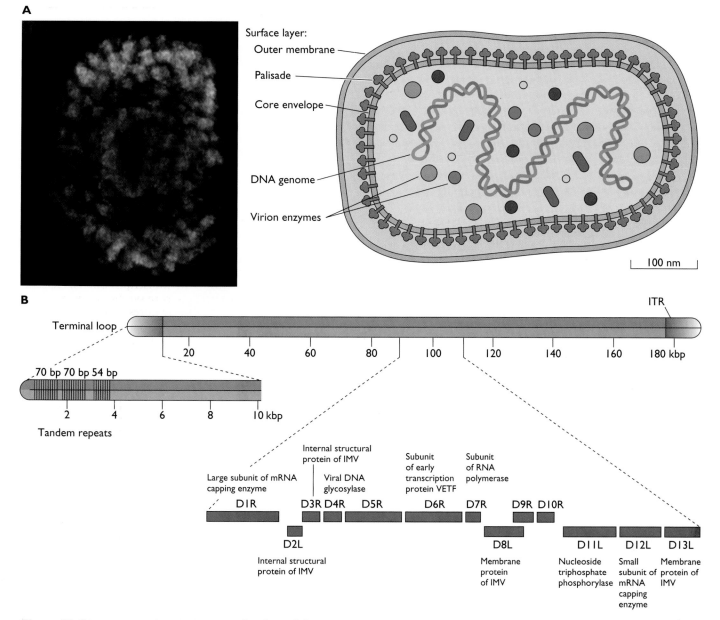

**A**

Surface layer:
Outer membrane
Palisade
Core envelope
DNA genome
Virion enzymes

100 nm

**B**

Terminal loop

ITR

20   40   60   80   100   120   140   160   180 kbp

70 bp 70 bp 54 bp

2   4   6   8   10 kbp

Tandem repeats

Large subunit of mRNA capping enzyme

Internal structural protein of IMV

Viral DNA glycosylase

Subunit of early transcription protein VETF

Subunit of RNA polymerase

D1R   D3R  D4R   D5R   D6R   D7R   D9R  D10R

D2L

D8L   D11L   D12L   D13L

Internal structural protein of IMV

Membrane protein of IMV

Nucleoside triphosphate phosphorylase

Small subunit of mRNA capping enzyme

Membrane protein of IMV

**Figure 20 Structure and genome organization of the poxvirus vaccinia virus.**

779

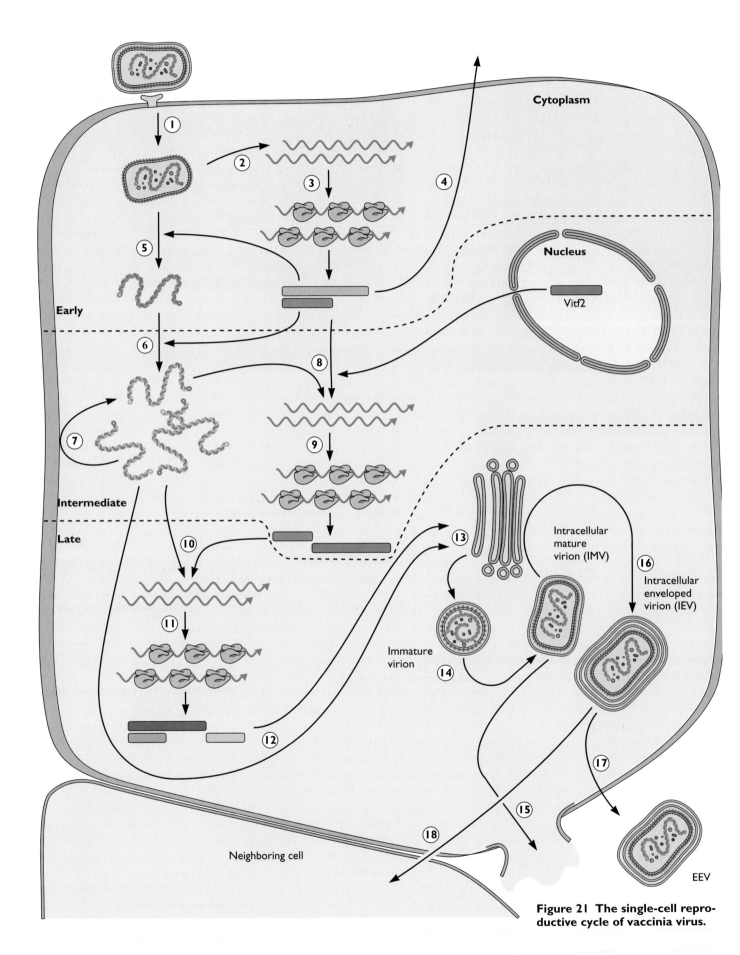

**Figure 21  The single-cell reproductive cycle of vaccinia virus.**

# Index